8th International Workshop DICE 2016: Spacetime-Matter-Quantum Mechanics

AA006918

Journal of Physics: Conference Series Volume 880

Castiglioncello, Italy
12 – 16 September 2016

ISBN: 978-1-5108-4722-4
ISSN: 1742-6588

Printed from e-media with permission by:

Curran Associates, Inc.
57 Morehouse Lane
Red Hook, NY 12571

Printed by Curran Associates, Inc. (2017)

For permission requests, please contact the Institute of Physics
at the address below.

Institute of Physics
Dirac House, Temple Back
Bristol BS1 6BE UK

Phone: 44 1 17 929 7481
Fax: 44 1 17 920 0979

techtracking@iop.org

Additional copies of this publication are available from:

Curran Associates, Inc.
57 Morehouse Lane
Red Hook, NY 12571 USA
Phone: 845-758-0400
Fax: 845-758-2633
Email: curran@proceedings.com
Web: www.proceedings.com

Table of contents

Volume 880

8th International Workshop DICE2016: Spacetime - Matter - Quantum Mechanics
12–16 September 2016, Castiglioncello, Italy

Accepted papers received: 9 August 2017
Published online: 21 August 2017

Preface

Invited Lectures

Contributed Papers

IOP Conf. Series: Journal of Physics: Conf. Series **880** (2017) 011001 doi:10.1088/1742-6596/880/1/011001

DICE 2016 : Spacetime Matter Quantum Mechanics

Presented in this volume are the Invited Lectures and the Contributed Papers of the *Eighth International Workshop on Decoherence, Information, Complexity and Entropy – DICE 2016*, held at **Castello Pasquini, Castiglioncello (Tuscany), September 12-16, 2016**.

These proceedings are meant to document the lively exchange of ideas during the conference for the interested public and the scientific community alike. – For further details on all aspects of the meeting, cf. the website at `http://www.df.unipi.it/~elze/DICE2016.html`.

Our intention has always been to unite leading researchers, advanced students, and renowned scholars from various areas, in order to stimulate new ideas and their exchange across the borders of specialization. The meetings in this series have successfully grown from DICE 2002,[1] followed by DICE 2004,[2] DICE 2006,[3] DICE 2008,[4] DICE 2010,[5] DICE 2012,[6] and DICE 2014.[7] About 120 participants representing almost 30 countries attended DICE 2016 this time.

It has been a great honour and inspiration having three eminent scholars presenting Keynote Lectures to all of us: Gabriele Veneziano (CERN & College de France, Paris), *"A quantum Universe before the Big Bang(s)?"*, in this volume – Bill Unruh (University of British Columbia, Vancouver), *"Measurement of Hawking radiation in the lab"* – Noam Chomsky (MIT, Cambridge MA), *"The Galilean challenge: architecture and evolution of language"*, in this volume.

Entitled *"Spacetime - Matter - Quantum Mechanics"*, the conference took place in the very pleasant atmosphere of Castello Pasquini – in beautiful surroundings, overlooking a piece of Tuscany's coast. The 5-day program addressed aspects of the following topics:

- Emergent spacetime, matter, and symmetries

[1] *Decoherence and Entropy in Complex Systems* ed H-T Elze *Lecture Notes in Physics* **633** (Berlin: Springer, 2004)
[2] *Proceedings of the Second International Workshop on Decoherence, Information, Complexity and Entropy – DICE 2004* ed H-T Elze *Braz. Journ. Phys.* **35**, 2A & 2B (2005) pp 205–529;
free access at: **www.sbfisica.org.br/bjp**
[3] *Proceedings of the Third International Workshop on Decoherence, Information, Complexity and Entropy – DICE 2006* eds H-T Elze, L Diósi and G Vitiello *Journal of Physics: Conference Series* **67** (2007);
free access at: **www.iop.org/EJ/toc/1742-6596/67/1**
[4] *Proceedings of the Fourth International Workshop on Decoherence, Information, Complexity and Entropy – DICE 2008* eds H-T Elze, L Diósi, L Fronzoni, J J Halliwell and G Vitiello *Journal of Physics: Conference Series* **174** (2009); free access at: **http://www.iop.org/EJ/toc/1742-6596/174/1**
[5] *Proceedings of the Fifth International Workshop on Decoherence, Information, Complexity and Entropy – DICE 2010* eds H-T Elze, L Diósi, L Fronzoni, J J Halliwell, E Prati, G Vitiello and J Yearsley *Journal of Physics: Conference Series* **306** (2011); free access at: **http://iopscience.iop.org/1742-6596/306/1**
[6] *Proceedings of the Sixth International Workshop on Decoherence, Information, Complexity and Entropy – DICE 2012* eds H-T Elze, L Diósi, L Fronzoni, J J Halliwell, E Prati, G Vitiello and J Yearsley *Journal of Physics: Conference Series* **442** (2013); free access at: **http://iopscience.iop.org/1742-6596/442/1**
[7] *Proceedings of the Seventh International Workshop on Decoherence, Information, Complexity and Entropy – DICE 2014* eds H-T Elze, L Diósi, L Fronzoni, J J Halliwell, C Kiefer, E Prati and G Vitiello *Journal of Physics: Conference Series* **626** (2015); free access at: **http://iopscience.iop.org/issue/1742-6596/626/1**

- Quantum coherence from micro- to macroscopic
- Physics of the oberserver
- New key experiments
- Quantum mechanics interfacing with gravity
- Quantum gravity / cosmology and high-energy physics

A Roundtable Discussion dedicated to the public under the theme *"La mente e la dinamica delle popolazioni di neuroni. Un filosofo e due fisici discutono di mente e cervello"* formed an integral part of the program. With the participation of A M Iacono (Pisa), E Prati (Milano), G Vitiello (Salerno), and moderated by H-T Elze (Pisa), the event attracted a large audience involved in the all-evening lively discussions. – Furthermore, Noam Chomsky gave an extended Press Conference during the workshop, which created much public interest and resulted in several articles (links on the website of the conference given above).

Three Best Posters were voted for by the participants and awards given by MDPI were presented to V.A. Bittencourt (Second Prize), L. Seveso (Second Prize) and A. Tilloy (First Prize) during a ceremony at the conference dinner.

The workshop has been organized by L Diósi (Budapest), H-T Elze (Pisa, chair), L Fronzoni (Pisa), J J Halliwell (London), C Kiefer (Cologne), E Prati (Milano) and G Vitiello (Salerno), with most essential help from our conference secretaries L Fratino (London), A Lampo (Barcelona), I Pozzana (Pisa) and F Vallone (Trento).

Several institutions and sponsors generously supported the workshop and their representatives as well as the citizens of Rosignano / Castiglioncello are warmly thanked for the kind hospitality:

- *Comune di Rosignano* – A Franchi (Sindaco di Rosignano), together with the Giunta Comunale di Rosignano, and special thanks to assessore L Montagnani.
- *Associazione Armunia* – Il Presidente e il Consiglio di Amministrazione / La direzione generale e la coordinazione artistica / Lo staff tecnico e organizzativo.
- *Solvay Italia SA, Rosignano* – A De Lorenzo (Site Communication & Total Quality Manager).

Funds made available by Università di Pisa, Domus Galilaeana (Pisa), Centro Interdisciplinare per lo Studio dei Sistemi Complessi – CISSC (Pisa), Dipartimento di Fisica (Università di Salerno), Istituto Italiano per gli Studi Filosofici – IISF (Napoli), Institute of Physics Publishing – IOP (Bristol), Springer Verlag (Heidelberg), MDPI AG (Basel) and by a grant from Foundational Questions Institute (FQXi) to C Kiefer (Cologne) are gratefully acknowledged.

We especially thank Laura Pesce (Vitrium Galleria – http://www.laurapesce.it) for exposing her artwork *"Complexity and Aesthetics"* dedicated to the conference at Castello Pasquini.

The proceedings have been edited by L Diósi, H-T Elze, L Fronzoni, J J Halliwell, C Kiefer, E Prati, and G Vitiello and the accepted papers follow essentially the order of presentation during the conference, separating, however, Invited Lectures and Contributed Papers. – Several Invited Lectures, cf. complete list at `http://www.df.unipi.it/~elze/DICE2016.html`, could not be reproduced here, partly for copyright reasons. – In the name of all participants, we would like to thank A Ashton, following S Toms, with collaborators at IOP Publishing (Bristol) for friendly advice and most valuable immediate help during the editing process and, especially, for their continuing efforts to make the Journal of Physics: Conference Series available to all.

Budapest, Pisa, London, Cologne, Milano, Salerno — June 2017

Lajos Diósi Hans-Thomas Elze Leone Fronzoni Jonathan Halliwell
Claus Kiefer Enrico Prati Giuseppe Vitiello

Peer review statement

All papers published in this volume of *Journal of Physics: Conference Series* have been peer reviewed through processes administered by the proceedings Editors. Reviews were conducted by expert referees to the professional and scientific standards expected of a proceedings journal published by IOP Publishing.

A Quantum Universe Before the Big Bang(s)?

Gabriele Veneziano

Theory Division, CERN, CH-1211 Geneva 23, Switzerland
and
Collège de France, 11 place M. Berthelot, 75005 Paris, France

E-mail: gabriele.veneziano@cern.ch

Abstract. The predictions of general relativity have been verified by now in a variety of different situations, setting strong constraints on any alternative theory of gravity. Nonetheless, there are strong indications that general relativity has to be regarded as an approximation of a more complete theory. Indeed theorists have long been looking for ways to connect general relativity, which describes the cosmos and the infinitely large, to quantum physics, which has been remarkably successful in explaining the infinitely small world of elementary particles. These two worlds, however, come closer and closer to each other as we go back in time all the way up to the big bang. Actually, modern cosmology has changed completely the old big bang paradigm: we now have to talk about (at least) two (big?) bangs. If we know quite something about the one closer to us, at the end of inflation, we are much more ignorant about the one that may have preceded inflation and possibly marked the beginning of time. No one doubts that quantum mechanics plays an essential role in answering these questions: unfortunately a unified theory of gravity and quantum mechanics is still under construction. Finding such a synthesis and confirming it experimentally will no doubt be one of the biggest challenges of this century's physics.

1. Outline

- Three revolutions 100 years ago
- The cosmological model of the 21st century
- The importance of QM in (modern) cosmology
- Which Big Bang?
- Which quantum gravity?
- String theory (and cosmology)
- Conclusions

2. Three revolutions 100 years ago

Three crucial revolutions in physics took place at the dawn of last century:

- 1900: Max Planck, puzzled by an ultraviolet divergence in black-body radiation, introduces his famous constant h and starts the *Quantum Revolution*, i.e. the *end of determinism*;
- 1905: Albert Einstein, building upon the finiteness and constancy of the speed of light c, introduces *Special Relativity*, i.e. the *end of absolute time*;

IOP Conf. Series: Journal of Physics: Conf. Series **880** (2017) 012001 doi:10.1088/1742-6596/880/1/012001

- 1915-16: Albert Einstein again, starting from the observed universality of free-fall, introduces a geometric theory of gravity, *General Relativity*. It marks the *end of an absolute geometry*, which is deformed by matter in a way proportional to the gravitational constant G.

Since then a lot of progress has been made in combining in various ways those three revolutions. The most significant of these are:

A. The Standard Model of Elementary Particles (SMEP)

B. The Standard Model of Gravitation and Cosmology (SMGC)

Together, they form our present:

Standard Model of Nature (SMN)

A. The SMEP is the result of combining the principles of the quantum and special relativity revolutions. We may say that it is the marriage between h and c, the two constants of Nature characterizing those two revolutions. Its successes have been amazing, last but not least the discovery in 2012 at CERN of the Higgs boson implementing the Brout-Englert-Higgs mechanism. For lack of time I will not go any further into describing this pillar of the SMN. Rather, I would like to recall the present status of the Standard Model of Gravity before turning to its cosmological applications.

B. As already mentioned the Standard Model of Gravity is based on Einstein's General Relativity, a beautiful theory that can be regarded as a synthesis between special relativity and the Newtonian theory of gravity, a marriage between c and G.

Its starting point is the so-called Equivalence Principle, deeply rooted in the observed universality of free fall. Tests of such universality are expected to reach soon incredible precision attaining the level of 1 part in 10^{15} in the coming few years from the already operative μscope experiment, with an even better precision foreseen from the future STEP experiment (if and when it will fly).

Besides, GR predicts well-verified corrections to Newtonian gravity (deflection of light, periastron precession, Lense-Thirring effect, ...). It also predicts qualitatively new phenomena, most notably:

- The existence and properties of black holes (BH)
- The existence of gravitational waves (GW)

Concerning BH, there has been for quite a while direct evidence for their existence, in particular for the presence, in the centre of most galaxies including our own, of huge black holes over a million times heavier than the sun.

Finally, as far as GW are concerned, the situation has changed dramatically this year when previous indirect evidence (through the observation of the slowing down of the revolution period of binary pulsars) was turned into direct observation in the LIGO interferometers by the LIGO/VIRGO collaboration. Amazingly, this confirmation of GR comes with further evidence for BH as well, this time as sources of GW when two of them (each one of tens of solar masses) merge.

Note that, so far, we have only discussed tests of classical General Relativity ($\hbar = 0$). As we will see in a moment, things change considerably once we extend the applications of GR to cosmology.

3. 21st century's cosmology

Let us start by recalling the cosmological model that prevailed in last century till the eighties. It was based on GR and was characterized by what we may call a *classical dogma* known under the resounding name of Hot Big Bang (HBB) cosmology. According to it:

> The Universe was born about 13.5 billion years ago
> in state of extremely high temperature and density.
> A singularity[1] prevents us from going beyond that moment
> in our past, therefore identified with the beginning of time.

Since then, an adiabatic expansion, very rapid initially, then slower and slower, cooled down the Universe. Today we see relics of that catastrophic event in the cosmic microwave background (CMB), a thermal spectrum of photons with a temperature of about $2.7K$.

The HBB scenario had several important successes. There is very little doubt that an event *like* a hot big bang must have taken place in our past. Unfortunately, its necessarily precise form poses some puzzling questions. For lack of time I will only mention a few of them.

1. The homogeneity problem

The universe is very homogeneous and isotropic on large scales: the CMB temperature varies by only one part in a hundred thousand (i.e. by tens of μK) depending of the direction it is coming from. Now, this property is put by hand (as initial condition) in the HBB scenario!

If, instead, we start from generic initial conditions and the expansion has always been a decelerating one, the Universe has always been too big (for its age!) for being able to become homogeneous in the short time elapsed since $t = 0$.

The problem can be visualized in Fig. 1, where the blue ellipses, representing the size of our observable Universe at a given time in our past, are compared to the red ones, representing the distance traveled by light from the Big Bang till that generic time. Clearly, throughout our past history the blue ellipses were larger than the red ones, showing that thermalization by causal processes (that cannot travel faster than light) is impossible.

2. The flatness problem

We can measure (e.g. by CMB experiments) how far the present geometry of space is from being exactly Euclidean. The result is that it can only have a very small spatial curvature (this is often stated by saying that $\Omega \sim 1$, or that the energy density is nearly critical). However, during a decelerating expansion spatial curvature increases meaning that, in order not to be larger than measured today, it had to be extremely small right after the HBB. Again an initial condition that one has to put in by hand.

3. Generating the large-scale structure of the Universe

The HBB model either predicts too much structure or none at all (i.e. no galaxies, clusters, stars). To get the right amount of structure one needs again a very fine-tuned initial state. All three problems can be ascribed to two basic features of HBB cosmology:

- The existence of a beginning of time;
- An always decelerating expansion (usually implied by the fact that ordinary matter resists the expansion).

[1] A moment at which several physical quantities, like, density and temperature become infinitely large.

IOP Conf. Series: Journal of Physics: Conf. Series **880** (2017) 012001 doi:10.1088/1742-6596/880/1/012001

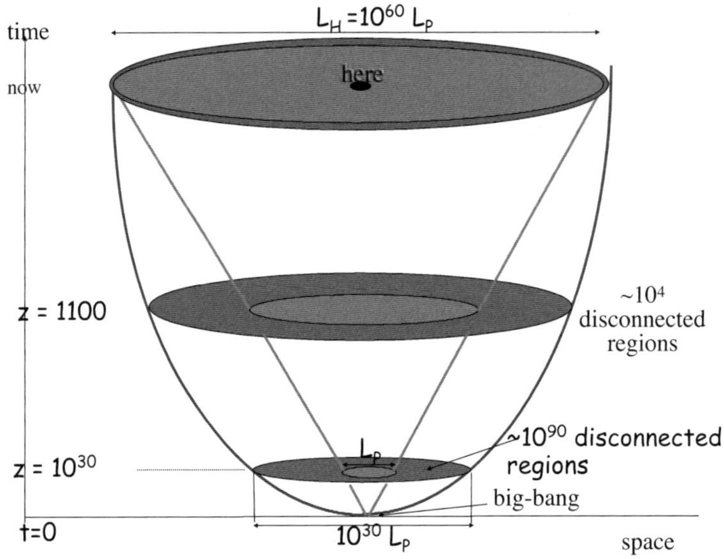

Figure 1. Time evolution (from bottom up) of the size of our presently observable Universe (blue ellipses) and of the distance traveled by light since the Big Bang (red ellipses) in the old HBB scenario. In our past the red ellipses were always smaller than the blue ones.

If, instead, a phase of accelerated expansion took place in the primordial Universe, then:
The homogeneity and flatness problems (see below for the 3rd one) are easily solved, since the Universe was so tiny during the inflationary phase that it could have easily become homogeneous by causal processes. This (a posteriori) very simple observation gave rise to an alternative to the HBB called the inflationary scenario (in which fig. 1 is replaced by fig. 2).

Eventually, in the nineties, cosmologists arrived to what is now considered the new standard model of cosmology, often dubbed as the "Concordance Model", a model which combines the inflationary paradigm with that of a dark sector in the energy budget of the Universe. The dark sector consists of two components:

- Dark Matter, whose existence is called for by the otherwise mysterious rotation curves of galaxies;
- Dark Energy, which is needed to explain the recent acceleration in the expansion.

So far, we are still talking classical physics/GR and everything looks fine. So why bother with the quantum?

4. The importance of QM in (modern) cosmology

There are, at least, two reasons why inflation without quantum mechanics does not solve the cosmological problems posed by HBB cosmology.

- The origin of large-scale structure

  Without QM inflation produces a perfectly homogeneous Universe. That's because any initial *classical* inhomogeneity (which, by definition, has to be on length scales larger than the Planck length) is stretched far beyond our present horizon during the 60 or more e-folds of expansion and thus effectively erased. In other words, without QM the structures we see today in the sky could not have formed.

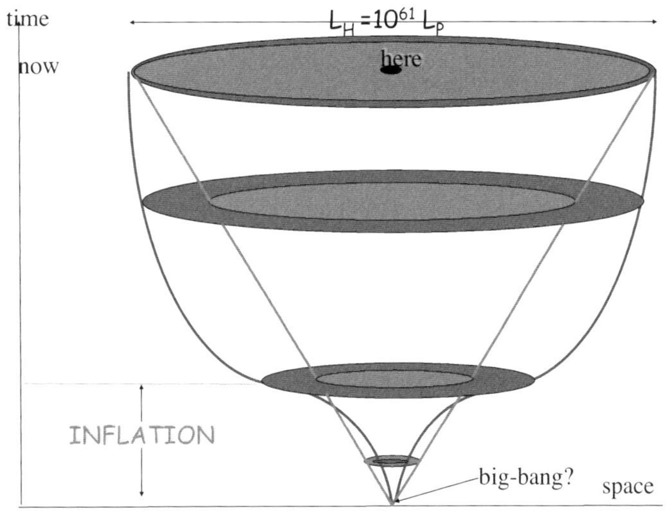

Figure 2. Time evolution (from bottom up) of the size of our presently observable Universe (blue ellipses) and of the distance traveled by light since the Big Bang (red ellipses) in the inflationary scenario. In our far past the red ellipses were larger than the blue ones.

Instead, with QM, classical initial inhomogeneities are erased, but they replaced by calculable quantum fluctuations produced, amplified and stretched throughout the inflationary epoch (see [1] for a recent review). Those created during the last 60 e-folds are still within our horizon today. And indeed the CMB fluctuations we observe today (e.g. via the Planck satellite) carry an imprint of those primordial quantum fluctuations.
Although, so far, we have only observed the so called density/curvature perturbations these, as the name indicates, can be either seen as matter or as geometry perturbations, the distinction between the two being coordinate(gauge)-dependent. Thus, I maintain that CMB observations are already[2] a clear indication that gravity needs to be quantized like we already know to be the case for the other fundamental interactions.

- The need to reheat after inflation

Inflation cools down the Universe (in case it was hot before inflation) to practically zero temperature. How can we generate a hot Universe *after* inflation? If we do not, there is no way to generate the CMB or to have the primordial nucleosynthesis of light elements such as Helium!
Note that, in inflationary cosmology, the Universe is still expanding after inflation. How can it keep expanding and yet reheat? This is where a second intervention of QM takes place by providing a dissipative, non adiabatic conversion of potential energy into a hot thermal soup of elementary particles. Indeed, one of the challenges for any inflationary model is to be able to generate sufficient reheating (or entropy production). It is hard, although perhaps not impossible, to find a classical mechanism achieving that goal but QM does it in a very natural way.

[2] The observation of cosmological tensor perturbation through the B-mode CMB polarization would be, of course, the ultimate smoking gun for the need to quantize gravity.

5. Which Big Bang?

An important consequence of our previous discussion is that the phase of reheating after inflation plays the role of the big bang of the old HBB cosmology, modulo some important differences:

1. It has no singularity (such as an infinite temperature) associated with it (if any it's non trivial to make it hot enough);
2. It has nothing to do with the beginning of time!

During many decades we have taught the general public the simple equation:

$$\text{Big Bang} = \text{Beginning of Time}$$

But, as just discussed, according to modern inflationary cosmology, we have to distinguish:

The *physical* non singular BB, at the end of inflation, the one that left measurable relics (the CMB, light nuclei) and:

A *theoretical* singular BB that could have preceded inflation whose relics have been washed out by inflation. In any case:

The BB we know something about has nothing to do with the beginning of time.

Thus the often shown picture depicted below is very misleading.

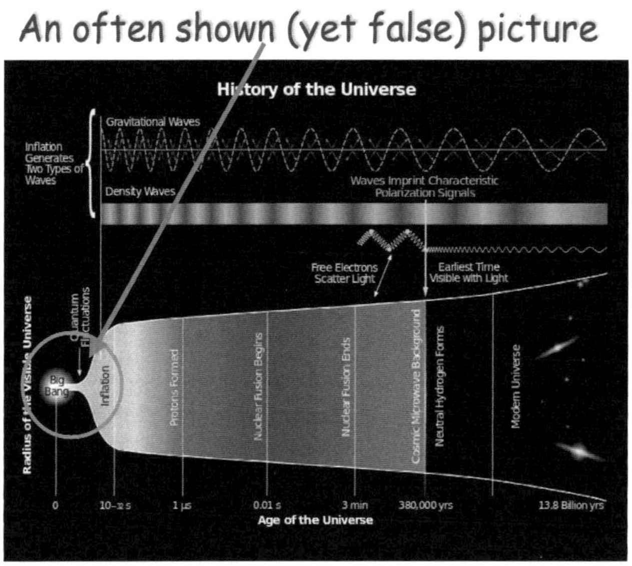

In conclusion, we are led to accept, on the basis of our discussion, a necessarily quantum cosmology taking place before and during the (true/physical) Big Bang!!

But what about the beginning of time? What preceded inflation? We do not know the answer to this question and standard inflationary cosmology is pretty agnostic about it. We only know that the answer lies in the answer to another question:

Which is the correct quantum theory[3] of gravity?

6. Which quantum gravity?

Quantizing General Relativity is a notoriously difficult task. Unlike in the SMEP, in the case of gravity the ultraviolet infinities that are present in any local quantum field theory (in four

[3] We are referring here to a full-fledged consistent quantum theory of gravity: cosmological perturbations can be treated semi-classically without such a complete quantum gravity formulation.

dimensions) cannot be lumped into a finite number of renormalization constants.

To this date, the most interesting attempts to quantum GR are those of Loop Quantum Gravity and of Asymptotic Safety. Both assume that one can make sense of quantum gravity by modifying the way to quantize GR, not GR itself. Their viability as theoretically consistent quantum theories is still to be demonstrated. The cosmological applications of LQG apparently reach conclusions similar to those of string theory. It is hard (at least for me) to assess the reliability of those claims.

By contrast, a drastically new approach to quantum gravity (and quantum field theory in general) is offered by (super)string theory thanks to a consistent non-local modification of fundamental interactions below a certain length scale. In order to introduce the basic concepts underlying string theory it is good to recall:

A lesson from particle physics

According to our present understanding, at the most microscopic quantum level, all fundamental interactions are transmitted by massless particles of spin 1 or 2. The first (e.g. the photon) give rise to non-gravitational interactions, while the latter (the graviton) is responsible for gravity.

Once the existence of such particles is assumed the basic theoretical structure of both gauge theories and general relativity (gauge invariance, general covariance) follows from consistency.

As we shall see in a moment, in string theory the existence of massless spinning particles is a consequence of Quantum Mechanics. Therefore Quantum String Theory (QST) represents a kind of Copernican Revolution in which quanta come first, quantum field theory comes next, and classical field theory (Maxwell, General Relativity,...) comes last. Let us see how.

7. String theory (and cosmology)

String theory: what's that? We do not quite have a definite answer to this question but, at least at a superficial level, the best reply seems to be:

String Theory is the theory of strings

Jokes apart, what that means is that one replaces the grand principles of gauge invariance and general covariance, which are at the basis of our SMN, by *just* the assumption that everything is made of:

Relativistic Quantum Strings

In other words, when the particles that are usually considered elementary and point-like are looked at with sufficient resolution they actually appear as little one-dimensional objects, open or closed strings (depending on whether or not they have ends). String theory thus combines two of the fundamental revolutions of a century ago (and two fundamental constants, $\hbar$ and c) but, rather than adding from the very beginning the third (General Relativity and G), simply adds the string postulate. As we shall argue below this will be enough thanks to some quantum magic. In one symbolic equation we can write:

$$\text{SR} + \text{QM} + \text{Strings} = \text{Magic Cocktail}$$

Let us start with two crucial quantum miracles:

7.1. Quantum magic I: emergence of a fundamental length

Classical relativistic strings have no characteristic length scale. Their tension T is simply a conversion factor associating a string size L with a string mass $M \sim TL$. Note the analogy with GR, where, through G and c we also associate a length (a Schwarzschild radius R) to a given mass $R = 2GMc^{-2}$.

IOP Conf. Series: Journal of Physics: Conf. Series **880** (2017) 012001 doi:10.1088/1742-6596/880/1/012001

By contrast, quantum strings have a minimal (optimal) size L_s given by $L_s^2 = \hbar/T$. This a truly fundamental scale analogous to the Bohr radius of an atom in non-relativistic QM. There is, again, an analogy with GR, which, at the quantum level, does have a length scale, the Planck length $L_P^2 = G\hbar c^{-3}$.

The string length scale L_s plays many different crucial roles in QST. In particular, all interactions are smeared over regions of order L_s and thus softened at distances shorter than L_s. This basic property of quantum strings cures the ultraviolet problems of conventional quantum gravity.

7.2. Quantum magic II: massless spinning particles

Classical string cannot have angular momentum without also having a finite size, and thus, through the string tension, a finite mass. Indeed one can prove a classical lower bound on the mass of a bosonic string with (orbital) angular momentum J:

$$M^2 \geq 2\pi TJ \tag{1}$$

This classical inequality is slightly violated at the quantum level and becomes (after a consistent regularization of the divergent sum):

$$M^2 \geq 2\pi T(J - a\hbar) \quad , \quad a = 0, \frac{1}{2}, 1, \frac{3}{2}, 2. \tag{2}$$

with the half-integer values of a occurring in the case of the superstring. This quantum miracle is obviously what leads to the already anticipated Copernican Revolution:

$M = 0, J = 1$ provides the photons and the other gauge bosons mediating non-gravitational forces

$M = 0, J = 2$ provides a graviton, mediating the gravitational force.

Unfortunately, the $M = J = 0$ case provides dangerous extra massless particles (and corresponding long-range forces, but this matter is beyond the scope of this lecture).

7.3. A Theory of everything?

In QST all elementary particles correspond to different vibrations of the same basic objects: open and closed strings! Thus, combining the two miracles we just described, provides:

A unified and finite theory of elementary particles, and of their gauge and gravitational interactions, not just compatible with, but based upon, Quantum Mechanics!

Quantizing strings gives rise, directly, to quantum versions of gauge and gravitational interactions whose classical limits are the conventional field theories traditionally seen as the starting point of QFTs. Furthermore, the first quantization of strings also provides short-distance corrections without which QFTs are plagued by UV divergences!

7.4. String Cosmology

We are now finally ready to ask the question: What does string theory tell us about the beginning of time [4]?

We have already stressed that QST implies short-distance modifications of GR. Not surprisingly these intervene as soon as the curvature-radius of the geometry (e.g. H^{-1}) becomes comparable to L_s, the minimal size of a quantized string.

Intuitively, this minimal size implies an upper limit to density, temperature, and curvature, all the quantities that supposedly blow-up at the putative beginning of time. Apparently, there

[4] See also a non-specialized article of mine in Scientific American [2].

is no place for the singular Big Bang of the old BB cosmology. But what is there to replace it? The answer is not known, but two possibilities emerge as most likely:

- A cosmology without a classical beginning.
  A very quantum "stringy" phase replaces the old big bang. The concepts of a classical space and time spring out of that phase as emerging concepts. This possibility is sketched in Fig. 3.
- A cosmology without a beginning.
  Here the "stringy" phase plays the role of a "quantum bridge" between our epoch and another classical one, seen as a sort of gravitational collapse from which the Universe bounced when the curvature radius became of order L_s (a Big Bounce). This possibility is sketched in Fig. 4.

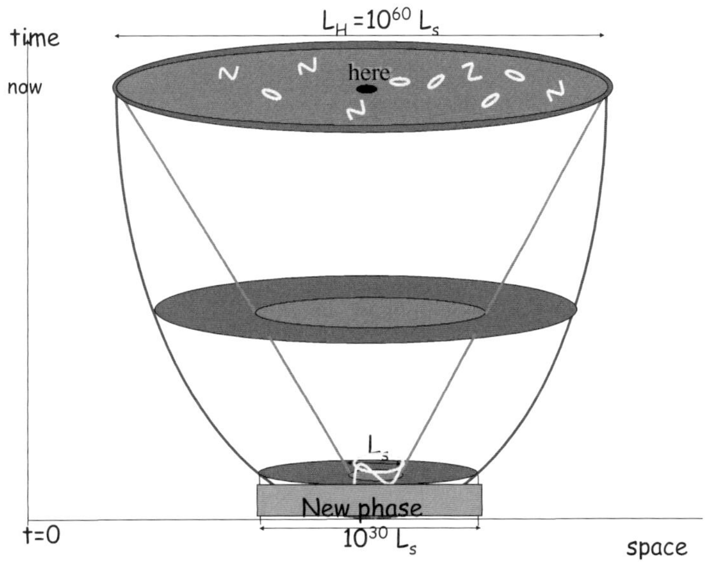

Figure 3. A cosmology without a classical beginning.

The pre-bounce phase could either prepare or just replace the inflationary one. In the latter case we may very well expect some:

Observable Cosmological Relics of a pre-big bang phase

The following ones are particularly noteworthy (see [3] and references therein):

- A stochastic background of gravitational waves, possibly observable in the next decade by advanced LIGO and/or VIRGO. A possible spectrum is illustrated in Fig. 5.
- A new source of adiabatic density/curvature fluctuations, with distinctive properties w.r.t. those of conventional inflation (e.g. with no primordial B-polarization in the CMB!)
- A new mechanism to generate seeds for the cosmic magnetic fields which are known to exist on galactic and inter-galactic scales and whose origin is mysterious in the absence of a mechanism producing appropriate seeds for them.

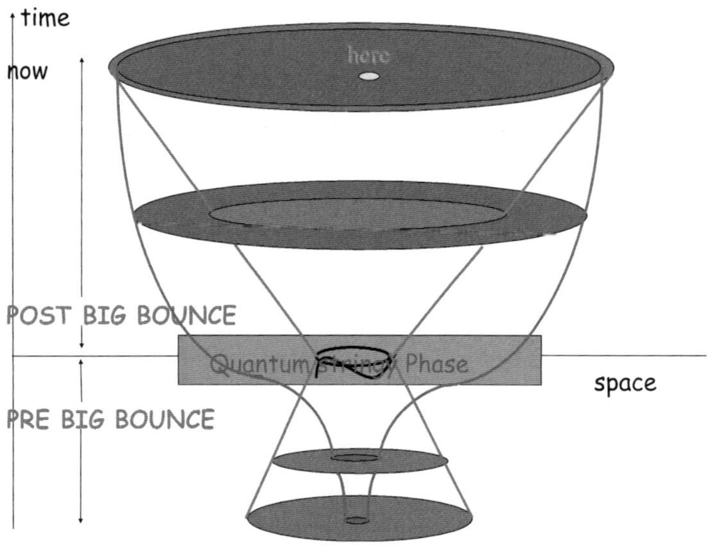

Figure 4. A cosmology without a beginning.

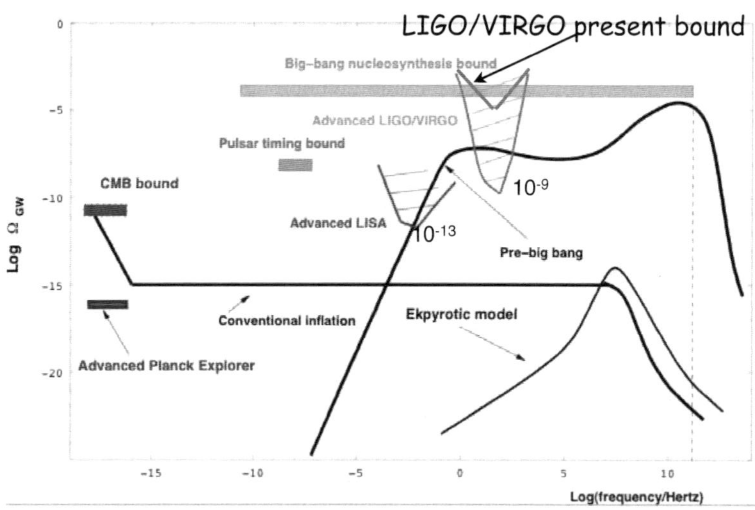

Figure 5. Different possible sources of a stochastic background of GW and the expected sensitivities for their detection.

8. Conclusions

Classical GR ($h = 0$) works very well in a large number of situations ... but:

- For its cosmological applications, and in particular in the inflationary scenario, GR needs QM.
- QM is at the origin of large scale structures in the Universe. It is also necessary to reheat it at the end of inflation.
- That moment of reheating plays the role of the traditional big bang but has nothing to do with the beginning of time.
- We don't know what preceded/started inflation. Answering such a question requires knowing the correct theory of quantum gravity and there are not so many options for it.
- In a string-theory context quantum gravity itself, as well as other QFTs, follows from QM, the classical BB singularity is (almost certainly) avoided, and the question of the beginning of time is well posed ...

... but not yet solved!

Acknowledgements
I am very grateful to the organizers of this Symposium, and in particular to Thomas Elze, for having kindly invited me to attend such an interesting event in such a beautiful setting.

References

[1] Mukhanov V 2005, *Physical Foundations of Cosmology* (Cambridge U. Press)
[2] Veneziano G 2014 in *A Matter of Time*, a collection of *Scientific American* articles
[3] Gasperini M and Veneziano G 2003, *Phys. Rept.* **373** 1 (*Preprint* arXiv:hep-th/0207130)

IOP Conf. Series: Journal of Physics: Conf. Series **880** (2017) 012002 doi:10.1088/1742-6596/880/1/012002

Quantum geometrodynamics of Einstein and conformal (Weyl-squared) gravity

Claus Kiefer and Branislav Nikolić

Institut für Theoretische Physik, Universität zu Köln, Zülpicher Straße 77, 50937 Köln, Germany

Abstract. We discuss the canonical quantization of general relativity and Weyl-squared gravity. We present the classical and quantum constraints and discuss their similarities and differences. We perform a semiclassical expansion and discuss the emergence of time for the two theories. While in the first case semiclassical time has a scale and a shape part, in the second case it only has a shape part.

1. Introduction

The quantization of gravity is perhaps the most important open problem in theoretical physics [1]. Among the oldest approaches is quantum geometrodynamics ("Wheeler-DeWitt equation"). This is still a promising approach because it is very conservative: its equations are found if one searches for quantum wave equations that directly lead to Einstein's equations in the semiclassical (WKB) limit [2]. In 1926, Erwin Schrödinger found his wave equation by formulating classical mechanics in Hamilton-Jacobi form and "guessing" a wave equation that leads to this form in the limit of geometric optics. We recall what he wrote in 1926 [3]:

> *We know today, in fact, that our classical mechanics fails for very small dimensions of the path and for very great curvatures.* Perhaps this failure is in strict analogy with the failure of geometrical optics . . . that becomes evident as soon as the obstacles or apertures are no longer large compared with the real, finite, wavelength. . . . Then it becomes a question of searching for an undulatory mechanics, and the most obvious way is by an elaboration of the Hamiltonian analogy on the lines of undulatory optics.[1]

The same procedure can be applied to Einstein's equations [2]. If written into Hamilton-Jacobi form, it can be "translated" into the Wheeler-DeWitt equation and the diffeomorphism (momentum) constraints. These equations are timeless, that is, there is no external time parameter or external spacetime background. One can introduce an *intrinsic time* by the structure of the Wheeler-DeWitt equation whose kinetic term is locally hyperbolic; this intrinsic time is related to $\sqrt{h}$, where h is the determinant of the three-metric h_{ab}. An *extrinsic time* is sometimes used in the alternative approach of reduced quantization; a typical choice is York's time (see e.g. [1], p. 148), which is related to the trace K of the extrinsic curvature.

[1] *Wir wissen doch heute, daß unsere klassische Mechanik bei sehr kleinen Bahndimensionen und sehr starken Bahnkrümmungen versagt.* Vielleicht ist dieses Versagen eine volle Analogie zum Versagen der geometrischen Optik . . . , das bekanntlich eintritt, sobald die 'Hindernisse' oder 'Öffnungen' nicht mehr groß sind gegen die wirkliche, endliche Wellenlänge. . . . Dann gilt es, eine 'undulatorische Mechanik' zu suchen – und der nächstliegende Weg dazu ist wohl die wellentheoretische Ausgestaltung des Hamiltonschen Bildes.

Starting with quantum geometrodynamics as the fundamental framework, one can derive the limit of quantum (field) theory in an external background spacetime by a Born-Oppenheimer type of approximation scheme with respect to the Planck mass [1, 2]. At leading order, the semiclassical or "WKB time" is recovered as an emergence concept from the timeless quantum equations.

In this contribution, we want to shed more light on the nature of emergent time. We will study the quantum geometrodynamics corresponding to general relativity (GR) and the quantum geometrodynamics corresponding to conformal (Weyl-squared) gravity, briefly called Weyl theory. The latter serves as a model for a gravitational theory devoid of scale. We will discuss the common features of and the differences between those theories.

For this purpose, we will employ a unimodular decomposition of the three-metric. In quantum GR (section 2), we can identify the scale and the shape part of the WKB time; the scale part is related to the intrinsic time in the vacuum Wheeler-DeWitt equation. So the WKB time needs for its "passing" both scale and shape.

In section 3, we present a model of quantum geometrodynamics based on Weyl-squared gravity. We find there equations similar to the equations of quantum GR, but with a different explicit structure.

In section 4, we discuss the semiclassical limit of these equations, which gives rise to a "Weyl-WKB time" (WWKB time). It has the property that the evolution of the scale degree of freedom does not contribute to it; this is because the WWKB time inherits the conformal symmetry of the Weyl-squared action from which the intrinsic time is absent. WWKB time is thus scale-less and conformally invariant; we call it the *shape time*. In section 5, we present our conclusions. Further details and references can be found in our papers [4, 5, 6].

2. Canonical quantum general relativity

Instead of the usual canonical formulation of GR and its quantization, we formulate the quantum geometrodynamics here in variables which manifestly reveal the conformal behaviour of the theory. We are interested, in particular, in identifying the three-metric volume $\sqrt{h}$ and the trace of the extrinsic curvature K as canonical variables carrying a single degree of freedom. This allows us to keep track of "intrinsic time" and "extrinsic time" by investigating what happens to these canonical degrees of freedom. These variables are called *unimodular-conformal variables* and are discussed in detail in [4, 5, 6]. To formulate GR in these variables, we will briefly review here section 2 from [7].

The extrinsic curvature and its trace are given by

$$K_{ij} = \frac{1}{2N}\left(\dot{h}_{ij} - 2D_{(i}N_{j)}\right), \quad K = h^{ij}K_{ij} = \frac{1}{N}\left(\frac{\dot{\sqrt{h}}}{\sqrt{h}} - D_iN^j\right), \tag{1}$$

where D_i is the covariant derivative with respect to the three-metric. We decompose these as well as the metric variables into scale and shape part, called unimodular-conformal decomposition:

$$a := (\sqrt{h})^{1/3}, \quad \bar{h}_{ij} := a^{-2}h_{ij}, \tag{2}$$

$$\bar{N}^i := N^i, \quad \bar{N}_i := a^{-2}N_i, \quad \bar{N} := a^{-1}N, \tag{3}$$

$$\bar{K}^{\mathrm{T}}_{ij} := a^{-1}K^{\mathrm{T}}_{ij} = \frac{1}{2\bar{N}}\left(\dot{\bar{h}}_{ij} - 2\left[\bar{D}_{(i}\bar{N}_{j)}\right]^{\mathrm{T}}\right), \quad \bar{K} := \frac{aK}{3} = \frac{1}{\bar{N}}\left(\frac{\dot{a}}{a} - \frac{1}{3}D_iN^i\right), \tag{4}$$

where $\bar{D}_i$ is the conformal part of the covariant derivative depending only on $\bar{h}_{ij}$. Note that intrinsic time $\sqrt{h}$ is here captured by the scale density a.

The Einstein-Hilbert (EH) action in its ADM formulation can be written in unimodular-conformal variables as follows:

$$\mathcal{S}^{\mathrm{E}} = \int \mathrm{d}t\, \mathrm{d}^3x\, \mathcal{L}^{\mathrm{E}},$$
$$\mathcal{L}^{\mathrm{E}} = \frac{1}{2\kappa} N\sqrt{h}\left({}^{(3)}R + K_{ij}K^{ij} - K^2\right) = \frac{1}{2\kappa}\bar{N}a^2\left(a^2\, {}^{(3)}R + \bar{K}^{\mathrm{T}2}_{ij} - 6\bar{K}^2\right), \tag{5}$$

where $\kappa = 8\pi G$ $(c = 1)$, and we keep in mind that ${}^{(3)}R$ can also be decomposed if needed (see the Appendix in [5]). Since the configuration variables are a and $\bar{h}_{ij}$, the conjugate momenta will be p_a and $\bar{p}^{ij}$, respectively,

$$p_a = \frac{\partial \mathcal{L}^{\mathrm{E}}}{\partial \dot{a}} = -\frac{6a}{\kappa}\bar{K}, \quad \bar{p}^{ij} = \frac{\partial \mathcal{L}^{\mathrm{E}}}{\partial \dot{\bar{h}}_{ij}} = \frac{a^2}{2\kappa}\bar{K}^{\mathrm{T}}_{ab}\bar{h}^{ia}\bar{h}^{jb}. \tag{6}$$

Using the definitions (2)–(4) and recalling that the ADM momentum is given by $p^{ij} = \sqrt{h}\left(K^{ij} - h^{ij}K\right)/(2\kappa)$, the momenta (6) can be related to the trace and traceless part of the ADM momentum, respectively, by

$$p_a = \frac{2}{a}\, p_{\mathrm{ADM}}, \quad \bar{p}^{ij} = a^2\, p^{ij\mathrm{T}}_{\mathrm{ADM}}, \tag{7}$$

so that $p^{ij}_{\mathrm{ADM}} = p^{ij\mathrm{T}}_{\mathrm{ADM}} + h^{ij}p_{\mathrm{ADM}}/3$. Performing a Legendre transformation, we see that the total Hamiltonian is simply a linear combination of first class constraints,

$$H^{\mathrm{E}} = \int \mathrm{d}^3x \left\{\bar{N}\mathcal{H}^{\mathrm{E}}_{\perp} + N^i\mathcal{H}^{\mathrm{E}}_i + \lambda_{\bar{N}}p_{\bar{N}} + \lambda^i p_i\right\}. \tag{8}$$

The momenta with respect to the lapse density $\bar{N}$ and shift N^i are the primary constraints $p_{\bar{N}} \approx 0$ and $p_i \approx 0$, respectively. They give rise to secondary constraints: the Hamiltonian constraint $\mathcal{H}^{\mathrm{E}}_{\perp} \approx 0$ and the momentum constraints $\mathcal{H}^{\mathrm{E}}_i \approx 0$, which in unimodular-conformal variables have the following form:

$$\mathcal{H}^{\mathrm{E}}_{\perp} = -\frac{\kappa}{12}p_a^2 + \frac{2\kappa}{a^2}\bar{h}_{ik}\bar{h}_{jl}\bar{p}^{ij}\bar{p}^{kl} - \frac{a^4}{2\kappa}\left({}^{(3)}R - 2\Lambda\right) \approx 0, \tag{9}$$

$$\mathcal{H}^{\mathrm{E}}_i = -2\bar{D}_k\left(\bar{h}_{ij}\bar{p}^{jk}\right) - \frac{1}{3}D_i\left(a\, p_a\right) \approx 0. \tag{10}$$

For completeness, we have added the cosmological constant Λ.

It is important to notice that the constraints explicitly depend on the scale density, which is related to intrinsic time $\sqrt{h}$, recall (2). It is this and *only* this variable that is responsible for the conformal non-invariance of GR [7].

Canonical quantization now proceeds by promoting the canonical variables into operators acting on wave functionals $\Psi\left[\bar{h}_{ab}(\mathbf{x}), a(\mathbf{x})\right]$:

$$\hat{\bar{h}}_{ab}(\mathbf{x})\Psi = \bar{h}_{ab}(\mathbf{x})\cdot\Psi, \qquad \hat{a}(\mathbf{x})\Psi = a(\mathbf{x})\cdot\Psi, \tag{11}$$

$$\hat{\bar{p}}^{cd}(\mathbf{x})\Psi = \frac{\hbar}{\mathrm{i}}\frac{\delta}{\delta\bar{h}_{cd}(\mathbf{x})}\Psi, \qquad \hat{p}_a(\mathbf{x})\Psi = \frac{\hbar}{\mathrm{i}}\frac{\delta}{\delta a(\mathbf{x})}\Psi. \tag{12}$$

From the classical Poisson brackets, one gets the standard commutators. The constraints are then promoted to operators annihilating the wave functional, leading to the Wheeler-DeWitt equation and the momentum constraints, respectively,

$$\hat{\mathcal{H}}_{\perp}\Psi = \left[+\frac{\kappa\hbar^2}{12}\frac{\delta^2}{\delta a^2} - \frac{2\kappa\hbar^2}{a^2}\bar{h}_{ik}\bar{h}_{jl}\frac{\delta^2}{\delta\bar{h}_{ij}\delta\bar{h}_{kl}} - \frac{a^4}{2\kappa}\left({}^{(3)}R - 2\Lambda\right) + \hat{\mathcal{H}}^{\mathrm{m}}_{\perp}\right]\Psi \approx 0, \tag{13}$$

$$\hat{\mathcal{H}}_i\Psi = 2\,\mathrm{i}\hbar\bar{D}_k\left(\bar{h}_{ij}\frac{\delta\Psi}{\delta\bar{h}_{jk}}\right) + \frac{\mathrm{i}\hbar}{3}D_i\left(a\frac{\delta\Psi}{\delta a}\right) + \hat{\mathcal{H}}^{\mathrm{m}}_a\Psi \approx 0. \tag{14}$$

IOP Conf. Series: Journal of Physics: Conf. Series **880** (2017) 012002 doi:10.1088/1742-6596/880/1/012002

We have added the parts of the Hamiltonian and momentum constraints $\hat{\mathcal{H}}_\perp^{\rm m}$ and $\hat{\mathcal{H}}_a^{\rm m}$ coming from quantized matter (non-gravitational fields), so that the wave functional now depends on matter fields, too, $\Psi \equiv \Psi\left[\bar{h}_{ab}, a, \phi\right]$.

It is evident from (13) that the Wheeler-DeWitt equation has an indefinite kinetic term, with a playing the role of "intrinsic time". This intrinsic time is, however, constructed from the three-metric, so the Wheeler-DeWitt equation is timeless in the sense of absence of spacetime. That spacetime has disappeared is in full analogy to the absence of classical trajectories in quantum mechanics. The absence of spacetime is symbolized in Figure 1 by the absence of the pointer in the clock. (But the scale is present, waiting to be measured in the semiclassical limit.)

Figure 1. Absence of time in full quantum gravity

The notion of spacetime can be recovered from full canonical quantum GR by a Born-Oppenheimer type of expansion with respect to the Planck-mass squared, $m_{\rm P}^2 = \hbar/G$. In this limit, we obtain the well known theories of quantum field theory in curved spacetime (see e.g. [1] for all technical details). We write the total wave functional in the form

$$\Psi[\bar{h}_{ab}, a, \phi] \equiv \exp\left(\frac{\rm i}{\hbar} S[\bar{h}_{ab}, a, \phi]\right), \tag{15}$$

where ϕ stands for 'matter field', and perform an expansion of S (which is a complex function) with respect to the Planck-mass squared,

$$S[\bar{h}_{ab}, a, \phi] = m_{\rm P}^2 S_0^{\rm E} + S_1^{\rm E} + m_{\rm P}^{-2} S_2^{\rm E} + \ldots .$$

We insert this expansion into the Wheeler-DeWitt equation and compare different orders of $m_{\rm P}^2$. At highest order, $m_{\rm P}^4$, we find that $S_0^{\rm E}$ is independent of ϕ,

$$\frac{\delta S_0^{\rm E}}{\delta \phi} = 0 \quad \Rightarrow \quad S_0^{\rm E} = S_0^{\rm E}[\bar{h}_{ij}, a]. \tag{16}$$

At order $m_{\rm P}^2$, one obtains an equation for $S_0^{\rm E}$:

$$-\frac{1}{12}\left(\frac{\delta S_0^{\rm E}}{\delta a}\right)^2 + \frac{2}{a^2}\bar{h}_{ik}\bar{h}_{jl}\frac{\delta S_0^{\rm E}}{\delta \bar{h}_{ij}}\frac{\delta S_0^{\rm E}}{\delta \bar{h}_{kl}} - \frac{a^4}{2}\left({}^{(3)}R - 2\Lambda\right) = 0\,. \tag{17}$$

Rewriting this in terms of the original variables and using the chain rule for the functional derivative with respect to the metric,

$$\frac{\delta}{\delta h_{ij}} = \frac{\delta a}{\delta h_{ij}}\frac{\delta}{\delta a} + \frac{\delta \bar{h}_{kl}}{\delta h_{ij}}\frac{\delta}{\delta \bar{h}_{kl}} = \frac{1}{6a}\bar{h}^{ij}\frac{\delta}{\delta a} + \frac{1}{a^2}\frac{\delta}{\delta \bar{h}_{ij}}, \tag{18}$$

one realizes that (17) is just the Einstein-Hamilton-Jacobi equation,

$$\frac{2\kappa}{\sqrt{\hbar}}\mathcal{G}_{ikjl}\frac{\delta S^{\mathrm{E}}}{\delta h_{ij}}\frac{\delta S^{\mathrm{E}}}{\delta h_{kl}} - \frac{\sqrt{h}}{2\kappa}{}^{(3)}R = 0, \tag{19}$$

for the Hamilton-Jacobi functional $S^{\mathrm{E}} := S_0^{\mathrm{E}}/\kappa$, where $\mathcal{G}_{ikjl} = (h_{ik}h_{jl} + h_{il}h_{jk} - h_{ij}h_{kl})/2$ is the inverse of the DeWitt supermetric [1]. Since this equation is equivalent to all Einstein equations [8], one has recovered classical GR at this order of semicalssical expansion.

At the next order, m_{P}^0, an equation for S_1^{E} is obtained that can be simplified by writing

$$\psi^{(1)} \equiv D[\bar{h}_{ij}, a]\exp\left(\frac{\mathrm{i}}{\hbar}S_1^{\mathrm{E}}\right) \tag{20}$$

and demanding for D the standard WKB prefactor equation to hold [1].

After some manipulations, one arrives at the following equation for $\psi^{(1)}$:

$$\mathrm{i}\hbar\left[-\frac{1}{24}\frac{\delta S_0^{\mathrm{E}}}{\delta a}\frac{\delta}{\delta a} + \frac{1}{a^2}\bar{h}_{ik}\bar{h}_{jl}\frac{\delta S_0^{\mathrm{E}}}{\delta h_{ij}}\frac{\delta}{\delta h_{kl}}\right]\psi^{(1)} \equiv \mathrm{i}\hbar a\mathcal{G}_{ijkl}\frac{\delta S_0^{\mathrm{E}}}{\delta h_{ij}}\frac{\delta}{\delta h_{kl}}\psi^{(1)} = \hat{\mathcal{H}}_{\perp}^{\mathrm{m}}\psi^{(1)}. \tag{21}$$

The form of this quation reminds one of the Tomonaga-Schwinger equation for $\psi^{(1)}$. Using again (18), the left-hand side of this equation has been rewritten in terms of the original variables (middle expression in (21)), which is well known from the literature [1].

To make the Tomonaga-Schwinger form explicit, we define

$$\frac{\delta}{\delta\tau(\mathbf{x})} := \mathcal{G}_{ijkl}\frac{\delta S_0^{\mathrm{E}}}{\delta h_{ij}}\frac{\delta}{\delta h_{kl}}, \tag{22}$$

thus introducing a local "bubble" (Tomonaga-Schwinger) time functional $\tau(\mathbf{x})$. We can thus write (21) in the form

$$\mathrm{i}\hbar\frac{\delta\psi^{(1)}}{\delta\tau} = \hat{\mathcal{H}}_{\perp}^{\mathrm{m}}\psi^{(1)}, \tag{23}$$

also known as Tomonaga-Schwinger equation. Note that $\hat{\mathcal{H}}_{\perp}^{\mathrm{m}}$ in (23) differs from $\hat{\mathcal{H}}_{\perp}^{\mathrm{m}}$ in (22) by a factor of a, which also induces the same relative rescaling of the bubble time. This is just a consequence of using unimodular-conformal variables.

It is of interest to have a closer look at the bubble time in unimodular-conformal variables,

$$\frac{\delta}{\delta\tau(\mathbf{x})} := -\frac{1}{24a}\frac{\delta S_0^{\mathrm{E}}}{\delta a}\frac{\delta}{\delta a} + \frac{1}{a^3}\bar{h}_{ik}\bar{h}_{jl}\frac{\delta S_0^{\mathrm{E}}}{\delta h_{ij}}\frac{\delta}{\delta h_{kl}}. \tag{24}$$

We see from this that there are two main components, a derivative along the direction (in functional configuration space) of the scale density a and a component along the five directions of the conformal (shape) part $\bar{h}_{ij}$. The scale part is the part that corresponds to intrinsic time in the full Wheeler-DeWitt equation (13). We will call the two *independent* contributions to the bubble time as "scale time" and "shape time", respectively. Figure 2 symbolizes the emergence of time (the pointer) in the semiclassical approximation. While the picture shows only scale time, the full semiclassical time contains, of course, also the shape part.

We finally note that τ is not a scalar function, because this would be in contradiction to the commutator of the matter Hamiltonian densities at different space point [9]. A functional Schrödinger equation can be obtained from the Tomonaga-Schwinger equation after choosing a particular foliation and integrating over space. The next order of the Born-Oppenheimer expansion, m_{P}^{-2}, leads to genuine quantum gravitational effects; for example, corrections to the power spectrum of the CMB anisotropies [10].

IOP Conf. Series: Journal of Physics: Conf. Series **880** (2017) 012002 doi:10.1088/1742-6596/880/1/012002

Figure 2. Emergence of semiclassical time

3. Quantum Weyl-squared gravity and the absence of scale

In this section, we address Weyl-squared gravity as a model of a conformally invariant gravitational theory. It is defined by the action

$$S^{\mathrm{W}} := -\frac{\alpha_{\mathrm{W}}\hbar}{4}\int \mathrm{d}^4x\sqrt{-g}\,C_{\mu\nu\lambda\rho}C^{\mu\nu\lambda\rho}, \tag{25}$$

where $C_{\mu\nu\lambda\rho}$ is the Weyl tensor. This action is invariant under the Weyl transformations

$$g_{\mu\nu}(x) \to \tilde{g}_{\mu\nu}(x) = \Omega^2(x)g_{\mu\nu}(x). \tag{26}$$

We have written this action in a form that is appropriate for discussing the semiclassical approximation of its quantum version: α_{W} is a dimensionless constant, and it is the quantity $S^{\mathrm{W}}/\hbar$ that is relevant for the semiclassical expansion. We do not expect the Weyl theory to be a classical alternative to GR, but it may be of relevance in the early universe and at the Planck scale, where scales may become irrelevant [11].

We note that (25) defines a theory with higher derivatives, which demands an enlargement of the configuration space (see [5] for details and references to earlier work). Here, we have to add the extrinsic curvature K_{ab} as an additional independent canonical variable besides the three-metric h_{ab}. This can be accomplished by adding a constraint to the original Lagrangian which takes care of the definition (1). In unimodular-conformal variables, the new Lagrangian becomes [5]

$$\mathcal{L}_c^{\mathrm{W}} = \mathcal{L}^{\mathrm{W}} - a^5\lambda^{ij\mathrm{T}}\left[2\bar{K}_{ij}^{\mathrm{T}} - \frac{1}{\bar{N}}\left(\dot{\bar{h}}_{ij} - 2\left[\bar{D}_{(i}\bar{N}_{j)}\right]^{\mathrm{T}}\right)\right] - 2a^3\lambda\left[\bar{K} - \frac{1}{\bar{N}}\left(\frac{\dot{a}}{a} - \frac{1}{3}D_aN^a\right)\right]\Bigg\}, \tag{27}$$

where "T" stands for the traceless part of the corresponding quantity. For the canonical momenta, we find from (27):

$$p_{\bar{N}} = \frac{\partial\mathcal{L}_c^{\mathrm{W}}}{\partial\dot{\bar{N}}} \approx 0\,, \quad p^i = \frac{\partial\mathcal{L}_c^{\mathrm{W}}}{\partial\dot{N}^i} \approx 0\,, \quad \bar{P} = \frac{\partial\mathcal{L}_c^{\mathrm{W}}}{\partial\dot{\bar{K}}} \approx 0\,, \tag{28}$$

$$\bar{p}^{ij} = \frac{\partial\mathcal{L}_c^{\mathrm{W}}}{\partial\dot{\bar{h}}_{ij}} = a^5\lambda^{ij\mathrm{T}}\,, \quad p_a = \frac{\partial\mathcal{L}_c^{\mathrm{W}}}{\partial\dot{a}} = 2a^2\lambda\,, \tag{29}$$

$$\bar{P}^{ij} = \frac{\partial\mathcal{L}_c^{\mathrm{W}}}{\partial\dot{\bar{K}}_{ij}^{\mathrm{T}}} = -\alpha_{\mathrm{W}}\hbar\,\bar{h}^{ia}\bar{h}^{jb}\bar{C}_{ab}^{\mathrm{T}}. \tag{30}$$

Note that the momenta $\bar{p}^{ij}$ and $\bar{P}^{ij}$ are traceless. The quantity $\bar{C}_{ab}^{\mathrm{T}}$ is the traceless part of the "electric" part of the Weyl tensor [5].

IOP Conf. Series: Journal of Physics: Conf. Series **880** (2017) 012002 doi:10.1088/1742-6596/880/1/012002

The constraint analysis proceeds in the standard way. In addition to the primary constraints with respect to lapse and shift (as known from GR), we have here an additional primary constraint given by

$$\bar{P} = \frac{\partial \mathcal{L}_c^{\mathrm{W}}}{\partial \dot{\bar{K}}} \approx 0.$$

The presence of this constraint suggests that $\bar{K}$ is arbitrary, in the same manner as $p_{\bar{N}} \approx 0$ and $p_i \approx 0$ suggest that $\bar{N}$ and N^i are arbitrary. The secondary constraints follow from the conservation of the primary constraints and read (attention is here restricted to the vacuum case)

$$\mathcal{H}_\perp^{\mathrm{W}} = -\frac{\bar{h}_{ik}\bar{h}_{jl}\bar{P}^{ij}\bar{P}^{kl}}{2\alpha_{\mathrm{W}}\hbar} + \left({}^{(3)}\bar{R}_{ij}^{\mathrm{T}} + \partial_i \bar{D}_j\right)\bar{P}^{ij} + 2\bar{K}_{ij}^{\mathrm{T}}\bar{p}^{ij} - \alpha_{\mathrm{W}}\hbar \bar{C}_{ijk}^2 \approx 0, \tag{31}$$

$$\mathcal{H}_i^{\mathrm{W}} = -2\bar{D}_k\left(\bar{h}_{ij}\bar{p}^{jk}\right) - 2\bar{D}_k\left(\bar{K}_{ij}^{\mathrm{T}}\bar{P}^{jk}\right) + \bar{P}^{jk}\bar{D}_i\bar{K}_{jk}^{\mathrm{T}} \tag{32}$$

$$\mathcal{Q}^{\mathrm{W}} = a p_a \approx 0. \tag{33}$$

A combination of the secondary constraint $\mathcal{Q}^{\mathrm{W}}$ and the primary constraint $\bar{P}$ generates conformal transformations (of a and $\bar{K}$) [12, 5]. Due to the enlargement of the configuration space, the structure of these constraints differs from those of GR, see (9) and (10). We note, in particular, that the intrinsic time in (9), which is related to the determinant of the three-metric, is absent here because the scale density is arbitrary due to conformal invariance. This will later be connected with the emergence of pure shape time in the semiclassical limit.

Canonical quantization now proceeds in a manner analogously to (11) and (12), taking into account the enlarged configuration space. Adding a (conformally coupled) matter part symbolized by ϕ, we arrive at a quantum wave functional $\Psi[\bar{h}_{ij}, a, \bar{K}_{ij}^{\mathrm{T}}, \bar{K}, \phi]$, which is subject to the following constraints:

$$\hat{\mathcal{H}}_\perp^{\mathrm{W}}\Psi = 0\,, \quad \hat{\mathcal{H}}_i^{\mathrm{W}}\Psi = 0\,, \quad \hat{\bar{P}}\Psi = 0\,, \quad \hat{\mathcal{Q}}^{\mathrm{W}}\Psi = 0. \tag{34}$$

The first equation is the "Weyl-Wheeler-DeWitt equation" (WWDW equation) and reads

$$\left[\frac{\hbar}{2\alpha_{\mathrm{W}}}\bar{h}_{ik}\bar{h}_{jl}\frac{\delta^2}{\delta\bar{K}_{ij}^{\mathrm{T}}\delta\bar{K}_{kl}^{\mathrm{T}}} - \mathrm{i}\hbar\left({}^{(3)}\bar{R}_{ij} + \partial_i\bar{D}_j\right)^{\mathrm{T}}\frac{\delta}{\delta\bar{K}_{ij}^{\mathrm{T}}} - 2\mathrm{i}\hbar\bar{K}_{ij}^{\mathrm{T}}\frac{\delta}{\delta\bar{h}_{ij}} - \alpha_{\mathrm{W}}\hbar\bar{C}_{ijk\perp}^2 + \hat{\mathcal{H}}_\perp^m\right]\Psi = 0\,. \tag{35}$$

The new quantum momentum constraints are given by

$$\mathrm{i}\hbar\left[2\bar{D}_k\left(\bar{h}_{ij}\frac{\delta\Psi}{\delta\bar{h}_{jk}}\right) + 2\bar{D}_k\left(\bar{K}_{ij}^{\mathrm{T}}\frac{\delta\Psi}{\delta\bar{K}_{jk}^{\mathrm{T}}}\right) - \bar{D}_i\bar{K}_{jk}^{\mathrm{T}}\frac{\delta\Psi}{\delta\bar{K}_{jk}^{\mathrm{T}}}\right] + \hat{\mathcal{H}}_i^{\mathrm{m}}\Psi = 0. \tag{36}$$

As for the classical constraints, the WWDW equation is structurally different from the Wheeler-DeWitt equation (13). Its dynamics is not only determined by the three-metric, but also by the extrinsic curvature (which classically corresponds to the time evolution of the three-metric). But there is no scale a and thus no intrinsic time. This is symbolized in Figure 3. We also note that in the vacuum case $\hbar$ drops out of these equations.

The third and fourth of the above constraints show that the wave functional does not depend on a and $\bar{K}$,

$$\frac{\delta\Psi}{\delta\bar{K}} = 0\,, \quad a\frac{\delta\Psi}{\delta a} = 0, \tag{37}$$

and is thus conformally invariant (apart possibly from a phase). This is the most important difference to GR, especially in the light of the semiclassical approximation which we will discuss in the next section.

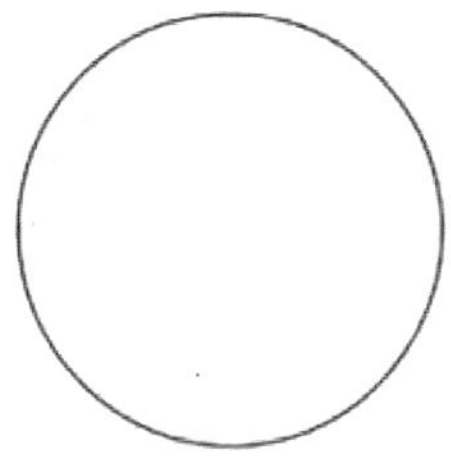

Figure 3. Absence of scale and absence of time

4. Recovery of (shape) time in the semiclassical limit

In this section, we will discuss the semiclassical approximation scheme for the WWDW equation [4]. While in quantum GR (section 2) m_{P}^{-2} was used as the expansion parameter, this role will be played here by α_{W}^{-1}.

Let us consider Weyl gravity with a conformally coupled matter field ϕ. We restrict ourselves to this case (massless vector or conformally coupled scalar field) because the constraints will remain first class and a and $\bar{K}$ will remain arbitrary. The general case is discussed in [6]. Writing

$$\Psi\left[\bar{h}_{ij}, \bar{K}^{\mathrm{T}}_{ij}, \phi\right] \equiv \exp\left(\frac{\mathrm{i}}{\hbar} S\left[\bar{h}_{ij}, \bar{K}^{\mathrm{T}}_{ij}, \phi\right]\right),$$

and performing an expansion of S with respect to the (dimensionless) variable α_{W}^{-1},

$$S = \alpha_{\mathrm{W}} \sum_{n=0}^{\infty} \left(\frac{1}{\alpha_{\mathrm{W}}}\right)^{n} S^{\mathrm{W}}_{n} \tag{38}$$

we find the following equations at consecutive orders of α_{W}:

- α_{W}^{2}: S^{W}_{0} is independent of ϕ, that is,

$$\left(\left(\hat{\mathcal{H}}^{m}_{\perp}\Psi\right)/\Psi\right)^{(2)} = 0 \quad \Rightarrow \quad \frac{\delta S^{\mathrm{W}}_{0}}{\delta\phi} = 0. \tag{39}$$

- α_{W}^{1}: This gives the Weyl-Hamilton-Jacobi equation for S^{W}_{0}. We first find

$$-\frac{1}{2\hbar}\bar{h}_{ik}\bar{h}_{jl}\frac{\delta S^{\mathrm{W}}_{0}}{\delta \bar{K}^{\mathrm{T}}_{ij}}\frac{\delta S^{\mathrm{W}}_{0}}{\delta \bar{K}^{\mathrm{T}}_{kl}} + \left({}^{(3)}\bar{R}_{ij} + \partial_i \bar{D}_j\right)^{\mathrm{T}}\frac{\delta S^{\mathrm{W}}_{0}}{\delta \bar{K}^{\mathrm{T}}_{ij}} + 2\bar{K}^{\mathrm{T}}_{ij}\frac{\delta S^{\mathrm{W}}_{0}}{\delta \bar{h}_{ij}} - \hbar\bar{C}^{2}_{ijk} = 0. \tag{40}$$

Relabelling $S^{\mathrm{W}} \equiv \alpha_{\mathrm{W}} S_0$, this is recognized as the Hamilton-Jacobi (HJ) equation for the Weyl theory after substituting

$$\bar{P}^{ij} \to \frac{\delta S^{\mathrm{W}}}{\delta \bar{K}^{\mathrm{T}}_{ij}}, \quad \bar{p}^{ij} \to \frac{S^{\mathrm{W}}}{\delta \bar{h}_{ij}} \tag{41}$$

into the Hamiltonian constraint (31). The classical solutions to the HJ equation are thus determined by the conformal parts of both the three-metric and the extrinsic curvature. It would be interesting to show (as expected) that the Weyl-HJ equation is equivalent to the Bach equations (equations of motion for the Weyl theory) in analogy to Gerlach's [8] work for GR, but we assume this equivalence here without proof.

- α_{W}^{0}: The equation for S_1^{W} can be simplified by introducing

$$\psi^{(1)} \equiv D[\bar{h}_{ij}, \bar{K}_{ij}^{\mathrm{T}}] \exp\left(\frac{\mathrm{i}}{\hbar} S_1^{\mathrm{W}}\right) \tag{42}$$

and demanding the "WKB prefactor equation" to hold for D, in analogy to the procedure in quantum GR. After the elimination of D, one is left with the following equation for $\psi^{(1)}$:

$$\mathrm{i}\hbar\left[-\frac{1}{2\hbar}\bar{h}_{ik}\bar{h}_{jl}\frac{\delta S_0^{\mathrm{W}}}{\delta \bar{K}_{kl}^{\mathrm{T}}}\frac{\delta}{\delta \bar{K}_{ij}^{\mathrm{T}}} + \left({}^{(3)}\bar{R}_{ij} + \partial_i \bar{D}_j\right)^{\mathrm{T}} \frac{\delta}{\delta \bar{K}_{ij}^{\mathrm{T}}} + 2\bar{K}_{ij}^{\mathrm{T}}\frac{\delta}{\delta \bar{h}_{ij}}\right]\psi^{(1)} = \hat{\mathcal{H}}_{\perp}^{\mathrm{m}}\psi^{(1)}, \tag{43}$$

which should be compared with (21) above. We can now define the following "bubble Weyl time"

$$\frac{\delta}{\delta \tau_{\mathrm{W}}(\mathbf{x})} := -\frac{1}{2\hbar}\bar{h}_{ik}\bar{h}_{jl}\frac{\delta S_0^{\mathrm{W}}}{\delta \bar{K}_{kl}^{\mathrm{T}}}\frac{\delta}{\delta \bar{K}_{ij}^{\mathrm{T}}} + \left({}^{(3)}\bar{R}_{ij} + \partial_i \bar{D}_j\right)^{\mathrm{T}} \frac{\delta}{\delta \bar{K}_{ij}^{\mathrm{T}}} + 2\bar{K}_{ij}^{\mathrm{T}}\frac{\delta}{\delta \bar{h}_{ij}}, \tag{44}$$

and one arrives at the "Weyl-Tomonaga-Schwinger" equation for $\psi^{(1)}$,

$$\mathrm{i}\hbar\frac{\delta \psi^{(1)}}{\delta \tau_{\mathrm{W}}} = \hat{\mathcal{H}}_{\perp}^{\mathrm{m}}\psi^{(1)}. \tag{45}$$

Comparing the Weyl bubble time τ_{W} with the τ in (22), we immediately recognize the absence of a and $\bar{K}$. But Weyl time has additional components compared with (22): time passes as a result of the dynamics in the extended configuration space. Since there is no scale part, the Weyl time is scale-less or, in other words, it is a pure shape time. This will be different if, for example, an R^2 term is added; the additional term leads to a scale time, so that the full action gives the total "higher-derivative" bubble time. The emergence of a scale-less time in the Weyl theory is symbolized in Figure 4. Here, time is only defined from the semiclassical *shape* degrees of freedom.

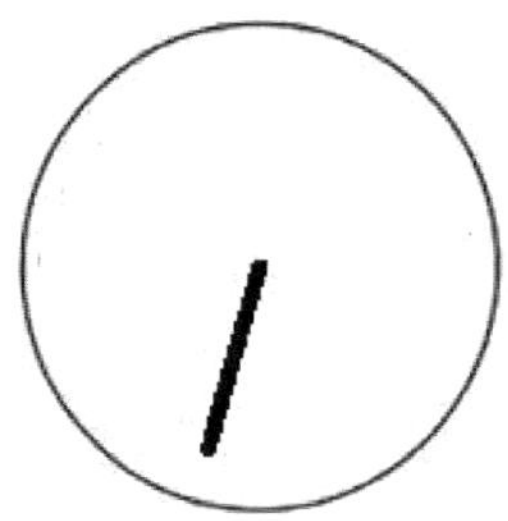

Figure 4. Emergence of scale-less time

We note that, as in the case of quantum GR, τ_{W} is not a scalar function [9]. We also note that $\psi^{(1)}$ is conformally invariant and that higher orders in α_{W}^{-1} give quantum gravitational correction terms.

An important consequence from these discussions is that semiclassical time is ambiguous with respect to the underlying theory. Both quantum GR and quantum Weyl gravity action can give rise (formally) to semiclassical time. The qualitative difference between the two times cannot be determined solely from the (functional) Schrödinger equation; one needs to go to higher orders for a (in principle) measurable distinction to emerge.

Another interesting question is if scale-ful and shape-ful times can emerge from a scale-less quantum gravity theory. This possibility would imply a breaking of conformal symmetry in the transition from Planckian to lower energy scales. One can represent this question visually as in Figure 5. This question will be discussed in [6].

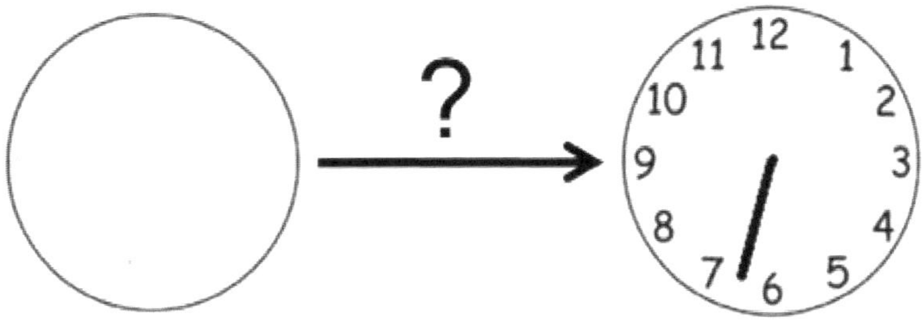

Figure 5. Emergence of full WKB time from a scale-less quantum theory of gravity?

5. Conclusion

Let us briefly summarize our main points as follows:

- The canonical quantization of Weyl gravity can be performed analogously to GR, but the structure of the configuration space is different: it consists of both the three-metric and the extrinsic curvature.
- We have performed a consistent decomposition of variables into scale part and conformally invariant part. The constraints can be rewritten in terms of the conformally invariant parts.
- We have displayed and discussed the Weyl-Wheeler-DeWitt equation and the Weyl diffeomorphism constraints. There are also new constraints which show that the wave functional is conformally invariant.
- We have discussed the semiclassical expansion and the recovery of time for both quantum GR and the quantum Weyl theory. Whereas in the former we get both a scale and a shape time, the latter only leads to a shape time.

References

[1] Kiefer C 2012 *Quantum Gravity* third edition (Oxford: Oxford University Press)
[2] Kiefer C 2009 *Gen. Rel. Grav.* **41** 877
[3] Schrödinger E 1926 *Ann. Phys. (Berlin)* **384** 489
[4] Kiefer C and Nikolić B 2017 in *Gravity and the Quantum* ed Bagla J S and Engineer S (Berlin: Springer)
[5] Kiefer C and Nikolić B 2017 *Phys. Rev. D* **95** 084018
[6] Kiefer C and Nikolić B 2017 Conformal and Weyl-Einstein gravity: Quantum geometrodynamics *in preparation*
[7] Nikolić B 2017 Treating the Einstein-Hilbert action as a higher derivative Lagrangian: revealing the missing information about conformal non-invariance *Preprint* gr-qc/1702.06337 *These Proceedings*
[8] Gerlach U H 1969 *Phys. Rev.* **177** 1929
[9] Giulini D and Kiefer C 1995 *Class. Quantum Grav.* **12** 403
[10] Brizuela D, Kiefer C and Krämer M 2016 *Phys. Rev. D* **94** 123527
[11] 't Hooft G 2015 *Int. J. Mod. Phys. D* **24** 1543001
[12] Irakleidou M, Lovrekovic I and Preis F 2015 *Phys. Rev. D* **91** 104037

Quantum Gravity and Cosmology: an intimate interplay

Mairi Sakellariadou

Theoretical Particle Physics and Cosmology Group, Department of Physics, King's College London, University of London, Strand, London, WC2R 2LS, U.K.

E-mail: mairi.sakellariadou@kcl.ac.uk

Abstract. I will briefly discuss three cosmological models built upon three distinct quantum gravity proposals. I will first highlight the cosmological rôle of a vector field in the framework of a string/brane cosmological model. I will then present the resolution of the big bang singularity and the occurrence of an early era of accelerated expansion of a geometric origin, in the framework of group field theory condensate cosmology. I will then summarise results from an extended gravitational model based on non-commutative spectral geometry, a model that offers a purely geometric explanation for the standard model of particle physics.

1. Introduction

Most of the fundamental questions in cosmology - which for a long time belonged to the realm of philosophy - can now be addressed within a physical theory and tested against increasingly accurate observational data. We live the golden age of cosmology. Still, fundamental puzzles remain. The most obvious is *what, if anything, happened before the big bang?* Another is *what is the universe made of?*

Standard cosmology is formulated within the framework of Einstein's general theory of relativity. Notwithstanding, general relativity is not adequate to explain the earliest stages of cosmic existence, and cannot provide an explanation for the big bang itself. The reason is that quantum mechanics must have played a rôle at sufficiently early times. The majority of early universe cosmological models are studied at a semi-classical level, hence they leave basic questions such as the beginning of the universe, and the origin of inflation (or of alternative models) unanswered. Modern early universe cosmology is in the need of a rigorous underpinning in quantum gravity. More precisely, quantum gravity offers the appropriate framework to address fundamental questions such as the origin of the space-time continuum and its effective dynamics, and the pictures of the earliest moments of the universe which emerge. These pictures must include an approximately homogeneous background and superimposed perturbations. In return, cosmological data represent the best (if not the only) chance for testing quantum gravity, thus guiding the formulation of the complete fundamental theory.

Quantum gravity proposals belong to two classes, they are either *top-down* or *bottom-up*. In the former class belong approaches like string/M-theory or a variant of non-perturbative approach to quantum gravity, like loop quantum gravity or group field theory. In the latter one belong the proposal of non-commutative (spectral) geometry, and that of asymptotic safety. String theory may provide the underlying theory required to resolve questions such as the initial

IOP Conf. Series: Journal of Physics: Conf. Series **880** (2017) 012003 doi:10.1088/1742-6596/880/1/012003

big bang singularity, the origin of an early inflationary era or of an alternative to inflation mechanism, and explain space-time dimensionality [1, 2]. Group field theory is a discrete quantum gravity approach which relies on pre-gerometric structures. Its hardest challenge is to retrieve, within an appropriate (semi-classical) limit, the geometry of continuum space-times and the corresponding dynamics as given by (classical) general relativity. A main tenet of non-commutative geometry is that above the Planck scale the whole concept of geometry breaks down and space-time is replaced by a non-commutative manifold. In this approach, below but close to the Planck energy scale, space-time is the product of a Riemannian spin manifold with a finite non-commutative space.

In what follows, I will highlight examples which show the intimate relation between the search for a theory of quantum gravity and the urge to find the appropriate theoretical framework to address questions about the origin and very early moments of our universe and unweave its fabric. I will first highlight a cosmological model based on a microscopic string theory proposal, called the *D-material universe*. I will then present the effective dynamics of a cosmological model, called the *group field theory condensate cosmology*, derived from group field theory.

2. A string theory inspired cosmological model: D-material universe

D-material universe [3] is a modified gravity model involving a fundamental vector field; it may appear as the low-energy limit of certain brane theories.

Consider a compactified (3+1)-brane propagating in a higher dimensional bulk, populated by point-like D0-branes, called D-particles. Since a brane is, by definition, the collection of end-points of open strings, this picture implies that open strings propagate in a medium of D-particles. Thus, brane-puncturing massive D-particles can be captured by electrically neutral matter (to be consistent with the fact that we are living in an electrically neutral universe) open strings. As a result, D-particles recoil and their velocity will locally break Lorentz invariance. The emerged vector-like excitations can lead to an early era of geometric inflation and contribute to large scale structure, enhancing dark matter perturbations [3, 4]. This is an example of a bi-metric theory, there is the sigma model background metric related to the Einstein-frame metric, and a metric describing the distortion of space-time surrounding D-particles [5].

The interaction between stringy matter (living in a brane-world of three longitudinal large dimensions) and a medium of recoiling D-particles, is given using the Dirac-Born-Infeld (DBI) action:

$$\begin{aligned} S_{\text{eff 4D}} &= \int \mathrm{d}^4x \Big[-\frac{\sqrt{-g}}{4} e^{-2\phi} \mathcal{G}_{\mu\nu}\mathcal{G}^{\mu\nu} - \frac{T_3}{g_{s0}} e^{-\phi} \sqrt{-\det(g + 2\pi\alpha' F)}(1-\alpha R(g)) \\ &\quad - \sqrt{-g}\frac{e^{-2\phi}}{\kappa_0^2}\tilde{\Lambda} + \sqrt{-g}\frac{e^{-2\phi}}{\kappa_0^2}R(g) + \mathcal{O}\Big((\partial\phi)^2\Big)\Big] + S_{\text{m}} \ , \end{aligned} \tag{1}$$

where $\mathcal{G}_{\mu\nu}$ is a flux gauge field, T_3 denotes the brane tension, ϕ is the dilaton field assumed to be constant, $g_{\text{s}} = g_{\text{s0}}e^{\phi}$ is the string coupling, which is just g_{s0} for a constant dilaton, g stands for the determinant of the gravitational field, $\tilde{\Lambda}$ is the cosmological constant, κ^{-2} is the four-dimensional bulk induced gravitational constant $\kappa^{-2} = \big(V^{(6)}/g_{\text{s0}}^2\big)M_{\text{s}}^2$ with M_{s} the string scale $M_{\text{s}} = 1/\sqrt{\alpha'}$. The vector field A_μ denotes the recoil velocity excitation during the string-matter with D-particle interactions; its field strength is

$$F_{\mu\nu} \equiv \partial_\mu A_\nu - \partial_\nu A_\mu \ . \tag{2}$$

The vector field A_μ satisfies the constraint

$$A_\mu A_\nu g^{\mu\nu} = -\frac{1}{\alpha'} \ . \tag{3}$$

In the low-energy weak approximation, one can expand the four-dimensional DBI action (1) and get, in the Einstein frame $g^{\rm E}_{\mu\nu} = e^{-2\phi} g_{\mu\nu}$, the effective action

$$S^{\rm E}_{\rm eff\ 4D} = \int {\rm d}^4x\sqrt{-g}\Big[-\frac{T_3 e^{3\phi_0}}{g_{s0}} - \frac{\tilde{\Lambda} e^{2\phi_0}}{\kappa_0^2} - \frac{\tilde{F}^2}{4}(1-\alpha e^{-2\phi_0}R) + \Big(\frac{\alpha T_3 e_0^{\phi}}{g_{s0}} + \frac{1}{\kappa_0^2}\Big)R$$
$$+\lambda\Big(\tilde{A}_\mu\tilde{A}^\mu + \frac{4\pi^2\alpha' T_3 e^{3\phi_0}}{g_{s0}}\Big)\cdots\Big] + S_{\rm m}\ , \qquad (4)$$

where $\tilde{F}_{\mu\nu}$ is the Maxwell field strength for the redefined vector field

$$\tilde{A}_\mu \equiv \Big(\frac{T_3(2\pi\alpha')^2}{g_{s0}} e^{3\phi_0}\Big)^{1/2} A_\mu\ , \qquad (5)$$

and $\alpha = \alpha'\zeta(2) = \alpha'\pi^2/6$. The last term in (4) implements the constraint (3) rewritten for the redefined vector field $\tilde{A}_\mu$ as

$$\tilde{A}_\mu\tilde{A}_\nu g^{\mu\nu} = -\frac{4\pi^2\alpha' T_3 e^{3\phi_0}}{g_{s0}}\ ; \qquad (6)$$

λ is a Lagrange multiplier.

Having the action in the low-energy weak-field approximation, therefore valid at late times, one can derive the graviton, vector field and dilaton equations of motion. The vector field equation of motion fixes the background value of the Lagrange multiplier field:

$$\langle\lambda(x)\rangle = \frac{e^{-3\phi_0} g_{s0}}{8\pi^2\alpha'|T_3|}\tilde{A}_\mu[\tilde{F}_{\nu\mu}(1-\alpha e^{-2\phi_0}R)]^{;\nu}\ , \qquad (7)$$

while considering galactic scales, the dilaton equation of motion fixes the value of the cosmological constant:

$$\Lambda_0 \simeq -\frac{T_3}{2g_{s0}} e^{3\phi_0}\ , \qquad (8)$$

which is negative for a positive brane tension T_3. Note that such anti-de Sitter type terms cancel against dilaton independent contributions to the brane vacuum energy, thus leading to a small positive cosmological constant

$$\Lambda^{\rm vac} \equiv \Lambda_0 + \frac{1}{8}\langle\mathcal{G}_{\mu\nu}\mathcal{G}^{\mu\nu}\rangle + \cdots > 0\ , \qquad (9)$$

during the galactic era.

At sufficiently late scales, so that the effective action (4) is valid, we will analyse gravitational lensing phenomenology. We consider a static spherically symmetric background

$$g_{\alpha\beta}{\rm d}x^\alpha{\rm d}x^\beta = -e^{\nu(\sqrt{x^2+y^2+z^2})}{\rm d}t^2 + e^{\zeta(\sqrt{x^2+y^2+z^2})}a^2(t)({\rm d}x^2+{\rm d}y^2+{\rm d}z^2)\ . \qquad (10)$$

Interactions of D-particles with open strings lead to recoil fluctuations of the former, corresponding to world-sheet deformations of gauge fields, as

$$A_i(\vec{x},t) \simeq \frac{1}{\alpha'} g_{ij}(\vec{x},t)u^j\Big(t\frac{a(t_{\rm c})^2}{a(t)^2} - t_{\rm c}\Big)\ ,\ t > t_{\rm c}\ , \qquad (11)$$

with $t_{\rm c}$ denoting the collision time, which is of the same order of magnitude as the Friedmann-Lemaitre-Robinson-Walker cosmic time, hence

$$t_{\rm c} \sim \frac{2}{H_0[1+(1+z)^2]}\ . \qquad (12)$$

for a galaxy at redshift z. The time of observation t is considered as the present time t_0. Using Eq. (11) with the constraint (3) and the definition (12) with $a(t_c) = a_0/(1+z)$, one finds that the *electric* type field strength components F_{ti},

$$F_{ti}(\vec{x},t) \sim \frac{1}{\alpha'} g_{ij}(\vec{x},t) u^j \left[\frac{1-3(1+z)^2}{(1+z)^2(1+(1+z)^2)} \right] , \tag{13}$$

associated with linear recoil momentum excitations, are much bigger than the *magnetic* type field strength components F_{ij}, corresponding to non-zero angular momentum of recoiling D-particles. Hence the latter may be neglected.

Considering late eras of the cosmological evolution, D-particles are assumed to have recoil velocities following gaussian stochastic statistics:

$$\langle\langle u^m u^n \rangle\rangle = \sigma_0^2(t)\delta^{mn} \ , \ \langle\langle u^m \rangle\rangle = 0 \ , \ \sigma_0^2(t) = [a(t)]^{-3}|\beta| \ , \tag{14}$$

so that macroscopically Lorentz invariance is maintained. The statistical fluctuations are proportional to the cosmic density of defects, with constant of proportionality the statistical variance $|\beta|$ of the recoil velocity:

$$|\beta| \sim \frac{1}{3}\frac{n_{\rm D}}{n_\gamma}\frac{\xi^2 |p_i^{\rm phys}|^2}{M_{\rm s}^2} g_{\rm s0}^2 \ , \tag{15}$$

where $n_{\rm D}$, n_γ denote respectively the number density of D-particles and radiation, ξ is a space-time local constant fudge factor, given by the microscopic theory and in general smaller than unity. Assuming scattering of D-particles is mainly with cosmic photons, p_i is the presently observed average energy density of cosmic microwave background (CMB) photons:

$$\sqrt{|p_i^{\rm physCMB}|^2} \sim 7 \times 10^{-4} {\rm eV} \ . \tag{16}$$

The string scale is $10^4{\rm GeV} \leq M_{\rm s} \leq 10^{18}{\rm GeV}$.

One then explores the parameter space of $|\beta|$, so that D-particles can play the rôle of dark matter candidates, while at the same time they can provide the seed for large scale structure. A detailed analysis in Ref. [3], combining lensing data with an estimate of the required densities so that the D-matter recoil-velocity fluid can mimic dark matter in galaxies, in the sense that its contribution to the energy density is of the same order as the mass density of a galaxy, led to the constraint

$$10^{-58} M_{\rm Pl}^4 \leq T_3 \leq 10^{-26} M_{\rm Pl}^4 \ . \tag{17}$$

At this point, let us emphasise that in this model we do not suggest a mechanism to replace dark matter, since we have for instance neutrinos, instead in our scenario the rôle of dark matter is enhanced by the vector field.

Moreover, neutrinos may appear as dark matter candidates that could be captured by D-particles. After the capture by the D-particle defect, the emerging stringy matter excitation may have a different flavor than its original one. Thus, D-particle populations in galaxies may act as a *medium* inducing flavor oscillations. Considering only $\nu_e \leftrightarrow \nu_\mu$ oscillations and computing the average of the neutrino stress tensor with respect to the flavor vacuum, the obtained [5] extra time-dependent dark matter energy contribution is

$$\Omega_\Lambda^{\nu_{\rm mixing}} \sim 0.24 \ . \tag{18}$$

D-particles may also induce a successful inflationary mechanism through condensation of their recoil velocity field. Such a mechanism is compatible with the latest Planck data provided the condensate fields are large ($\sigma(t) \gg 1$), namely we have dense D-particle populations in the early universe, which one expects to be sufficiently diluted today. This regime is appropriate for low string mass scales with respect to the hubble inflationary scale:

$$M_s \ll H_I \sim 10^{-5} M_{Pl} \ll M_{Pl} \ , \tag{19}$$

using current Planck data.

Using finite temperature formalism (Euclidean time) to account for the Hawking-Gibbons temperature (associated with an observer-dependent horizon) of a de Sitter space-time, one gets [3] the Euclidean DBI action and then through analytic continuation one recovers a Minkowski space. The Euclidean DBI action reads

$$S_{\rm eff\ 4D} \simeq \int d^4x \sqrt{\tilde{g}}\, \frac{1}{\kappa_0^2} \left[R(\tilde{g}) + \frac{1}{2} \partial_\mu \varphi\, \partial^\mu \varphi - \frac{\kappa_0^2}{4}\, \mathcal{G}_{\mu\nu} \mathcal{G}^{\mu\nu} - \frac{e^{-\sqrt{\frac{2}{3}}\varphi}}{\alpha} - \left(\tilde{\Lambda} - \frac{1}{\alpha} \right) e^{-2\sqrt{\frac{2}{3}}\varphi} \right] , \tag{20}$$

where we have redefined the metric as

$$\tilde{g}_{\mu\nu} = (1+\sigma) g_{\mu\nu} \ , \tag{21}$$

and φ denotes a canonically normalised scalar field:

$$\varphi = \sqrt{\frac{3}{2}} \ln(1 + \sigma(t)) \ . \tag{22}$$

Assuming the flux field condensates into a constant one, which contributes to the vacuum energy as

$$\frac{1}{4} \langle\langle \mathcal{G}_{\mu\nu}\, \mathcal{G}^{\mu\nu} \rangle\rangle \equiv \mathcal{D} \ , \tag{23}$$

the *Euclideanised* effective potential for the canonically normalised scalar field, reads

$$V^{\mathcal{E}} = -\kappa_0^2 \mathcal{D} - \frac{e^{-\sqrt{\frac{2}{3}}\varphi}}{\alpha} - \left(\tilde{\Lambda} - \frac{1}{\alpha} \right) e^{-2\sqrt{\frac{2}{3}}\varphi} \ . \tag{24}$$

Performing analytic continuation back to Minkowski, teh potential is approximated by

$$V(\varphi) \simeq \kappa_0^2 \mathcal{D} + \left(\tilde{\Lambda} - \frac{1}{\alpha} \right) e^{-2\sqrt{\frac{2}{3}}\tilde{\varphi}} + i\, \frac{e^{-\sqrt{\frac{2}{3}}\tilde{\varphi}}}{\alpha} \ , \tag{25}$$

with $\tilde{\varphi}$ real. The imaginary part in the effective potential above denotes an instability; it has the following important physical meaning. The field rolls down, the condensate becomes small, the imaginary part disappears, and consequently one can expand the square-root of the DBI action and recover the effective action valid at late eras, which we have used earlier to study gravitational lensing and growth of matter perturbations.

The real part of the effective potential (25) is [3]

$$\mathrm{Re}\, V(\tilde{\varphi}) = \tilde{\mathcal{D}} + \left(\tilde{\Lambda} - \frac{1}{\alpha} \right) e^{-2\sqrt{\frac{2}{3}}\tilde{\varphi}} \ , \quad \tilde{\mathcal{D}} \equiv \kappa_0^2 \mathcal{D} \ , \tag{26}$$

with $(\tilde{\Lambda} - 1/\alpha)$ negative relative to $\tilde{\mathcal{D}}$. The above potential (26) is of Starobinsky type, provided one can tune the flux-field condensate to be $\tilde{\mathcal{D}} > 0$ and such that the minimum of the potential occurs for the field value $\tilde{\varphi} = 0$ and corresponds to zero potential.

IOP Conf. Series: Journal of Physics: Conf. Series **880** (2017) 012003 doi:10.1088/1742-6596/880/1/012003

Fixing the spectral index to $n_s = 0.965$ we get $N = 57.7$ e-folds, leading to the slow-roll parameters ϵ, ξ, η [3]

$$\epsilon \simeq 5.6 \times 10^{-5} \ll 1 , \qquad \eta \simeq -1.7 \times 10^{-2} \ll 1 , \qquad \xi \simeq 3.0 \times 10^{-4} \ll 1 , \tag{27}$$

consistent with current CMB data. The CMB constraint on V/ϵ leads to

$$\mathcal{D} \simeq 3.2 \times 10^{-11} \, M_{\rm Pl}^4 \qquad \Leftrightarrow \qquad \tilde{\mathcal{D}} \simeq 3.2 \times 10^{-11} \, M_{\rm Pl}^2 \, . \tag{28}$$

It is worth noting that it is the gauge field flux condensate $\mathcal{G}_{\mu\nu}\mathcal{G}^{\mu\nu}$ that induces a de Sitter phase (positive, almost constant, vacuum energy), and hence inflation, but it is the recoiling D-particles velocity vector field that induces a slowly rolling scalar degree of freedom that allows exit of inflation.

3. A non-perturbative approach to QG: Group Field Theory Condensate Cosmology

3.1. The setup

Group Field Theory (GFT) [6, 7], a second quantisation formulation of loop quantum gravity, states that spacetime and geometry emerge through an effective description of the collective behaviour of pre-geometric degrees of freedom. This approach conjectures that the continuous macroscopic space-time we are familiar with, emerges through a condensation of bosonic GFT quanta, each representing an atom of space. Within this context, classical gravitational dynamics is recovered through an effective description of the GFT condensate dynamics.

A GFT model is defined by a choice of three ingredients: a group that plays the rôle of the local gauge group of gravity, a fundamental building block of space, and a field with its corresponding action encoding the dynamics. The main choices so far are respectively, the gauge group SU(2) of Ashtekar-Barbero gravity, a quantum tetrahedron created or annihilated by second quantised field operators, and a complex scalar field with an action given by a kinetic quadratic term and a sum of interaction polynomials weighted by corresponding coupling constants. The interaction terms implement the way the quanta of geometry are *glued* with each other.

The basic idea of GFT is that all data encoded in the fields is only of combinatorial and algebraic nature. Hence, GFT is a background-independent field theory. In the simplest case, the data, storing geometric information, is made of group elements attached to the edges of a graph. The data correspond to the holonomies of a connection and the fluxes of a triad field, which are the fundamental variables of loop quantum gravity. The group elements can be seen as parallel transports of a gravitational connection along the edges of a graph.

Being interested in building a four-dimensional quantum gravity, the complex scalar field φ lives on $d = 4$ copies of the G =SU(2) group: $\varphi(g_I) : G^d \to \mathbb{C}$, with $I = 1, \cdots, d$, where the group elements g_I correspond to parallel transports on the gravitational connection along a link. In addition, one imposes invariance of the field φ under the simultaneous right multiplication of all $g_I \in G$ by the same group element h, for all $h \in G$: $\varphi(g_1 h, \cdots, g_d h) = \varphi(g_1, \cdots, g_d), \forall h \in G$. The classical action can be written as

$$S[\varphi, \bar{\varphi}] = \int_G (dg)^d \int_G (dg')^d \bar{\varphi}(g_I) \mathcal{K}(g_I, g'_I) \varphi(g_I) + \mathcal{V}[\varphi, \bar{\varphi}] \, , \tag{29}$$

where $\mathcal{K}$ is the kinetic kernel and $\mathcal{V}$ represents the interaction term; the former is typically local whereas the latter is in general a nonlinear and nonlocal convolution of φ with itself. The above action leads to the classical equation of motion

$$\int_G (dg')^d \mathcal{K}(g'_I, g_I) \varphi(g_I) + \frac{\delta \mathcal{V}}{\delta \bar{\varphi}(g_I)} = 0 \, , \tag{30}$$

IOP Conf. Series: Journal of Physics: Conf. Series **880** (2017) 012003 doi:10.1088/1742-6596/880/1/012003

for the GFT field. In the quantum version of the theory, the dynamics is obtained from the partition function

$$Z = \int [\mathcal{D}\varphi][\mathcal{D}\bar{\varphi}]e^{-S[\varphi,\bar{\varphi}]} \ . \tag{31}$$

A no-space state, which therefore does not include any topological or quantum geometric information, is defined by the Fock vacuum:

$$\hat{\varphi}(g_I)|\emptyset\rangle = 0 \ . \tag{32}$$

The excitation of a GFT quantum over the Fock vacuum, written as

$$|g_I\rangle = \hat{\varphi}^\dagger(g_I)|\emptyset\rangle \ , \tag{33}$$

corresponds to the creation of a single open four-valent loop quantum gravity spin network vertex. One can thus construct N-particle states, which are needed to describe extended three-geometries, and then use second-quantised hermitian operators to obtain quantum geometric observable data for these states.

Group Field Cosmology (GFC) [8] is the quantum cosmological model based on GFT, in the sense that the condensate phase is identified with the continuum approximately smooth and spatially homogeneous space-time of our universe. In this context one applies mean field techniques for the corresponding GFT states and obtains an effective dynamics to which one attempts a cosmological interpretation. The geometric interpretation that one may have in mind is that of a gas of tetrahedra, each of them in the same microscopic quantum state, defined by the condensate mean field $\sigma(g_I)$ with

$$\langle\sigma|\hat{\varphi}|\sigma\rangle = \sigma \neq 0 \ . \tag{34}$$

Separate the field operator into the condensate mean field $\sigma(g_I)$ and the non-condensate contribution $\delta\hat{\varphi}(g_I)$ as

$$\hat{\varphi}(g_I) = \sigma(g_I) + \delta\hat{\varphi}(g_I) \ , \tag{35}$$

and consider the coherent (as eigenstate of the field operator) state

$$|\sigma\rangle = e^{-\frac{1}{2}\int (dg)^d|\sigma(g_I)|^2} \ e^{\hat{\sigma}}|\emptyset\rangle \ \ \text{with} \ \hat{\sigma} = \int (dg)^d\sigma(g_I)\hat{\varphi}^\dagger(g_I) \ , \tag{36}$$

constructed from quantum tetrahedra which encode the same quantum geometric information. The effective dynamics is obtained from the quantum cosmology equation

$$\frac{\delta S[\sigma,\bar{\sigma}]}{\delta\bar{\sigma}(g_I)} = \int (dg')^d\mathcal{K}(g_I,g'_I)\sigma(g'_I) + \frac{\delta\mathcal{V}}{\delta\bar{\sigma}(g_I)} = 0 \ , \tag{37}$$

with the action as in Eq. (29) but now written in terms of the condensate field σ, while the kinetic kernel is

$$\mathcal{K} = \delta(g'_I g_I^{-1})\delta(\phi' - \phi)\left[-(\tau\partial_\phi^2 + \sum_{I=1}^{4}\Delta_{g_I}) + m^2\right], \quad \tau > 0 \ , \tag{38}$$

where ϕ is a massless scalar field, playing the rôle of a relational clock [9]. Note that Eq. (37) is the analogue of Gross-Pitaevskii equation for real Bose condensates.

IOP Conf. Series: Journal of Physics: Conf. Series **880** (2017) 012003 doi:10.1088/1742-6596/880/1/012003

Consider first a static mean field $\sigma(g_I, \phi) = \sigma(g_I)$ and neglect all interactions setting $\mathcal{V} = 0$ as in Ref. [10]. The corresponding equation, in an isotropic restriction, can be expressed in terms of the angle parametrisation ψ as

$$-\Big[\frac{d^2}{d\psi^2} + 2\cot(\psi)\frac{d}{d\psi}\Big]\sigma(\psi) + 2\mu\sigma(\psi) = 0\ , \quad \psi \in \Big[0, \frac{\pi}{2}\Big]\ , \tag{39}$$

with $\mu \equiv m^2/12$. Assuming as a boundary condition, the *near-flatness* condition

$$\sigma(\psi)|_{\psi=\frac{\pi}{2}} = 0\ , \tag{40}$$

the eigensolutions of Eq. (39) are

$$\sigma_j(\psi) = \frac{\sin((2j+1)\psi)}{\sin(\psi)}\ , \quad \psi \in \Big[0, \frac{\pi}{2}\Big]\ , \tag{41}$$

with $j \in \frac{2\mathbb{N}_0+1}{2}$ corresponding to the eigenvalues $\mu = -2j(j+1)$. The probability density is concentrated around small ψ, namely around small curvature values [10], implying that the building blocks of geometry are almost flat. Studying the normalised spectrum of the loop quantum gravity volume operator

$$V = V_0 \sum_{m\in\mathbb{N}_0/2} |\sigma_{j;m}|^2 V_m \ \text{ with } \ V_m \sim m^{3/2}\ ,\ V_0 \sim l_{\rm Pl}^3\ , \tag{42}$$

(written in terms of Fourier coefficients $\sigma_{j;m}$) with respect to the eigensolutions $\sigma_j(\psi)$, we find that eigensolutions for smaller j have a bigger volume, while the volume remains finite for all j. Thus, a general solution of the quantum cosmology equation, decomposed in terms of eigensolutions, describes a finite space with the largest contributions from the low spin modes. The same conclusion is drawn studying the normalised spectrum of the area operator

$$A = A_0 \sum_{m\in\mathbb{N}_0/2} |\sigma_{j;m}|^2 A_m \ \text{ with } \ A_m \sim (m(m+1))^{1/2}\ ,\ V_0 \sim l_{\rm Pl}^2\ , \tag{43}$$

with respect to the eigensolutions $\sigma_j(\psi)$.

Let us now turn on the interaction term to study the impact of simplified interactions onto static GFT quantum gravity condensates which are thought to describe continuous three-geometries [10]. Consider the effective potential

$$V_{\rm eff}(\sigma) = \frac{\mu}{2}\sigma^2 + \frac{\kappa}{4}\sigma^4\ , \tag{44}$$

with $\mu < 0$ and $\kappa > 0$ in order to get a nontrivial (nonperturbative) vacuum with nonzero vacuum expectation value, $\langle\sigma\rangle \neq 0$, so that there is agreement with the condensate state ansatz. The system would settle into one of the two minima $\langle\sigma_0\rangle = \pm\sqrt{-\mu/\kappa}$ and describe a condensate. Considering the interaction term as a perturbation of the free case, one then analyses the effect of perturbations onto the spectra of geometric operators. Studying the probability density of the interacting mean field over ψ, we conclude that the finiteness of the free solutions at the origin is lost due to the interactions. However, the concentration of the probability densities around the origin can still be maintained, giving rise to nearly flat solutions, as long as $|\kappa|$ does not become too big. Studying the normalised spectrum of the volume and area operators with respect to the interacting mean field σ_j, we conclude that perturbations increase both the volume and the

IOP Conf. Series: Journal of Physics: Conf. Series **880** (2017) 012003 doi:10.1088/1742-6596/880/1/012003

area, but remain finite in the weakly nonlinear regime; these effects are more pronounced for small values of j. We have shown the validity of these results for

$$V_{\rm eff}(\sigma) = \frac{\mu}{2}\sigma^2 + \frac{\kappa}{4}\sigma^5 , \tag{45}$$

with either ($\mu < 0, \kappa > 0$) or ($\mu < 0, \kappa < 0$) in order to have a nontrivial nonperturbative vacuum. To investigate the condensate phase we have also studied the solutions around the nontrival minima, and conclude that the condensate consists of many discrete building blocks predominantly of the smallest nontrivial size. Summarising, a free condensate configuration in an isotropic restriction settles dynamically into a low-spin configuration of the quantum geometry. Hence, our analysis supports the proposal that an effectively continuous geometry may emerge from the collective behaviour of a discrete pre-geometric GFT substratum.

One may then lift the isotropic restriction to investigate anisotropic GFT condensate configurations. Studying the probability density of the mean field, we have found that anisotropies play an important rôle only at small values of the relational clock (i.e., at small volumes), whereas at late times the isotropic mode dominates [14].

3.2. Cosmological consequences through the effective Friedmann equation

The effective action to describe the dynamics of a GFT condensate reads

$$S = \int d\phi (A|\partial_\phi \sigma|^2 + \mathcal{V}(\sigma)) , \tag{46}$$

where σ is a complex scalar field representing the configuration of the Bose condensate of GFT quanta as a function of the massless scalar field ϕ that plays the rôle of relational time, and ρ_j is the modulus of the component of σ corresponding to the spin-j representation of SU(2). For convenience, we separate σ_j into its modulus ρ_j and phase θ_j as $\sigma(\phi) = \rho_j e^{i\theta_j}$.

Assuming the interactions between GFT quanta as being sub-dominant, the corresponding equations for the modulus and phase of σ_j read [10]

$$\begin{aligned} \rho_j'' - [m_j^2 + (\theta_j')^2]\rho_j &\approx 0 , \\ 2\rho_j'\theta_j' + \rho_j\theta_j'' &\approx 0 , \end{aligned} \tag{47}$$

leading to two constant charges

$$\begin{aligned} E_j &\approx (\rho_j')^2 + \rho_j^2(\theta_j')^2 - m_j^2\rho_j^2 , \\ Q_j &\approx \rho_j^2\theta_j' ; \end{aligned} \tag{48}$$

the former called the *GFT energy*.

The expectation values for the total volume and momentum of the massless scalar field operators with respect to ϕ, evaluated on isotropic GFT condensate states read:

$$\begin{aligned} V(\phi) &= \sum_{j\in\mathbb{N}_0/2} V_j[\rho_j(\phi)]^2 , \\ \pi_\phi(\phi) &= \frac{\hbar}{2i}\sum_j \left(\bar\sigma_j(\phi)\sigma_j(\phi)' - \bar\sigma_j(\phi)'\sigma_j(\phi)\right) . \end{aligned} \tag{49}$$

Since $\partial\pi_\phi/\partial\phi = 0$, which is the continuity equation on cosmology for the case of a massless scalar field, one concludes that the effective Friedmann and acceleration equations in the semi-classical limit read:

$$\frac{\partial_\phi V}{V} = 2\frac{\partial_\phi \rho}{\rho} \equiv 2g(\phi) , \tag{50}$$

$$\frac{\partial^2_\phi V}{V} = 2\left[\frac{\partial^2_\phi \rho}{\rho} + \left(\frac{\partial_\phi \rho}{\rho}\right)^2\right] , \tag{51}$$

respectively. Note that in Eq. (51) we have omitted the index j in ρ_j and hence wrote $V(\phi)$ as $V(\phi) = V_j\rho^2(\phi)$, since as we have shown earlier, the condensate field peaks on a particular representation j.

The evolution equation of the universe in the classical limit of GFT can be written as an effective Friedmann equation:

$$H^2 = \left(\frac{\partial_\phi V}{3V}\right)^2 \dot{\phi}^2 \equiv \frac{8}{9}g^2\epsilon , \tag{52}$$

where ϵ plays the rôle of an energy density. One can thus define an *effective gravitational constant* $G_{\text{eff}} = g^2/(3\pi)$, from the collective behaviour of space-time quanta.

The proposal of an effective gravitational constant is not new, and as it has been shown in the literature, it may need to interesting consequences [11]. In particular, quantum geometry effects may lead to stochastic fluctuations of the gravitational constant, which can be thus considered as a macroscopic effective dynamical quantity. In consequence, a time-dependent dark energy term in the modified field equation can be expressed in terms of a time-dependent dynamical gravitational constant. As a result, the late-time accelerated expansion of the universe may be ascribed to quantum fluctuations in the geometry of space-time rather than the vacuum energy from the matter sector [11].

From Eq. (50) it is clear that there is a value of ϕ, call it Φ, for which $g(\phi)$ vanishes, and hence at relational time Φ there is a *bounce*, where the energy density reaches a maximum value

$$\epsilon_{\text{max}} = \frac{1}{2}\frac{Q^2}{V^2_{\text{bounce}}} , \tag{53}$$

with V_{bounce} the volume at the bounce:

$$V_{\text{bounce}} = \frac{V_{j0}(\sqrt{E^2 + 12\pi G Q^2} - E)}{6\pi G} , \tag{54}$$

which gives the minimum (non-zero) value of the volume. Note that the singularity is always avoided for $E < 0$ and provided Q is non-zero, it is also avoided for $E > 0$.

Next issue we would like to address is whether one can get an early era of geometric-type inflation, meaning an early era of accelerated expansion through the different geometric background and not via the introduction of an inflaton field with an *ad hoc* potential. In standard (classical) cosmology, there is a scale factor a, leading to a volume $V = a^3$ and a *proper time* t. In the GFT context there is no metric, neither proper time, nevertheless we will be able to define an equivalent to $d^2a/dt^2 > 0$ condition, via the volume V. More precisely, within the cosmological context derived from GFT, we have

$$\frac{\ddot{a}}{a} = \frac{1}{3}\left(\frac{\pi_\phi}{V}\right)^2\left[\frac{\partial^2_\phi V}{V} - \frac{5}{3}\left(\frac{\partial_\phi V}{V}\right)^2\right] , \tag{55}$$

leading to the following condition for inflation

$$\frac{V''}{V} > \frac{5}{3}\left(\frac{V'}{V}\right)^2 , \tag{56}$$

IOP Conf. Series: Journal of Physics: Conf. Series **880** (2017) 012003 doi:10.1088/1742-6596/880/1/012003

which is valid even in the absence of space-time and the absence of proper time. Hence, within the GFT condensate cosmology, the condition for inflation reads [12]

$$4m^2 + \frac{2E}{\rho^2} > \frac{20}{3}g^2 \ , \tag{57}$$

and near the bounce the l.h.s. of the above equation is positive while the r.h.s. is zero. In conclusion, there is an early era of geometric inflation, namely an accelerated expansion in the absence of an inflaton field, for either $E < 0$ or $E > 0$; in the latter case the accelerated expansion is followed by a maximal deceleration [12].

Certainly the next question is to study whether such a geometric inflation can last sufficiently long, therefore one has to calculate the number of e-folds N defined here as

$$N = \frac{2}{3}\log\left(\frac{\rho_{\rm end}}{\rho_{\rm bounce}}\right) \ . \tag{58}$$

A careful analysis shows that in this case, where interactions have been neglected, the number of e-folds is $0.119 \leq N \leq 0.186$ for all possible values of m^2 and Q^2. Thus, GFT cosmology in the absence of interactions between building blocks cannot replace the standard inflationary scenario. In the following we will turn on the interactions, which is indeed the more natural and consistent scenario, as the quanta of geometry must be somehow glued with each other. Depending on our results we may also be able to give a prescription for the way one must build the GFT model and assign specific type of interactions in order to get at the semi-classical limit the desired properties for the space-time that must be then seen as the emergent 3-geometry of our homogeneous and isotropic universe [13].

Let us consider the effective action (46) for an isotropic GFT condensate, and take [13]

$$\mathcal{V}(\sigma) = B|\sigma(\phi)|^2 + \frac{2}{n}w|\sigma|^n + \frac{2}{n'}w'|\sigma|^{n'} \ , \text{with} \quad n' > n \ . \tag{59}$$

Such a choice is motivated from *spin foam* models for four-dimensional quantum gravity, which are mostly based on interaction terms of power 5 of the modulus field, as well as from *tensor* models, which are based on even powers of σ; the former are called *simplicial* and the latter *tensorial*.

A detailed study has shown [13] that the occurrence of a bounce and an early epoch of accelerated expansion found in the free theory, both hold in the interacting case. In the latter case, the solutions are cyclic motions describing oscillations around a stable minimum, hence interactions induce recollapse leading to cyclic cosmologies. Furthermore, in the interacting case, one can find the range of the parameter space for which inflation will last sufficiently long. Note that to have a successful inflationary scenario one will need to impose an extra condition, namely that there is no intermediate stage of deceleration between the bounce and end of inflation. Hence, one concludes that for an interacting potential, as in (59), with

$$\lambda < 0 \quad \text{and} \quad n \geq 5 \ (n' > n) \ ,$$

GFT cosmology can lead to an inflation-like era.

At this points let me make some remarks about the validity of the above results for a GFT model based on a real field. As analysed in Ref. [14], for real-valued condensate fields, solutions which avoid the singularity problem and grow exponentially after the bounce can only be found for negative GFT energy E. Studying the stability properties of an evolving isotropic system, it was shwon that the condensate models give rise to effectively continuous, homogeneous and isotropic three-geometries built from many smallest and almost flat building blocks of the quantum geometry. Finally, studying the stability properties of an evolving anisotropic system, it was shown the isotropisation for increasing values of the clock (i.e., the volume), and that the singularity avoidance is not altered by the occurrence of anisotropies.

IOP Conf. Series: Journal of Physics: Conf. Series **880** (2017) 012003 doi:10.1088/1742-6596/880/1/012003

4. The gravitational sector of Noncommutative Spectral Geometry

Here, one follows the approach that in order to construct a quantum theory of gravity coupled to matter, the gravity-matter interaction is the most important ingredient to determine the dynamics. The aim will be to guess the small-scale structure of space-time from our knowledge at the electro-weak scale, which is well-studied by collider experiments. We consider the Standard Model (SM) of particle physics, the most successful particle physics model we have at hand, as a phenomenological model which dictates the space-time geometry. Close but below the Planck scale, gravity and the SM fields are thought to be packaged into geometry and matter on a Kaluza-Klein non-commutative space-time, a four-dimensional space-time with an internal discrete zero dimensionality-space attached to each point. Thus, space is defined as the product of a four-dimensional compact Riemannian spin manifold $\mathcal{M}$ with an internal Kaluza-Klein space $\mathcal{F}$. Such simple non-commutative spaces $\mathcal{M}\times\mathcal{F}$, where the non-commutative algebra describing space is the algebra of functions over ordinary space-time, are called *almost commutative manifolds* [15].

The almost commutative manifold $\mathcal{M}\times\mathcal{F}$ is defined by the canonical spectral triple $(\mathcal{A}, \mathcal{H}, D)$, in terms of an algebra, a Hilbert space and a Dirac operator, as [16]:

$$\mathcal{M}\times\mathcal{F} := (C^\infty(\mathcal{M}, \mathcal{A}_\mathcal{F}), L^2(\mathcal{M}, S)\otimes\mathcal{H}_\mathcal{F}, \not{D}\otimes\mathbb{I}+\gamma_5\otimes\mathcal{D}_\mathcal{F}) ;$$

the set $C^\infty(\mathcal{M})$ of square-integrable spinors S on $\mathcal{M}$ forms an algebra under point-wise multiplication acting on the Hilbert space $L^2(\mathcal{M}, S)$ as multiplication operators, $\not{D}$ is the Dirac operator $-i\gamma^\mu\nabla^s_\mu$ acting as first order differential operator on the spinors, the operator γ_5 plays the rôle of a chirality operator. In addition one considers the antilinear isomorphism $J_\mathcal{M}$ as the charge conjugation operator on spinors. The internal discrete space $\mathcal{F}$ is defined by the spectral triple $(\mathcal{A}_\mathcal{F}, \mathcal{H}_\mathcal{F}, D_\mathcal{F})$; it encodes the internal degrees of freedom at each point in space-time. The Hilbert space $\mathcal{H}_\mathcal{F}$ is a 96×96 matrix, written in terms of 3×3 Yukawa mixing matrices and a real constant responsible for neutrino mass terms. To recover the SM, the algebra $\mathcal{A}_\mathcal{F}$ must be appropriately chosen. It was shown that [17]

$$\mathcal{A}_\mathcal{F} = M_a(\mathbb{H})\oplus M_k(\mathbb{C}) ,$$

with k an even number $k=2a$. The choice $k=4$ is the smallest possible value that produces correctly 16 fermions in each of the three generations.

It is important to note the physical meaning of the choice of such a non-commutative manifold. The spectral triple

$$(\mathcal{A}, \mathcal{H}, \mathcal{D}, J, \gamma) = (C^\infty(\mathcal{M}), L^2(\mathcal{M}, S), \partial\!\!\!/_\mathcal{M}, J_\mathcal{M}, \gamma_5)\otimes(\mathcal{A}_\mathcal{F}, \mathcal{H}_\mathcal{F}, \mathcal{D}_\mathcal{F}, J_\mathcal{F}, \gamma_\mathcal{F}) ,$$

defining $\mathcal{M}\times\mathcal{F}$, is the product

$$(\mathcal{A}_1, \mathcal{H}_1, \mathcal{D}_1, J_1, \gamma_1)\otimes(\mathcal{A}_2, \mathcal{H}_2, \mathcal{D}_2, J_2, \gamma_2) ,$$

with

$$\mathcal{A} = \mathcal{A}_1\otimes\mathcal{A}_2\ ,\ \mathcal{H} = \mathcal{H}_1\otimes\mathcal{H}_2\ ,\ \mathcal{D} = \mathcal{D}_1\otimes 1+\gamma_1\otimes\mathcal{D}_2\ ,\gamma = \gamma_1\otimes\gamma_2\ ,\ J = J_1\otimes J_2\ .$$

This *doubling* of the algebra is essential in order to lead to the SM physics. As we have shown in Refs. [18, 19] this doubling is intimately related to dissipation, gauge field structure, neutrino mixing, while it incorporates the seeds of quantisation.

To explore another physical meaning of the almost-commutative manifold, let us identify it with a pair of four-dimensional Minkowski space-times embedded in a five-dimensional Minkowski geometry. Following this approach, we have considered [20] fermions travelling within the light cone of the ambient five-dimensional space-time, and then derived the energy-momentum dispersion relation, thus arguing for its non-commutative geometry origin.

To write down the action we will use the spectral action principle, according which the bosonic action functional depends only on the spectrum of the fluctuated Dirac operator $\mathcal{D}_A$ (a Dirac operator with a gauge connection):

$$\mathcal{D}_A = \mathcal{D} + A + \epsilon' JAJ^{-1} , \tag{60}$$

with A a self-adjoint operator, J an anti-unitary operator and $\epsilon \in \{\pm 1\}$, and is of the form

$$\mathrm{Tr}(f(\mathcal{D}_A^2/\Lambda^2)) , \tag{61}$$

with f a cut-off function and Λ the cut-off energy scale at which the action holds. This action sums up all eigenvalues of the fluctuated Dirac operator $\mathcal{D}_A$ which are smaller than the cut-off energy scale Λ. We evaluate the trace using heat kernel techniques, in terms of the Seeley-de Witt coefficients a_n; its asymptotic expansion reads

$$\mathrm{Tr}(f(\mathcal{D}_A^2/\Lambda^2)) \sim 2f_4\Lambda^4 a_0(\mathcal{D}_A^2) + 2f_2\Lambda^2 a_2(\mathcal{D}_A^2) + f(0)a_4(\mathcal{D}_A^2) + \mathcal{O}(\Lambda^{-2}) . \tag{62}$$

Hence, the cut-off function f plays a rôle through only three of its momenta, f_4, f_2, f_0, which are respectively related to the cosmological constant, the gravitational constant and the coupling constants at unification. One must then add the fermionic part of the action:

$$(1/2)\langle J\Psi, \mathcal{D}_A\Psi\rangle \ ; \ \Psi \in \mathcal{H}^+ . \tag{63}$$

After a long calculation, the bosonic part of the spectral action can be schematically written as [16]

$$\begin{aligned} S_\Lambda \ = \ & \int d^4x\sqrt{g}\left(A_1\Lambda^4 + A_2\Lambda^2\left(\frac{5}{4}R - 2y_t^2H^2 - M^2\right)\right. \\ & +A_3\left(g_2^2 W^\alpha_{\mu\nu}W^{\alpha\ \mu\nu} + g_3^2 G^a_{\mu\nu}G^{a\ \mu\nu} + \frac{5}{3}g_1^2 B_{\mu\nu}B^{\mu\nu}\right) \\ & \left. +\text{other } \mathcal{O}(\Lambda^0) \ +\text{other } \mathcal{O}\left(\Lambda^{-2}\right)\right) , \end{aligned} \tag{64}$$

where y_t is the Yukawa coupling for the top quark; R is the curvature scalar; W, G, B are curvature tensors for the three interactions and M is (up to a numerical factor) a heavy Majorana right-handed neutrino mass. The A_i with $i = 1, 2, 3$ are constants which depend on the details of the cut-off function f and for typical choices of that function the three constants are not too different from unity. Finally, the g_1, g_2, g_3 are the corresponding gauge couplings of the three interactions; they satisfy $g_2^2 = g_3^2 = (5/3)g_1^2$, which hold in several Grand Unified Theories (GUTs), hence the cut-off scale Λ can be considered as the GUT scale, $\Lambda \sim (10^{14} - 10^{17})$ GeV.

It is worth noting that very few gauge groups fit into the spectral action model. The standard model does as well as the Pati-Salam group $\mathrm{SU}(2)_L\times$ $\mathrm{SU}(2)_R\times$ SU(4), but SO(10), of which the former group is an intermediate breaking stage, does not. The absence of larger groups with novel representation is a positive outcome since it prevents proton decay.

Using renormalisation group techniques one then derives predictions for the SM phenomenology; the results are in agreement with current particle physics data, including the Higgs mass [21].

The gravitational sector of the theory coupled to matter, in Euclidean signature, reads [16]

$$\begin{aligned} \mathcal{S}_\Lambda = \int \Big(& \frac{1}{2\kappa_0^2}R + \alpha_0 C_{\mu\nu\rho\sigma}C^{\mu\nu\rho\sigma} + \gamma_0 + \tau_0 R^\star R^\star + \frac{1}{4}G^i_{\mu\nu}G^{\mu\nu i} + \frac{1}{4}F^\alpha_{\mu\nu}F^{\mu\nu\alpha} \\ & +\frac{1}{4}B^{\mu\nu}B_{\mu\nu} + \frac{1}{2}|D_\mu\mathbf{H}|^2 - \mu_0^2|\mathbf{H}|^2 - \xi_0 R|\mathbf{H}|^2 + \lambda_0|\mathbf{H}|^4\Big)\sqrt{g}\ d^4x , \end{aligned} \tag{65}$$

where

$$\kappa_0^2 = \frac{12\pi^2}{96 f_2 \Lambda^2 - f_0 \mathfrak{c}} \quad , \quad \alpha_0 = -\frac{3 f_0}{10\pi^2} \, ,$$
$$\gamma_0 = \frac{1}{\pi^2}\left(48 f_4 \Lambda^4 - f_2 \Lambda^2 \mathfrak{c} + \frac{f_0}{4}\mathfrak{d}\right) \quad , \quad \tau_0 = \frac{11 f_0}{60\pi^2} \, ,$$
$$\mu_0^2 = 2\Lambda^2 \frac{f_2}{f_0} - \frac{\mathfrak{c}}{\mathfrak{a}} \quad , \quad \xi_0 = \frac{1}{12} \, ,$$
$$\lambda_0 = \frac{\pi^2 \mathfrak{b}}{2 f_0 \mathfrak{a}^2} \quad , \quad \mathbf{H} = (\sqrt{a f_0}/\pi)\phi \; ; \tag{66}$$

$\mathbf{H}$ a rescaling of the Higgs field ϕ to normalise the kinetic energy, and the momentum f_0 is physically related to the coupling constants at unification.

The action (65) is obtained from the *cut-off* bosonic spectral action (61). It contains the Einstein-Hilbert action with a cosmological term, a topological term, a conformal gravity term with the Weyl curvature tensor, as well as conformal coupling of the Higgs field to gravity. The coefficients of the gravitational terms depend upon the Yukawa parameters of the particle physics content. Given that the cut-off scale is $\Lambda \sim (10^{14} - 10^{17})$ GeV, the action (65) provides a natural framework to address early universe cosmology.

Since this action contains higher derivative terms, one may wonder whether this gravitational theory is plagued by linear instabilities, namely by the appearance of negative energy modes. We have shown [22] that there exists a class of almost commutative torsion geometry that leads to a Hamiltonian which is bounded from below and hence argued that the theory does not suffer from linear instabilities.

Studying the low-energy weak curvature regime, where one may neglect the non-minimal coupling term between the background geometry and the Higgs field, we have shown that any modifications to the background equation may appear at leading order only for anisotropic and inhomogeneous models [23]. As energies are approaching the Higgs scale, the non-minimal coupling of the Higgs field to the curvature can no longer be neglected. In the latter case, the coupling plays the rôle of an effective gravitational constant; alternatively, the coupling increases the self-interaction of the Higgs field. Note that one observes a similarity with compactified string models and with chameleon cosmology.

Exploring the cosmological rôle of the Higgs field [24, 25], which has conformal coupling to gravity, we have calculated the renormalisation of the Higgs self-coupling up to two-loops and shown that while the Higgs potential can lead to slow-roll inflation, the CMB constraints turn the predictions of such a scenario incompatible with the measured value of the top quark mass [25].

The gravitational sector of the bosonic spectral action may be seen as an extended theory of gravity. Studying its astrophysical consequences one may constrain f_0 (or equivalently $a_0 = -3f_0/(10\pi^2)$), one of the three free parameters.

Consider linear perturbations around Minkowski background [26] and study the propagation of gravitational waves. Comparing with data from binary pulsar systems, for which the rate of change of the orbital frequency is well-known, one finds [27]

$$\beta \gtrsim 7.55 \times 10^{-13}\mathrm{m}^{-1} \, .$$

This constraint is improved using Gravity Probe B [28]:

$$\beta \gtrsim 7.1 \times 10^{-5}\mathrm{m}^{-1} \, ,$$

or LARES data [29]:

$$\beta \gtrsim 1.2 \times 10^{-6}\mathrm{m}^{-1} \, . \tag{67}$$

However, the strongest constraint on β is obtained through torsion balance experiments [28]:

$$\beta \gtrsim 10^4 \mathrm{m}^{-1} . \tag{68}$$

To obtain the cut-off bosonic spectral action we have made an expansion in $1/\Lambda^2$. This expansion is however valid when the fields and their derivatives are small with respect to the cut-off scale Λ, hence the spectral action (62) is valid only in the weak-field approximation. To cure this problem, we have proposed a novel definition of the bosonic spectral action using zeta function regularisation:

$$S_\zeta \equiv \lim_{s\to 0} \mathrm{Tr}\, D^{-2s} \equiv \zeta(0, D^2) . \tag{69}$$

The ζ function is well defined and given by the a_4 heat kernel coefficient associated with the Laplace type operator D^2:

$$\begin{aligned} S_\zeta &= a_4\left[D^2\right] = \int d^4x\,\sqrt{g}\,L\ , \\ \text{with}\quad L(x) &= a_4(D^2, x)\ . \end{aligned} \tag{70}$$

Following this approach one can address issues of renormalisability and spectral dimensions. More precisely, in the zeta spectral action there are no operators of dimension higher than four, hence the theory is renormalisable. The zeta spectral action can be used up to the Planck scale where the very nature of space-time changes due to quantum gravitational effects. Moreover, the ζ spectral action exhibits viable spectral dimensions. More precisely, for Higgs scalars and gauge bosons the spectral dimensions coincide with the topological one and equal four, while in the gravitational sector the spectral dimension equals two, which implies improved ultraviolet behavior of the gravitational propagators.

5. Conclusions

To address remaining open questions in early universe cosmology we need a quantum gravity theory, while in return cosmology is the best (if not the only) way to test the validity of quantum gravity proposals. I have highlighted three cosmological models built upon three quantum gravity proposals from string/brane theory, group field theory and non-commutative geometry. I have shown how some issues may be resolved, in particular regarding an early era of accelerated expansion, growth of matter perturbations and resolution of the big bang singularity. I have then argued how one can get a purely geometric explanation for the standard model of particle physics, which still remains the best particle physics model we have at hand.

Given these promising results, one may further explore predictions of quantum gravity models which will allow to adjust their free parameters, modify them or even disprove them. A synergy between such proposals is required. For instance, one may investigate whether there is a semi-classical regime for which the effective dynamics of GFT can reduce to the dynamics of another non-perturbative approach to quantum gravity, like loop quantum cosmology. In addition, one may investigate the connection between non-perturbative approaches to quantum gravity, where the pre-geometric degrees of freedom are algebraic in nature, and non-commutative spectral geometry, which may help to quantise the latter.

References

[1] Brandenberger R and Vafa C 1989 *Nucl. Phys.* B **316** 391; Sakellariadou M 1986 *Nucl. Phys.* B **468** (1996) 319
[2] Durrer R, Kunz M and Sakellariadou M 2005 *Phys. Lett.* B **614** 125; Nelson W and Sakellariadou M 2008 *Phys. Rev.* D **78** 104012; Nelson W and Sakellariadou M 2009 *Phys. Lett.* B **674** 210
[3] Elghozi T, Mavromatos N, Sakellariadou M and Yusaf M F 2016 *JCAP* **1602** 060

[4] Mavromatos N, Sakellariadou M and Yusaf M F 2013 *JCAP* **1303** 015
[5] Mavromatos N and Sakellariadou M 2007 *Phys. Lett.* B **652** 97
[6] Freidel L 2005 *Int. J. Theor. Phys.* **44** 1769; Oriti D 2009 The Group field theory approach to quantum gravity, in Approaches to quantum gravity *Preprint* gr-qc/0607032
[7] Oriti D 2014 Group Field Theory and Loop Quantum Gravity *Preprint* gr-qc/1408.7112
[8] Gielen S, Oriti D and Sindoni L 2013 *Phys. Rev. Lett* **111** 031301; Gielen S, Oriti D and Sindoni L 2014 *JHEP* **1406** (2014) 013
[9] Oriti D, Sindoni L and Wilson-Ewing E 2016 *Class. Quant. Grav.* **33** 224001
[10] Pithis A, Sakellariadou M and Tomov P 2016 *Phys. Rev.* D **94** 064056
[11] de Cesare M, Lizzi F and Sakellariadou M 2016 *Phys. Lett.* B **760** 498
[12] de Cesare M and Sakellariadou M 2017 *Phys. Lett.* B **764** 49
[13] de Cesare M, Pithis A and Sakellariadou M 2016 *Phys. Rev.* D **94** 064051
[14] Pithis A and Sakellariadou M 2017 *Phys. Rev.* D **95** 064004
[15] Connes A 1994 *Noncommutative Geometry* (New York: Academic Press); Connes A and Marcolli M 2008 *Noncommutative Geometry, Quantum Fields and Motives* (India: Hindustan Book Agency)
[16] Chamseddine A H, Connes A and Marcolli M 2007 *Adv. Theor. Math. Phys.* **11** 991
[17] Chamseddine A H and Connes A 2007 *Phys. Rev. Lett.* **99** 191601
[18] Sakellariadou M, Stabile A and Vitiello G 2011 *Phys. Rev.* D **84** 045026
[19] Gargiulo M V, Sakellariadou M and Vitiello G 2014 *Eur. Phys. J.* C **74** 2695
[20] Sakellariadou M and Watcharangkool A 2017 *Phys. Rev.* D **95** 025027
[21] Stephan C A 2009 *Phys. Rev.* D **79** 065013; Chamseddine A H and Connes A 2012 *JHEP* **1209** 104; Devastato A, Lizz Fi and Martinetti P 2014 *JHEP* **1401** 042
[22] Sakellariadou M and Watcharangkool A 2016 *Phys. Rev.* D **93** 064034
[23] Nelson W and Sakellariadou M 2010 *Phys. Rev.* D **81** 085038
[24] Nelson W and Sakellariadou M 2009 *Phys. Lett.* B **680** 263
[25] Buck M, Fairbairn M and Sakellariadou M 2010 *Phys. Rev.* D **82** 043509
[26] Nelson W, Ochoa J and Sakellariadou M 2010 *Phys. Rev.* D **82** 085021
[27] Nelson W, Ochoa J and Sakellariadou M 2010 *Phys. Rev. Lett.* **105** 101602
[28] Lambiase G, Sakellariadou M and Stabile A 2013 *J. Cosm. & Astrop. Phys.* **12** 020
[29] Capozziello S, Lambiase G, Sakellariadou M, Stabile An and Stabile Ar 2015 *Phys. Rev.* D **91** 044012
[30] Kurkov M A, Lizzi F, Sakellariadou M and Watcharangkool A 2015 *Phys. Rev.* D **91** 065013

IOP Conf. Series: Journal of Physics: Conf. Series **880** (2017) 012004 doi:10.1088/1742-6596/880/1/012004

Fractal spacetimes in stochastic gravity? – *views from anomalous diffusion and the correlation hierarchy*

B L Hu

Maryland Center for Fundamental Physics and Joint Quantum Institute, University of Maryland, College Park, Maryland, 20742, USA

E-mail: blhu@umd.edu

Abstract. We explore in stochastic gravity theory whether non-Gaussian noises from the higher order correlation functions of the stress tensor for quantum matter fields when back-reacting on the spacetime may reveal hints of multi-scale structures. Anomalous diffusion may depict how a test particle experiences in a fractal spacetime. The hierarchy of correlations in quantum matter field induces the hierarchy of correlations in geometric objects via the set of Einstein-Langevin equations for each correlation order. This correlation hierarchy kinetic theory conceptual framework, aided by the characteristics of stochastic processes, may serve as a conduit for connecting the low energy 'Bottom-Up' approach with the 'Top-Down' theories of quantum gravity which predict the appearance of fractal spacetimes at the Planck scale.

1. Introduction

There are three levels of inquiry here: 1) stochastic gravity (in the restricted sense) [1, 2]: Gaussian noises associated with the second-order correlations in the stress energy tensor of quantum matter field backreacting on the spacetime induce metric fluctuations via the Einstein-Langevin equation [3]. 2) What has not yet been done, but doable in principle, is the higher order correlations which generate non-Gaussian noises in the quantum matter field and their backreaction on the spacetime dynamics. This is stochastic gravity theory (in a broader sense), as originally intended [4]. The new quest posed by the title question is: 3) Whether their backreaction effects near the Planck scale, at least in principle, may indicate / permit the existence of fractal (multi-scale) spacetimes?

Where would anomalous diffusion fit in? Recall the original formulation of stochastic gravity was inspired by quantum Brownian motion [5] describing normal diffusion, where Gaussian noise associated with the second-order correlations in the stress energy tensor of the quantum matter field can be defined exactly. Here, for non-Gaussian noises associated with the higher order correlations, we appeal to anomalous diffusion processes to see if spacetimes with multi-scale features may appear. This would entail a generalization of the Laplace-Beltrami operator governing normal wave propagation in a smooth curved manifold to some operator of more complex structure describing wave propagation in a fractal spacetime. Following this line of inquiry we give a quick description of each component in this conceptual scheme below. We then take a step back to look at the bigger picture and recap the kinetic theory [6] 'Bottom-Up' approach to quantum gravity [7] where correlation hierarchies both in the matter and the spacetime sectors take the center stage. We then examine dimensional reduction and fractal spacetime near the Planck scale, review the basics of anomalous diffusion, how it can act as a probe for fractal structures, then finally discuss the procedures from stochastic gravity.

1. Backreaction: Semiclassical Gravity and Stochastic Gravity

A. Backreaction of quantum matter fields – the vacuum expectation value of their stress energy tensor – on the background spacetime dynamics is the main theme in *semiclassical gravity* begun in the late 70s. The central equation is the semiclassical Einstein equation (SCEq).

B. Backreaction of the *fluctuations* of quantum matter fields – the vacuum expectation value of the stress energy bitensor $T_{mn}(x)T_{rs}(x')$ – on the spacetime dynamics is the main theme in *stochastic semiclassical gravity* initiated in 1994. The central equation is the Einstein-Langevin Equation (ELEq), with the two point function of T_{mn} of quantum fields, now living in the spacetime which is a solution of the SCEq, acting as the source driving the Einstein Equation.

2. Noise: Moments of Stress-Energy Tensor

A. Gaussian noise: For the two-point function of the stress-energy bitensor , one can use the Feynman-Vernon Gaussian functional identity [8] to express it as classical stochastic force: noise. Using them as source for the Einstein-Langevin equation gives *stochastic gravity in the restricted sense* (SG2, 2 for second correlation order)

For higher-point functions of the stress-energy tensor there is no identity, thus they cannot be expressed as simple noise. However, their innate qualities of fluctuations and correlations remain, and their importance unabated, in fact may increase, as we shall see.

B. Higher moments: As shown by Fewster, Ford, Roman (FFR) [9] and others, even at low energy, in Minkowski spacetime, under test field condition (no backreaction on spacetime) the higher moments of T_{mn} contribute significantly.

They may be expressed as *correlation noises* [10] which are predominantly non-Gaussian, i.e., noises associated with the higher order correlation functions in the correlation hierarchy (e.g., the BBGKY-Boltzmann hierarchy for classical gas, the Schwinger-Dyson hierarchy for interacting quantum fields). Considering the backreactions of the higher moments of T_{mn} extending to the whole correlation hierarchy is the task of *stochastic gravity theory in the broader sense* (SGn, n for nth correlation order).

3. Anomalous Diffusion:

A. non-Gaussian noise: Where do they arise? Nonlinearity.

Example: quantum Brownian motion (QBM) beyond the bilinear coupling order [11], interacting quantum fields [12], can be treated by perturbative techniques, both produce multiplicative colored noise.

B. Anomalous diffusion: A great variety of stochastic processes studied in various fields, see, e.g., [13]. They have very different spectral dimensions from normal (Brownian) diffusion and different small and large scale behaviors.

4. Spacetime near the Planck scale:

A. *Dimensional Reduction*: Spacetime at small length scales becomes effectively 2 dimensional [14]. See [15] for a lucid account of this feature from different angles. In the causal dynamical triangulation (CDT) program of Ambjørn, Jurkiewicz and Loll (AJL) et al [16] this is actually a more ubiquitous situation than the appearance of a large 4D spacetime. An oft-used way to define dimensionality of space is the spectral dimension from Brownian motion.

B. *Fractal Spacetime*: many 'Top-Down' theories (meaning, starting from some assumed microscopic structure of spacetime) contain this feature [17] The two earliest and longest running programs which reported on this behavior are CDT and asymptotic safety gravity (ASG) [18] program of Lauscher, Reuter et al [19]. This aspect is pursued in recent years with rigor by Calcagni and co-workers [20] . We shall draw on these work for inspirations in our pursuit. More on this later.

5. 'Bottom-Up' view: Does backreaction of non-Gaussian noise bring forth a modification of the wave operator reflecting possible fractal dimensionality in spacetime? We shall provide some background perspective for pursuing this in the 'Bottom-Up' approach (meaning, starting from the known and proven theories of spacetime and matter, namely, general relativity and quantum field theory) to see if this is possible. My present thinking leans towards the affirmative.

2. Perspective

I'd like to first recapitulate some statements about 1) the complementary functionality of the 'Top-Down' and the 'Bottom-Up' approaches, whose central tasks give meanings to the Emergent vs Quantum Gravity theories, respectively [7]. 2) The ubiquitous presence and importance of a stochastic regime between the quantum and the semiclassical regimes. 3) Why the 'Bottom-Up' approach, though admittedly difficult, even seemingly impossible, is necessary. I'll then summarize a conceptual scheme I proposed 15 years ago known as the "kinetic theory approach to quantum gravity" [6] where the correlation hierarchy plays a determinant role in unraveling the behavior of spacetime near the Planck scale, approached from low energy up. This way of thinking is enriched by interesting new results of Fewster Ford and Roman on the higher moments of the stress tensor.

1) In [7] I made the point that all candidate theories of Quantum Gravity, namely, theories for the microscopic structures of spacetime, should *look at their commonalities at the low energy – Planck scale – limit*, rather than their differences at the trans-Planckian scale, which is beyond present day experimental or observational verification capabilities. They should examine from their favorite theories of quantum gravity those features they all share, present definitive predictions at sub-Planckian energies and look for experiments and observations implementable at today's ultra-low energy which can imply, directly or indirectly, their Planck scale behavior. This transition region should serve also as the meeting ground for all 'Bottom-Up' approaches to compare their predictions with those from the 'Top-Down' theories.

2) *The existence and significance of a stochastic regime between the quantum and the semiclassical regimes* in many physical systems. Here, fluctuation phenomena play a pivotal role. An example we have seen played out in the last twenty years is (environment-induced) decoherence: a system's transition from quantum to classical is brought about by noise (in the environment). The role played by noise can be rephrased by correlations, as in a variant formulation of the fluctuation-dissipation theorem. Our prediction is that fluctuation phenomena should play an important role in the above-described transition region meeting ground. It is from this reasoning that we focus on effects such as the induced metric fluctuations ('spacetime foam') from the backreaction of quantum matter field fluctuations in stochastic gravity theory.

3) *The 'Bottom Up' approach is admittedly difficult, but not impossible.* After all, that is how physics has progressed over the centuries: from conceptual defects in a low energy theory or contradictory phenomena in its observational predictions, we conjecture, test, prove and establish the unknown higher energy theories and structures, which later become established laws and facts.

I'll mention a couple of prior examples where a proven low energy theory when examined closely and critically can provide insight into some new features of more viable theories at higher energies:

a) Renormalization (regularization) of the stress energy tensor *mandates* the appearance of quadratic curvature terms in the effective action.

IOP Conf. Series: Journal of Physics: Conf. Series **880** (2017) 012004 doi:10.1088/1742-6596/880/1/012004

b) Inclusion of (Gaussian) quantum field fluctuations *mandates* the appearance of stochastic components in the curved background spacetime – the induced metric fluctuations.

In light of this the question we are asking, it seems to me, is not unreasonable, or totally out of reach, namely, "Does the inclusion of non-Gaussian quantum matter field fluctuations ushers in fractal spacetime structures?" Let us recapitulate the structure of stochastic gravity and start from there.

Stochastic Gravity in the restricted sense (up to second order SG2) in relation to classical and semiclassical gravity can be represented by the following three levels of theoretical structures: *Classical Gravity* – Einstein equation with classical matter:

$$G_{\mu\nu}[g] = \kappa T_{\mu\nu}[g]$$

Semiclassical gravity (mean field theory) :

$$G_{\mu\nu}[g] = \kappa \left(T_{\mu\nu}[g] + \left\langle T^q_{\mu\nu}[g] \right\rangle \right)$$

where q denotes a quantum object and $\langle\ \rangle$ denotes taking its expectation value
Stochastic gravity (including quantum fluctuations):

$$G_{\mu\nu}[g+h] = \kappa \left(T_{\mu\nu}[g+h] + \left\langle T^q_{\mu\nu}[g+h] \right\rangle + \xi_{\mu\nu}[g] \right)$$

to linear order in perturbations h, where $\xi_{\mu\nu}$ is the stochastic force induced by the quantum field fluctuations with the correlation

$$\langle \xi_{\mu\nu}(x)\xi_{\alpha\beta}(y)\rangle_s = N_{\mu\nu\alpha\beta}(x,y)$$

where $\langle\ \rangle_s$ denotes taking a distribution average over stochastic realizations, and $N^{\mu\nu\alpha\beta}$ is the noise kernel

$$N^{\mu\nu\alpha\beta}(x,y) = \frac{1}{2}\left\langle \left\{ t^{\mu\nu}(x), t^{\alpha\beta}(y) \right\} \right\rangle$$

where $t^{\mu\nu}(x) = T^{\mu\nu}(x) - \langle T^{\mu\nu}(x)\rangle$.

2.1. Correlation hierarchy and the kinetic theory approach to quantum gravity

Thus noise carries information about the correlations of the quantum matter field stress tensor. One can further link *correlations* in quantum field stress tensors to *coherence* in quantum gravity. Stochastic gravity brings us closer than semiclassical gravity to quantum gravity in the sense that the correlations in the quantum matter field stress tensor and correlations in the induced geometric objects (such as the Riemann tensor correlator), which by its theoretical construct are fully present or accessible in quantum gravity, are partially retained in stochastic gravity. Because of the self-consistency condition required of the backreaction equations for the matter and spacetime sectors when solved simultaneously, the background spacetime has a way to tune in to the correlations of the quantum matter fields registered in the noise terms which manifest through the induced metric fluctuations from solutions of the Einstein-Langevin equations.

Viewed in this broader light the Einstein-Langevin equation is only a partial (lowest correlation order) representation of the more complete theory for the micro-structures of spacetime and matter. There, the quantum coherence in the geometry sector is locked in and related to the quantum coherence in the matter field, as the quantum description of the combined matter and gravity sectors should be given by a completely coherent wave function of both.

IOP Conf. Series: Journal of Physics: Conf. Series **880** (2017) 012004 doi:10.1088/1742-6596/880/1/012004

Semiclassical gravity forsakes all the quantum phase information in the gravity or geometry sector. Stochastic gravity captures only partial phase information or quantum coherence in the gravity sector by way of the correlations in the quantum matter fields. Since the degree of coherence can be represented by correlations, (putting aside entanglement issues for the problem at hand), the strategy for the stochastic gravity program (in the broader sense SGn) is to move up the hierarchy starting with the second order correlator (the variance) of the matter field stress energy tensor to the higher order correlations, and through their linkage with gravity sector provided by the Einstein-Langevin equations for each order, retrieve whatever quantum attributes (partial coherence) of the quantum gravity theory at trans-Planckian scales. Thus, as remarked in [4], in this vein, focusing on the noise kernel, the stress energy tensor two point function, is only the first step (beyond the mean field semiclassical gravity theory) towards reconstructing the full theory of quantum gravity. This is the conceptual basis for the so-called 'kinetic theory approach to quantum gravity' [6].

The kinetic theory approach to quantum gravity proposes to unravel the microscopic structure of spacetime by examining how the correlation functions of the geometric objects formed by the basic constituents of spacetime are driven by the correlation noises from the higher moments of the quantum matter stress energy tensor. This paradigm is structured around the so-called *Boltzmann-Einstein Hierarchy* of equations. For illustrative purpose, assuming that the micro-structure of spacetime can be represented by some interacting quantum field of micro-constituents, the Schwinger-Dyson hierarchy of n point correlation functions would be the quantum parallel to the BBGKY hierarchy, the lowest order being the Boltzmann equation. For any nth order correlation, if one can represent all the higher order correlations as noise, called 'correlation noise' in [10], this would give rise to a stochastic Boltzmann equation for the nth order correlation. At the second correlation order, the Einstein tensor correlator with induced metric fluctuations [21] has been calculated for the Minkowski space, so do the Weyl and the Riemann tensors [22] for the de Sitter space recently. The higher order correlations of these geometric objects with induced metric fluctuations can likewise be determined, albeit much harder, by solving the Einstein-Langevin equations with sources from the higher moments of the stress energy tensor of the corresponding orders. The combined set of correlators of all orders of the geometric objects is given the name of a 'Boltzmann-Einstein hierarchy' in [6] because it has the structure of a BBGKY hierarchy whose lowest order is the Boltzmann equation while in the present spacetime context the lowest order in this hierarchy is the Einstein equation. [1]

Stochastic Gravity in the broader sense SGn refers to the Boltzmann-Einstein hierarchy of equations in the spacetime sector *together with* the Einstein-Langevin equations for each order of the hierarchy (or levels of structure) in connecting the spacetime structure correlators to the matter field correlators. A figurative way suggested in [6] to understand the formal structure of these two inter-woven hierarchies of spacetime-matter relations is, if we assume the 'horizontal' dimension in this conceptual chart represents the E-L equations relating at each correlation order the spacetime correlators to the matter field correlators, then the B-E hierarchy of spacetime correlators occupy the 'vertical' dimension. What stochastic gravity does to reach quantum gravity is to 'climb up' the spacetime B-E hierarchy aided by the E-L equations which tap into the quantum matter sector at each level of structure.

In summary, viewed in the light of mesoscopic physics, stochastic gravity is the theory which enables one to probe into the higher correlations of quantum matter and spacetime. From

[1] This is *not* the Einstein-Boltzmann equation in classical general relativity and relativistic kinetic theory which frames the classical matter in the Boltzmann style as source of the Einstein equation. The Boltzmann-Einstein hierarchy refers to the spacetime sector alone.

IOP Conf. Series: Journal of Physics: Conf. Series **880** (2017) 012004 doi:10.1088/1742-6596/880/1/012004

the excitations of the collective modes in geometro-hydrodynamics one tries to deduce the kinetic theory of spacetime meso-dynamics and eventually the full theory of quantum gravity for spacetime micro-dynamics[2].

2.2. Probability distribution for quantum stress tensor fluctuations

The above is a grand scheme based on the importance of the fluctuations or noise of the quantum matter field manifested in the correlations of the stress energy tensor. The expectation values of the bi-tensor is the driving source of stochastic gravity theory (in the restricted sense). Going beyond, tackling the higher moments, is the start of this new journey. This topic has been explored by Fewster, Ford and Roman (FFR) [9] systematically in the past. Their recent findings, it seems to me, add more weight to the correlation hierarchy conceptual framework.

What FFR found was, for two-dimensional Minkowski space, the probability distribution for individual measurements of the stress-energy tensor for a conformal field in the vacuum state, smeared in time against a Gaussian test function, yields a shifted gamma distribution with the shift given by the optimal quantum inequality bound these authors found earlier. For small values of the central charge it is overwhelmingly likely that individual measurements of the sampled energy density in the vacuum give negative results. For 4D, the probability distribution of the smeared square field is also a shifted gamma distribution, but that the distribution of the energy density is not. There is a lower bound at a finite negative value, but no upper bound.

These results show that arbitrarily large positive energy density fluctuations are possible. Since they fall slower than exponentially, this may even allow for the dominance of vacuum fluctuations over thermal fluctuations. The implication of these findings for gravity and cosmology is that large passive geometry fluctuations are possible. (What Ford calls active and passive correspond to our intrinsic and induced). These findings testify to the importance of induced metric fluctuations from the backreaction of the stress-energy tensor correlations of higher orders, which is in the realm of stochastic gravity (in the broader sense).

3. Spacetime near the Planck scale

The findings of Fewster, Ford and Roman described above, that the higher moments of the stress energy tensor of quantum matter fields have nontrivial effects in Minkowski spacetime at today's ultra-low energy, also have implications on the ultra-short distance structure of spacetime near the Planck scale. One such effect is dimensional reduction [14].

3.1. Dimensional reduction

Many theories of quantum gravity contain the ingredients, or predict the occurrence, of dimensional reduction of spacetime at the Planck scale from 4D to 2D. Even at the classical level, within the theory of general relativity, it has been shown in the 60s-70s in the work of Belinsky, Khalatnikov, Lifshitz [25] and Misner [26] that the most general solutions to the Einstein equation near the cosmological singularity manifest a 'velocity-dominated' [27] behavior. This refers to the contribution of the extrinsic curvature (Kasner solution) dominating over the intrinsic curvature (mixmaster solution) near the singularity, or spacetime assuming an inhomogeneous Kasner solution at every point in space. Physically this means spatial points decouple, light cones strongly focus and shrink to timelike lines, or in a more poetic depiction,

[2] This paradigm was used by Mattingly [23] as an example of emergence of general relativity from quantum gravity, mirroring the 'general relativity as geometro-hydrodynamics' perspective [24], in analogy to hydrodynamics being an emergent theory from molecular dynamics.

IOP Conf. Series: Journal of Physics: Conf. Series **880** (2017) 012004 doi:10.1088/1742-6596/880/1/012004

spacetime becomes 'asymptotically silent'. For a description of further evidences for asymptotic silence as a generic Planck scale behavior, see, e.g., [15]

Along the line of reasoning of Planck scale focusing by non-Gaussian vacuum stress-energy fluctuations, Carlip, Mosna and Pitellis [28], using the results of [9] for the probability distribution of 2D conformal field theory, have recently shown that vacuum fluctuations of the stress-energy tensor in two-dimensional dilaton gravity lead to a sharp focusing of light cones near the Planck scale. Space is effectively broken into a large number of causally disconnected regions, thus adding to the evidence of spontaneous dimensional reduction at short distances. They also argued that these features should be present in four dimensions qualitatively.

It is of interest to see how the dimensionality of spacetime is defined. A common way is to use the spectral dimensions in Brownian motion, namely, the dimension of paths traversed by a random walker. Since going beyond Brownian motion is a main themes of this inquiry, it is perhaps worthy of a short description here [29].

The spectral dimension for a diffusion process is defined as follows: From the probability density $P(x; x'; \tau)$ of a diffusing particle on a background, one can define a return probability $\mathcal{P}_\tau = V^{-1} \int_x P(x; x; \tau)$. Here, x and x' denote coordinates on the Euclidean spacetime with volume V , and τ is an external diffusion time – 'external' here refers to what pertains only to the diffusion process and not related to the physical time in spacetime dynamics. The spectral dimension for a background with fixed dimensionality is then defined as

$$d_S = 2 \lim_{\tau \to 0} \frac{\partial ln \mathcal{P}(\tau)}{\partial ln \tau} \tag{1}$$

To allow for the spectral dimension to change as the scale length varies, one can generalize this definition to make $d_S(\tau)$ dependent on the diffusion 'time' τ by not taking the limit $\tau \to 0$.

The expectation value of the spectral dimension is relatively easy to evaluate numerically, since random walks are simple to model. Intuitively a random walker with more dimensions to explore will diffuse more slowly from a starting point, and will also take longer to return.

Quantitatively, a diffusion process on a d-dimensional manifold is described by a heat equation

$$\left(\frac{\partial}{\partial s} - \Delta_x\right) K(x, x'; s) = 0 \quad \text{with } K(x, x', 0) = \delta(x - x'), \tag{2}$$

with a short distance solution

$$K(x, x'; s) \sim (4\pi s)^{-d/2} e^{-\sigma(x,x')/2s} \left(1 + \mathcal{O}(s)\right) \tag{3}$$

where $\sigma(x, x')$ is Synge's world function, essentially the square of the geodesic distance. In particular, the return probability $K(x, x, s)$ is

$$K(x, x; s) \sim (4\pi s)^{-d/2}. \tag{4}$$

This relationship can extend to any space on which a diffusion process or random walk can occur. The spectral dimension is then defined as the coefficient corresponding to d in (4).

As Ambjørn, Jurkiewicz, and Loll discovered [30], and reconfirmed by Kommu [31] and others, the spectral dimension found by Causal Dynamical Triangulation is 4 for "long" random walks, but changes to 2 for 'short' random walks. The corresponding Green functions are those of four-dimensional fields at large scales, but those of two-dimensional fields at small scales. The cross over scale length is about 15 Planck lengths [32].

Obtained from a Mellin transform of the Green function, the *heat kernel* contains just as much physical information about the correlation functions for a quantum field. The heat kernel is a useful vehicle to extract the UV behavior of a quantum field in curved spacetime, under a small s Schwinger proper time expansion, as was used for the identification of UV divergences in several stress energy tensor regularization programs (see, e.g., [33]). The opposite limit of small s expansion gives the short distance behavior [34]. These are examples of how properties of quantum field theory in curved spacetime can provide useful hints for the spacetime structure at the Planck scale. However, beware that the Schwinger proper time s has nothing to do with physical time or Euclidean time. It is a fictitious construct introduced by Schwinger, thus the heat equation for quantum fields in ordinary 4D spacetime is written in 1+4 dimensions. The diffusion 'dynamics' measured by this fictitious 'time' has nothing to do with the diffusion of a physical particle in real 1+3D spacetime.

3.2. Fractal spacetime

That spacetime at the Planck scale can assume some fractional structure has been proposed from many independent considerations, such as in causal dynamical triangulation, asymptotically safe gravity, Harava-Lifshitz gravity, group field theory [17]. Take for example the results from the long running causal dynamical triangulation program [30], the prevailing geometries are 2D fractal spacetimes which evolve for a very short time. Long-lived smooth structures are rare, but they do exist, some even evolve to a 4D spacetime of large volume with homogeneity [32], desirable features like our universe. In fact after about 100 Planck time spacetime acquires semiclassical features.

One can get an idea about some properties of a fractal spacetime by watching how a test particle moves in it, much like geodesics in a smooth curved spacetime. Processes like anomalous diffusion (in a classical, flat Euclidean space) are of interest to the quantum gravity community from the following consideration: In a certain energy or lengthscale range, some qualitative features of the *quantum structure of spacetime* can be gleaned off from the diffusive motion of a probe particle. This type of motion is captured by a generalization of the Laplacian operator (Laplace-Beltrami, in curved spacetime) from the more familiar form defined for a smooth manifold to include stochastic features.

An important fact to bear in mind is that the effective metric 'seen' by a diffusing particle depends on the momentum of the probe. This can be captured by a renormalization group formulation, such as utilized in the asymptotic safety gravity (ASG) program [19]. There [29] it is found that in d dimensions, the spectral dimension d_S changes depending on the value of a parameter δ, which in turn depends on the probe length scale.

$$d_S = \frac{2d}{2+\delta} \tag{5}$$

One can identify three characteristic regimes where the spectral dimension is approximately constant over many orders of magnitude. At large distances one reaches the *classical regime* where $\delta = 0$ and the spectral dimension agrees nicely with both the Hausdorff and topological dimension of the spacetime, as required. At smaller distances one first encounters a *semiclassical regime* with $\delta = d$, before entering into the fixed-point *quantum gravity* regime with $\delta = 2$, signifying a dimensional reduction to 2. In this sense δ provides a measure of the quantum nature of spacetime.

3.3. Might we see these effects from 'Bottom-Up'?

With occurrence of dimensional reduction and appearance of fractal spacetime likely near the Planck scale, we now ask the question: might we see these effects from low energy up? By low energy we mean today's 4D spacetime with a smooth manifold structure, and it is believed that this smooth manifold structure will persist to around, but somewhat lower than, the Planck energy, say, $\approx 100\ell_{Pl}$, reached from below.

As mentioned earlier, in many physical systems one can find a stochastic regime in between the semiclassical and the quantum regimes. For spacetime structures we expect the same is true. Indeed, it is in the transition from the stochastic to the quantum regime where we would anticipate fractal spacetime to appear, possibly accompanying a transition from continuum to discrete structures (e.g., [35]) . As described earlier, for gravity there is a theory which can capture the essence of this stochastic regime, namely, stochastic gravity. This is the motivation for us to look for possible fractal structures in stochastic gravity as we move up in energy, or probing at a shorter scale length. The physical quantities of interest for this inquiry is the higher moments of the stress tensor giving rise to nonGaussian noise in the quantum matter fields. The stochastic processes to aid us visualize the features of such spacetimes are the anomalous diffusion processes, a much broader class than the quantum Brownian motion describing normal diffusion.

Let us separate the contexts of our inquiries into two kinds, one is easier to answer than the other.

Q1: Can a smooth manifold with a Laplace-Beltrami operator admit non-Gaussian noises from a quantum field? The answer is yes. For example, Ramsey et al in treating the nonequilibrium inflaton dynamics for reheating considered fermion production, which when expressed as noise, are multiplicative colored noises. The nonGaussian nature comes from the nonlinearity in Yukawa interaction between a scalar (inflaton) and a spinor (fermion) field. There are works in classical stochastic processes with multiplicative noise, such as the generalized Langevin equation of e.g., [36] and on-Gaussian noise in quantum stochastic processes, such as the nonlinear Langevin Equation of e.g., [37]

Q2: Consider non-Gaussian noises from a quantum field back-reacting on a spacetime. Does the requirement of self-consistency in the dynamics of spacetime and matter require / permit the possibility of a fractal spacetime? We think it is possible.

Note that in the above, we are talking about two different stochastic processes: Q1 concerns quantum matter (with non-Gaussian noise) moving in a given background space whereas Q2 refers to the classical spacetime dynamics driven by quantum noise via the Einstein- Langevin Equation. We explore the latter question here.

4. Anomalous diffusion

In this section I'll give a sketch of what anomalous diffusion is and in the next section, how it can serve as a probe into the quantum nature of spacetime [20]. I am not an expert in this topic so this is just sharing my learning process with you. The following contents are excerpted from the works of practitioners in these two (seemingly) disparate fields, statistical mechanics and quantum gravity, to prepare us for the journey we wish to embark upon.

4.1. Anomalous transport by fractional dynamics

'Anomalous' is in comparison to the 'normal' Brownian motion. In a classical diffusion process in d (embedding) spatial dimension, the deviation of the mean squared displacement is given by

$$\langle(\Delta r)^2\rangle \equiv \langle r^2\rangle - \langle r\rangle^2 = 2dK_\beta t^\beta \quad \text{(no summation on}\,\beta) \tag{6}$$

where t is the time in classical stochastic dynamics (τ when applied to the analysis of quantum spacetime structure), and K_β is the generalized diffusion constant with $\beta = 1$ being the normal process and $\beta \neq 1$ the anomalous processes. The case with $0 < \beta < 1$ are called subdiffusive (dispersive, slow), those with $\beta > 1$ the superdiffusive (enhanced, fast) processes. Usually the domain $1 < \beta \leq 2$ is considered, with $\beta = 2$ being the ballistic limit described by a wave equation, or its forward and backward components.

These processes are characterized by their PDFs: We mention several force-free processes whose PDFs share the form:

$$P(x', t) = (4\pi K_\beta t^\beta)^{-\frac{1}{2}} \exp\left(-x^2/(4\pi K_\beta t^\beta)\right) \tag{7}$$

The processes this encompasses are:
(a) Normal Brownian motion (BM): $\beta = 1$
(b) Fractional Brownian motion (FBM): $0 < \beta \leq 2$
(c) generalized Langevin equation (GLE) with power-law kernel $0 < \beta < 2$ with $\beta \neq 1$
One can find other commonly encountered processes, such as d) subdiffusion (SD), e) Levý flight (LF), f) Levý walk (LW) etc, described in e.g., Table 1 of the review by [38], which we follow in this subsection. The subdiffusive case below bears special significance for quantum spacetimes.

Note the qualitative differences between the first three (a-c) and the latter three (d-f) types of motion: The fractional dynamical equations corresponding to SD, LFs and LWs are highly non-local, and carry far-reaching correlations in time and/or space, represented in the integro-differential nature (with slowly decaying power-law kernels) of these equations. In contrast, FBM and GLE on the macroscopic level are local in space and time, and carry merely time- or space- dependent coefficients.

These generalizations can be obtained from Brownian motion (BM) by using the continuous time random walk (CTRW) model, which we describe below, following [38]. All of these models can be mapped onto the corresponding fractional equations, which we will describe afterwards.

1. In a standard random walk process each step is of a fixed length in a random direction at each tick of a system clock. A process having constant spatial and temporal increments, Dx and Dt, will give rise to the standard diffusion process in the long-time limit, i.e., $x(t) = N^{-\frac{1}{2}}\Sigma_{i=1}^{N} x_i$. After a sufficient number of steps the associated random variable where x_i is the position after the ith step will be distributed by a Gaussian due to the central limit theorem.

2 In a CTRW process the jump length and the waiting time are distributed according to two PDFs, $\lambda(x)$ and $\psi(t)$, respectively. The propagator for such a CTRW process in the absence of an external force is given in Fourier (k) - Laplace (u) space by

$$P(k, u) = \frac{1 - \psi(u)}{u[1 - \psi(k, u)]} \tag{8}$$

Subdiffusion is classically described in terms of a CTRW with a long-tailed inverse power-law waiting time:

$$\psi(t) \approx \tau^\beta / t^{1+\beta} \text{ for } 0 < \beta < 1 \tag{9}$$

A waiting time PDF of this form is obtained under the Laplace expansion $\psi(u) \approx 1 - (u\tau)^\beta$ for $u << \tau$. It includes normal diffusion at the limit $\beta = 1$, in which case $\psi(u) = e^{-u\tau} \approx 1 - u\tau$, and $\psi(t) = \delta(t - \tau)$. Now combine with the Fourier transform of the jump length $\lambda(x)$. After an analogous expansion of a short-range jump length PDF, $\lambda(k) \approx 1 - \mu k^2$ (for $k \to 0$) for the Fourier transform of λ, we obtain,

$$P(k, u) = \frac{1/u}{1 + u^{-\beta} K_\beta k^2} \tag{10}$$

where $K_\beta \equiv \mu/\tau^\beta$ is the anomalous diffusion constant. For Brownian motion $\beta = 1$, after using the differentiation and integration theorems for the Laplace and Fourier transforms, we obtain the normal diffusion equation:

$$\frac{\partial P(x, t)}{\partial t} = K_1 \frac{\partial^2 P(x, t)}{\partial x^2} \tag{11}$$

The subdiffusive cases ($0 < \beta < 1$) have a term of the form $u^{-\beta} f(u)$. Their PDFs obey the fractional diffusion equation

$$\frac{\partial P(x,t)}{\partial t} =_0 D_t^{1-\beta} K_\beta \frac{\partial^2 P(x,t)}{\partial x^2} \tag{12}$$

where ${}_0D_t^{1-\beta} \equiv \frac{\partial}{\partial t} {}_0D_t^{-\beta}$ and ${}_0D_t^{-\beta}$ is a Riemann-Liouville fractional differential operator defined by

$${}_0D_t^{-\beta} \equiv \frac{1}{\Gamma(\beta)} \int_0^t dt' \frac{f(t')}{(t-t')^{1-\beta}} \tag{13}$$

for any well-behaved function $f(t)$. It is important to keep track of the initial condition in this fractional diffusion equation. Noticing that ${}_0D_t^{-\beta} 1 = t^{-\beta}/\Gamma(1-\beta)$, we can rewrite Eq. (12) in the form

$${}_0D_t^{\beta} P(x,t) - t^{-\beta} P_0(x)/\Gamma(1-\beta) = K_\beta \frac{\partial^2 P(x,t)}{\partial x^2} \tag{14}$$

5. Diffusive processes and fractal spacetime

In this line of investigation the probability density function (PDF) contains more information than the spectral dimension, and is thus the focus of interest. The basic properties of PDF such as semi-positive definiteness need be checked for the diffusion process representative of quantum spacetime, case by case, for different theories of quantum gravity. This is discussed in e.g., [29] where it is shown several ways to construct diffusion equations which capture the quantum properties of spacetime while admitting solutions that are manifestly positive semi-definite.

Note again that the diffusion 'time' is not a physical time and thus the motion of the probe particle is not related to how matter moves in, affects, or is affected by the background spacetime. It carries no dynamical meaning itself beyond serving as a tool for 'imaging the topography' of the quantum spacetime structure.

One way suggested in [29] for obtaining a positive semi-definite PDF is to use nonlinear time. These authors use a renormalization group (RG)-improvement scheme where the scale k is related to the diffusion time τ with large diffusion times corresponding to the IR regime $k \to 0$ and short diffusion times to the UV regime $k \to \infty$. Assuming a power-law relation between the effective metrics at scale k and the reference scale k_0, $\langle g^{\mu\nu} \rangle_k \propto k^\delta \langle g^{\mu\nu} \rangle_{k_0}$; one gets

$$\left(\partial_\tau - k^\delta \langle g^{\mu\nu} \nabla_\mu \nabla_\nu \rangle_{k_0} \right) P(x, x', \tau) = 0 \,, \tag{15}$$

where k is the RG scale, k_0 is the IR reference scale and $g^{\mu\nu}$ is the fixed IR reference metric which is taken to be the flat Euclidean metric.

To encode the scaling effects in the diffusion time τ, one can multiply the equation with $k^{-\delta}$ so that the diffusion operator becomes a standard second-order Laplacian,

$$\left(k^{-\delta} \frac{\partial}{\partial \tau} - \nabla_x^2 \right) P(x, x', \tau) = 0 \,. \tag{16}$$

The relation between k and τ is then fixed on dimensional grounds. Since kx is dimensionless, Eq. (16) implies that, dimensionally, $\tau \sim k^{-\delta-2}$ which suggests the scale identification

$$k = \tau^{-\frac{1}{\delta+2}} \,, \tag{17}$$

where the proportionality constant has been absorbed into the diffusion time τ.

IOP Conf. Series: Journal of Physics: Conf. Series **880** (2017) 012004 doi:10.1088/1742-6596/880/1/012004

By changing the diffusion time variable from τ to τ^β with

$$\beta = \frac{2}{\delta + 2}, \tag{18}$$

the resulting equation can be cast into a diffusion equation in nonlinear time τ^β:

$$\left(\frac{\partial}{\partial \tau^\beta} - \nabla_x^2\right) P(x, x', \tau) = 0\,. \tag{19}$$

The probability density resulting from this diffusion equation is given by a Gaussian in $r = |x - x'|$:

$$P(r, \tau) = \frac{1}{(4\pi\tau^\beta)^{\frac{d}{2}}}\, e^{-\frac{r^2}{4\tau^\beta}}\,. \tag{20}$$

which is seen to be manifestly positive semi-definite. Moreover, the cutoff identification Eq. (17) implies that $P[r, \tau(k)] \propto k^d$ has the correct scaling behavior of a diffusion probability in d dimensions. Notice that for the classical regime value of $\delta = 0$ or $\beta = 1$, we obtain the same expression as the heat kernel given in Eq. (3) without the curvature correction terms, with (r, τ) here corresponding to (σ, s) there.

The spectral dimension resulting from Eq. (20) is independent of τ and given by Eq. (5) It was remarked in [29] that the spectral dimension obtained from the diffusion in nonlinear time, Eq. (19), and the one measured within CDT is identical to the one found in [39], thus offering useful grounds for the comparison of different approaches to quantum gravity, here in their stochastic dynamics representations. (See also [40].)

The averaged squared displacement of the test particle implied by Eq. (20) is found to be

$$\langle r^2 \rangle_{\text{nonlinear time}} = 2\, d\, \tau^\beta\,, \tag{21}$$

where angular brackets denote the expectation value with respect to the associated probability density function, $\langle f(x) \rangle = \int dx\, P(x, x', \sigma)\, f(x)$. For $\beta = 1$, this corresponds to a Wiener process of normal diffusion; the case $\beta < 1$ is subdiffusive. There are actually two possible stochastic processes underlying Eq. (19). One is scaled Brownian motion (SBM), i.e., a Wiener process which takes place in nonlinear time. The second is fractional Brownian motion (FBM), which is a stochastic process with correlated increments, thus non-Markovian. (See, e.g. [38] for details and references.) When the leading theories of quantum gravity are viewed in the anomalous diffusion light, whether the relevant stochastic processes are Markovian or non-Markovian has some special significance, because it may reveal the correlation and memory effects in the kinematics and dynamics of the basic microscopic constituents of spacetime, an essential goal of quantum gravity.

6. Fractal spacetimes in stochastic gravity?

In summary, anomalous diffusion has been used as a probe into the quantum nature of spacetime in several quantum gravity theories [20]. Here, what is studied is anomalous diffusion equations on a classical flat space, not even curved spacetime. The point is to get some ideas in how new qualitative features arise, such as fractal structure. A dynamical dimensional change is captured by a modification of the Laplacian operator appearing in the classical diffusion equation. The effective metric 'seen' by the diffusing particle depends on the momentum of the probe. Expressing this metric through a fixed reference scale leads to a modified diffusion equation providing an effective description of the propagation of the probe particle on the quantum gravity background. A multifractal structure with spectral dimension d such as given in Eq.(5) depends on the probed length scale whereby one sees a clear transition from classical to semiclassical to quantum regime. But as stressed before, the time in such processes is not the physical time in quantum gravity (time does not exist) and many technical and conceptual issues remain.

These reported activities in recent years are from 'Top-Down' theories. We now ask if new features like fractal spacetime may appear from 'Bottom-Up', namely, extrapolating from the low energy realm described by general relativity up in energy. We know the existence of the semiclassical and stochastic gravity regimes. Using the structural framework of stochastic gravity to probe at increasingly finer scales we now ask whether, and how, we can see some signs of fractal structure in spacetime. In the first part of my talk I motivated this feature from some long-running programs of quantum gravity theories, and in the second part I introduced some tools for examining these possibilities, amongst them anomalous diffusion processes driven by non-Gaussian noises. Let us now examine this pathway, try to identify new challenges and see what new measures we need to take to meet them.

First, calculate the vacuum expectation values of the higher correlation functions of the stress energy tensor of quantum matter fields and solve the corresponding order Einstein-Langevin equations. This is an extension of what has been done quite nicely in (Gaussian) stochastic gravity with Gaussian noises from the two point function of the stress energy tensor – the noise kernel – acting as the source. The solutions of the ELEq yield the second order correlations in the spacetime sector, as was done in [21] for the correlation function of the Einstein tensor in Minkowski spacetime, and recently for the Weyl and Riemann second order correlators [22] in de Sitter space which contain information of the induced metric fluctuations. Now, for the higher moments of the stress tensor, even though there are no obvious ways to identify them as noise, one can in principle solve the ELEq to obtain the higher order correlators of the geometric objects, e.g., the Riemann tensor correlator (call this the geometric route). One can nonetheless at first a) examine the range of weakly nonGaussian stress tensor, meaning, look at small departures from the second order correlations in stress tensor and b) carry out a perturbative calculation using as background spacetime the Gaussian second moment-induced solutions (Einstein or Riemmann tensors) mentioned above. The results of Ford et al and Verdaguer et al will be useful for these steps a) and b) respectively. The alternative route of viewing this as a stochastic process whereby one can conceptualize the physical picture easier (that's what led me to anticipate what kind of theory should lie beyond semiclassical gravity) and can borrow known techniques toward finding solutions of the stochastic equations (call this the stochastic route) requires identifying the non-Gaussian noises from the higher moments. The challenge here is, there is no clean separation between the real and imaginary parts of the influence action and there is no non-Gaussian functional integral identity whereby one can interpret a quantum object in terms of a classical stochastic variable. However, one can first consider a case of weak nonGaussianity (from the results for the third moment of the stress tensor by Fewster et al [9]) using perturbative methods off the known second moment results (where noise is well defined via the Feynman-Vernon identity). This is the route currently pursued with H T Cho [37].

Carrying out the above stated tasks is not easy, but a bigger challenge lies ahead. I see three demands both conceptual and technical: There may be a need to 1) create, figuratively speaking, new positions for newcomers, 2) anticipate the backreaction of non-Gaussian noise bringing forth a modification of the Laplace-Beltrami wave operator which contains the germs of a fractal spacetime 3) introduce new dynamical variables at shorter scales for the more fundamental constituents of a more basic theory and from them construct effective theories for collective variables which can match with the more familiar low energy theories.

On 1), recall an example given earlier: renormalization of the stress energy tensor for quantum matter fields requires the introduction of higher-order curvature terms. Thus in going from a test field theory, namely quantum field theory in curved spacetime, to semiclassical gravity with backreaction whereby the background spacetime dynamics is determined in a self-consistent manner, one needs to create new positions for the newcomers, namely, three seats for the ΔR, the Ricci curvature-squared and the Weyl curvature-squared terms (in 4D only two places are needed thanks to the Gauss-Bonnet theorem). I would imagine new places need be created in the spacetime sector (LHS of the ELEq) to accommodate the induced geometric fluctuations due to the backreaction of the higher-moments of the stress energy tensor. 2) may be the critical step, namely, the appearance of new structure like fractals, departing from the familiar smooth manifold structure at lower energies or probed with lesser resolutions. The example we gave from asymptotic safety quantum gravity program illustrates the change-over of dimensionality from the classical regime to the semiclassical to the quantum. The stochastic regime lying in between the quantum and the semiclassical should have a signature in this regard by itself. The challenge is to identify what types of non-Gaussian noise corresponding to what correlation order of the quantum matter source induce what kinds of generalized wave operators permitting what kinds of fractal structures. 3) is a task for all effective field theories but posed here in the reversed direction, namely, not merely constructing EFTs (nuclear physics, for example) from a proven-valid microscopic theory with known basic constituents (QCD, for the same example), but taking what is known in an effective theory to posit the more basic theories and even predict the more basic constituents and their dynamics. Phase transitions (e.g., [41]) in these interfaces add to the unpredictability but also the richness of possibilities. This is where cooperation between the 'Bottom-Up' and 'Top-Down' theories becomes necessary. What we have done here is to explore some useful ideas and tools for Task 2).

Acknowledgment I thank Prof. Thomas Elze and organizers of DICE2016 for their invitation, Profs. Steve Carlip, Larry Ford, Renate Loll and Enric Verdaguer for correspondences on this topic, and Prof. Hing-Tong Cho for discussions on the perturbative approach in treating non-Gaussian noises. Main themes of this talk were first presented at the Peyresq Physics Meeting in June 2016, supported partially by OLAM, Association pour la Recherche Fondamentale, Bruxelles. This essay contains materials from a part of the last chapter of [2]

[1] B. L. Hu and E. Verdaguer, "Stochastic gravity: Theory and Applications" Living Reviews in Relativity 11 (2008) 3 [arXiv:0802.0658]
[2] B. L. Hu and E. Verdaguer, *Semiclassical and Stochastic Gravity – Quantum Field Effects on Curved Spacetime* (Cambridge University Press, Cambridge, 2019)
[3] E. Calzetta and B.L. Hu, Phys. Rev. D 49, 6636 (1994); B. L. Hu and A. Matacz, Phys. Rev. D 51, 1577 (1995); B. L. Hu and S. Sinha, Phys. Rev. D 51, 1587 (1995); A. Campos and E. Verdaguer, Phys. Rev. D 53, 1927 (1996); F. C. Lombardo and F. D. Mazzitelli, Phys. Rev. D 55, 3889 (1997).
[4] B. L. Hu, "Stochastic Gravity" Int. J. Theor. Phys. 38, 2987 (1999); gr-qc/9902064.
[5] B. L. Hu, J. P. Paz and Y. Zhang, Phys. Rev. D45, 2843 (1992)
[6] B. L. Hu, "A Kinetic Theory Approach to Quantum Gravity" Int. J. Theor. Phys. 41 (2002) 2111-2138 [gr-qc/0204069]
[7] B. L. Hu, "Emergent /Quantum Gravity: Macro/Micro Structures of Spacetime" DICE2008 J. Phys. Conf. Ser. 174 (2009) 012015 arXiv:0903.0878
[8] R. Feynman and F. Vernon, Ann. Phys. (NY) 24, 118 (1963).
[9] C. Fewster, L. Ford and T. Roman, Probability distributions of smeared quantum stress tensors, Phys. Rev. D 81, 121901 (R) (2010). C. Fewster, L. Ford and T. Roman, Probability distributions for quantum stress tensors in four dimensions, Phys. Rev. D 85, 125038 (2012)
[10] E. Calzetta and B. L. Hu, "Correlations, Decoherence, Disspation and Noise in Quantum Field Theory", in Heat Kernel Techniques and Quantum Gravity, ed. S. Fulling (Texas A & M Press, College Station 1995); hep-th/9501040. E. Calzetta and B. L. Hu, Phys. Rev. D 61, 025012 (2000).
[11] B. L. Hu, J. P. Paz and Y. Zhang, Phys. Rev. D47, 1576 (1993)
[12] S. A. Ramsey, B. L. Hu, and A. M. Stylianopoulos, Phys. Rev. D57, 6003-6021 (1998).
[13] R. Metzler and J. Klafter, Phys. Rep. 339 1 (2000).
[14] G. 't Hooft, Dimensional reduction in quantum gravity, in Salamfestschrift, A. Ali, J. Ellis and S. Randjbar-

Daemi eds., World Scientific, Singapore (1993) [gr-qc/9310026]
J. Ambjørn, J. Jurkiewicz and R. Loll, Spectral dimension of the universe, Phys. Rev. Lett. 95 (2005) 171301
D. Benedetti and J. Henson, Spectral geometry as a probe of quantum spacetime, Phys. Rev. D 80 (2009) 124036 [arXiv:0911.0401]
S. Carlip, "The small scale structure of spacetime", in Foundations of Space and Time, G. Ellis, J. Murugan and A. Weltman eds., Cambridge University Press, Cambridge U.K. (2012) [arXiv:1009.1136]
S. Carlip, "Dimensional Reduction in Quantum Gravity", Talk at the "Shapes of Gravity" workshop, Nijmegen, the Netherlands April 2016

[15] S. Carlip, "Spontaneous dimensional reduction in short-distance quantum gravity?", AIP Conf. Proc. 1196 (2009) 72 [arXiv:0909.3329]

[16] J. Ambjørn, A. Goerlich, J. Jurkiewicz, R. Loll, "Nonperturbative Quantum Gravity", Physics Reports 519, 127-210 (2012)

[17] O. Lauscher and M. Reuter, Fractal spacetime structure in asymptotically safe gravity, JHEP 10 (2005) 050
Martin Reuter, Frank Saueressig, Asymptotic Safety, Fractals, and Cosmology, Lectures given at the Sixth Aegean Summer School on Quantum Gravity and Quantum Cosmology, Chora, Naxos (Greece), 2011.
G. Calcagni, Fractal universe and quantum gravity, Phys. Rev. Lett. 104 (2010) 251301 [arXiv:0912.3142]
L. Modesto, Fractal structure of loop quantum gravity, Classical Quant. Grav. 26 (2009) 242002
D. Benedetti, Fractal properties of quantum spacetime, Phys. Rev. Lett. 102 (2009) 111303
D. Benedetti and J. Henson, Spectral geometry as a probe of quantum spacetime, Phys. Rev. D80 (2009) 124036
T.P. Sotiriou, M. Visser and S. Weinfurtner, Spectral dimension as a probe of the ultraviolet continuum regime of causal dynamical triangulations, Phys. Rev. Lett. 107 (2011) 131303 [arXiv:1105.5646]
G. Calcagni, A. Eichhorn and F. Saueressig, Probing the quantum nature of spacetime by diffusion, Phys. Rev. D 87 (2013) 124028 [arXiv:1304.7247]
G. Calcagni and G. Nardelli, Spectral dimension and diffusion in multiscale spacetimes, Phys. Rev. D 88 (2013) 124025, [arXiv:1304.2709].
G. Calcagni, D. Oriti and J. Thürigen, Spectral dimension of quantum geometries, Class. Quant. Grav. 31 135014 (2014)

[18] Steven Weinberg, "Ultraviolet divergences in quantum theories of gravitation". In "General Relativity: An Einstein centenary survey", ed. S. W. Hawking and W. Israel. (Cambridge University Press 1979).

[19] O. Lauscher and M. Reuter, "Ultraviolet Fixed Point and Generalized Flow Equation of Quantum Gravity", Phys. Rev. D65 025013 (2002). M. Reuter and F. Saueressig, Lect. Notes Phys. 863, 185 (2013).

[20] Gianluca Calcagni, Astrid Eichhorn and Frank Saueressig, Probing the quantum nature of spacetime by diffusion, Phys. Rev. D87 (2013) 124028
G. Calcagni and G. Nardelli, Spectral dimension and diffusion in multiscale spacetimes, Phys. Rev. D 88 (2013) 124025.
Gianluca Calcagni, Daniele Oriti, and Johannes Thürigen, Spectral dimension of quantum geometries Class. Quantum Grav.31, 135014 (2014)
Johannes Thürigen, Discrete quantum geometries and their effective dimension arXiv:1510.08706 PhD thesis, Humboldt-Universität zu Berlin

[21] R. Martin and E. Verdaguer, Phys. Rev. D 60, 084008 (1999). Phys. Rev. D 61, 124024 (2000).

[22] M. B. Fröb, A. Roura, E. Verdaguer, JCAP 1208, 009 (2012). M. B. Fröb, A. Roura, E. Verdaguer, JCAP 1407, 048 (2014). M. B. Fröb, JCAP 1412, 010 (2014).

[23] James Mattingly, "Emergence of spacetime in stochastic gravity", Studies in History and Philosophy of Modern Physics Volume 44, Issue 3, August 2013, Pages 329-337

[24] B. L. Hu, "General Relativity as Geometro-Hydrodynamics" Invited talk at the Second Sakharov Conference, Moscow, May, 1996; gr-qc/9607070.

[25] Belinskii, V.A.; Khalatnikov, I.M.; Lifshttz, E.M. . "Oscillatory Approach to a Singular Point in the Relativistic Cosmology". Advances in Physics Vol. 19 (80): 525-573 (1970)

[26] Charles W. Misner, "Mixmaster Universe", Phys. Rev. Lett. 22, 1071 (1969).

[27] D. M. Eardley, R. K. Sachs and E. P-T. Liang, J. Math. Phys. 13, 99 (1972).

[28] S. Carlip, R. A. Mosna and J. P. M. Pitelli, Vacuum Fluctuations and the Small Scale Structure of Spacetime, Phys. Rev. Lett. 107, 021303 (2011)

[29] Gianluca Calcagni, Astrid Eichhorn and Frank Saueressig, Probing the quantum nature of spacetime by diffusion, PRD 87 (2013) 124028

[30] J. Ambjørn, J. Jurkiewicz and R. Loll, Spectral dimension of the universe, Phys. Rev. Lett. 95 (2005) 171301
J. Ambjørn, J. Jurkiewicz, and R. Loll, *Phys. Rev.* **D72**, 064014 (2005)

[31] R. Kommu, *Class. Quant. Grav.* **29**, 105003 (2012)

[32] J. Cooperman, "Scale-dependent homogeneity measures for causal dynamical triangulations", Phys. Rev. D 90, 124053 (2014)
[33] B. L. Hu and D. J. O'Connor, Phys. Rev. D 30, 743 (1984).
[34] S. Sinha and B. L. Hu, Phys, Rev. D 38, 2422 (1988).
[35] Astrid Eichhorn and Tim Koslowski, "Towards phase transitions between discrete and continuum quantum spacetime from the renormalization group", Phys. Rev. D 90, 104039 (2014)
[36] R. Mankin, K. Laas and A. Sauga, "Generalized Langevin equation with multiplicative noise: Temporal behavior of the autocorrelation functions" Phys. Rev. E 83, 061131 (2011).
[37] H. T. Cho and B. L. Hu, "Non-Gaussian noise and nonlinear Langevin Equation in Quantum Brownian Motion"
[38] Ralf Metzler and Joseph Klafter, "The restaurant at the end of the random walk: recent developments in the description of anomalous transport by fractional dynamics", J. Phys. A: Math. Gen. 37 (2004) R161-R208.
[39] M. Reuter and F. Saueressig, "Fractal space-times under the microscope: a renormalization group view on Monte Carlo data", J. High Energy Phys. 12 (2011) 012.
[40] D. N. Coumbe and J. Jurkiewicz. "Evidence for asymptotic safety from dimensional reduction in causal dynamical triangulations", Journal of High Energy Physics 03 (2015) 151.
[41] J. Ambjørn, S. Jordan, J. Jurkiewicz, R. Loll, "A second-order phase transition in CDT", Phys. Rev. Lett. 107, 211303 (2011)

Gravity theory through affine spheres

E. Minguzzi

Dipartimento di Matematica e Informatica "U. Dini", Università degli Studi di Firenze, Via S. Marta 3, I-50139 Firenze, Italy

E-mail: ettore.minguzzi@unifi.it

Abstract. In this work it is argued that in order to improve our understanding of gravity and spacetime our most successful theory, general relativity, must be destructured. That is, some geometrical assumptions must be dropped and recovered just under suitable limits. Along this line of thought we pursue the idea that the roundness of the light cone, and hence the isotropy of the speed of light, must be relaxed and that, in fact, the shape of light cones must be regarded as a dynamical variable. Mathematically, we apply some important results from affine differential geometry to this problem, the idea being that in the transition we should preserve the identification of the spacetime continuum with a manifold endowed with a cone structure and a spacetime volume form. To that end it is suggested that the cotangent indicatrix (dispersion relation) must be described by an equation of Monge-Ampère type determining a hyperbolic affine sphere, at least whenever the matter content is negligible. Non-relativistic spacetimes fall into this description as they are recovered whenever the center of the affine sphere is at infinity. In the more general context of Lorentz-Finsler theories it is shown that the lightlike unparametrized geodesic flow is completely determined by the distribution of light cones. Moreover, the transport of lightlike momenta is well defined though there could be no notion of affine parameter. Finally, we show how the perturbed indicatrix can be obtained from the perturbed light cone.

1. Introduction

In this work we are going to present some recent results on anisotropic gravity theories obtained in [1,2]. Although, ultimately the theory is of Lorentz-Finsler type we wish to introduce it from a different angle, emphasizing the role of affine differential geometry in its development.

The author was motivated by some general ideas and principles on how Physics should be expected to evolve. The history of Physics is intimately connected with that of Geometry and more, broadly, with that of Mathematics. With the work of Albert Einstein we learned that the gravitational phenomena are manifestations of a dynamical metric manifold. The special relativistic spacetime had to be replaced by a more dynamical structure which broke the, at the time undisputed, property of translational invariance.

Similarly, if our not so humble aim is that of trying to anticipate the next physical revolution, it seems a good idea to look for other tacit "rigidity" assumptions and to work to relax them. Some authors have focused on the topology of the manifold. Since one of the main goals remains the unification of the gravitational dynamics with that of quantum fields, and since the latter display peculiar phenomena of quantization in the observables, it has been suggested that, perhaps, the hidden assumption which has to be removed is contained in the very use of differentiable manifolds. Some authors suggested that manifolds should be replaced by graphs,

IOP Conf. Series: Journal of Physics: Conf. Series **880** (2017) 012005 doi:10.1088/1742-6596/880/1/012005

since graphs, being by their very nature quantized, are philosophically compatible with the hints coming from the quantum work. The most notable program in this direction is that of "Causal Set Theory", for which the spacetime manifold, with its causal structure is replaced by a directed graph [3]. The directed links provide a primitive notion of causality.

Although we shall not be concerned with quantum gravity, we wish to mention Causal Set Theory because it recognizes that two spacetime ingredients of general relativity should be preserved in the transition to the quantum world. They are the notion of *causality* and the notion of *spacetime volume*. There are good reason to do so; reasons that might be better understood by briefly recalling the geometrical assumptions behind general relativity.

We recall that in general relativity the spacetime is modeled by a Lorentzian time-oriented manifold (M, g). At each even $x \in M$ we have a Lorentzian bilinear form on T_xM as in special relativity

$$g_x = -(y^0)^2 + (y^1)^2 + (y^2)^2 + (y^3)^2.$$

where $\{y^\mu\}$ are coordinates induced on T_xM by the local coordinates $\{x^\mu\}$ on M. The cone of timelike vectors is, see Fig. 1,

$$\Omega = \{y \colon g_x(y, y) < 0,\ y^0 > 0\}$$

The set of lightlike vectors is

$$\partial\Omega = \{y \colon g_x(y, y) = 0, y \neq 0\}$$

The *velocity space* of massive particles/observers is

$$\mathbb{H} = \{y \colon g_x(y, y) = -1\}.$$

This is the usual *hyperboloid*. As a consequence on spacetime we have a distribution $x \to \Omega_x$ of timelike cones and a distribution $x \to \mathbb{H}_x$ of hyperboloids. A curve $x \colon t \mapsto x(t)$ is

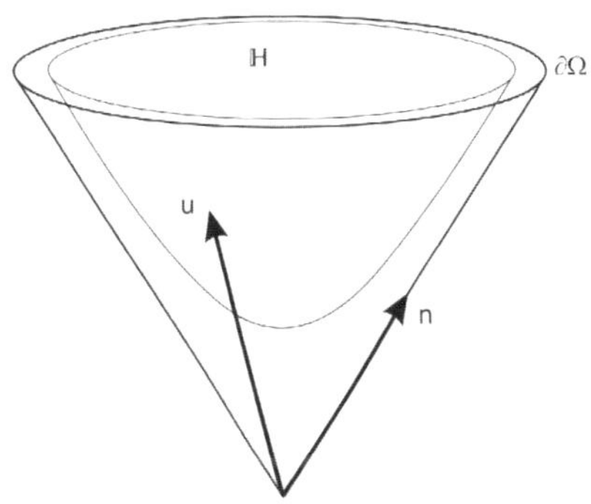

Figure 1. Light cone and velocity space in special and general relativity.

- Timelike: if $g(\dot{x}, \dot{x}) < 0$, (massive particles),
- Lightlike: if $g(\dot{x}, \dot{x}) = 0$, (massless particles).

It is *causal* if at every point it is timelike or lightlike. The *proper time* of a massive particle is

$$\tau = \int_{x(t)} \sqrt{-g(\dot{x}, \dot{x})}\mathrm{d}t.$$

A simple algebraic results tells us that any Lorentzian bilinear form is almost completely characterized by its light cone [4, App. D].

Proposition 1.1. *On a vector space of dimension* $n \geq 3$ *two Lorentzian bilinear forms* η_1, η_2 are proportional *if and only if they have the same timelike cone* Ω.

$$\Omega_1 = \Omega_2 \Leftrightarrow \exists a \in \mathbb{R} : \eta_1 = a^2\eta_2.$$

Since the volume form induced by the bilinear form is $\sqrt{-\det\eta}\,\mathrm{d}y^0 \wedge \cdots \wedge \mathrm{d}y^n$ we have

Corollary 1.2. *On a vector space of dimension* $n \geq 3$ *two Lorentzian bilinear forms* η_1, η_2 coincide *if and only if they share the same timelike cone and they induce the same volume form.*

So, introducing the x-dependence of general relativity we arrive at

IOP Conf. Series: Journal of Physics: Conf. Series **880** (2017) 012005 doi:10.1088/1742-6596/880/1/012005

Corollary 1.3. *Two spacetime metrics* g_1 *and* g_2 coincide *if and only if they induce the same distribution of timelike cones* $x \to \Omega_x$ *and the same volume form* $\mathrm{d}\mu = \sqrt{-\det g}\,\mathrm{d}x^0 \wedge \cdots \wedge \mathrm{d}x^n$.

The distribution of light cones defines the causal relation

$$J = \{(x, y) \in M^2 \colon \ x = y \text{ or there is a future directed causal curve from } x \text{ to } y\}$$

so as a minimal requirement the transition from the non-quantum to the quantum world must preserve the notions of causal relation and spacetime volume. Causal Set Theory has this feature. The spacetime volume is obtained from the number of vertices of the graph while the causal relation is obtained from the direction of the links. While we share the opinion that the pair

spacetime measure + causal order

is central and must be preserved we do not believe that the transition to the quantum world could be accomplished in one big step, by working directly with discretized objects such as graphs. In a sense, it would be like pretending to do Quantum Mechanics by working with the spectra of operators. In Quantum Mechanics the discretized features are a byproduct of the operator theory in Hilbert spaces, so the mathematics must evolve into something richer rather than simpler.

Furthermore, in this work we do not ask: what are the physical quantum theories which accommodate reasonable concepts which can be interpreted as causal relation and spacetime volume. Rather we stay in the non-quantum world and ask: what is the most general non-quantum theory which embodies these notions? So we are really looking for non-quantum modifications of general relativity.

In order to ask this question we have to look for some "rigidity" assumption, with the idea of dropping it. We shall focus on the rigidity assumption that light cones are round, namely have ellipsoidal sections, see Fig. 2. So the new dynamical variable will be the *shape of the light cones*. Since they are no more round, the speed of light in our theory could be (slightly) anisotropic.

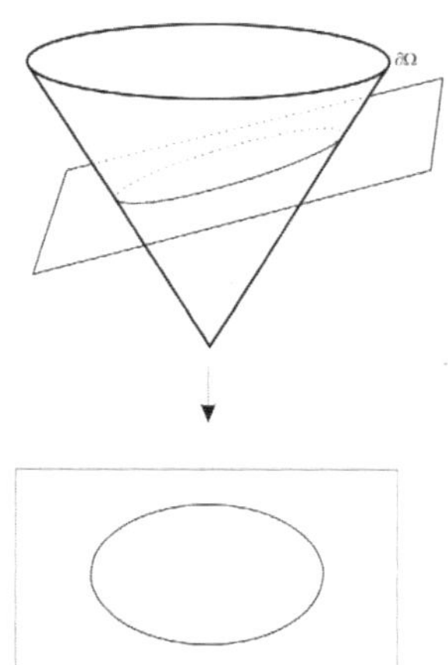

Figure 2. Usual light cones are round.

We have seen that the distribution of light cones in general relativity determines the metric up to a conformal factor. Moreover, unparametrized lightlike geodesics are conformally invariant so the specification of the light cone distribution fixes the motion of massless particles. There is reasonable hope that a distribution of non-isotropic light cones (not to count the specification of a volume form) could fix the motion of massless particles, namely a notion of lightlike geodesics (we shall see that this is indeed the case, cf. Sec. 3.3).

However, on spacetime we have also the massive particles. Since in general relativity the metric gets determined from the cone distribution and the spacetime volume, so does the hyperboloid $\mathbb{H}$ inside the timelike cone. In the anisotropic theory from the same ingredients we don't get a bilinear form on T_xM and so we don't get a velocity space for massive particles.

Here comes the main idea of our work on this problem: to use affine differential geometry to show that there is, in fact, a privileged hypersurface inside the timelike cone which, in the end, can be successfully interpreted as velocity space. So, let us recall some elements of affine differential geometry.

IOP Conf. Series: Journal of Physics: Conf. Series **880** (2017) 012005 doi:10.1088/1742-6596/880/1/012005

2. Affine differential geometry

Affine differential geometry was developed by Wilhelm Blaschke and his school in the first half of the XX century when he realized that non-degenerate (e.g. convex) hypersurfaces admit a natural notion of transverse. So let E be an affine space modeled on a $n+1$-dimensional vector space V endowed with a volume form ω. We stress that we don't have any scalar product on V and so no notion of angle or orthogonality. In order to fix the ideas let us consider a convex hypersurface S, namely a hypersurface bounding a strongly convex set.

Let us pick a point $p \in S$ and let us consider the plane P tangent to p and its translates P_t, $P_0 = P$, where t is a parameter. For every t the plane P_t cuts the hypersurface and selects a compact sector Q_t of finite volume inside S, see Fig. 3. The notion of barycenter (centroid) is affine, namely it is well defined in any affine space. So let G_t be the centroid of Q_t. As $t \to 0$ the point G_t draws a curve which ends at $G_0 = p$. The tangent to the curve at p selects a special direction, which is the direction of the affine normal. Of course, should V be endowed with a scalar product, the Euclidean normal need not coincide with the affine normal, though this turns out to be the case for spheres.

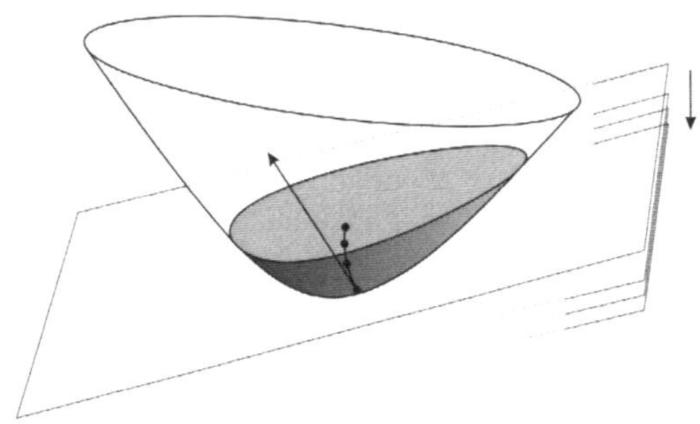

Figure 3. Geometrical construction of the affine normal.

Formally, the immersion is denoted $f\colon N \to E$, where $S = f(N)$. Next, though one does not know the privileged transverse one chooses one $\xi\colon N \to V$, so that $\xi(x)$ is transverse to S at $f(x)$. This choice allows one to split the derivatives, so for $X, Y \in TN$ we have, denoting with D the natural derivative one E

$$D_{f_*(X)} f_*(Y) = f_*(\nabla_X Y) + h(X,Y)\xi, \qquad \text{(Gauss equation)}$$
$$D_{f_*(X)} \xi = -f_*(S(X)) + \tau(X)\xi. \qquad \text{(Weingarten equation)}$$

which are used to define a torsionless connection ∇, a metric h, an endomorphism S and a one-form τ on N. Here the transverse field is determined by $n+1$ arbitrary parameters, so we shall need $n+1$ conditions to fix it.

Observe that the affine space has two natural structures, namely the torsionless (flat) connection D and the volume form ω which share the compatibility condition $D\omega = 0$ (the volume form is translationally invariant). So, one wants the same properties to hold on N where on this manifold we have already identified a torsionless connection ∇, while the volume form can be defined through $\theta = f^*(i_\xi \omega)$. By imposing the (equiaffine) condition $\nabla\theta = 0$ we remove some of the arbitrariness which resides in the choice of transverse ξ (we are imposing n equations, since the derivative takes n-directions). It turns out that this condition is equivalent to $\tau = 0$. A manifold endowed with a pair (∇, θ) satisfying the condition $\nabla\theta = 0$ is called *equiaffine space*. Its connection ∇ has necessarily a symmetric Ricci tensor. The notion of equiaffine space generalizes that of affine space with translationally invariant volume form.

Actually, here we can really define the volume form following a different route, namely, for any θ-positively oriented n-tuple, set

$$\theta'(X_1, \cdots, X_n) = |\det h(X_i, X_j)|^{1/2}.$$

So the remaining condition demands that there should be no ambiguity in the definition of volume, that is $\theta' = \theta$. In this way ξ is fixed unambiguously to the affine (Blaschke) normal, and so we obtain, as a consequence, the affine volume form θ, the affine connection ∇, the affine metric h, and the affine shape operator S.

From a privileged transverse direction we can introduce an important concept. Indeed, suppose that the transverse directions meet at the same point, see Fig. 4. Then the hypersurface is called *affine sphere* and the point of convergence is the *center* of the sphere.

It turns out that the are just three types of affine spheres: (a) the improper spheres, for which the the center is at infinity, namely all the privileged transverse directions are parallel, (b) the (proper) elliptic spheres, for which the center is inside the convex set bounded by the hypersurface, and (c) the (proper) hyperbolic spheres, for which the center is outside the convex set bounded by the hypersurface. Some deep theorems from affine differential geometry assure that the improper affine spheres are just the paraboloids while the elliptic affine spheres are just the ellipsoids.

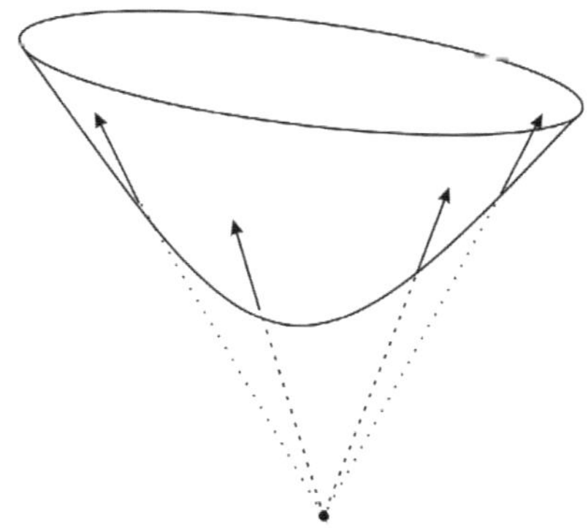

Figure 4. The definition of affine sphere.

What about the hyperbolic affine spheres? Eugenio Calabi [5] in 1971 conjectured that they are much more abundant, and that, in fact, up to homotheties there is a one-to-one correspondence between open sharp convex cones and hyperbolic affine spheres. Namely, every hyperbolic affine sphere is asymptotic to an open sharp convex cone and every such cone contains, up to homotheties, a unique hyperbolic affine sphere asymptotic to it, see Fig. 5. Calabi's conjecture was proved by Cheng and Yau in 1977, some steps being clarified by other mathematicians.

Suppose that we have been given on E a centroaffine hypersurface, namely a hypersurface transverse to the half-lines starting from some point $o \in E$. It is convenient to regard E as a vector space V with origin in o, to introduce linear coordinates $\{y^\alpha\}$ over V in such a way that the hyperplane $y^0 = 0$ does not intersect the closure of the cone generated by the hypersurface and o, saved for the origin. Moreover, it is convenient to parametrize the hypersurface using Klein's type coordinates ($v^i = y^i/y^0$) and regard the hypersurface as the embedding

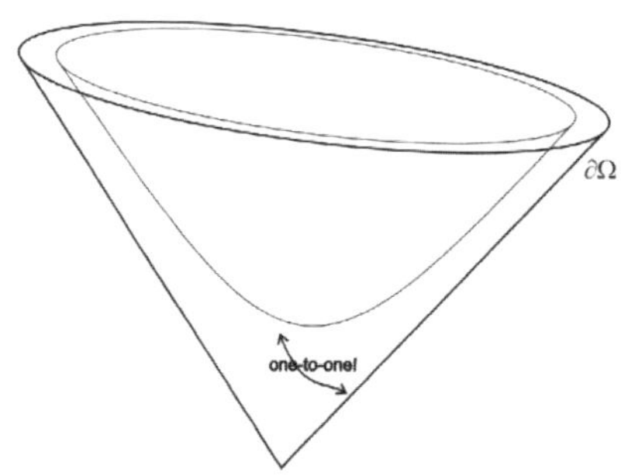

Figure 5. Calabi's conjecture.

$$f : \boldsymbol{v} \mapsto y(\boldsymbol{v}) = -\frac{1}{u(\boldsymbol{v})}(1, \boldsymbol{v})$$

so that $N = D \subset R^n$. Then the hypersurface is an affine sphere if u is convex, vanishes at the boundary of the convex set D and

$$\det u_{ij} = \left(-\frac{1}{u}\right)^{n+2}, \quad u|_{\partial D} = 0, \qquad \text{(affine sphere equation).} \tag{1}$$

Here n is the affine sphere dimension (we shall be interested in $n = 3$). The mathematical problem was precisely that of showing that this equations admits one and only one solution.

3. Physical interpretation

Now, we return to our physical problem. We needed a special hypersurface inside the non-round light cone so as to assign to it the role of velocity space of the observer. Our idea is now to let it coincide with the affine sphere of the light cone. In this way the distribution of light cones determines the velocity space of the observer, and hence the dispersion relation.

Observe that the coordinates $\{y^\alpha\}$ can be fixed as follows: first we choose a timelike direction and choose the y^0-axis along it, then the other axes are chosen so that the hyperplane $y^0 = 1$ is tangent to the hypersurface and the affine metric h is the identity at $y = (1, 0, \cdots, 0)$. The coordinates, up to rotation of the space coordinates $\{y^i\}$ are uniquely determined by the choice of timelike direction and are called *observer coordinates* relative to the observer selected by that direction. The coordinates $\boldsymbol{v}$ defined through $v^i = y^i/y^0$ are nothing else, in our physical interpretation, than the velocity of the particle as seen from the observer, while the domain D is nothing but the (observer dependent) shape of the domain of allowed velocities for massive particles. The remarkable fact is that the variables which most simplify the affine sphere PDE are precisely the variables which have physical meaning.

In our theory u is nothing but the classical Lagrangian for the particle (per unit mass). For instance, whenever the domain D is isotropic the solution of the Monge-Ampére equation is $u = -\sqrt{1 - \boldsymbol{v}^2}$, so that we recover the general relativistic kinematical space from the assumption of light speed isotropy. It turns out that in observer coordinates the function u has always the expansion

$$u = -1 + \tfrac{1}{2}\boldsymbol{v}^2 + o(v^2),$$

which is an equivalent form of expressing the fact that we are in observer coordinates. Furthermore, the Legendre transform $u^*(\boldsymbol{p})$ represents the classical Hamiltonian (per unit mass) for the theory and in observer coordinates it admits the expansion

$$E(\boldsymbol{p}) = u^*(\boldsymbol{p}) = 1 + \tfrac{1}{2}\boldsymbol{p}^2 + o(p^2),$$

it can be shown that $\boldsymbol{p} \mapsto (-u^*(\boldsymbol{p}), \boldsymbol{p})$ is an affine sphere dual to that on velocity space with which we worked so far.

It can be observed that we have obtained the velocity space on $E = T_xM$ by regarding it as an affine space endowed with a translationally invariant volume form and by assuming to have been given a timelike cone Ω_x on T_xM centered at 0. In order to have a translationally invariant volume form at each event we need a volume form on spacetime. So our spacetimes which we term *affine sphere spacetimes* can be equivalently characterized by two properties, namely as (a) a distribution of affine spheres with center in the zero section of TM or as (b) a pair given by a spacetime volume and a cone distribution. Thanks to the latter characterization the affine sphere spacetimes reflect the notions of measure and order, precisely the sort of objects we were looking for. Actually, there is a third characterization which will be mentioned in the next section.

3.1. Lorentz-Finsler geometry

In this section we clarify that our theory is a special type of Lorentz-Finsler theory.

In Finslerian generalizations of general relativity the spacetime is a $n+1$-dimensional manifold endowed with a Finsler Lagrangian $\mathscr{L} \colon \Omega \to \mathbb{R}$, $\Omega \subset TM\backslash\{0\}$, where Ω is an open sharp convex cone subbundle of the slit tangent bundle, $\mathscr{L}$ is positive homogeneous of degree two, that is $\forall s > 0, y \in \Omega_x$, $\mathscr{L}(x, sy) = s^2\mathscr{L}(x, y)$, $\mathscr{L}$ is negative on Ω and converges to zero at the boundary $\partial\Omega$, and finally, the fiber Hessian $g_{\mu\nu} = \partial^2\mathscr{L}/\partial y^\mu \partial y^\nu$ is Lorentzian. We shall not demand $\mathscr{L}$ to be differentiable at the boundary $\partial\Omega$, namely we adopt the *rough model* discussed in [6]. The set Ω_x represents the set of future directed timelike vectors at $x \in M$.

IOP Conf. Series: Journal of Physics: Conf. Series **880** (2017) 012005 doi:10.1088/1742-6596/880/1/012005

The indicatrix $\mathscr{I}_x \subset \Omega_x$ is the locus where $2\mathscr{L} = -1$ and it represents the velocity space of observers (this is the usual hyperboloid in general relativity). By positive homogeneity the Finsler Lagrangian can be recovered from the indicatrix as follows, for $y \in \Omega_x$

$$\mathscr{L}(x, y) = -s^2/2, \quad \text{where } s \text{ is such that } \ y/s \in \mathscr{I}_x. \tag{2}$$

By positive homogeneity the formulas $\mathscr{L} = \frac{1}{2} g_{\mu\nu} y^\mu y^\nu$, $\frac{\partial \mathscr{L}}{\partial y^\mu} = g_{\mu\nu} y^\nu$, hold true, where the metric might depend on y. If it is independent of y then we are in the quadratic case which corresponds to Lorentzian geometry and general relativity. The Cartan torsion is $C_{\mu\nu\alpha} = \frac{1}{2} \frac{\partial}{\partial y^\alpha} g_{\mu\nu}$. It is symmetric and annihilated by y^μ. The mean Cartan torsion is its contraction

$$I_\alpha := g^{\mu\nu} C_{\mu\nu\alpha} = \frac{1}{2} \frac{\partial}{\partial y^\alpha} \log |\det g_{\mu\nu}|. \tag{3}$$

In a series of recent works we have stressed the importance of the Lorentz-Finsler spaces for which $I_\alpha = 0$, see [1, 7, 8]. They are precisely the affine sphere spacetime which we met before. Observe that the condition $I_\alpha = 0$ implies that we have a well defined volume form on M given by the standard expression $\mathrm{d}\mu = |\det g_{\mu\nu}|^{1/2} \mathrm{d}x^0 \wedge \cdots \mathrm{d}x^n$. In general Lorentz-Finsler spaces there is no such volume form.

For a long time it was believed that Finslerian spacetimes which satisfy $I_\alpha = 0$ are not interesting. In fact a theorem by Deicke establishes that Finsler spaces which satisfy this property are Riemannian and so trivial. Previous researchers did not pay much attention to the signature of the metric. In the spacetime case Deicke's theorem does not really hold, in fact we can prove that the condition $I_\alpha = 0$ is really equivalent to the fact that the indicatrix is a hyperbolic affine sphere centered in the zero section.

The proof passes through the following symmetric tensor on N (Pick cubic form, the $\frac{1}{2}$ factor is due to a better correspondence with Finslerian objects)

$$c(X, Y, Z) = \frac{1}{2}[(\nabla_X h)(Y, Z) + \tau(X) h(Y, Z)]. \tag{4}$$

which in the context of Finsler geometry one evaluates on the indicatrix with respect to the centroaffine normal y (which is equiaffine, namely $\tau = 0$). It turns out that under the pullback of the immersion f, C corresponds to c and its trace I corresponds to $\mathrm{tr}_h c$, so $I = 0$ if and only if $\mathrm{tr}_h c = 0$. But the latter condition together with $\tau = 0$ really implies that y is proportional to the Blaschke normal, so that the affine normal directions meet at o and hence the indicatrix $\mathscr{I}$ is an affine sphere.

So $I = 0$ if and only if the indicatrix is an affine sphere centered at the origin of the tangent space $T_x M$. Once we have the indicatrix we can recover the Finsler Lagrangian from (2) which means that all the metrical properties follow from the indicatrix and hence from the light cone. The relationship between the classical Lagrangian u and the Finsler Lagrangian $\mathscr{L}$ is given by

$$\mathscr{L}((y^0, \mathbf{y})) = -\frac{1}{2}(y^0)^2 u^2(\mathbf{y}/y^0), \tag{5}$$

$$u(\mathbf{v}) = -\sqrt{-2\mathscr{L}((1, \mathbf{v}))}. \tag{6}$$

A remarkable fact concerning the concept of affine sphere spacetime is that it embodies in a nice way the non-relativistic limit [1]. In fact, non-relativistic spacetimes admit the same description of (relativistic) affine sphere spacetimes: the cotangent indicatrix is determined by a distribution of affine spheres which, in the non-relativistic case, have center at infinity. This result is connected with a theorem due to Jörgens, Pogorelov, Calabi, Cheng and Yau that establishes that these improper affine spheres are paraboloids. The algebraic ingredients entering the definition of paraboloid define a one-form field $\mathrm{d}t$, the classical time, and a space metric on its kernel, precisely the geometric ingredients needed to express the geometry of non-relativistic spacetimes [9].

IOP Conf. Series: Journal of Physics: Conf. Series **880** (2017) 012005 doi:10.1088/1742-6596/880/1/012005

3.2. Relativistic invariance

In [2] we have argued that the relativity principle holds at an event x if a linear subgroups of the endomorphisms of T_xM acts transitively on the indicatrix $\mathscr{I}_x$. In other words, the velocity space is homogeneous and so for an observer it becomes impossible to infer its position on velocity space from local measurements. One might ask what are the affine sphere spacetimes which satisfy this form of relativity principle. I have shown that there are just three possibilities connected with a classification of homogeneous cones due to Vinberg [10]. In these models, due to homogeneity, the shape of D is really independent of the observer.

The first model is isotropic, i.e. D is a ball, and has the usual Finsler Lagrangian

$$\mathscr{L} = \frac{1}{2}\big(-(y^0)^2 + \mathbf{y}^2\big). \tag{7}$$

The second model has a velocity domain D of tetrahedral shape, with Finsler Lagrangian

$$\begin{aligned}\mathscr{L}_C = -\frac{1}{2}\big[&(y^0 - y^1 - y^2 - y^3)^{1/2}(y^0 - y^1 + y^2 + y^3)^{1/2}\\ &(y^0 + y^1 - y^2 + y^3)^{1/2}(y^0 + y^1 + y^2 - y^3)^{1/2}\big],\end{aligned} \tag{8}$$

This model falls into a family investigated by Bogoslovsky and Goenner [11, 12] but from the geometric point of view was also considered by many other authors including Berwald and Moór [13, 14], Matsumoto [15] and Calabi [5].

The third model was not previously noticed. The Finsler Lagrangian is

$$\mathscr{L} = -\frac{2}{3^{3/4}}\Big(\frac{1}{2}y^0 + \frac{\sqrt{3}}{2}y^3\Big)^{1/2}\Big(\Big(\frac{\sqrt{3}}{2}y^0 - \frac{1}{2}y^3\Big)^2 - (y^1)^2 - (y^2)^2\Big)^{3/4}. \tag{9}$$

and the domain D is a cone in which the height is equal to the diameter of the base.

As can be seen the isotropic model is the only one with C^1 dependence of the speed of light on direction, so in the classical derivation of the Lorentz transformations the isotropy assumption could be replaced by a C^1 light velocity assumption.

For small velocities, namely for $y^i/y^0 \ll 1$ all the previous expressions reduce themeselves to (7), the special relativistic case. This is so because we wrote these models in observer coordinates.

I also considered general relativistic Lorentz-Finsler affine sphere spacetimes which in the analogous low-velocity limit reduce themselves to notable spacetimes of GR. For instance, with $r = \sqrt{z^2 + \rho^2}$ the next Lagrangian

$$\begin{aligned}\mathscr{L} = -\frac{2(1-\frac{2m}{r})}{3^{3/4}}&\left(\left(\tfrac{1}{2}\,\mathrm{d}t_S + \tfrac{\sqrt{3}}{2}\left(1-\tfrac{2m}{r}\right)^{-1}\frac{(1-2m\rho^2/r^3)\mathrm{d}z + 2mz\rho\,\mathrm{d}\rho}{\sqrt{1-2m\rho^2/r^3}}\right)^2\right)^{1/4}\\ &\left(\left(\tfrac{\sqrt{3}}{2}\,\mathrm{d}t_S - \tfrac{1}{2}\left(1-\tfrac{2m}{r}\right)^{-1}\frac{(1-2m\rho^2/r^3)\mathrm{d}z + 2mz\rho\,\mathrm{d}\rho}{\sqrt{1-2m\rho^2/r^3}}\right)^2 - \left(1-\tfrac{2m}{r}\right)^{-1}\left(\frac{\mathrm{d}\rho^2}{1-2m\rho^2/r^3} + \rho^2\mathrm{d}\varphi^2\right)\right)^{3/4}.\end{aligned}$$

is conic anisotropic at each event and reduces itself to the Schwarzschild metric in the said limit. Similarly, we obtained conic anisotropic versions of the Kerr-Schild, Kerr-de Sitter, Kerr-Newman, Taub, and FLRW spacetimes. These metrics have been deduced from their low-velocity limit (with respect to some conformal stationary observer) and from the requirement of relativistic invariance at every spacetime event, not from the imposition of dynamical field equation. However, the latter can perhaps be better identified if we regard these metrics as reasonable spacetime solutions.

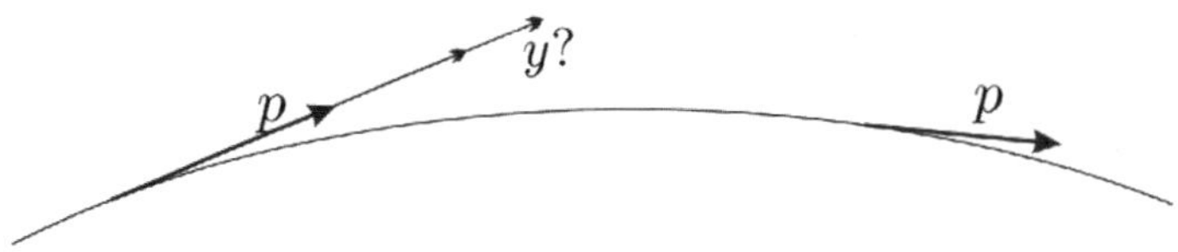

Figure 6. In Lorentz-Finsler theories the unparametrized lightlike geodesics and the transport of momenta make sense while the notion of affine parameter might not make sense.

3.3. Geodesic flow on the lightlike cotangent bundle

We have seen that saved for the isotropic case, the Lorentz-Finsler spaces which satisfy the relativity principle have a non smooth light cone. In general, we might regard these spacetimes as idealizations since they are considerably anisotropic. We are more interested in small deviations from isotropy and so in light cones which could be smooth. Unfortunately, even if the cone is smooth the function $\mathscr{L}$ determined from it has been proved [1] to be $C^{1,1/2}(\bar{\Omega}_x)$ and with non vanishing differential at the cone. This degree of differentiability is insufficient, for the geodesics solve Lagrange's equation

$$\frac{\mathrm{d}}{\mathrm{d}t}\frac{\partial \mathscr{L}}{\partial y^\mu}-\frac{\partial \mathscr{L}}{\partial x^\mu}=0, \qquad y^\mu=\frac{\mathrm{d}x^\mu}{\mathrm{d}t}$$

so that lightlike geodesics are well defined only if $\mathscr{L}(x,y)\in C^2$ for y belonging to the light cone. So it seem that we have a problem in the definition of lightlike geodesic.

This problem has a nice solution which shows that the unparametrized lightlike geodesics really depend only on the distribution of light cones, namely on $\mathscr{L}^{-1}(0)$, and not on the whole Finsler Lagrangian $\mathscr{L}$. Since we can always find another C^2 (subsidiary) Finsler Lagrangian $\hat{\mathscr{L}}$ with the same light cones, the unparametrized lightlike geodesics are indeed well defined provided we can show that they are independent of $\hat{\mathscr{L}}$. This is precisely what we proved in [1].

Indeed, since $\hat{\mathscr{L}}$ and $\hat{\mathscr{L}}'$ vanish on the light cone bundle $N=\cup_x\partial\Omega_x$, which is a codimension one hypersurface on $TM\backslash 0$, and are negative on Ω we must have $\mathrm{d}\hat{\mathscr{L}}=\phi\,\mathrm{d}\hat{\mathscr{L}}'$ for some positive function ϕ on N. This means that for $(x,y)\in N$,

$$\frac{\partial \hat{\mathscr{L}}}{\partial x^\mu}=\phi\frac{\partial \hat{\mathscr{L}}'}{\partial x^\mu}, \qquad \frac{\partial \hat{\mathscr{L}}}{\partial y^\mu}=\phi\frac{\partial \hat{\mathscr{L}}'}{\partial y^\mu}. \tag{10}$$

Let $x(t)=x'(t')$ be a curve in two different parametrization. Thanks to the identity

$$\Big(\frac{\mathrm{d}t}{\mathrm{d}t'}\Big)^2\phi^{-1}\Big(\frac{\mathrm{d}}{\mathrm{d}t}\frac{\partial \hat{\mathscr{L}}}{\partial y^\mu}-\frac{\partial \hat{\mathscr{L}}}{\partial x^\mu}\Big)=\Big(\frac{\mathrm{d}t'}{\mathrm{d}t}\Big)^{-1}\Big(\frac{\mathrm{d}}{\mathrm{d}t}\log\Big(\phi\frac{\mathrm{d}t'}{\mathrm{d}t}\Big)\Big)\frac{\partial \hat{\mathscr{L}}'}{\partial y^\mu}+\frac{\mathrm{d}}{\mathrm{d}t'}\frac{\partial \hat{\mathscr{L}}'}{\partial y^\mu}-\frac{\partial \hat{\mathscr{L}}'}{\partial x^\mu}$$

we have that every lightlike $\hat{\mathscr{L}}$-geodesic is a lightlike $\hat{\mathscr{L}}'$-geodesic for a parameter t' such that $\phi\frac{\mathrm{d}t'}{\mathrm{d}t}=cnst$. Thus unparametrized lightlike geodesics do not depend on the choice of $\hat{\mathscr{L}}$.

But there is much more to it. Notice that if $x(t)=x'(t')$ with $x(t)$ a $\hat{\mathscr{L}}$-lightlike geodesic and $x'(t')$ a $\hat{\mathscr{L}}'$-lightlike geodesic, then $\phi\frac{\mathrm{d}t'}{\mathrm{d}t}=cnst$. Given an unparametrized geodesic we say that a $\hat{\mathscr{L}}$-affine parameter t and a $\hat{\mathscr{L}}'$-affine parameter t' are syntonized if $\phi\frac{\mathrm{d}t'}{\mathrm{d}t}=1$. So if the two parameters are syntonized at one event then they are syntonized at every event. But this fact implies that the transport of momenta is well defined since we have the identity

$$\frac{\partial \hat{\mathscr{L}}}{\partial y^\mu}(x(t),\tfrac{\mathrm{d}x}{\mathrm{d}t}(t))=\phi\frac{\partial \hat{\mathscr{L}}'}{\partial y^\mu}(x(t),\tfrac{\mathrm{d}x}{\mathrm{d}t}(t))=\phi\frac{\mathrm{d}t'}{\mathrm{d}t}\frac{\partial \hat{\mathscr{L}}'}{\partial y^\mu}(x'(t'),\tfrac{\mathrm{d}x'}{\mathrm{d}t'}(t')). \tag{11}$$

Thus on the unparametrized lightlike geodesic $x(t) = x'(t')$ if $p_\mu(t_1) = p'_\mu(t'_1)$ at one event then $\phi \frac{\mathrm{d}t'}{\mathrm{d}t} = 1$ at the same event, namely t and t' are syntonized, which means $\phi \frac{\mathrm{d}t'}{\mathrm{d}t} = 1$ at any later event and so $p_\mu(t) = p'_\mu(t')$ at any later event. That is, the transport of lightlike momenta is well defined as independent of the subsidiary Lagrangian.

There is a similar Hamiltonian argument which allows us to reach the same conclusion. Let $\hat{\mathscr{L}} \colon \bar{\Omega} \to \mathbb{R}$ be a subsidiary Lagrangian and let $\hat{\mathscr{H}} \colon \bar{\Omega}^* \to \mathbb{R}$ be its Legendre transform. Here Ω^* is the polar cone of Ω, cf. [16, Sec. 2.5], $\hat{\mathscr{H}}$ vanishes on $N^* = \partial\Omega^*$, it has non-vanishing differential there and is negative in Ω^*.

The dynamics induced by a subsidiary Lagrangian $\hat{\mathscr{L}}$ can also be obtained from Hamilton's equations

$$i_X \omega = -\mathrm{d}\hat{\mathscr{H}},$$

where $\omega = \mathrm{d}p_\alpha \wedge \mathrm{d}x^\alpha$ is the symplectic form and X is a vector field on the slit cotangent bundle $T^*M \backslash 0$ defining the dynamical system on momentum space. Due to

$$L_X \omega = (\mathrm{d}i_X + i_X \mathrm{d})\omega = -\mathrm{d}\mathrm{d}\hat{\mathscr{H}} = 0,$$

the flow preserves the Hamiltonian and so it is tangent to $N^* = \hat{\mathscr{H}}^{-1}(0)$.

Any other subsidiary Hamiltonian $\hat{\mathscr{H}}'$ vanishes on N^*, thus there is a function $\varphi > 0$ such that $\hat{\mathscr{H}}' = \varphi \hat{\mathscr{H}}$ on N^*. Let X' be the field determined by $\hat{\mathscr{H}}'$, then since the integral curves stay in N^*

$$i_{X'} \omega = -\mathrm{d}\hat{\mathscr{H}}' = -\hat{\mathscr{H}} \mathrm{d}\varphi - \varphi \mathrm{d}\hat{\mathscr{H}} = -\varphi \mathrm{d}\hat{\mathscr{H}},$$

which implies, by the non degeneracy of the symplectic form, $X = \varphi X'$. In other words the integral lines coincide and so there is a well defined unparametrized flow on N^* determined solely by this codimension one hypersurface. Due to the positive homogeneity of the Hamiltonians, this flow is invariant under homotheties, a fact which for any given event and direction on the base M allows us to find a unique projected integral line passing through that event in that direction. This is the unparametrized lightlike geodesic on M.

The possibility of defining the transport of momenta without the need for an affine parameter is a remarkable feature, indeed the notion of affine parameter in general relativity is, for most part, an unnecessary concept quite detached from observations.

3.4. Perturbation of the isotropic theory

Since in practice the light cone is expected to be almost isotropic, we are interested in the perturbation of the isotropic affine sphere solution $u = -\sqrt{1 - v^2}$ as induced by a small deformation of the light cone. It is convenient to introduce the variable

$$w(\boldsymbol{v}) = -\frac{1}{2} u^2 = \mathscr{L}((1, \mathbf{v})). \tag{12}$$

where the coordinates $v^i = y^i / y^0$ correspond to the Klein's parametrization of $\mathbb{H}^n$. The metric of the hyperboloid is

$$h = h_{ij} \mathrm{d}v^i \mathrm{d}v^j = \frac{1}{(1 - v^2)^2} \{(1 - v^2)\delta_{ij} + v_i v_j\} \mathrm{d}v^i \otimes \mathrm{d}v^j.$$

Perturbing (1) we obtain

$$(\delta^{ij} - v^i v^j)(\delta w)_{ij} + 2v^j (\delta w)_j - 2\delta w = 0, \tag{13}$$

IOP Conf. Series: Journal of Physics: Conf. Series **880** (2017) 012005 doi:10.1088/1742-6596/880/1/012005

where the new domain D is obtained from the unit ball B through a displacement $-\delta w|_{\partial B}\hat{\mathbf{v}}$ of the boundary at $\hat{\mathbf{v}} \in \partial B = S^{n-1}$. If we define $\delta w = (1-v^2)f$ we can write it in the form

$$\Delta^{\mathbb{H}} f - 2(n+1)f = 0, \qquad \lim_{v\to 1}[(1-v^2)f(\boldsymbol{v})] = \delta w|_{S^{n-1}} \tag{14}$$

where there appears the Laplacian of the hyperbolic metric. The advantage of this equation stays in its coordinate invariance. Expanding the perturbation in spherical harmonics

$$\delta w = \sum_{\ell,\mu} a_{\ell\mu} U_\ell(v) Y_\ell^\mu(\theta,\phi), \tag{15}$$

where $a_{\ell\mu}$ are constants and $U_\ell(1) = 1$, and replacing in (13) we arrive at a hypergeometric differential equation whose solution is

$$U_\ell = n_\ell v^\ell \,\mathrm{Re}\left(F\left(\tfrac{\ell}{2}-1, \tfrac{\ell}{2}-\tfrac{1}{2}, \ell+\tfrac{n}{2}; v^2+i0^+\right)\right), \tag{16}$$

where n_ℓ is a normalization constant (not to be confused with the space dimension; of course we are mostly interested in the physical dimension $n = 3$).

Here we have taken the real part of the value obtained as the limit $z \to v^2$ for z converging to the real line from the upper half plane. We have done this because the hypergeometric function has a singularity at $z = 1$. Both Re and $+i0^+$ can be removed from the expression in the region $v \leq 1$.

Here we have taken the real part of the value obtained as the limit $z \to v^2$ for z converging to the real line from the upper half plane. We have done this because the hypergeometric function has a singularity at $z = 1$. Finally, the perturbation of the classical Lagrangian is obtained from $u+\delta u = \sqrt{-2(w+\delta w)} = u(1+\frac{1}{2}\frac{\delta w}{w}) + \ldots$, that is $\delta u = \delta w/\sqrt{1-v^2}$. Using this expansion we can relate the deformation of the light cone and the deformation of the indicatrix. We give an example in Figure 7.

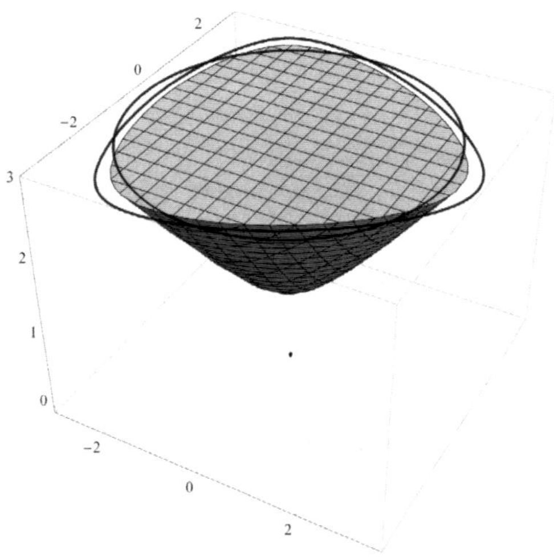

Figure 7. Perturbation of the 3-dimensional isotropic cone with the third spherical harmonic (black ring), and consequent deformation of the hyperboloid.

As it is well known, the Cosmic Microwave Background radiation presents some puzzling anisotropy feature at large angles (small ℓ). We believe that affine sphere relativity might provide the correct theoretical framework for the interpretation and explanation of such anomalies. The idea is simply that they are due to a deformation of the light cone in our spacetime neighborhood or in a spacetime neighborhood of the last scattering hypersurface. These considerations will be expanded in a future work.

4. Conclusions

We have presented a novel kinematical framework for a gravitational theory of Lorentz-Finsler type. This theory has several interesting features. It preserves the correspondence between the notion of spacetime and the pair given by (a) a cone distribution (i.e. a causal order) and (b) a spacetime measure. It also provides a dynamics for massive particles since these ingredients determine univocally a Finsler Lagrangian, and hence all the required metrical concepts of the theory. We have investigated the notion of relativity principle showing that the theories which satisfy it either coincide with GR or are highly anisotropic and so are not expected to be physically relevant in our spacetime neighborhood. Therefore, we have moved to a perturbative approach. Here we could only introduce some of the results we obtained in this direction. We believe that they are really promising since they seem to have the potential to explain some anomalies observed in the CMB radiation.

References

[1] Minguzzi E 2017 *Commun. Math. Phys.* **350** 749–801 arXiv:1702.06739
[2] Minguzzi E 2017 *Phys. Rev. D* **95** 024019 arXiv:1702.06745
[3] Bombelli L, Lee J H, Meyer D and Sorkin R D 1987 *Phys. Rev. Lett.* **59** 521–524
[4] Wald R M 1984 *General Relativity* (Chicago: The University of Chicago Press)
[5] Calabi E 1972 *Symposia Mathematica, Vol. X (Convegno di Geometria Differenziale, INDAM, Rome, 1971)* (Academic Press, London) pp 19–38
[6] Minguzzi E 2016 *Rep. Math. Phys.* **77** 45–55 arXiv:1412.4228
[7] Minguzzi E 2014 *Int. J. Geom. Meth. Mod. Phys.* **11** 1460025 erratum ibid 12 (2015) 1592001. arXiv:1405.0645
[8] Minguzzi E 2015 A divergence theorem for pseudo-Finsler spaces arXiv:1508.06053
[9] Künzle H P 1972 *Ann. Inst. H. Poincaré Phys. Theor.* **17** 337–362
[10] Vinberg È B 1963 *Trudy Moskov. Mat. Obšč.* **12** 303–358 [Trans. Mosc. Math. Soc. 12, 340403 (1963)
[11] Bogoslovsky G Y and Goenner H F 1998 *Physics Letters A* **244** 222–228
[12] Bogoslovsky G Y and Goenner H F 1999 *Gen. Relativ. Gravit.* **31** 1565–1603
[13] Berwald L 1939 *Compositio Math.* **7** 141–176
[14] Moór A 1954 *Acta Math.* **91** 187–188
[15] Matsumoto M and Shimada H 1978 *Tensor N.S.* **32** 275–278
[16] Minguzzi E 2015 *Commun. Math. Phys.* **334** 1529–1551 arXiv:1403.7060

IOP Conf. Series: Journal of Physics: Conf. Series **880** (2017) 012006 doi:10.1088/1742-6596/880/1/012006

Multi-time wave functions

Matthias Lienert[1], Sören Petrat[2] and Roderich Tumulka[1]

[1] Department of Mathematics, Rutgers University, 110 Frelinghuysen Road, Piscataway, NJ 08854-8019, USA

[2] Department of Physics, Jadwin Hall, Princeton University, Washington Road, Princeton, NJ 08544-0708, USA

E-mail: m.lienert@rutgers.edu, spetrat@princeton.edu, tumulka@math.rutgers.edu

Abstract. In non-relativistic quantum mechanics of N particles in three spatial dimensions, the wave function $\psi(\boldsymbol{q}_1, \ldots, \boldsymbol{q}_N, t)$ is a function of $3N$ position coordinates and one time coordinate. It is an obvious idea that in a relativistic setting, such functions should be replaced by $\phi((t_1, \boldsymbol{q}_1), \ldots, (t_N, \boldsymbol{q}_N))$, a function of N space-time points called a multi-time wave function because it involves N time variables. Its evolution is determined by N Schrödinger equations, one for each time variable; to ensure that simultaneous solutions to these N equations exist, the N Hamiltonians need to satisfy a consistency condition. This condition is automatically satisfied for non-interacting particles, but it is not obvious how to set up consistent multi-time equations with interaction. For example, interaction potentials (such as the Coulomb potential) make the equations inconsistent, except in very special cases. However, there have been recent successes in setting up consistent multi-time equations involving interaction, in two ways: either involving zero-range (δ potential) interaction or involving particle creation and annihilation. The latter equations provide a multi-time formulation of a quantum field theory. The wave function in these equations is a multi-time Fock function, i.e., a family of functions consisting of, for every $n = 0, 1, 2, \ldots$, an n-particle wave function with n time variables. These wave functions are related to the Tomonaga–Schwinger approach and to quantum field operators, but, as we point out, they have several advantages.

1. Introduction

Multi-time wave functions arise naturally when considering a particle-position representation of a quantum state in a relativistic setting. They were first introduced by Dirac in 1932 [4] and studied to some extent in the 1930s [5, 2], but not comprehensively. The basic idea is that, in a relativistic space-time, ordinary N-particle wave functions

$$\psi(\boldsymbol{q}_1, \boldsymbol{q}_2, \ldots, \boldsymbol{q}_N, t) \tag{1}$$

with $\boldsymbol{q}_j \in \mathbb{R}^3$ require the choice of a reference frame because they refer to the positions of several particles *at the same time t*. An alternative that does not require a choice of reference frame is to consider a wave function

$$\phi\Big((t_1, \boldsymbol{q}_1), (t_2, \boldsymbol{q}_2), \ldots, (t_N, \boldsymbol{q}_N)\Big) \tag{2}$$

that is a function of N space-time points $x_j = (t_j, \boldsymbol{q}_j)$ and thus of N time variables, called a *multi-time wave function.* We call an N-tuple of space-time points a *space-time configuration,* or simply a *configuration*. The function ϕ is a covariant object: It does not require the choice of any coordinate system on space-time $\mathscr{M}$ if we regard it as a function $\phi : \mathscr{M}^N \to S$, with S a suitable spin space (or $S = \mathbb{C}$ in the spinless case, or S a bundle of spin spaces if $\mathscr{M}$ is curved). More precisely, ϕ will often be defined only on the *spacelike configurations*, that is, on the set $\mathscr{S}_N$ of those N-tuples $(x_1, \ldots, x_N)$ of space-time points $x_j \in \mathscr{M}$ for which any two are spacelike separated or equal; see Figure 1. Note that $\mathscr{S}_N$ is also defined in a covariant way, and is also $4N$-dimensional.

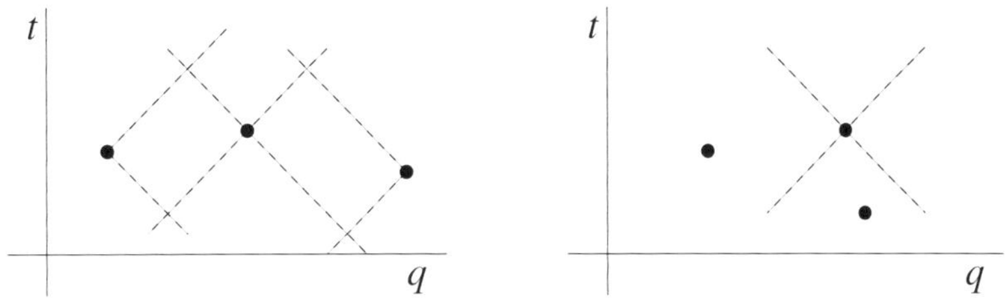

(a) (b)

Figure 1. Example of (a) a spacelike configuration of 3 points (shown as bullets), (b) a non-spacelike configuration. Both examples are shown in Minkowski space-time with light cones (dashed) drawn at 45°.

The relation between ψ and ϕ is simple: In the reference frame to which ψ refers, set all time variables in ϕ equal to obtain ψ,

$$\psi\Big(\boldsymbol{q}_1, \boldsymbol{q}_2, \ldots, \boldsymbol{q}_N, t\Big) = \phi\Big((t, \boldsymbol{q}_1), (t, \boldsymbol{q}_2), \ldots, (t, \boldsymbol{q}_N)\Big). \tag{3}$$

Put differently, this means that ψ is the restriction of ϕ to the *simultaneous configurations* relative to the chosen reference frame.

As we will explain in more detail below, ϕ is usually also directly related to detection probabilities according to the *curved Born rule*: If we place detectors along a spacelike hypersurface Σ, then the probability distribution on Σ^N of the detected configuration has density (relative to the volume defined by the 3-metric on Σ) given by

$$\rho(x_1, \ldots, x_N) = \big|\phi(x_1, \ldots, x_N)\big|^2 \tag{4}$$

for any $x_1, \ldots, x_N \in \Sigma$ and with $|\cdot|^2 = |\cdot|^2_\Sigma$ understood appropriately: for example, for Dirac wave functions with spin space $S = (\mathbb{C}^4)^{\otimes N}$,

$$|\phi|^2 = \overline{\phi}\left[\gamma^{\mu_1} n_{\mu_1}(x_1) \otimes \cdots \otimes \gamma^{\mu_N} n_{\mu_N}(x_N)\right] \phi\,, \tag{5}$$

where $n_\mu(x)$ is the future unit normal vector to Σ at $x \in \Sigma$. In words, the inner product in spin space (and thus the norm) depends on the Lorentz frame, and we need to use the local frame tangent to Σ.

A time evolution law for ϕ is a law that determines ϕ on its entire domain from initial data. The appropriate initial datum in a given reference frame specifies the values of ϕ at those configurations for which all $t_j = 0$ while the $\boldsymbol{q}_j$ are arbitrary; in other words, the initial datum is $\psi(t = 0)$. The kind of evolution analogous to the Schrödinger equation (we set $\hbar = 1$)

$$i\frac{\partial \psi}{\partial t} = H\psi \tag{6}$$

IOP Conf. Series: Journal of Physics: Conf. Series **880** (2017) 012006 doi:10.1088/1742-6596/880/1/012006

is a system of PDEs comprising one equation per time variable,

$$i\frac{\partial\phi}{\partial t_j} = H_j\phi\,, \tag{7}$$

called *multi-time (Schrödinger) equations*. The multi-time equations we are considering are linear equations, so that linear combinations of solutions are solutions. The chain rule and (3) then imply the single-time Schrödinger equation (6) with

$$H = \sum_j H_j \tag{8}$$

at space-time configurations with $t_1 = t_2 = \ldots = t_N$. A central issue about multi-time equations that does not arise for the ordinary Schrödinger equation is that the H_j need to fulfill a *consistency condition*, or else the equations (7) cannot be simultaneously satisfied, or can only for special initial conditions. Therefore, whenever we propose a system of multi-time equations, we need to prove their consistency. For non-interacting particles, the consistency condition is automatically fulfilled, and in fact a unique solution ϕ exists, not only on the spacelike configurations, but on all of $\mathscr{M}^N$ [24]. In contrast, to set up consistent multi-time equations with interaction is challenging. Apart from special examples [6, 7, 3, 30] (and early successes with fields [2, 28, 25], more below), this was successfully done only recently [20, 21, 12, 15]; we will elucidate below how. In all of these examples, the multi-time equations are remarkably simple, see Eqs. (27), (30), and (43) below.

In quantum field theory (QFT), the single-time wave function ψ can often be taken to be an element of Fock space, and thus a function on $\mathcal{Q} = \bigcup_{N=0}^{\infty}(\mathbb{R}^3)^N$ (with the union understood as a disjoint union), the configuration space of a variable number of particles. The corresponding multi-time wave function ϕ is defined on a subset of $\bigcup_{N=0}^{\infty}\mathscr{M}^N$ (with $\mathscr{M}$ the space-time, say $\mathscr{M} = \mathbb{R}^4$), viz., the set of spacelike configurations $\mathscr{S} = \bigcup_{N=0}^{\infty}\mathscr{S}_N$. The multi-time equations are then an infinite system of coupled partial differential equations of the type (7), with interaction implemented via creation and annihilation terms in the H_j. Since the N-particle sector $\phi^{(N)}$ of ϕ (i.e., the part of ϕ on $\mathscr{S}_N$) has N time variables, there are N equations for it; the creation and annihilation terms involve $\phi^{(N+1)}$ and $\phi^{(N-1)}$. In Section 6 we provide an explicit example of such a set of equations which has been shown to be consistent. There is also a simple connection of the multi-time wave function to an expression involving the field operators in the Heisenberg picture, and to the Tomonaga-Schwinger approach. In fact, under suitable conditions, all these three approaches can be translated into each other.

While our motivation comes from the wish for a manifestly covariant particle-position representation of the quantum state, we mention that Elze [9, 10, 11] has recently used multi-time wave functions for a different purpose in connection with certain discrete action principles called Hamiltonian cellular automata: Elze found that for an N-particle system with $N > 1$, such action principles for a multi-time wave function can yield a physically more reasonable time evolution (after setting all times equal) than for a single-time wave function.

Another application of multi-time wave functions finds particular use in multi-time equations *without interaction*: it concerns detection probabilities on a *timelike* hypersurface $\tilde{\Sigma}$, corresponding to detectors waiting for the particles to arrive. Since different particles can arrive at the detectors at different times, the joint distribution $\tilde{\rho}$ of the space-time points of detection naturally involves several time variables. While in the case with interaction, the computation of $\tilde{\rho}$ involves collapses of the wave function for any (attempted) detection, $\tilde{\rho}$ can be computed more easily in the case without interaction, in fact directly from the multi-time wave function ϕ, also at non-spacelike $(x_1 \ldots x_N)$, according to

$$\tilde{\rho}(x_1 \ldots x_N) = \overline{\phi(x_1 \ldots x_N)}\left[\gamma^{\mu_1}\tilde{n}_{\mu_1}(x_1) \otimes \cdots \otimes \gamma^{\mu_N}\tilde{n}_{\mu_N}(x_N)\right]\phi(x_1 \ldots x_N) \tag{9}$$

IOP Conf. Series: Journal of Physics: Conf. Series **880** (2017) 012006 doi:10.1088/1742-6596/880/1/012006

with $\tilde{n}_\mu(x)$ the outward unit normal vector to $\tilde{\Sigma}$ at x, at least in the following two cases: (i) for ideal hard detectors modeled by an absorbing boundary condition on $\tilde{\Sigma}$ [29]; and (ii) in the scattering regime [8], where detectors are placed along a very distant surface in space and stay there, so that the particles coming out of the scattering process do not interact because of the great distance.

The remainder of this article is organized as follows. In Section 2, we explain how the multi-time approach is related to a Hilbert space framework, and how the multi-time wave function relates to detection probabilities. In Section 3, we elucidate the need for and form of consistency conditions. In Section 4, we summarize results showing that interaction potentials make multi-time equations inconsistent. In Section 5, we describe a consistent model with zero-range interaction, and in Section 6 consistent models in QFT.

2. Hilbert spaces, unitarity, and detection probabilities

2.1. Hilbert spaces and unitarity

Unitarity plays a crucial role in the structure of quantum physics. Given a multi-time wave function ϕ on its natural domain $\mathscr{S}$, one cannot, however, insert N arbitrary time variables and expect that the integral of $|\phi|^2$ over the space variables yields unity. The reason is that $\phi(t_1, \boldsymbol{q}_1, \ldots, t_N, \boldsymbol{q}_N)$ is not defined for all configurations, but only for spacelike configurations. Instead, the integral of $|\phi|^2$ over a spacelike Cauchy hypersurface Σ yields unity. More precisely, let us define ϕ_Σ from ϕ through the appropriate "restriction to Σ," i.e., by considering only configurations on Σ:

$$\phi_\Sigma(q) := \phi(q), \quad q \in \Sigma^N. \tag{10}$$

Here N can be either fixed, in the case of a fixed number of particles, or take several values referring to different sectors of Fock space for a variable particle number. Now the integral of $|\phi_\Sigma|^2$ over $\mathcal{Q}_\Sigma = \Sigma^N$ (or $\mathcal{Q}_\Sigma = \bigcup_{N=0}^\infty \Sigma^N$) equals 1, as it must for the curved Born rule (4) to make sense. Thus, ϕ_Σ lies in the appropriate Hilbert space $\mathscr{H}_\Sigma$ associated with Σ, e.g., for N Dirac particles,

$$\mathscr{H}_\Sigma = \mathscr{H}_{N,\Sigma} = S_\pm L^2\big(\Sigma^N, (\mathbb{C}^4)^{\otimes N}\big) \tag{11}$$

with $S_\pm$ the (anti-)symmetrizer assuming the particles are bosons (fermions);[1] the inner product is

$$\langle f|g\rangle = \int_{\Sigma^N} d^3x_1 \cdots d^3x_N \, \overline{f(x_1 \ldots x_N)} \left[\gamma^{\mu_1} n_{\mu_1}(x_1) \otimes \cdots \otimes \gamma^{\mu_N} n_{\mu_N}(x_N)\right] g(x_1 \ldots x_N), \tag{12}$$

so that $\|f\|^2 = \langle f|f\rangle$ equals the integral of the probability density (5).

Since we can take ϕ_Σ as an initial datum, have the multi-time equations determine ϕ on all spacelike configurations, and then consider $\phi_{\Sigma'}$ on any other spacelike Cauchy hypersurface Σ', we obtain a time evolution operator

$$U_\Sigma^{\Sigma'} : \mathscr{H}_\Sigma \to \mathscr{H}_{\Sigma'}, \quad \phi_\Sigma \mapsto \phi_{\Sigma'} \tag{13}$$

that is unitary for the multi-time equations considered here. These unitaries satisfy the composition laws $U_{\Sigma'}^{\Sigma''} U_\Sigma^{\Sigma'} = U_\Sigma^{\Sigma''}$ and $U_\Sigma^\Sigma = I$. Furthermore, they provide the translation between the multi-time wave function ϕ and the Tomonaga-Schwinger equation, as we will discuss in Section 6.

[1] According to the spin-statistics connection, Dirac particles must be fermions, but for toy models we may equally well consider bosonic symmetry.

This family of unitaries is largely equivalent to the multi-time evolution. More precisely, if a family $(\phi_\Sigma)_\Sigma$ consisting of one element in each $\mathscr{H}_\Sigma$ is given, these functions fit together as a single function ϕ on the set $\mathscr{S}$ of spacelike configurations according to (10) if and only if

$$\phi_\Sigma(q) = \phi_{\Sigma'}(q) \quad \text{whenever } q \in \Sigma^N \cap (\Sigma')^N . \tag{14}$$

This relation is satisfied for $U_\Sigma^{\Sigma'}$ obtained from the Tomonaga-Schwinger equation in relevant examples.

In the case of a variable number of particles, the multi-time wave function becomes a function on $\bigcup_{N=0}^\infty \mathscr{S}_N \subset \bigcup_{N=0}^\infty \mathscr{M}^N$ called a "multi-time Fock function" [20]. It can be represented as a sequence of N-particle multi-time wave functions $\phi^{(N)}$,

$$\phi = \left(\phi^{(0)}, \phi^{(1)}, \phi^{(2)}, \ldots\right), \tag{15}$$

where $\phi^{(0)} \in \mathbb{C}$. We write $\phi(q) = \phi^{(N)}(q)$ if $q = (x_1, \ldots, x_N)$. ϕ_Σ then is an element of the Fock space

$$\mathscr{H}_\Sigma = \bigoplus_{N=0}^\infty \mathscr{H}_{N,\Sigma} \tag{16}$$

with

$$\|\phi_\Sigma\|^2 = \sum_{N=0}^\infty \|\phi_\Sigma^{(N)}\|^2 = 1 . \tag{17}$$

2.2. Detection probabilities and the curved Born rule

A full proof of the curved Born rule is the subject of work in progress [16]; here we briefly outline what needs to be proved, as well as prior results.

The unitarity of $U_\Sigma^{\Sigma'}$ entails that $|\phi_\Sigma|^2$ integrates up to 1 and thus qualifies as a probability distribution on $\mathcal{Q}_\Sigma = \Sigma^N$ or $\mathcal{Q}_\Sigma = \bigcup_{N=0}^\infty \Sigma^N$—it is the natural candidate for a curved Born rule. However, this rule cannot simply be postulated, because the usual Born rule in any one fixed Lorentz frame, together with the appropriate collapse rule, already determines the joint probability distribution of the detection events for detectors that we place at different times, including detectors that we place along any Σ. Specifically, if we approximate Σ in the given Lorentz frame by horizontal pieces of hypersurfaces as in Figure 2 with temporal discretization ε, then the usual Born and collapse rules apply to the horizontal pieces as a kind of iterated position measurements with repeated collapse, one after every attempted detection.

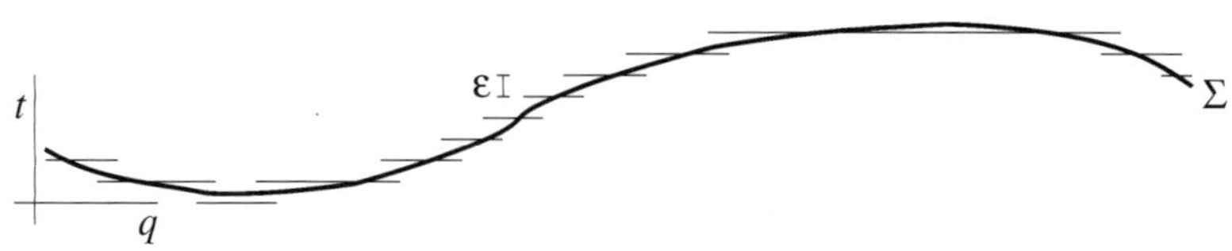

Figure 2. Construction for computing the detection probability distribution on a curved spacelike hypersurface Σ: Σ gets approximated by horizontal pieces with temporal distance ε. The length of the pieces is chosen such that every timelike curve intersects at least one piece.

In the limit $\varepsilon \to 0$ we obtain a distribution on $\mathcal{Q}_\Sigma$, and the claim is that this distribution coincides with $|\phi_\Sigma|^2$. Here it is relevant that wave functions do not propagate faster than light, and that interaction terms in the Hamiltonian do not provide faster-than-light interaction.

IOP Conf. Series: Journal of Physics: Conf. Series **880** (2017) 012006 doi:10.1088/1742-6596/880/1/012006

A preliminary result in this direction was already obtained by Bloch [2] (see also [14, Sec. 2.3] for a discussion in English): He derived the curved Born rule in the case that N particles are confined to spacelike separated regions as in Figure 3 (so they cannot interact), and Σ is horizontal within each region.

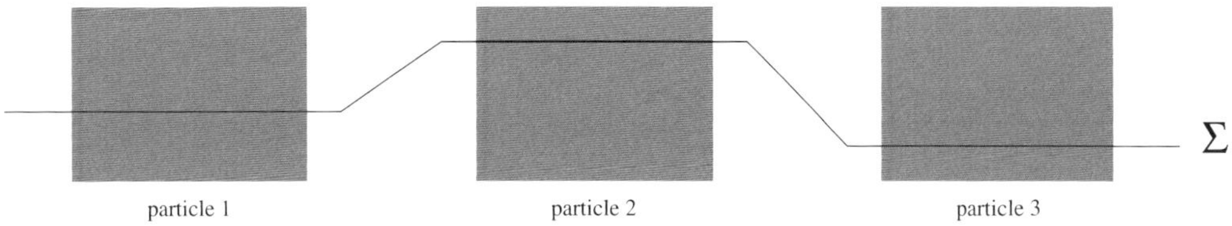

Figure 3. Bloch's [2] result concerns particles confined to spacelike separated regions (gray) and hypersurfaces that are horizontal in each region.

3. Consistency of evolution equations

Multi-time evolution equations are not necessarily consistent. One needs to ensure that the many simultaneous equations (7) do not contradict each other. Consider first the case in which the number N of particles (and thus of time variables) is fixed, and the H_j are time-independent self-adjoint operators on a Hilbert space $\mathscr{H}$. Regarding ϕ as a function $\mathbb{R}^N \to \mathscr{H}$, $(t_1, \ldots, t_N) \mapsto \phi(t_1, \ldots, t_N)$, and for given initial data

$$\phi\big(t_1 = 0, \ldots, t_N = 0\big) = \phi_0 \,, \tag{18}$$

the order of first time-evolving ϕ_0 in t_j and then in t_k or the other way around must be irrelevant, i.e., the following diagram has to commute:

$$\begin{array}{ccc} \phi(0,0) & \xrightarrow{e^{-iH_jt_j}} & \phi(t_j,0) \\ {\scriptstyle e^{-iH_kt_k}}\big\downarrow & & \big\downarrow{\scriptstyle e^{-iH_kt_k}} \\ \phi(0,t_k) & \xrightarrow{e^{-iH_jt_j}} & \phi(t_j,t_k). \end{array} \tag{19}$$

In words, $e^{-iH_jt_j}$ and $e^{-iH_kt_k}$ have to commute for all t_j and t_k, which happens if and only if the H_j commute (in the spectral sense) [22, thm. VIII.13]:

$$[H_j, H_k] = 0 \quad \forall j, k. \tag{20}$$

In that case,

$$\phi(t_1, \ldots, t_N) = e^{-iH_1t_1} \cdots e^{-iH_Nt_N} \phi(0, \ldots, 0) \,. \tag{21}$$

If the H_j depend on time, (20) has to be replaced by the following consistency condition [19]:

$$\Big[i\partial_{t_j} - H_j, i\partial_{t_k} - H_k\Big] = 0 \;\forall j, k. \tag{22}$$

This condition has been shown to be both necessary and sufficient in the case of bounded operators H_j on $\mathscr{H}$ [19] and some other cases [20]. To find a rigorous proof of necessity and sufficiency in general remains a task for future work. We conjecture that (22) is the appropriate consistency condition also when the operators H_j are general differential expressions, such that the multi-time equations remain first-order partial differential equations in the times t_j. This includes the cases of multi-time equations (7) (a) which are defined on a sub-domain of $\mathbb{R}^{4N}$, and (b) with a variable number of time coordinates. Case (a) occurs, e.g., as multi-time wave functions are naturally only defined on $\mathscr{S}$ (see the examples [20, 12]). Case (b) is the typical situation in quantum field theory when formulated in the particle-position representation [20]. Then, the expression for H_j is no longer an operator on Hilbert space; it still is a differential expression, as we will discuss below.

Furthermore, we conjecture that a given single-time dynamics (6) with finite propagation speed and local interactions can always be extended to yield a unique (consistent) multi-time evolution. This is supported by the examples [19, proof of thm. 8] and [20, Sec. 5.4] and shall be the subject of future work.

4. Inconsistency of interaction potentials

The consistency condition (22) is quite restrictive. Two of us obtained a no-go theorem about interaction potentials in [19] which was further extended in [17]. Here, "interaction potentials" are understood as arbitrary smooth matrix-valued functions $V_j(x_1, \ldots, x_N) : \mathbb{R}^{4N} \to (\mathbb{C}^4)^{\otimes N}$ in

$$H_j = H_j^0 + V_j(x_1, \ldots, x_N)\,, \tag{23}$$

where

$$H_j^0 = \sum_{k=1}^{3} i\gamma_j^0 \gamma_j^k \frac{\partial}{\partial x_j^k} + m\gamma_j^0 \tag{24}$$

is the free Dirac Hamiltonian (we set $c = 1$) acting on the coordinates and spin indices of the j-th particle, and γ_j^μ denotes the Dirac gamma matrix γ^μ acting on the j-th tensor factor of $(\mathbb{C}^4)^{\otimes N}$. (Note that the superscript 0 in γ^0 means the timelike component of γ^μ, whereas in H^0 it means something else: the free Hamiltonian.)

The combined results of [19, 17] then state that the only Poincaré invariant potentials V_j which satisfy the consistency condition (22) are $V_j \equiv 0$. Similar theorems for the cases $H_j^0 = -\Delta_j$ and H_j^0 given by arbitrary first-order differential operators can also be obtained [19]. If we drop the requirement about Poincaré invariance, then potentials satisfying (22) can be found, but these seem artificial and consequently only of mathematical interest [17]. Note that the proof in [19] was carried out for smooth potentials only, but we expect the result to hold for singular potentials as well, e.g., the Coulomb potential $|\boldsymbol{q}|^{-1}$.

This result raises the question: How can interaction be achieved in multi-time equations other than via potentials? Two answers, based on zero-range interaction and on particle creation/annihilation, will be provided in Sections 5 and 6. Other approaches have been suggested in [6, 7, 3, 30] (see also [13]); another notable approach is based on integral equations for multi-time wave functions [14, appendix A]; in fact, the well-known Bethe-Salpeter equation [23] belongs to this class.

5. Relativistic zero-range interactions

While the above no-go theorem excludes interaction potentials that are functions, it does not exclude δ potentials, also known as zero-range interactions. It is known [1] from non-relativistic quantum mechanics that zero-range interactions can be implemented rigorously by means of a boundary condition on the wave function at those configurations for which two particles meet. This clearly avoids the use of interaction potentials.

We now describe an example [12, 14] of a consistent multi-time evolution with zero-range interaction for two massless Dirac particles in 1+1 dimensions, $\mathscr{M} = \mathbb{R}^2$. The reasons for setting up the model in this way are the following. In the relativistic case, we need to choose a relativistic Hamiltonian such as the Dirac Hamiltonian. For the latter, it is known [26], however, that zero-range interactions exist in 1+1 but not in higher dimensions. In the massless case, the dynamics becomes particularly simple and even explicitly solvable such that new mathematical techniques not relying on the time-less functional analytic Hilbert space picture become available. These are needed for a manifestly covariant treatment of the model.

The multi-time wave function in this case is a map

$$\phi : \mathscr{S} \subset \mathbb{R}^2 \times \mathbb{R}^2 \to \mathbb{C}^2 \otimes \mathbb{C}^2 \simeq \mathbb{C}^4, \quad (t_1, z_1, t_2, z_2) \mapsto \phi(t_1, z_1, t_2, z_2) \tag{25}$$

and the multi-time equations are given by the free 1+1-dimensional Dirac equations with a boundary condition. The free equations read, with the notation $x_j = (t_j, z_j)$,

$$i\gamma_j^\mu \partial_{j,\mu} \phi(x_1, x_2) = 0, \quad j = 1, 2 \tag{26}$$

in covariant notation, or

$$\begin{aligned} i\frac{\partial}{\partial t_1}\phi(t_1, z_1, t_2, z_2) &= -i\,\sigma_3 \otimes 1_2 \frac{\partial}{\partial z_1}\phi(t_1, z_1, t_2, z_2), \\ i\frac{\partial}{\partial t_2}\phi(t_1, z_1, t_2, z_2) &= -i\,1_2 \otimes \sigma_3 \frac{\partial}{\partial z_2}\phi(t_1, z_1, t_2, z_2). \end{aligned} \tag{27}$$

in Hamiltonian form. Here, 1_2 stands for the 2×2 unit matrix,

$$\gamma^0 = \sigma_1 = \begin{pmatrix} 0 & 1 \\ 1 & 0 \end{pmatrix}, \quad \gamma^1 = \sigma_1\sigma_3 = \begin{pmatrix} 0 & -1 \\ 1 & 0 \end{pmatrix}, \tag{28}$$

and σ_i, $i = 1, 2, 3$ denote the Pauli matrices. The boundary conditions are prescribed (as limits) on the set of collision configurations,

$$\mathscr{C} := \{(t_1, z_1, t_2, z_2) \in \mathbb{R}^2 \times \mathbb{R}^2 : t_1 = t_2,\ z_1 = z_2\}, \tag{29}$$

and one particular example of a suitable boundary condition is given by (denoting the spin components of ϕ as ϕ_i, $i = 1, 2, 3, 4$)

$$\begin{aligned} \phi_2(t, z-0, t, z+0) &= e^{-i\theta}\phi_3(t, z-0, t, z+0), \\ \phi_2(t, z+0, t, z-0) &= e^{+i\theta}\phi_3(t, z+0, t, z-0), \end{aligned} \tag{30}$$

where $\theta \in (-\pi, \pi]$ is a phase.

Initial data are given at equal times as in (18); they have to satisfy the boundary condition as well. The main results are: the multi-time evolution is consistent and can be defined in a rigorous way, the model is interacting in the sense that a generic initial product wave function ϕ_0 becomes entangled with time, both multi-time equations as well as boundary conditions are Lorentz invariant, and the model is compatible with anti-symmetry for indistinguishable particles. Heuristically, the boundary conditions (30) correspond to a spin-dependent δ-potential $(\pi - \theta)\mathrm{diag}(0, 1, -1, 0)\delta(z_1 - z_2)$ at equal times [15]. Finally, the Dirac tensor current

$$j_\phi^{\mu\nu}(x_1, x_2) = \overline{\phi}(x_1, x_2)\gamma^\mu \otimes \gamma^\nu \phi(x_1, x_2) \tag{31}$$

is conserved,

$$\partial_{x_1^\mu} j^{\mu\nu}(x_1, x_2) = \partial_{x_2^\nu} j^{\mu\nu}(x_1, x_2) = 0\,, \tag{32}$$

which, together with the boundary conditions (30), ensures the unitarity of $U_\Sigma^{\Sigma'}$.

The model has also been extended to N particles in 1+1 dimensions in [15], and there are strong indications that non-zero masses will not change the results. An extension to higher dimensions, however, does not seem feasible as then the dimension of $\mathscr{C}$ is too low for boundary conditions to have impact on the dynamics. In conclusion, the results show that interacting dynamics for multi-time wave functions in one spatial dimension can be achieved in a rigorous and manifestly Lorentz invariant way.

IOP Conf. Series: Journal of Physics: Conf. Series **880** (2017) 012006 doi:10.1088/1742-6596/880/1/012006

6. Quantum field theory

Another way of implementing interaction in the multi-time framework is by particle creation and annihilation, i.e., by considering models from quantum field theory. We report here mainly about the results of [20, 21]. Interaction by particle creation naturally suggests itself when we are looking for relativistic theories and expect that interaction should not take place faster than light; besides, perhaps surprisingly, also the consistency condition of multi-time equations (relativistic or not) pushes us, since it excludes interaction through potentials, to considering particle creation. The formulation of models from QFT in terms of multi-time wave functions can be regarded as a new representation of QFT, a multi-time Schrödinger picture particle position representation. As such, it provides an alternative approach to fully relativistic formulations of QFT such as the Tomonaga-Schwinger formalism and quantum fields in the Heisenberg picture. We will later argue that these three pictures are in fact equivalent, in the sense that each can be translated into the others under suitable conditions. (Some of the statements in this direction we show in very general terms, while others we show for specific examples, but we conjecture to hold also for more general models.)

Nevertheless, an advantage of the multi-time framework is that as a mathematical object multi-time wave functions are simpler, since they are locally just functions of finitely many variables, and their evolution is determined by a coupled system of PDEs. Furthermore, it is possible to introduce a cut-off in the multi-time framework [18, 19], but not in the Tomonaga-Schwinger picture. This is so because there is no analogue of spacelike hypersurfaces with a cut-off, but there are versions of the set of all almost spacelike configurations that take into account a finite range of the interaction. On the other hand, the multi-time approach as we present it here has the limitation that it is tied to a particle-position representation of the involved particles (using Fock space), and cannot be applied to a field representation (in which ψ is a functional of a function on 3-space). In fact, Dirac, Fock, and Podolsky [5] considered a model of quantum electrodynamics in a particle representation for the electrons and a field representation for the radiation. As a multi-time formulation, they proposed one time variable for each electron and one time variable for the field. That led them to considering the field on a horizontal hyperplane, although it was a main motivation for multi-time wave functions to avoid being tied to configurations on horizontal hyperplanes; that is why we do not follow their suggestion here.

In the examples we consider here, we take all particles to be Dirac particles; photons could be included by taking a photon wave function to be a complexified Maxwell field [18]. We leave aside issues of the right choice of position observable and make no attempt to remove or redefine the negative energy solutions. We also leave aside the ultraviolet divergence problem and calculate in a non-rigorous way.

6.1. Multi-time equations with particle creation and annihilation

We describe in detail a simple toy model, the "emission-absorption model" [20]. It involves fermionic x-particles that can emit and absorb bosonic y-particles, so that the x-particle number is conserved and the y-particle number is not. We first define the single-time version of the model. Its Hilbert space is given by the tensor product of two Fock spaces,

$$\mathscr{H} = \mathscr{H}_x \otimes \mathscr{H}_y, \qquad \text{with } \mathscr{H}_{x/y} = \bigoplus_{N=0}^{\infty} S_{x/y} L^2(\mathbb{R}^3, \mathbb{C}^4)^{\otimes N}, \tag{33}$$

where S_x is the antisymmetrization operator, and S_y the symmetrization operator.[2] In other words, the wave function ψ in the (M, N)-particle sector takes values in the spinor space $(\mathbb{C}^4)^{\otimes M} \otimes (\mathbb{C}^4)^{\otimes N}$, i.e., it could be explicitly written as $\psi_{r_1 \dots r_M, s_1 \dots s_N}(x^{3M}, y^{3N})$, abbreviating $(x^{3M}, y^{3N}) = (\boldsymbol{x}_1, \dots, \boldsymbol{x}_M, \boldsymbol{y}_1, \dots, \boldsymbol{y}_N)$ with $\boldsymbol{x}_j, \boldsymbol{y}_k \in \mathbb{R}^3$ for all j, k. However, in a given expression, we usually only indicate the indices that other operators than the identity act on. The time evolution of the wave function $\psi_t \in \mathscr{H}$ is given by the Schrödinger equation

$$i\partial_t \psi_t = H\psi_t, \qquad \text{with } H = H_x + H_y + H_{\text{int}}. \tag{34}$$

In order to write down the interaction, we introduce the usual (spinor valued) creation and annihilation operators $a^\dagger, a$ for the x-particles and $b^\dagger, b$ for the y-particles. These satisfy the (anti-)commutation relations

$$\{a_r(\boldsymbol{x}), a^\dagger_{r'}(\boldsymbol{x}')\} = \delta_{rr'}\delta(\boldsymbol{x} - \boldsymbol{x}') \quad \text{and} \quad [b_s(\boldsymbol{y}), b^\dagger_{s'}(\boldsymbol{y}')] = \delta_{ss'}\delta(\boldsymbol{y} - \boldsymbol{y}'), \tag{35}$$

and all other combinations are zero (a and b operators commute). Explicitly written out in position space, they read

$$\big(a_r(\boldsymbol{x})\,\psi\big)(x^{3M}, y^{3N}) = \sqrt{M+1}\,(-1)^M\,\psi_{r_{M+1}=r}\big((x^{3M}, \boldsymbol{x}), y^{3N}\big) \tag{36}$$

$$\big(a^\dagger_r(\boldsymbol{x})\,\psi\big)(x^{3M}, y^{3N}) = \frac{1}{\sqrt{M}}\sum_{j=1}^{M}(-1)^{j+1}\,\delta_{rr_j}\,\delta^3(\boldsymbol{x}_j - \boldsymbol{x})\,\psi_{\widehat{r_j}}\big(x^{3M} \setminus \boldsymbol{x}_j, y^{3N}\big), \tag{37}$$

$$\big(b_s(\boldsymbol{x})\,\psi\big)(x^{3M}, y^{3N}) = \sqrt{N+1}\,\psi_{s_{N+1}=s}\big(x^{3M}, (y^{3N}, \boldsymbol{x})\big) \tag{38}$$

$$\big(b^\dagger_s(\boldsymbol{x})\,\psi\big)(x^{3M}, y^{3N}) = \frac{1}{\sqrt{N}}\sum_{k=1}^{N}\delta_{ss_k}\,\delta^3(\boldsymbol{y}_k - \boldsymbol{x})\,\psi_{\widehat{s_k}}\big(x^{3M}, y^{3N} \setminus \boldsymbol{y}_k\big), \tag{39}$$

where $\hat{\cdot}$ means that the variable is omitted and $(x^{3M} \setminus \boldsymbol{x}_j) := (\boldsymbol{x}_1, \dots, \boldsymbol{x}_{j-1}, \boldsymbol{x}_{j+1}, \dots, \boldsymbol{x}_M)$. Then the contributions to the Hamiltonian are

$$\begin{aligned} H_x &= \int d^3\boldsymbol{x}\, a^\dagger(\boldsymbol{x}) H^0 a(\boldsymbol{x}), \quad H_y = \int d^3\boldsymbol{x}\, b^\dagger(\boldsymbol{x}) H^0 b(\boldsymbol{x}), \\ H_{\text{int}} &= \int d^3\boldsymbol{x}\, a^\dagger(\boldsymbol{x})\Big(g^* \cdot b(\boldsymbol{x}) + g \cdot b^\dagger(\boldsymbol{x})\Big) a(\boldsymbol{x}), \end{aligned} \tag{40}$$

where H^0 is the free Dirac Hamiltonian as in (24), $g \in \mathbb{C}^4$ is a fixed spinor, and the summation over the spinor indices is implicitly understood in the above expressions ($\cdot$ denotes the inner product in spinor space). In order to compare better with the multi-time equation that we are going to introduce next, we write down the Hamiltonian in the position representation:

$$\begin{aligned} (H\psi)(x^{3M}, y^{3N}) &= \sum_{j=1}^{M} H^0_{x_j}\psi(x^{3M}, y^{3N}) + \sum_{k=1}^{N} H^0_{y_k}\psi(x^{3M}, y^{3N}) \\ &+ \sqrt{N+1}\sum_{j=1}^{M}\sum_{s_{N+1}=1}^{4} g^*_{s_{N+1}}\psi_{s_{N+1}}\big(x^{3M}, (y^{3N}, \boldsymbol{x}_j)\big) \\ &+ \frac{1}{\sqrt{N}}\sum_{j=1}^{M}\sum_{k=1}^{N} g_{s_k}\delta(\boldsymbol{y}_k - \boldsymbol{x}_j)\psi_{\widehat{s_k}}(x^{3M}, y^{3N} \setminus \boldsymbol{y}_k). \end{aligned} \tag{41}$$

[2] This choice is again contrary to the spin-statistics connection, since we take both x- and y-particles to be spin-$\frac{1}{2}$ Dirac particles. The spin-statistics *theorem* does not apply here, since our Hamiltonian (41) is unbounded from below.

In this setting, the multi-time wave function ϕ is a spinor-valued function on the set of all spacelike configurations $\mathscr{S} = \bigcup_{M,N=0}^{\infty} \mathscr{S}_{M,N}$, where $\mathscr{S}_{M,N}$ is the set of all spacelike configurations of M x- and N y-particles. Similar to before we write $(x^{4M}, y^{4N}) = (x_1, \dots, x_M, y_1, \dots, y_N)$ with $x_j, y_k \in \mathbb{R}^4$ for all j, k. Then the multi-time evolution equations are given by

$$i\frac{\partial \phi}{\partial x_j^0} = H_{x_j}\phi, \qquad i\frac{\partial \phi}{\partial y_k^0} = H_{y_k}\phi, \tag{42}$$

with

$$\begin{aligned} H_{x_j}\phi(x^{4M}, y^{4N}) &= H^0_{x_j}\phi(x^{4M}, y^{4N}) + \sqrt{N+1}\sum_{s_{N+1}=1}^{4} g^*_{s_{N+1}}\phi_{s_{N+1}}\big(x^{4M}, (y^{4N}, x_j)\big) \\ &\quad + \frac{1}{\sqrt{N}}\sum_{k=1}^{N} G_{s_k}(y_k - x_j)\phi_{\widehat{s_k}}(x^{4M}, y^{4N} \setminus y_k), \\ H_{y_k}\phi(x^{4M}, y^{4N}) &= H^0_{y_k}\phi(x^{4M}, y^{4N}), \end{aligned} \tag{43}$$

where $G : \mathbb{R}^4 \to \mathbb{C}^4$ is a Green's function, i.e., the solution to

$$i\frac{\partial G}{\partial y^0} = H^0_y G \quad \text{with } G(0, \boldsymbol{y}) = g\delta(\boldsymbol{y}) \tag{44}$$

for the fixed spinor $g \in \mathbb{C}^4$. One could also rewrite the multi-time equations in a covariant notation by multiplying by γ^0 and bringing the free Hamiltonian to the left-hand side; then they take the form

$$i\gamma^\mu_{x_j}\partial_{x_j^\mu}\phi + m_y\phi = \dots \quad \text{and} \quad i\gamma^\mu_{y_k}\partial_{y_k^\mu}\phi + m_y\phi = 0\,. \tag{45}$$

Note that for equal times $x_j^0 = y_k^0 = t$ for all j, k, the equations (42) indeed reduce to the single-time Schrödinger equation (34) with Hamiltonian (41). In fact, the multi-time equations are little more than the terms in the one-time H (41) grouped into terms associated with each particle—remarkably simple. Furthermore, the solution ϕ to (42) has the same permutation symmetry as the initial datum, but now as a permutation of *space-time* points:

$$\begin{aligned} \phi_{r_{\sigma(1)}\dots r_{\sigma(M)}, s_1\dots s_N}\big(x_{\sigma(1)}, \dots, x_{\sigma(M)}, y^{4N}\big) &= (-1)^\sigma \phi_{r_1\dots r_M, s_1\dots s_N}\big(x_1, \dots, x_M, y^{4N}\big), \\ \phi_{r_1\dots r_M, s_{\sigma(1)}\dots s_{\sigma(N)}}\big(x^{4M}, y_{\sigma(1)}, \dots, y_{\sigma(N)}\big) &= \phi_{r_1\dots r_M, s_1\dots s_N}\big(x^{4M}, y_1, \dots, y_N\big) \end{aligned} \tag{46}$$

for all permutations σ, with $(-1)^\sigma$ the sign of the permutation.

It was shown in [20] that also in this case with a variable number of time variables, the commutator condition (22) is necessary and sufficient for the consistency of the multi-time equations. It turns out that for the operators from (43) these commutators indeed vanish on all spacelike configurations (including collision configurations where some x_j or y_k are equal). Thus, on a non-rigorous level, the equations (42) possess a unique solution on $\mathscr{S}$, given initial conditions for equal times. In the two cases when $m_y = 0$ or $g^*\gamma^0 g = 0$, the commutators vanish on *all* configurations (also non-spacelike ones), but we believe these to be exceptional cases due to the simplicity of the model.

A few remarks about the multi-time model seem in order.

- When setting up multi-time equations starting from a single-time model with Hamiltonians such as (41), there is usually some freedom in how to distribute the different terms among the different multi-time equations. In (41), since the creation term contains a sum over all x- and y-particles, one can attribute its summands either to H_{x_j} or to H_{y_k}. That is, one could also set up the multi-time equations (42) with

$$H_{x_j}\phi(x^{4M},y^{4N}) = H^0_{x_j}\phi(x^{4M},y^{4N}) + \sqrt{N+1}\sum_{s_{N+1}=1}^{4} g^*_{s_{N+1}}\phi_{s_{N+1}}\big(x^{4M},(y^{4N},x_j)\big),$$
$$H_{y_k}\phi(x^{4M},y^{4N}) = H^0_{y_k}\phi(x^{4M},y^{4N}) + \frac{1}{\sqrt{N}}\sum_{j=1}^{M} G_{s_k}(y_k-x_j)\phi_{\widehat{s_k}}(x^{4M},y^{4N}\setminus y_k). \qquad (47)$$

However, it turns out that the multi-time equations with (47) are equivalent to the ones with (43) on the set $\mathscr{S}$ of spacelike configurations. This fact is rather surprising: How can two sets of equations that give different, non-equivalent expressions for certain partial derivatives of ϕ (viz., for $\partial\phi/\partial x_j^0$) be equivalent? This has to do with the set $\mathscr{S}$ of spacelike configurations: If ϕ were defined on the set $\bigcup_{M,N=0}^{\infty}\mathscr{M}^{M+N}$ of *all* configurations (spacelike or not), then the equations (47) would *not* be equivalent to (43) because one could simply compute $\partial\phi/\partial x_j^0$ and see whether it agrees with the right-hand side of the first equation in (43) or that in (47). However, for ϕ defined on $\mathscr{S}$ there are certain configurations where $\partial\phi/\partial x_j^0$ cannot be computed: the configurations where an x-particle and a y-particle meet, $x_j = y_k$. There, varying x_j^0 while keeping y_k^0 fixed would lead out of $\mathscr{S}$, so $\partial\phi/\partial x_j^0$ is not defined, whereas $(\partial/\partial x_j^0 + \partial/\partial y_k^0)\phi$ is. And the crucial term $G_{s_k}(y_k - x_j)$ vanishes in $\mathscr{S}_N$ *except at precisely those configurations* where $x_j = y_k$. At those configurations, the PDEs (42) are understood as determining the derivatives of ϕ that *are* defined, and that is why different choices of H_{x_j} and H_{y_k} can determine these derivatives in the same way, and thus define the same time evolution of ϕ.
- The model would also be consistent on $\mathscr{S}$ if the wave function had been chosen symmetric under exchange of x-particles, i.e., if the x-particles were bosonic. However, it is interesting to note that the model would *not* be consistent on $\mathscr{S}$ if the wave function was antisymmetric in the y-particles, i.e., if the y-particles were fermionic. (It would then not be consistent in the special cases $m_y = 0$ or $g^*\gamma^0 g = 0$ either; in fact, the commutators (22) for $x_i \neq x_j$ would be non-zero at *all* configurations.) In other words, the model is only consistent if the fermion number is conserved (which is believed to be one of the fundamental conservation laws of the Standard Model).
- The operator H_{x_j} from (43) or (47) is *not* an operator on a Hilbert space, because it involves changing a time variable, viz., setting $y^0_{N+1} = x^0_j$. It is a perfectly fine operator acting on ϕ, but cannot be understood as an operator on a Hilbert space. That is not surprising keeping in mind (i) that $H_{x_j}\phi$ provides $i\partial\phi/\partial x_j^0$ for a specific sector $\phi^{(M,N)}$ but depends on neighboring sectors $\phi^{(M,N\pm1)}$; and (ii) that the functions ϕ with fixed time coordinates and arbitrary space coordinates do not form a Hilbert space, as discussed in Section 2.1 above. The multi-time equations do define evolution operators $U_\Sigma^{\Sigma'}$ between Hilbert spaces $\mathscr{H}_\Sigma$ and $\mathscr{H}_{\Sigma'}$, as elucidated in Section 6.3 below.
- Initial data that determine ϕ can, in fact, be specified on any spacelike hypersurface Σ by specifying ϕ on $\mathcal{Q}_\Sigma = \bigcup_{M,N=0}^{\infty}\Sigma^{M+N}$, i.e., for all configurations on Σ.
- The multi-time equations (42) are not fully Lorentz invariant because they involve the choice of a fixed spinor g and the only Lorentz invariant 4-spinor is the zero spinor (which would make the model interaction-free). However, if the y-particles had integer spin, g could be replaced by a Lorentz-invariant object, and the equations would be fully covariant.

- As mentioned before, the equations (42) are actually not rigorously defined since they contain δ distributions in the creation terms via the Green's function G (which is a distribution). This is an instance of the ultraviolet divergence in QFT. The equations (42) could be defined in a rigorous sense by introducing an ultraviolet cut-off, i.e., replacing the δ distribution in (44) with some localized function φ. This is described in more detail in [18], where the evolution equations with such a cut-off are explicitly written down, and a proof for the consistency and existence and uniqueness of solutions on a (not Lorentz invariant) subset of the set of spacelike configurations is sketched. However, this obviously breaks the Lorentz invariance of the equations. Since the multi-time formalism is about finding a fundamentally relativistic invariant quantum theory, we chose to proceed with formal calculations. (However, note that for those equations where a renormalization scheme works, one could also apply this to the multi-time equations.) It will be of interest to explore whether and how multi-time equations can be set up for Hamiltonians for which creation and annihilation terms are defined by means of boundary conditions instead of δ functions [27].

6.2. Relation to field operators in the Heisenberg picture

There is a simple relation between the multi-time wave function ϕ, i.e., the solution to (42), and an expression involving creation and annihilation operators in the Heisenberg picture. In the latter, the state vector Ψ is fixed and only the operators are subject to the dynamics, i.e., we define for $x = (t, \boldsymbol{q})$ that

$$a(x) = e^{iHt}a(\boldsymbol{q})e^{-iHt}, \qquad b(x) = e^{iHt}b(\boldsymbol{q})e^{-iHt} \tag{48}$$

with H as in (41). Then one can show that

$$\phi(x^{4M}, y^{4N}) = \frac{(-1)^{M(M-1)/2}}{\sqrt{M!N!}}\langle\emptyset|a(x_1)\cdots a(x_M)b(y_1)\cdots b(y_N)|\Psi\rangle \tag{49}$$

on spacelike configurations, where $|\emptyset\rangle$ is the Fock vacuum. Equivalently, one can write using the field operators $\Phi_x = a + a^\dagger$, $\Phi_y = b + b^\dagger$,

$$\phi(x^{4M}, y^{4N}) = \frac{(-1)^{M(M-1)/2}}{\sqrt{M!N!}}\langle\emptyset|\Phi_x(x_1)\cdots\Phi_x(x_M)\Phi_y(y_1)\cdots\Phi_y(y_N)|\Psi\rangle \tag{50}$$

on *collision-free* spacelike configurations, i.e., those where none of the x_j's and y_k's are equal.

In fact, we could take (50) as the definition of ϕ on the collision-free spacelike configurations. Note that (50) would define some multi-time function $\tilde{\phi}$ also for configurations with collisions, and even for non-spacelike configurations. In the absence of interaction, $\tilde{\phi}$ agrees with ϕ, but in the presence of interaction the two differ at collision configurations, and ϕ is not defined at non-spacelike configurations; in that case, $\tilde{\phi}$ has the disadvantages that it does not necessarily satisfy any system of PDEs and that it is not related in a simple way to detection probabilities, as the curved Born rule (4) holds only on spacelike configurations. (In addition, in the Heisenberg picture the Hilbert space $\mathscr{H}$ and the state vector $\Psi \in \mathscr{H}$ refer to the initial time $t = 0$, and thus to a particular spacelike hypersurface. In contrast, ϕ and the PDEs (42) governing it are independent of any choice of hypersurface, so they provide a more fully covariant description.)

6.3. Relation to the Tomonaga-Schwinger picture

The Tomonaga-Schwinger approach associates a wave function $\tilde{\psi}_\Sigma$ with every spacelike hypersurface Σ. We have pointed out in (10) and (13) how multi-time equations define ϕ_Σ, here

$$\phi_\Sigma \in \mathscr{H}_\Sigma = \bigoplus_{M=0}^{\infty} S_x L^2(\Sigma, \mathbb{C}^4)^{\otimes M} \otimes \bigoplus_{N=0}^{\infty} S_y L^2(\Sigma, \mathbb{C}^4)^{\otimes N}, \tag{51}$$

and unitaries $U_\Sigma^{\Sigma'}$. They are closely related to the Tomonaga-Schwinger approach, except that the latter is formulated in the interaction picture, which arises if we would like to represent ϕ_Σ by vectors $\tilde{\psi}_\Sigma$ in a fixed Hilbert space $\tilde{\mathscr{H}}$. Then, we need to identify each $\mathscr{H}_\Sigma$ with $\tilde{\mathscr{H}}$. This can be done via the free time evolution

$$F_{\Sigma\to\Sigma'} = \left(\bigoplus_{M=0}^{\infty} F_{\Sigma\to\Sigma'}^{(1,x)\otimes M} \right) \otimes \left(\bigoplus_{N=0}^{\infty} F_{\Sigma\to\Sigma'}^{(1,y)\otimes N} \right), \tag{52}$$

where $F_{\Sigma\to\Sigma'}^{(1,x/y)}$ is the unitary operator obtained from solving the free Dirac equation with mass $m_{x/y}$. Then

$$\tilde{\psi}_\Sigma = F_{\Sigma\to\Sigma_0} \phi_\Sigma \tag{53}$$

is the wave function on Σ in the interaction picture, where Σ_0 is some fixed spacelike hypersurface connected to the choice of $\tilde{\mathscr{H}}$. The evolution of $\tilde{\psi}_\Sigma$ is given by the *Tomonaga-Schwinger equation*

$$i\left(\tilde{\psi}_{\Sigma'} - \tilde{\psi}_\Sigma\right) = \left(\int_\Sigma^{\Sigma'} d^4x H_I(x) \right) \tilde{\psi}_\Sigma, \tag{54}$$

for infinitesimally neighboring spacelike hypersurfaces Σ, Σ', where the integral is understood to be over the 4-dimensional volume enclosed between Σ and Σ', and where $H_I(x)$ is the Hamiltonian density in the interaction picture. For the model (40), it is given by

$$H_I(x) = e^{i(H_x+H_y)x^0} a^\dagger(\boldsymbol{x}) \Big(g^* \cdot b(\boldsymbol{x}) + g \cdot b^\dagger(\boldsymbol{x}) \Big) a(\boldsymbol{x}) e^{-i(H_x+H_y)x^0}. \tag{55}$$

The Tomonaga-Schwinger equation (54) has a solution for every initial datum if and only if the consistency condition

$$\big[H_I(x), H_I(y)\big] = 0 \quad \text{for all spacelike separated } x, y \tag{56}$$

holds. Furthermore, (54) is Lorentz invariant if $H_I(x)$ is a Lorentz scalar. Note that (55) is not a Lorentz scalar due to the choice of a fixed spinor g.

For the model (42), one can show that the $\tilde{\psi}_\Sigma$ obtained from ϕ through (53) indeed solves the Tomonaga-Schwinger equation (54) with interaction Hamiltonian density (55). Conversely, any given solution $\tilde{\psi}_\Sigma$ to the Tomonaga-Schwinger equation (54) with Hamiltonian density (55) is connected to a solution ϕ of the multi-time equations (42): (i) we can switch from the interaction picture to $\mathscr{H}_\Sigma$ by considering $\phi_\Sigma := F_{\Sigma_0\to\Sigma}\tilde{\psi}_\Sigma$; (ii) as mentioned before, Eq. (14) is the condition for the possibility of combining all the ϕ_Σ into a multi-time function ϕ; (iii) one can check [20] that (14) is indeed satisfied; and (iv) the multi-time evolution of ϕ agrees with (42) [20]. An analogous translation between Tomonaga-Schwinger equations for $\tilde{\psi}_\Sigma$ and multi-time equations for ϕ persists under rather weak assumptions on $\mathscr{H}_\Sigma$ and $H_I(x)$.

6.4. Other models

In [21], we set up a multi-time model involving three particle species x, y and z. In this model, x- and y-particles can annihilate each other and create a z-particle, and, conversely, a z-particle can decay into an x- and a y-particle. The interaction Hamiltonian is

$$H_{\text{int}} = \int d^3\boldsymbol{x} \Big(g a^\dagger(\boldsymbol{x}) b^\dagger(\boldsymbol{x}) c(\boldsymbol{x}) + g^* a(\boldsymbol{x}) b(\boldsymbol{x}) c^\dagger(\boldsymbol{x}) \Big), \tag{57}$$

where a, b, c are spinor-valued annihilation operators, and $g \in (\mathbb{C}^4)^{\otimes 3}$ (i.e., summation about spinor indices is implicitly understood). This model is inspired by electrons (x), positrons (y) and photons (z), but, as in the emission-absorption model, we take all three particles to be Dirac particles. The multi-time equations can be set up similarly as in (42) and (43). It turns out this model is consistent on spacelike configurations if and only if the fermion number is conserved, i.e., either none or two of the x, y, z particles are fermions. It can be shown that this model is related to quantum fields in the Heisenberg picture and the Tomonaga-Schwinger picture in the same way as described above.

We conjecture that multi-time equations can be set up consistently for many kinds of QFTs under some reasonable conditions, and that the equivalence to quantum fields in the Heisenberg picture and the Tomonaga-Schwinger approach still holds.

7. Conclusions

In this paper, we have given an overview of the theory of multi-time wave functions and its recent developments. It was elucidated that the consistency of the multi-time evolution is a restrictive condition that excludes the most common mechanism of interaction in non-relativistic quantum mechanics, i.e., potentials. We have described two alternative ways of constructing relativistic interactions in the multi-time picture: zero-range (or δ-potential) interactions and interactions via particle creation and annhilation in QFTs. We have also described the relations of the multi-time wave function to detection probabilities along spacelike hypersurfaces, to the field operators in the Heisenberg picture, and to the Tomonaga-Schwinger approach.

One striking trait of the multi-time approach lies in its parallels to non-relativistic quantum mechanics: in that the quantum state is represented by a *wave function*, that its time evolution is governed by PDEs, and that its modulus squared yields detection probabilities. And the results reported here suggest that it may be possible to formulate also more serious relativistic QFTs in terms of multi-time wave functions, which sets a goal for future research.

Acknowledgments

This project has received funding from the European Union's Framework for Research and Innovation Horizon 2020 (2014–2020) under the Marie Skłodowska-Curie Grant Agreement No. 705295. S.P. gratefully acknowledges support from the German Academic Exchange Service (DAAD).

References

[1] Albeverio S, Gesztesy F, Høegh-Krohn R and Holden H 1988 *Solvable Models in Quantum Mechanics* (Berlin: Springer-Verlag)
[2] Bloch F 1934 Die physikalische Bedeutung mehrerer Zeiten in der Quantenelektrodynamik *Physikalische Zeitschrift der Sowjetunion* **5** 301–5
[3] Crater H W and Van Alstine P 1983 Two-body Dirac equations *Ann. Phys. (NY)* **148** 57–94
[4] Dirac P A M 1932 Relativistic quantum mechanics *Proc. R. Soc. London* A **136** 453–64
[5] Dirac P A M, Fock V A and Podolsky B 1932 On quantum electrodynamics *Physikalische Zeitschrift der Sowjetunion* **2(6)** 468–79 Reprinted 1958 *Selected Papers on Quantum Electrodynamics* ed J Schwinger (New York: Dover)
[6] Droz-Vincent P 1982 Second quantization of directly interacting particles *Relativistic Action at a Distance: Classical and Quantum Aspects* ed J Llosa (Berlin: Springer-Verlag) pp 81–101
[7] Droz-Vincent P 1985 Relativistic quantum mechanics with non conserved number of particles *J. Geom. Phys.* **2** 101–19
[8] Dürr D and Teufel S 2004 On the exit statistics theorem of many particle quantum scattering *Multiscale Methods in Quantum Mechanics* ed P Blanchard and G Dell'Antonio (Basel: Birkhäuser)
[9] Elze H T 2016 Multipartite cellular automata and the superposition principle *Int. J. Quant. Info.* **14** 1640001 (*Preprint* `http://arxiv.org/abs/1604.04201`)
[10] Elze H T 2016 Quantum features of natural cellular automata *J. Phys.: Conf. Ser.* **701** 01201 (*Preprint* `http://arxiv.org/abs/1604.06652`)

[11] Elze H T 2017 in this volume
[12] Lienert M 2015 A relativistically interacting exactly solvable multi-time model for two mass-less Dirac particles in 1+1 dimensions *J. Math. Phys.* **56** 042301 (*Preprint* http://arxiv.org/abs/1411.2833)
[13] Lienert M 2015 On the question of current conservation for the Two-Body Dirac equations of constraint theory *J. Phys. A: Math. Theor.* **48** 325302 (*Preprint* http://arxiv.org/abs/1501.07027)
[14] Lienert M 2015 *Lorentz invariant quantum dynamics in the multi-time formalism* Ph.D. thesis, Mathematics Institute, Ludwig-Maximilians University, Munich, Germany
[15] Lienert M and Nickel L 2015 A simple explicitly solvable interacting relativistic N-particle model *J. Phys. A: Math. Theor.* **48** 325301 (*Preprint* http://arxiv.org/abs/1502.00917)
[16] Lienert M, Petrat S and Tumulka R 2017 Multi-time wave functions and detection probabilities *In preparation*
[17] Nickel L and Deckert D A 2016 Consistency of multi-time Dirac equations with general interaction potentials *J. Math. Phys.* **57** 072301 (*Preprint* http://arxiv.org/abs/1603.02538)
[18] Petrat S 2010 *Evolution equations for multi-time wavefunctions* Master's thesis, Rutgers, The State University of New Jersey http://dx.doi.org/doi:10.7282/T3SB45GJ
[19] Petrat S and Tumulka R 2014 Multi-time Schrödinger equations cannot contain interaction potentials *J. Math. Phys.* **55** 032302 (*Preprint* http://arxiv.org/abs/1308.1065)
[20] Petrat S and Tumulka R 2014 Multi-time wave functions for quantum field theory *Ann. Phys. (NY)* **345** 17–54 (*Preprint* http://arxiv.org/abs/1309.0802)
[21] Petrat S and Tumulka R 2014 Multi-time formulation of pair creation *J. Phys. A: Math. Theor.* **47** 112001 (*Preprint* http://arxiv.org/abs/1401.6093)
[22] Reed M and Simon B 1980 *Methods of Modern Mathematical Physics I: Functional Analysis* (Academic Press)
[23] Salpeter E E and Bethe H A 1951 A relativistic equation for bound-state problems *Phys. Rev.* **84**: 1232–1242
[24] Schweber S 1961 *An Introduction To Relativistic Quantum Field Theory* (Row, Peterson and Company)
[25] Schwinger J 1948 Quantum electrodynamics. I. A covariant formulation *Phys. Rev.* **74**: 1439–61
[26] Svendsen E 1981 The effect of submanifolds upon essential self-adjointness and deficiency indices *J. Math. Anal. Appl.* **80**: 551–565
[27] Teufel S and Tumulka R 2016 Avoiding ultraviolet divergence by means of interior-boundary conditions *Quantum Mathematical Physics* ed F Finster, J Kleiner, C Röken and J Tolksdorf (Basel: Birkhäuser) pp 293–311 (*Preprint* http://arxiv.org/abs/1506.00497)
[28] Tomonaga S 1946 On a relativistically invariant formulation of the quantum theory of wave fields *Progr. Theor. Phys.* **1(2)** 27–42
[29] Tumulka R 2016 Detection time distribution for the Dirac equation (*Preprint* http://arxiv.org/abs/1601.04571)
[30] Van Alstine P and Crater H W 1997 A tale of three equations: Breit, Eddington–Gaunt, and two-body Dirac *Found. Phys.* **27** 67–79

IOP Conf. Series: Journal of Physics: Conf. Series **880** (2017) 012007 doi:10.1088/1742-6596/880/1/012007

The discovery of gravitational waves: a gentle fight against noise

Giancarlo Cella
INFN Sez. Pisa – Largo B. Pontecorvo 3 56127 Pisa (Italy)
E-mail: giancarlo.cella@pi.infn.it

Abstract. The recent direct observation of gravitational waves coming from collisions between black holes is the fulfillment of a dream started more than sixty years ago with the pioneering experiments of J. Weber. This has been possible by reaching with interferometric detectors an unprecedented sensitivity, which requires the reduction of several technical and fundamental noise sources, thermal and quantum in particular. I discuss some of these, with some emphasis on quantum noise. I also briefly discuss the first detection results and their impact on fundamental physics, commenting about future perspectives and challenges.

1. Introduction

Albert Einstein elaborated general relativity motivated by the exigence of having a theory of gravitation not in conflict with special relativity. This was not the case of Newtonian theory, where gravity is an action at distance analogous to the Coulomb force in electromagnetism. Following this analogy, general relativity could be understood as a theory for a field which should play a similar role of electric and magnetic field in Maxwell's theory.

As shown by Einstein himself [1], the theory allows for wave-like solutions which are generated by accelerated mass motion (or, more generally, by energy momentum tensor as should be for a relativistic theory). Compared with electromagnetic interaction, the gravitational one has some peculiarities. By choosing appropriately the reference frame the gravitational effects can always be set to zero at a given event in space-time. This is the so-called equivalence principle which Einstein took as a cornerstone of his theory, and allows for a geometric interpretation: gravitational effects are due to space-time curvature, and the space-time metric tensor $g_{\mu\nu}$ play the role of the gravitational field.

As the source of gravitation is the energy-momentum tensor, it is expected that gravitational field should contain energy itself and must behave as a source for itself, in other words general relativity should be a nonlinear theory. All this is nicely described by the Einstein equation

$$G_{\mu\nu} = \frac{8\pi G}{c^4} T_{\mu\nu} \quad (1)$$

where $G_{\mu\nu}$ is a tensor which depends in a nonlinear way from the metric and its first and second derivatives, G the Newton's constant and c the speed of light.

The nonlinear character of the theory, and its covariance with respect to arbitrary changes of coordinates, triggered subtle controversies about the interpretation of their wave-like solutions, which can be obtained naively by adding a small perturbation $h_{\mu\nu}$ to the flat space solution described by the Minkowski metric $\eta_{\mu\nu}$

$$g_{\mu\nu} = \eta_{\mu\nu} + h_{\mu\nu} \tag{2}$$

and by linearizing Equation (1). In an appropriate coordinate system this gives an equation for waves which propagates at speed of light, with two independent polarizations.

At some point, Einstein himself was not convinced about the real existence of gravitational waves. He submitted together with Nathan Rosen a paper to Physical Review titled "Do the gravitational waves exist?". The paper was rejected, to great Einstein's disappointment, and then resubmitted [2] and published in another journal. After several decades of discussions a general consensus about the existence of gravitational waves was reached, and in particular about the back action effects of gravitational wave emission on the source. Appropriate approximation schemes for calculations were developed, starting from Einstein-Infeld-Hoffmann equations [3], which are appropriate for a source in slow motion.

An indirect experimental confirmation for the existence of gravitational waves came from the observation of the binary pulsar PSR B1913+16 by Hulse and Taylor [4]. The observed decrease of the orbital period of the system in very good agreement with the expectation of general relativity for the rate of energy loss.

The dominant contribution to a gravitational wave generated by this kind of source can be estimated as

$$h_{ij} \simeq \frac{1}{r}\frac{2G}{c^4}\ddot{Q}_{ij}^{TT} \tag{3}$$

where Q_{ij} is the transverse and traceless part of the quadrupole momentum of the source and h_{ij} the adimensional strain generated by it. The strain h_{ij} is a measurement of the effect of the wave, and can be understood as a relative change of length. Looking at the smallness of the prefactor, it can be realized that a $\ddot{Q}_{ij}^{TT}$ large enough to be directly detectable can be provided only by astrophysical (or cosmological) sources. Still, estimates of the expected h_{ij} give typically values $O\left(10^{-22}\right)$ or less.

A large enough sensitivity is currently reached by interferometric detectors such as Advanced LIGO [5] and Advanced VIRGO [6]. The basic detection principle is the accurate measurement of the difference between the lengths of the two arms of a Michelson interferometer (see Figure 1). A laser beam is injected in the system, and separated in two parts using a partially reflecting mirror (the beam splitter BS). The two beams obtained are then recombined after they traveled in the two arms, and a relative change of the arm lengths generate a relative phase shift which is detected at the output B1.

In order to increase the sensitivity a Fabry-Perot resonant cavity is built inside each arm. The length scale L of each arm is 4km for Advanced LIGO (3km for Advanced VIRGO): this means that the length sensitivity of the apparatus must be of the order of $Lh \simeq 4 \times 10^{-19}$, which is a fraction of the diameter of a proton. This is possible only by maintaining all the noise sources which couples to the detector at a low enough level.

2. Sources

There are several potential sources of detectable gravitational waves. A possible classification is based on the peculiarities of the expected signal: roughly speaking we can speak about *chirps, bursts, continuous signals and stochastic signals.*

A *chirp-like* signal is expected in compact binary coalescence events. They can be described as the last part of the evolution for a bound system of two compact objects of masses m_1 and m_2 (black holes or neutron stars). As in the case of the PSR B1913+16 binary system, the two orbiting bodies generates a $\ddot{Q}_{ij} \neq 0$, and radiates gravitational waves loosing energy. During this process the relative distance between the bodies decreases, and the orbital frequency increases. If

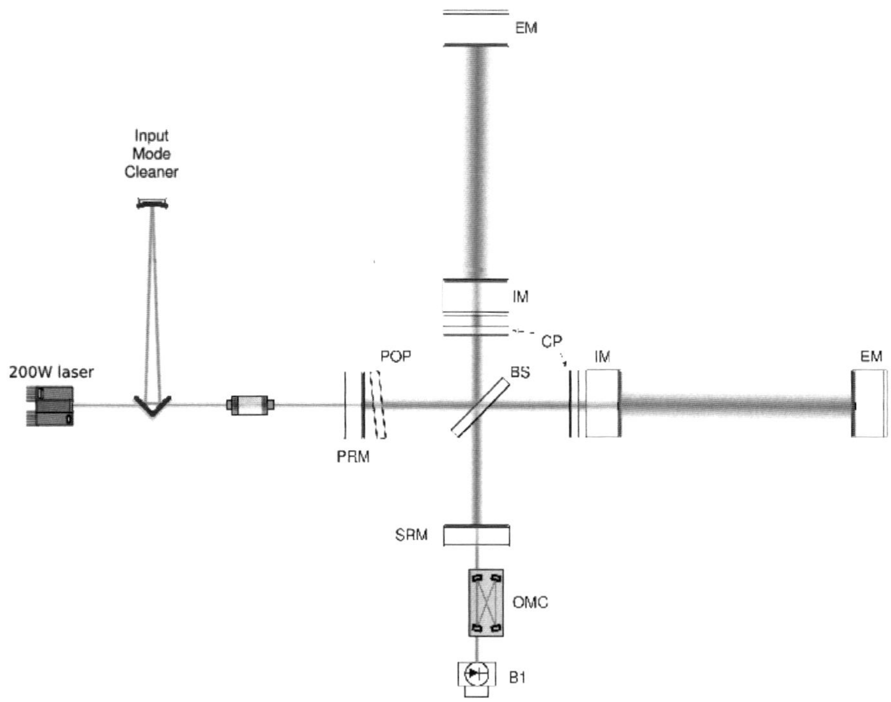

Figure 1. A simplified optical scheme of Advanced VIRGO [6]. The basic topology is the one of a Michelson interferometer. In each arm a resonant Fabry-Perot optical cavity is built (IM-EM mirrors). Power recycling (PRM) and signal recycling (SRM) mirrors are added to increase sensitivity.

we exclude the final part of the evolution we can neglect relativistic and strong field corrections: the relative energy loss for each orbit is small and we can describe the process in the adiabatic approximation. The result is a quasi-periodic signal, which can be written as

$$h_{ij} \sim \frac{4}{r}\left(\frac{GM_c}{c^2}\right)^{5/3}\left(\frac{\omega_{gw}(t)}{2c}\right)^{2/3}\cos\left[\Phi_0 + \int^t \omega_{gw}(t')dt'\right] \qquad (4)$$

neglecting an angular factor which depends on the polarization and on the source position in the sky. Here $M_c = (m_1 m_2)^{3/5}(m_1 + m_2)^{-1/5}$ is the chirp mass of the system and $\omega_{gw}(t)$ is the instantaneous angular frequency of the wave, which is two times the orbital angular frequency, owing to its quadrupolar nature. We see that both the frequency and the amplitude of the signal increases accordingly with the law

$$\omega_{gw}(t) = \frac{1}{4}\left(\frac{GM_c}{c^3}\right)^{-5/8}\left(\frac{5}{t_c - t}\right)^{3/8} \qquad (5)$$

When $t = t_c$ both the amplitude and the frequency diverges. Before that the relative distance between the two bodies become so small that they merge together. This process cannot be described using a simple analytical approximation but requires numerical simulations which fully take into account strong field effects in general relativity: anyway t_c is called coalescence time and can be used as an estimate for the time of merger. Equation (5) will be an acceptable approximation only until[1]

$$f_{gw} \lessapprox f_{\rm ISCO} = 2.2 \times 10^3 \text{Hz} \left(\frac{M_\odot}{m_1 + m_2} \right) \tag{6}$$

which we can take as an upper limit for the frequencies emitted by the source.

In the case of the Hulse-Taylor pulsar $\omega_{gw} \simeq 2 \times 10^{-4}$s: this means that the frequency of the gravitational wave emitted is too small to be detectable by VIRGO and LIGO, which as we will see have a good sensitivity in the frequency band between $O(10)$ Hz and $O(10^4)$ Hz. It means also that PSR B1913+16 will merge in $O(10^8)$ years: in the final phase for the system evolution the signal will enter in the detection bandwidth of the detectors, and will stay there until $f_{\rm ISCO}$ will be reached.

Note the inverse proportionality between $f_{\rm ISCO}$ and the mass of the system. This means that if the total mass of the system is large the merger will happen when the signal is inside the detection band. When the total mass increases, the time spent inside the detector's sensitivity window will decrease until it will disappear completely.

A *burst* signal is characterized by its small duration, typically of the $O(10^{-2})$s. They can be produced in violent explosive events, such as a *supernovae* explosions but also in some more exotic cases, for example in cosmic string dynamics.

In the case of *supernovae*, a requirement for the generation of gravitational wave is that the core collapse must be asymmetric, otherwise no quadrupole or higher time dependent moment can be generated. There is no clear consensus about the asymmetry one can expect, as the physics involved is quite intricate and large scale numerical simulations are needed. An *supernova* event in our galaxy is expected once in $10 - 100$y, so the chances of observing it are not very high. However if this will happen the study of the emitted signal could add a lot to the knowledge of this still poorly understood physical system.

Continuous signals can be produced by a rotating neutron star. This very compact object (a mass typically of about $1.4 M_\odot$ is confined in a region with a radius of few kilometers) can generate gravitational waves if a small deviation from sphericity is present. This can be allowed by elastic or magnetic effects in the crust, and the expected deviation are quite small. However a signal of this kind could help, if detected, in clarifying the nature of the matter inside the star, and in particular its equation of state [7].

Finally a *stochastic signal* can be generated by a superposition of several events, both of astrophysical or cosmological origin. It is called stochastic because, owing to the large overlap between different contributions, it is not possible to resolve the single event. As a consequence it is not possible to describe it using a deterministic theoretical waveform to be searched for, maybe parametrized by a set of unknown parameters as is the case for a *chirp* or for a continuous signal. Instead the signal must be modeled as a stochastic process, and is characterized by their statistical properties such as the time correlation functions.

One consequence of this is that in order to detect a stochastic background of gravitational waves more than one detector is needed. The basic strategy is to estimate the correlation between the output of a pair of detectors. Under the hypothesis that the two noises are uncorrelated (which should not be taken as granted, especially if the detectors are one near the other) this allows for a detection or, at least, to the definition of an upper limit.

An astrophysical stochastic background could be the cumulative superposition of many binary coalescence events, or many bursts, or many continuous signals. Its detection could give interesting information about the validity of stellar evolution models.

A cosmological stochastic background can be the result of processes happened in the first phase of the universe, such as inflation, phase transitions, cosmic string network evolution. They

[1] This estimate of $f_{\rm ISCO}$ is strictly valid only when one mass is much smaller than the other, and neglects spin effects.

are in principle very interesting because, owing to the weakness of gravitational interaction, the universe can be considered transparent for gravitational wave radiation since its beginning, so the gravitational waves generated can propagate unperturbed toward the present epoch. For this reason the detection of a cosmological stochastic background of gravitational waves could give a unique opportunity for getting information about the very young universe [8].

Currently there are three gravitational wave detectors of comparable sensitivity, the two of the LIGO Advanced project in USA, and the VIRGO Advanced in italy, near Pisa. Only the first two where running during the first direct detection, while VIRGO is expected to join in august 2017. The availability of several detectors is important from several point of view. As we saw in the case of stochastic background at least a pair of detector is mandatory.

Ideally the experimental noise can be described as a Gaussian stochastic process. If this is the case the detection probability can be increased also for a deterministic signal by combining together in a coherent way the measurements of all the available detectors. In a realistic situation, where the statistical characterization of the experimental noise is never completely under control, the request for a coincident detection increases the confidence on the result.

A signal from a source at a given position in the sky and with a given polarization couples in a different way to detectors with different location and orientation. Using several detectors at the same time allows for the determination of the source position, mainly by looking at the relative delays observed between the signals to do a triangulation. In this case, at least three detectors are needed to solve the problem and to determine the coordinates in the sky, with an accuracy which depends also, of course, on the signal to noise ratio. This is quite important because in many cases a non gravitational counterpart of the event is expected. For example, a binary coalescence between two neutron stars is expected to be a trigger for a gamma ray burst, and it is important to localize accurately the source, in a time as short as possible, to allow for a complementary electromagnetic observation.

Another interesting application of a multidetector analysis is the determination of the polarization of the gravitational wave, which allows for a more accurate determination of the source parameters and is an important test for the validity of general relativity. This is because extended models for theory of gravitation allows, quite generically, additional polarizations beside the canonical two foreseen by Einstein's theory.

3. A fight against the noise

The simplest characterization of the sensitivity of a gravitational wave detector can be given by its strain equivalent amplitude noise spectrum. The idea is to normalize the detector's output $s(t)$ in such a way that

$$s(t) = h(t) + n(t) \tag{7}$$

where $h(t)$ is the gravitational strain and $n(t)$ the noise. If the noise is stationary the power spectrum of $n(t)$ can be written as

$$\left\langle \tilde{n}(\omega)^* \tilde{n}(\omega') \right\rangle = 2\pi\delta\left(\omega - \omega'\right) S_n(\omega) \tag{8}$$

and the strain equivalent noise spectrum is just given by $\sqrt{S_n}$. The noise budget planned for Advanced VIRGO is plotted in Figure 2. The best sensitivity window is between 20Hz and 4×10^3kHz. We see also that the dominant source of noises are the thermal one (below 300Hz) and the quantum one, in the higher frequency region.

Seismic noise does not appear in the picture, but becomes the main limit below 10Hz. Special care is taken to attenuate it, the reason being that the typical spectral amplitude of the motion of the surface of the earth is at least nine orders of magnitude larger than the expected signal in the detection bandwidth. The principle used for seismic noise attenuation in VIRGO is based on a chain of oscillators which are interposed between the ground and the suspended mirror. They

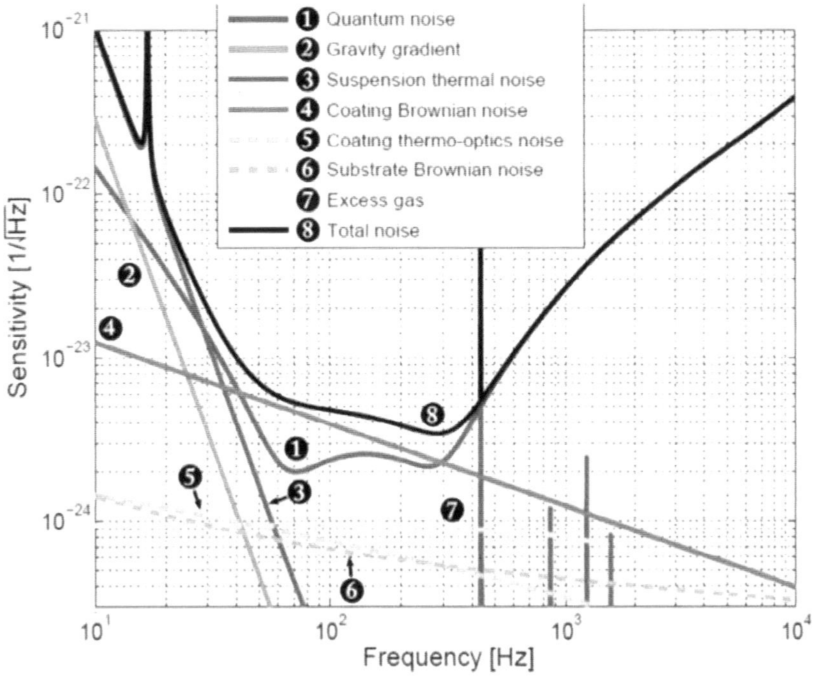

Figure 2. The strain equivalent spectral amplitudes $\sqrt{S_n(\omega)}$ of several noises coupled to VIRGO Advanced, as a function of the frequency. Most relevant contributions comes from quantum noise and from thermal noise of mirror coating and suspensions. Seismic noise dominates below 10Hz and is not represented.

are designed in such a way that the proper frequencies ω_p of the chain are below 1Hz. Above these the oscillator chain spectral transfer function is depressed by a factor $\left(\frac{\omega_p}{\omega}\right)^{2n}$, where n is the number of stages. With a large enough n the required attenuation factor can be reached. Currently $O(10^{-15})$ attenuation factors can be obtained in this way.

The main target for second generation (advanced) detectors was the improvement of the best sensitivity of one order of magnitude in h. This will be obtained (when the design sensitivity will be reached) mainly by reducing thermal and quantum noise. I will give come details about quantum noise in the next section, and I concentrate here on thermal noise.

Thermal noise is the unavoidable result of the coupling between the apparatus and the environment. When the system is in thermodynamical equilibrium its thermal fluctuations can be evaluated by using the fluctuation-dissipation theorem. This theorem told us that the spectral amplitude of thermal noise is proportional to dissipative effects.

In an interferometric detector we are interested in the phase fluctuations induced by the thermal agitation of the mirror surface on the laser beam. There are essentially three different possibilities for reducing them:

(i) The thermodynamical way. Thermal noise amplitude is proportional to the square root of the temperature, so it is possible to reduce it with a cryogenic setup. This is not easy, especially because the cryogenic infrastructure can reintroduce other kind of noises, such as mechanical vibration. Beside that, a non trivial problem is the elimination of the heat which is transferred by the laser to the mirrors. This is a small fraction of the total power, as a large effort is done to use mirrors with a very low absorption, but can become significant if the laser power is very high. Neither VIRGO Advanced nor LIGO Advanced are cryogenic experiments.

(ii) The statistical way. The total phase shift of the laser beam is an average of the mirror displacement weighted by the laser intensity profile. A direct way to reduce thermal fluctuations is to use larger beams, which average more. There are however some limitations, because the dimension of a beam is limited by the mirror size and it is not possible to increase this by a large factor.

(iii) The material improvement way. By optimizing the material and the mechanical construction we can reduce the source of dissipation, which are a measurement of the coupling between the system and the environment. The largest contribution to thermal noise comes currently from the optical coating of the mirrors, which is needed to obtain the requested reflectivities (very large or very small, as needed). But the mechanical properties of the materials used for the coating are not good enough, and there is an intense activity in searching for better materials and solutions.

One of the main difficulty in the thermal noise reduction program is about the strong interdependence between possible solutions and other requirements, such as mechanical and optical properties of the apparatus.

4. Quantum noise

Looking at Figure 2 we see above 300Hz the sensitivity is limited by the so-called quantum noise. In this high frequency region this is commonly called "shot noise", and is connected with the quantum nature of light.

The most direct way to decrease the shot noise is to increase the laser power. However this solution is not a free lunch: a more powerful laser can introduce problems connected with thermal lensing and power absorption which can be quite difficult to solve, such as the parametric resonance one [9]. An alternative is based on the use of squeezed light, as I will discuss next.

In future detectors thermal noise will be reduced further. The impact of quantum noise will be more and more important at intermediate frequencies, and its subtle nature will come out.

4.1. An example: optical cavity

A crude model for a laser beam, which however is good enough to understand several peculiarities of ponderomotive effects in interferometers, is a plane wave with some given frequency ω_0. When this plane wave interacts with optical elements such as a moving mirror or a beam splitter it is reflected and modulated. We will write the classical electric field at a given point as the first component of the bidimensional vector

$$\boldsymbol{E}(t) = \mathbb{R}\left(\omega_0 t\right) \boldsymbol{e}(t) \tag{9}$$

Here $\mathbb{R}(\theta)$ is a rotation matrix by an angle θ and $e_i(t)$ are functions with a slow variation on the timescale ω_0^{-1}. When modulations effect are small, we can write Eq. (9) in a slightly different way

$$\boldsymbol{E}(t) = \mathbb{R}\left(\omega_0 t\right) \left[\overline{\boldsymbol{e}} + \delta \boldsymbol{e}(t)\right] \tag{10}$$

with $|\delta \boldsymbol{e}| \ll |\overline{\boldsymbol{e}}|$. In this way we emphasize the separation between the ideal, monochromatic plane wave contribution $\overline{\boldsymbol{e}}$ and the small corrections $\delta \boldsymbol{e}$ which carry the relevant information coming from the detector, both the signal and the noise.

This is the basic idea beyond the graphical representation in Figure 3. The average part $\mathbb{R}\overline{\boldsymbol{e}}$ is represented by the black sticks, rotating with angular frequency ω_0, the "fluctuating" one $\mathbb{R}\delta\boldsymbol{e}(t)$ by the gray "balls". From this graphical representation it is easy to understand that the component of the fluctuating vector $\delta \boldsymbol{e}$ parallel to $\overline{\boldsymbol{e}}$ represents the amplitude fluctuation

IOP Conf. Series: Journal of Physics: Conf. Series **880** (2017) 012007 doi:10.1088/1742-6596/880/1/012007

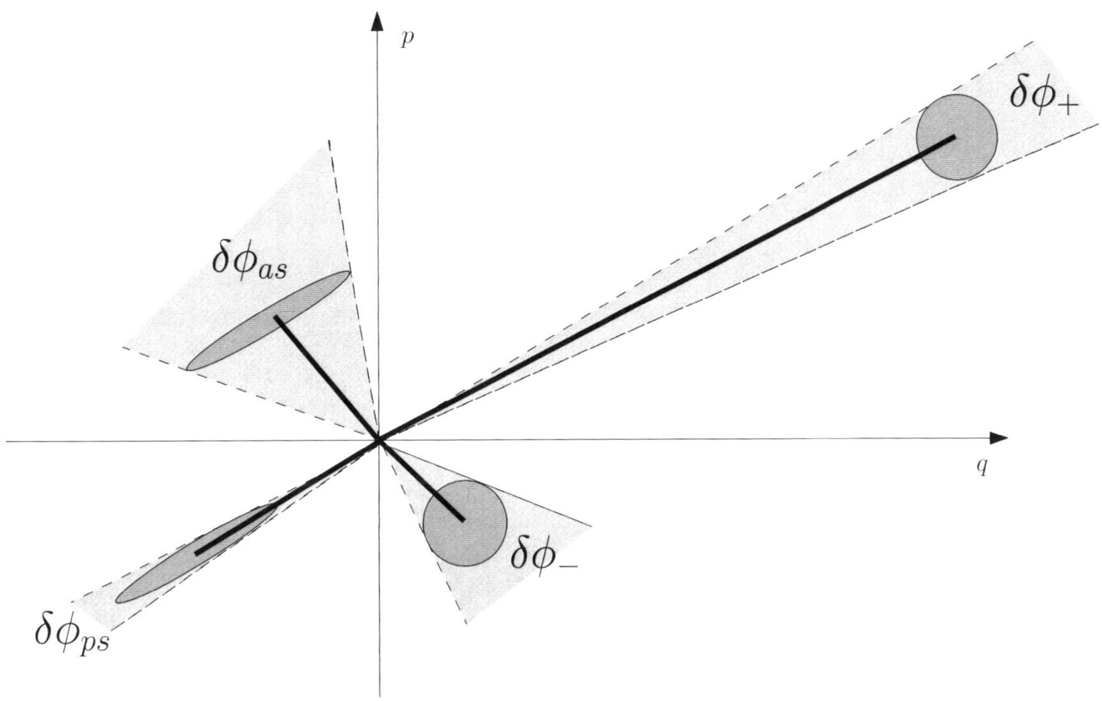

Figure 3. A graphic representation of the field described by Equation (10). The black stick represents the "average" part of the field (which is monochromatic with angular frequency ω_0). The dark gray areas represent the fluctuating parts of the field: they can be seen as the typical region where the field can be found. It follows that the light gray regions are estimates of phase fluctuation.

of the field, and the perpendicular component is proportional to the phase fluctuation, namely $\delta\phi(t) \sim |\overline{\boldsymbol{e}}|^{-2} |\delta\boldsymbol{e} \wedge \overline{\boldsymbol{e}}|$.

It is easy to understand how the modulation vectors which describe the field at different points are related. For example, after a free propagation over a length ℓ the field is simply delayed by ℓ/c, which means

$$\overline{\boldsymbol{e}}' + \delta\boldsymbol{e}'(t) = \mathbb{R}\left(\frac{\omega_0 \ell}{c}\right)\left[\overline{\boldsymbol{e}} + \delta\boldsymbol{e}\left(t - \frac{\ell}{c}\right)\right] \tag{11}$$

Note that the fluctuating part is not simply rotated, but it gets an additional delay.

The most important rule from our point of view is the one that gives the field reflected on a moving mirror. In the context of interferometric detectors the mirror's displacement from his reference position $\delta z(t)$ is slow and small. By assuming it of the same order of field's fluctuation in the linear approximation we obtain from (11)

$$\overline{\boldsymbol{e}}' + \delta\boldsymbol{e}'(t) = \mathbb{R}\left(\omega_0 \delta\tau\right)\left[\overline{\boldsymbol{e}} + \delta\boldsymbol{e}\left(t - \delta\tau\right)\right] \simeq \overline{\boldsymbol{e}} + \delta\boldsymbol{e}\left(t\right) + \frac{\omega_0 \delta\tau}{c}\mathbb{R}\left(\frac{\pi}{2}\right)\overline{\boldsymbol{e}} \tag{12}$$

Here $\delta\tau(t) = 2\delta z(t)/c$ is the additional delay of the field induced by the mirror displacement and we neglected $O(\delta^2)$ effects. Note that the average part is unchanged, while the fluctuating one gets a contribution proportional to the average field and perpendicular to it: as expected, the mirror movement generates phase fluctuations.

The mirror moves for several reasons: it can be affected by noises such as seismic or thermal, and hopefully it can respond to the tidal effect of a gravitational wave. The field itself act on the mirror with a radiation pressure force, which is proportional to the square of the field, whose fluctuating part will be given, after an average over ω_0^{-1} timescales, by

$$\delta F_{RP} = \chi \left[\overline{\boldsymbol{e}}^T \delta \boldsymbol{e}\,(t) + \overline{\boldsymbol{e}}'^T \delta \boldsymbol{e}'\,(t)\right] \tag{13}$$

Note that δF_{RP} is proportional to the field's fluctuations parallel to $\overline{\boldsymbol{e}}$, namely to amplitude fluctuations. The mirror will respond to external forces with something like

$$\delta z(t) = \mathcal{G} \star (\delta F_{RP} + \delta F)\,(t) = \mathcal{G} \star \left[2\chi \overline{\boldsymbol{e}}^T \delta \boldsymbol{e} + \delta F\right](t) \tag{14}$$

where $\mathcal{G}$ is the appropriate mechanical Green function. Together with Equation (12) this gives

$$\delta \boldsymbol{e}'(t) \simeq \delta \boldsymbol{e}(t) + \overline{\boldsymbol{e}} \frac{2\omega_0}{c} \mathbb{R}\left(\frac{\pi}{2}\right) \mathcal{G} \star \left(2\chi \overline{\boldsymbol{e}}^T \delta \boldsymbol{e} + \delta F\right) \tag{15}$$

The fields near the two mirrors inside an optical cavity are related by

$$\delta \boldsymbol{e}_{IM} = \mathbb{T}\,(\mathbb{I} + \mathbb{M}_B \star)\,\delta \boldsymbol{e}_{EM} + \mathbb{T} \boldsymbol{g}_{EM} \star \delta F \tag{16}$$

$$\delta \boldsymbol{e}_{EM} = t \mathbb{T} \delta \boldsymbol{e}_{IM} + r \mathbb{T}\,(\mathbb{I} - \mathbb{M}_A \star)\,\delta \boldsymbol{e}_{IM} \tag{17}$$

Here we introduced a matrix which describes the modulation introduced by the mirror J

$$\mathbb{M}_J \equiv \frac{4\chi\omega_0}{c} \mathbb{R}\left(\frac{\pi}{2}\right) \mathcal{G}_J \overline{\boldsymbol{e}}_J \overline{\boldsymbol{e}}_J^T \tag{18}$$

a vector which describes the coupling to an external force applied to the mirror

$$\boldsymbol{g}_J \equiv \frac{2\omega_0}{c} \mathbb{R}\left(\frac{\pi}{2}\right) \overline{\boldsymbol{e}}_J \mathcal{G}_J \tag{19}$$

and a time delay operator $\mathbb{T}$ defined by its action

$$\mathbb{T} \delta \boldsymbol{e}(t) \equiv \mathbb{R}_c \delta \boldsymbol{e}\left(t - \frac{L}{c}\right) \tag{20}$$

which represents a propagation from one mirror to the other ($\mathbb{R}_c = \mathbb{R}(\omega_0 L c^{-1})$).

The front mirror IM, which is used to inject an external field $\delta \boldsymbol{e}_{IN}$, is described by the reflection and transmission coefficients r, t. The end mirror EM is supposed to be completely reflective and an external force δF is applied to it. By solving for the fields we get

$$\delta \boldsymbol{e}_{IM} = t \mathbb{C}^{-1} \mathbb{T}\,(\mathbb{I} + \mathbb{M}_B \star)\,\mathbb{T} \delta \boldsymbol{e}_{IN} - \mathbb{C}^{-1} \mathbb{T} \boldsymbol{g}_{EM} \star \delta F \tag{21}$$

$$\delta \boldsymbol{e}_{EM} = t \mathbb{C}^{-1} \mathbb{T} \delta \boldsymbol{e}_{IN} + r \mathbb{C}^{-1} \mathbb{T}\,(\mathbb{I} - \mathbb{M}_A \star)\,\mathbb{T} \boldsymbol{g}_{EM} \star \delta F \tag{22}$$

where

$$\mathbb{C} = \mathbb{I} - r \mathbb{T}\,(\mathbb{I} + \mathbb{M}_B \star)\,\mathbb{T}\,(\mathbb{I} - \mathbb{M}_A \star) \tag{23}$$

and finally we can write the field exiting from the cavity as a function of the incoming one

$$\delta \boldsymbol{e}_{OUT} = \mathbb{C}^{-1} \left\{\left[\mathbb{T}\,(\mathbb{I} + \mathbb{M}_{EM} \star)\,\mathbb{T} - r\,(\mathbb{I} + \mathbb{M}_{IM} \star)\right] \delta \boldsymbol{e}_{IN} - t \mathbb{T} \boldsymbol{g}_{EM} \star \delta F\right\} \tag{24}$$

In order to interpret in the correct way these expressions we need also the values of the mean fields near the mirrors. These are given by a set of equations completely equivalent to (21) and (22), but with $\mathbb{T} = \mathbb{R}_c$, $\mathbb{M}_J = 0$ and $\delta F = 0$. The final result is

$$\overline{\boldsymbol{e}}_{IM} = t \left(1 - r \mathbb{R}_c^2\right)^{-1} \mathbb{R}_c^2 \overline{\boldsymbol{e}}_{\mathrm{IN}} \tag{25}$$

$$\overline{\boldsymbol{e}}_{EM} = t \left(1 - r \mathbb{R}_c^2\right)^{-1} \mathbb{R}_c \overline{\boldsymbol{e}}_{\mathrm{IN}} \tag{26}$$

This relatively compact expression contains a rich phenomenology. By choosing appropriately the reference frame we can couple the cavity to a gravitational wave writing

$$\delta F = \frac{1}{2} L \ddot{h} \tag{27}$$

where h is the strain induced in the direction of the cavity. In this way we can compare the noise and the signal of the cavity, by looking at the first and second term on right side of Equation (24). I will restrict to the particular case of a cavity with $\mathcal{G}_{IM} = \mathcal{G}_{EM} \equiv \mathcal{G}$ at the resonance (which means $\mathbb{R}_c = \mathbb{I}$). For simplicity I will consider also only the high finesse ($t \ll 1$) and quasistatic ($\Omega L/c \ll 1$) regime. In the frequency domain we get

$$\delta \tilde{\boldsymbol{e}}_{OUT} \simeq \delta \boldsymbol{e}_{IN} + \frac{4\omega_0}{c} \frac{\mathbb{R}\left(\frac{\pi}{2}\right) \overline{\boldsymbol{e}}_{\mathrm{IN}}}{1-r} \tilde{\mathcal{G}}(\Omega) \left(2\chi \overline{\boldsymbol{e}}_{\mathrm{IN}}^T \delta \tilde{\boldsymbol{e}}_{\mathrm{IN}} + \frac{L}{2} \Omega^2 \delta \tilde{h} \right) \tag{28}$$

A quantum description of the system can be obtained by promoting the fluctuating fields $\delta \boldsymbol{e}(t)$ to operators $\delta \hat{\mathbf{e}}(t)$ with appropriate commutation relations. As a consequence of these the two components of $\delta \hat{\mathbf{e}}(t)$ can't be measured with arbitrary precision at the same time: quantum fluctuations will always be present, will be characterized by the symmetrized expectation values

$$\mathbb{K}\left(t - t'\right) = \left\langle \delta \hat{\mathbf{e}}(t) \delta \hat{\mathbf{e}}(t') \right\rangle_{symm} \tag{29}$$

whose Fourier transform can be interpreted as a noise power spectrum.

The laser beam used in gravitational detectors can be modeled as a a single mode of the electromagnetic field in a coherent state. Coherent states are minimum uncertainty states, because they saturate the Heisenberg uncertainty relations. In particular they are a "fair" compromise between phase and amplitude indeterminacy of the field, and in this sense they can be seen as the "most classical" states of light. Whatever quadrature we choose to measure, we will obtain the same noise.

If we look at Equation (28) we see that the measurement which maximize the signal must be done in the direction perpendicular to $\overline{\boldsymbol{e}}_{\mathrm{IN}} = \overline{\boldsymbol{e}}_{\mathrm{OUT}}$, which is the direction associated to phase fluctuations. The terms proportional to $\delta \boldsymbol{e}_{\mathrm{IN}}$ are noise contributions. the first from the left comes from the input phase fluctuations, which are simply reflected by the cavity, and can be interpreted as shot noise. The second is generated by the back action of the radiation pressure on the mirror, and is called radiation pressure noise. The third term is the signal. Note that, as I said, the ratio between the signal and the shot noise can be reduced by increasing the amplitude of the input field. However when we do this we reduce the ratio between the signal and the radiation pressure noise, as this one is proportional to the second power of the input field. It seems that there is a fundamental limit (which is called *standard quantum limit)* set by quantum noise which cannot be evaded. As a matter of fact this is not true.

4.2. Squeezed states

If we look again at Equation (28) and we analyze the results of a measurement of a generic quadrature, we find a surprising result. It comes out that the quadrature which maximize the ratio between signal and quantum noise is not the phase one. The reason is that the output field $\delta \boldsymbol{e}_{\mathrm{OUT}}$ is not in a coherent state, but in a squeezed one.

There is a particular class of squeezed states whose members share with the coherent states the minimum uncertainty property. They can have, for example, a reduced phase uncertainty at the expense of a larger amplitude one (see Figure 3). In that case for a given value of the laser power the shot noise (connected with phase indeterminacy) will be reduced.

A qualitative representation of coherent and squeezed state in the quadrature space, which can be seen as the analogous of the classical phase space for an harmonic oscillator, is given in Figure (3). It comes out that fluctuation of coherent and squeezed states are Gaussian. For this reason the fluctuation of a coherent state is represented by a circle which indicates the uncertainty region of the field (the indeterminacy of both quadratures δp and δq are the same and are uncorrelated). A squeezed state is represented by an ellipse, with a fluctuation reduced along one quadrature and increased along the other. The phase fluctuation $\delta\phi_-$ can be reduced both by increasing the laser power ($\delta\phi_+ < \delta\phi_-$) or by appropriately squeezing the state ($\delta\phi_{ps} < \delta\phi_-$). The appropriate tool to give a well defined meaning to the error ellipse is the Wigner quasi-probability functions, which for coherent and minimum uncertainty squeezed states is Gaussian.

In the case of $\delta \boldsymbol{e}_{\mathrm{OUT}}$ the squeezing is generated by the back action of the radiation pressure, and is frequency dependent. In other words, if we evaluate the Fourier transform of the correlation matrix (29) we obtain a frequency dependent spectral covariance matrix. Typically when the mechanical response of the apparatus described by $\mathcal{G}$ is large (for example near a resonance or in the low frequency region for a free mass) back action (ponderomotive) effects become large and we obtain a strongly stretched ellipse.

The idea of using squeezed light to improve the sensitivity of an interferometric detector of gravitational waves is around since a long time [10], and the possibility of measuring an optimized quadrature is just one of the possible options.

Squeezed states of light can be produced routinely today, with increasing squeezing levels. The main difficulty is to produce squeezing in the frequency range of interest, which is the audio region for earth bound detectors. But progresses are quite fast.

By injecting a squeezed vacuum in an interferometer it is possible to evade the standard quantum limit [11, 12, 13], and the reduction of shot noise will be pursued by both LIGO and VIRGO using this technique. It is also possible in principle to completely eliminate radiation pressure effects by a combination of squeezed vacuum injection and output quadrature optimization [13].

4.3. Optomechanical effects

Several other techniques for evading the standard quantum limit have been demonstrated theoretically in several papers. It can be shown that in the low photon number regime the optimal states are the Fock ones [14, 15]. When the number of photons involved is high, the optimal approach is based on squeezed states [16].

A general observation is that the final sensitivity depends on both the optical degrees of freedom of the apparatus and the mechanical ones. These are coupled by radiation pressure, which acts in a dual way:

- The quantum state of the optical modes can be changed by the back action of the mechanical degrees of freedom. For example a mirror recoiling under the action of the radiation pressure fluctuation can induce a phase fluctuation on the reflected beam correlated with the first, generating a squeezed state (*ponderomotive squeezing*). In the general case, within the linearized approximation for a simple cavity, this effect is completely described by Equation (24), once the appropriate parameters of the cavity are inserted.
- The dynamic of the mirrors can be changed. For example the radiation present inside a slightly out of resonance (detuned) cavity can produce a considerable effective mechanical stiffness, the so-called *optical spring effect*. This effect is also implicitly contained inside Equation (24), but can be understood simply by looking at the poles of the transfer function, which are the solution of $\det \mathbb{C} = 0$ (see Equation (23)). These can be associated to the modes of the cavity, and in the general case will be determined by both mechanical and optical effects. In particular a force on the mirror IM can have an effect, mediated by the radiation pressure, on the mirror EM. Delay effects can also induce damping (or anti-damping) effects, and instabilities can appear.

The study of optomechanical effects is an active field of research, which is connected with fundamental questions about the foundations of quantum mechanics and quantum information. It is also expected to produce an increasing quantity of technological applications. Usually the most accessible examples of optomechanical devices are small scale ones (such as for example microcavities). A relevant point to be stressed is that in the case of gravitational waves detectors we deal with macroscopic masses, and in principle we have the possibility of studying quantum mechanical effects, such as entanglement [17], in this regime.

The direct observation of the ponderomotive squeezing effect is of great interest. Several attempts have been made, and a number of experiments are currently in various phases of their development. The effect has been observed on membranes [18] and using an optomechanical cavity coupled to a waveguide [19]. Direct evidence of ponderomotive squeezing is still lacking in macroscopic optical cavities.

There are other possible schemes for standard quantum limit evasion. Of particular interest are optical schemes based on the Sagnac effect [20] which are intrinsically unaffected by back action effects, and can be seen as examples of *Quantum Non Demolition* measurements. Other examples are schemes which exploit the *optical spring effect* previously mentioned [21, 22, 23]. Both approaches are starting to become technically feasible.

4.4. Squeezing and thermal noise

The interaction of the system with the environment can be modeled, via the fluctuation dissipation theorem, by introducing appropriate external stochastic forces. It is intuitive (and can be explicitly verified by looking at Equation (15)) that these external forces will introduce an additional phase noise. The interplay between optical noise and thermal noise is not completely trivial when squeezed light is used, and the optimal strategies to be used in order to improve the sensitivity must be carefully designed.

A point worth to be mentioned is that in presence of an optical spring the evaluation of thermal noise must be done with care. In particular, the mechanical response of the system that must be used to apply the fluctuation dissipation theorem must not take into account the optical stiffness. The reason for that can be understood intuitively because the source of optical stiffness is the laser electromagnetic field inside the cavity, which can't be seen as a system in thermal equilibrium with the environment.

This can be exploited to reduce thermal noise in the frequency domain of interest. For example it is possible to push the resonance of a system to higher frequencies adding an optical spring constant to the mechanical one. Modeling the mechanical damping with a loss angle ϕ we can write

$$K_{eff} = K_{mech}\left(1 + i\phi\right) + K_{opt} \tag{30}$$

and when $K_{opt} \gg K_{mech}$ we obtain a large Q factor by a kind of "dilution effect". As a side note, the application of the fluctuation dissipation effect in a situation which cannot be strictly defined of thermodynamic equilibrium is by itself an interesting subject in my opinion worth of further theoretical and experimental investigation.

4.5. Squeezing and losses

In order to use squeezed states of light a particular care should be used in order to avoid optical losses due to absorption or scattering. These deteriorate the squeezing level and the problem is particularly relevant because the implementation of standard quantum limit evasion requires often additional optical elements.

When there are optical losses inside the apparatus a quantum description becomes somewhat subtle. The main issue is that quantum mechanics must be unitary, and this is not compatible with a naive description. It is easy to understand the point using a simple model. Suppose that a field described by an operator $\hat{a}$ is transmitted and partially absorbed through a lens. One could be tempted to describe the transmitted field with an operator

$$\hat{b} = \sqrt{1-\eta^2}\hat{a} \tag{31}$$

where η parametrize the loss. However the commutation rules of this operator can't be canonical:

$$\left[\hat{b}, \hat{b}^\dagger\right] = \left(1-\eta^2\right)\left[\hat{a}, \hat{a}^\dagger\right] \neq \left[\hat{a}, \hat{a}^\dagger\right] \tag{32}$$

In the correct modelization we must look at the absorbed field of the mode $\hat{a}$ as to a scattered one in another mode $\hat{a}_s$. And the same mechanism which produces this scattering will be able to scatter a mode $\hat{b}_s$ into the mode $\hat{b}$. The correct description will be

$$\hat{b} = \sqrt{1-\eta^2}\hat{a} - \eta\hat{b}_s \tag{33}$$

$$\hat{a}_s = \eta\hat{a} + \sqrt{1-\eta^2}\hat{b}_s \tag{34}$$

and it is easy to verify that the commutation rules are preserved.

A practical consequence of this is that if $\hat{a}$ describes a carefully prepared squeezed state, after the loss the squeezing of the field $\hat{b}$ will be reduced. The reason is that $\hat{b}_s$ will describe a non squeezed vacuum state, and his fluctuations will add to the ones of $\hat{a}$.

5. The first detection

The first direct detection of a gravitational wave signal was made by the two LIGO interferometers on September 14, 2015 [24]. The signal where detected first by the LIGO Livingston Observatory, and then with a relative delay of 7ms by the LIGO Hanford Observatory. The signal to noise ratio was quite high, of about 24, and allowed for a determination of the main parameters of the event with a good confidence.

The source was identified as a binary coalescence of two black holes[25, 24], with masses of $36.2^{+5.2}_{-3.8} M_\odot$ and $29.1^{+3.7}_{-4.4} M_\odot$. The remnant of the coalescence is a Kerr black hole of $62.3^{+3.7}_{-3.1} M_\odot$. The spins of the final black hole was determined to be about $0.68^{+0.005}_{-0.06}$, while the determination of the spin of the two initial objects was much less accurate.

Looking at the initial and final masses, it comes out that the energy radiated by the system, almost completely in the last 0.5s of the coalescence, was of about $3^{+0.5}_{-0.4} M_\odot c^2$. For a system of two black holes the only expected mechanism for energy loss is toward emission of gravitational waves.

Using the observed delay between the H1 and L1 signals and the constraint about the consistence between the two waveforms it was possible to constrain the position in the sky of the source in a fraction of an annulus of about 230deg^2, which is quite large but it is the best that can be done with a pair of detectors. Finally, the event happened at a redshift of $z = 0.09^{+0.03}_{-0.04}$, which correspond to a distance of about 1.3×10^9 ly from us.

The total time spent inside the sensitivity band of the detectors was about 0.5s. The reason is that the mass of the coalescing objects was quite large, with an higher frequency cut-off quite low[2].

There is a good agreement between the observed signal and the prediction of general relativity. Owing to its good signal to noise ratio it was possible to make several tests, in particular by looking at the merger regime where nonlinear effects are expected to be relevant and in the final phase (the so-called ring down) which is dominated by the relaxation of a small number of excitation of the final Kerr black hole. These are under good theoretical control, and give a signature of the validity of the theory. During the first observing run a second black hole coalescence ($14.2M_\odot + 7.5M_\odot \rightarrow 21.8M_\odot$) was clearly observed.

[2] This can estimated by using Eq. (6), or just looking at Figure 4. Remember that Eq. (6) is strictly valid only in the test mass approximation.

IOP Conf. Series: Journal of Physics: Conf. Series **880** (2017) 012007 doi:10.1088/1742-6596/880/1/012007

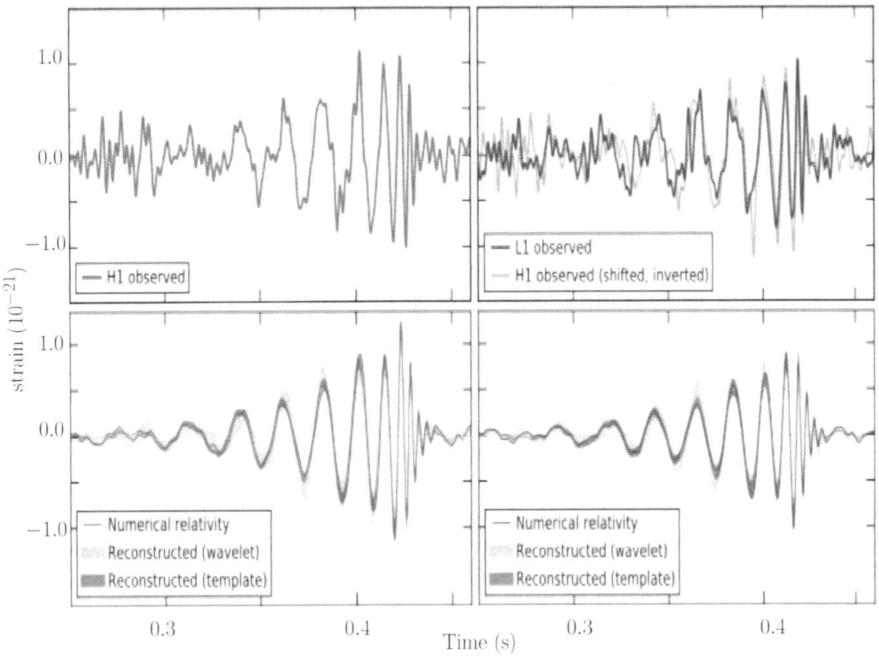

Figure 4. Plots of the signal of the first gravitational wave direct detection. On the top left, the signal measured at the LIGO Hanford Observatory (H1). Noise at frequencies different from the signal's ones have been filtered out. On the top right, the signal measured at the LIGO Livingston Observatory (L1) overimposed with the H1 one, after the compensation of the observed delay and of an expected sign flip. On the bottom, the measured signals compared with the prediction of numerical relativity. Adapted from [24].

6. Conclusions and perspectives

For the first time a gravitational wave signal has been directly detected. At present, a couple of binary black hole coalescences were seen, but this is just one example of a large set of possible sources. We are confident that in a short time a coalescence between a pair of neutron stars (or between a neutron star and a black hole) will be detected. This will open the possibility of studying the equation of state of the matter in the extreme conditions that can be found in the interior of a neutron star, which will give important information about fundamental physics. Other sources are expected, as I discussed, and maybe something completely unexpected can happen. In short, it can be said that we are at the beginning of the gravitational wave astronomy era.

If we look at the predictions about the future improvement of sensitivity of earth bound detectors, we can say that in the next few years with advanced detectors we will be able probably to detect $O(10^2)$ events similar to the one of the first detection. This can appear surprising, but we should remember that the volume of the universe explored is proportional to the third power of the sensitivity.

This year Advanced VIRGO will join the network of observatories, and is expected to give an important contribution, especially for the determination of the parameters of the sources. In 2018 KAGRA, a cryogenic Japanese detector, is expected to join also, and probably a fifth one (LIGO India) will start its operations after 2020.

Still larger improvements are foreseen for the so called *third generation detectors*, such as the *Einstein Telescope* [26]. In this case it is possible that optical noise will become the most

IOP Conf. Series: Journal of Physics: Conf. Series **880** (2017) 012007 doi:10.1088/1742-6596/880/1/012007

fundamental limit of sensitivity for the detector both at higher and lower frequencies [27]. Another important sensitivity limitation will become the so-called Newtonian noise. This could be described as the effect of direct gravitational coupling between the apparatus and environmental density fluctuations. It is possible that its effect could be seen in advanced detectors also (see Figure 2) and the study of possible methods for its mitigation is already started (see [28] for a review).

Newtonian noise can be seen as a fundamental limit that make difficult to push the detection window to lower frequencies for an earth bound experiment. The LISA project [29] for a space based observatory will be immune to seismic and Newtonian noise. Its long baseline will allow for a sensitivity window between 10^{-3}Hz and 1Hz. It will be complementary to earth bound detector, and it will allow to study very massive black holes coalescences and cosmological stochastic background in particular.

Acknowledgments

I wish to thank the organizers of DICE workshop for inviting me, and for their patience waiting the preparation of this paper.

References

[1] Einstein A 1916 Näherungsweise Integration der Feldgleichungen der Gravitation *Sitzungsberichte der Königlich Preussischen Akademie der Wissenschaften Berlin* **688**
[2] Einstein and Rosen N 1937 On Gravitational Waves *Journal of the Franklin Institute* **223** 43
[3] Einstein A, Infeld L and Hoffmann B 1938 The Gravitational Equations and the Problem of Motion *Annales of Mathematics* **39** 65
[4] Weisberg J M and Taylor J H 2005 The Relativistic Binary Pulsar B1913+16: Thirty Years of Observations and Analysis *In Binary Radio Pulsar, ASP Conference Series. Eds. F. A. Rasio and I. H. Stairs* **328** 25
[5] Aasi J et al. (The LIGO Scientific Collaboration) 2015 Advanced LIGO *Class. Quantum Grav.* **32** 074001
[6] Acernese F et al. (VIRGO Collaboration) 2015 Advanced Virgo: a second-generation interferometric gravitational wave detector *Class. Quantum Grav.* **32** 024001
[7] Jones D I 2002 Gravitational waves from rotating strained neutron stars *Class. Quantum Grav.* **19** 1255
[8] Maggiore M 2000 Gravitational wave experiments and early universe cosmology *Physics Reports* **331** 283
[9] Braginsky V B, Strigin S E and Vyatchanin S P 2001 Parametric oscillatory instability in Fabry–Perot interferometer *Phys. Lett. A* **287** 331
[10] Caves C M 1981 Quantum mechanical noise in an interferometer *Phys. Rev. D* **23** 1693
[11] Yuen H P 1983 Contractive States and the Standard Quantum Limit for Monitoring Free-Mass Positions *Phys. Rev. Lett.* **51** 719
[12] Jaekel M T and Reynaud S 1990 Quantum Limits in Interferometric Measurements *Europhys. Lett.* **13** 301
[13] Kimble H J, Levin Y, Matsko A B, Thorne K S, and Vyatchanin S P 2001 Conversion of conventional gravitational-wave interferometers into quantum nondemolition interferometers by modifying their input and/or output optics *Phys. Rev. D* **65** 022002
[14] Mitchell M W, Lundeen J S and Steinberg A M 2004 Super-resolving phase measurements with a multiphoton entangled state *Nature* **429** 161
[15] Rarity J G, Tapster P R, Jakeman E, Larchuk T, Campos R A, Teich M C, and Saleh B E A 1990 Two-photon interference in a Mach-Zehnder interferometer *Phys. Rev. Lett.* **65** 1348
[16] Demkowicz-Dobrzański R, Banaszek K and Schnabel R 2013 Fundamental quantum interferometry bound for the squeezed-light-enhanced gravitational wave detector GEO 600 *Phys. Rev. A* **88** 041802
[17] Müller-Ebhardt H, Rehbein H, Schnabel R, Danzmann K and Chen Y 2008 Entanglement of macroscopic test masses and the Standard Quantum Limit in laser interferometry *Phys. Rev. Lett.* **100** 013601
[18] Purdy T P et al. 2013 Observation of Radiation Pressure Shot Noise on a Macroscopic Object *Science* **339** 801
[19] Safavi-Naeini A H, Gröblacher S, Hill J T, Chan J, Aspelmeyer M and Painter O 2013 Squeezed light from a silicon micromechanical resonator *Nature* **500** 185
[20] Eberle T et al. 2010 Quantum Enhancement of the Zero-Area Sagnac Interferometer Topology for Gravitational Wave Detection *Phys. Rev. Lett.* **104** 251102
[21] Buonanno A and Chen Y 2002 Signal recycled laser-interferometer gravitational-wave detectors as optical springs *Phys. Rev. D* **65** 042001

IOP Conf. Series: Journal of Physics: Conf. Series **880** (2017) 012007 doi:10.1088/1742-6596/880/1/012007

[22] Sheard B S, Gray M B, Mow-Lowry C M, McClelland D E, and Whitcomb S E 2004 Observation and characterization of an optical spring *Phys. Rev. A* **69** 051801
[23] Corbitt T, Wipf C, Bodiya T, Ottaway D, Sigg D, Smith N, Whitcomb N and Mavalvala N 2007 Optical Dilution and Feedback Cooling of a Gram-Scale Oscillator to 6.9mK *Phys. Rev. Lett.* **99** 160801
[24] Abbott B P et al. (LIGO Scientific Collaboration and Virgo Collaboration) 2016 Observation of Gravitational Waves from a Binary Black Hole Merger *Phys. Rev. Lett.* **116** 061102
[25] Abbott B P et al. (LIGO Scientific Collaboration and Virgo Collaboration) 2016 Binary Black Hole Mergers in the First Advanced LIGO Observing Run *Phys. Rev. X* **6** 041015
[26] Punturo M et al. 2010 The third generation of gravitational wave observatories and their science reach *Class. Quant. Grav.* **27** 084007
[27] Hild S et al. 2010 A xylophone configuration for a third-generation gravitational wave detector *Class. Quantum Grav.* **27** 015003
[28] Harms J 2015 Terrestrial Gravity Fluctuation *Living Rev Relativ* **18** 3
[29] Jennrich O 2004 LISA: a mission to detect and observe gravitational waves *In Gravitational Wave and Particle Astrophysics Detectors, edited by James Hough, Gary H. Sanders, Proceedings of SPIE* **5500**

IOP Conf. Series: Journal of Physics: Conf. Series **880** (2017) 012008 doi:10.1088/1742-6596/880/1/012008

The atoms of spacetime and the cosmological constant

Thanu Padmanabhan
IUCAA, Pune University Campus, Ganeshkhind, Pune, India.
E-mail: paddy@iucaa.in

Abstract. I describe an approach which relates classical gravity to the quantum microstructure of spacetime. In this approach, the field equations arise from maximizing the density of states of the matter plus geometry. The former is identified using the thermodynamics of null surfaces. The latter arises from the existence of a zero-point length in the spacetime which associates an internal degree of freedom with each event in the spacetime, in the form of a fluctuating vector of constant norm. The density of states, as well as the resulting field equations, remain invariant under the shift $T^a_b \to T^a_b$+(constant)δ^a_b (arising from the addition of a constant to the matter Lagrangian, which is a symmetry of the matter sector). The cosmological constant (Λ) arises as an integration constant and renders the amount of cosmic information (I_c) accessible to an eternal observer finite. The relation between Λ and I_c allows us to determine the numerical value of (Λ) from the information content of the quantum spacetime, within the context of cosmology.[1]

1. Gravity from the atoms of space: Summary

It is possible to obtain the gravitational field equations from an extremum principle based on the density of states of the quantum geometry. I will motivate a suitable definition of the density of states of the quantum spacetime — which is the key new idea in this approach — that leads to deeper insights into the microscopic degrees of freedom of the spacetime and, as a bonus, provides [1] a fresh perspective on the numerical value of the cosmological constant. In this section, I will summarize the results and highlight the conceptual aspects.[2] Later sections have some mathematical details and I refer the reader to two recent reviews [1,2] for a more extensive discussion.

Let me begin by recalling that the physics of the fluids can be presented at two different levels, one more fundamental than the other. The first level is the continuum description which ignores the fact that there are microscopic degrees of freedom in the fluid in the form of atoms/molecules. This approach uses variables like density $\rho(x^i)$, pressure $P(x^i)$, temperature $T(x^i)$, fluid momentum $P^\mu(x^i)$, etc. and the dynamics is governed by the continuity equation and, say, the Navier-Stokes equation. Of course, the existence of temperature (proportional

[1] This is an updated version of the lecture I gave at the DICE 2016 meeting. I will concentrate here on the conceptual aspects of the approach. More technical details can be found in Ref. [1,2].

[2] The signature is $(-,+,+,+)$ and I use natural units with $c=1, \hbar=1$ and set $\kappa=8\pi G=8\pi L_P^2$ where L_P is the Planck length $(G\hbar/c^3)^{1/2}=G^{1/2}$ in natural units. Latin letters i,j etc. range over spacetime indices and the Greek letters α,β etc. range over the spatial indices. I will write x for x^i, suppressing the index, when no confusion is likely to arise.

to the random kinetic energy of the atoms) as well as the transport coefficients tells us that such a description is fundamentally inconsistent. The second, deeper, level of description uses the distribution function $f(x^i, p_j)$, which *counts* the number of atoms $dN = f(x^i, p_j)d^3xd^3p$ per unit phase space volume d^3xd^3p (with the constraint $p^2 = m^2$ making the phase space six-dimensional). This description, in terms of a distribution function, is remarkable because it allows us to use the continuum language and — at the same time — recognize the discrete nature of the fluid. (One could equivalently think of f as the number of *degrees of freedom* per unit phase space volume.) The new feature which arises at this level of description is the "internal" variable p^μ which allows us to describe atoms with different *microscopic* momenta co-existing at the same event x^i. The *macroscopic* momentum of the fluid $P^\mu(x^i)$ is given by the average value $\langle p^\mu \rangle = P^\mu(x^i)$. Similarly, we have $\langle p^\mu p^\nu \rangle = P^\mu(x^i)P^\nu(x^i) + \Sigma^{\mu\nu}$ where the second term arises from the dispersion of momentum.

In a similar manner, I want to introduce a function $\rho_g(x^i, \phi_A)$ to describe the density of states of the quantum spacetime. Here, ϕ_A (with $A = 1, 2, 3, ...$) denotes possible internal degrees of freedom (analogous to the momentum p_i for the distribution function for the molecules of a fluid) which exist as fluctuating *internal* variables at *each* event x^i. Their behaviour, at any event x^i, is determined by a probability distribution $P(\phi_A, x^i)$, the form of which depends on the microscopic quantum state of the spacetime. Of course, we cannot determine this function without knowing more about the quantum structure of spacetime, but — fortunately, as we will see — we only need some properties of the average values like $\langle \phi_A \rangle, \langle \phi_A \phi_B \rangle$ etc., for the purpose of obtaining and interpreting the classical field equations of gravity. The dependence of $\rho_g(x^i, \phi_A)$ on x^i arises only indirectly through the geometrical variables like the metric tensor, curvature tensor etc., (which I will collectively denote as $\mathcal{G}_N(x)$ with $N = 1, 2, 3, ...$), so that $\rho_g(x, \phi_A) = \rho_g(\mathcal{G}_N(x), \phi_A)$. Such a description is not exact but will be valid at some mesoscopic scales larger than L_P so that the variables $\mathcal{G}_N(x)$ can be defined. At these scales, the Planck length plays a role which is analogous to that of the mean-free-path in the kinetic theory.

It turns out that there is a natural way of defining $\rho_g(x^i, \phi_A)$, if we introduce discreteness into the spacetime through a zero-point length. Remarkably enough, this procedure *also* identifies the internal variable ϕ_A as a constant norm four-vector n^a, which can be thought of as a microscopic, fluctuating, quantum variable at each event x^i. In terms of this internal, vector degree of freedom, the $\rho_g(x^i, n_a)$ is given by

$$\ln \rho_g \propto \left[1 - \frac{L_P^2}{8\pi} R_{ab}(x) n^a n^b\right] \tag{1}$$

The fluctuations of n^a are governed by some probability functional $P[n^i(x), x]$, which is the probability that the quantum geometry is described by a vector field $n_a(x)$ at every x. As I said before the form of P is unknown at present; but, fortunately, we will need only two properties of this probability distribution $P[n^i(x), x]$ which can be derived: (i) It preserves the norm of n^i, which is unity in the Euclidean sector and zero in the Lorentzian sector; i.e $P[n^i(x), x]$ will have the form $F[n(x), x]\delta(n^2 - \epsilon)$ with $\epsilon = 1$ in the Euclidean space and zero in the Lorentzian spacetime. (ii) The average of n^a over the fluctuations (in a given quantum state of the geometry) gives,[3] in the Lorentzian spacetime, a null normal $\ell^a(x^i)$ to a patch of

[3] This is analogous to the average value of the microscopic momenta of fluid particles $\langle p^\mu \rangle = P^\mu(x^i)$ giving rise to the macroscopic momentum of the fluid. The key difference is that, in normal fluid mechanics, the distribution function itself is used to do the averaging while here the fluctuations are governed by some other probability distribution $P(x, n)$. Recall that the null normal ℓ_a also defines the tangent vector to the null geodesic congruence on the null surface; in this sense, it is indeed the momentum of the photons traveling along the null geodesics.

null surface; i.e., $\langle n^a \rangle = \ell^a(x^i)$. Such averages can be defined through the functional integral

$$\langle n_a \rangle = \int \mathcal{D}n \; n_a(x) P[n^a(x), x] = \ell_a(x) \tag{2}$$

where $P[n^a(x), x]$ is parametrized by some null vector field $\ell_a(x)$. (For example, P could be a sharply peaked Gaussian functional in the variable $[n^a(x) - \ell^a(x)]$.) Of course, different quantum states of the spacetime geometry will lead to different P with different null normals $\ell_a(x^i)$ as their mean values; so, in fact, the expectation value $\langle n_a \rangle$ actually leads to the *set of all null normals* $\{\ell_a(x^i)\}$ at an event x^i when we take into account all quantum states. Similarly, $\langle n_i n_j \rangle = \ell_i \ell_j + \sigma_{ij}$ where the second term σ_{ij} represents quantum gravitational corrections to the mean value, etc. Therefore, the mean value $\langle \ln \rho_g(x^i, n_a) \rangle$, in the continuum limit, is given by:

$$\langle \ln \rho_g(x^i, n_a) \rangle \propto 1 - \frac{1}{8\pi} L_P^2 R_{ab} \ell^a \ell^b + ... \tag{3}$$

where we have not displayed terms proportional to $R_{ab}\sigma^{ab}$ which are of higher order and independent of $\ell_a(x)$. Remarkably enough, the quantity

$$\mathcal{H}_g \equiv -\frac{1}{8\pi L_P^2} R_{ab} \ell^a \ell^b \tag{4}$$

which occurs here leads to a term quadratic in the *derivatives* $\nabla_a \ell_b$ (plus an ignorable total divergence) and has an interpretation as the gravitational contribution to the heating rate (per unit area) of the null surface to which ℓ_a is the normal. Its integral over the null surface, Q_g, can be interpreted as the gravitational contribution to the heat content of the null surface.

To complete the picture, we also need the corresponding expression for the matter sector. In the continuum limit, it is straightforward to show — using the concept of local Rindler horizons — that the corresponding quantity for matter is given by:

$$\langle \ln \rho_m \rangle \propto L_P^4 T_{ab} \ell^a \ell^b = L_P^4 \mathcal{H}_m \tag{5}$$

where — as we shall see — $\mathcal{H}_m$ can be interpreted as the heat density contributed by matter crossing a null patch. Taking into account both matter and spacetime, the total number of degrees of freedom, in the continuum limit — in a state characterized by the vector field $\ell_a(x)$ — will be

$$\langle \Omega_{\rm tot} \rangle_\ell = \prod_x \langle \rho_g \rangle \langle \rho_m \rangle = \exp \sum_x (\langle \ln \rho_g \rangle + \langle \ln \rho_m \rangle) \equiv \exp[S_{\rm grav}(\ell) + S_{\rm m}(\ell)] \tag{6}$$

to the lowest order (when we ignore the fluctuations, so that $\ln\langle \rho \rangle \approx \langle \ln \rho \rangle$). The proportionality constants in Eq. (1) and Eq. (5) can be taken to be the same factor, say, μ, by choosing the measure in this sum appropriately. The ℓ_a dependent part of the configurational entropy $S_{\rm tot} = S_{\rm grav} + S_{\rm m}$ is given by the functional

$$S_{\rm tot}[\ell(x)] = \int_{\mathcal{S}} d^3V_x \; \mu E^a_b \langle n_a n^b \rangle = \int_{\mathcal{S}} d^3V_x \; \mu \left(T^a_b(x) - \frac{1}{\kappa} R^a_b(x) \right) \ell_a(x) \ell^b(x) + ... \tag{7}$$

where, in the continuum limit, the sum over x is replaced by integration over the null surface $\mathcal{S}$ for which $\ell^a(x)$ is the normal, with the measure $d^3V_x = (d\lambda d^2x \sqrt{\gamma}/L_P^3)$ and the proportionality constant μ is introduced.

The gravitational field equations can be obtained by extremizing the expression for $\langle \Omega_{\rm tot} \rangle_\ell$ or, equivalently, the configurational entropy $S_{\rm tot} = S_{\rm grav} + S_{\rm m}$ over ℓ and demanding that the extremum condition holds for all ℓ_a. Since different quantum states of geometry will lead to different $\ell_a(x)$ at the same event x^i, this is equivalent to demanding the validity of the extremum condition for all quantum states of the geometry, which are relevant in the classical limit. The extremum of Eq. (7) with respect to $\ell^a \to \ell^a + \delta\ell^a$, subject to the constraint ℓ^2 = constant, will then lead to Einstein's equations, with a cosmological constant arising as an integration constant.[4] Moreover, we will see later that the integrand of Eq. (7) *itself* can be interpreted as the heating rate of the null surface. So, the classical limit makes perfect thermodynamic sense.

So we obtain the field equations *without* treating the metric as a dynamical variable. While it is unusual, there are strong reasons to believe that the metric tensor is probably not the correct dynamical variable to use in an extremum principle. I have described this aspect in detail in Ref. [1, 3] but let me briefly highlight this historical accident, in the construction of the field equations of general relativity, because of which an innocent geometrical object like g_{ab} acquired the status of a dynamical variable in an action principle.

Given the principle of equivalence and general covariance, you can conclude that: (a) Gravity is best described as the effect of curvature of spacetime by using a line interval $ds^2 = g_{ab}dx^a dx^b$. (b) The influence of gravity on other systems can be obtained by demanding the validity of special relativity in the local inertial frames and general covariance. (c) In the Newtonian limit, the only nontrivial metric coefficient is $g_{00} = -(1 + 2\phi)$. What Einstein was looking for at this stage was a generalization of the Newtonian field equation $\nabla^2(2\phi) = \kappa\rho$ which can be written as $-\nabla^2 g_{00} = \kappa T_{00}$ where ρ is identified with the time-time component T_{00} of the divergence-free, second rank symmetric energy momentum tensor T_{ab}. Since ∇^2 is not Lorentz invariant, you might think (alas, wrongly!) that you need to "generalize" the ∇^2 to $\Box^2$ so that the left hand side has second derivatives with respect to *both space and time* coordinates. The second derivatives of the metric tensor can be expressed covariantly in terms of the curvature tensor, which led Einstein to look for a divergence-free, second rank symmetric tensor to replace $\nabla^2 g_{00}$ in the left hand side. This leads to $G_{ab} = \kappa T_{ab}$.

But Einstein could have taken a different, and better, route! One can generalize the Newton's law of gravity $\nabla^2\phi \propto \rho$, *retaining the right hand side as it is and without introducing second time derivatives in the left hand side.* To do this, note that: (i) The energy density $\rho = T_{ab}u^a u^b$ in the right hand side, is observer dependent where u^i is the four velocity of an observer. There is no need to be afraid of u^i appearing in this equation. (ii) Consistency demands that we should generalize the left hand side, $\nabla^2\phi$, to some scalar which also *depends on the four-velocity u^i of the observer* bilinearly. (iii) It is perfectly acceptable for the left hand side *not* to have second *time* derivatives of the metric, in the rest frame of the observer, just as they do not occur in $\nabla^2\phi$.

To obtain such a scalar (which will replace $\nabla^2\phi$) containing *spatial* second derivatives that depend bilinearly on u^i, we first project the indices of R_{abcd} to the space orthogonal to u^i, using the projection tensor $P^i_j = \delta^i_j + u^i u_j$, thereby obtaining the tensor $\mathcal{R}_{ijkl} \equiv P^a_i P^b_j P^c_k P^d_l R_{abcd}$. The only scalar you can construct from $\mathcal{R}_{ijkl}$ is $\mathcal{R}^{-2} \equiv \mathcal{R}^{ij}_{ij}$ where $\mathcal{R}$ can be thought of as the radius of curvature of the space.[5] The natural generalization of $\nabla^2\phi \propto \rho$ is then given by $\mathcal{R}^{-2} \propto \rho = T_{ab}u^a u^b$. Working out the left hand side (see e.g., p. 259 of Ref. [4]), we get: $G_{ab}u^a u^b = \kappa T_{ab}u^a u^b$. If we demand that this relation holds for all observers, we get back $G_{ab} = \kappa T_{ab}$. A natural generalization of this approach is to demand $G_{ab}\ell^a\ell^b = \kappa T_{ab}\ell^a\ell^b$ for all *null* vectors ℓ_a which has the attractive feature that it is invariant under $T^a_b \to T^a_b + (\text{constant})\delta^a_b$, thereby preserving a symmetry respected by matter sector. This is exactly what we get in our

[4] This is straightforward: Demanding that the extremum condition holds for all ℓ_a leads to $R^a_b - \kappa T^a_b = f(x)\delta^a_b$. Taking the divergence of this equation and using $\nabla_a T^a_b = 0$ and $\nabla_a R^a_b = (1/2)\partial_b R$, you get $f(x) = (1/2)R+$ a constant, leading to Einstein's equations, with a cosmological constant arising as an integration constant.

[5] The $\mathcal{R}_{ijkl}$ and $\mathcal{R}$ should *not* to be confused with the curvature tensor ${}^3R_{ijkl}$ and the Ricci scalar ${}^3R$ of the 3-space orthogonal to u^i.

IOP Conf. Series: Journal of Physics: Conf. Series **880** (2017) 012008 doi:10.1088/1742-6596/880/1/012008

approach.

Even though we vary the null vector field ℓ_a in our extremum principle in Eq. (7), the vector field ℓ_a also cannot be thought of as a dynamical variable in the conventional sense. Recall that, normally, if you vary a dynamical variable q_i in an action principle, you get an evolution equation for that variable q_i. Here, we demand that the extremum condition should hold for all q_i and use this demand to constrain the background spacetime. Such an extremum condition is more similar to thermodynamic extremum principles (describing an equilibrium configuration) rather than to action principles (leading to dynamical evolution). While a bit unusual, all this is consistent with the idea that the gravitational field equations must be interpreted in a thermodynamic language.

I will now describe how the expression for ρ_g is obtained and how the internal degree of freedom n_a arises.

2. Area associated with a spacetime event

The two primitive geometrical constructs one can think of in a spacetime are the area and the volume. It is, therefore, natural to assume that the density of states of the quantum spacetime, $\rho_g(\mathcal{P})$, at an event $\mathcal{P}$, should be some function F of either the area $A(\mathcal{P})$ or the volume $V(\mathcal{P})$ that we can "associate with" the event $\mathcal{P}$. Further, the total degrees of freedom, $\rho_g(\mathcal{P})\rho_g(\mathcal{Q})$, associated with two events $\mathcal{P}$ and $\mathcal{Q}$ is multiplicative while the primitive area/volume elements are additive. Hence this function F should be an exponential.[6] In terms of area, for example, we then get

$$\ln\left\{\begin{array}{c}\text{density of states of the}\\ \text{quantum geometry at } \mathcal{P}\end{array}\right\} \propto \left\{\begin{array}{c}\text{area "associated with"}\\ \text{the event } \mathcal{P}\end{array}\right\} \tag{8}$$

That is, $\ln\rho_g(\mathcal{P}) \propto A(\mathcal{P})$.

We next need to give a precise meaning to the phrase, area (or volume) "associated with" the event $\mathcal{P}$. Any natural definition in standard Riemannian geometry will attribute zero volume and zero area to an event. But when we introduce the discreteness of the spacetime in terms of a zero-point length, we find that [18] the area associated with an event becomes nonzero but the volume will still remain zero. What is more, this approach will introduce an arbitrary, constant norm, vector n_a into the discussion. (Its norm is unity in the Euclidean sector and it will map to a null vector with zero norm in the Lorentzian sector.). The area "associated with" an event will still be a fluctuating, indeterminate variable depending on a quantum degree of freedom n^a. I will show how all these results arise, concentrating on the area.

Let me first introduce the notion of an equi-geodesic surface, which can be done either in the Euclidean sector or in the Lorentzian sector; I will work in the Euclidean sector. An equi-geodesic surface $\mathcal{S}$ is the set of all points at the same geodesic distance σ from some specific point P, which we take to be the origin [6–9]. I can now "associate" an area element with a point P in a fairly natural way by the following limiting procedure: (i) Construct an equi-geodesic surface $\mathcal{S}$ around a point P at a geodesic distance σ. (ii) Calculate the area element $\sqrt{h}$ from the induced metric h_{ab} on $\mathcal{S}$. (iii) Use the limit $\sigma \to 0$ to define the area element associated with the point P. In standard differential geometry, we can show [10] that, in the limit of $\sigma \to 0$, the area element, normalized to the flat spacetime value, is given by:

$$\frac{\sqrt{h}}{\sqrt{h_{flat}}} = \left(1 - \frac{1}{6}\sigma^2 R^a_b n_a n^b\right); \tag{9}$$

[6] This is an improvement over the discussion in my earlier works, like e.g., Ref. [1] wherein I approximated it as a linear function. The lowest order results remain the same in either case but $\ln\rho_g(\mathcal{P}) \propto A(\mathcal{P})$ makes better physical sense.

where $n_a = \nabla_a \sigma$ is the normal to $\mathcal{S}$. The second term involving $\mathcal{E} \equiv R^a_b n_a n^b$ gives the curvature correction to the area element of an equi-geodesic surface. In fact, Eq. (9) describes a textbook result in differential geometry and is often presented as a measure of the curvature at any event.

As you can readily see from Eq. (9), $\sqrt{h} \to 0$ when $\sigma \to 0$ since, in this limit $\sqrt{h_{flat}} \propto \sigma^3$. Even though the normal $n_a = \nabla_a \sigma$ becomes ill-defined in this limit, this ambiguity of the second term is irrelevant when $\sigma \to 0$ because of the σ^2 factor. This is, of course, to be expected. The existence of non-zero microscopic degrees of freedom of the spacetime requires some kind of discrete structure in the spacetime. They cannot arise if the spacetime is treated as a continuum all the way. (This is analogous to the fact that you can't associate a finite number of molecules of a fluid with an event P if the fluid is treated as a continuum all the way.) Classical differential geometry, which leads to Eq. (9), knows nothing about any discrete spacetime structure and hence cannot give you a nonzero ρ_g. To obtain a nonzero ρ_g, we need to know how the geodesic interval and the metric get modified in the quantum description of spacetime. In particular, we would expect to have a $\sqrt{h}$ which does not vanish in the coincidence limit in such a quantum description. I will now turn to the task of describing a spacetime metric which is modified by quantum gravitational effects from some general considerations.

There is a large amount of evidence (see e.g., [11–16]) which suggests that a primary effect of quantum gravity will be to introduce into the spacetime a zero-point length, by modifying the geodesic interval $\sigma^2(x, x')$ between any two events x and x' to a form like $\sigma^2 \to \sigma^2 + L_0^2$ where L_0 is a length scale of the order of the Planck length.[7] This allows us to determine a quantum corrected, effective metric along the following lines: Just as the original σ^2 is obtained from the original metric g_{ab}, we demand that the geodesic interval $S(\sigma^2)$ — which incorporates the effects of quantum gravity — arises from a corresponding, quantum gravity-corrected metric [6], which we will call the qmetric q_{ab}. Of course, no such local, non-singular q_{ab} can exist; this is because, for any such q_{ab}, the resulting geodesic interval will vanish in the coincidence limit, by definition of the integral. The $q_{ab}(x, x')$ will be a bitensor, which is singular at all events in the coincidence limit $x \to x'$. The fact that the pair $(q_{ab}, S(\sigma^2))$ should satisfy the same relationships as (g_{ab}, σ^2) is enough to determine [7, 8, 17] the form of q_{ab}. We can then compute the area element $(\sqrt{h}\, d^3x)$ of an equi-geodesic surface using the qmetric.

The computation is straightforward and — for $S(\sigma^2) = \sigma^2 + L_0^2$ in $D = 4$, though similar results [17, 18] hold in the more general case in D dimensions — we get:[8]

$$\frac{\sqrt{h}}{\sqrt{h_{flat}}} = \left[1 - \frac{1}{6}(R_{ab}n^a n^b)\left(\sigma^2 + L_0^2\right)\right]; \qquad L_0^2 = \frac{3L_P^2}{4\pi} \tag{10}$$

(It will turn out that the correct Newtonian limit of the theory requires $L_0^2 = 3L_P^2/4\pi$ which is the value we will use.) When $L_0^2 \to 0$, we recover the standard result in Eq. (9), as we should. Our interest, however, is in the coincidence limit $\sigma^2 \to 0$ evaluated with finite L_0. Something

[7] A more general modification will have the form $\sigma^2 \to S(\sigma^2)$ where the function $S(\sigma^2)$ satisfies the constraint $S(0) = L_0^2$ with $S'(0)$ finite. The results described here are insensitive to the explicit functional form of $S(\sigma^2)$. For the sake of illustration, I will use the function $S(\sigma^2) = \sigma^2 + L_0^2$.

[8] You might think that the result in Eq. (10) arises from the standard result in Eq. (9), just by the simple replacement of $\sigma^2 \to (\sigma^2 + L_0^2)$. This happens to be true but it turns out that this replacement trick does *not* work for the volume element $\sqrt{q}$ which actually vanishes [7, 8, 17] when $\sigma \to 0$. This is why each event has zero volume, but a finite area, associated with it! A further insight into this curious feature is provided by the following fact: The leading order dependence of $\sqrt{q}d\sigma \approx \sigma d\sigma$ leads to the volumes scaling universally as σ^2 while the area measure is finite. This, in turn, leads to the result [18] that *the effective dimension of the quantum-corrected spacetime tends to $D = 2$ close to Planck scales,* independent of the original D. Similar 'dimensional reduction' has been noticed by several people [19] in different, but specific, models of quantum gravity. The approach described here leads to this result in a fairly *model-independent* manner.

IOP Conf. Series: Journal of Physics: Conf. Series **880** (2017) 012008 doi:10.1088/1742-6596/880/1/012008

nice happens when we do this and we get a non-zero limit:

$$\lim_{\sigma\to 0}\frac{\sqrt{h}}{\sqrt{h_{flat}}}=\left[1-\frac{L_P^2}{8\pi}R_{ab}n^a n^b\right] \tag{11}$$

That is, the qmetric attributes to every point in the spacetime a finite area measure (but a zero volume measure)! So we define [20] the dimensionless density of states of the quantum spacetime, as:[9]

$$\ln\rho_g(x^i,n_a)\propto\lim_{\sigma\to 0}\frac{\sqrt{h(x,\sigma)}}{\sqrt{h_{flat}(x,\sigma)}}\propto\left[1-\frac{L_P^2}{8\pi}R_{ab}n^a n^b\right]=\mu\left[1-\frac{L_P^2}{8\pi}R_{ab}n^a n^b\right] \tag{12}$$

where μ is a dimensionless proportionality constant. As we mentioned earlier this is related to the choice of the measure; alternatively, we can set $\mu=1$ in Eq. (12) and absorb it in the measure.

We see that ρ_g, defined by the first equality in Eq. (12), depends on the extra, internal degree of freedom, n_a which could take all possible values (at a given x^i) except for the constraint that it has unit norm in the Euclidean space. *This quantity is a relic of the discrete nature of the spacetime.* This is analogous to the p_j which appears in the fluid distribution function $f(x^i,p_j)$ which can take all possible values at a given x^i — as a relic of the discrete nature of the fluid — except for the constraint that it has a constant norm $p^2=m^2$. In the case of the fluid, we have $\langle p^\mu\rangle=P^\mu(x^i)$ where $P^\mu(x^i)$ is the mean momentum of the fluid in the continuum description and $\langle p^\mu p^\nu\rangle=P^\mu(x^i)P^\nu(x^i)+\Sigma^{\mu\nu}$, where the second term arises from the dispersion of momentum. Similarly n^a is a microscopic, fluctuating variable such that its average over fluctuations gives some vector field $\langle n^a\rangle=\ell^a_E(x^i)$ of unit norm in the Euclidean continuum limit and $\langle n^i n^j\rangle=\ell^i_E\ell^j_E+\sigma^{ij}$ where the second term σ_{ij} represents higher corrections to the mean value.

So far, we have been working in the Euclidean sector with $n_a=\nabla_a\sigma$ being the unit normal to the equi-geodesic surface, $\sigma=$ constant. The limit $\sigma\to 0$ in the Euclidean sector makes the equi-geodesic surface shrink to the origin. *But, in the Lorentzian sector, the same limit leads to the null surface which acts as the local Rindler horizon around the chosen event.* Therefore, in this limit, we need to identify n^a and its mean value ℓ^a_E with Lorentzian null vectors. This is how an internal degree of freedom enters into the density of states of the quantum spacetime, $\langle\ln\rho_g\rangle$, in the form of a null vector ℓ^a.

To see this in some more detail, consider the Euclidean version of the local Rindler frame. There are two natural ways of extending the null surface and the Rindler observer off the TX plane in the Lorentzian sector. You can extend the null line (the 45 degree line in TX plane) to the null *plane* $T=X$ in the spacetime. Alternatively, you can extend the null line to the null *cone* by $R^2-T^2=0$ where $R^2=X^2+Y^2+Z^2$ and the trajectories of Rindler observers to the hyperbolas in the hyperboloid ($R^2-T^2=$ constant) which will go 'around' the null cone in the Lorentzian spacetime (see the left part of Fig. 1). Observers living on this hyperboloid will use their respective X axes obtained by rotation. When we analytically continue to the Euclidean sector, the hyperboloid $R^2-T^2=\sigma^2$ will become a sphere $R^2+T_E^2=\sigma_E^2$ (see the right half of Fig. 1). The light cone $R^2-T^2=0$, which transforms to $R^2+T_E^2=0$, collapses into the origin. The local Rindler observers, living on the hyperboloid $R^2-T^2=\sigma^2$, will perceive local patches of the light cone $R^2-T^2=0$ as their local Rindler horizon (see the left half of Fig. 1). Thus, taking the limit $\sigma_E\to 0$ in the Euclidean sector corresponds to approaching the local Rindler horizons in the Lorentzian sector with the hyperboloid degenerating into the light cones

[9] As you will see, the fact that the term involving R_{ab} comes with a *minus sign* in Eq. (12) is crucial for the success of our programme and we have *no* control over it!

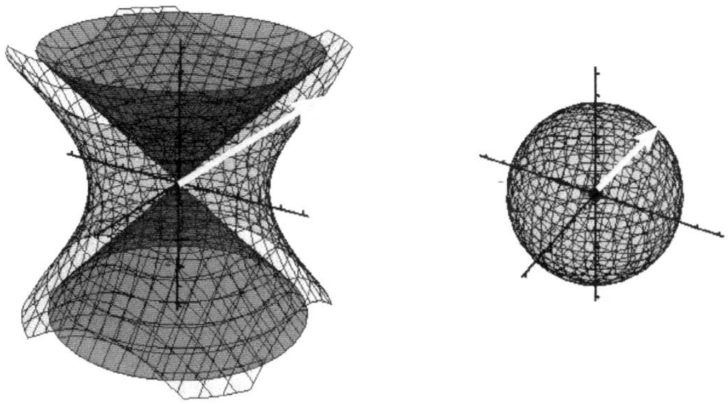

Figure 1. (a) Left: In the local inertial frame (in Lorentzian spacetime), the light cones originating from an event (taken to be the origin) are the null surfaces with $R^2 - T^2 = 0$ with a normal ℓ_a. The local Rindler observers on the hyperboloid $R^2 - T^2 = \sigma^2 =$ constant located around these light cones perceive a patch of the light cone as a local Rindler horizon with a non-zero temperature. The arrow denotes the normal to the hyperbola. (b) Right: In the Euclidean sector, the hyperboloid $R^2 - T^2 = \sigma^2$ becomes a sphere $R^2 + T_E^2 = \sigma_E^2$ and the normal to the hyperboloid becomes the normal to the sphere. The light cone $R^2 - T^2 = 0$ translates to $R^2 + T_E^2 = 0$, and hence collapses into the origin. The limit $\sigma_E \to 0$, viz., approaching the origin in the Euclidean sector, corresponds to approaching the Rindler horizon in the Lorentzian sector. In this limit, the hyperboloid becomes the two light cones emanating from $\mathcal{P}$. In the Euclidean sector, the direction of the normal to the sphere becomes ill-defined when the radius of the sphere tends to zero. In the Lorentzian sector, it corresponds to the normal to the null surface in the limit when the hyperboloid degenerates to the light cone. The dependence of ρ_g on the normal ℓ_a^E to equi-geodesic surface (in the Euclidean sector) is what translates into its dependence on the null normal ℓ_a in the Lorentzian sector.

emanating from the event $\mathcal{P}$. The normal ℓ_a^E to the Euclidean sphere can be now identified with the normal to the local Rindler horizon ℓ_a. The dependence of ρ_g on n_a in the Euclidean equi-geodesic surface translates into its dependence on the null normal ℓ_a to which it maps in the Lorentzian sector.

We can summarize the net result of the exercise as follows: Take any event x^i and a patch of any null surface — with normal $\ell_a(x)$ and passing through x^i — that can act as a horizon to the local Rindler observers. Identify $\langle n_a \rangle$ with $\ell_a(x)$ in the continuum limit. There are, of course, an infinite number of null patches (and normals) at any event. The quantum state of the microscopic geometry decides the expectation value $\langle n_a \rangle$ in a more fundamental description. Different quantum states will lead to different null normals $\ell_a(x^i)$. Thus, the expectation value $\langle n_a \rangle$ actually corresponds to the *set of* all null normals $\{\ell_a(x^i)\}$ at an event x^i when we consider all possible quantum states. Then the microscopic degrees of freedom of the spacetime associated with any event in the continuum spacetime is given by the mean value:

$$\langle \ln \rho_g(x^i, n_a) \rangle = \mu \left[1 - \frac{L_P^2}{8\pi} R_{ab} \ell^a \ell^b \right] + ... \tag{13}$$

where we have not displayed terms proportional to $R_{ab}\sigma^{ab}$ which are of higher order and independent of $\ell_a(x)$.

This approach brings to the center-stage the geodesic interval $\sigma^2(x, x')$ (rather than the metric) as the useful variable to describe spacetime geometry [17]. In the classical spacetime, both $\sigma^2(x, x')$ and $g_{ab}(x)$ contain the same amount of information and one is derivable from the other. But the geodesic interval $\sigma^2(x, x')$ is better suited to take into account quantum gravitational effects to a great extent. Introducing a zero-point length in the spacetime by the modification $\sigma^2 \to S(\sigma^2)$ (with a finite $S(0)$) allows us to derive all the results presented above.

This modification provides an operational principle for incorporating quantum gravitational effects at mesoscopic scales around any event $\mathcal{P}$. You can do this as follows: (a) Around the event $\mathcal{P}$, construct a local Euclidean version of the spacetime; this is almost always possible *locally*. (b) Construct a synchronous frame around this event in which the line element has the form $ds^2 = d\sigma^2 + h_{\alpha\beta}dx^\alpha dx^\beta$. (c) Modify this metric to a quantum corrected qmetric, $q_{ab}(x, \mathcal{P})$ such that the geodesic interval acquires a zero-point length [18, 29]. One can then explore the quantum corrected geometrical features using the bi-tensor $q_{ab}(x, \mathcal{P})$ in the neighbourhood of $\mathcal{P}$. In particular, this allows us to provide a natural definition of density of states of the quantum spacetime through Eq. (12).

The procedure outlined above associates a constant norm, randomly fluctuating vector field n_i with every event in spacetime which is a relic from quantum gravity. This vector field can take arbitrary values throughout the spacetime and hence is described by some probability functional $P[n_i(x), x]$. While one cannot determine the form of $P[n_i(x), x]$ with our current knowledge of quantum gravity, we can use it to obtain the semi-classical limit of geometry with some reasonable assumptions. The n_a gets mapped to the normal ℓ_a to the null surface when we take the limit $\sigma \to 0$ in the Euclidean sector. This mapping, in turn, depends on the fact that the condition $\sigma^2(x, y) = 0$ will lead to $x = y$ in the Euclidean space while it will be satisfied by all events connected by a null ray in the Lorentzian spacetime.

This expression in Eq. (13) has a natural interpretation as the gravitational contribution to the heating rate (per unit area) of the null surface of which ℓ_a is the normal. Its integral over the null surface, Q_g, can be interpreted as the gravitational contribution to the heat content of the null surface. These results arise because the term $R_{ab}\ell^a\ell^b$ is related to the concept of "dissipation without dissipation" [24] of the null surfaces. Let me explain briefly how this connection comes about.

Construct the standard description of a null surface by introducing the complementary null vector k^a (with $k^a\ell_a = -1$) and defining the 2-metric on the cross-section of the null surface by $q_{ab} = g_{ab} + \ell_a k_b + k_a \ell_b$. Define the expansion $\theta \equiv \nabla_a \ell^a$ and shear $\sigma_{ab} \equiv \theta_{ab} - (1/2)q_{ab}\theta$ of the null surface where $\theta_{ab} = q^i_a q^j_b \nabla_i \ell_j$. (It is convenient to take the null congruence to be affinely parametrized.) One can then show that [25]:

$$-\frac{1}{8\pi L_P^2} R_{ab}\ell^a\ell^b \equiv \mathcal{D} + \frac{1}{8\pi L_P^2}\frac{1}{\sqrt{\gamma}}\frac{d}{d\lambda}(\sqrt{\gamma}\theta) \tag{14}$$

where

$$\mathcal{D} \equiv \left[2\eta\sigma_{ab}\sigma^{ab} + \zeta\theta^2\right] \tag{15}$$

is the standard expression for the viscous heat generation rate of a fluid with shear and bulk viscous coefficients [26–28] defined[10] as $\eta = 1/16\pi L_P^2, \zeta = -1/16\pi L_P^2$. Ignoring the total divergence in Eq. (14), we can identify the integral of $R_{ab}\ell^a\ell^b$ with:

$$Q_g = -\frac{1}{8\pi L_P^2}\int \sqrt{\gamma}\, d^2x\, d\lambda\, R^a_b \ell_a \ell^b = \int \sqrt{\gamma}\, d^2x\, d\lambda\, \mathcal{D} = \int \sqrt{\gamma}\, d^2x\, d\lambda \left[2\eta\sigma_{ab}\sigma^{ab} + \zeta\theta^2\right] \tag{16}$$

[10] The fact that the null fluid has negative bulk viscosity coefficient is well-known in the literature [26–28], especially in the context of the black hole membrane paradigm. So I will not pause to discuss this aspect.

IOP Conf. Series: Journal of Physics: Conf. Series **880** (2017) 012008 doi:10.1088/1742-6596/880/1/012008

which represents the heat content of the null surface due to gravitational degrees of freedom and the integrand $\mathcal{D}$ is the rate of heating of the null surface due to gravitational degrees of freedom.

Of course, we do not want the null surfaces to exhibit heating or dissipation! This is ensured by the presence of matter which is needed if $R_{ab} \neq 0$. As we will see, the contribution to the heating from the microscopic degrees of freedom of the spacetime precisely cancels out the heating of any null surface by the matter, on-shell. In fact, this allows us to reinterpret the field equation, expressed as

$$-\frac{1}{8\pi L_P^2} R_{ab}\ell^a\ell^b + T_{ab}\ell^a\ell^b = \mathcal{H}_g + \mathcal{H}_m = 0 \tag{17}$$

as a zero-dissipation principle. I will now introduce the physics of the matter sector and show how this comes about.

3. Heat density of matter

I will next show how the combination $\mathcal{H}_m \equiv T_{ab}\ell^a\ell^b$ for any null normal ℓ_a can be thought of as the heat density contributed by matter crossing a null surface. This will eventually lead to the result in Eq. (5).

To acquire some preliminary insight, consider first the case of an ideal fluid, with $T^a_b = (\rho + p)u^a u_b + p\delta^a_b$. In this case, the combination $T^a_b \ell_a \ell^b$ is precisely the *heat density* $\rho + p = Ts$ where T is the temperature and s is the entropy density of the fluid. (The last equality follows from the Gibbs-Duhem relation. We have chosen the null vector such that $(\ell.u)^2 = 1$ for simplicity.) But $T^a_b u^b u_a$ is the energy density for *any* kind of T^a_b, not just for that of an ideal fluid. How can we interpret $T^a_b \ell_a \ell^b$ as the *heat* density in a *general* context when T^a_b could describe any source — not necessarily a fluid — for which concepts like temperature or entropy do not exist intrinsically? *Remarkably enough, you can do this!*. In any spacetime, around any event, we can introduce the local Rindler observers who will indeed interpret $T^a_b \ell_a \ell^b$ as the heat density contributed by the matter to a null surface which they perceive as a horizon. Let me describe this result:

We begin by introducing a freely falling frame (FFF) with coordinates $(T, \mathbf{X})$ in a region around some fiducial event $\mathcal{P}$ in the spacetime. Next, we transform from the FFF to a local Rindler frame (LRF) — with coordinates $(t, \mathbf{x})$ — using the transformations: $X = \sqrt{2ax}\cosh(at), T = \sqrt{2ax}\sinh(at)$ constructed using some acceleration a. (This transformation is valid for $X > |T|$ and similar ones exist for other wedges.) One of the null surfaces passing though $\mathcal{P}$, will now get mapped to the $X = T$ surface of the FFF and will act as a patch of horizon to the local Rindler observers with the trajectories $x =$ constant. This construction leads to a beautiful result [21, 22] in quantum field theory. The local vacuum state, appropriate for the freely-falling observers around $\mathcal{P}$, will appear to be a thermal state to the local Rindler observers, with a temperature proportional to their acceleration a:

$$k_B T = \left(\frac{\hbar}{c}\right)\left(\frac{a}{2\pi}\right) \tag{18}$$

(This acceleration a can, of course, be related to other geometrical variables of the spacetime in different contexts). The existence of the Davies–Unruh temperature tells us that around *any* event, in *any* spacetime, you will always have a class of observers who will perceive the spacetime to be hot.

Consider now the flow of energy associated with the matter, with some *arbitrary* T_{ab}, that crosses the null surface. Nothing strange happens when this phenomenon is viewed in the FFF by the locally inertial observer. But the local Rindler observer, who attributes a temperature T to the horizon, sees it as a hot surface. Therefore, she will interpret the energy ΔE, dumped on

IOP Conf. Series: Journal of Physics: Conf. Series **880** (2017) 012008 doi:10.1088/1742-6596/880/1/012008

the horizon (by the matter that crosses the null surface in the FFF), as energy deposited on a *hot* surface, thereby contributing a *heat* content $\Delta Q = \Delta E$ to the surface. (Recall that, in the case of a *black hole* horizon, an outside observer will find that any matter takes an infinite amount of time to cross the horizon, thereby allowing for thermalization to take place. In a similar manner, a local Rindler observer will find that matter takes a very long time to cross the local Rindler horizon.) It is easy to compute ΔE in terms of T^a_b. The LRF has an approximate Killing vector field, ξ^a, generating the Lorentz boosts in the FFF, which coincides with a suitably defined[11] null normal ℓ^a of the null surface in the appropriate limit. Using the heat current that arises from the energy current $T_{ab}\xi^b$, we can compute the total heat energy dumped on the null surface to be:

$$Q_m = \int \left(T_{ab}\xi^b\right) d\Sigma^a = \int T_{ab}\xi^b\ell^a\sqrt{\gamma}\, d^2x d\lambda = \int T_{ab}\ell^b\ell^a\sqrt{\gamma}\, d^2x d\lambda \tag{19}$$

where we have used the fact that $\xi^a \to \ell^a$ on the null surface. Therefore, the combination

$$\mathcal{H}_m \equiv \frac{dQ_m}{\sqrt{\gamma}d^2x d\lambda} = T_{ab}\ell^a\ell^b \tag{20}$$

is indeed the heat density (energy per unit area per unit affine time) of the null surface, contributed by matter crossing a local Rindler horizon. This interpretation is valid for *any* kind of T^a_b.

The entropy S_m associated with the heat Q_m in Eq. (19) is $S_m = Q_m/T_H$ where T_H is a temperature introduced essentially for dimensional purposes. (A natural choice will be to take it to be the temperature associated with the acceleration of the Rindler observers very close to the horizon. But, as to be expected, none of the results will depend on its numerical value if we choose the measure of integration appropriately.) We have

$$S_m(\ell) = \frac{1}{T_H}\int d\lambda\, d^2x\sqrt{\gamma}\,\mathcal{H}_m(x,\ell) = \mu\int \frac{d\lambda\, d^2x\sqrt{\gamma}}{L_P^3}\left(L_P^4\mathcal{H}_m(x,\ell)\right) \tag{21}$$

where we have introduced suitable factors of L_P to exhibit clearly the dimensionless nature of S_m and defined $\mu \equiv (1/L_P T_H)$. Replacing the integration by a summation over the spacetime events for conceptual clarity, we can write

$$S_m(\ell) = \sum_x L_P^4\mathcal{H}_m(x,\ell) \equiv \sum_x \ln\rho_m(x,\ell) = \ln\prod_x \rho_m(x,\ell) \tag{22}$$

so that $\ln\rho_m(x,\ell) = L_P^4\mathcal{H}_m(x,\ell)$ with a proportionality constant set to unity if we absorb the factor μ in the integration measure. (Alternatively, we could have set the proportionality constants in Eq. (1) and Eq. (5) to be μ and used the measure without the factor μ. The results will be the same.) This connects up with our discussion in Sec. 1 and, in particular, with the result in Eq. (5) in the continuum limit. Note that the total number of degrees of freedom is correctly given by

$$\Omega_m(\ell) = \exp S_m(\ell) = \prod_x \rho_m = \prod_x \exp(L_P^4\mathcal{H}_m) = \exp\left[\mu\int\frac{d\lambda\, d^2x\sqrt{\gamma}}{L_P^3}\left(L_P^4\mathcal{H}_m\right)\right] \tag{23}$$

[11] Since the null vectors have zero norm, there is an overall scaling ambiguity in any expression involving them. This can be resolved, in this case, by considering a family of hyperboloids $\sigma^2 \equiv X^2 - T^2 = 2ax =$ constant and treating the light cone as the (degenerate) limit $\sigma \to 0$ of these hyperboloids. We set $\ell_a = \nabla_a\sigma^2 \propto \nabla_a x$ and then take the corresponding limit.

IOP Conf. Series: Journal of Physics: Conf. Series **880** (2017) 012008 doi:10.1088/1742-6596/880/1/012008

The first equality is the standard relation between the entropy and the degrees of freedom, the second expresses the result as a product over the degrees of freedom associated with each event and the third equality presents it in terms of the variable $\mathcal{H}_m = T^{ab}\ell_a\ell_b$.

Since the matter stress tensor appears only through the combination $T^a_b\ell_a\ell^b$ in the theory, it is invariant under the shift $T^a_b \to T^a_b +$ (constant)δ^a_b arising, for example, by the addition of a constant to matter Lagrangian. This is a symmetry of the matter sector, since the equations of motion for matter are unaffected by the addition of a constant to the matter Lagrangian. But this symmetry is not respected by gravity in the conventional approaches, in which the metric is varied as a dynamical variable in an extremum principle. Any such approach is incapable of solving the cosmological constant problem because the shift $T^a_b \to T^a_b +$ (constant)δ^a_b will regenerate it. In the approach presented here, it is the heat density $\rho + p$ rather than the energy density ρ which couples to gravity and cosmological constant arises as an integration constant. As I will describe in Sec. 4, its value can also be determined within this approach in a self-consistent manner.

In the above discussion, we identified the average $\langle n_a\rangle = \ell_a(x)$ with the normal to the null surface and hence the heat density of matter can be written as:

$$T^{ab}\ell_a\ell_b = T^{ab}\langle n_a\rangle\langle n_b\rangle \approx T^{ab}\langle n_a n_b\rangle \tag{24}$$

when we ignore higher order fluctuations. This suggests that the expression $\langle \ln \rho_m\rangle$ again can arise as the average value of

$$\ln \rho_m \propto L_P^4 T_{ab} n^a n^b \tag{25}$$

which describes the (postulated) interaction of matter with with the microscopic degrees of freedom of the spacetime at these mesoscopic scales. The x^i dependence of $\rho_m(x^i, n_a)$ arises through a symmetric divergence-free energy momentum tensor $T_{ab}(x)$.

We can also obtain the results of Sec.1 directly using n_a itself, along the following lines: The total density of states of the quantum spacetime when the internal degrees of freedom are described by a configuration $n_a(x)$ can be written as:

$$\begin{aligned}\Omega_{\rm tot}[n_a(x)] &= \prod_x \rho_g(x^i, n_a)\,\rho_m(x^i, n_a) = \exp\sum_x (\ln\rho_g + \ln\rho_m) \\ &= \exp\mu\int dV_x\; L_P^4 E^{ab} n_a(x) n_b(x)\end{aligned} \tag{26}$$

with $E_{ab} \equiv T_{ab} - (1/\kappa)R_{ab}$. The average over n_a is given by:

$$\langle\Omega_{\rm tot}\rangle = \int \mathcal{D}n\; P[n_a(x)]\Omega_{\rm tot}[n_a(x)] = \int \mathcal{D}n\; P[n_a(x)]\exp\mu\int dV_x\; L_P^4 E^{ab} n_a(x) n_b(x) \tag{27}$$

The gravitational field equations arise on extremizing this expression for $\langle\Omega_{\rm tot}\rangle$. The n^a dependent part of the exponential reduces to one proportional to

$$\int dV_x E^a_b n_a n^b \equiv \int dV_x \left(T^a_b(x) - \frac{1}{\kappa}R^a_b(x)\right) n_a n^b; \qquad \kappa \equiv 8\pi L_P^2 \tag{28}$$

The extremum condition for this expression, with respect to $n^a \to n^a + \delta n^a$, subject to the constraint $n^2 =$ constant, leads to Einstein's equations, with a cosmological constant arising as an integration constant.

There is a more interesting interpretation which is possible for the expression in Eq. (28). Recall that $T^{ab}\ell_a\ell_b$ has an unambiguous physical meaning as the heat density contributed by matter while $\mu = 1/L_P T_H$ is the dimensionless Rindler horizon temperature. Therefore, one can think of $\exp -(1/T_H)(L_P^3 T^{ab}\ell_a\ell_b)$ as the Boltzmann factor for the energy in a Planck volume L_P^3 near the Rindler horizon. Of course, it doesn't make sense to talk about Planck volumes using ℓ_a but we can do it with n_a. This suggests the interpretation of

$$F \equiv \exp -\frac{1}{T_H}(L_P^3 E^a_b n_a n^b) = \exp -\frac{1}{T_H}\left(T^a_b(x) - \frac{1}{\kappa}R^a_b(x)\right) n_a n^b \tag{29}$$

as the the Boltzmann factor for the excitation of the internal variable n_a. The extremum principle can be thought of as maximizing this expression. I hope to discuss this interpretation in detail in a later work.

4. Cosmic information and the cosmological constant

A hallmark of any good paradigm is that it could throw light on issues which it was not originally designed to tackle. I will briefly explain how the approach presented above allows you to solve a major problem in theoretical physics, viz., the tiny value of cosmological constant. In fact, unlike any other approach to quantum gravity, this approach makes a falsifiable numerical prediction which, I believe, is vital for any formalism to be called science. I have described how this comes about in detail in previous works (see e.g., [1, 2]) and hence will be somewhat brief here.

Observations indicate that the evolution of our universe can be described by three different phases, viz., an inflationary phase very early on, followed by a radiation/matter dominated phase which lasted until recently, and an accelerated phase dominated by a small cosmological constant which has started in the near-past and will continue forever. These three phases can be characterized by three densities $\rho_{\rm inf}, \rho_{\rm eq}$ (which is the density of matter at the epoch when the matter and radiation densities were equal) and ρ_Λ. These three densities act as the signature of our universe, in the sense that the entire evolutionary history can be determined in terms of these numbers. Current observations determine $\rho_{\rm eq}$ and ρ_Λ fairly accurately as: $\rho_{\rm eq} = [(0.86 \pm 0.09)\ \mathrm{eV}]^4$ and $\rho_\Lambda = [(2.26 \pm 0.05) \times 10^{-3}\mathrm{eV}]^4$; we do not have a direct handle on $\rho_{\rm inf}$ but it is usually taken to be about $\rho_{\rm inf} \simeq (10^{15}\ \mathrm{GeV})^4$. So, in standard cosmology, these three densities have absolutely no relation with each other and they are widely different.

I invite you to form a strange dimensionless number I out of these three densities by the definition:

$$I = \frac{1}{9\pi}\ln\left(\frac{4}{27}\frac{\rho_{\rm inf}^{3/2}}{\rho_\Lambda\, \rho_{\rm eq}^{1/2}}\right) \tag{30}$$

and evaluate its numerical value. Surprisingly enough, you will find that

$$I \approx 4\pi\left(1 \pm \mathcal{O}\left(10^{-3}\right)\right) \tag{31}$$

That is, $I = 4\pi$ to an accuracy of one part in thousand for the standard values expected for our universe. This should make you wonder why the Eq. (30) leads to such a pleasing value as 4π, since it is not often that such strange things happen.

The paradigm described here provides an answer [30]. The right hand side of Eq. (30) can actually be interpreted, in a well-defined manner, as the amount of of cosmic information accessible to an eternal observer [1,2,30]; in fact it is this interpretation that leads to the rather strange combination in the first place. The reason this information content has a value 4π is related to the quantum microstructure of spacetime.[12] This, in turn, arises from the following facts: The effective dimension of the quantum-corrected spacetime becomes $D = 2$ close to

[12] It is an *observational fact* that I, defined via Eq. (30), indeed has a numerical value 4π for our universe. It is hard to believe such a result is not telling us anything about our universe and can be ignored as "just one of those things"!

Planck scales, independent of the original D. This result can be derived, in a fairly model-independent manner using a renormalized quantum effective metric [18]. Similar results have been established earlier by several other authors (for a sample, see e.g., [19]) in a number of approaches to quantum gravity). This, in turn, implies that [1, 18] the unit of information associated with a quantum gravitational 2-sphere of radius L_P is just $I_{QG} = 4\pi L_P^2/L_P^2 = 4\pi$.

This relation allows us to determine ρ_Λ in terms of $\rho_{\rm eq}$ and $\rho_{\rm inf}$ (both of which could eventually be determined from high energy physics):

$$\rho_\Lambda = \frac{4}{27}\frac{\rho_{\rm inf}^{3/2}}{\rho_{\rm eq}^{1/2}}\exp\left(-36\,\pi^2\right) \tag{32}$$

If we take the typical values $\rho_{\rm inf} = (1.2\times 10^{15}\ \text{GeV})^4, \rho_{\rm eq} = (0.86\ \text{eV})^4$, we get $\rho_\Lambda = (2.2\times 10^{-3}\ \text{eV})^4$ which agrees very well with the observed value! That is, the idea that the cosmic information content accessible to an eternal observer, I_c, is equal to the basic quantum gravitational unit of information $I_{QG} = 4\pi$, determines the numerical value of the cosmological constant correctly.

Theoretically, the value of ρ_Λ is the holy grail of cosmology. Observationally, however, we know the values of $\rho_{\rm eq}$ and ρ_Λ very well today but have no direct handle on $\rho_{\rm inf}$. Using Eq. (32), we can actually *predict* the value of $\rho_{\rm inf}$ in terms of the cosmologically determined parameters $\rho_{\rm eq}$ and ρ_Λ. We then find that $\rho_{\rm inf}^{1/4} = 1.2\times 10^{15}$ GeV. This is the *only* model with quantum gravitational inputs *that leads to a falsifiable prediction*[13].

More recently, we could improve this model in a significant manner and eliminate the inflationary phase altogether.[14] The new model is based on the idea that the universe made a transition from a quantum pre-geometric phase to the classical Friedmann phase at an energy scale $E_{QG} = \nu^{-1}E_{\rm Pl}$ where ν is a free parameter. We then make a single postulate: The total number of modes N which enter the Hubble radius throughout the history of the universe is equal to 4π. From this single assumption, one could obtain the following results: (a) The universe *must have* a late-time accelerated phase. (b) The cosmological constant, which could drive this late-time acceleration, is given by:

$$\rho_\Lambda L_P^4 = \frac{4}{27}\left(\frac{3}{8\pi}\right)^{3/2}\frac{1}{\nu^6(\rho_{\rm eq}L_P^4)^{1/2}}\exp\left(-36\pi^2\right) \tag{33}$$

(c) The primordial fluctuations (generated at the quantum to classical transitions) will be scale invariant and their amplitude is given by:

$$\mathcal{A} = \left[\frac{k^3P(k)}{2\pi^2}\right]^{1/2} = \frac{0.19c_1}{\nu} \tag{34}$$

where c_1 is a numerical factor of order unity and whose exact value can be determined by more detailed analysis. Thus, a *single* parameter ν determines the values of *both* ρ_Λ as well as $\mathcal{A}$. There is absolutely no reason why we should be able to find a value for ν which gives the correct numerical values for *both* these numbers. *Incredibly enough, it turns out to be possible.* Using the value of ν determined from Eq. (33), we find that $\mathcal{A}_{\rm theory} = 3.05c_1\times 10^{-5}$ which has to be compared with the observed value $\mathcal{A}_{\rm obs}\approx 4.69\times 10^{-5}$. We see that the results are remarkably consistent with $c_1 = 1.54 = \mathcal{O}(1)$.

[13] In obtaining Eq. (32), I have assumed instantaneous reheating; ambiguities in the reheating dynamics can change this result by a factor of a few, leading to the prediction $\rho_{\rm inf}^{1/4}\approx(1-5)\times 10^{15}$ GeV.
[14] The summary below is based on results obtained in collaboration with H. Padmanabhan [31].

These results highlight the notion of spacetime information and its role in gravitational dynamics, already seen in several other contexts [1]. It also strengthens the viewpoint, suggested in Refs. [1,2], that the universe should *not* be treated as a particular solution to the gravitational field equations but instead, be approached as a special dynamical system.

5. Highlights and Discussion

This approach, like any other approach to quantum gravity, involves both algebraic features and conceptual features. The algebraic aspects are extremely simple[15] and, in fact, fairly trivial. Consider a functional of a null vector field ℓ_a given by Eq. (7) (or for n_a given by Eq. (28)). Demand that (a) this functional should be an extremum for the variations of the vector field which maintain its norm and (b) the extremum condition should hold for all the null vector fields at any given event. It is trivial to show that these demands lead to Einstein's equations with the cosmological constant arising as an integration constant.

What is really non-trivial about this approach are the conceptual aspects and the somewhat unfamiliar flow of logic. In particular, we need to understand the following aspects: (a) Where does the null vector field n^a (or ℓ_a) spring from? (b) What is the physical meaning of $R_{ab}n^a n^b$ which appears in the extremum principle? (c) What is the physical meaning of $T_{ab}n^a n^b$? (d) Why do we demand that the extremum condition should hold for all null vector fields at every event?

I believe I have provided reasonably clear answers to all these questions. Just to summarize the previous discussion, the answers — in brief — are as follows: (a) The vector field n^a arises as a relic of quantum gravity when we introduce a zero point length to the spacetime. It measures the density of states of the quantum spacetime at any given event through Eq. (12). In fact, this is probably the most important new feature of this approach. (b) The physical meaning of $R_{ab}n^a n^b$ is provided by Eq. (12) when you think of n_a as a fluctuating quantum variable. If we replace $n^a n^b$ by $\langle n^a n^b\rangle \approx \ell^a \ell^b$, then this term acquires an interpretation as the "dissipation without dissipation" on the null surfaces through Eq. (16). (c) This is probably the most difficult issue. *The Einstein equation* $G_{ab} = \kappa T_{ab}$ *equates apples to oranges.* The left hand side is geometrical while the right hand side, viz. matter, is made of discrete, quantized degrees of freedom bearing no relation to spacetime geometry. A description of matter in terms of T_{ab} — or even in terms of the quantum expectation value $\langle T_{ab}\rangle$ — is probably not going to be useful at the mesoscopic scales.[16] We could couple both the spacetime geometry and matter to n_a through the terms $R^a_b n_a n^b$ and $T^a_b n_a n^b$ respectively, thereby leading to an effective coupling between them. But this idea requires further investigation; one possible way forward could be through Eq. (29). But all these ideas still treat apples and oranges as though they are the same. We need a more unified way of generating both matter and geometry from some other fundamental building blocks, which is currently lacking. It is possible that the zero-dissipation-principle of Eq. (17) tells us something important in this context but it is not clear what exactly is the message. So, I would consider the inclusion of matter into this formalism to be the major challenge at present. (d) This is easy. In this approach, the mean value of n^a maps to a null vector field ℓ^a. Different quantum states of the geometry will lead to different ℓ^a at a given event. The demand that the extremum condition should hold for all ℓ^a is equivalent to demanding that the extremum condition holds for all quantum geometries; rather, the semi-classical spacetime arises only when this condition holds for all quantum geometries which are relevant for such a spacetime.

[15] Unlike some others, I do not believe that the algebraic structure of quantum gravity has to be complicated for it to be correct!

[16] This is, in fact, a key issue not addressed satisfactorily in any approach to quantum gravity. We do not know how to describe matter fields close to Planck scales.

A real bonus, of course, is the fact that the approach provides a refreshingly different way of solving the cosmological constant problem.

Acknowledgments
I thank the organizers of DICE 2016 for excellent hospitality. Part of the discussion here is based on unpublished work [31] done in collaboration with Hamsa Padmanabhan. I thank Sumanta Chakraborty, Sunu Engineer and Dawood Kothawala for useful discussions.

References
[1] Padmanabhan T 2016 *Int. Jour. Mod. Phys.* D **25** 1630020 [arXiv:1603.08658]
[2] Padmanabhan T 2017 Do We Really Understand the Cosmos? *To appear in Comptes Rendue* [arXiv:1611.03505]
[3] Padmanabhan T 2016 *J. Phys (Conf. Series)* **701** 012018 [arXiv:1602.01474]
[4] Padmanabhan T 2010 *Gravitation: Foundations and Frontiers* (Cambridge University Press).
[5] Padmanabhan T 1998 *Phys. Rev. Lett.* **81** 4297 [hep-th-9801015]; Padmanabhan T 1999 *Phys. Rev.* D **59** 124012 [hep-th-9801138].
[6] Kothawala D and Padmanabhan T 2014 *Phys. Rev.* D **90** 124060.
[7] Kothawala D 2013 *Phys. Rev.* D **88** 104029.
[8] Stargen D J and Kothawala D 2015 *Phys. Rev.* D **92** 024046 [arXiv:1503.03793].
[9] Kothawala D and Padmanabhan T 2015 *Phys. Lett.* B **748** 67.
[10] Gray A 1974 *Mich. Math. J.* **20** 329.
[11] DeWitt B S 1964 *Phys. Rev. Lett.* **13** 114.
[12] Padmanabhan T 1985 *Gen. Rel. Grav.* **17** 215
[13] Padmanabhan T 1985 *Ann. Phys.* **165** 38
[14] Padmanabhan T 1997 *Phys. Rev. Lett.* **78** 1854 [hep-th/9608182]
[15] Garay L 1998 *Phys. Rev. Lett.* **80** 2508 [gr-qc/9801024]
[16] Garay L 1995 *Int. J. Mod. Phys.* A **10** 145.
[17] Chakraborty S and Padmanabhan T [work in progress].
[18] Padmanabhan T, Chakraborty S and Kothawala D 2016 *Gen. Rel. Grav.,* **48** 55 [arXiv:1507.05669].
[19] Carlip S, Mosna R and Pitelli J 2011 *Phys. Rev. Lett. 107* 021303; Ambjorn J, Jurkiewicz J and Loll R 2005 *Phys. Rev. Lett.* **95** 171301; Modesto L 2009 *Class. Quantum Grav.* **26** 242002; Husain V, Seahra S S and Webster E J 2013 *Phys. Rev.* D **88** 024014; Modesto L and Nicolini P 2010 *Phys. Rev.* D **81** 104040 [arXiv:0912.0220]
[20] Padmanabhan 2015 *Entropy* **17** 7420-7452 [arXiv:1508.06286]; also see Padmanabhan T Exploring the nature of gravity [arXiv:1602.01474]
[21] Davies P C W 1975 *J. Phys.* A **8** 609.
[22] Unruh W G 1976 *Phys. Rev.* D **14** 870.
[23] Majhi B R and Padmanabhan T 2012 *Phys.Rev.* D **86**, 101501 [arXiv:1204.1422]
[24] Padmanabhan T 2011 *Phys.Rev.* D **83** 044048 [arXiv:1012.0119]; Kolekar S and Padmanabhan T 2012 *Phys.Rev.* D **85** 024004 [arXiv:1109.5353]
[25] Padmanabhan T 2014 *Gen. Rel. Grav.* **46** 1673 [arXiv:1312.3253]
[26] Damour T 1979 *Th'ese de doctorat dEtat, Universite Paris.*
[27] Damour T 1982 Surface effects in black hole physics *Proceedings of the Second Marcel Grossmann Meeting on General Relativity.*
[28] Thorne K S, Price R H and MacDonald D A 1986 *Black Holes: The Membrane Paradigm* (Yale University Press)
[29] Kothawala D 2011 *Phys.Rev.* D **83** 024026 [arXiv:1010.2207]
[30] Padmanabhan T and Padmanabhan H 2014 *Int.Jour.Mod. Phys.* D **23**, 1430011 [arXiv:1404.2284]
[31] Padmanabhan T and Padmanabhan H 2017 Information paradigm for gravity: Cosmic Information determines the cosmological constant and the amplitude of primordial perturbations (unpublished).

Analogue gravity models of emergent gravity: lessons and pitfalls

Stefano Liberati

SISSA, Via Bonomea 265, 34136 Trieste, Italy, EU

E-mail: liberati@sissa.it

Abstract. Analogue models of gravity have provided a test bed for many classical and quantum field theory effects in curve spacetime. Here we present a review of some relevant results towards their extension as toy models of emergent gravity scenarios. From these models we shall try to draw general lessons about the emergent gravity tackle on the cosmological constant problem as well as about the characteristic phenomenological signatures they suggest. Finally, we shall discuss current constraints on these signatures and the field's future perspectives.

1. Introduction

General Relativity (GR) still stands strong after 100 years from its lay out by Einstein. However, in spite of its numerous successes we cannot say we fully understand it: many odd features of the theory still baffle us and suggest that only the tip of the iceberg has been uncovered. In particular, an incomplete list of puzzling facts may comprise

- The nature of the expected singularities of classical GR.
- Critical phenomena in gravitational collapse
- Horizon thermodynamics and their problems (information loss and transplanckian issues)
- Spacetime thermodynamics: Einstein equations as equations of state.
- Thermodynamics interpretation of Einstein equations (see Padmanabhans talk and papers)
- The dark ingredients of our universe
- Faster than light and Time travel solutions
- AdS/CFT duality, holographic behaviour
- Gravity/fluid duality

These are all features that we struggle to understand nowadays and it is not encouraging to realise that while new items have been added to this list none of them has been ticked away as solved. In response to this growing evidence of our lack of a deep understanding of the nature of gravity, several fundamental approaches have been devised. In particular a new framework has been suggested which goes under the general name of "emergent gravity" (EG).

The basic idea of EG is that gravity is not a fundamental interaction and that spacetime is a composite object approximately like a fluid is. In this scenario GR is seen as a sort of hydrodynamics emerging from a deeper theory of the fundamental constituents which are not quanta of spacetime but rather abstract mathematical entities. In this sense, all sub-Planckian physics has to be seen as low energy physics, a phase akin to a Bose–Einstein condensate of the fundamental constituents [1]. Within such a framework also singularities change interpretation, from limits of predictability of GR to phase transitions of the fundamental theory where the hydrodynamic limit comes about (big bang) or ceases to exists (black holes or big rip).

It is obvious that such concepts, as fascinating as they can be, mays sound more as buzz words rather that a solid proposal for the ultimate nature of gravity. Furthermore, how can we test them? As usual in physics, in order to move from abstract ideas to more concrete physical intuition, it is then appropriate to introduce relatively simple systems that can act as tests beds for our conjectures. In what follows we shall consider a possible route in this direction i.e. the Analogue Gravity (AG) proposal.

2. Analogue gravity: an example of emergent spacetime

Analogue gravity stemmed from the realisation that within several condensed matter systems there are regimes in which the effective degrees of freedom are represented by fields propagating over effective pseudo-Riemannian structures [2].

For instance, in the case of perfect, irrotational and barotropic fluids, it can be proved that the perturbations in the velocity potential (i.e. the scalar function θ whose gradient gives the velocity of the fluid, $\vec{v} \propto \vec{\nabla}\theta$) do obey a massless Klein–Gordon equation in a curved effective spacetime whose metric tensor is given by the so-called acoustic metric,

$$g_{\mu\nu} = \frac{\rho}{c_s^2} \begin{pmatrix} -(c_s^2 - v^2) & \vdots & v^i \\ \cdots & \cdot & \cdots \\ v^i & \vdots & \delta_{ij} \end{pmatrix}, \tag{1}$$

where ρ is the local density of the fluid, c_s it the (local) speed of sound and v_i is the velocity field of the fluid flow. Analogue models have been used to understand (and possibly to test in a laboratory) some peculiar aspects of physics in curved spacetimes, otherwise inaccessible (e.g. Hawking radiation). For a review of the subject see [2]. For the large majority, these analogue models for gravity offer the possibility of studying some kinematical aspects of physics of curved spacetimes, leaving aside the issue of dynamics. But there are some exceptions.

2.1. BEC as a prototype model

Bose–Einstein condensates (BECs) have become the subject of extensive study as possible analogue models of general relativity (see e.g. [3, 4, 5, 6]).

Let us start by very briefly reviewing the derivation of the acoustic metric for a BEC system, and show that the equations for the phonons of the condensate closely mimic the dynamics of a scalar field in a curved spacetime. In the dilute gas approximation, one can describe a Bose gas by a quantum field $\widehat{\Psi}$ satisfying

$$\mathrm{i}\hbar \frac{\partial}{\partial t}\widehat{\Psi} = \left(-\frac{\hbar^2}{2m}\nabla^2 + V_{\mathrm{ext}}(\mathbf{x}) + \kappa(a)\, \widehat{\Psi}^\dagger \widehat{\Psi} \right) \widehat{\Psi}. \tag{2}$$

Here κ parameterises the strength of the interactions between the different bosons in the gas. It can be re-expressed in terms of the scattering length a as $\kappa(a) = (4\pi a \hbar^2)/m$.

As usual, the quantum field can be separated into a macroscopic (classical) condensate and a fluctuation: $\widehat{\Psi} = \psi + \widehat{\varphi}$, with $\langle \widehat{\Psi} \rangle = \psi$. Then, by adopting the self-consistent mean field approximation $\widehat{\varphi}^\dagger\widehat{\varphi}\widehat{\varphi} \simeq 2\langle\widehat{\varphi}^\dagger\widehat{\varphi}\rangle\ \widehat{\varphi} + \langle\widehat{\varphi}\widehat{\varphi}\rangle\ \widehat{\varphi}^\dagger$ one can arrive at the set of coupled equations:

$$\mathrm{i}\hbar\,\frac{\partial}{\partial t}\psi(t,\mathbf{x}) = \left(-\frac{\hbar^2}{2m}\nabla^2 + V_{\rm ext}(\mathbf{x}) + \kappa\, n_c\right)\psi(t,\mathbf{x}) + \kappa\left\{2\tilde{n}\psi(t,\mathbf{x}) + \tilde{m}\psi^*(t,\mathbf{x})\right\}, \tag{3}$$

$$\mathrm{i}\hbar\,\frac{\partial}{\partial t}\widehat{\varphi}(t,\mathbf{x}) = \left(-\frac{\hbar^2}{2m}\nabla^2 + V_{\rm ext}(\mathbf{x}) + \kappa\, 2n_T\right)\widehat{\varphi}(t,\mathbf{x}) + \kappa\, m_T\,\widehat{\varphi}^\dagger(t,\mathbf{x}), \tag{4}$$

with $n_c \equiv |\psi(t,\mathbf{x})|^2$, $m_c \equiv \psi^2(t,\mathbf{x})$, $\tilde{n} \equiv \langle\widehat{\varphi}^\dagger\,\widehat{\varphi}\rangle$, $\tilde{m} \equiv \langle\widehat{\varphi}\,\widehat{\varphi}\rangle$, $n_T = n_c + \tilde{n}$, and $m_T = m_c + \tilde{m}$. The equation for the classical wave function of the condensate is closed only when the back-reaction effect due to the fluctuations are neglected. (This back-reaction is hiding in the parameters $\tilde{m}$ and $\tilde{n}$.) This is the approximation contemplated by the Gross–Pitaevskii equation. In general one will have to solve both equations simultaneously. Adopting the Madelung representation for the wave function of the condensate $\psi(t,\mathbf{x}) = \sqrt{n_c(t,\mathbf{x})}\ \exp[-\mathrm{i}\theta(t,\mathbf{x})/\hbar]$, and defining an irrotational "velocity field" by $\mathbf{v} \equiv \boldsymbol{\nabla}\theta/m$, the Gross–Pitaevskii equation can be rewritten as a continuity equation plus an Euler equation:

$$\frac{\partial}{\partial t}n_c + \boldsymbol{\nabla}\cdot(n_c\mathbf{v}) = 0, \tag{5}$$

$$m\frac{\partial}{\partial t}\mathbf{v} + \boldsymbol{\nabla}\left(\frac{mv^2}{2} + V_{\rm ext}(t,\mathbf{x}) + \kappa n_c - \frac{\hbar^2}{2m}\frac{\nabla^2\sqrt{n_c}}{\sqrt{n_c}}\right) = 0. \tag{6}$$

These equations are completely equivalent to those of an irrotational and inviscid fluid apart from the existence of the so-called quantum potential

$$V_{\rm quantum} = -\frac{\hbar^2}{2m}\frac{\nabla^2\sqrt{n_c}}{\sqrt{n_c}}, \tag{7}$$

which has the dimensions of an energy.

Apart from the wave function of the condensate itself, we also have to account for the (typically small) quantum perturbations of the system (4). These quantum perturbations can be described in several different ways, here we are interested in the "quantum acoustic representation"

$$\widehat{\varphi}(t,\mathbf{x}) = e^{-\mathrm{i}\theta/\hbar}\left(\frac{1}{2\sqrt{n_c}}\,\widehat{n}_1 - \mathrm{i}\,\frac{\sqrt{n_c}}{\hbar}\,\widehat{\theta}_1\right), \tag{8}$$

where $\widehat{n}_1, \widehat{\theta}_1$ are real quantum fields. By using this representation Equation (4) can be rewritten as

$$\partial_t\widehat{n}_1 + \frac{1}{m}\boldsymbol{\nabla}\cdot\left(n_1\,\boldsymbol{\nabla}\theta + n_c\,\boldsymbol{\nabla}\widehat{\theta}_1\right) = 0, \tag{9}$$

$$\partial_t\widehat{\theta}_1 + \frac{1}{m}\boldsymbol{\nabla}\theta\cdot\boldsymbol{\nabla}\widehat{\theta}_1 + \kappa(a)\,n_1 - \frac{\hbar^2}{2m}\,D_2\widehat{n}_1 = 0. \tag{10}$$

Here D_2 represents a second-order differential operator obtained from linearizing the quantum potential. Explicitly:

$$D_2\,\widehat{n}_1 \equiv -\frac{1}{2}n_c^{-3/2}\,[\nabla^2(n_c^{+1/2})]\,\widehat{n}_1 + \frac{1}{2}n_c^{-1/2}\,\nabla^2(n_c^{-1/2}\,\widehat{n}_1). \tag{11}$$

The equations we have just written can be obtained easily by linearizing the Gross–Pitaevskii equation around a classical solution: $n_c \to n_c + \widehat{n}_1$, $\phi \to \phi + \widehat{\phi}_1$.

From the previous equations for the linearised perturbations it is possible to derive a wave equation for $\widehat{\theta}_1$ (or alternatively, for $\widehat{n}_1$). All we need is to substitute in Equation (9) the $\widehat{n}_1$ obtained from Equation (10). This finally leads to a wave equation for θ_1 which can be easily rewritten as

$$\partial_\mu (f^{\mu\nu}\, \partial_\nu \widehat{\theta}_1) = 0. \tag{12}$$

Where the $f^{\mu\nu}$ are *differential operators* acting on space only.

Now, if we make a spectral decomposition of the field $\widehat{\theta}_1$ we can see that for wavelengths larger than $\hbar/mc_s$ (the so called "healing length" of the condensate), the terms coming from the linearization of the quantum potential (the D_2) can be neglected in the previous expressions, in which case the $f^{\mu\nu}$ can be approximated by (momentum independent) numbers, instead of differential operators. (This is the heart of the acoustic approximation.) Then, by identifying

$$\sqrt{-g}\; g^{\mu\nu} = f^{\mu\nu}, \tag{13}$$

the equation for the field $\widehat{\theta}_1$ becomes that of a (massless minimally coupled) quantum scalar field over a curved background

$$\Delta \theta_1 \equiv \frac{1}{\sqrt{-g}}\, \partial_\mu \left(\sqrt{-g}\; g^{\mu\nu}\, \partial_\nu\right) \widehat{\theta}_1 = 0, \tag{14}$$

with an effective metric of the form

$$g_{\mu\nu}(t, \mathbf{x}) \equiv \frac{n_c}{m\, c_s(a, n_c)} \begin{bmatrix} -\{c_s(a, n_c)^2 - v^2\} & \vdots & -v_j \\ \cdots\cdots\cdots\cdots & \cdot & \cdots\cdots \\ -v_i & \vdots & \delta_{ij} \end{bmatrix}. \tag{15}$$

Here the magnitude $c_s(n_c, a)$ represents the speed of the phonons in the medium $c_s^2(a, n_c) = \kappa(a)\, n_c/m$ With this effective metric now in hand, the analogy is fully established, and one is now in a position to start asking more specific physics questions.

2.2. Lorentz breaking in BEC models

It is interesting to consider the case in which the above "hydrodynamical" approximation for BECs does not hold. In order to explore a regime where the contribution of the quantum potential cannot be neglected we can use the so called *eikonal* approximation, a high-momentum approximation where the phase fluctuation $\widehat{\theta}_1$ is itself treated as a slowly-varying amplitude times a rapidly varying phase. This phase will be taken to be the same for both $\widehat{n}_1$ and $\widehat{\theta}_1$ fluctuations. In fact, if one discards the unphysical possibility that the respective phases differ by a time varying quantity, any time-constant difference can be safely reabsorbed in the definition of the (complex) amplitudes. Specifically, we shall write

$$\widehat{\theta}_1(t, \mathbf{x}) = \mathrm{Re}\,\{\mathcal{A}_\theta\, \exp(-i\phi)\}\,, \tag{16}$$

$$\widehat{n}_1(t, \mathbf{x}) = \mathrm{Re}\,\{\mathcal{A}_\rho\, \exp(-i\phi)\}\,. \tag{17}$$

As a consequence of our starting assumptions, gradients of the amplitude, and gradients of the background fields, are systematically ignored relative to gradients of ϕ. (Warning: What we are doing here is not quite a "standard" eikonal approximation, in the sense that it is not applied directly on the fluctuations of the field $\psi(t, \mathbf{x})$ but separately on their amplitudes and phases ρ_1 and ϕ_1.) We adopt the notation

$$\omega = \frac{\partial \phi}{\partial t}; \qquad k_i = \nabla_i \phi. \tag{18}$$

Then the operator D_2 can be approximated by the *function* $D_2 \to -n_c^{-1}k^2/2$.

As desired, this has the net effect of making $f^{\mu\nu}$ a matrix of (explicitly momentum dependent) numbers, not operators. The physical wave equation (12) now becomes a nonlinear dispersion relation

$$f^{00}\,\omega^2 + (f^{0i} + f^{i0})\,\omega\,k_i + f^{ij}\,k_i\,k_j = 0. \tag{19}$$

After substituting the approximate D_2 into this dispersion relation, rearranging, and introducing the speed of sound c_s one obtains

$$\omega = v_0^i\,k_i \pm \sqrt{c_s^2 k^2 + \left(\frac{\hbar}{2m}\,k^2\right)^2}. \tag{20}$$

Coming back to considering the form of (20) at this stage some observations are in order: It is easy to see that (20) actually interpolates between two different regimes depending on the value of the wavelength $\lambda = 2\pi/||k||$ with respect to the "acoustic Compton wavelength" $\lambda_c = h/(mc_s)$. In particular, if we assume $v_0 = 0$ (no background velocity), then for large wavelengths $\lambda \gg \lambda_c$ one gets a standard phonon dispersion relation $\omega \approx c||k||$. For wavelengths $\lambda \ll \lambda_c$ the quasi-particle energy tends to the kinetic energy of an individual gas particle and in fact $\omega \approx \hbar^2 k^2/(2m)$.

3. From emergent spacetime analogue to full emergent gravity

In order to see how some sort of gravitational dynamics is encoded in the BEC, a suitable framework must be set up in order to see how the quasiparticles backreact over the condensate.

This can be achieved by considering an improved version of the Gross–Pitaevski equation which consistently takes into account the effect of the particles out of the condensate. The Gross–Pitaevski (GP) equation is replaced by the so-called Bogoliubov–de Gennes (BdG) equation

$$i\hbar\frac{\partial}{\partial t}\psi = -\frac{\hbar^2}{2m}\nabla^2\psi - \mu\psi + \kappa|\psi|^2\psi + 2\kappa\mathfrak{n}\psi + \kappa\mathfrak{m}\psi^*, \tag{21}$$

where $\mathfrak{n}, \mathfrak{m}$ are given by the expectation values:

$$\mathfrak{n} = \langle\Xi|\widehat{\varphi}(x)^\dagger\widehat{\varphi}(x)|\Xi\rangle, \qquad \mathfrak{m} = \langle\Xi|\widehat{\varphi}(x)^2|\Xi\rangle, \tag{22}$$

where the state $|\Xi\rangle$ is the particular state one is considering. Notice that, if this state were the Fock vacuum state for particles, these expectation values would be identically zero. Notice also that one is implicitly taking a normal ordering in the particle operator, so that an unphysical (divergent) zero point energy is removed automatically.

When exploring the possibility of casting the equations for the BEC background in a gravitational form, it is clear that the non-relativistic nature of the latter implies that at most some form of Newtonian gravity could be expected to emerge. However, in order to construct some analogue of Newtonian gravity, we need massive particles as sources of the gravitational field (massless particles do not gravitate in Newtonian gravity). Therefore, the quasiparticles must not be Goldstone bosons, but instead pseudo-Goldstones: the $U(1)$ symmetry has to be broken explicitly at the level of the Hamiltonian.

In [7] an extra term was added to the Hamiltonian,

$$\widehat{H}_0 \to \widehat{H} = \widehat{H}_0 + \widehat{H}_\lambda, \qquad \widehat{H}_\lambda = -\frac{\lambda}{2}\int d^3x\left(\widehat{\Psi}(x)^2 + (\widehat{\Psi}^\dagger(x))^2\right), \tag{23}$$

where λ is a coupling constant having the dimensions of an energy in these choice of units. The breaking of the $U(1)$ symmetry has an obvious interpretation: the number of bosons is no longer a conserved charge (see [7] for concrete examples in which such system could be realized).

Preliminary analysis of the condensate wavefunction based on the extension of the GP equation to this case, leads to the homogeneous solution

$$\psi = \sqrt{n_c} e^{-i\theta/\hbar}, \qquad n_c = \frac{\mu + \lambda}{2}, \qquad \theta = 0, \tag{24}$$

where the phase $\theta = 0$ is fixed by stability of the condensate itself: for different values of θ the quasiparticles would show a tachyonic instability.

The analysis of the properties of the quasiparticles in the case of homogeneous background (see [7] for details) leads to the conclusion that the quasiparticles dispersion relation is

$$E = \left(\mathcal{M}^2 c_s^4 + p^2 c_s^2 + \frac{p^4}{4m^2} \right)^{1/2}, \tag{25}$$

where

$$c_s^2 = \frac{\mu + 2\lambda}{m}, \qquad \mathcal{M}^2 = 4\frac{\lambda(\mu + \lambda)}{(\mu + 2\lambda)^2} m^2. \tag{26}$$

In the case of small momenta, and when the condensate wavefunction is not exactly homogeneous (*i.e.* when $u(\mathrm{x}) \neq 0$), the Hamiltonian for the quasiparticles takes the shape

$$\widehat{H}_{q.p.} \approx \mathcal{M} c_s^2 - \frac{\hbar^2 \nabla^2}{2\mathcal{M}} + 2\frac{(\mu + \lambda)(\mu + 4\lambda)}{\mathcal{M} c_s^2} u(\mathrm{x}), \tag{27}$$

which leads to the identification of a "gravitational potential":

$$\Phi_{grav}(\mathrm{x}) = \frac{(\mu + 4\lambda)(\mu + 2\lambda)}{2\lambda m} u(\mathrm{x}). \tag{28}$$

3.1. The emergent gravitational system

Having presented the main ideas and required tools, we pass to the results. Consider the Hamiltonian with the $U(1)$ breaking term. In the limit in which the backreaction of the condensate is small, *i.e.* in the limit in which there are few quasiparticles, when the condensate is almost homogeneous, the Bogoliubov–de Gennes equation can be rewritten in terms of the above mentioned effective Newtonian potential as

$$\left(\nabla^2 - \frac{1}{L^2} \right) \Phi_{grav} = 4\pi G_N \rho_{\mathrm{matter}} + C_\Lambda, \tag{29}$$

where

$$G_N \equiv \frac{\kappa(\mu + 4\lambda)(\mu + 2\lambda)^2}{4\pi\hbar^2 m \lambda^{3/2} (\mu + \lambda)^{1/2}}, \qquad C_\Lambda \equiv \frac{2\kappa(\mu + 4\lambda)(\mu + 2\lambda)}{\hbar^2 \lambda} \left(\mathfrak{n}_\Omega + \frac{1}{2}\mathfrak{m}_\Omega \right), \tag{30}$$

$$L^2 \equiv \frac{\hbar^2}{4m(\mu + \lambda)}. \tag{31}$$

Notice the peculiar splitting of the source term. A detailed analysis [7] shows that the expectation values (22) always split into two contributions, one nonlocal term due to the quasiparticles (ρ_{matter}), and an unavoidable vacuum contribution, C_Λ, due to the inequivalence between the Fock vacuum for particles and the Fock vacuum for quasiparticles.

The reader will immediately realise that the would-be Poisson equation includes a term which makes the interaction short ranged. In particular, this range is set by the healing length L, which is an UV scale for the physics of the BEC (very much like the Planck scale in quantum gravity). This might have been guessed from the beginning, since that the healing length represents the typical scale for the dynamics of the condensate. Henceforth, this system is not an analogue for a realistic form of gravitational interaction, however it does offer some intriguing hints which we further develop in what follows.

IOP Conf. Series: Journal of Physics: Conf. Series **880** (2017) 012009 doi:10.1088/1742-6596/880/1/012009

3.2. A lesson for the cosmological constant in emergent gravity?

Note that the source term in the correct weak field approximation of Einstein equations is $4\pi G_N(\rho + 3p/c^2)$. For standard non-relativistic matter, p/c^2 is usually negligible with respect to ρ. However, it cannot be neglected for the cosmological constant, since $p_\Lambda/c^2 = -\rho_\Lambda$. As a consequence $C_\Lambda = -2c_s^2\Lambda$, where Λ would be the GR cosmological constant.

Indeed it can be shown via an explicit computation in an homogeneous background, no external potential, and with the condensate is at rest it equals [8]

$$\Lambda = -\frac{20m\, g\rho_0\,(g\rho_0 + 3\lambda)}{3\sqrt{\pi}\hbar^2\lambda}\sqrt{\rho_0 a^3}\, F_\Lambda\left(\frac{\lambda}{g\rho_0}\right), \tag{32}$$

where we used that in the above regime $\mu = g\rho_0 - \lambda$ and $F_\Lambda(\lambda/g\rho_0)$ is a monotonically decreasing function ranging from $F(0) = 1$ to $F(1) = 0$.

It is important to stress that if one compares the value of Λ either with the ground-state grand-canonical energy density h or to the ground-state energy density ϵ of of the BEC, Λ does not correspond to either of them [8]. Actually, since Λ is proportional to $\sqrt{\rho_0 a^3}$, it can even be arbitrarily smaller both than h and than ϵ, if the condensate is very dilute. Furthermore, Λ is proportional only to the subdominant second order correction of h or ϵ, which is strictly related to the depletion factor of the condensate.

Indeed, several scales show up in this system, in addition to the naive Planck scale computed by combining $\hbar$ and the emergent constants $G_{\rm N}$ and c_s:

$$L_{\rm P} = \sqrt{\frac{\hbar c_s^5}{G_{\rm N}}} \propto \left(\frac{\lambda}{g\rho_0}\right)^{-3/4} (\rho_0 a^3)^{-1/4} a. \tag{33}$$

For instance, the Lorentz-violation scale $L_{\rm LV} = \xi \propto (\rho_0 a^3)^{-1/2} a$ differs from $L_{\rm P}$, suggesting that the breaking of the Lorentz symmetry might be expected at scale much longer than the Planck length (energy much smaller than the Planck energy), since the ratio $L_{\rm LV}/L_{\rm P} \propto (\rho_0 a^3)^{-1/4}$ increases with the diluteness of the condensate.[1]

It is also instructive to compare the energy density corresponding to Λ to the Planck energy density:

$$\mathcal{E}_\Lambda = \frac{\Lambda c_s^4}{4\pi G_{\rm N}}, \quad \mathcal{E}_{\rm P} = \frac{c_s^7}{\hbar G_{\rm N}^2}, \quad \frac{\mathcal{E}_\Lambda}{\mathcal{E}_{\rm P}} \propto \rho_0 a^3 \left(\frac{\lambda}{g\rho_0}\right)^{-5/2}. \tag{34}$$

The energy density associated with the analogue cosmological constant is much smaller than the values computed from zero-point-energy calculations with a cut off at the Planck scale. Indeed, the ratio between these two quantities is controlled by the diluteness parameter $\rho_0 a^3$.

Taken at face value, this relatively simple model displays too many crucial differences with any realistic theory of gravity to provide conclusive evidences. However, it displays an alternative path to the cosmological constant from the perspective of a microscopic model. The analogue cosmological constant that we have discussed *cannot* be computed as the total zero-point energy of the condensed matter system, even when taking into account the natural cut-off coming from the knowledge of the microphysics [10]. Indeed, the value of Λ is related only to the (subleading) part of the zero-point energy proportional to the quantum depletion of the condensate.

The implications for gravity are twofold. First, there could be no *a priori* reason why the cosmological constant should be computed as the zero-point energy of the system. More properly, its computation must inevitably pass through the derivation of Einstein equations emerging

[1] Note also that $L_{\rm LV}$ scales with $\rho_0 a^3$ exactly as the range of the gravitational force, signalling that this model is too simple to correctly grasp all the desired features. However, in more complicated systems [9], this pathology can be cured, in the presence of suitable symmetries, leading to long range potentials.

IOP Conf. Series: Journal of Physics: Conf. Series **880** (2017) 012009 doi:10.1088/1742-6596/880/1/012009

from the underlying microscopic system. Second, the energy scale of Λ can be several orders of magnitude smaller than all the other energy scales for the presence of a very small number, nonperturbative in origin, which cannot be computed within the framework of an EFT dealing only with the emergent degrees of freedom (e.g. semiclassical gravity).

4. Quasi-particles and locality

An aspect of the physics of quasiparticles which is not often stressed concerns the issue of locality, *i.e.* whether the effective Lagrangian of the quasiparticles does obey the axioms of local quantum field theory.

In BEC, the particles ϕ and quasiparticles ω field operators are related by a Bogoliubov transformation, which have the following general structure:

$$\omega^A(\mathrm{k}) = M^A_B(\mathrm{k})\phi^B(\mathrm{k}), \tag{35}$$

when working in momentum space this relation is always linear. However, it is nonlocal in coordinate space as

$$\omega^A(\mathrm{x}) = \int d^3y\, K^A_B(\mathrm{x},\mathrm{y})\phi^B(\mathrm{y}), \tag{36}$$

where the kernel K is determined by the Bogoliubov coefficients:

$$K^A_B(\mathrm{x},\mathrm{y}) = \int d^3k\, M^A_B(\mathrm{k})e^{-i\mathrm{k}\cdot(\mathrm{x}-\mathrm{y})}. \tag{37}$$

This is the mathematical statement of the fact that quasi-particles are collective degrees of freedom. The structure of the transformation immediately implies that there is a mismatch between the notion of locality of the quasi-particle with respect to the notion of locality of the atoms. As one easily realizes, the two classes of operators ϕ, ω, separately, do obey canonical equal time commutation relations:

$$[\phi^A(\mathrm{x}), (\phi^B)^\dagger(\mathrm{y})] = \delta^{AB}\delta^3(\mathrm{x}-\mathrm{y}), \qquad [\omega^A(\mathrm{x}), (\omega^B)^\dagger(\mathrm{y})] = \delta^{AB}\delta^3(\mathrm{x}-\mathrm{y}), \tag{38}$$

which are a direct consequence of the fact that Bogoliubov transformations are preserving the algebra of the creation-annihilation operators. Therefore, as long as we use only one of the two families, there is no way in which a deviation from standard local quantum field theory can be manifest. However, the mixed commutators are nontrivial and it is straightforward to see that

$$[\phi^A(\mathrm{x}), (\omega^B)^\dagger(\mathrm{y})] = (K^{AB})^*(\mathrm{x},\mathrm{y}), \tag{39}$$

Of course, this fact becomes crucial when the effective Lagrangian describing the physics of the quasi-particles involves terms mixing particle and quasi-particle operators.

In the case of the BEC it is pretty clear how the underlying dynamics induces in the action for the quasi-particles an interaction term of the form

$$L_{\mathrm{int}} = -\frac{\kappa}{4}\left((\phi^A)^\dagger\phi_A\right)^2 \tag{40}$$

which explicitly involves the particle fields, rather than the quasi-particles. Hence, when computing the effects of the interaction terms, *e.g.* scattering processes between the quasi-particles, the nonlocality encoded in the kernel K will necessarily enter in the physical quantities.

IOP Conf. Series: Journal of Physics: Conf. Series **880** (2017) 012009 doi:10.1088/1742-6596/880/1/012009

5. Analogue gravity in Relativistic BEC

Bose–Einstein condensation can occur not only for non-relativistic bosons but for relativistic ones as well. The main differences between the thermodynamical properties of these condensates at finite temperature are due both to the different energy spectra and also to the presence, for relativistic bosons, of anti-bosons. These differences result in different conditions for the occurrence of Bose–Einstein condensation, which is possible, e.g., in two spatial dimensions for a homogeneous relativistic Bose gas, but not for its non-relativistic counterpart – and also, more importantly for our purposes, in the different structure of their excitation spectra.

In reference [11] an analogue model based on a relativistic BEC was studied. We summarize here the main results. The Lagrangian density for an interacting relativistic scalar Bose field $\widehat{\phi}(\mathbf{x},t)$ may be written as

$$\widehat{\mathcal{L}} = \frac{1}{c^2}\frac{\partial\widehat{\Psi}^\dagger}{\partial t}\frac{\partial\widehat{\Psi}}{\partial t} - \nabla\widehat{\Psi}^\dagger\cdot\nabla\widehat{\Psi} - \left(\frac{m^2c^2}{\hbar^2} + V(t,\mathbf{x})\right)\widehat{\Psi}^\dagger\widehat{\Psi} - U(\widehat{\Psi}^\dagger\widehat{\Psi};\lambda_i)\ , \tag{41}$$

where $V(t,\mathbf{x})$ is an external potential depending both on time t and position $\mathbf{x}$, m is the mass of the bosons and c is the light velocity. U is an interaction term and the coupling constant $\lambda_i(t,\mathbf{x})$ can depend on time and position too (this is possible, for example, by changing the scattering length via a Feshbach resonance [12]). U can be expanded as

$$U(\widehat{\Psi}^\dagger\widehat{\Psi};\lambda_i) = \frac{\lambda_2}{2}\widehat{\rho}^2 + \frac{\lambda_3}{6}\widehat{\rho}^3 + \cdots \tag{42}$$

where $\widehat{\rho} = \widehat{\Psi}^\dagger\widehat{\Psi}$. The usual two-particle $\lambda_2\widehat{\Psi}^4$-interaction corresponds to the first term $(\lambda_2/2)\widehat{\rho}^2$, while the second term represents the three-particle interaction and so on.

In this case it is convenient to use the representation

$$\widehat{\Psi} = \psi\,(1+\widehat{\chi})\ . \tag{43}$$

It is worth noticing now that the expansion in Eq. (43) can be linked straightforwardly to the previously discussed expansion in phase and density perturbations $\widehat{\theta}_1$, $\widehat{\rho}_1$, by noting that

$$\frac{\widehat{\rho}_1}{\rho} = \frac{\widehat{\chi}+\widehat{\chi}^\dagger}{2},\quad \widehat{\theta}_1 = \frac{\widehat{\chi}-\widehat{\chi}^\dagger}{2i}\ .$$

Setting $\chi \propto \exp[i\,(\mathbf{k}\cdot\mathbf{x}-\omega t)]$ one then gets from the equation of motion [11]

$$\left(-\frac{\hbar}{m}\mathbf{q}\cdot\mathbf{k} + \frac{u^0}{c}\omega - \frac{\hbar}{2mc^2}\omega^2 + \frac{\hbar}{2m}k^2\right)\left(\frac{\hbar}{m}\mathbf{q}\cdot\mathbf{k} - \frac{u^0}{c}\omega - \frac{\hbar}{2mc^2}\omega^2 + \frac{\hbar}{2m}k^2\right) - \left(\frac{c_0}{c}\right)^2\omega^2 + c_0^2k^2 = 0\ ,$$

where, for convenience we have defined the following quantities

$$u^\mu \equiv \frac{\hbar}{m}\eta^{\mu\nu}\partial_\nu\theta\ ,\quad c_0^2 \equiv \frac{\hbar^2}{2m^2}U''(\rho;\lambda_i)\rho\ ,\quad \mathbf{q} \equiv m\mathbf{u}/\hbar\ . \tag{44}$$

Here $\mathbf{q}$ is the speed of the condensate flow, c is the speed of light. For a condensate at rest ($\mathbf{q}=0$) one then obtains the following dispersion relation

$$\omega_\pm^2 = c^2\left\{k^2 + 2\left(\frac{mu^0}{\hbar}\right)^2\left[1+\left(\frac{c_0}{u^0}\right)^2\right] \pm 2\left(\frac{mu^0}{\hbar}\right)\sqrt{k^2 + \left(\frac{mu^0}{\hbar}\right)^2\left[1+\left(\frac{c_0}{u^0}\right)^2\right]^2}\right\}\ . \tag{45}$$

The dispersion relation (45) is sufficiently complicated to prevent any obvious understanding of the regimes allowed for the excitation of the system. It is much richer than the non-relativistic case. For example, it allows for both a massless/gapless (phononic) and massive/gapped mode, respectively for the ω_- and ω_+ branches of (45). Nonetheless, it should be evident that different regimes are determined by the relative strength of the the first two terms on the right hand side of (45) (note that the same terms enter in the square root). This can be summarized, in low and high momentum limits respectively, for k much less or much greater than

$$\frac{mu^0}{\hbar}\left[1+\left(\frac{c_0}{u^0}\right)^2\right] \equiv \frac{mu^0}{\hbar}(1+b), \tag{46}$$

where b basically encodes the relativistic nature of the condensate (the larger b the more the condensate is relativistic).

Finally, it is also possible to recover an acoustic metric for the massless (phononic) perturbations of the condensate in the low momentum limit ($k \ll mu^0(1+b)/\hbar$) [11]

$$g_{\mu\nu} = \frac{\rho}{\sqrt{1-u_\sigma u^\sigma/c_o^2}}\left[\eta_{\mu\nu}\left(1-\frac{u_\sigma u^\sigma}{c_0^2}\right)+\frac{u_\mu u_\nu}{c_0^2}\right]. \tag{47}$$

5.1. Emergent scalar gravity in Relativistic BEC

The above discussed relativistic BEC system albeit still unrealised in laboratory is an obvious candidate for further exploring the dynamics of the emergent spacetime this time in a fully relativistic context (see [13] for a complete derivation). In order to do so let us go back to the equation of motion of the BEC

$$\left(\Box - m^2\right)\Psi - 2\lambda|\Psi|^2\Psi = 0. \tag{48}$$

This was the starting equation in ref. [11] where the acoustic metric was first derived.

Let us again decompose $\widehat{\Psi}$ as $\widehat{\Psi} = \psi(1+\widehat{\chi})$, where ψ is the condensed part of the field ($\langle\Psi\rangle = \psi$), which we now take to be real, and $\widehat{\chi}$ is the fractional fluctuation.[2] Note that $\widehat{\chi}$ is instead complex and $\langle\widehat{\chi}\rangle = 0$. It can be written in terms of its real and imaginary parts $\chi = \chi_1 + i\chi_2$ (and from now on we drop the hat notation). Substituting this decomposition in eq. (48) and taking the expectation value we get the equation of motion for the condensate

$$(\Box - m^2)\psi - 2\lambda\psi^3 - 2\lambda\psi^3\left[3\langle\chi_1^2\rangle + \langle\chi_2^2\rangle\right] = 0, \tag{49}$$

where we have assumed that the cross-correlation of the fluctuations vanish, i.e., $\langle\chi_1\chi_2\rangle = 0$. This is justified a posteriori by equations (53), which show that χ_1 and χ_2 do not interact with each other at the order of approximation we are working. Eq. (49) determines the dynamics of the condensate taking into account the backreaction of the fluctuations. It is the relativistic generalization of the Gross–Pitævskii equation.

5.2. Dynamics of perturbations: acoustic metric

Having determined the dynamics of the condensate we now want to calculate the equations of motion for the perturbations themselves. To this end, we insert $\Psi = \psi(1+\chi_1+i\chi_2)$ in eq. (48) and expand it to linear order in ψ's. Using the Gross–Pitævskii equation to that order and separating the real and imaginary parts we get the equation of motion for χ_1 and χ_2,

$$\Box\chi_1 + 2\eta^{\mu\nu}\partial_\mu(\ln\psi)\partial_\nu\chi_1 - 4\lambda\psi^2\chi_1 = 0, \tag{50a}$$
$$\Box\chi_2 + 2\eta^{\mu\nu}\partial_\mu(\ln\psi)\partial_\nu\chi_2 = 0. \tag{50b}$$

[2] The reality of the condensate order parameter is the crucial assumption here. See the relevant discussion in [13].

We therefore see that χ_2 is the massless mode, which is the Goldstone boson of the broken $U(1)$ symmetry, while χ_1 is the massive mode with mass $2\psi\sqrt{\lambda}$. We now define a "acoustic" metric, which is conformal to the background Minkowski,

$$g_{\mu\nu} = \psi^2\, \eta_{\mu\nu}. \tag{51}$$

The relation between the d'Alembertian operators for $g_{\mu\nu}$ and $\eta_{\mu\nu}$ is given by,

$$\Box_g = \frac{1}{\psi^2}\,\Box + \frac{2}{\psi^2}\,\eta^{\mu\nu}\,\partial_\mu(\ln\psi)\,\partial_\nu. \tag{52}$$

Equations (50) can be written in terms of the d'Alembertian of $g_{\mu\nu}$ as

$$\Box_g\chi_1 - 4\lambda\chi_1 = 0, \tag{53a}$$
$$\Box_g\chi_2 = 0. \tag{53b}$$

We see from eqs. (53) that the fluctuations propagate on a curved metric, called the acoustic metric, which in this case is conformal to the background Minkowski space eq. (51). Note that in this derivation there was no low-momentum approximation needed in order to derive the acoustic metric. On the other hand, they back-react on the condensate through the relativistic generalization of the Gross-Pitaevskii equation (49). It is natural to ask if it is possible to have a geometric description of the dynamics of the condensate too.

The Ricci tensor of the acoustic metric (51) can be calculated to be

$$R_g = -6\,\frac{\Box\,\psi}{\psi^3} \tag{54}$$

Dividing the relativistic Gross-Pitaevskii equation by ψ^3, eq. (49) can be written as

$$R_g + 6\frac{m^2}{\varphi_0^2} + 12\lambda = \langle T_{\rm qp}\rangle, \tag{55}$$

where we have defined $\langle T_{\rm qp}\rangle := -12\lambda\left[\,3\,\langle\chi_1^2\rangle + \langle\chi_2^2\rangle\,\right]$ and the subscript "qp" reminds us that this quantity is determined by the quasi-particle excitations of the condensate.

Eq. (55) is evidently reminiscent of the Einstein–Fokker equation describing Nordström gravity [14, 15],

$$R + \Lambda = 24\pi\frac{G_{\rm N}}{c^4}\,T, \tag{56}$$

where R and T are, respectively, the Ricci scalar and the trace of the stress-energy tensor of matter. Unfortunately, the gravitational analogy of our equation is spoiled by the mass term. Therefore we will consider our system in the zero mass limit, something doable in the presence of a suitable chemical potential (see discussion in [13]). otice that this limit does not spoil the presence of a condensate or the uniqueness of the Lorentz group for constituents and excitations found in sec.

The striking resemblance of equations (55) with zero mass term and (56) is still not enough to draw conclusions. Indeed, the dimensions of the various quantities appearing in eq. (55) are not canonical and need to be fixed for such comparison to be meaningful. This is due to the fact that, as is usual in the analogue gravity literature, our acoustic metric is a dimensional quantity because ψ is dimensional. The fractional perturbations χ_1 and χ_2, on the other hand, are dimensionless.

One therefore needs to suitably rescale fields in order to have a dimensionless metric and (mass) dimension one scalar fields propagating on the curved metric. The upshot of this dimensional analysis is that we need to scale the field $\psi \to \frac{\mu}{\sqrt{c\hbar}}\psi$ and perturbation $\chi \to \frac{\sqrt{c\hbar}}{\mu}\chi$ [13]. Finally, using these rescaled quantities we can rewrite eq. (55) (with $m = 0$) in the form of eq. (56) as

$$R + \Lambda_{\text{eff}} = \langle T_{\text{qp}} \rangle, \tag{57}$$

and T_{qp} here and in the following is the same expression as in (55) but with the mass dimension one fields. Equations of motion of the quasi-particles (50) can also be rewritten in terms of the rescaled fields as

$$\Box_g \chi_1 - \frac{4\lambda\mu^2}{\hbar c}\chi_1 = 0, \tag{58a}$$

$$\Box_g \chi_2 = 0, \tag{58b}$$

where all quantities, including the $\Box_g$ operator, now pertain to those of the rescaled fields.

Remarkably, by defining the actual SET of the quasiparticle as the usual variation of their quadratic action w.r.t. the emergent metric one one finds [13] that this basically coincides with T_{qp} modulo a proportionality factor. Indeed one finds,

$$\langle T \rangle = -2\lambda \frac{\mu^2}{c\hbar}\left[3\langle \chi_1^2 \rangle + \langle \chi_2^2 \rangle\right] = \frac{1}{6}\frac{\mu^2}{c\hbar}\langle T_{qp} \rangle. \tag{59}$$

Due to this last expression one sees that the RHS of eq. (57) is actually given by $\frac{6c\hbar}{\mu^2}\langle T \rangle$ and hence our emergent Nordström gravity equation will be exactly of the form (56) with the identification $G_{\text{eff}} = \hbar c^5/(4\pi\mu^2)$. This value corresponds to an emergent analogue Planck scale $M_{\text{Pl}} = \mu\sqrt{4\pi}/c^2$.

We have thus succeeded in expressing the dynamics of the background for our rBEC analogue model in a geometric language

$$R + \Lambda_{\text{eff}} = 24\pi \frac{G_{\text{eff}}}{c^4}\langle T \rangle. \tag{60}$$

The acoustic metric itself is sourced by the expectation value of the trace of the stress-energy tensor of the perturbations of the condensate playing the role of the matter. These matter fields in turns propagate relativistically on a conformally flat acoustic metric (51) with equations (58).

A final comment is deserved by the emergent, positive, cosmological constant term Λ_{eff}. The quantity of interest for what concern the usual cosmological constant problem is the ratio between the energy density associated to the (emergent) cosmological constant $\epsilon_{\Lambda_{\text{eff}}} \sim \left(\frac{\Lambda_{\text{eff}} c^4}{G_{\text{eff}}}\right)$ and the emergent Planck energy density $\epsilon_{\text{pl}} \sim \frac{c^7}{\hbar G_{\text{eff}}^2}$. In our case this ratio is given by

$$\frac{\epsilon_{\Lambda_{\text{eff}}}}{\epsilon_{\text{pl}}} \simeq \frac{3\lambda\hbar c}{4\pi^2}. \tag{61}$$

As one can see the ratio is proportional to $\lambda\hbar$ and so is clearly pretty small due to the presence of Planck constant and of the natural assumption of a weakly interacting system. Of course in principle this term can be "renormalised" by the vacuum contribution of the matter fields (basically the vacuum expectation value $\langle T \rangle$).

6. Going beyond: the issue of background independence

The result presented in Eq. (60) is at the same time striking and disappoint. While we succeeded in obtaining a dynamical equation for the analogue metric which reproduce a well known theory of gravity, such a theory is just a scalar theory of gravity (obviously as our fundamental fields are scalar in nature) which admits only conformally flat metrics.

Indeed, Nordström gravity is not fully background independent as the the Minkowski metric can be see then as a background structure, what one may call a prior geometry. One may hence say that diffeomorphism invariance is somewhat of a weaker form in Nordström gravity with respect the one present in general relativity. In particular, while the essence of diffeomorphism invariance in GR is encoded in the associated Hamiltonian constraints, these are not defined in the present formulation of Nordström gravity.

Another way to see this point is to resort to the famous derivation of Einstein equations from the Clausius equation ($dS = \delta Q/T$) presented by Jacobson in his 1995 seminal paper [16]. In that work, one uses the equivalence principle to construct around an arbitrary point of spacetime a Rindler wedge as observed for some accelerated observer. The wedge provides a natural notion of entropy as the entanglement entropy of the vacuum (which will be locally the Minkowski one) and of temperature through the Unruh effect. Heat fluxes can be associated to matter fluxes through the wedge. The entanglement entropy is proportional to the area which can in turn be written in terms of the expansion of the null congruence associated to the wedge. The variation of the entropy due to some flux is then associated to the variation of the expansion which is determined by the Raychaudhuri equation. The latter contains the Ricci tensor. In the end by equating the integrands on both sides of the Clausius equation and by using the conservation of the stress energy tensor (SET) one gets the Einstein equation with a cosmological constant (which come about as an integration constant) [16].

Now, it is interesting to note that if one takes an ideal system such as a perfect fluid (inviscid and irrotational), then perturbations propagate on the acoustic metric at all scales (there is no scale to provide for Lorentz breaking in the equations). In such system, all of the above argument can be reproduced: the metric is locally flat and all of the effects of quantum field theory leading to entanglement entropy and Unruh temperature still holds for phonons of the fluid. So, should we conclude that the Einstein equation can be recovered even in an (albeit ideal) analogue system? Of course this would not be compatible with that fact that the real equations, governing the fluid dynamics, are the hydrodynamic ones!

After some thinking one can realize that there is a step that even in our idealised fluid system is not granted, that is the assumption that the null congruence bounding the Rindler wedge reacts to phonon fluxes as in the true gravitational setting, i.e. ultimately, we cannot assume that phonons gravitate (affect the background geometry) via their SET. Indeed we saw explicitly in BEC systems that phonons "gravitate" in rather different ways: at best with their energy density or with the trace of their SET. So in the end the impossibility of recovering the full Einstein equations in analogue models seems to be linked to the fact that phonons will not gravitate through the right object i.e. a conserved SET, which in turns is directly linked to the background independence issue.

This point seems also strictly connected to a theorem recently demonstrated in [17]. The starting point is that background independent gravitational theories, with universal coupling to energy, are characterised by Hamiltonians that are pure boundary terms on shell. Then it is shown that in order for this to be the low energy effective description of a field theory with *local* kinematics, all bulk dynamics must be frozen and thus irrelevant to the construction.

The result indeed implies that gravitational theories emergent along the analogue gravity framework, could be truly diffeomorphism invariant only if a different notion of micro-causality for the fundamental and emergent fields is present (something not realised e.g. in BEC).

In conclusion it seems that emergent theories of gravity would be characterised either by extra fields possibly leading to Lorentz breaking in the UV or by some degree of non-locality (which can appear or at the kinematical level or at the level of the interactions). In the last part of this contribution we shall discuss possible phenomenological consequences of these departures from standard physics and what constraints we can cast on them.

7. From analogue models to phenomenology

In closing this brief explorations on toy models of emergent gravity it is interesting to consider what we can say about the phenomenological consequences of the framework they seem to suggest. In this sense we have two main streams of investigation offered by analogue gravity ideas, i.e. we could generically expect or UV deviations from Lorentz invariance and/or intrinsic non-locality to appear both in interactions (as suggested by the phonon-atom duality in BEC) as well as in free propagation (if some for of fundamental non-locality has to be implemented as suggested by Marolf's theorem discussed above).

7.1. Lorentz breaking phenomenology

With regard of UV Lorentz breaking, both in the form of higher mass dimension dispersive or dissipative operators to be included the EFT for elementary particles, there is plenty we can say. A systematic investigation has been carried out mainly in the last 15 years based on the so called Standard Model Extension (i.e. an extension of the standard model of particle physics with Lorentz breaking operators added order by order in mass dimension and catalogues by their even or odd nature under CPT) which lead to a wealth of constraints (see e.g. [18, 19]). In particular, this approach leads to modified dispersion relations of the form

$$E^2 = p^2 + m_i^2 + \sum_{n=1}^{\infty} \eta_{i(n)} \frac{p^n}{M_{\rm Pl}^{n-2}} , \tag{62}$$

where we have put the low energy speed of light $c = 1$ and labelled the particle types by the i index and $M_{\rm Pl} = 1.22 \times 10^{19}$ GeV. In this parametrisation casting a strong constraints corresponds to show that the dimensionless coefficient η_i has to be much smaller than one. Most commonly the values of n considered are 3 and 4 (linear and quadratic Planck suppressed terms respectively).

It should not come as a surprise that constraints on this ansatz came mostly from high energy astrophysics and cosmology. Indeed the color dependence of the group velocity of photons can lead to observable differences of time of arrival for light emitted from very distance sources (e.g. gamma ray bursts). Putting an upper bound on color dependent delays can cast bounds on the size of the Lorentz breaking terms.

Similarly, modified dispersion relations can lead to depolarisation (vacuum birefringence), anomalous threshold reactions (e.g. vacuum Cherenkov or photon decay) as well as modify standard threshold reactions by shifting the standard threshold energy as well as by introducing "upper thresholds" (a maximum energy for which the reaction can happen). See [20, 21] for an overview.

All this new physics led to severe constraints on different particle species illustrated schematically in Table 1 (taken from [19] so updated to 2013). These seem indeed very tight constraints, but a caveat is in order: everywhere you see a (CR) label the constraints is performed using observations of Ultra High Energy Cosmic Rays (UHECR). The status of these observations and in particular the actual observation of the so called GZK cutoff is subject of intense debate nowadays and recent evidence from the leading experiment in the field, AUGER, seems to strongly hint in disfavour of this claim. This in particular makes the reliability of the constraints on the $n = 4$ Lorentz breaking corrections very weak and in strong need of further investigations.

IOP Conf. Series: Journal of Physics: Conf. Series **880** (2017) 012009 doi:10.1088/1742-6596/880/1/012009

Order	photon	e^-/e^+	Protrons	Neutrinos[a]
n=2	N.A.	$O(10^{-16})$	$O(10^{-20})$ (CR)	$O(10^{-8} \div 10^{-10})$
n=3	$O(10^{-16})$ (GRB)	$O(10^{-16})$ (CR)	$O(10^{-14})$ (CR)	$O(40)$
n=4	$O(10^{-8})$ (CR)	$O(10^{-8})$ (CR)	$O(10^{-6})$ (CR)	$O(10^{-7})^*$ (CR)

Table 1. Summary of typical strengths of the available constrains on the SME at different n orders for rotational invariant, neutrino flavour independent LIV operators. GRB=gamma rays burst, CR=cosmic rays. [a] From neutrino oscillations we have overwhelming constraints on the difference of LIV coefficients of different flavors. * Expected constraint from future experiments.

Even the possibility of dissipative dispersion relations has been considered (inspired by analogue gravity investigations for viscous fluids, see e.g. [22]). For example, we can conjecture a dispersion relation for photons of the form [23]

$$\omega^2 = c^2k^2 - i\sigma_2 c^2 \frac{k^3}{M_{\rm Pl}}, \tag{63}$$

where $\sigma_2 = (4\nu_2 M_{\rm Pl})/3c$ is the dimensionless coefficient controlling the magnitude of the Lorentz violation (LV) and ν is the "viscosity" coefficient of the spacetime fluid. Using the observed 80 TeV photons from the Crab nebula which is at a distance $D_{\rm Crab} \simeq 1.9$ Kpc one obtains$\sigma_2 \approx 1.3\times 10^{-26}$. similar strengths constrains can be obtained for neutrinos. If spacetime emergent as a fluid, it better has to be a superfluid.

It is worth stressing that higher order dissipative terms can and in principle should be considered. For example, nothing forbids such terms in superfluids (which have zero viscosity) to be non-zero. Similarly, if some fundamental, custodial, symmetry of the underlying, quantum gravitational system would forbid the above mentioned "spacetime viscosity" term still one could expect non-zero dispersive $O(k^4)$ and dissipative $O(k^5)$ terms to appear. These are sufficiently high energy modifications for which we do have relatively weak constraints on dispersion and basically no constraints on dissipation [23]. Casting strong constraints on this higher order dissipative terms would be very informative as some sort of dissipative effect should be expected at higher order even for superfluids.

7.2. Non-locality phenomenology

As said, an alternative element that seems to be required for a consistent emergent gravity scenario " a' la analogue gravity" appears to be some form of non-locality and indeed there are QG scenarios resorting to different forms of emergence where Lorentz invariance is held as a guiding principle, while spacetime is seen as emergent from more fundamental, planckian structures. Typical scenarios in this sense are those of String Field Theory (SFT) [24] and Causal Set Theory (CST) [25]. But what kind of equations of motions one should expect in this case?

A general approach to the problem could consist in considering for example the dynamics of a free scalar field in flat spacetime, i.e. the standard Klein-Gordon (KG) equation $(\Box+m^2)\phi(x) = 0$. The simplest modification to this equation, which preserves Lorentz invariance, is one which generalises the KG operator to some function thereof: $(\Box+m^2) \to f(\Box+m^2)$. Furthermore, in order to avoid Ostrogradsky-like [26] instabilities, which arise in general whenever the dynamics contains more than two time derivatives, such an equation must be non-local in both space *and* time, in the sense of possessing an infinite number of spatial and temporal derivatives. Generically, the definition of f will contain a characteristic, covariantly defined non-locality length scale l_k, which allows for a suitable power law expansion characterising the deviation from the standard local field equations.

Not surprisingly, the above expectations are fulfilled at least by two QG models which implements Lorentz invariance at the fundamental level, namely String Field and Causal Set theories. Indeed, in String Field theory one finds that the KG equation in four dimensions is modified to [27]

$$f(\Box + m^2) = (\Box + m^2)\, \exp\left[l_k^2(\Box + m^2)\right], \tag{64}$$

with f therefore an analytic function. On the other hand, in Causal Set Theory one finds that $f(\Box + m^2)$ is generally non-analytic, and in four dimensions can be expanded as [28, 29]

$$f(\Box + m^2) = (\Box + m^2) - \frac{3l_k^2}{2\pi\sqrt{6}}(\Box + m^2)^2 \left[3\gamma - 2 + \ln\left(\frac{3l_k^4(\Box + m^2)^2}{2\pi}\right)\right] + \dots \tag{65}$$

where γ is Euler-Mascheroni's constant.

It is important to note that the non-locality length scale l_k, in both models needs not to be related to the quantum/discreteness scale normally associated to QG, i.e. the Planck scale. In fact, this scale is a mesoscopic one possibly lying somewhere between the TeV scale and the Planck scale; a fact which is of particular relevance within the context of casting phenomenological constraints.

The phenomenology associated to this kind of theories is in general different depending on the form of the function f. For non-analytic forms, like the one emergent from studies in CAUSET theory, one gets in $4d$ Green functions for a massless scalar field with non-zero support inside the light cone, i.e. the theory fails to satisfy Huygens' principle, i.e. there is a "leakage" inside the light cone of the field emitted from a delta function source [28]. This could prove fruitful in testing the theory, since one can envisage performing high precision tests of radiation emitted by very localised sources to check if such afterglow is present.

Alternative the case of analytical f functions can be studied by looking at the non-relativistic limit of the Klein-Gordon equation. This provides a non-local Schrödinger evolution which can be solved in a perturbative way. In particular, it was shown in [30, 31] that the corresponding non-local evolution of opto-mechanical quantum oscillators is characterised by a spontaneous periodic squeezing that cannot be generated by environmental effects. Quite surprisingly future experiments (already under construction) will either see such effects or otherwise cast severe bounds on the non-locality scale of order $l_k \lesssim 10^{-26}m$ (well beyond the current limits set by the Large Hadron Collider $l_k \lesssim 10^{-19}m$).

8. Conclusions

In summary, we think that this brief explorations serves to show the rich landscape of scenarios and phenomenological implications offered by emergent gravity ideas. Analogue gravity models are probably not going to give us a definite answer about the actual viability of an emergent gravity framework for recovering general relativity. However, we have seen that they do provide a set of toy models that may serve as test beds for our conjectures and suggest testable predictions of emergent gravity scenarios. While the route to a consistent picture is still long, we do hope that the results presented here will encourage more researchers to explore this largely unbeaten path.

Acknowledgments

The author acknowledge the John Templeton Foundation for the supporting grant #51876.

References

[1] Hu B L 2005 *Int. J. Theor. Phys.* **44** 1785–1806 (*Preprint* gr-qc/0503067)
[2] Barcelo C, Liberati S and Visser M 2005 *Living Rev. Rel.* **8** 12 [Living Rev. Rel.14,3(2011)] (*Preprint* gr-qc/0505065)

[3] Garay L J, Anglin J R, Cirac J I and Zoller P 2000 *Phys. Rev. Lett.* **85** 4643–4647 (*Preprint* gr-qc/0002015)
[4] Garay L J, Anglin J R, Cirac J I and Zoller P 2001 *Phys. Rev.* **A63** 023611 (*Preprint* gr-qc/0005131)
[5] Barcelo C, Liberati S and Visser M 2001 *Class. Quant. Grav.* **18** 1137 (*Preprint* gr-qc/0011026)
[6] Fedichev P O and Fischer U R 2004 *Phys. Rev.* **A69** 033602 (*Preprint* cond-mat/0303063)
[7] Girelli F, Liberati S and Sindoni L 2008 *Phys. Rev.* **D78** 084013 (*Preprint* 0807.4910)
[8] Finazzi S, Liberati S and Sindoni L 2012 *Phys. Rev. Lett.* **108** 071101 (*Preprint* 1103.4841)
[9] Liberati S, Visser M and Weinfurtner S 2006 *Phys. Rev. Lett.* **96** 151301 (*Preprint* gr-qc/0512139)
[10] Volovik G E 2006 *AIP Conf. Proc.* **850** 26–33 (*Preprint* cond-mat/0507454)
[11] Fagnocchi S, Finazzi S, Liberati S, Kormos M and Trombettoni A 2010 *New J. Phys.* **12** 095012 (*Preprint* 1001.1044)
[12] Cornish S L, Claussen N R, Roberts J L, Cornell E A and Wieman C E 2000 *Phys. Rev. Lett.* **85** 1795–1798
[13] Belenchia A, Liberati S and Mohd A 2014 *Phys. Rev.* **D90** 104015 (*Preprint* 1407.7896)
[14] Deruelle N 2011 *Gen. Rel. Grav.* **43** 3337–3354 (*Preprint* 1104.4608)
[15] Giulini D 2008 *Stud. Hist. Phil. Sci.* **B39** 154–180 (*Preprint* gr-qc/0611100)
[16] Jacobson T 1995 *Phys. Rev. Lett.* **75** 1260–1263 (*Preprint* gr-qc/9504004)
[17] Marolf D 2015 *Phys. Rev. Lett.* **114** 031104 (*Preprint* 1409.2509)
[18] Mattingly D 2005 *Living Rev. Rel.* **8** 5 (*Preprint* gr-qc/0502097)
[19] Liberati S 2013 *Class. Quant. Grav.* **30** 133001 (*Preprint* 1304.5795)
[20] Mattingly D, Jacobson T and Liberati S 2003 *Phys. Rev.* **D67** 124012 (*Preprint* hep-ph/0211466)
[21] Baccetti V, Tate K and Visser M 2012 *JHEP* **03** 087 (*Preprint* 1111.6340)
[22] Visser M 1998 *Class. Quant. Grav.* **15** 1767–1791 (*Preprint* gr-qc/9712010)
[23] Liberati S and Maccione L 2014 *Phys. Rev. Lett.* **112** 151301 (*Preprint* 1309.7296)
[24] Siegel W 1988 *Adv. Ser. Math. Phys.* **8** 1–244
[25] Surya S 2011 (*Preprint* 1103.6272)
[26] Woodard R P 2015 *Scholarpedia* **10** 32243 (*Preprint* 1506.02210)
[27] Koshelev A S 2012 *Rom. J. Phys.* **57** 894–900 (*Preprint* 1112.6410)
[28] Belenchia A, Benincasa D M T and Liberati S 2015 *JHEP* **03** 036 (*Preprint* 1411.6513)
[29] Johnston S 2015 *Class. Quant. Grav.* **32** 195020 (*Preprint* 1411.2614)
[30] Belenchia A, Benincasa D M T, Liberati S, Marin F, Marino F and Ortolan A 2016 *Phys. Rev. Lett.* **116** 161303 (*Preprint* 1512.02083)
[31] Belenchia A, Benincasa D M T, Liberati S, Marin F, Marino F and Ortolan A 2017 *Phys. Rev.* **D95** 026012 (*Preprint* 1611.07959)

IOP Conf. Series: Journal of Physics: Conf. Series **880** (2017) 012010 doi:10.1088/1742-6596/880/1/012010

Non-linear dynamics in biological microtubules: solitons and dissipation-free energy transfer

Nick E Mavromatos

King's College London, Department of Physics, Strand, London WC2R 2LS, UK

E-mail: `Nikolaos.Mavromatos@kcl.ac.uk`

Abstract. I review some recent developments concerning soliton solutions in biological microtubules and their significance in transferring energy without dissipation. I discuss various types of soliton solutions, as well as 'spikes', of the associated non-linear Lagrange equations describing the dynamics of a 'pseudo-spin non-linear σ-model' that models the dynamics of a microtubule system with dipole-dipole interactions. These results will hopefully contribute to a better understanding of the functional properties of microtubules, including the motor protein dynamics and the information transfer processes. With regards to the latter we also speculate on the use of microtubules as 'logical' gates. Our considerations are classical, but the soliton solutions may have a microscopic quantum origin, which we briefly touch upon.

1. Introduction

The rôle of solitons in biological systems as providers of efficient (mostly dissipation free) energy transport, in analogy with the frictionless electric current transfer in superconductivity theories, has a long history. Already in 1968, H. Fröhlich [1] have suggested that macroscopic quantum coherent phenomena may be responsible for dissipation-free energy and signal transfer in biological systems through coherent excitations in the microwave region of the spectrum due to nonlinear couplings of biomolecular dipoles. The frequency with which such coherent modes are 'pumped' in biological systems was conjectured to be of order of the inverse of the time interval $t_{\text{coherence Froehlich}} \sim 10^{-11} - 10^{-12}$ s , which is known as Fröhlich's coherence time.

Soon after, A.S. Davydov [2], proposed that *solitonic excitation states* may be responsible for dissipation-free energy transfer along the α-helix self-trapped amide in a fashion similar to superconductivity: there are two kinds of excitations in the α-helix: deformational oscillations in the α-helix lattice, giving rise to quantized excitations ("phonons"), and internal amide excitations. The resulting non-linear coupling between these two types of excitations is a Davydov soliton, which traps the vibrational energy of the α-helix and thus prevents its distortion (solitons are classical field theory configurations with finite energy). In a rather different approach, F. Popp [3] suggested that studies of the statistics of counts of photons in ultra-weak bioluminescence in the visible region of the spectrum point towards the existence of a coherent component linked to the living state.

In the 1990's a suggestion on the rôle of quantum effects on brain functioning, and in particular on conscious perception, has been put forward by S. Hameroff and R. Penrose [4], who concentrated on the microtubules (MT) [5] of the brain cells. In particular, they noted that one may view the tubulin protein dimer units of the MT as a quantum two-state system, in coherent superposition. The model of [4] assumes, without proof, that sufficient environmental isolation occurs, so that the *in vivo* system of MT in the brain undergoes *self-collapse*, as a result of sufficient growth that allowed it to reach a particular threshold, namely a critical mass/energy, related to *quantum gravity* (orchestrated reduction method). This type of collapse should be distinguished from the standard *environmental decoherence* that physical quantum systems are subjected to [6]. In this way, the authors of [4] argue that decoherence times of order O(1 s), which is a typical time for conscious perception, may be achieved, thereby deducing that consciousness is associated with *quantum computations* in the mind.

Unfortunately, in my opinion, environmental decoherence, even for *in vivo* MT, cannot be ignored. Microtubules (MT) are fundamental constituents of most cells [5], not only the brain ones, and play a crucial rôle in the cell mytosis. They are cylindrically shaped cytoskeletal biopolymers. They are found in eukaryotic cells and are formed by the polymerization of heterodimers built of two globular proteins, alpha and beta tubulin. The MTs can grow up to 50 μm long (with an average length of 25 μm). Each MT is built of a set of macroscopic dipoles which generate dynamical electric fields, that prove crucial for an understanding of the functional properties of MT and their interactions.

One-spatial dimensional Solitons in ferreoelectric models of MT have been discussed from a rather phenomenological point of view in [7], and argued to provide efficient energy-transfer mechanisms. These solitons are kinks of an appropriate variable, on the main axis of the MT, which is associated with the appropriate projection of the dipole vectors of the MT. In a series of works [8, 9], we have developed a microsocpic *quantum electrodynamics cavity model* for MT, in which electromagnetic interactions between the electric dipole moments of the tubulin protein dimer units and the corresponding dipole quanta in the (thermally isolated) water interiors of the *in vivo* MT, are argued to be the dominant forces, inducing environmental entanglement and eventual decoherence [6] in at most $O(10^{-6} - 10^{-7})$ s. Such times are much shorter than the required time scale for conscious perception, but have been argued to be sufficient for *dissipation-less* energy transfer and signal transduction along moderately long MT of length sizes of order $\mu\text{m} = 10^{-6}$ m. As I will discuss below, the basic underlying mechanism is the formation of appropriate *solitonic* dipole states along the protein dimer walls of the MT, which are reminiscent of the quantum coherent states in the Fröhlich-Davydov approach. These dipoles states are classical, obtained after decoherence of quantum states, and correspond to solutions of the non-linear equations that describe the dynamics of the MT within certain models that take proper account of the dipole-dipole interactions. We have also speculated [9] that under sufficient environmental isolation, which however is not clear if it can be achieved in *in vivo* MT systems, these coherent states may provide the basis for an operation of the MT as quantum logic and information teleporting gates. In particular, it has been argued in [8] that sufficiently strong dipole-dipole interactions between ordered water molecules in the interior of the MT cavities and the dimers are responsible for providing thermal isolation, and thus sustainable conditions for the formation of coherent dipole solitonic states on the MT surface, which could then be responsible for dissipation-free energy and signal transfer. At any rate, our main concern in the above works was the search for, and modeling of, possible quantum effects in cell MT which may not be necessarily associated with conscious perception. In fact in this talk I will disentangle the latter from dissipation-free energy and signal transfer in biological matter, which I will restrict my attention to.

In this respect I will discuss some recent developments I was involved with in ref. [10], in which we discussed and classified interesting types of solitons in MTs, including one and two-spatial dimensional solutions, and in particular those with a helical shape, which have a chance of providing more realistic models for non-linear MT dynamics. I will not discuss the environment of ordered water molecules, however I will take into account dipole-dipole interactions between dimers on the MT surface and argue that they also can also provide the necessary non-linear

IOP Conf. Series: Journal of Physics: Conf. Series **880** (2017) 012010 doi:10.1088/1742-6596/880/1/012010

dynamics for the formation of solitonic dipole states. In this latter respect, the rôle of the ordered-water molecules via their dipole-dipole interactions with the MT dimers can be viewed as providing the necessary thermal isolation, according to the ideas of [8], so that the coherent solitonic states can be formed.

The structure of the talk is as follows: In the next section 2, I discuss the model for MT, which is based on the so-called pseudo-spin non-linear σ-model, and which includes dipole-dipole interactions that they were crucial in our microscopic approach [8, 9] in providing the necessary isolation from thermal losses. In the following section 3, I classify the various solitonic solutions arising from the non-linear dynamics of the model. I discuss one-, and two-dimensional representations of the solutions and their significance as far as energy transfer along the MTs is concerned. I also discuss 'spike configurations, and the properties of the associated electric fields produced by such configurations. In the concluding section, 4, I speculate on the rôle of physical systems of MTs with solitons as logical gates, which may be important for 'quantum computations' of biological matter, which in this way optimises signal and energy transduction, by appropriate decision taking processes that chose the most efficient path, minimising losses. We note that similar speculations have been made on mechanisms for light harvesting by marine algae [11].

2. The pseudo-spin σ-model for microtubules and the phase diagram

MTs are realized as hollow cylinders typically formed by 13 parallel protofilaments (PFs) covering the wall of MT [5, 12, 13]. Each PF represents a tubulin heterodimer with the electric dipole momentum, **P**. (See Fig. 1.) Due to their interaction with the complex biological

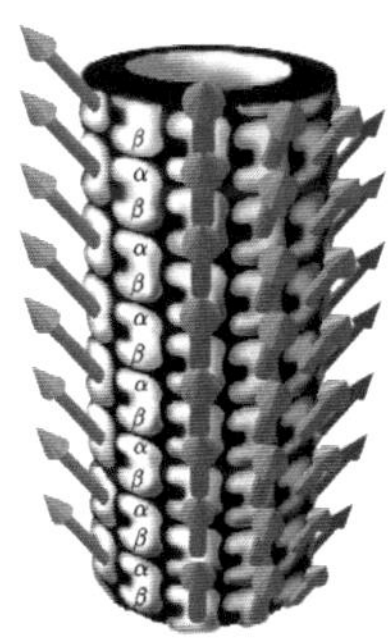

Figure 1. The structure of the cytoskeleton microtubule (MT). The arrows indicate the orientation of the permanent dipole moments of the tubulin heterodimers with respect to the MT surface. Picture from ref. [10].

environment (solvent) the MTs may experience a strong radial electrostatic field leading to the additional (radial) polarization of tubulins [14].

The tibulin heterodimer contains approximately 900 amino acid residues with the number of atoms about 14000. The total mass of the heterodimer can be estimated as, ($M \approx 1.84 \cdot 10^{-19}$g). Each heterodimer can be considered as effective electric dipole with α and β tibulin being as positive and negative side of dipole, respectively [15].

In our model [10] we treat each dipole as a classical pseudo-spin, $\mathbf{S}_i$, with a constant modulus. The potential energy of the system can be written as:

$$U_0 = \sum_{\langle i,j \rangle} J_{ij} \big(\mathbf{S}_i \cdot \mathbf{S}_j - 3(\mathbf{S}_i \cdot \mathbf{e}_{ij})(\mathbf{S}_j \cdot \mathbf{e}_{ij}) \big) - B \sum_i \mathbf{S}_i \cdot \mathbf{e}_r, \tag{1}$$

where $\mathbf{e}_{ij}$ is a unit vector parallel to the line connecting the dipoles, $\mathbf{S}_i$ and $\mathbf{S}_j$. The scalar product is understood as follows: $\mathbf{S}_i \cdot \mathbf{S}_j = S_i^1 S_j^1 + S_i^2 S_j^2 + S_i^3 S_j^3$. The first sum describes the dipole-dipole interaction, and the second one characterizes the effect of the transversal (radial) electrostatic field acting on the dipoles. The interaction coupling J_{ij} is given by $J_{ij} = \frac{1}{4\pi\epsilon\,\epsilon_0\, r_{ij}^3}$, where ϵ is the permittivity of the MT medium, ϵ_0 that of the vacuum and r_{ij} is the distance between sites i and j of the lattice model. This coupling expresses the well-known fact that dipole-dipole interactions in electromagnetism scale with the inverse of the cubic power of the distance between the dipoles.The quantity B denotes the amplitude of the effective electric field, produced by the solvent environment, along the radial direction $\mathbf{e}_r$.

Since the MTs may exhibit ferroelectric properties at room temperature, one can consider the MT as a ferroelectric system [13, 16]. To include into consideration the ferroelectric properties of the MT, we adopt the approach developed in [17]. In this case, the overall effect of the environment on the effective spin, $\mathbf{S}_i$, is described by the double-well quartic on-site potential,

$$V(\mathbf{S}_i) = P(\mathbf{S}_i \cdot \mathbf{e}_z)^2 + Q(\mathbf{S}_i \cdot \mathbf{e}_z)^4. \tag{2}$$

It is convenient to parameterize the pseudo-spin $\mathbf{S}_i$ by the unit vector $\mathbf{n}_i$, as: $\mathbf{S}_i = S\mathbf{n}_i$, where S is the module of $\mathbf{S}_i$. Then, the total potential energy of the system can be written as,

$$\begin{aligned} U = S^2 \sum_{\langle i,j \rangle} J_{ij}\big(\mathbf{n}_i \cdot \mathbf{n}_j - 3(\mathbf{n}_i \cdot \mathbf{e}_{ij})(\mathbf{n}_j \cdot \mathbf{e}_{ij})\big) \\ + \sum_i \big(PS^2(\mathbf{n}_i \cdot \mathbf{e}_z)^2 + QS^4(\mathbf{n}_i \cdot \mathbf{e}_z)^4 - BS\mathbf{n}_i \cdot \mathbf{e}_r\big). \end{aligned} \tag{3}$$

In what follows we shall use the local spherical coordinates (Θ_i, Φ_i) to define the orientation of the dipole, It is commonly accepted that coupling constants, J_{ij}, are nonzero only for the nearest-neighbor dipole moments. The system of MT dimers may be represented as a triangular lattice, as shown in Fig. 2, so that each spin has six nearest neighbors. We denote the constants of interaction between the central dipole in Fig. 2 and nearest neighbors as, $J_{0\alpha}$, and the distance between the central spin and its nearest neighbors as, d_α ($\alpha = 1, 2, \ldots, 6$), setting $d_{01} = d_{04} = a$, $d_{02} = d_{05} = b$, $d_{03} = d_{06} = c$. The corresponding angles (between the central dimer and others) are denoted as, θ_1, θ_2 and θ_3, so that: $\mathbf{e}_{01} \cdot \mathbf{e}_{01} = \cos\theta_1$, $\mathbf{e}_{01} \cdot \mathbf{e}_{02} = \cos\theta_2$, $\mathbf{e}_{01} \cdot \mathbf{e}_{06} = \cos\theta_3$.

Typical values of parameters known from the literature are: $a = 8\,\text{nm}$, $b = 5.87\,\text{nm}$, $c = 7.02\,\text{nm}$, $\theta_1 = 0$, $\theta_2 = 58.2^\circ$, $\theta_3 = 45.58^\circ$, $S = 1714$ Debye [18, 17] (See Fig. 2b.) The radius of the MT can be estimated as, $R \approx 11.2\,\text{nm}$ [16, 19]. The unit cell shown in Fig. 2 consists of the central spin surrounded by six neighbors. Its area is: $\Sigma_0 = 3ad = 120\,\text{nm}^2$.

From now on we shall work in the Continuum approximation, which suffices for the classification of the various solutions describing the non-linear dynamics of the system. The continuum limit is obtained by allowing the area per a site, Σ_0 to go to zero, keeping the total area, $N\Sigma_0$ fixed. In this limit, the summation is replaced by the integral over the MT surface: $\sum_{\langle ij \rangle} \to (1/2)\int_\Sigma d^2x$. The variable, $\mathbf{n}_i = \mathbf{n}(\mathbf{r}_i)$, should be replaced by a smooth function of the continuum coordinates: $\mathbf{n}(\mathbf{r}_i) \to \mathbf{n}(\mathbf{r})$.

In the cylindrical coordinates the metric on Σ can be written as,

$$ds^2 = R^2 d\varphi \otimes d\varphi + dz \otimes dz, \tag{4}$$

where R is the radius of the MT. In what follows, we use the abbreviation: $\nabla n^a \cdot \nabla n^b = g^{ij}\partial_i n^a \partial_j n^b$.

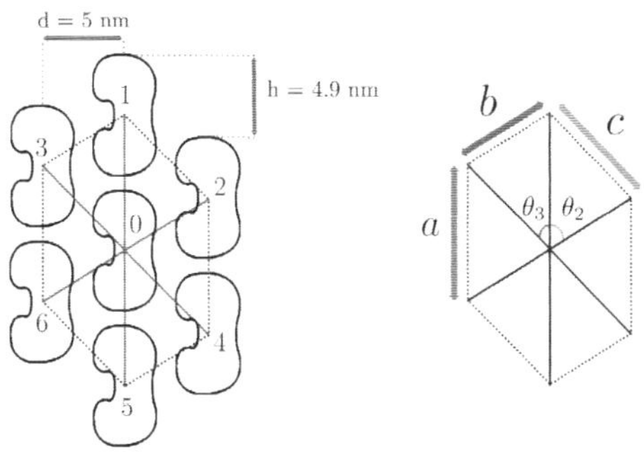

(a) (b)

Figure 2. (Color online) Tubulin neighborhood in the hexagonal unit cell of the microtubule. The distance between dimers is d. The heterodimer helix direction is defined by the height, h. The typical values of parameters are: $a = 8\,\text{nm}$, $b = 5.87\,\text{nm}$, $c = 7.02\,\text{nm}$, $d = 5\,\text{nm}$, $h = 4.9\,\text{nm}$, $\theta_1 = 0$, $\theta_2 = 58.2^o$, $\theta_3 = 45.58^o$ [18, 16, 17, 19] .

The metric in the intrinsic space of pseudo-spins is given by: $G_{ab} = \delta_{ab} - h_{ab}$ $(a, b = 1, 2, 3)$, where

$$h_{22} = \frac{6S^2}{J}\sum_{\alpha=1}^{3} J_{0\alpha}\sin^2\theta_\alpha,\ h_{23} = \frac{3S^2}{J}\sum_{\alpha=1}^{3}(-1)^\alpha J_{0\alpha}\sin\theta_\alpha\cos\theta_\alpha,\ h_{33} = \frac{6S^2}{J}\sum_{\alpha=1}^{3} J_{0\alpha}\cos^2\theta_\alpha\ . \tag{5}$$

The computation of the constants yields: $h_{22} = 1.55$, $h_{23} = 0.11$, $h_{33} = 1.45$.

Further, it is convenient to introduce the dimensionless coordinates, $\zeta = z/\sqrt{\Sigma_0}$ and $\tilde{R} = R/\sqrt{\Sigma_0}$. Now, the total action yielding the equations of motion can be written as,

$$S_{tot} = J\int dt \int_\Sigma \mathcal{L} d\Sigma + S_\lambda, \tag{6}$$

where $d\Sigma = \tilde{R} d\zeta d\varphi$ and

$$S_\lambda = J\int dt \int_\Sigma \lambda(\mathbf{n}\cdot\mathbf{n} - 1)\, d\Sigma. \tag{7}$$

The Lagrangian of the system is given by,

$$\mathcal{L} = \frac{\rho}{2}\left(\frac{\partial \mathbf{n}}{\partial t}\right)^2 + \frac{1}{2}G_{ab}\nabla n^a \cdot \nabla n^b - \mathcal{V}(\mathbf{n}), \tag{8}$$

where $\rho = I/J$ and

$$\mathcal{V}(\mathbf{n}) = G_{ab}n^a n^b + g_0(n^3)^2 + g_1(n^3)^4 - g_2 n^1. \tag{9}$$

Taking into account that $h_{23}, |h_{33} - h_{22}| \ll 1$, we neglect by contributions of these terms and keep only terms with h_{33}. This approximation transforms the Lagrangian (8) into the following [10]

$$\mathcal{L} = \frac{\rho}{2}\left(\frac{\partial \mathbf{n}}{\partial t}\right)^2 + \frac{1}{2}(\nabla\mathbf{n})^2 - \frac{h}{2}(\nabla n^2\cdot\nabla n^2 + \nabla n^3\cdot\nabla n^3) - \mathcal{W}(\mathbf{n}), \tag{10}$$

where $h = h_{33}$ and

$$\mathcal{W}(\mathbf{n}) = h(n^1)^2 + g_0(n^3)^2 + g_1(n^3)^4 - g_2 n^1. \tag{11}$$

As one can see, in the continuum limit the electric properties of the MT are described by the nonlinear anisotropic σ-model. The order parameter, $\mathbf{n}$, is the local polarization unit vector specified by a point on the sphere, S^2.

Using the local spherical coordinates (Θ, Φ) to define the orientation of the dipole: $\mathbf{n} = (\sin\Theta\cos\Phi, \sin\Theta\sin\Phi, \cos\Theta)$, the Lagrangian of the system can be recast as follows:

$$\begin{aligned}\mathcal{L} =& \frac{\rho}{2}((\partial_t\Theta)^2 + \sin^2\Theta(\partial_t\Phi)^2) + \frac{1}{2}\left((\nabla\Theta)^2 + (\nabla\Phi)^2\right) - \frac{h}{2}(\cos\Theta\sin\Phi\nabla\Theta + \sin\Theta\cos\Phi\nabla\Phi)^2 \\ &- \frac{h}{2}\sin^2\Theta(\nabla\Theta)^2 - \mathcal{W}(\Theta,\Phi),\end{aligned} \tag{12}$$

where

$$\mathcal{W}(\Theta,\Phi) = (g_0 - h)\cos^2\Theta + g_1\cos^4\Theta - h\sin^2\Theta\sin^2\Phi - g_2\sin\Theta\cos\Phi. \tag{13}$$

The Euler-Lagrange equations can be written as [10],

$$\begin{aligned}\rho\frac{\partial^2\Theta}{\partial t^2} &= \frac{\delta\mathcal{L}}{\delta\Theta}, \\ \rho\frac{\partial}{\partial t}\left(\sin^2\Theta\frac{\partial\Phi}{\partial t}\right) &= \frac{\delta\mathcal{L}}{\delta\Phi}.\end{aligned} \tag{14}$$

The ground state of the MT, yielding the permanent dipole moment with $\Phi = 0$, is defined by the minimum value of the energy,

$$E(u) = E_0 + J\int_\Sigma \mathcal{V}(u)\, d\Sigma, \tag{15}$$

where $u = \cos\Theta$,

$$E_0 = -Jg_1\int_\Sigma \sigma^2\, d\Sigma, \tag{16}$$

and

$$\mathcal{V} = g_1\left((\sigma - u^2)^2 - \kappa\sqrt{1-u^2}\right). \tag{17}$$

Here we set $\sigma = (h - g_0)/(2g_1)$ and $\kappa = g_2/g_1$. One can see that there are three critical points: $u_1 = 0$, and $u_{2,3}$ defined from the equation:

$$u^6 - (1+2\sigma)u^4 + \sigma(2+\sigma)u^2 + \kappa^2/16 - \sigma^2 = 0. \tag{18}$$

The behavior of the dimensionless energy density of the system, $w = \mathcal{V}/g_1$, as a function of u and parameters σ and κ is presented in Fig. 3.

First, we consider the case when the parameter $\kappa = 0$. In this case, the critical points of the Hamiltonian are given by

$$u_1 = 0, \qquad u_{2,3} = \pm\sqrt{\sigma}. \tag{19}$$

As one can see, if $\sigma < 0$, the ground state of the MT is paraelectric, $u_1 = 0$. It corresponds to the radial orientation of the permanent dipole moments of the tubulin dimers with respect to the surface of the MT (Fig. 1). For $\sigma > 0$, the homogeneous ground state is a doubly degenerate ferroelectric state. The dipole momentum of the tubulin dimer is given by $u_{2,3} = \pm\sqrt{\sigma}$ (see Fig. 3a).

It follows from the phase diagram of Fig. 4 that, when $\kappa > 4\sigma$, the ground state of the MT is *paraelectric*, corresponds to the radial orientation of the permanent dipole moments of the tubulin dimers with respect to the surface of the MT. When $\kappa < 4\sigma$, the ground state of the system is *ferroelectric*.

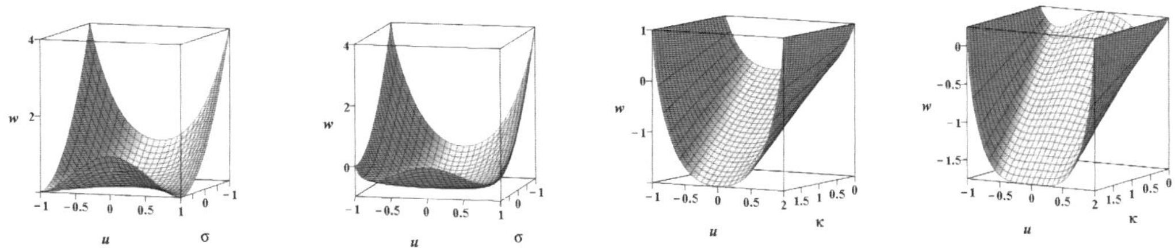

(a) (b) (c) (d)

Figure 3. Dimensionless energy density w as a function of u and parameters σ and κ. (a,b) w vs u and σ: (a) $\kappa = 0$, (b) $\kappa = 1$; (c,d) w vs u and κ: (c) $\sigma = 0$, (d) $\sigma = 0.5$.

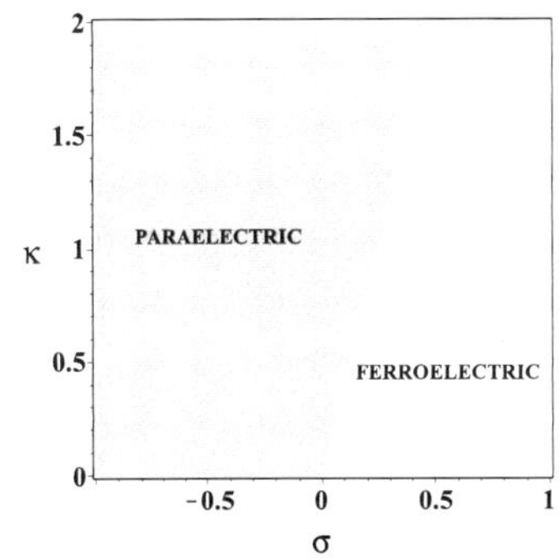

Figure 4. The phase diagram.

3. A classification of the non-linear solutions: Snoidal waves, Kinks, Helical Solitons and Spikes

In this section we proceed with a classification of the non-linear solutions of the equations of motion (14). We shall be brief in our description, referring the interested reader for details to our work in ref. [10].

In order to construct solutions for nonlinear waves moving along the MT with the constant velocity, we use the traveling wave ansatz. We assume that in cylindrical coordinates the field variables are functions of

$$\xi = \sqrt{\frac{2}{\eta p \Sigma_0}}(z + h_0\varphi/2\pi - vt), \tag{20}$$

where $\eta = h/g_1$ and $p = 1 + (h_0/2\pi R)^2$, the velocity of the wave being v. One can then show that the field equations possess the first integral of motion:

$$(u_0^2 - \cos^2\Theta)\left(\frac{d\Theta}{d\xi}\right)^2 + \sin^2\Theta\left(u_0^2 - \frac{1}{h}\cot\Theta - \sin^2\Phi\right)\left(\frac{d\Phi}{d\xi}\right)^2 + \frac{1}{2}\sin(2\Theta)\sin(2\Phi)\frac{d\Theta}{d\xi}\frac{d\Phi}{d\xi}$$
$$- (\sigma - \cos^2\Theta)^2 + \eta\sin^2\Theta\sin^2\Phi + \kappa\sin\Theta\cos\Phi = \text{const}, \tag{21}$$

where $u_0^2 = 1 - 1/h - \rho v^2/(hp\Sigma_0)$. This implies for the nonlinear wave propagation velocity v:

$$v = \sqrt{(\sigma_0^2 - u_0^2)\frac{hp\Sigma_0}{\rho}}, \tag{22}$$

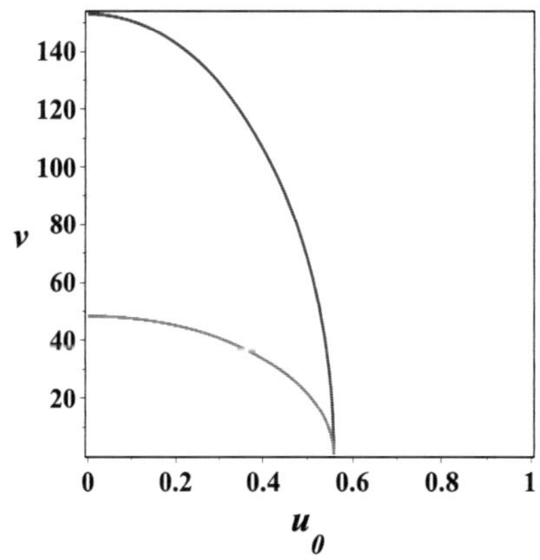

Figure 5. Velocity of the excitation (m/s): $M = 10^{-23}$ g (blue), $M = 10^{-22}$ g (red), $l = 2$ nm. From ref. [10].

where we set $\sigma_0^2 = 1 - 1/h$. The dependence of v on the parameter u_0 is depicted in Fig. 5, from which we conclude [10] that the velocity of the wave is bounded from above $v \leq v_0$, where $v_0 \approx 155$m/s.

3.1. Solutions with $\Phi = 0$

Here we discuss solutions of the system of non-linear equations (14) and (21) with $\Phi = 0$ and $\Theta = \Theta(\xi)$. For the function, $\Theta(\xi)$, we obtain the nonlinear differential equation,

$$(u_0^2 - \cos^2\Theta)\frac{d^2\Theta}{d\xi^2} + \frac{1}{2}\sin(2\Theta)\left(\frac{d\Theta}{d\xi}\right)^2 - \sin(2\Theta)(\sigma - \cos^2\Theta) + \frac{\kappa}{2}\cos\Theta = 0. \tag{23}$$

The phase portraits of the system $(\Theta(\xi), P_\Theta = d\Theta(\xi)/d\xi)$ is determined from (21), which in this case reads [10]

$$(u_0^2 - \cos^2\Theta)\left(\frac{d\Theta}{d\xi}\right)^2 - (\sigma - \cos^2\Theta)^2 + \kappa\sin\Theta = \text{const.}\ , \tag{24}$$

and are depicted in Figs. refPP1 and 7, for various parameters. One can observe the occurrence of the three elliptic points for $\sigma > u_0^2$ (Fig. 6a). When $\sigma < u_0^2$, two elliptic points disappear.

By substitution $u = \cos\Theta$ into Eq. (24), one can rewrite it as,

$$\frac{u_0^2 - u^2}{(1-u^2)}\left(\frac{du}{d\xi}\right)^2 d - (\sigma - u^2)^2 + \kappa\sqrt{1-u^2} = \text{const.} \tag{25}$$

which can be integrated to give

$$\left(\frac{du}{d\xi}\right)^2 + V(u) = 0, \tag{26}$$

where $-\varepsilon$ is the integration constant, and

$$V(u) = -\frac{((\sigma - u^2)^2 - \kappa\sqrt{1-u^2} - \varepsilon)(1-u^2)}{u_0^2 - u^2}. \tag{27}$$

Thus, the dynamics of the dipoles on the surface of the MT can be considered as the motion of the effective particle of mass $m = 2$ in the potential $V(u)$, with the total energy of the system being, $E = 0$. The phase portrait of the system (26) in the plane (Θ, P_u), with $P_u = du/d\xi$, is shown in Fig. 8.

IOP Conf. Series: Journal of Physics: Conf. Series **880** (2017) 012010 doi:10.1088/1742-6596/880/1/012010

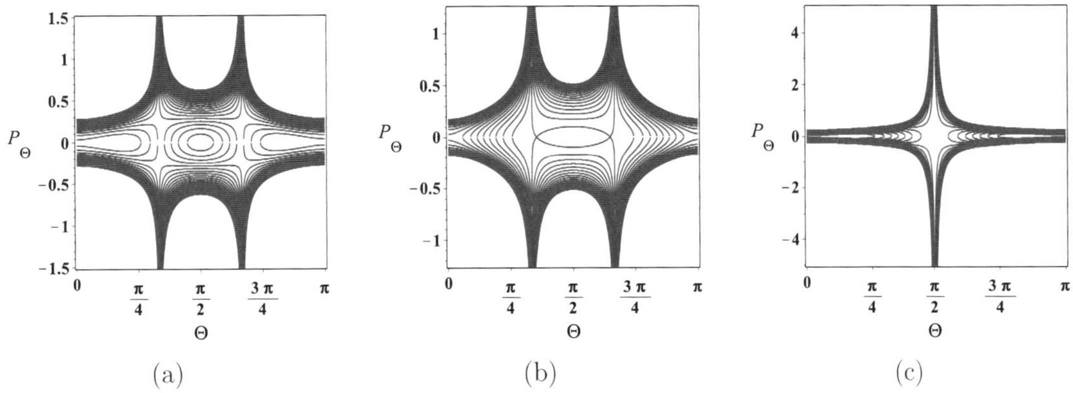

Figure 6. Phase portrait of the system (24). The momentum P_Θ is defined as, $P_\Theta = d\Theta/d\xi$. (a) $\sigma = 0.75$, $u_0 = 0.5$; (b) $\sigma = 0.2$, $u_0 = 0.5$; (c) $\sigma = 0.75$, $u_0 = 0$. In all cases: $\kappa = 0$. From ref. [10].

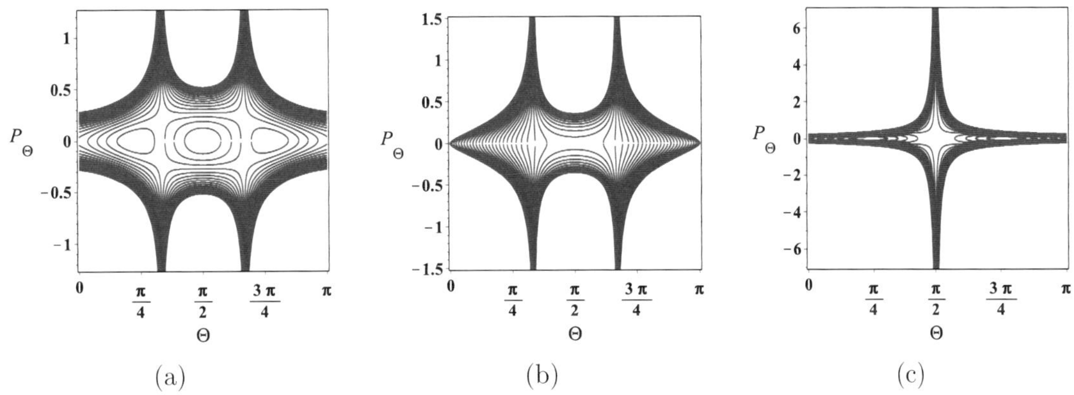

Figure 7. Phase portrait of the system (24). The momentum P_Θ is defined as, $P_\Theta = d\Theta/d\xi$. (a) $\sigma = 0.75$, $u_0 = 0.5$; (b) $\sigma = 0$, $u_0 = 0.5$, ; (c) $\sigma = 0.75$, $u_0 = 0$. In all cases: $\kappa = 0.5$. From ref. [10].

3.1.1. Snoidal waves and kinks: $\kappa = 0$. A sn-oidal function $\mathrm{sn}\, u$ is a particular Jacobi elliptic function. In particular, such functions are defined by means of the inverse of the following incomplete elliptic integral of the first kind [20]

$$u = \int_0^\varphi \frac{d\theta}{\sqrt{1 - m\sin^2\theta}} \Rightarrow \mathrm{sn}\, u = \sin\varphi\ , \tag{28}$$

where φ is called the amplitude, and the parameter $0 \le m \le 1$. Thus a sn oidal function, as all elliptic functions, are functions of two variables (φ, m). [1]

To construct *sn-oidal waves* in our case, we first assume $\kappa = 0$, which implies absence of the intrinsic radial electric field ($g_2 = 0$). Choosing the constant of integration in Eq. (24) as, $\varepsilon = (\sigma - u_0^2)^2$, we obtain,

$$\left(\frac{du}{d\xi}\right)^2 = (2\sigma - u_0^2 - u^2)(1 - u^2). \tag{29}$$

[1] One can define other functions, such as $\mathrm{cn}u = \cos\varphi$ (a cn-oidal function), or $\mathrm{dn}\, u = \sqrt{1 - m\sin^2\varphi}$ (dn-oidal function).

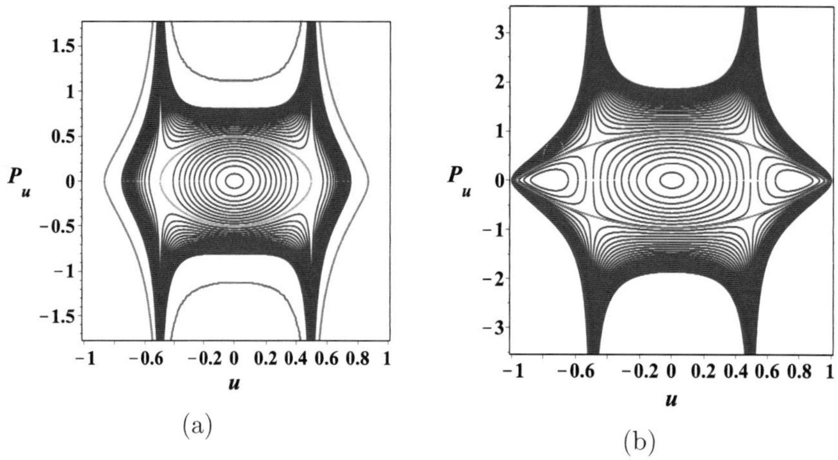

Figure 8. Phase portrait of the system (25) in the plane (u, P_u): (a) $\sigma = u_0^2 = 0.25$, $k = 0.5$; (b) $\sigma = 0.6$, $k = 0.975$. Parameters: $u_0 = 0.5$, $\kappa = 0$.

Assume $u_0^2 < 2\sigma < 1 + u_0^2$, then the analytical solution of this equation is given by a snoidal wave,

$$u = k\,\mathrm{sn}(\xi - \xi_0, k). \tag{30}$$

Here $k = \sqrt{2\sigma - u_0^2}$, and $\mathrm{sn}(z, k)$ is the Jacobi elliptic function. In Fig. 9 the static sn-solutions for different choices of the constant k are depicted. In Fig. 8a, the orbit for $k = 0.5$ is represented by the orange curve.

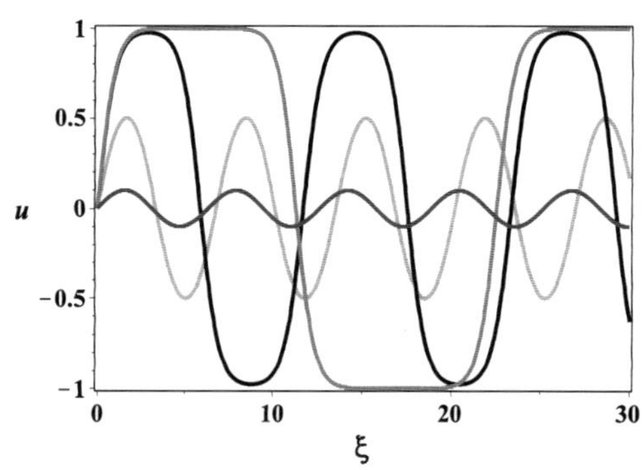

Figure 9. The sn-solution: $k = 0.1$ (blue), $k = 0.5$ (orange), $k = 0.975$ (black), $k = 0.9999$ (red).

The period of the sn-wave is given by $T = 4K$, where

$$K = \int_0^{\pi/2} \frac{d\varphi}{\sqrt{1 - k^2 \sin^2 \varphi}}, \tag{31}$$

is the complete elliptic integral of the first kind [20].

IOP Conf. Series: Journal of Physics: Conf. Series **880** (2017) 012010 doi:10.1088/1742-6596/880/1/012010

For $k^2 \ll 1$ and $k'^2 = 1 - k^2 \ll 1$, applying the Maclaurin Series in k^2 and k'^2 [20], we obtain (setting, for simplicity, $\xi_0 = 0$)

$$u = k \sin \xi - \frac{k^3}{4}(\xi - \sin \xi \cos \xi) \cos \xi + \mathcal{O}(k^5), \tag{32}$$

$$u = \tanh \xi - \frac{k'^2}{4}(\xi + \sinh \xi \cosh \xi)\mathrm{sech}^2 \xi \ . + \mathcal{O}(k'^4). \tag{33}$$

In particular, for $k = 0$, we obtain $u = 0$. This solution corresponds to the elliptic point located at the center of the phase space portrait of Fig. 8. When $k = 1$, the sn-waves become the *kink*

$$u = \tanh(\xi - \xi_0), \tag{34}$$

with the boundary conditions: $u(\pm\infty) = \pm 1$. In Fig. 8b, the corresponding orbit is presented by the separatrix (red curve).

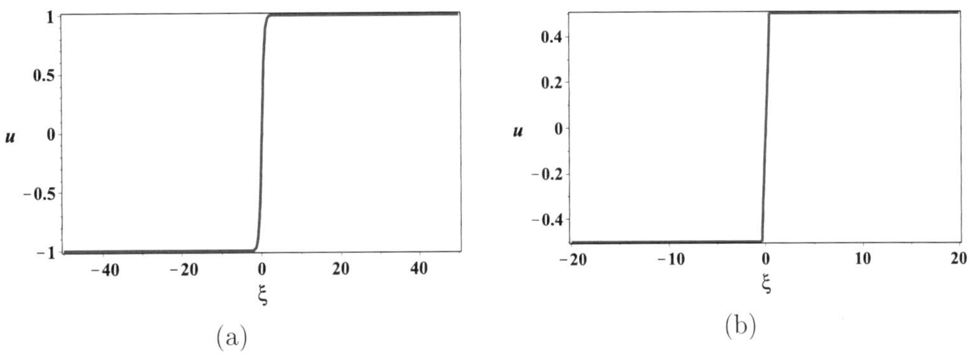

Figure 10. Kink. (a) analytical solution, $\sigma = 0.25$, $\varepsilon = -0.25$. (b) numerical solution, $\sigma = 0.625$, $\varepsilon = 0.141$. Parameters: $u_0 = 0.5$, $\kappa = 0$.

A topological classification of kinks is given in terms of homotopy group [21]. The topological charge, π_0, of kink is determined by the magnitude, n_z of the polarization vector at the ends of the MT:

$$\pi_0 = \frac{1}{2}(n_z(+\infty) - n_z(-\infty)). \tag{35}$$

To change the topological charge one needs to overcome the potential barrier, proportional to the size of the MT (formally, infinite potential barrier).

We remark at this point that such one-dimensional solitions have been considered in connection with dissipation-free energy and signal transduction in phenomenological one-dimensional models of MT in [7, 8]. here such solutions have been derived from realistic three-dimensional lattice models, entailing dipole-dipole interactions, whose physical importance has been stressed above and in [8].

3.1.2. Spikes: $\kappa = 0$ A spike solution can be obtained as excitation of the ground state, u_g. To estimate energy carried by spike, we approximate it by a step function. Using Eq. (15), we then obtain [10]

$$\Delta w_{sp} = w_g - w_{sp} = -\frac{Jg_1}{\Sigma_0}(u_{sp}^2 - u_g^2)^2, \tag{36}$$

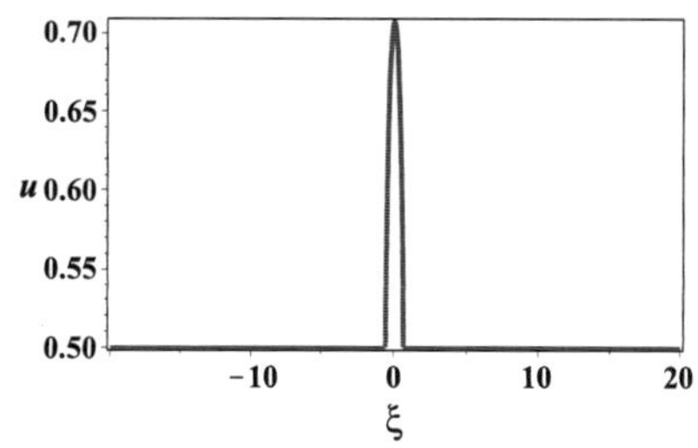

Figure 11. Spike: $\varepsilon = 0.25$, $u_0 = 0.5$, $\sigma = 0.25$, $\kappa = 0$.

where u_{sp} is the height of the spike, and $w_g = -Jg_1 u_g^4/\Sigma_0$ is the energy density of the ground state (see Eq.(16)).

The electric field produced by the spike can be estimated as [10], $\Delta E_z = E_z^{\max}(u_{sp}^2 - u_g^2)^2$, where $E_z^{\max} = \frac{Jg_1}{S}$ is the maximum value of the electric field due to the permanent dipoles, which is reached when all dipoles are aligned along the MT (in which $u_g = 1$) The maximum value of the electric field produced by spike $\Delta E_z \leq \Delta E_z^{\max}$, has been estimated in [10] as

$$\Delta E_z^{\max} = E_z^{\max}(1 - u_g^2)^2 = E_z^{\max} \cos^4 \Theta_0 \leq E_z^{\max}. \tag{37}$$

where Θ_0 denotes the angle between the permanent dipole and axis orthogonal to the surface of the MT. Notice that the maximum magnitude of the electric field produced by spike is bounded by $E_z^{\max}$. As discussed in the literature [16], in the ground state the orientation of the dipoles with respect to the surface of the MT can be defined by $\Theta_0 \approx 29^\circ$. Substituting these data into Eq. (37), we obtain the following estimation for the electric field produced by the spike: $\Delta E_z^{\max} \approx 0.6 E_z^{\max}$. To evaluate $E_z^{\max}$, we use data available for the electric field inside of the MT: $E_z \sim 10^5 \div 10^8$ V/m [7]. Then, we obtain the following estimate for the electric field produced by the spike: $\Delta E_z^{\max} \lesssim 0.6 \cdot (10^5 \div 10^8)$ V/m . The localized spike solution is presented in Fig. 11. In the phase space portrait of Fig. 8, the corresponding orbit is indicated by the red curve on the right. We note that the spike solutions can be important candidates for information transfer by the MTs.

3.2. Solutions with $\Theta = \pi/2$ - Chiral solitons

In this section, we study solutions related to the paraelectric ground state. We seek a solution of the non-linear Lagrange equations (14) in the form: $\Theta = \pi/2$. Substituting $\Theta = \pi/2$ into Eq. (21), we obtain

$$\left(u_0^2 - \sin^2 \Phi\right)\left(\frac{d\Phi}{d\xi}\right)^2 + \eta \sin^2 \Phi + \kappa \cos \Phi = \text{const}, \tag{38}$$

Introducing a new function, $u_\varphi = \sin \Phi$, one can recast this equation as,

$$\left(\frac{du_\varphi}{d\xi}\right)^2 + U(u_\varphi) = 0, \tag{39}$$

where

$$U(u_\varphi) = \frac{(\varepsilon - \eta u_\varphi^2 - \kappa\sqrt{1 - u_\varphi^2})(1 - u_\varphi^2)}{u_\varphi^2 - u_0^2}. \tag{40}$$

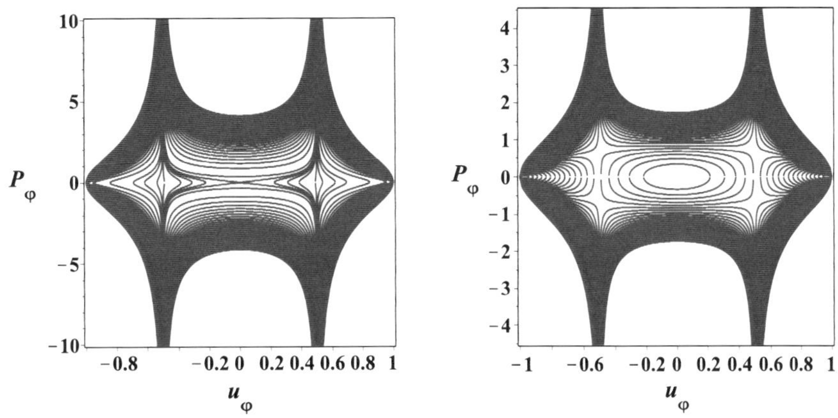

(a) (b)

Figure 12. Phase portrait of the system (39) in the plane (u_φ, P_φ): (a) $\eta = 0.1$, $\kappa = 0.75$; (b) $\eta = 0.75$, $\kappa = 0.25$. Parameters: $u_0 = 0.5$.

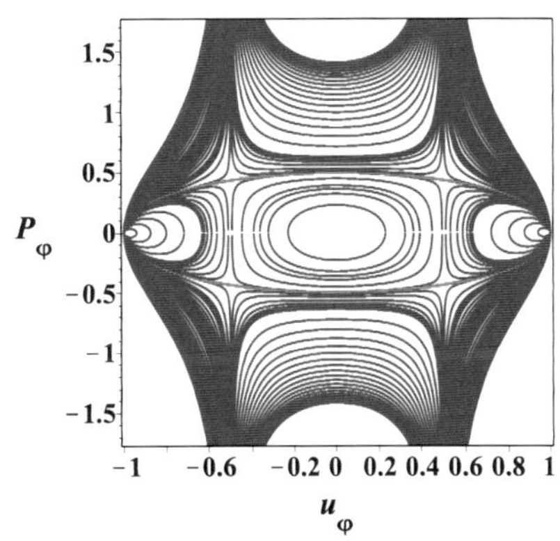

Figure 13. Phase portrait of the system (39) in the plane (u_φ, P_φ): $\eta = 0.25$, $\kappa = 0$, $u_0 = 0.5$.

We denote by ε the constant of integration in Eq. (12). "Chirality" is a topological charge described by the relative homotopy group and defined as [21]:

$$\chi = \frac{1}{\pi}\int_{-\infty}^{\infty} dz\, \mathbf{e}_z \cdot \left(\mathbf{n} \times \left(\left(\frac{\partial \mathbf{n}}{\partial z}\right)\right)\right) = \frac{1}{\pi}\int_{-\infty}^{\infty} dz\, \sin^2\Theta \frac{\partial \Phi}{\partial z}. \tag{41}$$

Taking into account that in our case $\Theta = \pi/2$, we obtain $\chi = \frac{1}{\pi}(\Phi(+\infty) - \Phi(-\infty))$, Chiral solitons in the phase space are presented by orbits located in the interval $(-u_0, u_0)$. (See Figs. 12 and 13.) We remark that Chiral solitons can produce quantized charge transport across the MT that is topologically protected and controllable by the soliton's chirality.

Suppose that $\kappa = 0$, then taking the constant of integration as, $\varepsilon = \eta u_0^2$, one can rewrite (12) as:

$$\left(\frac{du_\varphi}{d\xi}\right)^2 = \eta(1 - u_\varphi^2). \tag{42}$$

The analytical solution of this equation is given by

$$u_\varphi = \sin(\sqrt{\eta}(\xi - \xi_0)) \ , \tag{43}$$

which is a *chiral soliton*. The corresponding orbit is represented in Fig. 13 by the separatrix (red curve).

3.3. Two-dimensional solutions

The solutions presented in previous sections have the form: $\Theta = \Theta(z + \nu\varphi - vt)$ and $\Phi = \Phi(z + \nu\varphi - vt)$. Thus, they describe two-dimensional nonlinear waves propagating on the MT surface along the z-direction. However, in realistic cases one may have fully two-dimensional solutions propagating on the MT surface. In Fig. 14a,b, the static helicoidal sn-solution is depicted. In Fig. 14c, the helicoidal sn-wave is presented. Finally, in Fig. 15, the solution describing kink moving in the z-direction, is depicted. All parameters are given in the corresponding figure captions. In addition one has

$$(u_0^2 - \cos^2\Theta)\left(\frac{d\Theta}{d\xi}\right)^2 - (\sigma - \cos^2\Theta)^2 + \kappa \sin\Theta = \text{const}. \tag{44}$$

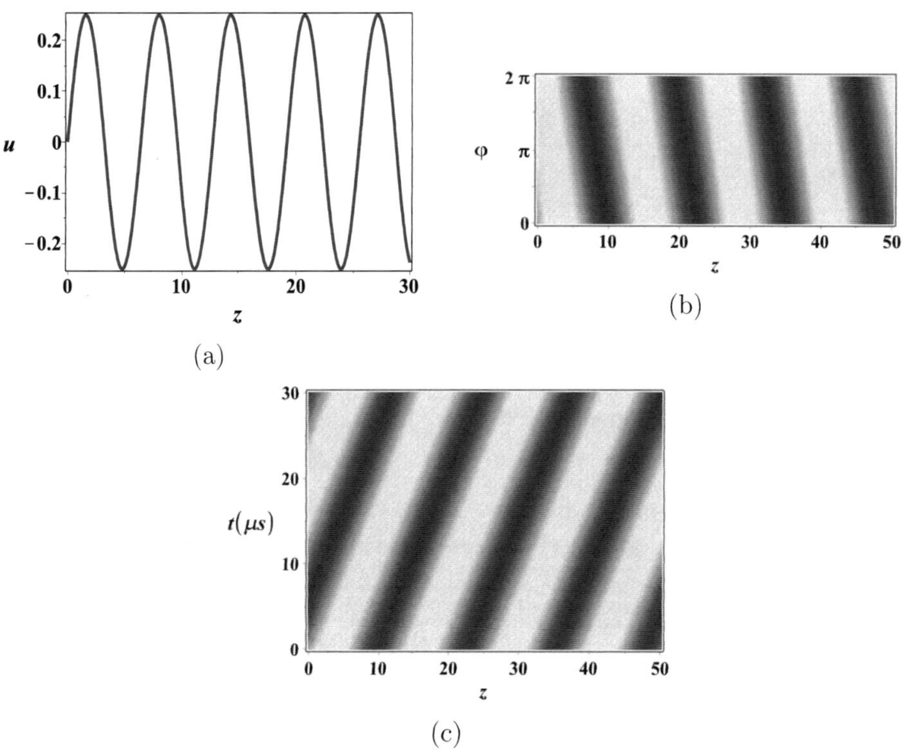

Figure 14. Sn-solutions. (a) u vs z ($v = \nu = 0$); (b) Density plot of the helicoidal static snoidal solution $v = 0$. Density plot of the propagating sn-wave along the MT ($\varphi =$ conts). Parameters: $v = 0.1\text{m/s}$, $\nu = 100\,\text{nm}$, $C = 0.5$, $k = 0.25$.

This completes the classiifcation of the soliton solutions arising from the non-linear dynamics of the pseudospin model of MT.

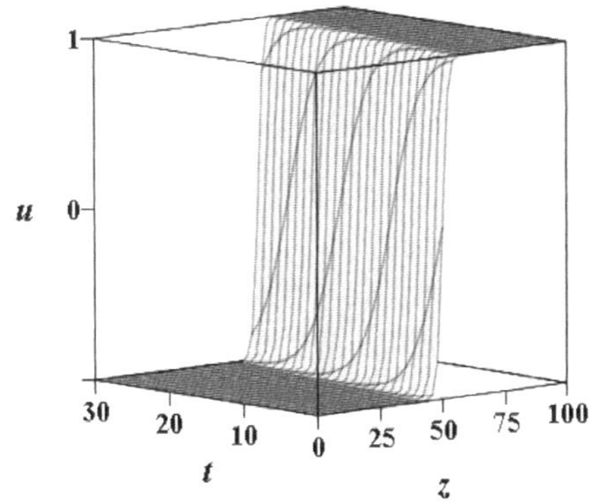

Figure 15. Propagating kink excitation. Parameters: $v = 0.1m/s$, $\nu = 100\,\mathrm{nm}$, $C = 0.5$, $k = 0.25$.

4. Conclusions and outlook

In this talk, I reviewed some recent work on soliton solutions arising in the non-linear dynamics of dimer dipoles in MicroTubular bio-systems modelled by pseudo spin non-linear σ-models. The presence of such solitons, if confirmed in realistic MT systems [22], would be an important step towards our understanding of energy and signal transduction by these biological entities. Solitons may transport energy and information in a dissipation-free and thus efficient way for the complex biological processes MT are associated with.

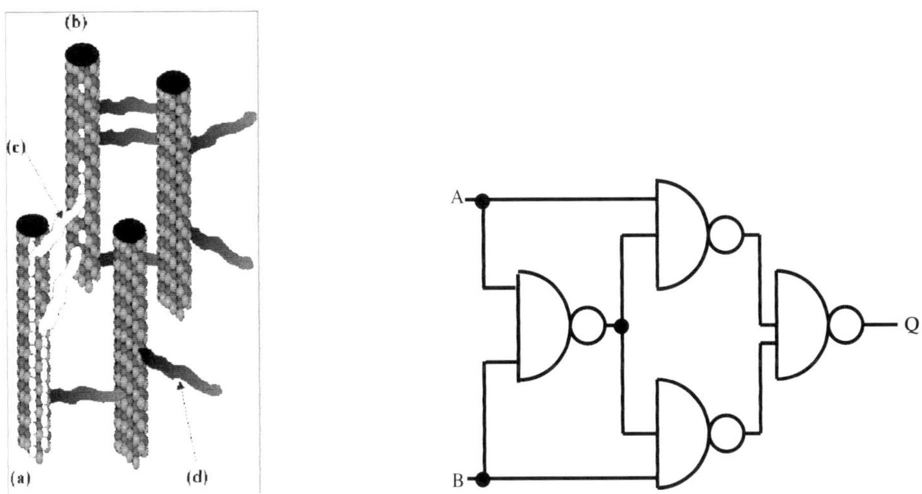

Figure 16. A MT arrangement in cell as a 'logic' XOR gate. Left panel: the biological arrangement. Right panel: the electronic XOR gate for comparison.

In addition to efficient energy transport, the presence of solitonic structures in MT, may imply their rôle as biological gates [9] (See Fig. 16, left panel). Although MT do not naturally branch, nevertheless the analogue of a 'logic' XOR gate (see Fig. 16, right panel) by MT arrangements in cells can be provided due to the existence of Microtubule Associated Proteins (MAP) that connect the various MTs in a network. Once a soliton is formed along one MT, an "active MAP" (yellow colour in left panel of Fig. 16) can transport it from one MT to another. In the MT arrangement depicted on the left panel of Fig. 16, an XOR logic gate can be realised provided the "0" entry is represented by the absence of a soliton and the "1" entry by the presence of a soliton. In this arrangement, MT (a) acts as the "Input" MT, whilst MT (b) is the "Output" MT. (c) is a MAP transmitting a soliton, while (d) represents a "quiet" MAP (green coloured MAPs). MT (a) has two solitons travelling (yellow colour), encountering two MAPs (yelow coloured MAPs) that transmit both solitons to MT (b). In this hypothetical scenario, the solitons arrive out of phase at MT (b) and cancel each other out. The truth table for XOR reads: $0,0 \to 0; 0,1 \to 1; 1,0 \to 1, 1,1 \to 0$. and in this case is realized by MTs if the MAPs are arranged in such a way that each can transmit a soliton independently but if they both transmit, the solitons cancel out.

The above ideas are of course speculative, but it is possible that represent reality in *in vivo* situations involving MT networks. Above we have treated the solitons as classical solutions, and in fact the XOR gate rôle of MT arrangements can be due to classical physics. However, we have already mentioned above that in ref. [8] the solitons have been viewed as *macroscopic quantum coherent states*, which may survive long enough so that certain processes such as energy and signal transport along moderately long MT of length of a few μm can take place. In [8] we have discussed the conditions under which such a situation can be realised in nature, provided sufficient isolation of the MT dimer system from thermal and in general environmental losses occurs. Strong dipole-dipole interactions between the dipole moments of water molecules in the interior of the MT and the electric dipoles of the MT surface dimers have been argued to provide such conditions. In view of the logic gate representation of MT arrangements, then, it is evident that such a binary information system can provide the basic substrate for quantum information processing inside a (not exclusively neural) cell. In a typical MT network, there may be about 10^{12} tubulin dimers. Such a number is macroscopic, and one is tempted to express doubt as to whether, in realistic biological situations, such macroscopic populations of 'particles' can be entangled quantum mechanically, with the entangled state being maintained for a relatively long period of time. However, in atomic physics the experiments of ref. [23] have demonstrated *experimentally* the existence of long-lived entangled states of *macroscopic* populations of Cs gas samples, each sample containing 10^{12} atoms. In such experiments entanglement is generated via interaction with pulses of light. Thus it is not impossible that *in vivo* one has, under certain circumstances specified above, similar entanglement of MT coherent quantum states.

Acknowledgments

I would like to thank the organisers of DICE2016 for their invitation to speak in a plenary session and for organising such an interesting and thought stimulating meeting. I also acknowledge interesting discussions with my co-authors in this research and thank them for the enjoyable collaborations.

References

[1] Fröhlich H 1968 *Int. J. Quantum Chem.* **2** 641; *do.* 1970 *Nature* **228** 1093; *do.* 1972 *Phys. Lett.* **39A** (2) 153; also in *Bioelectrochemistry* eds F Guttman and H Keyzer (New York: Plenum 1986)
[2] Davydov A S 1977 *J. Theoretical Biology* **66** (2) 379
[3] Popp F A 1979 *Electromagnetic Bio- Information* eds F A Popp, G Becker, H L König and W Peschka (Munich-Vienna-Baltimore: Urban & Schwarzenberg)
[4] Hameroff S and Penrose R 1995 *J. Consciousness Studies* **2** 98; *do.*1996 *ibid.* **3** 36; *do.* Mathematics and Computers in Simulation **40** 453
[5] See, *e.g*: Dustin Y 1984 *MicroTubules* (Berlin: Springer); Engleborghs Y 1992 *Nanobiology* **1** 97
[6] Zurek W H 2003 *Rev. Mod. Phys.* **75** 715, and references therein

[7] Satarić M V, Tuszyński J A and Žakula R B1993 *Phys. Rev.* E **48** 589
[8] Mavromatos N E and Nanopoulos D V 1997 *Int. J. Mod. Phys.* B **11** 851; *do.* 1998 *ibid.* **12** 517; *do.* 1998 *Adv. Struct. Biol.* **5** 283
[9] Mavromatos N E 1999 *Bioelectrochemistry and Bioenergetics* **48** 273; Mavromatos N E, Mershin A and Nanopoulos D V 2002 *Int. J. Mod. Phys.* B **16** 3623
[10] Nesterov A I, Ramírez M F, Berman G P and Mavromatos N E 2016 *Phys. Rev.* E **93** 062412
[11] Collini E, Wong C Y, Wilk K E, Curmi P M G, Brumer P and Scholes G D 2010 *Nature* **463** 644
[12] Tuszyński J A, Brown J A, Hawrylak P and Marcer P 1998 *Phil. Trans. R. Soc. Lond. A* **356** 1897
[13] Satarić M V and Tuszynski J A 2005 *J. Bio. Phys.* **31** 487
[14] Baker N A, Sept D, Simpson J, Holst M J and McCammon J A 2001 *Proc. Nat. Acad. Sci.* **98** 10037
[15] Satarić M V 2014 *Bulletin T. CXLVI Acad. Serb. Sci. et Art.* **39** 1
[16] Tuszyński J A, Brown J A, Crawford E, Carpenter E J, Nip M L A, Dixon J M and Satarić M V 2005 *Mathematical and Computer Modelling* **41** 1055
[17] Tuszyński J A, Hameroff S, Satarić M V, Trpisova B and Nip M L A 1995 *J. Theor. Biol.* **174** 1055
[18] Slyadnikov E E 2011 *Technical Physics* **56** 1699
[19] Tuszynski J A and Craddock T J A 2010 *J. Biol. Phys.* **36** 53
[20] Abramowitz M and Stegun I A 1964 *Handbook of Mathematical Functions with Formulas, Graphs, and Mathematical Table* (Dover Publications)
[21] Mermin N D 1979 *Rev. Mod. Phys.* **51**(3) 591
[22] Sahu S, Ghosh S, Hirata K, Fujita D and Bandyopadhyay A 2013 *Appl. Phys. Lett.* **102** 123701
[23] Julsgaard B, Kozhekin A and Polzik E 2001 *Nature* **413** 400

Why is the World four-dimensional? Hermann Weyl's 1955 argument and the topology of causation

Silvia De Bianchi

Department of Philosophy, Building B, Autonomous University of Barcelona, 08193 Bellaterra (Barcelona), Spain

E-mail: silvia.debianchi@uab.cat

Abstract. This paper approaches the question of space dimensionality by discussing a neglected argument proposed by Hermann Weyl in 1955. In *Why is the World Four-Dimensional?* (1955), Weyl offered a different argument from the one generally attributed to him and presented in *Raum-Zeit-Materie*. In the first sections of the paper, this new argument and its features are spelled-out, and in the last section, I shall develop some useful remarks on the concept of topology of causation that can still inform our reflection on the dimensionality of the world.

1. Introduction

In the last decade, Ptkov, among others, suggested that of the most fundamental questions of the 21st century is the search for an explanation of the dimensionality of the world [1]. In his view, by considering the analysis of the kinematical effects of special relativity one can find the key to answer this question. Indeed, he thinks that these effects and experiments which confirmed them would be impossible if the world were three dimensional. Therefore, one has to consider the world as four dimensional. The way in which Petkov touches the question raises at least two important problems. First, one should define what one means by the word 'World' in such a discussion [2]. Second, given that both three- and four-dimensional representations of the physical world are consistent and allowed in the relativistic picture,[1] it results that Petkov does not explain *why* the world is four dimensional, he just shows that it *is* four dimensional and that this dimensionality correctly reflects the dimensionality of the world.

In this paper, I shall discuss another possible approach, one that explores the search for an explanation of the necessity of the four dimensionality of the world. Thus, this approach assumes the four dimensionality of space-time as given[2] and aims at explaining *why* the world has such a configuration and not another one. To reach this goal, not only philosophy and physics are needed, but a fundamental role is also played by mathematics.[3] In 1955, a few months before his

[1] According to Ohanian, physics clearly points to the fact that the observable world is both 3D and 4D, because it can be considered as a manifold and as an embedded curved hypersurface [2].

[2] In other words, it already assumes it as a fact whose features are given by Minkowski's definition of the world.

[3] An approach that is similar (only methodologically) to the argument suggested by Weyl is offered by Tegmark [3]. Tegmark explored the possibilities entailed by PDEs solutions and spelled out how the problem of boundary conditions is of fundamental importance in judging the capacity of a scientific theory to explain the world-dimensionality problem.

death, in Washington D.C. Weyl gave a talk entitled *Why is the World four-dimensional?*[4]. The paper has been recently published in the Weyl collection *Levels of Infinity* edited by Peter Pesic and has been neglected for almost 60 years, even if in its title one can detect a timely question bringing with it the spirit of an old and complex problem, one that invests the foundations of our scientific theories. Both scientific and philosophical literature is replete of articles and books dealing with the problem of space dimensionality and the dimensionality of the world [1][2][5].[4]

Nowadays, the reflection on the question of the dimensionality of the world is often discussed in connection with the assumption of the Anthropic Principle, be it in its weak or strong form. However, I claim that from a rigourous philosophical standpoint one should avoid to ground the argument for the explanation of world dimensionality on the anthropic principle as this prevents us from endorsing a fruitful approach to this question. I shall discuss this point in the last sections of this paper. In offering a deep analysis of Weyl's arguement, I aim at generating an interest among physicists about the possibility of shaping the question of the world dimensionality in a way that is not entirely philosophical nor physical and foremost that is independent of any anthropic principle. Hermann Weyl's 1955 argument represents such an alternative and for this reason is worth being investigated. I will offer an enriched picture of his thought and will offer a new reading of his argument, at least new with respect to Barrow's and Tipler's interpretation of Weyls take on this question [8]. In *The anthropic cosmological principle*, they noticed:

> Weyl pointed out that only in (3+1) dimensional space-time can Maxwells theory be founded upon an invariant integral form of the action; only in (3+1) dimensions it is conformally invariant and this "does not only lead to a deeper understanding of Maxwells theory, but the fact that the world is four dimensional, which has hitherto always been accepted as merely accidental becomes intelligible through it" [8], 260-261).[5]

In what follows, I shall spell out the content of the 1955 argument and then show how it represents an alternative to other approaches to the explanation of the world-dimensionality that use the anthropic principle. Finally, I expound the notion of topology of causation emerging from Weyl's work and which sort of guiding questions it leaves open to future investigation.

2. Why is the World Four-dimensional? (1955)

Even if his paper starts with a question, Weyl immediately admits that he is unable to offer an answer and rather aims at showing why theoretical physicists have been unable to answer until then. First, he distinguishes the question of the dimensionality of the world, which includes time, from the question of the dimensionality of space. In order to do so he makes two premises. First, Weyl endorses Minskowski's definition of the world, namely the world is "the totality of space-time-points, of possible localizations in space and time", so that, having space 3 and time 1 dimension, one assumes the existence of the four-dimensional world (which is what we want to explain and therefore assume it as given). Then Weyl makes a second assumption, according to which there is a crucial difference between the representation of space-time in special relativity and that of space-time in general relativity:

> The metric field is revealed as a agent of formidable reality. Now something that acts upon matter in such forcible fashion has no claim to be enthroned in sphinx-like rigidity above the ever-changing world of matter; it should itself be flexible and respond to the changing distribution of matter; he who makes suffer must suffer itself ([4], 210).

[4] One of the most discussed examples of arguments on space dimensionality is Kant's [6][7]. The latter is to be considered as alternative to teleological approaches to this question and their adaptation into modern versions of the anthropic principle.
[5] The reference is to [9], 284.

Weyl thus immediately highlights that to look at the metrical structure of space-time is not enough to answer the question of the world-dimensionality. This is because the notion of Riemannian manifold is not bound to the special dimensionality 4, and as far as this metrical structure goes, any other dimension would do just as well. He then suggests to look at all possible laws of the physical phenomena that take place on this stage and asks whether they can be carried out in the same unambiguous way from 4 to any other number of dimensions. They are all such as to allow generalization to n dimensions and therefore, according to Weyl:

> There is no reason, until we do not discover other physical laws that work differently in specific dimensionalities, to think that there is something special in the Creator or in the Nature's choice of modelling a 4 dimensional world ([4], 211).

Precisely in this passage, Weyl discards the possibility to appeal to any anthropic principle to justify the dimensionality of the world. The four-dimensionality of the world appears to be something contingent. One could expect to see Weyl appealing to the causal structure of spacetime in order to argue for its necessity, but this is not the case. Surprisingly, in 1955, he depicts causation as a necessary but not sufficient condition to determine dimensionality. Causation is a conformal structure of the world prior to the dimensionality of space and this structure preserves the laws of physics invariant under the modification of the metrical field. However, not all the physical systems seem to be reducible to conformal structures. Therefore, even if for Weyl spacetime can be treated as a structure originated by the conformal or causal structure of the world, yet this cannot explain why the world is four dimensional. Consider the following passage, where Weyl shows in which terms the causal structure is a necessary condition:

> Locally, that is to say in the infinitesimal neighborhood of a world point P, the equation of the light cone
>
> $$ds^2 = \sum_{i,j} g_{ij} dx_i dx_j = 0$$
>
> describes the worlds causal structure. Knowing the causal structure, we do not know the g_{ij} themselves but only their ratios. The causal structure is not changed when the g_{ij} defining the metric are replaced by $\lambda \cdot g_{ij}$ with a positive factor of proportionality λ which is an arbitrary function of P. [...] It would not seem unreasonable to assume that the world is endowed not with a metric, but with a conformal or causal structure only, in other words that only lengths of line elements at the same point are comparable to each other. Then the laws of nature would not be affected by the modification $g_{ij} \to \lambda \cdot g_{ij}$ of the metrical field. Now it is a fact that the equations characterizing a harmonic linear tensor formula are invariant under this substitution if and only if $n = 2e$, i.e. if the dimensionality n is twice the rank e of the electromagnetic tensor f. If $r = 1$, $e = 2$ we would thus obtain the desired $n = 4$. ([4], 212-213)

At this point Weyl introduces the reason why conformal structure is a necessary but not sufficient condition for originating spacetime dimensionality:

> I wish I could close here, but I am afraid that it would be thoroughly dishonest. It is true that the laws of electromagnetic field in empty space presuppose only a conformal structure of the world, they are invariant with respect to the replacement of g_{ij} defining the metric by $\lambda \cdot g_{ij}$. But when we pass from the non-homogenous Maxwell equations which describe the generation of such fields by matter, or to Einsteins laws of gravitation even in empty space, this is no longer the case. The field laws are not conformally invariant neither for the gravitational field nor for the electromagnetic field in the presence of electric charge and current. ([4], 213)

Therefore, even if is certainly important to highlight the fact that Maxwell equations for the electromagnetic field in empty space, in particular the laws for the propagation of light, are conformal-invariant if and only if $n = 4$, still this does not lead very far. Weyl wants a deeper reason *why* the world has a Riemannian structure and in order to reach this goal, he reshaped the question of dimensionality as follows:

- Why is the orthogonal group among all groups of homogenous linear transformations the one that is characteristic of the local metric of the world?
- Why is this metric Pythagorean?

Weyl thus advances here the proposal of an alternative approach to arguments appealing to the anthropic principle or to conformal structure only. For him, the development of mathematics and topology could lead us to an answer for the question of the dimensionality of the world, even if he was unable to give an answer to such long-standing question.

3. Weyl's view of topology

Before spelling out in detail Weyl's argument for the methodology to search for the explanation of the world dimensionality, let us clarify why for him topology can guarantee the sufficient condition for such explanation. In his view, both conformal structure and topology together can justify the actual dimensionality of the world, but topology is assumed by Weyl to be the essential ingredient to reach this goal. Why? In *The mathematical way of thinking* (1940), Weyl clarified what is for him the causal structure of the universe: "At the ground of the words past, present and future referring to time we find something more tangible than time, namely the causal structure of the universe. Events are localized in space and time; an event of small extension takes place at a space-time or world point, a here-now" ([10], 70-71). According to Weyl, the topological scheme is what allows us to embrace both special and general relativity:

> A certain 4-dimensional scheme can be used for the localization of events, of all possible here-nows; physical quantities which vary in space and time are functions of a variable point ranking over the corresponding symbolically constructed 4-dimensional topological space. In this sense, the world is a four-dimensional continuum. The causal structure will have to be constructed within the medium of this 4-dimensional world, i.e. out of the symbolic material constituting our topological space. Incidentally, the topological viewpoint has been adopted on purpose, because only thus our frame becomes wide enough to embrace both special and general relativity theory. The special theory envisages the causal structure as something geometrical, rigid, given once and for all, while in the general theory it becomes flexible and dependent on matter in the same way as the electromagnetic field ([10], 77).

The topology of causation does not only embrace special and general relativity theories, but also preserves the distinction between them. Furthermore, there is another element to consider. General relativity theory establishes laws of nature that connect the flexible causal structure with other flexible physical entities, distribution of masses, or the electromagnetic field:

> These laws in which the flexible things figure as variables are in their turn constructed by the theory in an explicit a priori way. Of course the topological structure cannot be flexible as the causal structure is, but one must have a free outlook on all topological possibilities before one can decide by the testimony of experience which of them is realized by our actual world. To that end one turns to topology ([10], 81).

Topology can offer a solution to the question of the dimensionality of the world, in the sense that it is able to offer the necessary and sufficient condition for its justification, and this sufficient

condition is embodied by the fact that topological possibilities can be tested via experiments. In analyzing a continuum, like space, Weyl wants to proceed in "a more general manner than by measurement of coordinates and adopt the topological viewpoint, so that two continua arising one from the other by continuous deformation are the same to us" ([10], 74).
In other words, it means that one can construct the connection between here-now points and one can construct such a connection in view of its systematic exposition, which can lead us to express it in the form of a physical law.

4. Weyl's argument revisited

It is now clear that for Weyl topology has the special status of a field chiefly able of developing mathematical possibilities, but at the same time is the key-element that enables them to be experimentally checked. We have now more elements to understand and discuss Weyl's 1955 argument that goes as follows.

First, let us consider that to any group of transformations one can ascribe a structure, which is represented by the corresponding abstract group. The structure of the orthogonal group turns out to be quite different for different dimensions n. All orthogonal groups except that of dimension 4 have the structural property of being simple[6] and only to the 4-dimensional orthogonal group one can ascribe a more complex structure, being the direct product of two single groups.[7]
Weyl notices at this point of his argument that the orthogonal group for a positive definite quadratic form is closely related to the notion of sphere. Therefore, topology will reveal the elementary and highly significant differences in the behavior of spheres of different dimensions. Weyl was led to this insight by also considering Witold Hurewicz's studies on the homotopy group. By improving these studies, Weyl believed that important hints for the explanation of the world-dimensionality could come from the consideration that there are elementary and highly significant differences in the behaviour of spheres of different dimensions concerning the continuos mappings $S_n \to V$ of an n-dimensional sphere S_n into a given manifold or variety V. Such a mapping is considered trivial if one can deform it continuously into one that maps S into a single point of V. In the case where V is also a sphere, one obtains a sphere S_{n-d} of lower dimension $n - d$ than the first S_n. Now, the question is whether all mappings $S_n \to S_{n-d}$ are trivial or not and the answer depends only on the difference d of dimensions, at least as soon as $\geq 2(d-1)$. Therefore, one obtains the following answers:

- yes for $d = 4$ and 5
- no for $d = 1, 2, 3, 7$ (non-trivial mappings)
- no for $d = 2p - 3$ where p is a prime number.

That a simple property of a spherical space varies from dimension to dimension in a "strange way" and thus it is not insensitive to dimension points to the existence of a deeper level that pertains to mathematics, but that could also be expressed by a physical law. Unfortunately, Weyl notices that the research of this deepest level in which properties of spheres change from dimension to dimension is still unable to provide a definite answer. His hope was that:

> One day physics will discover laws of nature, the mathematical formulation of which takes into account of such structural features as are highly sensitive to dimension. Only this will allow us to explain and understand the specific character of the world actual dimensionality 4 ([4], 215).

[6] A simple group has no normal subgroups other than itself and the identity, thus it cannot be split into smaller elements.
[7] In discussing this point, Weyl ironically points out that this fact is against the trivial idea that nature is simple.

5. Weyl's argument and the alternative to the Anthropic Principle

Weyl's treatment of the problem of the world-dimensionality implies a genuine philosophical question. To recognize that among different possible metrics only the Pythagorean metric embodies real physical space means to attribute to it a necessary status. However, if this representation of space-time is necessary, there must be a deep reason for it. And Weyl thought that a clue could be offered by the understanding of the uniqueness of the Lorentz group.

Weyl admits the possibility of recognizing a double status to the Pythagorean metric, namely it can be necessary and contingent at the same time. It is an "essence" that could be replaced by another one. To attribute the status of contingent necessity or necessary contingency to the dimensionality of the world does not rule out in principle theological and teleological arguments for the explanation of its dimensionality, but can lead us to the possibility of rethinking the conditions under which natural laws "generate" dimensionality.

This is the most difficult point of Weyl's approach and it goes beyond any analysis of Weyl's or any other argument on dimensionality. I refer here to what I call the *Problem of Generation* (PoG). It arises when one wants to define the principle or the rules according to which, independently from the generalization of the laws of physics to n-dimensions, a certain index and thus the four-dimensionality of the world arises. Which characters do we have to attribute to such a dimensionality generation? Does this generation imply the passage from the mere possibility or mathematical reality to existence? The PoG reveals that there is a profound link between the way in which we shape the strategy to explain world-dimensionality and the content of our scientific theories. And once this link is disclosed, also the effectiveness of mathematics and the question about our ability to produce physical theories describing and explaining the world receive a more refined answer.

Let us now go back to the question oulined at the beginning of this section. How could, in Weyl's view, the four dimensionality described by necessary physical laws be also contingent? The answer is simple, but it has important consequences on our way of conceiving of physical theories. It might be found indeed that there is a physical theory that is able to use other than the Lorentz group to represent features of space-time.[8]

This new mathematical entity would possess some new properties that enable the effectiveness of new physical laws in their application to the world and in explaining phenomena (which would render the Lorentz group applied to physics contingent), but it also entails the conservation of former essential properties of the Lorentz group and therefore the capacity of representing fundamental properties of the world (which still renders this group necessary).

Since for Weyl the particular representation of important features of the world, i.e. four dimensionality expressed via physical theories arises from the capacity of the Lorentz group of representing and capturing these essential features, to use another formalism or mathematical entity would automatically make the feature of four dimensionality contingent with respect to the new formalism and the new physical theory.

Thus the contingent character of the four dimensionality of the world would be just epistemological rather than an ontological one. This, of course, originates an immediate association to some versions of the anthropic principle, and apparently would seem to fit nicely with them. We have seen in the introduction that Barrow and Tipler did not take into account Weyl's late view of the question of the dimensionality of the world. Nevertheless, we are accostumed to consider their study of the anthropic principle as being independent from arguments, such as Weyl's 1955. Let me however ask the following question: does Weyl's 1955 argument challenge Barrow's and Tipler's view of the anthropic principle?[9] Barrow and Tipler

[8] Notice that for Weyl mathematics is the realm of the open possibilities, whereas physics concers actuality. Mathematics is portrayed by Weyl as a constructive (creative) activity necessary to discover patterns and in this respect it is irreducible to physics.

[9] I shall not discuss the logical inconsistency of Barrow's and Tipler's argument, for such a discussion and its

considered different versions of the Anthropic Principle, and today the less problematic and acceptable one is called *Weak Anthropic Principle* (WAP):

> WAP: The observed values of all physical and cosmological quantities are not equally probable, but they take on values restricted by the requirement that there exist sites where carbon-based life can evolve and by the requirement that the Universe be old enough for it to have already done so ([8], 15).

Barrow and Tipler considered this as a restatement of the bare fact that in scientific practice it is essential to take into account the limitations of the measuring apparatus when interpreting observations ([8], 23). And, therefore, they considered our existence as generating a selection effect in assessing the probability of the observed features of the universe.

Weyl's 1955 approach to the question of the dimensionality of the world represents a challenge to WAP, and in a profound way. Indeed, the argument proceeds completely a priori by following a heuristic rule and what he called the topological scheme. Weyl's approach can be classified as relying on one premise that is very different from the postulate assumed by the anthropic principle. The latter presupposes taking the world as a whole [11], whereas we have seen that Weyl presupposes Minkowski's definition of the world as a totality of space-time-points, but meant to be only possible localizations in space and time. Starting from a completely different premise, Weyl's argument assumes an internalist perspective (internalist with respect to a scientific theory, namely relativity theory) and searches for an explanation that is to be found outside this theory by means of mathematics. He tried to assess whether the features of the Lorentz group governing the behaviour and the properties of the constitutive structure of spacetime can offer such an explanation, but he found that topology was not yet able to provide it, and therefore he had to limit his analysis in describing why scientists could not achieve an explanation for the four-dimensionality of the world.

Supporters of WAP devise a different strategy and methodology. Their argument is meant to justify observed values of physical and cosmological quantities and not to explain them. Therefore, basic features of the universe (shape, size, age etc.) must be observed to be of a type that allows the evolution of the observers, because our existence acts as a selection effect in assessing these features [8]. Therefore, given that on the ground of our existence and faculties we have to expect to observe a world possessing precisely three spatial dimensions, supporters of WAP are led to conclude that to search for an explanation of the world dimensionality is meaningless. One will notice that this argument is circular and fallacious. It is circular because the ground for the features to be observed as they are is also their consequence. It is fallacious and weak due to the necessary teleological link established between observed physical quantities and the evolution of the observers. Indeed, none of the physical theories used to interpret observations of the features of the universe needs the theory of evolution to be universally valid. More importantly, none of these theories has laws that only allow the observation of features or quantities directly connected to our evolution. What is then the justification for thinking that observed quantities should allow the evolution of observers? We can summarize the main tenets of WAP as follows:

- WAP assumes a premise according to which we cannot know the world-system as a whole in its totality on the base of observation only.
- WAP uses the existence of human beings and their intelligence to justify this premise, but in so doing, it pretends to exclude alternative ways of investigating basic features of the universe, including its dimensionality.
- WAP reverses the status of consequences and makes them appear as a ground and it does so without any justification.

formal reconstruction, see [11].

- WAP uses the existence of human beings to justify the observation of basic features of the universe by using principles that are valid outside the domain of the scientific theories that are supposed to explain these observed features.

The hidden premise of WAP is of a teleological nature and assumes that there must be a necessary and actual means-to-end relationship (that actually is just purely thought by us) between actual observations of physical quantities and biological evolution of the observers. This amounts to admit a sort of "driving force" linking the existence of intelligent beings and the features of the universe. This attitude inevitably leads to teleological arguments, one that assumes the formal characters of the kind used by Paley in the 19th century [12].

If one accepts to break the alleged mutual influence of the premises and consequences of WAP, due to the impossibility of explaining something existent on the ground of a possibility or a reality entailed in some scientific theories (not all of them), then one can easily see that the quest for the explanation of why and how does the world have such a dimensionality is still open and worth being investigated. One thus can and should possibly avoid to search for this explanation just on the ground of WAP.

6. When philosophy, mathematics and physics meet: Weyl's legacy

Weyl's 1955 argument in this sense provides a valuable hint as to the way to proceed. Furthermore, it shows that the problem is not to prove whether 3D and/or 4D are correct descriptions of the world or whether they are compatible with each other. As Weyl underlined, the crucial question is to explain the uniqueness of the four-dimensional Lorentz group, by looking for physical laws justifying it.[10] Weyl urged that the distinctive character of the four-dimensional Lorentz group, either as a group of linear transformations or as an abstract group, must have been clarified. Thus, when looking for an explanation of the dimensionality of the world, a first step can be a sort of search for a mathematical explanation. I say "sort of" because Weyl himself took it as something contingent, as something that should serve as a guiding principle in the investigation. The problem and the answer, indeed, are not only mathematical: Weyl asks whether another group can replace the four-dimensional Lorentz group in the construction of natural laws by also sharing with the known natural laws those features that are essential for the constitution of physical world and for describing the homogeneity of the four-dimensional world.

In other words, a second step in explaining the world dimensionality is necessary and cannot be considered as a mere analysis of the mathematical and formal properties of the group considered *per se*. This second step consists in finding out what is essential to the properties of the Lorentz group that allows physical laws of a scientific theory to be applicable to the actual world. This second step basically enables the transition from the level of pure possibility to the level of physical reality. And this is why for Weyl one cannot say to have an explanation for the dimensionality of the world if one cannot answer this fundamental question: "one cannot claim to have understood nature unless one can establish the uniqueness of the four-dimensional Lorentz group" ([13], 23). To answer this question also allows us to grasp which kind of features in group theory count in order to be able to recognize them in the future and to use them when giving birth to a new scientific theory.[11]

Before concluding, I would like to remark that, when elaborating his 1955 argument, Weyl was working at the Institute of Advanced Study in Princeton and could witness the debates surrounding difficulties, exhaltation and great results of both relativity and quantum mechanics. He was one of the few mathematical physicists who had access to the formalisms of both theories

[10] Weyl defined the four-dimensional Lorentz group as 'basic group pattern of the universe' ([13], 22).

[11] It would be interesting to compare Weyl's argument with current studies on emergence of a 4D World and causal quantum gravity [14].

and mastered them. He therefore had the privilege to describe at a deep level how scientific explanation is a complex process and moreover a process constituted by degrees or levels. The latter also depend on the intrinsic historical dimension of mathematics that is never complete. This is the reason why Weyl identified a proper heuristic role of mathematical explanation in the making of a scientific theory and as a step to be included within the process of scientific explanation. These questions, above all that on the nature of scientific explanation, are widely discussed among philosophers of science and mathematics, but somehow the debates focus more on the analysis of the scientific theories at hand, rather than on the processes at stake when constructing a theory. The latter is something left to historians of science for reconstruction. However, in more recent times, we witnessed an exception in the debates surrounding the attempts at constructing a theory of quantum gravity or when exhamining the search for heuristic tools relying on the study of the properties instantiated by mathematical objects. If one wants to apply Weyl's methodology to current philosophy of physics and to the epistemology and methodology of theory construction, then there appear to be some guiding questions:

- Is it possible to trace the topological and differential structures of spacetime back to something that could be interpreted as a physical law?
- Can we understand the role of mathematics without taking into account physics?
- How should we conceive of mathematics within physics when looking for explanation?

Especially this last question touches the methodology of natural science that Weyl conceived in terms of symbolic construction. I won't enter into details here, because it would be further the aims of this paper, but I would like to offer the following quote as a stimulous for further reflection on our current views of the relationship between mathematics and physics and on the different ways in which human beings search for explanations:

> Not content with an answer to the question 'How is it?' we wish to know 'How did it come to be so?' Man, wherever he awakens to ponder the riddles of existence, is prone to expect *evolution* to enlighten him about the essence of things. [...] The experience of science accumulated in her own history has led to the recognition that evolution is far from being the basic principle of world understanding; it is the end rather than the beginning of an analysis of nature. Explanation of a phenomenon is to be sought not in its origin but in its immanent law. Knowledge of the laws and of the inner constitution of things must be far advanced before one may hope to understand or hypothetically reconstruct their genesis [15], 285-286.

Therefore, if Weyl's 1955 argument for the dimensionality of the world is to be classified, it should be recognized as an argument endowed of an internalist perspective, one that looks at mathematics as the most appropriate source for the internal analysis of physical reality.

7. Conclusion

Even if it was not possible for him to offer an explanatory answer to the question of the world dimensionality, Weyl believed that it was just a matter of waiting for the development of mathematics, and of topology in particular, to enlarge our understanding of the essence of space-time structure. His reflection clearly invest the foundations of physics, encourage a dialogue between philosophers and physicists in reflecting upon the relationship between mathematics and physics, on the character of scientific explanation and above all it forces philosophy to rethink of old problems, such as that of the contingent necessity of a being that could lead to the formulation of new philosophical concepts and therefore 'tools' to understand and give a meaning to the world. Thus Weyl's approach to the problem of the dimensionality of the world should be first of all understood as a methodology to obtain a deeper understanding of the space-time structure according to physical laws, or better, by recalling Galileo's words, his approach aimed at encouraging us in spelling out the letters of the alphabet through which we can read the book of the universe.

Acknowledgments
I am indebted to Hans-Thomas Elze, Claus Kiefer and Gabriele Veneziano for fruitful comments and suggestions received during the conference DICE 2016. This contribution has been made possible thanks to the Marie Curie fellowship (Grant No. BP2013-100101 - EC-FP7 PEOPLE COFUND) and the Ramón y Cajal fellowship (Grant No. RYC-2015-17289). This research received support from the Catalan Government (AGAUR, 2014 SGR 1410) and the Spanish Ministry of Economy, Indystry and Competitiveness (MINECO, HAR2014-57776-P).

References

[1] Petkov V 2007 Relativity, dimensionality, and existence *Relativity and the Dimensionality of the World* ed V Petkov (Dordrecht: Springer) pp 115-135
[2] Ohanian H C 2007 The Real World and Space-Time *Relativity and the Dimensionality of the World* ed V Petkov (Dordrecht: Springer) pp 81-100
[3] Tegmark M 1997 On the dimensionality of spacetime *Class. Quantum Grav.* **14** L69
[4] Weyl H 2013 Why is the world four-dimensional? *Levels of infinity: selected writings on mathematics and philosophy* ed P Pesic (NY: Dover) pp 203-216
[5] Callender C 2005 *Stud. Hist. Philos. Sci.* **36** 113
[6] Kant I 2012 Thoughts on the true estimation of living forces *Immanuel Kant: Natural science* ed E Watkins (Cambridge: Cambridge University Press) pp 1-55
[7] De Bianchi S, Wells J D 2015 *Synthese* **192** 287
[8] Tipler F J, Barrow J 1986 *The anthropic cosmological principle* (NY: Oxford University)
[9] Weyl H 1952 *SpaceTimeMatter* (Translated from the 4th German edition) ed H Brose (NY: Dover)
[10] Weyl H 2013 The mathematical way of thinking *Levels of infinity: selected writings on mathematics and philosophy* ed P Pesic (NY: Dover)
[11] Craig W L 1988 Br. J. Philos. Sci. **39** 389
[12] Paley W 1802 *Natural theology* (Philadelphia: H Maxwell)
[13] Weyl H 1948 *Similarity and congruence: a chapter in the epistemology of science* ETH Bibliothek, Hochschularchiv Hs 91a:31 23 Bl.
[14] Ambjørn J, Jurkiewicz J, Loll R 2004 Phys. Rev. Lett. **93** 131301
[15] Weyl H 1949 *Philosophy of mathematics and natural science* (Princeton: Princeton University Press)

IOP Conf. Series: Journal of Physics: Conf. Series **880** (2017) 012012 doi:10.1088/1742-6596/880/1/012012

A few reflections on protective measurements and more.

M Genovese

INRIM, Strada delle Cacce 91, I-10135 Torino, Italy
INFN, Via P. Giuria 1, I-10125 Torino, Italy

E-mail: m.genovese@inrim.it

Abstract. Quantum measurement represents the most severe and charming problem of foundations of quantum mechanics. Several issues ranging from the collapse of the wave function to the ontic meaning of the wave function remain unclarified. Here we briefly summarize some recent ideas about this problem, with particular attention to protective measurements and their meaning.

1. Introduction

Quantum measurement represents the most severe and charming problem of foundations of quantum mechanics. Several issues ranging from the collapse of the wave function to the ontic meaning of the wave function remain unclarified [1].

The recent research in this sense has, on the one hand, pointed to define more precisely the formalism and, on the other hand, to look deeper in quantum measurement mechanisms for better understanding the meaning of the wave function and its collapse.

In the first line of research a very successful attempt is the one based on monoidal categories [2].

In summary, a symmetric monoidal category is defined from objects, A, forming a commutative monoid with the tensor product $\bigotimes$ and identity I. Objects are interpreted as system types, as qubits, more qubits, classical data, ... When every object has a dual (the adjoint) the category is compact.

Morphisms $I \rightarrow A$ represent states, through their preparation. Defined classical objects X, a non-demolitive measurement is represented by $M : A \rightarrow X \bigotimes A$. The point is how defining classical states. This is done defining a classical state as a state coming together with a copying operation $\delta : X \rightarrow X \bigotimes X$. Formally a classical object in a compact category is defined through a Frobenius algebra.

The concept of projector is captured by the one of idempotent morphism, i.e. by the fact that repeated projective measurements of the same observable on the same state give the same result, $A \underset{\rightarrow}{\overset{M}{=}} X \bigotimes A \underset{\rightarrow}{\overset{1_X \bigotimes M}{\text{——}}} X \bigotimes X \bigotimes A \underset{\leftarrow}{\overset{\delta \bigotimes 1_A}{\text{——}}} X \bigotimes A \underset{\leftarrow}{\overset{M}{=}} A$.

Nevertheless, describing a genuine quantum measurement requires a further step for including decoherence. This is achieved by defining a genuine quantum measurement

$$Meas := (1_B \bigotimes \delta \circ \delta \bigotimes 1_b) \circ (M \bigotimes M_*). \tag{1}$$

This can then be extended to demolitive measurements $m : A \to X$ by $(\delta \circ \delta) \circ (m \bigotimes m_*)$.

This new formulation of quantum mechanics in terms of monoidal categories can indeed provide a new compact definition of quantum mechanics, connecting it to the last developments in mathematics, but, anyway, does not provide any new physical insight on the physics of quantum measurements.

For achieving new insights on the physical meaning of quantum measurement, in the second line of research we mentioned, a significant research has been addressed to study quantum measurement in weak coupling regime.

In 1988 Aharonv, Albert and Vaidman introduced [3] weak value measurements that represent a new paradigm of quantum measurement where so little information is extracted from a single measurement that the state does not collapse. In little more detail, the weak value of an observable $\widehat{A}$ is defined as $\langle \widehat{A} \rangle_w = \frac{\langle \psi_f | \widehat{A} | \psi_i \rangle}{\langle \psi_f | \psi_i \rangle}$, where the key role is symmetrically played by the pre-selected ($|\psi_i\rangle$) and post-selected ($|\psi_f\rangle$) quantum states. In category calculus this requires an accordingly modification of Eq.1. When the pre- and post-selected states are equal, the weak value is just the expectation value of $\widehat{A}$.
If we consider a von Neumann coupling between the observable $\widehat{A}$ and a pointer observable $\widehat{P}$, according to the unitary transformation $\widehat{U} = \exp(-ig\widehat{A} \otimes \widehat{P})$, in the weak interaction regime the evolution of this system is

$$\langle \psi_f | e^{-ig\widehat{A} \otimes \widehat{P}} | \psi_i \rangle \simeq \langle \psi_f | \psi_i \rangle (\mathbf{1} - ig \langle \widehat{A} \rangle_w \widehat{P}). \tag{2}$$

Thus, weak values are measurable quantities and provide some information on the system. The interpretation of this provided information is anyway not trivial. Weak values being normalised transition amplitudes are neither conventional probabilities nor expectation values/ eigenvalues of (Hermitian) operators.

Furthermore, weak values present non-classical properties assuming anomalous values (imaginary, negative, unbounded values), i.e. "the result of a measurement of a component of a spin 1/2 particle can turn to be 100" [3], and when sequential weak measurements are considered one can find paradoxical situations where for two projection operators P_a and P_b one has $\langle P_a \rangle_{weak} > 0$ and$\langle P_b \rangle_{weak} > 0$, while $\langle P_a \cdot P_b \rangle_{weak} \leq 0$ or $\langle P_a + P_b \rangle_{weak} \leq 0$, a property now also demonstrated experimentally [6]. This situation represents a real conundrum [10] since, on the one hand, at variance with projective measurements, weak values (due to the weak coupling) do not collapse the wave function and they can be performed sequentially on the same systems and, on the other hand, for the previous description in terms of von Neumann coupling demonstrates the measurement of the projection operator P_a can trigger the meter of the pointer only if the system before postselection had a non zero probability corresponding to P_a.

Therefore, the attribution of a clear meaning to weak values is still missing. Eventually it can be searched in categorical formalism of quantum mechanics.

For these amazing properties, weak values in the recent years have been subject of a large interest since their non-classical properties allow investigating fundamental aspects of quantum mechanics [5, 6]. Furthermore, the possibility of amplifying the measurement of small parameters by exploing weak values paved the way to significant applications in quantum metrology & sensing [4] and other quantum technologies [7].

2. The protective measurements

Another interesting example of non conventional quantum measurement is provided by protective measurements. They represent a new paradigm of quantum measurement [8] allowing the possibility of measuring directly the expectation value of a quantum observable. In little more detail, let us consider the standard von Neumann quantum measurement of an observable A: it involves an interaction Hamiltonian coupling the observable A to the pointer momentum P

$$H_A = g(t)PA \tag{3}$$

In a standard impulsive measurement $g(t) \neq 0$ in a very short interval, leading to a shift of the pointer proportional to the eigenvalue a_n of the observable A corresponding to the eigenstate $|a_n\rangle$ in which the state is collapsed. Thus, disposing of a single state no information on the original state is acquired. On the other hand, in a protective measurement $g(t) \sim 1/T$ is smooth enough that the state is not changed (i.e. a "protection" of the state is realised). The evolution of the initial product state describing the system state (in the eigenvalue $|\nu\rangle$ of system Hamiltonian H_S) and the pointer state, $|\Psi(0)\rangle = |\nu\rangle|\phi(r_0)\rangle$, in terms of the system energy eigenstates $|\mu\rangle$ is

$$\begin{aligned} |\Psi(T)\rangle &= \Sigma_{n,\mu} e^{\frac{i}{\hbar}(P\langle A\rangle/T+\langle\mu|H_S|\mu\rangle)T}|a_n\rangle|\mu\rangle\langle\mu|\nu\rangle\rangle a_n|\phi(r_0)\rangle \simeq \\ &\simeq e^{\frac{i}{\hbar}(P\langle A\rangle+\langle\mu|H_S|\mu\rangle)}|\nu\rangle|\phi(r_0)\rangle = \quad (4) \\ &= e^{\frac{i}{\hbar}(\langle\mu|H_S|\mu\rangle}|\nu\rangle|\phi(r_0+\langle A\rangle)\rangle \quad (5) \end{aligned}$$

which is valid in the adiabatic limit when the energy eigenstates remain unchanged. Thus, at the end of the measurement the pointer shift is given by the average value $\langle A\rangle$ and the system, in its initial state $|\nu\rangle$, is not entangled with the pointer. The essential point is that the average value of an observable can be evaluated directly by a single system, eventually even not an eigenstate of the corresponding operator, and not as a statistical average of eigenvalues for an ensemble. By measuring the averages of a sufficiently large number of variables the wave function of the system can eventually be reconstructed to any desired precision.

In practice, the protection of the state can be performed in two ways. In a first case when it is a nondegenerate eigenstate of system Hamiltonian (as above). In a second case by exploiting Zeno effect, i.e. through frequent measurements of a variable for which the state is a nondegenerate eigenstate. Of course in both cases the protection of the state requires some "a priori" knowledge of the state itself [16]; in category language this means that we are now considering a map $X \bigotimes A \to X \bigotimes A$. Anyway, for practical applications one has to distinguish between the information available to the party protecting the state and the one measuring the state.

In particular, protective measurements represent an interesting element in the debate about the ontic vs. epistemic role of the wavefunction. The quantum wavefunction is the fundamental element of quantum mechanics. Originally Schrödinger attributed to the wavefunction an ontological meaning, i.e. that it really exists regardless of our knowledge. However, after Born promoted the epistemological view of the wavefunction, i.e. seeing it as a statistical tool depending on our present knowledge of the system, this opinion prevailed up to recent years when the discussion re-emerged [9].

The relevance of protective measurements in this debate stems from the fact that protective measurements would allow inferring the properties of a single quantum state of a single system and even the (protected) wavefunction itself only negligibly disturbing it during the process, while in general the state of an unknown system, according with the "no cloning" theorem, should remain undetectable.

Nevertheless, of course protective measurements require an "a priori" knowledge of the state (eventually regarding a subspace of Hilbert space if the Hamiltonian evolution in Eq. 3 does not involve the whole space), for casting the protection, and therefore their meaning in the previous discussion can be questioned [14, 15, 16, 17, 18, 19, 20].

However, protective measurements have found several interesting theoretical applications in foundations of quantum mechanics, among which we can mention exploring the meaning of the wavefunction [20], examining empiricist vs. realist aspects of quantum mechanics [11] or the discussion about Bohmian interpretation of quantum mechanics [12, 13, 21]. An application to quantum state measurements has also been proposed [22]: here the optimization of protective measurement is used for minimising the state disturbance when performing a protective measurement of a single system. As a further interesting study, we can mention the extension of the concept of protection to include protection in pre- and post-selected systems [23, 20].

Finally, protective measurements also represent an interesting quantum information protocol: Alice prepares a single particle in a well-defined (pure) state. She then sends this "particle" to Bob who has to find its state. Bob knows the no-cloning theorem, which renders this task impossible and hence asks Alice to send also a proper protection mechanism for the specific state she prepared. Alice agrees, sending him the protection apparatus in the form of a black box. Bob can nevertheless employ this black box for preforming his set of weak measurements to find what is the state of the particle without altering it. A protocol that can find applications, for instance, in testing state preparation, and that provides the most significant experimental motivation for this kind of measurement, that is currently being realised in INRIM (see Fig.1).

Figure 1. Picture of the experiment realising protective measurements on photon polarization. One can recognise, in foreground, a titanium-sapphire laser used for generating, through parametric down conversion, heralded single photons, that are then addressed to a seven stages protective measurements, in the background left, before reaching a cdd, in the background right.

3. Conclusions

Drawing some conclusion, the field of quantum measurement remains one of the most fascinating fields of investigation, involving the very basis of the most fundamental an s successful theory we dispose of. Protective measurements are an interesting tool in this debate and may also represent a useful protocol for quantum technologies.

Acknowledgments

This work has been supported by EMPIR-14IND05 "MIQC2".

References

[1] Genovese M 2010 *Adv. Sci. Lett.* **3** 249
[2] Coecke B and Pavlovic D 2006 *Preprint* arXiv:0608035
[3] Aharonov Y, Albert D and Vaidman 1988 *Phys. Rev. Lett.* **60** 1351
[4] Hosten O and Kwiat P 2008 *Science* **319** 787
[5] Piacentini F et al. 2016 *Phys. Rev. Lett.* **116** 180401
[6] Piacentini F et al. 2016 *Phys. Rev. Lett.* **117** 170402
[7] Troupe J and Farinholt J 2017 *Preprint* arXiv:1702.04836

[8] Aharonov Y and Vaidman L 1993 Phys. Lett. A **178** 38
[9] Pusey MF, Barrett J and Rudolph T O 2012 *Nat. Phys.* **8** 475
[10] Svennsson B 2017 *Found. of Phys.* 10.1007/s10701-017-0068-5
[11] Dickson M 1995 *Philos. Sci.* **62** 122
[12] Aharonov Y and Vaidman L 1996 About Position Measurements Which Do Not Show the Bohmian Particle Position, in *Bohmian Mechanics and Quantum Theory: An Appraisal* eds J T Cushing, A Fine and S Goldstein (Dordrecht: Kluwer 1996) pp. 141-154
[13] Aharonov Y, Englert B and Scully M 1999 *Phys. Lett. A* **263** 137
[14] Rovelli C 1994 *Phys. Rev. A* **50** 2788
[15] Unruh W 1994 *Phys. Rev. A* **50** 882
[16] D'Ariano G, Yuen H 1996 *Phys. Rev. Lett.* **76** 2832
[17] Aharonov Y et al. 1996 *Found. Phys.* **26** 117
[18] Dass N, Qureshi T 1999 *Phys. Rev. A* **59** 2590
[19] Uffink J 1999 *Phys. Rev. A* **60** 3474
[20] Gao S 2015 *Protective Measurement and Quantum Reality* (Cambridge University Press, UK)
[21] Holland P 2016 *Preprint* arXiv:1601.03993
[22] Schlosshauer M 2016 *Phys. Rev. A* **93** 012115
[23] Aharonov Y and Vaidman L 1996 *Preprint* arXiv:quant-ph/9602009

Unveiling the curtain of superposition: Recent gedanken and laboratory experiments

E Cohen[1,2] **and A C Elitzur**[2]

[1] H.H. Wills Physics Laboratory, University of Bristol, Tyndall Avenue, Bristol, BS8 1TL, U.K
[2] Iyar, The Israeli Institute for Advanced Research, POB 651, Zichron Ya'akov 3095303, Israel

Email: eliahu.cohen@bristol.ac.uk

Abstract. What is the true meaning of quantum superposition? Can a particle genuinely reside in several places simultaneously? These questions lie at the heart of this paper which presents an updated survey of some important stages in the evolution of the three-boxes paradox, as well as novel conclusions drawn from it. We begin with the original thought experiment of Aharonov and Vaidman, and proceed to its non-counterfactual version. The latter was recently realized by Okamoto and Takeuchi using a quantum router. We then outline a dynamic version of this experiment, where a particle is shown to "disappear" and "re-appear" during the time evolution of the system. This surprising prediction based on self-cancellation of weak values is directly related to our notion of Quantum Oblivion. Finally, we present the non-counterfactual version of this disappearing-reappearing experiment. Within the near future, this last version of the experiment is likely to be realized in the lab, proving the existence of exotic hitherto unknown forms of superposition. With the aid of Bell's theorem, we prove the inherent nonlocality and nontemporality underlying such pre- and post-selected systems, rendering anomalous weak values ontologically real.

1. Preface

(He looked under the bed, in the chimney, in the cupboard; – nobody. He could not understand how he got in, or how he escaped.

Hoffmann. – Nocturnal Tales)

Oh! how often have I heard and seen him, Scarbo, when at midnight the moon glitters in the sky like a silver shield on an azure banner strewn with golden bees.

How often have I heard his laughter buzz in the shadow of my alcove, and his fingernail grate on the silk of the curtains of my bed!

How often have I seen him alight on the floor, pirouette on a foot and roll through the room like the spindle fallen from the wand of a sorceress!

Do I think him vanished then? the dwarf grows between the moon and me like the belfry of a gothic cathedral, a golden bell shakes on his pointed cap!

But soon his body becomes blue, translucent like the wax of a candle, his face pales like the wax of a candle end – and suddenly he is extinguished.

Scarbo (translated to English), from Gaspard de la nuit / Aloysius Bertrand

IOP Conf. Series: Journal of Physics: Conf. Series **880** (2017) 012013 doi:10.1088/1742-6596/880/1/012013

2. Introduction

In a famous letter Einstein has replied to a bewildered letter from Louis de Broglie's PhD advisor. Paul Langevin did not know what to think about the youngster's idea which has generalized Einstein's earlier Nobel-winning discovery that light, hitherto regarded as a wave, was in fact composed of particles. Now came de Broglie and proposed that electrons and all other particles were in fact waves. Einstein was enthusiastic: "He has lifted a corner of the great veil." This marked the birth of the infamous "wave-particle duality" that underlies quantum uncertainty with all its consequences. Any particle, as long as not measured, resides it several locations or states. It is superposed. Einstein famously hated this term, yet he was honest enough to regard de Broglie's discovery as lifting Nature's veil.

In what follows we suggest that time is ripe for this veil to be lifted even further, strictly within present-day quantum theory. Based solely on ordinary quantum measurements, albeit under special post-selections, we show that what has been so far grudgingly accepted as "a particle existing in several locations" may in reality be a plethora of states, so far unnoticed. These states are sometimes very strange, *e.g.*, "a particle disappearing in one location and reappearing in another." It is post-selection, a simple method having a crucial role in the Two-State Vector Formalism (TSVF), which offers us a glimpse into this deeper reality beneath the quantum realm, with much more lying ahead.

This article's outline is as follows. In Secs. 3,4 we review the three-boxes paradox and its "shutter version", respectively. In Sec. 5 we present a novel version of this experiment where a particle seems to disappear and then re-appear in another place. In Sec. 6 this disappearing and re-appearing particle acts as a shutter, thereby making the paradox more tangible. Secs. 7,8 put these thought experiments in a broader context, relating them to some other ideas and Sec. 9 concludes the work.

3. The three-boxes paradox

The three-boxes paradox [1] is a simple and by now well-known paradox, but it already contains some intriguing features we shall rely on.
A system is prepared, i.e. pre-selected, at the superposition state:

$$|\psi(t=0)\rangle = \frac{1}{\sqrt{3}}\left(|1\rangle + |2\rangle + |3\rangle\right). \tag{1}$$

Later in time, it is (strongly) measured, i.e. post-selected, and found at the state:

$$|\phi(t=t_f)\rangle = \frac{1}{\sqrt{3}}\left(|1\rangle + |2\rangle + |3\rangle\right). \tag{2}$$

A surprising prediction arises as to measurements at intermediate times: If the particle is looked for in Box 1 at some $0<t'<t_f$, it will be found there with certainty. To understand why this is the case, let us assume that we look for the particle at Box 1 and don't find it there, meaning that the initial state has now collapsed to: $|\psi(t=t')\rangle = \frac{1}{\sqrt{2}}\left(|2\rangle + |3\rangle\right)$. Since this state is orthogonal to $|\phi(t=t_f)\rangle$, the latter would not occur in contrast to our assumption, hence, the particle must be in the first box. However, if we alternatively look for it in Box 2, we always find it there based on the same logic. But how can it reside in both? An analysis in terms of weak values accordingly reveals that $\langle \Pi_1 \rangle_w = \langle \Pi_2 \rangle_w = +1$, where $\Pi_i = |i\rangle\langle i|$ is a projection on the i-th Box, while intriguingly the weak value at the third box is found to be anomalous $\langle \Pi_3 \rangle_w = -1$, meaning that the sum of all three weak values is still 1, as it should be. Since the weak values of projections into the first two boxes coincides with the eigenvalues of these dichotomic operators, it follows [1] that they could have been found using a strong measurement as well. This is not true for the last box which contains a particle with negative properties, a notion that will become clearer in Sec. 5.

4. The "shutter version" and its recent realization

The three-boxes paradox is based on a counterfactual: "Had one measured the particle in Box 1, it would be there, and similarly for 2" This is an extravagant type of nonlocality, where the very act of measurement in a certain location seems to force the particle to "collapse" there. Yet it is a retrodiction, holding for a past state which is by definition no longer accessible. This seems to rob the paradox much of its acuity.

This obstacle was overcome by Aharonov and Vaidman [2] in the general N boxes case, with a scheme analogous to Hardy's delayed measurement [3]. This time it is a strong projective measurement of a particle operating as a "shutter." It has two stages: i) The shutter particle, after the pre-selection (1), is coupled with a superposed test particle. ii) Then, after the former's post-selection (2), the test particle is subjected to measurement which reveals the shutter particle's intermediate state. The retrodiction thus turns into a standard prediction: In all cases where the shutter's post-selection succeeds, the test particle is reflected from all the N-1 boxes, demonstrating the shutter's simultaneous existence in all of them.

Okamoto and Takeuchi [4] have recently tested this prediction of the TSVF for N=3 in a quantum optics setup employing a novel quantum router [5] which enables the shutter and the probe photon to indirectly interact. Their results show that the shutter did not randomly occupy one of the two boxes, but rather (within experimental accuracy limitations) reflected a photon from both.

5. The disappearing and reappearing particle

Whereas the three boxes experiment presents an intriguing *state* – one particle "collapsing" in whatever box inspected – our recent disappearing-reappearing particle gedankenexperiment [6] gives even more intriguing *evolution*: The particle is bound to be found in mutually-exclusive boxes at different times. And here again, the prediction is for strong (projective) measurements.

The pre-selection (in fact a preparation) is

$$\left|\psi(t=0)\right\rangle=\frac{1}{\sqrt{3}}\left(\left|1\right\rangle+i\left|2\right\rangle+\left|3\right\rangle\right), \tag{3}$$

and the post-selection

$$\left|\phi(t=t_f)\right\rangle=\frac{1}{\sqrt{3}}\left(-\left|1\right\rangle+i\left|2\right\rangle+\left|3\right\rangle\right). \tag{4}$$

We then introduce a time evolution $H=\varepsilon\sigma_x$ during $0\le t\le t_f$, allowing the particle to tunnel between Boxes 1 and 2.

Surprisingly, we now have three different predictions for three earlier instants during this interval:

$$\Pi_1(t_1)=1,\ \ \Pi_3(t_1)=1, \tag{5}$$

$$\Pi_1(t_2)=0,\ \ \Pi_2(t_2)=0,\ \ \Pi_3(t_1)=1, \tag{6}$$

$$\Pi_2(t_3)=1,\ \ \Pi_3(t_3)=1, \tag{7}$$

where the weak value notation is omitted since these the above values coincide with the eigenvalues of the projectors and hence could have been found in a strong measurement too [1].

It is intriguing nevertheless that $\left\langle\Pi_2(t_1)\right\rangle_w=\left\langle\Pi_1(t_3)\right\rangle_w=-1$, suggesting that the disappearance from Boxes 1 and 2 in Eq. 6 is a result of self-cancellation (see also Sec. 8). And moreover, we thus understand that the total number of particles within these boxes is understood to be zero throughout the course of the experiment, yet their sub-division is non-trivial.

It is time-symmetry that obliges this surprising prediction within orthodox quantum mechanics. Followed from past to future, this formulation gives a reasonable account: Once you have obtained the

pre-selection (3), then, *if* you get either (5), (6) or (7), your probability to get the post-selection (4) goes from 11% to 33%.
But followed conversely (as if pre- and post-selection are interchanged), the result is odd: If you did *not* perform any intermediate measurement, just obtained (3) and (4), then *all three retrodictions (5)-(7) are equally correct*!

This, of course, is a very extraordinary claim, even in comparison with other quantum effects. *How can the probe particle indicate, with certainty, the shutter particle's disappearance from one location and reappearance at another?* It would therefore be only natural to consider a more moderate, even trivial alternative: The pre- and post-selections give us an *ensemble*, termed "family of histories" by Griffiths [7], comprised of three *distinct* groups, each of which has its own two properties:

i) Particles that went to Box 1 and yield the post-selection (4) if measured at t_1;

ii) Particles that went to Box 3 and yield the post-selection (4) if measured at t_2;

iii) Particles that went to Box 2 and yield the post-selection (4) if measured at t_3.

No position changes, then, because each particle has only one history: If found in Box 1/2/3, at any time, then it has been there *all along*.

Disproving this alternative is straightforward. Being "all along" somewhere is by definition a *local hidden variable*. Here the abovementioned Bell-Hardy proof [3] becomes crucial: Our single particle's wave-function, split into three, is as nonlocal as an EPR pair or GHZ triplet. Here too, then, any hidden variable must be nonlocal, this time extremely so to the point of obliging the particle to abruptly jump from one trajectory to another.

6. The disappearing-reappearing particle made to act as a shutter

Based on the Okamoto-Takeuchi [4] realization of Aharonov and Vaidman's protocol [2], our proposal for a realization of the disappearing-reappearing particle naturally follows [8].

The main new element added to the experiment is a temporal superposition preceding the spatial one. Split an initial photonic beam into three equal-intensity beams (see *e.g.* [9]) and delay the 2nd and 3rd beams such that the three parts pass through the 3-boxes system at t_1, t_2 and t_3, respectively.

At each instant, direct the 1/3 ray to one or more boxes in accordance with the TSVF prediction (Fig. 1) about the shutter-particle's position at that instant:

i. At t_1, the first 1/3 beam is split again by a simple BS into 2 and goes to Boxes 1+3.

ii. At t_2, the second 1/3 goes only to 3.

iii. At t_3, the last 1/3, again split by simple BS, goes to 2+3.

Then, after passing all the boxes at all times, the photon is reunited in a precisely reverse manner to the above splits, first spatially and then temporally. Once all splits are completely undone, we measure the photon to see whether its initial quantum state is restored by interference.

Notice the required care: *Only this specific combination space and time paths* – (1+3)(t_1), 3(t_2), (2+3)(t_3) – will restore the photon's initial state.

In addition to this direct test of the shutter-particle's presence, there is a complementary test, easier but less decisive: *Send the photon through the boxes predicted to be* empty *at* t_2. This, conveniently, does not involve complicated momentum exchange between shutter and photon. Here too, the specific combination of (1+3)(t_2) – restores the photon's initial state (and probabilistically also the combination 2(t_1) and 1 (t_3)).

Earlier, with the *gedanken* version of the disappearing-reappearing particle, we used the Bell-Hardy proof against the "all along" local account. This involved a counterfactual: two parts of the shutter-particle's wave *could* as well interact with two atoms, which *would* then become entangled, manifesting a Bell-test violation.
The present setting offers us a stronger, *active* test of this kind. As Okamoto and Takeuchi [4] point out, the shutter and the test-particle become *EPR entangled*. Why, then, not subject them to a Bell-test? Violations of Bell's inequality are then bound to emerge, indicating that the "which path" outcome of each particle is nonlocally determined by the other particle's choice of measurement. Consequently, the "all along" alternative is ruled out in favor of the disappearance-reappearance account.

Fig. 1 shows a simplified test of this kind, where each particle's wave-function is split into two halves within an interferometer. The two interferometers cross each other on both arms. In 1/2 of the cases, each particle would turn out to have taken the path not crossed by the other, thereby detected by one of the detectors placed farther on its non-collision path in Fig. 1. Let these cases be selected out. The remaining cases, where the detectors remain silent, assure us, by Interaction-Free-Measurement [10], that *collision between the two particles has occurred, in superposition, on all four intersecting MZI paths*.
This is actually our post-selection. It gives rise to a full entanglement between the shutter and the test-particle:

$$|\psi\rangle = \frac{1}{\sqrt{2}}\left(|1\rangle|1\rangle + |2\rangle|2\rangle\right), \tag{8}$$

enabling an Alice and Bob to subject them to EPR-Bell measurements where each party randomly chooses between "which-path" and interference measurements. By Bell [11], each outcome is determined also by the type of measurement chosen for the other, distant particle. So much for the "all along" alternative.

7. From superposition to quantum entanglement – the quantum liar paradox

We recently suggested another paradox [12] belonging to the same family. Two atoms are prepared, one excited and one ground: $|\psi\rangle = |e\rangle|g\rangle$ We assume that a photon emitted from the first would necessarily excite the second. We wait the half-life time and now, apart from a few technicalities described in [12] we are sure that the atoms are entangled

$$|\psi\rangle = \frac{1}{\sqrt{2}}\left(|e\rangle|g\rangle + |g\rangle|e\rangle\right). \tag{9}$$

However, we can now run a Bell-test to verify the nonlocal correlation between them, but now a paradox ensues: In all cases where the first atom is still excited (i.e. oblivious Eq. 9), we believe that no photon was emitted, but in this case, how can they be entangled and exhibit nonlocal correlations? It turns out that the mere *possibility* of photons emission has the power of entangling the atoms and create nonlocal correlations between them. In [12] we continued in this line of thought to analyze the multi-particle scenarios (which are even more striking) and also discussed a verification using weak measurements.

IOP Conf. Series: Journal of Physics: Conf. Series **880** (2017) 012013 doi:10.1088/1742-6596/880/1/012013

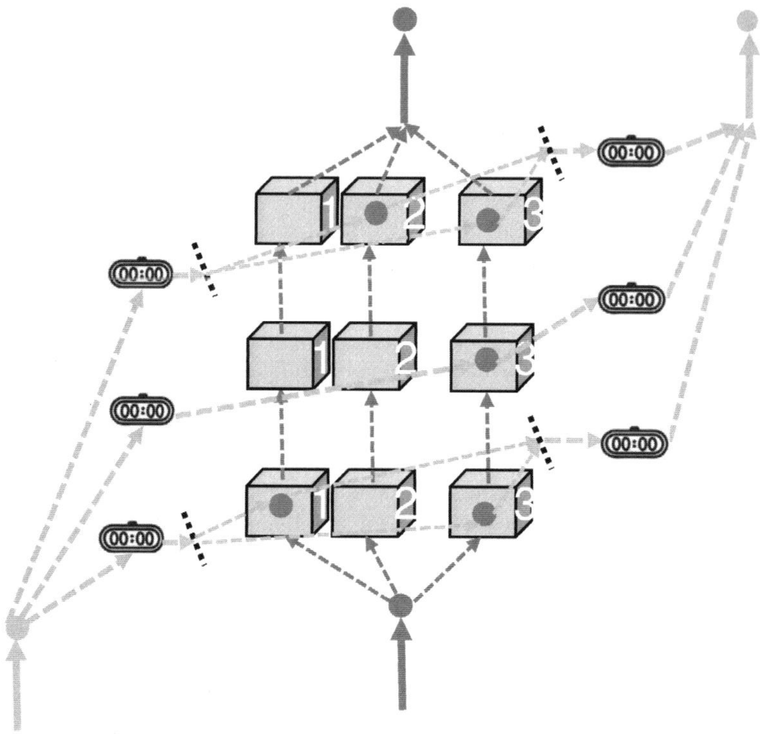

Figure 1. The measurement of the disappearing and re-appearing shutter is "strong" but delayed. The particle in red operates as a shutter while being superposed in Boxes 1-3. Its varying location is measured by a photon (orange) whose wave-function is split into five parts in both space and time: (1+3) at t_1, (3) at t_2 and (2+3) at t_3. Finally it is re-united in a reverse manner to test its interference-like pattern. This is guaranteed only with this precise combination of spacetime paths.

8. Quantum Oblivion

Our final task is to show that all the above effects share something in common. They all demonstrate the extreme reversibility governing quantum phenomena, to the point of events happening and then "unhappening." This is Quantum Oblivion [13], presented in greater detail DICE 2014 [14]. Moreover, it offers the very key for understanding numerous other quantum oddities.

The thought experiment in the last section demonstrates very lucidly this oblivion. The Quantum Liar paradox is based on couternfactuals, i.e. events that *could* have occurred, but eventually *haven't*. In terms of TSVF [15-17], this unique pair of happening and unhappening can be decomposed into a sum of self-cancelling *weak values* [18]. This mechanism has been shown to be ubiquitous in the quantum realm [13, 18], but here we can straightforwardly see its dynamics in action – the disappearance of a particle that should have resided in the first or second box, but eventually did not, is understood through the null weak value in these boxes which is a sum of earlier +1 and -1 weak values. The mathematical formalism leads us to recognize them as two particles, one with "positive properties" and one with "negative properties" (as can be inferred thorough weak measurements [19, 20]) which strictly cancel one another at t_2 thus representing a genuine disappearance, yet with an interesting underlying reality. To emphasize the latter point we suggest verification tests [6] sensitive to a non-zero modular momentum within the first and second box even in the absence of the particle.

Quantum oblivion has been shown [13] to underlie many well-known phenomena, from Interaction-Free-Measurement [10, 21] and the Quantum Zeno effect [22] to the Aharonov-Bohm effect [23]. They all turn out to share the common denominator of Nature herself totally "forgetting" a certain event, leaving an intriguing gap in causality and locality. Further work along this promising line of research are currently underway.

9. Discussion

We have seen a major progress in the gedanken and practical levels of quantum paradoxes. Recent pre- and post-selected paradoxes [6, 12, 21, 24] become more and more acute and can be examined using strong (projective) measurements, and not only via weak measurements. Moreover, during the last few months a laboratory demonstration [3] has achieved just that, and another one is currently under feasibility study [8]. All these broaden the notion of quantum superposition, so as to include exotic states, thus shedding light on the very foundations of quantum mechanics.

Acknowledgments
It is a pleasure to thank Yakir Aharonov for many helpful discussions. E.C. was supported by ERC - AdG NLST.

References

[1] Aharonov Y and Vaidman L 1991 *J. Phys. A: Math. Gen.* **24** 2315
[2] Aharonov Y and Vaidman L 2003 *Phys. Rev. A* **67** 042107
[3] Hardy L 1994 *Phys. Rev. Lett.* **73** 2279
[4] Okamoto R and Takeuchi S 2016 *Sci. Rep.* **6** 35161
[5] Shomroni I, Rosenblum S, Lovsky Y, Bechler O, Guendelman G and Dayan B 2014 *Science* **345** 903-906
[6] Aharonov Y, Cohen E, Landau A and Elitzur A C 2017 The case of the disappearing (and re-appearing) particle *accepted to Sci. Rep.* (*Preprint* arXiv:1610.09025)
[7] Griffiths R B 2016 *Phys. Rev. A* **94** 032115
[8] Aharonov Y, Cohen E, Elitzur A C, Okamoto R and Takeuchi S 2017 Nonlocal position changes of a particle under special pre- and post-selection, revealed by strong measurements and proved by Bell's theorem (forthcoming)
[9] Schnabel R, Bunkowski A, Burmeister O. and Danzmann K 2006 *Opt. Lett.* **31** 658-660
[10] Elitzur A C and Vaidman L 1993 *Found. Phys.* **23** 987
[11] Bell J S 1964 *Physics* **1** 195–200
[12] Aharonov Y, Cohen E, Elitzur A C and Smolin L 2016 Interaction-free effects between distant atoms *Preprint* arXiv:1610.07169
[13] Elitzur A C and Cohen E 2015 *Int. J. Quantum Inf.* **12** 1560024
[14] Cohen E and Elitzur A C 2015 *J. Phys. Conf. Ser.* **626** 012013
[15] Aharonov Y, Bergmann P G and Lebowitz J L 1964 *Phys. Rev.* **134** 1410-1416
[16] Aharonov Y and Vaidman L 1990 *Phys. Rev. A* **41** 11
[17] Aharonov Y, Cohen E., Gruss E and Landsberger T 2014 *Quantum Stud.: Math. Found.* **1** 133-146
[18] Elitzur A C and Cohen E 2016 *Philos. T. R. Soc. A* **374** 20150242
[19] Aharonov Y, Albert D and Vaidman L 1988 *Phys. Rev. Lett.* **60** 1351
[20] Aharonov Y, Cohen E and Elitzur A C 2014 *Phys. Rev. A* **89** 052105
[21] Aharonov Y, Cohen E and Popescu S 2015 A current of the Cheshire cat's smile: Dynamical analysis of weak values *Preprint* arXiv:1510.03087
[22] Sudarshan E C G and Misra B 1977 *J. Math. Phys.* **18** 756
[23] Aharonov Y and Bohm D 1959 *Phys. Rev.* **115** 485
[24] Aharonov Y, Colombo F, Popescu S, Sabadini I, Struppa D C and Tollaksen J 2016 *P. Natl. Acad. Sci. USA* **113** 532-535

IOP Conf. Series: Journal of Physics: Conf. Series **880** (2017) 012014 doi:10.1088/1742-6596/880/1/012014

Reflections on the information paradigm in quantum and gravitational physics

Philipp Andres Höhn

Institute for Quantum Optics and Quantum Information, Austrian Academy of Sciences, Boltzmanngasse 3, 1090 Vienna, Austria, and
Vienna Center for Quantum Science and Technology, Universität Wien, Boltzmanngasse 5, 1090 Vienna, Austria

E-mail: p.hoehn@univie.ac.at

Abstract. We reflect on the information paradigm in quantum and gravitational physics and on how it may assist us in approaching quantum gravity. We begin by arguing, using a reconstruction of its formalism, that quantum theory can be regarded as a universal framework governing an observer's acquisition of information from physical systems taken as information carriers. We continue by observing that the structure of spacetime is encoded in the communication relations among observers and more generally the information flow in spacetime. Combining these insights with an information-theoretic Machian view, we argue that the quantum architecture of spacetime can operationally be viewed as a locally finite network of degrees of freedom exchanging information. An advantage – and simultaneous limitation – of an informational perspective is its quasi-universality, i.e. quasi-independence of the precise physical incarnation of the underlying degrees of freedom. This suggests to exploit these informational insights to develop a largely microphysics independent top-down approach to quantum gravity to complement extant bottom-up approaches by closing the scale gap between the unknown Planck scale physics and the familiar physics of quantum (field) theory and general relativity systematically from two sides. While some ideas have been pronounced before in similar guise and others are speculative, the way they are strung together and justified is new and supports approaches attempting to derive emergent spacetime structures from correlations of quantum degrees of freedom.

1. Introduction

The basic discipline underlying the current information age, namely information theory, is not a physical theory (in contrast to thermodynamics ruling the age of steam engines). Nevertheless, the information paradigm is now also permeating (at least part of) physics, offering novel perspectives on old and new problems, specifically in quantum and gravitational physics, and thereby attaining direct physical relevance. After all, physics without information is not possible; a description of the world requires to gather information about it. The information paradigm in physics is by construction very operational. From an information-theoretic perspective, physical systems are information carriers which can be used to acquire, store and communicate information. This perspective is practically realised in the field of quantum information which takes quantum systems to perform all sorts of information-theoretic tasks with them.

Here we are less interested in practical applications within the limits of quantum information; rather, we wish to ask how far the information paradigm, and specifically the picture of physical systems as information carriers, can possibly lead us in understanding the physical content of quantum theory and general relativity from a new angle and what it may suggest for the construction of a new physical theory incorporating both. So the questions we will ask (and partially address) are much more basic than typically investigated in quantum information.

In particular, we will argue, among others, using an informational reconstruction of quantum theory, that its physics lies in the relation between the observer and observed systems; one can regard quantum theory as a law book which governs the observer's acquisition of information from physical systems in terms of *how* and *how much* information is accessible. In quantum theory the information flow is physical, encoded in physical systems that can undergo their own dynamics, such as electrons. This physical information flow is under the focus of the theory but the spatiotemporal stage on which such information flows is externally given.

By contrast, we will argue that much of the physics of general relativity lies in the communication relations among observers (or more generally systems); general relativity can be understood as a law book governing *where* and *when* information can be communicated. In general relativity the information flow among idealised observers, at least in terms of light signals, is, strictly speaking, external, while the stage on which this information flows, i.e. spacetime, is dynamical and the focus of the theory. Alice can send a light signal to Bob, depending on their causal relations, but the influence of this light signal on the dynamical spacetime they live in is ignored, no matter how energetic it is.

Therefore, in both quantum theory and general relativity the spatiotemporal structure and (at least part of) the information flow among observers live a life on their own in the sense that either one or the other is dynamical, but not both; they do not interact other than that the causal structure of spacetime determines from where to where information can be communicated. This division is, of course, perfectly legitimate for the purposes of either theory. For the purpose of quantum theory, to describe localised interactions with microphysical systems, and, likewise, for the purpose of general relativity, to describe the large scale structure of spacetime, the back-reaction of the localised information flow among observers on spacetime can be safely ignored.

But what is irrelevant to the large scale structure or to communication *within* it may turn out to be very relevant for the small scale spatiotemporal structure. A theory consistently incorporating both quantum theory and general relativity should eliminate the above division and any external ingredients; both the information flow among 'all degrees of freedom' and the spatiotemporal structure should be internal, governed by the laws of the theory. Does a quantum theory of gravity therefore include the fundamental physical laws which somehow govern the information exchange among 'all degrees of freedom' simultaneously in terms of *how* and *how much* information and *where* and *when* it can be communicated?[1]

To set our discussion into a broader context, we emphasise that here we will not delve into such depths of the information paradigm as inquiring into whether *"... all things physical are information-theoretic in origin ..."*, as in Wheeler's it-from-bit aphorism [1], or whether the opposite should be true and all information derives from physical entities (bit-from-it) [2]. Instead, we shall be pragmatic. At the level of our discussion, an informational perspective is largely universal in that it depends little on the precise (physical incarnation of the) underlying degrees of freedom. By itself it will not tell us much about the concrete physics of information carriers (is the qubit incarnated as an electron spin or a photon?, what are their interactions?,...). But physics clearly puts constraints on what and how information-theoretic processes can be physically implemented and, vice versa, informational properties can impose constraints on the physics. It is in this sense that an informational perspective, whether fundamental or not, can tell us something about the architecture of a physical theory and how to reason about the world.

The remainder of this manuscript is organised as follows. In sec. 2, we summarise how the formalism of quantum theory can be reconstructed from rules constraining an observer's acquisition of information about physical systems. We focus mostly on the conceptual discussion and its implications. In sec. 3, we offer a purely qualitative analysis – but with reference to examples – of how classical spacetime structure can, in principle, be extracted from communication relations among observers. Combining the informational insights from quantum

[1] To avoid a misunderstanding: quantum field theory on curved spacetimes provides a framework which incorporates quantum theory and general relativity. However, the external nature of communication among idealised observers on a classical geometry is the same as in general relativity. A fully internal picture of information flow and spatiotemporal structures might only arise in quantum gravity.

theory and general relativity, we conclude in sec. 4 with the proposal that the fundamental architecture of spacetime should be viewed as a locally finite network of systems acquiring and exchanging information with smooth geometry emerging from the information flow only in a coarse-grained limit. We further argue that this picture should be concretised into a systematic top-down approach to quantum gravity to complement extant bottom-up approaches and help close the scale gap between the unknown Planck scale physics and the familiar physics of quantum (field) theory and general relativity. The informational perspective on gravity is perhaps less established than that on quantum theory. But it has an equal potential to revolutionise our understanding of spacetime architecture – and plenty of developments in this direction are currently taking place, on some of which we comment.

2. Quantum theory from rules on information acquisition

In this section, we shall argue that quantum theory can be understood as a law book, governing an observer's acquisition of information from physical systems. The idea behind this is not new. It appears in various guises in various informational interpretations of quantum theory, such as Hartle's interpretation [3], Rovelli's *relational quantum mechanics* [4], the Brukner-Zeilinger interpretation [5–7], QBism [8], and many others. However, here we shall restrict attention to a reconstruction of quantum theory [9–11] which manifests this perspective perhaps most concretely.

The paradigmatic and a priori counter-intuitive phenomena of special relativity, such as relativity of simultaneity, time dilation and Lorentz contraction, are naturally explained by the seemingly innocent principle of relativity: "all the laws of physics are the same for all inertial observers." Over the years the question has arisen whether one can similarly explain the paradigmatic and a priori counter-intuitive phenomena of quantum theory (and its entire formalism) from equally natural physical principles. Starting with Hardy's seminal work [12], this has given rise to a whole wave of recent reconstructions of (finite dimensional) quantum theory from operational axioms [9–19]. While there are significant differences among these various quantum reconstructions as far as their ingredients and starting points are concerned, they all have contributed to answering Wheeler's question "how come the quantum?" [1].

Nevertheless, the situation with quantum theory is more complicated than with special relativity; all of these derivations have come short of providing an explanation of quantum theory as 'intuitive' as that of special relativity through the relativity principle. The reasons for this are manifold. While it is possible that we simply lack the 'right' idea, it is quite conceivable that there just isn't such an 'intuitive' explanation of quantum theory, possibly because dealing with probabilistic structures seems less intuitive than with velocities, positions and light signals. Moreover, of course, the level of 'intuitiveness' depends on which ingredients one takes for granted when formulating the principle(s).

For example, a lot of non-trivial assumptions go into the relativity principle: one presupposes a landscape of mechanical theories in which Newton's axioms make sense (the notion of *inertial* observer requires Newton's first law). In particular, the continuous three-dimensional Euclidean space, in which each inertial observer sees the dynamics taking place – and non-accelerated objects moving on straight lines –, as well as its symmetries are an *input*, not an output of the construction of special relativity.

By contrast, most quantum reconstructions arguably start from a much more puristic level. They do *not* presuppose other physical laws but make very basic statements about what sort of operations (e.g., which kinds of measurements, transformations and information-processing tasks) an observer can or cannot perform on physical systems. Crucially, these statements make *no* reference to any concrete physics and, in particular, to *what* physical properties the observer can measure. Such statements come *before* a concrete physical law and give rise to the theory rather as a universal framework to which all other, concrete physics of the systems has to be

IOP Conf. Series: Journal of Physics: Conf. Series **880** (2017) 012014 doi:10.1088/1742-6596/880/1/012014

subjected and by means of which the observer can reason about his world and whatever the concrete observed system. Yet, the dimension, continuity and geometry of state spaces in which the quantum dynamics takes place as well as the full set of transformations thereon are an *output*, not an input of quantum reconstructions.

So one could argue that quantum reconstructions come a substantially longer way and from more elementary assumptions than the derivation of special relativity from the relativity principle. It is thus perhaps not surprising that quantum reconstructions are less 'intuitive'. However, they are no less important.

In the remainder we now outline exclusively a reconstruction [9–11] which, first of all, makes the perspective explicit that quantum theory, at least for systems of arbitrarily many qubits, is a law book governing an observer's acquisition of information. Namely, it derives quantum theory (its formalism, state spaces, unitaries, projective measurements) from rules constraining an observer's acquisition of information from physical systems which we take as information carriers. Secondly, and in relation to the discussion above, it offers the hitherto most compelling operational explanations for paradigmatic and a priori counter-intuitive phenomena of quantum theory such as entanglement, monogamy, non-locality and generally quantum correlations.

Part of this reconstruction has been inspired by Rovelli's *relational quantum mechanics* [4] and the Brukner-Zeilinger informational interpretation [5,6] so that the end result can be regarded as a completion of such ideas for qubit systems. In particular, we follow the premise that we shall only speak about information that the observer has access to, resulting in the quantum state assuming the role of a state of information. However, we emphasise that, while the reconstruction has been inspired by these interpretations of quantum theory, it does not actually rely on them so this should not discourage a reader unsympathetic with these interpretations.

Our focus here will lie on explaining the physical intuition behind the reconstruction and on offering a conceptual overview over the most important reconstruction steps. Our aim here is not, however, to provide a technically rigorous account of the reconstruction as this is done elsewhere. For technical details and a precise formulation of the below, the reader is encouraged to consult the actual reconstruction in [9, 10] or the accessible review in [11].

The reconstruction proceeds in three main steps. First, we have to abandon the mathematical formalism of quantum theory (after all, we want to recover it) and construct a very general landscape of alternative operational theories containing quantum theory and ideally also classical probability theory. These alternative theories all have in common that they describe an observer's acquisition of information about physical systems in some logically conceivable worlds. This theory landscape is an analogue of the landscape of mechanical theories underlying the construction of special relativity. The second aim is to find physical statements, here as rules constraining an observer's acquisition of information, which single out quantum theory within this landscape. Third, one has to actually derive quantum theory from these operational axioms.

2.1. Landscape of alternative theories for generalised bit systems

We need a notion of information acquisition without reference to quantum theory. The basic idea is to formulate the acquisition of information of an observer O about an ensemble of systems $\{S_a\}_{a=1}^n$ in terms of an interrogation of the systems with questions. This encodes a general notion of measurement with answers representing apparatus-registered outcomes. The further idea is to develop an elementary calculus for these questions in terms of whether they provide (in)dependent information and whether O may know their answers simultaneously or not. This requires to introduce an operational notion of outcome probabilities for these questions. The theories in the landscape are formulated in terms of these questions, probabilities and in terms of how these probabilities may change in time; quantum theory is a special one of them.

As an aside, it is worth noting that as a crucial difference to various earlier reconstructions, such as [12–16], which use the framework of *general probabilistic theories* and derive quantum theory rather abstractly from properties of states, the reconstruction outlined here derives the most crucial quantum properties from the new question calculus. Through these questions and their relations it is more directly connected to what can be measured in the laboratory and thereby gives an operationally more compelling explanation of various quantum phenomena.

Every system S comes with a specific set of questions $\mathcal{Q}$ which can be meaningfully answered by it. Ultimately, since we are interested in the most elementary systems, generalised bits, we shall henceforth assume $\mathcal{Q}$ to only contain *binary*, i.e. yes-no questions.[2] But this assumption is also in line with the idea that all things physical derive their existence from apparatus-elicited yes or no answers as in Wheeler's it-from-bit paradigm [1] or von Weizsäcker's ur-theory [20].

We also need the notion of *composite* systems; S_A and S_B with question sets $\mathcal{Q}_A, \mathcal{Q}_B$ form a composite system S_{AB} if

$$\mathcal{Q}_{AB} = \mathcal{Q}_A \cup \mathcal{Q}_B \cup \tilde{\mathcal{Q}}_{AB}, \tag{1}$$

and $\tilde{\mathcal{Q}}_{AB}$ only contains compositions, via some logical connectives, of questions from $\mathcal{Q}_A$ and $\mathcal{Q}_B$. That is, a composite system admits inquiries about individual subsystem and composite properties. Composite systems with more than two subsystems can be defined recursively.

The systems are prepared in some preparation device. Each way of preparing the $\{S_a\}_{a=1}^n$ shall yield a specific statistics over the answers to the $Q \in \mathcal{Q}$ (for n sufficiently large) which O can record. Through his experiments O is assumed to have developed a theoretical model for $\mathcal{Q}$ and for the set Σ of all the possible answer statistics for all $Q \in \mathcal{Q}$ for all preparations. A belief updating, according to this model and any prior information on the way of preparation, enables O to assign, for the next S_a to be interrogated, a prior probability y_i that S_a's answer to $Q_i \in \mathcal{Q}$ will be 'yes'.

We assume the y_i encode everything O could possibly say about the future outcomes to arbitrary experiments on S_a ($\mathcal{Q}$ is tomographically complete in this regard). This makes it natural to identify O's 'catalogue of knowledge' about the given S_a, i.e. the collection of $\{y_i\}_{\forall Q_i \in \mathcal{Q}}$, with its (prior) state. Similarly, Σ constitutes the state space of S_a.

Which prior state would O begin with in a belief updating when he knows 'nothing' about the preparation of the $\{S_a\}$? To this end, we assume there to exist a distinguished state in Σ defined by $y_i = \frac{1}{2}, \forall Q_i \in \mathcal{Q}$. (This assumption is a constraint on the *pair* $(\mathcal{Q}, \Sigma)$.) This state corresponds to O's best guess that all outcomes are equally likely and will be referred to as the *state of no information*. Clearly, for this state to make operational sense, there must exist a preparation which yields completely random answer statistics for all Q_i in $\mathcal{Q}$.

The state $\{y_i\}_{\forall Q_i \in \mathcal{Q}}$ is the prior state for the next S_a of the ensemble to be interrogated (but also coincides with the state O assigns to the ensemble $\{S_a\}$). After the interrogation of S_a, O's information about this specific S_a as well as the ensemble $\{S_a\}$ may change. This has to be reflected in consistent rules by means of which O updates both the (now *posterior*) state of the specific S_a *and* the ensemble state of $\{S_a\}$ (which now is the prior state of the next system S_b from the ensemble to be interrogated).[3] We assume O to have such rules so that he is able to consistently update the states he assigns to the systems according to the received answers.

Now that O is equipped with the tools to update states, we are able to make sense of basic question relations: can O know the answers to various questions simultaneously, does one answer imply another? The notion of question independence requires a special state. $Q_i, Q_j \in \mathcal{Q}$ are

(maximally) independent if, after having asked Q_i to S in the state of no information, the posterior probability $y_j = \frac{1}{2}$. That is, if the answer to Q_i relative to the state of no information tells O 'nothing' about the answer to Q_j.

[2] E.g., in the case of quantum theory, such a question could read "is the spin of the electron up in x-direction?."
[3] This requires a distinction of single and multiple shot interrogation [9, 11].

dependent if, after having asked Q_i to S in the state of no information, the posterior probability $y_j \neq \frac{1}{2}$. (If $y_j = 0$ or 1 they are maximally dependent.) That is, if the answer to Q_i relative to the state of no information gives O at least partial information about the answer to Q_j.

(maximally) compatible if O may know the answers to both Q_i, Q_j simultaneously, i.e. if there exists a state in Σ such that y_i, y_j can be simultaneously 0 or 1.

(maximally) complementary if every state in Σ which features $y_i = 0, 1$ necessarily implies $y_j = \frac{1}{2}$. Notice that complementarity implies independence (but not vice versa).

These relations shall be symmetric; e.g. Q_i is independent of Q_j if and only if Q_j is independent of Q_i, etc. These concepts yield an elementary calculus for questions in terms of their (in)dependence and compatibility which is convenient for encoding many quantum structures [9,11]. Since all propositions must have operational meaning, O is allowed to apply classical rules of inference (in terms of Boolean logic) exclusively to sets of mutually compatible questions.

The parametrization of a state by $\{y_i\}_{\forall Q_i \in \mathcal{Q}}$ is unpractical. Is there a complete description of the systems in terms of a smaller, yet *informationally complete* set of questions which encodes all the independent information they possibly carry? We assume this to be the case; any set of pairwise independent questions which is maximal in the sense that no further question from $\mathcal{Q}$ can be added to it without destroying pairwise independence shall be informationally complete. Such a set will be countable, so the state of a system can be represented by a vector $\vec{y} = (y_1, y_2, \ldots, y_D)$. All such maximal sets have the same number D of elements [9].

We also need a suitable measure to quantify O's information; O's information about S_a's answer to Q_i shall be a function $\alpha(y_i)$ of the corresponding probability only with $0 \leq \alpha(y_i) \leq 1$ `bit` and $\alpha(y) = 0$ `bit` $\Leftrightarrow y = \frac{1}{2}$ and $\alpha(1) = \alpha(0) = 1$ `bit`. The precise form of $\alpha(y_i)$ follows from the quantum postulates. For O's total information about S_a we make the additive ansatz

$$I(\vec{y}) := \sum_{i=1}^{D} \alpha(y_i). \tag{2}$$

Finally, O may subject the systems to interactions so that the state $\vec{y}$ can change in time. Any legitimate time evolution must map the state space to itself. The set of all possible time evolutions to which O has access is denoted by $\mathcal{T}$ and forms part of O's theoretical world model.

2.2. Rules on O's acquisition of information which single out qubits

We now consider a composite system S_N of N generalised bits with $(\mathcal{Q}_N, \Sigma_N, \mathcal{T}_N)$. The aim is to formulate rules on O's acquisition of information which single out the quantum theory of N qubits within the theory landscape. Here we give all rules in colloquial form and only the first three also in their technical formulation. For further details on the rules, see [9–11].

Elementary systems admit only limited information to an observer: e.g., an electron allows him to only know a single binary proposition such as its spin in x-direction at a time, but nothing fully independent of it. The maximally accessible amount of information shall characterise S_N.

Rule 1. (Limited Information) "The observer O can acquire maximally $N \in \mathbb{N}$ *independent* `bits` of information about the system S_N at any moment of time."
There exists a maximal set Q_i, $i = 1, \ldots, N$, of N mutually maximally independent and compatible questions in $\mathcal{Q}_N$.

There is also Bohr's complementarity: systems admit more independent propositions than what the information limit allows them to answer at a time. The observer's ignorance is reflected in the unanswered properties being random. For example, an observer may also measure the spin of the electron in y-direction. The price is total ignorance about its spin in x- and z-directions. Being a composite system, complementarity shall exist at each subsystem level of S_N. For

IOP Conf. Series: Journal of Physics: Conf. Series **880** (2017) 012014 doi:10.1088/1742-6596/880/1/012014

example, suppose the N Q_i of the maximal set of rule 1 correspond to N questions about the individual gbits making up S_N. Then to each such Q_i there should exist a complementary Q'_i which is compatible with all $Q_{j\neq i}$. We require this property for any set abiding by rule 1.

Rule 2. (Complementarity) "The observer O can always get up to N *new* independent `bits` of information about the system S_N. But whenever O asks S_N a new question, he experiences no net loss in his total amount of information about S_N."
There exists another maximal set Q'_i, $i = 1, \dots, N$, of N mutually maximally independent and compatible questions in $\mathcal{Q}_N$ such that Q'_i, Q_i are maximally complementary and $Q'_i, Q_{j\neq i}$ are maximally compatible.

Rules 1 and 2 offer a mathematical implementation of earlier conceptual ideas of Rovelli [4] and Zeilinger and Brukner [5,6].

Next, O shall also not gain or lose information *without* asking questions.

Rule 3. (Information Preservation) "The total amount of information O has about (an otherwise non-interacting) S_N is preserved in-between interrogations."
$I(\vec{y})$ is constant *in time in-between interrogations for (an otherwise non-interacting) S_N.*

In this article, we will not need the mathematical formulation of the remaining rules [9–11] and only give them colloquially. We impose no further restrictions on time evolution of states other than that it be continuous and consistent with the other rules (and obviously the landscape).

Rule 4. (Time Evolution) "O's 'catalogue of knowledge' about S_N evolves *continuously* in time in-between interrogations and every consistent such evolution is physically realisable."

Probabilities for 'yes'-answers by S_N to any $Q \in \mathcal{Q}_N$ and in any state can be derived within the landscape [10]. O shall be allowed to ask S_N *any* question which 'makes (probabilistic) sense'.

Rule 5. (Question Unrestrictedness) "Every question which yields legitimate probabilities for every way of preparing S_N is physically realisable by O."

Remarkably, these five rules cannot distinguish complex and real numbers: qubit and rebit quantum theory (two-level systems over real Hilbert spaces) are the two solutions for the triple $(\mathcal{Q}_N, \Sigma_N, \mathcal{T}_N)$ surviving these rules [9,10]. But the following additional rule eliminates rebits.

Rule 6. (Tomographic Locality) "O can determine the state of the composite system S_N by interrogating only its subsystems (and doing statistics over the outcomes)."

As shown in [9,10] and reviewed in [11], rules 1–6 single out quantum theory in the landscape: they yield exactly the right state spaces, unitary transformations, projective measurements, etc.

Theorem. ([9, 10]) *The only solution to rules 1–6 is qubit quantum theory where*

- *Σ_N is the space of $2^N \times 2^N$ density matrices over $\mathbb{C}^{2^N}$,*
- *states evolve unitarily according to $\mathcal{T}_N \simeq \mathrm{PSU}(2^N)$ and the equation describing the state dynamics is the von Neumann evolution equation,*
- *$\mathcal{Q}_N$ is the set of projective measurements onto the $+1$ eigenspaces of N-qubit Pauli operators*[4] *and the probability for $Q \in \mathcal{Q}_N$ to be answered with 'yes' in some state is given by the Born rule for projective measurements.*

[4] A Hermitian operator on $\mathbb{C}^{2^N}$ is a Pauli operator iff it has two eigenvalues ± 1 of equal multiplicity.

2.3. A glimpse of the reconstruction steps

The reader only interested in the main flow of arguments can safely skip this more involved subsection as the subsequent discussion will not rely on it. Its purpose is to keep the article relatively self-contained by offering a taste of the informational insights into, e.g., entanglement, monogamy and quantum non-locality gained through the individual reconstruction steps.

The first steps are to construct informationally complete sets of questions for S_N, exploiting the independence, compatibility and complementarity structure on $\mathcal{Q}_N$ introduced in rules 1 and 2. The pairwise independent questions of an informationally complete set $\{Q_1, Q_2, \dots, Q_{D_1}\}$ for S_1 must be mutually complementary for otherwise O could break the 1 `bit` limit of rule 1. Also, $D_1 \geq 2$ by rule 2; this will become the dimension of the Bloch ball. Since S_N is composite for $N > 1$, one has to clarify how to build up composite questions with logical connectives from single system questions. Pairwise independence of an informationally complete set implies that one has to use the XNOR $\leftrightarrow$ (or equivalently its negation, the XOR) connective [9,11]. For instance, an informationally complete set for S_2 can be shown to be given by $\{Q_i, Q'_j, Q_{ij}\}_{i,j=1,\dots,D_1}$, where Q_i, Q'_j are the individual questions of the two informationally complete sets associated to the two single systems making up S_2 and

$$Q_{ij} := Q_i \leftrightarrow Q'_j \tag{3}$$

are correlation questions, standing for "are the answers to Q_i, Q'_j the same?."[5]

It is convenient to encode the compatibility relations graphically. Individual questions are represented as *vertices* and bipartite correlation questions as *edges* between them, e.g.

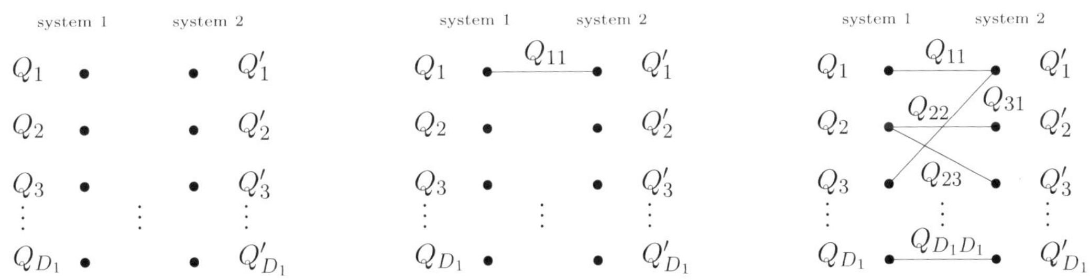

Rules 1 and 2 entail that vertices are compatible with edges if and only if they are vertices of the latter and edges are compatible if and only if they do not intersect in a vertex [9]; e.g., Q_1 is compatible with Q_{11} and complementary to Q_{22}, while Q_{11} and Q_{31} are complementary and Q_{11} and Q_{22} are compatible.

This implies *entanglement*: nonintersecting edges are independent, compatible and no individual question can be simultaneously compatible with multiple of them. Accordingly, the maximally accessible $N = 2$ *independent* `bits` about S_2 of rule 1 can be reached by O through correlation questions only, at the expense of any information about individual questions, yielding Schrödinger's notion of entanglement (*"...the best possible knowledge of a whole does not necessarily include the best possible knowledge of all its parts..."* [21]). For example, a state with $Q_{11} = Q_{22} =$ 'yes' corresponds in quantum theory to a Bell state with spins of qubits 1 and 2 correlated in x- and y-direction. This can be generalised to arbitrary entangled states [9]. Entanglement thereby follows directly from an information limit and complementarity.

But also *monogamy* follows compellingly from rules 1 and 2. Let S_A, S_B, S_C be three generalised bits. Suppose S_A and S_B are maximally entangled so that O has saturated the $N = 2$ *independent* `bit` limit about the bipartite composite S_{AB} inside the tripartite composite S_{ABC}.

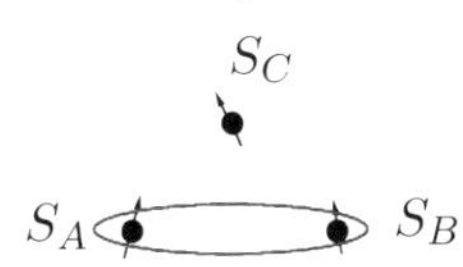

[5] $Q_{ij} =$'yes' if $Q_i = Q'_j$ and $Q_{ij} =$'no' otherwise. In quantum theory, Q_{ij} corresponds to "are the spins of qubit 1 in i- and of qubit 2 in j-direction correlated?" and $\leftrightarrow$ to the tensor product in $\sigma_i \otimes \sigma_j$ where σ_i is a Pauli matrix.

IOP Conf. Series: Journal of Physics: Conf. Series **880** (2017) 012014 doi:10.1088/1742-6596/880/1/012014

O can then reach the $N = 3$ *independent* bit limit for S_{ABC} only by inquiring individual information about S_C. The maximally entangled pair S_{AB} can thus *not* be entangled with any other system S_C. Using the question calculus and informational *monogamy inequalities* for $N \geq 3$, one can make this heuristic monogamy explanation rigorous [9].

Rules 1 and 2 also imply $D_1 \leq 3$ (and, together with rule 6, $D_1 = 3$) [9, 11]. The reason is essentially as follows: the diagonal correlation questions $Q_{11}, Q_{22}, \ldots, Q_{D_1 D_1}$ for S_2 are not only pairwise independent and pairwise compatible, but also mutually compatible so that O can know the answers to all D_1 such questions simultaneously [9]. But they cannot be also mutually independent for otherwise O could exceed the $N = 2$ *independent* bit limit of rule 1. So these mutually compatible questions must be logically related; e.g., it turns out [9, 11] that either

$$Q_{jj} = Q_{11} \leftrightarrow Q_{22}, \qquad \text{or} \qquad Q_{jj} = \neg(Q_{11} \leftrightarrow Q_{22}), \qquad j = 3, \ldots, D_1 \tag{4}$$

so that if $D_1 > 3$, pairwise independence of Q_{jj} would be violated for $j \neq 1, 2$.

Such line of reasoning also elucidates non-locality: similar arguments [9, 11] show that either

$$Q_{11} \leftrightarrow Q_{22} = Q_{12} \leftrightarrow Q_{21}, \qquad \text{or} \qquad Q_{11} \leftrightarrow Q_{22} = \neg(Q_{12} \leftrightarrow Q_{21}). \tag{5}$$

In terms of the individual questions Q_i, Q'_j, the first case (without relative negation) reads

$$(Q_1 \leftrightarrow Q'_1) \leftrightarrow (Q_2 \leftrightarrow Q'_2) = (Q_1 \leftrightarrow Q'_2) \leftrightarrow (Q_2 \leftrightarrow Q'_1). \tag{6}$$

Any distribution of simultaneous truth values over the Q_i, Q'_j satisfies (6). It is a *classical logical identity* and therefore compatible with *local* hidden variables for Q_i, Q'_j [9]. However, it contains complementary questions and therefore conflicts with the premise of section 2.1 which allows O to apply classical rules of inference exclusively to mutually compatible questions. Since one of the two cases (5) *must* hold and the first case of classical logic is ruled out, we infer that the second case $Q_{11} \leftrightarrow Q_{22} = \neg(Q_{12} \leftrightarrow Q_{21})$ must be correct. Indeed, it is consistent with the theory landscape and rules 1–6 and also precludes a local hidden variable interpretation [9].

In this manner one can reconstruct informationally complete sets and the correlation structure for arbitrarily many qubits [9,11]. In particular, one can show that an informationally complete set for S_N, and thereby the state space Σ_N, is $(4^N - 1)$-dimensional. This is the correct number of degrees of freedom in an N-qubit density matrix.

The next step consists in employing rules 3 and 4 which imply that O's total information $I(\vec{y})$ is a 'conserved charge' of time evolution and that any evolution consistent with the landscape and rules is physically realisable to show [9, 11]:

(a) $\mathcal{T}_N$ is a group and acts *linearly* on states represented as Bloch vectors $\vec{r} = 2\,\vec{y} - \vec{1} \in \mathbb{R}^{4^N-1}$

$$\vec{r}(\Delta t + t_0) = T(\Delta t)\,\vec{r}(t_0), \qquad T(\Delta t) \in \mathcal{T}_N \tag{7}$$

(b) the information measure is quadratic $\alpha(y_i) = (2\,y_i - 1)^2$. Thus, O's total information (2)

$$I_N(\vec{y}) = \sum_{i=1}^{4^N-1} (2\,y_i - 1)^2 = |\vec{r}|^2 \tag{8}$$

coincides with the square norm of the Bloch vector and *not* the Shannon entropy. A similar quadratic measure has earlier been proposed by Brukner and Zeilinger [6, 22, 23].

Together this entails $\mathcal{T}_N \subset \mathrm{SO}(4^{\mathrm{N}} - 1)$ since time evolution must be connected to the identity and, by rule 3, preserve $I_N(\vec{y})$.

IOP Conf. Series: Journal of Physics: Conf. Series **880** (2017) 012014 doi:10.1088/1742-6596/880/1/012014

We provide an exemplary glimpse of how to reconstruct the state space and set of unitaries for a single qubit S_1. As mentioned, it can be shown that $D_1 = 3$ so that an informationally complete set contains three questions $\{Q_1, Q_2, Q_3\}$. It is a maximal set of mutually complementary questions: no further question can be added without destroying mutual complementarity in the set. The information limit of rule 1 then implies [9, 11]

$$0\,\texttt{bit} \leq I_{N=1}(\vec{y}) = r_1^2 + r_2^2 + r_3^2 = (2\,y_1 - 1)^2 + (2\,y_2 - 1)^2 + (2\,y_3 - 1)^2 \leq 1\,\texttt{bit}. \tag{9}$$

The 1-bit-state $\vec{y}_* = (1, 0, 0)$ – a state of maximal information or pure state of S_1 – lies in Σ_1 thanks to rule 1. As noted above, $\mathcal{T}_1 \subset \mathrm{SO}(3)$. In fact, the full group SO(3) acting on $\vec{y}_*$, according to (7), produces the entire Bloch sphere of quantum pure states all of which are consistent with the rules and theory landscape. Accordingly, rule 4 entails $\mathcal{T}_1 = \mathrm{SO}(3) \simeq \mathrm{PSU}(2)$ and that all states on the Bloch sphere, as states of maximal information, are legitimate. States of non-maximal information and Σ_1 as the complete Bloch *ball* of qubit states, incl. the state of no information at the center, are obtained through operational convexity arguments [9, 11].

In a nutshell: *for pure states of S_1, the maximal mutually complementary set carries* exactly *1 bit of information and this is a conserved charge of time evolution (rule 3) which defines both the unitary group and the state space.* Remarkably, this generalises to S_2: the unitary group PSU(4) and set of density matrices for two qubits follow likewise from six conserved informational charges, each associated to a maximal complementarity set [10,11]. The reconstruction of $\Sigma_N, \mathcal{T}_N$ for $N > 2$ then exploits universality results from quantum computation [24,25], establishing that local and bipartite unitaries (which must be in $\mathcal{T}_N$ since S_N is composite) generate the entire unitary group $\mathrm{PSU}(2^{\mathrm{N}})$ [10].

Ultimately, one obtains quantum theory in its *adjoint* (i.e., Bloch vector) representation [9–11]: the derived set Σ_N of states, represented as Bloch vectors $\vec{r}(t) \in \mathbb{R}^{4^N-1}$, is equivalent to the hermitian representation of quantum states, namely the set of density matrices $\rho(t) = \frac{1}{2^N}(\mathbb{1} + \vec{r}(t)\cdot\vec{\sigma})$ over $\mathbb{C}^{2^N}$. Here $\vec{\sigma}$ is a vector of a basis of Pauli operators $\sigma_{\mu_1\cdots\mu_N} = \sigma_{\mu_1}\otimes\cdots\otimes\sigma_{\mu_N}$, $\mu_i = 0, 1, 2, 3$, except all $\mu_i = 0$ simultaneously, where $\sigma_0 = \mathbb{1}$ and $\sigma_1, \sigma_2, \sigma_3$ are Pauli matrices. Similarly, $T(t) \in \mathcal{T}_N = \mathrm{PSU}(2^{\mathrm{N}})$, acting on states according to (7), is equivalent to the set of unitary transformations $\rho(t) = U(t)\,\rho(0)\,U^\dagger(t)$, $U(t) = e^{-i\,H\,t} \in \mathrm{SU}(2^{\mathrm{N}})$ for some hermitian operator H. This, in turn, is equivalent to $\rho(t)$ satisfying the von Neumann evolution equation

$$i\frac{\partial\rho}{\partial t} = [H, \rho]. \tag{10}$$

Since we are only working with operationally accessible information it is not surprising that we recover the theory in its adjoint rather than Hilbert space formulation which contains operationally redundant information such as global phases or Hilbert space vector lengths.

We abstain from outlining how the questions in $\mathcal{Q}_N$ are equivalent to projective measurements of Pauli operators $\sigma_{\mu_1\ldots\mu_N}$ and how the Born rule is recovered. This is demonstrated in [10, 11].

2.4. Implications of the reconstruction

This reconstruction not only offers an informational explanation for the architecture of quantum theory but also unravels previously unknown 'informational charges' that characterise the unitary group and sets of pure states. It provides a compelling elucidation of its paradigmatic phenomena such as entanglement, monogamy and non-locality from limited accessible and existence of complementary information. Overall, it yields the quantum theory of qubits as a universal framework, a law book which governs the acquisition of information of an observer from elementary systems. The framework does not by itself specify what the information and the concrete interpretation is. The concrete physics has to be subjected to this universal framework, but provides the degrees of freedom and concrete interpretation. It identifies whether the yes-no questions correspond to, e.g., electron spin, polarization, the occupation of one of two energy levels, etc. The framework then governs how this information is accessible to the observer; e.g., how much information and which properties the observer can know at once.

The argumentation should be extended to systems with *continuous* measurement outcomes as in quantum *mechanics* or even quantum field theory before solid claims can be made that quantum theory can in general be understood as a law book governing an observer's acquisition of information about physical systems, incl. fields. But the reconstruction delivers a proof of principle and informational interpretations [3, 4, 7, 8] do not stop at qubit systems in any case. Alternatively, followers of more radical proposals such as Wheeler's it-from-bit paradigm [1] or von Weizsäcker's ur-theory [20] may argue such an extension to be unnecessary. After all, these proposals posit all continuous physics to be just an approximation to the real bit-based physics, embodied in apparatus-elicited yes or no answers to which all physics could be reduced.

Of course, classical probability theory in the form of classical bits, as any other theory in the landscape of alternative theories, likewise constitutes a framework governing an observer's acquisition of information. And for all these theories in the landscape, a system's state is a state of information, the observer's 'catalogue of knowledge' about it. However, there is a crucial difference between classical probability theory and quantum theory: While both account for an incomplete knowledge of the preparation of the system through mixed rather than pure states – a lack of information which could, in principle, be avoided through more accurate measurement – what really distinguishes the two theories is complementarity. Thanks to complementarity an observer will necessarily have 'incomplete' information; in the language of the above reconstruction, a system cannot answer all questions simultaneously. Even for a system in a pure state, i.e. a state of maximal knowledge about the preparation, the observer will always find a complete randomness of some properties. This inherent randomness – or 'missing' information – is responsible for the multitude of interpretations of the system's state and, if it is a state of information (not all interpretations agree on this!), of what exactly this information is about.

Before elaborating on this, it is worth recalling why such profound interpretational issues do not arise in classical probability theory or classical statistical physics. In classical physics a phase space point corresponds to a true state of the system, a state of reality, that can, in principle, be measured. Of course, in practice this state can often not be determined due to experimental limitations which is one of the reasons for the necessity of statistical physics. Statistical states as classical probability distributions over phase space are so-called epistemic states: they represent an observer's knowledge about the *real* state of the system, given the experimental limitations, but they are not real states themselves. Indeed, they can be determined using the maximum-entropy principle from information theory to estimate the least biased probability distribution with regard to the missing information [26]. The epistemic state can be taken as the observer's best bet for the real physics. Only in the unpractical scenario that the observer measures the system's state to arbitrary precision will the epistemic state, now as a delta-function on phase space, be in one-to-one correspondence with its real state. Such a delta-function state is a pure state, a state of maximal knowledge which encodes a complete description of the real physical state. Ignoring the usual debates about the meaning of probabilities (frequentist, Bayesian,...), the interpretation of a statistical state in classical physics and in classical probability theory as an epistemic state is uncontroversial. But for quantum theory the situation is more intricate.

It was hoped by various authors that a successful reconstruction of quantum theory would finally settle the debate about its interpretation, see e.g. [4, 5, 8]. After all, what could be better for manifesting an interpretation than deriving the theory from its specific perspective on it? However, it is safe to say that so far this hope has not materialised, very much thanks to complementarity which is at the heart of the whole interpretation debate. Where is the 'missing' information which is reflected in the inherent quantum randomness? Does it exist at

IOP Conf. Series: Journal of Physics: Conf. Series **880** (2017) 012014 doi:10.1088/1742-6596/880/1/012014

all? Informational interpretations [3–8], connecting with Bohr's ideas (*"It is wrong to think that the task of physics is to find out how Nature is. Physics concerns what we can say about Nature"* [27]), will hold that there simply is no 'missing' information because a system's properties are not observer-independent. Precisely that is disputed by realist interpretations, claiming there to be an observer-independent reality and appropriate hidden variables carrying the 'missing' information, e.g. see [28–32]. It is not evident that the question as regards the 'missing' information will ever be convincingly settled so long as quantum theory continues to accurately describe experiments and no deeper theory superseding it was discovered. In that case the 'missing' information is neither needed for the description of experiments nor accessible and the interpretation of its status remains exactly that – an interpretation.

For instance, the present quantum reconstruction is established in the spirit of Rovelli's relational interpretation [4] and the Brukner-Zeilinger informational interpretation [5,6]. Modulo a precise take on the origin of probabilities it is also generally compatible with ideas underlying QBism [8]. We followed the premise to only speak about the information accessible to the observer and the interpretation of the state as a state of information is thereby built-in by construction. By default we thus said nothing about whether or not some appropriate hidden variables could give rise to the information accessible to the observer by determining the systems' answers. This was simply not necessary and from the perspective of this reconstruction hidden variables can thereby be regarded as superfluous, but nothing fundamentally inhibits them. Clearly, this reconstruction, while close in spirit to the informational interpretations [4–8], is therefore also compatible with hidden variables[6] and so the issue of the interpretation is by no means settled by it (regardless of the precise form of the axioms).

For similar reasons, it is hard to foresee how any further reconstruction could single out the 'right' interpretation. The quantum formalism simply is open to a multitude of interpretations and there may be a similar number of conceptually inequivalent ways to reconstruct it.

Note, however, that stringent no-go theorems [33–36] severely constrain epistemic interpretations [31,32] of quantum states which view the quantum state, in analogy to classical statistical physics, as a state of knowledge over the 'real' physical state which is encoded in some hidden variables. To avoid confusion, we emphasise again that the informational interpretations [4–8], while also regarding the quantum state as a state of information, do *not* take it as epistemic, refuting the very idea of hidden variables.

In any case, we wish not to get entrenched in religious wars about the interpretation of quantum theory. The precise interpretation – other than that quantum theory can be understood as a law book governing the acquisition of information – will not be of relevance in the discussion below. And perhaps more telling about the nature of reality than any individual interpretation by itself is the fact that we have a multiplicity of incompatible, yet consistent interpretations of quantum theory. This raises the question whether reality, rather than being one unambiguously existing entity, is just a consistent interpretation of our interactions with the physical world.

3. Spacetime from communication relations

Information theory also hints at a novel perspective on gravity and spacetime. Indeed, many gravitational phenomena also admit a distinctly information-theoretic flavour; black hole or cosmological horizons as information barriers constitute the most prominent examples. More generally, a general relativistic spacetime can, in principle, be reduced to the information flow among all observers or systems contained in it – at least up to the conformal structure. This is perhaps an under-appreciated perspective.

[6] We recall from section 2.3 that *local* hidden variables are, of course, ruled out.

IOP Conf. Series: Journal of Physics: Conf. Series **880** (2017) 012014 doi:10.1088/1742-6596/880/1/012014

3.1. Classical spacetime from quantum correlations

Gravity is spacetime geometry and geometry tells matter how to move, thereby controlling the flow of information. In particular, the geometry's causal structure determines from where to where a signal can be sent. This standard perspective regards spacetime geometry as the primary notion and the information flow as secondary:

$$\text{spacetime geometry} \;\Rightarrow\; \text{causal structure} \;\Rightarrow\; \text{information flow.}$$

However, information and causality are in an intimate relation: the causal structure not only controls the propagation of information but, vice versa, from the full information flow one could, in principle, deduce the set of all causal relations in spacetime. Assuming spacetime to be given by a Lorentzian geometry (and relatively mild conditions such as stable causality[7]), the causal structure determines the geometry of spacetime up to conformal re-scalings [37, 38].[8] Thanks to this intimate relation, it becomes a matter of choice what we regard as primary and what as secondary; we are equally well entitled to take the information flow as primary and to deduce the conformal geometry as secondary from it:

$$\text{information flow} \;\Rightarrow\; \text{causal structure} \;\Rightarrow\; \text{spacetime geometry (up to scale)}$$

Since an observer can probe and access spacetime structure only indirectly through interactions with matter this converse logic is operationally more adequate.

Whichever of these perspectives one prefers, the structure of a general relativistic spacetime is encoded in the information flow among a continuum of observers. After the equivalence principle, which essentially implies spacetime to be a Lorentzian geometry, the essence of relativity can thus be said to lie in observers and their communication relations. General relativity can be regarded as a law book which governs *where* and *when* information can be communicated.

But how can one access the information flow in spacetime? Since all matter is described by quantum field theory the material information flow is encoded in the correlation structure of quantum fields. After all, the correlation structure encodes the matter dynamics and usually all the information about the quantum field theory. Hence, the correlation structure of quantum fields should also encode the information about spacetime geometry. Indeed, the strength of vacuum field correlations (e.g., in Minkowski space) typically depends only on the spacetime distance between two events; it decays polynomially (e.g., for massless scalar fields) or exponentially (e.g., for massive scalar fields) in spacelike directions and oscillates in timelike directions. Vacuum field correlations are special; e.g., they (a) contain no EPR entangled pair excitations which would alter locally the drop-off behaviour of quantum correlations; (b) usually satisfy area laws, i.e. the entanglement entropy of the field state for a subregion scales with the area of that region because it is dominated by the short-distance degrees of freedom near the interface [40–43]. Vacuum-like states are thus particularly convenient for reading out geometric properties, such as spatiotemporal distances and areas, directly from the strength of correlations. They seem to encode the actual geometry and not only the causal structure which determines conformal geometry. But also for more general states we anticipate that one should be able to start with the information flow, as encoded in a field's correlation structure, and use correlation strengths to *derive causal relations and distances*:

$$\text{correlation structure of fields} \;\Rightarrow\; \text{causal structure} \;\Rightarrow\; \text{spacetime geometry (up to scale?).}$$

[7] This means that opening up the light cones does not introduce closed timelike curves. The spacetime is thus not 'close' to causally pathological ones.

[8] It is necessary to assume spacetime to be given by a Lorentzian metric manifold because not all sets of events with causally well-defined relations can be embedded into a spacetime geometry. This is precisely one of the challenges of the causal set approch to quantum gravity [39].

However, for non-vacuum states, the entanglement of excitations may destroy a simple relation between correlation strength and spacetime distances. Yet, since entangled states must still respect causality with correlations behaving differently in spacelike and timelike directions, they should still encode the causal and thereby the conformal structure.

The perspective above has not been technically fully worked out yet in the context of general relativity. However, there are developments which are compatible with it and warrant further research in this direction. For example, in analogy to information theory which permits one to reproduce the continuous information of a continuous signal through a discrete sampling [44], one can probe continuous geometries through a discrete sampling of quantum field correlations [45, 46]. This requires inverse spectral geometry and can determine the geometry down to a cut-off scale determined by the sampling density. Unfortunately, the situation with Lorentzian signature spacetimes is technically intricate and most work has focused thus far on Euclidean signature geometries (but see [47]). This could, nevertheless, be useful for probing spatial geometries through field correlations in a canonical picture of general relativity. Another – remarkable – piece of evidence supporting our above perspective is Jacobson's derivation of the semiclassical Einstein equations from equilibrium conditions on vacuum entanglement entropy for small geodesic ball regions (at least for first-order variations) [48]. In line with our discussion above, the input to this derivation is a quantum field living on a spacetime given by some Lorentzian geometry. Through Jacobson's result even the dynamics of spacetime assumes an information-theoretic flavour: colloquially, and in view of our understanding of entanglement through the quantum reconstruction, the Einstein equations are essentially equivalent to maximising the composite information in the field, namely, to maximising the information contained in the vacuum correlations between small geodesic ball regions and their environment. Reading out spacetime structure from quantum field states has also been proposed in [49,50] via a spacetime-free formulation of quantum field theory borrowing concepts from local quantum physics [51].

For later purpose we note that, unless one considers semiclassical general relativity where the expectation value of the energy-momentum tensor of the quantum field sources the gravitational field, the material information flow is external to spacetime because it does not back-react. The dynamics of this information flow is not directly coupled with that of spacetime. This separation is legitimate as long as one is interested in the dynamics *in*, not of spacetime.

In the same vein, idealised observers with their reference frames of rods and clocks are not incorporated self-consistently into general relativity, classical or semiclassical alike. They are non-dynamical, extensionless and do not back-react, neither on spacetime nor on other fields, being external too. As Einstein pointed out: "*The theory.... introduces two kinds of physical things, i.e., (1) measuring rods and clocks, (2) all other things, e.g., the electro-magnetic field, the material point, etc. This, in a certain sense, is inconsistent; strictly speaking measuring rods and clocks would have to be represented as solutions of the basic equations..., not, as it were, as theoretically self-sufficient entities...*" [52]. The communication among idealised observers in terms of, e.g. light signals is external to spacetime in the same sense. What matters are the causal relations but they are oblivious to the signal's energy.

Of course, for the purpose of general relativity, namely, to provide a global picture of the large scale structure of spacetime, it is unproblematic to neglect back-reaction of observers and their communication and in this way assign an external role to them. Moreover, when discussing an observer's experiences, one is again interested in the dynamics *in* not *of* spacetime, as expressed by, e.g., geodesic equations as equations of motion. This external concept of observer and communication is what ultimately allows one to conceive of the idealised observer as an agent with free will whose actions and communications are not all pre-determined by the dynamical equations of spacetime and matter fields but who nevertheless is subject to their influence.

This separation between information flow and spacetime will be challenged in the next section.

3.2. *Spacetime and quantum communication relations*

It is worthwhile to note that there is also an intimate relation between abstract finite dimensional quantum theory and spacetime properties, complementing the perspective above. Indeed, one can read out elementary spacetime structure, *without* presupposing it, from communication relations and probability measurements on physical information carriers.

Specifically, what does the dimension of spacetime mean operationally and how can one read it out from information exchange processes? We noted in sec. 2 that the three-dimensionality of space is an input and not an output of special relativity and that the dimensionality of quantum state spaces is an output not an input of quantum reconstructions. Given the intimate relationship between the three-dimensional state space of a single qubit – the Bloch ball – and the set of spatial directions, can one reconstruct the three-dimensionality of physical space together with quantum theory from operational axioms? This question was essentially already asked by von Weizsäcker using his ur-theory by means of which he sought to build up all of physics from the elementary quantum alternative – the 'ur' (or qubit in modern language) – as the primordial ingredient of nature [20]. Von Weizsäcker argued heuristically that space is three-dimensional precisely because the 'ur' has a three-dimensional state space from which also the local rotational symmetry of space is directly inherited. While this may sound naïve at first sight, surprisingly, the three-dimensionality of physical space, *together* with abstract quantum theory, can indeed be derived (a) from operationally plausible conditions on the communication between two agents[9] [53], and (b) from elementary postulates on the 'classical limit' of a generalised probability theory [54]. These results rely on the assumption that physical space is locally given by d-dimensional Euclidean space (d unspecified), but underline the intimate intertwinement of the structure of abstract quantum theory and the three-dimensionality of space. Furthermore, the derivation in [53] also provides a recipe for observers to infer local spatial geometry from probability measurements, in line with the general perspective of this section.

But the relation between abstract quantum structures and spacetime goes further. The Lorentz transformations are normally derived within classical physics and quantum physics is adapted to them. However, assuming universality of quantum theory, they should have a quantum justification. In an attempt to deliver such a justification, it is instructive to consider an information-theoretic reference frame agreement protocol for observers in distinct laboratories who have never met but can communicate with quantum systems [55]. The relation between their frames can be defined as the 'least information-theoretic effort' required for them to synchronise their descriptions of local quantum physics. Under the assumption that there are sufficiently many observables which can be measured universally on several different quantum systems, one can show that their descriptions by different observers must be related by the orthochronous Lorentz group $\mathrm{O}^+(3,1)$ of correct dimension [55]. Crucially, this result neither presupposes *any* particular spacetime structure (incl. the dimension), nor introduces any apart from the group translating among different descriptions of abstract quantum states. Can one use this result nevertheless to reconstruct $3+1$-dimensional Minkowski space from an abstract quantum communication perspective? We also note that related work derives local Lorentz covariance in finite-dimensional local quantum physics from transformation properties of thermal states [50].

Using such arguments from quantum information and renormalization in quantum gravity, one can also entertain the surprising possibility that the local linearity of spacetime might ultimately be a consequence of the probabilistic linearity of a fundamental theory of nature [56].

The above results manifest a deep relation between elementary spacetime properties and the mere possibility of certain abstract communication tasks.

[9] The most crucial condition is that any spatial direction can be encoded in the most elementary system.

IOP Conf. Series: Journal of Physics: Conf. Series **880** (2017) 012014 doi:10.1088/1742-6596/880/1/012014

4. 'Information flow = spatiotemporal structure' in quantum gravity?

Equipped with these observations, we shall now take the liberty to speculate on an informational perspective on quantum gravity. The ensuing picture may ultimately lead nowhere, but the evidence in its favour is strong enough to attempt a systematic pursuit of this perspective. Furthermore, the information paradigm, which by construction is operational in nature, might be useful for injecting a somewhat missing operational sense into the problem of quantum gravity.

4.1. Spacetime architecture as an informational network

We have argued that quantum theory can be regarded as a law book governing how and how much information can be acquired from (or, in turn, encoded in) physical systems. We also argued that general relativity can be conceived of as a law book which governs from where to where information can be communicated. These observations invite one to speculate that a theory which incorporates both quantum theory and general relativity should somehow include the laws which govern fundamentally the information exchange among all degrees of freedom in terms of both how and how much information can be communicated and where and when.

The dynamics of (at least part of) the information flow and of spacetime are treated separately in quantum theory and general relativity with one or the other taking an external role according to the picture that the information flows *on* this spacetime with a negligible back-reaction on the latter's large scale structure. However, the details that are irrelevant for the large scale structure of spacetime will presumably become very relevant when attempting to unify both theories in a single framework which coherently describes the small scale structure of spacetime also. A self-consistent (more) fundamental theory of nature should not rely on any external structure but provide a fully internal picture of physics. It should lift the separation, treating neither information flow nor spacetime as external, but their dynamics jointly.

The diffeomorphism invariance of general relativity implies that objects cannot be localised with reference to a background, but only in relation to one another: *"Objects are not located in spacetime. They are located with respect to one another"* [57]. This is spacetime relationalism: localisation via coincidences of worldlines, gravitational and matter degrees of freedom, etc., but no external reference frame. The physical general relativistic spacetime *is* the continuum of such relations. This is essentially Mach's principle.[10]

Driving this to the extreme for a (more) fundamental framework encompassing both quantum theory and general relativity, there should be no separate spatiotemporal structure and no separate information flow. Are they rather fundamentally the same? After all, spacetime is about localisation via coincidences, information flow requires interaction and *"objects are where they interact, objects interact where they are"* [58].[11] At its most basic level, spatiotemporal structure should reduce to the most basic relations compatible with the essences of quantum theory and general relativity. What more basic relation is there for degrees of freedom than having information about one another (e.g. through correlation)? At its most basic level, spatiotemporal structure and the totality of such 'having-information-about-one-another'-relations among all degrees of freedom, i.e. ultimately the totality of the information flow, would – in an information-theoretic Machian sense – be one and the same. Familiar classical spacetime should only emerge from this information flow in some suitable (coarse-graining) limit. Such a picture would automatically be background and external frame independent.

A spatiotemporal architecture arising from 'having-information-about-one-another'-relations can be anticipated to be dramatically different than that of general relativity. From an operational perspective, the continuum of observers and the infinite amount of information they would have to communicate in general relativity in order to encode an even arbitrarily

[10] Strictly speaking, Mach's principle can be argued not to be fully implemented in general relativity and neither to exist in a single, generally agreed upon form [57]. However, here we shall not worry about such details.

[11] Thoughts in a related direction have also been expressed in sec. 5.6.4 of [57].

small spacetime region is dissatisfying, given that any physical information exchange is finite. Already Feynman was deeply concerned about the fact that it would take a computer an infinite register to compute what happens in no matter how small a patch of spacetime [59].

Indeed, in analogy to the signal band-width of Shannon's sampling theory [44], a meaningful quantification and sampling of information actually requires a scale in continuous theories. For example, in statistical physics, a sensible phase space measure for defining a probability density and an information measure (entropy) therefrom relies on a cell size $\hbar$ which only finds a natural justification in quantum theory. Without such a scale, any (from an operational viewpoint necessarily discrete) sampling of physical information would only yield an approximation to the 'true' physics.

For instance, as mentioned in sec. 3, classical spacetime geometry can also be probed through a Shannon-type discrete sampling of field correlations, and thereby of the spectra of suitable geometric operators (e.g., the Laplacian) [45,46]. Via inverse spectral geometry, this specifies the continuous spacetime geometry down to a 'band-width' scale determined by a cut-off eigenvalue of the geometric operator which, in turn, is determined by the sampling density. Below this scale, geometries become operationally indistinguishable through the considered network of sampling points *in* spacetime. Hence, the probed geometries are part of an isospectral equivalence class of geometries with a cut-off, i.e. of a class of geometries which have indistinguishable geometric spectra up to the cut-off.[12] Since one could always increase the sampling density, the extracted isospectral geometry is considered to only yield an approximation to the purported 'true' continuum physics. This is completely analogous to determining spatial geometry from a sampling of the cosmic microwave background.

By contrast a (more) fundamental theory implementing an 'information-theoretic Machian principle' should provide a (more) fundamental picture of spatiotemporal structure and not an approximation. In short, at the deepest level: information exchange/sampling, yes; approximation, no. In quantum gravity, we suspect that all physical information should be naturally quantifiable. Hence, this requires some universal scale – presumably the Planck scale – as a universal 'band-width' of spacetime. Then it is the continuum which is the approximation to the real physics. In support of this perspective, various arguments, e.g. holographic arguments from black hole entropy calculations, suggest that, in contrast to standard quantum field theory, there is only a locally finite amount of degrees of freedom in the universe [1,60–62].

These observations hint at a picture, viewing *the fundamental architecture of spacetime as a locally finite network of systems – constituting finite information registers – which acquire and communicate finite amounts of information.* A picture in which the clear distinction between systems and observers disappears, but which only has a very general notion of physical systems, comprised of some degrees of freedom, exchanging information.

As soon as a universal 'band-limit' of spatiotemporal structure – and especially so minuscule as the Planck scale – enters the picture, the challenge arises to connect physics at this universal scale to the known operational physics of quantum theory and general relativity at large scales. In particular, since the notion of system is an inherently effective one, depending on the considered degrees of freedom, the acquisition and exchange of information must be scale dependent. In such an informational network, this scale must ultimately be related to the level of complexity under consideration. Such a network picture could thus never be consistent without a bridge, over levels of complexity, from the micro- to the macrocosm. This requires the machinery of renormalization [63], however, now in an informational (e.g., see [64, 65]) and background independent (e.g., see [66]) incarnation, where 'scale' is related to the level of network complexity under consideration rather than an ordinary background energy scale.

[12] Given the arguments of sec. 3.1, one might even insist that all relevant operational information about a spacetime geometry is encoded in quantum correlations such that differences among members of a given isospectral equivalence class may be operationally inaccessible.

Specifically, it necessitates a consistent way of separating relevant from irrelevant information and coarse-graining finer into collective degrees of freedom in the network. In fact, coarse-graining is already implicit in the sampling picture: changing the sampling density changes the level of coarse-graining of the information under consideration.

4.2. An operational alternative to the 'wave function of the universe'?

If the fundamental distinction between observer and systems disappears but information exchange remains central, one might wonder whether an informational state interpretation, as in sec. 2, might carry over to such a network picture and ultimately to quantum gravity and cosmology. It might, but as proposed in [9] only in relative fashion. In this background independent context, any information acquisition by any register, assuming the role of observer, is *internal*, i.e. occurs within the network; a global observer outside the network is meaningless. Taking the state as the 'observer's catalogue of knowledge', as in sec. 2, the *self-reference problem* [67, 68] impedes a given register to infer the global state of the entire network (incl. itself) from its interactions with the rest. Accordingly, relative to any subsystem, one can assign a state to the rest of the network but, without external observer and reference frame, there should then be no global state (aka 'wave function of the universe') for the entire network at once – at least not fundamentally. This stands in stark contrast to most approaches to quantum gravity and cosmology, although absence of global states has been proposed before [61, 69] without reference to such an informational network picture.

Such a proposal must face up to the fact that on large scales observers seem to agree on an observer independent reality. Given that the large scale structure will presumably be the result of renormalization, an effective (not fundamental) global state describing the large scale structure as in quantum cosmology should emerge from coarse-graining. The informational state interpretation would thus have to be made consistent with coarse-graining several perspectives into one. This requires the possibility to change relative perspectives and to decide, in the absence of an objective external reference, when perspectives agree and can be merged. In analogy to a frame independent formulation of general relativity, this might require some perspective neutral theory rather than an external reference. The perspective neutral theory would by itself not have an immediate operational meaning because it would encode all perspectives at once, but should be boiled down to specific perspectives through additional choices. Presumably, this should be related to a primitive notion of symmetry akin to diffeomorphism invariance. Altogether, this could offer an operational alternative to the problematic concept of the 'wave function of the universe' which is ubiquitous in standard approaches to quantum cosmology.

4.3. A new informational top-down approach to the architecture of spacetime

One will rightly wonder what precisely the degrees of freedom in such an informational network should be. However, this is not something which an informational approach alone can tell us. As already discussed in sec. 2, an informational perspective is rather ignorant of the precise underlying degrees of freedom and their concrete physics; it is *universal* by focusing on properties which are independent of their precise physical incarnation. Instead, it is the task of bottom-up approaches to quantum gravity, e.g. loop quantum gravity and string theory, to specify the fundamental degrees of freedom at the bottom of (length) scales. The daunting challenge bottom-up approaches face is to start with such microscopic ingredients and to construct a theory valid over the entire scale gap between the unknown Planck scale physics and the known operational physics of quantum (field) theory and general relativity at familiar large scales.

This informational picture of spatiotemporal structure sketched here is thus not an attempt to invalidate, but to complement bottom-up efforts in quantum gravity. Given the enormous difficulties of bottom-up approaches, it would be injudicious to focus exclusively on closing the scale gap from the bottom up. It should be systematically closed from *both* sides; one should also start with the known physics, reformulate them in some suitable manner and go down in scale to help close this gap from the top down – but without too narrow a prejudice about the bottom of scales. Indeed, informational ideas as outlined here should be materialised into a *systematic* top-down approach to the architecture of spacetime which – thanks to its universal informational perspective – has the potential to elucidate how classical geometries may emerge at large scales largely independently from whatever the precise microphysics are. As such, it might even be compatible with multiple bottom-up approaches.

Developing, as suggested in sec. 3, a reconstruction of a classical spacetime geometry from field correlations (information flow) can be seen as such a top-down step. This would yield a reformulation of known general relativistic spacetime properties in terms of quantum correlations which could generalise to a network picture and lower scales. Given the arguments of subsection 4.1 that spatiotemporal structure should arise from information flow, we suspect it to likewise emerge from quantum correlations across such a network. Now a classical spacetime will not be given and instead distances and causal relations will have to be directly defined through correlation strength. Heuristically, the more correlated degrees of freedom are in vacuum-like states – or the more information they have about one another – the 'closer' they are. An emergent notion of locality will likewise have to be defined in terms of correlation strength.

Such an approach will thus rely on a thorough understanding of correlations in networks of quantum systems. Fortunately, the precise degrees of freedom in the network might be of secondary importance, thanks to the evidence that essential (coarse-grained) properties of quantum correlations are fairly model independent. For instance, fall-off properties of quantum correlation strengths are qualitatively similar for ground-like states in quantum many-body systems with local interactions and quantum field theory. Furthermore, area laws for entanglement entropy appear in quantum field theory and gravity [40–43, 70–73] as well as in many-body physics with gapped Hamiltonians [74, 75] for ground-like states, but not for general states. Independently of the precise degrees of freedom and interactions, these correlation properties are rooted in locality and causal structure (in quantum many-body physics coming from Lieb-Robinson bounds [76]). One situation is relativistic, the other non-relativistic. So what relates them? Since one cannot literally identify non-relativistic condensed matter systems with relativistic field theory, it must be that (coarse-grained) properties of quantum correlations in complex quantum systems with local interactions are *universal*. These universal features render condensed matter systems useful network toy models for simulating gravitational scenarios [77–80]. Similarly to how condensed matter insights into symmetry breaking once led to a breakthrough in particle physics, condensed matter insights into entanglement/geometry relations may lead to breakthroughs in gravitational physics.

To be sure, there are already developments which could be counted to such a top-down approach. In particular, Jacobson's derivation of the semiclassical Einstein equations [48] can be seen in this light. By assuming universal entanglement entropy area laws for vacuum states coming from the ultra-violet through some unknown coarse-graining, it makes assumptions about the underlying Planck scale physics which can ultimately only be justified with bottom-up quantum gravity approaches, but which lead back to the known dynamics of general relativity. Constraining the fundamental theory to produce emergent metric geometries with area laws seems to be an easier task than directly asking it to generate Einstein's equations, especially given the above mentioned evidence that area laws are quite generic for ground states of complex quantum systems.

Moreover, there are many exciting developments in the holographic context of the AdS/CFT correspondence. Most importantly, the entanglement entropy of a region in the boundary theory seems to be directly related to the area of an extremal surface in the semiclassical bulk geometry [70, 71]. These holographic area laws seem to have profound implications. Not only can they be used to derive the bulk Einstein equations perturbatively around a pure AdS background [81, 82], in some analogy to Jacobson's holography-free derivation [48]. But they can also be used to argue that entanglement essentially acts as the 'glue' which keeps a semiclassical spacetime together. Indeed, boundary regions which are entangled/non-entangled seem to correspond to connected/disconnected bulk regions [72, 73].[13] This is a compelling picture that is heuristically compatible with the information-theoretic Machian view of sec. 4.1 according to which spatiotemporal structure might at the deepest level be the same as the totality of the information flow among all degrees of freedom. Namely, without regions having information about one another, there is no spacetime; 'no entanglement = no spacetime'.

In harmony with the ideas sketched here, recently a model framework has been devised [83] which aims precisely at defining and extracting spatial geometries from correlations in a network of abstract finite-dimensional quantum systems – without a background. We have mentioned several times the special role that vacuum-like states assume for reading out geometric behaviour from correlation strengths. It would thus be highly desirable to have a characterisation of such a vacuum-like class of states in purely information-theoretic terms, *without* reference to a specific interaction or background geometry. The authors of [83] propose such a characterisation purely in terms of correlation properties, essentially as states with redundant information in higher correlations. Geometric information across the network is defined using mutual information, which upper bounds correlations of arbitrary quantum observables (see also [80]), and entanglement entropies of abstract regions. The focus lies on reproducing flat Euclidean geometries due to a lack of an obvious origin of the emergent dimension and signature in the correlation structure of the network. The question is whether derivations of local Lorentz covariance from finite dimensional quantum systems [50, 55], as discussed in sec. 3, can be adapted to such a network to open up the possibility to also discussing emergent 3+1-dimensional curved geometries. This should at least yield a valid picture in a large scale regime where the framework corresponds to a set of observers sampling quantum correlations, in the spirit of [45, 46], across a network.

Given that such a putative top-down approach should complement and connect with bottom-up efforts, it is also worthwhile to point out some developments in full (not semiclassical) quantum gravity which fit the general picture. Most of these have occurred in the context of the spin networks of loop quantum gravity. Beginning with the proposal [84, 85] to define emergent geometries therefrom through coarse-graining and correlation strengths, there are now systematic efforts to materialise these ideas [86, 87]. However, this proposal faces a great challenge: a proper notion of entanglement and physical correlations in quantum gravity requires a diffeomorphism invariant description of subsystems.[14] In a canonical language this is tantamount to finding commuting sets of degrees of freedom in the physical Hilbert space – in some analogy to the notion of localised subsystems in local quantum physics [88] – and is thus deeply intertwined with the observable problem in quantum gravity [57, 89–91]. In particular, diffeomorphism invariant observables are highly non-local in a spacetime manifold sense such that localisation of subsystems or subregions which can be entangled must be done in a relational manner, coming back to sec. 4.1 (see also the recent discussion in [92]). In the network picture proposed here, locality should in any case emerge through correlation strength.

Despite promising recent advances [92–98], subsystem localisation and entanglement in the context of diffeomorphism (and, more generally, gauge) symmetry remains to be far better understood. However, this challenge is at the same time a key asset: the diffeomorphism invariance at the fundamental level would ensure that there is indeed no separation between the information flow and spatiotemporal structure – in support of information-theoretic Machian

[13] The connecting extremal surface in the bulk vanishes without entanglement of the boundary regions.

[14] E.g., as exemplified by the topological case of 3D vacuum quantum gravity, constraint imposition can render kinematical entanglement physically irrelevant, given that the resulting physical Hilbert space is one-dimensional.

views. Spacetime – and with it locality – has to emerge from the correlations among the fundamental quantum geometric degrees of freedom without a separate spatiotemporal structure given at the outset.

The intention here is not to give the impression that all the above mentioned diverse developments will necessarily fit together into a single picture. Rather, we are discussing a new paradigm – the information paradigm in quantum and gravitational physics – which has the potential to forge a novel interdisciplinary research field. Given the infancy of this field, this requires to follow such general ideas systematically and to take inspiration from many of the developments coming currently together. Regardless of the involved details, these developments suggest that entanglement may play a crucial role in the emergence of spacetime structures from a deeper theory, rendering the tools from quantum information also invaluable in gravity.

Indeed, leaving the details of different approaches aside, we had argued that the network and its degrees of freedom should be locally finite. This would provide a natural ultra-violet cut-off – and thus a minimal scale for sampling information. Accordingly, whatever the fundamental degrees of freedom, a top-down sampling of correlations could never yield a unique classical spectral geometry. Any emergent geometry would thus *necessarily* be contained in a non-trivial isospectral equivalence class such that any macroscopic limit could not distinguish the microphysics and would thereby be compatible with a multitude of fundamental networks. This, in fact, would even leave open the possibility that several microscopic network theories could give rise to the same *universal* large scale behaviour through coarse-graining. The precise details of the physics at a fundamental scale might thus ultimately be of less significance thanks to universal coarse-graining and entanglement properties. The hope is that one can also explain in this light the universal entanglement area laws crucially going into Jacobson's derivation of the semiclassical Einstein equations [48]. A spacetime emerging from the informational network picture might therefore be an inherently effective one, being largely independent of the detailed microphysics.

Acknowledgments
I would like to thank S. Carrozza, B. Dittrich, S. Gielen, T. Koslowski, M. Müller and D. Rätzel for discussions and comments on an earlier draft version. The project leading to this publication has received funding from the European Union's Horizon 2020 research and innovation programme under the Marie Sklodowska-Curie grant agreement No 657661.

References

[1] Wheeler J 1990 in *Complexity, Entropy, and the Physics of Information*, Ed Zurek W (Addison-Wesley) 309
[2] Barbour J 2015 in *It From Bit or Bit From It?*, Ed Aguirre A, Foster B and Merali Z (Springer, The Frontiers Collection) 197
[3] Hartle J 1968 *Americ. J. Phys.* **36** 704
[4] Rovelli R 1996 *Int. J. Theor. Phys.* **35** 1637
[5] Zeilinger A 1999 *Found. Phys.* **29** 631
[6] Brukner C and Zeilinger A 2003 in *Time, Quantum and Information*, Eds Castell L and Ischebeck O (Springer)
[7] Brukner C 2016 in *Quantum (Un)speakables II*, Eds Bertlmann R and Zeilinger A (Springer, The Frontiers Collection) 95
[8] Fuchs C 2002 *Preprint* arXiv:quant-ph/0205039
[9] Höhn P 2014 to appear in *Quantum, Preprint* arXiv:1412.8323
[10] Höhn P and Wever C 2017 *Phys. Rev. A* **95** 012102 (*Preprint* arxiv:1511.01130)
[11] Höhn P 2017 *Entropy* **19**(3) 98 (*Preprint* arxiv:1612.06849)
[12] Hardy L 2001 *Preprint* quant-ph/0101012
[13] Dakic B and Brukner C 2011 in *Deep Beauty* Ed Halvorson H (Cambridge University Press) pp. 365
[14] Masanes L and Müller M 2011 *New J. Phys.* **13** 063001
[15] Barnum H, Müller M and Ududec C 2014 *New J. Phys.* **16** 123029
[16] de la Torre G, Masanes L, Short A and Müller M 2012 *Phys. Rev. Lett.* **109** 090403
[17] Chiribella G, D'Ariano G and Perinotti P 2011 *Phys. Rev. A* **84** 012311

[18] Goyal P 2010 *New J. Phys.* **12** 023012
[19] Appleby M, Fuchs C, Stacey B and Zhu H 2016 *Preprint* arXiv:1612.03234
[20] von Weizsäcker C 2006 *The Structure of Physics* (Springer)
[21] Schrödinger E 1935 *Math. Proc. Camb. Phil. Soc.* **31** 555
[22] Brukner C and Zeilinger A 1999 *Phys. Rev. Lett.* **83** 3354
[23] Brukner C and Zeilinger A 2001 *Phys. Rev. A* **63** 022113
[24] Bremner M et al 2002 *Phys. Rev. Lett.* **89** 247902
[25] Harrow A 2009 *Quant. Inf. Comput.* **9** 773
[26] Jaynes E 1957 *Phys. Rev.* **106** No. 4 620
[27] Bohr N 1962 *Atomic Theory and The Description of Nature* (Cambridge, Reprint)
[28] Einstein A, Podolsky B and Rosen N 1935 *Phys. Rev.* **47** 777
[29] Bohm D 1952 *Phys. Rev.* **85** 166
[30] Dieks D and Vermaas P (eds) 1998 *The Modal Interpretation of Quantum Mechanics* (Kluwer Academic Publishers)
[31] Ballentine L 1970 Rev. Mod. Phys. **42** 358
[32] Spekkens R 2007 *Phys. Rev.* **A 75** 032110
[33] Bell J 1966 *Rev. Mod. Phys.* **38** 447
[34] Kochen S and Specker E 1967 *J. Math. Mech.* **17** 59
[35] Pusey M, Barrett J and Rudolph T 2012 *Nature Physics* **8** 475
[36] Frauchiger D and Renner R 2016 *Preprint* arXiv:1604.07422
[37] Hawking S 1966 PhD Thesis (Cambridge)
[38] Geroch R and Horowitz G 1979 in *General Relativity: An Einstein centenary survey*, Eds Hawking S and Isreal W (Cambridge) 212
[39] Dowker F 2013 *Gen. Rel. Grav.* **45** 1651
[40] Bombelli L, Koul R, Lee J, and Sorkin R 1986 *Phys. Rev. D* **34** no 2 373
[41] Srednicki M 1993 *Phys. Rev. Lett.* **71** 666
[42] Casini H and Huerta M 2009 *J. Phys. A* **42** 504007
[43] Bianchi E and Myers R 2014 *Class. Quant. Grav.* **31** 214002
[44] Shannon C and Weaver W 1949 *The mathematical theory of communication* (University of Illinois Press)
[45] Kempf A 2009 *Phys. Rev. Lett.* **103** 231301
[46] Saravani M, Aslanbeigi S and Kempf A 2016 *Phys. Rev. D* **93** 045026
[47] Yazdi Y and Kempf A 2017 *Class. Quant. Grav.* **34** 094001
[48] Jacobson T 2016 *Phys. Rev. Lett.* **116** 201101
[49] Raasakka M 2017 *SIGMA* **13** 006
[50] Raasakka M 2017 *Preprint* arXiv:1705.06711
[51] Haag R 1996 *Local Quantum Physics: Fields, Particles, Algebras* (Springer)
[52] Einstein A 1949 in *Albert Einstein: Philosopher-Scientist* Ed Schilpp P (Open Court Illinois)
[53] Müller M and Masanes L 2013 *New J. Phys.* **15** 053040
[54] Dakic B and Brukner C 2016 in *Quantum Theory: Informational Foundations and Foils*, Eds Chiribella G and Spekkens R (Springer) 249
[55] Höhn P and Müller M 2016 *New J. Phys.* **18** 063026
[56] Müller M, Carrozza S and Höhn P 2017 *J. Phys. A: Math. Theor.* **50** 054003
[57] Rovelli C 2004 *Quantum Gravity* (Cambridge)
[58] Drieschner M 1974 *Lect. Notes Phys.* **29** 250
[59] Feynman R 1982 *Int. J. Theor. Phys.* **21** 467
[60] 't Hooft G 2000 *Preprint* arXiv:hep-th/0003004
[61] Markopoulou F 2002 *Preprint* arXiv:gr-qc/0210086
[62] Bao N, Carroll S and Singh A 2017 *Preprint* arXiv:1704.00066
[63] Wilson K 1975 *Rev. Mod. Phys.* **47(4)** 773
[64] Vidal G 2010 in *Understanding Quantum Phase Transitions*, Ed Carr L (Taylor & Francis, Boca Raton)
[65] Bény C and Osborne T 2013 *Preprint* arXiv:1310.3188
[66] Dittrich B 2017 in *Loop Quantum Gravity: The first 30 Years*, Eds Ashekar A and Pullin J (World Scientific) 153
[67] Breuer T 1993 *Phil. Sci.* **62** 197
[68] Dalla Chiara M 1977 *J. Phil. Log.* **6** 331
[69] Crane L 1995 *J. Math. Phys.* **36** 6180.
[70] Ryu S and Takayanagi T 2006 *Phys. Rev. Lett.* **96** 181602
[71] Hubeny V, Rangamani M and Takayanagi T 2007 JHEP **0707** 062
[72] van Raamsdonk M 2010 *Gen. Rel. Grav.* **42** 2323

[73] van Raamsdonk M 2016 *Preprint* arXiv:1609.00026
[74] Eisert J, Cramer M and Plenio M 2010 *Rev. Mod. Phys.* **82** 277
[75] Popescu S, Short A and Winter A 2006 *Nature Physics* **2** 754
[76] Kliesch M, Gogolin C and Eisert J 2014 in *Many electron approaches in physics*, Eds Site L and Bach V (Springer)
[77] Swingle B 2012 *Phys. Rev. D* **86** 065007
[78] Pastawski F, Yoshida B, Harlow D and Preskill J 2015 JHEP **1506** 149
[79] Hamma A and Markopoulou F 2011 *New J. Phys.* **13** 095006.
[80] Qi X 2013 *Preprint* arXiv:1309.6282
[81] Swingle B and van Raamsdonk M 2014 *Preprint* arXiv:1405.2933
[82] Faulkner T et al 2017 *Preprint* arXiv:1705.03026
[83] Cao C, Carroll S and Michalakis S 2017 *Phys. Rev. D* **95** 024031
[84] Girelli F and Livine E 2005 *Class. Quant. Grav.* **22** 3295
[85] Livine E and Terno D 2006 *Preprint* arXiv:gr-qc/0603008
[86] Chirco G, Mele F, Oriti D and Vitale P 2017 *Preprint* arXiv:1703.05231
[87] Bianchi E, Guglielmon J, Hackl L and Yokomizo N 2016 *Phys. Rev. D* **94** 086009
[88] Yngvason J 2014 *Preprint* arXiv:1401.2652
[89] Kuchar K 2011 *Int. J. Mod. Phys. D* **20** 3 (Reprint)
[90] Isham C 1993 in *Integrable Systems, Quantum Groups, and Quantum Field Theories*, Eds Ibort L and Rodriguez M (Kluwer Academic Publishers) 157
[91] Dittrich B, Höhn P, Koslowski T and Nelson M 2017 *Phys. Lett. B* **769** 554
[92] Donnely W and Giddings S 2017 *Preprint* arXiv:1706.03104
[93] Donnelly W and Freidel L 2016 JHEP **1609** 102
[94] Delcamp C, Dittrich B and Riello A 2016 JHEP **1611** 102
[95] Geiller M 2017 *Preprint* arXiv:1703.04748
[96] Speranza A 2017 *Preprint* arXiv:1706.05061
[97] Wieland W 2017 *Preprint* arXiv:1704.07301
[98] Wieland W 2017 *Preprint* arXiv:1706.00479

IOP Conf. Series: Journal of Physics: Conf. Series **880** (2017) 012015 doi:10.1088/1742-6596/880/1/012015

The Galilean Challenge: Architecture and Evolution of Language

Noam Chomsky
Linguistics & Philosophy, MIT, Cambridge MA 02139, USA
E-mail: nchomsky3@gmail.com

In the early days of the modern scientific revolution, Galileo and his contemporaries issued a crucial challenge to those concerned with the nature of human language, a challenge that was scarcely recognized until it was taken up in the mid-20th century and became the primary concern of much of the study of language. For short, I will refer to it as the Galilean challenge. These great founders of modern science expressed their awe and wonder at the fact that language permits us to construct "from 25 or 30 sounds an infinite variety of expressions, which although not having any resemblance in themselves to that which passes through our minds, nevertheless do not fail to reveal all of the secrets of the mind, and to make intelligible to others who cannot penetrate into the mind all that we conceive and all of the diverse movements of our souls."[1]

We can now see that the Galilean challenge requires some qualifications, but it is very real, and should I think be recognized as one of the deepest insights in the rich history of inquiry into language and mind in the past 2500 years.

The challenge was not entirely ignored. For Descartes, the human capacity for unbounded and appropriate use of language was a primary basis for his postulation of mind as a new creative principle. In later years, there is occasional recognition that language is a creative activity that involves "infinite use of finite means," in Wilhelm von Humboldt's formulation, that it provides "audible signs for thought," in the words of linguist William Dwight Whitney a century ago. There has also been awareness that these capacities are a species-property, shared by humans and unique to them, the most striking feature of this curious organism and a foundation for its remarkable achievements. But there was never much to say beyond a few phrases.

There is a good reason why the insights languished until mid-20th century: intellectual tools were not available for even formulating the problem in a clear enough way to address it seriously. That changed thanks to the work of Alan Turing and other great mathematicians who established the general theory of computability on a firm basis, showing in particular how a finite object like the brain can generate an infinite variety of expressions. It then became possible, for the first time, to address at least part of the Galilean challenge directly – although, regrettably, the earlier history was entirely unknown at the time.

[1] The quote is from Antoine Arnauld and Claude Lancelot, *Grammaire générale et raisonée* (1660). Galileo's earlier version, in his *Dialogo sopra i due massimi sistemi del mondo* (1632) was similar, except for referring to the alphabet rather than the sounds of language.

With these intellectual tools available, it becomes possible to formulate what we may call the Basic Property of human language: the language faculty of the human brain provides the means to construct a digitally infinite array of structured expressions, each of which has a semantic interpretation expressing a thought, and each of which can be externalized by means of some sensory modality. The infinite set of semantically interpreted objects constitutes what has sometimes been called a language of "thought": the system of thoughts that receive linguistic expression and that enter into reflection, inference, planning, and other mental processes, and when externalized can be used for communication and other social interactions. We may fairly assume that the language faculty is shared among humans. There are no known group differences in language capacity, and individual variation is at the margins.

We now know that although speech is the usual form of sensorimotor externalization, it can just as well be sign or even touch, discoveries that require a slight reformulation of the Galilean challenge. A more fundamental qualification has to do with the way the challenge is formulated: in terms of production of expressions. So formulated, the challenge overlooks some basic issues. Production, like perception, accesses the internal language but cannot be identified with it. We must distinguish the internalized system of knowledge from the actions that access it. The theory of computability enables us to establish the distinction, which is an important one, familiar in other domains.

Consider for example human arithmetical competence. In studying it, we routinely distinguish the internal system of knowledge from the actions that access it, like multiplying numbers in our head, an action that involves many factors beyond intrinsic knowledge; memory constraints, for example. The same is true of language. Production and perception access the internal language, but involve other factors as well, including again short-term memory, matters that began to be studied with some care as soon as the Galilean challenge was addressed in the 1950s.

There has been considerable progress in understanding the nature of the internal language, but its free creative use remains a mystery. That comes as no surprise. In a recent review of the state of the art concerning far simpler cases of voluntary action, two leading researchers, neuroscientists Emilio Bizzi and Robert Ajemian, write that we are beginning to learn something about the puppet and the strings, but the puppeteer remains shrouded in mystery.[2] That is even more dramatically true for such creative acts as the normal use of language, the unique human capacity that so impressed the founders of modern science.

The fundamental task of inquiry into language is to determine the nature of the Basic Property. To the extent that its properties are understood, we can seek to investigate particular internal languages, each an instantiation of the Basic Property, much as each individual visual system is an instantiation of the human faculty of vision. We can investigate how the internal languages are acquired and used, how the language faculty itself evolved, its basis in human genetics and the ways it functions in the human brain. This general program of research has been called the Biolinguistic Program. The theory of the genetically based language faculty is called Universal Grammar; the theory of each individual language is called its Generative Grammar.

Languages appear to be extremely complex, varying radically from one another. And indeed, a standard belief among professional linguists 60 years ago was that languages can vary in arbitrary ways and each must be studied without preconceptions. Similar views were held at the time about organisms generally. Many biologists would have agreed with molecular biologist

[2] Emilio Bizzi and Robert Ajemian, "A Hard Scientific Quest: Understanding Voluntary Movements," *Daedalus* 144.1 (2015).

IOP Conf. Series: Journal of Physics: Conf. Series **880** (2017) 012015 doi:10.1088/1742-6596/880/1/012015

Gunther Stent's conclusion that the variability of organisms is so free as to constitute "a near infinitude of particulars which have to be sorted out case by case."[3] When understanding is thin, we expect to see extreme variety and complexity.

A great deal has been learned since then. Within biology, it is now recognized that the variety of life forms is very limited, so much so that the hypothesis of a "universal genome" has been seriously advanced. My own feeling is that linguistics has undergone a similar development, along lines that I will outline.

The Basic Property takes language to be a computational system, which we therefore expect to observe general conditions on computational efficiency. A computational system consists of a set of atomic elements and rules to construct more complex ones. For generation of the language of thought, the atomic elements are word-like, though not words; for each language, the set of these elements is its lexicon. The lexical items are commonly regarded as cultural products, varying widely with experience and linked to extra-mental entities – an assumption expressed in the titles of standard works, such as W.V. O. Quine's influential study Word and Object. Closer examination reveals a very different picture, one that poses many mysteries. Let's put that aside for now, turning to the computational procedure.

Clearly, we will seek the simplest computational procedure consistent with the data of language, for reasons that are implicit in the basic goals of scientific inquiry. It has long been recognized that simplicity of theory translates directly to explanatory depth. A more specific version of this quest for understanding was provided by a famous dictum of Galileo's, which has guided the sciences since their modern origins: nature is simple, and it is the task of the scientist to demonstrate this, from the motion of the planets, to an eagle's flight, to the inner workings of a cell, to the growth of language in the mind of a child. Linguistics has an additional motive of its own for seeking the simplest theory: it must face the problem of evolvability. Not a great deal is known about evolution of modern humans, but the few facts that are well established, and others that have recently been coming to light, are rather suggestive, and conform well to the conclusion that the language faculty is near optimal for a computational system, the goal we should seek on purely methodological grounds.

One fact that does appear to be well established I have already mentioned: that the faculty of language is a true species property, invariant among human groups – and, furthermore, unique to humans in its essential properties. It follows that there has been little or no evolution of the faculty since human groups separated from one another. Recent genomic studies place this date not very long after the appearance of anatomically modern humans about 200,000 years ago, perhaps some 50,000 years later, when the San group in Africa separated from other humans. There is no evidence of anything like human language, or symbolic activities altogether, before the emergence of modern humans. That leads us to expect that the faculty of language emerged along with modern humans or not long after, a very brief moment in evolutionary time. It follows, then, that the Basic Property should indeed be very simple. The conclusion conforms to what has been discovered in recent years about the nature of language, a welcome convergence.

The discoveries about early separation of the San people are highly suggestive. They appear to share the general human language capacity, but have significantly different externalized languages. With irrelevant exceptions, their languages are all and only the languages with phonetic clicks, with corresponding adaptations in the vocal tract. The most likely explanation for these facts, developed in detail in current work by Dutch linguist Riny Huijbregts,[4] is that

[3] Gunther Stent, "From Probability to Molecular Biology," *Cell* 36 (1984).

[4] Riny Huijbregts, "Phonemic Clicks and the Mapping Asymmetry: How Language Emerged and Speech Developed" <http://www.sciencedirect.com/science/article/pii/S0149763416305450>, *Neuroscience & Biobehavioral Reviews* (2017).

possession of internal language preceded separation, which in turn preceded externalization, the latter in somewhat different ways in separated groups. Externalization seems to be associated with the first signs of symbolic behavior in the archaeological record, after the separation. Putting these observations together, it seems that we are reaching a stage in understanding where the account of evolution of language can perhaps be fleshed out in ways that were unimaginable until quite recently.

Returning to the Basic Property, as we have seen we have reason to believe that it may be quite simple. The challenge for research, then, is to show how the facts of language are accounted for in terms of the Basic Property: more fully, by the interaction of the Basic Property, specific experience, and language-independent principles, including principles of computational efficiency. The challenge is of particular interest and significance when it is clear that the child's experience provides little or no relevant evidence – a situation far more prevalent than commonly realized, so careful examination reveals, from acquisition of word meaning on to syntactic structures and the semantic properties of the generated language of thought.

Also of particular interest are the universal properties of the language faculty that began to come to light as soon as serious efforts were undertaken to construct generative grammars, including quite simple principles that had never been noticed, and that are quite puzzling. One crucial and puzzling principle is structure-dependence: the rules that yield the language of thought attend solely to structural properties, ignoring properties of the externalized signal, even such simple properties as linear order.

The property is illustrated by elementary examples. Consider the sentence "John and his father are tall" – not *is tall*, though the bigram frequency of *father-is* is much greater than of *father-are*, and the computation using linear order (adjacency) is far simpler than the computation that has to analyze the sentence into phrases and use phrasal locality.

To take an example of semantic construal, consider the sentence *birds that fly instinctively swim*. It is ambiguous: the adverb "instinctively" can be associated with the preceding verb (*fly instinctively*) or the following one (*instinctively swim*). Suppose now that we extract the adverb from the sentence, forming *instinctively, birds that fly swim*. Now the ambiguity is resolved: the adverb is construed only with the linearly more remote but structurally closer verb *swim*, not the linearly closer but structurally more remote verb *fly*. The only possible interpretation – *birds swim* – is the unnatural one, but that doesn't matter: the rules apply rigidly, independent of meaning and fact. What is puzzling, again, is that the rules ignore the simple computation of linear distance and keep to the far more complex computation of structural distance.

The principle of structure dependence holds for all constructions in all languages, and it is indeed puzzling. Furthermore, it is known without relevant evidence, as is clear in cases like the one I just gave and innumerable others. Experiment shows that children understand that rules are structure-dependent as early as they can be tested, by about age 3 – and are of course not instructed. We can be quite confident, then, that structure-dependence follows from principles of universal grammar that are deeply rooted in the human language faculty.

There is evidence from other sources that supports the conclusion that structure-dependence is a true linguistic universal, deeply rooted in language design. Research conducted in Milan a decade ago, initiated by Andrea Moro, showed that invented languages keeping to the principle of structure-dependence elicit normal activation in the language areas of the brain, but much simpler systems using linear order in violation of these principles yield diffuse activation, implying that experimental subjects are treating them as a puzzle, not a language.[5] Similar

[5] Mariacristina Musso et al., "Broca's Area and the Language Instinct" <http://www.nature.com/neuro/journal/v6/n7/abs/nn1077.html>, *Nature Neuroscience* 6.7 (2003).

results were found in work by Neil Smith and Ianthi Tsimpli in their investigation of a cognitively deficient but linguistically gifted subject. They also made the interesting observation that normals can solve the problem if it is presented to them as a puzzle, but not if it is presented as a language, presumably activating the language faculty.

The only plausible conclusion, then is that structure-dependence is an innate property of the language faculty. That raises the question why this should be so? There is only one known answer, and fortunately, it is the answer we seek for general reasons: the computational operations of language are the simplest possible ones. Again, that is the outcome that we hope to reach on methodological grounds, and that is to be expected in the light of the evidence about evolution of language already mentioned.

The simplest recursive operation, embedded in one or another way in all others, takes two objects already constructed, say X and Y, and forms a new object Z, without modifying either X or Y or adding any further structure. Accordingly, Z can be taken to be just the set $\{X, Y\}$. In current work, the operation is called Merge. Since Merge imposes no order, the objects constructed, however complex, will be hierarchically structured but unordered, and operations on them will necessarily keep to structural distance, ignoring linear distance. It follows that the linguistic operations yielding the language of thought will be structure-dependent, as indeed is the case, resolving the puzzle.

Externalization of language maps internal structures into some sensorimotor modality, usually speech. The sensorimotor system requires linear order; we cannot speak in parallel. But none of this enters the generation of the language of thought, which keeps to structural relations. More generally, externalization of language seems to be a peripheral aspect of language, not entering into its core function of providing a language of thought, contrary to a long tradition, including the formulation of the Galilean challenge.

Perception yields further evidence in support of this conclusion. The auditory systems of apes are quite similar to those of humans, even attuned to the phonetic features that are used in language. But the shared auditory-perceptual systems leave apes without anything remotely like the human faculty of language. Some have attributed this lack to deficiency of articulatory apparatus and vocal learning, but apes can gesture quite easily and as has been known for many years, sign language is virtually identical to spoken language in its basic properties and acquisition, and human groups (including children), under the right conditions, even invent normal sign languages with no linguistic input at all.

These results support the conclusion that internal language is independent of externalization, and that internal language evolved quite independently of the process of externalization.

The most powerful evidence, however, is what we learn from investigation of language design. One crucial example is the one just mentioned: the explanation for the puzzle posed by the linguistic universal of structure-dependence, which follows from the null hypothesis: that the computational system is optimal, and hence ignores linear order, the most elementary feature of externalization.

Not long ago it would have seemed absurd to propose that the operations of human language could be reduced to Merge, along with language-independent principles of computational efficiency. But work of the past few years has shown that quite intricate properties of language follow directly from this assumption, along with a few other quite simple ones. One important result has to do with the property of displacement, a ubiquitous and also quite puzzling property of language: phrases are heard in one position but interpreted both there and in some other position that is silent but where they could have occurred – a puzzling property, which is never built into artificial symbolic systems for metamathematics, programming, or other purposes.

For example, the sentence "which book will you read?" is interpreted as meaning roughly: "for which book x, you will read the book x," with the nominal phrase *book* heard in one position but interpreted in two positions.

I will not go into the details, but it is easy to show that Merge-based computation automatically yields displacement with copies; in this case, two copies of *which book*, yielding the correct semantic interpretation directly. The same process yields quite intricate semantic interpretations, and also has significant implications about the nature of language. To see why, consider for example the sentence "the boys expect to see each other" and the same sentence preceded by "which girls": "which girls do the boys expect to see each other." In the latter sentence, "each other" does not refer back to the closest antecedent, "the boys," as such phrases universally do, but rather to the more remote antecedent "which girls." The sentence means "for which girls the boys expected those girls to see each other." That is what reaches the mind, under Merge-based computation with automatic copies, though not what reaches the ear. What reaches the ear violates the locality condition of referential dependency. Deletion of the copy in externalization causes processing problems: such filler-gap problems, as they are called, can be become quite severe, and are among the major problems of automatic parsing and perception. If the copy were not deleted, the problem would not arise. Why then is it deleted? Again, because of principles of efficient computation that reduce what is computed to the minimum: at least one copy must appear or there is no evidence that displacement took place at all, so only the structurally most prominent one remains (with important qualifications strengthening the conclusion, which I will put aside), leaving a gap that must be filled by the hearer – a matter that can become quite intricate.

These examples illustrate a general phenomenon of some significance. Language design appears to maximize computational efficiency but disregards communicative efficiency. In fact, in every known case in which computational and communicative efficiency conflict, communicative efficiency is ignored. These facts argue against the common belief that communication is the basic function of language. They also further undermine continuity assumptions about language evolving from animal communication. And they provide further evidence that externalization, which is necessary for communication, is a peripheral aspect of language.

As I mentioned, there are methodological reasons and also some evolutionary reasons to expect that the basic design of language will be quite simple, perhaps even close to optimal. With regard to externalization of language, the same methodological arguments hold, but the evolutionary arguments do not apply. In fact, externalization of language may not involve evolution at all. The sensorimotor systems were in place long before the appearance of language. Mapping the internal language to some sensorimotor system for externalization is a hard cognitive problem, relating two systems that are unrelated: an internal system that may be highly efficient computationally, and a sensory modality unrelated to it. That would lead us to expect that the variety, complexity, and easy mutability of observed languages might lie primarily in externalization. Increasingly, it seems clear that that is the case. And in fact it should be expected, since the principles of the internal language are largely known by children without evidence, as, indeed, is a great deal more about language, including almost all semantic and most syntactic properties – a matter of contention, but solidly established, I think.

Let us return finally to the second component of a computational system, the atomic elements: for language, the lexical items. As I mentioned, the conventional view is that these are cultural products, and that the basic ones – those used for referring to the world – are associated with extra-mental entities. This representationalist doctrine has been almost universally adopted in the modern period. The doctrine does appear to hold for animal communication: a monkey's calls, for example, are associated with specific physical events. But the doctrine is radically false for human language, as was recognized as far back as classical Greece.

To illustrate, let's take the first case that was discussed in pre-Socratic philosophy, the problem posed by Heraclitus: how can we cross the same river twice? To put it differently, why are two appearances understood to be two stages of the same river? Contemporary philosophers have suggested that the problem is solved by taking a river to be a four-dimensional object, but that simply restates the problem: why this object and not some different one, or none at all?

When we look into the question, puzzles abound. Suppose that the flow of the river has been reversed. It is still the same river. Suppose that what is flowing becomes 95% arsenic because of discharges from an upstream plant. It is still the same river. The same is true of other quite radical changes in the physical object. On the other hand, with very slight changes it will no longer be a river at all. If its sides are lined with fixed barriers and it is used for oil tankers, it is a canal, not a river. If its surface undergoes a slight phase change and is hardened, a line is painted down the middle, and it is used to commute to town, then it is a highway, no longer a river. Exploring the matter further, we discover that what counts as a river depends on mental acts and constructions. The same is true quite generally of even the most elementary concepts: *tree*, *water*, *house*, *person*, *London*, or in fact any of the basic words of human language. Unlike animals, the items of human language and thought uniformly violate the representationalist doctrine.

For such reasons, incidentally, deep-learning approaches to object-recognition, whatever their interest, cannot in principle discover the meanings of words.

Furthermore, the intricate knowledge of the means of even the simplest words, let alone others, is acquired virtually without experience. At peak periods of language acquisition, children are acquiring about a word an hour, that is, often on one presentation. It must be, then, that the rich meaning of even the most elementary words is substantially innate. The evolutionary origin of such concepts is a complete mystery, one that may not be resolvable by means available to us.

Returning to the Galilean challenge, it has to be reformulated to distinguish language from speech, and to distinguish production from internal knowledge, the latter an internal computational system that yields a language of thought, a system that might be remarkably simple, conforming to what the evolutionary record suggests. Secondary processes map the structures of language to one or another sensorimotor system for externalization. These processes appear to be the locus of the complexity and variety of linguistic behavior, and its mutability over time.

There are suggestive recent ideas about the neural basis for the operations of the computational system, and about its possible evolutionary origins. The origin of the atoms of computation, however, remains a complete mystery, as does a major question that concerned those who formulated the Galilean challenge, the Cartesian question of how language can be used in the normal creative way, in a manner appropriate to situations but not caused by them, in ways that are "incited and inclined" but not "compelled," in Cartesian terms. The mystery holds for even the simplest forms of voluntary motion, as discussed earlier.

A great deal has been learned about language since the Biolinguistic Program was initiated. It is fair to say, I think, that during these years more has been learned about the nature of language, and about a very wide variety of typologically different language, than in the entire 2500 year history of inquiry into language. New questions have arisen, some quite puzzling. And there are some surprising answers, which lead us to revise what has long been believed about language and mental processes generally.

But as is familiar in the sciences, the more we learn, the more we discover what we do not know. And the more puzzling it often seems.

IOP Conf. Series: Journal of Physics: Conf. Series **880** (2017) 012016 doi:10.1088/1742-6596/880/1/012016

Quantum field theory and the linguistic Minimalist Program: a remarkable isomorphism

M Piattelli-Palmarini[1] **and G Vitiello**[2]

[1] Department of Linguistics, University of Arizona, Tucson, AZ, USA
[2] Dipartimento di Fisica "E.R.Caianiello" and INFN, Università di Salerno, Fisciano (SA) - 84084 Italy

E-mail: massimo@email.arizona.edu - vitiello@sa.infn.it

Abstract. By resorting to recent results, we show that an isomorphism exist between linguistic features of the Minimalist Program and the quantum field theory formalism of condensed matter physics. Specific linguistic features which admit a representation in terms of the many-body algebraic formalism are the unconstrained nature of recursive Merge, the operation of the Labeling Algorithm, the difference between pronounced and un-pronounced copies of elements in a sentence and the build-up of the Fibonacci sequence in the syntactic derivation of sentence structures. The collective dynamical nature of the formation process of Logical Forms leading to the individuation of the manifold of concepts and the computational self-consistency of languages are also discussed.

1. Introduction

In this paper, by closely following refs. [1], we present some recent results showing that some features of the so-called Minimalist Program [2] (MP) in linguistic studies are quite well suited to a mathematical representation in terms of algebraic methods and tools. The founder and main contributor of the linguistic domain called "Generative Grammar" (GG), has been, and still is, Noam Chomsky (see his contribution to these Proceedings [3]). Details of some specific linguistic features and of the quantum field theory (QFT) mathematical formalism can be found in [1]. For brevity we will omit to discuss the debate concerning the GG approach and its developments. For some basic references see e.g. [4] and [5] where a use-based explanation of language is offered; for counters see [6] and [7, 8]; for a statistical approach to syntax and language learning, see [9, 10, 11], for counters see [12, 13]. See [14, 15] and references therein for a broader approach to language and language evolution.

In our opinion, the results we present are very rewarding since, on the one hand, they allow to recognize the deep dynamical processes underlying the MP linguistic structures; on the other hand, they show the linguistic content of the many-body formalism.

The plan of the paper is the following. The MP and the basic Merge operation are presented in Section 2 (see also the Appendix A). In Sections 3 are discussed the X-bar tree, its self-similarity properties and the breakdown of time-reversal symmetry. The interfaces, the manifold of concepts and the copies of lexical elements, comments on the entropy and the computational self-consistency of language are presented in Section 4. Section 5 is devoted to final remarks. Finally, the definition and properties of the Fibonacci matrix are presented in the Appendix B.

IOP Conf. Series: Journal of Physics: Conf. Series **880** (2017) 012016 doi:10.1088/1742-6596/880/1/012016

2. The binary Merge and the third factors of language design

In the "The Minimalist Program" an increasing emphasis has been put in recent years on "third factors of language design", namely, principles that are basically minimal, strictly local search, minimal computation, not specific to language, nor specific to biological systems. In other words, the physics and the mathematics of language. The other two factors are genetic predispositions and peculiarities of the local language that the child has to learn [16].

In the early Eighties, the Generative Grammar theory developed in its refined form of the theory of Government and Binding (GB). The core of GB were several syntactic "modules", each taking care of a kind of syntactic operations (Case, Theta Criterion, Binding, Control etc.). See [17] for a basic introduction. It turned out that all Phrases had the recursive structure of a tree: the X-bar tree, or X-bar structure. An element of the structure, a node of the X-bar tree, can contain another X-bar structure, and so on recursively. The symbol X generically denotes all kinds of Phrases. In Minimalism, the picture has been drastically reduced, gaining in simplicity and depth of explanatory power [18].

The basic operation now is binary Merge: combine two elements of the lexicon, α and β, into a binary un-ordered set $\{\alpha, \beta\}$. This is, as a whole, of the same category as one of them (the head). For instance: $\{_\alpha\alpha, \beta\}$. *The smart man* is a man, *man* is the head, and we have a Noun Phrase; *Had been bought* is an event of buying (the verb *buy* in the past tense is the head), we have a Verb Phrase. This set, as a whole, is then Merged with a third element from the lexicon γ getting $\{_\gamma\gamma, \{_\alpha\alpha, \beta\}\}$. The new construct (Syntactic Object, SO) can have the category of γ.

Without going into further details, we observe that this binary Merge of larger and larger syntactic objects is again recursively repeated, in more complex sentences, with subordinates, relatives, embeddings, until the whole sentence is terminated [19]. The syntactic process, called "derivation", is similar to a proof ending when the sentence is terminated.

Merge can create binary sets by pairing different lexical items or whole constituents (e.g. $\{V, NP\}$ a verb and a noun phrase). This is called External Merge (EM). Internal Merge (IM) merges a set X with a term Y of X (where a term is a subset of X or a subset of a term of X), yielding $\{Y, X\}$ with two *copies* of Y, one the copy merged with X and one the term Y of X, which remains, because X is unchanged by the operation. This means that displacement yields copies, providing the basis for semantic interpretation ("reconstruction"). Syntactic movement is now reduced to the operation of IM. Merge leaves the items being merged unaltered (this is called No Tampering Condition; for instance, the sub-units that form a composite word like *undiagonalizable* (*un-, al,- ble*) are never treated separately by the derivation).

Note that there are intermediate cyclic points of derivational (computational) closure, called Phases. We will use upper case P for Phases in syntax and lower case p for phases in physics. Thus, the syntactic derivation (the specific mental computation) stops when a Phase is reached, and then a higher Phase is opened. The process continues inside-out, building higher and higher components in the syntactic hierarchy.

There is also "pair-merge", for asymmetric adjunction, when a sentence has some optional qualifications of the noun phrase (as in *the book of poems [with the glossy cover] [from Blackwell]*, where the expressions in square brackets are adjuncts). This Merge must be asymmetric. We cannot have *the book with the glossy cover from Blackwell of poems*. Finally, we have the Agree operation (*he goes, they go, he has gone, they have gone*) under conditions of minimal, strictly local search. Syntax has no other component than these, it is therefore called Narrow Syntax.

In the previous theory of GB [20, 21], syntax had many operators and modules. In MP, these are subsumed under the basic operation Merge. The head gives the name to the constituent it generates, nouns to Noun Phrases, verbs to Verb Phrases and so on. Generally, we have an $\{H, XP\}$ construction, a Head and a Phrase (the X in XP denoting any phrasal category).

Heads such as nouns, verbs, adjectives, prepositions, have been known in traditional linguistics; other ones, such as complementizer, inflection, tense, negation and more, are of

IOP Conf. Series: Journal of Physics: Conf. Series **880** (2017) 012016 doi:10.1088/1742-6596/880/1/012016

abstract kind and their identification has been not obvious. What were in GB called "empty categories", because they are not pronounced or written, are now simplified in terms of copies. Copies are elements already present in previous steps of the derivation, for instance items extracted from the lexicon.

We remark that the condition of "strict locality" applies to the structure of the sentence. The number and kind of nodes separating the affected elements in the syntactic tree are important, not the number of words separating the affected elements in the sentence. This is a central property of syntax and is called "structure dependence". It constitutes a sharp departure from anti-generativist approaches to language based on statistics or conventions of use.

3. The X-bar tree, the Fibonacci progression and the breakdown of time-reversal symmetry

In Generative Grammar we have a collection of binary entities. Lexical items are represented by convention as $(+/-)$; for instance Nouns as $(+N, -V)$, Verbs as $(-N, +V)$. This notation can be extended to Phrasal Heads $(+H, -C)$ and Complements $(-H, +C)$. In the syntactic derivation, we have Terminal nodes $(+T)$ and nonterminal nodes $(-T)$. Copies of lexical items, or of larger structures, in a sentence can be pronounced $(+Pr)$ or not-pronounced $(-Pr)$. Recursive applications of Merge may produce a Phase $(+Ph)$ or not $(-Ph)$. The basic syntactic operation, Merge, generates a binary set. Syntactic trees (the X-bar tree) have thus only two branches departing from each node, which is referred to as "binary branching" [22].

We formalize this in terms of vector space of states, the Pauli matrices σ_i, $i = 1, 2, 3$, and the unit matrix I. The vector space on which the matrices operate is built on the basis of the states $\begin{pmatrix} 0 \\ 1 \end{pmatrix}$ and $\begin{pmatrix} 1 \\ 0 \end{pmatrix}$, denoted by $|0\rangle$ and $|1\rangle$, respectively, with scalar product $\langle i|j\rangle = \delta_{ij}$, $i, j = 0, 1$. In full generality, they may represent two lexical elements or two levels of the same lexical element. They are fermion-like eigenstates of σ_3 with eigenvalue $\mp 1/2$, respectively. Thus we consider the $SU(2)$ group and the $su(2)$ algebra [23] with commutation relations which, written in terms of the matrices $\sigma^{\pm} = \sigma_1 \pm i\,\sigma_2$, are

$$[\sigma_3, \sigma^{\pm}] = \pm\sigma^{\pm}, \quad [\sigma^-, \sigma^+] = -2\sigma_3 \,. \tag{1}$$

The Pauli matrices σ_i, $i = 1, 2, 3$, and the unit matrix I form a basis for the $SU(2)$ group, so that any 2×2 matrix can be expressed in terms of superpositions of the σ_i, $i = 1, 2, 3$, and I. For example, the so-called Chomsky matrix C_1 is given by

$$C_1 \equiv \begin{pmatrix} 1 & 0 \\ 0 & -i \end{pmatrix} = \sigma_3 + \frac{I}{2} + i\left(\sigma_3 - \frac{I}{2}\right) = \sigma^+\sigma^- \; - \; i\,\sigma^-\sigma^+ \tag{2}$$

and its hermitian conjugate $C_2 = C_1^{\dagger} = \sigma^+\sigma^- \; + \; i\,\sigma^-\sigma^+$.

We will assume to have a collection of N two-level objects ("particles" or "lexical elements"). In a standard fashion they are represented for each i by the "ground states" $|0\rangle_i$ and "excited states" $|1\rangle_i$, $i = 1, 2, 3, ...N$. We also write $\sigma_{3i} = \frac{1}{2}(|1\rangle_{ii}\langle 1| - |0\rangle_{ii}\langle 0|)$, with eigenvalues $\pm\frac{1}{2}$, and $\sigma_i^+ = |1\rangle_{ii}\langle 0|$ and $\sigma_i^- = |0\rangle_{ii}\langle 1|$. In the algebra Eq. (1), we then have $\sigma^{\pm} = \sum_{i=1}^{N} \sigma_i^{\pm}$ and $\sigma_3 = \sum_{i=1}^{N} \sigma_{3i}$. We recall that the anticommutation relations of the σ's are $\{\sigma_i, \sigma_j\} = (1/2)\delta_{ij} I$, implying the antisymmetry under permutation of the fermion-like states.

In order to see how the "binary Merge" between two states is generated, consider first the two states $|0\rangle$ and $|1\rangle$. In the following we will consider generalization to the collection of N elements and restore the index i, which now for simplicity we omit. Start with $|0\rangle$. Here and in the following we do not consider the (trivial) possibility to remain in the fundamental state $|0\rangle$ (which is dynamically equivalent to "nothing happens"). The physical meaning of this is that we neglect fluctuations in the ground state, which can be described by $\sigma^- \sigma^+|0\rangle = 1|0\rangle$, i.e. $|0\rangle \to |0\rangle$. In the quantum formalism, this is achieved by considering the normal ordering or Wick product of the operators. The interesting possibility is the one of the excitation process from $|0\rangle$ to $|1\rangle$. This is obtained by applying σ^+ to $|0\rangle$: $|0\rangle \to \sigma^+|0\rangle = |1\rangle$, which, by associating $0 \leftrightarrow |0\rangle$ and $1 \leftrightarrow |1\rangle$, may be represented as $0 \to 1 \equiv |1\rangle$. As a first single step the state $|1\rangle$ has been singled out. After that we describe the "action" on that state by application of the sigmas.

Consider now that from the anticommutation relations above recalled $\sigma^+ \sigma^+ = 0 = \sigma^- \sigma^-$, which express the Pauli principle, and $\sigma^+ \sigma^- |1\rangle = 1|1\rangle$ and $\sigma^- \sigma^+ |0\rangle = 1|0\rangle$. Note that application of $\sigma^+\sigma^-$ is considered to produce a single step since it is equivalent to the application of the unit matrix I to $|1\rangle$. Note also that $|1\rangle \to |0\rangle$ describes the "decay process" of the excited state . $|1\rangle \to |1\rangle$ describes the "persistence" in the excited state, which represents a dynamically non-trivial possibility and thus we have to consider it. We thus may easily obtain the "tree"

$$\begin{array}{lll} & & \nearrow \; \sigma^-\sigma^+|0\rangle = |0\rangle \;\; \to \; \sigma^+|0\rangle \; = \; |1\rangle \qquad (3) \\ |0\rangle \;\; \to \;\; \sigma^+|0\rangle = |1\rangle \;\; \to & & \\ & & \searrow \; \sigma^+\sigma^-\sigma^+|0\rangle = |1\rangle \;\; \begin{array}{l} \nearrow \sigma^-\sigma^+|0\rangle \; = \; |0\rangle \\ \searrow \sigma^+\sigma^-\sigma^+|0\rangle \; = |1\rangle \end{array} \qquad (4) \end{array}$$

$$1 \qquad\qquad 1 \qquad\qquad\qquad 2 \qquad\qquad\qquad 3 \qquad (5)$$

and so on. At each step, new branching points $\begin{smallmatrix}\nearrow \\ \searrow\end{smallmatrix}$ (new nodes of the X-bar tree) are obtained and the X-bar tree is generated by the $SU(2)$ recursive dynamical process.

Note that in general, any power $(\sigma^+\sigma^-)^n \, \sigma^+|0\rangle = 1 \times \sigma^+|0\rangle$, for any n.

The conclusion at this point is that as a direct consequence of the fermion-like $su(2)$ algebra we have the "number of the states" in these first steps in the sequence: 1 1 2 3, starting with $|0\rangle$, [one state] , then [2 states], and (3) and (4) [3 states], (cf. Eq. (5)).

From here, from the two $|1\rangle$'s we will have in the next step two $|0\rangle$'s and two $|1\rangle$'s, and from the $|0\rangle$ we will get one single $|1\rangle$, in total 5 states: 1 1 2 3 5. We will get thus, in the subsequent steps, other states, and their numbers obtained at each step are in the Fibonacci progression ($\{F_n\}, F_0 \equiv 0$) with the ones obtained in previous steps. In general, suppose that at the step F_{p+q} one has p states $|0\rangle$ and q states $|1\rangle$; in the next step we will have: $(p+q)\,|1\rangle$ and $q\,|0\rangle$, $F_{q+(p+q)}$. In the subsequent step: $(p+2q)\,|1\rangle$ and $(p+q)\,|0\rangle$, a total of states $2p+3q = (q+p+q)+(p+q)$, i.e. the sum of the states in the previous two steps, according to the rule of the Fibonacci progression construction.

In conclusion, the X-bar tree (or F tree) is obtained as a result of the $SU(2)$ dynamics, and its recursivity or self-similarity properties turn out to be described by the Fibonacci progression.

It is interesting to remark that $(\sigma^+\sigma^-)^n \, \sigma^+|0\rangle = 1 \times \sigma^+|0\rangle$ can be thought of as a "fluctuating" process: σ^+ applied to the (ground) state $|0\rangle$ excites it to $|1\rangle$. Then σ^- brings it down to $|0\rangle$, and σ^+ again up to $|1\rangle$: $\sigma^+\sigma^-$ induces fluctuations $|1\rangle \leftrightarrow |0\rangle \leftrightarrow |1\rangle$ (through the "virtual" state $|0\rangle$), this is the meaning of the fact above observed that $\sigma^+\sigma^-$ is equivalent to 1 at any power n when operating on $\sigma^+|0\rangle$. Similar fluctuations can be obtained by considering $(\sigma^-\sigma^+)^n \, \sigma^-|1\rangle = 1 \times \sigma^-|1\rangle$, i.e. $|0\rangle \leftrightarrow |1\rangle \leftrightarrow |0\rangle$. This "fluctuating activity" corresponds, in the syntactic derivation, to successive applications of Merge. Simplifying a bit, when recursive Merge reaches the topmost node of a Phase, that is, a point of computational closure, everything underneath, in the tree, becomes off limit. The condition called "Phase Impenetrability Condition" [19],[24]-[26] specifies that nothing in a lower Phase is accessible to the syntactic operations that create the immediately higher Phase. The syntactic objects of the lower Phase and the lower Phase itself are dynamically "demoted" to a $|0\rangle$ state. The "fluctuating activity" is also much suggestive when one thinks of the processes (of milliseconds or so) in the selection of lexical items and the recursive Merge of these into syntactic objects.

3.1. Self-similarity, brain functional activity and breakdown of time-reversal symmetry

Note that in the construction of X-bar tree clarifying "how to go" from one step to the next one introduces the "dynamics". We have thus an "operatorial dynamics" by which multiplicities of states are generated. We observe that a Jaynes-Cummings-like model Hamiltonian can be constructed generating the full set of σ^+ and σ^- products of any order, in all possible orderings compatible with the $su(2)$ algebra (the products used above and leading, as we have seen, to the F progression). Thus, the X-bar tree, which plays so a crucial role in the MP, arises as a result of a *dynamical* model in linguistic, its recursive property being related to the self-similarity property of the Fibonacci progression. The paramount importance of the Fibonacci progression in language has been stressed in [27]-[29]. In the Appendix B we introduce the Fibonacci matrix and show its relation with the Chomsky C_1 matrix. We remark that according to known results in QFT [30],[31]-[33], fractal-like self-similar structures, such as the ubiquitous logarithmic spiral, the Fibonacci progression, the golden spiral and other self-similar structures signal the occurrence of underlying coherent dynamical processes, so that these fractal structures appear as the *manifestation* of the coherent states of the system (more specifically of squeezed or q-deformed coherent states). It is also remarkable that in neuroscience, the analysis of the brain functional activity also shows self-similar features [34, 35] again related to the coherent amplitude and phase modulated neuronal oscillatory patterns observed in the laboratory. It appears thus that our discussion of the recursive Merge in MP points to the emergence of a quite general phenomenon which is deeply rooted in the very brain physiological activity. In the following we comment more on this point.

We close this subsection by observing that at any given step of the X-bar tree (the F tree), the simple knowledge of the state $|0\rangle$ or $|1\rangle$ is not sufficient in order to know its parent state in the previous step; we should also know which one is the branch we are on. This in part corresponds to the Phase Impenetrability Condition mentioned above and to one of the major problems in all of contemporary linguistic theory. In speaking and reading we proceed left to right, from the "outside" (the main sentence), to the "inside" (subordinate sentence), but the syntactic derivation proceeds from right to left, from inside out. This creates a conflict, that presumably the construction of Phases, of periodic points of closure, solves [29, 36, 37].

While the tree construction (the "way forward") is fully determined by the σ's operations, the "way backwards" is not uniquely determined solely by the knowledge of the state $|0\rangle$ or $|1\rangle$. On the other hand, suppose one goes backward q steps starting from a given, say, $|1\rangle$ (or $|0\rangle$) . Then returning to such a specific state is no more guaranteed since at each branching point one has to chose which way to go (unless one keeps memory of its previous path, the Ariadne's thread...). In the syntactic derivation, "forward" consists in building further structure from the inside out, from right to left, proceeding upwards in the syntactic tree. The opposite, "backwards", consists in the derivation "looking down" to lower levels. The Phase Impenetrability Condition, as we have just seen, constrains this operation to a strict minimum. Omitting details, only the leftmost (and topmost) "edge" of the lower Phase is (quite briefly) still accessible to the operations building the next higher Phase.

The lesson is that, parameterizing by time the moving over the X-bar tree, time-reversal symmetry is broken. We therefore need to deal with dissipative formalism.

4. Semantics, copies of lexical elements and self-consistency of language

The conceptual interpretive (semantic) system (CI) and the sensory-motor system (articulation, auditory or visual perception) (SM) are assumed to be the two systems with which Narrow Syntax has to interface. Language cannot be simply restricted to "sounds" with meanings, it needs to include gestures as in Sign Language, also touch in the case of deaf-and-blind subjects [38, 39] (see also the analysis of American Sign Language [40] and the Tadoma method for deaf-and-blind subjects [38]).

Although in a highly nonlinear process, the CI "sees" all copies of the lexical elements, and interprets them; however, at the SM interface only one copy is pronounced (externalization) while the other copy (or copies) remains silent (is deleted at SM), as, for instance, in: *Which books did you read [books]?*. In most languages only the higher copy is pronounced, but there are languages in which the lower copy is pronounced and also languages in which all copies are pronounced. In the latter case, this applies to "short" elements (equivalent to the English "who", "which" and similar), never to whole Noun Phrases. We thus realize that in linguistics "copies" are important objects. We will show below how our modeling may account for this.

The present view is that Merge works without categorial constraints (a bit like Feynman's sum of all histories, before amplitudes give the wave function). With some simplifications, we can say that categories are needed only at the interface with CI for interpretation: which one is a verb, which one a noun, an adjective etc. (labeled heads). Thus, a minimal search process, or labeling process, *The Labeling Algorithm* [41]-[43], has to produce such a categorization at the CI interface. Ordering, implying non-commutativity, is, however, also important. There are two notions of order to be taken into account: ordering of the syntactic operations, which will be considered in the following, and ordering of the items in the externalized linguistic expression at the Sensory-Motor interface, namely, what to pronounce first, second etc. and what not to pronounce at all (deleted copies). Such an ordering, which needs to preserve anyway the interpretative categorization, is probably a reflex of the Sensory Motor system, not feeding narrow syntax or CI. Also in relation to previous views and analyses, see [44]-[48].

In summary, Third Factors (physics), underlying unconstrained Merge, the Labeling Algorithm and general principles of Minimal Computation (MC), are thought to account for all syntax properties. We show that this finds an explicit formalization in terms of QFT, with the absolutely not trivial result that linguistic features are actually manifestations of underlying dynamical processes.

4.1. From syntax to semantics. The manifold of concepts

Let us resume the discussion of the algebraic formalism. We restore the subscript i labeling the N elements introduced in Section 3. The associated states can be regarded as the ones at a given step, of high multiplicity, in the Fibonacci tree. Since N can be as large as one wants, we may always have a direct product of a large number (in principle an infinite number, hence one needs field theories) of factor states, $\Pi_{i=1,N}|s_i\rangle \equiv |s_1\rangle \otimes |s_2\rangle \otimes ...|s_i\rangle \otimes ... \equiv |s_1, s_2, ...s_i...\rangle$, with $s_i = 0$ or 1 for each $i = 1, 2, .., N$. The most general state, denote it by $|l\rangle$, is then a superposition of all states with l elements in $|1\rangle$ and $N - l$ elements in $|0\rangle$. The difference between the number of elements in $|1\rangle$ and the one of the elements in $|0\rangle$ is measured by σ_3 and is given by $\langle l|\sigma_3|l\rangle = l - \frac{1}{2}N$. This quantity is called the order parameter. Its being non-zero signals that the $SU(2)$ symmetry is broken (like, for example, in the case of the magnetic dipoles, where the order parameter provides the measure of the magnetization).

One can show that in the large N limit the $su(2)$ algebra of the σ matrices, represented in the space of the $|l\rangle$ states, for any l, and written in terms of $S^{\pm}$ and $S_3 \equiv \sigma_3$, where $S^{\pm} = \sigma^{\pm}/\sqrt{N}$, "contracts" ("rearranges") into the Weyl-Heisenberg algebra

$$[S_3, S^{\pm}] = \pm S^{\pm} , \quad [S^-, S^+] = 1 . \tag{6}$$

The result Eq. (6) is a central result. It expresses the "rearrangement" of the $su(2)$ algebra [Eq. (1)] in the $e(2)$ algebra [Eq. (6)], which is isomorph to the Weyl-Heisenberg algebra, with S_3 playing the role of the number operator and $S^{\pm}$ the role of *boson* ladder operators. Notice the dynamical transition from fermion-like degrees of freedom in the $su(2)$ algebraic frame above discussed to the Weyl-Heisenberg ($e(2)$) boson algebra. Such a *dynamical rearrangement of symmetry* and the contraction of the algebra for large N are well known *dynamical* processes in QFT [33],[49]-[52]. They occur when there is spontaneous breakdown of symmetry characterized by the order parameter, which behaves as a non-vanishing classical field. One can show [33, 49, 52, 53] that then the space of the states of the system splits into (infinitely many) unitarily inequivalent representations of the algebra Eq. (6), that is it undergoes a process of *foliation*, splitting in many physically inequivalent subspaces, each one labeled by a specific value assumed by the order parameter. Each of these subspaces is a well defined vector space and represents a possible *phase* in which our system can live. Moreover, each of these *phases* is characterized by collective, coherent waves, represented by the ladder operators $S^{\pm}$. More specifically, the theorem can be proven [54], which predicts the dynamical formation of long range correlation modes (the Nambu-Goldstone (NG) modes) when the symmetry gets spontaneously broken. In Generative Grammar the symmetry breaking phenomenon (the anti-symmetry of syntax, and the dynamic anti-symmetry of syntax) have been cogently argued for by Richard Kayne [55] and Andrea Moro [56]. This is in part why issues about the status of X-bar (as part of Narrow Syntax or as an emergent configuration of recursive binary Merge) have been recently debated [41, 42] (see also [27]). These NG collective modes $S^{\pm}$ are the carrier of the *ordering* information through the system volume [33, 49, 52, 53]. *Order* thus appears as a collective dynamical property of the system which manifests in the limit $N \gg l$. The order parameter provides indeed a measure of the system ordering. Different degrees of ordering corresponding to different values of the order parameter in a convenient range of variability self-consistently defined by the dynamics (different densities of the NG modes). Different values of the order parameter thus denote different, i.e. inequivalent, phases of the system [33, 49, 52, 53]. The dynamics of the system undergoing *phase transitions* is thus characterized by *criticality*, in much a similar way as the brain functional activity is characterized by critical transition through many regimes in a far from the equilibrium dynamics [34, 35, 57, 58].

We thus realize that our system has undergone a formidable dynamical transition, moving from the regime of being a collection of (fermion-like) *atomistic* components (lexical elements) to the regime of (boson-like) collective, coherent $S^{\pm}$ fields. Our main assumption at this point is to identify a specific *conceptual, meaningful* linguistic content (a Logical Form, LF [59]) with the collective coherent phase associated to a specific value of the order parameter. The semantic level (the "manifold of concepts") thus emerges as a dynamical process out of the syntactic background of lexical elements, in a way much similar (mathematically isomorph) to the one by which macroscopic system properties emerge as a coherent physical phase out of a collection of elementary components at a microscopic (atomistic) level in many-body physics [33, 52].

In conclusion, we can now give a quantitative characterization of the "interfaces" where the Narrow Syntax has to make contact with the Conceptual Interpretative (CI) system: interfaces are met when the spontaneous breakdown of symmetry is met, i.e. as the limit $N \gg l$ is approached. It is there that a specific meaning or "concept" arises by selecting out one representation of the algebra from many of them "unitarily inequivalent" among themselves (each corresponding to a different concept) [57]. The concept appears at that point as a collective mode, not a result of associative process pulling together bits and little lexical pieces, words etc.. Rather it binds all of them together in the specific explicit entanglement of which it is expression, as we will exhibit below. The collectiveness comes from the "phase coherence", whose carriers are the collective NG $S^{\pm}$ fields. It can be shown that coherence guaranties also the stability of the semantic content, its robustness against "defects" at the atomistic level. For instance, missing or partially corrupted, or deformed words, a non perfect grammatical construction, etc. barely affect the "meaning", provided however they occur far from a "criticality threshold". In the opposite case, even a slight change at the elementary level may trigger a phase transition process leading to a "dramatic" (i.e. unitarily inequivalent, in a formal sense) change in the semantic landscape. A measure of the departure from the "semantic stability" is provided by the consistency of the violation of the minimization of the free energy [1]. We also understand why "only at the interfaces the issues of ordering become relevant". Order indeed is lack of symmetry and it can only appear when this is spontaneously broken.

Categorization and non-commutativity (and order) are only necessary at the CI interface. Indeed, only at the large N limit CI needs labeled heads: which one is a verb, which one a noun, an adjective etc.. We have seen that the formal construction of the binary Merge does not require labeled structures (Noun, Verb, Adjective, Preposition etc.). The necessity of labeling (*The Labeling Algorithm*) only arises at the interface with meaning. Interpreting a Determiner Phrase (*the dog, many boys, most books*) or a Verb Phrase (*was going to Rome, had bought the book*) or a Complementizer Phrase (*that I know him, whether it's wise, who did that*) is a necessity for the Conceptual Intentional system, with the formal label of a syntactic object triggering different Intentional landscapes. Once the Narrow Syntax has made contact with CI system, through the action-perception cycle [57] of the cortex dynamics, the sensory-motor system (SM) gets also involved and therefore the linguistic structures can be externalized, allowing to communicate to other speakers all the required subtleties of meaning.

The formalism here presented thus endorses Chomsky's thesis that Merge is unconstrained, and that issues of labeling (headedness, categorization of lexical items) and ordering only arise at the interfaces of Narrow Syntax with the Conceptual-Intentional (CI) system and the Sensory-Motor (SM) system.

4.2. Copies of lexical elements

We now consider the feature of the copies of lexical elements in the Minimalist Program (MP).

At the end of Section 3.1, we have observed that time-reversal symmetry is broken moving along the X-bar tree. Time-reversal symmetry breakdown characterizes the dynamics of dissipative systems, which are systems open to the environment in which they are embedded. We extend therefore to the X-bar structures the formalism which is used for the study of dissipative systems. In such a mathematical formalism a characteristic feature is the doubling of the system degrees of freedom $A \to \{A, \tilde{A}\}$ [60], where A generically denotes the degree of freedom of the system and $\tilde{A}$ its "copy" or the doubled degree of freedom. They belong to two copies of the algebra $\mathcal{A}$. Thus the doubling is obtained by introducing the mapping $\mathcal{A} \to \mathcal{A} \times \mathcal{A}$ [61]. Typically A and $\tilde{A}$ are operators represented by some matrices, with some properties specified by use of a subscript $\mathbf{k}$, e.g. $A_{\mathbf{k}}$, which, however, we will omit for notational simplicity, as far as no misunderstanding occurs.

We also have the doubling of the state space $\mathcal{F} \to \mathcal{F} \times \tilde{\mathcal{F}}$. The operators A and $\tilde{A}$ act on $\mathcal{F}$ and $\tilde{\mathcal{F}}$, respectively. In the case of dissipative systems, $\tilde{A}$ represent the environment degrees of freedom and the system elements cannot be exchanged with the elements of the environment. In the present case of linguistics, at the syntactic and semantic levels, lexical elements and conceptual contents cannot be simply interchanged. This is described by the algebraic structure called the non-commutative Hopf algebra or q-deformed Hopf algebra, with q called the deformation parameter.

For notational simplicity from now on we will denote by A and $A^\dagger$ the operators S^- and S^+ in Eq. (6), respectively. Thus, the doubling process implies that correspondingly we also have $\tilde{S}^-$ and $\tilde{S}^+$, which will be denoted as $\tilde{A}$ and $\tilde{A}^\dagger$, respectively.

By adopting $q(\theta) = e^{\pm\theta}$, one can show that the system ground state is [33, 52, 60]

$$|0(\theta)\rangle_{\mathcal{N}} = e^{i\sum_{\mathbf{k}}\theta_\kappa G_{\mathbf{k}}}|0\rangle = \prod_{\mathbf{k}} \frac{1}{\cosh\theta_\kappa} \exp(\tanh\theta_\kappa A^\dagger_\kappa \tilde{A}^\dagger_\kappa)\,|0\rangle, \tag{7}$$

where the subscript $\mathbf{k}$ has been restored, and $|0\rangle \equiv |0\rangle \times |0\rangle$ is the state annihilated by A and $\tilde{A}$: $A|0\rangle = 0 = \tilde{A}|0\rangle$ (the vacuum state). However, A and $\tilde{A}$ do not annihilate $|0(\theta)\rangle_{\mathcal{N}}$. It is annihilated instead by the operators $A(\theta)$ and $\tilde{A}(\theta)$, obtained as

$A(\theta) = \exp(i\sum_{\mathbf{k}} \theta_\kappa G_{\mathbf{k}}) A \exp(-i\sum_{\mathbf{k}} \theta_\kappa G_{\mathbf{k}})$ and $\tilde{A}(\theta) = \exp(i\sum_{\mathbf{k}} \theta_\kappa G_{\mathbf{k}}) A \exp(-i\sum_{\mathbf{k}} \theta_\kappa G_{\mathbf{k}})$. The meaning of the subscript $\mathcal{N}$ is clarified below, θ denotes the set $\{\theta_\kappa, \forall\mathbf{k}\}$ and $|0(\theta)\rangle_\mathcal{N}$ is a well normalized state: ${}_\mathcal{N}\langle 0(\theta)|0(\theta)\rangle_\mathcal{N} = 1$. The vacuum $|0(\theta)\rangle_\mathcal{N}$ turns out to be a generalized $SU(1,1)$ coherent state of condensed couples of A and $\tilde{A}$ modes [23, 60], which are entangled modes in the infinite volume limit. The operator $\exp(i\sum_{\mathbf{k}} \theta_\kappa G_{\mathbf{k}})$, with $G_{\mathbf{k}} \equiv -i\,(A^\dagger_{\mathbf{k}} \tilde{A}^\dagger_{\mathbf{k}} - A_{\mathbf{k}} \tilde{A}_{\mathbf{k}})$, is the generator of the state $|0(\theta)\rangle_\mathcal{N}$ as shown in Eq. (7).

Incidentally, we observe that in language, in first approximation, the vacuum state is silence. Just like in the present algebraic formalism, there are many kinds of silence. Not only how a silence gap is interpreted in the unfolding of a conversation, but in a more specific and more technical sense. There is, literally, a syntax of silence [62] in linguistic constructions called ellipsis (*Mary bough a book and Bill did ____ too*) and sluicing (*Ann danced with someone but I do not know who ____*). Jason Merchant and other syntacticians and semanticists have shown that what can be omitted is never just an arbitrary "bunch of words", but an entire syntactic constituent (an entire Verb Phrase, most frequently). The syntax and semantics of several well defined but unpronounced elements has been part of the theory since the beginning of Generative Grammar [63].

One can show [60, 61, 64] that ${}_\mathcal{N}\langle 0|0(\theta)\rangle_\mathcal{N} \to 0$ and ${}_\mathcal{N}\langle 0(\theta')|0(\theta)\rangle_\mathcal{N} \to 0$, $\forall\theta \neq \theta'$, in the infinite volume limit $V \to \infty$. Thus we conclude that the state space splits in infinitely many physically inequivalent representations in such a limit, each representation labeled by a θ-set $\{\theta_{\mathbf{k}} = \ln q_{\mathbf{k}}, \forall\mathbf{k}\}$. This is the *$q(\theta)$-foliation* process of the state space. In the present case of linguistics this represents the process of generation of the manifold of concepts. It is a dynamical process since the generator $G_{\mathbf{k}}$ is essential part of the system Hamiltonian [60]. In our present case, "physically inequivalent" representations means that the "manifold of concepts" is made of "distinct", different spaces, each one representing a different "concept" (in language we have the Logical Forms (LFs) composing the global Logical Form of the entire sentence), here described as the coherent collective mode generated through the X-bar tree. These spaces (concepts) are protected against reciprocal interferences since the spaces are "unitarily inequivalent", i.e. there is no unitary operator (here is rooted the criticality of the dynamics) able to transform one space in another space [57, 58], which corresponds to the fact that syntactic Phases cannot be commingled, nor "reduced" one into the other. Phases are mutually impenetrable.

In practice, however, the unitary inequivalence is smooted out by realistic limitations, such as, for example, the impossibility to reach in a strict mathematical sense the $V \to \infty$ limit (i.e. the "infinite number" of lexical elements or the theoretically infinite number of choices for the co-referentiality indices in the logical form of even the simplest sentences). Thus, realistically, we may also move from concept to concept in a chain or trajectory going through the manifold of concepts [34, 35, 57],[65]-[67]. This corresponds to the compositionality of meanings, when the syntactic derivation proceeds "upwards" (that is: forward) from the lower Phases to the higher Phases, from local Logical Forms to the composition of more inclusive Logical Forms. Remarkably, these trajectories in the unitarily inequivalent spaces (the "manifold of concepts") have been shown to be classical chaotic trajectories [65], which again links these processes in language to the chaotic features of brain functional dynamics observed in neuroscience labs and discussed since long in the literature [68].

4.3. Self-consistency of language and its built-in self-reference

In order to better understand the role played by the "tilde copies", $\tilde{A}$, let's compute $N_{A_{\mathbf{k}}} = A^\dagger_{\mathbf{k}} A_{\mathbf{k}}$ in the state $|0(\theta)\rangle_\mathcal{N}$

$$\mathcal{N}_{A_{\mathbf{k}}}(\theta) \equiv {}_\mathcal{N}\langle 0(\theta)|A^\dagger_{\mathbf{k}} A_{\mathbf{k}}|0(\theta)\rangle_\mathcal{N} = {}_\mathcal{N}\langle 0(\theta)|\tilde{A}_{\mathbf{k}}(\theta)\tilde{A}^\dagger_{\mathbf{k}}(\theta)|0(\theta)\rangle_\mathcal{N} = \sinh^2\theta_{\mathbf{k}}. \tag{8}$$

From this we see that for any $\mathbf{k}$ the only non-vanishing contribution to the number of non-tilde modes $\mathcal{N}_{A_{\mathbf{k}}}(\theta)$ comes from the tilde operators, which can be expressed by saying that these last ones constitute the dynamic *address* for the non-tilde modes (the reverse is also true, the only non-zero contribution to $\mathcal{N}_{\tilde{A}_{\mathbf{k}}}(\theta)$ comes from the non-tilde operators).

This also shows that the physical content of $|0(\theta)\rangle_{\mathcal{N}}$ is specified by the $\mathcal{N}$-set $\equiv \{\mathcal{N}_{A_{\mathbf{k}}}(\theta), \mathcal{N}_{A_{\mathbf{k}}}(\theta) = \mathcal{N}_{\tilde{A}_{\mathbf{k}}}(\theta), \forall \mathbf{k}\}$, called the *order parameter*. The $\mathcal{N}$-set characterizes the vacuum; this also explains the meaning of the $\mathcal{N}$ subscript in $|0(\theta)\rangle_{\mathcal{N}}$.

All of this sheds some light on the relevance of "copies" in the MP. In some sense they are crucial in determining (indeed providing the address of) the whole conceptual content of the considered linguistic structure. They provide the *dynamic reference* for the non-tilde modes. Unpronounced copies, being silent, do not reach the Sensory Motor system, but they are crucially interpreted by the Conceptual Intentional system. They are necessary to the understanding of the meaning of what is actually pronounced. Remarkably, they are "built in" in the scheme here proposed; they are not imposed by hand by use of some constraint "external" to the linguistic system. It is in this specific sense that we speak of "self-consistency": our formal scheme is computationally (logically) self-contained [69]. Perhaps the real power of the linguistic tool available to humans consists in such a specific feature [1].

The entanglement between the modes A and $\tilde{A}$ is seen by expanding $|0(\theta)\rangle_{\mathcal{N}}$ in Eq. (7) as

$$|0(\theta)\rangle_{\mathcal{N}} = \left(\prod_{\mathbf{k}} \frac{1}{\cosh\theta_k}\right) \left[|0\rangle \otimes |0\rangle + \sum_{\mathbf{k}} \tanh\theta_k |A_{\mathbf{k}}\rangle \otimes |\tilde{A}_{\mathbf{k}}\rangle + \ldots\right] \tag{9}$$

and we see that it cannot be factorized into the product of two single-mode states $|A_{\mathbf{k}}\rangle$ and $|\tilde{A}_{\mathbf{k}}\rangle$. A and $\tilde{A}$, for any $\mathbf{k}$, are thus entangled modes. $|0(\theta)\rangle_{\mathcal{N}}$ can be also written as [52, 60]:

$$|0(\theta)\rangle_{\mathcal{N}} = \textstyle\sum_{n=0}^{+\infty} \sqrt{W_n}\,(|n\rangle \otimes |n\rangle),$$

with $W_n = \prod_{\mathbf{k}} \sinh^{2n_{\mathbf{k}}}\theta_k / \cosh^{2(n_{\mathbf{k}}+1)}\theta_k$ and n denoting the set $\{n_{\mathbf{k}}\}$, $0 < W_n < 1$, $\sum_{n=0}^{+\infty} W_n = 1$. The probability of having the component state $|n\rangle \otimes |n\rangle$ in the state $|0(\theta)\rangle_{\mathcal{N}}$ is thus W_n, which would be suppressed for large n in the case of a finite number of terms in the summation, since it is a decreasing monotonic function of n. However, this is not the case if the summation is over infinite terms. In such a limit there is no unitary generator able to disentangle the $A - \tilde{A}$ modes, $|0(\theta)\rangle_{\mathcal{N}}$ and $|0\rangle$ are unitarily inequivalent states. The entropy S is given by [33, 52, 60]

$${}_{\mathcal{N}}\langle 0(\theta)|S|0(\theta)\rangle_{\mathcal{N}} = \sum_{n=0}^{+\infty} W_n \, \log W_n \ . \tag{10}$$

Consistently with the breakdown of time-reversal symmetry (the *arrow of time*), one finds that variations of the entropy control the time evolution (the trajectories) in the manifold of concepts (the space of the *infinitely many* LF). In this formalism entropy is thus related with the semantic level of the LF, *meanings*, dynamically arising as collective modes out of the syntactic (atomistic) level of lexical elements. One can also exhibit the free energy functional and show that its minimization is related to the stability of the system dynamic regime (phase) [1].

5. Concluding remarks

In this paper we have reported some recent results showing that the many-body formalism may describe central features of the Minimalist Program in linguistics, with particular reference to the generation of the X-bar tree structure, its self-similarity properties and relation with the Fibonacci progression and the breakdown of time-reversal symmetry. We have discussed the role of the copies in the conceptual interpretative system CI and shown that they can be accounted by the Hopf algebra structure of the doubled tilde operators in the quantum field theory formalism. What emerges from our physical modeling of the Minimalist Program is that a truly dynamical structure underlies linguistic features. Among others, two specific results may be worth to be stressed. One is the dynamical emergence of Logical Forms, namely the dynamical transition from a numeration of lexical items to syntax and from syntax to the logical form (LF) of the sentence and to meaning. This has brought us to the identification of the "manifold of concepts", formally represented by the set of unitarily inequivalent state spaces in the foliation process of QFT. The formation of concepts appears then as the dynamical generation of collective coherent modes spanning as waves the lexical items entering the sentence or collection of sentences. These collective modes are associated to the ordering of the lexical items, but not strictly dependent on each of them. This aspect is expression of a stability principle, namely the robustness of meaning against corruption, alteration or even deletion of one or more, however below some critical threshold, of the lexical items. On the other hand, different ordering (different meanings), i.e. different values of the order parameter, denote inequivalent phases of the system [33, 49, 52, 53]. The undergoing dynamics of *phase transitions* is characterized by *criticality*: in a similar way as in critical dynamics of the brain functional activity [34, 35, 57, 58], we may have trajectories going through the manifold of concepts [34, 35, 57],[65]-[67] (move from concept to concept). These trajectories have been shown to be classical chaotic trajectories [65]. This again links these processes in language to the observed chaotic features of brain functional dynamics [68].

The other result is the linguistic computational self-consistency. This arises from the nonlinear interaction between the doubled entities (the non-tilde and the tilde modes representing the lexical items and their copies) entering the specific Hopf algebraic structure of the QFT formalism. The copies or tilde modes have been recognized to provide the *dynamical reference* (the "address") of the non-tilde modes. The result is the logical self-consistency, by inclusion of the *reference* terms, of languages. The resulting state is an entangled state of the non-tilde/tilde modes. Within such a scheme we can consistently define thermodynamic operators such as the entropy and the free energy. Languages appear thus to posses in their own structure, a "built in" "truth (or falsehood) evaluation function", a feature where perhaps resides, as we have observed, the real power of the linguistic tool available to humans.

Summing up, we have uncovered the isomorphism between the linguistic aspects of the Minimalist Program and the physics of many-body systems. As observed in [1], although the algebraic properties of the QFT formalism have been exploited by us, in our scheme the linguistic structures are "classical" ones. This should not be surprising, since [33, 52] QFT describes not only the quantum world of elementary particles and condensed matter physics, but also macroscopically behaving systems characterized by ordered patterns [33, 52].

Our analysis also suggests that the dynamics underlying the biochemical phenomenology of the brain behavior [34, 35, 57, 58, 67] may also provide the basic mechanisms of linguistics. As observed, it is remarkable that the brain functional activity also shows self-similar features [34, 35] related to coherent amplitude and phase modulated neuronal oscillatory patterns. This suggests that the recursive Merge in linguistic MP might be the manifestation of a quite general phenomenon, deeply rooted in the very same brain physiological activity.

Acknowledgments
We are grateful to Noam Chomsky for many illuminating discussions and suggestions. We are also thankful to the late Walter J. Freeman and Emilio Del Giudice for many insights. Very useful discussions with Gianfranco Basti are also acknowledged. Miur and INFN are acknowledged for partial financial support.

Appendix A. A little bit of history

(1) *Phrase Structure Grammars*

Until the mid-Fifties, a "grammar" was the result of applying some procedures (typically, segmentation and categorization) to a corpus of data[1]. Chomsky's first attempt at a new perspective was his master thesis [71, 72], but did not reach any audience until *Logical Structure of Linguistic Theory* (1955), his "Three models" paper in 1956 [73] and *Syntactic Structures* in 1957 [74, 75]. Before the advent of his "transformational grammar", the dominant models were Phrase Structure Grammars (PSG) a usually long list of general rules specifying how each kind of Phrase (each phrasal constituent) was to be expanded, all the way to the manifest expression of a complete sentence. The arrows indicate "rewriting rules", that is: what appears on the left has to be rewritten as what appears on the right. For instance (a very simple one) for a whole sentence (S), we typically have as constituents a Noun Phrase (NP), a Verb Phrase (VP) and (optionally) a Prepositional Phrase (PP). More concretely

NP1 → the cat
AUX → is
VP → (is) chasing
NP2 → a mouse
PP → in the garden
(S) = The cat is chasing a mouse in the garden.

(2) *Enter Transformational Grammar*

Chomsky did persuasively show that such grammars were radically insufficient to account for real languages. That there was a *kernel* sentence that can be *transformed* into variants that are intimately related, by permuting the order of the words in the phrases (obtaining passives, interrogatives, negations etc.). In his work *The Logical Structure of Linguistic Theory* LSLT [63], which really was also the first complete phrase structure grammar, kernel sentences were derived by obligatory transformations from a more abstract phrase structure. E.g., a transformation was required to convert [[be-ing] chase] to [be chasing] (and another similar one to yield is), what Haj Ross later called "Affix-hopping". Suitable mathematical analyses, in terms of the Theory of Automata and the theory of recursive functions, did show that Transformational Grammars (TG) constitute a more powerful kind of automata than Phrase Structure Grammars [76]. The only empirically significant concept in Generative Grammar is strong generative capacity. Canonical examples of transformations are like the following:

The cat is not chasing a mouse in the garden. NEGATION
A mouse is being chased by the cat in the garden. PASSIVE
What/who is the cat chasing in the garden? ACTIVE INTERROGATIVE
What/who is being chased by the cat in the garden? PASSIVE INTERROGATIVE
What/who is not being chased by the cat in the garden? NEGATIVE INTERROGATIVE

Instead of having many more ad hoc phrase structure rules, transformations took care of such cases[2]. The kernel, then was generalized into a deep structure distinct from the surface structure. The link with meaning was supposed to be the deep structure, not the surface structure. All speakers understand that the basic syntactic and semantic relations are preserved in sentences such as

He met his friend.
He succeeded in meeting his friend.
He failed to meet his friend.
He avoided meeting his friend.

Separate ad hoc phrase structure rules would have to be introduced, while basic transformational rules cover all such construction and explain why the basic meaning is invariant

[1] The most sophisticated development procedures were those of Zellig S. Harris (Chomsky's teacher, at the time) [70].

[2] Omitting inessential details and stressing that negation has a representation in formal logic, all these sentences have the same Logical Form (provided we insert or do not insert the logical symbol for negation, as required).

at the level of the deep structure[3]. Also, there are physically absent, but mentally present, syntactic elements that PSG[4] rules could not possibly deal with. Between square parentheses what is not pronounced (or written)

He saw the man standing at the bar.
He saw the man [the man was] standing at the bar.
They saw the prisoner escape from the prison.
They saw the prisoner [when they saw the prisoner he was] escap(ing) from the prison.

Importantly, there are verbs that block such constructions[5]. Ill-formed sentences are, by a long tradition, preceded by an asterisk.

They suspected that the prisoner would try to escape.
**They suspected the prisoner trying to escape.*

Contrast this with the perfectly OK sentence: *They saw the prisoner trying to escape.*

So, silent elements were introduced for the first time in linguistics. They are silent ("empty" in the technical terminology) but mentally present and very important in the syntactic structure and in the interpretation.

(3) *The Theory of Government and Binding*

In subsequent years, syntactic movement became a key notion, more general than transformations. Elements in a sentence can be moved to a different position, leaving a trace (an "empty category"). The interpretive system understands the meaning of the sentence by (mentally, silently) paying attention to the position of the trace and to what "governs" the trace: *Which books did you read?*

Books are the object of read and are so interpreted, as being to the right of read, where the trace is. Making the trace explicit by the letter t, we have: *Which books did you read t?*

Making syntactic movement even more explicit, we can write:

Which books did you read t?
↑←−−←−−←−−←−−↓

Several kinds of un-pronounced (and un-written) syntactic elements were introduced, called "empty categories" (empty because un-pronounced). But they had to be governed in a very similar way to the explicit elements of the sentence (this was called ECP, the Empty Category Principle).

Government was exactly defined, but we will not go into it here, because this notion has been subsumed by a deeper, more abstract notion in the current Minimalist version of the theory.

It's worth noticing that there are "barriers" to syntactic movement, that is: some instances of syntactic movement are blocked. There are "islands" to extraction and movement.

That the police would arrest several rioters was a certainty.
**Who was that the police would arrest a certainty?*

In the Theory of Government and Binding, these barriers for movement and the nature of "islands" were explained in terms of strict locality. Summarizing and simplifying drastically a number of acute analysis and several refinements, we can indicate here the so-called Minimal Link Condition [2]. Derived from general principles of derivational economy, it stated that movement can only occur to the closest potential landing site.

[3] We will not detail here how modifier verbs like "succeed" and "fail" are easily expressed in Logical Form.
[4] Phrase Structure Grammars
[5] Lexical semantics explains why these differences exist. Simplifying drastically, verbs like "suspect" (called Psych Verbs) cannot take a gerundive form as their direct object, while verbs of perception like "see" can.

IOP Conf. Series: Journal of Physics: Conf. Series **880** (2017) 012016 doi:10.1088/1742-6596/880/1/012016

A similar situation occurs, for instance, with agreement. In Italian there are masculine and feminine features for inanimates: *la bella sedia*, [FEM] *il grande coltello* [MASC], also plural/singular for determiners and adjectives (*le belle sedie, i grandi coltelli*). Once the noun gives the interpretable information singular (only one) or several (plural), these agreement features on the determiners and the adjectives are redundant (in fact English does not have them). For reasons of economy, Narrow Syntax must delete all un-interpretable features *before* the sentence reaches the Conceptual-Intentional interface. Basically, the process was the following: uninterpretable features must be *checked* (sic) to be deletable. Syntactic Objects with such features must "check" them with Syntactic Objects that have interpretable features. The search is *strictly local*, as local as possible. Best: inside the same constituent, at worst: inside the same Phase. The process is called probe-goal. When the probing Syntactic Object encounters the goal, the uninterpretable features are deleted and the result is sent to the Conceptual-Intentional interface. If the local search fails, we have an ungrammatical sentence (the derivation is said to "crash"). Uninterpretable features that have been successfully checked are deleted at the Conceptual-Intentional interface, but not necessarily (in fact usually not) also at the Sensory Motor interface. In fact we do see and hear them at the level of pronunciation and writing. In *Le molte belle figurine* (the many beautiful little cartoons) feminine and plural features are visible and audible in the determiner, the quantifier and the noun. This is called Agreement, the object of many, many linguistics studies [77, 78]. Agree is what makes determiners in languages such as Italian (*il, lo , la, molto, molti, poco, pochi* etc.) and adjectives (*bello, bella, belli*) agree with the noun. Also what makes auxiliaries (*have, be*) and past participles agree with the tense of the verb (*era stato visto, furono visti, saranno stati comprati* etc.). Agree must be local, very local, in a structural sense, not necessarily "local" in terms of proximity of words in the surface structure. I *libri di cui mi parlasti erano stati tutti comprati in libreria* (the books you had told me about have all been bought in a bookshop). *Libri* and *comprati* are not "near" on the surface, but they are near structurally, because the Prepositional Phrase *di cui mi parlasti*, which contains a Verb Phrase, veers off (so to speak) into a sideway structure (it's an adjunct). A similar constraint applies to the assignment of Case (nominative, accusative, genitive etc.) [77]. Case assigners (verbs and prepositions) must be structurally local to the nouns, adjectives etc. to which they assign case. Again, locality is structural, not necessarily on the surface. (Let's think of Latin, Russian etc. where Case is overtly expressed).

Appendix B. On the Fibonacci Matrix

We define the Fibonacci matrix [1] to be given by

$$F \equiv \frac{1}{2} I + \sigma_3 + 2\,\sigma_1 = \begin{pmatrix} 1 & 1 \\ 1 & 0 \end{pmatrix}$$

For the n-powers F^n of the F matrix, with $n \neq 0$, we have

$$F^n = \begin{pmatrix} F_{n+1} & F_n \\ F_n & F_{n-1} \end{pmatrix} = F_{n-1}\, I + F_n\, F, \qquad \mathrm{n} \neq 0$$

where the matrix elements F_{n+1}, F_n, F_n, F_{n-1}, with $F_0 \equiv 0$, for any $n \neq 0$, are the numbers in the Fibonacci progression $F_0 = 0$, $F_1 = 1$, $F_2 = 1$, $F_3 = 2$, $F_4 = 3$, $F_5 = 5$, $F_6 = 8$, $F_7 = 13, \ldots$ Moreover, also the coefficients of the matrices I and F in the last member on the r.h.s. of the above relation are the Fibonacci numbers F_{n-1} and F_n. We can indeed verify that

$$F^1 = \begin{pmatrix} 1 & 1 \\ 1 & 0 \end{pmatrix} = F, \qquad F^2 = \begin{pmatrix} 2 & 1 \\ 1 & 1 \end{pmatrix} = I + F,$$

$$F^3 = \begin{pmatrix} 3 & 2 \\ 2 & 1 \end{pmatrix} = I + 2\,F, \qquad F^4 = \begin{pmatrix} 5 & 3 \\ 3 & 2 \end{pmatrix} = 2\,I + 3\,F, \qquad \text{etc....}$$

We also have $C_1 + (1-i)2\sigma_1 - i2\sigma_3 = (1-i)F = \sqrt{2}\, e^{-i\frac{\pi}{4}} F$, which shows the relation between the Chomsky matrix C_1 and the Fibonacci matrix F.

References

[1] Piattelli-Palmarini M and Vitiello G 2015 *Biolinguistics* **9** 96-115
2015 Linguistics and some aspects of its underlying dynamics *Preprint* arXiv:1506.08663v1
2017 Third factors in language design: some suggestions from Quantum Field Theory in *The Cambridge Companion to Chomsky* (2nd Edition) ed J McGilvray (Cambridge, UK: Cambridge University Press) (in press) pp 134-152
[2] Chomsky N 1995 *The Minimalist Program* (Cambridge, MA: The MIT Press)
[3] Chomsky N 2017 Contribution to these Proceedings
[4] Tomasello M 2003 *Constructing a Language: A Usage-Based Theory of Language Acquisition* (Cambridge, MA: Harvard University Press)
[5] Bybee J 2007 *Frequency of Use and the Organization of Language* (Oxford: OUP)
[6] Wexler K 2002 Lenneberg's dream: Learning, normal language development and specific language impairment in *Language Competence across Populations: Towards a Definition of Specific Language Impairment* eds J Schaffer and Y Levy (Mahwah, NJ: Erlbaum) pp 11-60
[7] Crain S, Thornton R and Khlentzos D 2009 *Cognitive Linguistics* **20**(1) 145-155
[8] Pietroski P M 2003 *The character of natural language semantics Epistemology of Language* (Oxford: OUP)
[9] Reali F and Christiansen M H 2005 *Cognitive Science* **29** 1007-1028
[10] Perfors A, Tenenbaum J B and Regier T 2011 *Cognition* **118**(3) 306-338
[11] Christiansen M H and Chater N 2015 *Behavioral and Brain Sciences* **39** 1-72
[12] Berwick R C, Pietroski P, Yankama B and Chomsky N 2011 *Cognitive Science* **35**(7) 1207-42.
[13] Berwick R C, Chomsky N, Piattelli-Palmarini M 2013 Poverty of the Stimulus Stands: Why Recent Challenges Fail in *Rich Languages from Poor Inputs* eds M Piattelli-Palmarini, R C Berwick (Oxford: OUP) pp 19-42
[14] Perlovsky L and Sakai K L 2014 *Front Behav Neurosci* **8** 436
[15] Perlovsky L 2013 *Front Behav Neurosci* **7** 123
[16] Chomsky N 2005 *Inquiry* **36**(1) 1-22
[17] Carnie A 2007 *Syntax: A Generative Introduction* (2nd Edition) (Oxford: Wiley-Blackwell)
[18] Krivochen D G 2015 *The Linguistic Review* **32**(3) 515-572
[19] Chomsky N 2001 Derivation by Phase in *Ken Hale: A life in Language* ed M Kenstowicz (Cambridge MA: The MIT Press) pp 1-52
[20] Chomsky N 1981 *Lectures on Government and Binding. The Pisa Lectures* (Dordrecht: Foris Publications)
[21] Haegeman L 1991 *Introduction to Government and Binding Theory* (Oxford: Wiley-Blackwell)
[22] Kayne R S 1984 *Connectedness and Binary Branching* (Dordrecht: Foris Publications)
[23] Perelomov A M 1986 *Generalized coherent states and their applications* (Berlin: Springer Verlag)
[24] Chomsky N 2000 Minimalist inquiries: The framework in *Step by Step: Essays on Minimalist Syntax in Honor of Howard Lasnik* eds R Martin D Michaels, J Uriagereka (Cambridge MA: The MIT Press) pp 89-155
[25] Richards M D 2007 *Linguistic Inquiry* **38**(3) 563-572
[26] Gallego A J Ed 2012 *Phases: Developing the Framework* (Berlin: Walter de Gruyter)
[27] Medeiros D P 2008 *Biolinguistics* **2**(3) 152-195
[28] Idsardi W J and Uriagereka J 2009 *Biolinguistics* **3**(4) 404-406
[29] Piattelli-Palmarini M and Uriagereka J 2008 *Physics of Life Reviews* **5**(4) 207-224
[30] Vitiello G 2012. *Phys. Lett.* A **376** 2527-2532
[31] Vitiello G 2014 *Systems* **2** 203 - 216 doi:10.3390/systems2020203
[32] Vitiello G 2009 *New Math. and Natural Computation* **5** 245 - 264
[33] Blasone M, Jizba P, Vitiello G 2011 *Quantum field theory and its macroscopic manifestations* (London: ICP)
[34] Freeman, W. J. and Vitiello, G. 2006 *Physics of Life Reviews* **3** 93-118
[35] Freeman, W. J. and Vitiello, G. 2008 *J. Phys. A: Math. Theor.* **41** 304042
[36] Piattelli-Palmarini M and Uriagereka J 2004 The Immune Syntax: The Evolution of the Language Virus in *Variations and Universals in Biolinguistics* **4** ed L Jenkins (Amsterdam: North-Holland) pp 341-372
[37] Piattelli-Palmarini M and Uriagereka J 2005 *Lingue e Linguaggio* **IV**(1) 27-79
[38] Chomsky C 1986 *Journal of Speech and Hearing Research* **29**(3) 332-374
[39] Chomsky C 1969 *The Acquisition of Syntax in Children from 5 to 10* (Cambridge MA: The MIT Press)
[40] Klima E S and Bellugi U 1979 *The Signs of Language* (Cambridge MA: Harvard University Press)
[41] Chomsky N 2013 *Lingua* **130** (June) 33-49

IOP Conf. Series: Journal of Physics: Conf. Series **880** (2017) 012016 doi:10.1088/1742-6596/880/1/012016

[42] Chomsky N 2014 Problems of Projection: Extensions (Abstract) Department of Linguistics MIT
[43] Chomsky N 2015 Problems of projection: Extensions in *Structures Strategies and Beyond. Studies in honour of Adriana Belletti* eds E Di Domenico, C Hamann, S Matteini (Amsterdam: John Benjamins) in press
[44] Berwick R C, Friederici A D, Chomsky N and Bolhuis J J 2013 *Trends in Cognitive Sciences* **17**(2) 89-98
[45] Cecchetto C and Donati C 2010 *Syntax* **13**(3) 241-278
[46] van Gelderen E 2014 Problems of Projection: The Role of Language Change in *Labeling Paradoxes unpublished*
[47] Hornstein N 1999 *Linguistic Inquiry* **30**(1) 69-96
[48] Hornstein N, Nunes J and Grohmann K K eds 2005 *Understanding Minimalism* (Cambridge UK: Cambridge University Press)
[49] De Concini C and Vitiello G 1976 *Nucl. Phys.* B **116** 141-156
[50] Beige A, Knight P L and Vitiello G 2005 *New J. Phys.* **7** 96 - 9
[51] Inönü E and Wigner E P 1953 *Proc Nat Acad Sci* **39** 510-524
[52] Umezawa H 1993 *Advanced field theory: Micro macro and thermal physics* (New York: AIP)
[53] Shah M N Umezawa H and Vitiello G 1974 *Phys. Rev.* B **10** 4724-4736
[54] Goldstone J Salam A and Weinberg S 1962 *Phys. Rev.* **127** 965-970
[55] Kayne R S 1994 *The Antisymmetry of Syntax* (Cambridge MA: The MIT Press)
[56] Moro A 2000 *Dynamic Antisymmetry* (Cambridge MA: The MIT Press)
[57] Vitiello G 1995 *Int. J. Mod. Phys.* B **9** 973-989
[58] Vitiello G 2001 *My double unveiled* (Amsterdam: John Benjamins)
[59] May R 1985 *Logical Form: Its Structure and Derivation* (Cambridge MA: The MIT Press)
[60] Celeghini E, Rasetti M, Vitiello G 1992 *Annals Phys.* **215** 156-170
[61] Celeghini E, De Martino S, De Siena S, Iorio A, Rasetti M and Vitiello G 1998 *Phys. Lett.* A **244** 455-461
[62] Merchant J 2001 *The Syntax of Silence: Sluicing Islands and the Theory of Ellipsis* (Oxford: OUP)
[63] Chomsky N 1955 *The Logical Structure of Linguistic Theory* (Chicago: The University of Chicago Press, 1985) [part. ed. 1956, microfilm and pdf form; published in large part in 1975 with an explanatory introduction]
[64] Blasone M, Srivastava Y N, Vitiello G and Widom A 1998 *Annals Phys.* **267** 61-74
[65] Vitiello G 2004 *Int. J. Mod. Phys. B* **18** 785-792
[66] Vitiello G 2004 The dissipative brain in *Brain and Being. At the boundary between science philosophy language and arts* eds G G Globus K H Pribram, G Vitiello (Amsterdam: J. Benjamins Publ Co) pp 315-334
[67] Capolupo A, Freeman W J and Vitiello G 2013 *Physics of Life Reviews* **10** 85-94
[68] Freeman W J 1991 Scientific American **264**(2) 78-85
[69] Basti G, Capolupo A, Vitiello G 2017 Quantum Field Theory and Coalgebraic Logic in Theoretical Computer Science, *Preprint* arXiv:1701.00527
[70] Harris Z S 1951 *Methods in Structural Linguistics* (Chicago: University of Chicago Press)
[71] Chomsky N 1951 *Morphophonemics of Modern Hebrew* (MA thesis) (University of Pennsylvania)
[72] Chomsky N 1979 *Morphophonemics of modern Hebrew* (New York: Garland Pub)
[73] Chomsky N 1956 Three Models for the Description of Language in *Readings in Mathematical Psychology* **2** eds R D Luce R R Bush and E Galanter (New York: John Wiley and Sons) pp 105-124
[74] Chomsky N 1957 *Syntactic Structures* (The Hague: Mouton)
[75] Lasnik H, Depiante M, Stepanov A 2000 *Syntactic Structures Revisited* (Cambridge Ma: The MIT Press)
[76] Chomsky N 1959 *Information and Control* **2** 137-167
[77] Baker M C 2013 *Lingua* **130** (June) 14-32
[78] Miyagawa S 2010 *Why Agree? Why Move? Unifying Agreement-Based and Discourse-Configurational Languages* (Cambridge MA: The MIT Press)

Constraints on quantum information field and "human gain medium" making possible functioning of social laser

Andrei Khrennikov
International Center for Mathematical Modeling
in Physics and Cognitive Sciences
Linnaeus University, Växjö, S-35195, Sweden

E-mail: Andrei.Khrennikov@lnu.se

Abstract. Starting with the quantum-like paradigm on application of quantum information and probability outside of physics we proceed to the social laser model describing Stimulated Amplification of Social Actions (SASA). The basic components of social laser are the quantum information field carrying information excitations and the human gain medium. The aim of this note is to analyze constraints on these components making possible SASA. The soical laser model can be used to explain the recent wave of color revolutions as well as such "unpredictable events" as Brexit and election of Donald Trump as the president of the United States of America. The presented quantum-like model is not only descriptive. We shall list explicitly conditions for creation of social laser.

1. Introduction

The recent quantum information revolution has tremendous consequences not only for development of novel physical technologies, but even social ones. We know well about successes in quantum physical technologies related to quantum computing, cryptography, teleportation, and recently quantum stimulator: from philosophizing (Einstein-Podolsky-Rosen paradox, 1933) to real quantum technological projects (the recent EU call for 1 billion Euro). We want to point out to similar success in development of novel social technologies based, in fact, on quantum informational principles. These technologies exhibited they power in a series of color revolutions and recently in Brexit and the election of Donald Trump as the USA president.

The main consequence of the quantum information revolution is that nowadays quantum systems are treated as *carriers and processors of information.* A photon or an electron are just carriers of quantum bits. The physical structure of a system seems to be irrelevant for theoretical modeling. (Of course, it has to be taken into account in the process of the real physical realization.) Nowadays quantum mechanics (QM) has a few purely information and probabilistic interpretations, see, e.g., [1]-[9] and also appendix.

However, the majority of physicists (even those advertising the information interpretations) still think that carriers and processors of quantum information should follow very special physical laws. Thus, for them, it is difficult (and for many totally impossible) to accept that quantum information and probability theories can be applied outside of quantum physics.

In a series of works , see, e.g., [10]-[12], the author demonstrated that, in fact, the essence of quantum information processing is in sensitivity of systems to changes of surrounding contexts, adaptivity to environment.

These studies led to formulation of the **Quantum-like Paradigm** (Khrennikov):

The mathematical formalism of quantum information and probability theories can be used to model behavior not only of genuine quantum systems, but all context-sensitive systems, e.g., humans.

Nowadays quantum information and quantum probability are widely used in psychology, economics, finances, social and political sciences, see, e.g., [10]-[22] In particular, quantum probability relaxes some constraints on statistical data imposed by classical probability theory. One of such constraints is *the law of total probability and in general Bayesian inference,* see also appendix.

The use of quantum probability, instead of classical probability, can resolve some paradoxes of classical theory of decision making, economics, and game theory; e.g., the Elsberg and Machina paradoxes. In general the mathematical formalism of QM is useful to model a variety of psychological effects, e.g., the order, disjunction, and conjunction effects, see, e.g., the monographs [11], [15].

Recently the author of this paper presented a novel application of quantum theory outside of physics, the model of a kind of social laser [17, 18], describing *Stimulated Amplification of Social Actions* (SASA). Here one has to use more advanced methods of quantum theory comparing with the previous works about decision making and cognition. In particular, QM cannot describe functioning of lasers. The formalism of quantum field theory (QFT) has to be explored. The detailed presentation of this model can be found in open access paper [18]. However, considerations in both papers [17], [18] are rather longly and not well structured. In this paper we present the formulation of explicit conditions which should be imposed onto the two basic components of the social laser, the quantum information field and the human gain medium. The components satisfying these conditions can serve for design of the SASA process. These conditions will be listed in the compact form comparable with the corresponding conditions on the components of the physical laser.

The aim of the creation of the quantum-like model of social laser is to explain the recent wave of color revolutions (at the post-Soviet territory, Middle East) as well as such "unpredictable events" (from the viewpoint of the modern democratic system) as *Brexit and the election of Donald Trump as the president of the United States of America,* see section 3. The presented quantum-like model is not only descriptive. We shall list explicitly the conditions implying the possibility of initiation of SASA.

2. Laser

Since we hope that this paper will be readable for researchers from the fields of psychology, sociology, management, political science, we start with the brief recollection of the history of laser. Invention of lasers was one of the main technological outputs of quantum physics. The theoretical basis for laser-invention was implicitly present in works of Einstein on stimulated absorption and emission (1917). In 1939, Soviet physicist Fabrikant considered the theoretical possibility of stimulated amplification of microwaves. But only in 50th, Townes in cooperation with Gordon and Zeiger designed and realized at the lab level the first microwave amplifier, nowadays such devices are known as masers. Townes's maser was incapable of continuous output. In 1955, Prokhorov and Basov suggested optical pumping of a multi-level system as a method for obtaining the population inversion, later a main method of laser pumping.

For completeness of presentation, we present the picture illustrating absorption of photons by atoms and spontaneous and stimulated emissions, see Fig. 1.

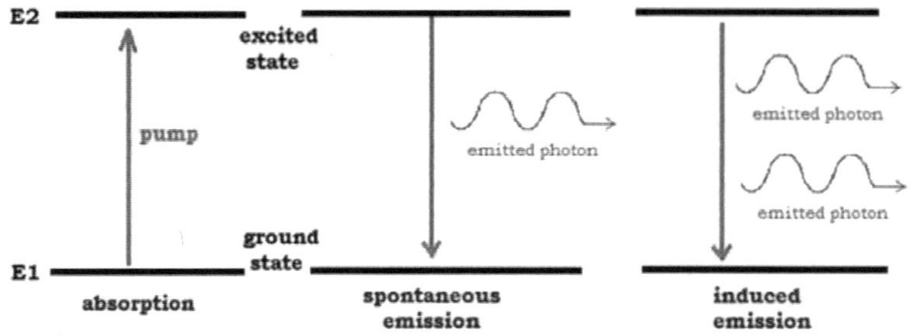

Figure 1. Absorption and spontaneous and stimulated emissions

For simplicity, we consider the two level atom. It has the ground and excited states with energies E_0 and E_1, respectively. The energy gap between these levels is given by $\Delta = E_1 - E_0$. Any atom in the ground state can absorb a photon with energy which is equal to this gap. An atom in the excited state can spontaneously emit a photon of energy Δ. If there is an ensemble of atoms, then spontaneously emitted photons are not coherent, i.e., they propagate in random directions and have random polarizations. Stimulated emission is a more interesting process. If an atom in the excited state interacts with a photon of energy Δ, then this atom emits a photon. Thus one photon produces two photons. The main feature of these two photons is that they are coherent, i.e., they have the identical states.

Now suppose that we were able to prepare an ensemble of atoms in the excited state. Suppose that one of them has emitted (e.g., spontaneously) a photon. Then interaction of it with some excited atom would lead to emission (stimulated by this photon) of new photon. These two photons by interacting with two atoms generate four photons. We have the cascade process of stimulated emission. The crucial difference of this process from spontaneous emission is that the output flow consists of coherent radiation, i.e., the states of emitted photons are identical.

To design laser one has to be able to create ensembles of atoms in which more than half of atoms are in the excited state - to approach the population inversion. This is the main problem and complexity of its solution explains why lasers were not created in 1920th. In fact, it is impossible to approach the population inversion for an ensemble of two level atoms; one has to operate with at least three level atoms (Prokhorov and Basov).

3. Recent "social revolutions": from post-Soviet states and Middle East to UK and USA

Color revolutions started at territory of the former Soviet Union, e.g., Georgia's Rose Revolution (2003), Ukraine's Orange Revolution (2004), Kyrgyzstan's Tulip Revolution (2005), Belarus' Jeans Revolution (2006), Moldova's Grape Revolution (2009), second Ukraine's Orange Revolution (2014). We can also mention Lebanon's Cedar Revolution (2005), Kuwait's Blue Revolution (2005), Myanmar's Saffron Revolution (2009). Recent years were characterized by the wave of color revolutions at the Middle East: Tunisia's Jasmine Revolution (2011), Egypt's Lotus Revolution (2011), Muslim military upraises in Iraq, Libya, Syria.

We emphasize that the diversity of opinions and theorizing about these social events presented in socio-political literature is really amazing, cf. [23]-[28]. There is no consistent and commonly acceptable theory of these phenomena. We shall explore the quantum-like laser model to describe the possibility of explanation of these events.

We point to the following strange features of all these color revolutions which are debated in political literature:

- No strong and bright leaders.
- No clear programs and aims.

Masses demonstrate coherent social actions (in the variety of forms, from massive protests and military upraises to referendums and elections) without leaders such as, e.g., Lenin and Trotsky or Atatürk. Such "revolutions" happen practically spontaneously and typically they are ended simply as relaxation of "energy of masses", i.e., without real constructive output.

4. Physical Laser: basic components and their features

Laser's operation is based on two components: *the quantum electromagnetic field and the gain medium* which is compound of atoms.

The quantum electromagnetic field interacts with matter with the aid of its excitations known as photons (carrying discrete portions of energy proportional to their frequencies). To simplify transition to the social modeling, we ignore polarization of photons. This field has the following features playing the fundamental role in functioning of laser:

- (F0) Addivity of physical energy carried by photons.
- (F1) Indistinguishability of excitations: two photons of the same energy cannot be distinguished.
- (F2) Bose-Einstein statistics of excitations: a few photons can occupy the same energy state.

Additivity of energy is so fundamental in physics that including of (F0) into the fundamental features of photons may seems strange. However, we are preparing the framework for the corresponding social considerations.

This is also a good place to make the following foundational remark. In contrast to a classical field, a quantum field cannot be imagined as a "physical field" propagating in space-time. This is the symbolic operational expression describing interactions of its excitations with matter (in the case of laser - with a gain medium). One may say that, in contrast to the classical electromagnetic field which can be imagined as continuously propagating in the space waves of energy, the quantum electromagnetic field seems to be nonphysical. However, physicists would not agree with such a viewpoint. Of course, believe in "reality" of the quantum electromagnetic field is often based on the rather naive picture of photons propagating in physical space. We stress that the latter is simply the relict of the classical thinking.

Laser's gain medium is composed of atoms (treated as quantum systems). Again, to simplify transition to the social modeling, we ignore electron's spin and degeneration of atom's levels. Atoms have the following features playing the fundamental role in functioning of laser:

- (G1) The discrete spectrum of energy.
- (G2) Fermi-Dirac statistics: only one electron in an atom can occupy the fixed energy state.
- (G3) An atom can absorb and emit only discrete portions of the energy, adapted to the differences between levels.
- (G4) Atoms in the gain medium have the same spectrum (the structure of energy levels).
- (G5) Stimulated emission.

5. Thermodynamics from combinatorics of state distribution

The combinatorial approach to thermodynamics, see, e.g., Schrödinger [29], provides the possibility to exclude physical components from reasoning leading to the basic thermodynamic laws. This approach is applicable to arbitrary systems (i.e., not only physical ones) which are characterized by the possibility to assign them states corresponding to the fixed values of some quantity E which is *additive with respect to elements of an ensemble of systems.* We can call it, e.g., *energy.*

In physics E is the "real" physical energy. In our considerations E will have the meaning of the *social energy.* Similarly to the physical energy, the social energy represents potentiality to perform social actions. A higher value of E can generate stronger action and its strength is proportional to E.

This is a good place to make another foundational remark. The classical treatment of the physical energy as the objective property of a physical system cannot be extended to quantum physics. We cannot consider "photon's energy" as its objective property. In quantum physics we can speak only about observables and observable quantities. Thus "photon's energy" is meaningful only as the result of measurement of energy which is performed, e.g., with the aid of photodectors. This quantum interpretation of energy (as well as other quantum quantities) is very important for justification of invention of its social analog. We do not claim that each individual "carries" some portion of social energy. The value of the social energy can be obtained through some measurement procedure. Of course, the specification and calibration of social measurement procedures is the separate and complex problem. But heuristically the social energy is not more mystical than the physical energy of a quantum system. This discussion is heavily based on treatment of humans as quantum-like systems, see introduction.

We proceed under the assumption about the possibility to assign the energy-like quantity E to systems under consideration (where "assign" is understood very generally - matching with quantum and quantum-like phenomenology). Then we obtain two different thermodynamical models:

- Thermodynamics of distinguishable systems.
- Thermodynamics of indistinguishable systems.

In physics, the first case corresponds to classical systems and the second one to quantum systems. In thermodynamics of indistinguishable systems, we distinguish two types of systems such that

- Any number of systems can occupy the same state (Bose-Einstein statistics).
- The fixed number, say q, of systems can occupy the same state (parastatistics).

Parastatistcis corresponding to $q = 1$ (the same state can be occupied by only one system) plays the most important role in quantum physics and it is known as Fermi-Dirac statistics. Only this kind of parastatistics can be found for real quantum systems. In social quantum-like modeling the question of possible types of parastatistics is open. However, to make life simpler, we proceed similarly to quantum physics.

6. Social Laser

6.1. Basic components and their features

The basic components of social laser, a structure performing stimulated amplification of social actions are

- Quantum information field, its excitations - quanta of information.
- Gain medium: humans.

The quantum information field was introduced in [17] and this paper contains the extended foundational discussion. This field provides the operational representation of information flows produced by mass-media, TV, and Internet. Excitations of this field carry information about some events. They can be considered as quanta of such field. Information carried by these information quanta encodes portions of social energy.

We list the basic features of the quantum information field making possible creation of social laser:

- (IF0) Addivity of social energy: energy transmitted to a person by excitations of the information field is integrated.
- (IF2) Indistinguishability of information excitations: the content of messages is not analyzed, we simply absorb social energy carried by them.
- (IF3) Bose-Einstein statistics of excitations: messages with totally different content can carry the same portion of social energy

We now list the basic features of human gain media making possible creation of social laser:

- (IG1) The discrete spectrum of the social energy.
- (IG2) Distribution of the social energy in individual's state space follows Fermi-Dirac statistics.
- (IG3) An individual can absorb and emit only discrete portions of the social energy, adapted to the differences between his/her levels.
- (IG4) Individuals belonging to the gain medium have the same spectrum.[1]
- (IG5) Stimulated emission of the social energy.

In this model a person "falling" from the energy level $\mathbf{E}_{i+1}$ to the lower level $\mathbf{E}_i$ cannot stop in between. This matches well the discrete social layer structure of the modern society. There is no place for those who does not match to the discrete structure of social layers.

6.2. *Human gain medium: comments*

In our model human being is treated as an *information transmitter*. He can absorb social energy carried by the information field; he can carry a portion of energy, and he can emit a portion of energy.

To be able to create a population G of individuals which would function as a kind of laser emitting waves of social energy, the processes of absorption and emission of social energy have to be discrete. *Individuals in G have to have the discrete structure of their energy levels.*

Consider the simplest case of the two level structure of individual's social energy. She, say Alice, can occupy either the ground state characterized by the minimal energy level $\mathbf{E}_0$ or the excited state with energy $\mathbf{E}_1$. She can absorb or emit only portions of energy $\Delta = \mathbf{E}_1 - \mathbf{E}_0$.

6.3. *Quantum information field: comments*

In our model an information excitation is characterized completely by its social energy. Thus from a variety of features characterizing an event we extract only one - the magnitude of the social energy carried by this event. Other its features are ignored (in this simplest model). Thus two events O_1 and O_2 belonging to the same type of excitations, i.e., having the same social energy, are indistinguishable.

The model can be generalized to take into account the content of an event. Similarly in reality an excitation of the quantum electromagnetic field (a photon) is characterized not only by its energy (which is proportional to the frequency), but also by its internal degree of freedom

[1] Of course, in reality it is impossible to prepare an ensemble of systems with precisely the same spectrum. In physics, variation of spectra follow the Gaussian law. One can expect this even in social systems.

- polarization. Thus social events can be assigned with some content characteristics, say politics, economics,....

In contrast to quantization in physics, the origin of social quantization is trivial. This is the ability of humans (and simpler cognitive systems) to represent reality as a chain of discrete events.

6.4. Indistinguishability of excitations of the quantum information field

Indistinguishability is the basic feature of information delivered to consumers by modern mass-media. In the modern society it is approached through creation of a massive flow of information such that, for people (receivers of information), it is difficult to inspect its content. They would "absorb information only to absorb information." Here the amount of the social energy carried by information communication is the subject of interest and not its content. People absorb information from TV, newspapers, Internet, mobiles. There is no neither time nor mental resources to analyze its content.

Such people overloaded by information differ crucially from people who leaved on the Earth say 30-50 years ago. We can speak about *post-human.* From the mental viewpoint, this is really a new being. Although biological differences are not (yet?) visible, behavioral differences are evident.

7. The probabilistic interpretations of quantum mechanics: the Växjö interpretation and Quantum Bayesianism

By the so-called Växjö interpretation of QM [2]-[5] QM is treated as a novel probabilistic formalism producing nonclassical probabilistic effects in the form of *violation of the formula of total probability.* The classical law of total probability is modified to take into account the interference term generated by incompatibility of observables. Quantum probabilities are interpreted as objective probabilities with emphasizing their frequency content.

Fuchs and Schack [6], [7] also proposed the probabilistic interpretation of QM known as Quantum Bayesianism (QBism). And they also pointed to violation of the classical law of total probability. Their modification of this law differs from the Växjö version and it is based on the use of so-called symmetric informationally complete positive operator valued measures (SIC POVMs). Fuchs and Schack interpret quantum probabilities as subjective probabilities and they stress the private agent perspective in observers analysis of experimental data.

I should honestly recognize that the subjective probability approach matches better applications of quantum probability to model cognition and the process of decision making. Thus QBism can be considered as the basic interpretation of the quantum-like approach to decision making [30]. However, the QBists' modification of the classical formula of total probability [6], [7] is hardly applicable for such purpose. Here we simply do not have SIC POVMs. Observables do not provide complete information about the belief state of a decision maker. Therefore the Växjö version of this formula [2]-[5] has to be in the use. It is based on the pair of incompatible observables (neither of them should be informationally complete).

However, in the social laser model we have to proceed with objective probabilities determining the population levels for the ground and excited states of individuals in human gain media. Thus in such modeling we have to use the Växjö interpretation of QM.

Acknowledgments

This work was partially supported by the grant "Mathematical Modeling of Complex Hierarchic Systems" of the Faculty of Technology of Linnaeus University. It was also supported by the EU-project "Quantum Information Access and Retrieval Theory" (QUARTZ), Grant No. 721321.

References

[1] Brukner C and Zeilinger A 1999 *Acta Physica Slovava* **49** 647
[2] Khrennikov A 1999 *Interpretations of Probability* (Utrecht/Tokyo: VSP Int. Sc. Publishers)
[3] Khrennikov A 2002 *Quantum Theory: Reconsideration of Foundations* ed Khrennikov A (Växjö: Växjö Univ. Press) p. 163
[4] Khrennikov A 2004 *Quantum Theory: Reconsideration of Foundations* ed A Khrennikov (Växjö: Växjö Univ. Press) p 323
[5] Khrennikov A 2009 *Contextual Approach to Quantum Formalism* (Berlin-Heidelberg-New York: Springer)
[6] Fuchs C A 2002 *Quantum Theory: Reconsideration of Foundations*, ed A Khrennikov (Växjö: Växjö University Press) p 443
[7] Fuchs C A and Schack R 2014 *Phys. Scr.* **90** 015104
[8] D' Ariano G M 2007 *Foundations of Probability and Physics-3* vol 889, ed Adenier et al. (Melville, NY: AIP.) p 79
[9] Plotnitsky A and Khrennikov A 2015 *Found. Phys.* **45** 269
[10] Khrennikov A 1999 *Found. Phys.* **29** 1065
[11] Khrennikov A 2010 *Ubiquitous Quantum Structure: from Psychology to Finances* (Berlin-Heidelberg-New York: Springer)
[12] Khrennikov A 2015 *Frontiers Psych.* **6** art. 997
[13] Aerts Aerts D, Sozzo S, Tapia J 2012 *Quantum Interaction. Lecture Notes in Computer Science* **7620** 48
[14] Asano M, Basieva I, Khrennikov A, Ohya M, Tanaka Y, Yamato I 2015 *Found. Phys.* **45** 1362
[15] Busemeyer J R and Bruza P D 2012 *Quantum Models of Cognition and Decision* (Cambridge: Cambridge University Press)
[16] Haven E and Khrennikov A 2009 *J. Math. Psych.* 53 378
[17] Khrennikov A 2015 *Entropy* **17** 6969
[18] Khrennikov A 2016 *Phil. Trans. R. Soc.* A **374** 20150245
[19] Khrennikova P, Haven E and Khrennikov A 2014 *Int. J. Theor. Phys.* **53**(4) 1346
[20] Khrennikova P and Haven E 2016 *Phil. Trans. Royal Soc. A* **374** 20150106
[21] Pothos E and Busemeyer J R 2009 *Phil. Trans. Royal Soc. B* **276** 2171
[22] Sozzo S 2014 *J. Math. Psych* 58(1) 1
[23] Mason P 2012 *Why It's Kicking off Everywhere: The New Global Revolutions* (London: Verso)
[24] Fukuyama F 2013 The Middle-Class Revolution. The Wall Street Journal, June 28
[25] Krastev I 2014 Democracy disrupted. The global politics of protest. Philadelphia: Penn Press
[26] Putin V 2017 We must stop a Ukraine-style 'coloured revolution' in Russia. *The Telegraph*, January 1.
[27] Travers T 2016 Why did people vote for Brexit? Deep-seated grievances lie behind this vote, *British Politics and Policy*, June 29.
[28] Fishwick C 2016 Why did people vote for Brexit? Deep-seated grievances lie behind this vote, *The Gardian*, November 9.
[29] Schrödinger E 1989 *Statistical Thermodynamics* (Dover Publications)
[30] Khrennikov A 2016 *Phil. Trans. Royal Soc.* A **374** 20150245

Quantum neuromorphic hardware for quantum artificial intelligence

Enrico Prati

Istituto di Fotonica e Nanotecnologie, Consiglio Nazionale delle Ricerche, Piazza Leonardo da Vinci 32, I-20133 Milano, Italy

E-mail: enrico.prati@cnr.it

Abstract. The development of machine learning methods based on deep learning boosted the field of artificial intelligence towards unprecedented achievements and application in several fields. Such prominent results were made in parallel with the first successful demonstrations of fault tolerant hardware for quantum information processing. To which extent deep learning can take advantage of the existence of a hardware based on qubits behaving as a universal quantum computer is an open question under investigation. Here I review the convergence between the two fields towards implementation of advanced quantum algorithms, including quantum deep learning.

1. Learning from the working principles of the brain

The invention of the modern hardware architecture of computers, the so called *Von Neumann architecture*, is often attributed to John Von Neumann. It was only in 1998 that Konrad Zuse was officially recognized by the Conference on the History of Computing in Padernborn as the inventor of such architecture as he developed and patented the Z1 model, based on floating point and binary digits in 1938 and the Z3 with the full modern architecture in 1941 [1], before Von Neumann [2]. In order to acknowledge both the two fathers of such an architecture, we should call it Von Neumann-Zuse architecture. Though, there is an additional reason for which Von Neumann should be acknowledged: when inventing the computer, he started with the aim to create an artificial brain, but in order to ensure control of the computation with the tools available at that time, he opted to impose two major simplifications, i.e. the system, contrarily to the unreliable and asynchronous building blocks of brain namely the real neurons, was based on deterministic and synchronous design. [2]. This enabled him to create the computer with telegraph relays and vacuum tubes, but he left open the alternative of pursuing the opposite strategy. As he was aware that brain processes information in parallel, and considered the brain a parallel and asynchronous computer, he implicitly also defined the characteristics of a kind of Non-Von Neumann (NVN) architecture. Indeed, he pointed out two prominent aspects of biological neurons: they process information in a parallel fashion and they are very unreliable (a synapse works in average once every 10 solicitations) and affected by several unavoidable sources of noise, ranging from atomic fluctuations across their membrane to potential fluctuation due to release of molecules by vesicles. Despite such sources of inaccuracy and indeterminism, neurons and synapses successfully carry information processing by redundancy and parallelism. As discussed later, there are several hardware approaches enabling device functionalities towards integrate and fire response of neurons and synaptic plasticity, which generally operate in low disturbance conditions. For researchers working in quantum information processing, where noise is a unavoidable source of loss of coherence and therefore of information, the fact that the brain perfectly works despite the noise may open a very tempting path. Furthermore, its working principles become surprising if one observes that the brain works not only despite the noise, but more precisely thanks to noise. There is a long list of useful and necessary effects of noise in order to achieve and improve brain functionalities in several cortical areas[4, 5, 6]. The deep learning algorithms [7, 8] inspired by the layered structure of visual cortex has pushed to the right direction the exploitation of such a parallelism and it has provided a new paradigm for the pattern recognition in complex data structures. In some cases, such as for Boltzmann machines, the statistical behavior reproduces the intrinsic variability of neuronal activity. Deep learning neural networks can be applied not only to image processing, but to any kind of data set, ranging from DNA research to economics to identify patterns and regularities often hidden to human intuition. In the next section the role of quantum mechanics in brain is discussed.

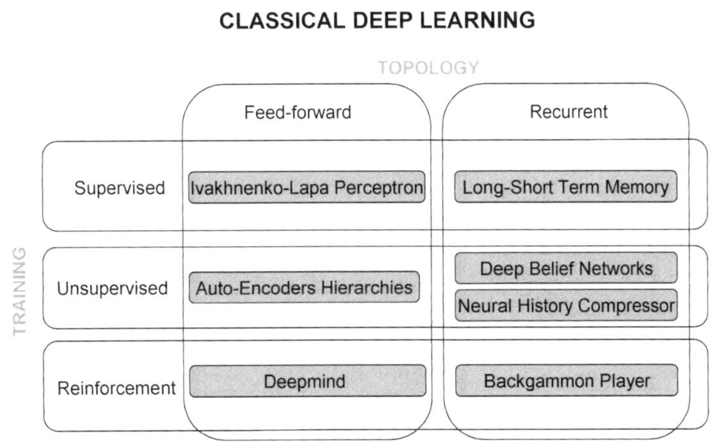

Figure 1. Classical deep learning architectures. The implementation depends on both the training method and on the topology (either feedforward and recurrent). Notice that Deepmind [25] consists of a special feed–forward topology called convolutional. See the main body of the text for the reference of each of the above implementations.

2. Quantum effects and information processing in the brain

There is a number of researches claiming that quantum mechanics plays a direct role in the information processing in the brain. By *direct role*, here we intend that the brain has been supposed to exploit quantum superposition of states, for instance in relatively large molecules such as chains in microtubules. Even if the topic is debated, there is a relatively large leap between the existence of quantum superposition effects in the brain, and their exploitation in the information processing. Quantum effects are normally present at the atomic scale and they are involved from instance at the scale of ion channel processes, where single Na, K and Ca ions move across the membrane of the neurons, to molecular dynamics and matching in the complex creation of proteins from genes as an expression of DNA. Several noise processes present in the brain are ultimately related to random quantum fluctuations.[9, 10, 11, 12] Despite a number of interesting claims about room temperature observable effects in biology of quantum mechanics [13], including some structures present in the brain [14], there is currently no evidence that such a superposition is used to perform the signaling process between neurons. Therefore, if one is interested to create a neuromorphic hardware exploiting quantum mechanics, one has two options. One may look either to include quantum fluctuations in the classical hardware, in order to natively implement sources of fluctuations which proved effective for the information processing of several cortical areas, or alternatively to export the organization structure of some cortical areas such as visual cortex, and to extend the deep learning architectures towards a quantum domain. Currently, there is no evidence that superposition of states is a key ingredient *per se* to support neuronal information processing. Among the neuromorphic classical hardware, we may include aluminum atomic bridges, CuS and AgS synapses, silicon CMOS platform, HfO_2 RRAMs, InGaZnO synapses, GST synapses, hybrid organic/silicon transistors [10]. Among them, only silicon devices and HfO_2 devices proved effective to cover both plasticity and spontaneous fluctuations [10, 15, 16, 17, 18]. Artificial neurons are currently built only by silicon devices. A prominent example of deep learning network designed on CMOS silicon hardware at circuit level has been provided in [19]. The opposite strategy, namely the exploitation of quantum hardware to extend the deep learning by quantum algorithms has been investigated. In the last section of this mini-review, the latter topic is outlined.

3. From machine learning to quantum artificial intelligence

In the years, three main methods have been developed to train both feedforward and recurrent neural networks: supervised learning (the training is based on correct input-output pairs), reinforcement learning (the correct input-output pairs are never presented), and unsupervised learning (contrarily to previous, the input data are unlabelled so there is no estimate of the accuracy of the output) and hybrid combinations. The implementation can be based on either *feedforward networks*, whose topology is such that input signal can only travel in one direction paths from the input neurons to the output neurons, or *recurrent networks* where loops in the topology are present. Deep learning approach consists in the separation of the nodes into groups providing multiple layers used as nonlinear processing units, which create a hierarchy from low-level to high-level features representation layers (Figure 1). Deep learning has been employed for deep supervised learning such as supervised deep feedforward multilayer perceptrons invented by Ivakhnenko and Lapa in 1965 [20] and the more recent *long short-term memory* recurrent neural networks [21] respectively, deep reinforcement learning [22, 23] applied to both recurrent (the backgammon player of [24]) and later feedforward (more specifically of convolutional kind, used to beat Atari2000 arcades and in AlphaGo [25]) neural networks respectively, and deep unsupervised learning in both versions based on feedforward (*autoencoders hierarchies* [26] since 1987) and recurrent (both *deep belief networks* based on restricted Bolzmann machines [27] and *neural history compressor* [28]) neural networks. It is worth mentioning at this point that restricted Boltzmann machines are a subclass of recurrent neural networks. They are a special case of a Hopfield network which is a neural network where all the nodes are symmetrically connected with the other nodes so that its weights matrix is symmetric. In a Boltzmann machine the internal connections of a Hopfield network define a bipartite network split between a visible layer and a hidden layer, and the units are stochastic and subject to thermalization according to the Boltzmann statistics. A restricted Bolzmann machine (RBM) consists in a subclass of Boltzmann machines for which no connections internally exist between nodes of the visible and of the hidden layers respectively. A deep RBM is composed by many RBM layers (see Figure 2). The prospect of employing spiking neuron in artificial neural networks has been summarized in the review of Schmidhuber [29], even if currently spiking neurons cannot yet compete with the best traditional deep NNs in practical applications. On the other hand, the advancement of scalable hardware qubits such as superconductive [30] and silicon

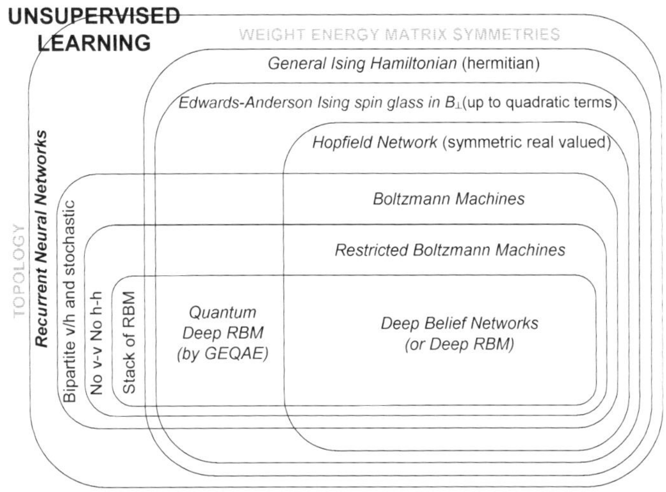

Figure 2. Hierarchy of different kind of Recurrent Neural Networks (RNN) including quantum domain for unsupervised learning. Starting from Hopfield networks (its energy cost function is based on a classical Ising hamiltonian, symmetric and real valued, corresponding to a symmetric operator acting on usual classical states), one can both restrict the topology to more sofisticated cases, such as bipartite (between visible v and hidden h layer) and stochastic, which corresponds to so called Boltzmann machines, and by further resctricting the connection so that there are neither v-v nodes connections, nor h-h nodes connections – the so called restricted Boltzmann machine (RBM) – which in turn can be used for deep belief networks by using a stack of them. On the other side, the Hopfield network can be implemented in a more general domain such that provided by quantum hardware for quantum annealing with Ising Edwards-Anderson Hamiltoninan or a general Ising Hamiltonian acting on quantum states. Quantum deep restricted Boltzmann machines, or more briefly quantum deep learning, corresponds to a Deep Belief Network exploited in a more extended class of Hamiltonians spanning a quantum space of vectors.

[31, 32] qubits make closer the implementation of quantum algorithms. Indeed, fault tolerant quantum information processing requires handling of the fragile states of physical qubit. A number of quantum error correction codes such as Steane code [36] and surface code [37] have been developed to assess logical error-free qubits. Such logical qubits are managed by the application layer for algorithm implementation with no care about the underlying hardware. Most significant architecture are classified among circuital (digital) quantum computing, based on universal sets of quantum logic ports, and quantum annealing (and adiabatic quantum computing). A major example of hybrid architecture has also been demonstrated [30], enabling the control of 5 logical qubits distilled by a virtual qubit layer constituted by 500 physical qubits. The quantum annealing corresponds to the minimization of a complex functional and is suitable for optimization problems by exploiting minimization search via quantum tunneling in a complex potential landscape mapping the problem to be solved. It appears that the analogy with minimization problem solved by deep learning algorithm in the regression operation is more than just qualitative. For instance, optimization problems like the four-queens on a chessboard has been mapped onto an adiabatic quantum computing framework by exploiting the analogy between Hopfield network and its quantum counterpart [33]. In the remaining of this section,

two examples of quantum algorithm for quantum artificial intelligence are reviewed. The first examples consists in the set of quantum deep learning algorithms provided by Microsoft in 2014 [34]. Such algorithms is based on the quantum mechanical extension of a deep unsupervised recurrent neural network where the recurrent neural network is organized as a deep restricted Boltmann machine. In its classical version, the latter is mathematically represented by an Ising model which is in thermal equilibrium. Let's clarify where the quantumness sets in. Usually the training of the network is carried by employing gradient descent methods. The energy of the system is expressed in terms of parameters called weights and biases respectively, which enable to define a gradient descent to find such weights and biases in order to maximize a likc-hood function during training by appropriately updating such values. The difficulty of computing the gradients for updating the parameters is exponentially hard by raising the number of both visible and hidden nodes, so two quantum algorithms have been proposed to speed up such process: the *Gradient Estimation via Quantum Sampling* (GEQS) and the *Gradient Estimation via Quantum Amplitude Estimation* (GEQAE) respectively. In general, the starting point when training the network is far from the Gibbs state which would maximize the probability of a given Boltzmann machine, so the quantum speedup is provided by its natural capability to rapidly converge thanks to qubits to a mean field state (a non-uniform prior distribution for the probabilities of each configuration) which is closer to the Gibbs state and therefore suitable for accelerating the search. GEQS consists in the preparation of a mean field state by exploiting the polynomial time required by qubits in order to achieve it. GEQAE algorithm is a new form of training, possible in cases where the training data is provided via a quantum oracle (a black box whose implementation is not specified in the algorithm), which allows access to the training data in superposition rather than sequentially. This method is preferable for large training sets of data. Finally, one may observe that advantages by employing even small scale of noisy physical qubits is sufficient to efficiently compute solutions even without applying quantum error correction codes. A five superconductive qubits called transmons chip has been fabricated and programmed to use it as a quantum oracle and to speed up a learning task, known as binary classification (to identify an unknown mapping between a set of bits onto 0 or 1).[35]

4. Conclusion

To conclude, the development of artificial intelligence based on the emulation of the functional organization of the visual cortex enabled efficient machine learning called deep learning. Such algorithm may be extended to its quantum version to be implemented on a quantum computer. Quantum deep learning by a quantum deep belief network is the first example of a class of quantum algorithms which has been proved to outperform classical algorithms. Such algorithm creates abstraction layers based on quantum superpositions which belong to a representations space not accessible to human brain understanding.

References

[1] Zuse H 2016 Konrad Zuse's Computer Z3 *http://www.zuse.de*
[2] Von Neumann J 1993 First Draft of a Report on the EDVAC *IEEE Annals of the History of Computing* **15** 4 27-75
[3] Von Neumann J 2000 *The Computer and the Brain* (New Haven: Yale University Press, 2nd Edition, Mrs. Hepsa Ely Sillman Memorial Lectures) *Applied Physics Letters* **102** 123701
[4] Burton R M and Mpitsos G J 1992 Event-dependent control of noise enhances learning in neural networks *Neural Networks* **5** 4 627–637
[5] Brascamp J W et al. 2006 The time course of binocular rivalry reveals a fundamental role of noise *Journal of vision* **6** 11 8
[6] Orlandi J G et al. 2013 Noise focusing and the emergence of coherent activity in neuronal cultures *Nature Physics* **9** 9 582–590
[7] LeCun Y et al 1998 Gradient-Based Learning Applied to Document Recognition *Proc of the IEEE* **86** 11 2276–2324

[8] Hinton G E and Salakhutdinov R R 2006 Reducing the Dimensionality of Data with Neural Networks *Science* **313** 504
[9] Richardson MJE and Gerstner W 2005 Synaptic shot noise and conductance fluctuations affect the membrane voltage with equal significance *Neural computation* **17** 4 923–947
[10] Prati E 2016 Atomic scale nanoelectronics for quantum neuromorphic devices: comparing different materials *Int. J. of Nanotech.* **13** 7 509–523
[11] Prati E 2016 Noise-assisted transmission of spikes in Maeda–Makino artificial neuron arrays *Int. J. of Paral., Emerg. and Distr. Syst.* doi:10.1080/17445760.2016.1189914
[12] Hasani R et al 2017 Control of the Correlation of Spontaneous Neuron Activity in Biological and Noise–activated CMOS Artificial Neural Microcircuits arxiv: 1702.07426v1
[13] Panitchayangkoona G et al 2010 Long-lived quantum coherence in photosynthetic complexes at physiological temperature *PNAS* **107** 29 1276612770
[14] Sahu S et al. 2013 Multi-level memory-switching properties of a single brain microtubule
[15] Prati E, Fanciulli M, Calderoni A, Ferrari G and Sampietro M 2007 Microwave irradiation effects on random telegraph signal in a MOSFET *Phys. Lett.* **A 370** 491-493
[16] Prati E, Fanciulli M 2008 Manipulation of localized charge states in n-MOSFETs with microwave irradiation *Phys. Lett.* **A 372** 3102-3104
[17] Prati E et al 2010 Measuring the Temperature of a Mesoscopic Electron System by means of Single Electron Statistics *Applied Physics Letters* **96** 113109
[18] Prati E et al 2016 Band transport across a chain of dopant sites in silicon over micron distances and high temperatures *Scientific Reports* **6** 19704
[19] Lu J et al 2015 A 1 TOPS/W analog Deep machine-learning engine with floating-gate storage in 0.13 μm CMOS *IEEE J. of Solid-State Circ.* **50** 1 1-12
[20] Ivakhnenko A G and Lapa V G 1965 *Cybernetic Predicting Devices* CCM Information Corporation
[21] Hochreiter S and Schmidhuber J 1997 Long Short-Term Memory *Neural Computation***9** 8 1735-1780
[22] Sun R and Sessions C 2000 Self-segmentation of sequences: Automatic formation of hierarchies of sequential behaviors *IEEE Transactions on Systems, Man, and Cybernetics: Part B Cybernetics* **30** 3 403-418
[23] Volodymyr M et al 2015 Human-level control through deep reinforcement learning *Nature* **518** 529533
[24] Tesauro G 1994 TD-gammon, a self-teaching backgammon program, achieves master-level play *Neural Computation***6** 2 215-219
[25] Mnih V et al. Playing Atari with deep reinforcement learning 2013 *Deepmind Technologies Technical Report* arXiv:1312.5602
[26] Ballard D H 1987 Modular learning in neural networks *Proc. AAAI* 279-284
[27] Hinton G and Salakhutdinov R 2006 Reducing the dimensionality of data with neural networks *Science* **313** 504-507
[28] Schmidhuber J 1992 Learning complex, extended sequences using the principle of history compression *Neural Computation* **4** 2 234-242
[29] Schmidhuber J 2015 Deep Learning in Neural Networks: An Overview *Neural Networks* **61** 85-117
[30] Barends R et al. 2016 Digitized adiabatic quantum computing with a superconducting circuit *Nature* **534** 222–226
[31] Rotta D et al. 2016 Maximum density of quantum information in a scalable CMOS implementation of the hybrid qubit architecture *Quantum Information Processing* **15** 6 2253–2274
[32] Rotta D, Sebastiano D, Charbon E and Prati E 2017 *npg Quantum Information* arXiv:1704.06365
[33] Kinjo M, Sato S, Nakamiya Y and Nakajima K (2005) Neuromorphic quantum computation with energy dissipation *Physical Review A* **72** 5 052328
[34] Wiebe N, Kapoor A and Svore K M 2014 Quantum deep learning arXiv:1412.3489
[35] Risté D et al. 2017 Demonstration of quantum advantage in machine learning *npj Quantum Information* **3** 16
[36] Steane A M 1996 Error correcting codes in quantum theory *Phys. Rev. Lett.* **77** 793
[37] Fowler A G et al. 2012 Surface codes: towards large scale quantum computation *Phys. Rev. A* **86** 032324

Newtonian self-gravity in trapped quantum systems and experimental tests

André Großardt

Zentrum für angewandte Raumfahrttechnologie und Mikrogravitation (ZARM), Universität Bremen, Am Fallturm 2, 28359 Bremen, Germany
Dipartimento di Fisica, Università degli Studi di Trieste, Strada Costiera 11, 34151 Miramare-Trieste, Italy

E-mail: andre.grossardt@zarm.uni-bremen.de

Abstract. No experimental evidence exists, to date, whether or not the gravitational field must be quantised. Theoretical arguments in favour of quantisation are inconclusive. The most straightforward alternative to quantum gravity, a coupling between classical gravity and quantum matter according to the semi-classical Einstein equations, yields a nonlinear modification of the Schrödinger equation. Here, effects of this so-called Schrödinger-Newton equation are discussed, which allow for technologically feasible experimental tests.

1. Introduction

The question how to reconcile the foundations of quantum theory with the dynamical space-time structure of general relativity is often reduced to problems arising at the Planck scale, e. g. in the physics of black holes or the early universe. Nonetheless, the status of the description of both gravity and quantum matter in a common framework is far from clear also at low energies: certainly from the point of view of experimental observations, but also from a theoretical perspective.

An illustrative example is the familiar double-slit experiment, if conducted with a massive[1] particle. Assume a high sensitivity gravitational probe (a test mass attached to a sophisticated force metre) could be placed behind the slit, sensing the gravitational field of the spatial superposition state prepared by the slit. What is the gravitational field it would observe?

In analogy to electrodynamics one is tempted to think that the answer is obvious: The gravitational field should be in a superposition state itself, which is entangled with the particle state. However, the analogy between gravity and electrodynamics must eventually fail; at the latest when it comes to the question of renormalisability. One may, therefore, ask whether this analogy is still appropriate for the case of macroscopic non-classical states at low energy. In general relativity, on the other hand, the gravitational field is an expression of the space-time structure. Matter acts as a source, determining the curvature of space time. In the presence of quantised matter fields, one would ask: how does this quantum matter in a non-localised state source the gravitational field?

[1] What quantifies a massive particle depends, of course, on the precise experimental context, specifically on feasible force sensitivities.

Keeping the classical space-time structure of general relativity, and treating quantum fields—which are then living on this curved space-time—as a curvature source, gives rise to the idea of (fundamentally) semi-classical gravity. The seemingly most straightforward way to incorporate quantum matter as a source into Einstein's equations is to replace the classical stress-energy tensor by the expectation value of the corresponding quantum operator:

$$R_{\mu\nu} - \frac{1}{2} g_{\mu\nu} R = \frac{8\pi G}{c^4} \langle\Psi|\hat{T}_{\mu\nu}|\Psi\rangle . \tag{1}$$

These are the semi-classical Einstein equations which have first been proposed in this context by Møller [1] and Rosenfeld [2].

Arguments have been invoked attempting to rule out this type of coupling between quantum matter and gravity solely on theoretical grounds. It has been pointed out that Eq. (1) is incompatible with both an instantaneous wave-function collapse and a no-collapse interpretation of quantum mechanics [3]. Nonetheless, this leaves open the possibility of a dynamical collapse in the spirit of collapse models [4]. More general assertions against any type of semi-classical coupling [5] were found inconclusive [6, 7, 8]. It seems that Rosenfeld was right when emphasising the necessity of experimental tests whether or not the gravitational field should be quantised: *"There is no denying that, considering the universality of the quantum of action, it is very tempting to regard any classical theory as a limiting case to some quantal theory. In the absence of empirical evidence, however, this temptation should be resisted. The case for quantizing gravitation, in particular, far from being straightforward, appears very dubious on closer examination."* [2]

It was Carlip [9] who first pointed out that an experimental test of semi-classical gravity could be feasible in the nonrelativistic, low-energy limit of laboratory quantum systems. While the proposed test in matter wave interferometry experiments turned out to still be five to six orders of magnitude beyond the state of the art [10], tests in optomechanical set-ups are at the edge of feasible observations [11, 12].

Here, the underlying physics behind these proposed tests is reviewed. In section 2 the nonrelativistic limit of Eq. (1) to the so-called Schrödinger-Newton (SN) equation is discussed, followed by a review of how one obtains an equation for the centre of mass of a many-particle system in section 3. Effects of this modification of the Schrödinger equation in a harmonic oscillator are analysed in section 4 with a discussion of the possibilities for testing them in section 5. Some open questions are addressed in the conclusion section 6.

2. Semi-classical gravity in the nonrelativistic limit

Starting with the *classical* Einstein equations for a *classical* Klein-Gordon (or Dirac) field, one can show that the gravitational self-coupling is promoted to the Schrödinger equation (as a differential equation for the evolution of a classical field) obtained in the nonrelativistic ($c \to \infty$) limit [13], where it yields the nonlinear one-particle SN equation:

$$\mathrm{i}\hbar\frac{\partial}{\partial t}\psi(t,\mathbf{r}) = \left(-\frac{\hbar}{2m}\nabla^2 - Gm^2 \int \mathrm{d}^3\mathbf{r}' \, \frac{|\psi(t,\mathbf{r}')|^2}{|\mathbf{r}-\mathbf{r}'|}\right)\psi(t,\mathbf{r}) . \tag{2}$$

For a *quantum* field, coupling to gravity according to Eq. (1), one can proceed in a similar fashion [14]. In the nonrelativistic, weak coupling limit, Einstein's equations yield the Poisson equation $\nabla^2 V = 4\pi G m \langle\Psi| \hat{\psi}^\dagger\hat{\psi} |\Psi\rangle$, while the field equations on curved space-time limit to the Fock space Schrödinger equation including the solution V of this Poisson equation as a nonlinear potential. The sectors belonging to different particle numbers separate in the nonrelativistic limit. In conclusion, one ends up with the N-particle equation [15]

$$i\hbar\frac{\partial}{\partial t}\Psi(t,\mathbf{r}_1,\cdots,\mathbf{r}_N) = \left(-\sum_{i=1}^{N}\frac{\hbar^2}{2m_i}\nabla_i^2 + V_{\text{ext}} + V_g[\Psi]\right)\Psi(t;\mathbf{r}_1,\cdots\mathbf{r}_N) \tag{3a}$$

$$V_g[\Psi](t,\mathbf{r}_1,\cdots,\mathbf{r}_N) = -G\sum_{i=1}^{N}\sum_{j=1}^{N} m_i m_j \int \mathrm{d}^3\mathbf{r}'_1\cdots\mathrm{d}^3\mathbf{r}'_N \frac{|\Psi(t;\mathbf{r}'_1,\cdots,\mathbf{r}'_N)|^2}{\left|\mathbf{r}_i-\mathbf{r}'_j\right|}, \tag{3b}$$

where V_{ext} stands for any additional non-gravitational potentials. Eq. (3) contains both the mutual gravitational interactions between particles and the gravitational self-interaction of each particle. In the one-particle case, $N = 1$, one obtains Eq. (2).

3. The Schrödinger-Newton equation for the centre-of-mass motion

Contrary to linear interaction potentials, the wave-function dependent potential (3b) does not separate into centre of mass and relative coordinates exactly. Note, however, that the electromagnetic forces which determine the shape of a complex quantum system like a large molecule are considerably stronger than the gravitational forces. One can, therefore, make use of an approximation similar to the Born–Oppenheimer approximation in atom physics [16]. The characteristic time scale of the relative motion, dominated by the electromagnetic interactions, is much shorter than that of the centre of mass motion, which is only affected by self-gravitational forces. The potential (3b) can then be averaged over the relative degrees of freedom, resulting in a potential that only depends on the centre of mass coordinate, and in the following centre of mass SN equation:

$$i\hbar\frac{\partial}{\partial t}\psi(t,\mathbf{r}) = \left(-\frac{\hbar^2}{2M}\nabla^2 + V_{\text{ext}} + V_g[\psi]\right)\psi(t,\mathbf{r}) \tag{4a}$$

$$V_g[\psi](t,\mathbf{r}) = -G\int \mathrm{d}^3\mathbf{r}'\,|\psi(t,\mathbf{r}')|^2\, I_{\rho_c}(\mathbf{r}-\mathbf{r}') \tag{4b}$$

$$I_{\rho_c}(\mathbf{d}) = \int \mathrm{d}^3\mathbf{u}\,\mathrm{d}^3\mathbf{v}\,\frac{\rho_c(\mathbf{u})\,\rho_c(\mathbf{v}-\mathbf{d})}{|\mathbf{u}-\mathbf{v}|}, \tag{4c}$$

where M is the total mass, ψ the centre of mass wave-function, and ρ_c is the (averaged) mass distribution of the quantum system, defined from the relative wave-function χ as

$$\rho_c(\mathbf{r}) = \sum_{i=1}^{N-1} m_i \int \mathrm{d}^3\mathbf{r}_1\cdots\mathrm{d}^3\mathbf{r}_{i-1}\,\mathrm{d}^3\mathbf{r}_{i+1}\cdots\mathrm{d}^3\mathbf{r}_{N-1}\,|\chi(\mathbf{r}_1,\ldots,\mathbf{r}_{i-1},\mathbf{r},\mathbf{r}_{i+1},\ldots,\mathbf{r}_{N-1})|^2\,. \tag{4d}$$

Now, if the size of the considered system (e. g. molecule) is small compared to the extent of its centre of mass wave-function, ρ_c can be approximated by a delta distribution, and Eq. (4) is approximately equivalent to the one-particle equation (2).

If the width of the wave-function becomes comparable to the size of the system then its structure, represented by the function I_{ρ_c}, must be taken into account. For a homogeneous spherical particle of radius R, for instance, one finds

$$I_{\rho_c}^{\text{sphere}}(d) = \frac{M^2}{R}\times\begin{cases}\frac{6}{5}-2\left(\frac{d}{2R}\right)^2+\frac{3}{2}\left(\frac{d}{2R}\right)^3-\frac{1}{5}\left(\frac{d}{2R}\right)^5 & \text{for } d\leq 2R\,,\\ \frac{R}{d} & \text{for } d\geq 2R\,.\end{cases} \tag{5}$$

A more realistic model of a molecule also takes into account the crystalline structure of the atoms. A spherical particle of radius R with atoms localised with a Gaussian distribution of

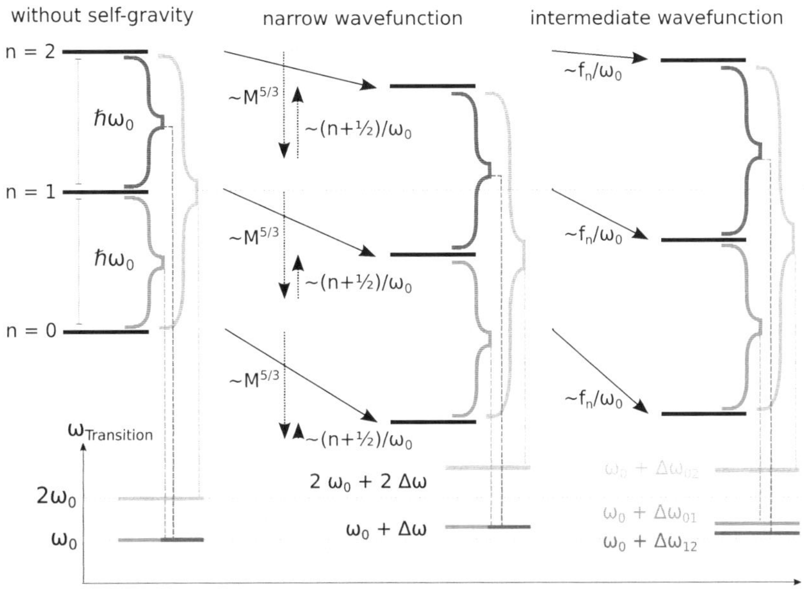

Figure 1. Schematic picture of the self-gravitational effect on the harmonic oscillator spectrum in the narrow and intermediate wave-function regime [12].

width σ is well approximated by the function [12, 17]

$$I_{\rho_c}^{\text{cryst}}(d) \approx \frac{6\,M^2}{5\,R} + \frac{M\,m_{\text{atom}}}{d}\,\text{erf}\left(\frac{d}{\sqrt{2}\,\sigma}\right), \tag{6}$$

as long as the wave-function has an extent small compared to the radius R of the whole particle. The first term stems from the mutual gravitational attraction of the atoms, while the second term describes the accumulated self-gravitation forces of each atom with its own marginal wave-function.

In the narrow wave-function regime, where the extent of the wave-function is small also compared to the localisation length σ of the atoms, the error function can be expanded around $d = 0$, yielding the quadratic approximation

$$I_{\rho_c}^{\text{narrow}}(d) \approx \frac{6\,M^2}{5\,R} + \frac{2\,M\,m_{\text{atom}}}{\sqrt{2\pi}\sigma}\left(1 - \frac{d^2}{6\,\sigma^2}\right). \tag{7}$$

4. Effects on harmonically trapped quantum systems

We assume that the potential V_{ext} is such that it strongly confines the wave-function in the y- and z-direction. Then the SN equation can be shown to become effectively one-dimensional [17]. If we assume a quadratic potential in the x-direction, $V_{\text{ext}} = M\,\omega_0^2\,x^2/2$, then for the narrow wave-function regime, using Eq. (7), one obtains the Hamiltonian

$$\begin{aligned} H^{\text{narrow}} &= -\frac{\hbar}{2\,M}\nabla^2 + \frac{M}{2}\omega_0^2\,x^2 - \frac{6\,G\,M^2}{5\,R} - \frac{2\,G\,M\,m_{\text{atom}}}{\sqrt{2\pi}\sigma}\left(1 - \frac{x^2 - 2\,x\,\langle x\rangle + \langle x^2\rangle}{6\,\sigma^2}\right) \\ &= -\frac{\hbar}{2\,M}\nabla^2 + \frac{M}{2}\omega_{\text{SN}}^2\,\langle x^2\rangle - M\,\omega_{\text{SN}}^2\,x\,\langle x\rangle + \frac{M}{2}\left(\omega_0^2 + \omega_{\text{SN}}^2\right)x^2 + \text{const.}, \end{aligned} \tag{8}$$

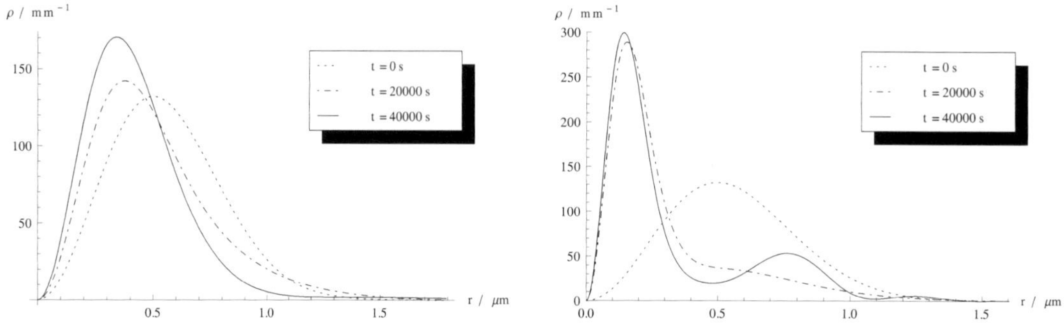

Figure 2. Inhibitions of the free dispersion of a wave packet. Plotted is the radial probability density $\rho = 4\pi\, r^2\, |\psi|^2$ for a mass of 7×10^9 u (left) and 10^{10} u (right) at different times.

with

$$\omega_{\mathrm{SN}} = \sqrt{\sqrt{\frac{2}{\pi}}\frac{G\, m_{\mathrm{atom}}}{3\,\sigma^3}}\,. \tag{9}$$

For a wave-function comparable to the atom localisation scale, the Hamiltonian becomes a much more complicated nonlinear functional. However, for the sake of determining the distortion of the energy spectrum one can simply use first order perturbation theory. The potential (4b) evaluated for the unperturbed eigenstates $\psi_n^{(0)}$ is then treated as a perturbation, yielding the energy shift

$$\Delta E_n = \langle \psi_n^{(0)} \mid V_g[\psi_n^{(0)}](\mathbf{r}) \mid \psi_n^{(0)} \rangle = -\frac{G\,\hbar\, m_{\mathrm{atom}}}{4\,\sigma^3\,\omega_0}\, f_n(\alpha)\,, \qquad \text{with} \qquad \alpha = 2\sigma\sqrt{\frac{M\omega_0}{\hbar}}\,. \tag{10}$$

In the case of the narrow wave-function, the function f_n depends on the quantum number n *linearly*, implying that the energy gap between eigenstates changes to $\Delta n\, \hbar\omega_0 (1 + \omega_{\mathrm{SN}}^2/\omega_0^2)$. Nonetheless, the transition energy only depends on the difference $\Delta n = (n_1 - n_2)$, not on n_1 and n_2 explicitly.

This changes in the regime of wider wave-functions, where f_n becomes nonlinear in n [12, 17]:

$$f_n(\alpha) \approx \text{const.} + \alpha^3 \sqrt{\frac{2}{\pi}} \int_0^\infty \mathrm{d}\zeta\, \exp\left(-\frac{\alpha^2\,\zeta^2}{2}\right) P_n(\alpha\,\zeta) \left(\frac{\mathrm{erf}\,(\sqrt{2}\,\zeta)}{2\,\zeta} - \sqrt{\frac{2}{\pi}}\right), \tag{11a}$$

$$P_n(z) = \frac{\mathrm{e}^{-z^2/2}}{\sqrt{2\pi}\,(2^n\, n!)^2} \int_{-\infty}^{\infty} \mathrm{d}\xi\, \mathrm{e}^{-2\xi^2}\, H_n\,(\xi)^2 \left(\mathrm{e}^{2\,z\,\xi}\, H_n\,(\xi - z)^2 + \mathrm{e}^{-2\,z\,\xi}\, H_n\,(\xi + z)^2\right), \tag{11b}$$

where H_n are the Hermite polynomials. Now the frequency gap between adjacent energy levels depends on n_1, n_2 explicitly, yielding a fine structure for the transition frequencies as depicted in Fig. 1.

Instead of the energy spectrum, one can study the effect of the self-gravitational potential on the dynamics of a quantum harmonic oscillator. For a squeezed coherent (i. e. Gaussian) state, the dynamics can be reduced to a set of two differential equations for the first and second moments:

$$0 = \frac{\mathrm{d}^2}{\mathrm{d}t^2}\langle x\rangle + \omega_0^2\langle x\rangle \tag{12a}$$

$$0 = \frac{\mathrm{d}^3}{\mathrm{d}t^3}\langle (x - \langle x\rangle)^2\rangle + 4\,\left(\omega_0^2 + \omega_{\mathrm{SN}}^2[\psi]\right)\,\frac{\mathrm{d}}{\mathrm{d}t}\langle (x - \langle x\rangle)^2\rangle\,. \tag{12b}$$

IOP Conf. Series: Journal of Physics: Conf. Series **880** (2017) 012019 doi:10.1088/1742-6596/880/1/012019

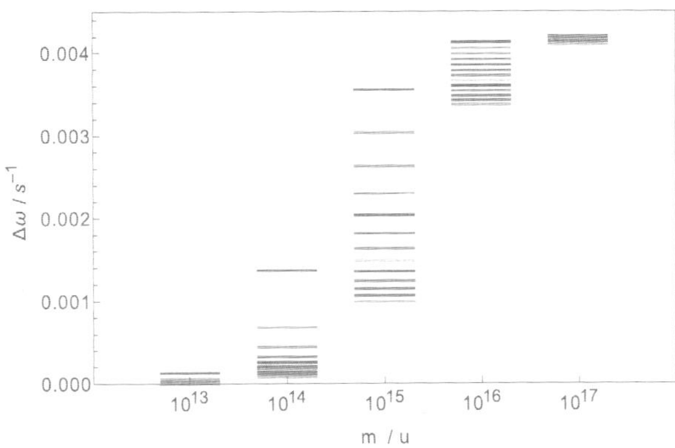

Figure 3. Resulting frequency spectrum of the transition frequency $\Delta\omega$ between adjacent eigenstates ($\Delta n = 1$) for osmium at $T = 100\,\mathrm{mK}$, $\omega_0 = 2\pi \times 10\,\mathrm{s}^{-1}$.

The first equation describes the usual oscillation of the centre of the wave-function with the unperturbed trap frequency ω_0. The second equation describes the internal oscillation, for which the frequency is shifted by a factor of $\sqrt{1 + \omega_{\mathrm{SN}}^2/\omega_0^2}$.

In the general case of a wider wave-function this frequency shift depends on the wave-function. However, it is strongest for a narrow wave-function, where ω_{SN} takes the value (9). For a wider wave-function, the self-gravitational effect only becomes less significant. The effect in the narrow regime was discussed by Yang et al. [11].

5. Experimental tests

First ideas to test the SN equation experimentally [9] were based on the loss of interference in matter-wave interferometry experiments [18]. Current experiments reach masses of up to 10^4 u and are clearly in the wide wave-function regime where Eq. (2) is valid. Numerical and analytical estimates [10] yield a required mass around 10^{10} u for a micrometre sized wave packet to show significant inhibitions of the free dispersion, see Fig. 2 for example plots of the evolution of a Gaussian wave packet. The required mass values seem infeasible on Earth, although they might be achievable in space experiments [19]. Note also that for the required mass scales, the effects of the internal structure of the particles are no longer negligible and, therefore, Eq. (2) is only of limited use for good predictions. However, approximation schemes exist [20, 21] which allow for a simple numerical calculation of the magnitude of the deviation between the nonlinear dynamics according to the SN equation and the ordinary, linear Schrödinger equation.

As far as ground based experiments are concerned, the two effects discussed here for particles trapped in quadratic potentials seem the most promising routes towards an experimental test. A proposal for a test of the spectral effect was presented in Ref. [12]. It consists of a levitated osmium disc in a Paul trap which is cooled to a temperature below 100 mK. At this temperature osmium becomes superconducting, avoiding decoherence due to blackbody radiation. The suggested experimental parameters are a particle mass of 10^{14} u trapped at a frequency of 10 Hz. At these values, a frequency resolution of 0.1 mHz would be sufficient for a detection of the spectral fine structure due to the SN equation. Fig. 3 shows a quantitative plot of the expected frequency spectrum for these values. At low masses, self-gravity becomes negligible. At high masses all spectral lines are shifted equally. The intermediate regime, where a significant splitting appears, spans about 3 orders of magnitude in mass.

The crucial parameter for the choice of material is the Debye–Waller factor, which determines the localisation σ of the atoms within the crystal. It takes an ideal value for osmium, even compared to materials of larger atomic mass such as gold; see Tab. 1 for a comparison of some materials. Levitation of a micrometre particle in a Paul trap has already been demonstrated experimentally [22], as have the low frequencies required [23], and the control of decoherence [24].

Table 1. Relevant experimental parameters for different materials [12].

Material	m_{atom} / u	ρ / g cm^{-3}	σ / pm	$\omega_{\text{SN}}^2/\omega_0^2$
Silicon	28.086	2.329	6.96	0.00246
Tungsten	183.84	19.30	3.48	0.128
Osmium	190.23	22.57	2.77	0.264
Gold	196.97	19.32	4.66	0.0574

Cooling to low energy states has not been demonstrated for the required masses, but the technological tools exist.

The dynamical effects [11] are somewhat disadvantageous to observe: they require both preparation of a squeezed state and a larger mass in order to reach the narrow wave-function regime. On the other hand, the required set-up is closer to existing ones, such as cooled mirrors (e. g. LIGO or LISA pathfinder). The fact that all of these systems are based on silicon rather than osmium, however, requires another two orders of magnitude better resolution compared to the same system based on osmium (cf. Tab. 1). At 10 Hz trap frequency, a quality factor better than 10^6 is required [11]. The feasibility for a test in several systems has been examined in Ref. [25].

6. Conclusion and open questions

From the above discussion we conclude that, although a realisation is still challenging, the SN equation seems testable with existing technology in the nearer future. What would we learn from such a test?

Obviously, if the predicted deviations from linear quantum mechanics would be observed, this would result in a complete paradigm shift of our view on both gravity and the foundations of quantum mechanics. If, on the other hand, standard quantum mechanics is confirmed and the SN equation ruled out by experiments (likely the outcome expected by a large majority of physicists) this would rule out the semi-classical Einstein equations (1) as a fundamental model. It would, however, not be conclusive evidence for the quantisation of gravity, since other types of coupling of quantum matter to a classical space-time could be possible, which do not yield the SN equation as a nonrelativistic limit. In fact, with stochastic gravity [26, 27] there is an example for such a model.

As the SN equation was first discussed by Diósi [15] and Penrose [28], it is often considered in context of the gravitational collapse model named after these two [29, 30, 31, 28, 32]. Although not directly related to it, one can discuss whether the SN equation could play a role in the collapse of the wave-function. On the one hand, the classicality of the gravitational force provides exactly the non-unitary ingredient needed for an objective collapse. On the other hand, the SN alone does not explain the emergence of classicality, since the SN "collapse" is completely deterministic. It does not recover the Born probability rule [14].

Probably the biggest concern against a fundamental validity of the SN equation is that measurements in the usual instantaneous collapse prescription allow for faster-than-light signalling when combined with a nonlinear dynamics [14]. It might be that a consistent description which makes use of semi-classical gravity as an explanation of the wave-function collapse can circumvent this problem. This description, however, has yet to be found.

Acknowledgments
The advances reviewed here resulted from collaborations with M. Bahrami, A. Bassi, J. Bateman, S. Donadi, D. Giulini, and H. Ulbricht. I gratefully acknowledge funding from the German Research Foundation (DFG) and the Italian National Institute for Nuclear Physics (INFN).

References
[1] Møller C 1962 *Colloques Internationaux CNRS* vol 91 ed Lichnerowicz A and Tonnelat M A (CNRS, Paris)
[2] Rosenfeld L 1963 *Nucl. Phys.* **40** 353–356
[3] Page D N and Geilker C D 1981 *Phys. Rev. Lett.* **47** 979–982
[4] Bassi A, Lochan K, Satin S, Singh T P and Ulbricht H 2013 *Rev. Mod. Phys.* **85** 471–527 (*Preprint* 1204.4325)
[5] Eppley K and Hannah E 1977 *Found. Phys.* **7** 51–68
[6] Mattingly J 2005 *Einstein Studies Volume 11. The Universe of General Relativity* Einstein Studies ed Kox A J and Eisenstaedt J (Boston: Birkhäuser) chap 17, pp 327–338
[7] Kiefer C 2007 *Quantum Gravity* 2nd ed (*International Series of Monographs on Physics* vol 124) (Oxford: Clarendon Press)
[8] Albers M, Kiefer C and Reginatto M 2008 *Phys. Rev. D* **78** 064051 (*Preprint* 0802.1978)
[9] Carlip S 2008 *Class. Quantum Grav.* **25** 154010 (*Preprint* 0803.3456)
[10] Giulini D and Großardt A 2011 *Class. Quantum Grav.* **28** 195026 (*Preprint* 1105.1921)
[11] Yang H, Miao H, Lee D S, Helou B and Chen Y 2013 *Phys. Rev. Lett.* **110** 170401 (*Preprint* 1210.0457)
[12] Großardt A, Bateman J, Ulbricht H and Bassi A 2016 *Phys. Rev. D* **93** 096003 (*Preprint* 1510.01696)
[13] Giulini D and Großardt A 2012 *Class. Quantum Grav.* **29** 215010 (*Preprint* 1206.4250)
[14] Bahrami M, Großardt A, Donadi S and Bassi A 2014 *New J. Phys.* **16** 115007 (*Preprint* 1407.4370)
[15] Diósi L 1984 *Phys. Lett. A* **105** 199–202
[16] Giulini D and Großardt A 2014 *New J. Phys.* **16** 075005 (*Preprint* 1404.0624)
[17] Großardt A, Bateman J, Ulbricht H and Bassi A 2016 *Sci. Rep.* **6** 30840 (*Preprint* 1510.01262)
[18] Arndt M and Hornberger K 2014 *Nat. Phys.* **10** 271–277
[19] Kaltenbaek R, Arndt M, Aspelmeyer M, Barker P F, Bassi A, Bateman J, Bongs K, Bose S, Braxmaier C, Brukner C, Christophe B, Chwalla M, Cohadon P F, Cruise A M, Curceanu C, Dholakia K, Diósi L, Döringshoff K, Ertmer W, Gieseler J, Gürlebeck N, Hechenblaikner G, Heidmann A, Herrmann S, Hossenfelder S, Johann U, Kiesel N, Kim M, Lämmerzahl C, Lambrecht A, Mazilu M, Milburn G J, Müller H, Novotny L, Paternostro M, Peters A, Pikovski I, Pilan-Zanoni A, Rasel E M, Reynaud S, Riedel C J, Rodrigues M, Rondin L, Roura A, Schleich W P, Schmiedmayer J, Schuldt T, Schwab K C, Tajmar M, Tino G M, Ulbricht H, Ursin R and Vedral V 2016 *EPJ Quant. Tech.* **3** 5 (*Preprint* 1503.02640)
[20] Großardt A 2016 *Phys. Rev. A* **94** 022101 (*Preprint* 1503.02622)
[21] Colin S, Durt T and Willox R 2016 *Phys. Rev. A* **93** 062102 (*Preprint* 1402.5653)
[22] Wuerker R F, Shelton H and Langmuir R V 1959 *J. Appl. Phys.* **30** 342–349
[23] Gerlich D 2003 *Hyperfine Interactions* **146-147** 293–306 ISSN 0304-3843
[24] Poulsen G, Miroshnychenko Y and Drewsen M 2012 *Phys. Rev. A* **86**(5) 051402
[25] Gan C, Savage C and Scully S 2016 *Phys. Rev. D* **93** 124049
[26] Hu B L and Verdaguer E 2008 *Living Rev. Relativity* **11** 3
[27] Anastopoulos C and Hu B L 2014 *New J. Phys.* **16** 085007 arXiv:1403.4921 (*Preprint* 1403.4921)
[28] Penrose R 1998 *Phil. Trans. R. Soc. A* **356** 1927–1939
[29] Diósi L and Lukács B 1987 *Ann. Phys. (Berlin)* **499** 488–492
[30] Diósi L 1989 *Phys. Rev. A* **40** 1165–1174
[31] Penrose R 1996 *Gen. Relativ. Gravit.* **28** 581–600
[32] Penrose R 2014 *Found. Phys.* **44** 557–575

IOP Conf. Series: Journal of Physics: Conf. Series **880** (2017) 012020 doi:10.1088/1742-6596/880/1/012020

Centre of mass decoherence due to time dilation: paradoxical frame-dependence

Lajos Diósi

Wigner Research Centre for Physics, Budapest 114, P.O.Box 49, H-1525 Hungary

E-mail: diosi.lajos@wigner.mta.hu

Abstract. The recently proposed centre-of-mass decoherence of composite objects due to gravitational time-dilation [Pikovski *et al*, Nat.Phys. 11, 668 (2015)] is confronted with the principle of equivalence between gravity and observer's acceleration. In the laboratory frame, a positional superposition $|x_1\rangle + |x_2\rangle$ can quickly decohere whereas in the free-falling frame, as I argue, the superposition can survive for almost arbitrary long times. The paradoxical result is explained by the so far unappreciated feature of the proposed model: the centre-of-mass canonical subsystem is ambiguous, it is different in the laboratory and the free-falling frames, respectively.

As long as the centre-of-mass motion of the composite object is non-relativistic, a simple Galilean-covariant Hamiltonian represents the Pikovski *et al* theory with exactly the same physical predictions. We emphasize the power of this Hamiltonian to understand essential features of the Pikovski *et al* theory and to moderate a few divergent statements in recent works.

1. Introduction[1]

Pikovski, Zych, Costa and Brukner (PZCB) have shown that the centre-of-mass of a composite object undergoes universal position-decoherence in the presence of gravitational field [1]. They proposed an approximate relativistic correction of order $1/c^2$ to the non-relativistic many-body Hamiltonian. It results in the coupling

$$-\frac{H_{\rm i}}{mc^2}\left(\frac{p^2}{2m} - mgx\right) \tag{1}$$

between the internal Hamiltonian $H_{\rm i}$ and the difference between kinetic and gravitational energies of the centre-of-mass. Here m is the total mass, x, p are the centre-of-mass canonical coordinate and momentum, resp., g is the gravitational acceleration. If the object is prepared at rest in superposition $|x_1\rangle + |x_2\rangle$ of two positions x_1, x_2 at large vertical difference Δx then the internal degrees of freedom will quickly decohere the said positional superposition. The universal mechanism, stemming from time-dilation in Earth gravity, decoheres composite systems into the position basis. This happens within standard physics, as emphasized by the authors.

[1] This section, apart from last paragraph, coincides with [2] rejected by *Nature Physics.*

IOP Conf. Series: Journal of Physics: Conf. Series **880** (2017) 012020 doi:10.1088/1742-6596/880/1/012020

I unearth that this decoherence is not independent of the frame of the observer. Changing the laboratory frame for the free-falling one, the coupling in question becomes

$$-\frac{H_\text{i}}{mc^2}\frac{p^2}{2m} \tag{2}$$

by the power of the equivalence principle. We can (though don't need to) take the same H_i in both the laboratory and the free-fall frames since its corrections would be relativistic to contain further $1/c^2$-factors. If I solve the Schrödinger equation in the free-falling frame, the wave function exhibits centre-of-mass kinetic energy decoherence, position decoherence is completely absent. In the room temperature example of the authors, a gram-scale object prepared in the positional superposition $|x_1\rangle + |x_2\rangle$ at $\Delta x = 10^{-6}$ cm gets decohered after about a millisecond — in the laboratory frame. It does not get decohered in the free-falling frame as long as the kinetic energy decoherence remains marginal. Indeed, the dispersion of p in the initial superposition must be much bigger than $\bar{h}/\Delta x$but otherwise it can be kept so small that the kinetic energy decoherence would remain ignorable for any conceivable period. Hence the superposition may survive practically forever in the free-falling frame. I find it fairly paradoxical. Although decoherence is frame-dependent relativistically, it is impossible that a superposition decays in one frame but it never decays in the other.

The reason of such paradoxical frame-dependence is simple but surprising. When we change from laboratory to free-fall frame, the laboratory momentum transforms like $p \Rightarrow p+(m+c^{-2}H_\text{i})gt+\mathcal{O}(1/c^4)$ — it drives out from the state space of the laboratory centre-of-mass subsystem! In other words, the split of the composite system into centre-of-mass and internal canonical subsystems is frame-dependent. The notion of *canonical* centre-of-mass, unlike the notion of centre-of-mass coordinate, is frame-dependent. Quantum mechanics uses the canonical notion. Although in the laboratory frame and, alternatively, in the free-falling frame we are supposed to measure the state of *the* centre-of-mass, the latter is defined differently in the two frames, allowing for the particular inequivalence of decoherence calculated in the two frames respectively. This is a remarkable feature encoded in the theory of PZCB.

For a related criticism on [1], see also ref. [3].

2. Debates, clarifications

In 2015, PZCB were neither aware nor interested in the behaviour of their theory in situations different from what they considered the important ones exclusively. Indeed, the claim [2] of paradoxical frame dependence originated from such applications rather than from — as the authors stated in [4] — conceptual oversight. In their concept, transformation between different frames means the laboratory and free-falling observers describe the same system and the same detector. No doubt then, presence or loss of interference fringes are invariant. Just frame dependence of interference fringes originates from another, not less sensible and standard concept of frame transformation: the laboratory observer applies laboratory detector while the free-falling observer applies free-falling detector. No doubt again, for a free-falling mass the laboratory detector sees PZCB decoherence but the free-falling detector shows perfect fringes. Using different detectors means different experiments, as [4] pointed it out correctly but that should not have been a reason to brush off the surprising outcomes. Recognition and resolution of the paradox could have been the *productive* alternative. For the unsupported (free-falling) mass two major points should have deserved attention.

First, the centre-of-mass reduced dynamics turned out to be unitary in the free-fall frame and heavily non-unitary in the laboratory frame. The offered resolution was the following [2]. The usual invariance of the split

$$\mathcal{H}_\text{tot} = \mathcal{H}_\text{cm} \otimes \mathcal{H}_\text{i} \tag{3}$$

of the total system into centre-of-mass and internal degrees of freedom is lost due to the coupling (1) between centre-of-mass and internal degrees of freedom. Without this coupling $\mathcal{H}_{\rm cm}$ is the same in both frames, q, p and centre-of-mass density matrices transform unitarily from one frame to the other. In the PZCB theory, however, $\mathcal{H}_{\rm cm}$ becomes different for laboratory and free-fall frames, respectively. Ref. [4], constantly disregarding the application of the theory in a different situation from ref.'s [1], failed to appreciate the point: the frame-dependence of the split (3) explains why the *reduced dynamics of the centre-of-mass are not unitary equivalent in both frames.*

Second, it turned out that measuring the same (identically prepared) object, a free-falling detector should see perfect fringes while a laboratory detector may see no fringes at all. That is paradoxical enough unless we consider a concrete model of detection. This happened finally with the lucid proof of Pang, Chen and Khalili [5]. These authors considered the superposition of two centre-of-mass wave packets of a free-falling object, prepared at respective vertical positions x_1, x_2 and at common horizontal momentum p. The two wave packets reach the vertical screen through distance L at arrival time Lm/p. If the screen is free-falling together with the mass, the wave packets form the following interference pattern on the screen:

$$\mathrm{const} \times \left(1 + \mathrm{V}\cos\left[\frac{\mathrm{p}(\mathrm{x}_1 - \mathrm{x}_2)/\mathrm{L}}{\hbar}\mathrm{x}_{\mathrm{scr}}\right]\right), \tag{4}$$

where visibility V reaches 1 in ideal experiment. (We set $x_{\rm scr} = 0$ to height $(x_1 + x_2)/2)$). Now let the srceen move at vertical speed $v_{\rm scr}$ relative to the mass. The fringes get shifted on the screen:

$$\mathrm{const} \times \left(1 + \mathrm{V}\cos\left[\frac{\mathrm{p}(\mathrm{x}_1 - \mathrm{x}_2)/\mathrm{L}}{\hbar}\left(\mathrm{x}_{\mathrm{scr}} - \mathrm{v}_{\mathrm{scr}}\frac{\mathrm{Lm}}{\mathrm{p}}\right)\right]\right). \tag{5}$$

This shift does not influence fringe visibility. Now let us insert the relativistic correction $H_{\rm i}/c^2$ of the mass. The dispersion ΔE of the internal energy $H_{\rm i}$ means dispersion of arrival times at the screen (i.e: *dispersion of time dilation* according to ref. [1]) resulting in reduction of fringe visibility by the factor

$$\exp\left(-\frac{1}{2}\left(v_{\rm scr}\frac{(x_1 - x_2)\Delta E}{\hbar c^2}\right)^2\right). \tag{6}$$

This decoherence effect is special relativistic, depends on the relative transverse velocity of the mass and the screen, but unrelated to gravity. If we insert $v_{\rm scr} = gt$ corresponding to the screen supported in the laboratory on Earth, and understand that $t = Lm/p$ is the (mean) time passed bye from preparation of the superposition until its detection then we expect the above decoherence factor coincide with the prediction of the PZCB theory:

$$\exp\left(-\frac{t^2}{2}\left(\frac{g(x_1 - x_2)\Delta E}{\hbar c^2}\right)^2\right). \tag{7}$$

In [7] PZCB literally distanced their decoherence effects from those found by ref. [5] on the moving screen (cf. Fig. 1).

3. Shortcut to the Pikovski *et al* theory

Let us start from standard non-relativistic quantum mechanics of a free composite object:

$$H = \frac{p^2}{2m} + H_{\rm i}. \tag{8}$$

IOP Conf. Series: Journal of Physics: Conf. Series **880** (2017) 012020 doi:10.1088/1742-6596/880/1/012020

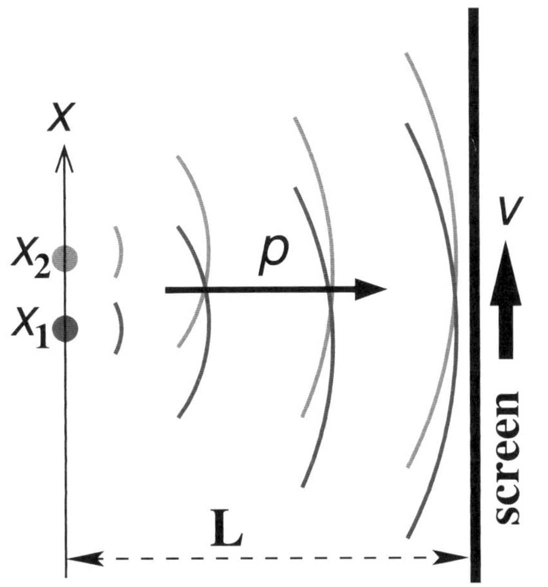

Figure 1. Interference of vertical superposition of two centre-of-mass wave packets in inertial motion, traveling horizontally at momentum p up to the screen which is in upward motion at velocity v_{scr}. Vertical shift of the fringes is $-v_{\text{scr}}$ times the arrival time. Since the arrival time Lm/p becomes $L(m+c^{-2}H_{\text{i}})/p$ due to special relativistic time dilation, fringe visibility becomes corrupted by the dispersion of the internal energy H_{i}.

The non-standard ansatz is the following: we take the relativistic correction H_{i}/c^2 of the mass into the account:

$$H = \frac{p^2}{2(m + c^{-2}H_{\text{i}})} + H_{\text{i}}. \tag{9}$$

This couples the centre-of-mass and internal degrees of freedom. It is important to emphasize that we ignore other possible relativistic corrections of the Hamiltonian. The obtained theory is a non-relativistic theory with a relativistic correction of the mass only. The Hamiltonian is Galilean-covariant, more precisely, it is covariant under the trivial extension of the Galilean group [6]. Moreover, it satisfies the Newtonian equivalence principle: an accelerated frame is equivalent with gravity. In particular, if we change the inertial (free-falling) reference frame for upward accelerated one (equivalent to laboratory frame) then, after canonical transformation, the Hamiltonian reads:

$$H = \frac{p^2}{2(m + c^{-2}H_{\text{i}})} + (m + c^{-2}H_{\text{i}})gx + H_{\text{i}}, \tag{10}$$

which, indeed, captures the presence of gravity g. The gravitating mass contains the same relativistic correction that we added to the inertial mass. The above non-relativistic theory, with its simple Galilean and Newtonian transformation properties, contains all physics of the PZCB theory as long as the centre-of-mass velocities can be ignored on the scale of light's velocity c, so that we can safely ignore the usual $\propto (p/mc)^2$ relativistic corrections of the Hamiltonian.

PZCB [1] proposed a fruitful relativistic correction to spatial dynamics of non-relativistic composite objects. The small parameter of usual relativistic corrections is $(p/mc)^2$. However, in ref. [1] there is a second small parameter (H_{i}/mc^2). *The two corrections are independent.*

The latter correction, without the usual ones, captures all the novelties due to relativistic time dilation of the internal 'clock' of the composite object. If the spatial motion is non-relativistic, that has been the case in most, if not all proposals so far, then the above Galilean-Newtonian Hamiltonian theory is equivalent with the PZCB theory [1].

Note that the PZCB coupling (1) corresponds to the Hamiltonian (10) expanded up to terms linear in the small parameter (H_i/mc^2). Although the relativistic consistency of higher orders of (H_i/mc^2) may be questionable, the full Hamiltonian (10) is convenient for its exact (extended) Galilean symmetry.

4. Epilogue

The Pikovski *et al* theory [1] is an *original, thoughtful and highly motivating* proposal for the effective relativistic coupling between centre-of-mass and internal degrees of freedom, with all its consequences like spatial gravitational decoherence, and with all its interpretations in context of special and general relativistic time dilation and of quantum foundations. The PZCB theory had been conceived as an *approximate Lorentz invariant* theory, up to order of $1/c^2$. This approach has long masked the fact that there is an *exact extended-Galilean-invariant* reduction of PZCB theory, fully incorporating relativistic time dilation of internal motion and being equivalent to PZCB theory *as long as the centre-of-mass motion is non-relativistic.* Interestingly, in ref. [8] Zych and Brukner were fully aware of this theory but, apparently, they did not accept it as a relevant option. Recognition would have waived desperate divergences in previous debates, would help ongoing discussions as well.

Bondar, Okon and Sudarsky [3] criticized the PZCB theory for it violates Bargman's mass superselection rule [9]. The PZCB refutal [4] was this: the theory is derived from a relativistic invariant theory, it is not Galilean-invariant hence not subject to mass superselection [9]. The precise resolution of the conflict is that mass superposition is allowed by a trivial extension of the Galilean symmetry, see [6], under which (10) is exact covariant. This was clear for the authors of [8] but, unfortunately, the refutal did not mention it.

A composite object supported at rest in the laboratory frame, rather than in free-fall, was the exclusive object of the original derivation of PZCB centre-of-mass decoherence [1]. A recent work [10] claimed that the decoherence effect becomes fully canceled by the supporting potential. Fortunately, PZCB [11] point out correctly that [10] constructs the mistaken supporting potential which turns out to be a gravitational field $-g$ just canceling with Earth's gravity g. Now, the mistake of [10] emerged from the desire to replace the non-relativistic supporting potential well $V(x)$ by some relativistically covariant supporting field. Had one started from the Hamiltonian (10), an additional supporting potential $V(x)$ would have been added consistently and the PZCB decoherence would have been derived correctly and in agreement with [1].

Acknowledgments

The author thanks Igor Pikovski, Magdalena Zych, Fabio Costa and Časlav Brukner for extended and useful discussions with each of them. This work was supported by EU COST Actions MP1006, MP1209, CA15220 and the Hungarian Scientific Research Fund under Grant No. 103917.

Appendix A. Derivation of PZCB Hamiltonian from special relativity

We start with the standard Lagrange function of classical pointlike mass in special relativity, in Minkowski coordinates:

$$L = -mc\sqrt{c^2 - \dot{x}^2} \tag{A.1}$$

yielding the Hamilton function:

$$H = c\sqrt{m^2c^2 + p^2}. \tag{A.2}$$

Consider the same motion in Rindler coordinates $\tilde{t}, \tilde{x}$ (without restriction of generality, we can take a single spatial dimension). Substituting

$$t = \frac{1}{c}(\tilde{x} + c^2/g)\sinh(g\tilde{t}/c), \quad x = (\tilde{x} + c^2/g)\cosh(g\tilde{t}/c) - c^2/g, \tag{A.3}$$

the Lagrange function reads:

$$\tilde{L} = -mc\sqrt{(c + g\tilde{x}/c)^2 - \dot{\tilde{x}}^2}, \tag{A.4}$$

yielding the Rindler Hamilton function:

$$\tilde{H} = c\sqrt{m^2(c + g\tilde{x}/c)^2 + \tilde{p}^2}. \tag{A.5}$$

The quantized motion of relativistic point-like masses needs field theory, but the quantum mechanical model exists at small velocities $p \ll mc$, with the following Hamiltonian in Minkovski coordinates:

$$H = mc^2 + \frac{p^2}{2m}\left(1 - \left(\frac{p}{2mc}\right)^2\right), \tag{A.6}$$

and in Rindler coordinates:

$$\tilde{H} = mc^2 + \frac{1}{2}\left\{\frac{\tilde{p}^2}{2m} + mg\tilde{x}, 1 - \left(\frac{\tilde{p}}{2mc}\right)^2\right\}. \tag{A.7}$$

If the mass has internal degrees of freedom of Hamiltonian H_{i} then we can introduce the relativistic correction H_{i}/c^2 to the inertial mass. If we retain terms linear in H_{i}, the above two Hamiltonians take the following forms:

$$H = mc^2 + H_{\mathrm{i}} + \frac{p^2}{2m}\left(1 - \left(\frac{p}{2mc}\right)^2 - \left(\frac{H_{\mathrm{i}}}{mc^2}\right)^2\right), \tag{A.8}$$

$$\tilde{H} = mc^2 + H_{\mathrm{i}} + \frac{p^2}{2m}\left(1 - \left(\frac{p}{2mc}\right)^2 - \left(\frac{H_{\mathrm{i}}}{mc^2}\right)^2\right) + \frac{1}{2}\left\{mg\tilde{x}, 1 - \left(\frac{\tilde{p}}{2mc}\right)^2 + \left(\frac{H_{\mathrm{i}}}{mc^2}\right)^2\right\}. \tag{A.9}$$

Appendix B. Reduced dynamics in laboratory and in free-fall frames[2]

We are going to determine the centre-of-mass reduced density matrix $\rho_{\mathrm{cm}}(t)$ using an uncorrelated initial state:

$$\rho_{\mathrm{cm}}(t) = \mathrm{Tr}_{\mathrm{i}}\left[\mathrm{e}^{-itH/\hbar}\left(\rho_{\mathrm{cm}}(0) \otimes \frac{\mathrm{e}^{-H_{\mathrm{i}}/k_B T}}{Z}\right)\mathrm{e}^{itH/\hbar}\right]. \tag{B.1}$$

Let our Hamiltonian be

$$H = \frac{p^2}{2(m + c^{-2}H_{\mathrm{i}})} + (m + c^{-2}H_{\mathrm{i}})gx + H_{\mathrm{i}} \tag{B.2}$$

which yields the coupling used by the authors in the leading order of $1/c^2$. We use this non-perturbative form just for convenience of calculations, without attributing any physical significance to the higher order terms. [The Hamiltonian of ref. [1] must contain a further

[2] As appeared in v2 of [2].

potential $V(x)$, to support the two wave packets that are superposed initially.] In the free-falling frame we determine $\rho_{\rm cm}(t)$, using the Hamiltonian

$$H = \frac{p^2}{2(m + c^{-2}H_{\rm i})} + H_{\rm i} \tag{B.3}$$

obtained from (B.2) in the free-fall coordinate $x - gt^2/2$. The laboratory and free-falling frames coincide at $t = 0$, we assume the same initial state, cf. (B.1), in both frames, respectively. It turns out that $\rho_{\rm cm}(t)$ may show quick positional decoherence in the laboratory frame and practically no decoherence ever in the free-falling frame.

We introduce the following centre-of-mass Hamiltonian in function of the internal energy eigenvalues E:

$$\begin{aligned} H(E) &= \frac{p^2}{2(m + c^{-2}E)} + (m + c^{-2}E)gx \\ &\equiv K(E) + (m + c^{-2}E)gx \end{aligned} \tag{B.4}$$

where $K(E)$ stands for the kinetic part. The reduced state of interest (B.1) can be expressed in this form:

$$\rho_{\rm cm}(t) = \sum_E \frac{{\rm e}^{-E/k_BT}}{Z} {\rm e}^{-itH(E)/\hbar} \rho_{\rm cm}(0) {\rm e}^{itH(E)/\hbar}. \tag{B.5}$$

The following useful identity holds for the operator part:

$$\begin{aligned} &\langle x_1|{\rm e}^{-itH(E)/\hbar}\rho_{\rm cm}(0){\rm e}^{itH(E)/\hbar}|x_2\rangle = \\ &= e^{-it(m+c^{-2}E)g(x_1-x_2)/\hbar} \times \\ &\quad \times \langle x_1{-}\tfrac{1}{2}gt^2|{\rm e}^{-itK(E)/\hbar}\rho_{\rm cm}(0){\rm e}^{itK(E)/\hbar}|x_2{-}\tfrac{1}{2}gt^2\rangle. \end{aligned} \tag{B.6}$$

We can ignore the unitary evolution $\exp[-itK(E)/\hbar]$ if both the coherent momentum and momentum uncertainty of the initial state $\rho_{\rm cm}(0)$ are suitably small. This can simply happen to a massive object initially at rest, as we see later. So we insert the identity (B.6) with $K(E) = 0$ into the expression (B.5) of $\rho_{\rm cm}(t)$; we obtain:

$$\begin{aligned} &\langle x_1|\rho_{\rm cm}(t)|x_2\rangle = \\ &= \sum_E \frac{{\rm e}^{-E/k_BT}}{Z} e^{-it(m+c^{-2}E)g(x_1-x_2)/\hbar} \times \\ &\quad \times \langle x_1{-}\tfrac{1}{2}gt^2|\rho_{\rm cm}(0)|x_2{-}\tfrac{1}{2}gt^2\rangle. \end{aligned} \tag{B.7}$$

We evaluate the pre-factor using the Gaussian approximation of thermodynamic fluctuations. The result is an explicit positional decoherence pre-factor:

$$\begin{aligned} &\langle x_1|\rho_{\rm cm}(t)|x_2\rangle = \\ &= e^{-\frac{1}{2}[c^{-2}\Delta E g(x_1-x_2)t/\hbar]^2} \times \\ &\quad \times e^{-it(m+c^{-2}\bar{E})g(x_1-x_2)/\hbar} \langle x_1{-}\tfrac{1}{2}gt^2|\rho_{\rm cm}(0)|x_2{-}\tfrac{1}{2}gt^2\rangle \end{aligned} \tag{B.8}$$

where $\bar{E}, \Delta E$ are the mean and the fluctuation, respectively, of the internal energy. We read out the decoherence time:

$$\tau_{\rm dec} = \frac{\hbar c^2}{\Delta E g|x_1 - x_2|}. \tag{B.9}$$

[This result is equivalent with the one obtained by [1] for the superposition resting in a potential $V(x)$.] The squared fluctuation is proportional with the heat capacity of the object:

$$(\Delta E)^2 = k_B T^2 \frac{d\bar{E}}{dT}. \tag{B.10}$$

Let us apply the result (B.8-B.10) to an initial superposition of two Gaussian wave packets of width σ each, standing at x_1 and x_2, respectively. For concreteness, we choose $\sigma = 10^{-8}$ m and $|x_1 - x_2| = 10^{-6}$ m, guaranteeing the perfect separation of the two wave packets. [This is a faithful physical representation of the symbolic superposition $|x_1\rangle + |x_2\rangle$.] Also we take $m = 1$ g and $T = 300$ K. The influence of the kinetic Hamiltonian $K(E)$ is completely ignorable as long as $t \ll m\sigma^2/\hbar \sim 10^{10}$ s. Hence, for any conceivable period, the results (B.8-B.10) are correct.

Guessing the heat capacity of the 1 g object by $(d\bar{E}/dT) \sim 1$ J/K, we get $\Delta E \sim 10^{-4} - 10^{-5}$ erg from eq. (B.10). Therefore eq. (B.9) yields $\tau_{\rm dec} \sim 0.1 - 0.01$ ms. [If one calculates ΔE from the microscopic model of ref. [1], eq. (B.9) recovers its decoherence time ~ 1 ms.]

Let us change the reference frame for the free-falling one. Using the Hamiltonian (B.3), the counterpart of eq. (B.7) reads:

$$\rho_{\rm cm}(t) = \sum_E \frac{e^{-E/k_B T}}{Z} e^{-itK(E)/\hbar} \rho_{\rm cm}(0) e^{itK(E)/\hbar}, \tag{B.11}$$

where $K(E) = \frac{1}{2}p^2/(m + c^{-2}E)$ and p is the momentum operator in the free-fall frame. We showed above that the chosen initial superposition with the "wide" standing wave packets remain practically static against the kinetic Hamiltonian $K(E)$. Accordingly, we can set $K(E) = 0$ for them and we end up with the trivial result:

$$\rho_{\rm cm}(t) = \rho_{\rm cm}(0). \tag{B.12}$$

The said superposition of the 1g object remains static practically forever in the free-fall reference frame.

References

[1] Pikovski I, Zych M, Costa F and Brukner Č 2015 *Nature Physics* **11** 668
[2] Diósi L 2015 Centre of mass decoherence due to time dilation: paradoxical frame-dependence *Preprint* arXiv:1507.05828v1
[3] Bonder Y, Okon E and Sudarsky D 2016 *Nature Physics* **12** 2 (2016)
[4] Pikovski I, Zych M, Costa F and Brukner Č 2015 Time Dilation in Quantum Systems and Decoherence: Questions and Answers *Preprint* arXiv:1508.03296v1
[5] Pang B, Khalili F Y and Chen Y 2016 *Phys. Rev. Lett.* 117, 090401
[6] Giulini D 1996 *Ann. Phys.* **249** 222
[7] Pikovski I, Zych M, Costa F and Brukner Č 2017 *New J. Phys.* **19** 025011
[8] Zych M, and Brukner Č 2015 Quantum formulation of the Einstein Equivalence Principle *Preprint* arXiv:1502.00971
[9] Bargman V 1954 *Ann. Math.* **59** 1
[10] Toroš M, Großardt A and Bassi A 2017 Quantum mechanics for non-inertial observers *Preprint* arXiv:1701.04298
[11] Pikovski I, Zych M, Costa F and Brukner Č 2017 Comment on "Quantum mechanics for non-inertial observers *Preprint* arXiv:1702.06670

IOP Conf. Series: Journal of Physics: Conf. Series **880** (2017) 012021 doi:10.1088/1742-6596/880/1/012021

Extensions of Born's rule to non-linear quantum mechanics, some of which do not imply superluminal communication

Bassam Helou and Yanbei Chen

Institute of Quantum Information and Matter, and Burke Institute of Theoretical Physics, M/C 350-17, California Institute of Technology, Pasadena, CA 91125, USA

E-mail: bhelou@caltech.edu

Abstract. Nonlinear modifications of quantum mechanics have a troubled history. They were initially studied for many promising reasons: resolving the measurement problem, formulating a theory of quantum mechanics and gravity, and understanding the limits of standard quantum mechanics. However, certain non-linear theories have been experimentally tested and failed. More significantly, it has been shown that, in general, deterministic non-linear theories can be used for superluminal communication. We highlight another serious issue: the distribution of measurement results predicted by non-linear quantum mechanics depends on the formulation of quantum mechanics. In other words, Born's rule cannot be uniquely extended to non-linear quantum mechanics. We present these generalizations of Born's rule, and then examine whether some exclude superluminal communication. We determine that a large class do not allow for superluminal communication, but many lack a consistent definition. Nonetheless, we find a single extension of Born's rule with a sound operational definition, and that does not exhibit superluminal communication. The non-linear time-evolution leading to a certain measurement event is driven by the state conditioned on measurements that lie within the past light cone of that event.

1. Introduction

Non-linear quantum mechanics (NLQM) has long been considered as a possible generalization of standard quantum mechanics (sQM) [1, 2, 3], for three main reasons. First, the measurement process is controversial, because if we assume that linear quantum mechanics explains all processes, then it is very difficult to explain wavefunction collapse [4]. Phenomenological non-linear stochastic, and experimentally falsifiable, extensions of quantum mechanics have been proposed to explain the measurement process [5], and upper bounds on the parameters of such theories have been obtained in [5, 6, 7]. Second, we would like to test the domain of validity of sQM. One possible feature to test is linearity. Experimental tests of certain non-linear theories have been performed in [8, 9, 10, 11, 12], all with negative results. Third, non-linear and deterministic theories of quantum mechanics have been proposed to unite quantum mechanics with gravity. For instance, the Schroedinger-Newton theory describes a classical spacetime which is sourced by quantum matter [13], and the correlated worldlines theory is a quantum theory of gravity, which postulates that gravity correlates quantum trajectories [14].

Despite these appealing reasons to study NLQM, there was a backlash after Gisin showed that deterministic NLQM could imply superluminal communication [15]. The no-signaling condition states that one cannot send information faster than the speed of light, and is a cornerstone of the special theory of relativity. The community regards the condition as being inviolable. Gisin's work was quickly followed by others with similar conclusions [16, 17]. Additional work then showed that under general conditions NLQM implies superluminal communication [18, 19]. In this article, we show that one can still satisfy the no-signaling condition by exploiting an ambiguity in Born's rule when it is extended to NLQM.

Along with the Schroedinger equation, Born's rule is a fundamental tenet of quantum mechanics. The former determines the time evolution of a wavefunction, while Born's rule provides the probabilistic interpretation of the wavefunction, which allows one to make predictions about the distribution of measurement results. Born's rule has, so far, passed all experimental tests, and so any non-linear theory must make predictions that become equivalent to Born's rule in sQM when the non-linearity vanishes.

In [6], we showed that NLQM suffers from a serious conceptual issue: Born's rule cannot be uniquely extended to NLQM. Specifically, there are multiple prescriptions of assigning the probability of a measurement outcome, that are equivalent in sQM, but become distinct in NLQM. Note that we treat Born's rule as a phenomenological prescription for determining the distribution of measurement results, and so there is no *a priori* reason to pick one extension over another.

We first present an ambiguity in the choice of the state we non-linearly evolve, by providing a simple example involving a single measurement. We then discuss a more elaborate setup, which is similar to that used to show that non-linear theories violate the no-signaling condition. We determine that there are no prescriptions based on this ambiguity that do not result in superluminal communication. In section 3, we present another type of ambiguity: the choice of wavefunction that drives the non-linear time evolution. We determine that there is an infinite number of prescriptions based on this ambiguity that do not violate the no-signaling condition. However, the majority do not have a sensible interpretation. In section 5, we present and discuss a particular prescription with a reasonable operational definition and that does not exhibit superluminal communication.

2. Ambiguity in choosing which state is non-linearly time evolved

In this section, we discuss one type of ambiguity that we encounter when extending Born's rule to NLQM: a choice in the state that we time-evolve. We illustrate this with a simple example, and then discuss whether the generated class exhibits superluminal signaling.

2.1. A simple example

According to Born's rule, if a system is initially in a state $|i\rangle$, the probability that a measurement device measures it to be in some eigenstate $|f\rangle$, after some period of evolution under a unitary operator $\hat{U}$, is

$$p_{i\to f} = \left|\left\langle f\middle|\hat{U}\middle|i\right\rangle\right|^2 . \tag{1}$$

The above formula can be written in a number of equivalent ways, such as

$$p_{i\leftarrow f} = \left|\left\langle i\middle|\hat{U}^\dagger\middle|f\right\rangle\right|^2 . \tag{2}$$

In NLQM, the time evolution operator depends on the state it acts on. As a result, Eqs. (1) and (2) become

$$p^{\rm NL}_{i\to f} = \left|\langle f|\mathcal{U}_i i\rangle\right|^2 , \qquad p^{\rm NL}_{i\leftarrow f} \propto \left|\left\langle i|\mathcal{U}^\dagger_f f\right\rangle\right|^2 , \tag{3}$$

where $|\mathcal{U}_i i\rangle$ is the time-evolved $|i\rangle$ and $\left|\mathcal{U}^\dagger_f f\right\rangle$ is the backwards time-evolved $|f\rangle$, under some non-linear dynamics. The superscript NL explicitly indicates that NLQM is being used. Moreover,

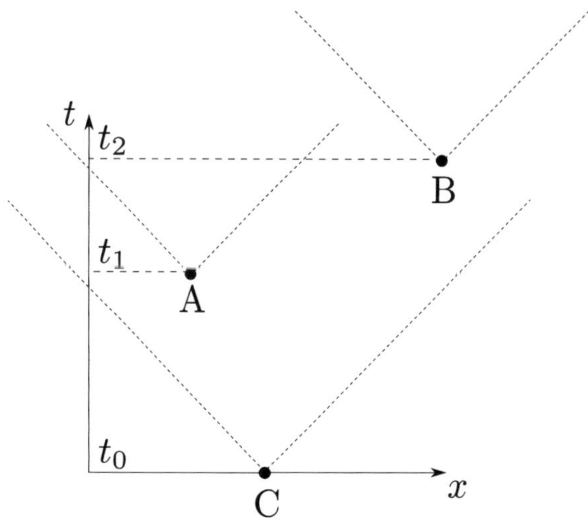

Figure 1. A setup with multiple measurement events. Event C describes the preparation of an ensemble of identical 2-particle states $|\Psi_{\text{ini}}\rangle$ by Charlie. Event A (B) describes Alice (Bob) measuring her (his) particles. The dashed lines show the light cone centered around each event.

the proportionality sign in $p^{\text{NL}}_{i\leftarrow f}$ follows from $\sum_f \left|\left\langle i|\mathcal{U}^\dagger_f f\right\rangle\right|^2$ being not, in general, normalized to unity. $p^{\text{NL}}_{i\rightarrow f}$ and $p^{\text{NL}}_{i\leftarrow f}$ are not necessarily equivalent, and so Born's rule cannot be uniquely extended to NLQM.

2.2. Multiple measurements

Having established the ambiguity of Born's rule, we move on to study more complicated setups with multiple measurements. Our aim is determining whether there exists an extension of Born's rule that does not exhibit superluminal communication, and so we will study the prototypical setup for showing that NLQM violates the no-signaling condition. As shown in Fig. 1, Charlie prepares a collection of identical *arbitrary* 2-particle states $|\Psi_{\text{ini}}\rangle$, and then sends them to Alice and Bob, such that they both hold one part of each of the states $|\Psi_{\text{ini}}\rangle$. Alice performs measurements on her ensemble of particles at t_1, and then Bob on his at a later time t_2. We assume that Alice's measurement events are space-like separated from Bob's. The ordering of t_1 and t_2, which depends on the reference frame that the experiment is viewed in, won't affect our analysis.

Denote the probability that Alice measures α in some basis z, and Bob measures β in some basis x by $p(\alpha,\beta)$. For example, if the particles were spins, z could be the $\hat{\sigma}_z$ eigenstates, $|\uparrow\rangle, |\downarrow\rangle$, and x could be the $\hat{\sigma}_x$ eigenstates $|\pm\rangle = (|\uparrow\rangle \pm |\downarrow\rangle)/\sqrt{2}$. We will first determine the probability distribution of different measurement events according to sQM, and then discuss the different ways of generalizing the probability distribution to NLQM.

In sQM, $p(\alpha,\beta)$ is given by

$$p(\alpha,\beta) = \left\langle \Psi_{f|\alpha,\beta} | \Psi_{f|\alpha,\beta} \right\rangle, \tag{4}$$

where $\left|\Psi_{f|\alpha,\beta}\right\rangle$ is the unnormalized joint quantum state of Alice and Bob at time t_2, conditioned on the measurement results α and β:

$$\left|\Psi_{f|\alpha,\beta}\right\rangle = \left(\hat{I}\otimes\hat{P}_\beta\right)\hat{U}_2\left(\hat{P}_\alpha\otimes\hat{I}\right)\hat{U}_1\left|\Psi_{\text{ini}}\right\rangle, \tag{5}$$

where $\hat{U}_1$ is the total time evolution operator (for both Alice and Bob's particles) from the initial time t_0 till t_1, and $\hat{U}_2$ is the total time evolution operator from t_1 till t_2. The projectors are $\hat{P}_\alpha = |\alpha\rangle\langle\alpha|$ and $\hat{P}_\beta = |\beta\rangle\langle\beta|$. Note that we chose to work with pure states instead of density matrices for clarity and simplicity. $|\Psi_{\text{ini}}\rangle$ can always be enlarged to include the initial state of the environment.

Substituting the expression of $\left|\Psi_{f|\alpha,\beta}\right\rangle$ into Eq. (4), we have

$$p(\alpha,\beta) = \left\langle \Psi_{\text{ini}} \middle| \hat{U}_1^\dagger \hat{P}_\alpha \hat{U}_2^\dagger \hat{P}_\beta \hat{U}_2 \hat{P}_\alpha \hat{U}_1 \middle| \Psi_{\text{ini}} \right\rangle. \tag{6}$$

$p(\alpha,\beta)$ can also be written in a more convenient form by writing the projection operators in the Heisenberg picture:

$$\hat{P}_\alpha^H \equiv \hat{U}_1^\dagger \hat{P}_\alpha \hat{U}_1; \qquad \hat{P}_\beta^H \equiv \hat{U}_{\text{tot}}^\dagger \hat{P}_\beta \hat{U}_{\text{tot}}, \quad \hat{U}_{\text{tot}} \equiv \hat{U}_2 \hat{U}_1. \tag{7}$$

We rewrite $p(\alpha,\beta)$ in terms of $\hat{P}_\alpha^H$ and $\hat{P}_\beta^H$:

$$p(\alpha,\beta) = \left\langle \Psi_{\text{ini}} \middle| \hat{P}_\beta^H \hat{P}_\alpha^H \middle| \Psi_{\text{ini}} \right\rangle, \tag{8}$$

where we have used that Alice and Bob's measurement events are space-like separated, and so their corresponding measurement operators commute. This concise way of writing $p(\alpha,\beta)$ has the advantage of being easily generalizable to more than just 2 measurements.

$p(\alpha,\beta)$ can be extended to uncountably many ways in NLQM. For instance, we can exploit the fact that for any projection operator $\hat{P}^\dagger \hat{P} = \hat{P}$, and that $\hat{I}$ can be decomposed as the product of any unitary operator and its hermitian conjugate (e.g. $\hat{I} = \hat{U}_1^\dagger \hat{U}_1$). Each unitary operator, when extended to NLQM, can act on the state to its left or to its right. Note that, in general, we will need to add a normalization factor to $p^{\text{NL}}(\alpha,\beta)$:

$$p^{\text{NL}}(\alpha,\beta) = \frac{1}{\mathcal{N}} \left\langle \Psi_{f|\alpha,\beta} | \Psi_{f|\alpha,\beta} \right\rangle, \tag{9}$$

where

$$\mathcal{N} \equiv \sum_{\gamma,\delta} \left\langle \Psi_{f|\gamma,\delta} | \Psi_{f|\gamma,\delta} \right\rangle. \tag{10}$$

Can we exploit this ambiguity to find a prescription that does not violate the no-signaling communication?

2.3. Imposing the no-signaling condition

Superluminal communication from Alice to Bob is not possible when

$$p(\beta) = \sum_\alpha p(\alpha,\beta) \tag{11}$$

is independent of the basis chosen by Alice. Otherwise, if $p(\beta)$ is influenced by Alice's choice of measurement basis in a deterministic way, then since Bob can easily estimate $p(\beta)$, he can determine the basis Alice chose for her measurement results, which can form the foundation of a communication strategy. For instance, both Alice and Bob can agree that a particular choice of Alice's measurement basis could be associated with sending the bit 0, while another choice could be associated with the bit 1.

IOP Conf. Series: Journal of Physics: Conf. Series **880** (2017) 012021 doi:10.1088/1742-6596/880/1/012021

In other words, to satisfy the no-signaling condition, we require that $p(\beta)$ depends only on β. In sQM, this is clearly satisfied, as

$$p(\beta) = \left\langle \Psi_{\text{ini}} \middle| \hat{P}^H_\beta \left(\sum_\alpha \hat{P}^H_\alpha \right) \middle| \Psi_{\text{ini}} \right\rangle = \left\langle \Psi_{\text{ini}} \middle| \hat{P}^H_\beta \middle| \Psi_{\text{ini}} \right\rangle, \tag{12}$$

because

$$\sum_\alpha \hat{P}^H_\alpha = \hat{U}_1^\dagger \left(\sum_\alpha |\alpha\rangle \langle\alpha| \right) \hat{U}_1 = \hat{I}. \tag{13}$$

We now analyze superluminal communication for general non-linear theories. We would like $\sum_\alpha p^{\text{NL}}(\alpha,\beta)$ to be independent of α, so $\sum_\alpha \langle \Psi_{f|\alpha,\beta} | \Psi_{f|\alpha,\beta} \rangle$ has to be independent of α for all β. Note that the normalization cannot, in general, help. A non-trivial measurement apparatus has at least two outcomes, so both the numerator and denominators of $p^{\text{NL}}(\alpha,\beta)$ consist of a sum of at least two terms. The only way that $\mathcal{N}$ gets rid of the dependence of each term in the numerator on the different αs is if each term is of the form $A(\{\alpha\})\, B(\beta)$, where $B(\beta)$ is a general function that depends only β, and $A(\{\alpha\})$ is a function identical among all the $\langle \Psi_{f|\alpha,\beta} | \Psi_{f|\alpha,\beta} \rangle$. Such a separable expression is not in general possible, because the initial state can be chosen to be anything, including entangled states, and the non-linear dynamics are arbitrary.

We can choose $p^{\text{NL}}(\beta)$ to be independent of the different αs. This can be seen by first expanding Eq. (8):

$$p(\alpha,\beta) = \left\langle \Psi_{\text{ini}} \middle| \hat{U}^\dagger_{\text{tot}} \hat{P}_\beta \hat{U}_2 \hat{P}_\alpha \hat{U}_1 \middle| \Psi_{\text{ini}} \right\rangle. \tag{14}$$

When extending $p(\alpha,\beta)$ to $p^{\text{NL}}(\alpha,\beta)$, let $\hat{U}^\dagger_{\text{tot}}$ act on $\langle \Psi_{\text{ini}}|$:

$$\langle \Psi_{\text{ini}} | \hat{U}^\dagger_{\text{tot}} \to \langle \mathcal{U}_{\text{tot}} \Psi_{\text{ini}} | . \tag{15}$$

Let $\hat{U}_2$ act the conditional state $\langle \mathcal{U}_{\text{tot}} \Psi_{\text{ini}} | \hat{P}_\beta$, which we will define as Ψ_c:

$$\hat{P}_\beta | \mathcal{U}_{\text{tot}} \Psi_{\text{ini}} \rangle \equiv |\Psi_c\rangle \tag{16}$$

$$\langle \Psi_{\text{ini}} | \hat{U}^\dagger_{\text{tot}} \hat{P}_\beta \hat{U}_2 \to \left\langle \mathcal{U}_2^\dagger \Psi_c \right|, \tag{17}$$

where $\mathcal{U}_2^\dagger$ denotes backwards time evolution. Finally, let $\hat{U}_1$ act on $|\Psi_{\text{ini}}\rangle$:

$$p^{\text{NL}}(\alpha,\beta) = \frac{1}{\mathcal{N}} \left\langle \mathcal{U}_2 \Psi_c \middle| \hat{P}_\alpha \middle| \mathcal{U}_1 \Psi_{\text{ini}} \right\rangle. \tag{18}$$

Notice that

$$p^{\text{NL}}(\beta) = \sum_\alpha p^{\text{NL}}(\alpha,\beta) = \frac{1}{\mathcal{N}} \left\langle \mathcal{U}_2 \Psi_c \middle| \left(\sum_\alpha \hat{P}_\alpha \right) \middle| \mathcal{U}_1 \Psi_{\text{ini}} \right\rangle \tag{19}$$

depends only on β because the sum over α can be brought over to inside the expectation value, where it acts solely on $\hat{P}_\alpha$ and gives $\hat{I}$.

We have ensured that Alice cannot communicate with Bob faster than the speed of light. However, Bob can superluminally communicate with Alice, because when Eq. (18) is summed over β, the sum cannot be brought over to $\hat{P}_\beta$, and so in general $\sum_\beta p^{\text{NL}}(\alpha,\beta)$ does not depend only on α. There is no way to prevent superluminal communication with the analysis performed above, as was concluded in [18]. Nonetheless, we have shown that some prescriptions can prevent one party from communicating to another.

In the next section, we exploit one more ambiguity to show that some extensions of Born's rule do not violate the no-signaling condition.

3. Ambiguity in the state driving the non-linear time evolution

We can take advantage of an unorthodox ambiguity to meet the no-signaling condition. We first present the ambiguity without regards to interpretation, and then investigate whether extensions of Born's rule that exploit this ambiguity have a reasonable interpretation, and a consistent operational definition.

3.1. The ambiguity

In general, the non-linear Schroedinger equation contains a nonlinear term $\hat{V}_{\text{NL}}$ and a linear term, $\hat{H}_L$:

$$i\hbar\partial_t |\psi\rangle = \left(\hat{H}_L + \hat{V}_{\text{NL}}(\psi)\right)|\psi\rangle, \tag{20}$$

where $\hat{V}_{\text{NL}}(\psi)$ is a shorthand for a non-linear potential that can depend arbitrarily on $\psi(z)$ for all times z. When Eq. (20) is used to predict distributions of measurement outcomes, $\hat{V}_{\text{NL}}$ does not have to act on ψ. Instead, it can act on a pre-determined wavefunction ϕ:

$$i\hbar\partial_t |\psi\rangle = \left(\hat{H}_L + \hat{V}_{\text{NL}}(\phi)\right)|\psi\rangle. \tag{21}$$

Notice that, unless ϕ is chosen to be ψ, Eq. (21) is a *linear* evolution equation, and associated to it is a unitary time evolution operator which we define as $\hat{U}(\phi)$. In the limit that $\hat{V}_{\text{NL}}$ vanishes, $\hat{U}(\phi)$ recovers the time evolution operator predicted by sQM. Consequently, in the same limit, extensions of Born's rule based on Eq. (21) recover the Born's rule in sQM.

3.2. Imposing the no-signaling condition

Consider the following extension of Eq. (8):

$$p^{\text{NL}}(\alpha,\beta) = \left\langle \Psi_{\text{ini}} \middle| \hat{P}_\beta^{H_\phi} \hat{P}_\alpha^{H_\phi} \middle| \Psi_{\text{ini}} \right\rangle, \tag{22}$$

where the superscript H_ϕ denotes a Heisenberg picture for the linear Schroedinger equation given by Eq. (21). For instance, $\hat{P}_\alpha^H$ becomes

$$\hat{P}_\alpha^{H_\phi} \equiv \hat{U}_1^\dagger(\phi)\,\hat{P}_\alpha \hat{U}_1(\phi). \tag{23}$$

As long as ϕ does not depend on α or β, the expression for $p^{\text{NL}}(\alpha,\beta)$ given by Eq. (22) does not violate the no-signaling condition.

3.3. Interpretation

Eq. (22) describes extensions of Born's rule that do not exhibit superluminal communication, but most do not have a reasonable operational definition. The issue that we immediately encounter is the choice of ϕ in Eq. (22). In [6], we gave two reasonable choices for ϕ. The first, which we termed *pre-selection*, takes ϕ to be the initial state of the experiment $|\Psi_{\text{ini}}\rangle$ and its time evolution under the full non-linear Schroedinger equation given by Eq. (20). The second, which we termed *post-selection*, takes ϕ to be the final state of the experiment $|\alpha,\beta\rangle$ and its (backwards) time evolution under Eq. (20).

These prescriptions have issues. Post-selection violates the no-signaling condition because all evolution operators in Eq. (22) depend on the measurement outcomes α and β. Pre-selection does not allow superluminal communication. However, choosing $|\Psi_{\text{ini}}\rangle$ is a delicate matter. Consider again the setup shown by Fig. 1. We chose that $|\Psi_{\text{ini}}\rangle$ be prepared by a third party: Charlie. In the analysis of the canonical EPR setup shown by Fig. 1, such a statement doesn't affect the calculation of the distribution of measurement results, but in pre-selection it does.

IOP Conf. Series: Journal of Physics: Conf. Series **880** (2017) 012021 doi:10.1088/1742-6596/880/1/012021

Charlie must have manipulated some state, which we call $\left|\Psi'_{\text{ini}}\right\rangle$, to prepare $|\Psi_{\text{ini}}\rangle$. Consequently, if pre-selection were to be applied, then it should be on $\left|\Psi'_{\text{ini}}\right\rangle$ instead of $|\Psi_{\text{ini}}\rangle$. This process could, in principle, be repeated *ad infinitum* back to the initial state of the universe, which is unknown to us. Even if it were known, working with the initial state of the universe would make the analysis of any experiment infeasible.

4. A sensible extension of Born's rule

In this section, we propose a sensible extension of Born's rule, where the state that drives the non-linear time evolution up to a certain measurement event is chosen to be the time evolved initial state conditioned on measurement results that lie within or on the past light cone. Our scheme is similar to Adrian Kent's proposal in [20]. He argued that if the non-linear evolution depends only on local states, defined by allowing only projective measurements in the past light cone of a particle and then tracing out the rest of the degrees of freedom, then superluminal communication is not possible.

4.1. A general example

To clarify and justify our prescription, we present a general example. Consider the setup shown in Fig. 2, which is a more elaborate version of Fig. 1. The thought experiment now includes two additional parties: Dylan, who performs a measurement in the past light cone of Charlie, and Eve who performs a measurement in the future light cone of Bob, and outside the future light cone of Alice. Dylan is included to demonstrate that our proposed prescription does not suffer from the same drawback as pre-selection, and Eve is included to show that even for a complicated configuration of measurement events, our prescription does not exhibit superluminal communication.

We begin our analysis with the predictions of sQM for the final state of the experiment conditioned on Charlie, Alice, Bob and Eve measuring C, α, β and ϵ with corresponding measurement eigenstates $|\Psi_{\text{ini}}\rangle$, $|\alpha\rangle$, $|\beta\rangle$ and $|\epsilon\rangle$, respectively. This conditional state is given by

$$|\psi_c\rangle \propto \hat{P}_\epsilon \hat{U}_3 \hat{P}_\beta \hat{U}_2 \hat{P}_\alpha \hat{U}_1 \hat{P}_C \hat{U}_0 \left|\Psi'_{\text{ini}}\right\rangle \propto \hat{P}_\epsilon \hat{U}_3 \hat{P}_\beta \hat{U}_2 \hat{P}_\alpha \hat{U}_1 |\Psi_{\text{ini}}\rangle , \tag{24}$$

where $\left|\Psi'_{\text{ini}}\right\rangle$ is the initial state prepared by Dylan, $\hat{U}_0$ denotes time evolution from the time of Dylan's measurement till t_0, $\hat{U}_1$ from t_0 till t_1, $\hat{U}_2$ from t_1 till t_2 and $\hat{U}_3$ from t_2 till t_3. The projection operators are defined by

$$\hat{P}_C = |\Psi_{\text{ini}}\rangle \langle\Psi_{\text{ini}}| , \quad \hat{P}_\alpha = |\alpha\rangle\langle\alpha| , \quad \hat{P}_\beta = |\beta\rangle\langle\beta| , \quad \hat{P}_\epsilon = |\epsilon\rangle\langle\epsilon| . \tag{25}$$

According to our proposed prescription, $|\psi_c\rangle$ extends to NLQM in the following way:

$$\begin{aligned} |\psi_c\rangle &\propto \hat{P}_\epsilon \hat{U}_3 \left(\hat{P}_\beta \Psi_{\text{ini}}\right) \hat{P}_\beta \hat{U}_2 (\Psi_{\text{ini}}) \hat{P}_\alpha \hat{U}_1 (\Psi_{\text{ini}}) \hat{P}_C \hat{U}_0 \left(\Psi'_{\text{ini}}\right) \left|\Psi'_{\text{ini}}\right\rangle & (26) \\ &\propto \hat{P}_\epsilon \hat{U}_3 \left(\hat{P}_\beta \Psi_{\text{ini}}\right) \hat{P}_\beta \hat{U}_2 (\Psi_{\text{ini}}) \hat{P}_\alpha \hat{U}_1 (\Psi_{\text{ini}}) |\Psi_{\text{ini}}\rangle , & (27) \end{aligned}$$

where we remind the reader that $\hat{U}(\phi)$ is the time evolution operator associated with Eq. (21). Charlie's measurement event lies in the past light cone of Alice, so time evolution up to Alice's measurement is driven by the conditional state $\hat{P}_C \hat{U}_0 \left(\Psi'_{\text{ini}}\right) \left|\Psi'_{\text{ini}}\right\rangle$ which is just $|\Psi_{\text{ini}}\rangle$. Bob's past light cone does not include Alice's measurement event, but includes Charlie's, so we have to use the conditional state at Charlie, $|\Psi_{\text{ini}}\rangle$, to drive the non-linear time evolution up to Bob's measurement. Finally, Eve's past light cone includes Bob's measurement event, so we use the conditional state $\hat{P}_\beta \hat{U}_1 (\Psi_{\text{ini}}) |\Psi_{\text{ini}}\rangle$ to drive the nonlinear time evolution between Bob and Eve.

Eq. (27) contains genuine non-linear time evolution, such as $\hat{U}_0 \left(\Psi'_{\text{ini}}\right) \left|\Psi'_{\text{ini}}\right\rangle$. This equation predicts dynamics that are identical to those predicted by the full non-linear Schroedinger equation given by Eq. (20) with an initial state of $\left|\Psi'_{\text{ini}}\right\rangle$. Moreover, notice that measurements within the past light cone of Charlie, like that of Dylan's, do not affect our analysis. Indeed, preparation events are always in the past light cone of the final measurements of an experiment, because the particles, whose speed is upper bounded by the speed of light, have to travel to where they will be measured. Consequently, our proposed scheme avoids the issue of pre-selection, which is the difficulty of choosing the initial state that will drive the non-linear time evolution.

We show that our proposed prescription does not violate the no-signaling condition by looking at the marginal probabilities, $p(\alpha|C)$, $p(\beta|C)$ and $p(\epsilon|C)$, conditioned on Charlie measuring $|\Psi_{\text{ini}}\rangle$. The probability of obtaining the measurement results α, β, ϵ, and that Charlie measures $|\Psi_{\text{ini}}\rangle$ is given by the norm of $|\Psi_c\rangle$ in Eq. (27):

$$\begin{aligned} & p^{\text{NL}}(\alpha, \beta, \epsilon, C) \\ = & \left|\left\langle \Psi_{\text{ini}} \middle| \hat{U}_0 \left(\Psi'_{\text{ini}}\right) \middle| \Psi'_{\text{ini}} \right\rangle\right|^2 / \mathcal{N} \times \\ & \left\langle \Psi_{\text{ini}} \middle| \hat{U}_1^\dagger (\Psi_{\text{ini}}) \hat{U}_2^\dagger (\Psi_{\text{ini}}) \hat{P}_\beta \hat{U}_3^\dagger \left(\hat{P}_\beta \Psi_{\text{ini}}\right) \hat{P}_\epsilon \hat{U}_3 \left(\hat{P}_\beta \Psi_{\text{ini}}\right) \hat{P}_\beta \hat{U}_2 (\Psi_{\text{ini}}) \hat{P}_\alpha \hat{U}_1 (\Psi_{\text{ini}}) \middle| \Psi_{\text{ini}} \right\rangle, \end{aligned}$$

where $\mathcal{N}$ is a normalization factor given by the sum of $p^{\text{NL}}(\alpha, \beta, \epsilon, C)$ over all possible measurement results. We calculate it to be unity. We have also simplified $p^{\text{NL}}(\alpha, \beta, \epsilon, C)$ by noting that $\hat{P}_\alpha$, which corresponds to an event space-like separated from Bob and Eve's measurements, commutes with $\hat{U}_2$, $\hat{U}_3$, $\hat{P}_\beta$ and $\hat{P}_\epsilon$. We are interested in $p^{\text{NL}}(\alpha, \beta, \epsilon|C)$, so we have to divide by $p^{\text{NL}}(C)$, which can be obtained by summing $p^{\text{NL}}(\alpha, \beta, \epsilon, C)$ over α, β and ϵ:

$$\begin{aligned} & p^{\text{NL}}(\alpha, \beta, \epsilon|C) \\ = & \left\langle \Psi_{\text{ini}} \middle| \hat{U}_1^\dagger (\Psi_{\text{ini}}) \hat{U}_2^\dagger (\Psi_{\text{ini}}) \hat{P}_\beta \hat{U}_3^\dagger \left(\hat{P}_\beta \Psi_{\text{ini}}\right) \hat{P}_\epsilon \hat{U}_3 \left(\hat{P}_\beta \Psi_{\text{ini}}\right) \hat{P}_\beta \hat{U}_2 (\Psi_{\text{ini}}) \hat{P}_\alpha \hat{U}_1 (\Psi_{\text{ini}}) \middle| \Psi_{\text{ini}} \right\rangle. \end{aligned}$$

4.2. Superluminal signaling is forbidden

Can Alice send signals to Bob or Eve, or vice versa? Let us first calculate $p^{\text{NL}}(\beta|C)$:

$$p^{\text{NL}}(\beta|C) = \left\langle \Psi_{\text{ini}} \middle| \hat{U}_1^\dagger (\Psi_{\text{ini}}) \hat{U}_2^\dagger (\Psi_{\text{ini}}) \hat{P}_\beta \hat{U}_2 (\Psi_{\text{ini}}) \hat{U}_1 (\Psi_{\text{ini}}) \middle| \Psi_{\text{ini}} \right\rangle, \tag{28}$$

which only depends on β. Next, let us look at Eve's distribution of measurement results. We obtain that $p^{\text{NL}}(\epsilon|C)$ depends on $\{\beta\}$ so Bob can send signals to Eve. This is acceptable as the distance between their measurement events is time-like. Finally, Alice cannot violate the no-signaling condition because

$$p^{\text{NL}}(\alpha|C) = \left\langle \Psi_{\text{ini}} \middle| \hat{U}_1^\dagger (\Psi_{\text{ini}}) \hat{P}_\alpha \hat{U}_1 (\Psi_{\text{ini}}) \middle| \Psi_{\text{ini}} \right\rangle \tag{29}$$

depends only on α.

The prescription described in this section has genuine non-linear time evolution, and yet does not violate the no-signaling condition. We do not contradict the conclusions of [18], because we violate one of the general conditions they stated are needed for deterministic NLQM to have superluminal communication. In particular, we break the assumption that states are represented by vectors in a Hilbert space. In Eq. (21), 2 states determine time evolution: ψ, and ϕ, which drives the non-linear time evolution under $\hat{V}_{\text{NL}}$. The choice of ϕ to use depends on the spacetime region the observer is in, as is depicted in Fig. 3.

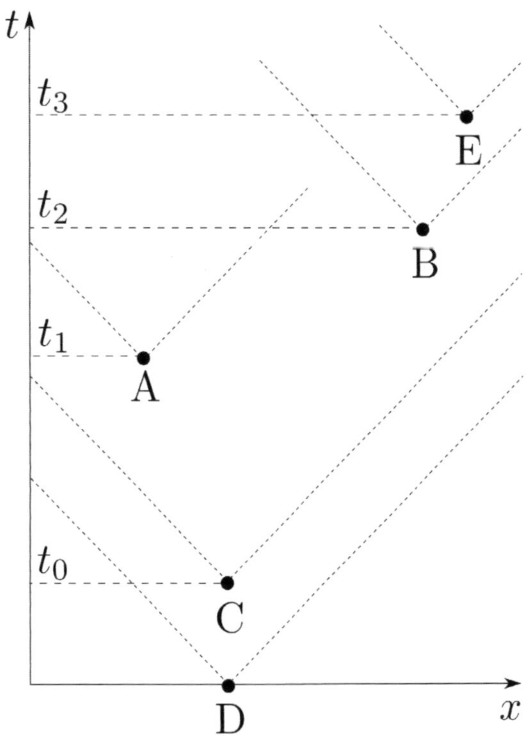

Figure 2. A setup similar to that described by Fig. 1, but more elaborate. Event D is Dylan preparing the state $\left|\Psi'_{\rm ini}\right\rangle$, Event C is Charlie measuring the eigenstate $|\Psi_{\rm ini}\rangle$. Event A (B) describes Alice (Bob) measuring her (his) particles. The dashed lines show the light cone centered around each event.

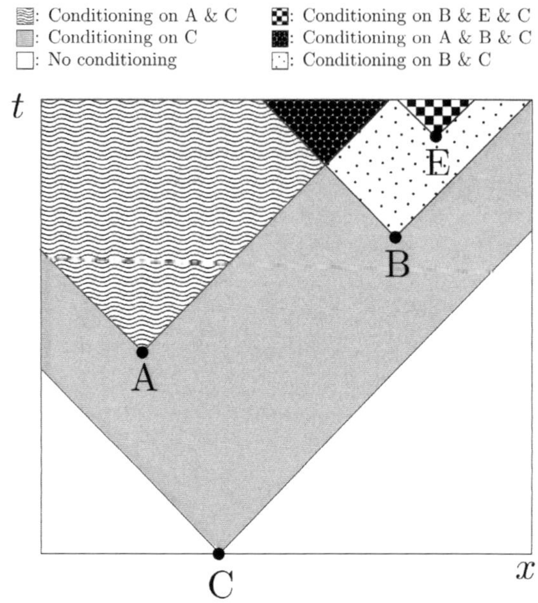

Figure 3. Partioning of spacetime into different regions according to which wavefunction drives the non-linear time evolution. There are 4 measurement events: C, A, B and E, arranged identically to the events in Fig. 2. We didn't include Event D to limit clutter. The 4 events result in 6 regions. In each, the wavefunction which drives the non-linear time evolution is the time-evolved initial state of the experiment conditioned on measurements which are presented in the legend at the top of the figure.

5. Conclusions

In summary, Born's rule cannot be uniquely extended to non-linear quantum mechanics. This ambiguity seems like another significant penalty for breaking linearity in sQM. However, when we demanded a formulation of NLQM that does not violate the no-signaling condition and that has a reasonable operational definition, we arrived at a single extension of Born's rule. Specifically, we chose that the non-linear time-evolution leading to a certain measurement event is driven by the time-evolved initial state of the experiment conditioned on measurements that lie within the past light cone of that event.

References

[1] Bialynicki-Birula I and Mycielski J 1976 Nonlinear wave mechanics *Annals of Physics* **100** 1 62.
[2] de Broglie L 1960 *Non-linear wave mechanics, A causal interpretation* (Elsevier Pub. Co.).
[3] Weinberg S 1989 Testing quantum mechanics *Annals of Physics* **194** 2 336.
[4] Bassi A and Ghirardi G 2000 A general argument against the universal validity of the superposition principle *Phys. Lett.* A **275** 373.
[5] Bassi A, Lochan K, Satin S, Singh T and Ulbricht H 2013 Models of wave-function collapse, underlying theories, and experimental tests *Rev. Mod. Phys.* **85** 2 471.

[6] Helou B, Luo J, Yeh H.-C., Shao C.-g., Slagmolen B. J., McClelland D. E. and Chen, Y 2016 Measurable signatures of quantum mechanics in a classical spacetime *Preprint* quant-ph/1612.06310.

[7] Vinante A, Bahrami M, Bassi A, Usenko O, Wijts G and Oosterkamp, T. H. 2016 Upper bounds on spontaneous wave-function collapse models using millikelvin-cooled nanocantilevers *Phys. Rev. Lett.* **116** 9 090402.

[8] Shull C. G., Atwood D. K., Arthur J and Horne M. A. 1980 Search for a nonlinear variant of the Schrodinger equation by neutron interferometry *Phys. Rev. Lett.* **44** 12 765.

[9] Bollinger J. J., Heinzen D. J., Itano W, Gilbert S. L. and Wineland D. J. 1989 Test of the linearity of quantum mechanics by rf spectroscopy of the $^9Be^+$ ground state *Phys. Rev. Lett.* **63** 10 1031.

[10] Chupp T. E. and Hoare R. J. 1990 Coherence in freely precessing ^{21}Ne and a test of linearity of quantum mechanics *Phys. Rev. Lett.* **64** 19 2261.

[11] Walsworth R. L., Silvera I. F., Mattison E. M. and Vessot R. F. C. 1990 Test of the linearity of quantum mechanics in an atomic system with a hydrogen maser *Phys. Rev. Lett.* **64** 22 2599.

[12] Majumder P. K., Venema B. J., Lamoreaux S. K., Heckel B. R. and Fortson E. N. 1990 Test of the linearity of quantum mechanics in optically pumped ^{201}Hg *Phys. Rev. Lett.* **65** 24 2931.

[13] Yang H, Miao H, Lee D, Helou B and Chen Y 2013 Macroscopic Quantum Mechanics in a Classical Spacetime *Phys. Rev. Lett.* **110** 17 170401.

[14] Stamp P. C. E. 2015 Rationale for a Correlated Worldline Theory of Quantum Gravity *New J. Phys.* **17** 6 065017.

[15] Gisin N 1990 Weinberg's non-linear quantum mechanics and supraluminal communications *Phys. Lett.* A **143** 1-2 1.

[16] Polchinski J 1991 Weinberg's nonlinear quantum mechanics and the Einstein-Podolsky-Rosen paradox *Phys. Rev. Lett.* **66** 4 397.

[17] Marek C 1991 Mobility and non-separability *Foundations of Physics Letters* **4** 4 351.

[18] Simon C, Bužek V and Gisin N 2001 No-signaling condition and quantum dynamics *Phys. Rev. Lett.* **87** 17 170405.

[19] Mielnik B 2001 Nonlinear quantum mechanics: a conflict with the Ptolomean structure? *Phys. Lett.* A **289** 1-2 1.

[20] Kent A 2005 Nonlinearity without superluminality *Phys. Rev.* A **72** 1 012108.

IOP Conf. Series: Journal of Physics: Conf. Series **880** (2017) 012022 doi:10.1088/1742-6596/880/1/012022

Cosmological implications of quantum mechanics parametrization of dark energy

Marek Szydłowski[1,2], **Aleksander Stachowski**[1] **and Krzysztof Urbanowski**[3]

[1] Astronomical Observatory, Jagiellonian University, Orla 171, 30-244 Krakow, Poland
[2] Mark Kac Complex Systems Research Centre, Jagiellonian University, Łojasiewicza 11, 30-348 Kraków, Poland
[3] Institute of Physics, University of Zielona Góra, Prof. Z. Szafrana 4a, 65-516 Zielona Góra, Poland

E-mail: marek.szydlowski@uj.edu.pl, aleksander.stachowski@doctoral.uj.edu.pl, k.urbanowski@if.uz.zgora.pl

Abstract. We consider the cosmology with the running dark energy. The parametrization of dark energy is derived from the quantum process of transition from the false vacuum state to the true vacuum state. This model is the generalized interacting ΛCDM model. We consider the energy density of dark energy parametrization, which is given by the Breit-Wigner energy distribution function. The idea of the process of the quantum mechanical decay of unstable states was formulated by Krauss and Dent. We used this idea in our considerations. In this model is an energy transfer in the dark sector. In this evolutional scenario the universe starts from the false vacuum state and goes to the true vacuum state of the present day universe. The intermediate regime during the passage from false to true vacuum states takes place. In this way the cosmological constant problem can be tried to solve. We estimate the cosmological parameters for this model. This model is in a good agreement with the astronomical data and is practically indistinguishable from ΛCDM model.

1. Introduction

We look for an alternative cosmological model, for the ΛCDM model, because we want to find a solution of the cosmological constant problem. Our proposition of the solution of the problem of the cosmological constant parameter assumes that the vacuum energy is given by the fundamental theory [1]. We apply quantum mechanics as a fundamental theory, which determines cosmological parameters and explain how cosmological parameters change during the cosmic evolution. The discussion about the cosmological constant problem can be found in papers [1]–[14].

Coleman *et al.* [15]–[17] showed that even if the early Universe is too cold to activate the energy transition to the minimum energy state then a quantum decay, from the false vacuum to the true vacuum, is still possible through a barrier penetration (by quantum tunneling). The discovery of the Higgs-like resonance at 125-126 GeV [18]–[21] gave the discussion about the instability of the false vacuum. In the neighborhood of the Planck epoch the Higgs mass $m_{\rm h} < 126$GeV causes that the electroweak vacuum is in a metastable state [19]. So the cosmological models of the early time Universe should respect the instability of the Higgs vacuum. The properties of the quantum mechanical decay process of metastable states should help to understand the properties of the observed universe[22] because the decay of the false vacuum is the quantum process [15, 16, 17].

We assume the Breit-Wigner energy distribution function as a model of the process of the energy transition from the false vacuum to the true vacuum. In consequence the parametrization of the dark energy has the following form

$$\rho_{\rm de} = E_0 + E_{\rm R}\frac{\alpha}{1-\alpha}\Re\left(\frac{J(t)}{I(t)}\right) \text{ or } \rho_{\rm de} = \Lambda_{\rm bare} + E_{\rm R}\left[1 + \frac{\alpha}{1-\alpha}\Re\left(\frac{J(t)}{I(t)}\right)\right], \quad (1)$$

where α, $E_{\rm R}$ are model parameters, which describe the variation from ΛCDM model and $\Lambda_{\rm bare} = E_0 - E_{\rm R}$. The values of α parameter belong to interval $(0,\ 1)$. When the α parameter or $E_{\rm R}$ is equal zero than this model is equivalent to the standard cosmological model. We assume that $8\pi G = c = 1$.

The functions $J(t)$ and $I(t)$ are described by the following formulas

$$J(t) = \int_{-\frac{1-\alpha}{\alpha}}^{\infty} \frac{\eta}{\eta^2+\frac{1}{4}} e^{-i\eta\tau} d\eta \text{ and } I(t) = \int_{-\frac{1-\alpha}{\alpha}}^{\infty} \frac{1}{\eta^2+\frac{1}{4}} e^{-i\eta\tau} d\eta. \quad (2)$$

The exact solutions of integrals $J(t)$ and $I(t)$ are given by

$$J(\tau) = \frac{1}{2}e^{-\tau/2}\left(-2i\pi + e^{\tau}\mathrm{E}_1\left(\left[\frac{1}{2} - \frac{i(1-\alpha)}{\alpha}\right]\tau\right) + \mathrm{E}_1\left(\left[-\frac{1}{2} - \frac{i(1-\alpha)}{\alpha}\right]\tau\right)\right), \quad (3)$$

$$I(\tau) = 2\pi e^{-\tau/2}\left(1 + \frac{i}{2\pi}\left(-e^{\tau}\mathrm{E}_1\left(\left[\frac{1}{2} - \frac{i(1-\alpha)}{\alpha}\right]\tau\right) + \mathrm{E}_1\left(\left[-\frac{1}{2} - \frac{i(1-\alpha)}{\alpha}\right]\tau\right)\right)\right), \quad (4)$$

where $\tau = \frac{\alpha(E_0 - \Lambda_{\rm bare})}{\hbar(1-\alpha)} V_0 t$ and V_0 is the volume of the Universe in the Planck epoch. We assume that $V_0 = 1$. The function $E_1(z)$ is the exponential integral and is given by the formula: $E_1(z) = \int_z^{\infty} \frac{e^{-x}}{x} dx$ (see [23]).

2. Preliminaries: unstable states

The parametrization of the dark energy transition from the false vacuum state to the true vacuum state following from the quantum properties of a such process will be used. This process is a quantum decay process. Quantum unstable systems are characterized by their survival probability. So if the system is in the initial unstable state $|\phi\rangle \in \mathcal{H}$, ($\mathcal{H}$ is the Hilbert space of states of the considered system), which was prepared at the initial instant $t_0 = 0$, then its survival probability (the decay law), $\mathcal{P}(t)$, of the unstable state $|\phi\rangle$ decaying in vacuum equals $\mathcal{P}(t) = |A(t)|^2$, where $A(t)$ is the probability amplitude of finding the system at the time t in the initial unstable state $|\phi\rangle$, $A(t) = \langle\phi|\phi(t)\rangle$, and $|\phi(t)\rangle$ is the solution of the Schrödinger equation for the initial condition $|\phi(0)\rangle = |\phi\rangle$, which has the following form

$$i\hbar\frac{\partial}{\partial t}|\phi(t)\rangle = \mathfrak{H}|\phi(t)\rangle. \quad (5)$$

Here $|\phi\rangle, |\phi(t)\rangle \in \mathcal{H}$, and $\mathfrak{H}$ is the total self-adjoint Hamiltonian for the system considered. The spectrum, $\sigma_c(\mathfrak{H})$, of $\mathfrak{H}$ is assumed to be bounded from below: $\sigma_c(\mathfrak{H}) = [E_{\rm min}, +\infty)$ of $\mathfrak{H}$). By using the basis in $\mathcal{H}$ formed by normalized eigenvectors $|E\rangle$, $E \in \sigma_c(\mathfrak{H})$ of $\mathfrak{H}$ and expanding $|\phi\rangle$ in terms of these eigenvectors one can express the amplitude $A(t)$ as the following Fourier integral

$$A(t) \equiv \int_{E_{min}}^{\infty} \omega(E)\, e^{-\frac{i}{\hbar}E\,t}\, dE, \quad (6)$$

where $\omega(E) > 0$ (see: [24]). Hence the amplitude $A(t)$, and thus the decay law $\mathcal{P}(t)$ of the unstable state $|\phi\rangle$, are completely determined by the density of the energy distribution $\omega(E)$ for the system in this state [24]. The amplitude $A(t)$ contains information about the decay law $\mathcal{P}_\phi(t)$ of the state $|\phi\rangle$: About the decay rate γ_ϕ^0 of this state, as well as the energy $\mathcal{E}_\phi^0$ of the system in this state. This information can be extracted from $A(t)$. It can be done by using the rigorous equation governing the time evolution in the subspace of unstable states, $\mathcal{H}_\parallel \ni |\phi\rangle_\parallel \equiv |\phi\rangle$. Such an equation follows from Schrödinger equation (5) for the total state space $\mathcal{H}$.

By using the Schrödinger equation (5) one can find that within the problem considered $i\hbar\frac{\partial}{\partial t}\langle\phi|\phi(t)\rangle = \langle\phi|\mathfrak{H}|\phi(t)\rangle$. From this relation a conclusion follows that the amplitude $A(t)$ satisfies the following equation

$$i\hbar\frac{\partial A(t)}{\partial t} = h(t)\, A(t), \tag{7}$$

where

$$h(t) = \frac{\langle\phi|\mathfrak{H}|\phi(t)\rangle}{A(t)} \equiv \frac{i\hbar}{A(t)}\,\frac{\partial A(t)}{\partial t}, \tag{8}$$

and $h(t)$ is the effective Hamiltonian governing the time evolution in the subspace of unstable states $\mathcal{H}_\parallel = \mathbb{P}\mathcal{H}$, where $\mathbb{P} = |\phi\rangle\langle\phi|$ (see [26] and also [27, 28, 29] and references therein). The subspace $\mathcal{H} \ominus \mathcal{H}_\parallel = \mathcal{H}_\perp \equiv \mathbb{Q}\mathcal{H}$ is the subspace of decay products. Here $\mathbb{Q} = \mathbb{I} - \mathbb{P}$. In general $h(t)$ is a complex function of time and in the case of $\mathcal{H}_\parallel$ of dimension two or more the effective Hamiltonian governing the time evolution in such a subspace it is a non-hermitian matrix $H_\parallel$ or non-hermitian operator. There is $h(t) = E_\phi(t) - \frac{i}{2}\varGamma_\phi(t)$, and $E_\phi(t) = \Re\,[h(t)]$, $\varGamma_\phi(t) = -2\,\Im\,[h(t)]$, are the instantaneous energy (mass) $E_\phi(t)$ and the instantaneous decay rate, $\varGamma_\phi(t)$ [26, 27, 28]. It is convenient to use relations (7), (8) when the density $\omega(E)$ is given and one wants to find the instantaneous energy $E_\phi(t)$ and decay rate $\varGamma_\phi(t)$: Inserting $\omega(E)$ into (6) one obtains the amplitude $A(t)$ and then using (8) one finds the $h(t)$ and thus $E_\phi(t)$ and $\varGamma_\phi(t)$. The simplest choice is to take $\omega(E)$ having the Breit-Wigner form: $\omega(E) \equiv \omega_{\rm BW}(E) \stackrel{\rm def}{=} \frac{N}{2\pi}\,\Theta(E - E_{\rm min})\,\frac{\varGamma_0}{(E-E_0)^2+(\frac{\varGamma_0}{2})^2}$, where N is a normalization constant and $\Theta(E) = 1$ for $E \geq 0$ and $\Theta(E) = 0$ for $E < 0$. The parameters E_0 and $\varGamma_0$ correspond with the energy of the system in the unstable state and its decay rate at the exponential (or canonical) regime of the decay process. Inserting $\omega_{BW}(E)$ into formula (6) for the amplitude $A(t)$ after some algebra one finds that

$$A(t) = \frac{N}{2\pi}\,e^{-\frac{i}{\hbar}E_0 t}\,I_\beta\left(\frac{\varGamma_0 t}{\hbar}\right), \tag{9}$$

where $I_\beta(\tau)$ is the integral given by (4) with $\beta = (1-\alpha)/\alpha \equiv (E_0 - E_{min})/\varGamma_0$. Here $\tau = \frac{\varGamma_0 t}{\hbar} \equiv \frac{t}{\tau_0}$, τ_0 is the lifetime. An explicit form of $I_\beta(t)$ is given by the formula (4). Next using this $A(t)$ given by the relation (9), and the relation (8) defining the effective Hamiltonian $h_\phi(t)$ one finds that within the Breit-Wigner model considered

$$h(t) = i\hbar\frac{1}{A(t)}\,\frac{\partial A(t)}{\partial t} = E_0 + \varGamma_0\,\frac{J_\beta(\frac{\varGamma_0 t}{\hbar})}{I_\beta(\frac{\varGamma_0 t}{\hbar})}, \tag{10}$$

where $J_\beta(\tau)$ is given by the formula (3) for $\beta = (1-\alpha)/\alpha$. It is important to be aware of the following problem: Namely from the definition of $J_\beta(\tau)$ one can conclude that $J_\beta(0)$ is undefined ($\lim_{\tau\to 0} J_\beta(\tau) = \infty$). This is because within the model defined by the Breit-Wigner distribution of the energy density, $\omega_{BW}(E)$, the expectation value of $\mathfrak{H}$, that is $\langle\phi|\mathfrak{H}|\phi\rangle$ is not finite. So all the consideration based on the use of $J_\beta(\tau)$ are valid only for $\tau > 0$. Note that simply $J_\beta(\tau) \equiv i\frac{\partial I_\beta(\tau)}{\partial \tau}$, which allows one to find analytical form of $J_\beta(\tau)$ having such a form for $I_\beta(\tau)$.

We need to know the energy $E_{\phi}(t) = \Re\,[h(t)]$ of the system in the unstable state $|\phi\rangle$ considered. Within the Breit-Wigner model one finds that

$$E_{\phi}(t) = E_0 + \varGamma_0\, \Re \left[\frac{J_{\beta}(\frac{\varGamma_0 t}{\hbar})}{I_{\beta}(\frac{\varGamma_0 t}{\hbar})} \right]. \tag{11}$$

There are cosmological scenarios that predict the possibility of decay of the Standard Model vacuum at an inflationary stage of the evolution of the universe (see eg. [31] and also [32] and reference therein) or earlier. Such scenario correspond with the hypothesis analyzed by Krauss and Dent [22]: The hypothesis that some false vacuum regions do survive well up to the cross–over time T or later was considered therein, (where T is the time when exponential and late time inverse power law contributions to the survival amplitude begin to be comparable). The fact that the decay of the false vacuum is the quantum decay process means that state vector corresponding to the false vacuum is a quantum unstable (or metastable) state. Therefore all the general properties of quantum unstable systems must also occur in the case of such a quantum unstable state as the false vacuum. This applies in particular to such properties as late time deviations from the exponential decay law and properties of the energy $E_0^{false}(t)$ of the system in the quantum false vacuum state at late times $t > T$: In [33] it was pointed out the energy of those false vacuum regions which survived up to T and much later differs from E_0^{false} [33]. Simply within the cosmological scenario in which the decay of false vacuum is assumed the unstable state $|\phi\rangle$ corresponds with the false vacuum state: $|\phi\rangle = |0\rangle^{\rm false}$. Then $|0\rangle^{\rm true}$ is the true vacuum state corresponding to the true minimal energy. In such a case $E_0 = E_0^{\rm false}$ is the energy of a state corresponding to the false vacuum measured at the canonical decay time (the exponential decay regime) and $E_0^{\rm true}$ is the energy of true vacuum (i.e., the true ground state of the system), so $E_0^{\rm true} \equiv E_{min}$. If one wants to generalize the above considerations to quantum field theory one should take into account among others a volume factor so that survival probabilities per unit volume should be considered and similarly the energies and the decay rate: $E \mapsto \rho(E) = \frac{E}{V_0}$, $\varGamma_0 \mapsto \gamma = \frac{\varGamma_0}{V_0}$, where $V_0 = V(t_0)$ is the volume of the considered system at the initial instant t_0, when the time evolution starts. It is easy to see that the mentioned changes $E \mapsto \frac{E}{V_0}$ and $\varGamma_0 \mapsto \frac{\varGamma_0}{V_0}$ do not change integrals $I_{\beta}(t)$ and $J_{\beta}(t)$. Similarly in such a situation the parameter $\beta = \frac{E_0 - E_{min}}{\varGamma_0}$ does not change. This means that the relation (11) can be replaced by corresponding relations for the densities $\rho_{\rm de}$ or Λ (see, eg., [29, 35, 36]).

3. Cosmological equations with $\rho_{\rm de} = \Lambda_{\rm bare} + E_{\rm R}\left[1 + \frac{\alpha}{1-\alpha}\Re\left(\frac{J(t)}{I(t)}\right)\right]$

In this paper, our model is introduced as the covariant theory from the following action

$$S = \int \sqrt{-g}(R + \mathcal{L}_{\rm m})\, d^4x, \tag{12}$$

where R is the Ricci scalar, $\mathcal{L}_{\rm m}$ is the Lagrangian for the barotropic fluid ($\mathcal{L}_{\rm m} = -\rho_{\rm tot}\left(1 + \int \frac{p_{\rm tot}(\rho_{\rm tot})}{\rho_{\rm tot}^2} d\rho_{\rm tot}\right)$, where $\rho_{\rm tot}$ is the total density of fluid and $p_{\rm tot}(\rho_{\rm tot})$ is the total pressure of fluid [37]) and $g_{\mu\nu}$ is the metric tensor. Here, the signature of the metric tensor is assumed as $(+,-,-,-)$. We consider the flat model (the constant curvature is zero).

We assume that the fluid consists of the baryonic matter $\rho_{\rm b}$, the dark matter $\rho_{\rm dm}$ and the dark energy $\rho_{\rm de}$. We treat the baryonic matter and the dark matter like dust ($p_{\rm b}(\rho_{\rm b}) = 0$ and $p_{\rm dm}(\rho_{\rm dm}) = 0$) and the equation of state for the dark energy is given by $p_{\rm de}(\rho_{\rm de}) = -\rho_{\rm de}$.

The Friedmann and acceleration equations can be find by variation action (12) by the metric $g_{\mu\nu}$. Then we get:

$$3H^2 = 3\frac{\dot{a}}{a}^2 = \rho_{\rm tot} = \rho_{\rm b} + \rho_{\rm dm} + \rho_{\rm de} \text{ and } \frac{\ddot{a}}{a} = -\frac{1}{6}(\rho_{\rm tot} + 3p_{\rm tot}(\rho_{\rm tot})) = \rho_{\rm b} + \rho_{\rm dm} - 2\rho_{\rm de}, \tag{13}$$

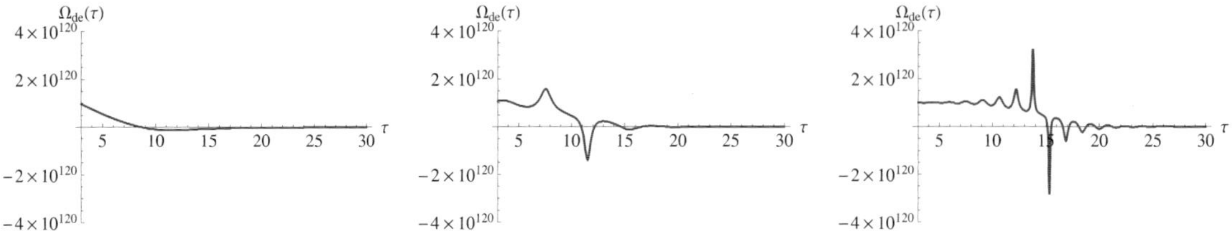

Figure 1. The dependence $\Omega_{\rm de}(\tau)$ for $\alpha = 0.2$ (right figure), $\alpha = 0.4$ (medium figure), $\alpha = 0.8$ (left figure) and $\frac{E_0}{3H_0^2} = 10^{120}$. The time τ for the left figure is given in unit $5.3 \times 10^{-145}s$, for the center figure is given in unit $2.0 \times 10^{-145}s$ and for the right figure is given in unit $3.3 \times 10^{-146}s$.

where $H = \frac{\dot{a}}{a}$ is the Hubble function,

From the above equations we can obtain the conservation equation

$$\dot{\rho}_{\rm tot} = -3H(\rho_{\rm tot} + p_{\rm tot}(\rho_{\rm tot})) \text{ or in the equivalent form } \dot{\rho}_{\rm m} = -3H\rho_{\rm m} - \dot{\rho}_{\rm de}, \tag{14}$$

where $\rho_{\rm m} = \rho_{\rm b} + \rho_{\rm dm}$.

If we introduce the interaction Q in the dark sector then Eq. (14) can be rewritten as

$$\dot{\rho}_{\rm b} = -3H\rho_{\rm b}, \ \dot{\rho}_{\rm dm} = -3H\rho_{\rm dm} + Q \text{ and } \dot{\rho}_{\rm de} = -Q. \tag{15}$$

If $Q > 0$ then the energy flows from the dark energy to the dark matter. If $Q < 0$ then the energy flows from the dark matter to the dark energy.

Fig. 1 shows the diagrams of the evolution $\Omega_{\rm de}(\tau) = \frac{\rho_{\rm de}(\tau)}{3H^2}$, where H_0 is the present value of the Hubble function, with respect of the time τ for different values of model parameter α ($\alpha = 0.2$, 0.4, 0.8) and $\frac{E_0}{3H_0^2} = 10^{120}$. We can see that the oscillations disappear for $\alpha > 0.4$ and the number of oscillations as well as the length of this intermediate phase decresess as the parameter α grows. On this figure we can see that the beginning value of the dark energy density, which is equal $\Omega_{\rm de} \approx 10^{120}$, is reduced to the present value of the dark energy density, which is $\Omega_{\rm de} \approx 0.7$. This final value of $\Omega_{\rm de}$ does not depend on the values of model parameters α and $\frac{E_0}{3H_0^2}$. This mechanism can be a trying to solve the cosmological constant problem. The characteristic of the intermidiate oscillatory phase is depended on the parameter α.

4. Statistical analysis

In statistical analysis, we use the following astronomical data: supernovae of type Ia (SNIa) (Union 2.1 dataset [39]), BAO data (Sloan Digital Sky Survey Release 7 (SDSS DR7) dataset at $z = 0.275$ [40], 6dF Galaxy Redshift Survey measurements at redshift $z = 0.1$ [41], and WiggleZ measurements at redshift $z = 0.44, 0.60, 0.73$ [43]), measurements of the Hubble parameter $H(z)$ of galaxies [55, 56, 57], the Alcock-Paczynski test (AP)[45, 46] (data from [47, 48, 49, 50, 51, 52, 42, 53, 54].) and measurements of CMB [58] and lensing by Planck, and low-ℓ polarization from the WMAP (WP).

We take the formula for likelihood function as $L_{\rm tot} = L_{\rm SNIa} L_{\rm BAO} L_{\rm AP} L_{H(z)} L_{\rm CMB+lensing}$. The likehood function $L_{\rm SNIa} = {\rm Exp}\left[-\frac{1}{2}[A - B^2/C + \log(C/(2\pi))]\right]$ is for SNIa, where $A = (\mu^{\rm obs} - \mu^{\rm th})\mathbb{C}^{-1}(\mu^{\rm obs} - \mu^{\rm th})$, $B = \mathbb{C}^{-1}(\mu^{\rm obs} - \mu^{\rm th})$, $C = {\rm Tr}\mathbb{C}^{-1}$ and $\mathbb{C}$ is a covariance matrix for SNIa, $\mu^{\rm obs}$ is the observer distance modulus and $\mu^{\rm th}$ is the theoretical distance modulus. The likehood function $L_{\rm BAO} = {\rm Exp}\left[-\frac{1}{2}\left(\mathbf{d}^{\rm obs} - \frac{r_s(z_d)}{D_V(\mathbf{z})}\right)\mathbb{C}^{-1}\left(\mathbf{d}^{\rm obs} - \frac{r_s(z_d)}{D_V(\mathbf{z})}\right)\right]$ is for BAO, where $r_s(z_d)$ is the sound horizon at the drag epoch [59, 44]. The likelihood function

Table 1. The best fit and errors for the estimated model with H_0 from the interval (66.0, 72.0), $\Omega_{\mathrm{m},0}$ from the interval $(0.27, 0.34)$. $\Omega_{\mathrm{b},0}$ is assumed as 0.048468.

parameter	best fit	68% CL	95% CL
H_0	68.82 km/(s Mpc)	$+0.61$ -0.55	$+0.98$ -0.92
$\Omega_{\mathrm{m},0}$	0.3009	$+0.0079$ -0.0084	$+0.0133$ -0.0134

$L_{H(z)} = \mathrm{Exp}\left[-\frac{1}{2}\sum_{i=1}^{N}\left(\frac{H(z_i)^{\mathrm{obs}}-H(z_i)^{\mathrm{th}}}{\sigma_i}\right)^2\right]$ is for measurements of the Hubble parameter $H(z)$ of galaxies. The likelihood function $L_{AP(z)} = \mathrm{Exp}\left[-\frac{1}{2}\sum_{i=1}^{N}\left(\frac{AP(z_i)^{\mathrm{obs}}-AP(z_i)^{\mathrm{th}}}{\sigma_i}\right)^2\right]$ is for AP, where $AP(z)^{\mathrm{th}} \equiv \frac{H(z)}{z}\int_0^z \frac{dz'}{H(z')}$ and $AP(z_i)^{\mathrm{obs}}$ are observational data. The likelihood function $L_{\mathrm{CMB+lensing}} = \mathrm{Exp}\left[-\frac{1}{2}(\mathbf{x}^{\mathrm{th}} - \mathbf{x}^{\mathrm{obs}})\mathbb{C}^{-1}(\mathbf{x}^{\mathrm{th}} - \mathbf{x}^{\mathrm{obs}})\right]$ is for CMB and lensing, where $\mathbb{C}$ is the covariance matrix with the errors, $\mathbf{x}$ is a vector of the acoustic scale l_A, the shift parameter R and $\Omega_b h^2$ where $l_A = \frac{\pi}{r_s(z^*)}c\int_0^{z^*}\frac{dz'}{H(z')}$ and $R = \sqrt{\Omega_{\mathrm{m},0}H_0^2}\int_0^{z^*}\frac{dz'}{H(z')}$, where z^* is the redshift of the epoch of the recombination [59].

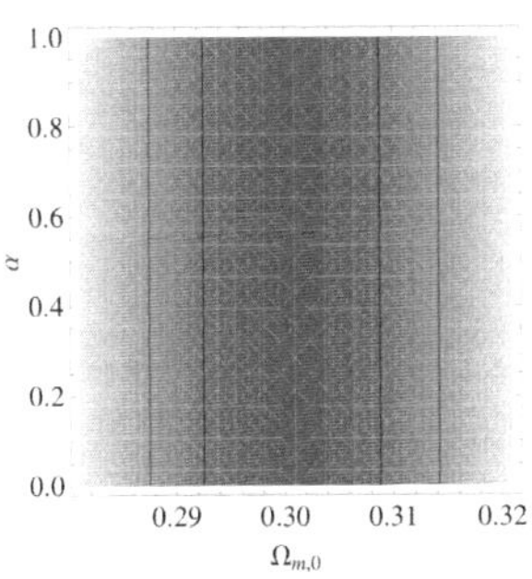

Figure 2. The diagram presents the intersection of the likelihood function of two model parameters $(\Omega_{\mathrm{m},0}, \alpha)$, with the marked 68% and 95% confidence levels. The plane of the intersection is the best fit of H_0 and $E_0/(3H_0^2) = 10^{120}$.

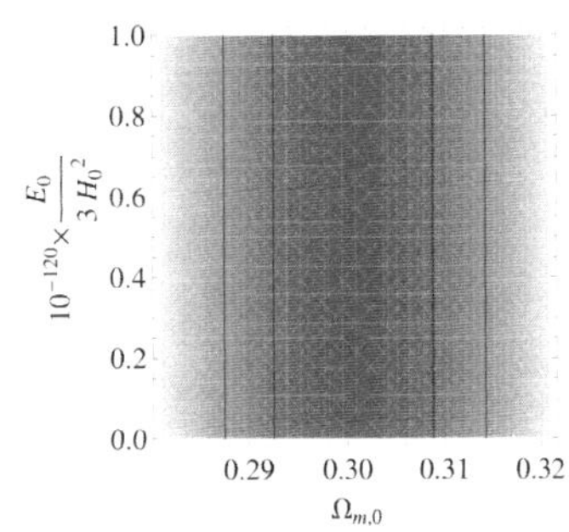

Figure 3. The diagram presents the intersection of the likelihood function of two model parameters $(\Omega_{\mathrm{m},0}, \frac{E_0}{3H_0^2})$, with the marked 68% and 95% confidence levels. The plane of the intersection is the best fit of H_0 and $\alpha = 0.1$.

We used our own code CosmoDarkBox in estimation of model parameter. This code uses the Metropolis-Hastings algorithm [60, 61].

In statistical analysis, we estimated four model parameters: H_0, $\Omega_{\mathrm{m},0} = \frac{\rho_{\mathrm{m},0}}{3H_0^2}$, α and E_0. Our statistical results are completed in Table 1. Figs. 2-3 show intersections of the likelihood function with 68% and 95% confidence level projections.

Changing of the values of α and E_0 parameter has not an influence for the values of the likelihood function. The changing of the values of the likelihood function are beyond abilities of numerical methods. We interprate this as the lack of sensitive of the present evolution of the universe for changing of α and E_0 parameter.

5. Conclusion

The main goal of our paper was to analyze the cosmological model with the running dark energy as well as matter in the dust form. We discussed the evolution of the dark energy with the respect that the decay of false vacuum to the true vacuum is a quantum decay process. In this case the cosmological model includes the interaction in the dark sector.

We found the intermediate phase of oscillations, in the early universe, between phases of the constant dark energy. There is a mechanism of jumping of the value of energy density of dark energy from the initial value of $E_0 = 10^{120}$ to present value of the cosmological constant of 0.7.

In this intermediate phase the amplitude of oscillatory increases, then there is a jump down followed by the decreasing oscillations. This type of oscillation appears for $0 < \alpha < 0.4$, where α is a model parameter. The number of oscillations as well as the length of this intermediate phase decresess as the parameter α grows. For $\alpha > 0.4$ the oscilations, in the intermediate phase, disappear and only the jump down of energy density of dark energy remains. The jump down mechanism can lead to solve the cosmological constant problem.

In this model, the energy transfer, in the early universe in the dark sector, is negligible because the energy density of dark energy is significantly lower than the energy density of dark matter.

From the statistical analysis, we found that the present value of the energy density of the dark energy is independent of the assumed values of the parameters α and E_0. The ΛCDM model is an attractor for the all models with different values of parameters α and E_0. The final interval of evolution for which we have data at dispose is identical for whole class of models. So it is impossible to find the best fit values of model parameters (degeneration problem).

References

[1] Polchinski J 2006 *The Quantum Structure of Space and Time: Proceedings of the 23rd Solvay Conference on Physics. Brussels, Belgium. 1–3 December 2005* 216–236 (*Preprint* arXiv:hep-th/0603249)
[2] Weinberg S 1989 *Rev. Mod. Phys.* **61** 1–23
[3] Carroll M S et al. 1992 *Ann. Rev. Astron. Astrophys.* **30** 499–54
[4] Dolgov A D 1997 *Phase transitions in cosmology. Proceedings, 4th Cosmology Colloquium, Euroconference, Paris, France, June 4-9, 1997* 161–175 (*Preprint* arXiv:astro-ph/9708045)
[5] Sahni V and Starobinsky A A 2000 *Int. J. Mod. Phys.* D **9** 373–444 (*Preprint* arXiv:astro-ph/9904398)
[6] Straumann N 1999 *Eur. J. Phys.* **20** 419–427 (*Preprint* arXiv:astro-ph/9908342)
[7] Weinberg S 2000 *Sources and detection of dark matter and dark energy in the universe. Proceedings, 4th International Symposium, DM 2000, Marina del Rey, USA, February 23-25, 2000* 18–26 (*Preprint* arXiv:astro-ph/0005265)
[8] Carroll S M 2001 *Living Rev. Rel.* **4** 1 (*Preprint* arXiv:astro-ph/0004075)
[9] Rugh S E and Zinkernagel H 2002 *Stud. Hist. Phil. Sci.* B **33** 663–705 (*Preprint* arXiv:hep-th/0012253)
[10] Padmanabhan T 2003 *Phys. Rept.* **380** 235–320 (*Preprint* arXiv:hep-th/0212290)
[11] Yokoyama J 2003 *Proceedings, 12th Workshop on General Relativity and Gravitation (JGRG12): Tokyo, Japan, November 25-28, 2002* (*Preprint* arXiv:gr-qc/0305068)
[12] Sarkar S 2005 *Curr. Sci.* **88** 2120–2128
[13] Copeland E J et al. 2006 *Int. J. Mod. Phys.* D **15** 1753–1936 (*Preprint* arXiv:hep-th/0603057)
[14] Szydlowski M 2015 *Phys. Rev.* D **91** 123538
[15] Coleman S R 1997 *Phys. Rev.* D **15** 2929–2936
[16] Callan C G and Coleman S R 1977 *Phys. Rev.* D **16** 1762–1768
[17] Coleman S R and De Luccia F 1980 *Phys. Rev.* D **21** 3305
[18] Kobakhidze A and Spencer-Smith A 2013 *Phys. Lett.* B **722** 130–134 (*Preprint* arXiv:1301.2846)
[19] Degrassi G et al. 2012 *J. High Energy Phys.* JHEP08(2012)098 (*Preprint* arXiv:1205.6497)
[20] Elias-Miro J et al. 2012 *Phys. Lett.* B **709** 222–228 (*Preprint* arXiv:1112.3022)
[21] Chao W et al. 2012 *Phys. Rev.* D **86** 113017 (*Preprint* arXiv:1210.0491)
[22] Krauss L M and Dent J 2008 *Physical Review Letters*, **100**, 171301
[23] Abramowitz M and Stegun I A 1964 *Handbook of Mathematical Functions*, (Washington, D.C.: Natl. Bur. Stand. Appl. Math. Ser. No 55)
[24] Krylov N S, Fock V A 1947 *Zh. Eksp. Teor. Fiz.* **17** p.p. 93–107 [in Russian]

[25] Khalfin L A 1957 *Zh. Eksp. Teor. Fiz. (USSR)* **33** p.p. 1371–1382 [in Russian]; [Sov. Phys. — JETP **6**, (1958), 1053]
[26] Urbanowski K 1994 *Phys. Rev.* A **50** 2847–2853
[27] Urbanowski K 2008 *Eur. Phys. J.* C **58** 151
[28] Urbanowski K 2009 *Cent. Eur. J. Phys.* **7** 696 – 703 DOI:10.2478/s11534–009–0053–5
[29] Urbanowski K 2017 *Theoretical and Mathematical Physics* **190** 458 (*Preprint* arXiv:1609.03382)
[30] Urbanowski K and Raczyńska K 2014 *Phys. Letters* B **731** 236–241
[31] Branchina V and Messina E 2013 *Physical Review Letters* **111** 241801
[32] Bezrukov F and Shaposhnikov M 2015 *Comptes Rendus Physique* **16** 994
[33] Urbanowski K 2011 *Phys. Rev. Lett.* **107** 209001
[34] Urbanowski K 2009 *Eur. Phys. J.* D **54** 25–29
[35] Urbanowski K and Szydlowski M 2013 *AIP Conf. Proc.* **1514** 143
[36] Szydlowski M, Stachowski A and Urbanowski K 2015 *J. Phys.: Conference Series* **626** 012033
[37] Minazzoli O and Harko T 2012 *Phys. Rev.* D **86** 087502 (*Preprint* arXiv:1209.2754)
[38] De Felice A and Tsujikawa S 2010 *Living Rev. Rel.* **13** 3 (*Preprint* arXiv:1002.4928)
[39] Suzuki N et al. 2012 *Astrophys. J.* **746** 85 (*Preprint* arXiv:1105.3470)
[40] **SDSS** Collaboration, W. J. Percival et al. 2010 *Mon. Not. Roy. Astron. Soc.* **401** 2148–2168 (*Preprint* arXiv:0907.1660)
[41] Beutler F et al. 2011 *Mon. Not. Roy. Astron. Soc.* **416** 3017–3032 (*Preprint* arXiv:1106.3366)
[42] Anderson L et al. 2013 *Mon. Not. Roy. Astron. Soc.* **427**, no. 4 3435–3467 (*Preprint* arXiv:1203.6594)
[43] Blake C et al. 2012 *Mon. Not. Roy. Astron. Soc.* **425** 405–414 (*Preprint* arXiv:1204.3674)
[44] Eisenstein D J and Hu W 1998 *Astrophys. J.* **496** 605 (*Preprint* astro-ph/9709112)
[45] Alcock C and Paczynski B 1979 *Nature* **281** 358–359
[46] Lopez-Corredoira M 2014 *Astrophys. J.* **781** no. 2 96 (*Preprint* arXiv:1312.0003)
[47] Sutter P M, Lavaux G, Wandelt B D and Weinberg D H 2012 *Astrophys. J.* **761** 187 (*Preprint* arXiv:1208.1058)
[48] Blake C et al. 2011 *Mon. Not. Roy. Astron. Soc.* **418** 1725–1735 (*Preprint* arXiv:1108.2637)
[49] Ross N P et al. 2007 *Mon. Not. Roy. Astron. Soc.* **381** 573–588 (*Preprint* astro-ph/0612400)
[50] Marinoni C and Buzzi A 2010 *Nature* **468** no. 7323 539–541
[51] da Angela J, Outram P J, and Shanks T 2005 *Mon. Not. Roy. Astron. Soc.* **361** 879–886 (*Preprint* astro-ph/0505469)
[52] Outram P J et al. 2004 *Mon. Not. Roy. Astron. Soc.* **348** 745 (*Preprint* astro-ph/0310873)
[53] Paris I et al. 2012 *Astron. Astrophys.* **548** A66 (*Preprint* arXiv:1210.5166)
[54] **SDSS** Collaboration, D. P. Schneider et al. 2010 *Astron. J.* **139** 2360–2373 (*Preprint* arXiv:1004.1167)
[55] Simon J, Verde L andJimenez R 2005 *Phys. Rev.* D **71** 123001 (*Preprint* astro-ph/0412269)
[56] Stern D, Jimenez R, Verde L, Kamionkowski M and Stanford S A 2010 *J. Cosmol. Astropart. Phys.* JCAP02(2010)008 (*Preprint* arXiv:0907.3149)
[57] Moresco M et al. 2012 *J. Cosmol. Astropart. Phys.* JCAP08(2012)006 (*Preprint* arXiv:1201.3609)
[58] Planck Collaboration, P. A. R. Ade et al. 2015 Planck 2015 results. XIV. Dark energy and modified gravity *Preprint* arXiv:1502.01590
[59] Hu W and Sugiyama N 1996 *Astrophys. J.* **471** 542–570 (*Preprint* astro-ph/9510117)
[60] Metropolis N, Rosenbluth A W, Rosenbluth M N, Teller A H and Teller E 1953 *J. Chem. Phys.* **21** 1087–1092
[61] Hastings W K 1970 *Biometrika* **57** 97–109

IOP Conf. Series: Journal of Physics: Conf. Series **880** (2017) 012023 doi:10.1088/1742-6596/880/1/012023

Quantum groups and noncommutative spacetimes with cosmological constant

A Ballesteros[1], I Gutiérrez-Sagredo[1], F J Herranz[1], C Meusburger[2] and P Naranjo[1]

[1] Departamento de Física, Universidad de Burgos, Plaza Misael Bañuelos s.n., E-09001 Burgos, Spain

[2] Department Mathematik, FAU Erlangen-Nürnberg, Cauerstr. 11, D-91058 Erlangen, Germany

E-mail: angelb@ubu.es, igsagredo@ubu.es, fjherranz@ubu.es, catherine.meusburger@math.uni-erlangen.de, pnaranjo@ubu.es

Abstract. Noncommutative spacetimes are widely believed to model some properties of the quantum structure of spacetime at the Planck regime. In this contribution the construction of (anti-)de Sitter noncommutative spacetimes obtained through quantum groups is reviewed. In this approach the quantum deformation parameter z is related to a Planck scale, and the cosmological constant Λ plays the role of a second deformation parameter of geometric nature, whose limit $\Lambda \to 0$ provides the corresponding noncommutative Minkowski spacetimes.

1. Introduction

Non-Abelian algebras play a prominent role in the Hamiltonian description of physical systems. For instance, nonrelativistic Quantum Mechanics is based on a 'noncommutative phase space' in which position and momenta operators generate the Lie algebra

$$[\hat{x}_i, \hat{p}_j] = i\hbar\,\delta_{ij}, \qquad [\hat{x}_i, \hat{x}_j] = 0, \qquad [\hat{p}_i, \hat{p}_j] = 0, \qquad i, j = 1, \ldots, N, \tag{1}$$

which is the direct sum of N copies of the Heisenberg–Weyl algebra. Here, noncommutativity is controlled by the fundamental constant $\hbar$, since the $\hbar \to 0$ limit of (1) leads to an Abelian algebra, and the classical limit of (1) is defined as $\{x_i, p_j\} := \lim_{\hbar\to 0} \frac{[\hat{x}_i, \hat{p}_j]}{i\hbar} = \delta_{ij}$. In this way we recover the symplectic structure of Classical Hamiltonian Mechanics, which can be properly said to be a Poisson-noncommutative theory.

On the other hand, spacetimes with nonvanishing cosmological constant Λ lead to noncommuting momenta in a natural way. For instance, the (3+1) (anti-)de Sitter (hereafter (A)dS) Lie algebra of isometries of the corresponding spacetimes can be written as

$$\begin{aligned}
&[J_i, J_j] = \epsilon_{ijk} J_k, && [J_i, P_j] = \epsilon_{ijk} P_k, && [J_i, K_j] = \epsilon_{ijk} K_k, \\
&[P_i, P_j] = -\omega\, \epsilon_{ijk} J_k, && [P_i, K_j] = -\,\delta_{ij} P_0, && [K_i, K_j] = -\,\epsilon_{ijk} J_k, \\
&[P_0, P_i] = \omega K_i, && [P_0, K_i] = -P_i, && [P_0, J_i] = 0,
\end{aligned} \tag{2}$$

where $i, j = 1, 2, 3$, $\epsilon_{123} = 1$ and $\omega = -\Lambda$. Here $\{J_i, P_0, P_i, K_i\}$ denote the generators of rotations, time translation, space translations and boosts. The algebra (2) is called the AdS_ω Lie algebra.

IOP Conf. Series: Journal of Physics: Conf. Series **880** (2017) 012023 doi:10.1088/1742-6596/880/1/012023

In this framework, when $\omega > 0$ we recover the (3+1) anti-de Sitter Lie algebra $so(3, 2)$, when $\omega < 0$ we find the (3+1) de Sitter Lie algebra $so(4, 1)$, and the flat limit $\omega \to 0$ gives rise to the (3+1) Poincaré algebra $iso(3, 1)$. Note that the commutation rules amongst the translation generators read

$$[P_i, P_j] = -\omega\, \varepsilon_{ijk} J_k, \qquad [P_0, P_i] = \omega K_i,$$

and the fundamental parameter that controls the noncommutativity of momenta is just the cosmological constant $\omega = -\Lambda$.

In this contribution we shall deal with a third type of noncommutativity: The one arising in different approaches to Quantum Gravity aiming to describe the 'quantum' structure of the geometry of spacetime at the Planck scale through a noncommutative algebra of 'quantum spacetime coordinates' [1, 2, 3, 4, 5]. In this approach the Planck length l_P (or energy E_p) is the parameter that governs the noncommutativity of the spacetime algebra, thus generating uncertainty relations between noncommuting coordinates that can be used in order to describe a 'fuzzy' or 'discrete' nature of the spacetime at very small distances or high energies [6, 7].

It is worth recalling that most of the 'quantum' spacetimes that have been introduced so far are noncommutative versions of the Minkowski spacetime [8, 9, 10, 11, 12]. As a consequence, the construction of noncommutative spacetimes with nonvanishing cosmological constant arises as a challenging problem in order to describe the interplay between the nonvanishing curvature of spacetime and quantum gravity effects, having in mind the possible cosmological consequences of Planck scale physics [13, 14, 15, 16, 17, 18].

In this paper we report on recent results [19, 20, 21, 22, 23, 24, 25] concerning the construction of noncommutative (A)dS spacetimes, which have been obtained by making use of the theory of quantum groups. We recall that the so-called 'quantum' deformations of Lie groups and algebras (see [26, 27, 28, 29, 30, 31, 32] and references therein) present many features that make them suitable to be considered in a Quantum Gravity scenario:

- They are Hopf algebra deformations of kinematical Lie groups in which the quantum deformation parameter, hereafter $z = \ln q$, can be related to a Planck scale parameter.
- They give rise to noncommutative spacetimes which are covariant under quantum group (co)actions. In this context, several notions in 'quantum kinematical geometry' can be rigorously generalized, like Poisson and quantum homogeneous spaces [33, 34, 35, 36, 37].
- Quantum groups can be thought of as Hopf algebra quantizations of Poisson–Lie (PL) groups, and the relevance of the latter in (2+1) gravity has been rigorously established (see [38, 39, 40, 41, 42, 43, 44] as well as the connections between dynamical r-matrices and gauge-fixing [45]).
- Deformed Casimir operators can be interpreted as modified dispersion relations of the same type that appear in several phenomenological approaches to Quantum Gravity [46, 47, 48].

Indeed, quantum group techniques can only be used to construct noncommutative analogues of spacetimes that can be obtained either as group manifolds or as homogeneous spaces, but Minkowski and (A)dS spacetimes fall into this class. Moreover, each Lie group admits a number of inequivalent quantum group deformations, and the quantum spacetime arising from each of them can be essentially different. The classification and explicit construction of such plurality of quantum geometries constitutes one of the main issues in the theory of quantum kinematical groups, which is far from being completed. In general, the results contained in the next sections show that noncommutative spacetimes with cosmological constant can be viewed as 'geometric' nonlinear deformations (with parameter Λ) of the noncommutative Minkowski spacetimes with quantum deformation parameter z related to either l_P or E_p.

The paper is organized as follows. In the next section we summarize the construction of noncommutative spacetimes from quantum groups. Section 3 is devoted to present two different noncommutative (2+1) (A)dS spacetimes that have been obtained by making use of two quantum deformations whose classical counterpart is known to be compatible with the Chern-Simons approach to (2+1) gravity, together with a brief summary of the results obtained so far in (3+1) dimensions.

2. Quantum groups, Lie bialgebras and noncommutative spacetimes

Quantum groups can be understood as noncommutative generalizations of algebraic groups, and all their properties are mathematically encoded within the Hopf algebra of functions on the group. Alternatively, they can also be interpreted as quantizations of PL groups, *i.e.*, of the Poisson-Hopf algebras of multiplicative Poisson structures on Lie groups. It is well known that PL structures on a (connected and simply connected) Lie group G are in one-to-one correspondence with Lie bialgebra structures (g, δ) on $g = \mathrm{Lie}(G)$ [49], where the skewsymmetric cocommutator map $\delta : g \to g \wedge g$ fulfils the two following conditions:

- i) δ is a 1-cocycle, *i.e.*,

$$\delta([X, Y]) = [\delta(X),\, Y \otimes 1 + 1 \otimes Y] + [X \otimes 1 + 1 \otimes X,\, \delta(Y)], \qquad \forall\, X, Y \in g.$$

- ii) The dual map $\delta^* : g^* \wedge g^* \to g^*$ is a Lie bracket on g^*.

Therefore, each quantum group G_z (with quantum deformation parameter $z = \ln q$) can be put in correspondence with a PL group G, and the latter with a Lie bialgebra structure (g, δ).

Moreover, in the same manner as Lie algebras provide the infinitesimal version of Lie groups around the identity, quantum algebras play the same role with respect to quantum groups. more explicitly, quantum algebras $U_z(g)$ are Hopf algebra deformations of universal enveloping algebras $U(g)$, and are constructed as formal power series in a deformation parameter z and coefficients in $U(g)$. The Hopf algebra structure in $U_z(g)$ is provided by a coassociative coproduct map $\Delta_z : U_z(g) \longrightarrow U_z(g) \otimes U_z(g)$, which is an algebra homomorphism, together with its associated counit ϵ and antipode γ mappings. It can be easily shown that the first-order deformation (in z) of the coproduct map

$$\Delta_z = \Delta_0 + z\,\delta + o[z^2],$$

is just the Lie bialgebra cocommutator map δ, where we have denoted the primitive (nondeformed) coproduct for $U(g)$ as $\Delta_0(X) = X \otimes 1 + 1 \otimes X$. Again, each quantum deformation is related to a unique Lie bialgebra structure (g, δ). If we consider a basis for g where

$$[X_i, X_j] = C_{ij}^k X_k,$$

any cocommutator δ will be of the form

$$\delta(X_i) = f_i^{jk}\, X_j \wedge X_k\,,$$

where f_i^{jk} is the structure tensor of the dual Lie algebra g^*, that will be given by

$$[\hat{\xi}^j, \hat{\xi}^k] = f_i^{jk}\, \hat{\xi}^i\,, \tag{3}$$

where $\langle \hat{\xi}^j, X_k \rangle = \delta_k^j$. In particular, if G is a group of isometries of a given spacetime (for instance, (A)dS or Poincaré), then X_i will be the Lie algebra generators and $\hat{\xi}^j$ will be the local coordinates on the group. We thus realize that if the cocommutator δ is non-vanishing (which means that we have a non-trivial deformation of $U(g)$) then, automatically, the commutator (3) among the spacetime coordinates associated to the translation generators of the group will be non-zero. This is just the way in which noncommutative spacetimes arise from quantum groups. Moreover, we stress that, as we will see in what follows, higher-order contributions to the noncommutative spacetime are obtained from dualizing the full quantum coproduct Δ_z.

In some cases the 1-cocycle δ is found to be coboundary

$$\delta(X) = [X \otimes 1 + 1 \otimes X, r], \qquad \forall X \in g, \tag{4}$$

where r (the classical r-matrix) $r = r^{ab} X_a \wedge X_b$, has to be a solution of the modified classical Yang–Baxter equation

$$[X \otimes 1 \otimes 1 + 1 \otimes X \otimes 1 + 1 \otimes 1 \otimes X, [[r,r]]\,] = 0, \qquad \forall X \in g,$$

where the Schouten bracket is defined as $[[r,r]] := [r_{12}, r_{13}] + [r_{12}, r_{23}] + [r_{13}, r_{23}]$, where $r_{12} = r^{ab} X_a \otimes X_b \otimes 1$, $r_{13} = r^{ab} X_a \otimes 1 \otimes X_b$, $r_{23} = r^{ab} 1 \otimes X_a \otimes X_b$ (recall that $[[r,r]] = 0$ is just the classical Yang–Baxter equation). An important application of the classical r-matrix associated to a given quantum deformation consists in the fact that the PL structure on G linked to δ is explicitly given by the so-called Sklyanin bracket

$$\{f, g\} = r^{ij}(X_i^L f\, X_j^L g - X_i^R f\, X_j^R g),$$

where X_i^L, X_i^R denote the right- and left-invariant vector fields on G. The linearization of this bracket in terms of the local coordinates in G is just the Poisson version of (3), and the quantization of the Sklyanin bracket gives the commutation relations that define the quantum group G_z in terms of noncommuting local coordinates. Moreover, if we compute the Sklyanin bracket for the spacetime coordinates we will obtain the 'semiclassical' version of the noncommutative spacetime associated to the quantum group G_z, and the main algebraic features induced from the spacetime noncommutativity can be straightforwardly appreciated at this Poisson level, such as, for instance, the role played by the cosmological constant. Moreover, if the classical spacetime is a homogeneous space $M = G/H$, where H is a certain subgroup of G, and the 1-cocycle δ fulfils the so-called coisotropy condition (see [25] and references therein)

$$\delta(h) \subset h \wedge g, \qquad h = \mathrm{Lie}(H), \quad g = \mathrm{Lie}(G), \tag{5}$$

then the canonical projection of the Sklyanin bracket to the M submanifold generates a Poisson subalgebra, which is just the Poisson homogeneous space associated to the chosen r-matrix, whose quantization will be the quantum homogeneous space associated to G_z.

Finally, it is is also worth mentioning that Lie bialgebras can alternatively be described as Drinfel'd double (DD) Lie algebras given by the relations

$$[X_i, X_j] = C_{ij}^k X_k, \qquad [\hat{\xi}^j, \hat{\xi}^k] = f_i^{jk}\, \hat{\xi}^i, \qquad [\hat{\xi}^i, X_j] = C_{jk}^i \hat{\xi}^k - f_j^{ik} X_k. \tag{6}$$

Moreover, for any DD Lie algebra $D(g)$ (6), its corresponding double Lie group $D(G)$ can be endowed with a PL structure generated by the canonical classical r-matrix

$$r = \sum_i x^i \otimes X_i \tag{7}$$

which is a solution of the classical Yang–Baxter equation $[[r,r]] = 0$. In fact, if a given even-dimensional Lie algebra can be written in the form (6), then the quantum deformation associated to the classical r-matrix (7) is called a DD one.

It can be proven that for the (A)dS and Poincaré algebras in (2+1) and (3+1) dimensions all 1-cocycles δ are coboundaries. Therefore, the classification problem for the quantum deformations of these Lie algebras is equivalent to finding all inequivalent (under automorphisms) solutions of the modified classical Yang–Baxter equation. This is, by no means, a simple task (see [50, 51, 52, 53]). In the sequel we will present the explicit construction and properties of some noncommutative spacetimes with nonvanishing cosmological constant that have been recently obtained through certain quantum (A)dS deformations of DD type.

3. AdS$_\omega$ noncommutative spacetimes from Drinfel'd doubles

In the specific case of (2+1)-quantum gravity, there are indications [14] that the perturbations of the vacuum state of a Chern–Simons (CS) quantum gravity theory with cosmological constant Λ are invariant under a certain quantum (A)dS algebra, whose zero-curvature limit is the κ-Poincaré quantum algebra [11]. Also, it is well-known that PL structures on the isometry groups of (2+1) spaces with constant curvature play a relevant role as phase spaces in the framework of a CS gauge theory. In this context, the gauge group is the isometry group, and the phase space of (2+1)-gravity coupled to point particles is related to the moduli space of flat connections, while the Poisson structure on the moduli space is a PL group (see [38, 39, 40, 41, 42, 43]).

Therefore, it seems reasonable to assume that quantum deformations of the (A)dS and Poincaré groups could play a relevant role in (2+1) quantum gravity. Nevertheless, a plethora of possible quantum deformations for these groups do exist (see [50, 51, 52, 53]), and some physically motivated criteria were needed in order to filter them. An answer to this question was presented in [19, 20], where it was proven that all the classical r-matrices coming from a DD structure of the (A)dS and Poincaré groups are compatible with the CS formalism, thus supporting the idea that natural PL structures for (2+1)-gravity are classical doubles [43, 44]. Moreover, all DD structures for the (A)dS Lie algebras ($so(3,1)$ and $so(2,2)$) were explicitly obtained in [20], and a similar classification for the Poincaré algebra has been also worked out [54]. As a result, two main candidates for quantum deformations of the (A)dS symmetries that would be suitable in a (2+1)-gravity CS setting were obtained. The main properties of these two quantum deformations and their associated noncommutative spacetimes will be sketched in what follows. In this approach, the cosmological constant will be included as a 'geometric deformation' parameter, and its flat (Poincaré) limit can be smoothly obtained.

To this aim we shall make use of the (2+1) AdS$_\omega$ algebra (which can be obtained from (2))

$$\begin{array}{lll} [J,P_i]=\epsilon_{ij}P_j, & [J,K_i]=\epsilon_{ij}K_j, & [J,P_0]=0, \\ [P_i,K_j]=-\delta_{ij}P_0, & [P_0,K_i]=-P_i, & [K_1,K_2]=-J, \\ [P_0,P_i]=\omega K_i, & [P_1,P_2]=-\omega J, & \end{array}$$

where $i,j=1,2$ and $\epsilon_{12}=1$. The anti-de Sitter algebra $so(2,2)$ corresponds to $\omega=-\Lambda>0$, the de Sitter algebra $so(3,1)$ to $\omega=-\Lambda<0$ and the Poincaré algebra $iso(2,1)$ to $\omega=\Lambda=0$.

3.1. A (2+1) anti-de Sitter noncommutative spacetime

The first type of anti-de Sitter classical r-matrix coming from a DD structure is given by [20]

$$r=\eta\, J\wedge K_1-\tfrac{1}{2}\left(-J\wedge P_0-K_2\wedge P_1+K_1\wedge P_2\right) \tag{8}$$

where $\Lambda=-\eta^2=-\omega<0$. The associated cocommutator map (4) reads

$$\begin{aligned} &\delta(J)=-\eta K_2\wedge J, \qquad \delta(K_2)=0, \qquad \delta(K_1)=-\eta K_2\wedge K_1, \\ &\delta(P_0)=\left(P_1\wedge P_2+\eta P_1\wedge J-\eta^2 K_1\wedge K_2\right), \\ &\delta(P_1)=\left(P_0\wedge P_2+\eta P_0\wedge J-\eta P_2\wedge K_1+\eta^2 K_1\wedge J\right), \\ &\delta(P_2)=\left(P_1\wedge P_0+\eta P_1\wedge K_1-\eta^2 J\wedge K_2\right). \end{aligned} \tag{9}$$

From it, the first-order noncommutative spacetime (3) is defined by the dual algebra g^*, namely

$$[\hat{x}_0,\hat{x}_1]=-\hat{x}_2, \qquad [\hat{x}_0,\hat{x}_2]=\hat{x}_1, \qquad [\hat{x}_1,\hat{x}_2]=\hat{x}_0, \tag{10}$$

where the $\hat{x}_i$ coordinates are dual to the translation generators through $\langle\hat{x}_j,P_k\rangle=\delta_{kj}$. The Lie algebra (10) is just $so(2,1)$, but we have to recall that higher-order contributions can arise from the full quantum coproduct. In fact, the full noncommutative Poisson structure on the homogeneous space $\mathrm{AdS}^{2+1} = \mathrm{SO}(2,2)/\mathrm{SO}(2,1)$ is obtained through the canonical projection of the Sklyanin bracket defined by the r-matrix (8). In terms of the AdS^{2+1} group coordinates x_a this Poisson noncommutative spacetime reads [22, 25]

$$\begin{aligned} \{x_0, x_1\} &= -\frac{\tanh \eta x_2}{\eta}\,\Upsilon(x_0,x_1) = -x_2 + o[\eta], \\ \{x_0, x_2\} &= \frac{\tanh \eta x_1}{\eta}\,\Upsilon(x_0,x_1) = x_1 + o[\eta], \\ \{x_1, x_2\} &= \frac{\tan \eta x_0}{\eta}\,\Upsilon(x_0,x_1) = x_0 + o[\eta], \end{aligned} \tag{11}$$

where $\Upsilon(x_0,x_1) = \cos\eta x_0(\cos\eta x_0 \cosh\eta x_1 + \sinh\eta x_1)$. The linearisation of this Poisson bracket is just the Lie bracket (10), and higher-order terms turn out to depend on the cosmological constant. Since the 1-cocycle (9) fulfils the coisotropy condition (5), then (11) is a Poisson homogeneous space, although its quantisation is by no means trivial, as it was discussed in [22].

3.2. A (2+1) twisted κ-AdS$_\omega$ noncommutative spacetime

Another relevant AdS_ω classical r-matrix coming from a different DD structure reads [20, 21]

$$r = z(K_1 \wedge P_1 + K_2 \wedge P_2) + \vartheta J \wedge P_0, \tag{12}$$

where ϑ is a second quantum deformation parameter associated to a twist (this is the 'time-like' realization of the r-matrix, for the 'space-like' one we refer to [21, 25]). Here δ reads

$$\begin{aligned} &\delta(P_0) = \delta(J) = 0, \\ &\delta(P_1) = z(P_1 \wedge P_0 - \omega K_2 \wedge J) + \vartheta(P_0 \wedge P_2 + \omega K_1 \wedge J), \\ &\delta(P_2) = z(P_2 \wedge P_0 + \omega K_1 \wedge J) - \vartheta(P_0 \wedge P_1 - \omega K_2 \wedge J), \\ &\delta(K_1) = z(K_1 \wedge P_0 + P_2 \wedge J) + \vartheta(P_0 \wedge K_2 - P_1 \wedge J), \\ &\delta(K_2) = z(K_2 \wedge P_0 - P_1 \wedge J) - \vartheta(P_0 \wedge K_1 + P_2 \wedge J), \end{aligned}$$

from which a first-order noncommutative spacetime is obtained:

$$[\hat{x}_0, \hat{x}_1] = -z\hat{x}_1 - \vartheta\hat{x}_2, \qquad [\hat{x}_0, \hat{x}_2] = -z\hat{x}_2 + \vartheta\hat{x}_1, \qquad [\hat{x}_1, \hat{x}_2] = 0.$$

As expected, when the twist parameter ϑ vanishes, we recover the well-known κ-Minkowski spacetime [8, 9, 10, 11]. Once again, these are first-order relations, and the cosmological constant $\Lambda = -\omega = -\eta^2$ will appear in higher-order corrections generated by the full quantum deformation. Namely, the full Poisson noncommutative spacetime obtained through the Sklyanin bracket from (12) reads

$$\begin{aligned} \{x_0, x_1\} &= -z\,\frac{\tanh \eta x_1}{\eta \cosh^2 \eta x_2} - \vartheta \cosh \eta x_1 \frac{\tanh \eta x_2}{\eta}, \\ \{x_0, x_2\} &= -z\,\frac{\tanh \eta x_2}{\eta} + \vartheta\,\frac{\sinh \eta x_1}{\eta}, \qquad \{x_1, x_2\} = 0. \end{aligned} \tag{13}$$

The coisotropy condition (5) is again fulfilled, and the quantization of (13) can be performed by asuming that $[\hat{x}_1, \hat{x}_2] = 0$. Therefore, the noncommutative twisted κ-AdS_ω spacetime is

$$\begin{aligned} [\hat{x}_0, \hat{x}_1] &= -z\left(\hat{x}_1 - \frac{1}{3}\omega\hat{x}_1^3 - \omega\hat{x}_1\hat{x}_2^2\right) - \vartheta\left(\hat{x}_2 + \frac{1}{2}\omega\hat{x}_1^2\hat{x}_2 - \frac{1}{3}\omega\hat{x}_2^3\right) + \mathcal{O}(\omega^2), \\ [\hat{x}_0, \hat{x}_2] &= -z\left(\hat{x}_2 - \frac{1}{3}\omega\hat{x}_2^3\right) + \vartheta\left(\hat{x}_1 + \frac{1}{6}\omega\hat{x}_1^3\right) + \mathcal{O}(\omega^2), \end{aligned}$$

where the cosmological constant generates nonlinear deformation terms that disappear after performing the Minkowski limit $\omega \to 0$.

3.3. A (3+1) noncommutative AdS$_\omega$ spacetime

Although there is no CS approach to gravity in (3+1) dimensions, it seems natural to explore whether DD quantum deformations do exist and can be related to the (2+1) ones. The answer is affirmative for the twisted κ-AdS$_\omega$ deformation, whose DD structure has been recently presented in [23, 24]. In the basis (2), the corresponding classical r-matrix reads

$$r = z(K_1 \wedge P_1 + K_2 \wedge P_2 + K_3 \wedge P_3 + \sqrt{\omega}\, J_3 \wedge J_1) + \vartheta\, P_0 \wedge J_2,$$

where the term $\sqrt{\omega}\, J_3 \wedge J_1$ (which does not exist in (12)) implies that the rotation subalgebra becomes quantum deformed when ω does not vanish. From this expression, it can be shown that the first-order noncommutative spacetime turns out to be

$$[\hat{x}^1, \hat{x}^0] = z\,\hat{x}^1 + \vartheta\,\hat{x}^2, \qquad [\hat{x}^2, \hat{x}^0] = z\,\hat{x}^2 - \vartheta\,\hat{x}^1, \qquad [\hat{x}^3, \hat{x}^0] = z\,\hat{x}^3,$$

together with $[\hat{x}^a, \hat{x}^b] = 0$, $(a, b = 1, 2, 3)$. This algebra is not isomorphic to the (3+1) κ-Minkowski spacetime whenever ϑ is not zero (see [23]). Again, a nonlinear deformation of this algebra in terms of the cosmological constant is obtained for higher orders [54], thus being the dual counterpart of the full (3+1) Hopf algebra deformation that has been explicitly given in [24].

A remarkable problem consists in finding the explicit form of the deformed version of the two AdS$_\omega$ Casimir operators, namely the quadratic one coming from the Killing–Cartan form

$$\mathcal{C} = P_0^2 - \mathbf{P}^2 + \omega\left(\mathbf{J}^2 - \mathbf{K}^2\right),$$

and the fourth-order invariant

$$\mathcal{W} = W_0^2 - \mathbf{W}^2 + \omega\,(\mathbf{J}\cdot\mathbf{K})^2,$$

where $W_0 = \mathbf{J}\cdot\mathbf{P}$ and $W_a = -J_a P_0 + \epsilon_{abc} K_b P_c$ are the components of the Pauli–Lubanski 4-vector. As it has been shown in [24], the quantum version of the former is

$$\begin{aligned}\mathcal{C}_z = {} & \frac{2}{z^2}\left[\cosh(zP_0)\cosh(z\sqrt{\omega}J_3) - 1\right] + \omega\cosh(zP_0)(J_1^2 + J_2^2)e^{-z\sqrt{\omega}J_3} \\ & - e^{zP_0}\left(\mathbf{P}^2 + \omega\mathbf{K}^2\right)\left[\cosh(z\sqrt{\omega}J_3) + \frac{z^2\omega}{2}(J_1^2 + J_2^2)e^{-z\sqrt{\omega}J_3}\right] \\ & + 2\omega e^{zP_0}\left[\frac{\sinh(z\sqrt{\omega}J_3)}{\sqrt{\omega}}R_3 + z\left(J_1R_1 + J_2R_2 + \frac{z\sqrt{\omega}}{2}(J_1^2 + J_2^2)R_3\right)e^{-z\sqrt{\omega}J_3}\right],\end{aligned} \tag{14}$$

where $R_a = \epsilon_{abc} K_b P_c$. Indeed, the κ-Poincaré Casimir is obtained as the $\omega \to 0$ limit of (14),

$$\mathcal{C}_z = \frac{2}{z^2}\left[\cosh(zP_0) - 1\right] - e^{zP_0}\mathbf{P}^2 = \frac{4}{z^2}\sinh^2(zP_0/2) - e^{zP_0}\mathbf{P}^2. \tag{15}$$

Therefore, the κ-Minkowski deformed dispersion relation coming from (15) turns out to be significantly modified when the cosmological constant does not vanish. This raises the question concerning the physical meaning of the quantum AdS$_\omega$ Casimir (14) in a cosmological context. Also, the classification of DD structures for the (3+1) (A)dS algebras is a challenging open problem, and the corresponding Poincaré case has been recently solved in [54]. Finally, we mention that the characterization of (3+1) (A)dS Poisson homogeneous spaces is currently under investigation by following the approach presented in [25].

Acknowledgments

This work has been partially supported by Ministerio de Economía y Competitividad (MINECO, Spain) under grants MTM2013-43820-P and MTM2016-79639-P, by Junta de Castilla y León (Spain) under grants BU278U14 and VA057U16 and by the Action MP1405 QSPACE from the European Cooperation in Science and Technology (COST). I.G-S. acknowledges a predoctoral grant from Junta de Castilla y León and the European Social Fund.

IOP Conf. Series: Journal of Physics: Conf. Series **880** (2017) 012023 doi:10.1088/1742-6596/880/1/012023

References

[1] Snyder H S 1947 *Phys. Rev.* **71** 38
[2] Podleś P and Woronowicz S L 1990 *Commun. Math. Phys.* **130** 381
[3] Doplicher S, Fredenhagen K and Roberts J E 1995 *Commun. Math. Phys.* **172** 187
[4] Szabo R 2003 *Phys. Rept.* **378** 207
[5] Heckman J J and Verlinde H 2015 *Nucl. Phys. B* **894** 58
[6] Garay L J 1995 *Int. J. Mod. Phys.* A **10** 145
[7] Maggiore M 1993 *Phys. Lett.* B **319** 83
[8] Maslanka P 1993 *J. Phys. A: Math. Gen.* **26** L1251
[9] Majid S and Ruegg H 1994 *Phys. Lett.* B **334** 348
[10] Zakrzewski S 1994 *J. Phys. A: Math. Gen.* **27** 2075
[11] Lukierski J, Nowicki A, Ruegg H and Tolstoy V N 1991 *Phys. Lett.* B **264** 331
[12] Ballesteros A, Herranz F J, del Olmo M A and Santander M 1995 *Phys. Lett. B* **351** 137
[13] Ballesteros A, Bruno N R and Herranz F J 2003 *Phys. Lett. B* **574** 276; ibid arXiv:hep-th/0401244
[14] Amelino-Camelia G, Smolin L and Starodubtsev A 2004 *Class. Quantum Grav.* **21** 3095
[15] Marcianó A, Amelino-Camelia G, Bruno N R, Gubitosi G, Mandanici G and Melchiorri A 2010 *J. Cosmol. Astropart. Phys. B* **06** 030
[16] Amelino-Camelia G, Barcaroli L, D'Amico G, Loret N and Rosati G 2016 *Phys. Lett. B* **761** 318
[17] Amelino-Camelia G, Gubitosi G, Palmisano G 2016 *Int. J. Mod. Phys. D* **25** 1650027
[18] Barcaroli L, Gubitosi G 2016 *Phys. Rev. D* **93** 124063
[19] Ballesteros A, Herranz F J and Meusburger C 2010 *Phys. Lett.* B **687** 375
[20] Ballesteros A, Herranz F J and Meusburger C 2013 *Class. Quantum Grav.* **30** 155012
[21] Ballesteros A, Herranz F J, Meusburger C and Naranjo P 2014 *Sym. Integ. Geom.: Meth. Applic.* **10** 052
[22] Ballesteros A, Herranz F J and Meusburger C 2014 *Phys. Lett. B* **732** 201
[23] Ballesteros A, Herranz F J and Naranjo P 2015 *Phys. Lett. B* **746** (2015) 37
[24] Ballesteros A, Herranz F J, Musso F and Naranjo P 2017 *Phys. Lett. B* **766** 205
[25] Ballesteros A, Meusburger C and Naranjo P *Preprint* arXiv:1701.04902
[26] Kulish P P and Reshetikhin N 1981 *Proc. Steklov Math. Inst.* **101** 101
[27] Drinfel'd V G 1987 *Quantum Groups* in *Proc. Int. Congress of Math. (Berkeley 1986)* ed A V Gleason (Providence RI: AMS) p 798
[28] Jimbo M 1985 *Lett. Math. Phys.* **10** 63
[29] Takhtajan L A 1990 *Introduction to quantum group and integrable massive models of Quantum Field Theory*, eds Mo-Lin Ge and Bao-Heng Zao (Singapore: World Scientific) p 69
[30] Reshetikhin N Y, Takhtadzhyan L A and Faddeev L D 1990 *Leningrad Math. J.* **1** 193
[31] Chari V and Pressley A 1994 *A Guide to Quantum Groups* (Cambridge: Cambridge University Press)
[32] Majid S 1995 *Foundations of Quantum Group Theory* (Cambridge: Cambridge University Press)
[33] Drinfel'd V G 1993 *Theor. Math. Phys.* **95** 524
[34] Dijkhuizen M S and Koornwinder T H 1994 *Geom. Dedicata* **52** 291
[35] Zakrzewszki S 1995 in *Quantum Groups* eds Lukierski J, Popowicz Z and Sobczyk J (Warszaw: PWN) p 629
[36] Reyman A G 1996 in *Quantum Groups and its applications in Physics* eds Castellani L and Wess J (Amsterdam: IOS) p 407
[37] Ciccoli N 1996 *J. Phys. A: Math. Gen.* **29** 1487
[38] Alekseev A Y and Malkin A Z 1995 *Commun. Math. Phys.* **169** 99
[39] Fock V V and Rosly A A 1992 Preprint ITEP-72-92; ibid 1999 *Am. Math. Soc. Transl.* **191** 67
[40] Achucarro A and Townsend P K 1986 *Phys. Lett. B* **180** 89
[41] Witten E 1988 *Nucl. Phys. B* **311** 46
[42] Meusburger C and Schroers B J 2008 *J. Math. Phys.* **49** 083510
[43] Meusburger C and Schroers B J 2009 *Nucl. Phys.* B **806** 462
[44] Papageorgiou G, Schroers B J 2009 *JHEP* **11** 009; ibid 2010 *JHEP* **11** 020
[45] Meusburger C and Schönfeld T 2011 *Class. Quantum Grav.* **28** 125008
[46] Amelino-Camelia G, Ellis J, Mavromatos N E, Nanopoulos D V and Sarkar S 1998 *Nature* **39** 763
[47] Amelino-Camelia G 2010 *Symmetry* **2** 230; ibid 2013 *Living Rev. Rel.* **16** 5
[48] Mattingly D 2005 *Living Rev. Rel.* **8** 5
[49] Drinfel'd V G 1983 *Soviet Math. Dokl.* **27** 68
[50] Ballesteros A, Gromov N A, Herranz F J, del Olmo M A and Santander M 1995 *J. Math. Phys.* **36** 5916
[51] Zakrzewski S 1997 *Comm. Math. Phys.* **185** 285
[52] Ballesteros A, Herranz F J and Musso F 2014 *J. Phys.: Conf. Series* **532** 012002
[53] Borowiec A, Lukierski J and Tolstoy V N 2016 *Phys. Lett. B* **754** 176
[54] Ballesteros A, Gutiérrez-Sagredo I and Herranz F J 2017, in preparation

IOP Conf. Series: Journal of Physics: Conf. Series **880** (2017) 012024 doi:10.1088/1742-6596/880/1/012024

Expanding space-time and variable vacuum energy

Claudio Parmeggiani
University of Milano-Bicocca - Italy (UNIMIB)
E-mail: c.parmeggiani@campus.unimib.it

Abstract. The paper describes a cosmological model which contemplates the presence of a vacuum energy varying, very slightly (now), with time. The constant part of the vacuum energy generated, some 6 Gyr ago, a deceleration/acceleration transition of the metric expansion; so now, in an aged Universe, the expansion is inexorably accelerating. The vacuum energy varying part is instead assumed to be eventually responsible of an acceleration/deceleration transition, which occurred about 14 Gyr ago; this transition has a dynamic origin: it is a consequence of the general relativistic Einstein-Friedmann equations. Moreover, the vacuum energy (constant and variable) is here related to the zero-point energy of some quantum fields (scalar, vector, or spinor); these fields are necessarily described in a general relativistic way: their structure depends on the space-time metric, typically non-flat. More precisely, the commutators of the (quantum field) creation/annihilation operators are here assumed to depend on the local value of the space-time metric tensor (and eventually of its curvature); furthermore, these commutators rapidly decrease for high momentum values and they reduce to the standard ones for a flat metric. In this way, the theory is "gravitationally" regularized; in particular, the zero-point (vacuum) energy density has a well defined value and, for a non static metric, depends on the (cosmic) time. Note that this varying vacuum energy can be negative (Fermi fields) and that a change of its sign typically leads to a minimum for the metric expansion factor (a "bounce").

1. Introduction. Energy-momentum tensors

On the space-time four-dimensional manifold are assigned a metric/gravitational tensor $\boldsymbol{g}$ and an energy-momentum (macroscopic) tensor $\boldsymbol{T}$; their local expressions, in a Friedmann *synchronous* chart $(t, r_1, r_2, r_3; U) = (t, \mathbf{r}; U)$ are ($i, j = 1, 2, 3$; U is an open subset of the space-time manifold):

$$g_{00}(t, \mathbf{r}) = c^2(t), \quad g_{ij}(t, \mathbf{r}) = -a^2(t)\gamma_{ij}(\mathbf{r}) \tag{1a}$$

$$T_{00}(t, \mathbf{r}) = \rho(t), \quad T_{ij}(t, \mathbf{r}) = -\sigma(t)a^2(t)\gamma_{ij}(\mathbf{r}) \tag{1b}$$

while $g_{0j}(t, \mathbf{r}) = T_{0j}(t, \mathbf{r}) = 0$; a (the *expansion factor*), g_{00} and T_{00} are always > 0. The curvature of the 3-dimensional metric expressed by $a^2(t)\gamma_{ij}$ is assumed (isotropy) to depend only on t; this "spatial" curvature will be denoted by $K(t)$ $(= K_0/a^2(t), -\infty < K_0 < +\infty)$. The purely spatial metric can be diagonalized: $\gamma_{ij}(\mathbf{r}) = \delta_{ij}/b^2(\mathbf{r})$, where $b(\mathbf{r}) = 1 + \frac{K_0}{4}\mathbf{r}^2$. Usually (and here) the chart is chosen in such a way that $c = 1$, but other options are available (for example $\rho = 1$); in any case, when $c = 1$, ρ can be identified to the (proper) energy density. See Friedmann [4] (Annex 1) or, for example, Einstein [2], Hawking and Ellis [6], Tolman [9], Weinberg [10].

IOP Conf. Series: Journal of Physics: Conf. Series **880** (2017) 012024 doi:10.1088/1742-6596/880/1/012024

The *energy conservation* law, $a(t)\rho'(t) + 3a'(t)(\rho(t) - \sigma(t)) = 0$ leads, assuming (in a first approximation) that σ is *constant*, to:

$$\rho(t) = \sigma + \frac{\mu}{a^3(t)}; \tag{2}$$

$\mu \; (= (\rho - \sigma)\, a^3)$ is an (universal) constant. If σ is > 0, ρ is *always* $> \sigma$ or $< \sigma$ according to the sign of μ, but when $\mu < 0$ the (weak) *energy condition* $\rho \geq 0$, $\rho \geq \sigma$ ([6], Chapter 4) is badly violated. For a $\sigma < 0$, μ has to be > 0 and a is bounded ($a^3 < \mu/|\sigma|$). Note that, because $\operatorname{sign}\rho' = -\operatorname{sign}(a'(\rho - \sigma))$, ρ increase (decrease) when a decrease (increase), if $\rho > \sigma$.

We are primarily interested to an expansion factor *strictly increasing* on some open, real line interval $]t_M, t_N[$ (and $t_M < t_0 < t_N$, if t_0 is the *present time*); typically $t_N = +\infty$ and $\lim_{t\to+\infty} a(t) = +\infty$ (so $K(+\infty) = 0$). Instead t_M can be $> -\infty$; in this case or $a_M = a(t_M)$ is a *minimum* of a and $a(t)$ decrease for $t < t_M$ or $a(t)$ is defined only for $t > t_M$ and $\lim_{t\to t_M^+} a(t) = 0$. In any case, for $t_M < t < t_N$, ρ can be expressed as a function of a, for example:

$$\rho = F\left(\tfrac{1}{a}\right) \quad \left(= \sum\nolimits_{n\geq 0} \frac{\rho^{(n)}}{a^n}\right) \tag{3}$$

that is $\rho = F \circ 1/a$; F is some differentiable (eventually analytical) function. Consequently $a\rho' = -(a'/a)\, F'\left(\tfrac{1}{a}\right) = -3a'(\rho - \sigma)$ and

$$\sigma = F\left(\tfrac{1}{a}\right) - \tfrac{1}{3a} F'\left(\tfrac{1}{a}\right) \quad \left(= \sum\nolimits_{n\geq 0} \left(1 - \tfrac{n}{3}\right) \frac{\rho^{(n)}}{a^n}\right). \tag{4}$$

Note that, for σ, the coefficient of a^{-3} is always $= 0$ and the one of a^{-4} is $= -\frac{1}{3}\rho^{(4)}$; this fact is, in our context, simply a consequence of the tridimensionality of the (ordinary) space. The energy condition $\rho \geq 0$, $\rho \geq \sigma$ will hold when $\sum_{n\geq 0} \frac{\rho^{(n)}}{a^n}$ and $\sum_{n\geq 0} n\frac{\rho^{(n)}}{a^n}$ are both ≥ 0.

If $\rho^{(ref)}$ is some constant, "reference" energy density and $\Omega^{(n)} := \rho^{(n)}/\rho^{(ref)}$,

$$\rho = \rho^{(ref)} \sum\nolimits_{n\geq 0} \frac{\Omega^{(n)}}{a^n}; \tag{5}$$

the traditional choice for $\rho^{(ref)}$ is $\frac{1}{\kappa}(H_0)^2$, the *critical density*, $\rho^{(crit)}$; $\kappa = \frac{8}{3}\pi G_{Newton}$, $\kappa\rho^{(n)} = (H_0)^2\,\Omega^{(n)}$, $H := a'/a$ is the Hubble-Lemaitre function ([4], Annex 1), $H_0 = H(t_0)$ and t_0 is always the present time. A better choice for $\rho^{(ref)}$, not related to some particular value of t, would be simply $\rho^{(0)}$ if $\rho^{(0)}$ is > 0. Observe that here $\Omega^{(3)}$ and $\Omega^{(4)}$ are only the coefficient of a^{-3} and a^{-4} in the energy density expansion: they do not necessary correspond to $\Omega^{(mat)}$ (non-relativistic matter) and to $\Omega^{(rad)}$ (relativistic matter, radiation): $\rho \; (= \rho^{(mat)} + \rho^{(rad)} + \rho^{(vac)})$ is the *total* energy density, from matter, radiation and some other sources that we call collectively *vacuum*.

2. Dynamics of expansion

The Einstein-Friedmann equation $a''/a = -\frac{\kappa}{2}(\rho - 3\sigma)$ can now be written, taking into account the energy conservation law, as

$$a'' = \kappa\left(aF\left(\tfrac{1}{a}\right) - \tfrac{1}{2}F'\left(\tfrac{1}{a}\right)\right) = H_0^2 \sum\nolimits_{n\neq 2} \left(1 - \tfrac{n}{2}\right)\Omega^{(n)} a^{1-n}; \tag{6}$$

this *ordinary*, *semi-linear* differential equation, supplemented by some initial conditions (for example $a(t_0) = 1$, $a'(t_0) = H_0$, if t_0 is the present time) has an unique solution, defined on some open (eventually unbounded, Section 1) interval of the real line. See, for some numerical example, the Section 5.

The "other" Einstein-Friedmann equation, $(a'/a)^2 + K_0/a^2 = \kappa\rho$, can be written as

$$(a')^2 = \kappa a^2 F\left(\tfrac{1}{a}\right) - K_0 = H_0^2 \sum\nolimits_{n\neq 2} \Omega^{(n)} a^{2-n} + \left(H_0^2\Omega^{(2)} - K_0\right); \quad (7)$$

it is, evidently, a *first integral* of the above, second order differential equation; the integration constant is related to the value of the spatial curvature K $(= K_0/a^2)$ at time t_0, if $a(t_0) = 1$. As expected, $\sum_{n\geq 0} \Omega^{(n)} = 1 + K_0/H_0^2$.

Assuming that only $\rho^{(0)}$, $\rho^{(3)}$ and $\rho^{(4)}$ are $\neq 0$, we arrive to the "classical" cosmological models, *old* (when $\rho^{(0)}$ is $= 0$ or < 0, see [10]) and *new* (when $\rho^{(0)}$ is > 0, see [12]); $\rho^{(3)}$ and $\rho^{(4)}$are always > 0. In the following we will consider a more general model where, eventually, all the $\rho^{(n)}$ are $\neq 0$, that is when we are in presence of a *variable* vacuum energy density. Note that if $\rho^{(vac)}$ *diverge negatively* faster than $\rho^{(mat)} + \rho^{(rad)}$ when $a \to a_L$ (for some $a_L \geq 0$) then $a(t)$ is typically defined for $-\infty < t < +\infty$ and has here a global *minimum* (and eventually two *flex*, between the minimum and the present time, see the Section 5).

3. Digression. Quantum fields in a connection space

Now the quantum fields structure depends on the space-time connection, typically non-flat.

3.1. Affine connection spaces

(X, ∇) is an *affine connection space*: X is a n-dimensional, orientable manifold, ∇ is its *absolute derivative* (here assumed to be torsion-less); $\mathcal{T}_x$ and $\mathcal{T}'_x$ (the dual of $\mathcal{T}_x$) are respectively the tangent and the cotangent vector spaces of X, at $x \in X$.

U is an open, *geodetically convex*, subset of X: for every couple of points of U, x and y, there is *one and only one* (fixed a parametrization) ∇-geodesic joining x to y and lying entirely in U; $v(x, y)$ is the vector of $\mathcal{T}_x$ tangent to the geodesic at x and such that $v(x, x) = 0_x$, $\exp_x(0_x) = x$ and $\exp_x(v(x, y)) = y$ ($\exp_x$ is the canonical geodesic map from $\mathcal{T}_x$ to X, generally defined only in a neighborhood of 0_x, V_x; see [6], Section 2.5 or any differential geometry textbook, for example Chern, Chen and Lam [1], Section 5.2). If $(e_1(x), \ldots, e_n(x))$ is a basis of $\mathcal{T}_x$ and $v(x, y) = \sum_i y^i e_i(x)$, the $y^1, \ldots, y^n$ are the so called (geodetic) *normal coordinates* of y, based on x. The image by $\exp_x^{-1}$ of a geodetic path crossing x is a (portion of a) straight line of $\mathcal{T}_x$, crossing 0_x.

In an *affine space* $(E, \mathcal{V})$ (for example in an Euclidean space or in a Galilean space-time) the vectors $v(x, y)$ joining points of E (a generic set) belong to a *fixed*, n-dimensional vector space, $\mathcal{V}$; they are usually denoted by $y - x$ (or $\overrightarrow{xy}$) and $\exp_x(v)$ is denoted by $x + v$; then $x + (v + w) = (x + v) + w$ while, fixed an $x \in E$, $v \mapsto x + v$ is a bijection from $\mathcal{V}$ to E. The canonical connection induced by D (the "ordinary" derivative) is now globally flat, but, on E, it is obviously possible to assign other, non-flat, derivatives. The more common space-time models are based on affine connection spaces, but sometimes are also considered affine spaces, with two (one flat, the other curved) physically recognizable connections. See Gupta [5] or Feynman [3]; it is possible, also in this case, to build quantum fields, perhaps more easily.

3.2. Quantum fields

Fixed a geodetically convex set U of (X, ∇) (the "observable" space-time), we want to build *local quantum fields*, defined in U. Now $\mathbf{A}$ is a $\mathbb{C}$-algebra, normed or semi-normed (and complete) with an involution $(^*)$ and an identity $(\mathbf{I})$; $\mathbf{a}_x$ and $\mathbf{a}_x^*$, the *local* annihilation and creation "operators", are, for every $x \in U$, $\mathbf{A}$-valued (Radon) measures defined on $\mathcal{T}'_x$: this means that $f \mapsto \mathbf{a}_x(f)$ and $f \mapsto \mathbf{a}_x^*(f)$ $\left(:= \left(\mathbf{a}_x\left(\overline{f}\right)\right)^*\right)$ are $\mathbb{C}$-linear and continuos maps, from a space of complex test functions $\mathcal{D}_x$ (functions defined on $\mathcal{T}'_x$) to $\mathbf{A}$; $\mathbf{a}_x(f)$ is usually written as $\int f(p).d\mathbf{a}_x(p)$ (or also $\int f(p)\,\mathbf{a}(x,p).dp$, with some abuse of notation). Moreover we have (Bose statistics):

$$[\mathbf{a}_x(g),\ (\mathbf{a}_x(f))^*]_- = B_x(f,g).\mathbf{I} = \int \overline{f(p)}g(p).d\mu_x(p).\mathbf{I} \tag{8}$$

where μ_x is some real and positive measure, generally depending on the connection of X (perhaps on its *Ricci tensor*). The sesquilinear form B_x assigns a pre-Hilbert structure on $\mathcal{D}_x$ so, by completion, it is possible to build a Hilbert space $\mathcal{H}_x$ (and its Fock space).

The scalar and neutral quantum field φ ($\varphi^* = \varphi$) is then defined by (here and in the following $h_{Plank} = 2\pi$):

$$\varphi_x(y) := \int \left(e^{-ip.v(x,y)}.d\mathbf{a}_x(p) + e^{+ip.v(x,y)}.d\mathbf{a}^*_x(p)\right) \tag{9}$$

or, setting $\phi(x,v) = \varphi_x(\exp_x(v))$,

$$\phi(x,v) := \int \left(e^{-ip.v}.d\mathbf{a}_x(p) + e^{+ip.v}.d\mathbf{a}^*_x(p)\right) = \mathbf{a}_x(\chi_v) + \mathbf{a}^*_x(\overline{\chi_v}). \tag{10}$$

In the above x and $y \in U$, $v \in V_x \subset \mathcal{T}_x$, $p \in \mathcal{T}'_x$; $\chi_v(p) := e^{-ip.v}$; $p.v(x,y) = \sum_i p_i.v^i(x,y)$ ($= \sum_i p_i y^i$, if the $y^1,\ldots,y^n$ are the normal coordinates of y, based on x); the measures $\mathbf{a}_x$ and $\mathbf{a}^*_x$ (as μ_x) are implicitly assumed to be *prolongable* to a space of continuos and bounded (in fact polinomially increasing) test functions. The $(y^1,\ldots,y^n) \mapsto \Phi_x(y^1,\ldots,y^n)$ $(= \varphi_x(y))$ are differentiable, $\mathbf{A}$-valued ordinary functions, defined on some open set of $\mathbb{R}^n$.

It is now possible to define a (curved) version of the *Pauli-Jordan* function (which is a true function):

$$\Delta(y,z) := \tfrac{i}{2}[\varphi_x(y), \varphi_x(z)]_- \tag{11}$$
$$= \int \sin(p.(v(x,y) - v(x,z))).d\mu_x(p) = \int \sin\left(\textstyle\sum_i p_i\left(y^i - z^i\right)\right).dm_x\left(p^1,\ldots,p^n\right).$$

In an *affine* space-time, equipped with only the canonical connection (see the previous Subsection), $v(x,y) - v(x,z) = (y-x) - (z-x) = y - z$ and μ_x do not depends on x; therefore $\Delta(y,z) = \int_{\mathcal{V}'} \sin(p.(y-z)).d\mu(p)$ depends exclusively on the vector joining y and z (but it is generally "singular"; see, for example, Pauli [7] or Weinberg [11]).

3.3. Metric tensors

In all physically interesting cases the manifold X is 4-dimensional and Riemannian, $\boldsymbol{g}$ is its metric tensor (of signature -2, as in Section 1) and $(X,\boldsymbol{g})$ is *time oriented* by $\boldsymbol{g}$. The measure μ_x is now concentrated on the *positive* (with respect to the local time orientation) M-mass hyperboloid of $\boldsymbol{g}_x$:

$$d\mu_x(p) = C_x(p).d\mu_M(p) \quad \left(d\mu_M(p) = \delta^{(+)}_{p^2_x - M^2}\right); \tag{12}$$

here p^2_x stands for $p.\boldsymbol{g}^{-1}_x.p$ $(= \sum_{ij} g^{ij}(x)p_i p_j)$, M is some positive parameter characterizing the field and C_x is a positive, rapidly decreasing (or also of compact support) continuos function. Chosen a $\boldsymbol{g}_x$-orthogonal and positive basis on $\mathcal{T}_x$ (and $\mathcal{T}'_x$), if $(\varepsilon, p_1, p_2, p_3) = (\varepsilon, \mathbf{p})$ are the components of p, then $\varepsilon > 0$, $\varepsilon^2 - \mathbf{p}^2 = M^2$ and $d\mu_M(p)$ can by replaced by $\frac{1}{2}\left(\mathbf{p}^2 + M^2\right)^{-1/2} d\mathbf{p}$.

Now it is quite reasonable to assume that the above $C_x(p)$ is built from $\boldsymbol{g}$ and from the Ricci tensor of the connection, $\boldsymbol{R}$: this is apparently the simplest, non trivial choice. And it is usually assumed that ∇ is the Levi-Civita connection of $\boldsymbol{g}$ (but ∇ and $\boldsymbol{g}$ can be totally or partially independent: ∇ can be, for example, the connection of $k^2\boldsymbol{g}$, where k is some scalar field). Anyway in the following we shall simply assume that ∇ is the connection of $\boldsymbol{g}$ and that C_x depends only on its Ricci tensor, $\boldsymbol{R}$; for example

$$C_x(p) = \mathcal{N}(x,p)\exp\left(-\frac{\lambda}{2M_P^2}\widehat{\boldsymbol{R}}_x(p,p)\right). \tag{13}$$

Here $\widehat{\boldsymbol{R}} := \boldsymbol{g}^{-1}.\boldsymbol{R}.\boldsymbol{g}^{-1}$ (so $\widehat{\boldsymbol{R}}(p,p) = \sum_{ijkl} R_{kl} g^{ik} g^{lj} p_i p_j$), M_P is a "very large" mass (the Plank mass, to say) and λ is some universal parameter, eventually dependent on the statistics (Bose or Fermi) of the quantum field. Obviously $\lambda\widehat{\boldsymbol{R}}_x(p,p)$ has to be always > 0 (when $p_x^2 = M^2$); $\mathcal{N}(x,p)$ is a *normalization factor* fixing the value of $C_x(p)$ on some chosen space-time point; or at a fixed time, in an "spatially homogeneous" space-time, as in Section 1. In this case we have $R_{00} = -3\left(H^2 + H'\right)$, $R_{ij} = \left(H' + 3H^2 + 2K\right)(a/b)^2\delta_{ij}$ and $R_{0j} = 0$ $(i,j = 1,2,3)$; therefore, on the M-mass hyperboloid

$$\tfrac{\lambda}{2}\widehat{\boldsymbol{R}}_x(p,p) = -\tfrac{3\lambda}{2}\left(H^2 + H'\right)M^2 + \lambda\left(K - H'\right)\mathbf{p}^2; \tag{14}$$

remember that $H(t) = a'(t)/a(t)$ and that $K(t) = K_0/a^2(t)$; $a''/a = H' + H^2$. Now $\lambda(K(t) - H'(t))$ has to be > 0, for every t $(> t_M)$: accordingly the quantum fields are "gravitationally" regularized (nothing diverge).

4. Vacuum energy density

The *zero-point* (vacuum) energy density of some (scalar, vector or spinor) quantum field ϕ, in a spatially homogeneous and *non-static* space-time, can now be written as (see, for example, Pauli [7], Section 9 or Zeldovich [13], Appendix 8, for the "flat" case):

$$\rho_\phi^{(vac)}(t) = \int (p_0)^2 C_x(p)\,.d\mu_{M_\phi}(p) \quad \text{(by definition)} \tag{15a}$$

$$= \mathcal{N}_\phi^{(0)}(t, M_C)\int_{\mathbb{R}^3}\exp\left(-\theta(t)\frac{\mathbf{p}^2}{M_C^2}\right)\sqrt{\mathbf{p}^2 + M_\phi^2}.d\mathbf{p} \quad \text{(Section 3)} \tag{15b}$$

$$\approx M_C^4\frac{\mathcal{N}_\phi^{(2)}}{\theta^2(t)} + M_C^2 M_\phi^2\frac{\mathcal{N}_\phi^{(1)}}{\theta(t)} + \ \ldots \quad (\text{for } M_C \to +\infty). \tag{15c}$$

In the above: the dimensionless numerical factors $\mathcal{N}_\phi^{(0)}, \mathcal{N}_\phi^{(1)}, \ldots$ depend on the field's mass and type (spin and statistics, they are *negative* for a spinor field); the *positive* function θ $\left(:= \frac{\lambda}{2}(M_C/M_P)^2(K - H')\right)$ depends on the space-time metric and it is normalized, fixing the value of M_C, in such a way that $\theta(t) \to 1$ when $t \to +\infty$ (or, more cautiously, in such a way that $\theta_0 = \theta(t_0) = 1$); M_C is again a cutoff mass and M_ϕ is the mass of the field ϕ. The next term in the asymptotic expansion ($M_C \to +\infty$) is a logarithmic one, here irrelevant, but it is the *main term* in the self-mass or self-charge calculations.

Consequently for the total vacuum energy, considering all the fields and retaining only the first two terms in the asymptotic expansion, we have ($\rho^{(vac)} := \sum_\phi \rho_\phi^{(vac)}$, $\mathcal{N}^{(2)} := \sum_\phi \mathcal{N}_\phi^{(2)}$, $\mathcal{N}^{(1)} := \sum_\phi (M_\phi/M_C)^2\mathcal{N}_\phi^{(1)}$, $M^{(crit)} := \sqrt[4]{\rho^{(crit)}} \cong 2.46 \cdot 10^{-12}$ *Gev*, [8], Astrophysical Constants.):

$$\frac{\rho^{(vac)}(t)}{\rho^{(crit)}} \cong \left(M_C/M^{(crit)}\right)^4\left(\frac{\mathcal{N}^{(1)}}{\theta(t)} + \frac{\mathcal{N}^{(2)}}{\theta^2(t)}\right) = \Omega^{(0)}\left(\frac{\eta^{(1)}}{\theta(t)} + \frac{\eta^{(2)}}{\theta^2(t)}\right). \tag{16}$$

When $t = +\infty$, $\rho^{(vac)} = \rho^{(0)}$ and $\theta = 1$ so $1 \cong \eta^{(1)} + \eta^{(2)}$ (or, having set $\theta(t_0) = 1$, $\frac{\rho^{(vac)}(t_0)}{\rho^{(0)}} \cong \eta^{(1)} + \eta^{(2)}$). Note that the signs of $\eta^{(1)}$ and $\eta^{(2)}$ can be different: this is, in fact, the more interesting case.

IOP Conf. Series: Journal of Physics: Conf. Series **880** (2017) 012024 doi:10.1088/1742-6596/880/1/012024

For a spatially homogeneous and *static* space-time ([2], end of Chapter 3; [9], Chapter 10) the spatial curvature $K(t)$ is a *constant*, as the vacuum energy density (which can be positive or negative, but reasonably $\neq 0$); unfortunately, in this case, the space-time "singularity" ($\lim_{t\to t_M^+} a(t) = 0$) seems unavoidable (see the Section 5). For a "generalized" de Sitter metric (the expansion parameter L_M is > 0 but $\neq K_M = K(t_M)$), $a(t) = a_M \cosh\sqrt{L_M}\,(t - t_M)$, $H^2/L_M = 1 - (a_M/a)^2$, $H'/L_M = (a_M/a)^2$; in this case $\lambda(K_M - L_M)$ has to be > 0 and, setting $\theta(t_0) = 1$, $\theta(t) = 1/a^2(t)$, physically quite unacceptable. In the Section 5 we shall consider the more general case, where

$$\theta(t) \cong 1 - \frac{\zeta}{a^2(t)}, \tag{17}$$

the parameter ζ is now > 0 but $\ll 1$, $a(t)$ will be $> \sqrt{\zeta}$.

5. Numerical Examples

It is convenient to measure the time in term of $1/H_0$ ($= 14.5\ Gyr$): in this section, we shall set $H_0 = a'(t_0) = 1$ (t_0 is always the present time and $a(t_0) = 1$). Now the $\Omega^{(n)}$ are related to K_0 and to a_0'' ($= a''(t_0)$, the *acceleration parameter*) by $\sum_{n\geq 0}\Omega^{(n)} = 1 + K_0$ and by $\sum_{n\geq 0}\left(1 - \frac{n}{2}\right)\Omega^{(n)} = a_0''$ (Section 2) hence, if only $\Omega^{(0)}$ and $\Omega^{(3)}$ are significantly $\neq 0$ and $K_0 \cong 0$,

$$\Omega^{(0)} = \tfrac{1}{3}\left(1 + 2a_0'' + K_0\right) \cong \tfrac{1}{3} + \tfrac{2}{3}a_0'' \tag{18a}$$

$$\Omega^{(3)} = \tfrac{2}{3}\left(1 - a_0'' + K_0\right) \cong \tfrac{2}{3} - \tfrac{2}{3}a_0''. \tag{18b}$$

The *curvature parameter* ([12], Section 1.5) is defined by $\frac{|K(t)|}{H^2(t)} = \frac{|K_0|}{(a'(t))^2}$, therefore, if $K_0 \neq 0$, $\frac{|K_0|}{(a')^2} = \left|1 - \frac{a^2}{K_0}\sum_{n\geq 0}\frac{\Omega^{(n)}}{a^n}\right|^{-1}$; then if $\Omega^{(0)} > 0$, $a'(t) \to +\infty$ and $|K_0|/(a')^2 \to 0$, for $t \to +\infty$. When $a'' = 0$ ($t = t_F$) the expansion factor has a *flex* and the curvature parameter has a maximum.

First at all we shall consider, as a reference, the case where the vacuum energy is constant and $\Omega^{(0)} = 0.685$, $\Omega^{(3)} = 0.315$ and $\Omega^{(4)} = 0.002$. Therefore $K_0 = 0.002$, R_0 ($:= 1/\sqrt{|K_0|}$, if $K_0 \neq 0$) $= 22.4$ ($= 3.25\cdot 10^{11}\ ly$), $a_0'' = 0.525$ and

$$t_0 - t_M = 0.944\ (13.70\ Gyr), \quad t_F - t_M = 0.529\ (7.67\ Gyr); \tag{19}$$

$a_F = 0.617$ while the curvature parameter at t_F is $= 0.0026$ ($= 0.002$ at t_0); $a(t) \to 0$ when $t \to t_M$, $t > t_M$. A smaller value of $\Omega^{(4)}$ increases the value of $t_0 - t_M$: if, for example, $\Omega^{(4)} = 0.001$, then $t_0 - t_M = 0.947$ ($13.73\ Gyr$).

But if the vacuum energy varies with time, for example if (see the Section 4),

$$\eta^{(1)} = 1.95, \quad \eta^{(2)} = -0.95, \quad \zeta = 0.001$$

things change dramatically: the $(t, a(t))$ curve has a *minimum* at t_M and *two flex* (between t_M and t_0), at t_F (as in the "classical" case) and at t_I (I as *inception*), then

$$t_0 - t_M = 0.939\ (13.62\ Gyr), \quad t_F - t_M = 0.523\ (7.58\ Gyr), \quad t_I - t_M = 3.9\ Myr. \tag{20}$$

Quite interesting is the behavior of the components of the energy density: the classical one ($\rho^{(mat)} + \rho^{(rad)}$) has a *maximum* at t_M then rapidly decrease, always respecting the energy conditions; the vacuum one ($\rho^{(vac)}$) has a *minimum* at t_M (where it is < 0) then rapidly increase and became > 0, never respecting the energy conditions; the total one (ρ) has a *minimum* at t_M (where $a' = 0$ as ρ' and $\rho_M \cong 0$) then increase, reaches a maximum (near t_M) and finally regularly decrease. Obviously everything happens because $\eta^{(1)} > 0$ and $\eta^{(2)} < 0$; generally the behavior of a and ρ is strongly influenced by the values of the parameters: for example if $\eta^{(1)}$ is too large the flex are lost and ρ_M is < 0; the flex are also lost if ζ is not enough small (and now the maximums and minimums are less sharp); for more, a realistic value of ζ (for our Universe) is perhaps much smaller than 0.001.

In any case, we can assume that different values of $\eta^{(1)}$, $\eta^{(2)}$ and ζ describe the contraction/expansion of some different massive object.

References

[1] Chern S S, Chen W H and Lam K S 2000 *Lectures on Differential Geometry* (Singapore: World Scientific)
[2] Einstein A 1955 *The Meaning of Relativity* (Princeton University Press)
[3] Feynman R P 1995 *Feynman Lectures on Gravitation* (Reading, Massachusetts: Addison-Wesley Publishing Company)
[4] Friedmann A and Lemaitre G 1997 *Essais de Cosmologie* (Paris: Editions du Seuil)
[5] Gupta S N 1954 Gravitation and electromagnetism *Physical Review* **96** 1683-1685
[6] Hawking S W and Ellis G F R 1973 *The Large Scale Structure of Space-time* (Cambridge University Press)
[7] Pauli W 1977 *Pauli Lectures on Physics: Selected Topics on Field Quantization* (Cambridge, Massachusetts: The MIT Press)
[8] Olive K A et al. (Particle Data Group) 2014 *Chinese Physics* C **38** 090001
[9] Tolman R C 1987 *Relativity, Thermodynamics and Cosmology* (New York: Dover Publications)
[10] Weinberg S 1972 *Gravitation and Cosmology: Principles and Applications of the General Theory of Relativity* (New York: John Wiley and Sons)
[11] Weinberg S 1995 *The Quantum Theory of Fields* I (Cambridge University Press)
[12] Weinberg S 2008 *Cosmology* (Oxford University Press)
[13] Zeldovich Ya B 1968 The cosmological constant and the theory of elementary particles *Soviet Physics Uspekhi* **11** 381-393

Probability of boundary conditions in quantum cosmology

Yasusada Nambu[1] and Hiroshi Suenobu[2,3]

[1,2] Department of Physics, Graduate School of Science, Nagoya University, Chikusa, Nagoya 464-8602, Japan

E-mail: [1]nambu@gravity.phys.nagoya-u.ac.jp

E-mail: [2]Suenobu.Hiroshi@ab.MitsubishiElectric.co.jp

Abstract. One of the main interest in quantum cosmology is to determine boundary conditions for the wave function of the universe which can predict observational data of our universe. For this purpose, we solve the Wheeler-DeWitt equation for a closed universe with a scalar field numerically and evaluate probabilities for boundary conditions of the wave function of the universe. To impose boundary conditions of the wave function, we use exact solutions of the Wheeler-DeWitt equation with a constant scalar field potential. We specify the exact solutions by introducing two real parameters to discriminate boundary conditions, and obtain the probability for these parameters under the requirement of sufficient e-foldings of the inflation. The probability distribution of boundary conditions prefers the tunneling boundary condition to the no-boundary boundary condition. Furthermore, for large values of a model parameter related to the inflaton mass and the cosmological constant, the probability of boundary conditions selects an unique boundary condition different from the tunneling type.

1. Introduction

Investigation of the very early period of the universe requires quantum treatment of gravity [1]. However, because we do not have the complete theory of quantum gravity yet in hand, simplified models with reduced dynamical degrees of freedom have been investigated to understand nature of canonical quantum gravity. This approach is the mini-superspace quantum cosmology (a general review of quantum cosmology is given by [2]). A quantum state of the model is represented by the wave function of the universe, which satisfies the Wheeler-DeWitt (WD) equation derived from the procedure of canonical quantization [3]. The wave function of the universe is represented as the path integral by summing over histories of the universe [4].

To obtain the wave function of the universe, we must impose boundary conditions of the WD equation. In the context of the quantum cosmology, there are two major candidates for the boundary condition, the tunneling proposal by Vilenkin [5, 6] and the no-boundary boundary condition proposal by Hartle and Hawking [7]. The former is given by the wave function only consisting of the outgoing mode at the asymptotic future of mini-superspace, and is analogous to the tunneling wave function in quantum mechanics. The latter is given by the path integral over Euclidean non-singular compact geometries with no-boundary. In the path integral representation of the wave function, a choice of the path integral contour corresponds to specifying a boundary condition of the WD equation [8, 9]. To predict the classical universe using quantum cosmology, we must derive a probability for classical observables from the wave function of the universe. The number of e-foldings of inflation is often used as a predictable observable and recent observational restriction requires this number must be greater than about 60. The amount of e-foldings predicted by quantum cosmology depends on models and boundary conditions. Thus, the main issue in quantum cosmology is to determine which type of boundary conditions are preferable to explain observational results.

In this research, we apply a numerical method to obtain predictions from wave functions of the universe. The main idea of our research is to represent boundary conditions of the wave

[3] Present affiliation: Mitsubishi Electric Corporation Information Technology R&D Center

IOP Conf. Series: Journal of Physics: Conf. Series **880** (2017) 012025 doi:10.1088/1742-6596/880/1/012025

function using exact solutions of the WD equation with a constant scalar field potential. This makes our problem of determining boundary conditions as the parameter estimation in space of boundary conditions. We aim to obtain a probability distribution of boundary conditions under the constraint of sufficient e-foldings of the inflation.

2. Mini-superspace model

2.1. Classical model and quantization

We consider the Einstein gravity with a cosmological constant and a minimally coupled massive scalar field as the inflaton. The action of the total system is given by

$$S = \frac{1}{16\pi G}\int d^4x\sqrt{-g}\,(R-2\Lambda) - \frac{1}{2}\int d^4x\sqrt{-g}\,[(\partial_\mu\Phi)^2 + m^2\Phi^2]. \tag{1}$$

We assume a homogeneous and isotropic closed universe. Then the geometry of the universe is given by the Friedmann-Robertson-Walker (FRW) metric with a scale factor. The metric is given by

$$ds^2 = \frac{3}{\Lambda}\left(-\frac{N^2}{q}d\lambda^2 + q\,d\Omega_3^2\right), \tag{2}$$

where λ is a dimensionless time parameter and N is a lapse function. We introduce a dimensionless field variable $\phi = (4\pi G/3)^{1/2}\,\Phi$ and its mass $\mu = (3/\Lambda)^{1/2}\,m$. In this model, dynamical variables are the scale factor $q(\lambda)$ and the inflaton field $\phi(\lambda)$. We represent them as coordinates of configuration space $q^A = (q^0, q^1) = (q, \phi)$. This configuration space is called mini-superspace. The total Hamiltonian of our mini-superspace model is given by

$$H_T = \frac{KN}{2}\left[\frac{1}{K^2}\left(-4p_q^2 + \frac{1}{q^2}p_\phi^2\right) - 1 + q\left(1+\mu^2\phi^2\right)\right] = NH. \tag{3}$$

where we introduced a constant $K \equiv 9\pi/(2G\Lambda)$. We obtain the Hamiltonian constraint $H = 0$ by taking variation of the lapse function N. Canonical quantization of the model is performed by replacing p_A in the Hamiltonian constraint by differential operator $\hat{p}_A = -i\partial/\partial q^A$, and imposing operator version of the Hamiltonian constraint $\hat{H}$ on a physical state $\Psi(q^A)$. It yields the Wheeler-DeWitt equation

$$\left[-\frac{1}{2K^2}G^{AB}\partial_A\partial_B + U(q)\right]\Psi(q) = 0, \quad U = -\frac{1}{2} + qV, \quad G^{AB} = \mathrm{diag}(-4, 1/q^2). \tag{4}$$

The wave function $\Psi(q^A)$ on mini-superspace is called the wave function of the universe. Although there is an operator ordering ambiguity in the quantization procedure, we choose the ordering which yields the equation (4) in our analysis.

2.2. Semi-classical approximation and probability

To extract predictions for the classical universe from the wave function, it must be expressed as the semi-classical form, which means the wave function behaves as the WKB solution of Eq. (4). We perform the WKB expansion of the wave function as $\Psi(q^A) = \sum_i C^{(i)}(q^A)e^{-\frac{1}{\hbar}I^{(i)}(q^A)}$ where sum is taken over the independent WKB components. For convergence of the path integral representation of the wave function, the contour of the path integral must be analytically continued in the complex plane. Thus, I and C would generally become complex functions. We write I as $I = I_R - iS$ where I_R and S are real functions. From the WD equation (4), we obtain a set of semi-classical equations for I, C:

$$O(\hbar^0): \ -\frac{1}{2K^2}(\nabla I)^2 + U(q^A) = 0, \quad O(\hbar^1): \ 2\nabla I\cdot\nabla C + C\nabla^2 I = 0, \tag{5}$$

where $(\nabla I)^2 = G^{AB}\partial_A I \partial_B I$, $(\nabla I)\cdot(\nabla C) = G^{AB}\partial_A I \partial_B C$, $\nabla^2 I = G^{AB}\partial_A \partial_B I$. Now, we consider a condition of the wave function to provide predictions for the classical universe. In the classical regime, the phase of the wave function must satisfies the Hamilton-Jacobi equation. This implies that the equation (5) corresponds to the Hamilton-Jacobi equation $1/(2K^2)(\nabla S)^2 + U = 0$ in the classical regime. Thus I_R and S should satisfy the condition $|\nabla I_R|^2/|\nabla S|^2 \ll 1$. This inequality is called the "classicality" condition [10, 11]. To predict the evolution of classical universes from the wave function, we expect that the probability can be obtained in the region of mini-superspace where the classicality condition is satisfied.

For the wave function Ψ satisfying the WD equation, we have the conserved current in mini-superspace $\mathcal{J}_A = (i/2)(\Psi^*\nabla_A\Psi - \Psi\nabla_A\Psi^*)$, $\nabla \cdot \mathcal{J}_A = 0$. In the classical region where the wave function has the WKB form, we can obtain the positive definite probability measure from $\mathcal{J}_A$. For each WKB components of the wave function, we define $J_A^{(i)} \equiv -|C^{(i)}|^2 \exp(-2I_R^{(i)})\nabla_A S^{(i)}$. They are conserved independently in the classical region $\nabla \cdot J^{(i)} = 0$. From the Hamilton-Jacobi equation, we can assign the canonical momentum in the classical region $p_A^{(i)} = \nabla_A S^{(i)} = \partial S^{(i)}/\partial q^A$. Here, we focus only on the components with $p_q^{(i)} = \partial_q S^{(i)} < 0$. From $dq^{(i)}/d\lambda \propto -p_q^{(i)} > 0$, these components correspond to expanding universes. Thus, we can introduce a conserved current corresponding to expanding universes as $J_A^+ \equiv -\sum_{p_q^{(i)}<0} |C^{(i)}|^2 \exp(-2I_R^{(i)})\nabla_A S^{(i)}$. Let us consider a surface Σ_c in mini-superspace which is spacelike with respect to the metric G_{AB} and has a unit normal n_A. We require the classicality condition is satisfied on this surface. Then the relative probability $\mathcal{P}(\Sigma_c)$ of classical histories passing through this surface is given by the component of the conserved current along the normal if it is positive. In the leading order in $\hbar$, this is $\mathcal{P}(\Sigma_c) \equiv J^+ \cdot n = -\sum_{p_q^{(i)}<0} |C^{(i)}|^2 \exp(-2I_R^{(i)})\nabla_n S^{(i)}$ where ∇_n means differentiation along the normal vector n_A. As a point on Σ_c is specified by the value of the scalar field, $\mathcal{P}(\phi) \equiv \mathcal{P}(\Sigma_c(\phi))$ provides the probability for the inflaton field to realize a value ϕ on Σ_c.

We can derive a probability for observables from the probability measure $\mathcal{P}(\phi)$. It can be given as the conditional probability [2] $P(s_0|s_1) = \int_{s_0} J \cdot d\Sigma_c / \int_{s_1} J \cdot d\Sigma_c$, $s_0 \subset s_1$ where s_1 is a subset of the hypersurface Σ_c defined by some theoretical constraints and s_0 is a subset of s_1 defined by restricting s_1 using observational constraints. By using the relation $p_A = \partial S/\partial q^A$, we can obtain classical trajectories starting from Σ_c. Namely, the probability measure on Σ_c with the classicality condition gives probability distribution of initial data (p_q, q, p_ϕ, ϕ) for the classical equation of motion. In our analysis, the number of e-foldings $\mathcal{N} \equiv \log(a(t_f)/a(t_i))$ is adopted as an observable. We define t_f as the end time of inflation driven by the scalar field potential. The number of e-foldings $\mathcal{N}$ is determined by the initial data and it is possible to translate the probability measure for ϕ on Σ_c to the probability measure for $\mathcal{N}(\phi)$. To introduce the conditional probability, we define an interval s_1 as $s_1 = [\phi_{\rm min}, \phi_{\rm pl}]$ where $\phi_{\rm min}$ is the lower bound of the interval and $\phi_{\rm pl} = 4\sqrt{2K}/(3\mu)$ is the value of the inflaton field corresponding to the Planck energy density $m_{\rm pl}^4$. Then an interval $s_0 \subset s_1$ is defined as $s_0 = [\phi_{\rm suf}, \phi_{\rm pl}]$ where $\phi_{\rm suf}$ corresponds to the number of e-foldings $\mathcal{N}_{\rm suf} \approx 60$ consistent with observations. Accordingly, the conditional probability to predict the universe with sufficient inflation becomes

$$P_{\rm suf} \equiv P(s_0|s_1) = \frac{\int_{\phi_{\rm suf}}^{\phi_{\rm pl}} d\phi\, \mathcal{P}(\phi)}{\int_{\phi_{\rm min}}^{\phi_{\rm pl}} d\phi\, \mathcal{P}(\phi)}. \tag{6}$$

These probability and expectation value depend not only on cosmological models but also on boundary conditions of the wave function. By calculating and comparing $P_{\rm suf}$ for given models and given boundary conditions, we can evaluate what type of models and boundary conditions are more suitable to explain observation of our universe.

IOP Conf. Series: Journal of Physics: Conf. Series **880** (2017) 012025 doi:10.1088/1742-6596/880/1/012025

wave function	parameter (a, b)	asymptotic form for $q \gg 1$
$\Psi_{\rm HH}$	$(\pi/4, 0)$	$\sim \exp(+K/(6V)) \cos S_0$
$\Psi_{\rm V}$	$(\pi/4, \pi/2)$	$\sim \exp(-K/(6V)) \exp(-iS_0)$
Ψ_1	$(\pi/2, \pi/2)$	$\sim \exp(-K/(6V)) \cos S_0$
Ψ_2	$(\pi/2, 0)$	$\sim \exp(+K/(6V)) \cos S_0$
Ψ_3	$(0, \text{any values})$	$\sim -\exp(-K/(6V)) \sin S_0$

Table 1. Typical wave functions and their parameters (a, b) and asymptotic behaviors. The phase function S_0 is defined by $S_0 = K/(6V)(2qV - 1)^{3/2} - \pi/4$.

3. Probability for boundary conditions

When we have some restriction on our models of inflationary universe from observations, we can investigate a probability which states preferable type of boundary conditions. It is possible to express this probability using Bayes' theorem:

$$P(B_i|S) = \frac{P(B_i)P(S|B_i)}{\sum_k P(B_k)P(S|B_k)}, \tag{7}$$

where $P(B_i|S)$ is a probability for B_i under S happened. Here, B_i is some candidate of a boundary condition of the wave function labeled by index i, and S means the universe with sufficient inflation, namely, $\mathcal{N} \geq 60$. Thus, $P(B_i|S)$ denotes the probability for B_i under the sufficiently inflated universe. On the contrary, $P(S|B_i)$ in the right hand side is the probability for sufficient inflation under the boundary condition B_i and is equivalent to $P_{\rm suf}$ defined in the previous section: $P(S|B_i) = P_{\rm suf}(B_i) = P(\mathcal{N} \geq 60\,|B_i)$. As we do not have any information on the prior probability $P(B_i)$, we assume that it is uniformly distributed. To represent different boundary conditions, we will introduce two parameters a, b in (10). Then the probability for the parameters a, b is given by

$$P(a, b|S) = \frac{P(S|a, b)}{\int da'db' P(S|a', b')}. \tag{8}$$

When we solve the WD equation, we have to impose some boundary condition on the wave function. For this purpose, we use exact solutions of the WD equation which are obtained when the scalar field potential $V(\phi)$ is constant. Based on the path integral representation of the wave function, for the constant scalar field potential case, the wave function corresponding to the no-boundary (Hartle-Hawking) and the tunneling (Vilenkin) type boundary conditions are expressed as [8] $\Psi_{\rm HH} = \Psi_2 + \Psi_3$, $\Psi_{\rm V} = \Psi_1 + i\,\Psi_3$, where

$$\Psi_1 \equiv (2V)^{-1/3}{\rm Ai}(z_0){\rm Ai}(z),\ \Psi_2 \equiv (2V)^{-1/3}{\rm Bi}(z_0){\rm Ai}(z),\ \Psi_3 \equiv (2V)^{-1/3}{\rm Ai}(z_0){\rm Bi}(z), \tag{9}$$

with $z = z(q) = (4V/K)^{-2/3}(1 - 2qV)$, $z_0 = z(0) = (4V/K)^{-2/3}$. For large values of the scale factor (classical region), $\Psi_{\rm HH}$ is superposition of expanding and contracting universes with amplitude $\exp(+K/(6V))$ which prefers small values of the potential. On the other hand, $\Psi_{\rm V}$ represents an expanding universe with amplitude $\exp(-K/(6V))$ which prefers large values of the potential. We can express more general type of wave functions introducing two real parameters a, b which represent boundary conditions of the wave function

$$\Psi_C = \tan a(\cos b\,\Psi_2 - i \sin b\,\Psi_1) + \Psi_3, \quad 0 \leq a, b \leq \pi/2. \tag{10}$$

Introduced parameters a, b distinguish boundary conditions of the wave function (Table 1). Solving the wave function and calculating probability $P_{\rm suf}$ for different values of (a, b), we can evaluate the probability for the parameters (a, b) using the relation (8).

IOP Conf. Series: Journal of Physics: Conf. Series **880** (2017) 012025 doi:10.1088/1742-6596/880/1/012025

4. Numerical simulation of the wave function

We solve the WD equation (4) numerically to obtain the probability of boundary conditions. The detail of our numerical method is explained in [12]. We prepare the initial surface $q = q_{\rm ini}$ in the Euclidean region of mini-superspace and impose the boundary condition $\Psi(q_{\rm ini}, \phi) = \Psi_C(q_{\rm ini}, \phi)$, $\partial_q\Psi(q_{\rm ini}, \phi) = \partial_q\Psi_C(q_{\rm ini}, \phi)$. As Ψ_C introduced by (10) is specified by two parameters (a, b), this boundary condition is also specified by these two parameters. We call Ψ_C the boundary wave function. For models with a constant scalar field potential, this boundary condition of course reproduces the exact solution Ψ_C. We fix the mass of the scalar field and consider two models with parameters $\mu = 0.2$ ($m^2 = 0.03$, $\Lambda = 2.25$) and $\mu = 3$ ($m^2 = 0.03$, $\Lambda = 0.01$). Different value of μ corresponds to different value of the cosmological constant in our analysis. The former choice results in slow roll inflation followed by over damped rolling of the inflaton field and the later results in inflation with slow rolling followed by oscillation of the inflaton about $\phi = 0$. Our simulation algorithm is as follows:

(i) Prepare an initial surface $q = q_{\rm ini}$ in the Euclidean region of mini-superspace close to $q = 0$. $q_{\rm ini}$ cannot be chosen too small because we must keep the Courant condition for stable numerical integration of the wave equation.

(ii) Solve the WD equation numerically from $q = q_{\rm ini}$ to $q_{\rm fin}$ with a given boundary wave function Ψ_C. We adopt the 5-step Adams-Bashforth method for numerical integration which has the 5-th order accuracy. We used 20000×200 grid size which covers $q_{\rm ini} \leq q \leq q_{\rm fin}, \phi_{\rm min} \leq \phi \leq \phi_{\rm max}$

(iii) We specify a hypersurface Σ_c on which the classicality condition is satisfied. We choose Σ_c as a constant S_0 surface. We numerically obtain the probability $\mathcal{P}(\phi)$ on Σ_c.

(iv) By integrating the classical equation of motion from Σ_c, we evaluate the number of e-foldings for each classical trajectories. Then calculate the probability measure of the e-foldings.

(v) Repeating step (ii) to step (iv) for different values of parameters (a, b), we obtain the probability of parameters (a, b) which specify boundary conditions of the wave function. We calculate the probability for 9×9 grid points in the parameter space (a, b) of boundary wave functions.

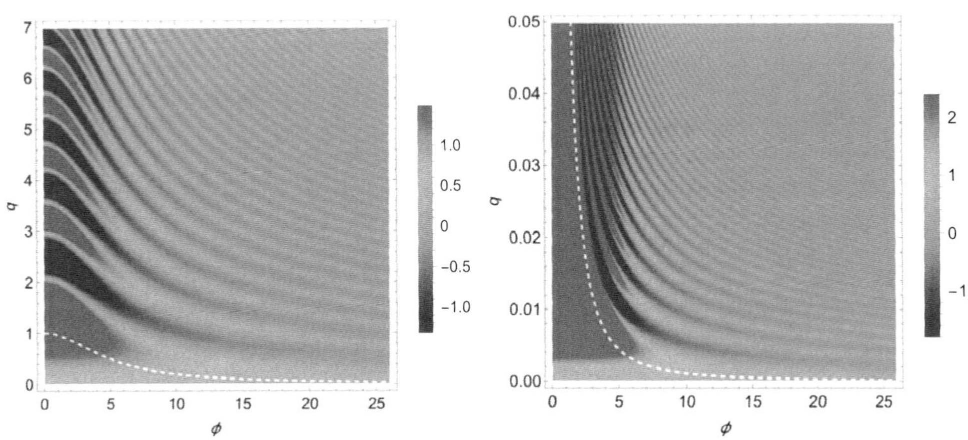

Figure 1. The density plot of wave functions with the no-boundary boundary condition (HH). For this boundary condition, wave functions are real. Left: $\mu = 0.2, q_{\rm ini} = 0.01$. Right: $\mu = 3, q_{\rm ini} = 0.0001$. The dashed line represents $2qV(\phi) = 1$ which is the boundary between the Euclidean region and the Lorentzian region in mini-superspace.

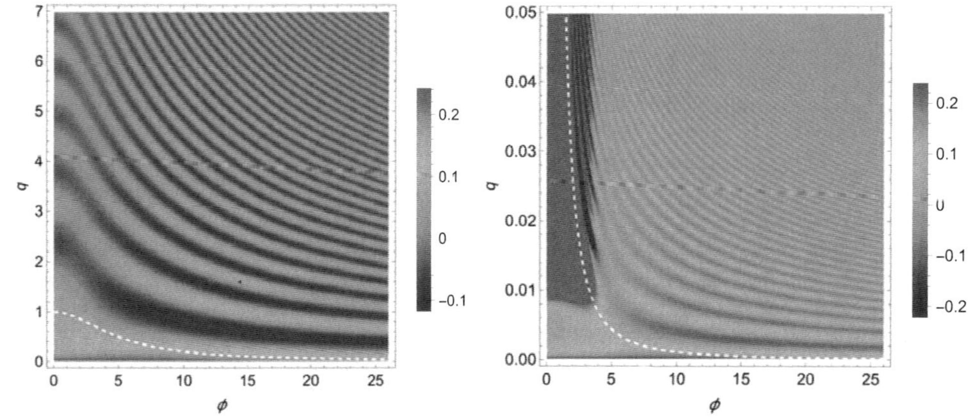

Figure 2. The imaginary part of the wave function with the tunneling boundary condition (V). Left: $\mu = 0.2$, $q_{\rm ini} = 0.01$, Right: $\mu = 3$, $q_{\rm ini} = 0.0001$. The dashed line represents the boundary between the Euclidean region and the Lorentzian region.

Fig. 1 shows wave functions with the boundary wave function $\Psi_{\rm HH}$ (the no-boundary boundary condition (HH)). Fig. 2 shows wave functions with the boundary wave function $\Psi_{\rm V}$ (the tunneling boundary condition (V)). Fig. 3 shows $\mathcal{P}(\phi)$ obtained from solutions of the WD equation with boundary wave functions Ψ_1, Ψ_2, Ψ_3 and $\Psi_{\rm HH}$, $\Psi_{\rm V}$. From now on, we denote wave functions with these boudary wave functions as $\Psi_1, \Psi_2, \Psi_3, \Psi_{\rm HH}, \Psi_{\rm V}$. This probability measure is not normalized because the conditional probability $P(S|a,b)$ can be obtained without normalizing $\mathcal{P}(\phi)$.

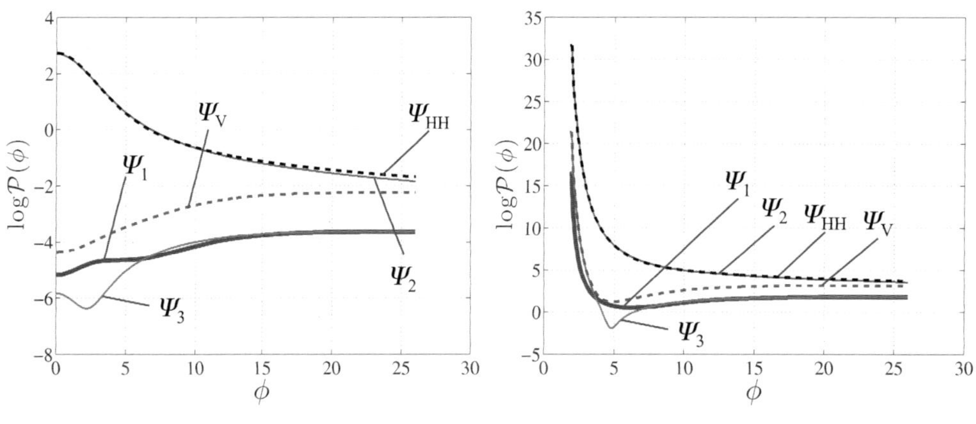

Figure 3. $\mathcal{P}(\phi)$ for $\mu = 0.2$ (left) and $\mu = 3$ (right). $\mathcal{P}(\phi)$ is not normalized.

For $\mu = 0.2$, wave functions Ψ_1, Ψ_3, $\Psi_{\rm V}$ prefer large values of ϕ while Ψ_2, $\Psi_{\rm HH}$ prefer small values of ϕ. This behavior of $\mathcal{P}(\phi)$ is the same as that obtained from the exact wave function Ψ_C. However, the distribution for small ϕ is slightly different from that obtained by Ψ_C. For $\mu = 3$, probabilities for Ψ_2 and $\Psi_{\rm HH}$ have the same behavior as these in the $\mu = 0.2$ model. However, probabilities for Ψ_1, Ψ_3 and $\Psi_{\rm V}$ show different behavior; The probabilities for small ϕ have significantly large values for the $\mu = 3$ model.

Using $\mathcal{P}(\phi)$, we obtain the probability measure for number of e-foldings $\mathcal{N}$ by numerical integrations of classical trajectories starting from Σ_c. Then, we evaluate the conditional probability for the sufficient inflation $P_{\text{suf}} = P(\mathcal{N} \geq 60)$. Finally, we obtain the probability $P(a, b)$ of boundary conditions using the relation (8) (Fig. 4). For $\mu = 0.2$, $P(a, b)$ has large

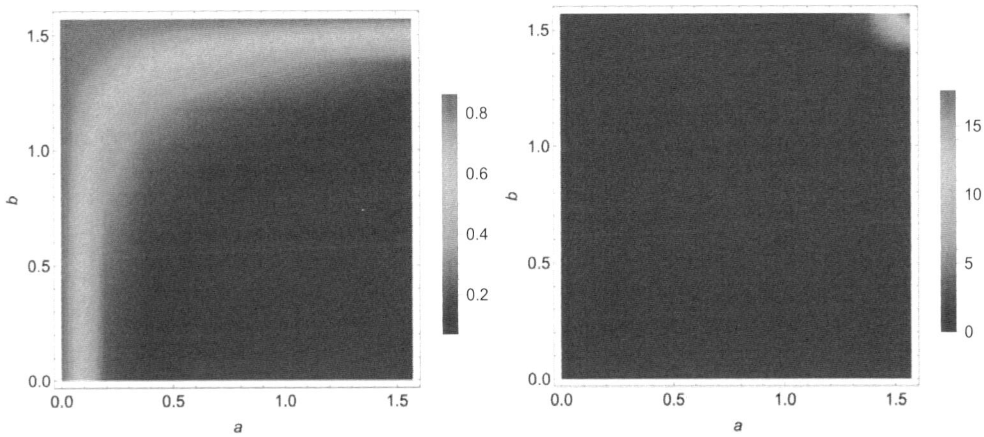

Figure 4. Density plot of $P(a, b)$ for $\mu = 0.2$ (left) and $\mu = 3$ (right). We evaluated this probability on 9×9 grid points in the parameter space of boundary conditions.

values on lines $a = 0$ and $b = \pi/2$. These two lines correspond to the wave functions Ψ_1, Ψ_3 and Ψ_{V}. $P(a, b)$ has small values on the line $b = 0$, corresponding to Ψ_2 and Ψ_{HH}. Consequently, Ψ_{V} is more preferable than Ψ_{HH} to realize large e-foldings, and this result is the same as one predicted by the wave functions for a constant scalar field potential. In contrast to that, for $\mu = 3$, $P(a, b)$ has large values only around the point $(a, b) = (\pi/2, \pi/2)$, which corresponds to the boundary wave function Ψ_1. This behavior is significantly different from the $\mu = 0.2$ case. Superiority of Ψ_{V} to Ψ_{HH} holds both for $\mu = 0.2$ and $\mu = 3$ models. We expect our results with parameter $\mu = 0.2$ and $\mu = 3$ are typical ones and $P(a, b)$ with $\mu < \mu_*$ behaves similar to the $\mu = 0.2$ case and $P(a, b)$ with $\mu > \mu_*$ behaves similar to the $\mu = 3$ case.

5. Conclusion

In this article, we considered boundary conditions for the wave function of the universe which lead to sufficient e-foldings of inflation. For this purpose, we adopted the exact solutions of the WD equation with a constant scalar field potential as the boundary condition of the wave function, and solved the WD equation numerically. This boundary condition is parametrized with two real parameters and includes both the tunneling and the no-boundary boundary conditions. We obtained the probability distribution function for these parameters under the condition of sufficient e-foldings of inflation. The parameters with large value of this probability determines the boundary condition of the wave function which predicts sufficient e-foldings of inflation. We found that the probability distribution of boundary conditions has two different behavior depending on the value of model parameter μ.

For small values of μ, the cosmological constant dominates and the inflaton field asymptotically approaches to zero without oscillation. In this case, ϕ dependence of the wave function is not so strong and the obtained wave function reproduces behavior of exact wave functions with a constant scalar field potential. Hence the probability of boundary conditions has large values for $\Psi_1, \Psi_3, \Psi_{\text{V}}$ and small values for Ψ_2, Ψ_{HH}. Thus, boundary conditions

Ψ_1, Ψ_3, Ψ_V are preferable to realize sufficient period of inflation and superiority among them is small. This behavior of the probability of boundary conditios can be expected from behavior of wave functions for a constant scalar field potential. On the other hand, for large values of μ, the slow roll inflation is followed by oscillation of the inflaton field. In this case, the derivative term of ϕ in the WD equation cannot be neglected and wave functions have large values about $\phi = 0$ for any boundary conditions. Owing to this behavior of the wave function, the probability of boundary conditions has large value about Ψ_1 (superiority of Ψ_V over Ψ_{HH} is kept as before). Thus, realistic inflationary models followed by oscillation of inflaton field select the boundary condition Ψ_1.

As an extention of analysis presented in this article, it is also possible to discuss probability for values of the model parameter μ. The probability distribution of boundary conditions has a sharp peak for the model with large μ and selects a specific boundary condition. This implies that a suitable boundary condition is automatically chosen for large values of μ (small values of the cosmological constant). If we assume that the probability of boundary conditions select an unique boundary condition, the parameter μ must acquire large value (the cosmological constant must be small).

Acknowledgments
YN was supported in part by JSPS KAKENHI Grant Number 15K05073 and 16H01094.

References
[1] Kiefer C 2012 *Quantum Gravity* (Oxford: Oxford University Press)
[2] Halliwell J J 2009 Introductory Lectures On Quantum Cosmology *Preprint* arXiv:0909.2566 [gr-qc]
[3] DeWitt B S 1967 Quantum Theory of Gravity 1. The Canonical Theory *Phys. Rev.* **160** 1113
[4] Halliwell J J 1988 Derivation of the Wheeler-DeWitt equation from a path integral for minisuperspace models *Phys. Rev. D* **38** 2468
[5] Vilenkin A 1982 Creation of universes from nothing *Phys. Lett. B* **117** 25
[6] Vilenkin A 1986 Boundary Conditions in Quantum Cosmology *Phys. Rev. D* **33** 3560
[7] Hartle J B and Hawking S W 1983 Wave function of the Universe *Phys. Rev. D* **28** 2960
[8] Halliwell J J and Louko J 1989 Steepest-descent contours in the path-integral approach to quantum cosmology I *Phys. Rev. D* **39** 2206
[9] Halliwell J J and Hartle J B 1990 Integration Contours for the No Boundary Wave Function of the Universe *Phys. Rev. D* **41** 1815
[10] Hartle J B, Hawking S W and Hertog T 2008 No-Boundary Measure of the Universe *Phys. Rev. Lett.* **100** 201301
[11] Hartle J B, Hawking W S and Hertog T 2008 Classical universes of the no-boundary quantum state *Phys. Rev. D* **77** 123537
[12] Suenobu H and Nambu Y 2017 Probability of Boundary Conditions in Quantum Cosmology *Gen. Relativ. Gravit.* **49** 19 (*Preprint* arXiv:1603.08172 [gr-qc])

Inflation, the Higgs field and the resolution of the Cosmological Constant Paradox

Francesco De Martini

Dipartimento di Fisica, Universita' La Sapienza, I- 00185 Roma, Italy

E-mail: francesco.demartini@uniroma1.it

Abstract. The nature of the scalar field responsible for the cosmological inflation, the "inflaton", is found to be rooted in the most fundamental concept of the Weyl's differential geometry: the parallel displacement of vectors in curved space-time. Within this novel dynamical scenario, the standard electroweak theory of leptons based on the $SU(2)_L \otimes U(1)_Y$ as well as on the conformal groups of spacetime Weyl's transformations is analyzed within the framework of a general-relativistic, co-covariant scalar-tensor theory that includes the electromagnetic and the Yang-Mills fields. A Higgs mechanism within a spontaneous symmetry breaking process is identified and this offers formal connections between some relevant properties of the elementary particles and the dark energy content of the Universe. An "Effective Cosmological Potential": V_{eff} is expressed in terms of the dark energy potential: $V_\Lambda \equiv M_\Lambda^2$ via the "mass reduction parameter": $\zeta \equiv \sqrt{\frac{|V_{eff}|}{|V_\Lambda|}}$, a general property of the Universe. The mass of the Higgs boson, which is considered a "free parameter" by the standard electroweak theory, by our theory is found to be proportional to the geometrical mean: $M_H \propto \sqrt{M_{eff} \times M_P}$ of the Planck mass, M_P and of the mass $M_{eff} \equiv \sqrt{|V_{eff}|}$ which accounts for the measured Cosmological Constant, i.e. the measured content of vacuum-energy in the Universe. The experimental result obtained by the ATLAS and CMS Collaborations at CERN in the year 2012: $M_H = 125.09(GeV/c^2)$ leads by our theory to a value: $M_{eff} \sim 3.19 \cdot 10^{-6}(eV/c^2)$. The peculiar mathematical structure of V_{eff} offers a clue towards the resolution of a most intriguing puzzle of modern quantum field theory, the "Cosmological Constant Paradox".

1. Introduction: Weyl geometry

A huge step forward in theoretical cosmology, and today a very important landmark of modern science, was the proposal by A. Starobinsky followed one year later by A.Guth, A. D. Linde, A. Albrecht and P.J. Steinhard [1, 2, 3, 4] of the *inflation*, an epoch of fast accelerating expansion of the early Universe that caused the Universe to expand through about 70 e-folds in a very small fraction of a second [5]. This expansion, driven by a scalar field called "inflaton", was originally argued to solve the problem of why the universe is so smooth at large scales. Moreover it later turned out to provide a consistent solution also to a host of different crucial problems among which the growth of structures in the Universe arising from magnified quantum fluctuations, the no observation of magnetic monopoles, the isotropy of the cosmic microwave background etc. [6, 7]. Over the years, the undeniable success of this idea was however somewhat questioned by the failure of finding the physical mechanism underlying the fundamental nature of the "inflaton" concept. In the present letter it is claimed that the fundamental nature of this scalar field is indeed *geometrical*, based on the conformal differential geometry introduced by Hermann Weyl two years after the publication of the first Einstein paper on General Relativity (GR) [8, 9]. This geometry rests on the following general statement:" *All Laws of Physics are invariant under* (A) *any change of coordinates including time and* (B) *under any change of calibration (i.e. gauge)*" The first part, (A) expresses the well known *covariance* of the GR based on the Riemann geometry. The second part (B), a significant addition by Weyl to Riemann's, expresses the *conformal* property of the metric space-time theory. Accordingly, the above statement is usually cast in a simpler form: "*All Laws of Physics are conformally-covariant*" (or *Weyl covariant* or, in short, *co-covariant*).

The rationale of the Weyl geometry can be outlined as follows. Taking Planck's constant ($\hbar$) and the velocity of light (c) to be constant by definition, any physical quantity X, e.g. represented by a scalar, a tensor, a spinor etc, can be assigned a unit that is a power of a physical length $\tilde{L}$: $X \to \tilde{L}^{W(X)}$. Furthermore, to that quantity can be assigned a transformation law $X \to e^{W(X)\lambda(x)}X$ under a conformal mapping. $W(X)$ is a (positive or negative) real number dubbed "*Weyl weight of* X" (or "dimensional number of X") and $\lambda(x)$ is a regular, real function of the space-time coordinates. Thus, the conformal mapping is a "unit transformation" amounting to a local space-time redefinition of the unity of length, i.e. of the "*calibration*". This concept is rooted into the most basic operation of the differential geometry i.e. the "parallel displacement" of any vector in a non-Euclidean manifold. In fact, according to Weyl, the parallel displacement from two infinitely nearby points P and $P+dP$ in spacetime acts on the length ℓ of any vector by inducing a calibration change $\delta\ell = \ell\phi_\rho dx^\rho$, where ϕ_ρ is a universal "*Weyl vector*", defined in the whole space-time. In summary, the metric structure of the Weyl geometry implies two fundamental forms: the *quadratic* Riemannian one, $g_{\rho\sigma}dx^\rho dx^\sigma$, being $g_{\rho\sigma} = g_{\sigma\rho}$ the metric tensor, and the Weyl *linear* one $\phi_\rho dx^\rho$. In Riemann's geometry it is always $\phi_\rho = 0$. The parallel displacement is *integrable* iff a scalar *Weyl potential* ϕ exists such as $\phi_\rho = \partial_\rho\phi$. As we shall see, the "*Weyl vector*" $\phi_\rho(x)$ and the corresponding scalar "Weyl potential" $\phi(x)$, both defined in the whole space-time spanned by the x^ρ coordinates play a basic role in the present work: indeed the "inflaton" is identified with ϕ [5].

We first note that under a physics perspective the Weyl geometry indeed consists of an abelian local scale-invariance gauge theory implying the following group of transformations

$$\begin{aligned} \phi_\rho &\to \phi_\rho + \partial_\rho\lambda(x) \\ g_{\rho\sigma} &\to e^{2\lambda} g_{\rho\sigma}. \end{aligned} \tag{1}$$

The inflaton may be considered a "gauge field" [8, 10]. The insightful perspective offered by this theory was supported by P.A.M. Dirac in a 1973 seminal paper [11]:
There is a strong reason in support of the Weyl's theory. It appears as one of the fundamental principles of Nature that the equations expressing basic laws should be invariant under the widest possible group of transformations. The confidence that one feels in Einstein GR theory arises because its equations are invariant under the wide group of transformations of curvilinear coordinates in Riemann space. The passage to Weyl geometry is a further step in the direction of widening the group of transformations underlying the physical laws. One has to consider transformations [...] which impose stringent conditions on them.
The stringent conditions alluded by Dirac imply in the first place the correct choice by definition of several constant units in terms of which the physical quantities are measured: these units must be mutually independent in the sense that a dimensionless number cannot be constructed with them. As in the relativistic quantum theory, it is conventional to take $\hbar, c, m_e$ (the electron mass) to be *constant by definition* [9, 10]. Other gauges, e.g. by replacing m_e by the gravitational constant G, lead in general to different theories which are mutually connected by

conformal mapping. Furthermore, the Lagrangian density L from which the dynamical equations are derived as well any measurable quantity X attained as result of the theory must be co-covariant, and then expressed by the zero value of the Weyl weight $W(L) = 0$, $W(X) = 0$. At the end of the calculation a careful application of the initial and boundary conditions must be undertaken in order to select the physically admissible results among the special solutions of the theory. The special solutions have the property that particle masses are independent of positions in space-time and the gravitational constant G is a true constant. Thus the statement that all electrons have the same mass, all hydrogen atoms have the same size etc. may be taken to be true by definition. All this circumvents the original criticism to the Weyl theory [8, 13, 14, 15]. As it will be shown below, the requirement of full covariance under conformal mapping is necessary only to determine the formal structure of the theory.

The implied meaning of the present work may be further enlightened by the fact that after 1970 several axiomatic approaches for deducing the projective and conformal structure of space-time were carried out by using basic concepts such as light rays and freely falling particles [13, 14, 15]. The highly remarkable fact is that all these investigations ended by assigning a Weylan, not a Riemannian structure to space-time. This raised a long lasting, and as yet unresolved interesting dilemma that could be understood by the possibility of the onset of a cosmological "phase transition" by which the now perceived Riemannian structure is induced via a dynamical process that breaks the Weyl invariance of the vacuum at some critical temperature T_c.

2. Inflation and the Weyl-Dirac Theory

All these concepts will be applied to a very general Weyl-Dirac conformal scalar-tensor theory involving the mass m_E of an elementary particle, e.g. an electron. In order to keep the conformal structure of the theory, the mass of the particle is expressed in form of a "mass field" $m_E \to k_E \cdot \mu(x)$, being the dimensionless coupling constant: k_E an intrinsic particle's property, and $\mu(x)$ a real scalar field, function of space-time, with weight $W(\mu) = -1$ [12]. The gravitational constant G appearing in the Einstein gravitational equation has $W(G) = +2$ and so it cannot be regarded as constant in the present approach. We define a dimensionless constant: α and assume that the ratio l_C/l_P between the particle's Compton length $l_C = \hbar/mc$ and the Planck length is independent of position in space and time. By applying the Dirac's Lagrange-multiplier method, the simplest form of the co-covariant Lagrangian density in $D = 4$ can be expressed in the scalar-tensor form [9, 11]:

$$L = \sqrt{-g}\left\{\alpha\mu^2\left[\bar{R} + 2V(T,\phi,\mu)\right] - D_\rho\mu D^\rho\mu - \frac{1}{4}\phi_{\rho\sigma}\phi^{\rho\sigma}\right\} \tag{2}$$

where $\bar{R} = R + R_W$ is the overall Riemann-Weyl curvature scalar and D_ρ is the Weyl co-covariant derivative [8, ?]. In particular, the Weyl curvature scalar in D=4 is: $R_W = 6[\phi_\rho\phi^\rho - \phi^\rho_{|\rho}]$. The generic potential $V(T,\phi)$, related to the cosmological constant Λ, and function of the temperature T, accounts for the self-interaction of the scalar field ϕ [28]. As we shall see immediately, the stringent conditions implied by the Weyl symmetry on each co-covariant addendum X appearing in the expression of L in Eq. (2), i.e. $W(L) = W(X) = 0$, impose a well defined exponential expression on the function $V(T,\phi)$.

In addition to the above considerations, we must further impose in the present context the general condition that the action: $I_\phi = \int dx^4\sqrt{-g}L_\phi$ is stationary respect to variations in ϕ, where: $L_\phi \equiv \left[\frac{1}{2}g^{\rho\sigma}\partial_\rho\phi\partial_\sigma\phi - V(T,\phi)\right]$ expresses the effect of the inflation field in the $D = 4$

space-time (Weinberg, 2008). This condition is expressed by:

$$\nabla_B \phi = -\frac{\partial V(\phi, T)}{\partial \phi} \tag{3}$$

being: $\nabla_B \phi \equiv \phi^{\rho}_{|\rho}$ the Laplace-Beltrami differential operator of the field: ϕ [?]. In the theory above the Lagrangian L, Eq. (2), is written in terms of the *geometrical* Weyl potential ϕ_W within a Weyl-Dirac scalar-tensor theory, e.g. according to the Texts by P.A.M. Dirac and E.A. Lord [11, 9] (Cfr. pages 410 and 197, respectively). On the other hand, the Lagrangian L_ϕ is written in terms of the *physical* "inflaton" field ϕ_I, e.g. according to the Texts by S. Dodelson and S. Weinberg (2008) [6, 7] (Cfr. pages 152 and 526, respectively). According to the basic conjecture of our theory the two fields should be considered as complementary aspects of the same entity: $\phi_W \equiv \phi_I \equiv \phi$. Accordingly, there is no conflict between the corresponding lagrangian theories but rather Eq. (3), i.e. the Euler-Lagrange result of the variation respect ϕ of L_ϕ, provides the necessary mathematical relation leading to the completion of the general theory in closed form, i.e. with no approximations.

Let us now make the simplifying assumption that the inflationary system is "trapped" in local minimum of the potential, a significant dynamical condition largely considered in the literature [6] :$V' \equiv \frac{\partial V(\phi,T)}{\partial \phi} = 0$. The variation of L respect to the relevant fields leads to the dynamical equations of the Weyl conformal theory. This one is selected to be integrable, so that $\phi_{\rho\sigma} = 0$. In particular, the variation respect $g_{\rho\sigma}$ leads for $\alpha = \frac{1}{7} \simeq 0.1428$ to the following Einstein equation [9]:

$$R_{\rho\sigma} - \frac{1}{2} g_{\rho\sigma} R = \left[\partial_\rho \phi \partial_\sigma \phi - \frac{1}{2} g_{\rho\sigma} \partial_\eta \phi \partial^\eta \phi + g_{\rho\sigma} V(T, \phi) \right] \tag{4}$$

being $R_{\rho\sigma}$ and R the Ricci tensor and the curvature scalar of the Riemann geometry. The expression within square brackets at the r.h.s. of Eq. (4) may be cast in the standard form: $K^2 T^*_{\rho\sigma} \equiv K^2 [\partial_\rho \phi \partial_\sigma \phi - g_{\rho\sigma} L_\phi]$ with: $K^2 = \frac{8\pi G}{c^4}$, by the rescaling: $\phi \rightarrow K\phi$ and: $V(\phi) \rightarrow K^2 V(K\phi)$ [i.e.:$\phi_{new} = K^{-1} \phi_{old}$]. The Eq. (4) is a somewhat suggestive result since it reproduces exactly the basic Einstein equation of the inflation theory reported in the standard texts on cosmology [6, 7, 28]. Note that our expression of $T^*_{\rho\sigma}$ is obtained, within the quoted simplifying assumption, by the formal application of the standard Euler-Lagrange variational procedures to the expression in Eq. (2) of the $D = 4$ Weyl's scalar curvature which is absent in Riemann's geometry [?]. Indeed, in modern texts an equation similar to Eq. (4) is attained via the introduction of an exotic "quintessence" object wich is assumed to represent artificially a modified matter model [16]. As said, by our novel interpretation of the inflaton field, the present work suggests the geometrical nature of the quintessence. All this may lead to several consequences of dynamical relevance, as for instance a straightforward, dynamically driven interchange between the weylan to riemannian symmetry conditions [30]. Note that all the variational calculations, e.g. leading to Eq. (4) are carried out by keeping the explicit spacetime dependence of the field: $\mu(x)$. Only after the completion of the variational process we have taken advantage of the conformal gauge invariance of the theory by choosing μ to be a *constant* field. An extended discussion on the conformal Weyl gauge theory is found in: [?, 9]. As anticipated, the Weyl symmetry imposes a restriction to the explicit form of the massive inflaton potential. In virtue of Eq. (1), since for any physical quantity: $X \rightarrow e^{\lambda(x) W(X)} X$, and because here: $W(V_\Lambda) = -2, W(M_P^2) = -2, W(\sqrt{-g}) = +4$, a possible expression to be inserted in Eq. (2) may be cast in the co-covariant form as an exponential: $V_\Lambda(T, \phi) \propto e^{-2K\phi}$ or as a superposition of exponentials. Another co-covariant solution could be: $V_\Lambda(T, \phi) \propto (\phi_\rho \phi^\rho)$ possibly associated with a superposition of exponentials. Another co-covariant solution apt to a superposition could be: $V_\Lambda(T,\phi) \propto (\phi^\rho_{|\rho})$. Etc. All these formal generalizations, by implying the definition of a finite set of positive or negative scalar parameters of the superpositions $C_n(T)$, lead in generakl to a $V_\Lambda(T,\phi)$ function showing an unlimited number of local maxima or minima for $K\phi$ ranging from zero to infinity [16]. We notice nevertheless that in general T, and hence any $C_n(T)$, is not constant within the evolution of the inflaton $K\phi$, e.g. during the rolling down of the Universe system towards levels of lower free energy in an expansion phase or, on the contrary, in a phase of re-heating. In spite of these complications, we may recognize that the theory above is able to account reasonably for the so far inscrutable paths of the Universe evolution.

As suggested by the title of this work, in the next Section we shall adopt a broader perspective by inquiring about the connections existing between the two relevant scalar fields that, albeit supposed to belong to distant theoretical domains, today play a major role in the vast scenario of modern physics. In order to accomplish this task we need to consider a more complete Lagrangian than the one given by Eq. (2). We make a connection with the electroweak theory of leptons [26, 27] and we consider the following:

$$\hat{L} = \sqrt{-g}\left\{\alpha\mu^2\left[\bar{R} + 2V(T,\phi,\mu)\right] - D_\rho\mu D^\rho\mu + |\lambda|\mu^4 - \frac{1}{4}f^{\rho\sigma}f_{\rho\sigma} - \frac{1}{4}F^l_{\rho\sigma}F^{l\rho\sigma}\right\} \tag{5}$$

where the skew - symmetric tensors are defined as follows: $f_{\rho\sigma} \equiv (\partial_\sigma A_\rho - \partial_\rho A_\sigma)$ for the $U(1)_Y$ gauge field and: $F^l_{\rho\sigma} \equiv (\partial_\sigma b^l_\rho - \partial_\rho b^l_\sigma + g\epsilon_{jkl}b^j_\rho b^k_\sigma)$ for the $SU(2)_L$ non-abelian gauge fields of the Yang-Mills theory. Here we introduce the gauge vector bosons: $\vec{A}_\rho$, i.e. the electromagnetic vector potential, for the group $U(1)_Y$ and the three components in the isospin space: $(b^1_\rho, b^2_\rho, b^3_\rho)$ of the gauge vector boson: $\vec{b}_\rho$ for $SU(2)_L$. Y is the "weak hypercharge" operator. The term proportional to $|\lambda|$ is added for dynamical stability against unbounded field oscillations. In the present more advanced theory the parameter α, assumed to be a true constant, cannot be considered a free parameter. In facts the starting point of a rigorous standard contruction of any lagrangian theory leading to the field equations of GR consists of the adoption of the following density: $\frac{1}{2K^2}\sqrt{-g}R + \tilde{L}$ where $\tilde{L}$ is the lagrangian density of all fields other than gravitation [9]. Accordingly, we shall set: $\alpha \cdot \mu^2 = \frac{c^4}{16\pi G} = \frac{M_P^2}{16\pi}$ within $\hat{L}$, henceforth. $M_P = 1.22 \times 10^{19}\frac{GeV}{c^2}$ is the Planck mass.

3. The Higgs field and the Vacuum Energy in the Universe

The wide conceptual scenario opened by the preceeding chapter and the structure of the Lagrangian $\hat{L}$, Eq. (5) offer the possibility of inquiring about the implications of the dark energy content of the Universe within some relevant aspects of the submicroscopic world of the elementary particles. The supposed mass generation properties of the Higgs field, and his pervasiveness that parallels analogous aspects of the inflationary field has already stimulated in recent years a wealth of research in the field referred to as: "Higgs inflation" [17, 18, 19]. In what follows we shall find that the connection between the two fields can be demonstrated in the context of the present conformally-covariant theory in which the "classical" GR approach based on the lagrangian Eq. (5) is associated with a spontaneously broken $SU(2)_L \otimes SU(1)_L$ gauge theory. We shall briefly consider this theory in the framework of the standard electroweak "*theory of leptons*" due to Steven Weinberg and Abdus Salam [24, 25]. Let us introduce a *complex* iso-doublet of scalar fields [26, 27]:

$$\vec{\mu} \equiv \begin{pmatrix} \mu^+ \\ \mu^0 \end{pmatrix} \tag{6}$$

that transforms as a $SU(2)_L$ doublet with heavy hypercharge: $Y_\mu = +1$ according to the Gell Mann-Nishijima relation for the electric charge: $Q = I_3 + \frac{1}{2}Y$. The weak-isospin projection I_3 and the weak-hypercharge are commuting operators: $[I_3, Y] = 0$ [26, 27]. Next, we introduce in the expression of$\hat{L}$, Eq. (5) the following replacement: $D_\rho \mu D^\rho \mu \rightarrow (D_\rho \vec{\mu})^\dagger \cdot (D^\rho \vec{\mu})$ where D_ρ expresses the Weyl-covariant, i.e. co-covariant, derivative, and the dot represents scalar vector multiplication in the isospin space. For the sake of simplicity we still keep the symbols: $\mu^2 \equiv (\vec{\mu}^\dagger \cdot \vec{\mu})$, and: $\mu^4 \equiv (\vec{\mu}^\dagger \cdot \vec{\mu})^2$ since these scalar quantities do not unsettle the *real* structure of the GR theory. The gauge co-covariant derivatives are:

$$D_\rho \vec{\mu} = [\partial_\rho + \phi_\rho]\,\vec{\mu} + \frac{i}{2}\left[g' Y A_\rho + g\hat{\tau} \cdot \hat{b_\rho}\right]\vec{\mu} \tag{7}$$

where the components of the vector $\hat{\tau}$: (τ_1, τ_2, τ_3) are the Pauli spin operators, g represents the coupling constant of the weak-isospin group $SU(2)_L$ and $\frac{g'}{2}$ represents the coupling constant of the weak-hypercharge group $U(1)_Y$. The presence of the inflaton field ϕ_ρ multiplied by $\vec{\mu}$ in the first *real* square bracket at the r.h.s. of the Eq. (7) is due to the weight $W(\vec{\mu}) = -1$ within the co-covariant derivative [?]. The Eq. 7 reproduces the spontaneous symmetry breaking theory of leptons [26, 27] by further introducing within the standard theory the formal change: $\partial_\rho \rightarrow [\partial_\rho + \phi_\rho]$. It will be shown that this change provides a contribution to the interaction between the Higgs and the inflaton fields we are now dealing with. We shall show that a larger contribution to this effect, proportional to α, is due to the $\phi_\rho \phi^\rho$ and $\phi^\rho_{\;|\rho}$ terms in the curvature R_W [?]. In summary, and most interesting, all that shows that it is precisely the conformally-covariant structure of the Weyl's geometry that establishes the connection between the two fields protagonists of the present analysis.
Let us briefly outline the standard theory of leptons on the basis of the classic texts [26, 27]. Consider in general terms the kinetic term of the dynamical equation for the scalar field $\vec{\mu}$:

$$\{(D_\rho \vec{\mu})^\dagger \cdot (D^\rho \vec{\mu}) - |\lambda|\mu^4\} = \{(\partial_\rho \vec{\mu})^\dagger \cdot (\partial^\rho \vec{\mu}) - \langle\mu\rangle^2 \mu^2 - |\lambda|\mu^4\} \tag{8}$$

If the mass term $\langle\mu\rangle^2$ is negative the continuous symmetry of the system's hamiltonian does not coincide with the symmetry of the vacuum and the condition of dynamical "spontaneous symmetry breaking" takes place. In virtue of a theorem [29] a possible Goldstone boson is associated with a generator of the gauge group that does not leave the vacuum invariant. We investigate this case by choosing the following vacuum expectation value of the scalar field Eq. (6):

$$\langle\phi\rangle_0 = \begin{pmatrix} 0 \\ \frac{v}{\sqrt{2}} \end{pmatrix} \tag{9}$$

where the vacuum field is: $v = \sqrt{\frac{-\langle\mu\rangle^2}{|\lambda|}}$. The vacuum is left invariant by any group generator G if: $G\langle\phi\rangle_0 = 0$. In our case we find all the $SU(2)_L$ and $U(1)_Y$ group generators: $\hat{\tau}_i, (i = 1, 2, 3)$ and Y operating on $\langle\phi\rangle_0$ break the symmetry of the vacuum. However the $U(1)_{EM}$ symmetry generated by the electric charge preserves the invariance since: $Q\langle\phi\rangle_0 \equiv \frac{1}{2}(\tau_3 + Y)\langle\phi\rangle_0 = \begin{pmatrix} 0 \\ 0 \end{pmatrix}$. The photon, therefore, remains massless and the other three gauge bosons will acquire mass. These are the heavy bosons: $W^\pm_\rho \equiv \frac{[b^1_\rho \mp ib^2_\rho]}{\sqrt{2}}$. and: $Z_\rho \equiv \frac{[-g' A_\rho + g b^3_\rho]}{\sqrt{g^2 + g'^2}}$. Expressed in very popular terms, this is the process referred to by Abdus Salam in a sentence reported in Ref. [12]: *"The massless Yang-Mills particles "eat" the Higgs particles (or fields) in order to gain weight, and the swallowed Higgs particles become ghosts"*. Upon expansion of the Lagrangian (8) about the shifted minimum of the Higgs potential we can investigate the small oscillations around the

IOP Conf. Series: Journal of Physics: Conf. Series **880** (2017) 012026 doi:10.1088/1742-6596/880/1/012026

vacuum v of of the "Higgs field", this one expressed by the field θ. This dynamics is expressed by the Lagrangian for small oscillations:

$$\bar{L} = \frac{1}{2}\Big[(\partial_\rho\theta)(\partial^\rho\theta) + 2\langle\mu\rangle^2\theta^2\Big] + \frac{v^2}{8}\left[g^2 \mid b_\rho{}^1 - ib_\rho^2 \mid^2 + (-g'A_\rho + gb_\rho^3)^2\right] + ... \qquad (10)$$

plus interaction terms. As shown by this equation, the Higgs field has acquired a $(mass)^2 \equiv (M_H)^2 = 2 \mid \langle\mu\rangle \mid^2$. The above sentences express in a very summary form the well known results of the standard electroweak theory.

We may now inquire about the effects on the theory of the change: $\partial_\rho \to [\partial_\rho + \phi_\rho]$ introduced in our present analysis by the co-covariant derivative (7) as well of the general relativistic structure of the overall co-covariant lagrangian $\hat{L}$ (5). Note first that nothing in the standard formulation of the electroweak theory based on the $SU(2)_L \otimes U(1)_Y$ group specifies the mass of the Higgs boson and none of the conceivable applications to the conventional processes depends in any way upon the value of M_H. It may therefore appear that M_H can indeed be considered a "free parameter" of the standard electroweak theory even if general constraints have been considered by some authors by imposing certain requirements of internal consistency [10].
This is not the case with our present theory. Since the Eq. (8) is formally included into the extended lagrangian *Eq.* (5), we can easily transfer to the last one all the physical conceptions and the theoretical considerations addressed so far to Eq. (8). In particular, the Riemann curvature R and the Weyl curvature R_W appearing in *Eq.* (5) are now involved in the spontaneous symmetry breaking scenario of the electroweak theory. On the other hand, they can also be expressed in terms of the relevant cosmological quantities: i.e. the inflation potential: $V_\Lambda(T, \phi)$, its ϕ-derivative: $V'_\Lambda(T, \phi)$ and the inflation vector field:ϕ_ρ. Consequently, and very important, the dynamical behaviour of all interactions affecting the properties of the elementary particles, including the Higgs field, is directly determined by these cosmological quantities, and then by the overall dynamics of the Universe. We believe that this is a striking result.
The present theory is very general and applies to all kinds of potential functions: $V_\Lambda(T, \phi)$. However, in view of what we believe to be the most interesting result of the present work, i.e. the resolution of the "Cosmological Constant Paradox" (referred to as: "Λ-Paradox", hereafter), in what follows we shall mostly deal with the exponential functions of ϕ: i.e. $V_\Lambda(T, \phi) \propto e^{-n\phi}$ and to the linear superpositions of these functions. There n is a "real number", i.e. either positive or negative, integer or fractional [16]. Within this restriction we claim the "universality" of the theory and then in what follows we shall consider only the case: n=2, as an example. For any other (n) the theoretical steps can be reproduced identically albeit the final numerical results of the calculations will be different.
Let us first define an "Effective Cosmological Potential":

$$V_{eff}(T, \phi) \equiv \left[V_\Lambda(T, \phi) + \frac{1}{2}\frac{\partial V_\Lambda(T, \phi)}{\partial\phi}\right] \qquad (11)$$

that can be either positive or negative depending on the sign and size of the proportionality parameters of $V_\Lambda(T, \phi)$ and/or of its ϕ-derivative. In case of a negative derivative a very small value of $|V_{eff}|$ may result from the sum of two very large contributions with opposite sign. As said, the contribution V_Λ is the vacuum energy, i.e. the dark-energy content of the Universe, conceptually connected with the Einstein's "Cosmological Constant" [28]. The quantity $|V_{eff}| \equiv M_{eff}^2$ represents the corresponding *measured* quantity. The above effect, expressed by the size of the "mass-reduction parameter": $\zeta \equiv \sqrt{\frac{|V_{eff}|}{|V_\Lambda|}} << 1$, indeed a general property of the Universe, can lead to a consequence of cosmological relevance since it represents a clue towards the resolution of the celebrated "Λ-Paradox" [20]. A conformally-covariant

solution of the Paradox based on Eq. (11) will be given later in this paper.
In agreement with our program and with the discussion above, we turn our attention to the evaluation of the mass of the Higgs boson, M_H. By considering Eq. (10), by collecting all terms proportional to μ^2 in $Eq.$ (5) and by expressing the Riemann curvature scalar in terms of $V_\Lambda(T,\phi)$, $V'_\Lambda(T,\phi)$ and: $\phi_\rho\phi^\rho$ [?], we find the expression:

$$M_H^2 = 24\alpha|V_{eff}| \tag{12}$$

We find that the spontaneous symmetry breaking condition implies: $V_{eff} < 0$. Remind the expression of α and make the simplifying assumption: $M_H^2 \cdot (2|\mu|^2) \to M_H^4$. This leads immediately to an expression of the Higgs mass:

$$M_H \simeq \left[\frac{3}{\pi}(M_{eff}^2 \cdot M_P^2)\right]^{\frac{1}{4}} \tag{13}$$

Interestingly enough, according to our theory the mass of the Higgs boson is proportional to the geometrical mean of the Planck mass and of the mass M_{eff} accounting for the *measured* cosmic vacuum-energy: $M_H = \left[\frac{3}{\pi}\right]^{\frac{1}{4}} \times \sqrt{M_P \cdot M_{eff}}$.
In conclusion, the present theory, which lies on the Weyl's geometrical foundations, prescribes that the value of the mass of any elementary particle belonging to the submicroscopic world depends on the average vacuum energy content of the Universe. This because it is precisely the Higgs field the source of the mass of all elementary quantum particles.
By taking into account the experimental datum measured by the joint ATLAS and CMS Collaborations at CERN in the year 2012:$M_H = 125.09(GeV/c^2)$ the corresponding size of the measurable dark-energy potential is found: $V_{eff} = 1.717 \cdot 10^{-12}(eV/c^2)^2$ [31, 32, 33].

The conceptual relevance of the "Effective Cosmological Potential" may be further enlightened by carring out the variational procedure respect $g_{\rho\sigma}$ on the Lagrangian $\hat{L}$ expressed by Eq. (5). For the sake of completeness we also include here the energy-momentum tensor $T_{\rho\sigma}$ due to external, unspecified matter and fields. The Euler-Lagrange equation consists of the Einstein's equation:

$$R_{\rho\sigma} - \frac{1}{2}g_{\rho\sigma}R - g_{\rho\sigma}\bar{\eta}\left[2V_{eff}(T,\phi) + \frac{1}{2}\partial_\rho\phi\partial^\rho\phi\right] = K^2 T_{\rho\sigma} \tag{14}$$

where:$\bar{\eta} \equiv [3 - (2\alpha)^{-1}]$. This equation is interesting because it shows that the generally very large "Inflation Potential" $|V_\Lambda(T,\phi)|$ appearing in the Lagrangian $\hat{L}$, Eq. (5) is replaced by the far smaller: $|V_{eff}(T,\phi)|$ within the corresponding Einstein equation. The expression within the square brackets in (14) is the *measured* "Cosmological Constant", Λ:

$$\Lambda \equiv \bar{\eta}\left[2V_{eff}(T,\phi) + \frac{1}{2}\partial_\rho\phi\partial^\rho\phi\right] \tag{15}$$

As we shall see: $|V_{eff}(T,\phi)| << |V_\Lambda(T,\phi)|$, and then: $\zeta << 1$. We stress here that the formal replacement $|V_\Lambda(T,\phi| \to |V_{eff}(T,\phi)|$ is precisely due to the application of the dynamical equation (3) to the expression of the Weyl-curvature R_W given in Ref. [?]. If the Weyl curvature is considered nonexistent in the source lagrangian $\hat{L}$, Eq. (5), as assumed in the context of Riemann's differential geometry (where: $R_W = 0$) the replacement $|V_\Lambda(T,\phi| \to |V_{eff}(T,\phi)|$ is not realized. This is precisely the origin and the real essence of the Λ-Paradox. Indeed, we shall see that the above replacement is the key argument for the resolution of the Paradox we shall propose later in this paper.

4. The Vector - Meson particle

If the "integrability condition" of the Weyl geometry, i.e. $\phi_\rho = \partial_\rho \phi$ and: $\phi_{\rho\sigma} = 0$, is relaxed, the quantities ϕ and ϕ_ρ must be considered independent variables of the theory and this one gets even richer. In particular, the inner structure of the Weyl geometry does not change, since the parallel displacement of vectors involves directly the vector-field ϕ^ρ, as said. Leaving aside all complications and deferring an exact analysis to future work, we may immediately apply the mathematical methods adopted in the previous Sections to the Lagrangian $\hat{L}$, Eq. (5). The variation respect to the *Weyl vector* $\phi_\rho(x)$ leads, via the gauge-fixing $\partial_\rho \phi^\rho = 0$, to the following Proca equation expressing the dynamics of a massive vector-meson, ϕ^ρ:

$$\left[\nabla_B + (\xi \frac{M_P c}{\hbar})^2\right] \phi^\rho = 0 \tag{16}$$

where M_P is the Planck mass and $\xi = \sqrt{2}(\beta)^{-1} \ll 1$ expresses the coupling of the particle to gravitons. It has been found that the massive meson ϕ^ρ, while strongly interacting with gravitons, does not interact with any $spin - \frac{1}{2}$ or $spin - 1$ elementary particle of the Standard Model. In other words, it is not coupled minimally to photons, hadrons and light or massive leptons [30, 34]. All these properties make this geometrical entity eligible for being considered an optimum candidate for Cold Dark Matter (CDM), the elusive object wich is now under (baffling) investigation in a large number of laboratories around the world. As it is well known, the (CDM) may consist of a Weakly Interacting Massive Particle (WIMP), a stable SUSY particle, a light neutralino with a mass of the order of $10^2(GeV/c^2)$ or even an "axion" particle with a mass as low as $10^{-5}(GeV/c^2)$ [37, 16]. Large concentrations of (CDM) have been detected in zones of the Universe characterized by a large inhomogeneous gravitational energy density. Indeed the CDM predominately clusters on the scale of galaxies. In summary, today the size of the mass and any other physical property of (CDM) is inferred from some sophisticated quantum theory based on an attributed "physical model".

Waiting for any novel, convincing experimental evidence, we refrain from considering further the putative identification of (CDM) with the ϕ^ρ-particle. We tend nevertheless to believe, according to a perspective biased by our "geometrical model", that the mass of this particle, i.e. the value of the parameter ξ in Eq. (16), must depend, once again, on a property of the Universe via some dynamical mechanism similar to the one already considered in this work. If the exact size of the mass of the ϕ^ρ-particle is presently out of reach, we are nevertheless able to determine the average size of this quantum excitation in the Universe. The Riemann curvature R is obtained by a standard variational procedure applied to $\hat{L}$ in order to find the Einstein equation.Then, both sides of this equation are multiplied by the metric tensor, by keeping in mind that, in D=4: $g_{\rho\sigma}g^{\rho\sigma} = 4$. It is expressed by:

$$R = 2\bar{\eta}\left[4|V_{eff}| - \phi_\rho \phi^\rho\right] \tag{17}$$

This equation is interesting since it shows the relative sizes of the two dominant contributions to the Universe curvature, i.e. the vacuum-energy, V_{eff} and the energy V_{DM} associated with the vector-meson ϕ^ρ. The mass of the ϕ^ρ field is:

$$M_{DM} \simeq \sqrt{\langle \phi_\rho \phi^\rho \rangle} \simeq 2\sqrt{|V_{eff}|} = 2.62 \cdot 10^{-6}(eV/c^2) \tag{18}$$

since the energy density of the Universe is "critical" and the Universe is "flat", i.e.$R \simeq 0$. This was indeed the main result of the BOOMERANG experiment carried out on (CMB) by Paolo De Bernardis and his group in the year 2000 [35]. These results were confirmed by recent measurement carried out by the PLANCK mission [36].

IOP Conf. Series: Journal of Physics: Conf. Series **880** (2017) 012026 doi:10.1088/1742-6596/880/1/012026

In conclusion, we propose here the following conjecture of cosmological relevance:
"The result of all experiments carried out on the cosmic vacuum-energy, i.e. dark-energy or zero-point-energy or cosmological-constant, by our terrestrial or extraterrestrial measurement apparata (i.e. carried by orbiting satellites or spatial vehicles) always consists of the reduced value of the effective cosmological potential, V_{eff} and never of the full value of V_Λ. This last quantity, a quantum-mechanical property of the Universe, is inaccessible to the human perception". The conjecture is proved by the results of the BOOMERANG, PLANCK and other recent astrophysical experiments. It is also consistent with our equations: (12), (17), (18).
As an alternative solution, the above conjecture may raise a profound epistemological and gnoseological issue whether the very *calibration* process within the parallel displacement of vectors in curved spaces determines at a very fundamental level all measurement processes, i.e. is *an intrinsic feature* of all measurement processes, and reveals itself when the experiments are carried out over cosmological spaces and times.

5. The Cosmological Constant Paradox and its solution

The following quote by Richard Feynman is enlightening [38]: *"..Such a mass density would, at first sight at least, be expected to produce very large gravitational effects which are not observed. It is possible that we are calculating in a naive manner, and, if all of the consequences of the general theory of relativity (such as the gravitational effects produced by the large stresses implied here) were included, the effects might cancel out; but nobody has worked all this out. It is possible that some cutoff procedure that not only yelds a finite energy density for the vacuum-state but also provides relativistic invariance may be found. The implications of such a result are at present completely unknown."*.

Let us analyze the Λ-Paradox in *"a naive manner"*, as in Feynman's words. The vacuum energy in the Universe E_{vac}, i.e. the "zero-point energy" of the quantum fields associated with all existing quantum particles is evaluated by (QFT). For simplicity we consider here only the "photon", which is the subject of (QED) (Quantum Electro-Dynamics), the chapter of (QFT) accounting for the electromagnetic (e.m.) phenomena. To carry out the calculation of E_{vac} we should first evaluate the spatial density ρ_{em} of the available $\vec{k} - modes$, i.e. of the spatial vectors over which the photons propagate in the free space. Afterwards, each mode, which is modelled by QED as a quantum-mechanical oscillator, is multiplied by 2-times (because of the two orthogonal polarizations) the oscillator's "zero-point energy", which is $(h\omega/2)$, where (h) is the Planck constant. The frequency ω is taken to range from zero to a cutoff that may be assumed to be determined by the Planck length, $l_P = 1.616 \times 10^{-33}(cm)$: $\omega_c = (2\pi c/l_P)$. At last, the resulting spatial energy-density must be multiplied by the volume of the Universe, W_U in order to get the final result: E_{vac} . We can amuse ourselves by carrying out the sequence of these operations, and more than that, by putting numbers at the end of every step of the calculation. I limit myself to give here the final result.The size of final vacuum-energy content is found: $E_{vac} = \frac{2\pi\hbar c}{L_P^4} \times W_U = \frac{2\pi c^7}{\hbar G^2} \times W_U = 2.9 \times 10^{98}(Joules/cm^3) \times W_U(cm^3)$ Since the linear size of our Universe evaluated by a (CMB) analysis is: $L_U \sim 124$ billions of light-years, the volume is: $W_U \sim (1.61 \cdot 10^{87})(cm^3)$. Then, the value of the content of the vacuum-energy, due only to photons, is: $E_{vac} \sim 5 \times 10^{185}(Joules) \sim 3 \times 10^{204}(eV)$. To be more conservative, the frequency cutoff ω_c could be determined by the size of the Compton wavelength of the proton: $\lambda_c \simeq 2 \cdot 10^{-14}$ cm. In this case all the numerical figures for the energies and for the energy densities given above should be multiplied by the factor $\sim 10^{-76}$. Even in this case the size of all numbers keeps being impressive ! We should remind here that in the vast realm of modern Science, the (QED) theory and, in particular, the "vacuum-field" concept were the most tested paradigms, ever. Uncountable and impressively precise experiments involving the atomic

physics, the optical spontaneous emission, the Casimir effect, the atomic spectra, the Lamb's shift, the electron's magnetic moment etc. were the landmarks of the great success of the XX century Physics. The calculation carried out for photons should now be extended to the other existing particles and the corresponding energy contributions should be added, without any reasonable chance of mutual cancellations.
Interestingly enough, the Λ-Paradox was also analyzed in the framework of the electro-weak theory of leptons by Chris Quigg (2013) [10].
The Λ-Paradox consists of the mysterious, humongous discrepancy existing between the enormous size of the overall *calculated*: $(E_{vac}/c^2)^2 = |V_\Lambda(T,\phi)|$ and the very small size of the "Cosmological Constant" Λ, Eq. (15), *measured* today.

Our present theory offers a conformally-covariant solution of the Λ-Paradox. Consider the following expression for the inflationary potential which accounts for the overall vacuum-energy content of the Universe calculated by the standard methods of (QFT) as shown above:

$$V_\Lambda(T,\phi) = \tilde{C}(T) \times \exp\left[2(\epsilon - 1)\right]\phi \tag{19}$$

Where $\tilde{C}(T)$ is assumed independent of $\mu(x)$ and of $g_{\rho\sigma}(x)$. We have already noted that, since for any physical quantity: $X \rightarrow e^{\lambda(x)W(X)}X$, and $W(\sqrt{-g}) = +4$ the above solution corresponding to $n = 2$ is indeed a conformally covariant solution. The quantity ϵ is a dimensionless real number: $\epsilon \approx 0$. Plugging Eq. (19) into Eq. (11) gives the ratio: $\frac{|V_{eff}|}{|V_\Lambda|} = \epsilon$. This is indeed a most drastic "mass-reduction" effect with the parameter: $\zeta = \sqrt{|\epsilon|} \sim 10^{(-210)}$, where this number only accounts for the e.m. field considered by the above numerical evaluation of E_{vac}. By this argument, the enormous content of vacuum-energy in the Universe, calculated by the (QFT) methods as shown above, and expressed by $|V_\Lambda(T,\phi)|$ is made consistent by our theory, i.e. by our Eqs. (11) and (19), with the very small value of $|V_{eff}(T,\phi)|$ and then, of the "effective" Cosmological Constant, Λ, Eq. (15), *measured* today [6, 7]. Therefore the Λ-Paradox is resolved by our theory for n=2.
As already stressed, for any real number (n) the resolution of the Λ-Paradox is achieved quite generally by our theory for any exponential potential function: $V_\Lambda(T,\phi) \propto e^{-n\phi}$ by inserting in the Lagrangian expressed by Eg. (5) the corresponding value of the parameter $\eta(n)$. In the case of a linear superposition: $\sum_j V_\Lambda^j(T,\phi)$ of exponential functions: $V_\Lambda^j(T,\phi) \propto e^{-n_j\phi}$, where n_j are generally uncorrelated real numbers, it suffices to plug within the square brackets of the Lagrangian $\hat{L}$, Eq. (5) the sum: $\sum_j \eta(n_j)V_\Lambda^j(T,\phi)$. For the sake of clarity, it is convenient to collect here all these results valid for the general case: $V_{eff} = [V_\Lambda + (n)^{-1}V_\Lambda']$, the Higgs boson mass: $M_H^2 = 2 \cdot \eta(n) \cdot \alpha|V_{eff}|$, the cosmological constant:$\Lambda = [\frac{\eta(n)}{2} + \frac{\bar{\eta}}{2}]$. $\eta(n) \equiv [(6 - \alpha^{-1}) \times n]$, $\bar{\eta} \equiv [3 - (2\alpha)^{-1}]$ and: n = any real number, positive or negative, integer or fractional. For $n \neq 2$ the co-covariance of the solution of the Λ-Paradox is attained by the multiplication of $\tilde{C}(T)$ with $(M_P)^\gamma$ with a suitable γ-exponent.
The Λ-Paradox cannot be resolved in the context of Riemann's differential geometry.

6. Conclusions

In conclusion, by our present work we have established several bridges between at least three cultural domains of the scientific endeavour that traditionally are rather disconnected: the foundations of the differential geometry, here focussed on the inspiring ideas by Hermann Weyl, the general relativistic cosmology, here focused on the inflation process, and the modern field-theory focused on some crucial interactive properties of the elementary quantum particles. Our aim was to search, by an unitarian perspective, for any possible conceptual and theoretical connection existing between the pervasive quantum fields that actively dwell the most remote corners of the Universe. We believe that precisely there can be found the key for undersanding the inner essence of Nature.

As a final comment to our somewhat unusual proposal let's identify here some key arguments of the underlying logical scenario. (A) The main result of our work is the expression of V_{eff}, Eq. (11) by the inclusion of the ϕ-derivative: V'. This is possible in virtue of Eq. (3), indeed the key-equation of our work. (B) Since Eq. (3) is the Euler-Lagrange equation of a variational procedure on the "physical" Lagrangian L_ψ, the quantity ϕ must be a "physical" field, and not a mere "geometrical" entity. Otherwise, Eq. (3) would be meaningless. (C) The same "physical" Eq (3) establishes a direct and necessary link from physics to geometry via the expression $\phi^{\rho}_{|\rho}$ appearing in the Weyl curvature R_W. All this implies a lucky intersection of several nicely interweaved physical-geometrical concepts. The Riemann's geometry, indeed a beautiful theory, proves to be *too simple* within our scenario. The lack in its foundational premises of the "calibration", i.e. of the "gauge" concept, moves it out of the game.

In summary, the geometrical mechanism proposed by the present work represents a unifying scenario by which a unique quantum field appears to play, by different routes and under different forms, an essential role in determining the evolution of the Universe "at large" as well as, at the microscopic level and via the dynamics of the Weyl scalar curvature R_W, of the everyday quantum phenomenology [39, 40].This appears to be a glimpse into quantum gravity.

Acknowledgments

Thanks are due to Enrico Santamato, Valeria Ferrari and Claus Kiefer for valuable advices and collaboration. Most of the content of the present work was first presented at the Conference "Quantum and Beyond" held in Vaxjo, Sweden, in the month of June 2016.

References

[1] Starobinsky A 1980 *Physics Letters* **91** 99
[2] Guth A 1981 *Physical Review D***23** 347
[3] Linde A D 1982 *Physics Letters B* **108** 389
[4] Albrecht A and Steinhardt P 1982 *Phys. Rev. Lett.* **48** 1220
[5] De Martini F and Santamato E 2016 *ArXiv:1609.01149[gr-qc]3044* URL `http://arxiv.org/abs/1609.01149`
[6] Dodelson S 2003 *Modern Cosmology* (New York : Academic Press)
[7] Weinberg S 2008 *Gravitation and Cosmology* (New York: Wiley)
[8] Weyl H 1952 *Space, Time, Matter* 4th ed (New York: Dover Publications, Inc.)
[9] Lord E A 1979 *Tensors, Relativity and Cosmology* (New Delhi, New York: Tata McGraw-Hill Co.)
[10] Quigg C 1983 *Theories of the Strong, Weak and Electromagnetic interactions* (Menlo Park: Benjamin Publishing Co.)
[11] Dirac P 1973 *Proceedings Royal Society of London A* **333** 403
[12] Jammer M 2000 *Concept of Mass* (Princeton: Princeton University Press)
[13] Ehlers J and Schild A 1973 *Commununications Mathematical Physics* **32** 119
[14] Trautman A 2012 *Gen. Rel. and Gravitation* **44** 1581
[15] Scholz E 2012 *ArXiv:1206.1559[gr-qc]* URL `http://arxiv.org/abs/1206.1559`
[16] Amendola L and Tsujikawa S 2010 *Dark Energy* (New York: Cambridge University Press)
[17] Bezrukov F L and Shaposhnikov M E 2014 *Physics Letters B* **734** 249
[18] Salvio A and Mazumdar A 2015 *Physics Letters B* **750** 194
[19] Hamada Y and Kawai H and Oda K 2015 *Physical Review D* **92** 4509
[20] Zee A 2013 *Einstein gravity in a nutshell* (New York: Princeton University Press)
[21] Higgs P 1964 *Phys. Rev. Letters* **13** 508
[22] Englert F and Brout R 1964 *Phys. Rev. Letters* **13** 321
[23] Guralnik G and Hagen C and Kibble T 1964 *Phys. Rev. Letters* **13** 585
[24] Weinberg S 1964 *Phys. Rev. Letters* **19** 1264

[25] Salam A 1964 *Elementary Particle Physics* (Stockolm: Almqvist, Wicksell)
[26] Weinberg S 2013 *The Quantum Theory of Fields, I and II* (Cambridge, New York: Cambridge University Press)
[27] Georgi H 2009 *Weak interactions and Modern Particle Physics* (New York: Dover Publications, Inc.)
[28] Penrose R and Hawking S *The nature of space and time* (New York: Princeton University Press)
[29] Goldstone J and Salam A and Weinberg S 1962 *Physical Review* **127** 965
[30] Hochberg D and Plunien G 1991 *Physical Review D* **43** 3358
[31] Aad G and ATLAS collaboration 2012 *Physics Letters B* **718** 1
[32] Chatrchyan S and CMS collaboration 2012 *Physics Letters B* **718** 30
[33] ATLAS and CMS collaborations 2015 *Physical Review Letters* **114** 191803
[34] Chen H 1988 *Physical Review Letters* **61** 2182
[35] De Bernardis P 2000 *Nature* **404** 955
[36] PLANCK 2015 *ArXiv:1502.01589v3[gr-qc]* URL `http://arxiv.org/abs/1502.01589v3`
[37] Sanders R H 2010 *The Dark Matter problem* (New York: Cambridge University Press)
[38] Feynman R P 1965 *Quantum Mechanics and Path Integrals* (New York: McGraw-Hill Publishing Co.)
[39] De Martini F and Santamato E 2014 *Physica Scripta* **T163** 014015
[40] Santamato E and De Martini F 2015 *Foundations of Physics* **45** 858

Treating the Einstein-Hilbert action as a higher derivative Lagrangian: revealing the missing information about conformal non-invariance

Branislav Nikolić

Institute for Theoretical Physics, Zülpicher Straße 77, Universität zu Köln, 50937 Köln, Germany

E-mail: nikolic@thp.uni-koeln.de

Abstract. The Hamiltonian formulation of conformally invariant Weyl-squared higher derivative theory teaches us that conformal symmetry is expressed through particular first class constraints related to the absence of the three-metric determinant and the trace of the extrinsic curvature from the theory. Any term depending on them which is added to this theory breaks conformal invariance and turns these constraints into second class ones. Such second class constraints are missing in the standard canonical formulation of the conformally non-invariant Einstein-Hilbert theory. It is demonstrated that such constraints do appear if the theory is treated as a higher derivative one: if the extrinsic curvature is promoted to an independent variable, the apparently missing information about conformal behavior is revealed.

1. Which variables encode conformal transformation in a theory?

Conformal (Weyl rescaling) symmetry refers to invariance of a Lagrangian under the following transformation of the metric and the involved fields,

$$g_{\mu\nu}(x) \quad \to \quad \tilde{g}_{\mu\nu}(x) = \Omega^2(x) g_{\mu\nu}(x) \,, \qquad \phi(x) \quad \to \quad \tilde{\phi}(x) = \Omega^n(x)\phi(x) \,, \tag{1}$$

where n is usually called the "conformal weight" of any kind of field ϕ. Conformal transformation and related invariance are important concepts. They are studied in many fields of physics, along with symmetry under a conformal transformations of *coordinates* (which we will not talk about here). In particular, while studying theories of gravity, it plays a crucial role e.g. in the transformation between Jordan and Einstein frames in scalar-tensor theories of gravity [1], in renormalization procedures while studying effective theories of gravity [2, 3], as well as in understanding the interplay between the scale and conformal invariance [4]. The list is far from complete, but it gives one a taste of the application and importance of the conformal transformation and invariance.

The usual way of studying conformal features of a theory is to directly transform all the involved objects under the conformal transformation (1). The necessary calculations may become quite cumbersome and can either lead to an appearance of non-vanishing terms depending on Ω and its derivatives, or can lead back to the very same form of the theory in question, in which case all dependence on Ω cancels (in general, this is true up to a total divergence which might appear in the Lagrangian after transformation). In the latter case, one concludes that the theory is invariant under conformal transformation (up to a total divergence).

Let us demonstrate this on the example of a non-minimally coupled scalar field (the dimension will be assumed to be four, unless otherwise specified):

$$\mathcal{L}^{\varphi}=-\frac{1}{2}\sqrt{-g}\left(g^{\mu\nu}\partial_{\mu}\varphi\partial_{\nu}\varphi+\xi R\varphi^{2}\right), \tag{2}$$

where ξ is a dimensionless non-minimal coupling constant. For the scalar field, we will assume that $n=-1$, since in that case the kinetic term is unchanged after the conformal transformation. Studying conformal properties of this theory requires looking at the transformation of R, which is induced by the transformation of 10 variables of the metric; this transformation can be found in the literature,

$$R \quad\rightarrow\quad \tilde{R}=\frac{1}{\Omega^{2}}\left(R-6\,\frac{\Omega^{-1}}{\sqrt{-g}}\partial_{\mu}\left(\sqrt{-g}g^{\mu\nu}\partial_{\nu}\Omega\right)\right). \tag{3}$$

The resulting transformed Lagrangian reads:

$$\mathcal{L}^{\varphi} \quad\rightarrow\quad \tilde{\mathcal{L}}^{\varphi}=-\frac{1}{2}\left[\sqrt{-g}\,g^{\mu\nu}\partial_{\mu}\varphi\partial_{\nu}\varphi+(1-6\xi)\Omega^{-1}\partial_{\mu}\left(\sqrt{-g}\,g^{\mu\nu}\partial_{\nu}\Omega\right)\varphi^{2}+\xi\sqrt{-g}R\varphi^{2}\right]$$
$$+\frac{1}{2}\partial_{\mu}\left(\Omega^{-1}\varphi^{2}\sqrt{-g}\,g^{\mu\nu}\partial_{\nu}\Omega\right). \tag{4}$$

The second term spoils conformal invariance. The piece proportional to -6ξ comes from the non-minimal coupling term $\xi R\varphi^{2}$, while the piece proportional to 1 comes from the kinetic term, using the Leibniz rule (giving rise to the last term). We see that the suggestive value $\xi=1/6$ exactly cancels the $\Omega-$dependence from the original kinetic term. This is called conformal coupling. Hence one says that the Lagrangian of the conformally coupled massless scalar field is conformally invariant up to a total divergence if the scalar field transforms as $\tilde{\varphi}=\Omega^{-1}\varphi$.

It will be demonstrated now that an appropriate decomposition of the variables suggests that *only one* of the resulting configuration variables is responsible for the conformal transformation of the Lagrangian — the determinant of the metric (raised to an appropriate power). This decomposition is deduced from the following reasoning. The metric has 10 independent elements, i.e., 10 degrees of freedom to start with. Its determinant is an object which carries only one degree of freedom — it is a scalar density. As a result of (1), the determinant transforms as

$$\sqrt{-g} \quad\rightarrow\quad \sqrt{-\tilde{g}}=\Omega^{4}\sqrt{-g} \tag{5}$$

in four dimensions. We can then say that a single power of Ω is produced from each $(\sqrt{-g})^{1/d}$. In this way, the conformal transformation of the metric can be thought of as if it results from a single degree of freedom only — the determinant. It is thus suggestive to decompose the metric into its determinant part and the remaining 9 degrees of freedom. This is achieved in the following way:

$$A:=\left(\sqrt{-g}\right)^{1/d}, \qquad g_{\mu\nu}=A^{2}\bar{g}_{\mu\nu}, \qquad g^{\mu\nu}=A^{-2}\bar{g}^{\mu\nu}, \qquad \det\bar{g}_{\mu\nu}=1 \tag{6}$$

in d dimensions, where $\bar{g}_{\mu\nu}$ is usually called "unimodular metric", due to its unit determinant (which is why it has $d-1$ degrees of freedom), or "conformal part of the metric", since it is invariant under conformal transformation. This is an *irreducible* decomposition under conformal (Weyl rescaling) transformation into its *scale density* (non-conformal) A and unimodular (conformal) $\bar{g}_{\mu\nu}$ part. Hence, it can now be understood that the conformal transformation of the metric results from the conformal transformation of its scale density (A) degree of freedom only. But nothing prevents one from extending this logic to *all* variables appearing in a Lagrangian. Namely, conformal transformation of the scalar field suggests the following decomposition:

$$\varphi = A^{-1}\chi\,, \tag{7}$$

where χ is a *conformally invariant scalar density of weight* $1/d$. Hence, the *geometric* degree of freedom (responsible for the conformal transformation) is separated from what one may call the "pure" matter degree of freedom, the scalar density χ.

With this decomposition, one may reinterpret the conformal behavior of the Lagrangian (4). Namely, decomposition of the Ricci tensor induced by the unimodular decomposition of the four-metric results in (see Appendix in [5])

$$\begin{aligned} R &= \frac{1}{A^2}\left(\bar{R} - \frac{2(d-1)}{A^2}\left[A\,\partial_\mu\left(\bar{g}^{\mu\nu}\partial_\nu A\right) + \frac{(d-4)}{2}\bar{g}^{\mu\nu}\partial_\mu A\,\partial_\nu A\right]\right) \\ &\overset{d=4}{=} \frac{1}{A^2}\left(\bar{R} - \frac{6}{A}\partial_\mu\left(\bar{g}^{\mu\nu}\partial_\nu A\right)\right)\,, \end{aligned} \tag{8}$$

where $\bar{R}$ is the part of the Ricci scalar determined only by the unimodular part of the metric, and is therefore conformally invariant. The conformal transformation of the Ricci scalar is now obtained simply by transforming $A \to \Omega A$ only. It is easy to see that

$$\frac{1}{A^3}\partial_\mu\left(\bar{g}^{\mu\nu}\partial_\nu A\right) \quad \to \quad \frac{1}{\Omega^2 A^3}\partial_\mu\left(\bar{g}^{\mu\nu}\partial_\nu A\right) + \frac{\Omega^{-3}}{\sqrt{-g}}\partial_\mu\left(\sqrt{-g}g^{\mu\nu}\partial_\nu\Omega\right) \tag{9}$$

indeed gives rise to the conformal transformation (3) already known from the literature (the last term is expressed in terms of original variables, for reference). Furthermore, applying the decomposition to (2), we obtain the Lagrangian

$$\mathcal{L}^\chi = -\frac{1}{2}\left[\bar{g}^{\mu\nu}\partial_\mu\chi\partial_\nu\chi + (1-6\xi)A^{-1}\partial_\mu\left(\bar{g}^{\mu\nu}\partial_\nu A\right)\chi^2 + \xi\,\bar{R}\chi^2\right] + \frac{1}{2}\partial_\mu\left(A^{-1}\chi^2\,\bar{g}^{\mu\nu}\partial_\nu A\right)\,, \tag{10}$$

reformulated for the non-minimally coupled scalar *density* field χ. An easy calculation shows that $A \to \Omega A$ reproduces the known conformal behavior (4). One can now understand the conformal features of this potentially conformally invariant Lagrangian in terms of the dependence on the scale density A. It is a simple matter to deduce that A is completely eliminated (up to a total divergence) only for the $\xi = 1/6$, which is the conformal coupling, resulting in the conformally invariant Lagrangian.

The lesson to take away from this introductory section is the following. Suppose a given theory is formulated in terms of the scale density and conformally invariant variables; it then follows that if the theory does not depend on scale density (up to a total divergence), it is conformally invariant.

2. Which constraints encode conformal features of a theory?

How can we learn about conformal behavior of theories in the Hamiltonian formulation? In studying the Hamiltonian formulation of metric theories of gravity, one usually uses the 3+1 decomposition of spacetime [6], such that the central role is taken over by the three-metric, instead of the four metric. Namely, the four-metric is decomposed as

$$g_{\mu\nu} = \begin{pmatrix} -N^2 + N_i N^i & N_i \\ N_i & h_{ij} \end{pmatrix}, \quad \sqrt{-g} = N\sqrt{h}\,, \tag{11}$$

IOP Conf. Series: Journal of Physics: Conf. Series **880** (2017) 012027 doi:10.1088/1742-6596/880/1/012027

where N and N^i are the lapse and the shift functions, and h_{ij} is the three-metric, the metric intrinsic to the 3-dimensional hypersurface, with its determinant denoted by h. Of importance is another tensor called the extrinsic curvature, defined by

$$K_{ij} = \frac{1}{2N}\left(\dot{h}_{ij} - 2D_{(i}N_{j)}\right), \quad K = h^{ij}K_{ij} = \frac{1}{N}\left(\frac{\dot{\sqrt{h}}}{\sqrt{h}} - D_iN^j\right), \tag{12}$$

where K is its trace (note that it represents time evolution of the three-volume $\sqrt{h}$), and D_i is the spatial covariant derivative. Equipped with these tensors, one can canonically formulate any metric theory of gravity via what is known as the ADM formalism (see e.g. [6]). However, for purposes discussed here, guided by the lesson from the previous section, let us go a step further and decompose the $3+1$ variables in such a way that their conformal transformation is completely determined only by the transformation of the *three-dimensional scale density* (the determinant of the three-metric $\sqrt{h}$), in analogy to the full four-dimensional case, discussed in the previous section. These variables are introduced in [5] (see also [8]) and are defined by

$$a := (\sqrt{h})^{1/3}, \quad \bar{h}_{ij} := a^{-2}h_{ij}, \quad \chi := a\varphi \tag{13}$$

$$\bar{N}^i := N^i, \quad \bar{N}_i := a^{-2}N_i, \quad \bar{N} := a^{-1}N, \tag{14}$$

$$\bar{K}^{\mathrm{T}}_{ij} := a^{-1}K^{\mathrm{T}}_{ij} = \frac{1}{2\bar{N}}\left(\dot{\bar{h}}_{ij} - 2\left[\bar{D}_{(i}\bar{N}_{j)}\right]^{\mathrm{T}}\right), \quad \bar{K} := \frac{aK}{3} = \frac{1}{\bar{N}}\left(\frac{\dot{a}}{a} - \frac{1}{3}D_iN^i\right), \tag{15}$$

where $\bar{h}_{ij}$ is now our unimodular part of the metric, and a is the three-dimensional scale density, in analogy with (6). Note that N^i is the only unchanged variable. Superscript "T" denotes a traceless object, $(i...j)$ represent symmetrization of indices, while $\bar{D}_i$ acts only with the conformal part of the three-dimensional covariant derivative. They may be called "unimodular-conformal variables". We have added the matter scalar density χ for completeness. It can be easily checked that only two of the above variables transform under a conformal transformation: the scale density a and the "trace density" $\bar{K}$. The trace density $\bar{K}$ is the only variable determined by the scale density a (through K, see (12)), and contains its temporal evolution, i.e. its velocity. Therefore, we expect that the conformal properties of any theory may be expressed in terms of the scale density a and the trace density $\bar{K}$, and so we keep an eye on them throughout the derivations.

The Einstein-Hilbert (EH) Lagrangian is usually given in the ADM form, but we further reformulate it in terms of variables (13)-(15) to obtain

$$\mathcal{L}^{\mathrm{E}} = \frac{1}{2\kappa}N\sqrt{h}\left({}^{(3)}R + K_{ij}K^{ij} - K^2\right) = \frac{1}{2\kappa}\bar{N}a^2\left(a^2\,{}^{(3)}R + \bar{K}^{\mathrm{T}2}_{ij} - 6\bar{K}^2\right), \tag{16}$$

where $\kappa \equiv 8\pi G$, $\bar{K}^{\mathrm{T}2}_{ij} \equiv \bar{K}^{\mathrm{T}}_{ij}\bar{h}^{ia}\bar{h}^{jb}\bar{K}^{\mathrm{T}}_{ab}$, and ${}^{(3)}R$ is left undecomposed for simplicity, but recall that it contains a dependence on a. It is now obvious that a and $\bar{K}$ are responsible for conformal transformations of the EH Lagrangian. It is well known [6] that the canonical EH theory is a *constrained* theory, meaning that it contains constraint equations. These constraints are the consequence of $\dot{N}$ and $\dot{N}^i$ not appearing in the Lagrangian, which gives rise to their vanishing momenta as primary constraints. Time preservation of these constraints gives rise to the secondary constraints called Hamiltonian $\mathcal{H}_\perp$ and momentum constraint $\mathcal{H}_i$, respectively. In unimodular-conformal variables, it is the absence of $\dot{\bar{N}}$ and $\dot{\bar{N}}^i$ which gives rise to the primary-secondary pair of constraints having the following form,

$$p_{\bar{N}} \approx 0 \quad \Rightarrow \quad \mathcal{H}_{\perp}^{\mathrm{E}} = -\frac{\kappa}{12}p_a^2 + \frac{2\kappa}{a^2}\bar{h}_{ik}\bar{h}_{jl}\bar{p}^{ij}\bar{p}^{kl} - \frac{a^2}{2\kappa}\left({}^{(3)}\bar{R} - \frac{4}{a^2}\left[a\,\partial_i\left(\bar{h}^{ij}\partial_j a\right) - \frac{1}{2}\bar{h}^{ij}\partial_i a\,\partial_j a\right]\right) \approx 0\,, \tag{17}$$

$$p_i \approx 0 \quad \Rightarrow \quad \mathcal{H}_i^{\mathrm{E}} = -2\partial_k\left(\bar{h}_{ij}\bar{p}^{jk}\right) + \partial_i\bar{h}_{jk}\bar{p}^{jk} - \frac{1}{3}D_i\left(a\,p_a\right) \approx 0\,, \tag{18}$$

where the momenta conjugate to the decomposed parts of the metric are

$$p_a = \frac{\partial\mathcal{L}^{\mathrm{E}}}{\partial\dot{a}} = -\frac{6a}{\kappa}\bar{K}\,, \quad \bar{p}^{ij} = \frac{\partial\mathcal{L}^{\mathrm{E}}}{\partial\dot{\bar{h}}_{ij}} = \frac{a^2}{2\kappa}\bar{K}^{\mathrm{T}}_{ab}\bar{h}^{ia}\bar{h}^{jb}\,, \tag{19}$$

and can be shown to be related to the trace and traceless parts of the ADM momentum $p^{ij} = \sqrt{h}\left(K^{ij} - h^{ij}K\right)/(2\kappa)$. The total Hamiltonian is given by

$$H^{\mathrm{E}} = \int \mathrm{d}^3x\,\left\{\bar{N}\mathcal{H}_{\perp}^{\mathrm{E}} + N^i\mathcal{H}_i^{\mathrm{E}} + \lambda_{\bar{N}}p_{\bar{N}} + \lambda^i p_i\right\}\,. \tag{20}$$

Note that we have decomposed the Ricci scalar above as well [5], in order to identify all terms where the scale density a resides. The sign "$\approx$" is Dirac's "weak equality", used in the constraint analysis to enable evaluation of the Poisson brackets. The only feature of canonical GR of importance for the present discussion is the fact that the constraints (17) and (18) are *first-class* constraints. This means that they are related to a certain symmetry of the theory, in this case the covariance of the Lagrangian. Whenever one encounters first class constraints, it is a signal that a symmetry is present in the theory, and vice versa. It is now important to keep in mind that these are *the only* constraints in the canonical EH theory.

Apart from GR, some further examples of Hamiltonian formulation of such theories can be found in e.g. [5, 9, 10, 11, 12], which contain curvature-squared terms $R^2, R^2_{\mu\nu}, R^2_{\mu\nu\rho\sigma}$. These theories are considered either as an extension of GR or as an alternative to it, from both the classical and quantum point of view, and are usually referred to as "higher derivative theories", because they are non-linear in second order time derivatives of the metric. Hence, unlike pure GR, the order cannot be reduced by simple partial integration and using a boundary term. One then usually promotes the extrinsic curvature to an *independent* variable, by adding a constraint to the Lagrangian,

$$\mathcal{L}(N, N^i, h_{ij}, \dot{h}_{ij}, \ddot{h}_{ij}) \quad \rightarrow \quad \mathcal{L}'(N, N^i, h_{ij}, K_{ij}, \dot{K}_{ij}) - N\sqrt{h}\lambda^{ij}\left[2K_{ij} - \frac{1}{N}\left(\dot{h}_{ij} - 2D_{(i}N_{j)}\right)\right]\,, \tag{21}$$

where $\mathcal{L}'$ is the same Lagrangian $\mathcal{L}$, but expressed in terms of new variables K_{ij}. This "hides" the second derivatives $\ddot{h}_{ij}$ and one can then proceed with the Hamiltonian formulation. The added constraint with Lagrange multiplier λ^{ij} follows from (12) and takes care of the additional number of degrees of freedom. It is important to realize that the momentum conjugate to the metric components is $p^{ij} = \sqrt{h}\lambda^{ij}$, i.e., is undetermined (λ^{ij} can be eliminated from the configuration space from the start, see [12]). Using unimodular-conformal variables, it is possible to further reformulate such theories, leading to a great insight into their conformal features expressed in terms of a and $\bar{K}$. As an example, we will take the Weyl-squared theory, whose Hamiltonian formulation is investigated in detail in [5], along with some of its quantum aspects in [8],

$$\mathcal{L}^{\mathrm{W}} := -\frac{\alpha_{\mathrm{W}}}{4}\int \mathrm{d}^4x\,\sqrt{-g}\,C_{\mu\nu\lambda\rho}C^{\mu\nu\lambda\rho}\,. \tag{22}$$

Here, $C^{\mu}{}_{\nu\lambda\rho}$ is the Weyl tensor, and $\alpha_{\rm W}$ is a positive dimensionless coupling constant. Due to covariance, the Hamiltonian and momentum constraints are present [11], as in GR, but they will not be discussed here (see [5] for more details). However, (22) enjoys conformal symmetry due to the conformal invariance of the Weyl tensor, and this richer structure is the reason why we choose this particular example. Consequentially, this theory generates *additional* first class constraints related to the generator of gauge conformal transformations, as first determined by [13]. They arise because a and $\bar{K}$ do not appear in the Weyl-squared theory — an important fact which seems not to have been completely realized in the literature until the theory was treated in unimodular-conformal vriables in [5]. These constraints read

$$\bar{P} = \frac{\partial \mathcal{L}^{\rm W}}{\partial \dot{\bar{K}}} \approx 0 \quad \Rightarrow \quad \dot{\bar{P}} = \mathcal{Q}^{\rm W} = a p_a \approx 0\,, \tag{23}$$

whose Poisson bracket trivially vanishes. The constraint $\bar{P} \approx 0$ appears because the Lagrangian does not depend on $\dot{\bar{K}}$, and as a consequence, $\mathcal{Q}^{\rm W}$ effectively states the same for $\dot{a}$, thanks to the fact that the Lagrangian is independent of $\bar{K}$. It is important to keep in mind that the appearance of these additional constraints is intimately related to the treatment of the extrinsic curvature as an *independent variable*, in contrast to the Lagrangian formulation of GR, where K_{ij} is not a configuration variable.

We are now interested in the change of constraints (23) if some non-conformal term is added to the Weyl-squared action. Since this is studied in detail in [5], we borrow here only the results. Namely, if the added term is the Lagrangian of the non-minimally coupled scalar field (2), the Hamiltonian analysis shows that the constraints (23) are turned into second class constraints and read

$$\bar{P} \approx 0 \quad \Rightarrow \quad \mathcal{Q}^{\rm W\chi} = a p_a + (1-6\xi)\chi p_\chi - 6\xi\,(1-6\xi)\,\bar{K}\chi^2 \approx 0\,, \tag{24}$$

with their non-weakly-vanishing Poisson bracket $\left\{\bar{P},\, \mathcal{Q}^{\rm W\chi}\right\} = 6\xi\,(1-6\xi)\,\chi^2$ (we omit the Dirac delta function). It is clear that the first class nature of these constraints is recovered only if $\xi = 1/6$, i.e., only if the scalar field is conformally coupled; see [5] for details. Looking more closely, we see that ξ controls the appearance of the trace density $\bar{K}$. Thus we can conclude that the appearance of a and $\bar{K}$ in the Lagrangian is responsible for the breaking of conformal symmetry. It follows that the information about conformal features is *not lost* upon breaking of symmetry — it is still contained in the *second-class* constraints.

If that is the case, then where is the information about the broken conformal symmetry in the canonical EH theory? Recall that there one has only first class constraints (17)-(18). Where are the second class constraints expressing its conformal non-invariance? To answer this question, one has to realize that any additional constraints would further reduce the number of degrees of freedom in GR, which we know is 2. Thus, additional constraints cannot appear unless additional variables are introduced to the theory. On the other hand, we have the Weyl-squared theory whose configuration space necessarily has to be expanded by additional variables, i.e. the extrinsic curvature, and we have seen that these variables ($\bar{K}$ in particular) give rise to additional constraints related to the conformal (non-)invariance. Let us then combine these two observations and propose to treat the EH theory as if it were a higher derivative theory, and see if we can learn something about its conformal behavior.

3. General Relativity treated as a higher derivative theory: manifest absence of conformal non-invariance

The EH Lagrangian does not contain second order time derivatives of the three-metric (after the partial integration), and it is not necessary to introduce the extrinsic curvature as an independent variable. But it is not harmful, either. Let us take for the Lagrangian $\mathcal{L}$ in (21) the EH Lagrangian (16) and call it $\mathcal{L}^{\mathrm{EH}}$. After some calculation similar to that in [5], the total Hamiltonian can be put in the following form,

$$H^{\mathrm{EH}} = \int \mathrm{d}^3x \left\{ \bar{N}\mathcal{H}_{\perp}^{\mathrm{EH}} + N^i\mathcal{H}_i^{\mathrm{EH}} + (2\bar{N}\bar{K}_{ij}^{\mathrm{T}})\mathcal{T}_{\mathrm{EH}}^{ij} + \left(\bar{N}\bar{K}\right)\mathcal{Q}^{\mathrm{EH}} + \lambda_{\bar{N}}p_{\bar{N}} + \lambda^i p_i + \Lambda_{ij}\bar{P}^{ij} + \Lambda_{\bar{K}}\bar{P} \right\} , \tag{25}$$

where $\bar{N}, N^i, (2\bar{N}\bar{K}_{ij}^{\mathrm{T}}), \left(\bar{N}\bar{K}\right), \lambda_{\bar{N}}, \lambda^i, \Lambda_{ij}, \Lambda_{\bar{K}}$ are Lagrange multipliers, and the following are constraints:

$$p_{\bar{N}} \approx 0 \quad \Rightarrow \quad \mathcal{H}_{\perp}^{\mathrm{EH}} = -\frac{a^2}{2\kappa}\left({}^{(3)}\bar{R} - \bar{K}_{ij}^{\mathrm{T}2} + 6\bar{K}^2 \right) + \frac{2}{\kappa}\left[a\,\partial_i\left(\bar{h}^{ij}\partial_j a\right) - \frac{1}{2}\bar{h}^{ij}\partial_i a\,\partial_j a \right] \approx 0 , \tag{26}$$

$$p_i \approx 0 \quad \Rightarrow \quad \mathcal{H}_i^{\mathrm{EH}} = -2\partial_k\left(\bar{h}_{ij}\bar{p}^{jk}\right) + \partial_i\bar{h}_{jk}\bar{p}^{jk} - \frac{1}{3}D_i\left(a\,p_a\right) \approx 0 , \tag{27}$$

$$\bar{P} \approx 0 \quad \Rightarrow \quad \mathcal{Q}^{\mathrm{EH}} \equiv a p_a + \frac{6a^2}{\kappa}\bar{K} \approx 0 , \tag{28}$$

$$\bar{P}^{ij} \approx 0 \quad \Rightarrow \quad \mathcal{T}_{\mathrm{EH}}^{ij} \equiv 2\bar{p}^{ij} - \frac{a^2}{\kappa}\bar{K}_{ab}^{\mathrm{T}}\bar{h}^{ai}\bar{h}^{bj} \approx 0 . \tag{29}$$

Compared to the constraints (17) and (18) from standard canonical GR, there are a few striking differences (apart from the momentum constraint (27), which is the same as (18)). First of all, there is no kinetic term in the Hamiltonian constraint (26). The momentum conjugate to the three-metric (and its decomposed parts) is undetermined through the Lagrange multiplier λ^{ij} from (21); see discussion that followed. Even though this might seem alarming, it is still the EH Lagrangian we are dealing with and no choice of variables should affect the dynamical structure of the theory. This leads us to investigate the second observation. Namely, there are six additional constraints (28) and (29), which turn out to be pairs of *second-class constraints*, because they do not commute,

$$\left\{\bar{P},\, \mathcal{Q}^{\mathrm{EH}}\right\} = -\frac{6a^2}{\kappa} , \quad \left\{\bar{P}^{ij},\, \mathcal{T}_{\mathrm{EH}}^{mn}\right\} = \frac{a^2}{\kappa}\left(\bar{h}^{im}\bar{h}^{jn} - \frac{1}{3}\bar{h}^{ij}\bar{h}^{mn} \right) . \tag{30}$$

What is the reason for this? Observe that the constraints (28) arise due to the absence of $\dot{\bar{K}}$ from the EH Lagrangian. A similar situation was encountered in the Weyl-squared theory, where the absence of $\dot{\bar{K}}$ resulted in first class constraints (23). However, the difference here is that $\bar{K}$ is *not* absent from the EH Lagrangian as in the Weyl-squared theory [5], which is why (28) are second class constraints. Also, the additional constraints (29) are not met in the Weyl-squared theory, because $\bar{K}_{ij}^{\mathrm{T}}$ is a dynamical variable there. These constraints are second class for a similar reason as why is second-class (28): $\bar{K}_{ij}^{\mathrm{T}}$ cannot be eliminated from the theory.

It seems that our new formulation of GR is contaminated with non-dynamical degrees of freedom. However, not all is lost. It turns out that $\dot{\mathcal{Q}}^{\mathrm{EH}} \approx 0$ and $\dot{\mathcal{T}}_{\mathrm{EH}}^{ij} \approx 0$ determine $\Lambda_{\bar{K}}$ and Λ_{ij}, respectively, but their explicit form is tedious to calculate and of no direct relevance for our discussion. The important thing is that the second-class constraints (28) and (29) can now be implemented, which has two remarkable consequence: firstly, it reduces the total Hamiltonian to the following form

$$H^{\mathrm{EH}} = \int \mathrm{d}^3x \left\{ \bar{N}\mathcal{H}_{\perp}^{\mathrm{EH}} + N^i\mathcal{H}_i^{\mathrm{EH}} + \lambda_{\bar{N}}p_{\bar{N}} + \lambda^i p_i \right\} , \tag{31}$$

which is of the same familiar form as (20); and secondly, the seemingly arbitrary $\bar{K}$ and $\bar{K}_{ij}^{\mathrm{T}}$ and seemingly arbitrary p_a and $\bar{p}^{ij}$ turn out to be actually related,

$$p_a = -\frac{6a}{\kappa}\bar{K} , \quad \bar{p}^{ij} = \frac{a}{2\kappa}\bar{K}_{ab}^{\mathrm{T}}\bar{h}^{ai}\bar{h}^{bj} . \tag{32}$$

Remarkably, this relationship is nothing than the definitions of the momenta in (19) in the standard ADM treatment of GR! Eliminating $\bar{K}^{\mathrm{T}}_{ij}$ and $\bar{K}$ from (26) via (32) in terms of p_a and $\bar{p}^{ij}$ recovers (17), which completes the equivalence of the two formulations. Moreover, one may check that the number of physical degrees of freedom is still 2: 16 original configuration variables $\bar{N}, N^i, \bar{h}_{ij}, a, \bar{K}^{\mathrm{T}}_{ij}$ and $\bar{K}$, 8 first class constraints (26) and (27), and 6 pairs of second class constraints (28) and (29) give $16-8-6=2$ physical degrees of freedom, as expected.

4. Conclusion

We learn from the Weyl-squared theory that the constraints $\bar{P}$, $\mathcal{Q}^{\mathrm{W}}$ are first class only if a and $\bar{K}$ — the only conformally variant variables — are absent from the theory. Any dependence of the Lagrangian on a and $\bar{K}$ induces a breaking of conformal symmetry and turns the constraints into second class ones. In the standard formulation of the EH theory, which we know to be conformally non-invariant, there are no second-class constraints which may give information about the obviously broken conformal symmetry. However, if extrinsic curvature is promoted to an independent configuration variable, the momentum conjugate to variable $\bar{K}$ turns out to vanish. This gives rise to *second-class* constraints (28), similarly to the Weyl-squared theory with non-minimally coupled scalar field (24). They express the missing information about the broken conformal symmetry of the EH theory. Similarly, the absence of $\dot{K}^{\mathrm{T}}_{ij}$ from the EH Lagrangian gives rise to second-class constraints (29).

Thus, one may say that broken conformal symmetry turns scale density (the volume) into a *dynamical* variable. The implementation of the second-class constraints ensures the equivalence with the standard ADM formulation, by recovering the ADM momentum (32). The question remains, however, what is the meaning of the second class constraints (29) which give rise to the dynamics of the unimodular part of the metric $\bar{h}_{ij}$, but we leave this for another occasion.

Acknowledgments

The author would like to thank Claus Kiefer for his encouragement to write this paper and for reading the manuscript. He is also thankful to Anirudh Gundhi, Nick Kwidzinski and David Wichmann for helpful comments.

References

[1] Capozziello S and Faraoni V 2011 *Beyond Einstein Gravity*, Fundamental Theories of Physics 170 (Springer, Dordrecht).
[2] Buchbinder I L, Odintsov S D and Shapiro I L 1992 *Effective Action In Quantum Gravity* (IOP Publishing Ltd, Bristol and Philadelphia).
[3] Steinwachs C 2014 *Non-minimal Higgs Inflation and Frame Dependence in Cosmology* (Springer, Heidelberg)
[4] Nakayama Y 2015 *Phys. Rep.* **569** 1
[5] Kiefer C and Nikolić B 2017 *Phys. Rev.* D **95** 084018
[6] Padmanabhan T 2010 *Gravitation – Foundation and Frontiers* (Cambridge University Press, Cambridge)
[7] Kiefer C and Nikolić B 2017 in *Gravity and the Quantum* edited by J. Singh, Baghla and S. Engineer (Springer, Cham) p. 127–143
[8] Kiefer C and Nikolić B 2017 Quantum geometrodynamics of conformal gravity (Contribution to these proceedings)
[9] Kaku M 1982 *Nucl. Phys.* B **203** 285
[10] Boulware D 1983 in *Quantum Theory of Gravity*, edited by S M Christensen (Adam Hilger Ltd, Bristol) p. 267–294
[11] Deruelle N, Sasaki M, Sendouda Y, and Yamauchi D 2010 *Prog. Theor. Phys.* **123** 169
[12] Klusoň J, Oksanen M and Tureanu A 2014 *Phys. Rev.* D **89** 064043
[13] Irakleidou M, Lovrekovic I and Preis F 2015 *Phys. Rev.* D **91** 104037

IOP Conf. Series: Journal of Physics: Conf. Series **880** (2017) 012028 doi:10.1088/1742-6596/880/1/012028

Multi-layer statistical gravity on the boundary

Pierre A Mandrin

Department of Physics, University of Zurich,
Winterthurerstrasse 190, 8057 Zürich, Switzerland

E-mail: pierre.mandrin@uzh.ch

Abstract. Starting from an important research path, we consider gravity as a collective phenomenon governed by statistical mechanics. While previous studies have focussed on the thermodynamic heat flow across a 2d-horizon as perceived by a single, accelerated observer, we evaluate here the number of microscopic states arising for multiple observers perceiving multiple horizons within foliations of the boundary of a space-time region. This yields a temperature-independent, Boltzmann-type "entropy" which is equivalent to the boundary action and which we call m-entropy. According to its statistical interpretation, the m-entropy distribution as a function of the gravitational field is maximum when Einstein's Field Equations hold. However, if the number of "atoms of space" is small, Einstein's Equations do not hold and no sharp geometry can be defined. On the other hand, the transition probability of microstates can be computed and can be interpreted as processes of a (alternative) model of quantum space-time.

1. Introduction

The discovery of the thermodynamic properties of black holes by Bekenstein [1] and Hawking [2] has initiated an important research path, according to which the gravitational field equations acquire the same status as e.g. the equations of fluid mechanics, see former work by Jacobson [3] and Padmanabhan [4]. The 2-surface on the null-horizon of the Rindler wedge is considered to be equivalent to a black hole horizon. Imposing the stationarity condition allows the definition of a local, observer-dependent temperature $T = \kappa/(2\pi)$ with surface gravity κ, as well as an entropy $S_{BHJP} = A/(4L_p^2)$ with associated horizon surface area A, where L_p is the Planck length.

In the present article, we consider an extension to a statistical model of the microscopic degrees of freedom. The statistical model is not restricted by (local) thermal equilibrium or by classical laws. Because the model is necessarily incompatible with a temperature concept, we have to introduce another type of "entropy" which we call the m-entropy S, it gathers missing information from multiple observers. The "temperature"-independence makes S very different from S_{BHJP}. As we will see, the underlying model can be interpreted as a quantum model of space-time on a dimensionally reduced space and can be used to compute transition probabilities of microscopic states.

Before presenting our model, we reformulate the action of general relativity so that it only acts on the boundary of a space-time region, and we then relate this action with the thermodynamic description of Rindler horizons. These preliminary steps are required in order to recognise later the statistical model capable of reproducing GR and, at the same time, identify a kind of alternative quantum model of space-time.

IOP Conf. Series: Journal of Physics: Conf. Series **880** (2017) 012028 doi:10.1088/1742-6596/880/1/012028

2. The boundary action of general relativity (GR)

Consider a compact space-time region V with piece-wise smooth boundary $\partial V = \sum_A B_A$, where every component B_A is non-null (time- or space-like). As we know, different expressions for the action equally describe GR. The Einstein-Hilbert action S_H and, to first order, the "covariant Einstein action" S_1 are defined as (omitting the cosmological constant for simplicity) [5]

$$S_H = \frac{c^4}{16\pi G} \int_V \sqrt{-g}\; R\; d^4x + S_m, \tag{1}$$

$$S_1 = \frac{c^4}{16\pi G} \Big[\int_V \sqrt{-g}\; R\; d^4x - \sum_A 2\varepsilon_A \int_{B_A} \sqrt{|\gamma|}\; K\; d^3x \Big] + S_m, \tag{2}$$

where S_m is the matter action, $K = K_{ab}\gamma^{ab}$, γ_{ab} is the induced metric on ∂V with determinant γ, the exterior curvature on ∂V with unit normal vector n^a is $K_{ab} = -\frac{1}{2}\mathcal{L}_\perp \gamma_{ab} = -g^{cd}\gamma_{da}\nabla_c n_b$, $\mathcal{L}_\perp = \mathcal{L}_n$, and $\varepsilon_A = n^a n_a$ on B_A. From now on, we ignore the matter term. Variation of (1) and (2) is known to yield a bulk term (leading to Einstein's field equations) and a boundary term

$$\delta S_H|_{\partial V} = \frac{c^4}{16\pi G} \sum_A \varepsilon_A \int_{B_A} \gamma_{ab}\; \delta N^{ab} d^3x, \quad \delta S_1|_{\partial V} = -\frac{c^4}{16\pi G} \sum_A \varepsilon_A \int_{B_A} N^{ab}\; \delta\gamma_{ab}\; d^3x, \tag{3}$$

respectively, where $N_{ab} = K_{ab} - K\gamma_{ab}$. We now combine

$$S_{\partial 1} = \frac{1}{2}(S_H - S_1) = \frac{c^4}{16\pi G} \sum_A \varepsilon_A \int_{B_A} \sqrt{|\gamma|}\; K\; d^3x, \qquad S_{\Sigma 1} = \frac{1}{2}(S_H + S_1) \tag{4}$$

with the variations ($S_{\partial 1}$ only acts on ∂V, and γ_{ab} and N^{ab} are now varied)

$$\delta S_{\partial 1} = \frac{c^4}{32\pi G} \sum_A \varepsilon_A \int_{B_A} \delta(N^{ab}\gamma_{ab})\; d^3x, \tag{5}$$

$$\delta S_{\Sigma 1} = \frac{c^4}{32\pi G} \sum_A \varepsilon_A \int_{B_A} \gamma_{ab}\; \delta N^{ab} - N^{ab}\; \delta\gamma_{ab}\; d^3x, \tag{6}$$

$$\delta S_{\partial 1}|_{\gamma_{ab}} = \delta S_{\Sigma 1}|_{g_{ab},\gamma_{ab}}, \quad \delta S_{\partial 1}|_{N^{ab}} = -\delta S_{\Sigma 1}|_{g_{ab},N^{ab}}, \tag{7}$$

where the suffix $|_Y$ means that Y is fixed at its saddle point value of GR, and $S_{\partial 1}$ has its extrema at the same values of γ_{ab} and N^{ab} as S_1 and S_H, respectively. Moreover, by using

$$2\delta S_{\partial 1}|_{\gamma_{ab}} = \frac{c^4}{16\pi G} \int_V \sqrt{-g}\; g_{ab}\; \delta R^{ab}\; d^4x, \tag{8}$$

the general bulk action ansatz $S_{\text{bulk}} \sim \int_V \sqrt{-g}\; Q^{ab}\; g_{ab}\; d^4x$ with

$$\delta S_{\text{bulk}} \sim \int_V \sqrt{-g}\; (Q^{ab} - \frac{1}{2}Q^c_c\; g^{ab})\; \delta g_{ab}\; d^4x + \int_V \sqrt{-g}\; g_{ab}\; \delta Q^{ab}\; d^4x \tag{9}$$

and the condition $\delta S_{\text{bulk}}|_{g_{ab}} = 2\delta S_{\partial 1}|_{\gamma_{ab}}$ forces $R^{ab} \sim Q^{ab}$, so that the theory defined by the "boundary action" $S_{\partial 1}$ is equivalent to GR.

3. Interpretation of the boundary action using multiple horizon thermodynamics

There have been arguments suggesting that the microscopic degrees of freedom of gravity live on 2-surfaces and there is a third dimension along which the microscopic state of the 2-surfaces is allowed to change. A well known reasoning at the root of the holographic principle is given by 't Hooft [6] and is based on general principles of quantum systems and on the properties of black holes. On the other hand, space-time may be described thermodynamically [3, 4]. An accelerated observer associates a temperature T and an entropy S_{BHJP} to the perceived Rindler horizon. T and S_{BHJP} are well-defined at the limit of stationarity (or local equilibrium) of the horizon. As we see next, the product TS_{BHJP} evaluated on a sequence of narrow null-horizon strips yields yet a different, Boltzmann-type entropy S (we call it m-entropy), and $S \sim S_{\partial 1}$.

Consider a compact space-time region V with non-null boundary ∂V. The boundary action $S_{\partial 1}$ (4) is equivalent to the boundary term in the notation of [4],

$$S_{\partial 1} = \frac{c^4}{16\pi G}\int_{\partial V} N^c_{ab}\, f^{ab}\, d^3\Sigma_c, \qquad f^{ab} = \sqrt{-g}\, g^{ab}, \qquad N^c_{ab}\, d^3\Sigma_c = N^{ab}\, d^3x, \tag{10}$$

N^c_{ab} is the momentum conjugate to f^{ab} and $d^3\Sigma_c$ is the 3-surface element covector.
We now cut the boundary ∂V into tiny cuboids $\Delta\partial V \subset \partial V$ and choose the length L_e of each edge e of $\Delta\partial V$ to satisfy $L_p \ll L_e <$ variation scale of γ_{ab} and K_{ab} on ∂V. We rewrite

$$S_{\partial 1} = \frac{c^4}{16\pi G}\sum_{\Delta\partial V \subset \partial V} S_{\partial 1\ \Delta\partial V} \tag{11}$$

with

$$S_{\partial 1\ \Delta\partial V} = \frac{c^4}{16\pi G}\int_{\Delta\partial V} N^c_{ab}\, f^{ab}\, d^3\Sigma_c = \frac{c^4}{32\pi G}\int_{\Delta\partial V} d^3x\, \sqrt{|\gamma|}\, \gamma^{ab}\, \mathcal{L}_\perp \gamma_{ab}. \tag{12}$$

Without loss of generality, we restrict ourselves to the gauge $g_{0a} = g_{00}\delta^0_a$. Because γ_{ab} is symmetric, it can be diagonalised using an orthogonal transformation, $x^a \to u^a_b x^b$, so that $\gamma_{ab} \to u^c_a \gamma_{cd}(u^T)^d_b$. The transformation u^a_b is held fixed within $\Delta\partial V$ as g_{ab} does not change significantly. If $e^I_{(a)}$ are the triads induced on ∂V, we write $\gamma_{ab} = e^I_{(a)}\eta_{IJ}e^J_{(b)}$, and $e^I_{(a)}$ is diagonal. The transformation u^a_b has unit Jacobian determinant and (12) transforms to

$$S_{\partial 1\ \Delta\partial V} = \frac{c^4}{16\pi G}\int_{\Delta\partial V} d^3x\, \sqrt{|\gamma|}\, \sigma, \qquad \sigma = \frac{1}{2}u^a_c\gamma^{cd}(u^T)^d_b\, \mathcal{L}_\perp u^e_b\gamma_{ef}(u^T)^f_a = \frac{1}{2}\gamma^{ab}\, \mathcal{L}_\perp\gamma_{ab}. \tag{13}$$

Therefore, γ_{ab} can be assumed to be diagonal without loss of generality. We reexpress (13) in terms of null coordinates $(\tilde{x}^a) = (\tilde{x}^1 = u, \tilde{x}^2 = v, x^3, x^4)$ instead of $(x^a) = (x^\perp, x^{||}, x^3, x^4)$, where $x^{||}$ is tangent to ∂V and x^3, x^4 are both spatial components. Furthermore, while restricting ourselves to $\Delta\partial V$, γ_{ab} is nearly constant and we are free to change to local coordinates satisfying $\gamma_{\perp\perp} = \gamma_{||||} = 1$ and change back to more general (null) coordinates when concluding the computations restricted to $\Delta\partial V$. Moreover, the range of integration in (13) approximately splits into an interval $\mathcal{I} = [x^{||}_-, x^{||}_+]$ and a piece of 2-surface $A(x^{||})$:

$$\int_{\Delta\partial V} d^3x = \int_{A(x^{||})} d^2x \int_{\mathcal{I}} dx^{||}. \tag{14}$$

If we assume that the classical spatial resolution of $x^{||}$ is no better than $\Delta x^{||}$, (13) can be written in terms of a sum of n strips of width $\Delta x^{||}$ with $n = (x^{||}_+ - x^{||}_-)/\Delta x^{||}$:

$$S_{\partial 1\ \Delta\partial V} = \frac{c^4}{16\pi G}\int_{\mathcal{I}} dx^{||} \int_{A(x_k^{||})} d^2x\ \sqrt{|\gamma|}\ \sigma = \frac{c^4}{16\pi G}\sum_{k=1}^{n}\int_{A(x_k^{||})} d^2x\ \sqrt{|\gamma|}\ \Delta x^{||}\ \sigma, \quad (15)$$

where $x_k^{||} = x_-^{||} + k\Delta x^{||}$ and $A(x_k^{||})$ is the intersection of $\Delta\partial V$ with the (x^3, x^4)-surface at $x^{||} = x_k^{||}$. Because the triads are diagonal, e_a^I also are diagonal, and we can rewrite (with $A = 3, 4$)

$$\begin{aligned} \sigma &= e_I^{(a)}\eta^{IJ}e_J^{(b)}e_{(b)}^K\eta_{KL}\mathcal{L}_\perp e_{(a)}^L = e_I^{(a)}\eta^{IJ}e_J^b e_b^K\eta_{KL}\mathcal{L}_\perp e_{(a)}^L = e_I^{(a)}\mathcal{L}_\perp e_{(a)}^I = e_I^A\mathcal{L}_\perp e_A^I + e_I^{||}\mathcal{L}_\perp e_{||}^I \\ &= e_I^A(\frac{\partial v}{\partial x^\perp}\ \mathcal{L}_v e_A^I + \frac{\partial u}{\partial x^\perp}\ \mathcal{L}_u e_A^I) + (e_I^u\ \mathcal{L}_v + e_I^v\ \mathcal{L}_u)(e_u^I\frac{\partial u}{\partial x^{||}} + e_v^I\frac{\partial v}{\partial x^{||}}) \\ &= e_I^A(\frac{\partial u}{\partial x^{||}}\ \mathcal{L}_v e_A^I + \frac{\partial v}{\partial x^{||}}\ \mathcal{L}_u e_A^I)\ + e_I^u\ \mathcal{L}_v e_u^I\frac{\partial u}{\partial x^{||}} + e_I^v\ \mathcal{L}_u e_v^I\frac{\partial v}{\partial x^{||}} + e_I^u\ \mathcal{L}_{||}e_v^I + e_I^v\ \mathcal{L}_{||}e_u^I. \end{aligned} \quad (16)$$

The last two terms of (16) sum up to zero because $e_I^v\ \mathcal{L}_{||}e_u^I = -e_u^I\ \mathcal{L}_{||}e_I^v = e_I^u\ \mathcal{L}_{||}e_v^I$ (by repeatedly using Leibnitz' rule and using $e_I^v e_u^I = \delta_u^v$, $e_I^v e_v^J = \delta_I^J$). With $\Delta x^{||} = \Delta u\frac{\partial x^{||}}{\partial u} = \Delta v\frac{\partial x^{||}}{\partial v}$, we obtain

$$\begin{aligned} \sigma\ \Delta x^{||} &= \Delta u\ (e_I^A\ \mathcal{L}_v e_A^I + e_I^u\ \mathcal{L}_v e_u^I) + \Delta v\ (e_I^A\ \mathcal{L}_u e_A^I + e_I^v\ \mathcal{L}_u e_v^I) \\ &= \Delta u\ e_I^{(a)_u}\ \mathcal{L}_v e_{(a)_u}^I + \Delta v\ e_I^{(a)_v}\ \mathcal{L}_u e_{(a)_v}^I, \end{aligned} \quad (17)$$

where $e_{(a)_w}^I$ are the triads induced on the hypersurface w = constant with $w = u, v$. We diagonalise $e_{(a)_w}^I$, expand back to metric notation, and finally change back to general null coordinates, to obtain

$$\begin{aligned} S_{\partial 1\ \Delta\partial V} &= \frac{c^4}{32\pi G}\sum_{k=1}^{n}\Big[\int_{A(x_k^{||})} d^2\tilde{x}\ \sqrt{|\gamma_{(u)}|}\ \Delta u\ \gamma_{(u)}^{ab}\ \mathcal{L}_v\gamma_{ab}^{(u)} + \int_{A(x_k^{||})} d^2\tilde{x}\ \sqrt{|\gamma_{(v)}|}\Delta v\ \gamma_{(v)}^{ab}\ \mathcal{L}_u\gamma_{ab}^{(v)}\Big] \\ &= \frac{c^4}{32\pi G}\sum_{k=1}^{n}\Big[\int_{\Delta u}\int_{A(x_k^{||})} d^3\tilde{x}\ \sqrt{|\gamma_{(u)}|}\ \gamma_{(u)}^{ab}\ \mathcal{L}_v\gamma_{ab}^{(u)} + \int_{\Delta v}\int_{A(x_k^{||})} d^3\tilde{x}\ \sqrt{|\gamma_{(v)}|}\gamma_{(v)}^{ab}\ \mathcal{L}_u\gamma_{ab}^{(v)}\Big], \end{aligned} \quad (18)$$

where $\gamma_{ab}^{(w)}$ and $\gamma_{(w)}^{ab}$ are the metric and inverse metric induced on the hypersurface w = constant, respectively. To link this result with horizon thermodynamics, we need to introduce locally flat coordinates $(\bar{t}, \bar{x}, \ldots)$ as well. The flat null coordinates $\bar{v}$, $\bar{u}$ are affine null parameters on the null surfaces $\bar{u}$ = constant and $\bar{v}$ = constant, respectively, which can locally be approximated by the null surfaces defined by the null strips. Then, for any general null-surface Σ, we have the identity [7]

$$\frac{c^4}{16\pi G}\int_\Sigma N_{ab}^c\ f^{ab}\ d^3\Sigma_c = \int_\Sigma T\ s\ d\lambda\ d^2x, \quad (19)$$

where s is the surface density of the entropy S_{BHJP} and T the temperature of the horizon parametrised by $\lambda = \bar{u}$ or $\bar{v}$, as seen by the corresponding Rindler observer. We consider two groups of equally accelerated observers, one observer per strip, one group for the strips $\bar{u}$ = constant observes the temperature T_u and one group for the strips $\bar{v}$ = constant observes the temperature T_v, respectively. Then, (18) can also be written as

$$\begin{aligned} S_{\partial 1\ \Delta\partial V} &= \sum_{k=1}^{n}\Big[\int_{\Sigma_{(u)}(x_k^{||})} T_u\ s\ du\ d^2x + \int_{\Sigma_{(v)}(x_k^{||})} T_v\ s\ dv\ d^2x\Big] \\ &= \sum_{k=1}^{n}\int_{A(x_k^{||})} [T_u\ s\ \Delta u + T_v\ s\ \Delta v]\ d^2x. \end{aligned} \quad (20)$$

IOP Conf. Series: Journal of Physics: Conf. Series **880** (2017) 012028 doi:10.1088/1742-6596/880/1/012028

What is the interpretation of (20) from the point of view of two neighbouring Rindler observers A and B separated by the vector $(-\Delta\bar{u}, \Delta\bar{v}, 0, 0)$? The Rindler wedges of A and B are given by $\bar{x}_A > |\bar{t}_A|$ and $\bar{x}_B > |\bar{t}_B|$, respectively. We then define the past horizons $\mathcal{H}_{A-}$ and $\mathcal{H}_{B-}$ ($\bar{v} =$ constant), the future horizons $\mathcal{H}_{A+}$ and $\mathcal{H}_{B+}$ ($\bar{u} =$ constant), and the 2-surfaces $\mathcal{P}_A : \bar{t}_A = \bar{x}_A = 0$ and $\mathcal{P}_B : \bar{t}_B = \bar{x}_B = 0$. If the distance $\Delta\bar{x}^{||}$ between $\mathcal{P}_A$ and $\mathcal{P}_B$ is time-like and A precedes B, $\mathcal{H}_{A+}$ and $\mathcal{H}_{B-}$ intersect precisely on a 2-surface $\mathcal{P}_I$ at an affine parameter distance $\Delta\bar{v}$ from $\mathcal{P}_A$ and at $\Delta\bar{u}$ from $\mathcal{P}_B$. (If $x^{||}$ is space-like, we would have to consider the horizons delimiting the time-like wedges $\bar{t}_A > |\bar{x}_A|$ and $\bar{t}_B > |x_B|$, accordingly.) Therefore, the extended physical system made of both observers A and B must see twice as much horizon area along $\mathcal{H}_{A-}$ and $\mathcal{H}_{B-}$ for values $\bar{u} < \Delta\bar{u}$ as would a single observer A along $\mathcal{H}_{A-}$. For this reason, we need to sum up the entropies for all the null-strips contained in (20) in order to describe the thermodynamics of the full multiple-observer system of the region $\Delta\partial V$.

We shall now have a closer look at the (observer dependent) parameters T_u, T_v. Depending on these values, the observers will have very different records:

(i) If $\Delta\bar{v} > (\sqrt{2}T_v)^{-1}$, $\mathcal{H}_{B-}$ masks the preceding observer A during its motion in the past to $\mathcal{P}_A$. Therefore, B has no chance to notice the presence of $\mathcal{H}_A$ before $\mathcal{H}_B$ peels off. On the other hand, we can squeeze the sequence of wedges along $x^{||}$ up to $\Delta\bar{v} = (\sqrt{2}T_v)^{-1}$ without unmasking $\mathcal{H}_B$ and therefore increase the number of wedges and observers accordingly. However, the spacings below the initial $\Delta x^{||}$ are to small for being resolved, by assumption, and not all horizon surfaces can be distinguished from each other. For wedges with space-like spacings $\Delta x^{||}$, the logic of masking horizons and of resolution works in a similar manner.

(ii) If $\Delta\bar{v} < (\sqrt{2}T_v)^{-1}$, the masking of observers in every wedge fails. In order to recover it, we have to stretch up the sequence of wedges to $\Delta\bar{v} = (\sqrt{2}T_v)^{-1}$, and the spacings of the wedges remain well resolved.

The conditions

$$\Delta\bar{u} = (\sqrt{2}T_u)^{-1} \quad \Delta\bar{v} = (\sqrt{2}T_v)^{-1} \tag{21}$$

are therefore relevant when we ask how much entropy is made inaccessible to any of the observers. In fact, imposing (21) dramatically simplifies (20) to the following identity:

$$S_{\partial 1} = \sqrt{2}\sum_{k=1}^{n} S_{BHJP}\big|_{x_k^{||};\ \bar{u}<\Delta\bar{u}=(\sqrt{2}T_u)^{-1};\ \bar{v}<\Delta\bar{v}=(\sqrt{2}T_v)^{-1}}. \tag{22}$$

(22) tells us that $S_{\partial 1}$ reduces to a sum of stacked entropies S_{BHJP} under stationary condition and for observers simultaneously resolving the loci of the stacked surfaces and maximising the amount of inaccessible information. For normal space, $S_{\partial 1}$ provides an extension of S_{BHJP} to multiple-observer systems as given by the usual boundary of an arbitrary compact space-time region V. Note that $S_{\partial 1}$ also differs from S_{BHJP} in that its value does not depend on the choice of a specific horizon temperature. In order to use an unambiguous terminology, we shall call $S_{\partial 1}$ the "m-entropy" (for multiple observer entropy), in contrast to the single-observer entropy.

4. Statistical interpretation of the boundary action

Now that we have found an interpretation of $S_{\partial 1}$ as a generalised form of entropy of the space-time geometry, we may look for a general statistical description of it. According to the above findings, the region $\Delta\partial V$ can be foliated into a certain maximum number n of pieces of space-like 2d-surfaces of aria A_k restricted on $\Delta\partial V$ ($k = 1, \ldots, n$). The number n is given by the best resolution we can achieve in a classical frame-work. The 2d-surfaces satisfy $x^b =$ constant ($b = 2, 3$ or 4). Each 2d-surface corresponds either to $\mathcal{P}_A$ (where A is an observer) or to $\mathcal{P}_I$.

Because the difference $\Delta x^{\perp}$ between $\mathcal{P}_A$ and the nearest $\mathcal{P}_I$ can hardly be resolved and is cancelled in the step to the following $\mathcal{P}_B$, we can neglect $\Delta x^{\perp}$ (simpler structure). Moreover, $\Delta\partial V$ is a cuboid, all A_k have the same size and we shorten $A_k \to A$.

We shall first consider the case for which $\Delta\partial V$ is space-like. Then, there is no preferred orientation b across which to foliate. Rather, we have to consider all three dimensions, $b = 2, 3, 4$. All orientation-dependent quantities inherit an orientation index, e.g. $A \to A_b$. To work on a similar basis as in the previous part (flat space coordinates for non-accelerated frame), we change to Minkowski index representation $x^I = e^I_a x^a$ so that we explicitly have $x^K = e^K_{b=K} x^{b=K}$ for the orientation K of the foliation. We also change $A_b \to A_K$ (K is not a Minkowski index here).

Due to Bekenstein's interpretation of a horizon in terms of entropy, A_K represents n_K space-time atoms which individually take on one out of n_q possible states $|q\rangle$, yielding a total of

$$p_K = n_q^{n_K} \tag{23}$$

possible microscopic states per layer. All states $|q\rangle$ have the same probability to occur, and the number p_K of microstates is related to a Boltzmann-type entropy σ_K per layer of area A_K, via

$$\sigma_K = \ln p_K, \quad A_K = 4L_p^2 \ln p_K. \tag{24}$$

If we foliate $\Delta\partial V$ across orientation K into N_K layers, the number of microscopic states is

$$\Omega_K(\Delta\partial V) = p_K^{N_K} = \exp s_K, \qquad s_K = N_K \sigma_K = N_K \ln p_K \tag{25}$$

is the Boltzmann-type entropy per volume $\Delta\partial V$, i.e. an "entropy density". As previously mentioned, the distance Δx^K between two consecutive layers is the smallest accessible spatial resolution. However, one expects Δx^K to have a fundamental lower bound of the order of the Planck length L_P. N_K can be estimated using the relation $N_K \approx k_K l^K$, where l^K is the size of $\Delta\partial V$ along x^K and the "narrowness parameter" k_K is a constant of order $\leq L_p^{-1}$.

We thus foliate $\Delta\partial V$ across all 3 orientations, independently from each other, and the total number of microscopic states is the product $\Omega = \Omega_1\Omega_2\Omega_3$:

$$\Omega(\Delta\partial V) = \prod_{K=1}^{3} \exp s_K = \exp \sum_{K=1}^{3} s_K. \tag{26}$$

We can rewrite s_K to a form which is equivalent to the integrand of (12):

$$s(\Delta\partial V) = \sum_{K=1}^{3} s_K = \sum_{K=1}^{3} k_K l^K \ln p_K = \sum_{K=1}^{3} k_K l^j e_j^K \ln p_K = \sqrt{\gamma}\kappa^j_K e_j^K, \tag{27}$$

with the definition $\sqrt{\gamma}\kappa^j_K = k_K l^j \ln p_K$ and reintroducing the Einstein summation convention in the last expression of (27). Finally, $\sqrt{2}$ times the total m-entropy

$$S = \sum_{\Delta\partial V} \sqrt{\gamma}\kappa^j_K e_j^K, \quad \sqrt{2}\kappa^j_K = K^j_K\, d^3x, \quad K^j_K = K^{ji} e^I_i \eta_{IK} \tag{28}$$

coincides with $S_{\partial 1}$ in (4), as far as the space-like parts of the boundary ∂V are concerned.

If $\Delta\partial V$ is time-like, the foliation method straight-forwardly applies to the time-like orientation $b = 2$ or $K = 2$. However, to reproduce $S_{\partial 1}$ (4) for time-like parts of ∂V as well, we need to include all 3 orientations, $K = 2, 3, 4$, so that time-like layers would represent space-time atoms as well. In fact, we can support even this hypothesis. While crossing a Rindler horizon, Rindler time becomes a spatial Rindler coordinate, and vice-versa. We arrive in a time-like wedge and transform to "new" Minkowski-coordinates. We then choose a different spatial direction for the acceleration of an observer, then cross the horizon again, and finally recover

Minkowski coordinates with a different time direction than before, but fully equivalent to the initial Minkowski coordinates. We can, however, recover the initial horizon area which is now time-like, while the Rindler horizons can be recovered via a substitution of the form $\bar{x}^1 \to i\bar{x}^1$. With this concept, time-like volumes $\Delta\partial V$ can be foliated across all 3 orientations as well.

To summarise, the boundary action $S_{\partial 1}$ is proportional to the total m-entropy S of the gravitational degrees of freedom on ∂V. $S_{\partial 1}$ can be interpreted as a multiple entropy of layers of microscopic degrees of freedom of space-time.

By virtue of the above construction, the microstates of gravity are given by arbitrary states $|q\rangle$ on every layer in every orientation, and they are not restricted to states satisfying the field equations of GR. Nevertheless, in order to obtain the macroscopic state with the highest probability, we maximise the distribution of m-entropy S as a function of e^I_a, i.e. we set the variation of S to zero with respect to e^I_a. This is equivalent to applying Hamilton's variation principle to the Einstein-Hilbert action and yields Einstein's Equations. Some kind of analogy between the gravitational action and the notion of "entropy" has been suggested earlier in [8] or [9]. For quantum mechanics, such a relationship also has been conjectured earlier [10].

In order to generate a non-vacuum space-time geometry, we have to constrain the space of admissible microstates. We shortly sketch one method which consists in constraining the macroscopic states. We illustrate this procedure for two different sources of information about the space-time geometry.

(i) We impose sources of gravitational field with unknown space-time distribution and specify how they determine the gravitational field. These are the matter fields which are described by a Lagrangian $\mathcal{L}_m(a^j, f_j)$ with a set of free constants a^j and fixed functions $f_j(e^\Delta_\mu, x^\mu)$.

(ii) We impose part of the space-time distribution of the gravitational field (or matter fields) directly. This corresponds to the information we could obtain from local measurements.

In both cases, the Lagrange multiplier method can be used so that we must add to S an additional term, we call it $\mu(x^\mu)$:

$$S_t = \sqrt{2}S + \int_V \mathrm{d}^4x\ \sqrt{g}\ \mu \tag{29}$$

If the expression S_t is maximised instead of S ($\delta S_t = 0$), it yields non-trivial classical solutions of GR under specific conditions and/or with non-vanishing matter.

5. The alternative quantum treatment

The above statistical interpretation of the boundary action has lead us to evaluate a discrete number of possible microstates. We can consider two limits.

The classical limit is $N_K, p_k \to \infty$, and the classical dynamics are obtained by maximising $S_t(N_K(x^\mu), p_k(x^\mu))$ as a function of the gravitational field $e^I_\nu(x^\mu; N_K(x^\mu), p_k(x^\mu))$. The classical limit is characterised by a sharp maximum of the entropy as a function of e^I_ν.

The quantum limit occurs for N_K, p_k not much greater than 1. The distribution of S does not have a sharp maximum as a function of the gravitational field. Neighbouring field values yield similar magnitudes. Each field point $\mathcal{G}$ is a set of many parameters (or discretised functions)

$$\mathcal{G} = (e^K_k(x^\mu), N^K(x^\mu), p^K(x^\mu)) \tag{30}$$

At a point $\mathcal{G}$ for which a parameter function π differs at some location x^μ by a small difference D_π from the maximum point $\mathcal{G}_{\max}$, we have

$$S_t(\mathcal{G}) - S_t(\mathcal{G}_{\max}) \sim -D_\pi^2 + \mathcal{O}(D_\pi^3). \tag{31}$$

We can thus approximate the distribution $\Omega(\mathcal{G})$ around the maximum by a Gaussian distribution, and the standard deviation of π determines the uncertainty of π. In this way, e_k^K acquires an uncertainty, and this affects any space-time distance. In this way, the statistical interpretation of gravity leads to a quantum uncertainty of the geometry.

Similarly to the n-point function computations of quantum field theory, we can evaluate the transition probability between n local quantum geometries $\mathcal{G}_m$ at the 3d-locations given by $\Delta\partial V_m$, $m = 1 \ldots n$. For example, let every number p_k^m of microscopic states per layer of $\mathcal{G}_m$ at $\Delta\partial V_m$ be restricted by the quantum channel

$$p_k^{m \pmod n +1} = p_k^m + \Delta p_k^{m \pmod n +1} \tag{32}$$

with Δp_k^m being integers not far from 0 (given for this channel) and $\sum_m \Delta p_k^m = 0$. Then, considering all possible configurations $\mathcal{G}_k$, we can compute the transition probability as

$$p(\Delta\partial V_m, \Delta p_k^m) = \frac{\sum_k \Omega(\mathcal{G}_k)|_{p_k^m}}{\sum_k \Omega(\mathcal{G}_k)}, \qquad \Omega(\mathcal{G}) = expS_t|_{\mathcal{G}}, \tag{33}$$

where the terms in the numerator are restricted by (32).

Expressions of the form (33) are the starting point for predictions on outcome rates of quantum processes. Notice that the denominator of (33) cannot be interpreted as a "partition function" because no relaxation can take place in 3d-quantum space.

6. Conclusions

By considering a statistical interpretation of gravity on the boundary of a space-time region, a connection between the classical and a quantum formulation of gravity has been identified. The statistical interpretation has been obtained from a multiple observer perspective of Rindler horizon thermodynamics. Within the statistical frame-work, the microscopic degrees of freedom can have arbitrary states, but not every macroscopic state (or geometry) is likely to occur. The geometries with highest probability satisfy the classical theory (GR). In the quantum regime (small numbers of degrees of freedom), the geometric parameters are not sharp. The transition probabilities of microscopic states can be computed as a weighted sum over microscopic states.

Acknowledgments

I would like to thank Gino Isidori for fruitful discussions and Philippe Jetzer for hospitality at University of Zurich.

References

[1] Bekenstein J D 1973 *Phys. Rev.* D **7** 2333
[2] Hawking S W 1975 *Commun. math. Phys.* **43** 199
[3] Jacobson T 1995 *Phys. Rev. Lett.* **75** 1260
[4] Padmanabhan T 2014 *Gen. Rel. Grav.* **46** 1673
[5] York J W 1986 *Found. Phys.* **16** 249
[6] 't Hooft G 1993 Dimensional reduction in quantum gravity *Preprint* gr-qc/9310026
[7] Majhi B R and Padmanabhan T 2013 *Eur. Phys. J.* C **73** 2651
[8] Mandrin P A 2014 *Int. J. Theor. Phys.* **53** 4250
[9] Munkhammar J 2015 On non-equilibrium thermodynamics of space-time and quantum gravity *Preprint* physics.gen-ph/1507.07793
[10] Lisi A G 2006 Quantum mechanics from a universal action reservoir *Preprint* physics.pop-ph/0605068

On the Einstein-Podolsky-Rosen paradox using discrete time physics

Roland Riek

Laboratory of Physical Chemistry, ETH Zurich, Switzerland

E-mail: roland.riek@phys.chem.ethz.ch

Abstract. The Einstein-Podolski-Rosen paradox highlights several strange properties of quantum mechanics including the super position of states, the non locality and its limitation to determine an experiment only statistically. Here, this well known paradox is revisited theoretically for a pair of spin $\frac{1}{2}$ systems in a singlet state under the assumption that in classical physics time evolves in discrete time steps Δt while in quantum mechanics the individual spin system(s) evolve(s) between the eigenstates harmonically with a period of $4\Delta t$. It is further assumed that time is a single variable, that the quantum mechanics time evolution and the classical physics discrete time evolution are coherent to each other, and that the precision of the start of the experiment and of the measurement time point are much less than Δt. Under these conditions, it is demonstrated for a spin $\frac{1}{2}$ system that the fast oscillation between the eigen states spin up $|\uparrow>$ and spin down $|\downarrow>$ reproduce the expected outcome of a single measurement as well as ensemble measurements without the need of postulating a simultaneous superposition of the spin system in its quantum state. When this concept is applied to a spin $\frac{1}{2}$ system pair in a singlet state it is shown that no entanglement between the two spins is necessary to describe the system resolving the Einstein-Podolski-Rosen paradox.

1. Introduction

The Einstein-Podolsky-Rosen paradox (EPR) [1] is a stimulating Gedankenexperiment in quantum mechanics challenging quantum mechanics as a complete description of physical reality. Motivated by the EPR paradox Schrödinger introduced the term entanglement [2]. An entangled system is thereby defined as a quantum state which cannot be described by a product of states of its individual constituents. A prototypical example thereof is the singlet state of two particles with spin $\frac{1}{2}$ $\hbar$(normalized in the following) having together spin zero [3]. Because the total spin is zero, whenever the first particle is measured to have spin up on some axis i $<\uparrow_1 |S_i| \uparrow_1>\equiv<\uparrow_1 | \uparrow_1>_i$, the other (when measured on the same axis) is always observed to have spin down $<\downarrow_2 | \downarrow_2>_i$ even upon long distance separation and vice versa. It is the paradox that before the measurement particle 1 is superimposed and thus no decision has been made yet whether at the measurement it will have spin up or down, but once the measurement is performed the state of the entire entangled system collapses instantaneously so that particle 2 has the orthogonal spin state of particle 1 without requiring any information transfer (which requires time) between the particles. Because such a behavior is regarded to violate causality, Einstein and others concluded that the established formulation of quantum mechanics must be incomplete [1]. A possible solution to this dilemma appeared to be the introduction of hidden variables, which, while not accessible to the observer, determine the future outcome of the spin measurements before the separation of the two particles. With other words, each particle carries the necessary information with it upon separation, and thus no transmission of information transfer from one particle to the other is required at the time of measurement. However, Bell's inequality theorem [4] rules out local hidden variables as an explanation of the mentioned paradox. Since the Bell inequalities have been supported by experiments [5] [6], EPR is considered no longer to be a paradox and the non local nature of quantum mechanics is thus widely accepted. It is however important to note that Bell's theorem still permits the existence of non-local hidden variables such as the Bohm interpretation [7] or the Conformal Quantum Geometrodynamics [8], while Accardi and coworkers showed recently that the locality condition in the Bell inequality is criticisable [9]. Alternatively one may request that there is a non-local connection of some sort between the system under study and the measurement devices as well as between the measurement apparatus themselves.

Here, the latter concept is realized by the assumption, that time is a single variable that is continuous in quantum mechanics but in classical physics evolves in discrete steps in a coherent fashion between the two frames. This rational guarantees that both measurement devises as well as the system under study are in clock and thus non-locally connected. Furthermore, it is assumed that in the quantum mechanical description of the system there is no super-position of quantum states, but rather a fast oscillation with a periodicity proportional to the classical time step size. This modification to quantum mechanics allows not only to describe a quantum system without super position, another paradox illustrated by Schrödinger's cat [3], but resolves the EPR paradox because the system behaves deterministically as we shall see.

After the introduction of a discrete dynamical time in classical physics (2.1) as well as time resolution considerations of setting up and measuring an experiment (2.1), the super position of two quantum states is replaced by a fast harmonic oscillation between the states followed by measuring it classically under the assumption of a discrete time (2.2). In 2.3 the introduced concepts of discrete time and the fast oscillation between eigenstates are applied to the singlet state of two particles with spin $\frac{1}{2}$ without the request of entanglement resolving thereby the EPR paradox, followed in paragraph 3 by a discussion.

2. Theory

2.1. The discreteness of time in classical physics

It is assumed that in contrast to quantum mechanics with a continuous time t, in classical physics time evolves in very small discrete time steps Δt of constant nature (for example Δt could be the Planck time $\Delta t_p = \sqrt{\frac{h\,G}{2\pi\,c^5}} = 5.4\,10^{-44}$ s with h the Planck constant, c the light velocity and G the gravitational constant). This assumption is nourished on the finding that the arrow of time and entropy can be derived by the introduction of a discrete time [10] and on the request of a dual relationship between energy and time from the corresponding uncertainty principle $\Delta E\,\Delta t > h/2$ (i.e. since in quantum mechanics energy is quantized time is (allowed to be) continuous and vice versa, since in classical physics energy is continuous time is requested to be discrete). Thus, by introducing a discrete time, any time-dependent observable $A(t)$ is represented by a sequence of discrete values [10],[11]:

$$(A_0,\, t_0 = 0),\, (A_1,\, t_1 = 1\,\Delta t), \ldots\ldots,\, (A_n,\, t_n = n\,\Delta t), \ldots\ldots,\, (A_{N+1},\, t_{N+1} = [N+1]\,\Delta t) \qquad (1)$$

with $(A_0,\, t_0)$ the initial and $(A_{N+1},\, t_{N+1})$ the final measurement and with n being an element of the natural numbers including 0. Furthermore, the following relationship between the continuous time in quantum mechanics t and the discrete time in classical physics holds:

$$t_n = n\,\Delta t \qquad (2)$$

denoting t_n to be the quantum mechanical time at classical time point n. It is further assumed, that time is a single variable (unlike space coordinates) and started at the beginning of the

IOP Conf. Series: Journal of Physics: Conf. Series **880** (2017) 012029 doi:10.1088/1742-6596/880/1/012029

universe and thus any object whether it is the system under study or the detector used are with the discrete time steps in tune/coherent to each other. In addition, it is important to mention that with state of the art technologies the time precision of experiments and thus the time point of measurement as well as the start of the experiment are many orders of magnitude less accurate than Δt (ca. $10^{26}\,\Delta t$). Thus, each experiment is somewhat different and it is expected that only an ensemble averaging over many experiments will give a valuable result. With other words already because of these experimental limitations the time-dependent deterministic quantum mechanics described by the Schrödinger equation [3] can only in average calculate the outcome of an experiment. This argumentation suggests that quantum mechanics is well calculating a single outcome, but is meaningless since the experimental set up is not of sufficient time resolution. But upon ensemble averaging quantum mechanics and experiments fit to each other.

2.2. Modifying the super position of quantum states by a time-resolved fast oscillation between the states

Let us study a particle with spin $\frac{1}{2}$. In the standard description of quantum mechanics [3], the eigenstates of a spin $\frac{1}{2}$ along the z-axis are the spin up $|\uparrow>_z$ state and the corresponding orthogonal spin down state $|\downarrow>_z$. When the particle state is not defined (i.e. is not in one of the eigen states), it is said that the particle is in both states, it is super imposed. This can be described by the following wave function using the Dirac notation:

$$|\Psi(t)>= \frac{1}{\sqrt{2}}|\downarrow>_z + \frac{1}{\sqrt{2}}|\uparrow>_z \tag{3}$$

While superimposed upon a single measurement, it is either in one of the two states, and upon ensemble averaging of many measurements it is 50% in the spin up state and 50% in the spin down state, expressed as follows:

$$<\Psi(t)|\Psi(t)>_z = \frac{1}{2} <\downarrow|\downarrow>_z + \frac{1}{2} <\uparrow|\uparrow>_z \tag{4}$$

It is now proposed here to modify the super position concept by suggesting that the spin $\frac{1}{2}$ system oscillates between the two states harmonically (Figure 1) with either

$$|\Psi^{\downarrow}_{hd}(t)> = \frac{1}{\sqrt{2}} cos(\frac{\pi\, t}{2\Delta t})|\downarrow>_z + \frac{1}{\sqrt{2}} sin(\frac{\pi\, t}{2\Delta t})|\uparrow>_z \tag{5}$$

if at starting time the quantum state was in a spin down $|\downarrow>_z$ state, or, with

$$|\Psi^{\uparrow}_{hd}(t)> = \frac{1}{\sqrt{2}} sin(\frac{\pi\, t}{2\Delta t})|\downarrow>_z + \frac{1}{\sqrt{2}} cos(\frac{\pi\, t}{2\Delta t})|\uparrow>_z \tag{6}$$

if at the start of the experiment the quantum system was in a spin up state $|\uparrow>_z$. It is evident that with these wave function descriptions denoted $\Psi_{hd}(t)$ with hd for harmonic and discrete, there is no superposition because the system oscillates forth and back between the two states with a periodicity of $4\,\Delta t$. As discussed above in measuring a single experiment one needs to consider the imprecision in the starting and measurement time. Thus, it is for a single measurement unknown whether the spin $\frac{1}{2}$ system starts with spin up $|\uparrow>_z$ or spin down $|\downarrow>_z$. Correspondingly, the measurement is done at $t_n = n\,\Delta t$ with n being either an even or odd number. An averaging over many experiments needs to take into account this limitation. Under these considerations the experiment may start with the spin $\frac{1}{2}$ in the spin down state $|\downarrow>_z$ measuring

$$< \Psi^{\downarrow}_{hd}(t)|\Psi^{\downarrow}_{hd}(t) >_{z,n} = \frac{1}{2}\, cos^2(\frac{\pi\, n\Delta t}{2\Delta t}) <\downarrow|\downarrow>_z + \frac{1}{2}\, sin^2(\frac{\pi\, n\Delta t}{2\Delta t}) <\uparrow|\uparrow>_z \tag{7}$$

(note the "bra" part may be regarded as the expectation value of the measurement device) which results by measuring at a time t_n with an even $n = 2\,m$ (with m being an element of the natural numbers including 0) to

$$< \Psi^{\downarrow}_{hd}(t)|\Psi^{\downarrow}_{hd}(t) >_{z,2m} = \frac{1}{2}\, cos^2(\frac{\pi\, 2m\Delta t}{2\Delta t}) <\downarrow|\downarrow>_z = \frac{1}{2} <\downarrow|\downarrow>_z \tag{8}$$

and for a time with an odd $n = 2\,m + 1$ to

$$< \Psi^{\downarrow}_{hd}(t)|\Psi^{\downarrow}_{hd}(t) >_{z,2m+1} = \frac{1}{2}\, sin^2(\frac{\pi\,(2m+1)\Delta t}{2\Delta t}) <\uparrow|\uparrow>_z = \frac{1}{2} <\uparrow|\uparrow>_z \tag{9}$$

Thus, the spin $\frac{1}{2}$ system that has started with a spin down state $|\downarrow>$ is always detected as a spin up or a spin down state although by our modified quantum mechanical description there is an oscillation between the two states.

When ensemble averaged over many measurements with either a measurement at even or odd time steps

$$\begin{aligned} < \Psi^{\downarrow}_{hd}(t)|\Psi^{\downarrow}_{hd}(t) >_z &= \frac{1}{2} < \Psi^{\downarrow}_{hd}(t)|\Psi^{\downarrow}_{hd}(t) >_{z,2m} + \frac{1}{2} < \Psi^{\downarrow}_{hd}(t)|\Psi^{\downarrow}_{hd}(t) >_{z,2m+1} \\ &= \frac{1}{2} <\downarrow|\downarrow>_z + \frac{1}{2} <\uparrow|\uparrow>_z \end{aligned} \tag{10}$$

and thus 50% of the case the system is in spin up and in 50% of the measurements it is in the spin down state, respectively. Correspondingly, if the spin $\frac{1}{2}$ starts with the spin up state $|\uparrow>$ it is detected upon averaging as

$$\begin{aligned} < \Psi^{\uparrow}_{hd}(t)|\Psi^{\uparrow}_{hd}(t) >_z &= \frac{1}{2} < \Psi^{\uparrow}_{hd}(t)|\Psi^{\uparrow}_{hd}(t) >_{z,2m} + \frac{1}{2} < \Psi^{\uparrow}_{hd}(t)|\Psi^{\uparrow}_{hd}(t) >_{z,2m+1} \\ &= \frac{1}{2} <\uparrow|\uparrow>_z + \frac{1}{2} <\downarrow|\downarrow>_z \end{aligned} \tag{11}$$

However, an ensemble measurement must take into account both the time variation-induced variation in the starting state as well as the measurement at even or odd time steps yielding

$$< \Psi_{hd}(t)|\Psi_{hd}(t) >_z = \frac{1}{2} <\downarrow|\downarrow>_z + \frac{1}{2} <\uparrow|\uparrow>_z = < \Psi(t)|\Psi(t) >_z \tag{12}$$

As demonstrated in eq. 12. the same result as in the conventional quantum mechanics approach is obtained but without requesting a simultaneous superposition of the two states generating a modified quantum mechanics that is upon exact measurement deterministic, while under current experimental time accuracies predicts statistically the outcome of an experiment identical to standard quantum mechanics.

2.3. The Einstein-Podolsky-Rosen paradox of a singlet state studied under discrete time physics

An entangled quantum system is discussed composed of two spins $\frac{1}{2}$ particles in a singlet state. This means that since the sum of the spin is 0, either of the particles is in the spin up state and the other in the spin down state, or vice versa. This singlet state can be described using the standard quantum mechanics formulation as follows:

$$|\Psi(t) > = \frac{1}{\sqrt{2}}|\uparrow\downarrow>_z - \frac{1}{\sqrt{2}}|\downarrow\uparrow>_z \tag{13}$$

IOP Conf. Series: Journal of Physics: Conf. Series **880** (2017) 012029 doi:10.1088/1742-6596/880/1/012029

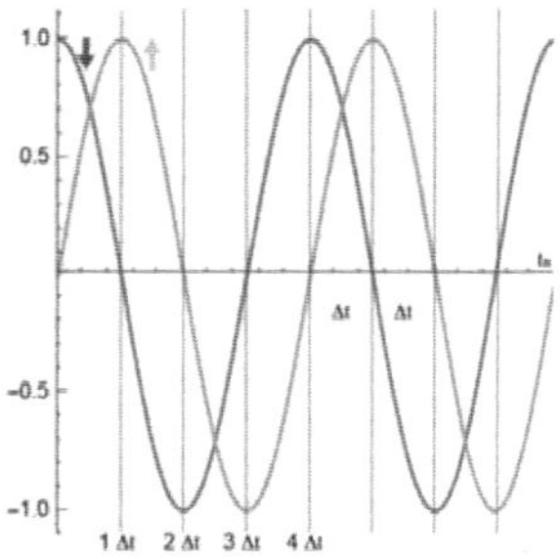

Figure 1. The time-resolved harmonic oscillation between spin down $\mid \downarrow>_z$ and spin up $\mid \uparrow>_z$ of a spin $\frac{1}{2}$ system is illustrated. With grey lines are the possible time points of the measurement indicated. They are in time steps of Δt as labeled.

Please note, that because of the entanglement the wave function of the singlet state is not the product of the two individual ones $|\Psi_i(t)> = \frac{1}{\sqrt{2}}|\downarrow_i>_z + \frac{1}{\sqrt{2}}|\uparrow_i>_z \; with\, i = 1, 2$.

If the entangled state is made observable along the z-axis

$$< \Psi(t)|\Psi(t)>_z = \frac{1}{2} <\uparrow\downarrow \mid \uparrow\downarrow>_z + \frac{1}{2} <\downarrow\uparrow \mid \downarrow\uparrow>_z \tag{14}$$

$$= \frac{1}{2} <\uparrow_1 \mid \uparrow_1>_z <\downarrow_2 \mid \downarrow_2>_z + \frac{1}{2} <\downarrow_1 \mid \downarrow_1>_z <\uparrow_2 \mid \uparrow_2>_z$$

which means that if particle 1 is upon measurement in the spin up state, particle 2 is simultaneously in the spin down state and vice versa. After many measurements, it is further the finding that in 50% of the cases particle 1 is in the spin up state and particle 2 is in the spin down state, while in the other 50% of measurements particle 1 is in the spin down state and particle 2 is in the spin up state, respectively.

For a more comprehensive analysis of the system under study, particle 1 is first detected along the z-axis followed by the measurement of particle 2 along an arbitrary angle $\varphi$to the z-axis [3, 4]. If the first measurement results in $<\uparrow_1 \mid \uparrow_1>_z$ the spin component of the wave function of particle 2 is given by $\mid \downarrow>_z = sin\, \frac{\varphi}{2}| \uparrow_2>_\varphi + cos\, \frac{\varphi}{2}| \downarrow_2>_\varphi$. The probability that the result of the second experiment is also positive (i.e. $<\uparrow_2 \mid \uparrow_2>_\varphi$) is therefore given by $P_{++}(\varphi) = sin^2\frac{\varphi}{2}$ and correspondingly the probability that spin 2 is measured as a spin down state $<\downarrow_2 \mid \downarrow_2>_\varphi$ is $P_{+-}(\varphi) = cos^2\frac{\varphi}{2}$. [3]. Accordingly, if the first measurement is $<\downarrow_1 \mid \downarrow_1>_z$ the probability that spin two is also negative is given by $P_{--}(\varphi) = sin^2\frac{\varphi}{2}$ and correspondingly $P_{-+}(\varphi) = cos^2\frac{\varphi}{2}$ [3].

In contrast to standard quantum mechanics described above, by using the harmonic oscillation and discrete time concept introduced above, the singlet state can be described by the product of the individual states $|\Psi_{i,hd}(t)>$ as follows:

$$\mid \Psi_{hd}(t)^{\downarrow\uparrow} >= |\Psi^{\downarrow}_{1,hd}(t)> * |\Psi^{\uparrow}_{2,hd}(t)> \tag{15}$$

if particle 1 started with spin down and with

$$\mid \Psi_{hd}(t)^{\uparrow\downarrow} >= |\Psi^{\uparrow}_{1,hd}(t)> * |\Psi^{\downarrow}_{2,hd}(t)> \tag{16}$$

if particle 1 started with spin up, respectively. Using the formulations of eqs. 5 and 6 these equations can be rewritten to

$$|\Psi_{hd}(t)^{\downarrow\uparrow} >= \frac{1}{2}[cos(\frac{\pi t}{2\Delta t})| \downarrow_1>_z + sin(\frac{\pi t}{2\Delta t})| \uparrow_1>_z] * [sin(\frac{\pi t}{2\Delta t})| \downarrow_2>_z + cos(\frac{\pi t}{2\Delta t})| \uparrow_2>_z] \tag{17}$$

$$= \frac{1}{2}cos^2(\frac{\pi t}{2\Delta t})| \downarrow_1>_z \mid \uparrow_2>_z + \frac{1}{2}sin^2(\frac{\pi t}{2\Delta t})| \uparrow_1>_z \mid \downarrow_2>_z + \frac{1}{2}cos(\frac{\pi t}{2\Delta t})\, sin(\frac{\pi t}{2\Delta t})| \downarrow_1>_z \mid \downarrow_2>_z$$

IOP Conf. Series: Journal of Physics: Conf. Series **880** (2017) 012029 doi:10.1088/1742-6596/880/1/012029

$$+\frac{1}{2}sin(\frac{\pi\,t}{2\Delta t})\,cos(\frac{\pi\,t}{2\Delta t})|\uparrow_1>_z|\uparrow_2>_z$$

and

$$\Psi_{hd}(t)^{\uparrow\downarrow}>=\frac{1}{2}[cos(\frac{\pi\,t}{2\Delta t})|\uparrow_1>_z+sin(\frac{\pi\,t}{2\Delta t})|\downarrow_1>_z]*[sin(\frac{\pi\,t}{2\Delta t})|\uparrow_2>_z+cos(\frac{\pi\,t}{2\Delta t})|\downarrow_2>_z]$$

$$|\Psi_{hd}(t)^{\uparrow\downarrow}>=\frac{1}{2}[cos^2(\frac{\pi\,t}{2\Delta t})|\uparrow_1>_z|\downarrow_2>_z+\frac{1}{2}[sin^2(\frac{\pi\,t}{2\Delta t})|\downarrow_1>_z|\uparrow_2>_z \quad (18)$$

$$+\frac{1}{2}cos(\frac{\pi\,t}{2\Delta t})\,sin(\frac{\pi\,t}{2\Delta t})|\uparrow_1>_z|\uparrow_2>_z+\frac{1}{2}sin(\frac{\pi\,t}{2\Delta t})cos(\frac{\pi\,t}{2\Delta t})|\downarrow_1>_z|\downarrow_2>_z$$

Please note, that the last terms with the mixed *sin* and *cos* functions are never observable (for any $t_n = n\,\Delta t$) because $sin(\frac{\pi t}{2\Delta t})cos(\frac{\pi t}{2\Delta t}) = \frac{1}{2}sin(\frac{\pi t}{\Delta t})$. These terms although present are not detectable because they oscillate twice as fast and thus at any measurement time are 0. It is important to notice that at the origin of the doubling of the oscillation is the request for a single time variable: both particles are evolving with the same time and are thus connected to each other, while not entangled. Thus, if a measurement is made on particle 1 at a given time point it has also consequences for particle 2. For example, if particle 1 started with spin down (eq. 20) and in the measurement is observed to have spin down $<\downarrow_1|\downarrow_1>_z$ (i.e. $n = 2m$).

$$<\downarrow_1|\Psi_{hd}(t)^{\downarrow\uparrow}>_{z,2m}=\frac{1}{2}<\downarrow_1|\downarrow_1>_z|\uparrow_2>_z \quad (19)$$

for particle 2 the spin up state $<\uparrow_2|\uparrow_2>_z$ will and must be detected.

The same results is obtained if particle 1 started with spin up (eq. 18) and the measurement is performed at a time point t_n with $n = 2m + 1$

$$<\downarrow_1|\Psi_{hd}(t)^{\uparrow\downarrow}>_{z,2m+1}=\frac{1}{2}<\downarrow_1|\downarrow_1>_z|\uparrow_2>_z \quad (20)$$

Alternatively, if particle 1 is started with spin down (i.e. eq. 17) and the measurement of particle 1 is performed at a time point t_n with $n = 2m + 1$ the spin up state $<\uparrow_1|\uparrow_1>_z$ is observed

$$<\uparrow_1|\Psi_{hd}(t)^{\downarrow\uparrow}>_{z,2m+1}=\frac{1}{2}<\uparrow_1|\uparrow_1>_z|\downarrow_2>_z \quad (21)$$

and thus for particle 2 the spin down state $<\downarrow_2|\downarrow_2>_z$ will be detected at any time point later.

The same result is obtained if particle 1 is started with spin up (i.e. eq. 17) and the measurement of particle 1 is performed at a time point t_n with $n = 2m$ yielding the spin up state $<\uparrow_1|\uparrow_1>_z$: $<\uparrow_1|\Psi_{hd}(t)^{\uparrow\downarrow}>_{z,2m}=\frac{1}{2}<\uparrow_1|\uparrow_1>_z|\downarrow_2>_z$

In analogy the corresponding results are obtained if first spin 2 is detected.

Finally, if averaged over many measurements in 50% of the measurements particle 1 is found in the spin up state and particle 2 in the down state, and vice versa in the other 50% of the measurements particle 1 is found in the spin down state and particle 2 in the spin up state:

$$<\Psi_{hd}(t)|\Psi_{hd}(t)>_z=\frac{1}{2}<\uparrow_1|\uparrow_1>_z<\downarrow_2|\downarrow_2>_z+\frac{1}{2}<\downarrow_1|\downarrow_1>_z<\uparrow_2|\uparrow_2>_z=<\Psi(t)|\Psi(t)>_z \quad (22)$$

Thus, under the assumption of a single coherent time variable, which is discrete in classical physics and by replacing the quantum super position by a fast oscillation between the states the detection along the z-axis of a singlet state can be described without the request of quantum entanglement. Actually, the quantum entanglement as exemplified here by the singlet state can be explained as two individual systems that are coherent in time and thus non-locally connected since it is assumed that there is only one single time variable.

This finding holds also if after the measurement of particle 1 along the z-axis at time point t_n, the second particle is measured at an angle φ to the z-axis at a later time point $t_{n+k} = t_n + k\,\Delta t$ (with k being an element of the natural numbers including 0). It is important to notice that the request for a single time variable is thereby considered. Let us first study the case that particle 1 is measured at time point t_n to have spin down $<\downarrow_1 \mid \downarrow_1>_z$ (i.e. either eqs. 19 or 20). For preparing the measurement of particle 2 along the φ axis a time interval $\delta_k = k\,\Delta t$ later the following expression is obtained

$$<\downarrow_1 |\Psi_{hd}(t)>_z = \frac{1}{2} <\downarrow_1 \mid \downarrow_1>_z [cos(\frac{\pi\,\delta_k}{2\Delta t})\, cos\,\frac{\varphi}{2}| \uparrow_2>_\varphi + sin(\frac{\pi\,\delta_k}{2\Delta t})\, sin\,\frac{\varphi}{2}| \downarrow_2>_\varphi] \tag{23}$$

The measurement of particle 2 at a time interval $\delta_k = k\,\Delta t$ later with an even $k = 2\,m'$ (with m' being an element of the natural numbers including 0).

$$< \Psi_{hd}(t)|\Psi_{hd}(t)>_{z,\varphi,2m'} = \frac{1}{2} <\downarrow_1 \mid \downarrow_1>_z cos^2\,\frac{\varphi}{2} <\uparrow_2 \mid \uparrow_2>_\varphi \tag{24}$$

and correspondingly the measurement at $\delta_k = k\,\Delta t$ later with an odd $k = 2\,m' + 1$ yields

$$< \Psi_{hd}(t)|\Psi_{hd}(t)>_{z,\varphi,2m'+1} = \frac{1}{2} <\downarrow_1 \mid \downarrow_1>_z sin^2\,\frac{\varphi}{2} <\downarrow_2 \mid \downarrow_2>_\varphi \tag{25}$$

The corresponding results are obtained if the particle 1 is measured at time point t_n to have spin up $<\uparrow_1 \mid \uparrow_1>_z$ (i.e. eq. 24 and thereafter) and the measurement of particle 2 along the φ axis a time interval $\delta_k = k\,\Delta t$ later is prepared

$$<\uparrow_1 |\Psi_{hd}(t)>_z = \frac{1}{2} <\uparrow_1 \mid \uparrow_1>_z [sin(\frac{\pi\,\delta_k}{2\Delta t})\, sin\,\frac{\varphi}{2}| \uparrow_2>_\varphi + cos(\frac{\pi\,\delta_k}{2\Delta t})\, cos\,\frac{\varphi}{2}| \downarrow_2>_\varphi] \tag{26}$$

The measurement of particle 2 at $\delta_k = k\,\Delta t$ later with an $k = 2\,m'$ yields

$$< \Psi_{hd}(t)|\Psi_{hd}(t)>_{z,\varphi,2m'} = \frac{1}{2} <\uparrow_1 \mid \uparrow_1>_z cos^2\,\frac{\varphi}{2} <\downarrow_2 \mid \downarrow_2>_\varphi \tag{27}$$

and correspondingly if measured at $\delta_k = k\,\Delta t$ later with $k = 2\,m' + 1$ yields

$$< \Psi_{hd}(t)|\Psi_{hd}(t)>_{z,\varphi,2m'+1} = \frac{1}{2} <\uparrow_1 \mid \uparrow_1>_z sin^2\,\frac{\varphi}{2} <\uparrow_2 \mid \uparrow_2>_\varphi \tag{28}$$

When these individual measurements are ensemble averaged over many experiments the same probabilities are obtained as calculated by standard quantum mechanics:

$$< \Psi_{hd}(t)|\Psi_{hd}(t)>_{z,\varphi} = \frac{1}{2}cos^2\,\frac{\varphi}{2}[<\uparrow_1 \mid \uparrow_1>_z<\downarrow_2 \mid \downarrow_2>_\varphi + <\downarrow_1 \mid \downarrow_1>_z<\uparrow_2 \mid \uparrow_2>_\varphi] \tag{29}$$

$$+\frac{1}{2}sin^2\,\frac{\varphi}{2}[<\uparrow_1 \mid \uparrow_1>_z <\uparrow_2 \mid \uparrow_2>_\varphi + <\downarrow_1 \mid \downarrow_1>_z <\downarrow_2 \mid \downarrow_2>_\varphi] =< \Psi(t)|\Psi(t)>_{z,\varphi}$$

3. Discussion

In addition to a quantized energy, quantum mechanics differs from classical physics by several counter intuitive findings such as the super position of quantum states, the non local nature of quantum mechanics, and the lack of the possibility to calculate a single measurement. The presented approach of combining the presumed existence of a discrete time in classical physics coherently connected to a single continuous time variable in quantum mechanics and the fast oscillation of the quantum system between its eigenstates is able to resolve some of these phenomena. It localizes the problem of measuring a quantum state to the issue of exact timing and explains why quantum mechanics is not able to calculate single experimental outcomes at current time resolutions, but calculates them correctly in average. It further resolves the causality issue as outlined in the EPR paradox [1]. At the root of the causal nature of the presented modified quantum mechanics theory is thereby the request of a single time variable, which is believed to be a logic axiom for a theory that guarantees causality.

While the presented modification of quantum mechanics for a two spin system may be regarded sound, the generalization - how it can be translated into a system with many eigen states - remains to be established. One may speculate that the system goes periodically through all the eigen states (which may be obtained by step functions or Lie algebra and may require a reformulation of the Schrödinger equation). Furthermore, finding experimental support for the presence of a discrete time and the modified quantum mechanics appear to be difficult since the current experimental time resolution is ca 26 orders of magnitude away from the Planck time Δt_p. Nonetheless, it can be stated, that the better the time resolution of the experiment (and eventually also the faster the processes under study) the more deviation from quantum mechanics towards a classical/deterministic behavior of the system is expected. Alternatively, the nature of the break down of a quantum mechanical system into its classical analog may give valuable hints in favor or against the presented theory. For example if a quantum system oscillates through all the eigen states in steps of $4\,\Delta t_p$ as speculated above, the system may behave classically or/and deterministic if the time the system needs to go through all significant eigen states is longer than the experimental time resolution while it would behave quantum mechanically if the time resolution of the experiment would be less (please note at the current time resolution this requests the study of a system with ca 10^{26} significantly populated eigen states).

It is evident that the concept of a discrete time is at the root of the presented interpretation. While not popular upon energy quantization in quantum mechanics the introduction of a discrete time appears to be obvious [11-17]. With the recent success in the derivation of a microscopic entropy under the assumption of a discrete time the hypothesis that time in classical physics may be discrete [10] is strengthend and worth to be disputed further.

Acknowledgments

I would like to thank Drs. G. Blatter, A. Sobol, and W. Kwiatkowski for helpful discussions.

References

[1] Einstein A, Podolsky B and Rosen N 1935 *Phys. Rev.* **47** 777–80
[2] Schrödinger E 1935 *Naturwissenschaften* **23** 807–12; 823–28
[3] Greiner W 1989 *Quantenmechanik* (Frankfurt am Main: Verlag Harri Deutsch)
[4] Bell J S 1964 *Physics* **1** 195–200
[5] Aspect A 1999 *Nature* **398** 189–90
[6] Hensen B 2015 *et al. Nature* **526** 682–6
[7] Bohm D 1952 *Phys. Rev.* **85** 166–79
[8] De Martin F and Santamato E 2014 arXiv:1406.2970
[9] Accardi L and Regoli M 2000 arXiv:quant-ph/0007005
[10] Riek R 2014 *Entropy* **16** 3149–72
[11] Lee T D 1983 *Physics Letters* B **122** 217–20
[12] Levi R 1927 *Journal de Physique et le Radium* **8** 182–98
[13] Caldirola P 1953 *Supplemento al Nuovo Cimento* **10** 1747–52
[14] Jaroszkiewicz G and Norton K 1997 *J. Phys. A: Math. Gen.* **30** 3115–44
[15] Rovelli C 2011 *PoS* **QGQGS2011, 3** arXiv:1102.3660
[16] Elze H-T 2014 *Phys. Rev. A* **89** 012111 arXiv:1312.1655
[17] Elze H-T 2013 *EPJ Web of Conferences* **58** 01013 arXiv:1310.2862

IOP Conf. Series: Journal of Physics: Conf. Series **880** (2017) 012030 doi:10.1088/1742-6596/880/1/012030

Quantum entanglement, Wheeler's delayed choice experiment and its explanation on the basis of quantization of fields

Alexey V Melkikh

Ural Federal University, Yekaterinburg, Russia

Email: melkikh2008@rambler.ru

Abstract. The result of Wheeler's delayed choice experiment is a natural consequence of the entanglement of moving photons and particles (atoms, molecules) of the slit through which they move. The inclusion of quantum fields (taking into account that speed of virtual particles is not limited) self-consistently explains why interaction's propagation velocity after closing one slit is larger than the speed of light.

1. Introduction

One manifestation of entanglement is Wheeler's delayed-choice experiment. Wheeler's thought experiment [1] is a variant of an experiment using two static slits, in which one of the slits may be closed after a photon passes through both slits, but before the photon has reached the screen. Practically, this applies a delayed choice to the structure of the system with which a photon interacts. How does a particle "know" that one of the slits was closed? How fast is the "knowing" communication? To answer these and other questions, a more detailed discussion of entanglement that explicitly takes into account the quantization of fields is required.

2. Wheeler's thought experiment and concepts of entanglement in quantum mechanics

Currently Wheeler's delayed-choice experiment has been experimentally implemented, although in a somewhat different form (see, for example, [2-5]). A schematic of Wheeler's delayed choice experiment is shown in Figure 1.

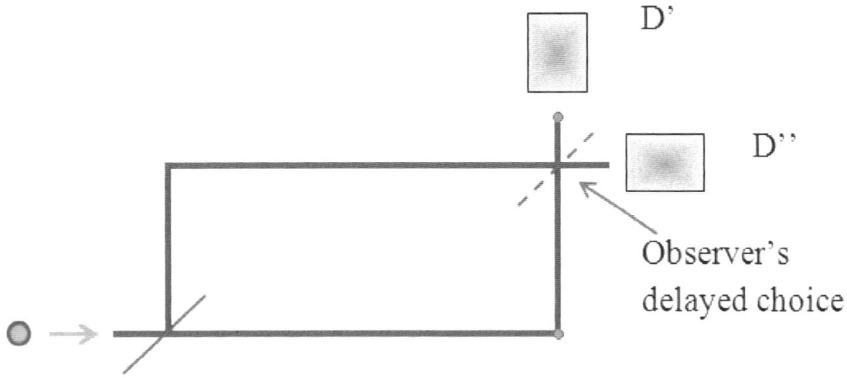

Figure 1. A schematic of Wheeler's delayed choice experiment in Mach-Zehnder interferometer.

A photon is sent to the Mach-Zehnder interferometer and is split into the superposition of two states, which move along two arms of the interferometer. By inserting the second splitter, either particle or wave behavior may be detected by the detector D' or D''. This is much like the photon passing through slits. If there is one slit, the screen image has a maximum corresponding to the photon's position (i.e., the photon shows particle properties). If there are two slits, an interference pattern is observed on the screen (i.e., the photon demonstrates wave properties).

Experiments have shown that the act of measurement fully determines how particles behave in the past, including moments, when the position of the particle is unachievable even with the speed of light. A variant of this experiment would be when the particle passes (for example, in outer space) very long distances before being recorded with the apparatus (such as a telescope).

Another variant of Wheeler's experiment is the quantum eraser [6, 7]. If the experimenter wishes to delete the information of slit through which the particle passed, then, anywhere it is situated, the interference pattern appears on the screen.

However, in the framework of quantum mechanics, including relativistic cases, it is difficult to explain this experiment. What is entangled with the photon? What conditions are needed for this entanglement? What is the mechanism of propagation for action at a distance? In Wheeler's experiment, why is this interaction's propagation velocity larger than the speed of light? To answer these and other questions, let us consider a more detailed picture, in which the quantization of fields is taken into account.

3. Explanation of entanglement on the basis of explicitly accounting for the quantization of fields

An explanation of entanglement may be given by explicitly accounting for quantum fields. Let's call the version of quantum mechanics, in which the fields are considered classically, "quantum mechanics of particles", distinct from one considering particles and fields. Quantum mechanics of particles and fields can explain entanglement as an interaction through virtual particles. Virtual particles are not subject to the limit imposed by the speed of light, so interaction at an arbitrary distance is natural for them. As mentioned earlier in connection with the solution of the problem of measurement [8], the quantum mechanics of particles is not a closed science - only the quantum mechanics of particles and fields together can be considered closed. Only such a comprehensive consideration avoids the measurement paradox while simultaneously solving the problem of the collapse of the wave function.

The system of equations for particles and fields (see, for example, [9, 10]) is as follows:

$$\left\{\gamma_\mu\left(\frac{\partial}{\partial x_\mu}-ieA_\mu(x)\right)+m\right\}\psi(x)=0 \quad , \tag{1}$$

$$\left\{\gamma_\mu^T\left(\frac{\partial}{\partial x_\mu}+ieA_\mu(x)\right)-m\right\}\bar{\psi}(x)=0 \quad , \tag{2}$$

$$\Box A_\mu(x)=-j_\mu(x) \quad , \tag{3}$$

where $\psi(x)$ is the wave function (operator) and $j_\mu(x)$ is the 4-density of electron current, which is equal to

$$j_\mu(x)=\frac{ie}{2}\left(\bar{\psi}(x)\gamma_\mu\psi(x)-\bar{\psi}^c(x)\gamma_\mu\psi^c(x)\right).$$

The subscript "c" indicates the change of the charge sign. A_μ is the potential of the electromagnetic field, γ_μ are the Dirac matrices, e is electron charge, and $\Box$ is

$$\Box\equiv\frac{\partial^2}{\partial t^2}-\frac{\partial^2}{\partial x^2}-\frac{\partial^2}{\partial y^2}-\frac{\partial^2}{\partial z^2}.$$

For a virtual particle, the well-known relation between the momentum and energy does not hold:

$$E^2\neq p^2c^2+m_0^2c^4.$$

As a consequence, virtual particles cannot be observed by classical devices. Is it possible to measure the virtual particles at all?

We may be able to speak about the measurement of the totality of the virtual particles. When an electron in a hydrogen atom jumps from one level to another, a photon emission of a certain length occurs. By measuring the properties of the photon, the energy levels of the atom can be obtained. In particular, we can determine the Lamb's shift of the levels. That is, obtained in the measurement (which is inevitably connected with the emission or annihilation of particles) are not only the properties of the electron at a certain level, but the properties of the "electron + virtual particles". Other experiments that demonstrate the indirect presence of virtual particles are the Casimir effect and dynamic Casimir effect [11].

Thus, measurements show agreement with a theory in which the virtual particles exist, but not with the theory in which fields are classical.

The same process involving virtual particles may look different. The larger virtuality of the particle, the shorter time and distance, in which the virtual process takes place. The speed of the virtual particles, determined by the momentum and energy, can be arbitrarily large.

There are various interpretations of quantum field theory. In particular, the "partial" interpretation was discussed in [12, 13].

The movement of quantum particles is associated with the group velocity, which is defined as

$$v_g=\frac{d\omega}{dk}.$$

This velocity cannot be larger than speed of light. For the phase velocity

$$v_p=\frac{\omega}{k}$$

such a restriction is absent.

It is believed that the phase velocity is not physically observable for elementary particles, so its use does not make practical sense. This, however, does not apply to virtual particles. Although they are not directly observable, as noted above, the effect of the totality of the virtual particles can be experimentally observed. Because in these experiments on particle entanglement, phase plays an

IOP Conf. Series: Journal of Physics: Conf. Series **880** (2017) 012030 doi:10.1088/1742-6596/880/1/012030

important role, the phase velocity, which can be attributed to the virtual particle, is fundamentally important.

4. Wheeler's experiments explanation on the basis of accounting for field quantization

The result of Wheeler's delayed choice experiment is also a natural consequence of the entanglement of moving photons and molecules (atoms, particles) of the slit through which they move. That is, a photon (particle A), flying through the two slits, was entangled with the atoms of the diaphragm through the exchange of virtual particles with the atoms. The most powerful entanglement is between the photon and the particles that actually form the slit because they alter the trajectory of the photon most significantly. With the more distant particles, the interaction is much weaker, so their condition will affect the state of the photon weakly.

How far would the photon flew, it continued to exchange virtual particles with atoms of the diaphragm. The rate of this exchange via virtual particles is not limited.

Suppose now that one slit is closed for a negligible time compared to the total time of flight of the photon. However, because the entanglement speed is not limited, the photon is now entangled with a modified diaphragm with one slit. As a result, its phase will change over the characteristic time of closing the slit; it will behave as if there were originally only one slit and the interference pattern on the screen disappears. This is the same, if as a result of the measurement we would receive information on which slit the photon has passed.

Thus, the slit closing may be considered a measurement-like process in a quantum sense; both processes create the same state event, wave emission by one slit. Indeed, the measurement determining a photon's (particle A) passage through a certain slit destroys its coherent state and a new wave is re-emitted (Figure 2).

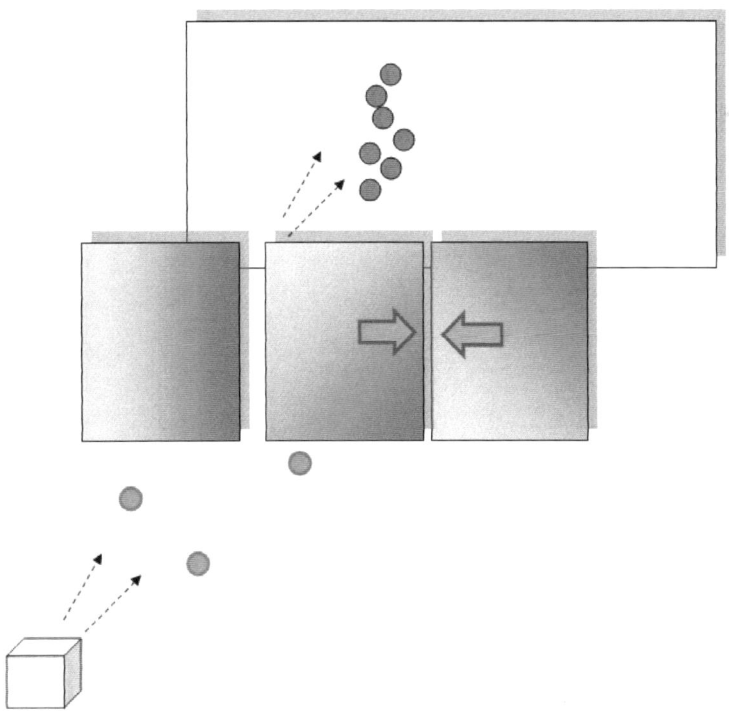

Figure 2. Photon is entangled with a modified diaphragm with one slit

In the erasing experiment the emitted photon can be absorbed (deleted). At the same time, the uncertainty generated by the measurement also disappears. If this is done arbitrarily late, particle A will be affected because it remains to be connected with emitted particles by means of virtual particles.

Thus, Wheeler's delayed choice experiment can be explained in terms of the quantum mechanics of particles of fields.

References

[1] Wheeler J A 1978, in *Mathematical Foundations of Quantum Mechanics*, ed A R Marlow (New York: Academic) pp 9–48
[2] Jacques V, Wu E, Grosshans F, Treussart F, Grangier P, Aspect A, Roch J-F 2007 *Science* **315** 966
[3] Kaiser F, Courdeau T, Milman P, Ostrowsky D B, Tanzilli S 2012 *Science* **338** 637
[4] Peruzzo A, Shadbolt P, Brunner N, Popescu S, O'Brien J L 2012 *Science* **338** 634
[5] Ma X S, Zotter S, Kofler J, Ursin R, Jennewein T, Brunker C, Zeilinger A 2012 *Nature Physics* **8** 479
[6] Yoon-Ho Kim, Yu R, Kulik S P, Shih Y H, Scully M 2000 *Phys. Rev. Lett.* **84** 1
[7] Walborn S P, Terra Cunha M O, Pádua S, Monken C H 2002 *Phys. Rev. A* **65** 033818
[8] Melkikh A V 2015 *Comm. Theor. Phys.* **64** issue 1 47
[9] Akhiezer A I, Berestetskii V B 1965 *Quantum electrodynamics* (New York: John Wiley & Sons)
[10] Berestetskii V B, Lifshits E M, Pitayevskii L P 1982 *Relativistic Quantum Theory* (Oxford: Pergamon Press)
[11] Wilson C M, Johansson G, Pourkabirian A, Simoen V, Johansson J R, Duty T, Nori F, Delsing P 2011 *Nature* **479** 376
[12] Teller P 1995 *An Interpretive Introduction to Quantum Field Theory* (Princeton: Princeton University Press)
[13] Bain J 2011 *Studies in History and Philosophy of Modern Physics* **42** 98

IOP Conf. Series: Journal of Physics: Conf. Series **880** (2017) 012031 doi:10.1088/1742-6596/880/1/012031

Many worlds in perspective

Heinrich Päs

Fakultät für Physik, Technische Universität Dortmund, 44221 Dortmund, Germany

E-mail: heinrich.paes@tu-dortmund.de

Abstract. A minimal approach to the measurement problem and the quantum-to-classical transition assumes a universally valid quantum formalism, i.e. unitary time evolution governed by a Schrödinger-type equation. As had been pointed out long ago, in this view the measurement process can be described by decoherence which results in a "Many-Worlds" or "Many-Minds" scenario according to Everett and Zeh. A silent assumption for decoherence to proceed is however, that there exists incomplete information about the environment our object system gets entangled with in the measurement process. This paper addresses the question where this information is traced out and - by adopting recent approaches to model consciousness in neuroscience - argues that a rigorous interpretation results in a perspectival notion of the quantum-to-classical transition. The information that is or is not available in the consciousness of the observer is crucial for the definition of the environment (i.e. the unknown degrees of freedom in the remainder of the Universe). As such the Many-Worlds-Interpretation, while being difficult or impossible to probe in physics, may become testable in psychology.

While it is general consensus that our experience has both objective and subjective aspects, apart from clearly subjective elements such as taste or value, there exist „perspectival" elements which depend on the observer's perspective onto the world albeit being also objectively determined by it. The identification and elimination of such perspectival elements has initiated some of the greatest scientific revolutions: For example, in the Copernican paradigm shift the apparent revolution of celestial bodies around the earth was resolved as consequence of the choice of a specific (and not very useful) coordinate frame with the Earth (and thus anthropic perspective) defining it's origin. Similarly in Special and General Relativity the only apparently observer-independent notions of space, time and gravity were eliminated and integrated into the truly observer-independent concepts of curved spacetime and general coordinate invariance. In a similar spirit, Huw Price [1] has argued that that the asymmetry of causation is anthropocentric in origin and that one should adopt a atemporal point-of-view to fully understand quantum mechanics. On the other hand, recent developments in quantum physics and in particular in quantum cosmology seem to require a generalization of the kind of perspectivism which is related to different local coordinate frames to include non-local perspectives as well. As we will see, the identification of this kind of perspectivism will in turn have interesting consequences for our understanding of time.

The problem of how to understand the quantum-to-classical transition dates back to the early years of quantum mechanics. Among the various interpretations of the quantum measurement process the "Many Worlds" or "Many Minds" interpretations (MWI) [2, 3] are recently becoming increasingly popular [4]. This scenario is attractive as it does not introduce any new elements into the formalism of quantum mechanics which go beyond the unitary evolution governed by a Schrödinger-type equation, and can in this sense be understood as "minimal" or "conservative".

In the following we briefly review the measurement process and the role of decoherence in the MWI, before we concentrate on the crucial importance of the observer's perspective.

In quantum mechanics, the state vector $|\Psi\rangle$ describes the complete knowledge about a system and evolves deterministically according to a Schrödinger-type equation

$$i\frac{d}{dt}|\Psi(t)\rangle = H|\Psi(t)\rangle. \tag{1}$$

The state vector does not, however, unambiguously determine the outcomes of measurements. What happens during a measurement can be easily illustrated by adopting a spin-1/2 or Q-bit object system $|\Psi\rangle \sim \{|\uparrow\rangle, |\downarrow\rangle\}$ which is coupled to a measurement apparatus with states $\{|0\rangle, |+\rangle, |-\rangle\}$. In the measurement process the object system gets entangled with the apparatus and unitary evolution yields a state vector $|\Psi_{\rm tot}\rangle$ for the complete (object + apparatus) system

$$\begin{aligned} |\Psi_{\rm tot}\rangle &\sim |\uparrow 0\rangle \rightarrow |\uparrow +\rangle \\ |\Psi_{\rm tot}\rangle &\sim |\downarrow 0\rangle \rightarrow |\downarrow -\rangle. \end{aligned} \tag{2}$$

In general our object system will be in a superposition $|\Psi\rangle \sim \alpha_\uparrow|\uparrow\rangle + \alpha_\downarrow|\downarrow\rangle$ which in the measurement process then evolves via

$$|\Psi\rangle \sim \alpha_\uparrow|\uparrow,0\rangle + \alpha_\downarrow|\downarrow,0\rangle \rightarrow |\Psi\rangle \sim \alpha_\uparrow|\uparrow,+\rangle + \alpha_\downarrow|\downarrow,-\rangle. \tag{3}$$

The latter two terms then correspond to the infamous Everett branches, "Many Worlds" or "Many Minds": if the apparatus is understood as observer the experimentalist will find herself within one of the branches and will not experience any alternative realities. The understanding of the measurement process as a coupling of object system, measurement apparatus and observer is known as the von-Neumann chain [5], and it leads immediately to a new problem: where in this chain is the quantum-to-classical transition, also known as "Heisenberg cut", happening? In principle there could always be an outside observer ("Wigner's friend") who would experience her experimentalist buddy in a quantum superposition or "Schrödinger cat state". All that seemed to be known for sure is that the mind of the observer is in a well-defined state ("psycho-physical parallelism" [5]). In this situation Wigner advocated a quantum collapse triggered in the experimentalist's consciousness ("consciousness as the last observer") and this understanding is known as the Wigner-von-Neumann interpretation [6].

Wigner later on changed his mind, however, when H.D. Zeh discovered decoherence ([3], see also [7, 8, 9, 10]): the phenomenon that entanglement with the environment leads to an extremely fast suppression of interference terms/quantum superpositions. Decoherence can be understood most easily in terms of reduced density matrices $\rho^{\rm r}$. If we consider the density matrix of a quantum system $\rho = |\Psi\rangle\langle\Psi|$ in the basis $|a\rangle$, the matrix element $\rho_{a'a}$ is given by the product of wave functions

$$\rho_{a'a} = \langle a'|\Psi\rangle\langle\Psi|a\rangle \equiv \psi^*(a')\psi(a) \tag{4}$$

and a total system consisting of two entangled sub-systems is described by

$$\rho_{a'b'ab} = \psi^*(a',b')\psi(a,b). \tag{5}$$

Considering now an observable acting only on the a-subsystem $L_{a'b'ab} \equiv L_{a'a}\delta_{b'b}$, one can calculate the expectation value as

$$\langle L\rangle = Tr(\rho L) = \sum_{a'b'ab} \psi^*(a',b')L_{a'a}\delta_{bb'}\psi(a,b) \equiv \sum_{a'a} \rho^{\rm r}_{a'a}L_{a'a} \tag{6}$$

with

$$\rho^{\rm r}_{a'a} = \sum_{b'} \psi^*(a',b')\psi(a,b'). \tag{7}$$

IOP Conf. Series: Journal of Physics: Conf. Series **880** (2017) 012031 doi:10.1088/1742-6596/880/1/012031

By looking only at the a-subsystem we trace out or average over the uninteresting degress of freedom of the remainder of the system. The resulting reduced density matrix ρ^{r} looks then exactly like a mixed state.

While this process provides an elegant and minimal explanation of the quantum measurement process, it also has two disturbing consequences:

- "Many Minds": the observer "splits" into multiple copies observing each possible outcome.
- Classical reality is a consequence of perspective. It results from our ignorance about the exact state of the environment. It is "emergent", not fundamental.

While heated debates have been fought out over the first of these points, this paper argues that the real "elephant-in-the-room" is the second consequence: Classical reality is a consequence not only of a measurement system being coupled to an environment but also of the incomplete knowledge about this environment, which is of course a consequence of the local observer who simply cannot have all possible information about the exact state of the entire Universe. This local perspective has been dubbed "frog perspective" (local, classical) by Tegmark and Zeh in contrast to the "bird perspective" (non-local, quantum) in which the entire quantum system would be observed and no quantum-to-classical transition takes place: The quantum-to-classical transition is perspectival!

Thus in principle there are two possible kinds of quantum systems:

- Isolated (typically microscopic) systems with no interaction with the environment. While all quantum systems we have experience with are of this type, this is naturally always an approximation.
- The entire quantum Universe: global, encompassing, with no external environment and thus not subject to decoherence. It is this latter system which constitutes the only true fundamental quantum state which can be experienced only in the non-local bird perspective.

As a side remark, let me mention that the quantum Universe as a pure quantum state has a vanishing von-Neumann entropy. It thus seems likely that the quantum Universe by itself is timeless, as described by the Wheeler-DeWitt equation [11]. It has been argued by Zeh [12] and Kiefer [13] that time itself then could be an emergent property of classical spacetime as a consequence of decoherence related to averaging out irrelevant gravitational degrees of freedom such as gravitational waves or tiny density fluctuations. In this case, not only the classical world but even time would be perspectival (compare the recent work by Rovelli [14] which argues along these lines).

The main open question of this approach seems to be what defines the "frog perspective" or - in Tegmark's words - the "factorization" [15] into subject, object and environment. Obviously this perspective is a consequence of the observer's consciousness being confined to her cerebral cortex. But what defines the boundaries of this "cognitive self" without assuming a classical description and preferred basis beforehand? In the following we argue that the "frog perspective" may be a crucial prerequisite of consciousness itself.

Of course the phenomenon of consciousness is far from being understood. In the recent years however an interesting approach called "Integrated Information Theory" [16] has been developed by neuroscientist Giulio Tononi, which pursues a mathematical framework for evaluating the quality and quantity of consciousness based on properties of the corresponding information processing such as the irreducibility into subsets or "integration of information". While such a property reminds of entangled quantum systems, Max Tegmark has provided two important results which suggest that consciousness at least in the IIT framework can most probably be no quantum process:

- In [17] Tegmark estimated decoherence times of neurons and microtubules within the human brain and found that quantum superpositions decay on extremely fast time scales of the order of $10^{-13} - 10^{-19}$ seconds.
- More recently Tegmark applied IIT to quantum systems and found that due to the free choice of the Hilbert space basis only an insufficient maximum integrated information of 0.25 bits can be obtained [15].

While these results are not unchallenged, we nevertheless thus adopt as a working hypothesis that consciousness should be understood as a classical algorithm operating in the cerebral cortex and defining the factorization into subject/conscious self, object and environment. In fact, taking Tononi and Tegmark seriously, consciousness seems to be emergent itself and only possible within a classical perspective. Consciousness may actually be a by-product or even the trigger of the quantum-to-classical transition. It should be stressed here that there is not only one possible definition of "self". For example, biologist Francisco Varela has argued that organisms have to be understood as a "mesh of virtual selves" [18], including the cognitive or conscious self, the immune or body self and so on. As we claim here, it is of crucial importance to understand which self or selves define(s) the local perspective giving rise to decoherence and classical reality. Indeed, situations where the various selves do not coincide seem to be particularly interesting in this regard. Apart from autoimmune diseases in which Varela was mainly interested, such situations include altered states of consciousness such as hallucinogen intoxication (for example under the influence of lysergic acid diethylamide (LSD-25)) where probands report at least from a subjective perspective a dissolution of the mental self and subjective time within a non-local experience [19]. It seems that these impressions result from a distortion of the information filtering usually proceeding within the thalamus as a consequence of the intoxication. These aspects of hallucinogen experience exhibit interesting (anti-)parallels to the adoption of the local "frog perspective" triggering the quantum-to-classical transition: Conversely, here a classical, local self and possibly even time itself emerges, as a consequence of the neglect of information about the environment. Given these parallels it seems not to be too far-fetched to speculate whether the local algorithm constituting the conscious self gets so strongly coupled with the environment as a consequence of the hallucinogen intoxication that it is lifted to a less local perspective and in this way is able to experience some kind of "quantum holism". One should seriously scrutinize whether such considerations can easily be disregarded as "New-Age-Bullshit".

More concretely, the question proposed to study here is what happens in mental states where consciousness fades away and is experienced as "dissolving". As Tegmark has argued in [17], classical systems exist in some kind of intermediate or critical regime between too little interaction (no decoherence) and too much interaction with the environment, where in the latter case it is not meaningful to consider the system as an independent entity anymore. If the "conscious self" defines the local perspective that leads to the perception of a classical world, it is the local and classical property of this "self" which is crucial for the quantum-to-classical transition. What is proposed here is to study the conditions where consciousness looses its independence by increasing the information flux into the algorithm assumed to be its physical correlate.

By adopting these ideas one could come to the fascinating conclusion that while the interpretation of quantum mechanics in general and the MWI in particular are notoriously difficult (if not impossible) to test in physics experiments, such tests may be possible in psychology. A simple setup could for example employ probands under the influence of LSD-25 performing quantum measurements (such as spin-up versus spin-down) on a computer screen, while an equally prepared control group deals with an equally looking interface connected to a classical simulation based on a random number generator. It is conceivable that the first group experiences quantum superpositions while the control group does not.

Obviously, the details of this idea are extremely vague, which is however to some extent a consequence of our limited knowledge about what the physical correlates of consciousness are and how they work. Of course it may very well be the case that the algorithm responsible for consciousness remains local even under influence of hallucinogens and that the non-local experience is only an illusion. Moreover, there are open questions about the experimental setup: Is a local perspective already defined by the measurement device or the data processing? Moreover, even a classical random number generator is a quantum physical object on a fundamental level. Can the difference between this object and e.g. a quantum-spin really be exploited this way? Finally, since even Time itself may cease to exist in this change of perspective, this process is barely understood even from a pure physics perspective. These questions cannot easily be answered by theoretical considerations. They may be resolved by the proposed experiment, though, in case of a positive result.

Such a result could be understood as a strong evidence for the MWI and would dramatically affect the way we think about quantum mechanics and the quantum-to-classical transition, and even about reality itself.

Acknowledgments
The author thanks H.D. Zeh, Claus Kiefer, Mathias Becker, Sinan Zeissner and Marc Wittmann for useful discussions.

References

[1] Price H 1996 *Time's arrow and Archimedes' point* (New York: Oxford University Press)
[2] Everett H 1957 *Rev. Mod. Phys.* **29** 454
[3] Zeh H D 1970 *Found. Phys.* **1** 69
[4] Tegmark M 1998 *Fortsch. Phys.* **46** 855 [quant-ph/9709032]
[5] von Neumann J 1932 *Mathematische Grundlagen der Quantenmechanik* (Berlin: Springer)
[6] Wigner E P 1963 *Am. J. Phys.* **31** 6
[7] Joos E and Zeh H D 1985 *Z. Phys.* **B 59** 223
[8] Zurek W H 1991 *Phys. Today* **44 N10** 36
[9] Schlosshauer M 2004 *Rev. Mod. Phys.* **76** 1267 [quant-ph/0312059]
[10] Schlosshauer M 2007 *Decoherence and the Quantum-to- Classical Transition* (Berlin: Springer)
[11] DeWitt B S 1967 *Phys. Rev.* **160** 1113
[12] Zeh H D 1986 *Phys. Lett.* **A 116** 9
[13] Kiefer C 2009 *Preprint* arXiv:0909.3767 [gr-qc]
[14] Rovelli C 2015 *Preprint* arXiv:1505.01125 [physics.hist-ph]
[15] Tegmark M 2015 *Chaos Solitons Fractals* **76** 238 [arXiv:1401.1219 [quant-ph]]
[16] Tononi G 2004 *BMC Neuroscience* **45** 42
[17] Tegmark M 2000 *Phys. Rev.* **E 61** 4194 [quant-ph/9907009]
[18] Varela F 1995 *The Emergent Self* in: *The Third Culture* ed John Brockman (New York: Simon & Schuster)
[19] Wittmann M 2015 *Consciousness and Cognition* **38** 172

IOP Conf. Series: Journal of Physics: Conf. Series **880** (2017) 012032 doi:10.1088/1742-6596/880/1/012032

Quantum inertia stops superposition: Scan Quantum Mechanics

Beatriz Gato-Rivera[1,2,3]

[1] Instituto de Física Fundamental, IFF-CSIC, Serrano 123, 28006 Madrid, Spain
[2] Nikhef, Science Park 105, 1098 XG Amsterdam, The Netherlands

E-mail: t38@nikhef.nl

Abstract. Scan Quantum Mechanics is a novel interpretation of some aspects of quantum mechanics in which the superposition of states is only an approximate effective concept. Quantum systems scan all possible states in the superposition and switch randomly and very rapidly among them. A crucial property that we postulate is quantum inertia, that increases whenever a constituent is added, or the system is perturbed with all kinds of interactions. Once the quantum inertia I_q reaches a critical value I_{cr} for an observable, the switching among its different eigenvalues stops and the corresponding superposition comes to an end, leaving behind a system with a well defined value of that observable. Consequently, increasing the mass, temperature, gravitational strength, etc. of a quantum system increases its quantum inertia until the superposition of states disappears for all the observables and the system transmutes into a classical one. Moreover, the process could be reversible. Entanglement can only occur between quantum systems because an exact synchronization between the switchings of the systems involved must be established in the first place and classical systems do not have any switchings to start with. Future experiments might determine the critical inertia I_{cr} corresponding to different observables, which translates into a critical mass M_{cr} for fixed environmental conditions as well as critical temperatures, critical electric and magnetic fields, etc. In addition, this proposal implies a new radiation mechanism from astrophysical objects with strong gravitational fields, giving rise to non-thermal synchrotron emission, that could contribute to neutron star formation. Superconductivity, superfluidity, Bose-Einstein condensates, and any other physical phenomena at very low temperatures must be reanalyzed in the light of this interpretation, as well as mesoscopic systems in general.

1. Introduction

Since its formulation in the 1920s and 1930s, the foundations of quantum mechanics have provided an endless source of intense and passionate debate and inspiration. The conventional Copenhagen interpretation, widely accepted for practical purposes, leaves however too many questions unanswered. Measurements of the microscopic entities by classical devices are supposed to produce the collapse of the wave functions into one of the possible states. That is, from all possible states in the superposition only one is realized in the measurement process, whereas the others just evaporate, as if macroscopic classical devices have the magical power to collapse wave functions. Moreover, the dividing line between macroscopic classical and

[3] Also known as B. Gato

microscopic quantum objects is completely unknown, even though many experiments have been performed in the last years trying to approach this frontier. These and some other questions have given rise to a plethora of different interpretations of quantum mechanics over the years.

Recently we have presented a novel interpretation of some aspects of quantum mechanics [1]. Our proposal, *Scan Quantum Mechanics* (SQM), is not radically different from conventional quantum mechanics since we let the mathematical formalism unmodified, as an effective correct description of the quantum systems. However, we put forward a mechanism underlying the superposition of quantum states, and a criterion for deciding the validity of the quantum description versus the classical description of the physical systems. To be precise, we postulate a property of all systems, quantum and classical alike, *quantum inertia* I_q, in such a way that quantum behaviour only manifests itself for values of I_q below some critical ones I_{cr} (one critical value for each observable). As a result, quantum inertia with its critical values marks the dividing line between the quantum and the classical worlds.

In the light of this interpretation, the answers to several long-standing crucial questions and paradoxes are most intuitive. This applies especially to the questions related to: the measurement problem, the superposition of quantum states, the elements of reality, the lack of entanglement between quantum systems and macroscopic objects, and the dividing line between quantum and classical behaviour, as we just mentioned. Other less intuitive but unavoidable features of this interpretation, like in conventional quantum mechanics, are the lack of well defined physical properties for quantum systems, randomness in the output of measurements and nonlocality, not only for quantum correlations but also for positions in space (i.e. quantum jumps instead of classical continuous trajectories).

Nevertheless, SQM can be tested with a number of laboratory experiments and by astrophysical observation, since it implies a new radiation mechanism in strong gravitational fields. This mechanism could contribute to neutron star formation and gives rise to non-thermal emission, especially γ-ray synchrotron radiation.

In what follows, as the purpose of this presentation is to emphasize the physical concepts of SQM, we will consider only pure states keeping the mathematical formalism to a minimum.

2. Scan Quantum Mechanics

Scan Quantum Mechanics consists basically of an interpretation of the wave function in which the superposition of the quantum states, at any instant, is only an approximate effective concept. The mathematical formalism is the same as in conventional quantum mechanics because there is no need to propose any modifications. For simplicity, let us consider a quantum system with only one observable. To be precise, we assume that the system is in a normalized superposition $\psi = \sum_k c_k |a_k\rangle$ of orthonormal eigenvectors $|a_k\rangle$ of an hermitian operator A, with eigenvalues a_k, that represents the observable. Then, the measurement is supposed to collapse the initial state, or wave function ψ, of the system to a unique well defined state, say $|a_m\rangle$, with the particular value a_m of the observable, the probability of encountering this value being given by the Born rule, i.e. $|c_m|^2$. Our proposal differs from conventional quantum mechanics in essentially two postulates:

1. The quantum systems scan the eigenstates of the observables, described by the wave function ψ, and switch among them randomly at very high speed so that, effectively, due to insufficient time resolution, the systems appear to be in a superposition of the different eigenstates $|a_k\rangle$, with eigenvalues a_k, at any instant. The relative time the quantum system spends in a particular state $|a_m\rangle$ coincides with the probability $|c_m|^2$ of finding that state during a measurement.

2. The quantum switching depends on a crucial property shared by all physical systems, *quantum inertia* I_q, that increases whenever a constituent is added, or the system is perturbed with all kinds of interactions and energies. If the value of I_q is below the critical one for the observable, $I_q < I_{cr}$, then the quantum switching among the available states is at work and the wave function ψ describes the system exactly as in conventional quantum mechanics. If, on the contrary, the value of I_q reaches the critical one, $I_q \geq I_{cr}$, then the system is unable to switch among the different eigenvalues of the observable, the quantum switching freezes and the superposition of states disappears leaving behind a classical system with a well defined value of the observable. As a consequence, the system's wave function ψ stops providing the correct description of the system. Thus for SQM the origin of classicality, the dividing line between quantum and classical behaviour, is simply described by the relation $I_q \geq I_{cr}$.

The generalization of postulate 2 for quantum systems with more than one observable, A, B, C,... is straightforward. Obviously, once the quantum inertia I_q reaches the critical values I^A_{cr}, I^B_{cr}, I^C_{cr},...., for all the observables, the quantum superpositions stop and the system becomes classical, with well defined values of all the observables. The situation in between, when $I_q \geq I_{cr}$ for some observables but not for all, leads to hybrid systems where quantum and classical behaviour should co-exist. These hybrid systems should exist, especially at very low temperatures and weak gravitational fields, as it actually happens in mesoscopic systems.

Notice that, as long as $I_q < I_{cr}$ for a system, SQM is indistinguishable from conventional quantum mechanics for all practical purposes although, in principle, it provides a more accurate description of the quantum systems at very small time scales for which we lack experimental resolution. Moreover, since the quantum systems are also evolving unitarily with time, as dictated by the Schrödinger equation, the rôle of time in SQM is twice as prominent as it is in conventional quantum mechanics. This is no obstacle, however, in order to transfer the essential concepts of our proposal to relativistic Quantum Field Theory.

For $I_q \geq I_{cr}$, however, the wave function ψ rather than collapsing goes 'out of order', in the sense that it does not provide the correct description of the physical systems once the switching among the quantum states has stopped. In other words, we regard the wave function as only a mathematical description of the system, in agreement with conventional quantum mechanics, in the same way that a parabolic trajectory is a description of the motion of a bullet, without further physical substance.

A different issue is whether Schrödinger's equation should be modified, by adding a non-linear term in the Hamiltonian, in order to account for the effect of the quantum inertia on the quantum systems, eventually leading to the end of the superposition of states. For us this is not at all obvious because our impression is that there are two levels of physical reality: the quantum level and the spacetime level, the former being more fundamental than the latter. In our views quantum inertia stays deep inside the quantum level, not showing up in the spacetime dynamics explicitly, and its major (or unique) rôle is to affect the ability of the system to switch among the available quantum states, providing as a consequence a criterion for the validity of the wave function and the applicability of the Schrödinger equation.

The masses of the particles, the interactions among them and many other perturbations and interactions coming from the environment (temperature, gravitation, electric and magnetic fields, collisions,...) all must contribute to the load of quantum inertia of a system I_q, apart from their contributions to the system's Hamiltonian, either to the kinetic or to the potential energy. Consequently, if due to these contributions I_q reaches the critical value I_{cr}, then the system stabilizes to only one state, say $|a_m\rangle$ with eigenvalue a_m. This state will be observed as a well defined physical state with value a_m by the measuring devices of the classical world.

An important feature of the quantum switching versus quantum inertia is its plausible *reversibility*. That is, if the load of quantum inertia I_q of a quantum system is pushed up until or above the critical value I_{cr} but it can be decreased back below I_{cr}, by lowering the temperature, the magnetic field, the gravitational strength, or by any other means, then as soon as $I_q < I_{cr}$ the quantum switching resumes activity leading to the apparent superposition of states again, and the wave function ψ comes back to work giving a correct effective description of the system, as a result. For the purpose of visualization, one could regard the quantum available states as located at degenerate minima of a potential, the quantum system oscillating very rapidly among these minima, like a special type of quantum tunneling. The load of quantum inertia I_q raises the barriers among the different states until the switching finally stops, when $I_q \geq I_{cr}$, and the system gets trapped in one of the minima as a consequence. The other way around, lowering the potential barriers below I_{cr}, by decreasing I_q, will immediately resume the frantic oscillations of the system among the minima of the potential.

One may wonder whether the quantum switching is really random or it follows a specific pattern. However, there is no reason a priori for the quantum switching to follow a specific pattern because the only requirement is that the times spent in the different states must be proportional to their probabilities as given by the corresponding coefficients in $|\psi|^2$. Moreover, a specific pattern for the quantum switching would require the existence of hidden variables that SQM does not need otherwise and that, in addition, could conflict with experimental results, namely with the violation of the Leggett inequalities [2] that discard many interpretations of quantum mechanics that make use of nonlocal hidden variables[4]. Consequently, we postulate that the patterns followed by the quantum switchings must be random.

Positions in space and momenta are treated as any other observable. Namely, we propose that quantum systems will be switching among available positions in space, without passing through intermediate points, as long as $I_q < I_{cr}$, where I_{cr} is the critical quantum inertia that stops the quantum jumps. Therefore, as soon as $I_q \geq I_{cr}$, the systems will stop the quantum switching among positions and will follow classical continuous trajectories like a tiny bullets. Quantum jumps are therefore the rule in SQM, and the possibility that they become continuous classical trajectories, whenever $I_q \geq I_{cr}$, opens up a world of new effects that have to be taken into account for the study of many physical phenomena. For example, if this happens to electrons in atoms due to, for example, the action of very strong gravitational fields, then the atoms will look like tiny planetary systems, with the electrons emitting γ-ray synchrotron radiation while falling into the nuclei. This non-thermal emission from astrophysical objects with strong gravitational fields, like neutron stars or black holes, should also be taken into account as an efficient mechanism for the conversion of atoms into neutrons in neutron stars and in accretion discs of some black holes, producing sudden explosions of gamma rays in all directions (less intense, however, than the 'canonical' gamma ray bursts). Since the gravitational strength of white dwarfs is not enough to produce this mechanism, as these stars would turn very quickly into neutron stars otherwise, it provides a lower bound for the critical gravitational strength necessary to ignite this mechanism. That is, $G_{cr} > G_{wd}$, where G_{wd} stands for the gravitational strength of white dwarfs.

The foregoing discussion is most helpful in order to understand what is going on when the position of a particle is given by a superposition of two trajectories, like in interferometry experiments, either by means of a double slit or a beam splitter. The particle then has 50% probability to follow one route or the other. In conventional quantum mechanics, due to particle-wave duality the particle follows both routes, showing its wave-like nature, until it is detected. Then magically the wave-like nature transmutes into a particle-like nature because in the measurement devices nobody sees a wave but just an impact of a particle. In contrast, for SQM the particle is a wave-like corpuscle equipped with some wave-like behaviour, which bears

[4] We remind the reader that the experimental violation of the Bell inequalities discards only the local hidden variable interpretations of quantum mechanics, i.e. it does not discard the nonlocal ones.

little resemblance with classical corpuscles and even less resemblance with classical waves, and follows the two routes jumping very fast between them, without passing through intermediate positions. If there is a double-slit setting, the particle actually passes through the two slits unless the thickness of the double-slit screen is extremely thin, as the particle will be able to switch many times trajectories while crossing through the slits. If one or two detectors are placed just behind the two slits, in order to determine which slit the particle came across, then obviously the particle will impact on only one of the detectors since it will stop switching trajectories (and often even existing) afterwards.

Due to its relationship with the quantum probabilities, the time spent by the systems in each available state must be quantized in terms of a time unit, that we call *quantum switching time* t_S. To be consistent, t_S must depend on the quantum inertia I_q. The simplest guess is:

$$t_S = C \, \frac{h}{I_{cr} - I_q} \, , \quad I_q < I_{cr} \, , \tag{1}$$

where h is the Planck constant and C is just a dimensionless proportionality constant. In this expression the quantum inertia has dimensions of energy or mass, as we set $c = 1$. The classical world corresponds to $I_q \geq I_{cr}$ where $t_S \to \infty$. Observe that the minimum value of t_S, given by $t_{Smin} = C\ h/I_{cr}$ is obtained when the quantum inertia vanishes, unless the latter can reach negative values, $I_q < 0$, a possibility that we do not consider. Let us remark that a true superposition corresponds to $t_S = 0$, since then the system would be in all the states simultaneously, but this can only occur in the limit $I_{cr} \to \infty$.

In the spirit of conventional quantum mechanics, it is tempting to propose as well a lower bound relating the indeterminacy on the quantum inertia ΔI_q with the corresponding indeterminacy on the switching time Δt_S, such as:

$$\Delta I_q \ \Delta t_S \geq h/4\pi. \tag{2}$$

Notice that if different observables have different values of the critical quantum inertia I_{cr}, these result in different values of the quantum switching time t_S. This we find most natural since, for example, switching among positions in space is a very different process than switching among spin states, or photon polarizations.

The relation $I_{cr} = C\ h/t_{Smin}$ provides interesting insight into the values of I_{cr} as function of the minimum switching time t_{Smin}. For example, if we set $C = 1$, for t_{Smin} equal to the Planck time $t_{Pl} = 5.39 \times 10^{-44}$ s, one gets for the critical quantum inertia $I_{cr} = 7.67 \times 10^{19}$ GeV $= 1.367 \times 10^{-7}$ kg. This value is far too large for a critical mass M_{cr} in order to mark the dividing line between the quantum and classical worlds, and therefore should be discarded.

In Table 1 some other values of I_{cr} are given as function of t_{Smin}, setting $C = 1$. We can compare those values with the masses of molecules currently used in interferometry experiments aiming to find the dividing line with respect to the mass and size [3]. For this it is useful to keep in mind that the proton mass is $m_p = 0.938272$ GeV while the value of an atomic mass unit, used in atomic and molecular physics, is: 1 amu = 0.931494 GeV. At present, the largest molecules for which quantum mechanical behaviour has been confirmed are in the mass range of 10^{-22} kg (10^5 amu), whereas the smallest objects known to behave according to classical mechanics have a mass of order 10^{-9} kg (10^{18} amu). As a result, there are 12 orders of magnitude to search for a critical mass M_{cr}, the corresponding values of t_{Smin} ranging from 10^{-41} to 10^{-30} s, as follows from $t_{Smin} = h/I_{cr}$. This means 12 orders of magnitude to find the dividing line where the center of mass motion of an object obeys quantum mechanics or classical mechanics instead.

Observe also in Table 1 that the values of t_{Smin} in unambiguous quantum territory, like 10^{-27} s corresponding to 4407 protons (4439 amu) are orders of magnitude smaller than the smallest

IOP Conf. Series: Journal of Physics: Conf. Series **880** (2017) 012032 doi:10.1088/1742-6596/880/1/012032

Table 1. Some values of the critical quantum inertia I_{cr}, expressed in GeV and kg, as function of the minimum switching time t_{Smin} in seconds, using the relation $I_{cr} = h/t_{Smin}$.

t_{Smin}(s)	I_{cr}(GeV)	I_{cr}(kg)
10^{-42}	4.135×10^{18}	7.37×10^{-9}
10^{-40}	4.135×10^{16}	7.37×10^{-11}
10^{-38}	4.135×10^{14}	7.37×10^{-13}
10^{-36}	4.135×10^{12}	7.37×10^{-15}
10^{-34}	4.135×10^{10}	7.37×10^{-17}
10^{-32}	4.135×10^{8}	7.37×10^{-19}
10^{-31}	4.135×10^{7}	7.37×10^{-20}
10^{-29}	4.135×10^{5}	7.37×10^{-22}
10^{-27}	4.135×10^{3}	7.37×10^{-24}

lifetime of particle decay. This provides a consistency check for our proposal since values of t_{Smin} larger than the lifetime of particles could lead to conflict.

As regards *quantum entanglement*, SQM requires exact synchronization between the switchings of the quantum subsystems involved with respect to the quantum time in the Hilbert space. Once one partner stops switching, for example due to a measurement, the switching stops instantly for all the entangled partners, without the need of any signals. The reason is that the resulting quantum system acts as a whole, independently of the physical locations of its parts in spacetime. In other words, the gain of quantum inertia I_q by any of the subsystems adds to the total load of I_q for the complete system. This produces the freezing of the entangled property for all the partners involved, but not forever, for as soon as they could release enough quantum inertia, the partners would resume the switching activity, liberating themselves from the *entanglement bound*.

A rather cumbersome aspect of quantum entanglement is its supposed proliferation and ubiquity, contrary to our daily experience of the world around, giving rise to very long *Von Neumann chains*. However, in SQM quantum systems cannot get entangled with classical systems by interacting with them, unless the quantum inertia I_q of the latter is very close to the critical value and, somehow, it can be lowered even more through the contact with the quantum system. Therefore we conclude that in SQM the Von Neumann chains are very short, as generically they only involve quantum systems that were already in superpositions before the interactions took place. For example, the minimal version of the Schrödinger's cat chain: quantum system $-->$ poison bottle $-->$ cat, reduces to only the quantum system, and neither the poison bottle nor the cat are in superpositions of any kind, as they are macroscopic classical objects. In other words, neither the bottle is in the superposition of $|broken\rangle$ and $|unbroken\rangle$ states, nor the cat is in the superposition of $|alive\rangle$ and $|dead\rangle$ states, in agreement with the conventional interpretation of quantum mechanics.

3. Experiments to probe Scan Quantum Mechanics

In [1] we reviewed, in the light of the more intuitive description offered by SQM, the most relevant past and present experiments that have been performed in order to test the conventional quantum mechanics. Apart from the classical interferometry and entanglement experiments, we also considered the behaviour of polarized light passing through a tourmaline crystal, as described by Dirac [4]. This provides another beautiful example where SQM improves our intuition and understanding about what is actually going on in quantum mechanical experiments. Let us say a few words about it.

Tourmaline crystals have the property of letting through only light plane-polarized perpendicular to its optic axis. As a result, if the incident light is polarized perpendicular to the optic axis, it will all go through; if parallel to the axis, none of it will go through; while if polarized at an angle α to the axis, a fraction $sin^2\alpha$ will go through. The description offered by SQM to the passing of light through a tourmaline crystal differs substantially from the picture held by conventional quantum mechanics. First of all, when a photon passes through a crystal its quantum inertia will necessarily increase, the exact increment depending on a number of factors such as the crystal structure and pattern, crystal chemical composition, temperature, incident angle of the photon and its polarization. Second, if the polarization of the incident photon split into a superposition of two perpendicular polarization states then SQM predicts that the photon will switch very fast and randomly between the two, increasing its quantum inertia I_q in the process, until it eventually reaches a critical value I_{cr}. At that instant the switching between polarization states stops and the photon continues its way in a well defined polarization state. If the photon is polarized parallel to the optic axis, then the crystal absorbs the photon and nothing is observed on the back side, whereas if the switching stops when the photon is polarized perpendicular to the optic axis, then the photon reaches the back side of the crystal.

In SQM finding the dividing line between quantum and classical behaviour amounts to finding the values of the critical quantum inertia I_{cr} for all the observables of a system. For this reason in [1] we also proposed some experiments to be performed in order to determine the critical values I_{cr} for some observables, that translate into critical masses M_{cr}, critical temperatures T_{cr}, critical electric or magnetic fields, etc. The critical values are experimental input in SQM, like the masses of elementary particles in the Standard Model of Particle Physics (no theory has been able to predict them so far). In addition, those experiments can also test the reversibility of the dividing line between the quantum and the classical worlds whenever this is possible: adding or substracting constituents to the system, increasing or decreasing the temperature, etc.

The contribution of the masses of the constituents to the quantum inertia I_q is very intuitive and is the major reason behind the dividing line between quantum and classical systems, since carefully increasing the number of constituents of a quantum system will eventually lead to its transmutation into a classical one when reaching a critical mass M_{cr}. To be more precise, for fixed environmental conditions, adding constituents to a quantum system until I_q reaches the corresponding critical value, $I_q \geq I_{cr}$, results in the cessation of the system's quantum jumps between the two (or more) trajectories in interferometry experiments: the system has become classical and follows a well defined trajectory. As a consequence, no interference pattern can be created by repeating the experiment with a large number of identical systems, unless one or more of the external conditions change, for example a decrease in temperature. We conclude, therefore, that it should be technically possible to determine (or estimate at least) the critical quantum inertia I_{cr} associated to the mass of physical systems, i.e. a critical mass M_{cr} for given external conditions, by performing interferometry experiments probing large molecules, like nanoparticle interferometry with complex organic molecules and inorganic clusters, as reported in ref. [3], or probing nanocrystals to which the experimentalists could add or subtract atoms one by one.

Apart from the masses of the constituents, according to SQM the many perturbations and interactions coming from the environment must contribute as well to the quantum inertia I_q of the physical systems. Thus one has to consider: temperature, gravitational forces, electric and magnetic fields and collisions with all kinds of matter and radiation. One has to take into account that these perturbations affect different observables in different ways and also that different observables have, in principle, different values for the critical inertia I_{cr}.

Once the dividing line had been found between quantum and classical behaviour regarding the mass of the physical systems, for fixed environmental conditions, the next interferometry experiments must be performed using borderline systems near the critical mass M_{cr}, varying slightly one of those conditions.

To start, one has to perform experiments increasing and decreasing the system's temperature to test its effect on the quantum inertia I_q, as well as testing the reversibility of the process, with the possibility of finding a critical temperature T_{cr}. A technical complication will always appear: the difficulty to distinguish between the influence of temperature on the quantum inertia and the thermal fluctuations, which also spoil quantum coherence. Although thermal noise is ubiquitous and some decoherence mechanisms might be at work to some extent, SQM offers a complementary explanation in order to understand what is its influence on physical systems. Namely, apart from the thermal noise, above certain critical temperature T_{cr} the quantum inertia I_q of the system surpasses the critical value I_{cr} and, as a result, the system becomes purely classical (with no tunneling possible, for example). Below the critical temperature $T < T_{cr}$ the system recovers its quantum behaviour, such as the tunneling ability. For this reason, we are convinced that all known experimental effects and phenomena that make use of very low temperatures must be reanalyzed in the light of SQM, i.e., taking into account the possibility of a critical temperature T_{cr} corresponding to a critical quantum inertia I_{cr}.

Unfortunately we cannot follow the same procedure increasing and decreasing the gravitational strength. Gravitation has been considered already for a long time as a possible cause for the collapse of the wave function or for decoherence [5], [6], [7] . According to SQM, increasing gravitational fields must increase the load of I_q with the corresponding increase in the switching time t_S. It is therefore possible that the quantum switching of a system stops altogether provided the gravitational strength reaches a strong enough value. This has very interesting implications for astrophysical objects with strong gravitational fields, providing an efficient mechanism for the capture of electrons by the protons in the nuclei, turning the atoms into neutrons, as we argued before.

The effect of magnetic and electric fields on I_q should also be tested in interferometry experiments using borderline systems, and one can consider also other tests, like the Stern-Gerlach experiment.

As a matter of fact, many experiments have been performed, or proposed, in the last years by a number of groups in order to test the quantum-to-clasical transition [3], [8], [9], [10]. Although these experiments were designed to test decoherence theory or collapse models, they can be applied to test SQM too, i.e., to determine (or estimate) the values of the critical quantum inertia I_{cr}, via a critical mass M_{cr} for fixed environmental conditions, or via critical temperatures T_{cr}, or critical values of other physical parameters.

To finish, an important observation about the possible effects of quantum inertia I_q on experiments is that in some experimental settings the researchers involved do not plan to destroy the system, usually a particle, in the measurement process, but only to get some information about it, like the trace it leaves in a bubble chamber, for example. Some of these procedures are called *weak measurements* and the experiments usually involve a weak measurement for the momentum of the particle followed by a normal (*strong*) measurement for the position of the particle, crashing it against a screen or a detector (see for example [11]). In our opinion, the weak measurements could disrupt a quantum system much more than the experimenters are aware of via I_q, for example stopping the quantum jumps among the available positions, so that the system continues its way as a tiny bullet, following a continuous classical trajectory that can be confused with a bohmian trajectory (since this is also a continuous classical trajectory).

4. Conclusions and final remarks

SQM puts forward a mechanism underlying the superposition of quantum states, and postulates a property of all systems, quantum inertia I_q, such that quantum behaviour only manifests itself for values of I_q below some critical ones I_{cr}. Therefore I_q with its critical values marks the dividing line between the quantum and the classical worlds.

SQM is not radically different from conventional quantum mechanics, but it provides a more accurate description of the quantum systems at very small time scales for which we lack experimental resolution. As a matter of fact, as long as $I_q < I_{cr}$ for all the observables of a physical system, SQM is essentially indistinguishable from conventional quantum mechanics.

One may wonder what causes the quantum switching in SQM, in the first place. In our opinion, the switching among the available states is a natural fact of the quantum world, without the need for any specific cause, neither a quantum potential nor any other forces. It only stops when the quantum inertia I_q reaches a critical value I_{cr}. More meaningful in this case is turning the question upside down: What causes the classical systems not to switch among different states? Not to have superpositions and present well defined properties instead? The answer given by SQM is quantum inertia I_q surpassing a critical value I_{cr} for each observable.

Regarding entanglement, observe that in principle it allows us to distinguish SQM from conventional quantum mechanics as SQM predicts the end of the entanglement bounds due to the release of I_q. In practice, however, to test this effect might be an impossible task if very long distances (interstellar or larger) were involved.

Let us stress again that mesoscopic systems in general as well as physical phenomena at very low temperatures, such as superconductivity, superfluidity, Bose-Einstein condensates, etc. must be reanalyzed in the light of SQM, i.e. taking into account the possibility of existence of quantum inertia I_q with critical values I_{cr} for the observables, giving rise to critical values for the mass, the temperature, electric and magnetic fields, etc. Similarly, non-thermal radiation by neutron stars and accretion discs of black holes must be reanalyzed taking into account the possibility of the emission of γ-ray synchrotron radiation, as described by SQM, turning atoms into neutrons.

Finally, it is interesting to notice also that, despite the asymmetry with respect to space and time, the essential ideas of SQM can be transferred straightforwardly to Quantum Field Theory, in which case the probabilities are implemented by switchings and quantum jumps of the particles, the quantum inertia giving rise to natural ultraviolet cutoffs. This brings about a very intriguing possibility: does quantum gravity really exist? For if strong enough gravitation disables quantum mechanics via quantum inertia, then there is no much room left for quantum gravity.

Acknowledgments
I thank the organizers of DICE 2016, especially Hans-Thomas Elze, for such an interesting and inspiring conference on fundaments of physics. This research was partially supported by the project FIS2012-38816, from the Spanish Ministerio de Economía y Competitividad.

References
[1] Gato-Rivera B 2015 Scan Quantum Mechanics: Quantum Inertia Stops Superposition *Preprint* arXiv:1512.03093
[2] Leggett A J 2003 Nonlocal Hidden-Variable Theories and Quantum Mechanics: An Incompatibility Theorem *Found. Phys.* **33** 1469
[3] Hornberger K, Gerlich S, Haslinger P, Nimmrichter S and Arndt M 2012 Colloquium: Quantum Interference of Clusters and Molecules *Rev. Mod. Phys.* **84** 157
[4] Dirac P A M 1930 *The Principles of Quantum Mechanics* (The International Series of Monographs on Physics) (Oxford: Clarendon Press)
[5] Penrose R 1996 On gravity's role in quantum state reduction *Gen. Rel. and Grav.* **28** 581
[6] Singh T P 2015 Possible role of gravity in collapse of the wave-function: a brief survey of some ideas *J. Phys.:Conf. Ser.* **626** 012009
[7] Diósi L 1984 Gravitation and quantummechanical localization of macroobjects *Phys. Lett.* A **105** 199
[8] Hornberger K, Uttenthaler S, Brezger B, Hackermüller L, Arndt M and Zeilinger A 2003 Collisional decoherence observed in matter wave interferometry *Phys. Rev. Lett.* **90** 160401

[9] Bassi A, Lochan K, Satin S, Singh T P and Ulbricht H 2013 Models of wave-function collapse, underlying theories, and experimental tests *Rev. Mod. Phys* **85** 471
[10] Bahrami M and Bassi A 2015 Proposing new experiments to test the quantum-to-classical transition *J. Phys.: Conf. Ser.* **626** 012006
[11] Kocsis S, Braverman B, Ravets S, Stevens M J, Mirin R P, Shalm L K and Steinberg A M 2011 Observing the Average Trajectories of Single Photons in a Two-Slit Interferometer *Science* **332** 1170

IOP Conf. Series: Journal of Physics: Conf. Series **880** (2017) 012033 doi:10.1088/1742-6596/880/1/012033

Quantum jumps: from foundational research to particle physics

Ignazio Licata[1,2] **and Leonardo Chiatti**[3]

[1]ISEM Institute for Scientific Methodology, Palermo (Italy)
[2]School of Advanced International Studies on Applied Theoretical and Non Linear Methodologies in Physics , Bari (Italy)
[3]ASL VT Medical Physics Laboratory, Viterbo (Italy)

E-mail: ignazio.licata@ejtp.info, leonardo.chiatti@asl.vt.it

Abstract. Since 1986 a vast body of experimental evidence has been accumulated of direct observation of quantum jumps in many physical systems. We can therefore assume that quantum jumps are genuine physical phenomena. On the other hand, substantial identity of "quantum jumps" and "collapses" induced by measurements can be admitted, both being represented by self-conjugate projection operators related to specific non-Hamiltonian aspects of micro-interactions. On this basis a model of quantum jump involving a single particle is discussed, and some consequences concerning hadronic physics (Hagedorn temperature, Regge trajectories) and quantum gravity are briefly sketched.

1. Introduction

Although the existence of quantum jumps (QJs) was rarely questioned by Bohr's time, the physics of these processes has never received much attention from researchers in foundations of quantum mechanics (QM). Despite of this, the direct observation of individual atomic QJs has become experimentally accessible since 1986 [1]. This observation has clearly evidenced that a QJ occurs as a discontinuous phenomenon whose duration is negligeable, not to be confused with the atomic transition (evolution of the atomic state vector from the initial state to the final state) ended by it. The transition, in fact, takes place in a finite time, which defines the process linewidth. A large experimental literature currently exists about QJs observation in many multi-level quantum systems, even macroscopic. These experimental facts compel us to consider in a new light the controversial "collapse of the wave function" associated with the measurement of a quantum observable. In fact, this collapse is also a discontinuous transformation represented by a projection operator on the final state of the process. It seems there are not essential differences between, for example, an atomic QJ and a collapse. As the atomic QJ is the projection of the time-dependent atomic orbital (solution of the time-dependent Schrödinger equation) on the final orbital, so the collapse is the projection of the time-dependent state vector of the system "apparatus + micro-object" on the final state associated with the actual result of the measurement. In other words, the collapse appears to be only a variant of QJ, a particular case of QJ. If this is true then, in accordance with the experimental results, the collapse must be a genuine physical phenomenon like the other QJs. So we have two consequences: 1) the objectivity of the collapse, which is a particular case of QJ and therefore a real physical phenomenon; 2) the cause of the collapse is the same as that of each other QJ: the coupling with an external field.

IOP Conf. Series: Journal of Physics: Conf. Series **880** (2017) 012033 doi:10.1088/1742-6596/880/1/012033

The very next question is what is the nature of the causal connection between the QJs. We can restrict ourselves to consider the connection between the QJ associated with the "preparation" of a quantum system and that, successive in time, relative to its "detection" in the final state. Cramer [2-4] suggested to see this pair of events as the two ends of a *transaction*, interpreting each of them as the emission of a wave signal Ψ towards the future and the concomitant emission of the conjugate signal Ψ^* toward the past. It is noteworthy that this is exactly the structure of the projector on the final state Ψ, which suggests an a-spatial description of the transaction as a loop connecting the two projectors associated with two events, we say $|\Psi\rangle\langle\Psi|$ and $|\Phi\rangle\langle\Phi|$. The statistics of these loops is then described by the transition amplitude $\langle\Psi|S|\Phi\rangle$ and its conjugate $\langle\Phi|S|\Psi\rangle$, where S is the time evolution operator [5]. This a-spatial concept of transaction, first formulated in [6], was promoted by Kastner [7] which underlined the difference from the original "spatiotemporal" reading proposed by Cramer. In our opinion, however, the existence of an additional level of the problem has to be considered. In fact, if the connection between the ends of a transaction is not attributable to the exchange of wave signals towards both the time directions within the light-cone, but is embodied in those time-simmetric a-spatial relations that we call loops, it is necessary considering that the latter are ultimately also a-temporal and indeed connected to an emergency process of time [5]. Transactions and QJs which represent their ends are thus structures that emerge from an a-spatial and a-temporal background through a *formal causation* process. The dynamic causality and temporal ordering, both essential aspects of the transactions, are emergent properties of this formal causation, rather than fundamental. It is then possible understand clearly that local interactions emerge from an *universal non-locality* which is much more extensive and radical than that involved in entangled states, which simply represents a particular manifestation of it [8]. In this context, the classical macroscopic reality can be seen as emergent from networks of QJs. In a cubic centimeter of ordinary matter at room temperature, a very high number of these events occur each second (think of the thermal noise and the constant interactions between atoms); and this is exactly what determines the appearance of that portion of matter as an object in the classical sense. In other words, the macro-objectification is an endogenous phenomenon. The border between classical and quantum is then defined by the fact that the average properties of large sets of QJs (the persistent "objects") can be described classically, while individual QJs or their direct connections require a more fundamental quantum description.

2. Quantum jumps

This section briefly summarize the QJ model we discussed elsewhere, referring the reader to the original papers for the necessary insights [9,10,11,12]. Let us consider a QJ in a system consisting of a single elementary particle, following which such a particle is released in the state Ψ. As we have seen, from the transactional point of view this QJ is represented by the action of the operator $|\Psi\rangle\langle\Psi|$ which, acting on a generic state $|\Psi\prime\rangle$, produces a new state proportional to $|\Psi\rangle$. A natural interpretation of the projector $|\Psi\rangle\langle\Psi|$ is the following: the shutdown of the component $|\Psi\rangle$ of $|\Psi\prime\rangle$ (section $\langle\Psi|$ of the projector) can be seen as the switching off of its de Broglie phase factor or, in other words, as the transfer of the particle rest energy to the vacuum: the component $|\Psi\rangle$ enters a condition in which "there is not motion in time." This condition is, from the point of view of the external time of the laboratory, a-temporal and then a-spatial, as a result of the relativistic relationship between spatial and temporal coordinates. The restart of the component $|\Psi\rangle$ (section $|\Psi\rangle$ of the projector) must then be interpreted as the restoration of the de Broglie phase factor or the transfer of an amount of energy (equal to the particle rest energy) from the vacuum to the particle. This component exits from the a-spatial and a-temporal condition to reappear as the particle state vector, undergoing the dynamical evolution in laboratory time. In this sense in a QJ we have the annihilation of the state Ψ followed by its re-creation. All the factors of the state vector $|\Psi\rangle$ other than the de Broglie phase factor (which express the dependence on the position, spin, and so on) remain unchanged. An important detail is that since only the QJ is observable (because it involves an energy transfer from one field to another through the vacuum), the component $|\Psi\prime\rangle - |\Psi\rangle$ of $|\Psi\prime\rangle$ remains counterfactual. The projector $|\Psi\rangle\langle\Psi|$ can act on its right transforming the component $|\Psi\rangle$ in $|\Psi\rangle$, or on its left transforming the component $\langle\Psi|$ in $\langle\Psi|$. In other words it can be seen as the concomitant emission of the state Ψ towards the future and its conjugated Ψ^* towards the past, according to Cramer's interpretation. Therefore, the algebraic or topological transformation that represents the QJ has to preserve the information associated with the de Broglie temporal phase factor of the state Ψ, recoding it in a form "frozen" with respect to the flow of external time t. We assume that in the background condition the information associated with the state Ψ is encoded in a kind of "internal wavefunction" inaccessible by direct observation. Let us introduce a background "internal time" which will be denoted by $\tau\prime$. The internal wavefunction associated with the particle will contain a factor $\Phi(\tau\prime)$, real and harmonic in $\tau\prime$, null at the boundary and outside the interval $[-\theta_0/2, +\theta_0/2]$. The reality condition must be satisfied for the absence of a temporal direction, which implies $\Phi = \Phi^*$. The following equation is a consequence of this postulate:

$$-\hbar^2 \frac{\partial^2}{[\partial(2\pi\tau')]^2}\Phi = (M_{sk}c^2)^2\Phi \tag{1}$$

with the condition $\Phi = 0$ for $\tau' \leq -\theta_0/2$, $\tau' \geq +\theta_0/2$, where M_{sk} is the original (skeleton) component of the particle mass. From Eq. (1) it follows that:

$$M_{sk}c^2 = n'\frac{\bar{}h}{\theta_0} \tag{2}$$

where $n\prime = 0,, 1, 3/2, ...$ is an integer for odd solutions, a half-integer for even solutions. A second postulate states that each oscillating solution of Eq. (1) is globally characterized by a second variable T (which is dimensionally a temperature) such that the not normalized probability of a given value of that variable is expressed by

$$\exp\left(-\frac{\bar{}h/\theta}{kT}\right) \tag{3}$$

for $kT \geq \bar{}h/\theta$, 0 otherwise, where k is the Boltzmann constant and $T \geq 0$. Assuming $\tau\prime\prime = \hbar/kT$ this probability becomes the square modulus of a second factor of the internal wavefunction:

$$\Lambda = \exp\left(-\frac{\tau''}{2\theta_0}\right) \tag{4}$$

If we assume that while in the background condition the particle is in a state of superposition of different values of $\tau\prime\prime$ we have:

$$-\imath\hbar\frac{\partial}{[\partial(\imath\tau'')]}\Lambda = \frac{\bar{}h}{2\theta_0}\Lambda \tag{5}$$

for $0 \leq \tau\prime\prime \leq \theta_0$, $\Lambda = 0$ otherwise. The creation of the wavefunction associated with the particle state Ψ exiting from the background can be considered as the passage from Eq. (3) to a probability identically equal to 1 for any value t of the "external" time accessible to the observer. This probability will be the square modulus of a factor that we can write as:

$$\Lambda \propto \exp\left(-\imath\frac{\tilde{\tau}}{2\theta_0}\right) \tag{6}$$

IOP Conf. Series: Journal of Physics: Conf. Series **880** (2017) 012033 doi:10.1088/1742-6596/880/1/012033

We can imagine this factor to derive from Eq. (4) by the Wick rotation $\tau'' \to \imath\tilde{\tau}$; simultaneously to this transformation, the factor $\Phi(\tau')$ must disappear. The quantity $\tilde{\tau}/2\theta_0$ can be interpreted as the external time t measured in units of the oscillation period of the old factor Φ, i.e. θ_0/n'. Therefore:

$$\frac{\tilde{\tau}}{2\theta_0} = \pm\frac{t}{\left(\frac{\theta_0}{n'}\right)} \Rightarrow \tilde{\tau} = \pm 2n't \tag{7}$$

And thus:

$$\Lambda \to \exp\left(\pm\frac{-\imath n't}{\theta_0}\right) = \exp\left(\mp\imath M_{sk}c^2\frac{t}{\hbar}\right) \tag{8}$$

i.e. the de Broglie's temporal phase factor. Annihilation is represented by the inverse process. It is immediately clear that (1) and (5) are two wave equations, respectively, in the real component τ' and the imaginary component $\imath\tau''$ of a complex internal time $\tau = \tau' + \imath\tau''$, which thus becomes the "precursor" of time t measured in the laboratory; this last physical quantity is thus emerging. The frequency of the "hidden" oscillation Φ becomes the frequency of the de Broglie's phase factor, experimentally accessible as the particle mass. The transformation (8) is, at the same time, algebraic (because it involves a Wick rotation) and topological (because it involves, on the complex plane of the variable τ, the transformation of a rectangular domain in a circumference [11]). Clearly, the entire model dissolves in the limit $\theta_0 \to 0$. The interval θ_0 must be a time scale characterizing the realm of elementary particles, and this leads to the conclusion that it is $c\theta_0 \approx 10^{-13}$ cm, and therefore $\hbar/\theta_0 \approx$ 50-100 MeV. Consideration should be given to the fact that this time scale is not a minimum time interval, nor does $c\theta_0$ define a minimum spatial interval. An estimation of θ_0 can be derived from the observation that the particle mass m should consist of the skeleton mass renormalized by perturbative effects (self-interaction) limited to the interaction vertex where the state Ψ is created [11,13]. This vertex coincides naturally with the QJ. If $n' = 0$, only the term of self-interaction survives in m. Only a fraction of the energy $\hbar/\theta_0$ needed to locate the particle in a temporal extension θ_0 is used, expressed by the dimensionless self-coupling constant $(g^2/\hbar c)$. This energy is therefore $g^2/(c\theta_0)$, and the particle is delocalized to a dressing region $\hbar c/g^2$ times larger than $c\theta_0$. This is the situation of lighter particles, i.e. electrons and neutrino mass eigenstates, whose masses are certainly lower than $\hbar/\theta_0$. With electrons, the self-interaction will be essentially electrostatic and therefore the self-interaction energy will be $e^2/c\theta_0$, where e is the elementary electric charge. Equating this expression to the rest energy of the electron, we obtain $c\theta_0$ is the classical radius of the electron. As a result, the fundamental skeleton mass interval $\hbar/\theta_0$ in Eq. (2) amounts approximatively to 70 MeV. Consequently, the interval θ_0 coincides (unless a factor 2/3) with the "chronon" introduced by Caldirola in its classical electron theory [14]. To our knowledge, this is the only classical electron theoretical description which is not affected by troubles with self-energy divergence or runaway solutions or violations of causality (pre-accelerations).

3. Mini big-bangs

The square modulus of Eq. (4):

$$|\Lambda(\tau'')|^2 = \exp\left(-\frac{\hbar/\theta_0}{kT}\right) = \exp\left(-\frac{\tau''}{\theta_0}\right) \tag{9}$$

describes the background as a set of thermostats at different absolute temperatures $T \in [\hbar/k\theta_0, \infty)$. When a particle with rest energy Mc^2 interacts with this background in a QJ, each thermostat gives a contribution:

IOP Conf. Series: Journal of Physics: Conf. Series **880** (2017) 012033 doi:10.1088/1742-6596/880/1/012033

$$dQ = Mc^2 \frac{\exp(-\frac{\tau''}{\theta_0})d\tau''}{\int_0^{\theta_0} \exp(-\frac{\tau''}{\theta_0})\, d\tau''} \tag{10}$$

to this energy. In the case of a contact interaction, that is confined within the QJ, as the strong interaction between hadrons is, each particle "sees" a single equivalent thermostat, with which it exchanges a thermal contribution equal to the sum of the thermal contributions exchanged with the various thermostats, and an entropy equal to the sum of the entropies exchanged with the various thermostats. Such exchanges are reversible due to the time simmetry of the QJ; the temperature T_H of the equivalent thermostat is thus defined by the relation:

$$\int_{\tau''=0}^{\tau''=\theta_0} dQ/T = \frac{\int_{\tau''=0}^{\tau''=\theta_0} dQ}{T_H} \tag{11}$$

This relation provides $kT_H = 160$ MeV, which is exactly the Hagedorn temperature [15,16]. It is also possible to see that the mass spectrum of the hadronic states implied by (9) is the same as that derived from the Hagedorn statistical bootstrap model, when the deconfinement temperature of the charge centres (quarks) is taken as T_H [15 ,16]. We can conclude by observing that the value of T_H is defined by θ_0, and $c\theta_0$ is the maximum value of the confinement radius of charge centres (de Sitter radius of the micro-particle universe [11]). The Hagedorn temperature is then connected to confinement. The "thermal" description of QJ outlined in this section is very similar to that proposed by us for the big bang [17,18]. The expression "mini big bangs" in use today with reference to interaction vertices involving real particles seems therefore particularly significant.

4. Regge trajectories

The introduction of the chronon θ_0 allows the definition of a moment of inertia of the particle of the order of $\hbar\theta_0$. This moment of inertia is not associated with a geometric extension of the particle or a mass distribution in the classical sense. It is simply a factor converting the frequency j/θ_0 (where j is the eigenvalue of particle total spin) in the spin j^-h. The "granularity" of the particle temporal localization in a QJ, expressed by the chronon, pesents some observable consequences in the case of hadrons. Hadrons can be generated in strong interactions whose range is of the order of $c\theta_0$. As we can see from Eq. (2), when a hadron exits from the interaction sphere of radius $c\theta_0$ where it has been formed, it has a definite skeleton mass m, where $\theta_0/n\prime = \hbar/mc^2$ (the effective mass will also contain contributions from internal degrees of freedom and perturbative effects). The newly formed mass m is therefore associated with a moment of inertia $mc^2\theta_0^2$ with respect to the centre of the sphere. The natural angular pulsation of the new-born hadron is $n\prime/\theta_0$, or more generally its integer submultiple $n\prime/n\theta_0$), so that the quantity of angular momentum transferred from the interaction vertex to the hadron is given by:

$$\Delta J = \Delta[(mc^2\theta_0^2)(\frac{n\prime}{n\theta_0})] = \Delta[(mc^2\theta_0^2)(\frac{mc^2}{n\hbar})] = \Delta[(mc^2)^2(\frac{\theta_0^2}{n\hbar})] \tag{12}$$

Bearing in mind that $J = j^-\hbar$we obtain, for n constant, the Regge trajectory:

$$(mc^2)^2 = jn(\hbar/\theta_0)^2 + \sigma \tag{13}$$

Here σ, which can be positive, zero or negative, represents the intercept of the trajectory [19]. Pulsations which are greater than $n\prime/\theta_0$ obviously have no relevance, because the hadron temporal localization makes the higher harmonics lose their physical meaning. Pulsations which are lower than $n\prime/\theta_0$ should be its submultiples, because angular motion is sampled by strong

IOP Conf. Series: Journal of Physics: Conf. Series **880** (2017) 012033 doi:10.1088/1742-6596/880/1/012033

interactions with a maximum frequency given by the inverse of the hadron temporal extension. In other words, n/θ_0 becomes a Nyquist frequency. Eq. (13) defines a rectilinear Regge trajectory whose slope is quantized, because n is an integer. For a discussion about experimental evidences of quantization the reader should consult ref [20].

5. Side effects on quantum gravity

The skeleton mass is the energy associated with the localization of the particle in an interaction vertex as a real particle incoming or outgoing that vertex. Experimentally, the entities that propagate from a QJ to another are the particles, not their charge centres (for example hadrons, not quarks). As we know from the equivalence principle, the gravitational charge of a particle is proportional to its inertial mass which in turn is derived from the skeleton mass. Consequently, this charge, as the skeleton mass, is a global property of the particle instead of individual charge centres. Now, the particle is coupled with external non gravitational fields through its charge centres (which, when coinciding with quarks, are also the sources of internal color field) and we know that the energy exchanged in such couplings does not present some known upper limit. This suggests that charge centres are pointlike. The particle that contains them has instead a finite radius $c\theta$ (with $\theta \leq \theta_0$). The gravitational field is however coupled with the particles through their mass. There is therefore an upper limit to the energy E that the gravitational field can exchange with the particles in a single interaction: when E exceeds $\hbar/\theta$ the gravitation "does not see" the particle, but only a fraction $(\hbar/\theta E)^3$ of it. For $E \gg \bar{h}/\theta$the particle is completely decoupled from gravitation. For example, for an electron is $\theta = e^2/mc^3$, where m is the electron mass [11-13], thus the electron decoupling starts at 70 MeV. For other particles the decoupling begins to higher energies. In gravitational collapse leading to the formation of a black hole, nucleons and electrons overlap and at a certain moment in a volume of radius $\approx c\theta_0$ will be concentrated a mass M whose gravitational self-energy is of the order of 70 MeV. At about this point the decoupling of matter by the gravitational field begins. This occurs at the critical density ρ_0 defined by the equation:

$$\frac{GM^2}{c\theta_0} = \frac{G[\rho_0(c\theta_0)^3]^2}{c\theta_0} \approx 70MeV \tag{14}$$

where G is the gravitational constant. When this density is reached the collapse continues due to inertia, but it is slowed by the pressure and eventually reverses; of consequence, the density decreases. When the density again reaches the critical value, the gravitation reappears. Oscillations around a stable point likely occur. The singularity is not formed, remaining an object of density comparable to that critical. From (14) we have $\rho_0 \approx 10^{32}$ g/cm^3, a value that is 62 orders of magnitude lower than the Planck density, given by the ratio between the mass of Planck and the cube of the Planck length: $M_{Planck}/(l_{Planck})^3 \approx 10^{-5}$ g/$(10^{-33}$ cm$)^3 = 10^{94}$ g/cm^3. The density of the collapsed object exceeds that of a neutron star of about 17 orders of magnitude. We must therefore assume that this object has gone through its own Schwarzschild radius so becoming a black hole. Ultimately, it is conceivable that a finite chronon could prevent the formation of singularities in gravitational collapse.

6. Conclusions

Our intent was to elucidate some possible connections between the foundational research in QM and current topics in particle physics. We can not hide that part of our motivation was to oppose the current trend of progressive isolation of this kind of research in an epistemological and philosophical scope removed from the confrontation with the cutting-edge research. The main results presented here can be summarized for points as follows:

(i) The external (laboratory) time is an emergent property at the particle scale;

(ii) the particle bare (skeleton) mass is finite;
(iii) it is also quantized;
(iv) the quantum of skeleton mass is $\hbar/\theta_0$, where θ_0 is a new constant of Nature;
(v) derivation (from θ_0) of the Hagedorn temperature $T_H = 160$ MeV;
(vi) Regge trajectories do exist, with quantized slopes;
(vii) θ_0 (probably) prevents the formation of singularities in gravitational collapse.

The derivation of these results and their relevant connections becomes possible only when we cease to see the dynamical causality as fundamental, and start to conceive it as an emerging aspect of a formal causation connecting the phenomenal domain to an a-spatial and a-temporal "background". The quantum formalism then becomes, how correctly guessed by Bohm many years ago [21], a (still partial) mathematical description of this causation. We have tried to enrich this description by adding a minimal model of QJ. In this model a time interval appears, the chronon, already known from the Caldirola work, which plays an essential role in the derivation of results listed above. The top-down structure of this approach, the exact opposite to that, bottom-up, of the conventional dynamical approach [22], expresses the fundamental nature of the non-local aspect of the physical world, of which the entanglement phenomena only represent the emerging peak.

Acknowledgments

The authors whish to thank Giuseppe Vitiello and Thomas Elze for their encouragement and stimulating atmosphere of DICE 2016, and Elinor Ruth Kastner and Erasmo Recami for fruitful discussions over the years.

References

[1] Bergquist J C, Hulet R G, Itano W M and Wineland D J 1986 Observations of quantum jumps in a single atom *Phys. Rev. Lett.* **57**(14) 1699-1702
[2] Cramer J G Generalized absorber theory and the Einstein-Podolsky-Rosen paradox 1980 *Phys. Rev.* D **22**(2) 362-376
[3] Cramer J G The Transactional interpretation of quantum mechanics 1986 *Rev. Mod. Phys.* **58** 647-687
[4] Cramer J G 2015 The transactional interpretation of quantum mechanics and quantum nonlocality *Preprint* arXiv:1503.0039 [quant-ph]
[5] Chiatti L 2013 The transaction as a quantum concept *Int. Journ. Res. App. Sci.* **16**(4) 28-47 (*Preprint* arXiv:1204.6636 [gen-ph])
[6] Chiatti L 2005 *The archetypal structures of physical world* (Roma: Di Renzo)
[7] Kastner R E 2013 *The Transactional interpretation of quantum mechanics: the reality of possibility* (Cambridge: Cambridge University Press)
[8] Licata I 2014 Transaction and non locality in quantum field theory *EPJ Web of Conferences* **70** 00039
[9] Chiatti L and Licata I 2017 Fluidodynamic representation and quantum jumps *Quantum structural studies; classical emergence from the quantum level*, eds R E Kastner, J Jeknic-Dugic and G Jaroszkiewicz (Singapore:World Scientific) pp 201-224
[10] Chiatti L 2016 Is Bohr challenge still relevant? *Beyond peaceful coexistence: the emergence of space, time and quantum*, ed I Licata (Singapore:World Scientific) pp. 545-557
[11] Licata I and Chiatti L 2015 Timeless approach to quantum jumps *Quanta* **4**(1) 10-26
[12] Chiatti L and Licata I 2016 Particle model from quantum foundations *Quant. Stud. Math. Found.* DOI 10.1007/s40509-016-0094-6
[13] Chiatti L 2012 A possible model for the mass spectrum of elementary particles *Phys. Ess.* **25**(3) 374-386 (*Preprint* arXiv:0905.3794)
[14] Caldirola P 1980 The introduction of the chronon in the electron theory and a charged-lepton mass formula *Lett. Nuovo Cim.* **27**(8) 225-228
[15] Hagedorn R and Rafelski J 1981 From hadron gas to quark matter I *Statistical mechanics of quarks and hadrons*, ed H Satz (Amsterdam:North Holland) pp.237-251

[16] Rafelski J and Hagedorn R 1981 From hadron gas to quark matter II *Statistical mechanics of quarks and hadrons*, ed H Satz (Amsterdam:Noorth Holland) pp.253-272
[17] Licata I and Chiatti L 2010 Archaic universe and cosmological model: big-bang as nucleation by vacuum *Int. Jour. Th. Phys.* **49**(10) 2379-2402
[18] Licata I and Chiatti L 2009 The archaic universe: big bang, cosmological term and the quantum origin of time in projective cosmology *Int. Journ. Theor. Phys.* **48**(4) 1003-1018
[19] Collins P D B 1977 *An introduction to Regge theory and high energy physics* (Cambridge:Cambridge University Press)
[20] Chiatti L 2014 A simple quantization rule for the slopes of Regge trajectories *Phys. Ess.* **27** 143-145
[21] Bohm D, Hiley B J and Stuart A E G 1970 On a new mode of description in physics *Int. Jour. Theor. Phys.* **3** 171-183
[22] Hiley B J 2016 Some remarks on the evolution of Bohm's proposals for an alternative to standard quantum mechanics (*Preprint* http://www.bbk.ac.uk/tpru/BasilHiley/History_of_Bohm_s_QT.pdf; retrieved 16/11/2016)

Nonclassical face of quantum decay processes

Krzysztof Urbanowski
University of Zielona Góra, Institute of Physics,
ul. Prof. Z. Szafrana 4a, 65–516 Zielona Góra, Poland.

E-mail: K.Urbanowski@if.uz.zgora.pl

Abstract. A detailed analysis of the quantum decay processes shows the survival probability $\mathcal{P}(t)$ can not take that exponential form at any time interval including times smaller than the lifetime τ. We show that for times $t \sim \tau$ and for the times later than τ the form of $\mathcal{P}(t)$ looks as a composition of an oscillating and exponential functions. The amplitude of these oscillations is very small for $t \ll \tau$ and grows with increasing time and depends on the model considered. We also study the survival probability of moving relativistic unstable particles with definite momentum $\vec{p} \neq 0$: It turns out that late time deviations of the survival probability of these particles from the exponential-like form of the decay law should occur much earlier than it follows from the classical standard approach resolving itself into replacing time t by t/γ (where γ is the relativistic Lorentz factor) in the formula for the survival probability $\mathcal{P}(t)$.

1. Introduction

Since the discovery of radioactivity and the formulation of the radioactive decay law by Rutheford and Sody [1] there is a conviction that the decay law has the exponential form, $N(t) = N_0 \exp[-\lambda t]$, where $\lambda > 0$ is a constant, $N(t)$ is the number of atoms of the radioactive element at the instant $t \geq 0$ and $N_0 = N(0)$. Wesisskopf–Wigner theory of spontaneous emission [2] has extended this belief on the quantum decay processes: They found that to a good approximation the quantum mechanical non–decay probability of the exited atomic levels is a decreasing function of time having exponential form. Further theoretical studies of the quantum decay process [3, 4] showed that such processes seem to have three stages: the early time (initial), exponential (or "canonical"), and late time having inverse–power law form [5]. Results of these theoretical studies were the reason that there is rather widespread belief that a universal feature of the quantum decay process is the presence of these three time regimes. In this situation, each experimental evidence of oscillating decay curve at times of the order of life times is considered as an anomaly: The so–called GSI–anomaly [6, 7] is an example. The question arises, if indeed in the case of one component quantum unstable systems such oscillations of the decay process at the "exponential" regime are an anomaly. The another question is: Whether and how such a possible oscillations depend on the motion of the unstable quantum system. So we need also to know how to describe the decay process of unstable quantum systems in motion.

Studying text books one finds that if the decay law of the unstable particle in rest has the following form $\mathcal{P}(t) = \exp[-\frac{\Gamma_0 t}{\hbar}] \equiv \mathcal{P}_c(t)$, then in the case of the moving particle with momentum $p \neq 0$ its decay law looks as follows $\mathcal{P}(t) \equiv \mathcal{P}(t;p) = \exp[-\frac{\Gamma_0 t}{\hbar \gamma}] = \mathcal{P}_c(t/\gamma)$, where Γ_0 is the decay rate (time t and Γ_0 are measured in the rest reference frame of the particle) and γ is the relativistic Lorentz factor, $\gamma \equiv 1/\sqrt{1-\beta^2}$, $\beta = v/c$, $v = |\vec{v}|$ is the velocity of the particle, $\vec{v} = c\vec{p}/\sqrt{\vec{p}^2 + m_0^2 c^2}$ and m_0 – is the rest mass. People are convinced that this equality being classical physics relation is valid also for any t in the case of quantum decay processes. The problem seems to be extremely important because from some theoretical studies it follows that in the case of quantum decay processes this relation is valid to a sufficient accuracy only for not more than a few lifetimes $\tau_0 = \hbar/\varGamma_0$ [8, 9, 10]. All these problems require a deeper analysis, elements of which will be presented below.

2. Properties of quantum unstable systems

Quantum unstable systems are characterized by their survival probability. We will assume that the reference frame $\mathcal{O}_0$ is the common inertial rest frame for the observer and for the unstable system. Now if the system in the rest frame is in the initial unstable state $|\phi\rangle \in \mathcal{H}$, ($\mathcal{H}$ is the Hilbert space of states of the considered system) prepared at the initial instant $t_0 = 0$, then the survival probability (the decay law), $\mathcal{P}_0(t)$, of the state $|\phi\rangle$ decaying in vacuum is given by the following formula $\mathcal{P}_0(t) = |a_0(t)|^2$, where $a_0(t)$ is the probability amplitude of finding the system at the time t in the rest frame $\mathcal{O}_0$ in the initial unstable state $|\phi\rangle$, $a_0(t) = \langle\phi|\phi(t)\rangle$. and $|\phi(t)\rangle$ is the solution of the Schrödinger equation for the initial condition $|\phi(0)\rangle = |\phi\rangle$,

$$i\frac{\partial}{\partial t}|\phi(t)\rangle = H|\phi(t)\rangle, \tag{1}$$

(We use units $\hbar = c = 1$). Here $|\phi\rangle, |\phi(t)\rangle \in \mathcal{H}$, and H is the total self–adjoint Hamiltonian of the system considered. We have $|\phi(t)\rangle = U(t)|\phi\rangle$, where $U(t) = \exp[-itH]$ is unitary evolution operator and $U(0) = \mathbb{I}$ is the unit operator.

The rest reference frame $\mathcal{O}_0$ is defined using common eigenvectors of H and of the momentum operator $\mathbf{P}$:

$$\mathbf{P}|\mu; p\rangle = \vec{p}|\mu; p\rangle, \quad \text{and,} \quad H|\mu; p\rangle = E'(\mu, p)\,|\mu; p\rangle, \tag{2}$$

where $\mu \equiv E'(\mu, 0) \in \sigma_c(H)$ and $\sigma_c(H)$ is the continuous part of the spectrum of H. Operators H and $\mathbf{P}$ act in the state space $\mathcal{H}$. We have (see [8, 9, 11, 12]),

$$E'(\mu, p) \equiv \sqrt{\mu^2 + (\vec{p})^2}. \tag{3}$$

One obtains the rest reference frame $\mathcal{O}_0$ of the quantum unstable system assuming that $\vec{p} = 0$. Then $|\mu; 0\rangle = |\mu; p = 0\rangle$ and

$$\mathbf{P}|\mu; 0\rangle = 0, \qquad H|\mu; 0\rangle = \mu\,|\mu; 0\rangle, \tag{4}$$

where $\mu \in \sigma_c(H)$. Eigenvectors $|\mu; 0\rangle$ are normalized as usual: $\langle 0; \mu|\mu'; 0\rangle = \delta(\mu - \mu')$.

The unstable system in the rest frame $\mathcal{O}_0$ is modeled as the following wave–packet $|\phi_0\rangle \equiv |\phi_{\vec{p}=0}\rangle \stackrel{\text{def}}{=} |\phi\rangle$,

$$|\phi_0\rangle \equiv |\phi\rangle = \int_{\mu_0}^{\infty} c(\mu)\,|\mu; 0\rangle\, d\mu, \tag{5}$$

where μ_0 is the lower bound of the spectrum $\sigma_c(H)$. The expansion coefficients $c(\mu)$ are functions of the mass parameter μ, (i.e. of the rest mass μ). The state $|\phi_0\rangle$ is normalized as follows:

$$\int_{\mu_0}^{\infty} |c(\mu)|^2\, d\mu = 1. \tag{6}$$

The expansion (5) and relation (4) allow one to find the amplitude $a_0(t)$ [4, 13],

$$a_0(t) \equiv \int_{\mu_0}^{\infty} \omega(\mu)\, e^{-i\,\mu\,t}\, d\mu, \tag{7}$$

where $\omega(\mu) \equiv |c(\mu)|^2 > 0$.

Note that the use of the Schrödinger equation (1) allows one to find that within the problem considered.

$$i\frac{\partial}{\partial t}\langle\phi|\phi(t)\rangle = \langle\phi|H|\phi(t)\rangle. \tag{8}$$

This relation leads to the conclusion that the amplitude $a_0(t)$ satisfies the following equation

$$i\frac{\partial a_0(t)}{\partial t} = h(t)\, a_0(t), \tag{9}$$

where

$$h(t) = \frac{\langle\phi|H|\phi(t)\rangle}{a_0(t)}, \tag{10}$$

and $h(t)$ is the effective Hamiltonian governing the time evolution in the subspace of unstable states $\mathcal{H}_{||} = P\mathcal{H}$, where $P = |\phi\rangle\langle\phi|$ (see [14] and also [15, 16] and references therein). The subspace $\mathcal{H} \ominus \mathcal{H}_{||} = \mathcal{H}_{\perp} \equiv Q\mathcal{H}$ is the subspace of decay products. Here $Q = \mathbb{I} - P$. If $\langle\phi|H|\phi\rangle$ exists then using unitary evolution operator $U(t)$ and projection operators P and Q the relation (10) can be rewritten as follows

$$h(t) = \langle\phi|H|\phi\rangle \ + \ \frac{\langle\phi|HQ\,U(t)|\phi\rangle}{a_0(t)}. \tag{11}$$

This effective Hamiltonian $h(t)$ can be also derived starting from the Schrödinger equation for the total state space $\mathcal{H}$ and and looking for the rigorous evolution equation for a distinguished subspace of states $\mathcal{H}_{||} \subset \mathcal{H}$ [14].

In general in the case of one–dimensional $\mathcal{H}_{||}$ there is

$$h(t) = \mu_{\phi}(t) - \frac{i}{2}\gamma_{\phi}(t), \tag{12}$$

and $\mu_{\phi}(t) = \Re\,[h(t)]$, $\gamma_{\phi}(t) = -2\,\Im\,[h(t)]$, are the instantaneous mass (energy) $\mu_{\phi}(t)$ and the instantaneous decay rate, $\gamma_{\phi}(t)$, [14, 15, 16]. The relations (9) and (10) are convenient when the density $\omega(\mu)$ is given and one wants to find the instantaneous mass $\mu_{\phi}(t)$ and decay rate $\gamma_{\phi}(t)$: Inserting $\omega(\mu)$ into (7) one obtains the amplitude $a_0(t)$ and then one can find $h(t)$ and thus $\mu_{\phi}(t)$ and $\gamma_{\phi}(t)$.

Note that the state vector $|\phi\rangle$ of the form (5) describing a quantum unstable system can not be an eigenvector of the Hamiltonian H, otherwise it would be that $\mathcal{P}_0(t) = |\langle\phi|\phi(t)\rangle|^2 = |\langle\phi|\exp{[-itH]}\phi\rangle|^2 \equiv 1$ for all times t. So the mass (energy) of such a system is not defined. Simply the mass can not take the exact constant value in this state $|\phi\rangle$. In such a case quantum systems are characterized by the mass (energy) distribution density $\omega(\mu)$ and the average mass $< m >= \int_{\mu_0}^{\infty} \mu\,\omega(\mu)\,d\mu$ or by the instantaneous mass (energy) $\mu_{\phi}(t)$ but not by the exact value of the mass.

As it was mentioned in Sec. 1 one of the aims of these considerations is to check whether the decay law $\mathcal{P}_0(t)$ may be of the pure exponential form at some time intervals. The simplest way to compare the decay law $\mathcal{P}_0(t)$ with the exponential (canonical) decay law $\mathcal{P}_c(t)$, where $\mathcal{P}_c(t) = |a_c(t)|^2$ and

$$a_c(t) = \exp{[-i\frac{t}{\hbar}(m_{\phi} - \frac{i}{2}\Gamma_{\phi}]}, \tag{13}$$

is to analyze properties of the following function [17]:

$$\zeta(t) \stackrel{\text{def}}{=} \frac{a_0(t)}{a_c(t)}. \tag{14}$$

It is because $|\zeta(t)|^2 = \frac{\mathcal{P}_0(t)}{\mathcal{P}_c(t)}$. One can easy find that

$$\frac{\partial \zeta(t)}{\partial t} = \frac{i}{\hbar}\,(m_\phi - \frac{i}{2}\,\varGamma_\phi)\,\zeta(t)\; - \frac{i}{\hbar}\,h(t)\,\zeta(t), \tag{15}$$

where $h(t)$ is the effective Hamiltonian defined by relations (9), (10). Let us assume now that $\langle\phi|H|\phi\rangle$ exists and such instants $0 < t_1 < t_2 < \infty$ of time t exist that for any $t \in (t_1, t_2)$ there is

$$\zeta(t) = \zeta(t_1) = \zeta(t_2) = const \stackrel{\text{def}}{=} c_\phi \neq 0. \tag{16}$$

It can occur only if $\frac{\partial \zeta(t)}{\partial t} = 0$ for all $t \in (t_1, t_2)$. By definition $\zeta(t) \neq 0$ and therefore from (15) we conclude that it is possible only and only if

$$h(t) \; - \; (m_\phi - \frac{i}{2}\,\varGamma_\phi) = 0, \quad \text{for } \; t_1 \leq t \leq t_2, \tag{17}$$

that is, if and only if

$$h(t_1) = h(t) = h(t_2) = const \stackrel{\text{def}}{=} c_h \neq 0 \quad \text{for } \; t_1 \leq t \leq t_2. \tag{18}$$

From (11) the conclusion follows that the equality $h(t_1) = h(t) = c_h$ can take place if

$$a_0(t)\,\langle\phi|HQ\,U(t_1)|\phi\rangle = a_0(t_1)\,\langle\phi|HQ\,U(t)|\phi\rangle. \tag{19}$$

We have that $a_0(t) \neq 0$, $a_0(t_1) \neq 0$ and $|a(t)| < \infty$. Hence $\lambda(t, t_1) \stackrel{\text{def}}{=} a_0(t)/a_0(t_1)$ is a complex function such that $0 < |\lambda(t, t_1)| < \infty$. Using this $\lambda(t, t_1)$ one can replace the relation (19) by

$$\langle\phi|HQ\,U(t_1)\,\Big[\,\lambda(t, t_1)|\phi\rangle \; - \; W(t, t_1)|\phi\rangle\Big] = 0, \tag{20}$$

where $W(t, t_1) = U^{+}(t_1)\,U(t)$ is the unitary operator.

As it is seen the condition (20) can be satisfied in two cases: The first one is

$$W(t, t_1)|\phi\rangle \; - \; \lambda(t, t_1)|\phi\rangle \; = \; 0, \tag{21}$$

and the second one takes place when $[\,\lambda(t, t_1)|\phi\rangle \; - \; W(t, t_1)|\phi\rangle] \neq 0$, and

$$(\langle\phi|H)^{+} = H|\phi\rangle \;\; \perp \;\; Q\,U(t_1)\,[\lambda(t, t_1)|\phi\rangle \; - \; W(t, t_1)|\phi\rangle]. \tag{22}$$

The first case implies that $h(t_1) = h(t) = c_h = const$ (which by (18) means that $\frac{\partial \zeta(t)}{\partial t} = 0$) if and only if the vector $|\phi\rangle$ representing an unstable state of the system is an eigenvector for the unitary operator $W(t, t_1)$. On the other hand if the condition (16) is satisfied then,

$$\lambda(t, t_1) = \frac{a_0(t)}{a_0(t_1)} \equiv \frac{a_0(t)}{a_c(t)}\,\frac{a_c(t)}{a_0(t_1)} \equiv \exp\,[\frac{i}{\hbar}(m_\phi - \frac{i}{2}\varGamma_\phi)(t - t_1)], \tag{23}$$

and thus for $t > t_1$ one finds that in such a case there must be $|\lambda(t, t_1)| < 1$. This conclusion means that the equation (21) has no solution when the condition (16) holds: Eigenvalues $\lambda(t, t_1)$ of any unitary operator must satisfy the condition $|\lambda(t, t_1)| = 1$.

Let us consider now the second case: The definition of the projectors P and Q suggests that this case can be realized only if the vector $H|\phi\rangle$ is proportional to the vector $|\phi\rangle$: $H|\phi\rangle = \alpha_\phi|\phi\rangle$. So $\frac{\partial \zeta(t)}{\partial t} = 0$ if and only if the vector $|\phi\rangle$ representing the unstable state of the system considered is an eigenvector for the total Hamiltonian H, which is in clear contradiction with the condition that the vector $|\phi\rangle$ representing the unstable state cannot be the eigenvector for H.

Implications of the above conclusions for possible realizations of the relation (20) allow one to deduce that the supposition that such time interval $[t_1, t_2]$ can exist that $h(t_1) = h(t) = c_h = const$ for $t \in (t_1, t_2)$ and thus $\zeta(t) = const = \zeta(t_1) = \zeta(t_2)$ for $t \in (t_1, t_2)$ is false. Hence from the definition of $\zeta(t)$ the conclusion follows: Within the approach considered in this paper for any time interval $[t_1, t_2]$ the decay law can not be described by the exponential function of time. This conclusion is the general one. It does not depend on models of quantum unstable states.

3. The Breit–Wigner (BW) model

In this part of the paper analytical results presented in the previous section will be illustrated graphically. The typical form of the survival probability $\mathcal{P}_0(t)$ is shown in Fig (1). The calculations were made for the distribution of the mass (energy) density $\omega(\mu)$ having the Breit–Wigner (BW) form $\omega(\mu) \equiv \omega_{BW}(\mu)$,

$$\omega_{BW}(\mu) = \frac{N}{2\pi}\,\Theta(\mu - \mu_0)\,\frac{\Gamma_0}{(\mu - m_0)^2 + (\frac{\Gamma_0}{2})^2}, \tag{24}$$

where N is a normalization constant and $\Theta(\mu)$ is a step function: $\Theta(\mu) = 0$ for $\mu \leq 0$ and $\Theta(\mu) = 1$ for $\mu > 0$. The form of the decay curves depend on the ratio $s_R = \frac{m_R}{\Gamma_0}$, where $m_R = m_0 - \mu_0$: The smaller s_R, the shorter the time when the late time deviations from the exponential form of $\mathcal{P}_0(t)$ begin to dominate.

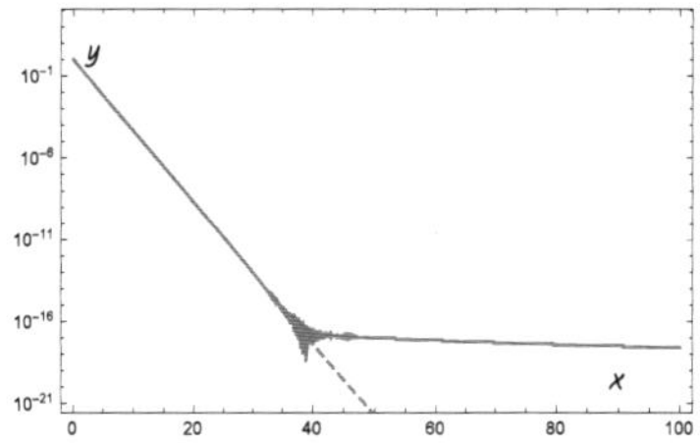

Figure 1. Decay curves obtained for $\omega_{BW}(E)$ given by Eq. (24). Axes: $x = t/\tau_0$ — time t is measured in lifetimes τ_0, y — survival probabilities on a logarithmic scale (The solid line: the decay curve $\mathcal{P}_0(t) = |a_0(t)|^2$; The dotted line: the canonical decay curve $\mathcal{P}_c(t) = |a_c(t)|^2$). The case $s_R = \frac{m_R}{\Gamma_0} = 1000$.

Within the considered model the standard canonical form of the survival amplitude $a_c(t)$ is given by formula (13) with m_0 and Γ_0 replacing m_ϕ and Γ_ϕ, (where Γ_0 is the decay rate and $\frac{\hbar}{\Gamma_0} \equiv \frac{1}{\Gamma_0} = \tau_0$ is the lifetime within the assumed system of units $\hbar = c = 1$: Here time t and Γ_0 are measured in the rest reference frame of the particle), and thus $\mathcal{P}_c(t) = |a_c(t)|^2 \equiv \exp\,[-\frac{\Gamma_0}{\hbar}\,t]$, is the canonical form of the survival amplitude.

The situation when $\omega(\mu) = \omega_{BW}(\mu)$ is the typical case discussed in numerous papers and used therein to model decay processes. Hence the need for analysis of the real form of the decay curves obtained using $\omega(\mu) = \omega_{BW}(\mu)$ and this is why we consider this case here.

As already noted the function $\zeta(t)$ (see (14)) is very helpful in achieving this goal. Analysis of properties of $|\zeta(t)|^2$ allows one to visualize all the more subtle differences between $\mathcal{P}_0(t)$ and $\mathcal{P}_c(t)$. The function $\zeta(t)$ was used to find numerically $|\zeta(t)|^2$ for $\omega(m) = \omega_{BW}(m)$. Results of numerical calculations are presented in Figs (2) and (3): One can see that in the case considered the form of $|\zeta(t)|^2$ also depend on the ratio $s_R \stackrel{\text{def}}{=} \frac{m_R}{\Gamma_0} \equiv \frac{m_0 - \mu_0}{\Gamma_0}$ [17].

From results presented in Figs (2), (3) one can see that the oscillating decay curves of one component unstable system can not be considered as something extraordinary or as anomaly: It seems to be a universal feature of the decay process [17]. Results of the model calculations presented in Figs (2) and (3) shows that at the initial stage of the "exponential" (or "canonical") decay regime the amplitude of these oscillations may be much less than the accuracy of detectors. Then with increasing time the amplitude of oscillations grows (see Fig. (3)), which increases the chances of observing them. So the conclusion that this is a true quantum picture of the decay process at the so–called "exponential" regime of times seems to be justified.

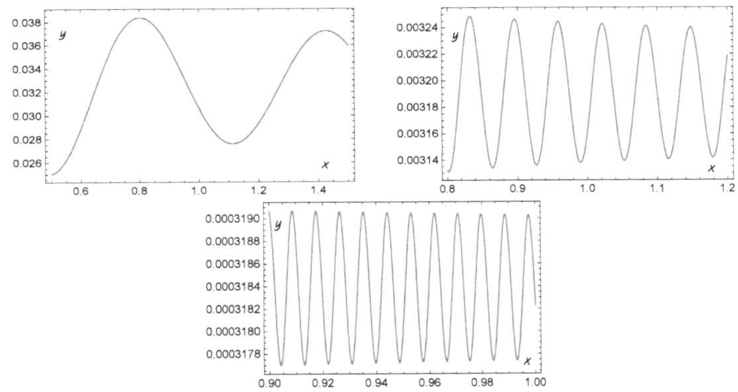

Figure 2. A comparison of decay curves obtained for $\omega_{BW}(\mu)$ given by Eq. (24) with canonical decay curves. Axes: $x = t/\tau_0$ — time t is measured in lifetimes τ_0, y — The function $f(t) = (|\zeta(t)|^2 - 1) = \frac{\mathcal{P}_0(t)}{\mathcal{P}_c(t)} - 1$, where $\zeta(t)$ is defined by the formula (14). **Left panel**: $s_R = 10$. **Right panel**: $s_R = 100$. **Lower panel**: $s_R = 1000$.

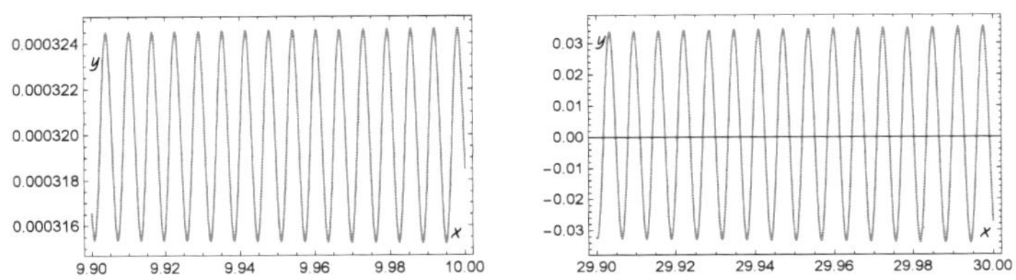

Figure 3. A comparison of decay curves obtained for $\omega_{BW}(\mu)$ given by Eq. (24) with canonical decay curves. **Both panels** – Axes: $x = t/\tau_0$ — time t is measured in lifetimes τ_0, y — The function $f(t) = (|\zeta(t)|^2 - 1) = \frac{\mathcal{P}_0(t)}{\mathcal{P}_c(t)} - 1$, where $\zeta(t)$ is defined by the formula (14), $\mathcal{P}_0(t) = |a_0(t)|^2$, $\mathcal{P}_c(t) = |a_c(t)|^2$. The case $s_R = 1000$.

4. Moving quantum unstable systems

Let us analyze now some properties of the moving quantum unstable systems. In this case we need the probability amplitude $a_p(t) = \langle\phi_p|\phi_p(t)\rangle$, which defines the survival probability $\mathcal{P}_p(t) = |a_p(t)|^2$. We have $|\phi_p(t)\rangle \stackrel{\text{def}}{=} \exp[-itH]\,|\phi_p\rangle$ in $\hbar = c = 1$ units. As it is seen we need the vector $|\phi_p\rangle$ and eigenvalues $E'(\mu, p)$ solving Eq. (2). Vectors $|\phi\rangle$ which is defined by the relation (5) and $|\phi_p\rangle$ are elements of the same state space $\mathcal{H}$ connected with the coordinate rest system of the observer $\mathcal{O}$: We are looking for the decay law of the moving particle measured by the observer $\mathcal{O}$. If to assume for simplicity that $\mathbf{P} = (P_1, 0, 0)$ and that $\vec{v} = (v_1, 0, 0) \equiv (v, 0, 0)$ then there is $\vec{p} = (p, 0, 0)$ for the eigenvalues $\vec{p}$ of the momentum operator $\mathbf{P}$ and $|\vec{p}| = p$. Hence (see [8, 9, 11, 12]),

$$H|\mu; p\rangle = \sqrt{\mu^2 + p^2}\,|\mu; p\rangle \tag{25}$$

which replaces Eq. (2).

We can model the moving quantum unstable particle ϕ with constant momentum, $\vec{p}$, analogously as the quantum unstable system in the rest frame (when $\vec{p} = 0$), that is as the following wave–packet $|\phi_p\rangle$,

$$|\phi_p\rangle = \int_{\mu_0}^{\infty} c(\mu)\,|\mu; p\rangle\, dm. \tag{26}$$

Here coefficients $c(\mu)$ are functions of the mass parameter μ, (in other words of the rest mass μ), which is Lorentz invariant and therefore the scalar functions $c(\mu)$ of μ are also Lorentz invariant and are the same as in the rest reference frame $\mathcal{O}_0$.

The final relation for for the amplitude $a_p(t)$ results from (25) and form the equation (26), (see [8, 9, 12]),

$$a_p(t) \equiv \int_{\mu_0}^{\infty} \omega(\mu)\, e^{-i\sqrt{\mu^2 + p^2}\; t}\, d\mu. \tag{27}$$

Results of numerical calculations are presented in Figs (4), (5), where calculations were performed for $\omega(\mu) = \omega_{BW}(\mu)$ and $\mu_0 = 0$, $E_0/\varGamma_0 \equiv m_0/\varGamma_0 = 1000$ and $cp/\varGamma_0 \equiv p/\varGamma_0 = 1000$. Values of these parameters correspond to $\gamma = \sqrt{2}$, which is very close to γ from the experiment performed by the GSI team [6, 7] and this is why such values of them were chosen in our considerations. According to the literature for laboratory systems a typical value of the ratio $m_0/\varGamma_0$ is $m_0/\varGamma_0 \geq O(10^3 - 10^6)$ (see eg. [18]) therefore the choice $m_0/\varGamma_0 = 1000$ seems to be reasonable minimum. Decay curves obtained numerically are presented in Fig (4) (see also [10]).

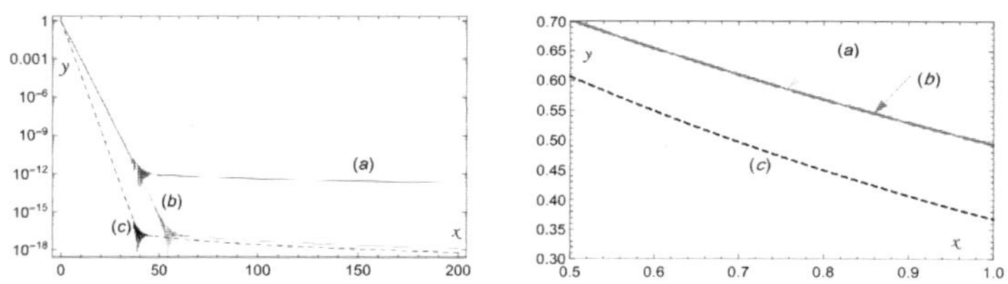

Figure 4. Decay curves obtained for $\omega_{BW}(m)$ given by Eq. (24). Axes: $x = t/\tau_0$ — time t is measured in lifetimes τ_0, y — survival probabilities (**Left panel**: the logarithmic scales, (a) the decay curve $\mathcal{P}_p(t)$, (b) the decay curve $\mathcal{P}_0(t/\gamma)$, (c) the decay curve $\mathcal{P}_0(t)$; **Right panel**: (a) – $\mathcal{P}_p(t)$, (b) – $\mathcal{P}_0(t/\gamma)$, (c) – $\mathcal{P}_0(t)$).

The ratio $\mathcal{P}_p(t)/\mathcal{P}_c(t/\gamma)$ in the case of moving particles can be also calculated similarly as it was in the case of quantum unstable systems in rest. Results of numerical calculations of this ratio are presented in Fig. (5), and calculations were performed for $\omega(\mu) = \omega_{BW}(\mu)$ and for $\mu_0 = 0$, $m_0/\varGamma_0 = 1000$, $cp/\varGamma_0 \equiv p/\varGamma_0 = 1000$ and $\gamma = \sqrt{2}$.

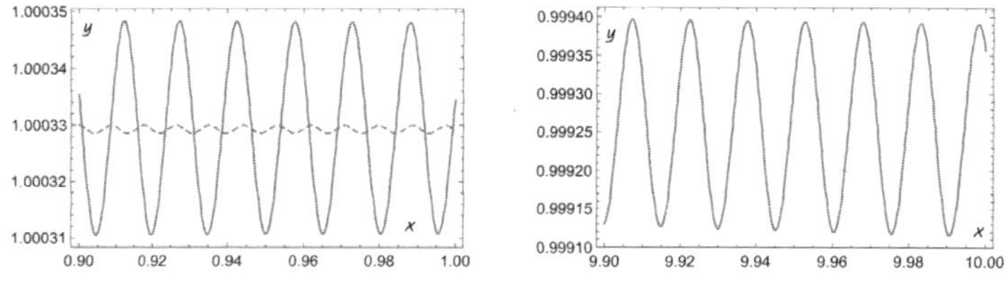

Figure 5. Left panel. Axes: $x = t/\tau_0$ — time t is measured in lifetimes τ_0, y — Ratio of probabilities — Solid line: $\mathcal{P}_p(t)/\mathcal{P}_c(t/\gamma)$; Dashed line $\mathcal{P}_0(t/\gamma)/\mathcal{P}_c(t/\gamma)$. **Right panel.** Axes: $x = t/\tau_0$ — time t is measured in lifetimes τ_0, y — Ratio of probabilities: $\mathcal{P}_p(t)/\mathcal{P}_c(t/\gamma)$.

5. Summary

As it was pointed out earlier the mass (energy) of the system in the unstable state $|\phi\rangle$ is not defined: It can not take the exact value. Unstable system can be characterized by the

IOP Conf. Series: Journal of Physics: Conf. Series **880** (2017) 012034 doi:10.1088/1742-6596/880/1/012034

mass distribution $\omega(\mu)$, the average mass $< m >= \int_{\mu_0}^{\infty} \mu\,\omega(\mu)d\mu$ and by instantaneous mass (energy) $\mu_\phi(t)$ but not by the mass. Also it was shown that there is no any time interval in which the survival probability (decay) law could be a decreasing function of time of the purely exponential form: Even in the case of the BW model in so–called "exponential regime" the decay curves are oscillatory modulated with smaller or large amplitude of oscillations depending on the parameters of the model. The another observation was that in the case of moving relativistic quantum unstable system moving with constant momentum $\vec{p}$, when unstable systems are modeled by the BW mass distribution $\omega_{BW}(\mu)$, only at times of the order of lifetime τ_0 it can be $\mathcal{P}_p(t) \simeq \mathcal{P}_0(t/\gamma)$ to a better or worse approximation. At times longer than a few lifetimes the decay process of moving particles observed by an observer in his rest system is much slower that it follows from the classical physics relation $\mathcal{P}_p(t) \stackrel{?}{=} \exp\,[-\frac{t}{\gamma}\,\varGamma_0]$: There is $\mathcal{P}_p(t) > \mathcal{P}_0(t/\gamma), \quad \text{for} \quad t \gg \tau_0$. Therefore there is a need to test the decay law of moving relativistic unstable system for times much longer than the lifetime. It was shown also that in the case of moving relativistic quantum unstable system moving with constant momentum $\vec{p}$ decay curves are also oscillatory modulated but the amplitude of these oscillations is higher than in the case of unstable systems in rest.

References

[1] Ruthheford E 1900 *Philosophical Magazine* **XLIX** 1, 161
[2] Weisskopf V F and Wigner E T 1930 *Z. Phys.* **63** 54; **65** 18
[3] L. A. Khalfin L A 1957 *Zh. Eksp. Teor. Fiz.* **33**, p.p. 1371; [1958 *Sov. Phys. — JETP* **6** 1053]
[4] Fonda L, Ghirardii G C and A. Rimini 1978 *Rep. on Prog. in Phys.* **41** 587
[5] Peshkin M, Volya A and Zelevinsky V 2014 *Europhysics Letters*, **107** 40001
[6] Litvinov Yu A *et al.* 2008 *Phys. Lett.* B **664** 162
[7] Kienle P *et al.* 2013 *Phys. Lett.* B **726** 638
[8] Stefanovitch V E 1996 *International Journal of Theoretical Physics* **35** 2539
[9] Shirokov M 2004 *International Journal of Theoretical Physics* **43** 1541
[10] Urbanowski K 2014 *Phys. Lett.* B **737** 346
[11] Gibson W M and Polard B R 1976 *Symmetry principles in elementary particle physics* (Cambridge)
[12] Khalfin L A 1997 Quantum theory of unstable particles and relativity (*Preprint* PDMI PREPRINT–6/1997: St. Petersburg Department of Steklov Mathematical Institute, St. Petersburg, Russia)
[13] Krylov N S and Fock V A 1947 *Zh. Eksp. Teor. Fiz.* **17** 93
[14] Urbanowski K 1994 *Phys. Rev.* A **50** 2847
[15] Urbanowski K 2008 *Eur. Phys. J.* C **58** 151
[16] Urbanowski K 2009 *Cent. Eur. J. Phys.* **7** 696
[17] Urbanowski K 2016 The true quantum face of the "exponential" decay law (*Prerint* quant–ph/1606.0730)
[18] Krauss L M and Dent J 2008 *Phys. Rev. Lett.* **100** 171301

IOP Conf. Series: Journal of Physics: Conf. Series **880** (2017) 012035 doi:10.1088/1742-6596/880/1/012035

Quantum friction imprints on the geometric phase of a moving atom in front of a dielectric plate

Fernando C. Lombardo and Paula I. Villar

Departamento de Física *Juan José Giambiagi*, FCEyN UBA and IFIBA CONICET-UBA, Facultad de Ciencias Exactas y Naturales, Ciudad Universitaria, Pabellón I, 1428 Buenos Aires, Argentina

Abstract. We compute the non-unitary geometric phase for the moving atom under the presence of the vacuum field and a dielectric mirror, analytically and numerically. We consider the atom (represented by a two-level system) moving in front of a dielectric plate, and study how decoherence of the particle's internal degrees of freedom can be found in the corrections to the geometric phase accumulated by the atom. We consider the particle to follow a classical, macroscopically-fixed trajectory and by integrating over the vacuum field and the microscopic degrees of freedom of the plate we may calculate friction effects. We find a velocity dependance in the correction to the unitary geometric phase due to quantum frictional effects. We also show in which cases decoherence effects could, in principle, be controlled in order to perform a measurement of the geometric phase using standard interferometry procedures.

1. Introduction

A quantum system can retain the information of its motion when it undergoes a cyclic evolution in the form of a geometric phase (GP), which was first put forward by Pancharatman in optics [1] and later studied explicitly by Berry in a general quantal system [2]. Since the work of Berry, the notion of geometric phases (GPs) was shown to have important consequences for quantum systems. Berry demonstrated that quantum systems could acquire phases that are geometric in nature. He showed that, besides the usual dynamical phase, an additional phase that was related to the geometry of the space state was generated during an adiabatic evolution. Since then, great progress has been achieved in this field. As an important evolvement, the application of the geometric phase has been proposed in many fields, such as the geometric quantum computation. Due to its global properties, the geometric phase is propitious to construct fault tolerant quantum gates. In this line of work, many physical systems have been investigated to realise geometric quantum computation, such as NMR (Nuclear Magnetic Resonance) [3], Josephson junction [4], Ion trap [5] and semiconductor quantum dots [6]. The quantum computation scheme for the GP has been proposed based on the Abelian or non-Abelian geometric concepts, and the GP has been shown to be robust against faults in the presence of some kind of external noise due to the geometric nature of Berry phase [7, 8, 9]. It was therefore seen that the interactions play an important role for the realisation of some specific operations. As the gates operate slowly compared to the dynamical time scale, they become vulnerable to open system effects and parameters fluctuations that may lead to a loss of coherence. Consequently, study of the GP was soon extended to open quantum systems. Following this idea, many authors have analysed the correction to the GP under the influence of an external environment using different approaches (see [10, 11, 12, 13] and references therein).

In this framework, we propose to track evidence of vacuum fluctuations in the geometric phase acquired by a neutral particle moving in front of an imperfect mirror. By measuring the interference pattern of the particle, it could be possible to find a dependence of the correction to the unitary geometric phase upon the velocity of the particle. The pattern obtained can be an indirect proof of the existence of a quantum frictional force [14]. We shall consider a neutral particle coupled to a vacuum field, which is also in interaction with the internal degrees of freedom of a dielectric plate. The particle's trajectory will be, along this paper, kept as an externally-fixed variable. We shall assume the particle moving at a constant velocity v, as is the most popular scenario in the literature [15, 16]. As we are interested in the dynamics of the internal degree of freedom of the particle, we will consider the neutral particle as a two-level quantum (a qubit) system (regarding it as a simple model for the atom) coupled to the vacuum field. We shall use a simple model for the microscopic degrees of freedom of the mirror, as we have done in a previous work [17]: a set of uncoupled harmonic oscillators, each of them also interacting locally with the vacuum field. In Fig.1 there is a picture of the global system and the interactions among the subsystems.

In order to consider how the relative motion between the particle and the plate affects the geometric phase acquired (and therefore consider the effects of quantum friction in the GP), we shall follow the procedure presented in previous works [17, 18], where we have used the in-out effective action approach, to account for the dissipative effects. Decoherence effects have been studied in Ref. [18], evaluating the influence action of composite environment (vacuum field + plate) over the particle [19].

2. Non-unitary geometric phase

We consider the moving particle to be a two-level quantum system (spin-1/2) coupled to a composite environment formed by the vacuum field and the dielectric mirror. Therefore, the main system bare Hamiltonian is given by

$$H_{\rm sys} = \Delta\sigma_z, \tag{1}$$

which simply represents a cyclic evolution with period $\tau = 2\pi/\Delta$ if isolated.

For simplicity, we are only considering a dephasing spin–bath interaction, neglecting relaxation. We take a product initial state for the complete spin-bath system as

$$\rho(0) = |\psi_0\rangle\langle\psi_0| \otimes |\varepsilon(0)\rangle\langle\varepsilon(0)|, \tag{2}$$

where $|\psi_0\rangle = \cos(\theta/2)|0\rangle + \sin(\theta/2)|1\rangle$ is a superposition initial state for the two–level system, and $|\varepsilon(0)\rangle$ is a general initial state of the composite environment. To compute the global phase gained during the evolution, one can use the Pancharatnam's definition [1], which has a gauge dependent part (i.e a *dynamical* phase $\Phi_d = \pi\cos\theta$) and a gauge independent part, commonly known as *geometric* phase $\Phi_g = \pi(1 + \cos\theta)$.

It is widely known that, when coupled to a bath, the reduced density matrix of the particle system satisfies a mater equation of the form (see for example [11] and references therein):

$$\dot{\rho}_{\rm r} = -i\Delta[\sigma_z, \rho_{\rm r}] - \mathcal{D}(t)[\sigma_z, [\sigma_z, \rho_{\rm r}]]. \tag{3}$$

This model describes a purely decoherent mechanism, solely containing a diffusion term $\mathcal{D}(t)$, where no energy exchange between system and the composite bath is present. The diffusion coefficient $\mathcal{D}(t)$ can be evaluated following Ref.[18], and its explicit form is not relevant at this point. Therefore, the coupling to the bath affects the system such that its reduced density matrix at a time t (after integrating out all the bath degrees of freedom) is

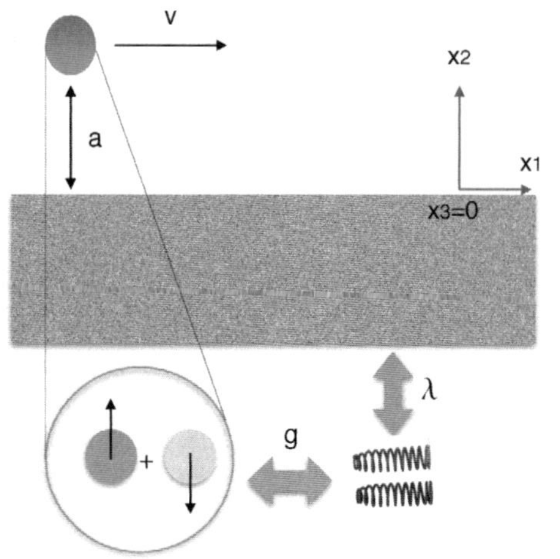

Figure 1. (Color online) We present a simple diagram of the system under consideration. That vacuum field is a massless scalar field $\phi(x)$ and the internal degrees of freedom of the plate are $\psi(x)$. The internal degrees of freedom of the particle will be considered as a two-level system σ_z.

$$\begin{aligned}\rho_{\rm r}(t) &= \cos^2(\theta_t)|00\rangle + \sin^2(\theta_t)|11\rangle \\ &+ \sin(\theta_t)\cos(\theta_t)e^{-i\Omega t}|01\rangle + \sin(\theta_t)\cos(\theta_t)e^{i\Omega t}|10\rangle,\end{aligned}$$

where we have used

$$\sin(\theta_t) = \frac{2(\epsilon_+ - \cos^2(\frac{\theta}{2}))}{\sqrt{|r(t)|^2\sin^2(\theta) + 4(\epsilon_+ - \cos^2(\frac{\theta}{2}))^2}} \tag{4}$$

$$\cos(\theta_t) = \frac{|r(t)|\sin(\theta)}{\sqrt{|r(t)|^2\sin^2(\theta) + 4(\epsilon_+ - \cos^2(\frac{\theta}{2}))^2}}. \tag{5}$$

Here, $r(t)$ is the decoherence factor induced by the composite environment (vacuum field and dielectric mirror). The eigenvalues of the above reduced density matrix are easily calculated, yielding:

$$\varepsilon_\pm(t) = \frac{1}{2} \pm \frac{1}{2}\sqrt{\cos^2(\theta) + |r(t)|^2\sin^2(\theta)}. \tag{6}$$

The phase Φ acquired by the open system after a period τ is defined as [10]:

$$\Phi = \arg\left[\sum_k \sqrt{\epsilon_k(\tau)\epsilon_k(0)}\langle\Psi_k(0)|\Psi_k(\tau)\rangle e^{-\int_0^\tau dt\langle\Psi_k(t)\frac{\partial}{\partial t}|\Psi_k(t)\rangle}\right], \tag{7}$$

where $|\Psi_k(t)\rangle$ and $\epsilon_k(\tau)$ are respectively the instantaneous eigenvectors and eigenvalues of $\rho_{\rm r}(t)$. Here, k refers to the two modes ($+$ and $-$) of the qubit model we are dealing with. The central result of Eq. (7) is to extract the purification independent part of the phase – which can be defined as a geometric phase because it is gauge invariant and reduces to the known results in the limit of a unitary evolution. In order to estimate the geometric phase, we only need to consider the eigenvector $|\Psi_+(t)\rangle$ since $\varepsilon_-(0) = 0$. Therefore, in this case, only the '+ mode" contributes to the GP. By the use of the several expressions obtained above, one can reach the final formula for the GP

$$\Phi = \Delta \int_0^\tau dt \cos^2(\theta_t). \tag{8}$$

The GP can be straightforwardly computed if we have the definition of $r(t)$. The decoherence factor $r(t)$ induced by this kind of composite environment has been already calculated in Ref.[18] but for the case of a harmonic oscillator internal degree of freedom for the moving particle. In the present case, we obtain the the decoherence factor induced upon the two-level system as:

$$r(t) = \exp\left\{-\frac{\gamma_0 t}{2}\left(1 + \frac{2}{3}v^2 + \tilde{\lambda}^2 v \frac{e^{-\frac{2\tilde{\Omega}}{v}\sqrt{1-v^2}}}{1-v^2}\right)\right\}, \tag{9}$$

by taking the limit $\Omega \gg \omega_0$ in [18], which means that the bare frequency of the harmonic oscillator is negligible compared with the natural frequency Ω of the dielectric harmonic oscillator degrees of freedom in the mirror. In the process, we have considered the internal degrees of freedom of the dielectric plate to be an infinite set of uncoupled harmonic oscillators of frequency Ω. Each of these oscillators is interacting locally in position with the vacuum field through a coupling constant λ. The two-level system (modeled by $\hat{\sigma_z}$ operator), also interacts linearly and locally with the vacuum field, but through a coupling constant named g. We have set dimensionless coupling constant to the dielectric plate $\tilde{\lambda}^2 = \lambda^2/\Omega^3$ and $\gamma_0 = g^2$.

It is important to note that the decoherence factor contains a term proportional to the coupling between the vacuum field and the dielectric mirror ($\sim \gamma_0\tilde{\lambda}^2$), in addition to the expected term corresponding to the coupling to the vacuum field, proportional to γ_0. The dependence of the decoherence factor upon the velocity v is also of remarkable importance. The contribution coming from the presence of the dielectric plate becomes less important when $v \to 0$ and grows for large values of v.

In Fig.2 we plot the ratio between the total geometric phase from Eq.(8) and the unitary phase $\Phi_g = \pi(1+\cos\theta)$, as a function of the initial angle θ and the tangential velocity v for fixed parameters of $\tilde{\lambda}$, $\tilde{\Omega}$ and γ_0. Therein, we can see that for small values of the initial angle (i.e. a spin very similar to $|\uparrow\rangle$) and very low values of the velocity, the GP obtained for this system is very similar to the one obtained for an isolated quantum system (i.e. a spin 1/2 particle evolving freely). The bigger difference in the open GP and the isolated one is obtained for bigger angles and bigger values of v.

The correction to the unitary GP is relevant for big values of v. What is important to remark, is that the mere presence of a velocity contribution to the phase, is an indication of the friction effect over the quantum degree of freedom of the atom. In this sense, the measurement of the geometric phase could, in principle, be an alternative way to find out quantum friction in a laboratory, even though the considered velocities in experiments are still far away from a relativistic case with $v \to 1$.

Finally, we can perform a series expansion in γ_0 and $\tilde{\lambda}$ (up to first non-trivial orders) in Eq.(8) in order to obtain an analytical expression for the GP acquired by the particle:

$$\begin{aligned} \Phi_{\rm approx} &\approx \pi(1+\cos\theta) + \frac{\pi^2}{3}\frac{\gamma_0}{(1-v^2)}\cos\theta\sin^2\theta \\ &\times \left[\left(3 - v^2 - 2v^4\right) + 3v\tilde{\lambda}^2 e^{-\frac{2\tilde{\Omega}}{v}\sqrt{1-v^2}}\right]. \end{aligned} \tag{10}$$

In the particular case in which the coupling between the atom and the dielectric plate is switched off, $\tilde{\lambda} = 0$, the correction to the unitary phase is given by $\delta\Phi \sim \pi^2\gamma_0(1 +$

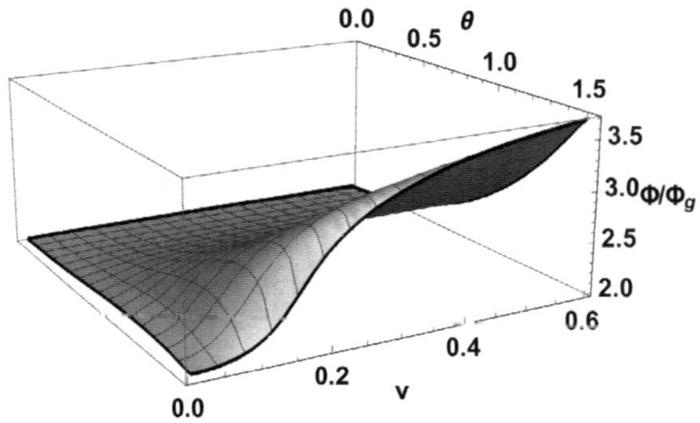

Figure 2. (Color online).We present the geometric phase normalised (Φ/Φ_g) by the unitary geometric fase $\phi_g = \pi(1 + \cos(\theta))$ as function of θ and v. Parameters used: $\tilde{\lambda} = 10$, $\gamma_0 = 0.05$, $\tau = 4\pi/\Delta$ and $\tilde{\Omega} = 0.03$.

$2/3v^2)] \cos\theta \sin^2\theta$ which agrees with the result obtained for a two-level system coupled to an environment composed by an infinite set of harmonic oscillators at equilibrium with $T = 0$ [11, 12, 20, 21, 22, 23, 24]. However, we can see that in this case the result is enhanced by a factor $1 + 2/3v^2$. This situation corresponds to the atom only coupled to the vacuum field. Up to the lowest perturbative order, the same result can be obtained in the limiting case of $v \to 0$.

3. Conclusions

We have considered the effects of quantum vacuum fluctuations on a particle moving parallel to an imperfect mirror. We have presented a simple model in which the vacuum field is a massless, real, scalar field coupled to the microscopic degrees of freedom of the mirror and the internal degree of freedom of the particle (following analysis done previously in Refs.[17, 18]). The plate is formed by uncoupled unidimensional harmonic oscillators, each of them interacting locally in position with the vacuum field. The macroscopic trajectory of the particle is externally fixed, and its internal degree of freedom is considered as a simple two-level system, also coupled to the scalar field. Using previous results for decoherence effects found in Refs.[17, 18], we have estimated the decoherence factor when the internal degree of freedom of the particle moving with parallel velocity v is a two-state system, as a toy model for an atom travelling over the dielectric plate. With the expression of the decoherence factor, it was possible to calculate the corrections to the geometric phase of the quantum states of the atom, induced by the interaction with the composite environment.

We have found that the phase acquired by the moving particle is different to the one the particle would have acquired if it has evolved freely. By measuring the correction to the unitary geometric phase, one can get an insight of the dependance of the phase on the parameters. In this way, we have seen that the bigger the velocity of the particle, the more correction to the phase. It is also noticeable that the effect of noise is bigger for initial states near the equator of the Bloch sphere.

Finally, we have obtained an approximate expression for the phase acquired and compared this result to the exact geometric phase. In this case, it is possible to see that the expression gives an accurate result for small values of the coupling constants (as expected), as well as small angles. It is also possible to note that in the case in which the coupling with the mirror can be neglected, the result obtained for the correction to the geometric phase would correspond to the one obtained in the case a qubit is solely in interaction with an infinite set of harmonic oscillators at thermal equilibrium with $T = 0$. Finally, it is remarkable that, in the case of the moving atom, there is a factor proportional to v^2 which enhances the correction to the phase, even though we are in the small velocity case $v < 1$.

We expect that in a interference experiment, the parameters of our model could be chosen in a way that would maximise the decoherence effects. By increasing the decoherence effect, the unitary geometric phase results in a major correction. In this sense, as quantum friction has not been measured in labs yet, we expect that an indirect evidence could be obtained from measuring the environmental induced corrections to the geometric phase. The dependance of the correction on the velocity v would be an indirect way to finally measure quantum friction.

Acknowledgments
This work was supported by UBA, CONICET and ANPCyT - Argentina. We would like to H. Thomas Elze for the organization of DICE 2016. FCL acknowledges to B. Lok Hu by the very interesting conversations about quantum friction, decoherence, and geometric phases.

References

[1] Pancharatnam S 1956 Generalized Theory of Interference, and its Applications. Part I. Coherent Pencils *Proc. Indian Acad. Sci. A* **44**, 247
[2] Berry M Y 1984 Quantal Phase Factors Accompanying Adiabatic Changes Proc. R. Soc. Lond. A **392** 45
[3] Jones J A, Vedral V, Ekert A and Castagnoli G 2000 Geometric quantum computation using nuclear magnetic resonance *Nature* **403** 869
[4] Faoro L, Siewert J and Fazio R 2003 Non-Abelian Holonomies, Charge Pumping, and Quantum Computation with Josephson Junctions *Phys. Rev. Lett.* **90** 028301
[5] Duan L M et. al. 2001 Geometric Manipulation of Trapped Ions for Quantum Computation *Science* **292** 1695
[6] Solinas P, Zanardi P, Zangh N and Rossi F 2003 Semiconductor-based geometrical quantum gates *Phys. Rev.* B **67** 121307
[7] Zanardi P, Rasetti M 1999 Holonomic Quantum Computation *Phys. Lett.* A **264**, 94
[8] Xiang-Bin W and Keiji M 2001 Nonadiabatic Conditional Geometric Phase Shift with NMR *Phys. Rev. Lett.* **87**, 097901
[9] Sjoqvist E, Tong D M, Andersson M L, Hessmo B, Johansson M, and Singh K 2012 Non-adiabatic holonomic quantum computation *New J. Phys.* **14**, 103035
[10] Tong D M, Sjoqvist E, Kwek L C, and Oh C H 2004 Kinematic Approach to the Mixed State Geometric Phase in Nonunitary Evolution *Phys. Rev. Lett.* **93**, 080405; see also Tong D M, Sjoqvist E, Kwek L C, and Oh C H 2005 Erratum: Kinematic Approach to the Mixed State Geometric Phase in Nonunitary Evolution [Phys. Rev. Lett. 93, 080405 (2004)] *Phys. Rev. Lett.* **95**, 249902
[11] Lombardo F C and Villar P I 2006 Geometric phases in open systems: A model to study how they are corrected by decoherence *Phys. Rev.* A **74**, 042311
[12] Lombardo F C and Villar P I 2008 Environmentally induced corrections to the geometric phase in a two-level system *Int. J. of Quantum Information (IJQI)* **6**, 707713
[13] Villar P I 2009 Spin bath interaction effects on the geometric phase *Phys. Lett.* A **373**, 206
[14] Pendry J B 1997 Shearing the vacuum - quantum friction *J. Phys. Condens. Matter* **9**, 10301; Pendry J B 2010 Quantum frictionfact or fiction? *New J. Phys.* **12**, 033028; Pendry J B 2010 Reply to comment on 'Quantum frictionfact or fiction? *New J. Phys.* **12**, 068002; Philbin T G and Leonhardt U 2009 No quantum friction between uniformly moving plates *New J. Phys.* **11**, 033035; Leonhardt U 2010 Comment on 'Quantum FrictionFact or Fiction? *New J. Phys.* **12**, 068001; Volokitin A I and Persson B N J 2007 Near-field radiative heat transfer and noncontact friction *Rev. Mod. Phys.* **79**, 1291
[15] Hoye J S and Brevik I 2010 Casimir friction force and energy dissipation for moving harmonic oscillators *Europhys. Lett.* **91**, 60003; Barton G 2010 On van der Waals friction: I. Between two atoms *New J. Phys.* **12**, 113044; Barton G 2010 On van der Waals friction. II: Between atom and half-space *New J. Phys.* **12**, 113045
[16] Intravaia F, Behunin R O, and Dalvit D A R 2014 Quantum friction and fluctuation theorems *Phys. Rev.* A **89** 050101
[17] Farías M B, Fosco C D, Lombardo F C, Mazzitelli F D, Rubio López A E 2015 Functional approach to quantum friction: Effective action and dissipative force *Phys. Rev.* D **91**, 105020

[18] Farías M B and Lombardo F C 2016 Dissipation and decoherence effects on a moving particle in front of a dielectric plate *Phys. Rev.* D **93**, 065035

[19] Feynman R and Vernon F 1963 The Theory of a General Quantum System Interacting with a Linear Dissipative System *Ann. Phys. (N.Y.)* **24**, 118; Lombardo F C, Mazzitelli F D, and Rivers R J 2003 Decoherence in Field Theory: General Couplings and Slow Quenches *Nucl. Phys.* B **672**, 462; Lombardo F C, Mazzitelli F D, and Rivers R J 2001 Classical behaviour after a phase transition *Phys. Lett.* B **523**, 317; Lombardo F C, Rivers R J, and Villar P I 2007 Decoherence of domains and defects at phase transitions *Phys. Lett.* B **648**, 64

[20] Cucchietti F M, Zhang J F, Lombardo F C, Villar P I and Laflamme R 2010 Geometric Phase with Nonunitary Evolution in the Presence of a Quantum Critical Bath *Phys. Rev. Lett.* 105 240406

[21] Villar P I and Lombardo F C 2011 Geometric phases in the presence of a composite environment *Phys. Rev.* A **83**, 052121

[22] Lombardo F C and Villar P I 2013 Nonunitary geometric phases: A qubit coupled to an environment with random noise *Phys. Rev.* A **87**, 032338

[23] Lombardo F C and Villar P I 2015 Correction to the geometric phase by structured environments: The onset of non-Markovian effects *Phys. Rev.* A **91**, 042111

[24] Lombardo F C and Villar P I 2015 Geometric phase and quantum correlations for a bipartite two-level system *Journal of Physics: Conference Series* **626** 012043.

Measurements of entanglement over a kilometric distance to test superluminal models of Quantum Mechanics: preliminary results.

B Cocciaro, S Faetti and L Fronzoni

Department of Physics Enrico Fermi, Largo Pontecorvo 3, I-56127 Pisa, Italy

E-mail: b.cocciaro@comeg.it, sandro.faetti@unipi.it, leone.fronzoni@unipi.it

Abstract. As shown in the *EPR* paper (Einstein, Podolsky e Rosen, 1935), Quantum Mechanics is a non-local Theory. The Bell theorem and the successive experiments ruled out the possibility of explaining quantum correlations using only local hidden variables models. Some authors suggested that quantum correlations could be due to superluminal communications that propagate isotropically with velocity $v_t > c$ in a preferred reference frame. For finite values of v_t and in some special cases, Quantum Mechanics and superluminal models lead to different predictions. So far, no deviations from the predictions of Quantum Mechanics have been detected and only lower bounds for the superluminal velocities v_t have been established. Here we describe a new experiment that increases the maximum detectable superluminal velocities and we give some preliminary results.

1. Introduction

The non local character of Quantum Mechanics (QM) has been object of a great debate starting from the famous Einstein-Podolsky-Rosen (EPR) paper [1]. Consider, for instance, a quantum system made by two photons a and b that are in the polarization entangled state

$$|\psi>= \frac{1}{\sqrt{2}} \left(|H,H> + e^{i\phi}|V,V>\right) \tag{1}$$

where H and V stand for horizontal and vertical polarization, respectively, and ϕ is a constant phase coefficient. The two entangled photons are created at point O, propagate in space far away one from the other (see Fig.1) and reach at the same time points A (Alice) and B (Bob) that are equidistant from O as schematically drawn in Fig.1. Two polarizing filters P_A and P_B lie at points A and B, respectively.

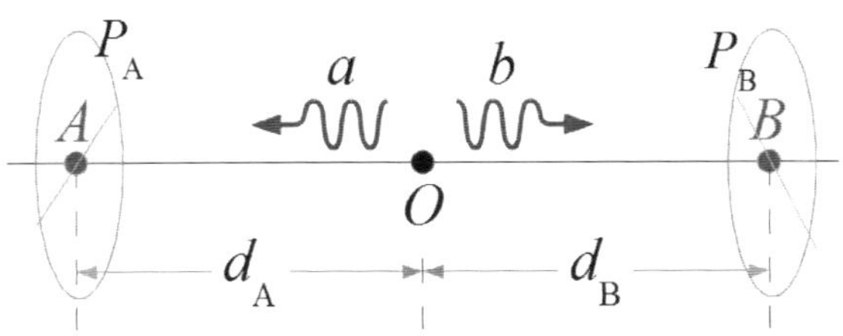

Figure 1: O: source of entangled photons (a and b); A(Alice) and B(Bob): points equidistant from O ($d_A = d_B$); P_A and P_B: Polarizing filters centered at points A and B, respectively.

Suppose, now, that the polarizers axes are aligned along the horizontal direction. According to *QM*, the passage of photon a (or b) through polarizer P_A (or P_B) leads to the collapse of the entangled state to $|H, H>$ everywhere, then, also photon b (or a) collapses to the horizontal polarization. This behaviour suggests the existence of a sort of "action at a distance" between entangled particles in complete disagreement with any other classical physic phenomenon (Electromagnetism , Gravity ....). According to Gisin [2, 3], classical correlations between far events have always due to two possible mechanisms: Common Cause or Communications. The Bell theorem[4] and many successive *EPR* experiments[5–14] demonstrated that correlations cannot be due only to a common cause (*hidden variables theories*) or to common cause + subluminal communications. According to Bell, "*in these EPR experiments there is the suggestion that behind the scenes something is going faster than light*"[15]. Models of *QM* based on the presence of superluminal communications (tachyons) have been proposed [16, 17]. Tachyons are known to lead to causal paradoxes (see, for instance, pages 52-53 in [18]), but no causal paradox arises if tachyons propagate isotropically in a preferred frame (*PF*) with velocity $v_t = \beta_t c$ $(\beta_t > 1)$[19–23].

Suppose, now, that quantum correlations are due to superluminal communications and that an *ideal experiment* is performed in the tachyon preferred frame S' where two polarizing filters lie at the same optical distances $d'_A = d'_B$ from source O. Photons a and b get the polarizers at the same time and no communication is possible. Then, correlations between entangled particles should differ from the predictions of *QM* and should satisfy the Bell inequality. However, from the experimental point of view, equality $d'_A = d'_B$ can be only approximatively verified within a given uncertainty $\Delta d'$. Consequently, photons a and b could get the polarisers at two different times $(\Delta t' = {}^{\Delta d'}/_{c})$ and could communicate if the tachyon velocity exceeds a lower bound $v_{t,min} = c\,(d'_{AB}/\Delta d')$ where d'_{AB} is the distance between polarizers P_A and P_B in the *PF*. Two are the possible experimental results: *i)* a lack of quantum correlations is observed; *ii)* quantum correlations are always satisfied. In the first case (*i)*) one should conclude that quantum correlations are due to exchange of superluminal messages with velocity lower than $v_{t,min}$. In the second case (*ii)*), due to the experimental uncertainty $\Delta d'$, one cannot invalidate the superluminal model of *QM* but can only establish a lower bound $v_{t,min} = c\,(d'_{AB}/\Delta d')$ for the superluminal velocities. It has been recently demonstrated an important theorem [24, 25]: if *QM* correlations are due to superluminal signals with finite velocity v_t, then also a *macroscopic superluminal signalling* becomes possible provided that states of three or four entangled particles are involved. This means that the superluminal signals do not remain hidden but they could lead to macroscopic superluminal communications. In conclusion, there are two possible alternative situations both involving some upheaval of the common thought: *a*) Nature is intrinsically non local and far events can be correlated without any common cause or communication (orthodox *QM*); *b*) Nature is local but, in this case, macroscopic superluminal signalling is possible (superluminal models). Physics is an experimental Science and, thus, we think that only the experiments can decide between these two alternatives.

The correlations between entangled particles can be experimentally tested measuring the number of coincidences $N(\alpha_A, \alpha_B)$ of photons passing through polarizers P_A and P_B for different values of the polarizers angles α_A and α_B with respect to the horizontal axis. In particular, two correlations parameters S_{max} and S_{min} can be measured (see equations (33) and (34) in reference [26]):

$$S_{max} = \frac{N(45°,67.5°) - N(0°,67.5°) - N(45°,112,5°) - N(90°,22.5°)}{N} \tag{2}$$

and

$$S_{min} = \frac{N(135°,202.5°) - N(0°,202.5°) - N(135°,157,5°) - N(90°,67.5°)}{N}, \tag{3}$$

where N is the number of coincidences with no polarizers (tacking into account for the polarizers transmission) that can be written as:

$$N = N(0°,0°)+N(0°,90°)+N(90°,0°)+N(90°,90°) \tag{4}$$

Eqs. (2), (3) and 4 have been obtained from equation (33) in reference [26] using the equalities:

$$\begin{aligned} N(a',\infty) &= N(a',b) + N(a',b+90°) \\ N(\infty,b) &= N(a,b) + N(a+90°,b) \end{aligned} \tag{5}$$

Quantum Mechanics predicts $S_{max} = 0.207$ and $S_{min} =$ - 1.207, respectively, whilst local theories must satisfy the inequalities $S_{max} \leq 0$ and $S_{min} \geq$ -1. Then, the measurement of one of these parameters makes possible a direct test of the superluminal models.

So far we considered an ideal experiment performed in the preferred frame, but the PF is unknown. A more complex EPR experiment can be still performed in the Earth if A and B are aligned along the Est-West axis and are equidistant (in the Earth frame) from the photons source at O. Of course, the entangled photons get simultaneously polarizers P_A and P_B in the Earth reference frame but not in the PF. However, according to Relativity, these events become simultaneous also in the PF if the velocity vector $\vec{V} = \vec{\beta}c$ of the PF is orthogonal to the A-B axis (see, for instance, the appendix in[27]). If the A-B axis coincides with the East-West direction, due to the Earth rotation around its axis, there are always two times t_1 and t_2 during each sidereal day where vector $\vec{V}$ becomes orthogonal to the A-B axis. If the A-B axis makes an angle $\gamma \neq 0$ with the Est-West axis, vector $\vec{V}$ becomes orthogonal to the A-B axis only if angle θ between vector $\vec{V}$ and the Earth polar axis lies in the interval $[\gamma, \pi - \gamma\ ]$. If this condition is satisfied, a loss of Quantum correlations should be observed at two given unknown times t_1 and t_2 each day if the tachyon velocity v_t is lower than the maximum detectable velocity $v_{t,min}$. However, there is an other important feature that can reduce the maximum detectable tachyons velocities in the Earth experiment. In fact, tachyons get simultaneously the polarizers also in the PF only at the two well defined times t_1 and t_2 but the measure of the coincidences numbers $N(\alpha_A, \alpha_B)$ is not instantaneous and requires a finite acquisition time δt. This produces a further uncertainty on the equalization of the optical paths that is an increasing function of the acquisition time and the reduced velocity β of the PF. Using the Relativity theory, it has been shown [28, 29] that the lower limit of the detectable tachyon velocities in a Earth experiment is:

$$\beta_{t,min} = \sqrt{1 + \frac{(1-\beta^2)\left[1-\bar{\rho}^2\right]}{\left[\bar{\rho} + \beta \sin\chi \sin\frac{\pi\delta t}{T}\right]^2}}, \tag{6}$$

where $\bar{\rho} = {}^{\Delta d}/_{d_{AB}}$, Δd is the uncertainty on the equalization of the optical paths in the Earth frame, T is the duration of the sidereal day, δt is the acquisition time, χ is the polar angle between the North-South axis and velocity $\vec{V}$ of the PF and β is the reduced PF velocity ($\beta = {}^{V}/_{c}$). In typical experimental conditions [28–30], the acquisition time δt is much smaller than the sidereal day T and $\beta_{t,min}$ is a decreasing function of both $\bar{\rho}$ and δt that reaches a minimum value if $\chi = {}^{\pi}/_{2}$. $\beta_{t,min}$ is also a decreasing function of β that assumes its maximum value $\beta_{t,min} = \frac{1}{\bar{\rho}}$ for $\beta = 0$ and approaches the minimum value $\beta_{t,min} = 1$ for $\beta \to 1$. Our following considerations and figures will be restricted to $\delta t \ll T$ and to the most unfavourable condition $\chi = {}^{\pi}/_{2}$. The typical plot of function $\beta_{t,min}$ versus the reduced velocity β of the PF for $\chi = {}^{\pi}/_{2}$ and for some values of $\bar{\rho}$ and δt is drawn in Fig.2.

Experiments of this kind have been performed by some groups in the last years[28–30]. In all these experiments no loss of QM correlations has been observed and, thus, only lower bounds $\beta_{t,min}$ for the tachyons reduced velocities have been established. Recently[27] we proposed a new experiment to increase the maximum detectable tachyons velocities by about two orders of

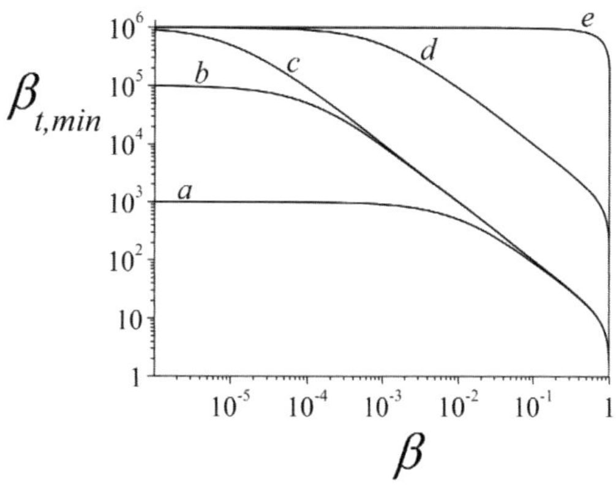

Figure 2: Function $\beta_{t,min}$ versus β for the unfavourable case $\chi = \pi/2$ and for some values of the experimental parameters $\bar{\rho}$ and δt. Curves a, b and c correspond to the fixed acquisition time $\delta t = 10^{-1} \times \frac{T}{\pi}$ and to decreasing values of $\bar{\rho}$ ($a : \bar{\rho} = 10^{-3}$, $b : \bar{\rho} = 10^{-5}$, $c : \bar{\rho} = 10^{-6}$). Curves c, d and e correspond to a fixed value $\bar{\rho} = 10^{-6}$ and to decreasing values of δt ($c : \delta t = 10^{-1} \times \frac{T}{\pi}$, $d : \delta t = 10^{-3} \times \frac{T}{\pi}$, $e : \delta t = 10^{-7} \times \frac{T}{\pi}$).

magnitude. Here we describe our improved experimental apparatus and we report some very preliminary experimental results. The main features of the experiment are discussed in Section 2. The preliminary experimental results are in Section 3, whilst the conclusions are in Section 4.

2. The experimental apparatus and the main sources of error.

2.1. Production and detection of entangled photons.

The main goal of our experiment is to make parameters $\bar{\rho} = \Delta d/d_{AB}$ and δt as smaller as possible to increase the lower bound $\beta_{t,min}$. Small values of $\bar{\rho} = \Delta d/d_{AB}$ ($\bar{\rho} \approx 1.8 \times 10^{-7}$) are obtained using a large distance d_{AB} ($d_{AB} = 1200\,\text{m}$) and a small uncertainty Δd ($\Delta d < 220\,\mu\text{m}$). A high intensity source of entangled photons provides a high coincidences rate (15000 coinc/s) and, thus, a small minimum acquisition time that is estimated to be $\delta t \approx 0.1\,\text{s}$. The experiment is performed in the "East-West" gallery of the European Gravitational Observatory (*EGO*) of Cascina that hosts the *VIRGO* experiment on the detection of gravitational waves[31]. Unfortunately, this gallery makes an angle $\gamma = 18°$ with the actual East-West axis. Then, vector $\vec{V}$ becomes orthogonal to the gallery axis at two times t_1 and t_2 only if angle θ between vector $\vec{V}$ and the Earth polar axis lies in the interval $[\gamma, \pi - \gamma]$. This means that we do not look at the entire celestial sphere but only at a $\approx 95\%$ fraction of it. In fact, the excluded solid angle is $\Omega = 2\int_0^\gamma 2\pi \sin\theta d\theta \approx 5\%$ of the 4π total solid angle.

The experimental apparatus is schematically shown in figure 3. A 220 mW diode laser beam ($\lambda = 406.5\,\text{nm}$) is polarized (polariser P_0) and the polarization axis can be rotated by a motorized $\lambda/2$ plate. All the measurements reported in this paper have been performed with the polarization axis making a 45° angle with the horizontal axis. The beam passes through a Babinet-Soleil compensator and impinges at normal incidence on two thin (0.56 mm) adjacent non-linear optical crystals (*BBO*) cut for type-I phase matching [32]. The beam is focused on the *BBO* plates with the beam waist having a 0.6 mm diameter. The optic axes of the *BBO* plates are tilted at the angle 29.05° and lie in planes perpendicular to each other with the first plane that is horizontal. The pump beam induces down conversion at the wavelength $\lambda = 813\,\text{nm}$ in each crystal[32] with maximum emission at the two symmetric angles $\gamma_A = -\gamma_B = 2.42°$ with respect to the pump laser beam. Suitable optical diaphragms select the entangled beams that are emitted within cones of aperture 0.8° centred at the maximum emission angles. The down converted photons are created in the maximally entangled state $\left(|H, H> + e^{i\phi}|V, V>\right)/\sqrt{2}$, where phase ϕ can be changed moving the motorized Babinet-Soleil compensator. Plates C, C_A and C_B in figure 3 are suitable compensating plates that provide a compensation of spurious effects due to the poor coherence of the pump beam (C) and to the anisotropy of the *BBO*

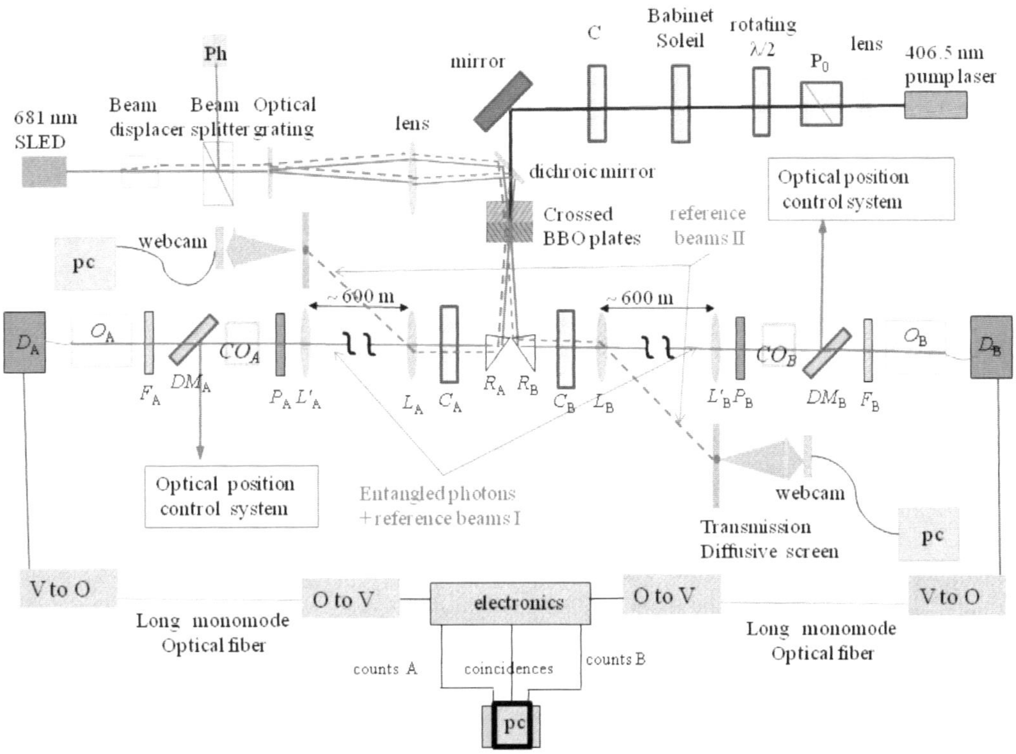

Figure 3: Schematic view of the experimental apparatus. To simplify the drawing some details that will be discussed below have not be inserted in the figure. The pump beam (blue in the figure) is polarized by the polarizer P_0 and the $\lambda/2$ plate. The Babinet-Soleil compensator introduces a variable optical dephasing between the horizontal and vertical polarizations. C , C_A and C_B are anisotropic compensator plates used to increase the fidelity of the entangled state. The pump beam is focused on two adjacent *BBO* plates where the entangled photons are generated and emitted at the angles $\pm 2.42°$ with respect to the pump laser beam. L_A, L_B, L'_A and L'_B are achromatic lenses aligned along the *EGO* gallery and having a 6 m focal length and a 15 cm diameter. Note that the figure is not to scale and, in particular, the distances between lenses L_A and L_B and L'_A and L'_B are very large (≈ 600 m). P_A and P_B are polarizing filters. O_A , O_B , CO_A and CO_B are systems of lenses. DM_A and DM_B are dichroic mirrors, F_A and F_B are optical filters, D_A and D_B are photon counting detectors. The superluminous diode (*SLED*), the beam displacer and the optical grating are used to produce two reference beams in each arm of the *EGO* gallery as discussed below (full and broken red lines). V to O denote electronic systems that transform the output voltage pulses produced by the photon counting detectors into optical pulses, whilst O to V transform the optical pulses into voltage pulses.

plates(C_A and C_B)[33–35]. With these compensating plates we obtain a high intensity source of entangled photons with high fidelity. All the components described above lie on a central optical table that is entirely enclosed in an insulating box. One of the lateral internal walls is made by a 50 cm × 150 cm aluminium plate (5 mm – thickness) in thermal contact with copper tube coils where a paraflu fluid circulates. Two 80 W fans inside the box move the air and homogenize the temperature everywhere. In this way, the internal temperature can be maintained fixed better than ±0.1 °C. Two couples of specially designed achromatic lenses L_A, L_B, L'_A and L'_B (diameter = 15 cm, focal length = 6 m) allow us to obtain two 1:1 images of the entangled photons source on two thin near infrared polarizing films (*LPNIR*, Thorlabs) P_A and P_B that

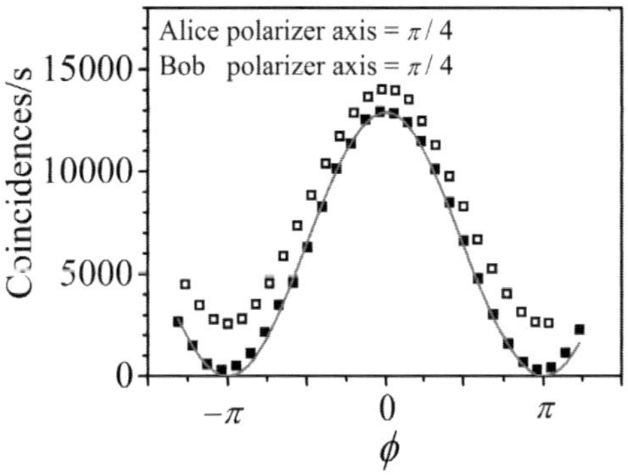

Figure 4: Dependence of the coincidences rate on phase ϕ when all the polarizers angles are fixed at 45 degrees. Full squares correspond to the experimental values with subctracted statistical coincidences. The open squares are the experimental results with subctracted statistical coincidences but whithout using the Kwiatt compensators C_A, C_B and C. The full line correspond to the prediction of Quantum mechanics for the entangled state of eq.(1).

lie at a 600 m distance from the source. The entangled photons pass through polarizers P_A and P_B and through the optical sets CO_A and CO_B that will be described below. Then, they are transmitted (98% transmission) by dichroic mirrors DM_A and DM_B (Chroma T760lpxr) and by two Chroma Techn. Corp. filtering sets F_A and F_B each composed by a bandpass filters ET810/40m ($\lambda = 810\,\text{nm} \pm 20\,\text{nm}$) and two low pass ET765lp filters ($\lambda_c = 765\,\text{nm}$). Finally, two identical optical lenses O_A and O_B focus the entangled photons on two Thorlabs multi mode optical fibres having a large diameter core (200 μm) and high numerical aperture (0.39). The ends of fibres are connected to the inputs of the single photons counters D_A and D_B (Perkin Elmer SPCM-AQ4C) that generate output voltage pulses with a 25 ns width. The voltage pulses are transformed into optical pulses by LCM155EW4932-64 modules of Nortel Networks (V to O module in fig. 3) that propagate in single mode optical fibres up to the central optical table where they are converted into electric pulses (O to V module in fig.3) and sent to an electronic monostable circuit that provides output squared voltage pulses together with coincidences pulses. Before starting the measurements we have measured the light spectral absorption due to air and we have verified that the adsorption in the wavelengths interval [790 nm–830 nm] is essentially due to water vapour. The total adsorbed light in this interval is a fraction lower than 3% of the incident light for a 45% air relative humidity. The coincidence rate measured by counters versus phase ϕ is shown in figure 4. Note the satisfactory contrast of the fringes that is obtained using the Kwiat et al. compensating plates[33–35].

2.2. Compensation of the beam deflections.
An interferometric method is used to equalize the optical paths from the source of the entangled photons to polarizers P_A and P_B. The method exploits two reference beams (beams I in figure 3) of wavelength $\lambda = 681\,\text{nm}$ and coherence length $L_A = 28.1\,\mu\text{m}$ produced by a super luminous diode ($SLED$ in figure 3). Due to the occurrence of vertical temperature gradients up to 3 °C/m in the EGO gallery produced by sunlight, it has been needed to use two couples of different reference beams (beams I and II) in each arm of the interferometer. It can be easily shown that an uniform vertical temperature gradient generates a vertical gradient of the air refractive index that produces the same effect as a diffused optical prism leading to a continuous deviation of the optical beams I and II up to about 1 m at a 600 m distance.The full curve in Figure 5(b) shows the average trajectory of beam I when a vertical temperature gradient occurs. A parabolic shape of the trajectory is predicted if the vertical temperature gradient is everywhere constant in the gallery. Furthermore, the small non uniformity of the vertical gradient of the air refractive index simulates a diffused cylindrical lens that leads to astigmatism of the images.

The accurate compensation of these effects is needed to collect a great number of entangled photons on the photon counting detectors. Beam I follows the same optical path of the entangled

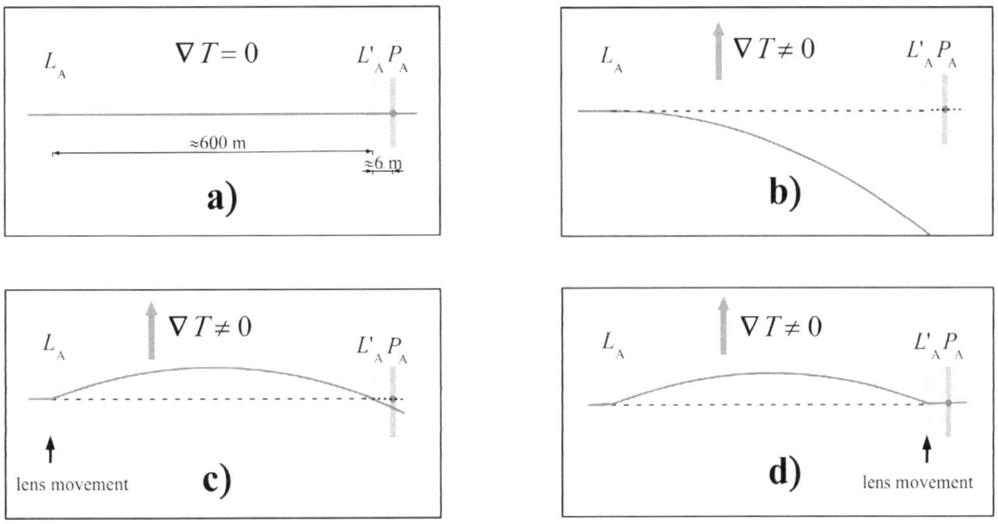

Figure 5: The average trajectory of beam I in an arm (the Alice arm) of the EGO gallery with and without vertical temperature gradient. The figure is not to scale but two characteristic dimensions are given in figure (a). (a): The vertical temperature gradient is not present: the average beam trajectory is a straight horizontal line and the beam impinges at the centre of polarizer P_A. (b) A vertical temperature gradient is present: the reference beam I exhibits a parabolic trajectory. (c) A suitable movement of lens L_A deviate the transmitted beam and leads to a new parabolic trajectory that passes through the centre of lens L'_A. Note that, now, the incidence angle of the beam on lens L'_A is not zero and the transmitted beam does not impinge at the centre of polarizer P_A. (d) A suitable movement of lens L'_A deviate the transmission beam toward the centre of polarizer P_A.

photons and provides an interferometric signal whilst beam II is horizontally displaced with respect to beam I and allows us to compensate the vertical deflections of the beams (up to 1 m at a 600 m distance) produced by the air refractive index gradients. The method to generate the reference beams has been greatly improved with respect to that proposed in [27]. Here we obtain the reference beams I and II using the beam displacer (Thorlabs BDY12U) in figure 3 to split the incident $SLED$ beam into two parallel beams at a 1.2 mm horizontal distance. The two beams pass through a beam splitter and are focused on a transmission phase grating that produce +1 and -1 order diffracted beams with 35% intensity with respect to the incident beam and at the average diffraction angles $+2.43°$ and $-2.43°$ that are virtually coincident with the maximum emission angles of the entangled photons ($\pm 2.42°$). The beam waists of the two reference beams spots on the optical grating have a 0.3 mm diameter and behave as two sources localized on the grating at a 1.2 mm horizontal distance. The optical rays emitted by these sources at the angles $+2.43°$ and $-2.43°$ pass through an achromatic lens having a 150 mm focal length, are reflected by a 565 nm short pass dichroic mirror (Chroma T565spxe) and produce 1:1 images of the grating spots on the BBO plates. Using a suitable optical method, the image of the reference source I on the BBO plates is centred with respect to the spot of the pump beam where the entangled photons are generated. The procedure above ensures that the reference beams I outgoing from the BBO plates are initially in phase and are superimposed to the entangled photons. This provides the easy alignment of the optical apparatus and the control of the optical paths of the entangled photons. Achromatic lenses L_A, L_B, L'_A and L'_B have been

IOP Conf. Series: Journal of Physics: Conf. Series **880** (2017) 012036 doi:10.1088/1742-6596/880/1/012036

built to have the same 6000 mm focal length (within ±10 mm) for the pump laser and for the *SLED*. 1:1 images of the reference source I and of the entangled photons source are produced by lenses L_A, L_B, L'_A and L'_B on polarizers P_A and P_B at a 600 m distance from the source. Reference beams II are horizontally deflected by lenses L_A and L_B and produce two spots at a horizontal distance of 12 cm from the centres of lenses L'_A and L'_B on two diffusing screens horizontally adjacent to the lenses (see figure 3). Two optical objectives collect the diffused beams and produce images of the spots on two webcams. All lenses L_A, L_B, L'_A and L'_B can be moved horizontally and vertically using Sigma Koki *PC* controlled motors. A labview program measures the position of the beams spots on the webcams and produces feedback signals that move lenses L_A and L_B to maintain the spots positions fixed. In such a way also the reference beams I (and the entangled photons) always remain fixed at the centre of lenses L'_A and L'_B (see Figure 5(c)). A 1 cm movement of lenses L_A and L_B produces a 1 m displacement of the beams spots at a 600 m distance. The feedback procedure leads to a complete control of the slow drifts of the beams but it cannot eliminate the rapid changes of the beam trajectories occurring within a few seconds time. This leads to a residual fluctuation of the beam spots at the centres of lenses L'_A and L'_B that is lower than ±30 mm. These residual displacements are appreciably smaller than the radius of the lenses (75 mm), then all the entangled photons are collected by them. The beams impinge at the centres of lenses L'_A and L'_B but the incidence angles change with time, due to the vertical refractive index gradient. Then, the images of the source of beams I (and of the entangled photons) that occur on polarizers P_A and P_B do not remain fixed at the centre of the polarizers (see figure 5c)). To stabilize these images at the centre of the polarizers, the reference beams I transmitted by the polarizers pass through two systems of cylindrical lenses CO_A and CO_B, are reflected by the 760 nm long pass dichroic mirrors DM_A and DM_B (Chroma T760lpxr) and impinge on two optical position control systems. Each optical position control system produces two 1:1 images of the beam spots occurring on the polarizers: one image is collected by a position sensing detector (Thorlabs PDP90A) and the other by a webcam. Labview feedback programs read the output of the position sensing detectors and move lenses L'_A and L'_B to maintain fixed the position of the beam spots at the centre of the two polarizers (see Figure 5(d)). In this case, too, slow drifts are completely removed but not the rapid displacements of the spots from the polarizers centres. These latter residual fluctuations remain always restricted below ±0.4 mm. Other labview feedback programs acquire the images of the webcams and measure the astigmatism of the images induced by non uniformities of the refractive index gradients. Each system of cylindrical lenses CO_A and CO_B is composed by a fixed cylindrical lens and a movable cylindrical lens that provide an effective cylindrical lens with a variable focal length. Suitable feedback signals generated by the labview program move the motorized cylindrical lenses to correct the astigmatism of the images. These procedures ensure that the spots of the reference beams I remain virtually fixed at the centre of polarizers P_A and P_B with a circular shape having a $\approx$ 0.3 mm diameter. Two images of the spot of the reference beam on polarizer P_A are shown in figures 6a) and 6b). Figure 6a) shows the spot for moderate sunlight with feedback *OFF*, whilst figure 6b) shows the same image with feedback *ON*. It must be remarked that our method stabilizes the spots of the 681 nm reference beams but the wavelengths of the entangled photons (790 nm – 830 nm) are different from those of the sled beam. However, the differences of the air refractive indices corresponding to the reference beams and to the entangled photons are very small and it can be shown that also the spots of the entangled photons on the polarizers always remain very close to the centres of the polarizers within 0.4 mm. In conclusion, our compensation procedure maintains the spot of the entangled photons restricted to a circular region close to the centre of the polarizers with a small diameter ($\approx$ 0.6 mm) and ensures that virtually all the entangled photons passing through the polarizers are collected by the photon counting detectors also in conditions of great sunlight.

a) **b)**

Figure 6: (a): The beam spot of the reference beam I on polarizer P_A in conditions of moderate sunlight and with feedback *OFF*. (b): The same spot with feedback *ON*. The diameter of the circular spot in 6b) is 0.3 mm.

2.3. Equalization of the optical paths.

To equalize the optical paths we exploit the reflections of the reference beams I from polarizers P_A and P_B. The reflected beams come back on the same path forward and impinge at the angles $+2.43°$ and $-2.43°$ on the optical phase grating where diffraction occurs again. The output beams that are diffracted orthogonally to the grating are reflected by the beam splitter and impinge on photodetector *Ph* where interference occurs. Air density fluctuations inside the *EGO* gallery induce oscillations of the optical path difference and, thus, an oscillating output voltage of the photodetector. The variations of the optical path differences are always greater than the optical wavelength and the output voltage oscillates from a minimum value (destructive interference) toward a maximum value (constructive interference). The peak to peak amplitude V_{pp} is measured by a simple electronic circuit. V_{pp} is maximized when the path difference is zero whilst V_{pp} tends to vanish if the path difference becomes greater than the *SLED* coherence length $L_c = 28.1\,\mu$m. Polarizer P_B is moved by a precision linear motorized stage (Physik Instruments M-406.22s) that is controlled by a *PC* through a labview program that generates a sweep of the P_B position and acquires the corresponding V_{pp} values. The typical dependence of V_{pp} on the polarizer position x during a summer night is shown in figure 7 (a) whilst the dependence during a summer day at the maximum sunlight is shown in figure 7(b). Note that the curve in figure 7(b) has a two bells profile. This behaviour can be explained assuming that the path difference oscillates with time around the average value Δd_{av} with a mean oscillation amplitude A. In these conditions, the typical Gaussian behaviour due to the finite coherence length of the *SLED* is expected to split into two nearly Gaussian profiles at distance $2A$. The central point between the two Gaussian peaks corresponds to the position of polarizer P_B where the average optical path difference is zero whilst the semi-distance between the two Gaussian Maxima corresponds to the average amplitude A of the fluctuations of the paths difference. The Full lines in figures 7(a) and 7(b) are the labview best fits of the experimental results with two Gaussians having the width $w = 0.020$ mm that characterizes the *SLED* source. From these best fits we deduce that the main amplitude of the oscillations of the path difference is smaller than $10\,\mu m$ during night but it becomes $33\,\mu m$ at the maximum sunlight.

The labview feedback program operates in this way: first of all a large amplitude sweep is made to localize the central point x_c between the two Gaussian, then the sweep amplitude is reduced to $200\,\mu m$ around x_c and the new x_c value is memorized and plotted. This latter procedure with a $200\,\mu m$ sweep is repeated continuously each 15 s for the entire measurement time (24 hours) and, thus, the difference between the optical paths remains restricted to $\Delta d = \pm 100\,\mu$m at each time. The time-dependence of x_c due to the temperature variations for an entire Summer day is shown in figure 8.

Note that the complete interference pattern is somewhat more complex than the small portion

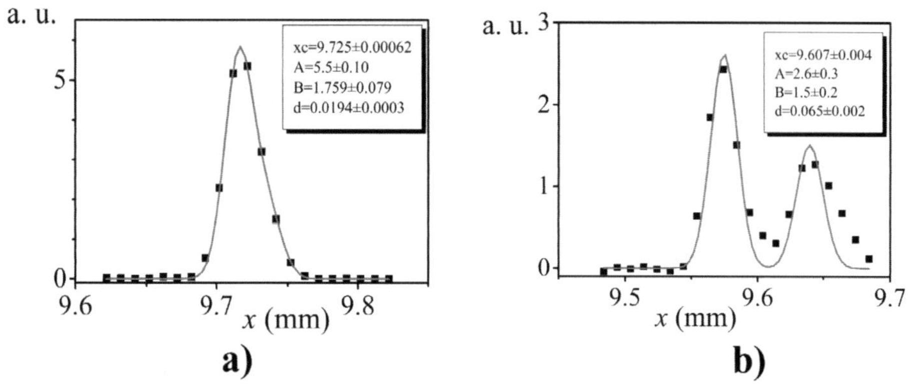

Figure 7: (a): The peak to peak of the output voltage of photodetector *Ph* during a Summer night versus the position x of the polarizer. Similar results are obtained in conditions of covered sky. The full line is the best fit with two slightly shiften Gaussian profiles $y = A\exp\left\{[(x - x_c - d/2)/w]^2\right\} + B\exp\left\{[(x - x_c + d/2)/w]^2\right\}$ where w is fixed to the value $w = 0.020$ mm that characterizes the *SLED* source and A, B,x_c and d are fitting parameters. x_c and d are measured in millimeters whilst A and B are in arbitrary unities. (b): The same dependence but in conditions of maximum sunligth. The two fitting Gaussians show a much greater separation due to the much greater amplitude of the optical paths fluctuations and the quality of the fit is poor due to noise effects induced by the strong air tourbolence.

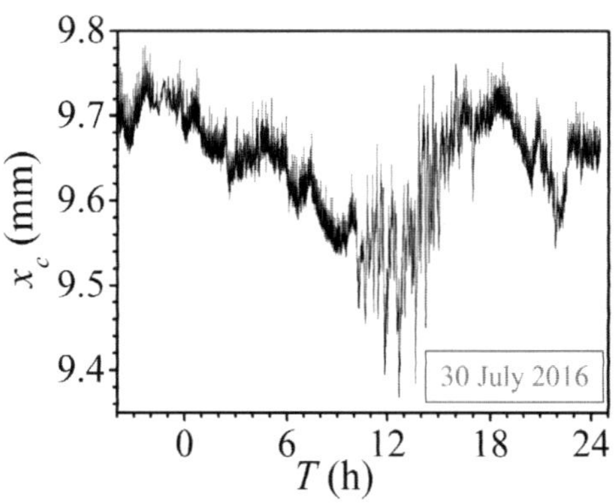

Figure 8: Variation of the equalization position x_c of the motorized polarizer during an entire Summer day. The maximum variation during the entire day is about 0.4 mm. Note the sharp and high amplitude variations due to the air turbulence during the maximum sunlight hours (from 10 h to 16 h).

shown in figures 7 since the *LPNIR* Thorlabs polarizers are made by a thin polarizing film (280 μm thickness) sandwiched between two glass plates. We have measured with accuracy the thickness of the glasses and we have found that their values are $904\,\mu\text{m} \pm 5\,\mu\text{m}$. Due to the sandwich shape of the polarizers, there are many interfaces giving reflected beams that can interfere. Also in the optimal night conditions we observe five main interference peaks that are separated each from the other by a distance about 1.5 mm. After a careful analysis, the central peak has been clearly identified as that which corresponds to the equalization of optical paths from the source to the polarizing thin layers. Then, our analysis has been restricted only to the central peak region corresponding to figures 7.

As shown in Section 1, one of the most important parameters of our experiment is the uncertainty Δd on the equalization of the optical paths. Here we resume the main contributions to this uncertainty (see also[27]) :

a) the above described uncertainty due to the motor sweep $\Delta d_{sweep} = 100\,\mu$m that

IOP Conf. Series: Journal of Physics: Conf. Series **880** (2017) 012036 doi:10.1088/1742-6596/880/1/012036

corresponds to a half of the total sweep excursion.

b) the uncertainty Δd_{pol} due to the finite thickness (280 μm) of the *LPNIR* Thorlabs polarizers layers. Since the extinction ratio of these polarizers at the entangled photons wavelengths is greater than 10^5, then we can estimate that 99% of photons with orthogonal polarization are adsorbed in a layer having a thickness $\approx$ 120 μm and we can assume the corresponding uncertainty value $\Delta d_{pol} = 120\,\mu$m.

c) The optical paths are equalized using the reference beams I that have not the same wavelength of the entangled photons. If the temperature would be uniform in the *EGO* gallery and the thickness of lenses L_A, L_B, L'_A and L'_B would be the same, also the entangled photons paths would be automatically equalized. This is no more true if there is a average temperature difference ΔT between the two arms of the interferometer or if there are differences between the thicknesses of the lenses. Calculations of these effects are somewhat complicate and need the knowledge of the temperature dependence of the refractive indices of the air and of the lenses at different wavelengths and the knowledge of the thickness differences between lenses L_A, L_B, L'_A and L'_B. The lenses thickness differences have been measured to be smaller than 0.1 mm and they do not affect appreciably the uncertainty. It results from the calculations that an average temperature difference $\Delta T = 1$°C between the two arms of the interferometer produces an optical paths difference slightly smaller than 10 μm. Since the horizontal temperature differences in the gallery are always smaller than 2-3 degrees, we get $\Delta d_{\Delta T} < 30\,\mu$m.

d) In our experiment we detect entangled photons with wavelengths from 790 nm toward 830 nm. Due to the optical dispersion, photons of different wavelengths see different optical paths in air and in the lenses, although this latter contribution is negligible. The difference of the optical paths due to the air optical dispertion is given by $\Delta d_{disp} = \frac{\partial n}{\partial \lambda}\Delta\lambda d$, where d is the distance from the source to polarizers (600 m) and $\Delta\lambda = 40$ nm is the bandwidth of the bandpass filters and n is the air refractive index. Substituting in the expression of Δd_{disp} the value $\frac{\partial n}{\partial \lambda} = 5.87 \times 10^{-9} nm^{-1}$ calculated using the Ciddor equation [36] at room conditions and with humidity = 50% and CO_2 = 450 micromol/mol, we get $\Delta d_{disp} = 144\ \mu m$. The resulting uncertainty in the optical paths differences is, then

$$\Delta d = \sqrt{\Delta d^2_{sweep} + \Delta d^2_{pol} + \Delta d^2_{\Delta T} + \Delta d^2_{disp}} = 215\,\mu m. \tag{7}$$

3. Preliminary experimental results.

In this Section we report some very preliminary experimental results concerning the *EPR* measurements. The main objective of these measurements is to verify that the experimental method provides accurate measurements of the *EPR* correlations with very small acquisition times of the coincidences. As shown in the Introduction, the presence of superluminal communications can be detected looking at the time dependence of the two correlation parameters S_{max} and S_{min}. Counts N_A and N_B of photons transmitted by polarizers P_A and P_B are detected and back-ground counts due to unwanted external light and to dark noise of the detectors are subtracted. Furthermore, the statistic spurious coincidences $N_{sp} = N_A N_B \delta' t$ are subtracted from the measured coincidences, where $\delta' t$ denotes the pulse duration $\delta' t = 25$ ns of the output pulses generated by the photon counting modules. Numbers $N(\alpha_A, \alpha_B)$ of the measured coincidences that appear in the expressions of the correlation parameters S_{max} and S_{min} are affected by the time-variations of the transmission coefficients in the two arms. The main causes of a variation of the transmission coefficients are: the occurrence of residual displacements of the transmitted beams that are not completely eliminated by the feedback method that reduce the collection efficiency of the entangled photons; the variation of the air relative humidity that induces a change of the light adsorption. In fact, counts $N_A(\alpha_A)$ and $N_B(\alpha_B)$ and coincidences $N(\alpha_A, \alpha_B)$ are related to the transmission coefficients according to

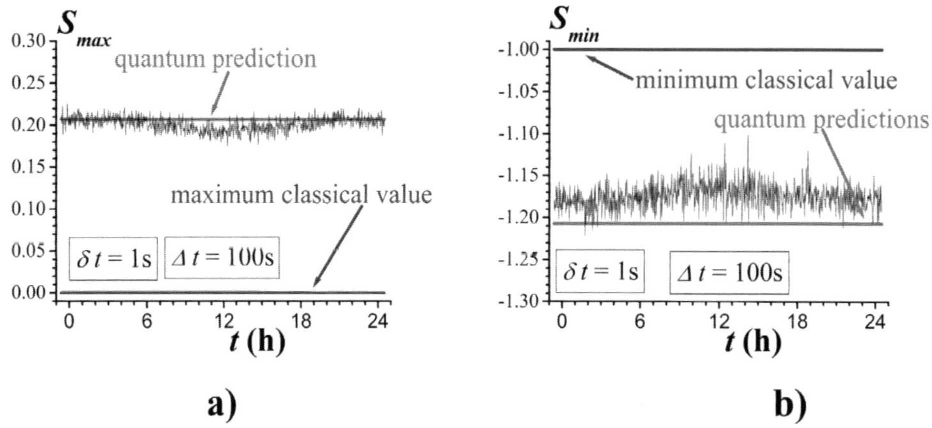

Figure 9: (a): Correlation parameter S_{max} versus the sidereal time. The full red line indicates the dependence predicted by Quantum Mechanics, whilst the blue lower line represents the maximum value allowed by local models of *QM*. (b): The same dependence but for the other parameter S_{min}. Note that parameter S_{min} is much more noisy because it is more sensitive to fluctuations of the coincidences numbers. Note the presence of a systematic effect of the sunlight in the interval between 9 h and 18 h.

relations (8):

$$\begin{aligned} N_A(\alpha_A) &= N\tau_A(\alpha_A)\epsilon_A(\alpha_A)p_A(\alpha_A) \\ N_B(\alpha_A) &= N\tau_B(\alpha_B)\epsilon_B(\alpha_B)p_B(\alpha_B) \\ N(\alpha_A,\alpha_B) &= N\tau_A(\alpha_A)\epsilon_A(\alpha_A)\tau_B(\alpha_B)\epsilon_B(\alpha_B)p(\alpha_A,\alpha_B) \end{aligned} \tag{8}$$

where N is the number of generated entangled photons, $\tau_A(\alpha_A)$ and $\tau_B(\alpha_B)$ are the transmission coefficients, $\varepsilon_A(\alpha_A)$ and $\varepsilon_B(\alpha_B)$ are the efficiencies of photon counting detectors, $p_A(\alpha_A)$ and $p_B(\alpha_B)$ are the probabilities that photons a and b pass through polarizers P_A and P_B (they are $p_A(\alpha_A) = p_B(\alpha_B) = 1/2$ for the entangled state) and $p(\alpha_A,\alpha_B)$ is the joint probability. We see that dividing the coincidences counts $N(\alpha_A,\alpha_B)$ for the product $N_A(\alpha_A)N_B(\alpha_B)$ and multiplying for the product of the average values of $\bar{N}_A(\alpha_A)$ and $\bar{N}_A(\alpha_A)$ one obtains a coincidences number that is no more affected by changes of the transmission coefficients and of the photodetectors efficiencies. Here below we will indicate by $N(\alpha_A,\alpha_B)$ the coincidences corrected according to the procedure outlined above. The correlations parameters S_{max} and S_{min} are obtained repeating measurements of coincidences $N(\alpha_A,\alpha_B)$ with the proper values of angles α_A and α_B that appear in equations (2) and (3). According to equations (2) , (3) and (4), 12 different couples of values α_A and α_B have to be selected rotating the motorized polarizers P_A and P_B. The rotations of polarizers P_A and P_B are controlled by a *PC* through a labview program that operates in this way: a couple of angles α_A and α_B is set (for instance the first angles 45° and 67.5° of the first contribution in equation (2)), then the polarizers axes are rotated until they reach the setted angles. The corresponding numbers of coincidences $N(\alpha_A,\alpha_B)$ in the acquisition time δt are measured. Then, angles are changed according to equations (2), (3) and (4) and the corresponding values of the coincidences are measured. When all the 12 values of coincidences that are needed to calculate parameters S_{max} and S_{min} have been measured, the labview program calculates these correlation parameters. Unfortunately, the average time that is needed to rotate the polarizers is of the order of 8 seconds and, thus, the duration of a single measurement of S_{max} and S_{min} requires a time $\Delta t \approx 100\,\mathrm{s}$ that is much larger than the acquisition time δt of coincidences. Then, the maximum superluminal velocity $\beta_{t,min}$ that can be detected in the present experiment is not limited by the acquisition time δt of the coincidences but by the much larger effective acquisition time $\Delta t = 100$ s. Figures 9a) and 9b) show the

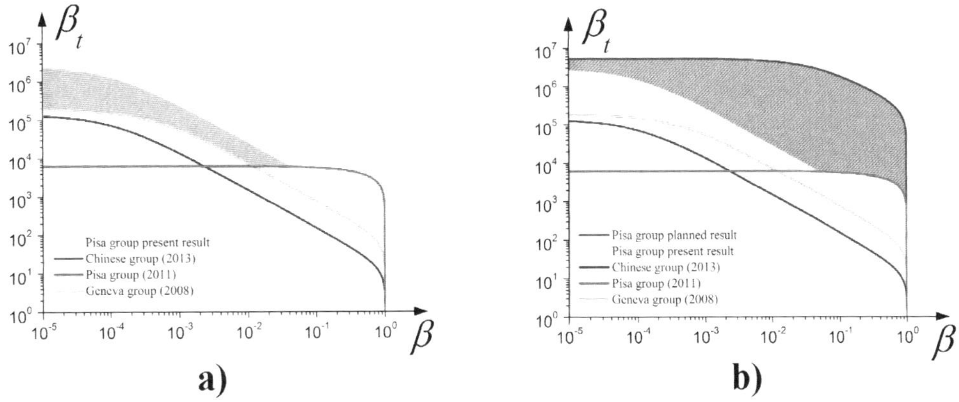

Figure 10: (a): The reduced superluminal velocity β_t versus the reduced velocity β of the preferred frame. The curves in the figure represent the lower bounds $\beta_{t,min}$ that have been established in the previous experiments and in the present one. The filled region represents the new region of superluminal velocities that are investigated here. (b) The filled region represents the new region of superluminal velocities that should be explored in the final experiment with the acquisition time $\delta t = 0.1$ s.

values of the correlation parameters S_{max} and S_{min} versus time during a sidereal day when the coincidences acquisition time was $\delta t = 1$ s but the effective acquisition time was $\Delta t = 100$ s. The full horizontal lines in figures 9a) e and 9b) correspond to the prediction of QM and to the maximum (figure 9a)) and minimum (figure 9b)) values allowed by local theories. According to the Introduction, parameter S_{max} would become lower than 0 and S_{min} would become greater than - 1 at two times each sidereal day if the superluminal signals have velocities lower than $\beta_{t,min}$. This behaviour is not observable in figures 9a) and 9b) and, thus, we can conclude that, if superluminal signals are responsible for QM correlations, then the superluminal velocities are greater than the maximum measurable values ($\beta_t > \beta_{t,min}$). The results in figures 9a) and 9b) were obtained using the coincidences acquisition time $\delta t = 1$ s but we have verified that sufficiently accurate results are also obtained using the much smaller acquisition time $\delta t = 0.1$ s where the relative statistical noise increases by a factor $\sqrt{10}$. Notice that parameter S_{min} exhibits much greater fluctuations than S_{max} and, thus, this latter parameter provides a much more accurate test of the EPR correlations. This behaviour is probably due to the fact that the absolute value of S_{max} is about six times lower than that of S_{min}. For this reason the planned final measurements will be made using the S_{max} parameter alone that is affected by a much smaller noise. Furthermore, it is important to remark that the measurements shown in figures 9a) and 9b) were obtained in a July day (30 July 2016) with very strong sunlight. The residual noisy effects due to sunlight are evident looking at the experimental points between times $t =$ 9 h and $t = 18$ h in the figures. All these effects are absent in conditions of fully covered sky.

Substituting the effective acquisition time $\Delta t = 100$ s in place of δt in equation (6) with the uncertainty $\Delta d = 215\,\mu$m, we obtain the lower bound $\beta_{t,min}$ that corresponds to our preliminary results. In figure 10 a) we show the lower bounds already found in some previous experiments [28–30] together with that obtained here. The filled region represents the new region of superluminal velocities investigated here. In figure 10 b) we show also the planned values of $\beta_{t,min}$ that should be obtained in our final experiment with a 0.1 s effective acquisition time. The filled region in figure 10 b) corresponds to the new region of superluminal velocities that will become accessible in the final experiment. The experimental method that will be used to bypass the problems related to the polarizers movement will be briefly outlined in the Conclusions below.

4. Conclusions

In the present paper we have developed an accurate and stable method to equalize the optical paths of the entangled photons over a kilometric distance. Due to vertical gradients of the air refractive index in the *EGO* gallery induced by sunlight it has been needed to greatly modify the experimental apparatus proposed in[27] introducing a complex feedback procedure to correct the deviations of the beams and the astigmatism of the images. In such a way we were able to obtain two virtually stable 1:1 images of the source of entangled photons at the centre of two polarizers lying at a distance of ≈ 600 m from the source. This ensures that virtually all the entangled photons transmitted by the polarizers are collected by the photon counting detectors also in the unfavourable conditions of maximum sunlight. The interference method used to equalize the optical paths exploits two reference beams reflected by the polarizers. The reference beams follow the average paths of the entangled photons. The method to produce the reference beam has been greatly improved with respect to the original project[27] thanks to the use of a suitable optical grating. The new method automatically ensures that the reference beams are superimposed to the entangled ones and that they have the same phase at the *BBO* plates without using the complex equalization procedure outlined in our original paper[27]. Finally, a suitable design of the optical components and the use of the compensation procedure developed by the Kwiat group[33–35] provides a great number of measured coincidences and makes possible to use a very short acquisition time $\delta t = 0.1$ s. Using this experimental apparatus we have continuously measured the correlation parameters S_{max} and S_{min} for an entire sidereal day to obtain some very preliminary results. Our experimental results are greatly affected by the long time that is needed to rotate the polarizers that leads to an effective acquisition time $\Delta t = 100$ s much greater than the minimum acquisition time of coincidences $\delta t = 0.1$ s. For this reason, the new explored region of the velocities of the superluminal signals investigated here (see figure 10 a)) is much smaller than the planned one (see figure 10 b)). Due to the strong vertical temperature gradients it has not been possible to perform the experiment along the true East-West direction that was proposed in our previous paper. In the present experiment the measures have been performed in the so called "East-West" gallery of *EGO* that makes the angle $\gamma = 18°$ with the actual East-West axis. Then, only a 95% portion of the celestial sphere is accessible with our experimental apparatus.

In order to become insensitive to the long time needed to rotate the polarizers and to reach an effective acquisition time $\delta t = 0.1$ s, it is needed to fully change the acquisition method. With this improved method we should obtain the planned results in figure 10 b). Here we describe only the main idea of the new method. Using a NTP+PTP GPS Network Time Server (TM2000A) and knowing the values of the difference *UTC-UT1* provided in the Web by *IERS* [37] we will be able to synchronize the measurements of the coincidences with the Earth rotation angle with respect to the fixed stars within a few milliseconds uncertainty. That accurate synchronization of the measurements cannot be obtained using a *PC* but requires the use of a real time acquisition. This will be obtained using a *Compact DAQ* (National Instruments 9132) in place of the *PC* to acquire the coincidences and a Real Time Labview program to control any aspect of the acquisition. In this way, will be possible to acquire the coincidences in successive days at the same Earth rotation times. The measurement method will follow the steps below: i) the real time labview program rotates polarizers to reach the angles α_A and α_B that correspond to the first contribution $N(45°, 67.5°)$ in the expression of S_{max} in equation (2). ii) When the Earth rotation angle reaches a well defined value, the acquisition of coincidences starts and 2^{20} values of coincidences are acquired in a full Earth rotation day. The successive day, the real time labview program sets the polarizers angles to the values corresponding to the second term in the expression of S_{max} and the acquisition of coincidences will start with sidereal synchronism with the measurements of the previous day. The same procedure will be repeated until all the contributions that are present in the expression of S_{max} have been obtained. With this procedure, the effective acquisition time coincides with the coincidences acquisition time δt and the planned region of superluminal velocities (filled region in figure 10 b)) should become accessible. If the revolution motion of the Earth around the Sun and the precession and nutation of the Earth axis would be absent, the losses of quantum correlations should occur exactly at the same earth rotation angles each sidereal day and, thus, one could utilize the coincidences $N(\alpha_A, \alpha_B)$ measured in successive days at the same earth rotation angles to calculate the time dependence of the correlation parameter S_{max} using equations (2) and (4). Unfortunately, the analysis of the experimental data will be much more complex due to the revolution motion of the earth around the Sun and to the precession and nutation motions. In fact, the losses of quantum correlations should occur when the relative velocity $\vec{V}$ of the preferred frame with respect to the Earth frame becomes orthogonal to the A-B axis. Due to the revolution motion of the earth around the Sun and to the precession and nutation motions, the angle between the relative velocity $\vec{V}$ of the preferred frame and the A-B axis is not a true periodic function having the period of the Earth rotation and, thus, the orthogonality condition will be not exactly satisfied at the same Earth rotation angles in different days. Then, the analysis of the experimental results will need more complex procedures that will be not discussed here.

Acknowledgements
We acknowledge Marco Bianucci for the realization of many electronic devices and for a great number of helpful and fundamental suggestions. We also thank the European Gravitational Observatory of Cascina (Italy) that host our experiment and, in particular, the director Federico Ferrini and Franco Carbognani and Stefano Cortese for their valuable support and for their kindness. We also thank the Istituto di tecnologie della comunicazione, dell'informazione e della percezione (S. Anna) for giving us two LCM1555EW4932-64 0f Nortel Networks. This work was supported by La Fondazione Pisa.

References
[1] Einstein A, Podolsky B and Rosen N 1935 *Phys. Rev.* **47**(10) 777–780
[2] Gisin N 2014 *Quantum Correlations in Newtonian Space and Time: Faster than Light Communication or Nonlocality* (Milano: Springer Milan) pp 185–203 URL `http://dx.doi.org/10.1007/978-88-470-5217-8_12`
[3] Gisin N 2014 *Quantum Chance - Nonlocality, Teleportation and Other Quantum Marvels* (Springer International Publishing)
[4] Bell J S 1964 *Physics* **1** 195–200
[5] Freedman S J and Clauser J F 1972 *Phys. Rev. Lett.* **28**(14) 938–941
[6] Aspect A, Dalibard J and Roger G 1982 *Phys. Rev. Lett.* **49**(25) 1804–1807
[7] Zeilinger A 1986 *Phys. Lett.* A **118** 1–2
[8] Tittel W, Brendel J, Zbinden H and Gisin N 1998 *Phys. Rev. Lett.* **81**(17) 3563–3566
[9] Weihs G, Jennewein T, Simon C, Weinfurter H and Zeilinger A 1998 *Phys. Rev. Lett.* **81**(23) 5039–5043
[10] Aspect A 1999 *Nature* **398** 189–190
[11] Pan J, Bouwmeester D, Daniell M, Weinfurter H and Zeilinger A 2000 *Nature* **403** 515–519
[12] Grangier P 2001 *Nature* **409** 774–775

[13] Rowe M A, Kielpinski D, Meyer V, Sackett C A, Itano W M, Monroe C and Wineland D J 2001 *Nature* **409** 791–794

[14] Matsukevich D N, Maunz P, Moehring D L, Olmschenk S and Monroe C 2008 *Phys. Rev. Lett.* **100**(15) 150404

[15] Davies P C W and Brown J R 1993 *The ghost in the atom: a discussion of the mysteries of quantum physics* canto ed. ed (Cambridge; New York: Cambridge University Press)

[16] Eberhard P H 1989 A realistic model for quantum theory with a locality property *Quantum theory and pictures of reality: foundations, interpretations, and new aspects* ed Schommers W (Berlin; New York: Springer-Verlag)

[17] Bohm D and Hiley B J 1991 *The undivided universe: an ontological interpretation of quantum mechanics* (Routledge)

[18] Møller C 1955 *The theory of relativity* (Oxford: Clarendon)

[19] Kowalczyński J 1984 *Int. J. Theor. Phys.* **23**(1) 27–60

[20] Reuse F 1984 *Ann. Phys.* **154** 161–210

[21] Caban P and Rembieliński J 1999 *Phys. Rev.* A **59**(6) 4187–4196

[22] Maudlin T 2001 *Quantum non-locality & relativity* (Oxford: Blackwell)

[23] Cocciaro B 2013 *Physics Essays* **26** 531–547

[24] Bancal J, Pironio S, Acín A, Liang Y, Scarani V and Gisin N 2012 *Nat. Phys.* **8** 867–870

[25] Barnea T J, Bancal J D, Liang Y C and Gisin N 2013 *Phys. Rev. A* **88**(2) 022123

[26] Aspect A 2002 Bell's theorem: The naive view of an experimentalist *Quantum (Un)speakables: From Bell to Quantum Information* ed Bertlmann R A and Zeilinger A (Springer)

[27] Cocciaro B, Faetti S and Fronzoni L 2013 *Journal of Physics: Conference Series* **442** 012005

[28] Salart D, Baas A, Branciard C, Gisin N and Zbinden H 2008 *Nature* **454** 861–864

[29] Cocciaro B, Faetti S and Fronzoni L 2011 *Phys. Lett.* A **375** 379–384

[30] Yin J, Cao Y, Yong H L, Ren J G, Liang H, Liao S K, Zhou F, Liu C, Wu Y P, Pan G S, Li L, Liu N L, Zhang Q, Peng C Z and Pan J W 2013 *Phys. Rev. Lett.* **110**(26) 260407

[31] The Virgo collaboration 1997 *Final Design Report* (Virgo Technical Note VIR-TRE-DIR-1000-13)

[32] Kwiat P G, Waks E, White A G, Appelbaum I and Eberhard P H 1999 *Phys. Rev. A* **60**(2) R773–R776

[33] Altepeter J, Jeffrey E and Kwiat P 2005 *Opt. Express* **13** 8951–8959

[34] Akselrod G M, Altepeter J B, Jeffrey E R and Kwiat P G 2007 *Opt. Express* **15** 5260–5261

[35] Rangarajan R, Goggin M and Kwiat P 2009 *Opt. Express* **17** 18920–18933

[36] Ciddor Equation URL `http://emtoolbox.nist.gov/wavelength/ciddor.asp`

[37] International Earth Rotation and Reference Systems Service URL `https://www.iers.org/IERS/EN/Publications/Bulletins/bulletins.html`

IOP Conf. Series: Journal of Physics: Conf. Series **880** (2017) 012037 doi:10.1088/1742-6596/880/1/012037

Quantum analysis and thermodynamic operator relations in stochastic energetics

Tomoi Koide

Instituto de Física, Universidade Federal do Rio de Janeiro, C.P. 68528, 21941-972, Rio de Janeiro, Brazil

E-mail: `tomoikoide@gmail.com`, `koide@if.ufrj.br`

Abstract. We introduce a model of the quantum Brownian motion coupled to a classical heat bath by using the operator differential proposed in the quantum analysis. We then define the heat operator by adapting the idea of the stochastic energetics. The introduced operator satisfies the relations which are analogous to the first and second laws of thermodynamics.

1. Introduction

The accelerating development in the nanotechnology enables us to access individual thermal random processes at microscopic scales. We however cannot directly apply thermodynamics to analyze the behaviors of such a system because the typical scale of the systems is very small and the effect of thermal fluctuations is not negligible. There is no established theory to describe general fluctuating systems thermodynamically. On the other hand, such a system is often modeled as a Brownian motion [1] and then the behaviors can be interpreted thermodynamically by using the stochastic energetics [2].

In this theory, energy, work and heat are represented by the variables of the Brownian particles, and we can show that the energy balance is satisfied and the expectation value of the heat flux has an upper bound. The former corresponds to the first law and the latter the second law in thermodynamics, respectively [2]. The prediction of the stochastic energetics is experimentally confirmed by analyzing extracted works from a microscopic heat engine [3]. Although this theory is generalized to relativistic systems [4] and the Poisson noise [5], the applications are still limited to classical systems [6].

In this work, we discuss the application of the idea of the stochastic energetics to quantum systems. For this purpose, we introduce the operator differential proposed in the quantum analysis [7, 8, 9]. The quantum analysis was proposed to expand the functions of operators systematically and has been applied to various problems in quantum mechanics and quantum statistical mechanics. For example, the expansion of the S matrix, the Baker-Campbell-Hausdorff formula, and the linear response theory can be regarded as the operator Taylor expansion. Using this, we introduce the model of the quantum Brownian motion coupled to a classical heat bath. Then we show that the heat operator can be defined satisfying operator relations which are analogous to thermodynamics [10].

2. Quantum analysis

In this section, we summarize the definition of the properties of the operator differential proposed in the quantum analysis [7, 8, 9].

Let us consider $f(\hat{A})$ where $f(x)$ is a smooth function of x. Then the operator differential with respect to $\hat{A}$ is expressed by $(df/d\hat{A})$, and introduced through the following equation,

$$f(\hat{A}+h\hat{C}) - f(\hat{A}) = \left(\frac{df}{d\hat{A}}\right) h\hat{C} + O(h^2), \tag{1}$$

where h is a small c-number and $\hat{C}$ is another operator which is in general not commutable with $\hat{A}$, $[\hat{A},\hat{C}] \neq 0$. Note that the value of the differential depends on the operator $\hat{C}$ and thus $(df/d\hat{A})$ is a hyper operator.

In the quantum analysis, this operator differential is defined by

$$\left(\frac{df}{d\hat{A}}\right) \equiv \int_0^1 d\lambda f^{(1)}(\hat{A} - \lambda\delta_A), \tag{2}$$

where $\delta_A = [\hat{A},\ \ ]$.

For example, let us consider the case of $f(x) = x^2$. Substituting this into the left hand side of Eq. (1), we find

$$(\hat{A}+h\hat{C})^2 - \hat{A}^2 = h(\hat{A}\hat{C} + \hat{C}\hat{A}). \tag{3}$$

Here we ignored the contribution of $O(h^2)$. The quantity on the right hand side can be reexpressed as

$$h\int_0^1 d\lambda 2(\hat{A} - \lambda\delta_A)\hat{C}. \tag{4}$$

This is nothing but our definition of the operator differential.

As the advantage of this definition, we can express the operator Taylor expansion in the following simple form,

$$f(\hat{A}+\hat{C}) = f(\hat{A}) + \sum_{n=1}^{\infty}\frac{1}{n!}\left(\frac{d^n f}{d\hat{A}^n}\right)\hat{C}^n, \tag{5}$$

where

$$\left(\frac{d^n f}{d\hat{A}^n}\right) = n!\int_0^1 d\lambda_1 \cdots \int_0^{\lambda_{n-1}} d\lambda_n f^{(n)}(\hat{A} - \sum_{i=1}^{n}\lambda_i\delta_A^{(i)}), \tag{6}$$

$$\delta_A^{(i)}\hat{C}^n = \hat{C}^{n-i}(\delta_A\hat{C})\hat{C}^{i-1}. \tag{7}$$

As an example of the application, we apply this operator Taylor expansion to the unitary time evolution operator in quantum physics $f = e^{i(\hat{H}_0+g\hat{V})t}$ where g is a coupling constant,

$$\begin{aligned}
e^{i(\hat{H}_0+g\hat{V})t} &= e^{i\hat{H}_0 t} + \sum_0^{\infty}\int_0^1 d\lambda_1\cdots\int_0^{\lambda_{n-1}} d\lambda_n e^{i(\hat{H}_0 - \sum_{i=1}^n \lambda_i\delta_{H_0}^{(i)})t}(g\hat{V})^n \\
&= e^{i\hat{H}_0 t} + \sum_0^{\infty} e^{i\hat{H}_0 t}\int_0^1 d\lambda_1\cdots\int_0^{\lambda_{n-1}} d\lambda_n e^{-i\sum_{i=1}^n \lambda_i\delta_{H_0}^{(i)} t}(g\hat{V})^n \\
&= e^{i\hat{H}_0 t} + \sum_0^{\infty} e^{i\hat{H}_0 t}\int_0^t dt_1\cdots\int_0^{t_{n-1}} dt_n e^{-i\sum_{i=1}^n t_i\delta_{H_0}^{(i)}}(g\hat{V})^n \\
&= e^{i\hat{H}_0 t} + \sum_0^{\infty} e^{i\hat{H}_0 t}\int_0^t dt_1 g\hat{V}(t_1)\cdots\int_0^{t_{n-1}} dt_n g\hat{V}(t_n),
\end{aligned} \tag{8}$$

where $\hat{V}(t)$ is the operator in the interaction picture,

$$\hat{V}(t) = e^{-it\delta_{H_0}}\hat{V} = e^{-i\hat{H}_0 t}\hat{V}e^{i\hat{H}_0 t}. \tag{9}$$

This is the well-known perturbative expansion in the S matrix. From the first to the second line, we used

$$\hat{A}(\delta_A \hat{C}) = \delta_A \hat{A}\hat{C}. \tag{10}$$

Besides this, there are several useful relations for δ_A,

$$f(\hat{A}-\delta_A)\hat{C} = \hat{C}f(\hat{A}), \quad \delta_A \hat{C} = -\delta_C \hat{A}, \quad e^{a\delta_A}\hat{C} = e^{a\hat{A}}\hat{C}e^{-a\hat{A}}. \tag{11}$$

And when $\hat{A}_t$ is a function of a c-number t,

$$\frac{df(\hat{A}_t)}{dt} = \frac{df(\hat{A}_t)}{d\hat{A}_t}\frac{d\hat{A}_t}{dt} = \int_0^1 d\lambda f^{(1)}(\hat{A}-\lambda\delta_A)\frac{d\hat{A}_t}{dt}. \tag{12}$$

3. Reexpression of quantum mechanics in quantum analysis

Before considering a quantum dissipative system, we investigate the possible role of the operator differential in the formulation of quantum mechanics.

3.1. Poisson bracket

We consider one-particle system which is characterized by the following Hamiltonian operator,

$$\hat{H} = \frac{\hat{p}_t^2}{2m} + V(\hat{x}_t), \tag{13}$$

where m is the mass of the particle. Then the Heisenberg equations of the position and momentum operators are, respectively,

$$\frac{d}{dt}\hat{x}_t = \frac{i}{\hbar}[\hat{H},\hat{x}_t] = \frac{\hat{p}_t}{m}, \tag{14}$$

$$\frac{d}{dt}\hat{p}_t = \frac{i}{\hbar}[\hat{H},\hat{p}_t] = -V^{(1)}(\hat{x}_t), \tag{15}$$

where $V^{(n)}(x) = \partial_x^n V(x)$.

With the help of the quantum analysis, we propose to define the Poisson bracket operator by

$$[\hat{A},\hat{B}]_{(\hat{x},\hat{p})} = \frac{\partial\hat{A}}{\partial\hat{x}_t}\frac{\partial\hat{B}}{\partial\hat{p}_t}1 - \frac{\partial\hat{A}}{\partial\hat{p}_t}\frac{\partial\hat{B}}{\partial\hat{x}_t}1. \tag{16}$$

Then the above Heisenberg equations can be cast into the following forms,

$$\frac{d}{dt}\hat{x}_t = [\hat{x}_t,\hat{H}]_{(\hat{x},\hat{p})} = \frac{1}{m}\int_0^1 d\lambda(\hat{p}_t-\lambda\delta_p)1 = \frac{\hat{p}_t}{m}, \tag{17}$$

$$\frac{d}{dt}\hat{p}_t = [\hat{p}_t,\hat{H}]_{(\hat{x},\hat{p})} = -\int_0^1 d\lambda V^{(1)}(\hat{x}-\lambda\delta_x)1 = -V^{(1)}(\hat{x}_t). \tag{18}$$

We confirmed that the Poisson bracket operator is applicable to general functions of $\hat{x}_t$ and $\hat{p}_t$, and the Heisenberg equation is reproduced [11]. The analysis with the operator differential may shed new light on the relation between classical and quantum physics.

3.2. Wigner function

In the usual formulation of quantum mechanics, the time evolution is characterized by the Hamiltonian operator. This is one of the origins of the difficulties in introducing dissipation in quantum systems [14]. On the other hand, if the formulation of the quantum analysis is consistent with quantum mechanics, we need not necessarily use the Hamiltonian operator to calculate expectation values. For example, suppose that we do not know the Hamiltonian of our system but the equations of the position and momentum operators are known to be given by Eqs. (14) and (15), respectively. Then we can still calculate the Wigner function.

As is well-known, the Wigner function is defined by

$$\rho_W(x,p,t) = \frac{1}{2\pi\hbar}\int dq\psi^*(x+q/2,t)\psi(x-q/2,t)e^{ipq/\hbar}, \tag{19}$$

where $\psi(x,t)$ is the wave function of our system. This definition is reexpressed as

$$\rho_W(x,p,t) = \langle\psi|\delta(x-\hat{x}_t+\delta_{x_t}/2)\delta(p-\hat{p}_t)|\psi\rangle. \tag{20}$$

Here $|\psi\rangle$ is an initial wave function. Using the latter expression, the time derivative of the Wigner function can be calculated directly from the Heisenberg equations as

$$\begin{aligned}
\partial_t\rho_W(x,p,t) &= \partial_t\int dkdq e^{i\hbar kq/2}e^{ikx}e^{iqp}\langle\psi|e^{-ik\hat{x}_t}e^{-iq\hat{p}_t}|\psi\rangle \\
&= -i\int dkdq e^{i\hbar kq/2}e^{ikx}e^{iqp}\langle\psi|\left[ke^{-ik\hat{x}_t}\int_0^1 ds\left(e^{iks\delta_{x_t}}\frac{d\hat{x}_t}{dt}\right)e^{-iq\hat{p}_t}\right. \\
&\quad \left. +qe^{-ik\hat{x}_t}e^{-iq\hat{p}_t}\int_0^1 dse^{iqs\delta_{p_t}}\frac{d\hat{p}_t}{dt}\right]|\psi\rangle \\
&= \left[-\frac{p}{m}\partial_x + V^{(1)}(x)\partial_p\right]\rho_W(x,p,t) + \Sigma(x,p,t),
\end{aligned} \tag{21}$$

where

$$\Sigma(x,p,t) = \sum_{l=1}^{\infty}\frac{V^{(2l+1)}}{(2l+1)!}\left(-\frac{\hbar}{4}\right)^l \partial_p^{2l+1}\rho_W(x,p,t) \tag{22}$$

This is the well-known equation satisfied for the Wigner function [12].

The Wigner function is not positive definite and thus cannot be interpreted as a probability density. In stead it should be considered as the measure of the integral. As a matter of fact, we can reexpress the expectation values of operators by the phase-space integral with this measure. As we observed, we constructed the above evolution equation of the Wigner function without using the form of the Hamiltonian operator. That is, the formulation based on the quantum analysis may be useful to consider the quantization of systems which do not have Hamiltonian, such as dissipative systems.

4. Application to quantum Brownian motion

In this section, we introduce the model of the quantum Brownian motion with the help of the quantum analysis.

Let us consider a quantum system which is characterized by the following equations [10],

$$d\hat{x}_t = \frac{\hat{p}_t}{m}dt, \tag{23}$$

$$d\hat{p}_t = -V^{(1)}(\hat{x}_t,\lambda_t)dt - \frac{\nu}{m}\hat{p}_t dt + \sqrt{2\nu k_B T}dB_t, \tag{24}$$

with $d\hat{A}_t = \hat{A}_{t+dt} - \hat{A}_t$. Here ν and T are the dissipative coefficient and the temperature, respectively, and B_t is the standard Wiener process satisfying

$$E[dB_t] = 0 \qquad E[(dB_t)^2] = 0, \tag{25}$$

where $E[\ ]$ denotes the ensemble average for the Wiener process. Other second-order correlations vanish. Note that λ_t represents the parameter which controls the form of the external potential V. That is, the change of the volume in this system is interpreted as that of the form of the potential through λ_t.

This system is dissipative and we cannot write down not only the Lagrangian but also the Hamiltonian [13]. However, we can still calculate the Wigner function using the above two operator equations by applying the quantum analysis [10]. The evolution equation is given by

$$\partial_t \rho_W(x,p,t) = \left[-\frac{p}{m}\partial_x + V^{(1)}(x)\partial_p + \frac{\nu}{m}\partial_p p + \frac{\nu}{\beta}\partial_p^2 \right] \rho_W(x,p,t) + \Sigma(x,p,t), \tag{26}$$

where

$$\Sigma(x,p,t) = \sum_{l=1}^{\infty} \frac{V^{(2l+1)}}{(2l+1)!} \left(-\frac{\hbar^2}{4} e^{-\nu t/m} \right)^l \partial_p^{2l+1} \rho_W(x,p,t). \tag{27}$$

The expectation values of the operators are calculated by the phase-space integrals with this integration mesure.

In this derivation, we used Ito's lemma for the stochastic operator. For example, we consider the operator characterized by the following stochastic differential equation of the operator,

$$d\hat{A}_t = \hat{L}_t dt + \sqrt{2\nu k_B T} dB_t. \tag{28}$$

Then for an arbitrary function of $\hat{A}_t$, we can show by using the operator Taylor expansion

$$df(\hat{A}_t) = \left[\int_0^1 d\lambda f^{(1)}(\hat{A}_t - \lambda\delta_{A_t})\hat{L}_t + \nu k_B T f^{(2)}(\hat{A}_t) \right] dt + \sqrt{2\nu k_B T} f^{(1)}(\hat{A}_t) \circ_i dB_t \tag{29}$$

$$= \left(\frac{df(\hat{A}_t)}{d\hat{A}_t} \right) \circ_s d\hat{A}_t + O(dt^{3/2}), \tag{30}$$

where

$$f(\hat{A}_t) \circ_i dB_t = f(\hat{A}_t)(B_{t+dt} - B_t), \tag{31}$$

$$f(\hat{A}_t) \circ_s dB_t = f(\hat{A}_{t+dt/2})(B_{t+dt} - B_t). \tag{32}$$

This is Ito's lemma for the stochastic operator $\hat{A}_t$ [10].

Exactly speaking, we need to show the existence of the appropriate probability space (σ algebra) for stochastic operators, but we simply assume it in this work.

One can see that the stationary solution of the above differential equation is given by

$$\lim_{t\to\infty} \rho_W(x,p,t) = \frac{1}{Z(\lambda)} e^{-\beta H(x,p,\lambda)}, \tag{33}$$

where

$$H(x,p,\lambda) = \frac{p^2}{2m} + V(x,\lambda), \tag{34}$$

$$Z(\lambda) = \int dx dp e^{-\beta H}. \tag{35}$$

IOP Conf. Series: Journal of Physics: Conf. Series **880** (2017) 012037 doi:10.1088/1742-6596/880/1/012037

It should be noted that the term Σ disappears in the classical limit and the differential equation of the Wigner equation is reduced to the Kramers equation which describes the thermal relaxation process of classical dynamics. That is, the Wigner function coincides with the classical phase-space distribution in the classical limit,

$$\lim_{\hbar\to 0} \rho_W(x,p,t) = \rho_{KR}(x,p,t), \tag{36}$$

where $\rho_{KR}(x,p,t)$ is the solution of the Kramers equation.

5. Quantum stochastic energetics

Various microscopic fluctuating systems are known to be modeled as the Brownian motion and then it is possible to define the heat operator by adapting a theory called stochastic energetics [2].

In this theory, the heat absorbed by a Brownian particle is defined as the work exerted by the heat bath on the Brownian particle. On the other hand, the interaction between the particle and the bath is represented by the dissipative term and the noise term. The heat absorbed from the heat bath is equivalent to the work exerted by the heat bath on the Brownian particle, which is, thus, defined by the product of a force and an induced displacement [2].

Extending this argument to our model, the heat operator can be defined by [10]

$$d\hat{Q}_t \equiv \left(d\hat{x}_t - \frac{1}{2}\delta_{dx_t} \right) \circ_s \left(-\frac{\nu}{m}\hat{p}_t + \sqrt{2\nu k_B T}\frac{dB_t}{dt} \right). \tag{37}$$

The operator δ_{dx_t} symmetrizes the order of the force and the displacement operators. Then we find the following operator relation is satisfied,

$$dH(\hat{x}_t, \hat{p}_t, \lambda_t) = d\hat{Q}_t + d\hat{W}_t, \tag{38}$$

where the work operator exerted by an external force is defined by [2]

$$d\hat{W}_t \equiv \partial_\lambda V(\hat{x}_t, \lambda_t) \circ_s d\lambda_t, \tag{39}$$

because the external force changes the form of V through its λ_t dependence. This energy balance corresponds to the first law of thermodynamics. Note that the energy balance is satisfied not for ensemble averages but for operators.

We can further show that the expectation value of the heat operator has an upper bound [10],

$$T\frac{d\sigma}{dt} \geq \langle\psi| E\left[\frac{d\hat{Q}_t}{dt}\right] |\psi\rangle. \tag{40}$$

Here we introduced the new function σ, defined by

$$\sigma = S_{SH} + S_{ME}, \tag{41}$$

where

$$S_{SH} = -k_B \int d\Gamma \rho_W(x,p,t) \ln|\rho_W(x,p,t)|, \tag{42}$$

$$S_{ME} = k_B \int^t ds \int d\Gamma \Big[\Sigma(x,p,s) \ln|\rho_W(x,p,s)| \\ -\beta\nu\delta^{(\hbar)} \rho_W(x,p,s) \left\{ \frac{p}{m} + \beta^{-1}\partial_p \ln|\rho_W(x,p,s)| \right\}^2 \Big]. \tag{43}$$

Note that $\delta^{(\hbar)}\rho_W(x,p,t) = \rho_W(x,p,t) - \rho_{KR}(x,p,t)$, which represents the modification of the phase-space distribution by quantum fluctuations. The first term $S_{SH}(t)$ is the Shannon entropy calculated by using the Wigner function instead of a probability distribution. The second term $S_{ME}(t)$ contains the memory effect and thus the behavior of $\sigma(t)$ depends on the hysteresis of the evolution. Note that $S_{ME}(t)$ is induced by quantum fluctuations and thus vanishes in the classical limit, leading to $S(t) = S_{SH}(t)$. That is, the above inequality reproduces the result of the classical stochastic energetics in the classical limit.

This corresponds to the second law of thermodynamics. In fact, $\sigma(t)$ can be regarded as the thermodynamic entropy in equilibrium because

$$\sigma|_{\rho_W=\rho_{eq}} = S_{SH}|_{\rho_W=\rho_{eq}} = \frac{\langle\psi|E[H]|\psi\rangle}{T} + k_B \ln Z. \tag{44}$$

6. Concluding remarks

We defined the heat operator for the model of the quantum Brownian motion coupled to a classical heat bath by introducing the definition of the operator differential in the quantum analysis and extending the idea of the stochastic energetics. Then we showed that the energy balance is satisfied (first law) and there exists the upper bound of the expectation value of the heat operator (second law). The detailed discussion for the thermodynamic laws observed in this model is discussed in Ref. [10].

Note that there is no Lagrangian which reproduces our dissipative system and thus there is no reason to consider that $\hat{p}_t$ is the canonical momentum of $\hat{x}_t$ in general. Therefore it is not obvious whether the commutation relation between them should be given by $[\hat{x}_t, \hat{p}_t] = i\hbar$. As a matter of fact, there is no established method to find the "canonical" variable for $\hat{x}_t$ in such a dissipative system. See also the discussion in Ref. [14] and references therein.

Acknowledgments

This work is financially supported by Conselho Nacional de Desenvolvimento Científico e Tecnológio (CNPq), project 307516/2015-6.

References

[1] Hänggi P and Marchesoni F 2009 Artificial Brownian motors: Controlling transport on the nanoscale *Rev. Mod. Phys.* **81** 387
[2] Sekimoto K 2010 *Stochastic Energetics* (Berlin: Springer-Verlag)
[3] Blickle V and Bechinger C 2012 Realization of a micrometre-sized stochastic heat engine *Nat. Phys.* **8** 143
[4] Koide T and Kodama T 2011 Thermodynamic laws and equipartition theorem in relativistic Brownian motion *Phys. Rev. E* **83** 061111
[5] Kanazawa K, Sagawa T and Hayakawa H 2012 Stochastic Energetics for Non-Gaussian Processes *Phys. Rev. Lett.* **108** 210601
[6] However, there is an attempt to apply SE to a quantum transport without changing the framework, Ghosh P K and Ray D S 2006 Stochastic energetics of quantum transport *Phys. Rev. E* **73** 036103
[7] Suzuki M 1997 Quantum analysis — Non-commutative differential and integral calculi *Commun. Math. Phys.* **183** 339
[8] Suzuki M 1999 General formulation of quantum analysis *Rev. Math. Phys.* **11** 243
[9] Suzuki M 2006 Refined formulation of quantum analysis, q-derivative and exponential splitting *J. Phys. A: Math. Theor.* **39** 5617
[10] Koide T 2016 Memory effect in the upper bound of the heat flux induced by quantum fluctuations *Phys. Rev. E* **94** 042140
[11] Koide T *in preparation*
[12] Wigner E P 1932 On the quantum correction for thermodynamic equilibrium, *Phys. Rev.* **40** 749
[13] We need to extend the usual frame work to describe dissipative phenomena in analytical mechanics. See, for example, Koide T and Kodama T 2012 Navier-Stokes, Gross-Pitaevskii and generalized diffusion equations using the stochastic variational method *J. Phys. A: Math. Theor.* **45** 255204

[14] Schuch D 1997 Nonunitary connection between explicitly time-dependent and nonlinear approaches for the description of dissipative quantum systems *Phys. Rev. A* **55** 935

Topological view of quantum tunneling coherent destruction

Alex E. Bernardini and Mariana Chinaglia

Departamento de Física, Universidade Federal de São Carlos, PO Box 676, 13565-905, São Carlos, SP, Brasil.

E-mail: alexeb@ufscar.br

Abstract. Quantum tunneling of the ground and first excited states in a quantum superposition driven by a novel analytical configuration of a double-well (DW) potential is investigated. Symmetric and asymmetric potentials are considered as to support quantum mechanical zero mode and first excited state analytical solutions. Reporting about a symmetry breaking that supports the quantum conversion of a zero-mode stable vacuum into an unstable tachyonic quantum state, two inequivalent topological scenarios are supposed to drive stable tunneling and coherent tunneling destruction respectively. A complete prospect of the Wigner function dynamics, vector field fluxes and the time dependence of stagnation points is obtained for the analytical potentials that support stable and tachyonic modes.

1. Introduction

The phase-space Wigner-Weyl formalism provides a suitable description of quantum mechanics. In the context of the dynamical evolution of quantum states, the Wigner function helps one to identify highlighting non-classical features revealed, for instance, by quantum tunneling [1, 3, 2]. The investigation of single-particle tunneling in double-well (DW) potentials can be found in several relevant scenarios [4, 5, 6, 7, 8, 9, 10]. For DW potentials, the maximum point ($\partial^2 V/\partial s^2 < 0$) between the two minima plays the role of a barrier that drives the quantum tunneling exhibited by the quantum superposition of ground and first excited states, when their energy are smaller than that from the hump. The quantum tunneling for a function generated by a two-wave superposition supported by novel symmetric and asymmetric DW potential configurations, when described in the phase-space, is investigated in this letter. In particular the dynamics of tachyonic modes supported by modified topological scenarios [11, 12], different from those which holds stable tunneling, is discussed. A symmetry breaking that supports the quantum conversion of a zero-mode stable vacuum into an unstable tachyonic quantum state is assumed [2] as to identify two inequivalent topological scenarios which drive stable tunneling and coherent tunneling destruction, respectively. In this scenarios, the Wigner flow stagnation points can be found where the potential is force free. They depend deeply on the potential critical points and bring on different effects whether they are a minimum, maximum or saddle points identified by winding numbers defined through [1] $\omega = \frac{1}{2\pi}\int d\varphi$, where φ is the angle between the positive position axis and the Wigner flow vectors. After a complete discussion of the tunneling phenomenon where the Wigner functions and their flow vector fields are computed, one identifies the non-classical features regarding the behavior of the stagnation points.

To summarize, proceeding is organized as following: in Sec. 2, the framework for the calculation of the ground and first excited states for DW potential configurations is presented, and results for a kink-like deforming function [11, 13, 12] are obtained; in Sec. 3 the quantum tunneling and Wigner flow for quantum superposition involving zero-mode stable and tachyonic states are calculated and discussed for symmetric and asymmetric DW configurations; the conclusions are drawn in Sec. 4.

2. Ground and first excited states in a DW potential

Given the ground state, ψ_0, of a *DW* potential, well-stablished quantum mechanical results, as those from [3], allow one to calculate the corresponding first excited, ψ_1, state through a multiplier function that relates ψ_0 with ψ_1 as [1]

$$\psi_1 = \alpha\psi_0. \tag{1}$$

Through the substitution of ψ_1 by $\alpha\psi_0$ into the Schrödinger equation $\psi''_n(s)+[E_n-V(s)]\psi_n(s) = 0$ for $n=1$ and subtracting the result from the Schrödinger equation for $n=0$ multiplied by α, one can find the quantity χ,

$$\chi(s) = \frac{\alpha'' + \Delta E\alpha}{2\alpha'} = -\frac{\psi'_0}{\psi_0} = -\frac{d\ln\psi_0}{ds}, \tag{2}$$

where ΔE is the energy difference between the states and henceforth shall be conveniently denoted by $\Delta E = 1/n$. Considering that α is an *ansatz* multiplier function, for a ground state described by,

$$\psi_0(s) = \mathcal{N}_0 \exp\left(-\int_0^s \chi(s')ds'\right), \tag{3}$$

where $\mathcal{N}_0$ is a normalization constant, the first excited state can be calculated through Eq. (1). Once a solution is found, the potential can be straightforwardly calculated through

$$V(s) = \frac{\psi''_0}{\psi_0} + E_0 = \chi^2 - \chi' + E_0. \tag{4}$$

An *ansatz* multiplier function set as $\alpha_1 = \beta + \tanh(s)$ [1, 2] which is, up to an additive constant, the solution for the differential equation driven by the $\lambda\phi^4$ model. This multiplier function is an input for both set of potentials: symmetric ($\beta = 0$) and asymmetric ($\beta \neq 0$) ones. Once one has chosen α, it is possible to find χ through Eq. (2) and proceed with the calculations for ψ_0 and ψ_1 by means of Eq. (3) and Eq. (1), respectively. The potential can be calculated either by means of χ or ψ_0 through the Eq. (4).

Once ψ_0 and ψ_1 have been established, two inequivalent topological scenarios that supports the quantum conversion of a zero-mode stable vacuum into an unstable tachyonic quantum state can be found. As to see the connection between their energies, one assumes a scalar field equation of motion, $\frac{\partial^2\phi}{\partial s^2} = \frac{dU}{d\phi}$, derived from a time independent Lagrangian in $1D$, from which, through quantum perturbations, by an increment of ξ, it is possible to find the Schrödinger-like equation,

$$\left(-\frac{d^2}{ds^2} + V_0^{(QM)}(s)\right)\psi_n(s) = w_n^2\psi_n(s), \tag{5}$$

with $V_0^{(QM)}(s) = d^2U(s)/d\xi^2(s)$, i. e. the quantum mechanical potential analogue. The eigenenergy, when real, i.e. for $w^2 > 0$, leads to a quantum stable solution. When it is imaginary, i.e. for $w^2 < 0$, one deals with unstable functions: the tachyonic modes.

IOP Conf. Series: Journal of Physics: Conf. Series **880** (2017) 012038 doi:10.1088/1742-6596/880/1/012038

By identifying $U(\xi) = \frac{z_\xi^2}{2}$, with $z_\xi = dz/d\xi = d\xi/ds = \xi'$, it is straightforward to show that Eq. (5) can be written as,

$$\left(-\frac{d^2}{ds^2} + \xi'''/\xi'\right)\psi_n(s) = w_n^2\psi_n(s). \tag{6}$$

Note that if the ψ_1 energy, w_1, is converted to zero, one has a novel scenario where the Schrödinger-like equation is

$$\left(-\frac{d^2}{ds^2} + V_1^{(QM)}(s)\right)\psi_1(s) = 0, \tag{7}$$

with $V_1^{(QM)}$ the novel quantum potential [2]. Now, there is a state that cross the origin producing a node. Eq. (1) thus implies that,

$$\left(-\frac{d^2}{ds^2} + V_1^{(QM)}(s)\right)\psi_0(s) = -w_1^2\psi_0. \tag{8}$$

Once one has the zero mode ψ_0 calculated through Eq. (3) and the potential found in Eq. (4), it is possible to compare it with $V_0^{(QM)}$ and identify that $\psi_0 \propto \xi'$. So there is an identification with the corresponding topological scenario that supports quantum perturbations for this $V_0^{(QM)}$ potential [2]. In fact, the zero-mode stable vacuum (ψ_0) can work as a mathematical tool for obtaining an unstable quantum state, which behaves like a tachyonic mode. Therefore, in this case, ψ_1 becomes the novel zero-mode stable solution in the tachyonic system causing the appearance of instabilities, as $\psi_0 + e^{-it/n}\psi_1$ is converted into $e^{-t/n}\psi_0 + \psi_1$, namely, an unstable configuration. For a DW potential as in (4) built from the ground state wave function, one identifies

$$\psi_0 = \mathcal{N}_0\, 2^{1/n}\exp\left(-\frac{(\beta + 2\arctan(e^s) - \frac{\pi}{2})\sinh[s]}{2n}\right)\cosh[s]^{\frac{n+1}{2n}}, \tag{9}$$

as well as the multiplier function, $\alpha = \beta + 2\arctan(e^s) - \frac{\pi}{2}$, where $\beta \neq 0$ drives the asymmetric profiles, as obtained from Eq. (4). The corresponding first excited states are obtained from Eq. (1), also for symmetric and asymmetric DW configurations. Fig. 1 shows the ground and first excited states along with both potentials, symmetric and asymmetric ones.

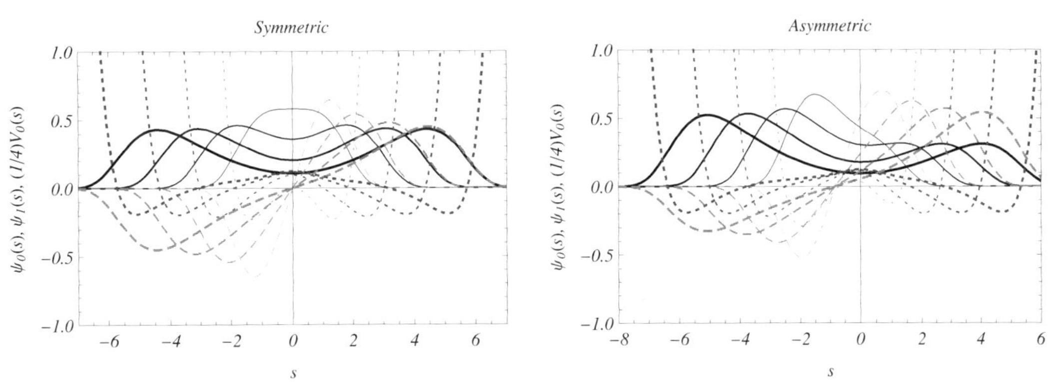

Figure 1. (Color online) Ground (solid black lines) and first excited (dashed red lines) states for symmetric ($\beta = 0$) (first plot) and asymmetric ($\beta = 3/4$) (second plot) QM potentials (dotted blue lines). The plots are for $\Delta E = 1/n$, with $n = 1$ (thickest), 4, 16, and 64 (thinnest).

IOP Conf. Series: Journal of Physics: Conf. Series **880** (2017) 012038 doi:10.1088/1742-6596/880/1/012038

3. Quantum tunneling and Wigner flow analysis

In order to investigate the quantum tunneling one needs the superposition function of the ground, ψ_0, and first excited, ψ_1, states:

$$\Psi(s,t) = \sin(\theta) exp\left(\frac{-iE_0 t}{\hbar}\right)\psi_0(s) + \cos(\theta) exp\left(\frac{-iE_1 t}{\hbar}\right)\psi_1(s), \tag{10}$$

which will be set into the Wigner function:

$$W(s,p,t) = \frac{1}{\pi\hbar}\int_{-\infty}^{\infty} dy \Psi^*(s+y,t)\Psi(s-y,t)e^{\frac{2i}{\hbar}py}, \tag{11}$$

where θ denotes the weight of ψ_0 and ψ_1 in the superposition function. As a quasi-probability function, the Wigner function attains only real values they can also be negative. The modulus of the normalized stable-unitary (black lines), and unstable→stationary (red lines) density probabilities are depicted in Fig. 2 for different times, ranging from $t = 0$ to $t = \pi/2$ (first row) and from $t = 5\pi/8$ to $t = \pi$ (second row). It is evidenced through the images that the normalized function completes the tunneling cycle going from *right to left* for both potentials while the tachyonic cycle is converted into a stationary state after the coherent destruction of the unstable mode. The Wigner functions squared for both scenarios relative to symmetric and

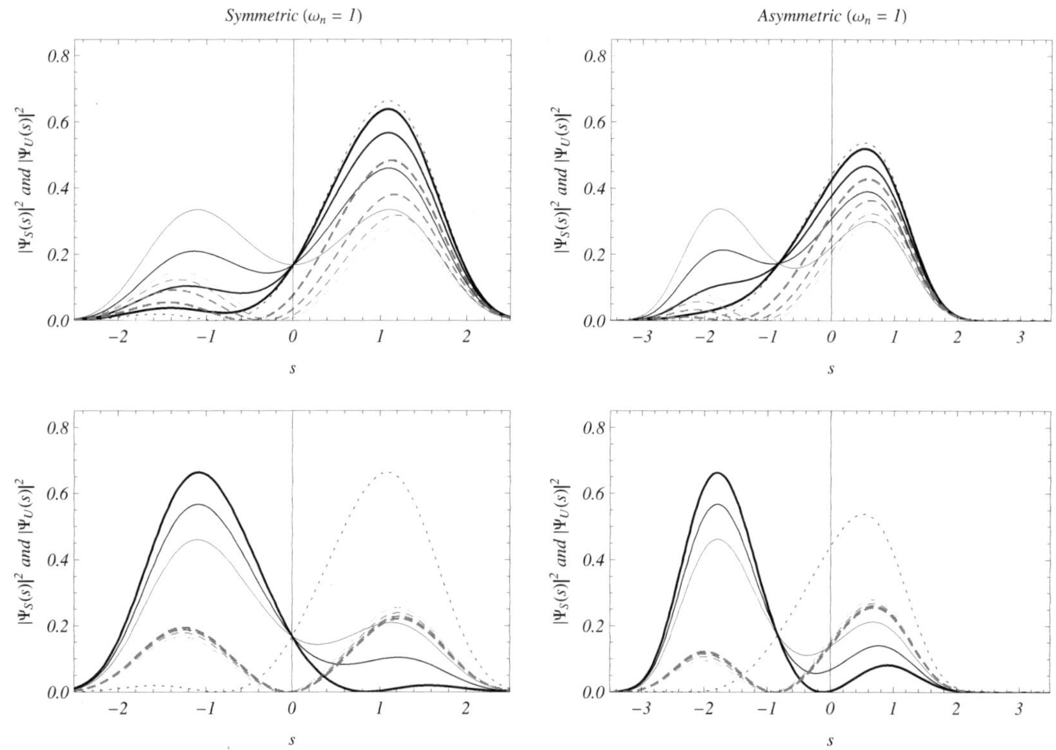

Figure 2. (Color online) Time evolution of probability densities for stable (solid black lines) and unstable (dashed red lines) composite states, Ψ_S and Ψ_U, in DW symmetric (first row) and asymmetric (second row) configuration. The curves are for times running from $t = \pi/8$ (thickest line) to $t = \pi/2$ (thinnest line) in the first plot, and from $t = 5\pi/8$ (thinnest line) to $t = \pi$ (thickest line) in the second plot, with steps $\Delta t = \pi/8$. Dotted blue lines correspond to $t = 0$ for both, stable and unstable tunneling. The results are for $n = 1$.

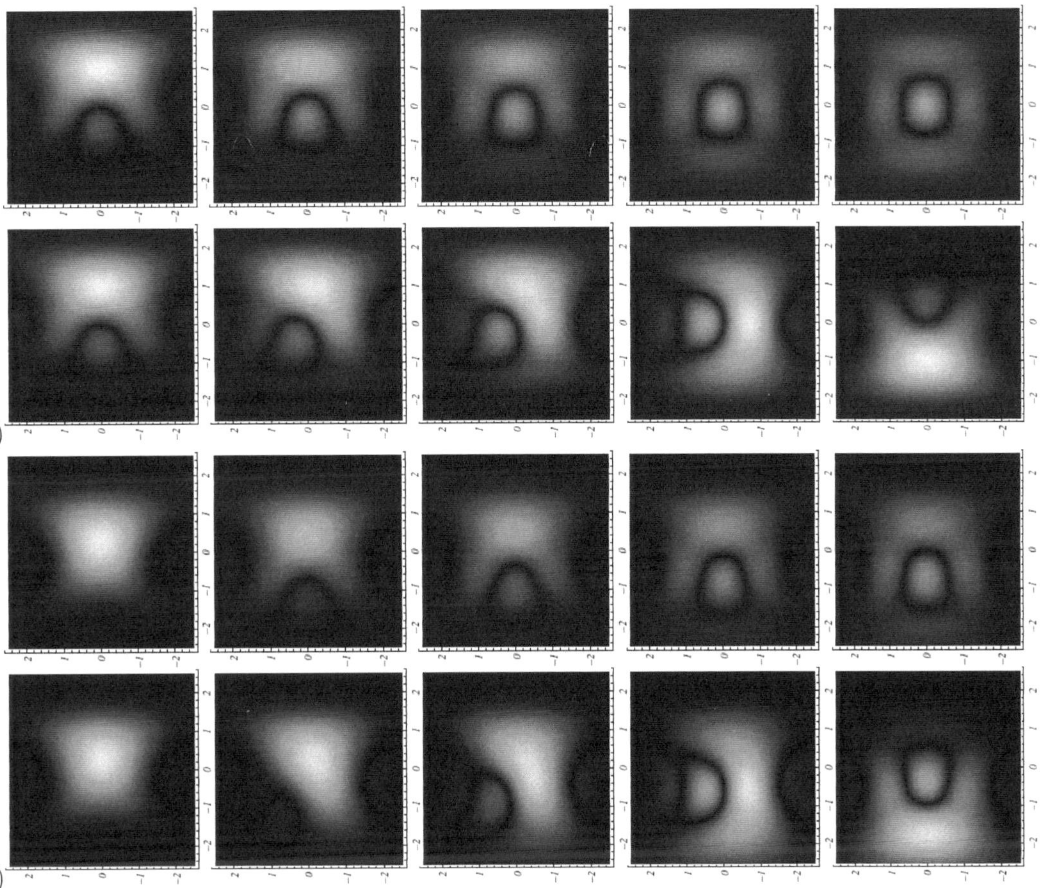

(a)

(b)

Figure 3. (Color online)Time evolution of Wigner amplitudes relative to symmetric (a) and asymmetric (b), unstable (first and third rows) and stable (second and forth rows) wave function configurations. The pictures are shown for time evolution from left $(t = 0)$ to right $(t = \frac{\pi}{2})$ for time intervals of $\frac{\pi}{8}$. Here again it was considered $n = 1$. Vertical and horizontal axes measure positions and momenta respectively.

asymmetric potentials can be seen in Fig. 3. Analyzing the image it is clear, as was previously seen with Fig. 2, that the tunneling occurs for both functions. The difference appears when one recognizes that for the normalized wave function the tunneling process is invertible whereas for tachyonic functions a limit state is reached. The invertible behavior appears because of the small difference between the energy of the states that causes a beat period, $T = 2\pi^{-}h/\Delta E$ The complete beating, that is when the function goes from one well to another and returns, is not presented in the tachyonic case because it is not normalizable. This latter smoothly attains an equilibrium lying in equal proportions in the two wells. The next step corresponds to the calculation of the Wigner flow,

$$\begin{pmatrix} J_s \\ \\ J_p \end{pmatrix} = \begin{pmatrix} \frac{p}{m} W(s,p,t) \\ \\ -\sum_{l=0}^{l=\infty} \frac{(\frac{i\hbar}{2})^{2l}}{(2l+1)!} \frac{\partial^{2l} W(s,p,t)}{\partial_p^{2l}} \frac{\partial^{2l+1} V(s)}{\partial^{2l+1} s} \end{pmatrix}. \tag{12}$$

which, in the classical limit, i.e. when one has $\hbar \to 0$ or $V(s)$ powers ≤ 2, leads to an equivalence between Wigner and classical Liouville flow. By convenience, the Wigner flow vector fields

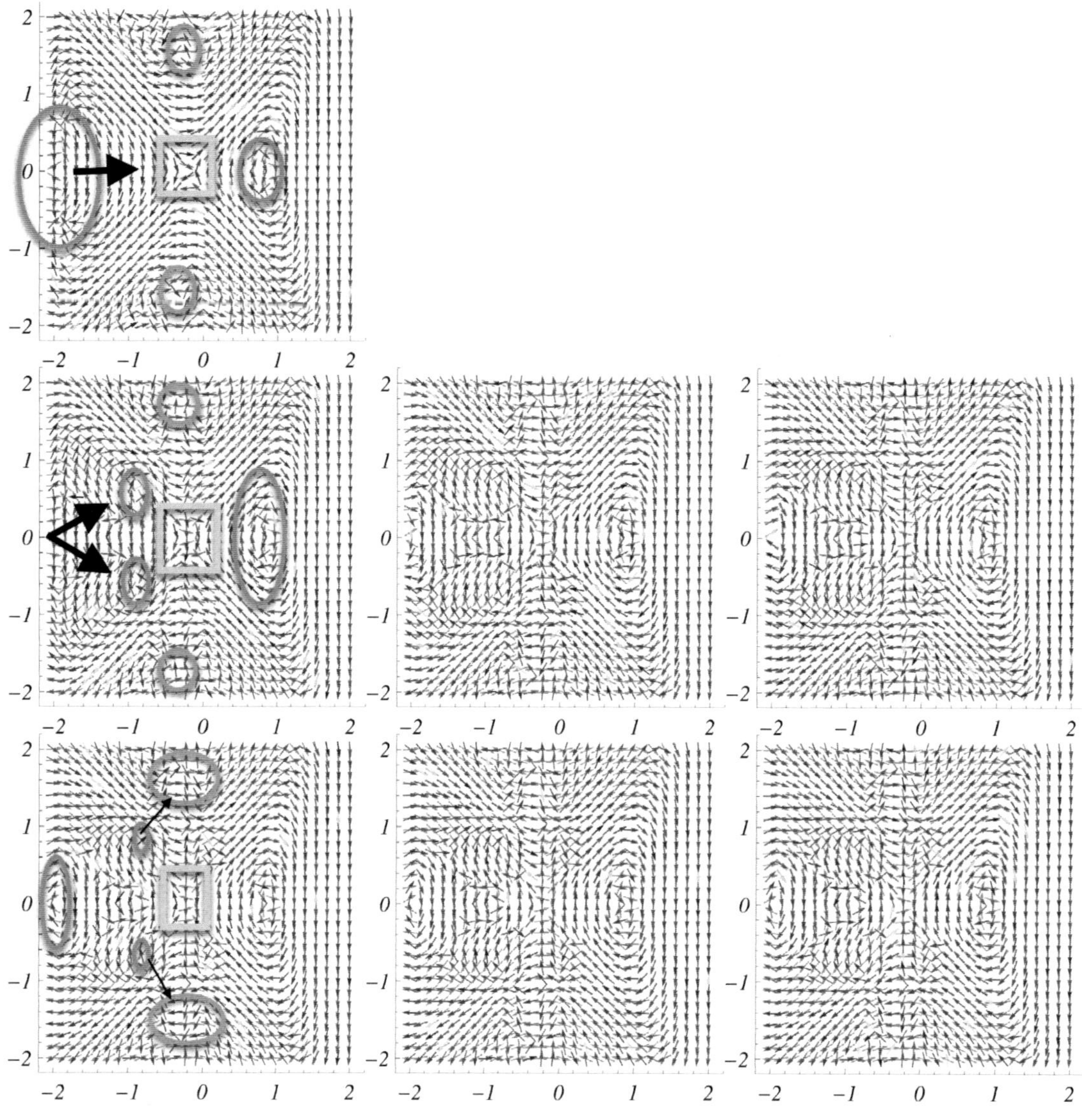

Figure 4. (Color online) First line portrait corresponds to $t = 0$. Last two lines represent the time evolution of the stable wave function for the asymmetric potential (last row of Fig. 3) up to half tunneling period. Red and green circles represent a CW and CCW vortex respectively, therefore both contributes with $\omega = +1$. Yellow rectangle represents a saddle point and contributes with $\omega = -1$. Vertical and horizontal axes measure p and x, respectively

relative to the stable wave functions driven by the asymmetric potential is considered in Fig. 4 for different times along the tunneling evolution. The first feature that is highlighted through the vector field is the difference between the classical (red) and quantum (blue) vector directions. The larger this difference, the greater the degree of non-classicality of the system [1]. While passing through a minimum of the potential, the flow spins towards this point and one has the winding number $\omega = +1$. Passing through a maximum of the potential, the flow also spins generating $\omega = +1$, but it spreads instead of concentrating. If it is a saddle point it elongates the flow making it slower and slower near the central point, through which one has $\omega = -1$.

Another feature is the flux reversion that corresponds to the regions where the Wigner function attains negative values. In order to investigate this feature ally with the stagnation points position, one can imagine a top-down vertical line around $s = -0.25$ in the first frame of Fig. 4. Following this line it is possible to first recognize one counterclockwise (CCW) vortex (green circle), then a flow reversal followed by a saddle point (yellow square), around $p = 0$. Moving to negative values of p one has a CCW vortex, causing again the flow inversion. This behavior is a further indication of the non-classical character of the system.

In order to see the stagnation points dynamics and calculate ω, the integration loop is depicted in each frame of Fig. 4. Note that the initial total winding number (first frame) is $\omega = +2$, correspondent to one clockwise (CW) vortex (red circle; $\omega = +1$), two CCW vortices (green circles; $\omega = +1$) and one saddle point (yellow square); $\omega = -1$). In the next frame the saddle point starts to crumble while another CW vortex appears in the left. Besides, one can no longer recognize the CCW vortices, which results again in a total winding number $\omega = +2$. Considering all subsequent frames one sees that the saddle point completely disappears engendering sequentially $\omega = +2$. Consequently, charge conservation is attained as conjectured [1]. The novelty is that an opposite conclusion appears when considering the unstable case for which one has no charge conservation. For the unstable function however this lack of conservation is not totally a surprise, once it is predicted only for systems with unitary quantum dynamics. Finally, the stagnation points also work as a tool for the recognition of tunneling. The function Ψ, which is found initially only in the right minimum of the potential, appears in the form of a CW vortex near $s = 0.8$ in Fig. 4. After time evolution Ψ coexists in both minima, thus causing the appearance of another CW vortex in the left-side in some subsequent frames. This behavior reveals the tunneling through the stagnation points.

4. Conclusions

Quantum tunneling analysis has shown that one has a reversibility of the process for the stable wave function whereas the tachyonic function attains a limit coexisting in both wells for any subsequent time. Thereat one can say that the tunneling undergoes a process of coherent destruction. It was also noticed that the Wigner vector field can be used to study non-classical features since it evinces the differences between quantum and classical vector flux profiles as well as the flow reversibility. The latter occurs where one has negative values for the Wigner function related to the stagnation points. Lastly, while one finds the expected charge conservation for stable wave functions, one draws attention to the fact that this conservation is not in principle expected for unstable functions. In fact, it is depicted from the vector flow analysis for the tachyonic case.

Acknowledgments
AEB thanks for the financial support from the Brazilian Agency FAPESP (grant 15/05903-4).

References

[1] Steuernagel O, Kakofengitis D and Ritter G 2013 *Phys. Rev. Lett.* **110** 030401
[2] Bernardini A E and Chinaglia M 2015 *Mod. Phys. Lett.* **A30** 1550118
[3] Caticha A 1995 *Phys. Rev.* **A 51** 4264
[4] Domcke W, Hänggi P and Tannor D 1997 *Chem. Phys.* **217** 117
[5] Schulz M D, Dusuel S, Schmidt K P and Vidal J 2013 *Phys. Rev. Lett.* **110** 147203
[6] Casetti L, Cohen E G D and Pettini M 1999 *Phys. Rev. Lett.* **82** 4160
[7] Bernardini A E and Bertolami O 2013 *Phys. Lett.* **B726** 512
[8] Bernardini A E and da Rocha R 2012 *Phys. Lett.* **B717** 238
[9] Bernardini A E and da Rocha R 2013 *AHEP* **2013** 304980
[10] Bernardini A E, Chinaglia M and da Rocha 2015 *Eur. Phys. J Plus* **130** 97
[11] Bernardini A E and da Rocha R 2016 *Phys. Lett.* **A380** 2279
[12] Bernardini A E, Chinaglia M and da Rocha R 2016 *Int. J. Theor. Phys.* **55** 4605
[13] Bernardini A E and da Rocha R 2016 *AHEP* **2016** 3650632

An information theory model for dissipation in open quantum systems

David M. Rogers

Department of Chemistry, University of South Florida, Tampa, FL 33620, USA

E-mail: davidrogers@usf.edu

Abstract. This work presents a general model for open quantum systems using an information game along the lines of Jaynes' original work. It is shown how an energy based reweighting of propagators provides a novel moment generating function at each time point in the process. Derivatives of the generating function give moments of the time derivatives of observables. Aside from the mathematically helpful properties, the ansatz reproduces key physics of stochastic quantum processes. At high temperature, the average density matrix follows the Caldeira-Leggett equation. Its associated Langevin equation clearly demonstrates the emergence of dissipation and decoherence time scales, as well as an additional diffusion due to quantum confinement. A consistent interpretation of these results is that decoherence and wavefunction collapse during measurement are directly related to the degree of environmental noise, and thus occur because of subjective uncertainty of an observer.

1. Introduction

This paper presents a new, simple ansatz for adding dissipation to arbitrary stochastic forcing of a quantum dynamical system. For Gaussian random forces, its Lindblad equation is identical to the Caldeira-Leggett model [1, 2, 3] up to order β^2, where the same term is present with a different prefactor. The system-centric, phase space picture here shows that this term represents a quantum confinement effect.

This ansatz is significant because nearly all other models for dissipation rely on the Feynman-Vernon influence functional approach to modeling a linearly coupled environment [4]. Because those approaches are based on a model of the environment containing infinitely many harmonic oscillators, they are complicated to carry over to other settings – for example the reconstruction problem in quantum field theory [5].

Work on the problem of quantum decoherence showed that there is a fundamental connection ($\tau/\theta = (\Delta q/\lambda_{db})^2$) between the timescale of energy dissipation, τ, and the decoherence timescale, θ [6, 7, 8]. Those two scales are separated by the squared distance per de Broglie wavelength, $\lambda_{db} = \hbar\sqrt{\beta/4m}$. However, these connections can be modified by changing length scales and models for the environment [9, 10]. The proper representation of the environment is not just an academic question, since the form of environmental coupling is related to different mechanisms for decoherence in laboratory experiments [11, 12], and even give information on the early universe and gravitational background [7, 13].

Investigation of the crossover from quantum to classical chaotic dynamics has been difficult, since classical constraints on energy flows do not have direct translations in the quantum theory [14]. Zurek used the Moyal bracket to relate the two and found that classical noise causes decoherence that limits the severity of quantum corrections [15]. The maximum entropy approach introduced here provides an alternative, straightforward route to derive force/flux relationships of the Onsager-Machlup type – via a 'kinetic' partition function for average fluxes. The associated fluctuation-dissipation theorem links energy exchange with the system to classical and quantum information loss from the system.

This work introduces the energy-based reweighting ansatz for modeling dissipation in several steps. Section 2 begins with a Bayesian framework for analyzing stochastic quantum dynamics. Sec. 3 shows that the forward-backward propagator, applied to the Wigner distribution, has a close connection to the classical Langevin equation. By applying a pre-averaging procedure, we illustrate how stochastic dynamics is caused by loss of phase information. That section breifly recalls key results of the classical moment generating function derived from maximum transition entropy [16]. The energy-based reweighting is finally introduced in Sec. 4 and its associated Wigner propagator is derived in Sec. 5. There, the moment generating function is found to provide an explicit definition for the local quantum temperature.

2. Information Game

For a closed quantum system, the full wavefunction is determined via the Schrödinger equation for all times from a known initial condition and a Hamiltonian. However, probability enters the description of the dynamics in an essential way at the point where the system interacts with its external environment.

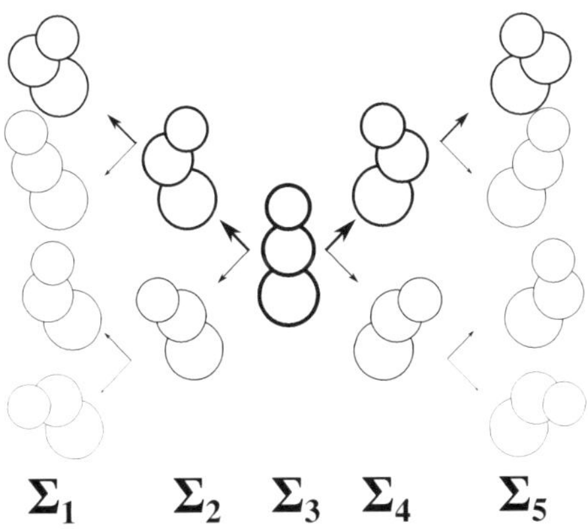

Figure 1. Set of possible histories as inferred from an observer with complete knowledge of the state at surface Σ_3. The action of a random environment is represented by two possibilities for the propagator at each step (up and down arrows, respectively). The rules of classical probability lead to decoherence.

In the information game model, every possible "state" of an open quantum system is assigned a probability using the rules of classical, Bayesian statistics [17]. In theory, it makes no difference whether the system "state" is defined indirectly as a set of local observables or as a wavefunction covering a complete spacelike (Cauchy) surface at a single instant in some timeframe. The important aspect of the model is that all predictions are made in a consistent way from classical information known to the observers. Each observer uses their information on the quantum state along with a quantum-mechanical Green's function to generate a classical probability distribution over the set of histories.

To fix these ideas, suppose that five different observers are each able to completely determine the density matrix of a single particle on a different spacelike Cauchy surface, Σ, and that they do so without interfering with the particle's dynamics. Order the observers by their separation in time, so that $\Sigma_1 < \Sigma_2 < \ldots$. Assume further that all observers are able to compute the full time evolution according to Schrödinger's equation and know the particle's Hamiltonian as well as the locations of all five surfaces. In this case, each observer is able to fully determine the information about the particle available to every other observer.

Now suppose that, in addition to everything above, a random environment is also present. The observers can no longer fully predict one another's measurements. Instead, observer i can only predict a probability distribution over the density matrix, ρ_j that will be available to observer j. The prediction process is straightforward if the information available to observer i is sufficient to define a measure, $(dP[G])$, assigning a probability to all possible propagators, G.

That measure can be used to construct the measure over the space of possible density matrices on Σ_j as a sum of point measures. The machinery of quantum mechanics appears only as a way to transform information about G into its image on the set of possible density matrices. Logical consistency is guaranteed by correct use of inference [18].

Figure 1 illustrates the situation above from the perspective of the third observer. Time evolution of the state is represented by motion of the figure as time progresses along the horizontal axis. In this schematic drawing, there are two possible Hamiltonians, giving rise to slightly different motions. The lighter shading of the lower figures at each timestep illustrates the situation where the corresponding Hamiltonian has a smaller classical probability ($dP[G_{\text{lower}}]$).

The sections that follow examine the inferred process, ρ_t, as viewed from a single fixed observer, and study the average, $\rho(x^1, x^2, t) = \int dP[G]\, G(t) * \rho_0$, which is appropriate since quantum mechanical operators are linear in ρ.

3. Incipient Classical Langevin Equation

The Wigner distribution [19] is defined as the Weyl symbol of the density matrix, ρ,

$$W_t(x,p) \equiv \mathcal{W}[\rho_t] = h^{-d} \iint_\Sigma dx_1 dx_2\; \rho_t(x_1, x_2) \delta(x - \tfrac{x_1+x_2}{2}) e^{ip(x_2-x_1)/\hbar}, \tag{1}$$

where d is the dimension of the position vector, $x \in \mathbb{R}^d$. Recognizing that $\mathcal{W}$ is a Fourier transform, we denote its inverse by $\mathcal{W}^{-1}$.

For compact notation, we use an underline to cast an operator, A, on Hilbert space into a superoperator, $\underline{A}$ (acting on other operators, B) as $\underline{A}[B] \equiv ABA^\dagger$, where $A^\dagger$ denotes the conjugate transpose. Note that if A, B are bounded, and B has real, non-negative eigenvalues, then $\underline{A}[B]$ does also.

It is well-known that the propagator for the wavefunction, $U^t \equiv e^{-it\hat{H}/\hbar}$, can be expressed as a path integral [20]. The time-slicing procedure for summation over histories has been rigorously defined for Cartesian and for general covariant spatial coordinates, and has been extended to field and point-particle systems in special relativity [21]. More refined derivations of continuous-time path integrals have also been given for particles in curved spacetime [22]. For simplicity in introducing the ideas here, our treatment will be confined to the Cartesian form.

The propagator for the density matrix is given by the forward-backward path integral [23],

$$\rho(x_N^1, x_N^2) = \prod_{n=0}^{N-1} \left[C^{-d} \iint dx_n^1 dx_n^2 \right] e^{\frac{i}{\hbar}(\mathcal{A}[x_0^1 \to x_N^1] - \mathcal{A}[x_0^2 \to x_N^2])} \rho(x_0^1, x_0^2)$$

$$C \equiv \epsilon h/m \tag{2}$$

$$\mathcal{A}[x_0 \to x_N] = \sum_{n=1}^{N} \frac{m(x_n - x_{n-1})^2}{2\epsilon} - \epsilon V\left(\frac{x_n + x_{n-1}}{2}\right) \tag{3}$$

over all pairs of paths, x^1, and x^2, connecting points on $x_0^1, x_0^2 \in \Sigma'$ and $x_N^1, x_N^2 \in \Sigma$. This goes over to continuous time in the limit $N \to \infty$ with $\epsilon N \to t$.

We use the path integral and the definition of the Wigner distribution [Eqs. (2) and (1)] to write the 'Wigner propagator' from (x', p') to (x, p) in the form,

$$-iL \equiv \lim_{t \to 0^+} t^{-1} \left(\mathcal{W} \underline{U}^t \mathcal{W}^{-1} - \underline{I} \right)$$

$$= \lim_{\epsilon \to 0^+} \epsilon^{-1} \left(C^{-d} \iint dy dy'\, e^{-\frac{i}{\hbar}\phi_0} - \underline{I} \right) \tag{4}$$

$$\phi_0 = \mathcal{A}[x' - \tfrac{y'}{2} \to x - \tfrac{y}{2}] - \mathcal{A}[x' + \tfrac{y'}{2} \to x + \tfrac{y}{2}] + py - p'y'. \tag{5}$$

In the limit, the action consists only of a single time-step, so that

$$\phi_0 = -\frac{m\Delta x\Delta y}{\epsilon} + \epsilon(V(\bar{x}+\bar{y}/2) - V(\bar{x}-\bar{y}/2)) + \Delta p\bar{y} + \Delta y\bar{p}, \tag{6}$$

where we have used the obvious notations, $\bar{x} = (x + x')/2$ and $\Delta x = x - x'$, etc. and noted that the transformations factor as $py - p'y' = \Delta p\bar{y} + \Delta y\bar{p}$.

The integral over Δy can be performed immediately. The short-time expansion of the evolution operator is thus,

$$e^{-iL\epsilon} = h^{-d}\delta\left(\Delta x - \tfrac{\epsilon\bar{p}}{m}\right)\int d\bar{y}\; e^{-\frac{i}{\hbar}\phi_1} \tag{7}$$
$$\phi_1(\bar{x},\bar{y}) \equiv \bar{y}\Delta p + \epsilon V(\bar{x}+\bar{y}/2) - \epsilon V(\bar{x}-\bar{y}/2)$$

This form shows the equivalence of the Wigner propagator with the Moyal bracket [24]. It was introduced by Schmidt [25].

The physical content of this equation can be seen assuming y is near zero [25] or using a "pre-averaging" method [26]. Pre-averaging makes a computational prescription out of the method of stationary phase by applying the identity $\int f(y)\,dy = \int (f * g)(y)\,dy$ [when $\int dy\, g(y) = 1$] in the form,

$$\int dy\; e^{-\frac{i}{\hbar}\phi_1(x,y)} = (2\pi l^2)^{-d/2}\iint dydz\; e^{-\frac{i}{\hbar}\phi_1(x,y-z)-z^2/2l^2}, \tag{8}$$

which holds for any width, l. If l is small enough, the exponential weighting allows expanding about $z = 0$ to good approximation.

$$\phi_1(x,y-z) = \phi_1(x,y) - z\cdot\delta_x\bar{\mathcal{A}} - \frac{1}{2}z\cdot\partial_y z\cdot\delta_x\bar{\mathcal{A}} + O(z^3), \tag{9}$$
$$\delta_x\bar{\mathcal{A}} \equiv \frac{1}{2}\left(\frac{\delta\mathcal{A}(x+y/2)}{\delta x} + \frac{\delta\mathcal{A}(x-y/2)}{\delta x}\right)$$

Retaining the term linear in z exhibits exponential fall-off away from Newton's law,

$$\int dy\; e^{-\frac{i}{\hbar}\phi_1(x,y)} \simeq \int dy\; e^{-\frac{i}{\hbar}\phi_1(x,y)}e^{-\frac{l^2(\delta_x\bar{\mathcal{A}})^2}{2\hbar^2}}. \tag{10}$$

Note that the factor $(\delta_x\bar{\mathcal{A}})^2$ generates the Langevin equation in the classical limit, $\bar{y} \to 0$.

$$\mathrm{P}\left(x,p|x',p'\right) = \mathcal{Z}_{\text{kin}}^{-1}(x',p')\delta(\Delta x - \epsilon\bar{p}/m)\exp\left\{-\frac{(\delta_x\mathcal{A})^2}{2\sigma^2\epsilon} + \frac{\beta}{2m}\bar{p}\cdot\delta_x\mathcal{A}\right\} \tag{11}$$

Eq. (11) clearly has a maximum entropy structure, with constraints on on $\langle(\delta_x\mathcal{A})^2|x',p'\rangle$ and $\langle dE|x',p'\rangle = \langle -\Delta x/\epsilon\cdot\delta_x\mathcal{A}\rangle$ – a connection that was first introduced in Ref. [16]. That derivation identified a normalization factor $\mathcal{Z}_{\text{kin}}(x',p')$ which is termed the kinetic partition function. Averages obtained by differentiating ($\log\mathcal{Z}_{\text{kin}}$) are identical to the results that would be obtained from the usual Stratonovich form of the stochastic calculus for the Langevin equation.

The structure of the Langevin equation is thus embedded in quantum mechanics itself. Quantum mechanics is still reversible because of the phase factor. Classically, however, the loss of the phase information should be modeled with random jumps in momentum on the scale of $\hbar/l$. Since z^3 and higher terms depend on 3rd order and higher derivatives of the potential. $l = l(x,y)$ can be safely set to the largest length over which the action functional remains quadratic [25, 27].

IOP Conf. Series: Journal of Physics: Conf. Series **880** (2017) 012039 doi:10.1088/1742-6596/880/1/012039

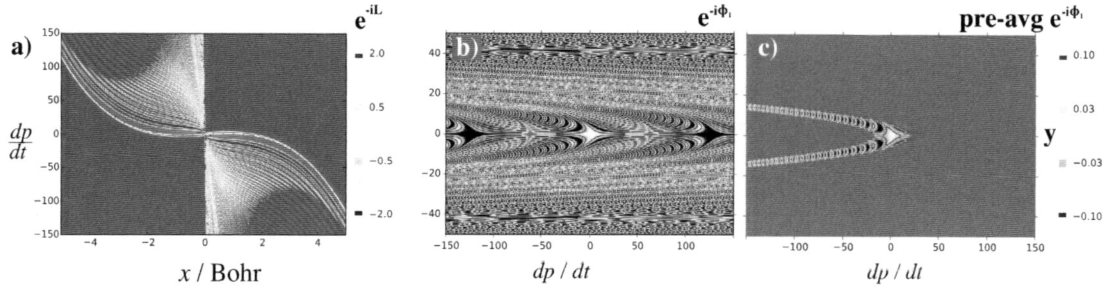

Figure 2. Wigner propagator (a), direct integrand (b), and pre-averaged integrand (c) for the quartic potential energy landscape $V(x) = x^4/4 - x^2$. The white dotted line in (a) shows the force, $-V'(x)$. The pre-averaged integrand in (c) was computed as the convolution of Eq. (8) using the expansion of Eq. (9) up to z^2.

Figure 2a plots the Wigner propagator for the symmetric quartic potential, $V(x) = x^4/4 - x^2$ at $\epsilon = 10^{-1}$ in atomic units ($\hbar$ = electron mass = 1). Although the full propagator is 4-dimensional, Eq. (7) shows that it only depends on the combination $\bar{x}$ and $\Delta p/\epsilon$. The integrals over $\bar{y}$ were carried out numerically using 200-point Gauss-Hermite quadrature after pre-averaging (integrating over z in Eq. (8)) using the expansion of Eq. (9) up to z^2 (rather than the first-order Eq. (10)). It was verified that the result was not affected by the particular choice of width, l (for which we used 1 Bohr). The dramatic effect of pre-averaging is shown in Fig. 2b and c, which plot the integrand before and after.

The phase structure visible in Airy-function like oscillations of Δp (vertical slices of Fig. 2a) is physically smoothed by the uncertainty principle. Although it is not shown here, propagating a coherent state (a minimum uncertainty wavepacket) is identical to smearing the propagator with a Gaussian function. Further smearing over $\Delta p/\epsilon$ is achieved by averaging over random forces (introduced in Sec. 4).

4. The quantum moment generating function

Following the classical weighting of ΔE, we introduce the following ansatz for the stochastic Wigner propagator,

$$-iL_M \equiv \lim_{t\to 0^+} t^{-1} \left(\int dP(k)\, \mathcal{W}\, \underline{\underline{M[U(k)^t]}}\, \underline{\mathcal{Z}_{\text{kin}}^{-1/2}} \mathcal{W}^{-1} - \underline{I} \right). \tag{12}$$

The super-operator $\mathcal{W}$ denotes the transformation of an operator into its Weyl symbol, and $\underline{M}$ is some maximum entropy weighting on the propagators, $U(k) = e^{-i\hat{H}_k/\hbar}$, that accounts for the environment's tendency to dissipate energy from the system. By analogy with the exponential weighting of energy changes in the classical case, we introduce dissipation using $M \equiv e^{-\beta\hat{H}/4}$, where $\beta = 1/k_B T$, and T is the temperature of the interacting thermal environment.

Each observer uses their information on $dP(k)$ and T to estimate the time-evolution as a series of applications of Eq. (12). This perspective makes a whole class of paradoxes disappear. The measurement of an observable at a future moment in time does not cause collapse of the wavefunction. Instead, both the preparation of the initial state and the measurement give distinct pieces of information on the possible influence of the environment, through the set of possible propagators (here $dP(k)$ and $\hat{H}, T$).

The exponential dependence on $\Delta\hat{H}$ makes sense in view of the fact that transitions are only generated by random deviations of $U(k)$ away from $e^{-i\hat{H}/\hbar}$. The energy weighting of M parallels the KMS condition on the symmetry of the corresponding transitions in a thermal bath [28].

In order for the propagator to be normalized we must have,

$$\mathcal{Z}_{\text{kin}} = \lim_{t\to 0^+} \int dP(k)\, \underline{\underline{M}}^\dagger[U(k)^{-t}][I]. \tag{13}$$

The dependence of the kinetic partition function on the starting configuration in the classical case translates to an operator valued $\mathcal{Z}_{\text{kin}}$ in the quantum case. Its derivatives generate moments,

$$\left.\frac{\partial^n \mathcal{Z}_{\text{kin}}}{\partial(\beta/2)^n}\right|_{\beta=0} = \sum_{m=0}^{n} \binom{n}{m} \sum_{j=0}^{m} 2^{-m} \binom{m}{j} \lim_{t\to 0^+} \int dP(k)\, \hat{H}^j(-\hat{H}_k(t))^{n-m} \hat{H}^{m-j} \tag{14}$$

where $\hat{H}_k(t) \equiv U(k)^{-t}\hat{H}U(k)^t$. When $\hat{H}$ commutes with the density matrix, ρ, the sums can be carried out to give

$$\frac{\partial^n \operatorname{tr}[\mathcal{Z}_{\text{kin}}\rho]}{\partial(-\beta/2)^n} = \int dP(k)\, \operatorname{tr}\left[(H_k(t) - H(0))^n \rho\right]. \tag{15}$$

To make an explicit comparison to the Langevin equation, we let $k \in \mathbb{R}$ be the strength of a random impulsive force so that the set of possible propagators for the system is described by a Wiener-process distribution for the forces,

$$U(k) = e^{-\frac{i}{\hbar}(\hat{H}-k\hat{x})}, \qquad dP(k) = (2\pi\sigma^2/\epsilon)^{-d/2} e^{-k^2\epsilon/2\sigma^2} dk. \tag{16}$$

With this choice, it can be shown that $-iL_M$ is the generator of a quantum dynamical semigroup with Lindblad form,

$$\begin{aligned} \mathcal{W}^{-1}[-iL_M[\mathcal{W}\rho]] = &-\tfrac{i}{\hbar}[\hat{H},\rho] - \frac{\sigma^2}{4\hbar^2}[\underline{M}[\hat{x}^2] - \underline{M}^\dagger[\hat{x}^2], \rho] \\ &- \frac{\sigma^2}{\hbar^2}\left(\underline{M}[\hat{x}]\rho\underline{M}^\dagger[\hat{x}] - \tfrac{1}{2}\{\underline{M}^\dagger[\hat{x}]\underline{M}[\hat{x}], \rho\}\right). \end{aligned} \tag{17}$$

The proof is by expanding $e^{-i\hat{H}_k\epsilon/\hbar}$ and retaining all terms of order ϵ, and noting that $\mathcal{Z}_{\text{kin}}^{-1/2} = I - \frac{1}{2}(\mathcal{Z}_{\text{kin}} - I) + O(\epsilon^2)$. Expanding $\underline{M}[\hat{x}]$ in powers of β proves the $O(\beta^2)$ equivalence with the Caldeira-Leggett master equation mentioned earlier.

5. The stochastic Wigner propagator

We proceed to simplify the Wigner propagator for the stochastic quantum mechanical evolution defined in Section 4. From its connection to the Caldeira-Leggett equation, it is expected that this propagator will recover the Langevin form at high temperature. Our purpose is to show how this comes about and to investigate the unique properties of the kinetic partition function.

We follow the derivation of Ref. [25] starting from the unnormalized propagator,

$$-iL'_M = \lim_{t\to 0^+} t^{-1}\left(\left(\int dP(k)\, \mathcal{W}\, \underline{\underline{M}}[U(k)^t]\mathcal{W}^{-1} - \underline{I}\right)\right) \tag{18}$$

$$\begin{aligned} \underline{M}[U(k)^t] &= e^{-\beta\hat{H}/4} e^{-it(\hat{H}+k\hat{x})/\hbar} e^{\beta\hat{H}/4} \\ &= e^{-it(\hat{H}+k\underline{M}[\hat{x}])/\hbar}. \end{aligned} \tag{19}$$

To proceed, we expand $\underline{M}[\hat{x}]$ as a power series in β and drop terms higher than $(\beta\hbar)^2$,

$$\begin{aligned} \underline{M}[\hat{x}] &= \hat{x} - \frac{\beta}{4}[\hat{H},\hat{x}] + \frac{\beta^2}{32}[\hat{H},[\hat{H},\hat{x}]] - O((\beta\hbar)^3) \\ &= \hat{x} + \frac{i\beta\hbar}{4m}\hat{p} + \frac{(\beta\hbar)^2}{32m}\hat{V}' - O((\beta\hbar)^3) \end{aligned} \tag{20}$$

Note that truncating the series at any order still leads to a propagator that is completely positive. This parallels the procedure by Diósi of truncating the expansion of $\beta\hbar\omega/2\coth(\beta\hbar\omega/2)$ [2]. The ratio of $\hat{x}$ and $\hat{p}$ coefficients in Eq. (20) shows the origin of the factor λ_{DB}^{-2} in the decoherence time-scale.

The presence of $i\hat{p}$ in the effective Hamiltonian technically requires us to re-derive the momentum part of the action [21]. However, the next step shows the same result even if $\hat{p}$ is directly substituted for $m\Delta x/\epsilon$ in writing down the action.

Using this damped Hamiltonian in the action functional, we note that the integral over Δy (see Ref. [25]) is unaffected. Thus, to order $(\beta\hbar)^2$,

$$e^{-iL'_M\epsilon} = h^{-d}\delta\left(\Delta x - \tfrac{\epsilon\bar{p}}{m}\right)\int d\bar{y}\int (2\pi\sigma^2/\epsilon)^{-d/2}dk\, e^{-\frac{i}{\hbar}\phi_k - \frac{k^2\epsilon}{2\sigma^2}} \tag{21}$$

$$\phi_k \equiv \bar{y}\Delta p + \epsilon V(\bar{x}+\bar{y}/2) - \epsilon V(\bar{x}-\bar{y}/2) + k\epsilon\left[\bar{y} + \frac{i\beta\hbar}{2\epsilon}\Delta x + \frac{(\beta\hbar)^2}{32m}(V'(\bar{x}+\bar{y}/2) - V'(\bar{x}-\bar{y}/2))\right]$$

Performing the Gaussian integral over k leads to,

$$e^{-iL'_M\epsilon} = h^{-d}\delta\left(\Delta x - \tfrac{\epsilon\bar{p}}{m}\right)\int d\bar{y}\, e^{-\frac{i}{\hbar}\phi_2 - \frac{\sigma^2\epsilon}{2\hbar^2}\Theta(\bar{x},\bar{y})^2 + \frac{\beta^2\sigma^2}{8\epsilon}\Delta x^2} \tag{22}$$

$$\phi_2 = \bar{y}\Delta p + \epsilon V(\bar{x}+\bar{y}/2) - \epsilon V(\bar{x}-\bar{y}/2) + \frac{\beta\sigma^2}{2}\Delta x\Theta(x,\bar{y})$$

$$\Theta(x,y) = y + \frac{(\beta\hbar)^2}{32m}(V'(x+y/2) - V'(x-y/2))$$

We derive corrections to the Langevin equation (Eq. (27)) by expanding in powers y (which scales as $\hbar$).

$$V'(x+y/2) - V'(x-y/2) = yV''(x) + \tfrac{y^3}{2^2 3!}V^{(4)}(x) + O(y^5) \tag{23}$$

$$\Theta(x,y) = y\left(1 + \frac{(\beta\hbar)^2}{32m}(V''(x) + \tfrac{y^2}{2^2 3!}V^{(4)}(x))\right) + O(y^5) \tag{24}$$

Neglecting the terms of order y^3 and higher, $\Theta \simeq \bar{y}u$ in

$$e^{-iL'_M\epsilon} \simeq \delta\left(\Delta x - \tfrac{\bar{p}\epsilon}{m}\right)(2\pi\sigma^2u^2\epsilon)^{-d/2}$$
$$\exp\left\{\frac{\beta^2\sigma^2}{8\epsilon}\Delta x^2 - \frac{1}{2\sigma^2u^2\epsilon}\left(\Delta p + \epsilon V'(\bar{x}) + \frac{\beta\sigma^2 u}{2}\Delta x\right)^2\right\} \tag{25}$$

$$u(\bar{x}) = 1 + \frac{(\beta\hbar)^2}{32m}V''(\bar{x}) \tag{26}$$

The normalization and the counter-term canceling Δx^2 in constructing $\beta dE/2$ both come out exactly. This is equivalent to the Langevin equation,

$$\Delta p = -\left(V'(x) + \frac{\beta\sigma^2 u}{2}\frac{\Delta x}{\epsilon}\right)\epsilon + \sigma u(\bar{x})k(t)\epsilon, \tag{27}$$

interpreted in the Stratonovich sense.

Even to first order in β, there is a clear difference from the classical case due to the appearance of the Hessian, V'', in $u(x)$. This can be seen as effectively an expansion term countering quantum confinement. The confinement term becomes important when the Hessian is on the order of $1/\hbar^2$. It then multiplies any noise present from the environment, increasing the

IOP Conf. Series: Journal of Physics: Conf. Series **880** (2017) 012039 doi:10.1088/1742-6596/880/1/012039

diffusion rate. Ref. [1] (Eq. (9.33) of Ref. [29]) identifies a tunneling contribution to crossing an energy barrier of $(\beta\hbar)^2(V''(\min) + V''(\text{barrier}))/24m$. This compares remarkably with the present derivation of $(\beta\hbar)^2V''/32m$ from a minimal model not based on specific properties of the environment.

This term also compares with the $[\hat{p}, [\hat{p}, \rho]]$ term put forward in Ref. [2]. There, expansion of $\coth(\beta\hbar\omega/2)$ and integration over an Ohmic frequency distribution showed that the coefficient should be the decoherence timescale times an extra factor of $\frac{4}{3}\lambda_{DB}^4\hbar^{-2} = \beta^2\hbar^2/12m^2$. To connect this to the Hessian, note that the $\hat{p}^2$ operator would effect an energy change on the order of $-mV''\delta x^2$. A similar expansion was noted in Ref. [3].

Because u appears in Eq. (27) as a scale for the temperature, it affects the low-temperature stationary distribution. Thus, the present work falls in-line with Ref. [30], which noted that a completely positive, translation invariant environment cannot achieve the Gibbs distribution at low temperature. We do not view this as a defect, since there is strong evidence that the unattainability of absolute zero is a real physical effect caused by continuous interaction with the environment [31, 32]. All of the previous works that find [33, 2, 30] and use [29] terms of this order include this effect.

Rather than analyze the dynamics as a classical or quantum Langevin equation, it is simpler still to study the properties of the normalization constant. Defining the normalization as in Eq. (13), it must have a Weyl symbol,

$$\mathcal{Z}_{\text{kin}}(x', p') = \iint dpdx\, e^{-iL'_M\epsilon} = \left(1 + \frac{\epsilon\beta\sigma^2 u}{2m}\right)^{-d/2} \exp\left(\frac{\epsilon\beta^2\sigma^2 p'^2}{8m^2}\right) \tag{28}$$

The energy exchange is,

$$\langle dE|x', p'\rangle = -\frac{\partial \log \mathcal{Z}_{\text{kin}}}{\partial\beta/2} = \frac{\sigma^2\epsilon}{m}\left(\frac{ud}{2} - \beta\frac{p'^2}{2m}\right). \tag{29}$$

In deriving Eq. (28), it is important to expand $\bar{p} = p' + \Delta p/2$ and retain all terms which contribute to order ϵ. The final result supports the choice of M from Eq. (12) and identifies the local quantum temperature with a scaled kinetic energy. Specifically, if the expected energy exchange during propagation is zero, then the observer can infer that the environment has an energy scale of T satisfying $k_BTu = \langle\frac{p^2}{m}\rangle$. In this formalism, temperature is a subjective, statistical property of the environment rather than an objective property of the system.

6. Conclusions

We have developed a maximum entropy formalism for describing the statistics of open quantum systems. The emergence of statistics has been framed in the context of multiple observers who have incomplete information on the time course of the system's Hamiltonian. This ansatz reproduces the Caldeira-Leggett equation, the connection between dissipative and decoherence timescales, and adds a quantum correction to the classical Langevin equation. It demonstrates how the density matrix of the system can collapse as a result of the inference process. This provides a method for attacking several open problems such as the reconstruction problem for relativistic theories of quantum mechanics [5]. Explicitly, local observables emerge because averaging over the noise process destroys coherences.

The analysis of Eq. (17) shows separate origins for decoherence and dissipation in the quantum Langevin equation. Decoherence comes about from random environmental noise, and this uncertainty carries an inherent time-scale. It plays the major role in determining the pointer states as those that commute with the environmental coupling [6]. However, dissipation comes from a separate source – the overall tendency of an infinite environment to take energy away from the system [7]. The dissipation used here (M) is characterized by an energy rather than a time-scale. Without random forces, M commutes with the dynamics, and has no opportunity to act. When randomness is present, both decoherence and dissipation work together to remove memory from the system. The decoherence timescale is much shorter because the uncertainty in the force destroys the phase structure (as a function of the momentum jumps, Δp) in the propagator.

This work can be adapted to more specific models for the stochastic environmental forces, including history-dependent processes. Our derivation offers a system-centric alternative to derive the fluctuation-dissipation theorem (as second derivatives of $\log \mathcal{Z}_{\text{kin}}$), and highlights the essential, physical features of the FDT and the relation between information and decoherence.

References

[1] Hanggi P, Grabert H, Ingold G L and Weiss U 1985 *Phys. Rev. Lett.* **55** 761–4
[2] Diósi L 1993 *Europhys. Lett.* **22** 1–3
[3] Kleinert H 2006 *J. Phys.: Conf. Series* **35** 25–38
[4] Feynman R P and Vernon F L 1963 *Ann. Phys.* **24** 118–73
[5] Carlip S 2001 *Rep. Prog. Phys.* **64** 885–942
[6] Zurek W H 1986 *Proc. NATO ASI Frontiers of Nonequilibrium Quantum Statistical Physics,* vol B135 ed G T Moore and M O Scully pp 145–9
[7] Unruh W G and Zurek W H 1989 *Phys. Rev.* D **40** 1071–94
[8] Diósi L 2011 *A Short Course in Quantum Information Theory* 2nd ed. (Lecture Notes in Physics vol 827) (Heidelberg: Springer)
[9] Cohen D 1997 *Phys. Rev.* E **55** 1422–41
[10] Schlosshauer M A 2007 *Decoherence and the Quantum-To-Classical Transition* (Heidelberg: Springer)
[11] Turchette Q A, Kielpinski D, King B E, Leibfried D, Meekhof D M, Myatt C J, Rowe M A, Sackett C A, Wood C S, Itano W M, Monroe C and Wineland D J 2000 *Phys. Rev.* A **61** 063418
[12] Turchette Q A, Myatt C J, King B E, Sackett C A, Kielpinski D, Itano W M, Monroe C and Wineland D J 2000 *Phys. Rev.* A **62** 053807
[13] Bera S, Donadi S, Lochan K and Singh T 2015 *J. Phys.: Conf. Series* **626** 012040
[14] Gallavotti G 2008 *Eur. Phys. J.* B **61** 1–24
[15] Zurek W H and Paz J P 1994 *Phys. Rev. Lett* **72** 2508–12
[16] Rogers D M, Beck T L and Rempe S B 2011 *J. Stat. Phys.* **145** 385–409;
Rogers D M and Rempe S B 2012 *J. Phys., Conf. Ser.* **402** 012014
[17] Jaynes E T 1957 *Phys. Rev.* **108** 171–90
[18] Jaynes E T 2003 *Probability Theory: the Logic of Science* (Cambridge: Cambridge Univ. Press)
[19] Wigner E 1932 *Phys. Rev.* **40** 749–59
[20] Feynman R 2005 *Feynman's Thesis: A New Approach to Quantum Theory* (Singapore: World Scientific)
[21] Kleinert H 2008 *Path Integrals in Quantum Mechanics, Statistics, Polymer Physics, and Financial Markets* (Singapore: World Scientific)
[22] DeWitte-Morette C, Maheshwari A and Nelson B 1977 *Gen. Relativity and Gravitation* **8** 581–93
[23] Shao J and Makri J 1999 *J. Phys. Chem.* A **103** 7753–6
[24] Moyal J E 1947 *Math. Proc. Cambridge Phil. Soc.* **45** 99–124
[25] Schmid A 1982 *J. Low Temp. Phys.* **49** 609–26
[26] Sabo D, Doll J D and Freeman D L 2002 *J. Chem. Phys.* **116** 3509–20
[27] Zurek W 2002 *Los Alamos Science* **27** 2–25
[28] Spohn H and Lebowitz J L 1978 *Adv. Chem. Phys.* vol 38 ed S A Rice (New York: Wiley) pp 109–42
[29] Hänggi P, Talkner P and Borkovec M 1990 *Rev. Mod. Phys.* **62** 251–342
[30] Kohen D, Marston C C and Tannor D J 1997 *J. Chem. Phys.* **107** 5236–53
[31] Gardas B and Deffner S 2015 *Phys. Rev.* E **92** 042126
[32] Rogers D M 2017 *Phys. Rev.* E **95** 012149
[33] Ford G W, Lewis J T and O'Connell R F 1988 *Phys. Rev.* A **37** 4419–28

IOP Conf. Series: Journal of Physics: Conf. Series **880** (2017) 012040 doi:10.1088/1742-6596/880/1/012040

A fluid/gravity prescription of the post-Newtonian parameter

Roldão da Rocha

Centro de Matemática, Computação e Cognição, Universidade Federal do ABC - UFABC, 09210-580, Santo André, Brazil.

E-mail: roldao.rocha@ufabc.edu.br

Abstract. The KSS bound for the shear viscosity-to-entropy density ratio is employed in the fluid/gravity correspondence setup, to provide a theoretical bound for the post-Newtonian parameter in the Casadio-Fabbri-Mazzacurati black string solution that matches the observational bound [1]. This derived bound is here shown to vary according to the black string temperature, and, although the post-Newtonian parameter reaches bigger values as the temperature increases for fixed black string masses, it is still into the observational/experimental bound range. The Casadio-Fabbri-Mazzacurati-AdS black brane is shown to converge to the Schwarzschild-AdS black brane, for sufficiently high order expansion of the dual boundary energy-momentum tensor.

1. Introduction

Extra-dimensional generalisations of the general relativity comprise extended models with unexpected consequences in the physics of black hole [2, 3] and ramifications into astrophysics [4, 5]. Particularly, black strings represent physical solutions of the brane effective Einstein equations [6, 7, 8, 9]. On the other hand, fluid dynamics can be thought of describing QFT, in the long wavelength regime [10]. At low energies, the associated hydrodynamics describe physical systems near the equilibrium and the transport parameters, for example conductivity and viscosity, encrypt the information on the perturbations and their propagation along the fluid. Such a paradigm relies on the AdS/CFT correspondence setup [11, 12], whose long wavelength regime establishes the interplay between black holes solutions in (asymptotically) anti-de Sitter (AdS) spacetime and hydrodynamics on the boundary, associated to a field theory in a strongly coupled regime [13, 14]. A prominent feature in fluid/gravity correspondence is given by the shear viscosity-to-entropy density ratio [10], whose universality, under certain conditions, has been explored in various contexts [13, 14, 15, 16, 17, 18].

Black strings and black branes, thus, have hydrodynamical aspects beyond entropy and temperature, encoded by transport coefficients as vorticity, viscosity, shear, diffusion parameters and rates. Black strings and black branes solutions have the shear viscosity-to-entropy density ratio η/s that is bounded [15] [1],

$$\frac{\eta}{s} \gtrapprox \frac{1}{4\pi}\,, \tag{1}$$

[1] Units with $\hbar = c = k_{\rm B} = 1$ shall be employed; spacetime 4D indexes run in the range $\mu = 0, 1, 2, 3$ and 5D indexes run in the range $A = 0, 1, 2, 3, 4$.

IOP Conf. Series: Journal of Physics: Conf. Series **880** (2017) 012040 doi:10.1088/1742-6596/880/1/012040

which is called the Kovtun-Starinets-Son (KSS) result [15]. In the case of Einstein gravity or any theory in the limit $N_c \to \infty$, the KSS result is an equality $\eta/s = 1/4\pi$. This result is derived in the context of SUGRA in a $\mathcal{N} = 4$ SUSY $SU(N_c)$ gauge theory, in the large 't Hooft coupling constant regime $g^2 N_c$ [17]. The AdS bulk dynamics corresponds to vacuum states of the boundary Yang-Mills setup, implying that AdS-Schwarzschild solutions are equivalent to boundary gauge theory states [19]. Other solutions of the Einstein equations [20, 1] correspond to a different thermal state in the boundary gauge theory.

The fluid/gravity correspondence was used to bound the post-Newtonian parameter via the black string [1], using the CFM black strings [4] associated with the effective Casadio-Fabbri-Mazzacurati (CFM) solutions [21, 22]. CFM metrics represent solutions of the bulk Einstein equations projected onto the brane, having a parametrised post-Newtonian parameter β corresponding to the Eddington-Robertson-Schiff parameter, bounded by experimental/observational data [23, 24]. In general, one can construct a dual fluid stress tensor by solving the bulk Einstein equations in asymptotically AdS spacetime and taking the long wavelength limit, but there are only a few black hole solutions that admit such a hydrodynamic dual on the boundary, even asymptotically. CFM solutions were studied in the context of the fluid/gravity correspondence [1]. A 5D CFM black string was proposed to have a 4D dual fluid dynamical prescription, and the fluid/gravity correspondence can impart a bound for the parametrised post-Newtonian parameter, through the universal KSS bound (1). After studying the bound for β, derived by the KSS result using the CFM-AdS black string, here such bound shall be further analysed, and its variation according to the black brane temperature shall be scrutinised.

The procedure to reported in this paper is, after reviewing the CFM solutions, to use the Green-Kubo formula to bound the parametrised post-Newtonian parameter through the shear viscosity-to-entropy density ratio by including up to the second-order hydrodynamics terms, discussing the unicity of the AdS-Schwarzschild solution.

2. A Casadio-Fabbri-Mazzacurati solution

In brane-world models, gravity can propagate into the bulk, originating a 5D Weyl fluid term that is projected onto the brane-world, via the Gauss-Codazzi method. Such a term must be taken into account in the Einstein effective equation. A by-product of the brane equations is the CFM metrics [22], which are originally vacuum solutions of the Einstein field equations, presenting a parametrised post-Newtonian parameter β. Ref. [21], for example, reported a specific bound for the parametrised post-Newtonian parameter, $\beta \approx 1$, by experimental/observational data [?]. Indeed, classical tests of general relativity provides the bound $|\beta - 1| \lesssim 0.003$ [23, 24], bringing data about the vacuum brane self-energy [2, 25, 26, 21].

The 4D Einstein effective equations can be derived when the 5D bulk Einstein equations are projected onto the 4D brane, by the Gauss-Codazzi method (the convention $8\pi G = c = 1 = \hbar$ is going to be fixed[2], where $G = \hbar c/M_{\rm pl}^2$ and $M_{\rm pl}$ denotes the Planck scale, and $\mu, \nu, \rho, \sigma = 0, 1, 2, 3$), yielding [27]

$$R_{\mu\nu} - \frac{1}{2} R g_{\mu\nu} + \Lambda g_{\mu\nu} - \mathrm{T}_{\mu\nu} = 0, \tag{2}$$

where $R_{\mu\nu}$, R, and Λ are, respectively, the Ricci tensor, the Ricci scalar, and the 4D cosmological constant. The effective stress tensor in Eq. (2) can be split into a sum, $\mathrm{T}_{\mu\nu} = T_{\mu\nu} + \mathcal{E}_{\mu\nu} + \sigma^{-1} S_{\mu\nu}$, where the first component $T_{\mu\nu}$ denotes the brane matter stress tensor, $\mathcal{E}_{\mu\nu}$ inscribes high-energetic corrections from the 5D Weyl fluid. The tensor $S_{\mu\nu}$ encrypts Kaluza-Klein imprints from the bulk onto the brane [27, 2]. Static, spherically symmetric, compact distributions on the

[2] Obviously, the precise values of all involved parameters shall be suitably taken into account, in the calculations in Sect. III.

brane, read

$$g_{\mu\nu}\, dx^\mu\, dx^\nu = -N(r)\, dt^2 + B(r)\, dr^2 + r^2\, d\Omega^2\,. \tag{3}$$

The CFM metric is derived by dismissing the condition $N(r) = B^{-1}(r)$ [21]. The CFM I solution comprises Eq. (3), with

$$N(r) = 1 - \frac{2\,G\,M}{r} \qquad \text{and} \qquad B(r) = \frac{1 - \frac{3\,G\,M}{2\,r}}{\left(1 - \frac{2\,G\,M}{r}\right)\left[1 - \frac{G\,M}{2\,r}(4\beta - 1)\right]}\,. \tag{4}$$

The solution (4) relies on the parameter β, being the Minkowski solution recovered when $M \to 0$. The horizon $r = R$ on the brane is derived by solving the algebraic equation $B^{-1}(R) = 0$. The sign of $(\beta - 1)$ determines whether the CFM solution is either colder or hotter than the Schwarzschild solution [21]. Moreover, the 4D Kretschmann scalar $R_{\mu\nu\rho\sigma}R^{\mu\nu\rho\sigma}$ diverges for $r = 0$ and $r = \frac{GM}{2}(4\beta - 1)$.

3. Fluid at the boundary, dual to the CFM black brane/black string solution

Hydrodynamics is the low-energy effective description for the low-momentum regime of correlation functions. Following the results derived in Ref. [1], the Green-Kubo formula plays a prominent role in the linear response theory, where the shear viscosity represents a fluid transport coefficient arising in response to a perturbation in the fluid stress-energy tensor.

The Kubo formula can be derived by coupling 4D gravity $g_{\mu\nu}$ to the fluid and determine the way how the fluid energy-momentum tensor responds to gravitational perturbations. In particular, for small metric perturbations which vary slowly in space and time, $g_{\mu\nu} = \bar{g}_{\mu\nu} + h_{\mu\nu}$ with $\|h_{\mu\nu}\| \ll 1$, the stress tensor can be expanded in the metric fluctuation $h_{\mu\nu}$ and in gradients [28]. This procedure was reported in Ref. [1], yielding

$$\begin{aligned}\langle T^{\mu\nu}(x)\rangle &= \langle T^{\mu\nu}\rangle - \frac{1}{2}\int d^4y\, G_{\rm R}^{\mu\nu|\rho\sigma}(x;y)h_{\rho\sigma}(y) + \frac{1}{8}\int d^4y\int d^4z\, G_{\rm R}^{\mu\nu|\rho\sigma|\tau\zeta}(x;y,z)\, h_{\rho\sigma}(y)\, h_{\tau\zeta}(z) + \ldots \\ &\equiv \langle T^{\mu\nu}_{(0)}\rangle + \langle T^{\mu\nu}_{(1)}\rangle + \langle T^{\mu\nu}_{(2)}\rangle + \cdots\,, \end{aligned} \tag{5}$$

where $\langle T^{\mu\nu}\rangle$ corresponds to $h = 0$ and $G_R^{\mu\nu|\cdots}$ are retarded n-point correlators. The fluid response to perturbations is derived via the conservation law $\nabla_\mu T^{\mu\nu} = 0$ and the conformal condition $T^\nu{}_\nu = 0$.

At zero order in derivatives, the energy momentum tensor reads

$$T^{\mu\nu}_{(0)} = (\epsilon + P)\, u^\mu\, u^\nu + P\, \bar{g}^{\mu\nu}\,, \tag{6}$$

where u^μ is the fluid four-velocity, ϵ denotes its energy density, P the pressure and $\bar{g}_{\mu\nu}$ is the unperturbed metric on the 4D boundary where the fluid lives. The response of the two-point function of the associated perturbed metric can then be written as the first order term $\langle T^{(1)}_{\mu\nu}\rangle$ in the expansion (5), $\langle T^{\mu\nu}(x)\rangle \sim \int dy\, G_R^{\mu\nu|\rho\sigma}(x;y)\, h_{\rho\sigma}(y)$, where the retarded Green function $G_R^{\mu\nu|\rho\sigma} = \langle T^{\mu\nu}(x)\, T^{\rho\sigma}(y)\rangle$ is employed [12]. Hence, from the gravity point of view of the correspondence, the above perturbation should arise from metric fluctuations $h_{\mu\nu}$ of the appropriate black brane metric.

The first-order expansion of the hydrodynamical stress-energy tensor, also known as the constitutive equation, includes dissipative terms, as well as shear and bulk viscosities. In curved space-times, the first order contribution is given by [29]

$$T^{\mu\nu}_{(1)} = -P^{\mu\rho}P^{\nu\sigma}\left[\left(\nabla_{(\rho}u_{\sigma)} - \frac{2}{3}\,\bar{g}_{\rho\sigma}\,\nabla_\tau u^\tau\right)\eta + \zeta\,\bar{g}_{\rho\sigma}\nabla_\tau u^\tau\right]\,, \tag{7}$$

where η is the shear viscosity, ζ the bulk viscosity, and ∇_μ represents the covariant derivative in the space-time metric $\bar{g}_{\mu\nu}$. In addition, $P^{\mu\nu} = \bar{g}^{\mu\nu} + u^\mu u^\nu$ is the projection tensor along spatial directions. The contribution to the shear viscosity η comes only from the component $h_{xy} = h_{xy}(t)$ of $h_{\mu\nu}$, and $h_{\mu\nu} = 0$, otherwise [1]. Such fluctuations around thermal equilibrium are small, making to assert the uniform fluid temperature $\mathcal{T}(x^\mu) = \mathcal{T}_0$, at rest in the chosen frame, $u^\mu = (1, u^i = 0)$.

The shear viscosity of the dual theory can be computed from gravity [13], by the Kubo formula, relating the viscosity to equilibrium correlation functions.

$$\delta\langle T^{xy}\rangle = -G_R^{xy|xy}(q)\,h_{xy}\ , \tag{8}$$

where $G_R^{xy|xy}(q) = -i\int d^4x\, e^{-i\,q_\mu\,x^\mu}\,\theta(x^0)\,\langle T^{xy}(x^\mu)\,T^{xy}(0)\rangle$. The Green-Kubo formula reads,

$$\eta = -\lim_{\omega\to 0}\frac{\Im\, G_R^{xy|xy}(\omega, 0)}{\omega}\ , \tag{9}$$

where $\Im$ denotes the imaginary part.

The CFM black string associated to the CFM I solution are related to black branes in AdS [1, 4], corresponding to a late time behaviour of their dual fluid. CFM black branes in AdS can be also obtained as a perturbative approach of CFM black holes on the brane [1, 4]. The 5D metric reads

$$g_{AB}\,dx^A\,dx^B = -N(r)\,dt^2 + B(r)\,dr^2 + \frac{r^2}{\ell^2}\,d\vec{x}^2\ , \tag{10}$$

where $d\Omega^2 = \ell^{-2}\,d\vec{x}^2$, the length ℓ is related to the AdS curvature [20] and the metric coefficients N and B equal the CFM expression (4). The entropy of these black strings can be computed by using the Hawking-Bekenstein formula, $S = \frac{A}{4\,G_5} = \frac{R^3\,V_3}{4\,G_5\,\ell^3}$, where R is the event horizon radius and $V_3 = \ell\int \ell^{-2}\,d\vec{x}^2$. Moreover, the volume density of entropy $s = S/V_3$ must be finite, which requires that $\beta < 5/4$ in Eq. (4). Introducing the variable $u = R/r$, and defining $\bar{M} = G\,M$, yields

$$B(u) = \frac{1 - \frac{3\,\bar{M}}{2\,R}\,u}{\left(1 - \frac{2\,\bar{M}}{R}\,u\right)\left[1 - \frac{\bar{M}}{2\,R}\,u\,(4\beta - 1)\right]} \tag{11}$$

and the metric finally reads

$$g_{AB}\,dx^A\,dx^B = -N(u)\,dt^2 + B(u)\,\frac{R^2}{u^4}\,du^2 + \frac{R^2}{u^2}\,d\Omega^2 \equiv g_{uu}\,du^2 + g_{\mu\nu}\,dx^\mu\,dx^\nu\ . \tag{12}$$

The value for β was derived in [1] for the fixed black string temperature, from the KSS result (1). The aim here is to analyse how the KSS behaves as the temperature increases. The background metric (12) can be perturbed as $g_{AB} \mapsto g_{AB} + h_{AB}$ [13, 15]. As mentioned before, the contribution to the shear viscosity comes from a single component of the metric fluctuations. Denoting this term by $\phi = \phi(t, u, \vec{x})$, the wave equation reads, after taking $\phi \simeq e^{i\,\omega\,t}\,\Phi(u)$ [15]

$$\partial_u\left(\sqrt{-g}\,g^{uu}\partial_u\Phi\right) - \sqrt{-g}\,g^{tt}\,\omega^2\,\Phi \approx 0\ , \tag{13}$$

which can be specialised for both CFM I metric, provided by Eqs. (4), that is

$$\frac{d^2\Phi}{du^2} + \frac{V}{u}\frac{d\Phi}{du} + \left(1 - \frac{2\,\bar{M}}{R}\,u\right)\omega^2\,\Phi = 0\ , \tag{14}$$

where V is given by Eqs. (29) in Ref. [1]. The retarded Green function [13, 14], $G_R(\omega, \vec{0}; \beta) = -\sqrt{-g}\, g^{uu}\, \Phi^* \frac{d\Phi}{du}\big|_{u\to 0}$, for both cases in Eqs. (29, 30) in Ref. [1] and then derive the shear viscosity from the Green-Kubo formula (9). The resulting propagator yields

$$\frac{\eta}{s} = -\frac{1}{s} \lim_{\omega\to 0} \frac{\Im G_R(\omega, k=0; \beta)}{\omega} = \frac{1}{4\pi}. \tag{15}$$

where we remarked that both s and η do depend on β. Eq. (15) was obtained in the linear hydrodynamics regime, corresponding to the first two terms in Eq. (5), that is Eqs. (6) and (7). Including the second-order regime of a dual hydrodynamic prescription [10, 19, 30, 28], the parametrised post-Newtonian parameter β bound can be refined.

Moreover, $\sigma^{\alpha\beta}$ denotes the shear tensor and $\Omega^{\alpha\beta}$ is the vorticity tensor. The second order dissipative part in the expansion of the stress-energy tensor (5) reads

$$\begin{aligned} T^{\mu\nu}_{(2)} &= \eta\,\tau_\Pi \left[\langle g_{\rho\alpha} u^\alpha \nabla_\rho \sigma^{\mu\nu}\rangle_\perp + \frac{1}{2}\sigma^{\mu\nu}(\nabla_\rho u^\rho)\right] + \kappa(g_{\rho\zeta} - 2u_\rho u_\zeta)\left(\frac{1}{2}P_\alpha^{\ \mu} P_\gamma^{\ \nu} R^{\rho(\alpha\gamma)\zeta}\right. \\ &\left. -\frac{1}{3}P_{\alpha\gamma} P^{\mu\nu} R^{\rho\alpha\gamma\zeta}\right) + \frac{1}{2}P_\alpha^{\ \mu} P_\gamma^{\ \nu}\left(\lambda_1\, \sigma_\zeta^{(\alpha}\, \sigma^{\gamma)\zeta} + \lambda_2\, \sigma_\zeta^{(\alpha}\, \Omega^{\gamma)\zeta} + \lambda_3\, \Omega_\zeta^{(\alpha}\, \Omega^{\gamma)\zeta}\right) \\ &-\frac{1}{3}P_{\alpha\gamma} P_{\mu\nu}\left(\lambda_1\, \sigma_\zeta^\alpha\, \sigma^{\gamma\zeta} + \lambda_2\, \sigma_\zeta^\alpha\, \Omega^{\gamma\zeta} + \lambda_3\, \Omega_\zeta^\alpha\, \Omega^{\gamma\zeta}\right), \end{aligned} \tag{16}$$

where $\langle\ \cdot\ \rangle_\perp$ denotes the traceless component that is orthogonal to the fluid velocity. The coefficients in (16) have different interpretations [28]. It is worth mentioning that the second-order hydrodynamic transport coefficients κ and λ_i have been calculated from the leading order AdS/CFT correlators [16, 28, 30]to , corresponding to the black brane horizon response properties. For instance, $\lambda_3 = -4 \lim_{\substack{k_1\to 0\\ k_2\to 0}} \frac{\partial}{\partial_{k_1}} \frac{\partial}{\partial_{k_2}} \lim_{\substack{\omega_1\to 0\\ \omega_2\to 0}} G^{xy|0x|0y}$. The obtained expressions are given by Eq. (37) in Ref. [1], irrespectively of the transport coefficients in the range in Refs. [13, 19, 31].

If one now takes, as usual in the literature, $h_{xy} = h_{xy}(t, z)$, Eq. (16) yields

$$T_{(1)xy} + T_{(2)xy} = -P\, h_{xy} - \eta\, \dot{h}_{xy} + \eta\, \tau_\Pi\, \ddot{h}_{xy} - \frac{\kappa}{2}\left[\ddot{h}_{xy} + \partial_z^2 h_{xy}\right], \tag{17}$$

where dots denote derivatives with respect to the time coordinate. The retarded Green function reads

$$G_R^{xy|xy}(\omega, k) = P - i\,\eta\,\omega + \eta\,\tau_\Pi\,\omega^2 - \frac{\kappa}{2}\left[\omega^2 + k^2\right], \tag{18}$$

providing the analogue of the Kubo formula with second-order derivative terms. The kinetic parameters τ_Π and κ are the coefficients of the ω^2 and k^2 terms in the low-momentum expansion of the retarded propagator. With the above expressions, and repeating the previous analysis, one finds bounds – temperature dependent – for the parametrised post-Newtonian parameter β, shown in Fig. 1. All terms in Eq. (16) are included, with transport coefficients provided by Eq. (37) in Ref. [1]. The KSS bound is illustrated in Fig. 2, as a function of the mass and the temperature. Although the KSS bound implies higher values for β as the temperature increases, these values still match the experimental/observational bounds [23].

Refined expressions involving more transport coefficients, like in Ref. [32], could be used. However, such contributions in the viscosity-to-entropy ratio would amount at most to order 10^{-5} corrections in the bounds for β. These preliminary results including second-order terms are thus fully compatible with the observational bound $|\beta - 1| \lesssim 0.003$ provided by the perihelion shift [23]. Nevertheless, the main result here is that, since the parametrised post-Newtonian

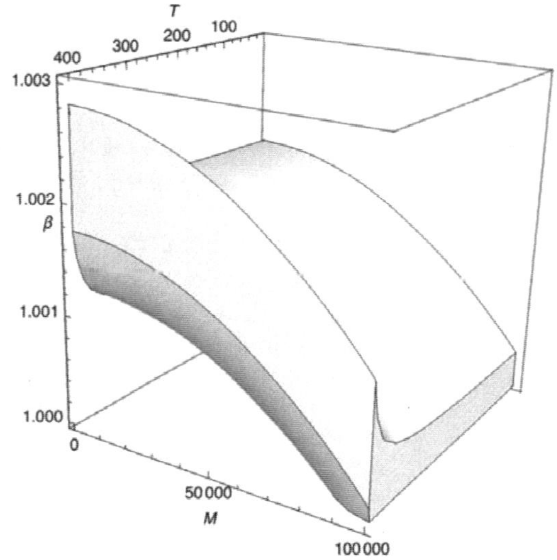

Figure 1. Allowed range for the parametrised post-Newtonian parameter β as a function of the mass M (in Solar mass units), for the CFM I case including second-order corrections.

parameter is bound to be close to the unity also in five dimensions, the CFM black branes must effectively reduce to their Schwarzschild counterparts. In fact, the CFM metric reduces to the Schwarzschild metric in the limit $\beta \to 1$. Here, we proved that the KSS result forces the CFM black branes to reduce to Schwarzschild black branes. For the boundary energy-momentum expansion (5), the first order approximation (7) provides $|\beta - 1| \lesssim 10^{-1}$, whereas the second order approximation (7) provides $|\beta - 1| \lesssim 3 \times 10^{-2}$. When the higher order term $\langle T^{\mu\nu}_{(3)} \rangle$ is included in Eq. (5), one finds $|\beta - 1| \lesssim 10^{-4}$, suggesting that $\beta \to 1$ when sufficiently high-order terms $\langle T^{\mu\nu}_{(n)} \rangle$ are taken into account in Eq. (5). This is in agreement with the conjecture of Ref. [33], that the Schwarzschild black brane [34, 35, 36] is the unique static black hole solution of vacuum gravity with a negative cosmological constant. In fact, Schwarzschild-AdS black branes can be recovered from CFM-AdS black branes, providing fundaments for the proof of a conjecture [33], that asserts that the Schwarzschild-AdS black hole is the unique, static, asymptotically AdS, black-hole solution of gravity with a negative cosmological constant.

4. Conclusions

The shear viscosity-to-entropy ratio has been obtained with respect to the parametrised post-Newtonian parameter of CFM-AdS black branes. The KKS result (1) yields the bounds on β, for different black brane masses. For fixed values of the black brane mass, β is shown to increase with the black brane temperature, still in agreement with observational bounds. The KSS bound yields the CFM black branes to converge to Schwarzschild black branes, when sufficiently high-order terms in the boundary energy-momentum tensor (5) are taken into account, in agreement with the conjectured unicity of the Schwarzschild black brane as a static black-hole solution of vacuum gravity with a negative cosmological constant [33].

Acknowledgements

RdR thanks to CNPq, grant No. 303293/2015-2, and to FAPESP, grant No. 2015/10270-0, for partial financial support.

IOP Conf. Series: Journal of Physics: Conf. Series **880** (2017) 012040 doi:10.1088/1742-6596/880/1/012040

References

[1] Casadio R, Cavalcanti R T and da Rocha R 2016 *Eur. Phys. J.* C **76** 556
[2] Maartens R and Koyama K 2010 *Living Rev. Rel.* **13** 5
[3] Antoniadis I, Arkani-Hamed N, Dimopoulos S and Dvali G R 1998 *Phys. Lett.* B **436** 257
[4] da Rocha R, Piloyan A, Kuerten A M and Coimbra-Araujo C H 2013 *Class. Quant. Grav.* **30** 045014
[5] Casadio R and da Rocha R 2016 *Phys. Lett.* B **763** 434
[6] Seahra S S, Clarkson C and Maartens R 2005 *Phys. Rev. Lett.* **94** 121302
[7] Casadio R, Ovalle J and da Rocha R 2014 *Class. Quant. Grav.* **31** 045016
[8] Bazeia D, Hoff da Silva J M and da Rocha R 2013 *Phys. Lett.* B **721** 306
[9] Coimbra-Araujo C H, da Rocha R and Pedron I T 2005 *Int. J. Mod. Phys.* D **14** 1883
[10] Bu Y and Lublinsky M *J. High Energy Phys.* JHEP11(2014)064
[11] Maldacena J M 1999 *Int. J. Theor. Phys.* **38** 1113 [Adv. Theor. Math. Phys. **2** 231 (1998)]
[12] Aharony O, Gubser S S, Maldacena J M, Ooguri H and Oz Y 2000 *Phys. Rept.* **323** 183
[13] Policastro G, Son D T and Starinets A O 2001 *Phys. Rev. Lett.* **87** 081601
[14] Policastro G, Son D T and Starinets A O *J. High Energy Phys.* JHEP09(2002)043
[15] Kovtun P, Son D T and Starinets A O 2005 *Phys. Rev. Lett.* **94** 111601
[16] Grozdanov S and Starinets A O *J. High Energy Phys.* JHEP03(2015)007
[17] Buchel A, Liu J T and Starinets A O 2005 *Nucl. Phys.* B **707** 56
[18] Cremonini S 2011 *Mod. Phys. Lett.* B **25** 1867
[19] Bhattacharyya S, Hubeny V E, Minwalla S and Rangamani M *J. High Energy Phys.* JHEP02(2008)045
[20] Sadeghi M and Parvizi S 2014 (*Preprint* `1411.2358`)
[21] Casadio R, Fabbri A and Mazzacurati L 2002 *Phys. Rev.* D **65** 084040
[22] Casadio R and Mazzacurati L 2003 *Mod. Phys. Lett.* A **18** 651
[23] Will C M 2014 *Living Rev. Rel.* **17** 4
[24] Cavalcanti R T, da Silva A G and da Rocha R 2016 *Class. Quant. Grav.* **33** 215007
[25] Casadio R and Harms B 2002 *Int. J. Mod. Phys. A* **17** 4635
[26] Casadio R and Harms B 2001 *Phys. Rev.* D **64** 024016
[27] Shiromizu T, Maeda K i and Sasaki M 2000 *Phys. Rev.* **D62** 024012 (*Preprint* `gr-qc/9910076`)
[28] Arnold P, Vaman D, Wu C and Xiao W *J. High Energy Phys.* JHEP10(2011)033
[29] Son D T and Starinets A O 2007 *Ann. Rev. Nucl. Part. Sci.* **57** 95
[30] Moore G D and Sohrabi K A *J. High Energy Phys.* JHEP11(2012)148
[31] Nakayama Y 2012 *Int. J. Mod. Phys.* A **27** 1250125
[32] Finazzo S I, Rougemont R, Marrochio H and Noronha J *J. High Energy Phys.* JHEP02(2015)051
[33] Emparan R and Reall H S 2008 *Living Rev. Rel.* **11** 6
[34] Abdalla E, Cuadros-Melgar B, Pavan A B and Molina C 2006 *Nucl. Phys.* B **752** 40
[35] Abdalla E, Piedra O P F and de Oliveira J 2010 *Phys. Rev. D* **81** 064001
[36] Duff M J, Khuri R R and Lu J X 1995 *Phys. Rept.* **259** 213

Working directly with probabilities in quantum field theory

R Dickinson[1], J Forshaw[1] and P Millington[2]*

[1]Consortium for Fundamental Physics, School of Physics and Astronomy, University of Manchester, Manchester M13 9PL, United Kingdom
[2]School of Physics and Astronomy, University of Nottingham, Nottingham NG7 2RD, United Kingdom

E-mail: robert.dickinson-2@manchester.ac.uk, jeff.forshaw@manchester.ac.uk, p.millington@nottingham.ac.uk

Abstract. We present a novel approach to computing transition probabilities in quantum field theory, which allows them to be written directly in terms of expectation values of nested commutators and anti-commutators of field operators, rather than squared matrix elements. We show that this leads to a diagrammatic expansion in which the retarded propagator plays a dominant role. As a result, one is able to see clearly how faster-than-light signalling is prevented between sources and detectors. Finally, we comment on potential implications of this approach for dealing with infra-red divergences.

1. Introduction

The utility of scattering-matrix theory has biased the development of techniques in quantum field theory towards the calculation of transition *amplitudes*. However, isolated transition amplitudes are not physical observables and they often suffer from artefacts that are eliminated only after we combine individual amplitudes into transition *probabilities*. For instance, the soft and collinear infra-red (IR) divergences occurring in gauge amplitudes are cancelled by the interference of virtual and real emissions by means of the Bloch-Nordsieck [1] and Kinoshita-Lee-Nauenberg theorems [2, 3] (in the case of spin-1) or the Weinberg soft graviton theorem [4] (in the case of spin-2). In non-Abelian gauge theories, unphysical polarisation states are removed by the contributions from Faddeev-Popov ghosts [5]. It seems reasonable therefore to pursue means of directly calculating transition probabilities that bypass the amplitude level altogether, the hope being that such artefacts never appear explicitly.

Another (not unrelated) artefact of dealing with amplitudes, and one that we will focus on in this note, is apparent a-causal behaviour, i.e. behaviour seemingly at odds with Einstein causality and the forbiddance of faster-than-light signalling. The archetypal example of such a signalling process is the famous Fermi two-atom problem [6]. Fermi considered two point-like atoms, A and B, separated by a distance R. Atom A is initially prepared in an excited state, and atom B is initially prepared in its ground state. Fermi calculated the probability that, after a time T, atom A would be found in its ground state, and atom B would be found in an excited state, following the exchange of a photon. Fermi claimed to prove that this probability

*Talk presented by P Millington.

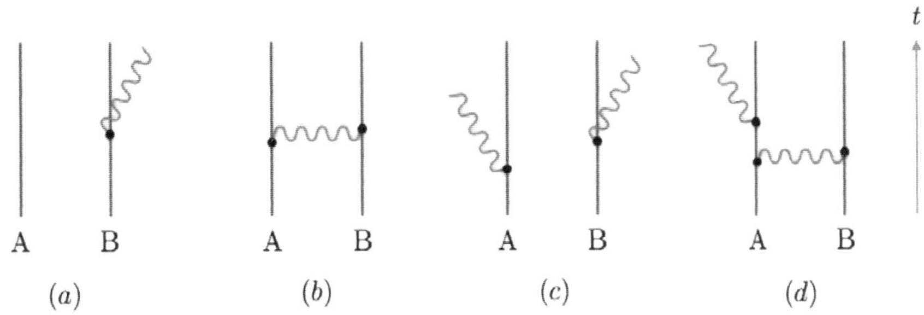

Figure 1. The four amplitudes relevant to the Fermi problem. Fermi considered diagram (b) in isolation, and it can be shown that $|(b)|^2 \neq 0$ for $T < R/c$. It is only in the combination $|(b)|^2 + |(c)|^2 + 2\,\mathrm{Re}[(a)(d)^*]$ that the a-causal parts cancel. Notice that this cancellation relies on the interference of diagrams involving only *unobserved* emissions.

is strictly zero for all times $T < R/c$. Fermi was in fact wrong, and one should expect a non-zero probability *for all* T, since the simultaneous measurement of the states of atoms A and B does not constitute a local measurement. Instead, we should ask for the probability that atom B is found in an excited state after a time T without making any restriction on the states of atom A or the electromagnetic field [7, 8, 9]. (A recent introduction to the Fermi problem and an overview of its history can be found in Ref. [10].) As we will see, this local measurement is independent of the state of atom A for all times $T < R/c$, therefore being manifestly causal in the weak sense (see, e.g., Ref [11]).

The local measurement described above is fully inclusive over the states of atom A and the electromagnetic field. As such, we must account for unobserved emission of photons in the final state, and the four amplitudes relevant to the Fermi problem are shown in Fig. 1. By working directly at the level of the probability, we will find that we do not need to sum explicitly over these unobserved emissions [10]. This suggests that one may be able to write down semi-inclusive transition probabilities in which the Bloch-Nordsieck cancellation is applied implicitly, potentially having a significant impact on the way in which we deal with IR divergences in gauge theories. As a step towards this, we will conclude the present note by describing an example of a Fock-space projection operator for (semi-)inclusive observables (see Ref. [12]).

2. Causality

Causality is built into quantum field theory *ab initio* through the vanishing of the equal-time commutator of field operators, in the case of bosons, or anti-commutator of field operators, in the case of fermions. This is the so-called *microcausality condition.* For example, given a real scalar field $\phi_x \equiv \phi(x)$, we have

$$\big[\phi_x,\, \phi_y\big] \;=\; 0 \qquad \forall\,(x-y)^2 \;<\; 0 \qquad \text{(space-like)}\,. \tag{1}$$

However, it is the Feynman propagator

$$\Delta^{(\mathrm{F})}_{xy} \;\equiv\; \Delta^{(\mathrm{F})}(x,y) \;=\; \frac{1}{2}\,\mathrm{sgn}(x_0-y_0)\,\langle\big[\phi_x,\,\phi_y\big]\rangle + \frac{1}{2}\,\langle\big\{\phi_x,\,\phi_y\big\}\rangle \;=\; \int\frac{\mathrm{d}^4k}{(2\pi)^4}\,\frac{ie^{-ik\cdot(x-y)}}{k_0^2-E_{\mathbf{k}}^2+i\epsilon} \tag{2}$$

that is ubiquitous in S-matrix theory and it has support over space-like separations due to the anti-commutator term $\langle\big\{\phi_x,\,\phi_y\big\}\rangle$. In Eq. (2), $E_{\mathbf{k}} = \sqrt{\mathbf{k}^2+m^2}$ is the on-shell energy (for a field of mass m), and $\mathrm{sgn}(x_0-y_0)$ is the signum function, which may be expressed in terms of unit-step functions as $\mathrm{sgn}(x_0-y_0) \equiv \Theta(x_0-y_0) - \Theta(y_0-x_0)$.

The fact that causality is not manifest at the amplitude level can be understood directly in terms of the Feynman prescription ($k_0^2 \to k_0^2 + i\epsilon$, $\epsilon = 0^+$), since this choice of analytic continuation allows us to Wick rotate between Minkowski space-time, which has causal structure, and four-dimensional Euclidean space, which does not. It is perhaps even more disconcerting that the a-causal part of the Feynman propagator actually corresponds to *on-shell* propagation, that is the part that we would normally associate with *real* particles. If causality is to be manifest then it is surely the retarded and advanced propagators

$$\Delta_{xy}^{(\mathrm{R})} \;=\; \Delta_{yx}^{(\mathrm{A})} \;=\; \frac{1}{i}\,\Theta(x_0-y_0)\,\langle[\phi_x,\,\phi_y]\rangle \;=\; \int\frac{\mathrm{d}^4k}{(2\pi)^4}\,\frac{e^{-ik\cdot(x-y)}}{k_0^2-E_{\mathbf{k}}^2\pm i\epsilon\,\mathrm{sgn}(k_0)} \tag{3}$$

that should play the dominant role, since these have support only over light- and time-like separations. Note that neither the retarded ($k_0^2 \to k_0^2 + i\epsilon\,\mathrm{sgn}(k_0)$, $\epsilon = 0^+$) nor advanced ($k_0^2 \to k_0^2 - i\epsilon\,\mathrm{sgn}(k_0)$, $\epsilon = 0^+$) pole prescriptions permit us to Wick rotate, as we should expect. In the next section, we will show that the retarded and advanced propagators do in fact play this dominant role but only at the level of *probabilities*, and their presence will ensure that faster-than-light signalling is forbidden. Moreover, we will see that they emerge through the interplay of time- and anti-time-ordered products of field operators.

Time-ordering of the two-point function corresponds to the Feynman prescription ($k_0^2 \to k_0^2 + i\epsilon$), which places the positive- and negative-frequency poles in the second and fourth quadrants of the complex plane, respectively. Instead, anti-time-ordering corresponds to the Dyson prescription ($k_0^2 \to k_0^2 - i\epsilon$), which places the positive- and negative-frequency poles in the first and third quadrants, respectively. It is through combinations of terms involving these two prescriptions that both poles can appear in the lower-half complex plane to give the retarded propagator or the upper-half complex plane to give the advanced propagator.

The role of both time orderings can also be illustrated by means of the Bogoliubov-Shirkov causality condition [13]. We begin with the familiar S-matrix operator

$$S[\lambda] \;=\; \mathrm{T}\exp\biggl[-i\int\mathrm{d}^4z\,\mathcal{H}_z^{\mathrm{int}}(\lambda)\biggr]\,, \tag{4}$$

where T is the time-ordering operator and the domain of integration over z_0 is $]-\infty,\infty[$. Taking a local interaction Hamiltonian density $\mathcal{H}_z^{\mathrm{int}}(\lambda) = \lambda\,\phi_z^4/4!$, we imagine promoting the coupling λ to one of two space-time dependent functions $\lambda_z \equiv \lambda(z)$ or $\lambda'_z \equiv \lambda'(z)$. If the coupling functions λ_z and λ'_z coincide for all z_0 earlier than some time t, the product $S[\lambda']S^\dagger[\lambda]$ must be independent of the behaviour of these functions for all times $z_0 < t$ by unitarity. Letting $\lambda'_z = \lambda_z + \delta\lambda_z$, where the infinitesimal variation $\delta\lambda_z$ is only non-zero for $z_0 > t$, we expand

$$S[\lambda']S^\dagger[\lambda] \;=\; S[\lambda]S^\dagger[\lambda] + \frac{\delta S[\lambda]}{\delta\lambda_y}\,\delta\lambda_y\,S^\dagger[\lambda] \;=\; \mathbb{I} + \frac{\delta S[\lambda]}{\delta\lambda_y}\,\delta\lambda_y\,S^\dagger[\lambda]\,, \tag{5}$$

in which the integral over y (for $y_0 > t$) in the first functional variations is left implicit. If this product is to be consistent with causality, Eq. (5) must still be independent of the behaviour of the coupling function for all times $x_0 < t < y_0$ and, by Lorentz covariance, for all x and y that are space-like separated. We therefore require that the quantity

$$\frac{\delta}{\delta\lambda_x}\biggl[\frac{\delta S[\lambda]}{\delta\lambda_y}\,S^\dagger[\lambda]\biggr] \tag{6}$$

vanish for all $x \precsim y$, i.e. for all x causally preceding or space-like separated from y. This gives the Bogoliubov-Shirkov causality condition [13]:

$$\frac{\delta^2 S[\lambda]}{\delta\lambda_x\,\delta\lambda_y}\,S^\dagger[\lambda] + \frac{\delta S[\lambda]}{\delta\lambda_y}\,\frac{\delta S^\dagger[\lambda]}{\delta\lambda_x} \;=\; 0 \qquad \forall\, x \precsim y\,, \tag{7}$$

which can be satisfied only through the cancellation of terms originating from S (time-ordered) and $S^\dagger$ (anti-time-ordered).

Making use of the explicit form of the S-matrix operator in Eq. (4), Eq. (7) can be written

$$\mathrm{T}\left[\frac{\partial\mathcal{H}_x^{\mathrm{int}}}{\partial\lambda_x}\frac{\partial\mathcal{H}_y^{\mathrm{int}}}{\partial\lambda_y}S\right]S^\dagger - \mathrm{T}\left[\frac{\partial\mathcal{H}_y^{\mathrm{int}}}{\partial\lambda_y}S\right]\bar{\mathrm{T}}\left[\frac{\partial\mathcal{H}_x^{\mathrm{int}}}{\partial\lambda_x}S^\dagger\right] = 0 \qquad \forall\, x \precsim y\,, \tag{8}$$

where $\bar{\mathrm{T}}$ is the anti-time-ordering operator. At lowest order in the coupling function, this requires

$$\Theta(x_0 - y_0)\left[\frac{\partial\mathcal{H}_x^{\mathrm{int}}}{\partial\lambda_x},\,\frac{\partial\mathcal{H}_y^{\mathrm{int}}}{\partial\lambda_y}\right] = 0 \qquad \forall\, x \precsim y\,, \tag{9}$$

and, on substituting for the interaction Hamiltonian density, we obtain the constraint

$$\frac{i}{(3!)^2}\,\Delta_{xy}^{(\mathrm{R})}\phi_x^3\phi_y^3 + \frac{1}{(2!)^3}\,[\Delta_{xy}^{(\mathrm{R})}]^2\phi_x^2\phi_y^2 - \frac{i}{3!}\,[\Delta_{xy}^{(\mathrm{R})}]^3\phi_x\phi_y - \frac{1}{4!}[\Delta_{xy}^{(\mathrm{R})}]^4 = 0 \qquad \forall\, x \precsim y\,. \tag{10}$$

This is automatically satisfied due to the presence of the retarded propagators,† which have originated from the commutator in Eq. (9). Proceeding to higher orders in the coupling, the combination of time- and anti-time-ordered products will give rise to a series of nested commutators of the Hamiltonian density that is again consistent with causality. In the next section, we will show that this series can be obtained straightforwardly at the level of expectation values by application of the Baker-Campbell-Hausdorff formula.

We remark that the Bogoliubov-Shirkov causality condition [Eq. (7)] can be expressed in terms of the largest time equation [14], of which Eq. (10) is an example. This hints at the fundamental relationship between unitarity and causality. This connection has recently been emphasised [15] by means of the tree-loop duality relations [16, 17], and it can be made explicit (see Ref. [18]) through the Kobes-Semenoff unitarity cutting rules [19, 20, 21] of the Schwinger-Keldysh closed-time-path (or in-in) formalism [22, 23].

3. Sources and detectors

We now turn out attention to the signalling between sources and detectors. For simplicity, we introduce a scalar analogue of the Fermi two-atom problem [10], which consists of a real scalar field ϕ that interacts with two static "atoms" S and D, fixed at positions $\mathbf{x}^S$ and $\mathbf{x}^D$, respectively. Working in the interaction picture, the free part of the Hamiltonian is

$$H_0 = \sum_n \omega_n^S\,|n^S\rangle\langle n^S| + \sum_n \omega_n^D\,|n^D\rangle\langle n^D| + \int \mathrm{d}^3\mathbf{x}\,\left(\tfrac{1}{2}(\partial_t\phi)^2 + \tfrac{1}{2}(\nabla\phi)^2 + \tfrac{1}{2}m^2\phi^2\right), \tag{11}$$

where the states $\{|n^X\rangle\}$ form a complete set of bound states for atom $X \in \{S, D\}$. The interaction Hamiltonian is

$$H_{\mathrm{int}}(t) = M^S(t)\,\phi(\mathbf{x}^S, t) + M^D(t)\,\phi(\mathbf{x}^D, t)\,, \tag{12a}$$

$$M^X(t) \equiv \sum_{m\neq n} \mu_{mn}^X\,e^{i\omega_{mn}^X t}\,|m^X\rangle\langle n^X|\,, \tag{12b}$$

†We have made use of the result (valid for all integer $n, m \geq 1$)

$$[\phi_x^n,\,\phi_y^m] = \sum_{k=1}^{\min(n,m)} (-1)^{k+1}\,\frac{n!m!}{k!(n-k)!(m-k)!}\,[\phi_x,\,\phi_y]^k\,\phi_x^{n-k}\phi_y^{m-k}\,.$$

IOP Conf. Series: Journal of Physics: Conf. Series **880** (2017) 012041 doi:10.1088/1742-6596/880/1/012041

where $\omega^X_{mn} \equiv \omega^X_m - \omega^X_n$ and μ^X_{mn} are monopole moments.

At time $t = 0$, the system is prepared with atom D in its ground state ($|g^D\rangle$), the field ϕ in its vacuum state ($|0^\phi\rangle$) and atom S in an incoherent superposition of its ground and excited states ($|g^S\rangle$ and $|p^S\rangle$). This corresponds to an initial density operator

$$\rho_0 \;\equiv\; \rho(0) \;=\; \gamma\,|i_p\rangle\langle i_p| \,+\, (1-\gamma)\,|i_g\rangle\langle i_g| \;, \qquad |i_{p(g)}\rangle \;=\; |p(g)^S\rangle \otimes |g^D\rangle \otimes |0^\phi\rangle \;, \tag{13}$$

where $\gamma \in \mathbb{R}$ parametrises the admixture of the initial states of the source atom S. The measurement (at some later time $t = T$) is effected by an operator E, which projects into the subspace of states in which the detector atom D is in its excited state ($|q^D\rangle$) but with no restriction on the states of atom S or the field ϕ:

$$E \;\equiv\; E^S \otimes E^D \otimes \mathcal{E} \;=\; \mathbb{I}^S \otimes |q^D\rangle\langle q^D| \otimes \mathbb{I}^\phi \,. \tag{14}$$

The probability of the local measurement outcome "atom D excited" is then

$$\mathbb{P}(D^*) \;=\; \mathrm{Tr}\,(E\rho_T) \;=\; \mathrm{Tr}\,(U^\dagger_{T,0}EU_{T,0}\rho_0) \;, \tag{15}$$

where

$$U_{T,0} \;=\; \mathrm{T}\exp\left[-i\int_0^T \mathrm{d}t\; H_{\mathrm{int}}(t)\right] \tag{16}$$

is the time-evolution operator. It is important to draw a clear distinction between the probability of the measurement outcome in Eq. (15) and the sensitivity of the detector to the initial preparation of the system. The latter tells us about our ability to signal between the source and detector atoms, and it is given by

$$\sigma_{pg} \;\equiv\; \frac{\mathrm{d}\,\mathbb{P}(D^*)}{\mathrm{d}\gamma} \;=\; \mathbb{P}_p(D^*) \,-\, \mathbb{P}_g(D^*) \;, \qquad \mathbb{P}_{p(g)}(D^*) \;\equiv\; \langle i_{p(g)}|U^\dagger_{T,0}EU_{T,0}|i_{p(g)}\rangle \;. \tag{17}$$

By applying the Baker-Campbell-Hausdorff formula to Eq. (17), the combination of time-ordered (from U) and anti-time-ordered exponentials (from $U^\dagger$) gives rise to an infinite series of nested commutators [10, 18, 24, 25]. This can be written in the form [10, 18]

$$\mathbb{P}_{p,g} \;=\; \sum_{j=0}^{\infty}\int_0^T \mathrm{d}t_1\,\mathrm{d}t_2\ldots\mathrm{d}t_j\;\Theta_{12\ldots j}\,\langle i_{p,g}|\,\mathcal{F}_j\,|i_{p,g}\rangle \;, \tag{18}$$

where $\mathcal{F}_0 \equiv E$,

$$\mathcal{F}_j \;=\; \tfrac{1}{i}\Big[\mathcal{F}_{j-1},\, H_{\mathrm{int}}(t_j)\Big] \;=\; \tfrac{1}{i}\Big[\mathcal{F}_{j-1},\, M^S_j\phi^S_j \,+\, M^D_j\phi^D_j\Big] \;, \tag{19}$$

and $\Theta_{ijk\ldots}$ is equal to 1 for $t_i > t_j > t_k > \ldots$ and 0 otherwise. We use the short-hand notation $M^X_j \equiv M^X(t_j)$ and $\phi^X_j \equiv \phi(\mathbf{x}^X, t_j)$. By defining

$$E^X_{\ldots k} \;\equiv\; \tfrac{1}{i}\big[E^X_{\ldots},\, M^X_k\big] \;, \qquad E^X_{\ldots\underline{k}} \;\equiv\; \big\{E^X_{\ldots},\, M^X_k\big\} \;, \tag{20a}$$

$$\mathcal{E}^{\ldots X}_{\ldots k} \;\equiv\; \tfrac{1}{i}\big[\mathcal{E}^{\cdots}_{\ldots},\, \phi^X_k\big] \;, \qquad \mathcal{E}^{\ldots X}_{\ldots\underline{k}} \;\equiv\; \big\{\mathcal{E}^{\cdots}_{\ldots},\, \phi^X_k\big\} \;, \tag{20b}$$

we can introduce a convenient "undercircle" notation [10]

$$E_{\underset{\circ\circ}{\underline{kl}}}\mathcal{E}_{\underset{\bullet\bullet}{\underline{kl}}} \;\equiv\; E_{kl}\mathcal{E}_{\underline{k}\underline{l}} \,+\, E_{k\underline{l}}\mathcal{E}_{\underline{k}l} \,+\, E_{\underline{k}l}\mathcal{E}_{k\underline{l}} \,+\, E_{\underline{k}\underline{l}}\mathcal{E}_{kl} \;, \tag{21}$$

which allows us to write the n-th order operator appearing in Eq. (18) as

$$\mathcal{F}_n \;=\; 2^{-n} \sum_{a=0}^{n} E^{S}_{(\underset{\circ}{\underline{1}}\dots\underset{\circ}{a}} \, E^{D}_{\underset{\circ}{a+1}\dots\underset{\circ}{n})} \, \mathcal{E}^{(S\dots S\;D\dots D)}_{(\underset{\bullet}{\underline{1}}\,\dots\,\underset{\bullet}{\underline{a}}\,\underset{\bullet}{a+1}\dots\underset{\bullet}{\underline{n}})} \,. \tag{22}$$

The parentheses indicate a summation over all unique ordered permutations of the indices, e.g., $E^S_{(12}E^D_{3)} = E^S_{12}E^D_3 + E^S_{13}E^D_2 + E^S_{23}E^D_1$.

In the effect operator [Eq. (14)], we have made no restriction on the states of atom S or the field ϕ, fixing $E^S = \mathbb{I}^S$ and $\mathcal{E} = \mathbb{I}^\phi$. We may then show that:

(i) $E^S_{k\dots} \;=\; 0$, such that the first index on $E^S_{\dots}$ must always be underlined, with $E^S_{\underline{k}} = 2M^S_k$.

(ii) $\mathcal{E}^{X\dots}_{1\dots} = 0$ and $\mathcal{E}^{XYZ\dots}_{\underline{1}23\dots} = 0$, such that the '1' index (the latest time) is never underlined on $E^X_{1\dots}$ and, more generally, any $\mathcal{E}^{\dots}_{\dots}$ operator vanishes when its first k indices consist of more non-underlined than underlined indices for any k.

These restrictions allow us to eliminate a large proportion of the terms in Eq. (22).

The first non-vanishing contributions to Eq. (17) arise from the fourth-order operator

$$\begin{aligned} \mathcal{F}_4 \;=\; \tfrac{1}{16}\Big(& E^D_{12\underset{\circ}{\underline{3}}\underset{\circ}{4}}\mathcal{E}^{DDDD}_{\underline{1}\underline{2}\underset{\bullet}{3}\underset{\bullet}{4}} + E^D_{1\underline{2}\underset{\circ}{3}4}\mathcal{E}^{DDDD}_{\underline{1}2\underline{3}\underset{\bullet}{4}} + E^D_{1\underline{2}\underset{\circ}{3}}E^S_{\underline{4}}\,\mathcal{E}^{DDDS}_{\underline{1}\underline{2}\underset{\bullet}{3}4} + E^D_{1\underline{2}3}E^S_{\underline{4}}\,\mathcal{E}^{DDDS}_{\underline{1}2\underline{3}4} \\ & + E^D_{1\underline{2}\underset{\circ}{4}}E^S_{\underline{3}}\,\mathcal{E}^{DDSD}_{\underline{1}\underline{2}3\underset{\bullet}{4}} + E^D_{1\underline{3}\underset{\circ}{4}}E^S_{\underline{2}}\,\mathcal{E}^{DSDD}_{\underline{1}2\underline{3}\underset{\bullet}{4}} + E^D_{1\underline{2}}E^S_{\underline{3}\underset{\circ}{4}}\mathcal{E}^{DDSS}_{\underline{1}\underline{2}3\underset{\bullet}{4}} + E^D_{1\underline{3}}E^S_{\underline{2}\underset{\circ}{4}}\mathcal{E}^{DSDS}_{\underline{1}2\underline{3}\underset{\bullet}{4}} \\ & + E^D_{1\underset{\circ}{\underline{4}}}E^S_{\underline{2}3}\mathcal{E}^{DSSD}_{\underline{1}2\underline{3}\underset{\bullet}{4}} + E^D_1 E^S_{\underline{2}3\underset{\circ}{4}}\mathcal{E}^{DSSS}_{\underline{1}2\underline{3}\underset{\bullet}{4}}\Big) \,. \end{aligned} \tag{23}$$

Substituting this expression into Eq. (17), the leading-order result for the sensitivity of atom D to the preparation of the system is [10]

$$\begin{aligned} \frac{\mathrm{d}\,\mathbb{P}(D^*)}{\mathrm{d}\gamma} \;=\; 2\sum_n \int_{t_1>t_2>t_3>t_4} & |\mu^S_{pn}|^2 |\mu^D_{qg}|^2 \\ \times \Big[& \cos(\omega^D_{qg}t_{12})\Big(\sin(\omega^S_{pn}t_{34})\Delta^{DS(\mathrm{H})}_{24} + \cos(\omega^S_{pn}t_{34})\Delta^{DS(\mathrm{R})}_{24}\Big)\Delta^{DS(\mathrm{R})}_{13} \\ & + \cos(\omega^D_{qg}t_{12})\Big(\sin(\omega^S_{pn}t_{34})\Delta^{DS(\mathrm{H})}_{14} + \cos(\omega^S_{pn}t_{34})\Delta^{DS(\mathrm{R})}_{14}\Big)\Delta^{DS(\mathrm{R})}_{23} \\ & + \cos(\omega^D_{qg}t_{13})\Big(\sin(\omega^S_{pn}t_{24})\Delta^{DS(\mathrm{H})}_{34} + \cos(\omega^S_{pn}t_{24})\Delta^{DS(\mathrm{R})}_{34}\Big)\Delta^{DS(\mathrm{R})}_{12} \\ & + \sin(\omega^S_{pn}t_{23})\Big(\cos(\omega^D_{qg}t_{14})\Delta^{SD(\mathrm{H})}_{34} + \sin(\omega^D_{qg}t_{14})\Delta^{SD(\mathrm{R})}_{34}\Big)\Delta^{DS(\mathrm{R})}_{12}\Big] + \cdots \,, \end{aligned} \tag{24}$$

where $t_{ij} \equiv t_i - t_j$, and we have defined the retarded and Hadamard propagators

$$\Delta^{XY(\mathrm{R})}_{jk} \;\equiv\; \tfrac{1}{i}\,\Theta_{jk}\,\langle[\phi^X_j\,,\,\phi^Y_k]\rangle \,, \tag{25a}$$

$$\Delta^{XY(\mathrm{H})}_{jk} \;\equiv\; \langle\{\phi^X_j\,,\,\phi^Y_k\}\rangle \,. \tag{25b}$$

Most importantly, we see that the latest time on the source atom is always connected to a later time on the detector atom by a retarded propagator. As a result, Eq. (24) is strictly zero if the source and detector atoms are space-like separated, and faster-than-light signalling is manifestly prohibited. The expectation value of $\mathcal{F}_4$ is illustrated diagrammatically in Fig. 2. A full exposition of the diagrammatic rules for this system is presented in Ref. [10]; this includes results for the expectation values of general nestings of commutators and anti-commutators of field operators, as well as the rules for computing the expectation values of the atom operators.

We remark that the above approach can be generalised to include additional fields, as well as their mutual and self-interactions; local interactions that involve spatial and temporal derivatives of the fields; and spatially extended sources and detectors (see Ref. [10]). In all of these cases, the manifest causality of local measurements persists.

IOP Conf. Series: Journal of Physics: Conf. Series **880** (2017) 012041 doi:10.1088/1742-6596/880/1/012041

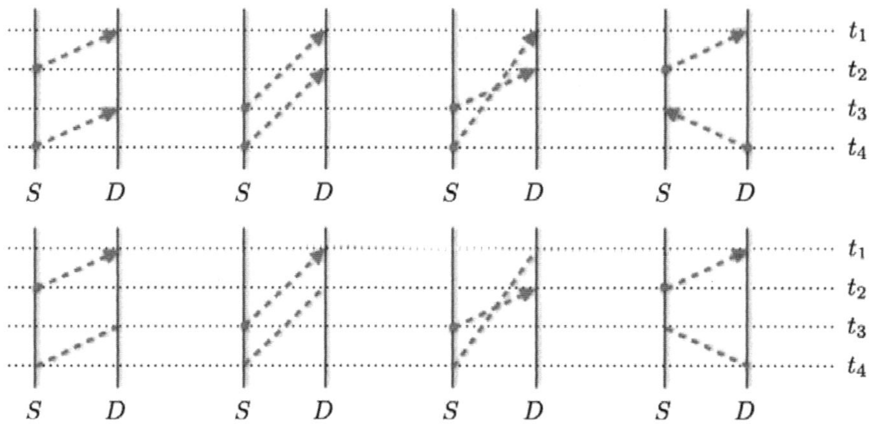

Figure 2. Diagrammatic representation of $\mathcal{F}_4$, describing the evolution of the density operator forwards in time due to the exchange of a ϕ quantum. The initial time $t = 0$ is at the bottom of the diagrams and the time of the measurement $t = T$ is at the top of the diagrams. The solid vertical lines track the source (blue) and detector (green) atoms. Orange dashed lines with arrows are associated with retarded propagators $\Delta_{ij}^{XY(\mathrm{R})}$, and those without arrows are associated with Hadamard propagators $\Delta_{ij}^{XY(\mathrm{H})}$. Each vertex is associated with a monopole moment μ_{mn}^{X}.

4. Implications for (semi-)inclusive observables

In order to obtain a causal result in Sec. 3, it was necessary to sum inclusively over the unobserved final states of the would-be photon field ϕ (and source atom S). Doing otherwise would have required us to sample the state of the field everywhere simultaneously, which would not constitute a local measurement. However, if we compare Figs. 1 and 2, we notice that the unobserved emissions necessarily accounted for in the amplitude-level calculation never appeared explicitly in the probability-level calculation. This can be traced back to the simple fact that, at the probability level, the effect operator corresponding to a fully inclusive measurement is just the unit operator, which trivially commutes with everything else, dropping out of the calculation entirely. The import of this observation is the following: by working directly with probabilities, we do not have to keep track of and calculate the individual amplitudes for all possible unobserved emissions in the final state.

Notwithstanding the potential implications of this for treating IR divergences, probability-level calculations allow us to introduce projection operators for (semi-)inclusive measurements, which do not have simple analogues at the amplitude level [12]. As an example, the effect operator that sums inclusively over final-state radiation with momenta below some scale μ is

$$E_\mu = :e^{-N_\mu}: \,, \qquad N_\mu = \int \frac{\mathrm{d}^3\mathbf{k}}{(2\pi)^3}\,\frac{1}{2E_{\mathbf{k}}}\,\theta(|\mathbf{k}|-\mu)\,a_{\mathbf{k}}^\dagger a_{\mathbf{k}} \,. \tag{26}$$

The colons indicate normal ordering, and $a_{\mathbf{k}}^\dagger$ and $a_{\mathbf{k}}$ are the usual scalar creation and annihilation operators. In the limit $\mu \to \infty$, the operator is fully inclusive over final-state radiation:

$$\lim_{\mu\to\infty} E_\mu = \mathbb{I} \,, \tag{27a}$$

$$\lim_{\mu\to\infty} \langle 0|\phi(x_1)\phi(x_2)E_\mu\phi(x_3)\phi(x_4)|0\rangle = \langle 0|\phi(x_1)\phi(x_2)\phi(x_3)\phi(x_4)|0\rangle \,, \tag{27b}$$

where the appearance of the four-point amplitude is a consequence of the optical theorem. In the limit $\mu \to 0$, the operator projects out the vacuum exclusively (see, e.g., Ref. [26]):

$$\lim_{\mu \to 0} E_\mu = |0\rangle\langle 0| \ , \tag{28a}$$

$$\lim_{\mu \to 0} \langle 0|\phi(x_1)\phi(x_2)E_\mu\phi(x_3)\phi(x_4)|0\rangle = \langle 0|\phi(x_1)\phi(x_2)|0\rangle\langle 0|\phi(x_3)\phi(x_4)|0\rangle \ . \tag{28b}$$

For finite μ, Eq. (26) defines an operator form of the Sudakov factor [27]. Such an object could not be written down at the amplitude level.

5. Concluding remarks

In this note, we have outlined a method for directly calculating transition probabilities in quantum field theory in terms of the expectation values of nested commutators and anti-commutators of field operators. These transition probabilities are manifestly causal and have the interesting feature that unobserved emissions are accounted for *implicitly* in (semi-)inclusive measurements. The ability to sum implicitly over soft emissions may have important implications for how we deal with IR divergences in gauge theories, and this motivates the further development of technologies for evaluating these probabilities. Such technology may, for instance, exploit the connection between causality and unitarity to which we have alluded, making use of the unitarity cutting rules of the closed-time-path formalism and building upon ideas presented in Ref. [18].

Acknowledgments

The work of PM is supported by STFC grant ST/L000393/1. PM would like to thank the organisers and participants of DICE2016 for their hospitality, questions and comments.

References

[1] Bloch F and Nordsieck A 1937 *Phys. Rev.* **52** 54–9
[2] Kinoshita T 1962 *J. Math. Phys.* **3** 650–77
[3] Lee T D and Nauenberg M 1964 *Phys. Rev.* **133** B1549–62
[4] Weinberg S 1965 *Phys. Rev.* **140** B516–24
[5] Faddeev L D and Popov V N 1967 *Phys. Lett.* B **25** 29–30
[6] Fermi E 1932 *Rev. Mod. Phys.* **4** 87–132
[7] Shirokov M I 1966 *Yad. Fiz.* **4** 1077 [1967 *Sov. J. Nucl. Phys.* **4** 774]
[8] Ferreti B 1968 in *Old and New Problems in Elementary Particles* ed Puppi G (New York: Academic Press)
[9] Power E A and Thirunamachandran T 1997 *Phys. Rev.* A **56** 3395–408
[10] Dickinson R, Forshaw J and Millington P 2016 *Phys. Rev.* D **93** 065054
[11] Hegerfeldt G C 1998 *Annalen Phys.* **7** 716–25
[12] Dickinson R, Forshaw J and Millington P 2017 *Preprint* arXiv:1702.04131 [hep-th]
[13] Bogoliubov N N and Shirkov D V 1959 *Introduction to the Theory of Quantized Fields* (New York: Interscience) pp 200–4
[14] 't Hooft G and Veltman M J G 1974 *NATO Sci. Ser.* B **4** 177–322
[15] Tomboulis E T 2017 *Preprint* arXiv:1701.07052 [hep-th]
[16] Catani S, Gleisberg T, Krauss F, Rodrigo G and Winter J-C 2008 *J. High Energy Phys.* 09 (2008) 065
[17] Bierenbaum I, Catani S, Draggiotis P and Rodrigo G 2010 *J. High Energy Phys.* 10 (2010) 073
[18] Dickinson R, Forshaw J, Millington P and Cox B 2014 *J. High Energy Phys.* 06 (2014) 049
[19] Kobes R L and Semenoff G W 1985 *Nucl. Phys.* B **260** 714–46
[20] Kobes R L and Semenoff G W 1986 *Nucl. Phys.* B **272** 329–64
[21] Kobes R 1991 *Phys. Rev.* D **43** 1269–82
[22] Schwinger J S 1961 *J. Math. Phys.* **2** 407–32
[23] Keldysh L V 1964 *Zh. Eksp. Teor. Fiz.* **47** 1515–27 [1965 *Sov. Phys. JETP* **20** 1018–26]
[24] Cliche M and Kempf A 2010 *Phys. Rev.* A **81** 012330
[25] Franson J D and Donegan M M 2002 *Phys. Rev.* A **65** 052107
[26] Louisell W H 1973 *Quantum Statistical Properties of Radiation* (New York: John Wiley & Sons) p 158
[27] Sudakov V V 1956 *Sov. Phys. JETP* **3** 65–71 [*Zh. Eksp. Teor. Fiz.* **30** 87–95]

IOP Conf. Series: Journal of Physics: Conf. Series **880** (2017) 012042 doi:10.1088/1742-6596/880/1/012042

Van der Waals and resonance interactions between accelerated atoms in vacuum and the Unruh effect

M Lattuca[1], J Marino[2], A Noto[1], R Passante[1,3], L Rizzuto[1,3], S Spagnolo[1] and W Zhou[1,4]

[1] Dipartimento di Fisica e Chimica, Università degli Studi di Palermo, Via Archirafi 36, 90123 Palermo, Italy

[2] Institut fur Theoretische Physik, Universität zu Köln, Zülpicher Straße 77, 50937 Cologne, Germany

[3] INFN, Laboratori Nazionali del Sud, 95123 Catania, Italy

[4] Center for Nonlinear Science and Department of Physics, Ningbo University, Ningbo, Zhejiang 315211, China

E-mail: `roberto.passante@unipa.it`

Abstract. We discuss different physical effects related to the uniform acceleration of atoms in vacuum, in the framework of quantum electrodynamics. We first investigate the van der Waals/Casimir-Polder dispersion and resonance interactions between two uniformly accelerated atoms in vacuum. We show that the atomic acceleration significantly affects the van der Waals force, yielding a different scaling of the interaction with the interatomic distance and an explicit time dependence of the interaction energy. We argue how these results could allow for an indirect detection of the Unruh effect through dispersion interactions between atoms. We then consider the resonance interaction between two accelerated atoms, prepared in a correlated Bell-type state, and interacting with the electromagnetic field in the vacuum state, separating vacuum fluctuations and radiation reaction contributions, both in the free-space and in the presence of a perfectly reflecting plate. We show that nonthermal effects of acceleration manifest in the resonance interaction, yielding a change of the distance dependence of the resonance interaction energy. This suggests that the equivalence between temperature and acceleration does not apply to all radiative properties of accelerated atoms. To further explore this aspect, we evaluate the resonance interaction between two atoms in non inertial motion in the coaccelerated (Rindler) frame and show that in this case the assumption of an Unruh temperature for the field is not required for a complete equivalence of locally inertial and coaccelerated points of views.

1. Introduction

A remarkable consequence of quantum field theory in accelerated frames is that the concept of vacuum is observer-dependent [1, 2, 3]. When viewed from a uniformly accelerated observer, the vacuum of a scalar field appears as a *thermal* bath of quanta at a temperature proportional to its acceleration, $T_U = \hbar a/(2\pi c k_B)$, a being the proper acceleration and k_B the Boltzmann constant. This phenomenon, which is the analogous in a flat space-time of the Hawking radiation from a black hole, is known as Davies-Fulling-Unruh effect, and its essence is that the particle content of a quantum field is not invariant for all coordinate transformations, but it depends on the reference frame [3, 4]. This conceptually subtle result, on the border between quantum field theory and general relativity, has been object of intense investigations in the last decades, and some controversial aspects concerning its physical meaning are still debated [5, 6, 7]. Several experimental schemes to detect the Unruh effect have been proposed in the literature [4, 8, 9, 10, 11], but a direct verification of this effect is still a challenge for experimentalists, due to the very high accelerations necessary to reveal a Unruh temperature of even a few Kelvin. For example, an acceleration of about 9.8 m/s^2, the gravitational acceleration experienced by a detector near the Earth's surface, gives a Unruh temperature of the order of 10^{-20} K, which is very difficult to detect.

Even if the absence of any direct experimental confirmation of the Unruh effect has led some physicists to question the very existence of this effect, it has been pointed out that the Unruh effect is a consequence of quantum field theory necessary to keep its coherency in accelerated systems [12]. A direct or indirect detection of the effect could thus allow to settle a fundamental question in quantum field theory.

A different although related question recently addressed in the literature, is whether the effects of a relativistic acceleration are strictly equivalent to a thermal field and what is the physical meaning of this equivalence [7, 12]. It has been recently discussed, for example, that nonthermal features associated with a uniform acceleration appear in the radiative properties of single accelerated atoms [13, 14], as well as in the van der Waals/Casimir-Polder interaction between atoms in non inertial motion [15], or in the entanglement generation between two accelerating atoms [16, 17].

Motivated by these intriguing issues, in this paper we investigate and review the effect of a uniformly accelerated motion on the van der Waals and resonance interactions between two atoms in vacuum, with the aim to explore different physical manifestations of the Unruh effect, as well as the possibility to observe signatures of such effect on the radiation-mediated interaction between accelerating atoms.

Van der Waals and Casimir-Polder dispersion forces are long-range interactions between atoms or neutral objects due to zero-point fluctuations of the quantum electromagnetic field and exchange of virtual photons between the atoms [18, 19]. These interactions, of purely quantum origin, have been measured with accuracy [20, 21]. When one of the two atoms is in an excited state, a real resonant photon can be exchanged between the atoms. These interactions are usually a fourth-order effect in the matter-field coupling, and thus they are very tiny. Resonance interactions can occur at second order in the coupling when two identical atoms are prepared in a correlated (symmetric or antisymmetric) state, with one atom excited and the other in the ground state. In this case the excitation is delocalized over the two atoms. Since such interaction is a second-order effect, it is usually much more intense than dispersion interactions and, moreover, can be of very long range, scaling asymptotically as R^{-1} (R being the interatomic distance) [19, 22].

In this paper, we explore the effect of atomic acceleration on both van der Waals/Casimir-Polder and resonance interactions between two atoms moving with the same uniform acceleration in the vacuum, in the free space or in the presence of a reflecting plane boundary. We show that the atomic acceleration significantly affects the van der Waals interaction between the two accelerated atoms, resulting in a different scaling of the interaction with the distance compared to inertial atoms and in an explicit time dependence of the interaction. When sufficiently long times are taken, we show that the change in the interaction energy due to the acceleration may become significant even for reasonable values of the acceleration [23]. This suggests that van der Waals interactions could be used as a probe to obtain an indirect signature of the Unruh effect, without necessity of extremely high accelerations [23]. We then investigate the effect of atomic acceleration on the resonance interaction between two entangled identical atoms accelerating in the vacuum space. By separating vacuum fluctuations and radiation reaction contributions, we find that the resonance interaction is exclusively due to the radiation reaction contribution, while vacuum fluctuations do not contribute [24, 25]. Accordingly, Unruh thermal

IOP Conf. Series: Journal of Physics: Conf. Series **880** (2017) 012042 doi:10.1088/1742-6596/880/1/012042

fluctuations, which are basically related to the vacuum field correlations in the locally inertial frame of both atoms, do not affect the resonance interaction between the accelerated atoms. Neverthless, the features of the interaction changes qualitatively due to acceleration, showing the presence of purely nonthermal effects of acceleration in the resonance interaction (i.e. not related to the equivalence between acceleration and temperature). In order to further address this point, we have investigated the resonance interaction between accelerated atoms also from the point of view of a coaccelerated frame. We show that, although the vacuum state in the comoving (i.e. locally inertial) and coaccelerated frames is different due to the Unruh effect, the resonance interaction between two atoms as measured by observers in these frames is the same, without the assumption of an Unruh temperature for the quantum field in the comoving frame [26]. This result is at variance with what has been obtained for other radiative effects, such as the spontaneous emission, where vacuum fluctuations are relevant [27]. Thus we conclude that the necessity to introduce an Unruh temperature in the coaccelerated frame depends on the specific physical process considered. Finally, we briefly discuss how the presence of a perfectly reflecting plate affects the resonance interaction between two accelerating atoms, showing the appearance of new features that could be relevant for detecting signatures of an accelerated motion in quantum field theory.

The paper is structured as follows. In Section 2 we investigate the van der Waals interaction between two uniformly accelerating atoms in vacuum. In Section 3 we discuss the effect of a uniform acceleration on the resonance interaction between two accelerated atoms in vacuum, both in the comoving and in the coaccelerated frame. Finally, in Section 4 we briefly discuss the effect of a perfectly reflecting mirror on the resonance interaction between two accelerated atoms. Last Section is devoted to our conclusive comments.

2. The van der Waals interaction between two uniformly accelerated atoms in vacuum

Let us consider two atoms, A and B, uniformly accelerating in the vacuum with the same acceleration. We assume the acceleration orthogonal to their distance, so that their separation is constant. The atoms interact with the quantum electromagnetic field in the vacuum state. We wish to evaluate the van der Waals/Casimir-Polder interaction energy between the two accelerating atoms. In order to do this, we exploit the heuristic model in [18], according to which the atom-atom Casimir-Polder interaction originates from the (classical) interaction between the instantaneous dipole moments induced and correlated by the spatially correlated zero-point fluctuations of the quantum electromagnetic field, and extend this model to accelerated atoms. For two ground-state atoms at rest separated by a distance $\mathbf{R} = \mathbf{r}_A - \mathbf{r}_B$, the Casimir-Polder interaction is thus obtained as

$$\Delta E_{CP} = \sum_{\mathbf{k}j} \alpha_A(k)\alpha_B(k)\langle 0 \mid E_\ell(\mathbf{k}, j; \mathbf{r}_A) E_m(\mathbf{k}, j; \mathbf{r}_B) \mid 0\rangle (V_{\ell m}(\mathbf{k}, \mathbf{R})) \tag{1}$$

where $\alpha(k)$ is the atomic dynamical polarizability, related to the instantaneous dipole moments by $\boldsymbol{\mu}_\ell(k) = \alpha(k)E_\ell(\mathbf{k}, j; \mathbf{r})$, $\mathbf{E}(\mathbf{k}, j; \mathbf{r})$ the Fourier component of the electric field operator, and $V_{\ell m}(\mathbf{k}, \mathbf{R})$ is the usual tensor potential [19].

We now sketch the main elements of our calculation. In order to generalize the method described above to uniformly accelerating atoms [23], we first calculate the electric field due to the accelerating dipole A at the position of the other dipole B, at the retarded time $t_r = t - \rho(t_r)/c$ [28]. Here $\rho(t_r)$ is an *effective* distance describing the distance traveled by the virtual photons exchanged between the two moving atoms. For atoms at rest, it coincides with the interatomic distance, that is $\rho(t_r) = R$; if the two atoms are accelerating, their effective distance depends on time because of the time-dependence of the Lorentz factor $\gamma = (1 - v^2/c^2)^{-1/2}$. We then calculate and Lorentz-transform the electric and magnetic field generated by the atomic dipoles to the comoving reference frame, where the two atoms are instantaneously at rest. The interaction energy of the two atoms is thus evaluated in the comoving frame of the accelerating atoms, and finally expressed in terms of quantities measured in the laboratory frame. After some lengthy calculations, assuming a non relativistic motion, we finally get [23]:

$$\Delta\tilde{E} = \Delta E^{\mathrm{r}} + \frac{a^2 t}{2c^3}\frac{\hbar c}{\pi R^3}\int_0^\infty \alpha_A(iu)\alpha_B(iu)\left(3+\frac{4}{uR}+\frac{2}{u^2R^2}\right)u^2\exp[-2uR]\,du$$
$$+\frac{a^2t^2}{6c^2}\frac{\hbar c}{\pi R^2}\int_0^\infty \alpha_A(iu)\alpha_B(iu)\left(-1+\frac{4}{uR}+\frac{8}{u^2R^2}+\frac{8}{u^3R^3}+\frac{4}{u^4R^4}\right)u^4\exp[-2uR]\,du \quad (2)$$

where ΔE^{r} is the well-known van der Waals potential energy for atoms at rest, scaling as R^{-6} in the near zone $R \ll \lambda$ (λ being the main atomic transition wavelength) and as R^{-7} in the far zone $R \gg \lambda$ [18, 29]. The other two terms in (2) are the changes in the potential energy due to the acceleration. These two terms depend explicitly on time and are responsible of a qualitative change of the distance-dependence of the interaction energy. In fact, in the near zone a term proportional to R^{-5} adds to the usual R^{-6} term for atoms at rest, while in the far zone a term proportional to R^{-6} adds to the usual R^{-7} behaviour. The presence of the term as R^{-6} in the far-zone is reminiscent of the Casimir-Polder interaction between two atoms at rest at finite temperature T [30], thus suggesting a connection with the Unruh-thermal effect. Also, the acceleration-dependent corrections are proportional to a^2t^2 or a^2t. This is a quite relevant feature of our result (2), since it suggests that modifications to the interaction energy between the two atoms could be observed also for reasonable values of the acceleration, provided that sufficiently long times are considered [23]. All this could be of relevance for a possible indirect detection of the Unruh effect through the van der Waals force between two accelerated atoms.

3. Resonance interaction between two accelerated atoms in vacuum

The physical arguments illustrated in the previous section, suggest to investigate if similar effects of acceleration manifest also in other quantum-electrodynamical processes, for example in the resonance interaction between two uniformly accelerated entangled identical atoms. In particular, in this Section we investigate if the effect of a uniform acceleration is equivalent to an Unruh *thermal* effect. In this context, it has been recently shown that the Casimir-Polder interaction energy between two uniformly accelerating atoms exhibits a qualitative change from a short-distance thermal behaviour to a long-distance nonthermal behaviour, with respect to a reference length identified with $z_a = c^2/a$, where a is the proper acceleration of the two atoms [15]. In this case the effect is due to the contribution of vacuum fluctuations to the interaction energy. Thus, effects of acceleration beyond the acceleration-temperature equivalence could be expected also in other radiative processes.

We consider here the resonance interaction between two uniformly accelerated identical atoms, one excited and the other in the ground state, prepared in a correlated (symmetric or antisymmetric) state and interacting with the electromagnetic field in the vacuum state. Since resonance interactions are usually much more intense than dispersion interactions, being a second-order effect, they could be another suitable candidate for observing the effect of accelerated motion in quantum field theory. Our aim is also to explore physical situations where the effects of the acceleration could be exclusively nonthermal [24]. We model the two atoms A and B as two-level systems with energy $\mp\frac{1}{2}\hbar\omega$, associated with the states $|g\rangle$ and $|e\rangle$, respectively. We assume that the two atoms interact with the electromagnetic field in the vacuum state, and accelerate with the same uniform acceleration along two parallel trajectories in the x direction, $x_A(\tau)$ and $x_B(\tau)$, maintaining constant their distance. In the comoving frame, their trajectory is [31]

$$t(\tau) = \frac{c}{a}\sinh\frac{a\tau}{c},\ x_{A/B}(\tau) = \frac{c^2}{a}\cosh\frac{a\tau}{c},\ y_{A/B}(\tau) = 0,\ z_A(\tau) = z_A,\ z_B(\tau) = z_B, \quad (3)$$

IOP Conf. Series: Journal of Physics: Conf. Series **880** (2017) 012042 doi:10.1088/1742-6596/880/1/012042

where τ is the proper time. We use the multipolar coupling scheme in the dipole approximation; in the instantaneous inertial frame of the two atoms, the Hamiltonian is [19, 24]

$$H = H_A + H_B + \sum_{\mathbf{k}j} \hbar\omega a^{\dagger}_{\mathbf{k}j} a_{\mathbf{k}j} \frac{dt}{d\tau} - \boldsymbol{\mu}_A(\tau) \cdot \mathbf{E}(x_A(\tau)) - \boldsymbol{\mu}_B(\tau) \cdot \mathbf{E}(x_B(\tau)), \tag{4}$$

where $j = 1, 2$ is the polarization index, $\boldsymbol{\mu} = e\mathbf{r}$ the atomic dipole moment operator, $a_{\mathbf{k}j}$, $a^{\dagger}_{\mathbf{k}j}$ the bosonic field operators, and $\mathbf{E}(x(\tau))$ the electric field operator. H_A and H_B are the Hamiltonians of the two atoms.

We assume the two identical atoms prepared in their symmetric or antisymmetric state

$$| \psi_{\pm}\rangle = \frac{1}{\sqrt{2}} (| g_A, e_B\rangle \pm | e_A, g_B\rangle) . \tag{5}$$

In these states the atomic excitation is delocalized between the two atoms. In the Dicke model, the symmetrical (antisymmetric) superposition is the well known superradiant (subradiant) state, because its spontaneous decay is enhanced (inhibited) [32].

To evaluate the resonance interaction between the two atoms, we exploit a procedure proposed in [33], which allows to separate at second order in perturbation theory, the contributions of vacuum fluctuations and radiation reaction to the energy shift of the two-atom system. This approach has been used to investigate the effect of the atomic acceleration on radiative properties of single atoms [13, 14, 34, 35, 36] and has been also recently generalized at fourth order in the coupling to investigate dispersion interactions between uniformly accelerated atoms [15, 37]. Following this procedure, we obtain the resonance interaction between two atoms ($\ell, m = x, y, z$ and the sum over repeated indices is assumed)

$$\delta E^{res} = -\frac{e^2}{2} \int_{\tau_0}^{\tau} d\tau' \chi^F_{\ell m}(x_A(\tau), x_B(\tau')) C^{A,B}_{\ell m}(\tau, \tau') + (A \leftrightarrow B \,\text{terms}) , \tag{6}$$

where $\chi^F(x(\tau), x(\tau'))$ and $C^{A,B}(\tau, \tau')$ are the field susceptibility and the atomic symmetric correlation function, respectively [13, 33]. From (6) it follows that the resonance energy shift of the two-atom system, is entirely due to the self-reaction contribution, involving only the field susceptibility, $\chi^F(x_A(\tau), x_B(\tau'))$, and the atomic symmetric correlation function, $C^{A,B}(\tau, \tau')$. The vacuum fluctuations term does not contribute. This is indeed expected on a physical basis, because the resonance interaction originates from an exchange of a real or virtual photons between the two correlated atoms, and therefore it is entirely due to the field radiated by the atoms (source fields). This feature has relevant consequences when we consider accelerating atoms, yielding that, as we now show, resonance interactions do not display any signature of the Unruh *thermal* effect. This situation should be compared with the dispersion interaction between two uncorrelated atoms discussed in the previous section, where the interaction is basically related to the spatial correlations of vacuum fluctuations, and thus thermal effects of acceleration do manifest in the interaction.

Using the appropriate field and atomic statistical functions (susceptibility and symmetric correlation function, respectively), taking the limits $\tau_0 \to -\infty$, $\tau \to \infty$ in Eq. (6), after some algebra we obtain the following expression for the resonance interaction between two accelerated atoms [24]

$$\delta E^{res} = \pm(\mu^A_{eg})_\ell(\mu^B_{ge})_m \left\{ \frac{1}{z^3} \widetilde{V}_{\ell m}(a, z, \omega_0) + \frac{a}{2z^2c^2} \widetilde{W}_{\ell m}(a, z, \omega_0) \right\} , \tag{7}$$

where + and − signs refer to the symmetric or antisymmetric state, respectively.

IOP Conf. Series: Journal of Physics: Conf. Series **880** (2017) 012042 doi:10.1088/1742-6596/880/1/012042

Expression (7) is valid for any value of az/c^2, $z = |z_B - z_A|$ being the interatomic distance. The functions $\widetilde{V}_{\ell m}(a, z, \omega_0)$ and $\widetilde{W}_{\ell m}(a, z, \omega_0)$, whose explicit expression we do not report here, are a generalization of the stationary tensor potential to the case of accelerated atoms, and are responsible of the qualitative change of the distance dependence of the resonance interaction energy we are going to discuss (see Ref. [24] for more details). For $za/c^2 \ll 1$, we recover the well-known inertial resonance interaction [19], which scales as z^{-3} in the near-zone $z \ll \lambda$ and as z^{-1} in the far-zone $z \gg \lambda$ and ($\lambda = c/\omega_0$ being the reduced atomic transition wavelength). On the other hand, at higher orders in az/c^2, Eq. (7) shows that the atomic acceleration leads to a change of the distance dependence of the resonance interaction, yielding a scaling with a different power law. For example, if both dipole moments are oriented along the same direction (x,y, or z), it can be shown that $\widetilde{W}_{\ell m}(a, z, \omega_0) = 0$, and only the term containing $\widetilde{V}_{\ell m}(a, z, \omega_0)$ contributes in (7), and for $za/c^2 \gg 1$ we obtain

$$\delta E \simeq \pm(\mu_{eg}^A)_\ell(\mu_{ge}^B)_m \frac{1}{z^3}\Bigg\{(\delta_{\ell m} - q_\ell q_m - 2n_\ell n_m)\left[\frac{2\omega_0 z}{c}\sin\Big(\frac{2\omega_0 c}{a}\ln\left(\frac{az}{c^2}\right)\Big) - \frac{\omega_0^2 z^2}{c^2}\left(\frac{2c^2}{za}\right)\right.$$
$$\left.\times\cos\Big(\frac{2\omega_0 c}{a}\ln\left(\frac{az}{c^2}\right)\Big)\right] + q_\ell q_m\left(\frac{8c^2}{az}\right)\cos\Big(\frac{2\omega_0 c}{a}\ln\left(\frac{az}{c^2}\right)\Big)\Bigg\}. \tag{8}$$

Here, $\mathbf{n} = (0, 0, 1)$ is the unit vector along the z direction and $\mathbf{q} = (1, 0, 0)$ is the unit vector along the direction of acceleration, x. On the other hand, when the dipoles are orthogonal to each other, with one of them along z and the other in the (x, y) plane, only the term with $\widetilde{W}_{\ell m}(a, z, \omega_0)$ contributes to the resonance interaction, being $\widetilde{V}_{\ell m}(a, z, \omega_0) = 0$. Since such a non-diagonal term is present only for $a \neq 0$ and the resonance interaction for stationary atoms vanishes in this specific configuration, its contribution is a sharp signature of the accelerated motion. Our results clearly show that the non-inertial motion yields a new scaling of the interaction with the distance between the atoms and, since it is not related to vacuum field fluctuations, this qualitative change is not a *thermal* effect and goes beyond the usual temperature-acceleration equivalence [24].

The nonthermal nature of the effect of acceleration on the resonance interaction induces us to question at which extent the equivalence between acceleration and temperature is valid [26]. Recent works have concerned with the problem of the equivalence of physical predictions in coaccelerated and comoving frames, and it was argued that a complete equivalence of the two different frames requires assuming in the Rindler frame an Unruh temperature proportional to the acceleration for the quantum fields [27, 12]. These investigations seem to suggest that the existence of the Unruh effect is necessary for the coherence of the results obtained in the comoving and coaccelerated frames.

We now address the question if this assumption is indeed required for all radiative process, or if there is some physical effect for which comoving and coaccelerated observers give the same prediction, without the additional assumption of an Unruh temperature. We consider the resonance interaction between two accelerated atoms from the point of view of a coaccelerated (Rindler) observer, and discuss the relation with the result obtained in the comoving frame. As before, we suppose the two atoms accelerating with the same acceleration a along the x direction. The trajectories of the two atoms in the Rindler coordinates (τ, ξ, y, z) are [31]

$$\tau_A = \tau\,, \quad \xi_A = y_A = 0\,, \quad z_A = z_1, \quad \tau_B = \tau\,, \quad \xi_B = y_B = 0\,, \quad z_B = z_2\,. \tag{9}$$

Again, we suppose the two atoms interacting with the quantum electromagnetic field in the vacuum state and prepared in one of the correlated states (5). Following the scheme in [38], we quantize the electromagnetic field in the Weyl gauge. In the multipolar coupling scheme and in the dipole approximation, we then calculate the resonance interaction energy following the same procedure developed in the previous section. From Eq.(6), after lengthy algebraic calculations, we recover the same result given in (7) [26]. This shows the complete equivalence between the two different, locally inertial and coaccelerated, frames, as far as the resonance interaction is concerned. This agrees with the mentioned fact that the resonance interaction is entirely due to the radiation reaction field, so that nonthermal effects are essentials, while vacuum fluctuations are not relevant. Therefore, although the vacuum in the comoving frame is not equivalent to the vacuum in the coaccelerated frame, the resonance interaction between the two uniformly accelerated atoms is the same in both frames without the additional assumption of an Unruh temperature for the field in the coaccelerated frame. We conclude that the assumption of an Unruh temperature in a coaccelerated frame is not a requirement for all radiative processes.

4. Resonance interaction between two accelerated atoms in the presence of a perfectly conducting plate

In this section we briefly illustrate preliminary results on the resonance interaction between two atoms moving with a uniform acceleration parallel to a perfectly conducting plate [25]. The presence of the reflecting plate changes the field modes and it is of great interest to investigate its effect on the resonance interaction between the two atoms. The atom-wall Casimir-Polder interaction between a uniformly accelerated atom and a perfectly reflecting plate has been already investigated in the literature [14, 39], and it has been shown that the presence of the boundary affects both vacuum fluctuations and radiation reaction contributions. We thus expect that also the resonance interaction should be modified by the presence of the reflecting plate.

Let us consider two atoms near a perfectly reflecting plate, placed at $z = 0$, and interacting with the electromagnetic field in the vacuum state. We suppose the atoms, separated by a distance L along the direction perpendicular to the wall, located at $z_A = z$ and $z_B = z + L$. They have a uniform acceleration along x, parallel to the plate. Again, we evaluate the resonance interaction between the two accelerating atoms, given by the radiation reaction contribution, in the presence of a conducting plate. After some algebra and using units such that $\hbar = c = 1$, we find [25]

$$\delta E = -\frac{\pi}{4}\left[\delta_{\ell m}(\mu_{ge}^A)_\ell(\mu_{eg}^B)_m P_{\ell m}(a,\mathcal{R},\omega_0) + ((\mu_{ge}^A)_x(\mu_{eg}^B)_z + (\mu_{ge}^A)_z(\mu_{eg}^B)_x)P_{xz}(a,\mathcal{R},\omega_0)\right] \quad (10)$$

where $\mathcal{R} = L + 2z$ and $P_{\ell m}(a, \mathcal{R}, \omega_0)$ is a function, that we do not report here, which modulates the resonance energy interaction with the distance $\mathcal{R}$ and the atomic acceleration a. Comparing with the case of atoms in free-space previously discussed, we find the emergence of a new behaviour originating from the presence of the boundary and related to the vectorial nature of the electromagnetic field. In fact, from Eq. (10) it follows that the effect of atomic acceleration on the resonance interaction can be enhanced or inhibited through an appropriate choice of the orientation of the two dipole moments, as well as of the distance of the two atoms from the plate. This gives additional possibilities to tailor and control the resonance interaction between accelerated atoms, and it could be also relevant for a possible detection of *nonthermal* effects of acceleration.

5. Conclusions

We have discussed and reviewed several physical effects related to the uniform acceleration of a system of two atoms interacting with the electromagnetic field in the vacuum state. We have first discussed the van der Waals/Casimir-Polder dispersion interaction between two accelerated ground-state atoms, and shown that the accelerated motion yields a new scaling of the interaction with the distance and an explicit time dependence of the interaction between the atoms. Exploiting this result, we have argued that our two-atom system could be a suitable setup for an indirect detection of the Unruh effect, even for a reasonable value of the acceleration. We have then investigated the resonance interaction between two identical atoms, one excited and the other in the ground-state and prepared in a *Bell-type* state, moving with uniform acceleration, both in free-space and in the presence of a perfectly reflecting plate. We have shown that the atomic acceleration significantly changes the resonance interaction between the atoms, resulting in a different scaling of the interaction with the distance and in a sharp manifestation of nonthermal effects of the acceleration for specific configurations of the two-atom system.

Acknowledgments

W.Z. would like to thank financial support from the National Natural Science Foundation of China (NSFC) under Grants No. 11405091 and China Scholarship Council (CSC). J.M. acknowledges support from the Alexander von Humboldt Foundation.

References

[1] Fulling S A 1973 *Phys. Rev. D* **7** 2850
[2] Davies P C W 1975 *J. Phys. A: Math. Gen.* **8** 609
[3] Unruh W G 1976 *Phys. Rev. D* **14** 870
[4] Crispino L C B, Higuchi A and Matsas G E A 2008 *Rev. Mod. Phys.* **80** 787
[5] Fedotov A M, Mur V D, Narozhny N B, Belinskii V A and Karnakov B M 1999 *Phys. Lett. A* **254** 126
[6] Ford G W and O'Connell R F 2006 *Phys. Lett. A* **350** 17
[7] Buchholz D and Solveen C 2013 *Class. Quantum Grav,* **30** 085011
[8] Yablonovitch E 1989 *Phys. Rev. Lett.* **62** 1742
[9] Schutzhold R, Schaller G and Habs D 2006 *Phys. Rev. Lett.* **97** 121302
[10] Cozzella G, Landulfo A G S, Matsas G E A and Vanzella D A T 2017 *Preprint* quant-ph/1701.03446
[11] Retzker A, Cirac J I, Plenio M B and Reznik B 2008 *Phys. Rev. Lett* **101** 110402
[12] Matsas G E A and Vanzella D A T 2002 *Int. J. Mod. Phys. D* **11** 1573
[13] Passante R 1998 *Phys. Rev. A* **57** 1590
[14] Rizzuto L and Spagnolo S 2009 *Phys. Rev. A* **79** 062110
[15] Marino J, Noto A and Passante R 2014 *Phys. Rev. Lett.* **113** 020403
[16] Hu J and Yu H 2015 *Phys. Rev. A* **91** 012327
[17] Benatti F and Floreanini R 2004 *Phys. Rev. A* **70** 012112
[18] Power E A and Thirunamachandran T 1993 *Phys. Rev. A* **48** 4761
[19] Craig D P and Thirunamachandran T 1998 *Molecular Quantum Electrodynamics* (Mineola, New York: Dover Publications)
[20] Lamoreaux S K 2005 *Rep. Prog. Phys.* **68** 201
[21] Buhmann S Y 2012 *Dispersion Forces I and II* (Heidelberg: Springer)
[22] Salam A 2010 *Molecular Quantum Electrodynamics* (Hoboken, New Jersey: John Wiley and Sons)
[23] Noto A and Passante R 2013 *Phys. Rev. D* **88** 025041
[24] Rizzuto L, Lattuca M, Marino J, Noto A, Spagnolo S, Zhou W and Passante R 2016 *Phys. Rev. A* **94** 012121
[25] Zhou W, Rizzuto L and Passante R 2017 *Preprint in preparation*
[26] Zhou W, Passante R and Rizzuto L 2016 *Phys. Rev. D* **94** 105025
[27] Zhu Z and Yu H 2007 *Phys. Lett. B* **645** 459
[28] Power E A and Thirunamachandran T 2001 *Proc. R. Soc. Lond. A* **457** 2757
[29] Casimir H B G and Polder D 1948 *Phys. Rev.* **73** 360
[30] Barton G 2001 *Phys. Rev. A* **64** 032102
[31] Birrell N D and Davies P C W 1982 *Quantum fields in curved space* (Cambridge: Cambridge University Press)
[32] Dicke R H 1954 *Phys. Rev.* **93** 99
[33] Dalibard J, Dupont-Roc J and Cohen-Tannoudji C 1982 *J. Physique* **43** 1617
[34] Audretsch J and Muller R 1995 *Phys. Rev. A* **52** 629
[35] Zhu Z, Yu H and Lu S 2006 *Phys. Rev. D* **73** 107501
[36] Rizzuto L 2007 *Phys. Rev. A* **76** 062114
[37] Noto A, Marino J, Zhou W and Passante R 2017 *Preprint in preparation*
[38] Lenz F, Ohta K and Yazaki K 2008 *Phys. Rev. D* **78** 065026
[39] Marino J, Noto A, Passante R, Rizzuto L and Spagnolo S 2014 *Phys. Scr.* **T160** 014031

IOP Conf. Series: Journal of Physics: Conf. Series **880** (2017) 012043 doi:10.1088/1742-6596/880/1/012043

Flavor mixing transformations for a uniformly accelerated observer

M Blasone[1,2], G Lambiase[1,2] and G G Luciano[1,2]

[1] Dipartimento di Fisica, Università degli Studi di Salerno, Via Giovanni Paolo II, 132 I-84084 Fisciano (SA), Italy

[2] INFN Sezione di Napoli, Gruppo collegato di Salerno, Italy

E-mail: blasone@sa.infn.it, lambiase@sa.infn.it, g.luciano9@studenti.unisa.it

Abstract. We study flavor mixing transformations for a uniformly accelerated observer (Rindler observer) in the simplest case of two charged scalar fields with different masses. This is obtained starting from an analysis of boson field mixing in inertial frame (Minkowski frame) in the hyperbolic basis. As a consequence of the non-trivial interplay between the Bogolubov transformation related to flavor mixing and the one arising from the Rindler spacetime structure, the Fulling-Rindler condensate of the inertial vacuum gets deeply modified, losing its original thermal nature.

1. Introduction

Since Pontecorvo's pioneering work [1], the issue of flavor mixing has turned out to be one of the hottest topics in particle-physics phenomenology. Despite the great progress coming from the theoretical and experimental developments of the two last decades, anyway, it remains one of the most puzzling phenomena. Many features, in fact, are still intriguing, among which, for instance, its origin within Standard Model, the related problem of the generation of neutrino masses and the non-trivial condensate structure exhibited by the vacuum for the mixed fields. The latter aspect, in particular, has been extensively explored, starting from the initial observation of the unitary inequivalence between mass and flavor vacua in QFT [2, 3]. In particular, oscillation formulas have been derived, exhibiting corrections with respect to the usual ones [4] and the entanglement properties of mixed particles have been recently investigated [5].

An equally important question dealing with flavor mixing and oscillations is the way in witch they are affected by a gravitational field [6, 7, 8, 9]. In this paper, we do a first step in this direction, by analyzing the QFT of two mixed scalar fields from the viewpoint of the Rindler observer, i.e. a uniformly accelerated observer in the Minkowski spacetime. It is worth stressing that till now the QFT of mixed fields has been studied only in inertial frame and in the plane-wave basis. Here we focus on the hyperbolic representation, where the boost generator is diagonal, which is a convenient starting framework for the analysis in the accelerated frame. The formalism we elaborate may serve as basis for studying flavor oscillations in curved background: once extended to neutrino fields, indeed, it can be compared with the other approaches existing in literature, such as the WKB approximation [6], the plane-wave method [7] or geometric treatments [8].

IOP Conf. Series: Journal of Physics: Conf. Series **880** (2017) 012043 doi:10.1088/1742-6596/880/1/012043

Analyzing field mixing in an accelerated frame, anyway, is not just a tool for studying the influence of gravitation on flavor oscillation formulas. There are, in fact, a number of theoretical problems appearing in such a context. For instance, it has been recently highlighted that the inverse β-decay rates of accelerated protons in the inertial and comoving frames disagree when neutrino mixing is taken into account [10]. A more detailed analysis of this and other aspects will be discussed elsewhere [11].

The paper is organized as follows: in Section 2 the standard QFT of two mixed scalar fields in the plane-wave expansion is briefly reviewed. Section 3 is devoted to the comparison of such a framework with the alternative hyperbolic basis; in particular, the physical equivalence between the two representations is shown. In Section 4, after a short discussion about the Rindler spacetime structure, the Rindler-Fulling quantization scheme is analyzed, both for free and mixed fields. Conclusions are summarized in the last Section.

2. Field mixing for an inertial observer: plane-wave representation

Let us start by discussing some of the major results about flavor mixing in the simplest case of two complex scalar fields. We review, in particular, the quantization of mixed fields for an inertial observer in the usual plane-wave basis. Such a topic has been widely analyzed in the last two decades, first in the case of Dirac fermions [2, 12] and later for other fields [3].

As it is well known, mixing transformations in a simplified two flavor model are defined according to the following equations

$$\phi_A(x) = \phi_1(x)\,\cos\theta + \phi_2(x)\,\sin\theta, \tag{1}$$

$$\phi_B(x) = -\phi_1(x)\,\sin\theta + \phi_2(x)\,\cos\theta, \tag{2}$$

where $\phi_j(x)$, $j = 1, 2$ are two free charged scalar fields of masses m_j, $\phi_\chi(x)$, $\chi = A, B$ are the mixed fields of "flavor" χ and θ is the mixing angle. In terms of the complete sets of plane-waves $\left\{U_{\mathbf{k},j}\,, U^*_{\mathbf{k},j}\right\}$, the expansion of $\phi_j(x)$ reads[1]:

$$\phi_j(x) = \int d^3\mathbf{k}\,\left\{a_{\mathbf{k},j}\,U_{\mathbf{k},j}(x) + \bar{a}^\dagger_{\mathbf{k},j}\,U^*_{\mathbf{k},j}(x)\right\}, \quad j = 1, 2, \tag{3}$$

where

$$U_{\mathbf{k},j}(x) = \left[2\omega_{\mathbf{k},j}(2\pi)^3\right]^{-\frac{1}{2}}\,e^{i\left(\mathbf{k}\cdot\mathbf{x}-\omega_{\mathbf{k},j}t\right)} \tag{4}$$

are solutions of the Klein-Gordon equation in the Minkowski coordinates $(t, x^1, \vec{x})$ for the field of mass m_j. It is not difficult to show that these mode functions are orthonormal with respect to the KG product

$$(\phi_\alpha, \phi_\beta) = i\int d^3\mathbf{x}\,\left[\phi^*_\beta(x)\,\overleftrightarrow{\partial_t}\,\phi_\alpha(x)\right], \tag{5}$$

where the integration is performed on a hypersurface of constant t.

The ladder operators $a^\dagger_{\mathbf{k},j}$ and $a_{\mathbf{k},j}$ ($\bar{a}^\dagger_{\mathbf{k},j}$ and $\bar{a}_{\mathbf{k},j}$) in Eq.(3) create and annihilate a particle (antiparticle) of momentum $\mathbf{k}$ and frequency $\omega_{\mathbf{k},j} = \sqrt{m_j^2 + \mathbf{k}^2}$, respectively. They are assumed to satisfy the canonical commutation relations. In addition, since we are dealing with two free fields, the vacuum state they annihilate is accordingly defined as $|0_M\rangle \equiv |0_M\rangle_1 \otimes |0_M\rangle_2$, where $|0_M\rangle_j$ is the vacuum for the field of mass m_j. We will refer to such a state as the "Minkowski inertial vacuum".

[1] The metric $\eta_{\mu\nu} = \mathrm{diag}(+1, -1, -1, -1)$ and natural units will be used throughout all the paper. Moreover, the following shorthand notation will be adopted for 4-, 3- and 2-vectors: $x \equiv (t, x^1, x^2, x^3)$, $\mathbf{x} \equiv \{x^1, x^2, x^3\}$ and $\vec{x} \equiv \{x^2, x^3\}$.

The completeness of plane-waves Eq.(4) allows us to adopt the following free fields-like expansions for the "flavor fields"

$$\phi_\ell(x) = \int d^3\mathbf{k}\Big\{a_{\mathbf{k},\ell}(t)\, U_{\mathbf{k},s}(x) + \bar{a}^\dagger_{\mathbf{k},\ell}(t)\, U^*_{\mathbf{k},s}(x)\Big\}, \tag{6}$$

where $(\ell, s) = (A, 1), (B, 2)$. The time dependent flavor operators $a_{\mathbf{k},\ell}$ are given by[2]

$$a_{\mathbf{k},\ell} = (\phi_\ell, U_{\mathbf{k},s})\,. \tag{7}$$

By using Eq.(1), for instance, we obtain for $a_{\mathbf{k},A}$ the following transformation

$$a_{\mathbf{k},A} = \cos\theta\, a_{\mathbf{k},1} + \sin\theta\, \left(\rho^{\mathbf{k}\,*}_{12}\, a_{\mathbf{k},2} + \lambda^{\mathbf{k}}_{12}\, \bar{a}^\dagger_{-\mathbf{k},2}\right), \tag{8}$$

where the Bogolubov coefficients $\rho^{\mathbf{k}}_{12}$ and $\lambda^{\mathbf{k}}_{12}$ are such that

$$\rho^{\mathbf{k}}_{12} \equiv \int d^3\mathbf{k}'\Big(U_{\mathbf{k},1}, U_{\mathbf{k}',2}\Big) = |\rho^{\mathbf{k}}_{12}|\, e^{i(\omega_{\mathbf{k},2}-\omega_{\mathbf{k},1})t}, \qquad \lambda^{\mathbf{k}}_{12} \equiv \int d^3\mathbf{k}'\Big(U^*_{\mathbf{k}',2}, U_{\mathbf{k},1}\Big) = |\lambda^{\mathbf{k}}_{12}|\, e^{i(\omega_{\mathbf{k},1}+\omega_{\mathbf{k},2})t}, \tag{9}$$

with

$$|\rho^{\mathbf{k}}_{12}| \equiv \frac{1}{2}\left(\sqrt{\frac{\omega_{\mathbf{k},1}}{\omega_{\mathbf{k},2}}} + \sqrt{\frac{\omega_{\mathbf{k},2}}{\omega_{\mathbf{k},1}}}\right), \qquad |\lambda^{\mathbf{k}}_{12}| \equiv \frac{1}{2}\left(\sqrt{\frac{\omega_{\mathbf{k},1}}{\omega_{\mathbf{k},2}}} - \sqrt{\frac{\omega_{\mathbf{k},2}}{\omega_{\mathbf{k},1}}}\right) \tag{10}$$

(a similar expression can be derived for $a_{\mathbf{k},B}$). Starting from Eqs.(8) and (9), one can easily prove that flavor operators, just as the corresponding mass operators, obey the canonical commutation relations (at equal times).

The flavor vacuum $|0(\theta,t)\rangle_{A,B}$ at time t can be now defined as

$$a_{\mathbf{k},\chi}(t)\; |0(\theta,t)\rangle_{A,B} = \bar{a}_{\mathbf{k},\chi}(t)\; |0(\theta,t)\rangle_{A,B} = 0\,, \qquad \chi = A, B\,. \tag{11}$$

By inverting the set of equations for the flavor operators with respect to the mass ones, it is possible to show that the vacuum Eq.(11) exhibits a condensate structure of particle-antiparticle pairs, both with the same and different masses. The condensation density for any t is given by

$${}_{A,B}\langle 0(\theta,t)|a^\dagger_{\mathbf{k}',j}\, a_{\mathbf{k},j}|0(\theta,t)\rangle_{A,B} = \sin^2\theta\; |\lambda^{\mathbf{k}}_{12}|^2\, \delta^3(\mathbf{k}-\mathbf{k}'), \qquad j = 1, 2\,. \tag{12}$$

It is worth noting that, in the infinite volume limit, flavor and mass vacua are orthogonal to each other due to the infinite number of degrees of freedom. This shows the fundamental result that the Hilbert space for the mixed field states is unitarily inequivalent to the Hilbert state space where the free-field operators are defined (see Ref.[2, 3] for the details).

3. Field mixing for an inertial observer: hyperbolic representation

The foregoing discussion is completely within the usual plane-wave basis. According to Lorentz covariance, anyway, there is no reason to prefer such a representation to those which diagonalize the other generators of the Lorentz group. We may wonder, for instance, how the above formalism appears if we quantize the fields in the hyperbolic representation, that is, the representation in which the Lorentz boost generator is diagonal[3]. The introduction of the latter quantization scheme, however, is not a purely mathematical exercise of change of basis. As we

[2] For simplicity, from now on we will omit the time dependence of flavor operators and vacuum when it is not strictly necessary.

[3] We will consider, in particular, the generator of a Lorentz boost along the x^1-axis.

shall see in the next Section, indeed, the hyperbolic expansion turns out to be an useful tool for analyzing the field quantization from the viewpoint of an accelerated observer [13].

In order to answer our question, let us expand the free fields $\phi_j(x)$, $j = 1, 2$ in terms of the complete and orthonormal sets of boost modes $\left\{\widetilde{U}^{(\sigma)}_{\kappa,j}, \widetilde{U}^{(\sigma)*}_{\kappa,j}\right\}$, related to the plane-waves by[4]

$$\widetilde{U}^{(\sigma)}_{\kappa,j}(x) \equiv \int_{-\infty}^{+\infty} dk_1\, p^{(\sigma)*}_{\Omega,j}(k_1)\, U_{\mathbf{k},j}(x) = \frac{e^{\sigma\pi\Omega/2}}{2\sqrt{2}\,\pi^2}\, K_{i\sigma\Omega}(\mu_{\vec{k},j}\xi)\, e^{i\left(\vec{k}\cdot\vec{x}-\sigma\Omega\eta\right)}, \tag{13}$$

where the hyperbolic coordinates (η, ξ) are defined according to the following equations

$$t = \xi \sinh\eta\,, \qquad x^1 = \xi\cosh\eta. \tag{14}$$

The subscript κ in Eq.(13) stands for $(\Omega, \vec{k})$, $K_{i\sigma\Omega}$ is the modified Bessel function of second kind and imaginary order and $\mu_{\vec{k},j} = \sqrt{m_j^2 + |\vec{k}|^2}$. The convolution function $p^{(\sigma)}_{\Omega,j}(k_1)$ is given by

$$p^{(\sigma)}_{\Omega,j}(k_1) = \frac{1}{\sqrt{2\pi\omega_{\mathbf{k},j}}}\left(\frac{\omega_{\mathbf{k},j}+k_1}{\omega_{\mathbf{k},j}-k_1}\right)^{i\sigma\Omega/2}, \quad j = 1, 2. \tag{15}$$

It is easy to show that the sets of these functions are both complete and orthonormal (see Ref.[14] for the details).

Strictly speaking, Eq.(13) holds only in the region $x^1 > |t| \cup x^1 < -|t|$. The correct global functions, namely the Gerlach's Minkowski Bessel modes, can be obtained by analytically continuing the solutions Eq.(13) across $x_1 = \pm t$ [15]. For our purpose, nevertheless, it is enough to consider the modes as above defined.

The free fields $\phi_j(x)$, $j = 1, 2$ can be thus expanded as follows

$$\phi_j(x) = \sum_\sigma \int_0^{+\infty} d\Omega \int d^2\vec{k}\, \left\{d^{(\sigma)}_{\kappa,j}\, \widetilde{U}^{(\sigma)}_{\kappa,j}(x) + \bar{d}^{(\sigma)\dagger}_{\kappa,j}\, \widetilde{U}^{(\sigma)*}_{\kappa,j}(x)\right\}, \tag{16}$$

where the role of the integration on k_1 in the plane-wave expansion Eq.(3) has been now replaced by the sum on σ and the positive integration on Ω. It naturally arises the question how the "hyperbolic" ladder operators in Eq.(16) are related to the corresponding ones in the plane-wave representation Eq.(3). In this regard, by equating the two alternative field expansions Eqs.(3) and (16) on a hyperplane of constant η and forming the KG inner product of both the sides with the boost mode $\widetilde{U}^{(\sigma)}_{\kappa,j}$, one immediately obtains

$$d^{(\sigma)}_{\kappa,j} = \int d^3\mathbf{k}' \left\{a_{\mathbf{k}',j}\left(U_{\mathbf{k}',j}, \widetilde{U}^{(\sigma)}_{\kappa,j}\right) + \bar{a}^\dagger_{\mathbf{k}',j}\left(U^*_{\mathbf{k}',j}, \widetilde{U}^{(\sigma)}_{\kappa,j}\right)\right\}, \quad j = 1, 2. \tag{17}$$

Since the boost modes $\widetilde{U}^{(\sigma)}_{\kappa,j}$ are linear combinations of positive frequency plane-waves alone (see Eq.(13)), the coefficient of $\bar{a}^\dagger_{\mathbf{k}',j}$ in Eq.(17) vanishes. Therefore, by exploiting the orthonormality condition of $\left\{U_{\mathbf{k},j}, U^*_{\mathbf{k},j}\right\}$, $j = 1, 2$, it follows that (similarly for $\bar{d}^{(\sigma)}_{\kappa,j}$)

$$d^{(\sigma)}_{\kappa,j} = \int_{-\infty}^{+\infty} dk_1\, p^{(\sigma)}_{\Omega,j}(k_1)\, a_{\mathbf{k},j}. \tag{18}$$

[4] The physical meaning of $\sigma = \pm 1$ and of the positive parameter Ω in Eq.(13) will be clear once the Rindler-Fulling quantization scheme for an accelerated observer is introduced.

In Ref.[14] such operators are explicitly shown to diagonalize the x^1-boost generator.
From Eq.(18) we find out that the vacuum annihilated by d-operators is the same as the usual Minkowski vacuum $|0_M\rangle$. Additionally, by using the orthonormality and completeness conditions of the sets $\left\{p^{(\sigma)}_{\Omega,j}\right\}$ and the commutation relations of $a_{\mathbf{k},j}$ and $a^{\dagger}_{\mathbf{k},j}$, $j = 1, 2$, it is possible to show that the transformation Eq.(18) is canonical. Combining both these results, we can definitely state that the quantum formalisms above introduced, the plane-wave and the hyperbolic constructions, are totally equivalent from the point of view of free fields.

To understand if such an equivalence still holds when mixing is involved, let us extend the hyperbolic quantization to the flavor fields Eqs.(1) and (2). Retracing the same steps of the plane-wave representation, we expand the above fields as follows

$$\phi_\ell(x) = \sum_\sigma \int_0^{+\infty} d\Omega \int d^2\vec{k} \left\{ d^{(\sigma)}_{\kappa,\ell}\, \widetilde{U}^{(\sigma)}_{\kappa,s}(x) + \bar{d}^{(\sigma)\dagger}_{\kappa,\ell}\, \widetilde{U}^{(\sigma)*}_{\kappa,s}(x) \right\}, \tag{19}$$

where $(\ell, s) = (A, 1), (B, 2)$ and $d^{(\sigma)}_{\kappa,\ell}$ are the flavor operators in the hyperbolic basis.
In order to establish whether the quantization formalism of mixed fields is somehow affected by such a change of basis, let us focus on flavor vacuum in this new representation. For this purpose, following the approach used in Ref.[13], we could express flavor d-operators in Eq.(19) in terms of the corresponding mass operators in Eq.(16). Exploiting Eq.(18), the relations thus obtained could be recast in terms of mass a-operators and, in the last step, of flavor a-operators by using the inverse of Eq.(8). Such a procedure, however, would be rather laborious. Here we propose a more straightforward approach, yielding the same result.

By directly comparing Eqs.(6) and (19) for the flavor fields and performing the same calculations as in absence of mixing, we readily obtain

$$d^{(\sigma)}_{\kappa,\ell} = \int_{-\infty}^{+\infty} dk_1\, p^{(\sigma)}_{\Omega,s}(k_1)\, a_{\mathbf{k},\ell}, \tag{20}$$

where $(\ell, s) = (A, 1), (B, 2)$. Therefore, the relation Eq.(18) between d- and a- mass operators also holds for the corresponding flavor operators[5].
Eq.(20) shows a significant result: since flavor d-annihilators are linear combinations of flavor a-annihilators alone, the change of representation from plane-wave to hyperbolic modes does not affect the flavor vacuum $|0(\theta, t)\rangle_{A,B}$ in Eq.(11) and, consequently, the Hilbert space structure for the mixed fields.

4. Field mixing for a uniformly accelerated observer: Fulling-Rindler quantization

In this Section we show how the hyperbolic quantization previously introduced turns out to be an helpful construction in the extension of QFT to an accelerated frame of reference.
As it is well known, the most suitable framework to describe the motion of an accelerated observer in the Minkowski spacetime is the Rindler metric[6]. The analysis of flavor mixing from the viewpoint of such an observer, therefore, requires a preliminary discussion about the properties of this background. In doing so, we also briefly review the Rindler-Fulling scheme [16], which is the natural way for quantizing the fields in an accelerated frame.

As a first step, let us recall that the line element $ds^2 = \eta_{\mu\nu} dx^\mu dx^\nu$ in the hyperbolic coordinates Eq.(14) takes the form

$$ds^2 = (dt)^2 - (dx^1)^2 - \sum_{j=2}^{3} (dx^j)^2 \underset{hyperb.\ coord.}{\longrightarrow} ds^2 = \xi^2 d\eta^2 - d\xi^2 - \sum_{j=2}^{3} (dx^j)^2 . \tag{21}$$

[5] This implies that the flavor operators $d^{(\sigma)}_{\kappa,\chi}$, $\chi = A, B$, just as the corresponding operators in the plane-wave basis, obey the canonical commutation relations (at equal times).
[6] As usual, we will refer to such an observer as the "Rindler observer".

The world line of a uniformly accelerated observer with proper acceleration a is thus given by

$$\xi(\tau) = \text{const} \equiv a^{-1}, \quad \vec{x}(\tau) = \text{const}, \tag{22}$$

where τ is the proper time measured along the line. From Eq.(21), we have $\eta(\tau) = a\tau$, that is, the proper time τ measured by an observer with proper acceleration a is the same as the Rindler time η, up to the scale factor a.

By using the Minkowski coordinates, one can easily verify that the line Eq.(22) describes an hyperbola in the (t, x^1) plane with asymptotes $t = \pm x^1$. Such a trajectory is parametrized by the proper acceleration a: for $a > 0$, the observer is confined within the right wedge $R_+ = \{x|x^1 > |t|\}$, for $a < 0$, instead, his motion occurs in the left wedge $R_- = \{x|x^1 < -|t|\}$. The two regions R_+ and R_-, moreover, are causally separated from each other; there is no way for an accelerated observer in R_+ to exchange information with one in R_- [17].

Keeping these properties in mind, the Rindler-Fulling procedure is now introduced as opposed to the more familiar Minkowski quantization. To this purpose, following the approach used in Ref.[14], we expand the free fields as follows

$$\phi_j(x) = \sum_\sigma \int_0^{+\infty} d\Omega \int d^2\vec{k} \left\{ b^{(\sigma)}_{\kappa,j}\, u^{(\sigma)}_{\kappa,j}(x) + \bar{b}^{(\sigma)\dagger}_{\kappa,j}\, u^{(\sigma)*}_{\kappa,j}(x) \right\}, \quad j = 1, 2. \tag{23}$$

The Rindler modes $u^{(\sigma)}_{\kappa,j}$ of positive frequency Ω with respect to the time η are defined by

$$u^{(\sigma)}_{\kappa,j}(x) = \theta(\sigma\xi) \left[2\Omega(2\pi)^2\right]^{-\frac{1}{2}} h^{(\sigma)}_{\kappa,j}(\xi)\, e^{i(\vec{k}\cdot\vec{x} - \sigma\Omega\eta)}, \tag{24}$$

where $\sigma = \pm 1$ refers to the right/left wedges $R_\pm$. They are solutions of the KG equation in Rindler coordinates for the field of mass m_j (see Ref.[14] for the mathematical details). As in Eq.(13), the function $h^{(\sigma)}_{\kappa,j}$ is given by the modified Bessel mode of second kind and imaginary order, up to a normalization factor. According to our previous considerations, the Heaviside step function $\theta(\sigma\xi)$ has been inserted in Eq.(24) in order to restrict such modes to only one of the two causally separated wedges (R_+ for $\sigma = +1$ and R_- for $\sigma = -1$). In addition, by expressing the KG inner product Eq.(5) in the Rindler coordinates and using Eq.(24), it is possible to show that the modes above defined form a complete orthonormal set.
As it is known, the ladder operators $b^{(\sigma)\dagger}_{\kappa,j}$ and $b^{(\sigma)}_{\kappa,j}$ ($\bar{b}^{(\sigma)\dagger}_{\kappa,j}$ and $\bar{b}^{(\sigma)}_{\kappa,j}$) in Eq.(23) can be interpreted as creators and annihilators of a Rindler particle (antiparticle), respectively. They are assumed to satisfy the canonical commutation relations. The Rindler vacuum is accordingly defined as $|0_R\rangle \equiv |0_R\rangle_1 \otimes |0_R\rangle_2$, where $|0_R\rangle_j$ is the vacuum for the field of mass m_j.

We are now interested in the relation between the free-field quantization in the inertial frame and the Rindler-Fulling scheme. To this end, let us remind that, as shown in the previous Section, the plane-wave and the hyperbolic constructions are completely equivalent from the viewpoint of an inertial observer. Therefore, in our comparison, we can equivalently consider the field expansions Eqs.(3) or (16) for such an observer. To simplify the calculations, we choose the latter. In such a way, by matching Eqs.(16) and (23) on a hypersurface of constant η and multiplying both the sides for the Rindler mode $u^{(\sigma)}_{\kappa,j}$, $j = 1, 2$ it follows that

$$b^{(\sigma)}_{\kappa,j} = \sqrt{1 + N_R(\Omega)}\, d^{(\sigma)}_{\kappa,j} + \sqrt{N_R(\Omega)}\, \bar{d}^{(-\sigma)\dagger}_{\tilde{\kappa},j}, \quad j = 1, 2, \tag{25}$$

where $\tilde{\kappa}$ stands for $(\Omega, -\vec{k})$ and

$$N_R(\Omega) = \frac{1}{e^{2\pi\Omega} - 1} \tag{26}$$

IOP Conf. Series: Journal of Physics: Conf. Series **880** (2017) 012043 doi:10.1088/1742-6596/880/1/012043

is the Bose-Einstein condensate of the Minkowski vacuum $|0_M\rangle$[7]

$$\langle 0_M| b^{(\sigma)\dagger}_{\kappa,j}\, b^{(\sigma)}_{\kappa',j}|0_M\rangle = N_R(\Omega)\,\delta^3(\kappa-\kappa'), \quad j=1,2. \tag{27}$$

From Eq.(27) we find out that the inertial vacuum appears to be not empty from the viewpoint of an accelerated observer; just as the flavor vacuum with respect to free fields, indeed, it exhibits a condensate structure in terms of the Rindler particles. The crucial point is that such a non-trivial result, both for mixing and Rindler condensation densities, mathematically arises from a Bogolubov transformation involving two sets of creators and annihilators: mass and flavor ladder operators in the former case (see Eq.(8)), Minkowski and Rindler ones for the latter effect (Eq.(25)).

Up to now we have separately analyzed the physics hiding under such Bogolubov transformations. It arises thus the natural question how the thermal condensate in Eq.(26) gets modified if we extend the Rindler-Fulling scheme to the mixed fields Eqs.(1) and (2).
For this purpose, all we need is to recast the mixing transformation Eq.(8) in terms of the Rindler b-operators. Once again, by exploiting the completeness of the Rindler modes in Eq.(24), we take for the flavor fields the following expansion

$$\phi_\ell(x) = \sum_\sigma \int_0^{+\infty} d\Omega \int d^2\vec{k}\left(b^{(\sigma)}_{\kappa,\ell}\, u^{(\sigma)}_{\kappa,s}(x) + \bar{b}^{(\sigma)\dagger}_{\kappa,\ell}\, u^{(\sigma)*}_{\kappa,s}(x)\right), \tag{28}$$

where $(\ell,s)=(A,1),(B,2)$ and $b^{(\sigma)}_{\kappa,\ell}$ are the flavor operators for the Rindler observer.
The interplay between the mixing and thermal Bogolubov transformations can be now easily investigated by comparing Eqs.(19) and (28). By multiplying both the sides for the Rindler mode $u^{(\sigma)}_{\kappa,j}$, $j=1,2$, one can show that the relation Eq.(25) between b- and d- mass operators also holds for the corresponding flavor operators [13]

$$b^{(\sigma)}_{\kappa,\chi} = \sqrt{1+N_R(\Omega)}\, d^{(\sigma)}_{\kappa,\chi} + \sqrt{N_R(\Omega)}\, \bar{d}^{(-\sigma)\dagger}_{\tilde{\kappa},\chi}, \quad \chi=A,B. \tag{29}$$

However, since our final aim is to calculate the Rindler distribution of mixed particles in the vacuum $|0_M\rangle$, it is useful to recast Eq.(29) in terms of the ladder operators $a_{\mathbf{k},j}$ and $\bar{a}_{\mathbf{k},j}$ in Eq.(3). The spectrum of mixed Rindler particles in the inertial vacuum takes thereby the form

$$\begin{aligned}\langle 0_M|\, b^{(\sigma)\dagger}_{\kappa,\chi}\, b^{(\sigma)}_{\kappa',\chi}\,|0_M\rangle \;=\; & \\ N_R(\Omega)\,\delta(\kappa-\kappa') \;+\; \sin^2\theta & \Bigg[\left(\sqrt{N_R(\Omega)\,N_R(\Omega')} + \sqrt{(1+N_R(\Omega))\,(1+N_R(\Omega'))}\right) N_{\lambda\lambda} \\ & + \sqrt{1+N_R(\Omega)}\,\sqrt{N_R(\Omega')}\,N_{\rho\lambda} + \sqrt{N_R(\Omega)}\,\sqrt{1+N_R(\Omega')}\,N^*_{\rho\lambda}\Bigg]\delta^2(\vec{k}-\vec{k}')\end{aligned} \tag{30}$$

where $\chi = A, B$ and

$$N_{\lambda\lambda} = \int_{-\infty}^{+\infty} dk_1\; p^{(\sigma)*}_{\Omega',j}(k_1)\, p^{(\sigma)}_{\Omega,j}(k_1)\,|\lambda^{\mathbf{k}}_{12}|^2, \qquad N_{\rho\lambda} = \int_{-\infty}^{+\infty} dk_1\; p^{(\sigma)*}_{\Omega',j}(k_1)\, p^{(\sigma)}_{\Omega,j}(-k_1)\,\lambda^{\mathbf{k}*}_{12}\,\rho^{\mathbf{k}}_{12}, \tag{31}$$

with $\rho^{\mathbf{k}}_{12}$ and $\lambda^{\mathbf{k}}_{12}$ defined in Eq.(9).

[7] One can recognize in Eqs.(25) and (26) the well-known Unruh effect: a uniformly accelerated observer feels in the Minkowski vacuum a heat bath with a temperature proportional to the magnitude of his acceleration [18].

Therefore, for mixed fields, the "usual" Rindler condensate of the inertial vacuum Eq.(26) gets non-trivially modified, losing its original thermal nature. Anyway, it can be readily seen that the standard value is recovered for $\theta \to 0$ and in the limit $m_1 \to m_2$ (see Eq.(10)), as one would expect in absence of mixing.
It is worth noting that the expectation value Eq.(30) is non-diagonal with respect to the parameters Ω and Ω'. Since such a result has been obtained without using any approximation, we can state that the factorization of the hyperbolic Bogolubov coefficients into the product of a Dirac delta distribution of Ω and Ω' with suitable functions adopted in Ref.[13] is not quite correct. It is our purpose to discuss this and other aspects in a forthcoming paper.

5. Conclusions

The issue of flavor mixing from the point of view of a uniformly accelerated observer has been discussed in the case of two complex scalar fields with different masses. In spite of such a minimal setting, the Bogolubov transformation built in field mixing has been found to non-trivially combine with the thermal Bogolubov transformation associated to the Rindler spacetime structure, leading to a meaningful modification of the "usual" Fulling-Rindler spectrum. The result obtained Eq.(30), in fact, shows that the particle distribution detected by an accelerated observer in the inertial vacuum loses its thermal nature when mixing is taken into account.

We stress that the formalism developed in this paper represents an useful springboard for studying flavor oscillations in the context of Quantum Field Theory on curved background. Indeed, once extended to the fermionic case and, in particular, to neutrino fields, the results derived within this framework can be compared with those obtained through the other approaches dealing with such a problem [6, 7, 8]. Additionally, it may be taken as starting point for investigating Lorentz invariance breakdown in the context of mixed neutrinos [19]. More work is inevitably required along these lines.

References

[1] Bilenky M S and Pontecorvo B 1978 *Phys. Rep.* **41** 225
[2] Blasone M and Vitiello G 1995 *Annals Phys.* **244** 283
[3] Blasone M, Henning P A and Vitiello G 1996 *La Thuile, Results and perspectives in particle physics* 139-152; Blasone M, Capolupo A, Romei O and Vitiello G 2001 *Phys. Rev.* D **63** 125015; Blasone M and Palmer J 2004 *Phys. Rev.* D **69** 057301
[4] Blasone M, Pires Pacheco P and Wan Chan Tseung H 2003 *Phys. Rev.* D **67** 073011; Blasone M, Henning P A and Vitiello G 1999 *Phys. Lett.* B **451** 140
[5] Blasone M, Dell'Anno F, De Siena S and Illuminati F 2014 *Europhys. Lett.* **106** 30002; 2009 *Europhys. Lett.* **85** 50002
[6] Stodolsky L 1979 *Gen. Rel. Grav.* **11** 391; Wudka J 2001 *Phys. Rev.* D **64**; Visinelli L 2015 *Gen. Rel. Grav.* **47** 62
[7] Fornengo N, Giunti C, Kim C W and Song J 1997 *Phys. Rev.* **D** 56; Konno K and Kasai M 1998 *Prog. Theor. Phys.* **100** 1145; Fornengo N, Giunti C, Kim C W and Song J 1999 *Nucl. Phys. Proc. Suppl.* **70** 264
[8] Cardall C Y and Fuller G M 1997 *Phys. Rev.* D **55** 7960; Zhang Y H and Li X Q 2016 *Nucl.Phys.* B911
[9] Lambiase G, Papini G, Punzi and Scarpetta G 2005 *Phys.Rev.* D**71** 073011
[10] Ahluwalia, D V, Labun L and Torrieri G 2016 *Eur. Phys. J. A* **52** 189
[11] Blasone M, Lambiase G and Luciano GG *work in progress*
[12] Blasone M,Gargiulo M V and Vitiello G 2016 *Phys. Lett.* B **761** 104
[13] Blasone M, Lambiase G and Luciano G G 2015 *J. Phys. Conf. Ser.* **631** 012053
[14] Takagi S 1986 *Prog. Theor. Phys. Suppl.* **88** 1
[15] Gerlach U H 1988 *Phys. Rev.* D **38** 514
[16] Fulling S A 1973 *Phys. Rev.* D **7** 2850
[17] Rindler W 1966 *Am. J. Phys.* **34** 1174
[18] Unruh W G 1976 *Phys. Rev.* D **14** 870
[19] Blasone M, Di Mauro M and Lambiase G 2005 *Acta Phys. Pol.* B **36** 3255; Blasone M, Magueijo J and Pires Pacheco P 2005 *Europhys. Lett.* **70** 600

Quantum corrections to newtonian potential and generalized uncertainty principle

Fabio Scardigli[a b], **Gaetano Lambiase** [c d], **and Elias Vagenas** [e]

[a] Dipartimento di Matematica, Politecnico di Milano, Piazza Leonardo da Vinci 32, 20133 Milano, Italy

[b] Department of Applied Mathematics, University of Waterloo, Ontario N2L 3G1, Canada

[c] Dipartimento di Fisica "E.R. Caianiello", Universita' di Salerno, 84084 Fisciano (Salerno), Italy

[d] INFN Gruppo Collegato di Salerno, Italy

[e] Theoretical Physics Group, Department of Physics, Kuwait University, P.O. Box 5969, Safat 13060, Kuwait

E-mail: fabio@phys.ntu.edu.tw, lambiase@sa.infn.it, elias.vagenas@ku.edu.kw

Abstract. We use the leading quantum corrections to the newtonian potential to compute the deformation parameter of the generalized uncertainty principle. By assuming just only General Relativity as theory of Gravitation, and the thermal nature of the GUP corrections to the Hawking spectrum, our calculation gives, to first order, a specific numerical result. We briefly discuss the physical meaning of this value, and compare it with the previously obtained bounds on the generalized uncertainty principle deformation parameter.

1. Introduction

Gravitation plays a pivotal role in the generalization of Heisenberg uncertainty principle (HUP), from the early attempts [1], to the more modern proposals, as string theory, loop quantum gravity, deformed special relativity, and studies of black hole physics [2, 3, 4, 5, 6, 7].

The dimensionless deforming parameter of the generalized uncertainty principle (GUP), usually denoted by β, is not (in principle) fixed by the theory, although it is generally assumed to be of order one (this happens, in particular, in some models of string theory, see for instance Ref. [2]). Therefore, many studies try to set bounds on β. For instance, in Refs. [8], [9], [10] a specific (in general, non linear) representation of the operators in the deformed fundamental commutator is utilized [1]

$$\left[\hat{X}, \hat{P}\right] = i\,\hbar \left(1 + \beta\, \frac{\hat{P}^2}{m_{\rm p}^2}\right) \tag{1}$$

in order to compute corrections to quantum mechanical quantities, such as energy shifts in the spectrum of the hydrogen atom, or to the Lamb shift, the Landau levels, Scanning Tunneling

[1] We shall work with $c = k_B = 1$, but explicitly show the Newton constant $G_{\rm N}$ and Planck constant $\hbar$. We also recall that the Planck length is defined as $\ell_{\rm p}^2 = G_{\rm N}\,\hbar/c^3$, the Planck energy as $\mathcal{E}_{\rm p}\,\ell_{\rm p} = \hbar\, c/2$, and the Planck mass as $m_{\rm p} = \mathcal{E}_{\rm p}/c^2$, so that $G_{\rm N} = \ell_{\rm p}/2\, m_{\rm p}$ and $\hbar = 2\,\ell_{\rm p}\, m_{\rm p}$.

Microscope, charmonium levels, etc. The bounds so obtained on β are quite stringent, ranging from $\beta < 10^{21}$ to $\beta < 10^{50}$.

In Refs. [11] and [12], a different path is proposed in order to find bounds on β. A deformation of classical newtonian mechanics is introduced by modifying the standard Poisson brackets in a way that resembles the quantum commutator

$$[\hat{x}, \hat{p}] = i\,\hbar\left(1 + \beta_0\,\hat{p}^2\right) \;\Rightarrow\; \{X, P\} = \left(1 + \beta_0\,P^2\right) \tag{2}$$

where $\beta_0 = \beta/m_{\rm p}^2$. However, in the limit $\beta \to 0$, Ref. [11] recovers *only* the newtonian mechanics but *not* General Relativity, and GR corrections must be added as an extra structure. The physical relevance of this approach (and the bound they get for β) remains therefore questionable.

Finally, in Refs. [13], [14], [15], authors consider the gravitational interaction when evaluating bounds on β. In the first two papers, a covariant formalism is used, which firstly is defined in Minkowski space, with the metric $\eta_{\mu\nu} = {\rm diag}(1, -1, -1, -1)$, and then easily generalized to curved space-times via the standard procedure $\eta_{\mu\nu} \to g_{\mu\nu}$. However, these two papers (as the previous ones) start from a deformation of classical Poisson brackets, although posited in covariant form. They obtain interesting consequences, like a β-deformed geodesic equation, which leads to a violation of the Equivalence Principle. This formalism remains anyway covariant when $\beta \to 0$ and it reproduces the standard GR results in the limit $\beta \to 0$ (unlike papers as Ref. [11]). In the third paper, Ref. [15], on the contrary, Poisson brackets and classical newtonian mechanics remain untouched. Therefore, the Equivalence Principle is preserved, and the equation of motion of a test particle is still given by the standard geodesic equation. Additionally, also here GR and standard quantum mechanics are recovered, when $\beta \to 0$. The bounds on β proposed by papers which take into account gravity range from $\beta < 10^{19}$, for those papers admitting a violation of Equivalence Principle, to $\beta < 10^{69}$ for the papers preserving the aforesaid principle.

In the present paper, following our main work in Ref. [16], we arrive directly to compute the value of β by comparing two different low energy (first order in $\hbar$) corrections for the expression of the Hawking temperature. The first is due to the GUP, and therefore involves β. The second correction, instead, is obtained by including the deformation of the metric due to quantum corrections to the newtonian potential. We require the two corrections to match (at the first order), and therefore we get a specific numerical value for β. It results to be of order of unity, in agreement with the general belief and with some particular models of string theory.

2. GUP-deformed Hawking temperature

One of the most common forms of deformation of the HUP (as well as the form of GUP that we are going to study in this paper) is

$$\Delta x\,\Delta p \geq \frac{\hbar}{2}\left(1 \,+\, \beta\,\frac{4\,\ell_{\rm p}^2}{\hbar^2}\,\Delta p^{\,2}\right) = \frac{\hbar}{2}\left[1 + \beta\left(\frac{\Delta p}{m_{\rm p}}\right)^2\right] \tag{3}$$

which, for mirror-symmetric states (with $\langle\hat{p}\rangle^2 = 0$), can be equivalently written in terms of commutators as

$$[\hat{x}, \hat{p}] = i\hbar\left[1 + \beta\left(\frac{\hat{p}}{m_{\rm p}}\right)^2\right] \tag{4}$$

since $\Delta x\,\Delta p \geq (1/2)|\langle[\hat{x}, \hat{p}]\rangle|$.

As is well known from the argument of the Heisenberg microscope [17], the size δx of the smallest detail of an object, theoretically detectable with a beam of photons of energy E, is roughly given by

$$\delta x \simeq \frac{\hbar}{2\,E} \tag{5}$$

since larger and larger energies are required to explore smaller and smaller details. From the uncertainty relation (3), we see that the GUP version of the standard Heisenberg formula (5) is

$$\delta x \simeq \frac{\hbar}{2\,E} + 2\,\beta\,\ell_{\rm p}^2\,\frac{E}{\hbar} \tag{6}$$

which relates the (average) wavelength of a photon to its energy E [2]. Conversely, using relation (6), one can compute the energy E of a photon with a given (average) wavelength $\lambda \simeq \delta x$. To compute the thermal GUP corrections to the Hawking spectrum, we follow the arguments of Refs. [18, 19, 20, 21, 22, 23, 24], and we arrive to the expression

$$T = \frac{\hbar}{8\pi G M}\left(1 + \frac{\beta\, m_{\rm p}^2}{4\pi^2\, M^2} + \ldots\right) . \tag{7}$$

where it is evident that to zero order in β, we recover the usual well known Hawking formula. We stress that we are here assuming that the correction induced by the GUP has a thermal character, and, therefore, it can be cast in the form of a shift of the Hawking temperature. Of course, there are also different approaches, where the corrections do not respect the exact thermality of the spectrum, and thus need not be reducible to a simple shift of the temperature. An example is the corpuscular model of a black hole of Ref. [25]. In this model, the emission is expected to gain a correction of order $1/N$, where $N \sim (M/m_{\rm p})^2$ is the number of constituents, and it becomes important when the mass M approaches the Planck mass.

3. Leading quantum correction to the newtonian potential

After early results by Duff [26], the leading quantum correction to the newtonian potential has been computed by Donoghue, by assuming General Relativity as fundamental theory of Gravity. In a series of beautiful papers (see for instance Ref. [27]) he reformulated General Relativity as an effective field theory, and, in particular, he considered two heavy bodies close to rest. The leading quantum correction derived from this model shows a long-distance *quantum* effect. More recently, Donoghue and other authors found that the gravitational interaction between the two objects can be described by the potential energy [28]

$$U(r) = -\frac{GMm}{r}\left(1 + \frac{3G(M+m)}{rc^2} + \frac{41}{10\pi}\frac{\ell_P^2}{r^2}\right) . \tag{8}$$

The first correction term does not contain any power of $\hbar$, so it is a classical effect, due to the non-linear nature of General Relativity. However, the second correction term, i.e., the last term of (8), is a true quantum effect, linear in $\hbar$. The potential generated by the mass M reads

$$V(r) = -\frac{GM}{r}\left(1 + \frac{3GM}{r}(1 + \frac{m}{M}) + \frac{41}{10\pi}\frac{\ell_P^2}{r^2}\right) . \tag{9}$$

4. Metric mimicking the quantum corrected newtonian potential

The effective newtonian potential [3] produced by a metric of the very general class

$$ds^2 = F(r)dt^2 - g_{ik}(x_1, x_2, x_3)dx^i dx^k \tag{10}$$

[2] Here, the standard dispersion relation $E = pc$ is assumed.

[3] The effective newtonian potential is produced by the metric given in (10) for a point particle which moves slowly, in a stationary and weak gravitational field, i.e., quasi-Minkowskian far from the source, $r \to \infty$.

IOP Conf. Series: Journal of Physics: Conf. Series **880** (2017) 012044 doi:10.1088/1742-6596/880/1/012044

where $r = |\mathbf{x}| = (x_1^2 + x_2^2 + x_3^2)^{1/2}$, and x_1, x_2, x_3 are the standard Cartesian coordinates, can be obtained with well known procedures [29], and is of the form

$$V(r) \simeq \frac{1}{2}(F(r) - 1) \tag{11}$$

or, equivalently,

$$F(r) \simeq 1 + 2V(r) . \tag{12}$$

We can therefore write down the metric which is able to mimic the quantum corrected newtonian potential proposed by Donoghue. Recalling (9), we have

$$F(r) \simeq 1 + 2V(r) = 1 - \frac{2GM}{r} - \frac{6\,G^2M^2}{r^2}\left(1 + \frac{m}{M}\right) - \frac{41}{5\pi}\frac{G^3M^3}{r^3}\left(\frac{\ell_P}{GM}\right)^2 .$$

Let us now define

$$\epsilon(r) = -\frac{6\,G^2M^2}{r^2}\left(1 + \frac{m}{M}\right) - \frac{41}{5\pi}\frac{G^3M^3}{r^3}\left(\frac{\ell_P}{GM}\right)^2 . \tag{13}$$

Therefore, $F(r)$ will now be of the form

$$F(r) = 1 - \frac{2GM}{r} + \epsilon(r) \tag{14}$$

and it is evident that when r is large, then $|\epsilon(r)| \ll 2GM/r$.

5. Temperature from a deformed Schwarzschild metric

We can legitimately wonder what kind of deformed Hawking temperature can be inferred from a deformed Schwarzschild metric as in (14) [4] The deformation (14) makes sense when $|\epsilon(r)| \ll GM/r$. We can also introduce a regulatory small parameter ε and, thus, we can write $\epsilon(r) \equiv \varepsilon\phi(r)$. At the end of the calculation, ε can go to unity. Of course, we look for the lowest order correction in the dimensionless parameter ε. The horizon's equation, i.e., $F(r) = 0$, now reads

$$r - 2\,G\,M + \varepsilon\, r\, \phi(r) = 0 . \tag{15}$$

Such equations can be solved, in a first approximation in ε, as follows. First, we formulate (15) in a general form

$$x = a + \varepsilon f(x) . \tag{16}$$

It is obvious that if ε is set equal to zero, then the solution will be $x_0 = a$. If ε is slightly different from zero, then we can try a test solution of the form $x_0 = a + \eta(\varepsilon)$ where $\eta(\varepsilon) \to 0$ for $\varepsilon \to 0$. Substituting the aforesaid test solution in (16), we get $x_0 = a + \varepsilon f(x_0)$ which means $\eta = \varepsilon f(a + \eta)$. To first order in η, we have $\eta = \varepsilon[f(a) + f'(a)\eta]$ from which we obtain $\eta = \varepsilon f(a)/[1 - \varepsilon f'(a)]$. Therefore, to first order in ε, the general solution of (16) reads

$$x_0 = a + \frac{\varepsilon f(a)}{1 - \varepsilon f'(a)} .$$

Applying this formula to (15), we get the solution

$$r_H = a - \frac{\varepsilon\, a\, \phi(a)}{1 + \varepsilon\,[\phi(a) + a\,\phi'(a)]} \tag{17}$$

[4] Recently, it was argued that in the special case in which $\epsilon(r) \sim 1/r^2$, the specific metric (14) could have some drawbacks in the context of GUP formalism [30]. However, none of those drawbacks appear here and, thus, there is no problem to employ (14) in our present study.

where $a = 2GM$.
The Hawking temperature is given by

$$T = \frac{\hbar}{4\pi} F'(r_H) \ . \tag{18}$$

Using Eq. (14) for $F(r)$, and Eq. (17) for r_H, and expanding in ε, one arrives to the deformed Hawking temperature as

$$T = \hbar \frac{F'(r_H)}{4\pi} = \frac{\hbar}{4\pi a} \left\{1 + \varepsilon \left[2\phi(a) + a\phi'(a)\right] + \varepsilon^2 \phi(a) \left[\phi(a) - 2a\phi'(a) - a^2\phi''(a)\right] + \dots\right\} \ . \tag{19}$$

It is noteworthy that the only function $\phi(r)$ that annihilates the first-order in ε temperature correction term is the solution of the differential equation $2\phi(r) + r\phi'(r) = 0$, namely $\phi(r) = A/r^2$, where A is an arbitrary constant. In particular, for the function $\phi(r) = G^2M^2/r^2$, the coefficient of ε in (19) is zero, and the coefficient of ε^2 is $-1/16$. It is also interesting to investigate what kind of function will eliminate the second-order in ε correction term. This function will be the solution of the differential equation $r^2\phi''(r) + 2r\phi'(r) - \phi(r) = 0$ which is an Euler equation. Its characteristic equation is of the form $\lambda^2 + \lambda - 1 = 0$ with roots $\lambda_1 = (-1 - \sqrt{5})/2$ and $\lambda_2 = (-1 + \sqrt{5})/2$. So, the functions which remove the ε^2-correction term are $\phi_1(r) = r^{-|\lambda_1|}$ and $\phi_2(r) = r^{\lambda_2}$.

6. Computing β

We are now in the position to compute the temperature generated by the metric (14), by simply employing (19). Therefore, the metric-deformed Hawking temperature is of the form [5]

$$T = \frac{\hbar}{4\pi a} \left\{1 + \left[2\epsilon(a) + a\epsilon'(a)\right] + \dots\right\} \tag{20}$$

while the GUP-deformed Hawking temperature reads

$$T = \frac{\hbar}{8\pi GM} \left(1 + \frac{\beta\, m_{\rm p}^2}{4\pi^2 M^2} + \dots\right) \ . \tag{21}$$

By comparing the two respective first-order correction terms in the two previous expansions, we obtain

$$\beta = \frac{4\pi^2 M^2}{m_{\rm p}^2} \left[2\epsilon(a) + a\epsilon'(a)\right] \ . \tag{22}$$

Using now expression (13) for $\epsilon(r)$, we get

$$2\epsilon(r) + r\epsilon'(r) = \frac{B}{r^3} \tag{23}$$

with $B = \frac{41\ G^3M^3}{5\pi}\left(\frac{\ell_P}{GM}\right)^2$. Therefore,

$$2\epsilon(a) + a\epsilon'(a) = \frac{B}{8G^3M^3} \tag{24}$$

and using (22), the parameter β will get the value

$$\beta = \frac{4\pi^2 M^2}{m_{\rm p}^2} \frac{41}{40\pi} \left(\frac{\ell_P}{GM}\right)^2 = \frac{82\pi}{5} \ . \tag{25}$$

[5] Notice that from (13) we have $\epsilon(r) \sim 1$ for $r \sim a$, so this would seem to spoil the expansion (20) when $r \sim a$. On the contrary, we can always imagine to first expand the temperature $T(r) = \hbar F'(r)/4\pi$ for $r \gg a$, when $\epsilon(r)$ is small. Then, the term in $1/r^2$ disappears from the expansion of $T(r)$ because of the condition $2\phi(r) + r\phi'(r) = 0$. Finally, we take the limit $r \to a$, and this yields (20).

7. Conclusions

In this work we have computed the value of the deformation parameter β of the GUP. We obtain this result by computing in two different ways the Hawking temperature for a Schwarzschild black hole.

The first way consists in using the GUP (in place of the standard HUP) to compute the Hawking formula. We get an expression of the temperature containing a correction term depending on β, i.e., the GUP-deformed Hawking temperature (7).

The second way involves the consideration of the quantum correction to the newtonian potential, computed years ago by Donoghue and others. The corrections to the newtonian potential imply naturally a quantum correction to the Schwarzschild metric. Therefore, the Hawking temperature computed through this quantum corrected Schwarzschild metric result to get corrections in respect to the standard Hawking expression, i.e., the metric-deformed Hawking temperature (20).

The request that the first-order corrections of the two different expressions of Hawking temperature must coincide, fixes unambiguously the numerical value of β to be $82\pi/5$.

Finally, a couple of comments are in order here. First, this numerical value is of order one, as expected from several string theory models, and from versions of GUP derived through gedanken experiments. In particular, this is the first time, to our knowledge, that a specific value is obtained for β by starting from the minimal assumptions we made. Second, as we know, in the last years much research has focused on the experimental bounds of the *size* of β, and several experiments have been proposed to test GUPs in the laboratory. In fact, it has been shown that one does not need to reach the Planck energy scale to test GUP corrections. Among the more elaborated proposals, where conditions can be created in a lab, are those of the groups of Refs. [31, 32, 33]. However, it is also worth of note that the best bounds on β presented in the literature are still by far much larger than the value computed here. This could require, presumably, a big leap in the experimental designs and techniques in order to search this region for the parameter β.

References

[1] Snyder H S 1947 *Phys.Rev.* **71** 38
Yang C N 1947 *Phys. Rev.* **72** 874
Mead C A 1964 *Phys. Rev.* **135** B 849
Karolyhazy F 1966 *Nuovo Cim.* A **42** 390
[2] Amati D, Ciafaloni M and Veneziano G 1987 *Phys. Lett.* B **197** 81
Gross D J and Mende P F 1987 *Phys. Lett.* B **197** 129
Amati D, Ciafaloni M and Veneziano G 1989 *Phys. Lett.* B **216** 41
Konishi K, Paffuti G and Provero P 1990 *Phys. Lett.* B **234** 276
Veneziano G 1990 Proceedings of PASCOS-90 *Quantum Gravity near the Planck scale* (Boston) 486
Capozziello S, Lambiase G and Scarpetta G 2000 *Int. J. Theor. Phys.* **39** 15
[3] Maggiore M 1993 *Phys. Lett.* B **304** 65
[4] Kempf A, Mangano G and Mann R B 1995 *Phys. Rev.* D **52** 1108
Bojowald M and Kempf A 2012 *Phys. Rev.* D **86** 085017
[5] Scardigli F 1999 *Phys. Lett.* B **452** 39
[6] Adler R J and Santiago D I 1999 *Mod. Phys. Lett.* **A14** 1371
[7] Scardigli F and Casadio R 2003 *Class. Quantum Grav.* **20** 3915
[8] Brau F 1999 *J. Phys.* A **32** 7691
[9] Das S and Vagenas E C 2008 *Phys. Rev. Lett.* **101** 221301
Ali A F, Das S and Vagenas E C 2011 *Phys. Rev.* D **84** 044013
[10] Pedram P, Nozari K and Taheri S H 2011 *JHEP* **1103** 093
[11] Benczik S, Chang L N, Minic D, Okamura N, Rayyan S and Takeuchi T 2002 *Phys. Rev.* D **66** 026003
[12] Nozari K and Akhshabi S 2008 *Chaos Solitons Fractals* **37** 324
Pedram P 2012 *Phys. Lett.* B **718** 638
[13] Ghosh S 2014 *Class. Quant. Grav.* **31** 025025

IOP Conf. Series: Journal of Physics: Conf. Series **880** (2017) 012044 doi:10.1088/1742-6596/880/1/012044

[14] Pramanik S and Ghosh S 2013 *Int. J. Mod. Phys.* A **28** 27, 1350131
Pramanik S 2014 *Phys. Rev.* D **90** 024023
[15] Scardigli F and Casadio R 2015 *Eur. Phys. J.* C **75** 425
[16] Scardigli F, Lambiase G and Vagenas E C 2017 *Phys. Lett.* B **767** 242
[17] Heisenberg W 1927 *Zeitschrift für Physik* **43** 172
[18] Scardigli F 1995 *Nuovo Cim.* B **110** 1029
[19] Adler R J, Chen P and Santiago D I 2001 *Gen. Rel. Grav.* **33** 2101
[20] Cavaglia M and Das S 2004 *Class. Quant. Grav.* **21** 4511
[21] Cavaglia M, Das S and Maartens R 2003 *Class. Quant. Grav.* **20** L205
[22] Susskind L and Lindesay J 2005 *An Introduction to Black Holes, Information, and the String Theory Revolution* (World Scientific, Singapore). See chapter 10.
[23] Nouicer K 2007 *Class. Quant. Grav.* **24** 5917
[24] Scardigli F, *Glimpses on the micro black hole planck phase*, [arXiv:0809.1832]
[25] Dvali G and Gomez C, *Black Hole's Quantum N-Portrait*, [arXiv:1112.3359]
[26] Duff M J 1974 *Phys. Rev.* D **9** 1837
[27] Donoghue J F 1994 *Phys. Rev. Lett.* **72** 2996
Donoghue J F 1994 *Phys. Rev.* D **50** 3874
[28] Khriplovich I B and Kirilin G G 2002 *J.Exp.Theor.Phys.* **95** 981
Bjerrum-Bohr N E J, Donoghue J F and Holstein B R 2003 *Phys.Rev.* D **68** 084005; 2005 Erratum-ibid.D **71** 069904
Bjerrum-Bohr N E J, Donoghue J F and Holstein B R 2003 *Phys.Rev.* D **67** 084033; 2005 Erratum-ibid.D **71** 069903
Akhundov A and Shiekh A 2008 *EJTP* 5 No.17 1
[29] Weinberg S 1972 *Gravitation and Cosmology* (John Wiley & Sons, New York)
[30] Farag Ali A, Khalil M M and Vagenas E C 2015 *Europhys. Lett.* **112** no.2 20005
[31] Brukner C et al. 2012 *Nature Physics* **8** 393
[32] Marin F et al. 2013 *Nature Physics* **9** 71
[33] Bawaj M et al. 2015 *Nature Commun.* **6** 7503 [arXiv:1411.6410]

IOP Conf. Series: Journal of Physics: Conf. Series **880** (2017) 012045 doi:10.1088/1742-6596/880/1/012045

Underground tests of quantum mechanics. Whispers in the cosmic silence?

C Curceanu[1,2,9], S Bartalucci[1], A Bassi[3], M Bazzi[1], S Bertolucci[4], C Berucci[1,5], A M Bragadireanu[1,2], M Cargnelli[5], A Clozza[1], L De Paolis[1], S Di Matteo[6], S Donadi[3], J-P Egger[7], C Guaraldo[1], M Iliescu[1], M Laubenstein[8], J Marton[5], E Milotti[3], A Pichler[5], D Pietreanu[1,2], K Piscicchia[9,1], A Scordo[1], H Shi[5], D Sirghi[1,2], F Sirghi[1,2], L Sperandio[1], O Vazquez Doce[10], J Zmeskal[5]

[1]INFN, Laboratori Nazionali di Frascati, CP 13, Via E. Fermi 40, I-00044, Frascati (Roma), Italy
[2]"Horia Hulubei"National Institute of Physics and Nuclear Engineering, Str. Atomistilor no. 407, P.O. Box MG-6, Bucharest - Magurele, Romania
[3]Dipartimento di Fisica, Università di Trieste and INFN– Sezione di Trieste, Via Valerio, 2, I-34127 Trieste, Italy
[4]University and INFN Bologna, Via Irnerio 46, I-40126, Bologna, Italy
[5]The Stefan Meyer Institute for Subatomic Physics, Boltzmanngasse 3, A-1090 Vienna, Austria
[6]Institut de Physique UMR CNRS-UR1 6251, Université de Rennes1, F-35042 Rennes, France
[7]Institut de Physique, Université de Neuchâtel, 1 rue A.-L. Breguet, CH-2000 Neuchâtel, Switzerland
[8]INFN, Laboratori Nazionali del Gran Sasso, S.S. 17/bis, I-67010 Assergi (AQ), Italy
[9]Museo Storico della Fisica e Centro Studi e Ricerche "Enrico Fermi", Roma, Italy
[10]Excellence Cluster Universe, Technische Universität München, Garching, Germany

E-mail: `catalina.curceanu@lnf.infn.it`

Abstract. By performing X-rays measurements in the "cosmic silence" of the underground laboratory of Gran Sasso, LNGS-INFN, we test a basic principle of quantum mechanics: the Pauli Exclusion Principle (PEP), for electrons. We present the achieved results of the VIP experiment and the ongoing VIP2 measurement aiming to gain two orders of magnitude improvement in testing PEP. We also use a similar experimental technique to search for radiation (X and gamma) predicted by continuous spontaneous localization models, which aim to solve the "measurement problem".

1. Introduction

To the best of our knowledge, there are 2 spin-separated classes of particles: fermions, with half-integer spin, and bosons, with integer spin. The Pauli Exclusion Principle (PEP), only valid for fermions, states that two fermions can not be in the same quantum state [1]. PEP is a fundamental principle in physics, for which an intuitive explanation is still missing [2]. The violation of PEP would certainly be related to new physics, beyond the Standard Model. A thorough test of the PEP is than necessary for all types of particles.

IOP Conf. Series: Journal of Physics: Conf. Series **880** (2017) 012045 doi:10.1088/1742-6596/880/1/012045

In the past, very few experiments have tested the PEP. Some of them put stringent limits to the probability of the violation of the Pauli Exclusion Principle in fermionic systems, for example the experiment performed by the DAMA collaboration [3], investigating transitions of stable systems from a non-Pauli violating state to a Pauli violating state. These transitions violate the Messiah-Greenberg superselection rule [4], which forbids the change of the symmetry of the state of a stable system. To bypass this rule, the VIP experiment introduces "new" electrons in the system by means of an electric current.

In the next sections we describe the experimental method, the VIP setup and the results obtained by running VIP at the underground Gran Sasso National Laboratory (LNGS). We present also the VIP2 setup, a major upgrade of VIP, which is using fast Silicon Drift Detectors (SDD) and a veto system, aiming to gain two orders of magnitude in PEP violation study.

We will then conclude the paper by presenting some ideas to use similar experimental techniques to perform measurements of X-rays spontaneously emitted by electrons and protons in matter, predicted by the collapse models [5, 6, 7, 8]. The collapse models deal with the "measurement problem" in quantum mechanics, by introducing a new physical dynamics that naturally collapses the state vector and make predictions which differ from those of standard quantum mechanics [9]. In this context, one of the most exciting tasks is to perform cutting-edge experiments, in order to asses whether quantum mechanics is exact, or an approximation of a deeper level theory.

2. The experimental method and VIP results

In the VIP and VIP2 experiments "new" electrons are introduced into a copper target by circulating an electric current. The VIP setup consisted of a copper cylindrical target, 45 mm radius, 50 μm thickness, and 88 mm height, where a current of 40 A was circulated, surrounded by 16 equally spaced Charge Coupled Devices (CCDs). The CCDs were at a distance of 23 mm from the copper cylinder, and paired one above the other. For the VIP2 setup, the cooper target consists of 2 strips with 25 μm thickness and 6 cm length, where an electric current of 100 A is circulated.

The "new" electrons introduced into the copper target have a certain probability to interact with and be captured by copper atoms. In the course of this interaction, the electrons might form a new symmetry state with the electrons of the atom. This process is the reason why the VIP experiment does not violate the Messiah-Greenberg superselection rule, since the current electrons are "new" to the electrons in the atom, i.e. they have no predefined symmetry of the state. With a certain probability which should be determined by the experiment, the newly formed symmetry state might have a symmetric component and the electron could undergo transitions from the 2p level to the 1s one, with the 1s already occupied by 2 electrons, which transition obviously violates the PEP, Figure 1.

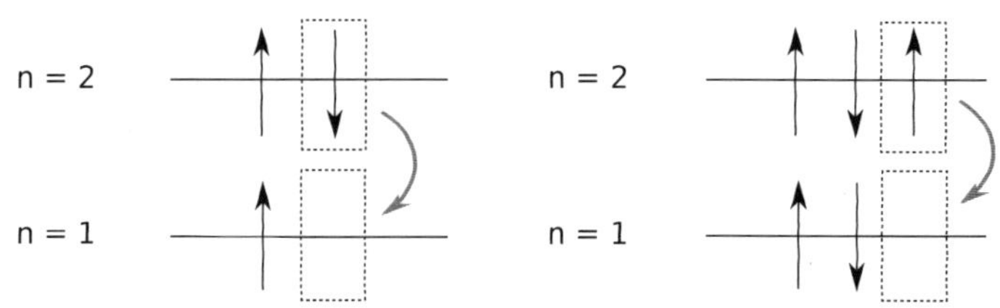

Figure 1. Normal 2p to 1s transition with an energy around 8 keV for Copper (left) and Pauli-violating 2p to 1s transition with a transition energy around 7,7 keV in Copper (right).

The non-Paulian transition has about 300 eV lower energy with respect to the allowed one. The energy of the X-rays which are emitted during the cascading process are recorded by X-

IOP Conf. Series: Journal of Physics: Conf. Series **880** (2017) 012045 doi:10.1088/1742-6596/880/1/012045

ray detectors, which are mounted close to the copper target. The recorded spectrum is then analyzed looking for an excess of counts over the background in the energy range of the forbidden transition, which is around 7,7 keV. The excess of counts or the lack of them will either signal discovery of the violation of the PEP or, else, determine a new upper bound for its violation.

The VIP experiment was installed at LNGS-INFN in Spring 2006 and was taking data until 2010, alternating period with current on (signal) to periods with current off (background). The data analysis resulted in an upper limit for the PEP violation probability for electrons of $\frac{\beta^2}{2} < 4.7 \times 10^{-29}$ [5, 7].

3. The VIP2 experiment

In the VIP2 experiment the CCDs were replaced by Silicon Drift Detectors (SDD), which have a superior energy resolution and offer timing capability. Six SDDs units, with a total active area of 6 cm^2 are mounted close to the Cu target, giving an acceptance which is about ten times as large as the acceptance of VIP CCDs. Moreover, an active shielding system (veto) was implemented, to reduce the background in the energy region of the forbidden transition. These systems will have an important contribution to improve the limit for the violation of the PEP by two orders of magnitude with new data which are presently comming by running the VIP2 experiment at LNGS [10].

In November 2015, the VIP2 setup was installed at Gran Sasso [11]. In the year 2016, data with the complete VIP2 detector system at the LNGS without shielding were taken. In Figure 2 the VIP2 setup as installed at LNGS is shown.

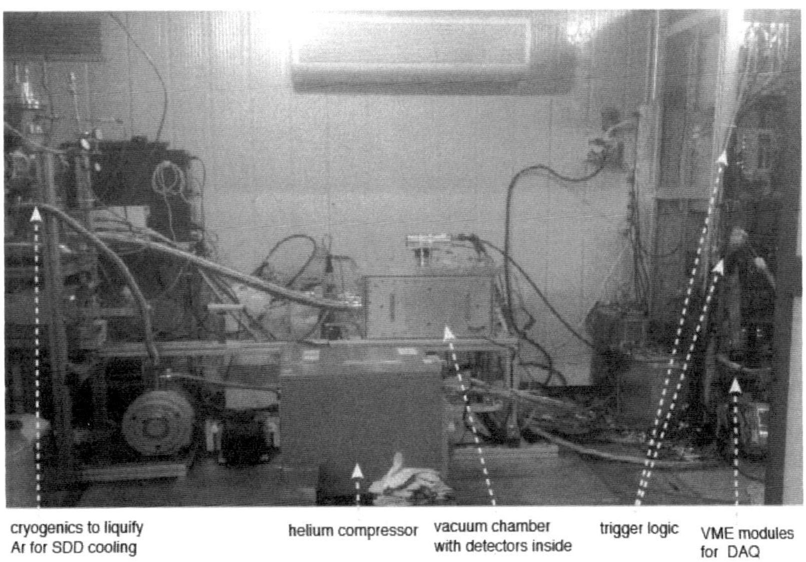

Figure 2. A picture of the VIP-2 setup in the barrack at LNGS .

Data with 100 Ampere DC current applied to the copper strip for 34 days was collected. Together with the data collected for 28 days without current, an analysis of the two spectra to determine a new value of the upper limit for PEP violation is on going. Figure 3 shows the summation of the data from 2016 of all the SDDs; the preliminary analysis shows the energy resolution of the summed spectra at 8 keV is less than 190 eV FWHM.

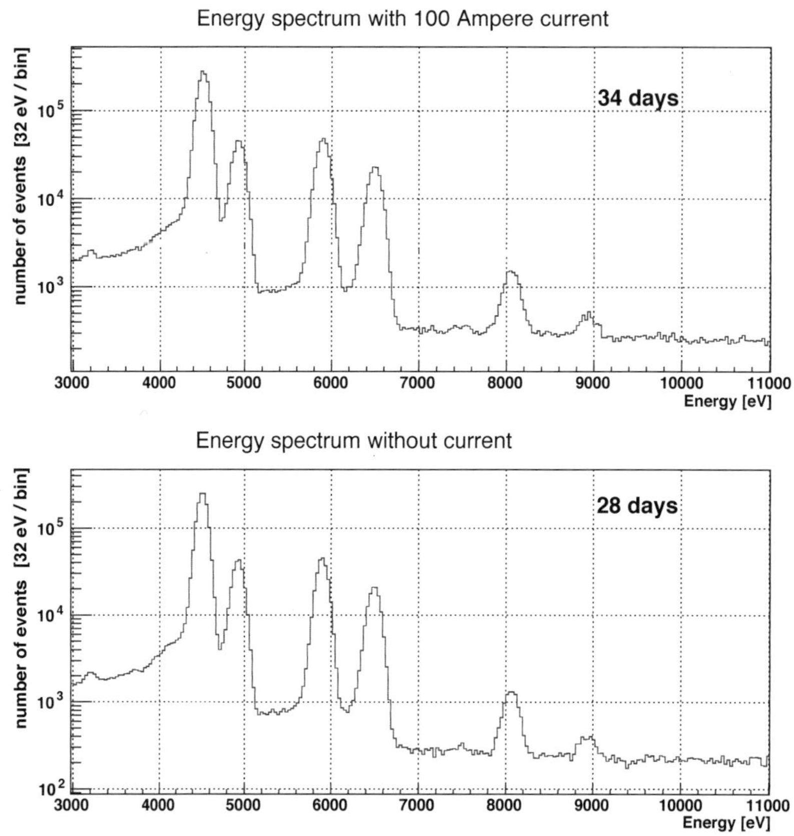

Figure 3. Upper figure: the energy spectrum obtained by VIP2 in 34 days with current (I=100 A); Lower figure: the energy spectrum obtained by VIP2 in 28 days without current.

In Table 1 [10] the improvement factors of VIP2, with respect to VIP, are presented. The overall improvement factor refers to the gain in the probability of the PEP violation ($\beta^2/2$) for electrons, obtained by applying the procedure described in [5, 6, 7] and taking into account the various expected gain factors shown in the table. In the coming 3 years of run we will gain two orders of magnitude in the PEP violation probability or will have signals of a violation.

4. X-ray measurements for testing the dynamical reduction models

The aim of the Dynamical Reduction Models (DRM) is to solve the so-called "measurement problem" in Quantum Mechanics (QM). The linear and unitary nature of the Shrödinger equation allows, in principle, the superposition of macroscopic states, but such superpositions are not observed in the measurement process, which is intrinsically non-linear and stochastic [12, 13]. The measurement problem led to the introduction of the wave packet reduction principle which, nevertheless, does not predict the scale at which the quantum-to-classical transition occurs, nor explains the collapse mechanism.

The work of Ghirardi, Rimini and Weber [14] lead to the development of a consistent DRM known as Quantum Mechanics with Spontaneous Localization (QMSL). According to the QMSL model each particle of a macroscopic system of n distinguishable particles experiences sudden spontaneous localizations, on the position basis, with a mean rate $\lambda = 10^{-16}$ s^{-1}, and a correlation length $a = 10^{-7}$ m. Between two localizations particles evolve according

Table 1. The improvement factors for VIP2 in comparison to VIP (in parenthesis).

Changes in VIP2	value VIP2 (VIP)	expected gain
acceptance	12 % ($\sim$ 1 %)	12
increase current	100 A (40 A)	> 2
reduced length	3 cm (8.8 cm)	1/3
total linear factor		8
energy resolution	170 eV (320 eV) @ 8 keV	4
reduced active area	6 cm ^{2} (114 cm ^{2})	20
better shielding and veto		5-10
higher SDD efficiency		1/2
background reduction		200 - 400
overall improvement (see text)		> 120

to the Shrödinger dynamics. The model ensures, for the macroscopic object, the decoupling of the internal and Center of Mass (CM) motions. The internal motion is not affected by the localization, whereas the CM motion is localized with a rate $\lambda_{macro} = n\,\lambda$.

Subsequently, the theory was developed in the language of the non-linear and stochastic Shrödinger equation [15, 16], where besides the standard quantum Hamiltonian, two other terms induce a diffusion process for the state vector, which causes the collapse of the wave function in space. In its final version [17] the model is known as the mass proportional Continuous Spontaneous Localization (CSL).

The value of the mean collapse rate is presently argument of debate. According to CSL λ should be of the order of 10^{-17} s^{-1}, whereas a much stronger value $10^{-8\pm2}$ s^{-1} was proposed by S. L. Adler [19] based on arguments related to the latent image formation and the perception of the eye.

DRM posses the unique characteristic to be experimentally testable, by measuring the (small) predicted deviations with respect to the standard quantum mechanics. The conventional approach is to generate spatial superpositions of mesoscopic systems and examine the loss of interference, while environmental noises are, as much as possible, under control. The present day technology, however, does not allow to set stringent limits on λ by applying this method. The most promising testing ground, instead, is represented by the search for the spontaneous radiation emitted by charged particles when interacting with the collapsing stochastic field [18]. A measurement of the emitted radiation rate thus enables to set a limit on the λ parameter of the models.

The radiation spectrum spontaneously emitted by a free electron, as a consequence of the interaction with the stochastic field, was calculated by Q. Fu [18] in the framework of the non-relativistic CSL model, and it is given by:

$$\frac{d\Gamma(E)}{dE} = \frac{e^2\lambda}{4\pi^2 a^2 m^2 E} \tag{1}$$

in eq. (1) m represents the electron mass and E is the energy of the emitted photon. In the mass proportional CSL model the stochastic field is assumed to be coupled to the particle mass density, then the rate is to be multiplied by the factor $(m/m_N)^2$, with m_N the nucleon mass. Using the measured radiation appearing in an isolated slab of Germanium [20] corresponding to an energy of 11 KeV, and employing the predicted rate eqn. (1), Fu obtained the following upper limit for λ (non-mass poportional model):

$$\lambda < 0.55 \cdot 10^{-16} s^{-1}. \tag{2}$$

In eq. (2) the QMSL value for a ($a = 10^{-7}$ m) is assumed and the four valence electrons were considered to contribute to the measured X-ray emission, since the binding energy is ~ 10 eV in this case, and they can be considered as *quasi-free*. Recent re-analyses of Fu's work [19, 21] corrected the limit in Eq. 4 to $\lambda < 2 \cdot 10^{-16} s^{-1}$.

We improved the limit on the collapse rate [22] by analysing the data collected by the IGEX (International Germanium EXperiment) experiment [23]. IGEX is a low-background experiment based on low-activity Germanium detectors dedicated to the $\beta\beta 0\nu$ decay research. We performed a fit of the published X-ray emission spectrum [24], which refers to an 80 kg day exposure, in the energy range $\Delta E = 4.5 \div 48.5$ KeV $\ll m$. The energy interval is compatible with the non-relativistic assumption of the model (Eq. (1)).

A Bayesian model was adopted to calculate the χ^2 variable minimized to fit the X ray spectrum, assuming the predicted (Eq. (1)) energy dependence:

$$\frac{d\Gamma(E)}{dE} = \frac{\alpha(\lambda)}{E}. \tag{3}$$

The obtained values for λ are:

$$\lambda \leq 2.5 \cdot 10^{-18} s^{-1}, \tag{4}$$

if no mass dependence is considered, and

$$\lambda \leq 8.5 \cdot 10^{-12} s^{-1}, \tag{5}$$

in the mass proportional CSL assumption. This analysis improves the previous limit [18] of two orders of magnitude.

By using a similar method, we are presently performing a dedicated experiment at LNGS which will allow for 1 - 2 orders of magnitude further improvement on the collapse rate parameter λ.

5. Conclusions

We presented the VIP experiment searching for the "impossible atoms", i.e., atoms where the Pauli Exclusion Principle (PEP) is violated. We have set a limit on PEP violation probability for electrons as $\frac{\beta^2}{2} < 4.7 \times 10^{-29}$, and we will improve this limit by two orders of magnitude, or find the PEP violation, in the coming 3 years by running the VIP2 setup at the LNGS underground laboratory. By using a similar experimental technique we are exploring the collapse theories, looking for spontaneously emitted radiation, to set stringent limits on the λ parameter characterizing these models. X-rays measurements performed in the "cosmic silence" of the LNGS-INFN laboratory, are a very important tool to test quantum mechanics.

Acknowledgments

We thank H. Schneider, L. Stohwasser, and D. Stückler from Stefan-Meyer-Institut for their fundamental contribution in designing and building the VIP2 setup.

We acknowledge the very important assistance of the INFN-LNGS laboratory staff during all phases of preparation, installation and data taking.

The support from the EU COST Action CA 15220 is gratefully acknowledged.

We thank the Austrian Science Foundation (FWF) which supports the VIP2 project with the grant P25529-N20 and Centro Fermi for the grant "Problemi aperti nella meccania quantistica".

Furthermore, this paper was made possible through the support of a grant from the Foundational Questions Institute, FOXi ("Events" as we see them: experimental test of the collapse models as a solution of the measurement problem) and a grant from the John Templeton Foundation (ID 58158). The opinions expressed in this publication are those of the authors and do not necessarily reflect the views of the John Templeton Foundation.

References

[1] Pauli W 1925 *Z. Phys.* **31** 765
[2] Feynman R *et al* 1963 *The Feynman Lectures on Physics*
[3] Bernabei R *et al* 2009 New search for processes violating the Pauli exclusion principle in sodium and in iodine *Eur. Phys. J. C* **62** 327
[4] Messiah A M I and Greenberg O W 1964 *Phys. Rev. B* **136** 248
[5] Curceanu (Petrascu) C *et al* 2012 *Journal of Physics: Conference Series* **361** 012006
[6] Curceanu (Petrascu) C *et al* 2011 *International Journal of Quantum Information* **9** 145
[7] Curceanu (Petrascu) C *et al* 2011 *Journal of Physics: Conference Series* **306** 012036
[8] Curceanu (Petrascu) C *et al* 2011 *Physics Proceedings* **17** 40
[9] Adler S L and Bassi A 2009 *Science* **325** 275
[10] Marton J *et al* 2013 *Journal of Physics: Conference Series* **447** 012070
[11] Shi H *et al* 2016 *Journal of Physics: Conference Series* **718 (4)** 042055
[12] Von Neumann J 1932 Mathematische Grundlagen der Quantummechanik (Springer-Verlag). English translation 1955: Mathematical Foundations of Quantum Mechanics (Princeton University Press)
[13] Bassi A and Ghirardi G C 2003 *Phys. Rep.* **379** 257
[14] Ghirardi G C, Rimini A and Weber T 1986 *Phys. Rev. D* **34** 47
[15] Pearle P 1989 *Phys. Rev. A* **39** 2277
[16] Ghirardi G C, Pearle P and Rimini A 1990 *Phys. Rev. A* **42** 78
[17] Pearle P and Squires E 1994 *Phys. Rev. Lett.* **73** 1
[18] Fu Q 1997 *Phys. Rev. A* **56** 1806
[19] Adler S L 2007 *Journal of Physics A* **40** 2935
[20] Miley H S *et al* 1990 *Phys. Rev. Lett.* **65** 3092
[21] Mullin J and Pearle P 2014 *Phys. Rev. A* **90** 052119
[22] Curceanu C, Hiesmayer B C and Piscicchia K 2015 *Journal of Advanced Physics* **4** 1
[23] Aalseth C E *et al* 1999 *Phys. Rev. C* **59** 2108
[24] Morales A *et al* 2002 *Phys. Lett. B* **532** 814

IOP Conf. Series: Journal of Physics: Conf. Series **880** (2017) 012046 doi:10.1088/1742-6596/880/1/012046

Internal clocks in timeless universe

Przemysław Małkiewicz
National Centre for Nuclear Research, Poland
E-mail: przemyslaw.malkiewicz@ncbj.gov.pl

Abstract. The Hamiltonian formalism of general relativity involves a Hamiltonian constraint. Attempts at quantisation of the Hamiltonian constraint formalism face an obstacle associated with the lack of predetermined time parameter whose existence is assumed in usual quantisation prescriptions. A way to deal with it is to employ an arbitrary internal degree of freedom as the internal clock to describe the evolution of a gravitational system. We use the so called reduced phase space approach in which the choice of internal clock is made prior to quantisation. We discuss the construction of reduced phase spaces based on the essential role played by internal clocks. Then we introduce the so called extended transformations which extend the well-known notion of canonical transformations. The extended transformations elucidate the relation between the canonical structure of the reduced phase space and the internal clock. Quantisation of reduced phase spaces and respective Hamiltonians reveals the relation between quantum dynamics and the choice of internal clock. Finally, we discuss concrete examples of canonical formalisms of the Friedmann-Lemaitre universe and Bianchi type I universe. We find that many physical features of the quantum dynamics depend on the choice of internal clock. In the Conclusions we speculate about the possible physical meaning of the result.

1. Introduction

In this short contribution we summarise our previous work [1–3] on the structure of timeless quantum mechanics. This kind of quantum mechanics does not involve a fixed, predetermined time. Instead, it involves clocks which are internal degrees of freedom in a given system and which are chosen ambiguously to parametrise the evolution. The immediate motivation to study such a formalism comes from attempts at quantisation of spacetime. The distinctive feature of general relativistic models of spacetime is the lack of the concept of predetermined time. They undergo an evolution but it is often expressed in terms of an auxiliary and ambiguous parameter. Attempts at quantisation of general relativity lead to an unresolved conundrum called the problem of time.

We will make use of the canonical formalism and canonical quantisation to investigate this problem. The canonical formalism for general relativistic models was formulated by Arnowitt, Deser and Misner [4] and its restriction to cosmological model can be found e.g. in [5]. The Hamiltonians of gravitational systems turn out to be constraints (we discuss this point below). In general, the Hamiltonian constraint of general relativity involves four terms and each of them corresponds to a generator of spacetime diffeomorphisms. In cosmological models, the underlying spatial symmetry enables to disregard spatial diffeomorphisms and formulate the spacetime dynamics in terms of a mechanical system with a single-term Hamiltonian constraint, a generator of time-like diffeomorphisms. Our analysis is going to be confined to this relatively simple case.

2. Hamiltonian constraint

The canonical formalism of general relativity involves a Hamiltonian which is a constraint. In other words, the Hamiltonian's role is twofold in this formalism. Namely, the Hamiltonian constraint, C, generates the motion through the Poisson bracket:

$$\frac{\mathrm{d}\mathcal{O}}{\mathrm{d}\tau} = \{\mathcal{O}, \mathrm{C}\},$$

where $\mathcal{O}$ is any observable and it constrains the space of physical states to a sub-manifold in the phase space:

$$\mathrm{C} = 0.$$

It can be seen that rescaling the Hamiltonian constraint with any non-vanishing function N,

$$\frac{\mathrm{d}\mathcal{O}}{\mathrm{d}\tau'} = \{\mathcal{O}, N\mathrm{C}\} = N \cdot \{\mathcal{O}, \mathrm{C}\} + \mathrm{C} \cdot \{\mathcal{O}, N\},$$

does not change the dynamics since the term $\mathrm{C} \cdot \{\mathcal{O}, N\} = 0$ vanishes. Nevertheless, it redefines the time parameter, $\tau \mapsto \tau'$, in such a way that:

$$N\mathrm{d}\tau' = \mathrm{d}\tau.$$

The parameters τ and τ' are auxiliary and have no physical meaning. The phase space on which the Hamiltonian constraint C is defined is called the kinematical phase space. The Hamiltonian constraint confines the physical motion to an odd dimensional surface $\mathrm{C} = 0$ in the kinematical phase space. Some of the functions on the constraint surface could be employed to measure the evolution. Contrary to time parameters, constraint surface functions carry physical meaning as they are parameters in the sub-manifold of physical states. A simple counting shows that the remaining physical parameters in the constraint surface must form an even dimensional space. As we will see they are endowed with a natural canonical structure and form the so-called reduced phase space.

Following Isham [6] there are three approaches to quantise Hamiltonian constraint systems: (I) time is introduced before quantisation, (II) time is introduced after quantisation or (III) time is not introduced at all and a timeless interpretation of quantum theory is proposed. Our basic tenet is that a selected degree of freedom is needed to parametrise the evolution of such systems, which is in compliance with the approaches (I) and (II). They both, however, suffer from the ambiguity in the choice of evolution parameter. To emphasise the identification of the evolution parameter with an internal degree of freedom, we will use the word 'clock' or 'internal clock' rather then 'time'. The approach (II) is based on the following constraint operator equation:

$$\hat{\mathrm{C}}\Psi = 0,$$

where $\hat{\mathrm{C}}$ is a quantised constraint. The physical state $\Psi(q_1, \ldots, q_n)$ has no time-dependence. However, it is assumed that one of its independent variables $q_1, \ldots, q_n$ (or, a combination of them) can be set as internal clock to describe a unitary evolution of Ψ understood as a wave-function of the remaining variables, e.g.:

$$\hat{\mathrm{C}}\Psi = \left(i\frac{\partial}{\partial q_1} - \hat{\mathrm{H}}\right)\Psi = 0,$$

by means of the Schrödinger equation based on a new Hamiltonian $\hat{\mathrm{H}}$ that generates evolution with respect to q_1.

IOP Conf. Series: Journal of Physics: Conf. Series **880** (2017) 012046 doi:10.1088/1742-6596/880/1/012046

The aim of our analysis is to understand how much predictions of a quantum theory depend on the employed clock. To address this problem we need to study the quantum dynamics of a given system with respect to various clocks. For the sake of simplicity we will employ the approach (I) which we describe in detail below. Although it may produce quantitatively different results from the approach (II), the addressed problem concerns them both and we do not expect to find any qualitative differences in this regard between these two approaches. The approach (I) entails first defining the reduced phase space equipped with a non-vanishing Hamiltonian generating physical dynamics with respect to a selected clock and next quantising it by usual procedures.

3. Ambiguities in quantum dynamics

The first explanation of the origin of ambiguous predictions produced by quantum theories based on different clocks was provided by Isham [6] and Kuchar [7]. They claimed that the relation between classical formalisms based on different internal clocks can be understood in terms of canonical transformations in the kinematical phase space. The latter relate different canonical coordinate systems and each of them includes a distinguished clock coordinate. Since many canonical transformations do not have a unitary representation they concluded that quantum theories based on different clocks must be in general unitarily inequivalent. Thus, they must produce, to some extent, different predictions.

It is known now that the above reasoning is mistaken. The real reason for unitary non-equivalence between different clock-based quantum theories lies elsewhere, namely in canonical non-equivalence of different reduced phase spaces associated with different clocks. This idea can be already found in the paper of Hajicek [8]. He shows by an explicit example that a gravitational model reduced and parametrised with different clocks entails different Poisson brackets between some metric components of the field. We will follow this observation and give it a more precise formulation below. The basic tool for this purpose is the idea of the so-called extended transformations that extend the notion of canonical transformations and which was already considered by Arnowitt, Deser and Misner in [9]. Some interesting considerations of explicit, though very limited, transformations of clock variable in the case of a spherically symmetric dust shell can be found in [10].

The notion of extended transformations (called pseudo-canonical transformations therein) was thoroughly investigated in the paper [3]. The extended transformations are defined in the reduced phase space rather than the kinematical phase space. They transform both canonical variables and internal clock. Although there is no longer Hamiltonian constraint in the reduced phase space, the role of extended transformations is to reflect the fact that the reduced phase space is descendent from the kinematical phase space and that its definition is tied to the ambiguous definition of the clock.

4. Internal clock

In this section we define the reduced phase space and show how it is related to the choice of clock. We introduce the extended transformations and discuss how they break at the quantum level.

4.1. Reduced phase space

Let us discuss how the reduced phase space is obtained. Solving a given Hamiltonian constraint $\mathrm{C} = 0$ in the kinematical phase space $\mathbf{P}$ of dimension $2n+2$ and equipped with a symplectic form Ω leads to a $2n+1$-dimensional sub-manifold $\mathcal{M}_{\mathrm{C}} \subset \mathbf{P}$ equipped with a degenerate two-form $\Omega|_{\mathrm{C}=0}$ induced from Ω. A very important property of $\Omega|_{\mathrm{C}=0}$ is that a field of its null vectors $\{v_{\mathrm{C}} \in T^*\mathcal{M}_{\mathrm{C}} : \Omega|_{\mathrm{C}=0}(v_{\mathrm{C}}) = 0\}$ generates a $2n$-dimensional space $\mathcal{S}$ of curves in $\mathcal{M}_{\mathrm{C}}$. These curves represent physical motion of the model. The key element of the construction is the choice of internal clock $t : \mathcal{M}_{\mathrm{C}} \mapsto \mathbf{R}$ which is a real function on the constraint surface and whose level sets $\{p \in \mathcal{M}_C : t(p) = t_{fixed}\}$ cross each curve in $\mathcal{M}_C$ at most once. Therefore, the internal clock provides a foliation of $\mathcal{M}_C$,

$$(t, \mathcal{S}) \mapsto \mathcal{M}_C. \tag{1}$$

At each level set of t a (non-degenerate) symplectic form may be introduced,

$$\Omega|_{C=0,t=t_{fixed}} \tag{2}$$

induced from $\Omega|_{C=0}$. The pair

$$\{(t_{fixed}, \mathcal{S}), \Omega|_{C=0,t=t_{fixed}}\} \tag{3}$$

is called the reduced phase space. $-\Omega|^{-1}_{C=0,t=t_{fixed}} = \{\cdot,\cdot\}_{phys}$ is the physical Poisson bracket in $(t_{fixed}, \mathcal{S})$. It can be shown that the degenerate two-form can be decomposed as follows:

$$\Omega|_{C=0} = \Omega|_{C=0,t=t_{fixed}} - \mathrm{d}t\mathrm{dH}, \tag{4}$$

where H is a function on $\mathcal{M}_C$ called the Hamiltonian. Furthermore it can be shown that the vector field

$$-\Omega|^{-1}_{C=0,t=t_{fixed}}(\cdot, \mathrm{H}) = \{\cdot, \mathrm{H}\}_{phys} \tag{5}$$

generates the physical curves $\mathcal{S}$ in $\mathcal{M}_C$ parametrised by t. In other words, given any observable $\mathcal{O} : \mathcal{M}_C \mapsto \mathbf{R}$, its evolution is given by means of the Hamilton equations:

$$\frac{\mathrm{d}\mathcal{O}}{\mathrm{d}t} = \frac{\partial \mathcal{O}}{\partial t} + \{\mathcal{O}, \mathrm{H}\}_{phys} \tag{6}$$

Clearly, the reduced phase space formalism is a reduction of the Hamiltonian constraint formalism to the unconstrained canonical formalism which includes a Hamiltonian H, a time t and a Poisson bracket. The construction is depicted in figure (1).

The key difference between the reduced phase space formalism and the unconstrained canonical formalism is that the choice of internal clock is ambiguous whereas the time parameter of the usual canonical formalism is predetermined and completely fixed. The consequences are immediately seen in the above construction. Namely, the choice of $\bar{t}$ instead of t as an internal clock leads to another foliation of the constraint sub-manifold, $(\bar{t}, \mathcal{S}) \mapsto \mathcal{M}_C$, and another decomposition of $\Omega|_{C=0}$,

$$\Omega|_{C=0} = \Omega|_{C=0,\bar{t}=\bar{t}_{fixed}} - \mathrm{d}\bar{t}\mathrm{d}\bar{\mathrm{H}}, \tag{7}$$

where $\Omega|_{C=0,\bar{t}=\bar{t}_{fixed}} \neq \Omega|_{C=0,t=t_{fixed}}$ are inequivalent symplectic forms which, when inverted, give inequivalent physical Poisson brackets. In other words, different choices of internal clocks establish canonically inequivalent reduced phase spaces. The same physical curves are obtained with the respective Hamilton equations, though they are now parametrised by $\bar{t}$.

4.2. Extended transformations

Let us introduce the so called extended transformations. By (q, p) we denote the canonical coordinates which always exist by the virtue of Darboux theorem,

$$\Omega|_{C=0,t=t_{fixed}} = \mathrm{d}q\mathrm{d}p. \tag{8}$$

We first recall the concept of canonical transformations. They are defined as such (time-dependent) transformations of canonical coordinates,

$$t \times (q, p) \mapsto (\bar{q}, \bar{p}), \tag{9}$$

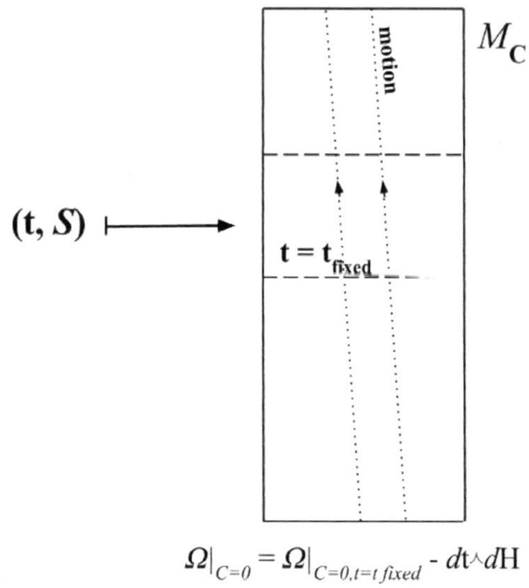

Figure 1. A foliation of the constraint surface $(t, \mathcal{S}) \mapsto \mathcal{M}_{\mathrm{C}}$ provides a decomposition of the degenerate form $\Omega|_{\mathrm{C}}$ into a symplectic term and another term proportional to the Hamiltonian. In this way a canonical formalism can be established for any choice of clock t. However, different choices of clock lead to different symplectic terms and canonically inequivalent formalisms.

which preserve the time coordinate t and the form of the symplectic form,

$$\Omega|_{\mathrm{C}=0,t=t_{fixed}} = \mathrm{d}\bar{q}\mathrm{d}\bar{p}. \tag{10}$$

The extended transformations are such transformations of canonical coordinates and time coordinate,

$$(q, p, t) \mapsto (\bar{q}, \bar{p}, \bar{t}), \tag{11}$$

which preserve the form of the degenerate form $\Omega|_{\mathrm{C}}$,

$$\Omega|_{\mathrm{C}} = \mathrm{d}q\mathrm{d}p - \mathrm{d}t\mathrm{d}H = \mathrm{d}\bar{q}\mathrm{d}\bar{p} - \mathrm{d}\bar{t}\mathrm{d}\bar{H}, \tag{12}$$

but in general they do not preserve the symplectic form,

$$\Omega|_{\mathrm{C},t=t_{fixed}} = \mathrm{d}q\mathrm{d}p \neq \mathrm{d}\bar{q}\mathrm{d}\bar{p} = \Omega|_{\mathrm{C},\bar{t}=\bar{t}_{fixed}} \tag{13}$$

The above transformations do not only transform a canonical coordinate system of a given reduced phase space but they can also transform a given reduced phases space into another one. Canonical transformation are indeed a subgroup of extended transformations.

4.3. Quantum level

At the classical level, the many inequivalent reduced phase space formalisms are related by extended transformations. They constitute a symmetry of the canonical formalism with respect to which the physical motion represented by physical curves in $\mathcal{M}_{\mathrm{C}}$ is invariant. This is no longer true at the quantum level. The reason is that quantisation imposes uncertainty on canonical variables and the notion of canonical variables is tied to the choice of clock. Therefore, uncertainties are imposed differently for different internal clocks and the classical symmetry

IOP Conf. Series: Journal of Physics: Conf. Series **880** (2017) 012046 doi:10.1088/1742-6596/880/1/012046

breaks down, that is, the quantised motion cannot be invariant with respect to the choice of clock. How can we see this more closely?

The problem is that quantisation of different reduced phase spaces and the respective Hamiltonians may produce different results not because the canonical structure underlying quantisation is different but just because some usual quantisation ambiguities occur (choice of basic variables, orderings...). To ensure that any dissimilarities between quantised reduced phase spaces and respective dynamics are induced by differences in employed clocks rather then usual quantisation ambiguities we should quantise all reduced phase spaces in a unique way in some sense. The precise prescription of how to achieve it was given in [1–3]. In what follows we just sketch its main points.

The dimensionality of the reduced phase space is equal to the number of constants of motion (we assume an explicitly integrable system). It can be easily proven that the Poisson algebra of the constants of motion is independent of the choice of clock. Thus, a quantisation of constants of motion can be fixed for all reduced phase space formalisms. However, the Poisson algebra of dynamical observables differs from one clock to another and their quantum representation must differ from one clock to another too. Nevertheless, it can be shown that quantisation of all constants of motion determines quantisation of all observables including dynamical observables as well. Therefore, quantisation of both non-dynamical (constants of motion) and dynamical observables in all reduced phase spaces can be fixed with a unique quantum representation of constants of motion only.

One of the consequences of the described prescription is that constants of motion are given exactly the same operators for all choices of clock. Any state in the Hilbert space can be then decomposed in terms of invariant eigenstates of self-adjoint constants of motion operators. In other words, the physical interpretation of a state as a wave-function on the spectrum of quantised constants of motion is unique. On the other hand, the quantum representation of dynamical observables depends on the choice of clock. A fixed dynamical quantity is given a different operator for different clocks. Therefore, assuming they are self-adjoint, a state is given a different wave-function on the spectrum of a fixed quantised dynamical quantity for different clocks. In other words, the way in which quantum evolution is interpreted depends on the employed clock and we will see this in the examples discussed below.

5. Review of results

In this section we review some of the previous results on the relation between quantum theory and clocks, which rely on the notion of extended transformations which break down at the quantum level.

In the work [3] the definition of extended transformations (called therein pseudo-canonical transformations) and their relation to canonical transformations in the kinematical phase space were established. This work includes a careful discussion on the separation of usual quantisation ambiguities from clock-induced effects found in quantised reduced phase spaces. It was shown by a simple example that spectra of dynamical operators can change their character from continuous to discrete upon transforming the internal clock.

In the work [2], a quantum theory of the Friedmann model of universe was studied. Initially, a variable associated with the perfect fluid filling the universe was set as the internal clock and a quantisation was performed. The quantum dynamics was shown to undergo a bounce which replaced the classical singularity. Then, extended transformations were applied and more reduced phase spaces were obtained and quantised. For simplicity, a method of phase space portraits based on coherent states was used. The approximate description of quantum dynamics in terms of phase space variables turned out to be enough to identify very strong dissimilarities between different clock-based quantum theories. They included the scale of the bounce, the number of bounces or the extent to which dynamics was symmetric with respect to the bounce.

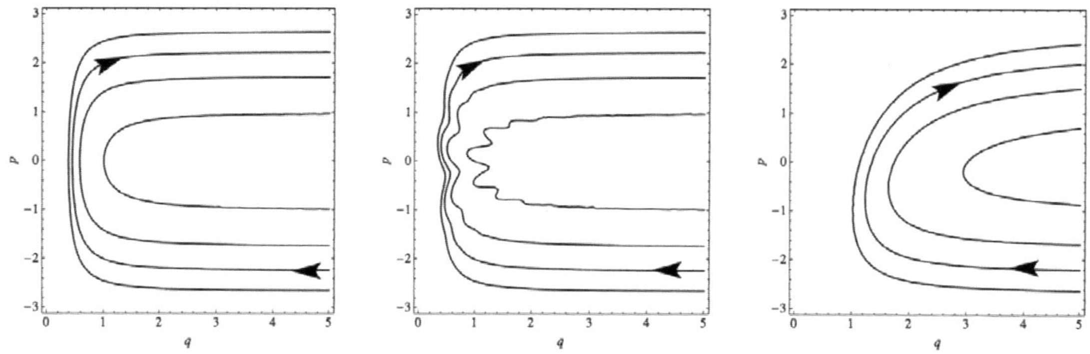

Figure 2. Three phase space portraits of quantum dynamics of the Friedmann universe with a bounce based on coherent states (see [2] for details). The portraits are given in a fixed phase space parametrised by q, which denotes the volume of the universe, and p, which is a canonically conjugate variable. Each of the three portraits was derived with a distinct clock and presents a distinct behaviour. The existence of a bounce which replaces the classical singularity and the classical behaviour exhibited by the portraits away from the bounce are universal features which occur for all choices of clock. Whereas, the scale of the bounce, the number of bounces and the symmetry between expansion and contraction clearly depend on the employed clock.

The only confirmed invariant feature of the quantum dynamics was the existence of the bounce itself and of the classical limit for both the expanding and contracting branches. Figure (2) illustrates the effect of the choice of clock on the semiclassical dynamics of the Friedmann universe.

In the work [1], the quantum dynamics of the anisotropic Bianchi type I model with a perfect fluid was studied. This model includes the Friedmann model described above in the isotropy limit. The justification for extending the reduced phase space to include anisotropy variables was to investigate the classical limit of quantum theory in more detail than it was possible in [2]. Recall that classical solutions are parametrised by a collection of conserved quantities, including the Hamiltonian. The key question addressed in this work was following: does the asymptotic classical solution of a given quantum solution, expressed in terms of a complete set of conserved quantities, is the same for both the expanding and contracting phases and is independent of the employed clock? It was shown that indeed an anisotropy state in the quantum Bianchi type I model with a bounce is unique for a given quantum solution and independent of the employed clock if registered far enough from the bounce. Then a general analysis showed that the classical asymptotic limits, if they exist, must be invariant with respect to transformations of underlying reduced phase spaces.

6. Conclusions

By timeless quantum mechanics we mean quantum mechanics of systems without a fixed, predetermined time parameter which quantifies their evolution. In such a theory, we set a dynamical variable as the internal clock which measures the evolution of the remaining variables. This procedure is shown to establish a canonical formalism for the underlying classical theory. Different choices of internal clock lead to different underlying canonical formalisms. This ambiguity imprints on the effects of imposition of uncertainties on canonical variables when a given system is quantised. The ambiguity in the notion of quantum dynamics for such theories is unavoidable. However, it was shown that when a system becomes in some sense classical the mere choice of clock does not influence the dynamics. Let us speculate about the possible physical meaning of the result.

There are two important cases of states of a given system without a fixed notion of time. The first case is a state in which there are no degrees of freedom which could be assumed classical. It seems impossible then to apply the usual understanding of quantum mechanics. For example, given a fixed dynamical observable, the choice of internal clock determines the spectral decomposition of that state, i.e. the form of the wave-function in the given representation. This undermines the physical interpretation of that state. The choice-of-clock-invariant piece of information about the state can be read only from its spectral decomposition with respect to a conserved quantity. Therefore, no information about the evolution of the state can be provided.

The second case is a state in which some degrees of freedom can be assumed classical. Setting one of them as the internal clock seems to be a preferred choice. Moreover, extended transformations, which transform the formalism from one clock to another clock which can be also assumed classical in a given state, do not change the spectral representation of that state for any dynamical observable. This seems to be the case of the usual quantum mechanics and laboratory systems which the usual quantum mechanics describes. These systems are indeed only subsystems of a global system which encompasses both the system and its environment. Indeed, the measuring apparatus, including clocks, belongs to the classical environment. Therefore, we see that timeless quantum mechanics includes and extends the usual quantum mechanics.

When we deal with local quantum systems in laboratories, the time standard is provided by external degrees of freedom in the form of apparatus that measures the progress of classical 'time'. But what should be the time standard for the evolution of global degrees of freedom like the ones we deal with in cosmological models? Clearly, there are no external degrees of freedom in this case and the only option seems to be choosing one of the global degrees of freedom as the internal clock. Then, however, anything we say about the quantum evolution of such a system strongly depends on the choice of clock we make. The only available and invariant information is time-independent. The quantum universe becomes truly timeless.

Acknowledgments

The author gratefully acknowledges support from Narodowe Centrum Nauki by decision No. DEC-2013/09/D/ST2/03714.

References

[1] Małkiewicz P 2016 Clocks and dynamics in quantum models of gravity *Preprint* arXiv:1601.04857 [gr-qc]
[2] Małkiewicz P 2015 What is Dynamics in Quantum Gravity? *Preprint* arXiv:1505.04730
[3] Małkiewicz P 2015 Multiple choices of time in quantum cosmology *Class. Quantum Grav.* **32** 135004 (*Preprint* arXiv:1407.3457)
[4] Arnowitt R Deser S Misner C W 1962 The dynamics of General Relativity *Gravitation: an introduction to current research* ed Louis Witten (Wiley) chapter 7 pp 227–265
[5] Uggla C (1997) Hamiltonian Cosmology *Dynamical Systems in Cosmology* ed J Wainwright and G F R Ellis (Cambridge University Press)
[6] Isham C J 1992 Canonical Quantum Gravity and the Problem of Time *Preprint* arXiv:gr-qc/9210011
[7] Kuchar K 1988 The problem of time in canonical quantization of relativistic systems *Conceptual Problems of Quantum Gravity* ed A Ashtekar and J Stachel (Proc. of Osgood Hill Conf.)
[8] Hajicek P 1999 Choice of Gauge in Quantum Gravity *Preprint* gr-qc/9903089
[9] Arnowitt R Deser S and Misner C W 1961 Heisenberg Representation in Classical General Relativity *Il Nuovo Cimento* Vol. XIX N. 4 p. 668-681
[10] Hajicek P Kijowski J 2000 Spherically symmetric dust shell and the time problem in canonical relativity *Phys. Rev.* D **62** 044025
[11] Klauder J R 2012 Enhanced Quantization: A Primer *J. Phys. A: Math. Theor.* **45** 285304-1-8

Quantum statistical gravity: time dilation due to local information in many-body quantum systems

Dries Sels[1,2] and Michiel Wouters[1]

[1]TQC, Universiteit Antwerpen, Universiteitsplein 1, B-2610 Antwerpen, Belgium
[2]Department of Physics, Boston University, Boston, MA 02215, USA

E-mail: dsels@bu.edu; michiel.wouters@uantwerpen.be

Abstract. We propose a generic mechanism for the emergence of a gravitational potential that acts on all classical objects in a quantum system. Our conjecture is based on the analysis of mutual information in many-body quantum systems. Since measurements in quantum systems affect the surroundings through entanglement, a measurement at one position reduces the entropy in its neighbourhood. This reduction in entropy can be described by a local temperature, that is directly related to the gravitational potential. A crucial ingredient in our argument is that ideal classical mechanical motion occurs at constant probability. This definition is motivated by the analysis of entropic forces in classical systems.

1. Introduction

After almost one century of coexistence, the relation between Einstein's theory of general relativity and quantum physics is still not well understood. At the same time, the precise relation between microscopic quantum physics and macroscopic classical physics has not been completely demystified. There are some suggestions that these two problems are related [1, 2, 3, 4, 5].

Two fundamentally different strategies are used to relate the quantum to the classical. The first one is based on the wave-particle duality and is most succinctly expressed in the path integral formulation. The correspondence between the classical and the quantum is here mathematically very direct, because it is the same action that appears in both theories. It can therefore be used to go from the classical to the quantum and vice versa. A familiar example is electromagnetism: Maxwell's equations are derived from the quantummechanical path integral by means of the stationary phase approximation.

The second strategy, statistical physics, is fundamentally different. From a quantum theory, thermodynamic relations can be computed, but those thermodynamic relations cannot be quantised. Still, it is the only method suited for the description of complex macroscopic systems, about which we only have thermodynamic and hydrodynamic information [6, 7].

The main efforts to find a quantum mechanical description of gravity have been based on the first method [8, 9] but also the second strategy has been explored [10, 11, 12]. The latter efforts go under the name of thermodynamic or entropic gravity. Conceptually, this strategy seems preferable, since the physics of gravity deals with macroscopic objects, that have nonzero entropy and that are coupled to environments [13, 14]. In thermodynamic gravity, the gravitational interaction is seen as emergent rather than as an explicit ingredient in the microscopic theory.

Gravity being the most universal force in the universe, it would ideally emerge in any complex quantum theory. In this paper, we will argue that this might be the case.

The first ingredient in our argument is that the presence of matter at some place in the universe constitutes information, defined as missing entropy [15, 16]. Within the framework of quantum mechanics, when knowledge about a particular realisation of the system is available, the incompatible part of the wave function has to be projected out. When the quantum system has entanglement, a local projection also influences the probability distribution in its vicinity. We will show that the effect of local information on its surroundings can be described by a position dependent 'entanglement' temperature [17, 18], defined by approximating the local reduced density matrix by a Gibbs state. We then demonstrate that the inhomogeneity of the entanglement temperature is reflected in a spatial variation of the magnitude of the energy fluctuations.

This mechanism is most easily illustrated in the EPR setting. When the first qubit of a Bell state $|\psi\rangle = \frac{1}{\sqrt{2}}(|\uparrow\uparrow\rangle + |\downarrow\downarrow\rangle)$ is not measured, the reduced density matrix of the second qubit is maximally mixed, equivalent to infinite temperature. On the other hand, when the first one is measured in the $\uparrow, \downarrow$ basis, the second qubit is in a pure state, which corresponds to a zero temperature density matrix.

It is actually well known that entanglement allows to make the connection between pure quantum states and statistical density matrices [19, 20]. Tracing over environment degrees of freedom leaves the system typically in a canonical Gibbs state, even if the composite state is pure [21]. We make a natural extension of this work by considering what happens if the environment is not fully traced over.

The main assumption in our quantum statistical approach to gravity is that the magnitude of energy fluctuations in a region sets the energy scale for the local physics. This follows from the fundamental characterisation of phenomena by their probability. We will demonstrate that it is intimately related to the equivalence principle. 'Gravitational red shifts' then appear because the entanglement temperature is lower closer to a source of information. We will also show that a position dependent entanglement temperature, implies the acceleration of semi-classical wave packets.

In our scenario, gravitational forces emerge as an interplay between quantum measurement and the statistical meaning of classical mechanics. Usually, we do not think about mechanical systems in terms of probability distributions, except in systems where 'entropic' forces occur [22, 23]. We therefore start our paper with a short description of entropic forces in Sec. 2. It is illustrated that the entropic attraction between two objects is due to correlations in their joint probability distribution.

In a quantum system, the role of the probability distribution of local states is played by the reduced density matrix. Correlations between regions are reflected in the quantum mutual information [24, 25]. An entropy and a temperature can be associated to the reduced density matrix. We show that local measurements can affect the temperature of the regions that are correlated with it. Our general arguments about local information in quantum systems are illustrated by calculations on a one-dimensional non-interacting fermion model in Sec. 3. From the requirement of probability conservation of a wave packet, we show in Sec. 4 that a temperature gradient leads to acceleration.

2. Classical entropic forces

Theoretical classical mechanics deals with isolated systems, that satisfy energy conservation. This excludes systems that are coupled to baths and therefore it is only an approximate description of real physical systems, where friction due to coupling with an environment is always present. In addition to friction, environments can also exercise forces. These are known as entropic forces [23] and are essential in e.g. depletion forces between colloidal particles in

IOP Conf. Series: Journal of Physics: Conf. Series **880** (2017) 012047 doi:10.1088/1742-6596/880/1/012047

fluids, osmotic pressure and the elasticity of polymers.

Let us start with an elementary analysis of entropic forces. It is the simplest setting in which one can see that probability conservation is a generalisation of energy conservation. We consider the classic example of two spheres immersed in a fluid. Due to repulsive interactions between the sphere and the fluid, there is a depletion region around each sphere. The presence of the spheres therefore lowers the entropy of the fluid (it reduces the available volume for each molecule). When the depletion regions of the two spheres overlap, the entropy of the fluid increases. The probability in terms of their positions $\mathbf{x}_{1,2}$ and momenta $\mathbf{p}_{1,2}$ reads

$$p = \frac{1}{Z} e^{S(\mathbf{x}_1 - \mathbf{x}_2) - \frac{p_1^2}{2m_1 T} - \frac{p_2^2}{2m_2 T}}, \tag{1}$$

where T is the temperature (we set $k_B = 1$) and Z a normalisation constant. In the absence of the entropic term S in the probability (1), energy conservation is equivalent to probability conservation and the temperature T is irrelevant.

In the presence of S on the other hand, probability and energy conservation are no longer the same. Energy conservation remains unaltered, but probability conservation requires

$$-T \ln p = \frac{p_1^2}{2m_1} + \frac{p_2^2}{2m_2} - TS(\mathbf{x}_1 - \mathbf{x}_2) \tag{2}$$

to be constant. A term $-TS$ is added to the energy, which gives rise to 'entropic forces'

$$\frac{d\mathbf{p}_i}{dt} = T \nabla_{\mathbf{x}_i} S. \tag{3}$$

The reason why the temperature usually does not appear in classical mechanics is that one implicitly requires that the mechanical energy E is much larger than the thermal energy. This means that classical mechanics is only concerned with statistically unlikely events, with $p \sim e^{-E/T} \ll 1$. This may seem contradictory to the standard classical to quantum correspondence, where the classical paths are the most likely, but corresponds to classical objects carrying a large amount of information. The meaning of the deterministic motion is that, conditional on being at position x at time t, there is unit probability to be at a place x' at time t'. In quantum physics, where probabilities are the most elementary quantities, it is then natural to elevate these implicit aspects to the definition of a classical mechanical object: it is an unlikely excitation that evolves at constant probability.

3. Local information and temperature in quantum systems

After these preliminary remarks on classical mechanics, we turn our attention to quantum systems. The central object in quantum mechanics of closed systems is its wave function. We will consider quantum systems that are defined on a lattice. A typical example is the Hubbard model. At first sight, one would have little hope for the emergence of relativistic physics out of such systems, but Lieb and Robinson showed in their seminal work that causality emerges in such systems under very general conditions [26].

In order to make our discussion more concrete, we perform some calculations on a specific model. The simplest systems from a theoretical point of view are the ones with quadratic Hamiltonians, that describe free quasi-particles. We will here consider the ground state of the fermionic hopping Hamiltonian of the form

$$\hat{H} = -\sum_i \left[\left(t\hat{c}_{i+1}^\dagger \hat{c}_i + h.c. \right) - \mu \hat{c}_i^\dagger \hat{c}_i) \right], \tag{4}$$

where t is the hopping amplitude and μ is the chemical potential. Apart from a Hamiltonian, we need to specify the quantum state of the system. Here, we will consider the ground state of the Hamiltonian, but the analysis could be extended to excited states as well.

All expectation values within a subregion A are described by the reduced density matrix $\rho_A = \mathrm{Tr}_{S\setminus A}\rho$, where the trace is over all sites except the ones in A, which is of the form $\hat{\rho}_A = \exp(-\hat{H}_A)$, where the 'modular' or 'entanglement' Hamiltonian can be written as

$$\hat{H}_A = \sum_{i,j} h_{ij}\hat{a}_i^\dagger \hat{a}_j. \tag{5}$$

An example for $N_A = 10$ sites is shown in Fig. 1. The structure of the original Hamiltonian is clearly visible in the entanglement Hamiltonian. When comparing entanglement Hamiltonians for different subsystem sizes, one finds that the main effect is that the magnitude of the matrix elements increases. We thus find that the entanglement Hamiltonian is of the form $h_A = \beta_A \tilde{h}_A$. We will call this scale dependent temperature T_A the 'entanglement' temperature [17, 18]. Fig. 1b shows the dependence of β_A on system size, where $\tilde{h}_A$ is normalised by the largest tunneling matrix element. A linear increase of the effective inverse temperature with system size is apparent, in agreement with analytical calculations exploiting the adS/CFT correspondence [27].

This behavior can be understood from the fact that, unlike the entropy of a Gibbs state, the entropy is not extensive. For 1-D free fermions it is well know that the ground state entanglement entropy is $S = 1/3\log(L)$. The entanglement temperature must thus decrease such that the entropy of a Gibbs state of the local (smooth) Hamiltonian equals the entanglement entropy.

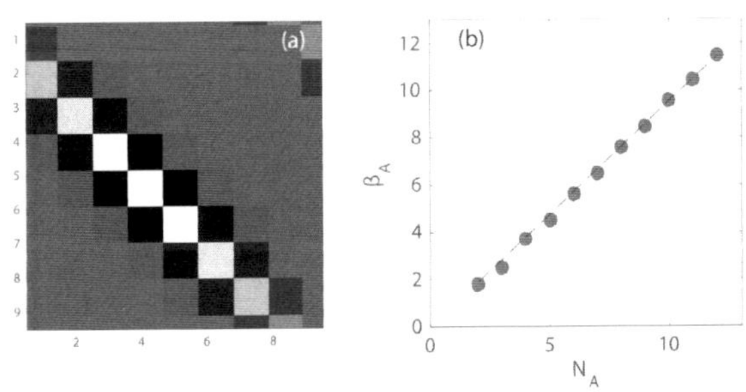

Figure 1. (a) The entanglement Hamiltonian h_{ij} (left) for the fermionic Hamiltonian (4) on a subregion of 10 sites. The structure of the original Hamiltonian is clearly visible, with hopping matrix elements next to the diagonal. (b) The entanglement temperature increases as a function of the system size.

Let us now consider a subregion AB that consists of two disconnected parts A and B, separated by a distance R. In terms of the entanglement Hamiltonian, there is a difference between h_{AB} and the direct sum $h_A \oplus h_B$. Because more information is present in the compound subsystem AB, larger matrix elements are found in h_{AB}: the joint system is at a lower temperature than the individual systems. Physically, this means that the uncertainty about the state A is reduced when information is obtained about system B.

It is instructive to look at the limiting cases $R = \infty$ and $R = 0$ when A and B have the same number of sites. When A and B are infinitely far apart they should not be entangled and the total entanglement Hamiltonian is just the direct sum entanglement Hamiltonian, hence the entanglement temperature is the same as for the individual systems. However, when the two systems touch they simple form one system that is twice as big. It was already shown above that the entanglement temperature is only half of that of the separate subsystems.

This is illustrated in Fig. 2, where we show the entanglement Hamiltonian consisting of two subregions of 10 sites that are separated by 15 sites. The left hand panel shows the full matrix, where one can clearly identify the direct sum of uncoupled entanglement Hamiltonians and some off-diagonal couplings. The right hand panels show the tunneling matrix elements and compares them to the case of a single subregion (dashed lines). Two features stand out. First, the tunneling matrix elements are larger, corresponding to a larger inverse temperature: $\beta_{AB} > \beta_A$. Secondly, there is also an asymmetry in the matrix elements, which could be interpreted as a temperature gradient.

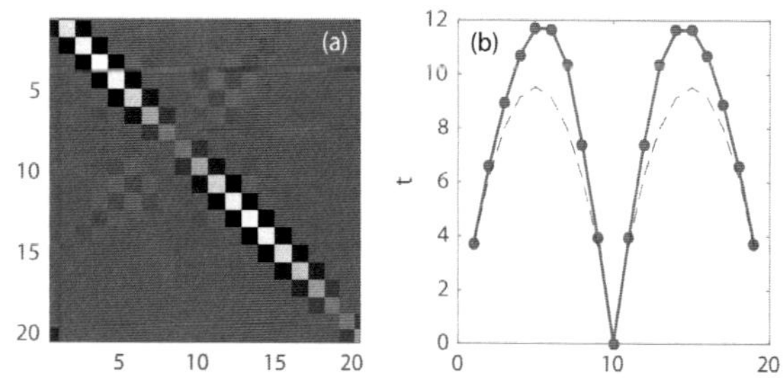

Figure 2. (a) The entanglement Hamiltonian h_{ij} for a subsystem consisting of two disjoint regions (10 sites each) separated by a distance of 15 sites. (b) Tunneling matrix elements (first off-diagonal) are compared with the case of a single region of 10 sites (dashed lines).

The entropy of the total system AB is also different from the sum of the entropies of A and B. The difference between the two is the mutual information

$$I_{AB} = S_A + S_B - S_{AB}, \tag{6}$$

which is always positive. Fig. 3 shows the mutual information as a function of the distance for our free fermion toy system (blue line). The mutual information is seen to decay slowly, with a power law behavior at large distance. This can be attributed to the fact that the system is gapless [28]. The red line shows the effective temperature β_{AB} of the joint density matrix $\hat{\rho}_{AB}$, which shows the same behavior at large distances. This shows a clear connection between the increase in mutual information and the decrease in effective temperature when the two subregions are brought closer together.

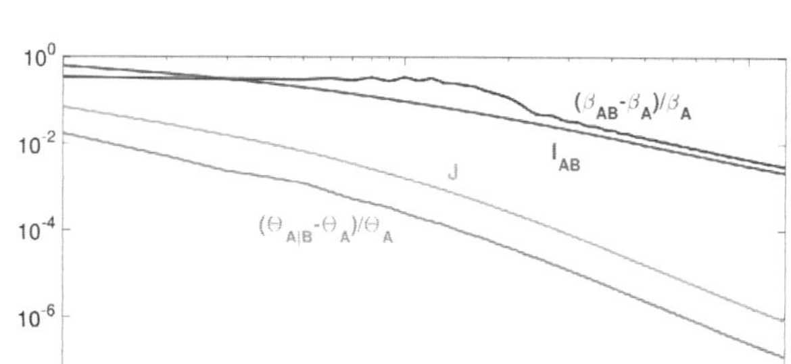

Figure 3. The mutual information I_{AB} as a function of the distance between two subregions of sizes $N_A = N_B = 10$ (blue), the effective temperature of the density matrix $\hat{\rho}_{AB}$ (red), the average conditional entropy J for projection on the Schmidt basis in region A (orange) and the energy average fluctuation width after projection on the Schmidt basis (green).

For classical systems, one can rewrite the mutual information in terms of conditional entropies as:

$$I_{AB} = S(A) - \sum_{x_B} p(x_B) S(A|x_B), \tag{7}$$

where the conditional entropy $S(A|x_B)$ is defined in terms of the conditional probability distribution as $S(A|x_B) = -\sum_{x_A} p(x_A|x_B) \ln p(x_A|x_B)$.

For quantum systems unfortunately, the situation is more complicated, because conditional probabilities are no longer simply defined in terms of joint probabilities, but by means of projection operators. It has been found that for quantum systems, the second definition of the mutual information is always smaller than the first one. This has led Zurek to introduce the notion of 'quantum discord' as [29]

$$D = I_{AB} - \max_{\Pi_j^A} J_{\Pi_j^A}(\rho), \tag{8}$$

where J is the conditional entropy that depends on the set of projective measurements Π_j^A. Unfortunately, the optimisation problem over measurements has been proven to be NP complete [30].

If one is interested in the physics of a region A, one would want to know the reduced density matrix ρ_A. Imagine now that a measurement in region B is performed. When there are correlations (quantified by the mutual information) between these two regions, the result of this measurement will affect the density matrix in region A. In general, the conditional entropy $S_{A|B}$ will be lower than S_A, with the mutual information as an upper bound on the average entropy reduction for a certain type of measurement. In view of the relation between entropy and temperature, this means that the temperature of the region A depends on information about region B. In other words, a measurement in B results in a position dependent temperature in its surroundings. Regions close to B will be more affected than regions farther away, as indicated by the behavior of the mutual information.

The simplest class of projections to be performed on this system is a projection on the Schmidt basis in region B. While this is clearly not the best basis in the sense of Eq. (8), it gives at least a lower bound on the average amount of information that can be gained with a projective measurement. From Fig. 3, it is seen that conditional entropy J is lower than the mutual information and decays faster as a function of the distance, but it still shows a power law type decay.

4. Energy statistics and mechanical acceleration

In order to make contact with fundamental physics, one should think about the density matrix before measurement as the 'vacuum'. An unlikely measurement outcome then corresponds to the detection of matter. If matter is measured to be present in region B, this measurement then affects the statistics in region A, depending on the distance between A and B.

The principal ingredient that is responsible for the entropic force is the spatial variation of the energy fluctuations. As we have argued in the previous section, these can arise from local information about the system. As an example, we plot in Fig. 3 the energy variance after projection on the Schmidt basis (as for the computation of J) with a green line. The spatial dependence of the energy fluctuations follows the behavior of the conditional entropy, establishing a link between entropy and energy fluctuations. This suggests that the energy fluctuations Θ are proportional to the entanglement temperature. We will therefore call Θ the 'local temperature' in the following.

As in the classical case, we can now obtain the entropic force by requiring constant probability for fluctuations. A massive fluctuation is characterized by its internal energy E_{int}. When the local temperature Θ is position dependent, the internal energy has to vary in space in order to keep the ratio E_{int}/Θ constant, so to ensure a constant probability for the excitation. From this condition one finds immediately that there is a gradient in the internal energy, corresponding to a force

$$F = -\frac{dE_{int}}{dx} = -E_{int}\frac{d\ln\Theta}{dx}. \tag{9}$$

If we interpret the energy E_{int} in Eq. (9) as the relativistic rest energy $E_{int} = mc^2$ of the excitation in the sense that $p = mv$, we obtain the acceleration

$$a = -c^2\frac{d\ln\Theta}{dx}. \tag{10}$$

The local temperature can then be identified with the Newtonian potential as

$$\phi = c^2 \ln \Theta. \tag{11}$$

Let us recapitulate how in our view gravity emerges from quantum mechanics

(i) Local properties of quantum systems are described by a reduced density matrix, characterised by a local temperature.
(ii) Local information leads to a spatial dependent temperature.
(iii) In order to conserve their probability, excitations have to change their internal energy when they move to a region with a different temperature, which leads to acceleration.

Looking back at the ingredients that led to the (weak) equivalence principle, the most crucial step is the requirement that $E_{int}/\Theta(x)$ is constant, motivated by constant probability. This is our translation of the physics of ideal mechanical motion into the language of statistics.

Einstein translated the physics of ideal mechanical motion (free falling objects) into the language of classical field theory as 'general covariance': the form of the equations should be independent of the coordinates. This principle has turned out to be hard to implement in a quantum setting. The reason could be the fact that quantum mechanics is fundamentally a statistical theory, requiring a direct formulation of physical processes in a probabilistic language.

5. Conclusions and outlook

In the present work, we have suggested a mechanism for time dilation that emerges out of a quantum system through a statistical analysis. It is clear that a large effort will be needed before our suggestion can become a viable candidate for a description of real quantum gravity.

As we have discussed, information should be extracted from the quantum system through slow observables, such that the obtained information is stable. Technically, finding these observables is a difficult problem: the search takes place in an exponentially large Hilbert space.

A further issue concerns the range the gravitational potential. In the simple model that we investigated, we found a power law decay for the mutual information in one-dimensional noninteracting fermions. An obvious question is to find the conditions for which one obtains a $1/r$ gravitational potential at large distances in three dimensional systems.

So far, we have only provided arguments for one aspect of Einstein gravity to emerge from a statistical analysis of quantum systems, namely time dilation in a static situation. If gravity is really emerging according to our mechanism, one should be able to derive the full set of geodesic and Einstein equations from statistical considerations, with far richer phenomenology than Newtonian gravity. It should ultimately include cosmology from the perspective of closed quantum systems.

In conclusion, we conjecture that gravity appears as a deformation of statistics due to local information. This type of locally deformed quantum states are usually not considered in calculations on many body quantum systems, that are concerned with small numbers of excitations. It would actually seem implausible that large fluctuations in entangled quantum systems do not influence their surroundings. Two possibilities then remain if quantum mechanics is a fundamental theory. Either this influence is gravity or it is another, unknown effect. Our arguments are encouraging for the gravity scenario to be correct. The description of large fluctuations will be mathematically challenging, but a thorough understanding seems unavoidable if we want to grasp how the physical world is related to our quantum models.

6. Acknowledgements

We thank Jacques Tempere and Jozef Devreese for encouraging discussions. D.S. acknowledges support of the FWO as post-doctoral fellow of the Research Foundation - Flanders.

IOP Conf. Series: Journal of Physics: Conf. Series **880** (2017) 012047 doi:10.1088/1742-6596/880/1/012047

References

[1] Penrose R 2014 *Found. Phys.* **44** 557–75
[2] Aguirre A and Tegmark M 2011 *Phys. Rev. D* **84** 105002
[3] Joos E, Zeh H D, Kiefer C, Giulini D J, Kupsch J and Stamatescu I O 2013 *Decoherence and the appearance of a classical world in quantum theory* (Springer Science & Business Media)
[4] Gell-Mann M and Hartle J B 1993 *Phys. Rev. D* **47** 3345–82
[5] Hu B L 2014 *J. Phys.: Conf. Series* **504** 012021
[6] Zubarev D, Mozorov V and Röpke G 1996 *Statistical Mechanics of Nonequilibrium Processes. Volume 1: Basic Concepts, Kinetic Theory* (Akadamie Verlag, Berlin)
[7] Grandy W 2012 *Entropy and the Time Evolution of Macroscopic Systems* (OUP Oxford)
[8] Polchinski J 1998 *String Theory* (Cambridge University Press)
[9] Rovelli C 2004 *Quantum Gravity* (Cambridge University Press)
[10] Jacobson T 1995 *Phys. Rev. Lett.* **75** 1260
[11] Padmanabhan T 2005 *Phys. Rep.* **406** 49 – 125
[12] Verlinde E 2011 *J. High Energy Phys.* **2011** 1–27
[13] Zurek W H 2003 *Rev. Mod. Phys.* **75** 715–775
[14] Schlosshauer M A 2007 *Decoherence: and the quantum-to-classical transition* (Springer Science & Business Media)
[15] Schrödinger E 1944 *What is life?: the physical aspect of the living cell; based on lectures delivered under the auspices of the Inst. at Trinity College, Dublin, in Feb. 1943* (University Press)
[16] Brillouin L 1953 *J. Appl. Phys.* **24** 1152–1163
[17] Wong G, Klich I, Zayas L A P and Vaman D 2013 *J. High Energy Phys.* **2013** 1–24
[18] Kumar S S and Shankaranarayanan S 2015 *preprint arXiv:1504.00501*
[19] Popescu S, Short A J and Winter A 2006 *Nat. Phys.* **2** 754–758
[20] Lloyd S 2013 *preprint arXiv:1307.0378*
[21] Goldstein S, Lebowitz J L, Tumulka R and Zanghì N 2006 *Phys. Rev. Lett.* **96** 050403
[22] Onsager L 1949 *Annals of the New York Academy of Sciences* **51** 627–659
[23] Asakura S and Oosawa F 1954 *J. Chem. Phys.* **22** 1255–1256
[24] Vedral V 2002 *Rev. Mod. Phys.* **74** 197–234
[25] Wolf M M, Verstraete F, Hastings M B and Cirac J I 2008 *Phys. Rev. Lett.* **100** 070502
[26] Lieb E H and Robinson D W 1972 *Comm. Math. Phys.* **28** 251–257
[27] Bhattacharya J, Nozaki M, Takayanagi T and Ugajin T 2013 *Phys. Rev. Lett.* **110** 091602
[28] Eisert J, Cramer M and Plenio M B 2010 *Rev. Mod. Phys.* **82** 277–306
[29] Zurek W H 2003 *Phys. Rev. A* **67** 012320
[30] Huang Y 2014 *New J. Phys.* **16** 033027

IOP Conf. Series: Journal of Physics: Conf. Series **880** (2017) 012048 doi:10.1088/1742-6596/880/1/012048

Model of the physical space from quantum mechanics

Otto C W Kong

Department of Physics and Center for Mathematics and Theoretical Physics, National Central University, Chung-li, Taiwan 32054

E-mail: otto@phy.ncu.edu.tw

Abstract. The physical world is quantum. However, our description of the quantum physics still relies much on concepts in classical physics and in some cases with 'quantized' interpretations. The most important case example is that of spacetime. We examine the picture of the physical space as described by simple, so-called non-relativisitic, quantum mechanics instead of assuming the Newtonian model. The key perspective is that of (relativity) symmetry representation, and the idea that the physical space is to be identified as the configuration space for a free particle. Parallel to the case of the phase space, we have a model of the quantum physical space which reduces to the Newtonian classical model under the classical limit. The latter is to be obtained as a contraction limit of the relativity symmetry.

1. Conceptual introduction

Newtonian mechanics is our first comprehensive model of a fundamental particle dynamics. Most of our theoretical conceptual notions in physics are tied to its formulation. As such, they are classical concepts. The physical world is quantum. Its description should be based on quantum concepts each of which may be very different from its classical counterpart, if there is at all a counterpart. Most if not all of our fundamental conceptual notions in physics as an intuitive, common sense, origin. However, such intuitive concepts do not fix their possible mathematical formulation which is an abstraction necessary for the analysis in a physical theory with precision. The formulation of such intuitive concepts into the concepts of classical physics is what physicists are familiar with, but not at all *a priori* more intuitive than their quantum counterparts. The unfortunate common miss-statement of quantum physics being less than intuitive has however hindered our appreciation of quantum physics with the more appropriate quantum formulation of the fundamental conceptual notions. Our model of space and time is the most central problem at hand. A common theme in the classical concepts is that the corresponding physical quantities, such as the position of a particle in space, have real number values. Quantum physics indicates otherwise, at least not necessarily by a finite number of real numbers. More generally, one should consider such quantities as having values modeled by some appropriate types of abstract algebraic entities. The system of real numbers as one of the latter has no preferred role apart from being our first order approximation to the physical nature as in the classical theory. We can summarize it all by saying that the main culprit in the limitation of the classical concept is the real number, which the philosopher W. Quine once called a 'convenient fiction'; with quantum physics, the old fiction is not so convenient any more.

Let us focus on the notion of the physical space as a collection of all possible positions, either of a free particle or equivalently as the center of mass for a system of particles. The first mathematical model is given by the three dimensional Euclidean geometry, which had been the only available model for centuries till beyond Newton's time. And it was of course adopted by Newton in his mechanics. Again, any mathematical model is beyond naive intuition. When, as it stand today, we have more than one such model, which one is the correct, or simply better, model for the physical space is a physics question. The answer is to be seek from matching the corresponding theories to experimental result. The limited success of Newtonian mechanics says that the Newtonian model may be not good enough. We are now familiar with the Einstein models, for his theories of special and general relativity. However, the Einstein models are still limited to real number geometry. Like the Newtonian model, they are classical (physical) models. We want to look at quantum models. A simple natural guess based on modern mathematics would be a kind of topological space which likely has some noncommutative geometry [1]. The idea of quantum geometry is not new in the subject matter related to deep microscopic structure of spacetime and quantum gravity. However, we present here a model of the physical space as behind simple quantum mechanics and illustrate its having the correct classical limit [2, 3].

At this point, some readers may feel uncomfortable about our suggestion of going beyond the notion of a three dimensional manifold calling it nothing counter-intuitive. The following comment is in order. The notion of a three 'dimensional' space, or four 'dimensional' spacetime, is no doubt quite intuitive. The Chinese term for the universe as a totality of all spacetime, for example, has two characters the first of which is the space part which literally means extending along the three principle axes. However, it is more tri-axial than three dimensional. Nothing in our intuition points towards identifying a definite position along any axis, or equivalently a distance with the abstract mathematical entity called a real number. One can see that the notion of being sort of tri-axial stays in our model.

2. The role of relativity symmetries

The 'particle' is the key concept for an ideal physical object in the Newtonian theory, one that still play a key role in simple quantum mechanics as a quantized version of the latter. To analyze the logic behind the Newtonian formulation, the first question is 'why particle?'. The particle is matter (with a single characteristic called mass) which has an unambiguous position in the Euclidean space. The Newton's laws are really definitions of notions like states (of motion), valid frames of reference, force, and mass (as inertia). Note that time is also modeled on an (one dimensional) Euclidean geometry. All the notions put together gives a model of physical phenomena the success of which obviously breaks down at the atomic scale and beyond. We argue that the modified theory of quantum mechanics has all the notions modified including those of position in space, though time is not touched so long as the 'non-relativistic' theory is concerned. We stick to the setting here.

In Ref.[2], we presented the perspective of taking the relativity symmetry as the starting point to look at the Newtonian model as well as formulating the quantum model and its classical limit. It is a perspective that is behind a big program aiming at eventually constructing the quantum spacetime model and its dynamics at the deep microscopic level [4, 5]. Symmetry is the single most important theoretical structure in modern physics. A relativity symmetry is the key symmetry behind a theory of mechanics. It is the group of reference frame transformation, and the symmetry of the spacetime model in the theory. The configuration space and phase space of a free particle are all natural homogeneous spaces as representations of the symmetry. The algebra of observables also carries a representation of it.

For the Newtonian theory, the relativity symmetry is given by the Galilei group. Newtonian space-time, as well as the single particle configuration and phase spaces can all be identified as coset spaces of the Lie group. The coset for the space-time is given by the quotient of the Galilei group $G(3)$ factored by the $ISO(3)$ subgroup of rotations and boosts. The corresponding case for the Einstein theory is the Minkowski spacetime $M^4 = ISO(1,3)/SO(1,3)$, the coset space of the Poincaré group factored by the Lorentz group. In fact, the Newtonian limit is exactly to be recovered by a contraction of $ISO(1,3)$ to $G(3)$ as $c \to \infty$ [6, 7]. What we have, as shown below, is much of a parallel picture for the quantum versus classical case. The action of $G(3)$ on the coset is given by the familiar expressions

$$\begin{pmatrix} t' \\ x'^i \\ 1 \end{pmatrix} = \begin{pmatrix} 1 & 0 & B \\ V^i & R^i_{\ j} & A^i \\ 0 & 0 & 1 \end{pmatrix} \begin{pmatrix} t \\ x^j \\ 1 \end{pmatrix} = \begin{pmatrix} t+B \\ V^i t + R^i_{\ j} x^j + A^i \\ 1 \end{pmatrix} , \tag{1}$$

and the more convenient infinitesimal form

$$\begin{pmatrix} dt \\ dx^i \\ 0 \end{pmatrix} = \begin{pmatrix} 0 & 0 & b \\ v^i & \omega^i_{\ j} & a^i \\ 0 & 0 & 0 \end{pmatrix} \begin{pmatrix} t \\ x^j \\ 1 \end{pmatrix} = \begin{pmatrix} b \\ v^i t + \omega^i_{\ j} x^j + a^i \\ 0 \end{pmatrix} . \tag{2}$$

Kinematical structure alone is enough to look at the configuration and phase spaces on which generic dynamics can be described with a time translation generated by a Hamiltonian with a nontrivial potential. Hence, we first restrict our analysis the subgroup, $G(3)$ with the time translations T taken out; denoted by $G(3)_s$. The Newtonian space can be identified with $G(3)_s/ISO(3)$ which is the same as $G(3)/ISO(3)\times T$. For convenience in matching with the familiar quantum picture, we use X_i generators in place of the boost generators $K_i = mX_i$. The Newtonian space then has $dx^i = \omega^i_{\ j} x^j + \bar{x}^i$, and for the phase space (p^i, x^i), we have also $dp^i = \omega^i_{\ j} p^j + \bar{p}^i$. The phase space is given by $G(3)_s/SO(3) = G(3)/SO(3)\times T$. The position and momentum coordinates x^i and p^i both transform as a three vector under rotations and have their own independent translations generated by P_i and X_i, respectively.

3. The quantum phase space is the physical space

The quantum symmetry differs from the classical one. We have the Heisenberg commutation relation which corresponds to having the Lie algebra to the Galilean symmetry modified by

$$[X_i, P_j] = i\delta_{ij} I \ , \tag{3}$$

where I is an extra central generator that commutes with all others. It is a bigger group $\tilde{G}(3)$ which is a $U(1)$ central extension of $G(3)$ [8]. Without the time translations, it is a Heisenberg-Weyl group supplemented with the rotations denoted by $H_R(3)$. In the setting, one has naively the space coset $H_R(3)/ISO(3)$ with

$$\begin{pmatrix} dx^i \\ d\theta \\ 0 \end{pmatrix} = \begin{pmatrix} \omega^i_j & 0 & \bar{x}^i \\ \bar{p}_j & 0 & \bar{\theta} \\ 0 & 0 & 0 \end{pmatrix} \begin{pmatrix} x^j \\ \theta \\ 1 \end{pmatrix} = \begin{pmatrix} \omega^i_j x^j + \bar{x}^i \\ \bar{p}_j x^j + \bar{\theta} \\ 0 \end{pmatrix} , \tag{4}$$

and the phase space coset $H_R(3)/SO(3)$ with

$$\begin{pmatrix} dp^i \\ dx^i \\ d\theta \\ 0 \end{pmatrix} = \begin{pmatrix} \omega^i_j & 0 & 0 & \bar{p}^i \\ 0 & \omega^i_j & 0 & \bar{x}^i \\ -\frac{1}{2}\bar{x}_j & \frac{1}{2}\bar{p}_j & 0 & \bar{\theta} \\ 0 & 0 & 0 & 0 \end{pmatrix} \begin{pmatrix} p^j \\ x^j \\ \theta \\ 1 \end{pmatrix} = \begin{pmatrix} \omega^i_j p^j + \bar{p}^i \\ \omega^i_j x^j + \bar{x}^i \\ \frac{1}{2}(\bar{p}_j x^j - \bar{x}_j p^j) + \bar{\theta} \\ 0 \end{pmatrix} . \tag{5}$$

Note that $H_R(3)$ is the central extension of $G(3)_s$ above. The funny cosets with a θ coordinate apparently do not serve our purpose. However, we know how to construct the quantum phase space from the second coset. That is the formulation of the Hilbert space of canonical coherent states [9].

IOP Conf. Series: Journal of Physics: Conf. Series **880** (2017) 012048 doi:10.1088/1742-6596/880/1/012048

The phase space coset is isomorphic to the Heisenberg-Weyl subgroup $H(3)$. A unitary representation of the latter is given by taking the linear span of the canonical coherent states each of which can be identified as a point in the coset. They are given by

$$e^{i\theta} \left| p^i, x^i \right\rangle = U(p^i, x^i, \theta) \left| 0 \right\rangle \tag{6}$$

where

$$U(p^i, x^i, \theta) \equiv e^{i\frac{x_i p^i}{2}} e^{i\theta \hat{I}} e^{-ix^i \hat{P}_i} e^{ip^i \hat{X}_i} = e^{i(p^i \hat{X}_i - x^i \hat{P}_i + \theta \hat{I})} , \tag{7}$$

and $|0\rangle \equiv |0,0\rangle$ is a fiducial normalized vector, $\hat{X}_i$ and $\hat{P}_i$ are representations of the generators X_i and P_i as Hermitian operators, and $\hat{I}$ is the identity operator representing the central generator I. As a basis, the set of coherent states is overcomplete. Wavefunction $\phi(p^i, x^i)$ for each of the state is a symmetry minimal uncertainty Gaussian centered at the state label, which correspond to expectation values rather than eigenvalues of the $\hat{P}_i$ and $\hat{X}_i$ observables. The operators generates translations of x^i and p_i exactly as described by the coset representation, with the nontrivial phase transformation. The $|0,0\rangle$ is the same as the ground state for a harmonic oscillator. The coherent states are like the classical state described in quantum mechanics.

What is particular interesting to note is that one can have the parallel construction starting from the (configuration) space coset of Eq.(4).The relevant subgroup is the one generated by P_i and I. We have a unitary representation as the span

$$e^{i\theta} \left| x^i \right\rangle = U'(x^i, \theta) \left| 0 \right\rangle \tag{8}$$

where

$$U'(x^i, \theta) \equiv e^{i\theta \hat{I}} e^{-ix^i \hat{P}_i} , \tag{9}$$

$|0\rangle$ is the fiducial normalized vector, and $\hat{P}_i$ and $\hat{I}$ the Hermitian operators generating translations in x^i and the phase rotation. Checking from the coset space action, one can see that the $|x^i\rangle$ states are position eigenstates. We have hence a Hilbert space as the one usually presented in basic quantum mechanics textbook, which is of course unitary equivalent to the previous one from the coherent states. It looks very much like it is the quantum space we have, and the quantum configuration space for a particle is the same as its phase space.

The real quantum phase space is not the Hilbert space but its projective counterpart. Each one dimensional subspace as a ray corresponds to a distinct physical state. The projective Hilbert space is known to have the structure of an infinite dimensional symplectic manifold and Kähler manifold [10]. We can look at it through an expansion of each state in terms of the Fock state basis as

$$|\phi\rangle = \sum (q_n + ip_n) |n\rangle \tag{10}$$

for real homogeneous coordinates q_n and p_n of the projective space. The Schrödinger equation

$$i\hbar \frac{d}{dt} |\phi\rangle = \hat{H} |\phi\rangle \tag{11}$$

for the state is actually equivalent to the set of Hamilton equations

$$\begin{aligned} \frac{d}{dt} q_n &= \frac{\partial}{\partial p_n} H(p_n, q_n) , \\ \frac{d}{dt} p_n &= -\frac{\partial}{\partial q_n} H(p_n, q_n) , \end{aligned} \tag{12}$$

IOP Conf. Series: Journal of Physics: Conf. Series **880** (2017) 012048 doi:10.1088/1742-6596/880/1/012048

with the Hamiltonian function $H(p_n, q_n) = \frac{2}{\hbar}\left\langle \phi|\hat{H}|\phi\right\rangle$. The basic fact is unfortunately much less appreciated than it should be. The projective Hilbert space also has a metric. It has no problem serving as a model of the physical space either. The countable infinite set of complex coordinates $q_n + ip_n$ are just $\langle n|\phi\rangle$. We can also think of $\phi(x^i) = \langle x^i|\phi\rangle$ as a set of complex coordinates. A functional analog of the Hamilton equations can be obtained for the real and imaginary parts of the function. The real and imaginary parts of each $\langle x^i|\phi\rangle$ then serves as a set of real coordinates. It is obviously essentially the same for the coherent state wavefunction $\phi(p^i, x^i) = \langle p^i, x^i|\phi\rangle$. There are uncountable infinite number of them, hence a set with much redundancy. The picture says a quantum wavefunction is a full and definite description of the position (and momentum) of the particle. The space of all such positions a free particle can possibly have, *i.e.* the full (projective Hilbert space is a model of the quantum space. Our familiar position operator $\hat{X}_i$ describe a notion of position which is obviously the one in the classical model. The quantum particle does not have a unique position in the classical model of space for the main reason that the latter fails as a good model of the space. We can also see that it is the phase transformations generated by the central charge that makes it no longer possible to have a reduced notion of the (configuration) space from the phase space. The Lagrangian submanifold of a symplectic manifold of a classical system, as the configuration space, can be taken as the real part of the symplectic manifold with a specific complex structure. A phase transformation that mixes the real and imaginary parts is a canonical transformation which mixes the configuration and momentum variable and not a symmetry transformation of the (configuration) space itself. With the quantum symmetry, we have phase transformations within the relativity symmetry. They are hence symmetry transformation of (configuration) space. The feature is essentially dictated by the Heisenberg commutation relation.

4. Classical limit as an approximation

We have our quantum model of the physical space. A nontrivial check that the idea is admissible is that it can explain the success of the corresponding classical picture in the proper limit. The latter is given by a symmetry contraction. More specifically, one should take the original representation which describes the quantum physics to the required limit rather than directly building the classical physics description from the contracted symmetry. We will see that the contraction of the representation does indeed give a representation of the contracted symmetry [4, 5]. Classical mechanics should be seen as such an approximation to quantum mechanics [11]. We require the quantum space model to reduce to the classical one as sort of the $\hbar \to 0$ limit through tracing the impact of the contraction on the Hilbert space representation(s) described. We are talking about the infinite dimensional curved space as the projective Hilbert space approximated by a finite dimensional Euclidean space. It works !

The symmetry contraction is given on the Lie algebra as the $k \to \infty$ with $X_i^c = \frac{1}{k}X_i$ and $P_i^c = \frac{1}{k}P_i$. Naively, one can take $\frac{1}{k^2}$ as $\hbar$. The nonzero commutator within $H(3)$, now in terms of X_i^c, P_i^c and I, goes as

$$[X_i^c, P_j^c] = \frac{i}{k^2}\delta_{ij} I \to 0 \,. \tag{13}$$

Hence, symmetry (sub)group, as well as the full $\tilde{G}(3)$ reduces to a trivial central extension of the classical symmetry. The I generator completely decoupled from the rest and becomes quite irrelevant. One can see that the transformation properties of the coset space representations in Eqs.(4) and (5) reduce to the classical analog as

$$H_R(3)/SO(3) \to G_s(3)/SO(3) \times U(1)$$

and

$$H_R(3)/ISO(3) \to G_s(3)/ISO(3) \times U(1) \,.$$

IOP Conf. Series: Journal of Physics: Conf. Series **880** (2017) 012048 doi:10.1088/1742-6596/880/1/012048

Explicitly, we have at the $k \to \infty$ limit

$$\begin{aligned} d\theta &= \bar{\theta} \,, \\ dx_c^i &= \omega_j^i x_c^j + \bar{x}_c^i \,, \\ dp_c^i &= \omega_j^i p_c^j + \bar{p}_c^i \,. \end{aligned} \tag{14}$$

For the coherent state picture, we first have the $|p^i, x^i\rangle$ states relabeled as $|\tilde{p}_i^c, \tilde{x}_i^c\rangle$ where $\tilde{p}_i^c = \sqrt{^-h}p$ and $\tilde{x}_i^c = \sqrt{^-h}x$ $(p_i = p^i,\ x_i = x^i)$. The new label correspond to the expectation values of $\hat{X}_i^c$ and $\hat{P}_i^c$. We have then

$$\begin{aligned} \langle \tilde{p}_i'^c, \tilde{x}_i'^c | \, \hat{X}_i^c \, | \tilde{p}_i^c, \tilde{x}_i^c \rangle &= \frac{(\tilde{x}_i'^c + \tilde{x}_i^c) - i(\tilde{p}_i'^c - \tilde{p}_i^c)}{2} \langle \tilde{p}_i'^c, \tilde{x}_i'^c | \tilde{p}_i^c, \tilde{x}_i^c \rangle \,, \\ \langle \tilde{p}_i'^c, \tilde{x}_i'^c | \, \hat{P}_i^c \, | \tilde{p}_i^c, \tilde{x}_i^c \rangle &= \frac{(\tilde{p}_i'^c + \tilde{p}_i^c) + i(\tilde{x}_i'^c - \tilde{x}_i^c)}{2} \langle \tilde{p}_i'^c, \tilde{x}_i'^c | \tilde{p}_i^c, \tilde{x}_i^c \rangle \,, \end{aligned} \tag{15}$$

with

$$\langle \tilde{p}_i'^c, \tilde{x}_i'^c | \tilde{p}_i^c, \tilde{x}_i^c \rangle = \exp\left[i \frac{\tilde{x}_i'^c \tilde{p}_i^c - \tilde{p}_i'^c \tilde{x}_i^c}{2\hbar} \right] \exp\left[-\frac{(\tilde{x}'^c - \tilde{x}^c)^2 + (\tilde{p}'^c - \tilde{p}^c)^2}{4\hbar} \right] , \tag{16}$$

$$\langle \tilde{p}_i^c, \tilde{x}_i^c | \tilde{p}_i^c, \tilde{x}_i^c \rangle = 1 \,. \tag{17}$$

The contraction result correspond to the $k \to \infty$ limit, hence $\hbar \to 0$. One can see that the last exponential factor hence the full result in Eq.(16) goes to 0. The basis states becomes mutually orthogonal. The equation above then say that the $\hat{X}_i^c$ and $\hat{P}_i^c$ operators become diagonal on the basis set. The conclusion there is that the original Hilbert space representation becomes a reducible representation. It reduces to a simple sun of the one dimensional spaces for the $|\tilde{p}_i^c, \tilde{x}_i^c\rangle$ states. But such rays of the original Hilbert space each correspond to a single state. They are the only states remain in the classical limit. All the nontrivial linear superpositions are eliminated. That is to say, the original projective Hilbert space reduces to the classical six dimensional phase space.

What about the physical space or the configuration space? We can actually do better than simply saying the latter is naturally a Lagrangian subspace of the (classical) phase space. We can start from looking at the $|x^i\rangle$ basis states we have in constructing the quantum Hilbert space as the model of the physical space and trace its contraction limit. The basis states are to be relabeled as $|x_i^c\rangle$, *i.e.* by eigenvalues of the $\hat{X}_i^c$. Now, $\hat{P}_i^c$ and hence all operators, say, as polynomials of $\hat{X}_i^c$ and $\hat{P}_i^c$ commute. We have the same conclusion for the reduction of the Hilbert space now seeing as going to the simple sum of the rays of $|x_i^c\rangle$. The projective Hilbert space is of course the three dimensional Euclidean space.

5. Concluding remarks

In physics, we are supposed to learn from experiments what constitutes a good/correct theoretical/mathematical model of any physical concept, and physical space should not be an exception. As physics advances, we see our understanding of all the intuitive concepts keeps deepening. The abstract mathematical contents to be given to such concepts changes as we replace a fundamental theory with a better one. Spacetime, or space and time, is the most basic concepts in our all discourse on physics. Einstein's relativity theories advanced our understanding of spacetime. Looking at it from the proper point of view, quantum mechanics does the same. In a way, it would be quite surprising if it does not. Our talk here offers exactly such a perspective.

A notion about spacetime being somehow emergent rather than fundamental has been getting popular lately. In many cases, it is not clear what the author(s) exactly meant about that. The way we see it: The familiar classical picture and all the related features are almost definitely not at all fundamental. They are like our classical approximate description of spacetime. It should be quite clear nowadays that the deep microscopic picture of spacetime has to be very different. We sure expect any good model of quantum spacetime model to give back the classical model as an approximation under proper limits. As such, one can certainly call the classical model a emergent one. However, to have a fundamental theory of physics built completely without some notion of spacetime in it, we think, would be quite impossible. There has to be some dynamical variables, whatever unfamiliar mathematical objects needed to described them, in the theory. For the part of such variables necessary to give a picture of the spacetime as we have in the known classical theories at the proper limit, it is only fair to call the collection of all their admissible values in the most general physical setting the spacetime model of the theory.

We would also like to point out that quantum field theory understood correctly gives another very interesting perspective about spacetime, a feature we expect to stay as a part of any model of the deep microscopic (quantum) spacetime. Quantum field theory says that spacetime is the only physical entity. The various quantum fields are not independent dynamical 'objects' but only dynamic degrees of freedom the spacetime has. Any state, says the state which corresponds to a single muon, involves all the degrees of freedom. With enough precision, we can see the role of any degree of freedom in any other basic properties of the state. The magnetic dipole moment of the muon has been used to reveal information about possible superparticles — postulated new quantum fields the discovery of which is one of the key target of the Large Hadron Collider experiments. We know already from quantum entanglement that the full system is more than the sum of its parts. There are no truly independent subsystems.

The above leads naturally to the question of gravitation or quantum gravity. We believe *'Noncommutative Geometry is to Quantum Gravity as NonEuclidean Geometry is to Gravity'*. One should build quantum gravity as a *geometrodynamics of quantum spacetime*. Our quantum space model presented here is of course very far from the final quantum spacetime model. We need to first go from the $\tilde{G}(3)$ setting to one that incorporates Poincaré symmetry. Actually, we have a basic perspective which goes beyond that to a full stable symmetry that cannot be the contraction limit of another symmetry [4]. Our basic stable symmetry have completely noncommutative X_μ and P_μ. We need also to see how the above perspective of quantum field theory. We need full dynamical pictures which includes analysis of the corresponding algebras of observables [11]. We have an exciting long way to go.

Let us have a last word on basic quantum mechanics. Starting with our quantum space model, quantum mechanics can be seen from a very different light. For instance, it begs the question of why not sees as physical observables all functions of the (quantum) position and momentum, as coordinates of the quantum space/phase space. The usual notion of physical observables as essentially functions as operator $\hat{X}_i$ and $\hat{P}_i$ may be too limited. They are only observables which have a clear classical analog. And as such, there is no reason to restrict possible knowledge about the operator/observables on a state to their expectation values either. All the other moments of the distributions are conceptually as measurable as the expectation values. On a complementary point of view, the noncommutative algebra of quantum observables (operators) can be thought of as all continuous function $C(X)$ of a topological space X is the general setting of noncommutative geometry [2]. Thinking about the quantum space as having noncommutative coordinates, essentially $x_i\star$ and $p_i\star$ (kind of $\hat{X}_i$ and $\hat{P}_i$) [11] is an alternative picture the infinite number of real number coordinates here. We are familiar with thinking about physical quantities as real number valued. That is no more than an assumption. The notion of physical quantities as having values in some other algebraic entity may be the way to go. We are familiar with the Bohr argument about all measurement being classical. That has to be the case only if we restrict results of measurements to be real numbers. A generic measurement should only to a process to extract information from a physical system. The system being quantum says that we may want to extract from it quantum information rather than the familiar classical one. We are still learning to manipulate quantum information well. But we can do that, we can deal with true quantum measurements.

Dealing with physical quantities as beyond real number values and quantum space(time) as beyond real number geometries is like a new Copernican revolution for our generation to push on!

Acknowledgments
The author is partially supported by research grants MOST 105-2112-M-008-017- from the MOST of Taiwan.

References

[1] Connes A 1994 *Noncommutative Geometry* (Academic Press);
van Suijlekom W D 2015 *Noncommutative Geometry and Particle Physics* (Springer)
[2] The results discussed here are based on Chew C S, Kong O C W and Payne J 2016 *Preprint NCU-HEP-k066, arXiv:1603.05371 [quant-ph]*
[3] A conceptual discussion in a boarder context has been presented in the earlier article in Chinese — Kong O C W 2015 *General Relativity : A Centennial Perspective* (Taiwan: Locus Publishing) p 185
[4] Kong O C W 2008 *Phys. Lett. B* **665** 58;
see also Kong O C W 2008 *Preprint arXiv:0906.3581 [gr-qc]*
[5] Cho D N and Kong O C W 2014 *Ann. Phys.* **351** 275
[6] Gilmore R 2005 *Lie Groups, Lie Algebras, and Some of Their Applications* (Dover);
Tung W K 1985 *Group Theory in Physics* (World Scientific)
[7] For a detailed pedagogical presentation, see Kong O C W and Payne J *Preprint NCU-HEP-k067*
[8] de Azcárraga J A and Izquierdo J M 1995 *Lie Groups, Lie Algebras, Cohomology and Some Applications in Physics* (Cambridge University Press)
[9] Peremolov A M 1977 *Sov. Phys. Usp.* **20** 103;
Zhang W-M, Feng D H and Gilmore R 1990 *Rev. Mod. Phys.* **62** 867;
Klauder J R and Sudarshan E C 1968 *Fundamentals of Quantum Optics* (Benjamin);
Klauder J R 2010 *A Modern Approach to Functional Integration* (Birkhäuser)
[10] Bengtsson I and Życzkowski K 2006 *Geometry of Quantum States* (Cambridge University Press)
[11] Chew C S, Kong O C W and Payne J 2017 *Preprint NCU-HEP-k070*, in preparation, gives a full presentation of the dynamical picture

IOP Conf. Series: Journal of Physics: Conf. Series **880** (2017) 012049 doi:10.1088/1742-6596/880/1/012049

Scale invariant non linear sigma model at finite temperatures

Marcus Benghi Pinto[1], Gabriel N Ferrari[1], Jean-Loïc Kneur[2] and Rudnei O Ramos[3]

[1] Departamento de Física, Universidade Federal de Santa Catarina, 88040-900 Florianópolis, Santa Catarina, Brazil

[2] Laboratoire Charles Coulomb (L2C), UMR 5221 CNRS-Université de Montpellier 2, 34095 Montpellier, France

[3] Departamento de Física Teórica, Universidade do Estado do Rio de Janeiro, 20550-013 Rio de Janeiro, RJ, Brazil

E-mail: marcus.benghi@ufsc.br

Abstract. The recently proposed renormalization group improved optimized perturbation theory is employed to evaluate the pressure of the two dimensional non linear sigma model at finite temperatures. We explicitly show how this powerful resummation method can turn the lowest (one loop) perturbative contribution to the pressure, which is not RG invariant, into a non perturbative quantity exhibiting scale invariance .

1. Introduction

The development of reliable approximation techniques to solve the equations of motion describing a physical system has always been considered a problem of fundamental importance in all branches of physics since most realistic situations are described by non linear interactions. In practice this means that an exact solution may not be easily attainable and theoretical predictions require that the evaluation of a physical quantity be performed within an appropriate approximation framework. Sometimes, when a problem cannot be exactly solved one can attempt to separate it into two parts with one of them representing an exactly solvable case. Then, if the underlying dynamics allows, one can treat the other piece as a "perturbation" and recur to the well established perturbation theory (PT) to approximately solve the problem by successively evaluating the terms which appear in a series written in powers of a (small enough) parameter which characterizes the perturbative interaction. Being relatively easy to implement, PT has been successfully employed to treat various widely different problems. However, the same approximation does not have the same capability in the case of quantum chromodynamics (QCD), which is the fundamental theory of strong interactions, since in this case the interaction parameter ($4\pi\alpha_s$) is not small at the energy scales relevant for nuclear physics. Today, the development of powerful computers offers the possibility to solve these non perturbative problems by employing the numerical methods of the so-called lattice field theory (LFT), which has been very successful in the description of the QCD phase transitions at finite temperatures and near vanishing baryonic densities [1]. However, the numerical sign problem, which arises at finite chemical potentials, has not yet been completely solved, preventing that the method be successfully used to describe the complete QCD phase diagram. A possible analytical alternative is to employ a variational approximation where the result of a related solvable case is rewritten in terms of a variational parameter. This strategy is adopted in the optimized perturbation theory (OPT) [2, 3] where, for scalar theories, a harmonic term $(1-\delta)m^2$ is added to the potential energy density while the anharmonic coupling, λ, is rescaled as $\lambda \to \delta\lambda$ with δ being formally treated as a (small) interaction parameter. Having evaluated a physical quantity to the desired order in δ one sets $\delta = 1$, and fixing the mass parameter, m, in a variational fashion. The OPT has been successfully used in physical situations relevant for hadronic [4] and condensed matter physics [5, 6, 7], while a similar technique, known as screened perturbation theory (SPT) [8], has been extended in order to cope with gauge theories. The resulting approximation, which provides a resummation of hard thermal loop perturbation theory (HTLpt) [9], has already been applied to QCD. Unfortunately, all these variational methods predict that scale invariance is not observed by thermodynamical physical observables such as the pressure, as many applications to scalar theories show [8, 10]. In the case of QCD the results obtained in Ref. [9] show a good agreement between HTLpt and LFT simulations at the "central" arbitrary renormalization scale value, $M = 2\pi T$, but the agreement is lost when the scale is varied even by a moderate amount. An alternative, combining OPT with renormalization group (RG) invariance, has been recently proposed [11]. This technique, known as renormalization group optimized perturbation theory (RGOPT), was originally used within the framework of the Gross-Neveu model and QCD at vanishing temperatures. More recently, the RGOPT has been extended to the finite temperature domain in an application to the scalar $\lambda\phi^4$ theory producing encouraging results [12]. Here, we illustrate how the method works by evaluating the pressure of the non linear sigma model (NLSM) in 1+1 dimensions which, despite of being a simple model, shares some common features with QCD such as asymptotic freedom, trace anomaly and the generation of a mass gap. At finite temperatures this model has been first analyzed in Ref. [13] and more recently in Refs. [14, 15]. For our present purpose we only need to compute the lowest order (one loop) contribution to the pressure. A more complete discussion (up to the two-loop level) will be given elsewhere [16]. The work is organized as follows. In the next section we recall the two dimensional NLSM, presenting the one loop perturbative result in Sec. 3, while the RGOPT is applied in Sec. 4 and some numerical results illustrated in Sec. 5. Our conclusions are given in Sec. 6.

2. The NLSM in 1+1-dimensions

The two-dimensional NLSM Lagrangian density, for π_i fields with $N-1$ components, can be written as [17]

$$\mathcal{L}_0 = \frac{1}{2}(\partial\pi_i)^2 + \frac{g_0(\pi_i\partial\pi_i)^2}{2(1-g_0\pi_i^2)} - \frac{m_0^2}{g_0}\left[\left(1-g_0\pi_i^2\right)^{1/2} - 1\right], \tag{1}$$

where the last parcel represents a constant and m_0 represents a mass parameter whose physical role will be specified later. Expanding he above theory to zeroth-order yields

$$\mathcal{L}_0 = \frac{1}{2}\left[(\partial\pi_i)^2 + m_0^2\pi_i^2\right] + \mathcal{E}_0 + \mathcal{O}(g_0), \tag{2}$$

where we have introduced a field independent term, $\mathcal{E}_0$, which corresponds to an infinite renormalization of the "zero-point" energy [3].

Within the imaginary time formalism the propagator is $1/(\omega_n^2 + \mathbf{p}^2 + m^2)$ where $\omega_n = 2\pi n T$ represent the bosonic Matsubara frequencies ($n = 0, \pm 1, \pm 2 \cdots$) and T is the temperature. In this work, the divergent integrals are regularized using dimensional regularization (within the minimal subtraction scheme $\overline{\rm MS}$) which, at finite temperature, and $d = 2-\epsilon$ dimensions can be easily implemented by using [18]

$$\int \frac{d^2p}{(2\pi)^2} \to T \sum\!\!\!\!\!\!\!\int_{\mathbf{p}} \equiv T \left(\frac{e^{\gamma_E} M^2}{4\pi} \right)^{\epsilon/2} \sum_{n=-\infty}^{+\infty} \int \frac{d^{1-\epsilon}\mathbf{p}}{(2\pi)}, \tag{3}$$

where γ_E is the Euler-Mascheroni constant and M is the $\overline{\rm MS}$ arbitrary regularization energy scale. Before performing the finite temperature evaluations we recall that in two-dimensions no spontaneous symmetry breaking of the global $O(N)$ symmetry can take place (at any coupling value) [19].

3. Perturbative pressure and scale invariance

If one considers the (one-loop) zeroth-order contribution to the pressure only the free gas type of term contributes yielding [13]

$$P = -\frac{(N-1)}{2} I_0(T) + \mathcal{E}_0, \tag{4}$$

where

$$I_0(T) = T \sum\!\!\!\!\!\!\!\int_{\mathbf{p}} \ln\left(\omega_n^2 + \omega_{\mathbf{p}}^2\right), \tag{5}$$

with the dispersion $\omega_{\mathbf{p}}^2 = \mathbf{p}^2 + m^2$.

Then, performing the sum over the Matsubara's frequencies within the $\overline{\rm MS}$ scheme one obtains

$$I_0(m,T) = \frac{m_0^2}{2\pi} \left\{ \frac{1}{\epsilon} + \left[\frac{1}{2} - \ln\left(\frac{m}{M}\right) \right] \right\} + T^2 \frac{2}{\pi} J_0(m/T). \tag{6}$$

In the above expression the thermal integral, $J_0(T)$, reads

$$J_0(m/T) = \int_0^\infty dz \ln\left(1 - e^{-\omega_z}\right), \tag{7}$$

where we have defined the dimensionless quantity $\omega_z^2 = z^2 + y^2$ with $z = |\mathbf{p}|/T$ and $y = m/T$. Then, to this lowest perturbative order the renormalized pressure can be obtained by setting $m_0 = m$, $g_0 = g$, where m is renormalized mass and g is the renormalized coupling. The only divergence can be eliminated by the zero point subtraction term, $\mathcal{E}_0 = (N-1)m^2/(4\pi\epsilon)$, so that the finite pressure is simply

$$P = -\frac{(N-1)}{2} I_0^{\rm r}(T), \tag{8}$$

where

$$I_0^{\rm r}(m,T) = \frac{m^2}{2\pi} \left[\frac{1}{2} - \ln\left(\frac{m}{M}\right) \right] + T^2 \frac{2}{\pi} J_0((\frac{m}{T})). \tag{9}$$

Next, let us consider the complete renormalization group (RG) operator defined by

$$M\frac{d}{dM} \equiv M\frac{\partial}{\partial M} + \beta\frac{\partial}{\partial g} - \frac{n}{2}\zeta + \gamma_m m\frac{\partial}{\partial m}. \tag{10}$$

Since the pressure is represented by a zero point Green's function ($n = 0$) we only need to consider the β and γ_m functions at the one-loop level [20] $\beta = -b_0 g^2$ and $\gamma_m = -\gamma_0 g$ where we have defined the RG coefficients in our normalization: $b_0 = (N-2)/(2\pi)$ and $\gamma_0 = (N-3)/(8\pi)$.

Let us now apply the RG operator to the zeroth-order pressure. Since β and γ_m are at least of order-g and $J_0(T)$ is scale independent, one gets

$$M\frac{dP}{dM} = -(N-1)\frac{m^2}{4\pi}\,, \tag{11}$$

which explicitly shows that the one loop pressure is not scale-invariant. Following Refs. [11, 12] one can fix this problem by adding a finite field independent term $m^2/g\sum s_k g^k$, where the coefficient s_k is fixed so that the pressure becomes RG invariant as we now demonstrate. Employing this RGOPT prescription one can then write the zeroth-order pressure as

$$P = -\frac{(N-1)}{2} I_0^{\rm r}(T) + \frac{m^2}{g} s_0\,. \tag{12}$$

Then, the RG equation, $MdP/dM = 0$, leads to

$$-(N-1)\frac{m^2}{4\pi} + \left(\beta\frac{\partial}{\partial g} + \gamma_m m\frac{\partial}{\partial m}\right)\frac{m^2 s_0}{g} = 0\,, \tag{13}$$

fixing $s_0 = 1$. In the next section we show how this RG invariant perturbative pressure can be interpolated in order to generate non perturbative results.

4. RG improved optimized perturbation theory

The next step is to perform the replacements [11, 12] $m \to (1-\delta)^a m$ and $g \to \delta g$, into the renormalized and RG invariant perturbative pressure that was obtained in the previous section and then fix the RGOPT exponent, a. It is important to note that within the standard OPT this parameter is fixed in an *ad hoc* way (for example, $a = 1/2$ in the case of scalar theories [2, 3, 10]). Note also that in the case of massless theory, such as the case studied here, m can also be interpreted as an infra red regulator. To fix the arbitrary mass parameter the RGOPT adopts the same variational criterion used within the standard OPT. Namely [21],

$$\frac{\partial P^{\rm RGOPT}}{\partial m}|_{m=\bar{m}} = 0\;, \tag{14}$$

which implies that the RG operator is *reduced* to

$$\left(M\frac{\partial}{\partial M} + \beta\frac{\partial}{\partial g}\right) P^{\rm RGOPT} = 0\,. \tag{15}$$

Performing the replacements, re-expanding to the zeroth order and using $s_0 = 1$, one can finally write

$$P^{\rm RGOPT} = -\frac{(N-1)}{2} I_0^{\rm r}(T) + \frac{m^2}{g}(1-2a). \tag{16}$$

Next,to fix the RGOPT exponent a one requires that the pressure given by Eq. (16) satisfies the *reduced* RG relation, Eq. (15), obtaining

$$a \equiv \frac{\gamma_0}{b_0} = \frac{(N-3)}{4(N-2)}, \tag{17}$$

which is also the value found in the context of the scalar $\lambda\phi^4$ theory [12]. Using $a = \gamma_0/b_0$ and the definitions for b_0 and γ_0 one can finally write

$$P^{\rm RGOPT} = -\frac{(N-1)}{2} I_0^{\rm r}(T) + (N-1)\frac{m^2}{(4\pi)g(M)b_0}. \tag{18}$$

where, at the one loop level, the running coupling is given by

$$g(M) = g(M_0)\left[1 + g(M_0) b_0 \ln\left(\frac{M}{M_0}\right)\right]^{-1} . \tag{19}$$

Before presenting some numerical results let us quickly illustrate how the RGOPT is able to produce non perturbative results already at this lowest order. Applying the optimization criterion, Eq. (14), to the RGOPT pressure, Eq. (18), yields a non trivial gap equation for $\bar{m}$ which, at $T=0$, leads to following the result for $\bar{m}$,

$$\bar{m}(0) = M \exp\left(-\frac{1}{g(M) b_0}\right) . \tag{20}$$

Moreover, as usual [22] the large-N result can be exactly reproduced when the $N \to \infty$ limit is appropriately taken within our formalism.

5. Numerical results

In order to make contact with the usual SPT/HTLpt results [8, 9] let us first define a temperature dependent scale, $M = 2\pi T\,\alpha$, where the case $\alpha = 1$ determines the "central" value. To obtain the pressure at high temperatures one can consider the high-T expansion for $J_0(T)$ [18]

$$J_0(m/T) = -\frac{\pi^2}{6} + \frac{\pi m}{2T} + \left(\frac{m}{2T}\right)^2 \left[\ln\left(\frac{m e^{\gamma_E}}{4\pi T}\right) - \frac{1}{2}\right] + \mathcal{O}(m^4/T^4) , \tag{21}$$

Then, using the Stefan-Boltzmann result for the NLSM [14], $P_{SB} = (N-1)\pi T^2/6$, one obtains

$$\frac{P^{\rm RGOPT}(\alpha)}{P_{SB}} = 1 - 6\hat{m}(\alpha) - 6\hat{m}^2(\alpha)\left[L_T(\alpha) - \frac{1}{b_0 g(\alpha,T)}\right] + O(\hat{m}^4) , \tag{22}$$

where, again in accordance with Refs. [8, 9], we have defined $\hat{m} = m/(2\pi T)$, $L_T(\alpha) = \ln(\hat{M}(\alpha) e^{\gamma_E}/2)$, and $\hat{M}(\alpha) = M/(2\pi T) = \alpha$. Setting $M_0 = 2\pi T_0$, where T_0 is a reference temperature one easily obtains $g(\alpha, T)$ from Eq. (19). Next, the optimized mass can be easily obtained by applying Eq. (14) to Eq. (22) which gives

$$\hat{m}(\alpha) = \frac{1}{2}\left(\frac{1}{b_0 g(\alpha,T)} - L_T(\alpha)\right)^{-1} . \tag{23}$$

Note that using the previous optimized mass gap solution, Eq. (23), within Eq. (22), the latter takes a much simpler expression (in the high-T limit here considered):

$$\frac{P^{\rm RGOPT}(\alpha)}{P_{SB}} = 1 - 3\hat{m}(\alpha) + O(\hat{m}^4) . \tag{24}$$

In order to compare the RGOPT and ordinary PT results, one can obtain the latter directly from Eq. (8) in the massless limit, which after straightforward algebra gives

$$\frac{P^{\rm PT}(\alpha)}{P_{SB}} = 1 - \frac{3}{2}\gamma_0\, g(\alpha, T). \tag{25}$$

On the other hand, the purely perturbative thermal "screening" mass at one-loop order can be obtained starting from the self-energy [17]:

$$\Gamma^{(2)}(p) = p^2(1 + g_0 I_1) + m_0^2\left[1 + \frac{(N-1)}{2} g_0 I_1\right] + \mathcal{O}(g^2) , \tag{26}$$

IOP Conf. Series: Journal of Physics: Conf. Series **880** (2017) 012049 doi:10.1088/1742-6596/880/1/012049

where I_1 is the basic (Euclidean) one-loop integral, in $\overline{\rm MS}$ renormalization scheme,

$$I_1 \equiv T \sum\!\!\!\!\!\!\!\int_{\mathbf{p}} \frac{1}{\omega_n^2 + \mathbf{p}^2 + m^2} = \frac{1}{2\pi}\left(\frac{1}{\epsilon} - \ln\frac{m}{M} - 2J_1(m/T)\right), \tag{27}$$

with the thermal part $J_1(m/T)$ having the high-T expansion (compare Eq. (21))

$$J_1(m/T) \equiv -\frac{T^2}{m}\frac{\partial J_0(\frac{m}{T})}{\partial m} \simeq -\frac{\pi}{2}\frac{T}{m} - \frac{1}{2}\ln\left(\frac{m\,e^{\gamma_E}}{4\pi T}\right) + \mathcal{O}(m^2/T^2). \tag{28}$$

Taking thus the pole mass $p^2 \equiv -m_D^2$ in Eq. (26) after mass renormalization, gives, in the massless limit $m \to 0$ relevant for the pure thermal one-loop mass:

$$m_D(\alpha) = \frac{N-3}{8}\,g(\alpha,T)\,T \equiv \pi\,\gamma_0\,g(\alpha,T)\,T, \tag{29}$$

which may be compared with the RGOPT nonperturbative one-loop result (23).

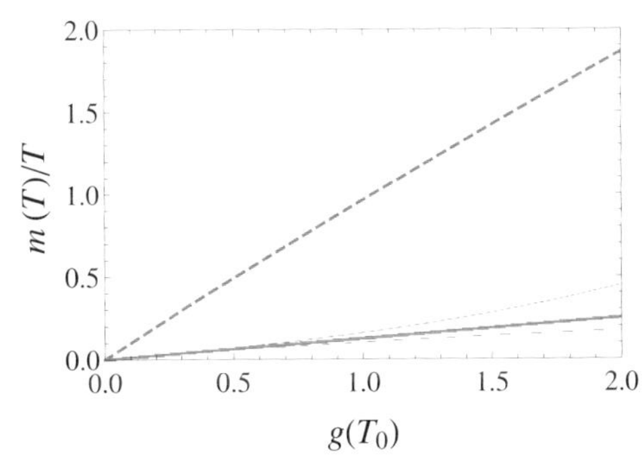

Figure 1. The thermal mass $m(T)/T$ for fixed $T = T_0$ as a function of the reference coupling, $g(T_0)$ for $\alpha = 0.5, 1$, and 2. Dashed line (RGOPT), upper continuous line (PT at $\alpha = 0.5$), middle continuous line (PT at $\alpha = 1$) and lower continuous line (PT at $\alpha = 2$).

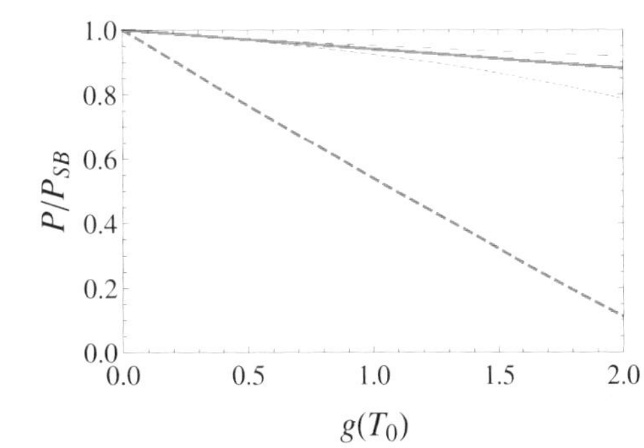

Figure 2. Normalized pressure P/P_{SB} as a function of $g(T_0)$ at $T = T_0$ for $\alpha = 0.5, 1$, and 2. Dashed line (RGOPT), upper continuous line (PT at $\alpha = 2$), middle continuos line (PT at $\alpha = 1$) and lower continuous line (PT at $\alpha = 0.5$).

Choosing $N = 4$, $\alpha = 0.5, 1, \text{and}\, 2$ we can now investigate different thermal quantities. Fig. 1 shows the PT and RGOPT (thermal) masses $m(T)/T$ for fixed $T = T_0$ as a function of the reference coupling, $g(T_0)$. The RGOPT mass displays exact scale invariance, in contrast to the PT mass (29). Next, Fig. 2 shows P/P_{SB} as a function of $g(T_0)$, also illustrating the exact scale invariance of the RGOPT pressure, and exhibiting a behavior similar to the one found for the $\lambda\phi^4$ model [12]. In particular, one should not be surprised by the different behavior (slope) as function of the coupling, with respect to PT one-loop results, of the RGOPT mass and pressure in Figs. 1 and 2 respectively. First, in contrast with the physical one-loop Debye (pole) mass, the optimized mass (23) is only an intermediate unphysical quantity in the pressure optimization procedure. Second, while comparing the different pressures as function of the running coupling Eq. (19) is sensible as concerns the scale dependence, at this stage the coupling value is not a physical quantity, being not fixed at some well-defined input scale and therefore essentially arbitrary in both PT and RGOPT cases. However, the overall physical consistency of our RGOPT pressure result can be checked by noting that, when the *arbitrary* mass in Eq. (24) is replaced with the *physical* thermal mass m_D Eq. (29), one consistently recovers the standard perturbative PT pressure as function of the coupling Eq. (25) (see Ref. [12] for a detailed discussion of similar results for the scalar ϕ^4 model).

Finally Fig. 3 shows P/P_{SB} as a function of T/T_0 for the fixed reference coupling, $g(T_0) = 1$, clearly displaying again the exact scale invariance of the nonperturbative RGOPT result. Although it is not obvious from the figure, note that the RGOPT pressure tends consistently towards the Stefan-Boltzmann limit for $T/T_0 \to \infty$, while the PT pressure reaches this limit more rapidly. This difference for a given reference coupling, and in particular the fact that the RGOPT pressure is substantially smaller than PT for moderate and small T/T_0, indicates that the RGOPT captures (resums) more essentially nonperturbative content.

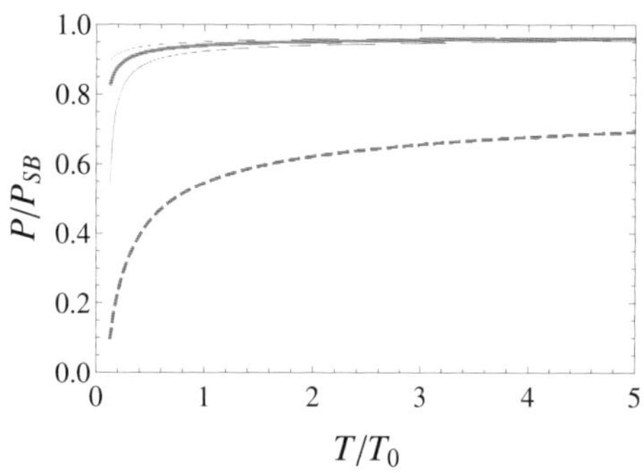

Figure 3. P/P_{SB} as a function of T/T_0 for the fixed reference coupling, $g(T_0) = 1$, $\alpha = 0.5, 1$, and 2. Dashed line (RGOPT), upper continuous line (PT at $\alpha = 0.5$), middle continuous line (PT at $\alpha = 1$) and lower continuous line (PT at $\alpha = 2$)

6. Conclusions

Using a simple model which shares common features with QCD, such as asymptotic freedom, we have briefly described how the RGOPT technique can generate scale invariant non perturbative results from the lowest order contribution to the pressure. As we have shown, within this framework one starts by requiring that the ordinary "free gas" type of contribution be RG invariant, which implies that an extra (field independent) term should be added to the action. Next, the Lagrangian density is deformed by the interpolating term, $(1-\delta)^a m$, and the arbitrary parameters, a and m, fixed by combining the RG equations with an optimization variational criterion. Circumventing the severe scale dependence issues observed within related variational methods (such as OPT, SPT, and HTLpt) the RGOPT method described here stands as a new potential alternative to LFT concerning the description of compressed strongly interacting matter.

Acknowledgments

MBP thanks the DICE 2016 organizers for the invitation. This work was partially supported by CNPq, CAPES and FAPERJ.

References

[1] Aoki Y, Endrodi G, Fodor Z, Katz S D and Szabo K K 2006 *Nature* **443** 675
Aoki Y, Fodor Z, Katz S D and Szabo K K 2006 *Phys Lett* B **643** 46
[2] Okopińska A 1987 *Phys Rev* B **35** 1835
Duncan A and Moshe M 1988 *Phys Lett* B **215** 352
[3] Chiku S and Hatsuda T 1998 *Phys Rev* D **58** 076001
[4] Kneur J L, Pinto M B and Ramos R O 2010 *Phys Rev* C **81** 065205
Ferroni L, Koch V and Pinto M B 2010 *Phys Rev* C **82** 055205

IOP Conf. Series: Journal of Physics: Conf. Series **880** (2017) 012049 doi:10.1088/1742-6596/880/1/012049

Restrepo T E, Macias J C, Pinto M B and Ferrari G N 2015 *Phys Rev* D **91** 065017
[5] Kleinert H 2003 *Mod Phys Lett* B **17** 1011
Kastening B 2003 *Phys Rev* A **68** 061601
Kneur J L, Neveu A and Pinto M B 2004 *Phys Rev* A **69** 053624
[6] Caldas H, Kneur J L, Pinto M B and Ramos R O 2008 *Phys Rev* B **77** 205109
[7] Kneur J L, Pinto M B and Ramos R O 2013 *Phys Rev* D **88** 045005
[8] Karsch F, Patkos A and Petreczky P 1997 *Phys Lett* B **401** 69
Andersen J O, Braaten E and Strickland M 2001 *Phys Rev* D **63** 105008
Andersen J O and Strickland M 2005 Ann Phys **317** 281
[9] Andersen J O, Braaten E and Strickland M 1999 *Phys Rev Lett* **83** 2139
Andersen J O, Leganger L E, Strickland M and Su N 2011 *JHEP* **1108** 05
Haque N, Bandyopadhyay A, Andersen J O, Mustafa M G, Strickland M and Su N 2014 *JHEP* **1405** 027
[10] Farias R L S, Krein G and Ramos R O 2008 *Phys Rev* D **78** 065046
[11] Kneur J L and Neveu A 2010 *Phys Rev* D **81** 125012
Kneur J L and Neveu A 2012 *Phys Rev* D **85**, 014005
Kneur J L and Neveu A 2013 *Phys Rev* D **88** 074025
Kneur J L and Neveu A 2015 *Phys Rev* D **92** 074027
[12] Kneur J L and Pinto M B 2015 *Phys Rev* D **92** 116008
Kneur J L and Pinto M B 2016 *Phys Rev Lett* **116** 031601
[13] Dine M and Fischler W 1981 *Phys Lett* B **105** 207
[14] Seel E, Smith D, Lottini S and Giacosa F 2013 *JHEP* **1307** 010
[15] Andersen J O, Boer D and Warringa H J 2004 *Phys Rev* D **69** 076006
[16] Ferrari G N, Kneur J L, Pinto M B and Ramos R O *in preparation*
[17] Zinn-Justin J 2002 *Quantum field theory and critical phenomena* (Oxford: Clarendon Press)
[18] Kapusta J I 1985 *Finite-Temperature Field Theory* (Cambridge: University Press)
[19] Mermin N D and Wagner H 1966 *Phys Rev Lett* **17** 1133
[20] Hikami S and Brézin E 1978 *J Phys* A **11** 1148
[21] Stevenson P M 1981 *Phys Rev* D **23** 2916
[22] Gandhi S K, Jones H F and Pinto M B 1991 *Nucl Phys* B **359** 429

IOP Conf. Series: Journal of Physics: Conf. Series **880** (2017) 012050 doi:10.1088/1742-6596/880/1/012050

Connections between 't Hooft's beables and canonical descriptions of dissipative systems

Dieter Schuch[1] **and Massimo Blasone**[2]

[1] Institut für Theoretische Physik, J.W. Goethe-Universität Frankfurt am Main, Max-von-Laue-Str. 1, D-60438 Frankfurt am Main, Germany

[2] Dipartimento di Fisica, Università di Salerno, Via Giovanni Paolo II, 132, 84084 Fisciano (SA), Italy and Istituto Nazionale di Fisica Nucleare (INFN), Gruppo Collegato di Salerno

E-mail: Schuch@em.uni-frankfurt.de,mblasone@unisa.it

Abstract. According to a proposal by 't Hooft, information loss introduced by constraints in certain classical dissipative systems may lead to quantization. This scheme can be realized within the Bateman model of two coupled oscillators, one damped and one accelerated. In this paper we analyze the links of this approach to effective Hamiltonians where the environmental degrees of freedom do not appear explicitly but their effect leads to the same friction force appearing in the Bateman model. In particular, it is shown that by imposing constraints, the Bateman Hamiltonian can be transformed into an effective one expressed in expanding coordinates. This one can be transformed via a canonical transformation into Caldirola and Kanai's effective Hamiltonian that can be linked to the conventional system-plus-reservoir approach, for example, in a form used by Caldeira and Leggett.

1. Introduction

In the attempt to find a deterministic description of quantum systems, 't Hooft[1]-[4] proposed the idea that deterministic degrees of freedom could operate at very high energy scales (e.g. Planck scale) and ordinary quantum mechanics (QM) would appear as a result of an information loss process. This would imply the existence of "beables", i.e. ontological (commuting) operators which, after information loss, would give rise to the usual quantum (non-commuting) observables.

It has been shown in Ref.[5] that in a model by Bateman [6] describing a damped harmonic oscillator, where the environment absorbing the energy is simply represented by one additional environmental degree of freedom, the corresponding Hamiltonian can be brought into a form that belongs to the same class as the one considered by 't Hooft. Then, by means of an appropriate constraint[5], the Hamiltonian for the quantum harmonic oscillator is obtained.

In this paper, we analyze the connections of Bateman's model with other models for the description of dissipative systems, thus providing the basis for alternative realizations of 't Hooft's scheme.

In Section 2 a short review of 't Hooft's method is outlined. In Section 3 the dissipative Bateman model is presented and its transformation into a form fitting 't Hooft's requirements is given. In Section 4, by removing the environmental degree of freedom via some constraint (that is not uniquely defined), the Bateman model is transformed into an effective model for the dissipative system alone.

In this effective model, position and momentum variables are not identical to the physical position and momentum (and not even related to them via a *canonical* transformation); but the model provides the correct equations of motion for the physical variables (in this case, a damped harmonic oscillator with linear velocity dependent friction force) and the Hamiltonian function is a constant of motion.

As the canonical position variable of this effective model is different from the physical position variable (it is expanding exponentially with respect to the physical position variable), the transformation from the beables to the observables, at least in the case of the position variable, is not yet completed.

However, this can be achieved by an addition *canonical* transformation of the expanding system into one proposed by Caldirola [7] and Kanai [8] where the position variable is identical to the physical one. The canonical effective model of Caldirola and Kanai can also be derived from the conventional system-plus-reservoir approach (see, e.g., Caldeira and Leggett [9, 10, 11]). This has been shown by Yu and Sun [12, 13]. In this sense, the transition from the beables, at least in the case of the observable position variable, can be achieved.

2. 't Hooft's model

In his model 't Hooft considered a class of systems which evolve deterministically even after quantization. This happens for a dynamics of the form [2, 5]

$$\dot{q}_i = f_i(q) = \{q_i, H_{\mathrm{tH}}\} \tag{1}$$

where $\{\ ,\ \}$ denotes the Poisson brackets and H_{tH} the Hamiltonian of the system, given by

$$H_{\mathrm{tH}} = \sum_i p_i f_i(q) \tag{2}$$

giving indeed $\dot{q}_i = \frac{\partial}{\partial p_i} H_{\mathrm{tH}} = f_i(q)$.

Once quantized, for such a system there exists a complete set of operators, the q_i, commuting at all times, which are indeed the beables. However the above Hamiltonian is not bounded from below. To fix this a constraint must be imposed. The lower bound is emerging during a coarse-graining of the beables to arrive at the observable degrees of freedom.

For this purpose, one can split the Hamiltonian according to

$$H_{\mathrm{tH}} = H_1 - H_2 \tag{3}$$

with

$$H_1 = \frac{1}{4\varrho}(\varrho + H_{\mathrm{tH}})^2 \quad \text{and} \quad H_2 = \frac{1}{4\varrho}(\varrho - H_{\mathrm{tH}})^2 \tag{4}$$

and $\varrho = \varrho(q_i)$= positive, H_1 and H_2: positively (semi) definite, with $\{H_1, H_2\} = \{\varrho, H_{\mathrm{tH}}\} = 0$.

The *dynamics* of the deterministic system can, *mathematically*, be formulated in terms of unitary evolution operators acting on a Hilbert space of states $|\Psi\rangle$ and be expressed *formally* by a Schrödinger-like equation.

A lower bound for the (operatorial) Hamiltonian can be obtained by the constraint condition onto the Hilbert space

$$H_2|\Psi\rangle = 0\ , \tag{5}$$

projecting out the states responsible for the negative part of the spectrum.

3. Bateman's model

It has been shown in Ref.[5] that a different model [6] connecting an apparently (formally) conservative system with a (physically) dissipative system involves a Hamiltonian that is, at first sight, different from 't Hooft's but can be shown to belong to the same class representing an explicit realization of 't Hooft's mechanism.

For this purpose, the two equations of motion

$$\ddot{x} + \gamma\dot{x} + \omega^2 x = 0 \quad \text{(damped harmonic oscillator)} \tag{6}$$

$$\ddot{y} - \gamma\dot{y} + \omega^2 y = 0 \quad \text{(accelerated harmonic oscillator)} \tag{7}$$

are considered that can be derived from the Hamiltonian

$$H_{\mathrm{B}} = \frac{1}{m}\, p_x p_y + \frac{\gamma}{2}\,(y p_y - x p_x) + m\left(\omega^2 - \frac{\gamma^2}{4}\right) xy \tag{8}$$

with

$$p_x = m\left(\dot{y} - \frac{\gamma}{2}y\right) \;,\; p_y = m\left(\dot{x} + \frac{\gamma}{2}x\right) \tag{9}$$

that is a constant of motion and was proposed by Bateman [6].

The system's degree of freedom x dissipates energy due to the friction force $-m\gamma\dot{x}$ whereas, in the equation of motion for the environmental degree of motion y, the accelerating force $+m\gamma\dot{x}$ replaces the friction force. Therefore, it seems reasonable that the energy dissipated by the x-system is gained by the y-system and, therefore, H_{B} is a constant of motion. However, if y would really be a physical position variable, it would be accelerated infinitely by the force $+m\gamma\dot{x}$ and for $t \to \infty$ the energy would diverge whereas the one of the x-system would vanish.

Therefore, the variable y cannot be considered a position variable like x because H_{B} would not be a constant of motion in this case. So, the interpretation of y as a second "position variable" (and the corresponding momentum), must be considered carefully in order to avoid unphysical results (for further details, see also [14, 15]).

To show that H_{B} belongs to the same class as H_{tH}, new variables are introduced [5]

$$x_1 = \frac{1}{\sqrt{2}}(x+y) \;,\; x_2 = \frac{1}{\sqrt{2}}(x-y) \tag{10}$$

$$p_1 = m\left(\dot{x}_1 + \frac{\gamma}{2}x_2\right) \;,\; p_2 = -m\left(\dot{x}_2 + \frac{\gamma}{2}x_1\right) \tag{11}$$

and further

$$x_1 = r\cosh u,\; x_2 = r\sinh u \;\rightarrow\; r^2 = x_1^2 - x_2^2 = 2xy\;. \tag{12}$$

Consequently, the Bateman Hamiltonian H_B can be rewritten in 't Hooft's form as

$$H_{\mathrm{B}} = \sum_{i=1}^{2} p_i f_i(q) \tag{13}$$

with

$$f_1(q) = 2\Omega \;,\; f_2(q) = -2\Gamma \quad \text{where} \quad \Omega = \sqrt{\omega^2 - \frac{\gamma^2}{4}} \;,\; \Gamma = \frac{\gamma}{2} \quad \text{and} \quad p_1 = \mathcal{C} \;,\; p_2 = \mathcal{J}_2 \tag{14}$$

where

$$\begin{aligned} \mathcal{C} &= \frac{1}{4\Omega m}\left[\left(p_1^2 - p_2^2\right) + m^2\Omega^2(x_1^2 - x_2^2)\right] \\ &= \frac{1}{4\Omega m}\left[\left(p_r^2 - \frac{1}{r^2}p_u^2\right) + m^2\Omega^2 r^2\right] = \text{const. of motion}\;, \end{aligned} \tag{15}$$

$$\mathcal{J}_2 = \frac{m}{2}\left[(\dot{x}_1 x_2 - \dot{x}_2 x_1) + \Gamma r^2\right] = \frac{1}{2}p_u \ . \tag{16}$$

Following the scheme outlined in Section 2, H_B can be split into two positive Hamiltonians according to

$$H_{\rm B} = H_{\rm I} - H_{\rm II} \tag{17}$$

with

$$H_{\rm I} = \frac{1}{2\Omega\mathcal{C}}\,(2\Omega\mathcal{C} - \Gamma\mathcal{J}_2)^2 \ , \quad H_{\rm II} = \frac{\Gamma^2}{2\Omega\mathcal{C}}\,\mathcal{J}_2^2 \ . \tag{18}$$

Implementing the constraint

$$\mathcal{J}_2|\Psi\rangle = 0 \ , \tag{19}$$

thus defining the physical states, the Bateman Hamiltonian leads to

$$H_{\rm B}|\Psi\rangle = H_{\rm I}|\Psi\rangle = \Omega\mathcal{C}|\Psi\rangle = \left(\frac{1}{2m}p_r^2 + \frac{m}{2}\Omega^2 r^2\right)|\Psi\rangle \ , \tag{20}$$

i.e., a Hamiltonian of a harmonic oscillator with frequency $\Omega = \sqrt{\omega^2 - \frac{\gamma^2}{4}}$ and the corresponding equation of motion

$$\ddot{r} + \Omega^2 r = 0 \ . \tag{21}$$

This equation of motion is formally identical to the one of a damped harmonic oscillator expressed in a coordinate system expanding exponentially as discussed in the next section.

4. Damped harmonic oscillator in an expanding coordinate system

It is possible to formulate the damped harmonic oscillator in a form that looks like an undamped oscillator, only with a reduced frequency $\Omega = \sqrt{\omega^2 - \frac{\gamma^2}{4}}$. In this case, the canonical Hamiltonian[1] [14, 16]

$$\hat{H}_{exp} = \frac{1}{2m}\,\hat{P}^2 + \frac{m}{2}\Omega^2\hat{Q}^2 \tag{22}$$

is a constant of motion providing the equation of motion

$$\ddot{\hat{Q}} + \Omega^2\,\hat{Q} = 0 \tag{23}$$

for the canonical variables

$$\hat{Q} = x\,e^{\frac{\gamma}{2}t} \ , \quad \hat{P} = m\,\dot{\hat{Q}} = m\left(\dot{x} + \frac{\gamma}{2}x\right)e^{\frac{\gamma}{2}t} \ . \tag{24}$$

Expressed in terms of the physical variables x and $p = m\dot{x}$, the Hamiltonian (22) obtains the form

$$\hat{H}_{exp} \mathrel{\hat{=}} \frac{m}{2}\left[\dot{x}^2 + \gamma\dot{x}x + \omega^2 x^2\right]e^{\gamma t} = \text{const.} \ , \tag{25}$$

and the equation of motion for the physical position variable x has the form

$$\ddot{x} + \gamma\dot{x} + \omega^2 x = 0 \tag{26}$$

which is identical to the equation of motion for the x-variable of the Bateman system.

[1] In the following, a hat denotes a canonical quantity, i.e., a quantity that obeys the rules of the canonical Hamiltonian formalism. In the dissipative case, these quantities are usually not identical to their physical counterparts like position and momentum and connected with these via non-canonical transformations (see below).

It should be noted that the connection between the expanding variables $\hat{Q}$ and $\hat{P}$ and the physical variables x and p is given by a $non-canonical$ transformation.

As Eq.(26) and Eq.(6) are identical and both corresponding Hamiltonians, $\hat{H}_{\mathrm{B}}$ as given in (8) and $\hat{H}_{exp}$ as given in (25), are constants of motion, the assumption that both Hamiltonians are identical seems to be legitimate. In other words, due to $\frac{d}{dt}\hat{H}_{\mathrm{B}} = \frac{d}{dt}\hat{H}_{exp}$, up to a constant factor, it can be assumed that

$$\hat{H}_{\mathrm{B}} \mathrel{\hat{=}} \hat{H}_{exp} \ . \tag{27}$$

To show this connection explicitly, both Hamiltonians must be expressed in terms of the same *physical* variables x and $\dot{x}$. The Bateman Hamiltonian can be written as

$$\hat{H}_{\mathrm{B}} = \left[\frac{1}{m}\, p_x p_y \ + m\left(\omega^2 - \frac{\gamma^2}{4}\right) xy\right] \ + \ \left\{\frac{\gamma}{2}\,(y p_y \ - \ x\hat{p}_x)\right\} \ = \ [\hat{H}_\Omega] + \{\hat{D}\} \ . \tag{28}$$

Comparison with $\hat{H}_{exp}$, using $p_y \ = \ m\ (\dot{x} \ + \ \frac{\gamma}{2}x)$ yields

$$\hat{H}_{exp} \mathrel{\hat{=}} \frac{m}{2}\mathrm{e}^{\gamma t}\left[\dot{x}^2 \ + \ \gamma\dot{x}x \ + \ \omega^2 x^2\right] \ = \ \hat{H}_{\mathrm{B}} \mathrel{\hat{=}} p_x\dot{x} \ + \ m\frac{\gamma}{2}\, y\dot{x} \ + \ m\omega^2 xy \ . \tag{29}$$

The lhs of (29) depends only on x and $\dot{x}$ whereas, on the rhs, y and p_x still appear. Therefore, y and p_x must be expressed in terms of x and $\dot{x}$.

For this purpose, the ansatz [15]

$$p_x \ = \ \mathrm{e}^{\gamma t}(a\ \dot{x} \ + \ b\ x) \quad \text{and} \quad y \ = \ \mathrm{e}^{\gamma t}(c\ \dot{x} \ + \ d\ x) \tag{30}$$

is applied.

Comparison of $\hat{H}_{\mathrm{B}}$ and $\hat{H}_{exp}$ provides $d = \frac{1}{2}$, but since there are only three equations for four parameters, there is still one parameter free of choice. Therefore, the remaining three parameters a, b and c are related via the two equations

$$a = \frac{m}{2}\ (1 \ - \ \gamma c) \ , \ b = m\ \left(\frac{\gamma}{4} \ - \ \omega^2\ c\right) \ . \tag{31}$$

For the particular choice $c = 0$, leading to $a = \frac{m}{2}$ and $b = m\frac{\gamma}{4}$, one obtains for p_x and y

$$\hat{p}_x \ = \ \frac{m}{2}\left(\dot{x} + \frac{\gamma}{2}\ x\right)\mathrm{e}^{\gamma t} \ = \ \frac{1}{2}\ \hat{P}\ e^{\frac{\gamma}{2}t} \quad \text{and} \quad \hat{y} = \frac{1}{2}x\ e^{\gamma t} \ = \ \frac{1}{2}\hat{Q}\ e^{\frac{\gamma}{2}t} \ . \tag{32}$$

Inserting this into $\hat{H}_B$ yields

$$\hat{H}_{\mathrm{B}} \ = \ \frac{1}{m}\ p_x\ p_y \ + \ m\left(\omega^2 - \frac{\gamma^2}{4}\right) x\ y \ = \ \hat{H}_\Omega \tag{33}$$

$$\hat{D} = \frac{\gamma}{2}\ (y\ p_y \ - \ x\ p_x) \ = \ 0\ . \tag{34}$$

Expressing $\hat{D}$ in terms of x, y, $\dot{x}$ and $\dot{y}$ leads to

$$\hat{D} = \frac{m}{2}\gamma(\dot{x}y - x\dot{y}) + \frac{m}{2}\gamma^2 xy \ , \ \text{ i.e.,} \ \ \hat{D} = \gamma\mathcal{J}_2 \ . \tag{35}$$

Therefore, the constraint $c = 0$ leading to $\hat{D} = 0$ is equivalent to the constraint $\mathcal{J}_2 = 0$. Consequently, $\hat{H}_{exp}$ is equivalent to $\hat{H}_I$ of the split Bateman Hamiltonian,

$$\hat{H}_{exp} \ = \ \frac{1}{2m}\ \hat{P}^2 \ + \ \frac{m}{2}\Omega^2\hat{Q}^2 = \hat{H}_{\mathrm{I}} = \frac{1}{2m}p_r^2 + \frac{m}{2}\Omega^2 r^2 \tag{36}$$

IOP Conf. Series: Journal of Physics: Conf. Series **880** (2017) 012050 doi:10.1088/1742-6596/880/1/012050

provided the following relations are fulfilled:

$$r = x\, \mathrm{e}^{\frac{\gamma}{2}t} = \hat{Q} \quad , \quad p_r = m\left(\dot{x} + \frac{\gamma}{2}x\right)\mathrm{e}^{\frac{\gamma}{2}t} = \hat{P}. \tag{37}$$

That means the dissipative system can be described within the canonical formalism but the price is a *non−canonical* transformation between the *physical* variables (x, p) and the *canonical* ones $(\hat{Q} = r, \hat{P} = p_r)$.

There is still the problem that the system obtained from the Bateman Hamiltonian (corresponding to the level of beables) via elimination of environmental information leading to the expanding canonical system and therefore to something on the level of the observables, depends on a position variable that is *not* identical to the physical (observable) position. However, this can be fixed by an additional *canonical* transformation on the canonical level, as will be shown in the next section.

5. Connection between different canonical descriptions of dissipative systems

In an attempt to describe dissipative systems with a linear velocity dependent friction force, like the one in Eq. (6), Caldirola [7] and Kanai [8] proposed the explicitly time-dependent Lagrangian

$$\hat{L}_{\mathrm{CK}} = \left[\frac{m}{2}\dot{x}^2 - V(x)\right]\mathrm{e}^{\gamma t} \tag{38}$$

with the *canonical* momentum

$$\hat{p} = \frac{\partial}{\partial \dot{x}}\hat{L}_{\mathrm{CK}} = m\dot{x}\,\mathrm{e}^{\gamma t} = p\,\mathrm{e}^{\gamma t} \tag{39}$$

where $p = m\dot{x}$ is the *physical* momentum and the canonical position variable $\hat{x}$ is identical to the physical one, $\hat{x} = x$. Due to the relation between the two momenta, the transition between the physical variables (x, p) and the canonical ones ($\hat{x} = x, \hat{p} = p\,\mathrm{e}^{\gamma t}$ is a *non − canonical* transformation (the Jacobian determinant is different from 1).

With the canonical momentum (39), the Hamiltonian corresponding to the Lagrangian (38) can be expressed as

$$\hat{H}_{\mathrm{CK}} = \frac{1}{2m}\mathrm{e}^{-\gamma t}\,\hat{p}^2 + \frac{m}{2}\omega^2 x^2 \mathrm{e}^{\gamma t}\,. \tag{40}$$

The Hamiltonian $\hat{H}_{\mathrm{CK}}$ is explicitly time-dependent, *not* a constant of motion and not equivalent to the energy of the dissipative system but related to it via

$$\hat{H}_{\mathrm{CK}} = \hat{H}_{\mathrm{CK}}(t) = E\,\mathrm{e}^{\gamma t}\,. \tag{41}$$

Expressing the canonical equations of motion obtained from the Hamiltonian (40) in terms of the physical position variable x, one obtains again the damped harmonic oscillator with linear friction, as in Eq. (6).

It can be shown straightforwardly that the dissipative canonical system of Caldirola and Kanai and the one in expanding coordinates are connected via a canonical transformation (that does not change the physical properties). In particular, the canonical variables $\hat{x} = x$ and $\hat{p} = p\,\mathrm{e}^{\gamma t}$ of the Caldirola–Kanai system are related to the ones of the expanding system via

$$\hat{Q} = \hat{x}\, e^{\frac{\gamma}{2}t} \quad , \quad \hat{P} = \hat{p}\, e^{-\frac{\gamma}{2}t} + m\,\frac{\gamma}{2}\hat{x}\, e^{\frac{\gamma}{2}t}\,. \tag{42}$$

The explicitly time-dependent generating function $\hat{F}_2(\hat{x}, \hat{P}, t)$ connecting the corresponding Hamiltonians via

$$\hat{H}_{\mathrm{exp}} = \hat{H}_{\mathrm{CK}} + \frac{\partial}{\partial t}\hat{F}_2 \tag{43}$$

is given by

$$\hat{F}_2(\hat{x},\hat{P},t) \;=\; \hat{x}\hat{P}\,e^{\frac{\gamma}{2}t} \;-\; m\frac{\gamma}{4}\hat{x}^2\,e^{\frac{\gamma}{2}t}\,, \tag{44}$$

turning the time-dependent Hamiltonian $\hat{H}_{\mathrm{CK}}$ into the constant of motion $\hat{H}_{\mathrm{exp}}$.

Furthermore, it has been shown by Yu and Sun [12, 13] that, starting from the conventional system-plus-reservoir approach where the dissipative system of interest plus the energy-absorbing environment together are regarded as a closed (conservative) Hamiltonian system, it is possible to arrive at the same Hamiltonian (40) as Caldirola and Kanai. For this purpose, they started with the approach of Caldeira and Leggett [9, 10, 11] where the system of interest is coupled to a bath of harmonic oscillators via

$$\hat{H}_{\mathrm{CL}} = H_{\mathrm{S}} + H_{\mathrm{R}} + H_{\mathrm{SR}}\;. \tag{45}$$

That is, the Hamiltonian $\hat{H}_{\mathrm{CL}}$ of Caldeira and Leggett is the sum of the Hamiltonians describing the system alone (H_{S}), the environment (reservoir) alone (H_{R}) and the interaction between them (H_{SR}) via

$$H_{\mathrm{S}} = \frac{1}{2m}p^2 + \frac{m}{2}\omega^2x^2\,,\quad H_{\mathrm{R}} = \sum_{i=1}^{N}\left(\frac{p_i^2}{2m_i} + \frac{m}{2}\omega_i^2q_i^2\right)\,,\quad H_{\mathrm{SR}} = -x\,\sum_i c_iq_i\;. \tag{46}$$

After taking the limit of N to infinity, eliminating the environmental degrees of freedom, assuming an Ohmic spectral density, etc. (for details, see, e.g., [17]), one finally obtains Eq. (6) as equation of motion for the physical position variable of the system of interest and, as effective Hamiltonian, one that is identical to that of Caldirola and Kanai,

$$\hat{H}_{\mathrm{CL,eff}} \;=\; \frac{1}{2m}\mathrm{e}^{\gamma t}\,p^2 + \frac{m}{2}\omega^2x^2\mathrm{e}^{\gamma t} \;\hat{=}\; \hat{H}_{\mathrm{CK}}\;. \tag{47}$$

Therefore, the connection from Bateman's canonical level (corresponding to the beables) to a *formal* canonical level (as in $\hat{H}_{exp}$) can be achieved and on this level further to a *formal* canonical description with the usual *physical* position variable (as in $\hat{H}_{CK}$). Even further, via the relation between the Hamiltonian $\hat{H}_{\mathrm{CK}}$ and the system-plus-reservoir approach also the connection to the conventional *physical* canonical level can be established. All these relations are depicted in Figure 1.

6. Conclusions

In this work the Bateman model, describing a system in contact with an energy-absorbing environment and providing an example of 't Hooft's model for deterministic QM, has been considered and its relations with other dissipative quantum models have been studied.

Applying some constraint (not uniquely defined) to the Bateman Hamiltonian, the remaining environmental degree of freedom can be eliminated and an effective description can be obtained in terms of one set of canonical position and momentum variables of the dissipative system alone (without any environmental degree of freedom). Such a canonical description involves a *canonical* position variable that is related to the *physical* position variable via an exponentially-time-dependent factor. This factor can be removed via the canonical transformation (42)-(44), leading to the approach by Caldirola and Kanai where the position variable is identical to the physical one, thus completing the transformation from the beables to the observables, at least for the position variable.

It is also possible to show a connection between the Caldirola–Kanai model and the conventional system-plus-reservoir approach (where the system with its *physical* position and

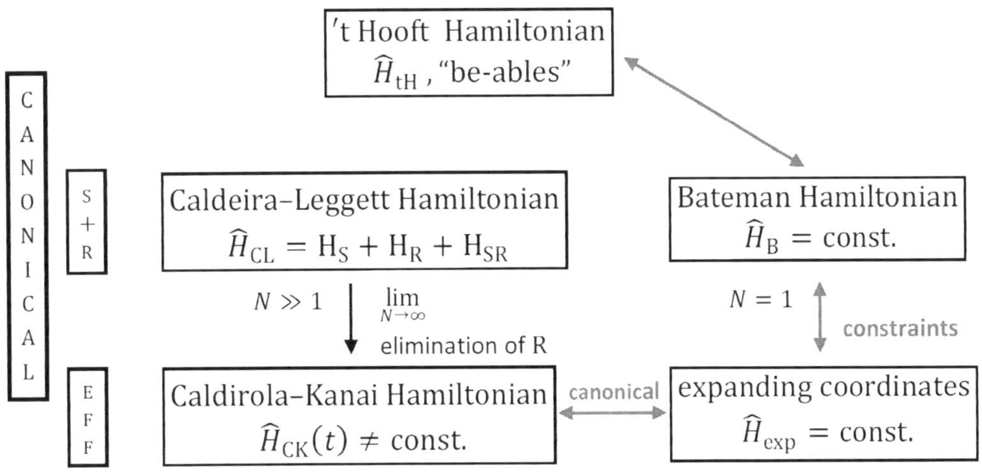

Figure 1. Relations between different descriptions of dissipative systems on the canonical level.

momentum variables is coupled to a large set of environmental degrees of freedom), according to Caldeira and Leggett. However, this model has some shortcomings as, for the corresponding density operator, it leads to a dissipative Hamiltonian that does not possess the Kossakowski–Lindblad form [17] and can therefore have negative values for the density matrix. A connection between 't Hooft's idea and a master equation with a dissipative Kossakowski–Lindblad term is mentioned in [18].

The problem to find the transition from the beables to a system that is entirely expressed in terms of *physical* position and momentum variables can be resolved by a *non – canonical* transformation between the variables of the Caldirola–Kanai (or expanding-coordinate) model, corresponding to a non-unitary transformation in the quantum mechanical case. This will be discussed elsewhere.

References

[1] 't Hooft G 1999 *Class. Quantum Grav.* **16** 3263
[2] 't Hooft G 1999 in: *Basics and Highlights of Fundamental Physics* Erice, hep-th/0003005
[3] 't Hooft G 1988 *J. Statist Phys.* **53** 323
[4] 't Hooft G 2016 *The Cellular Automaton Interpretation of Quantum Mechanics* (Springer Int. Publ.)
[5] Blasone M, Jizba P and Vitiello G 2001 *Phys. Lett. A* **287** 205
[6] Bateman H 1931 *Phys. Rev.* **38** 815
[7] Caldirola P 1941 *Nuovo Cimento* **18** 393
[8] Kanai E 1948 *Progr. Theor. Phys.* **3** 440
[9] Caldeira A O and Leggett A J 1981 *Phys. Rev. Lett.* **46** 211
[10] Caldeira A O and Leggett A J 1983 *Ann. Phys. (N.Y.)* **149** 374
[11] Caldeira A O and Leggett A J 1983 *Ann. Phys. (N.Y.)* **153** 445(E)
[12] Yu L H and Sun C P 1994 *Phys. Rev.* A **49** 592
[13] Sun C P and Yu L H 1995 *Phys. Rev.* A **51** 1845
[14] Schuch D 1998 *Symmetries in Science X* ed B. Gruber and M. Ramek (Plenum Press, New York)
[15] Schuch D, Guerrero J, López-Ruiz F F and Aldaya V 2015 *Phys. Scr.* **90** 045209
[16] Blasone M, Graziano E, Pashaev O K and Vitiello G 1996 *Annals Phys.* **252**, 115
[17] Weiss U 1993 *Quantum Dissipative Systems* (Singapore: World Scientific)
[18] Elze H-T 2009 *J. Phys.: Conf. Ser.* **174** 012009

Generating functional for the Green's functions of a two-flavor bosonic model

M Blasone[1], P Jizba[2] and L Smaldone[1]

[1] Dipartimento di Fisica, Università di Salerno, Via Giovanni Paolo II, 132 84084 Fisciano, Italy & INFN Sezione di Napoli, Gruppo collegato di Salerno, Italy
[2] FNSPE, Czech Technical University in Prague, Břehová 7, 115 19 Praha 1, Czech Republic

E-mail: blasone@sa.infn.it; p.jizba@fjfi.cvut.cz; lsmaldone@sa.infn.it

Abstract. We examine a simple two-flavor scalar model with a non-diagonal mass matrix. We argue that the conventional definition of the QFT partition function does not allow to evaluate the Green's functions with respect to the flavor vacuum as it favors only the mass vacuum. By using functional-integral techniques, we derive new generating functional for Green's functions on the flavor vacuum.

1. Introduction

The particle mixing mechanism is an important topic in the physics of Standard Model and beyond. Quark mixing was discovered theoretically in order to find an explanation for the suppression of some particular mesonic and hadronic processes involving the non conservation of quantum numbers like strangeness and flavor [1, 2]. Later on the concept of mixing was extended [3] so that it could also accommodate the discovery of CP violation in experiments on K^0-$\overline{K}^0$ or B^0-$\overline{B}^0$ oscillations [4, 5, 6].

A first viable mechanism for flavor oscillations was laid down in Pontecorvo [7, 8] and Bilenky [9] seminal works that allowed to explain quantitatively well the results of the Homestake experiment [10]. Phenomena such as neutrino or K^0-$\overline{K}^0$ oscillations are experimentally well accepted [11] nowadays. On the other hand, the theoretical aspects of oscillations still await a deeper understanding. The standard theoretical approach to this subject is based on quantum mechanics (QM) or on perturbative quantum field theory (QFT) [12]. However, in this last case the construction of neutrino's Fock space is done only in the ultra-relativistic limit.

A non-perturbative QFT approach to mixing was first developed in Ref. [13]. In consequence, phenomenologically relevant corrections to the standard Pontecorvo oscillation formula were found by using Green's functions evaluated on the flavor Fock vacuum [15]. A general review of these facts can be found in Ref. [14]. Most of developments in this direction were achieved by using operatorial techniques. They include, e.g., a deeper understanding of the mathematical structure of the mixing generator [16] or corrections to the usual Unruh radiation formula due to the presence of flavor mixing [17].

An interesting standpoint is to view the fermion mixing as a dynamical phenomenon, arising as a consequence of the spontaneous symmetry breaking in a some fermionic theory beyond the standard model [18]. This line of research was also recently pursued by the present authors who used a one-loop effective action for the composite Hubbard-Stratonovich boson field, to find a set of gap equations describing the dynamical generation of both masses and mixing [19]. On the other hand, a richer structure of the gap equations must exist. This was, on a general ground, shown in Ref. [18]. In fact, for each class of unitarily inequivalent Fock spaces one should expect a different class of gap equations [20, 21, 22]. We claim that the reason for this discrepancy can be retraced to the fact that the standard generating functional for the Green's functions is blind to the vacuum choice (or, perhaps better, to the choice of the particular representation of the Fock space).

Our aim here is to derive the generating functional for the Green's functions constructed both on mass and on flavor vacuum for a simple two-flavor charged scalar model. To set up the stage, we present in Section 2 some essentials of operatorial approach to scalar-field mixing. In doing so we follow Ref. [23]. In Section 3, we calculate the standard generating functional for the Green's functions (the mass vacuum one) and show how this can be extended in order to accommodate also the flavor-vacuum states. In order to produce a tractable result, we assume that the mixing angle θ is small so that the partition functional can be evaluated only to the first order in θ.

2. Boson field mixing

Let us consider the Lagrange density

$$\mathcal{L} \;=\; \partial_\mu \hat{\varphi}_f^\dagger \partial_\mu \hat{\varphi}_f \;-\; \hat{\varphi}_f^\dagger M \hat{\varphi}_f \,, \tag{1}$$

where

$$\hat{\varphi}_f \;=\; \begin{bmatrix} \hat{\varphi}_A \\ \hat{\varphi}_B \end{bmatrix}, \qquad M = \begin{bmatrix} m_A^2 & m_{AB}^2 \\ m_{BA}^2 & m_B^2 \end{bmatrix}, \tag{2}$$

which describes the dynamics of two coupled (mixed) scalar fields that we will call flavor fields, in analogy with the terminology used in quark and neutrino physics. However, flavor charge has to be interpreted as a generic quantum number. In fact, in experiments we observe, oscillations of different quantities such as strangeness or isospin. For this reason we consider a complex scalar field doublet, despite the fact that in the experiments are always observed oscillations of neutral mesons.

The Lagrange density (2) can be diagonalized thanks to the following transformation:

$$\hat{\varphi}_f \;=\; U \hat{\varphi}_m \,, \qquad U = \begin{bmatrix} \cos\theta & \sin\theta \\ -\sin\theta & \cos\theta \end{bmatrix}. \tag{3}$$

With this, $\mathcal{L}$ becomes

$$\mathcal{L} \;=\; \partial_\mu \hat{\varphi}_m^\dagger \partial_\mu \hat{\varphi}_m \;-\; \hat{\varphi}_m^\dagger M_d \hat{\varphi}_m \,, \tag{4}$$

where

$$\hat{\varphi}_m \;=\; \begin{bmatrix} \hat{\varphi}_1 \\ \hat{\varphi}_2 \end{bmatrix}, \qquad M \;=\; \begin{bmatrix} m_1^2 & 0 \\ 0 & m_2^2 \end{bmatrix}, \tag{5}$$

Lagrange density (4) thus describes two free scalar particles with masses m_1 and m_2. The above transformation can be rewritten by introducing the mixing generator:

$$\hat{G}_\theta(t) \;=\; \exp\left\{\theta \left[\hat{S}_+(t) - \hat{S}_-(t)\right]\right\}, \tag{6}$$

where

$$\hat{S}_+(t) \;=\; \hat{S}_-^\dagger(t) \;=\; -i \int \mathrm{d}^3\mathbf{x} \, \left[\hat{\pi}_1(x)\hat{\varphi}_2(x) \;-\; \hat{\varphi}_1^\dagger(x)\hat{\pi}_2^\dagger(x)\right]. \tag{7}$$

Here $\hat{\pi}_i$, $i = 1,2$ are conjugate momenta, so that equal-time canonical commutation relations (ETCR)

$$[\hat{\varphi}_i(x), \hat{\pi}_i(y)]_{t_x=t_y} \;=\; \delta^3(\mathbf{x}-\mathbf{y})\,, \tag{8}$$

are satisfied, with the other commutators equal to zero. We can thus write

$$\hat{\varphi}_f(x) \;=\; \hat{G}_\theta^{-1}(t)\,\hat{\varphi}_m(x)\,\hat{G}_\theta(t)\,. \tag{9}$$

This transformation provides us with the mapping between two representations of ETCR. Representation spaces (i.e., respective Fock spaces) are connected by the relation

$$|a(t)\rangle_f \;=\; \hat{G}_\theta^{-1}(t)|a\rangle_m\,. \tag{10}$$

In particular, the vacuum states in these two representations are related by

$$|0(t)\rangle_f \;=\; \hat{G}_\theta^{-1}(t)|0\rangle_m\,. \tag{11}$$

The notation used, put in evidence that the unitary generator (6) has the structure of a $SU(2)$ generalized (Perelomov) coherent state. The $SU(2)$ algebra is closed by the operators:

$$[\hat{S}_+(t), \hat{S}_-(t)] \;=\; 2\hat{S}_3(t)\,, \qquad [\hat{S}_3(t), \hat{S}_\pm(t)] \;=\; \pm\hat{S}_\pm(t)\,, \tag{12}$$

where

$$\hat{S}_3(t) \;=\; -\frac{i}{2}\int d^3\mathbf{x}\,\Big[\hat{\pi}_1(x)\hat{\varphi}_1(x) \;-\; \hat{\varphi}_1^\dagger(x)\hat{\pi}_1^\dagger(x) \;-\; \hat{\pi}_2(x)\hat{\varphi}_2(x) \;+\; \hat{\varphi}_2^\dagger(x)\hat{\pi}_2^\dagger(x)\Big]\,. \tag{13}$$

One can prove that these two Fock spaces are orthogonal (i.e., inequivalent) in the large V limit [13, 14]. In fact, by using the Gaussian decomposition

$$\hat{G}_\theta(t) \;=\; \exp\Big[-\tan\theta\hat{S}_+(t)\Big]\exp\Big[-2\log\cos\theta\hat{S}_3(t)\Big]\exp\Big[\tan\theta\hat{S}_-(t)\Big]\,, \tag{14}$$

we can derive the important result:

$${}_m\langle 0|0\rangle_f \;=\; \prod_{\mathbf{k}}{}_m\langle 0|G^{-1}_{\mathbf{k}\,\theta}(t)|0\rangle_m = \exp\left[-\frac{V}{(2\pi)^3}\int d^3\mathbf{k}\,\log\left(1 \;+\; \sin^2\theta|V_\mathbf{k}|^2\right)\right]\,, \tag{15}$$

where

$$V_\mathbf{k} \;=\; \frac{1}{2}\left(\sqrt{\frac{\omega_{\mathbf{k}\,1}}{\omega_{\mathbf{k}\,2}}} \;-\; \sqrt{\frac{\omega_{\mathbf{k}\,2}}{\omega_{\mathbf{k}\,1}}}\right)e^{i(\omega_{\mathbf{k}\,1}\,+\,\omega_{\mathbf{k}\,2})t}\,, \qquad \omega_{\mathbf{k}\,i} \;=\; \sqrt{|\mathbf{k}|^2 \,+\, m_i^2}\,,\quad i=1,2\,. \tag{16}$$

In the limit $V\to\infty$, as said, the RHS of Eq. (15) goes to zero: flavor and mass Fock states are orthogonal and so flavor and mass fields belong to unitarily inequivalent representations of ETCR (8). In other words, in this limit, the unitary generator (6) does not exist as a consequence of the second Schur's lemma [24]. It is thus important to decide which representation is the one on which we should work. It was argued in Ref. [23] that, in order to have a well defined form of the oscillation formula which is independent from an arbitrary mass parametrization [25], one has to choose the flavor Fock space. The exact oscillation formula reads in this case

$$\mathcal{Q}_{\mathbf{k}\,A}(t) \;=\; 1-\sin^2 2\theta\left[|U_\mathbf{k}|^2\sin^2\left(\frac{\omega_{\mathbf{k}\,1}-\omega_{\mathbf{k}\,2}}{2}\right)t - |V_\mathbf{k}|^2\sin^2\left(\frac{\omega_{\mathbf{k}\,1}\,+\,\omega_{\mathbf{k}\,2}}{2}\right)t\right]\,, \tag{17}$$

$$\mathcal{Q}_{\mathbf{k}\,B}(t) \;=\; \sin^2 2\theta\left[|U_\mathbf{k}|^2\sin^2\left(\frac{\omega_{\mathbf{k}\,1}\,-\,\omega_{\mathbf{k}\,2}}{2}\right)t \;-\; |V_\mathbf{k}|^2\sin^2\left(\frac{\omega_{\mathbf{k}\,1}\,+\,\omega_{\mathbf{k}\,2}}{2}\right)t\right]\,, \tag{18}$$

$$U_\mathbf{k} \;=\; \frac{1}{2}\left(\sqrt{\frac{\omega_{\mathbf{k}\,1}}{\omega_{\mathbf{k}\,2}}} \;+\; \sqrt{\frac{\omega_{\mathbf{k}\,2}}{\omega_{\mathbf{k}\,1}}}\right)e^{i(\omega_{\mathbf{k}\,1}\,+\,\omega_{\mathbf{k}\,2})t}\,, \qquad |U_\mathbf{k}|^2 \;-\; |V_\mathbf{k}|^2 \;=\; 1\,, \tag{19}$$

where $\mathcal{Q}_{\mathbf{k}\,\sigma}(t) \equiv \langle\nu_\sigma|\hat{Q}_{\mathbf{k}\,A}(t)|\nu_\sigma\rangle$, $\sigma = A, B$ is the expectation value of the flavor charge, for each family, on the single particle flavor state, in the flavor representation.

3. Mass and Flavor partition function

In this section we evaluate explicitly the generating functional for the Green's functions (known also as the QFT partition function) both on mass and on flavor vacuum. In doing so, we point out how the standard definition of partition function favors only the first case. The generating functional for the Green's functions on the flavor vacuum has to be determined in a completely new way.

Let us consider the standard Hamiltonian-based partition function:

$$\mathcal{Z}[J_\varphi, J_\pi] = \int \mathcal{D}\pi_f \mathcal{D}\varphi_f \mathcal{D}\pi_f^\dagger \mathcal{D}\varphi_f^\dagger \, e^{i\int d^4x \left(\pi_f^T \partial_0 \varphi_f + \partial_0 \varphi_f^\dagger \pi_f^* - \mathcal{H}_J\right)}, \tag{20}$$

where the Hamilton density is

$$\mathcal{H}_J = \int d^3\mathbf{x} \left(\pi_f^T \pi_f^* + \nabla\varphi_f^\dagger \cdot \nabla\varphi_f + \varphi_f^\dagger M \varphi_f - J_\varphi^\dagger \varphi_f - \varphi_f^\dagger J_\varphi - J_\pi^\dagger \pi_f^* - \pi_f^T J_\pi\right). \tag{21}$$

Performing the change of variables

$$\pi_f = \tilde{\pi}_f^* + \partial_0 \varphi_f^* + J_\pi^*, \qquad \pi_f^* = \tilde{\pi}_f + \partial_0 \varphi_f + J_\pi, \tag{22}$$

we have

$$\mathcal{Z}[J_\varphi, J_\pi] = \mathcal{N} e^{i\int d^4x J_\pi^\dagger J_\pi} \int \mathcal{D}\varphi_f^* \mathcal{D}\varphi_f \, e^{iS_J}, \tag{23}$$

where the action S_J reads

$$S_J = \int d^4x \left(\partial_\mu \varphi_f^\dagger \partial_\mu \varphi_f - \varphi_f^\dagger M \varphi_f + J_\varphi^\dagger \varphi_f + \varphi_f^\dagger J_\varphi + J_\pi^\dagger \partial_0 \varphi_f + \partial_0 \varphi_f^\dagger J_\pi\right). \tag{24}$$

This expression can be simplified by integrating by parts:

$$S_J = \int d^4x \left(\partial_\mu \varphi_f^\dagger \partial^\mu \varphi_f - \varphi_f^\dagger M \varphi_f + J_f^\dagger \varphi_f + \varphi_f^\dagger J_f\right), \tag{25}$$

where $J_f(x) = J_\varphi(x) - \partial_0 J_\pi(x)$ and $J_f^*(x) = J_\varphi^*(x) - \partial_0 J_\pi^*(x)$. Performing now the transformation

$$\varphi_f = U\varphi_m, \qquad U = \begin{bmatrix} \cos\theta & \sin\theta \\ -\sin\theta & \cos\theta \end{bmatrix}, \tag{26}$$

we obtain

$$J_m = U^T J_f, \qquad J_m^\dagger = J_f^\dagger U. \tag{27}$$

The solution can be written as:

$$\mathcal{Z}[J_m, J_\pi] = e^{i\int d^4x J_\pi^\dagger J_\pi} e^{-i\int d^4x \int d^4y \, J_m^\dagger(x) \Delta_F(x-y) J_m(y)}, \tag{28}$$

where

$$\Delta_F(x-y) = \begin{bmatrix} \Delta_F^1(x-y) & 0 \\ 0 & \Delta_F^2(x-y) \end{bmatrix}, \tag{29}$$

and $\Delta_F^i(x-y)$ is the Feynman propagator of the i-th mass field

$$\Delta_F^i(x-y) = \int_{\mathcal{C}_F} d^4p \, \frac{e^{-ip(x-y)}}{p^2 - m_i^2 + i\varepsilon}. \tag{30}$$

IOP Conf. Series: Journal of Physics: Conf. Series **880** (2017) 012051 doi:10.1088/1742-6596/880/1/012051

If $m_1 = m_2$, it follows, from Eq. (27) that the partition function (28) is the same in terms of both mass and flavor currents.

First we consider the case $J_\pi = 0$. In this case Eq. (28) becomes

$$\mathcal{Z}[J_m] \;=\; e^{-i\int d^4x \int d^4y\, J_m^\dagger(x)\Delta_F(x-y)J_m(y)}\,. \tag{31}$$

This is the generating functional of mass Green's functions on the mass vacuum[1]:

$${}_m\langle 0|\varphi_i(y)\varphi_j^\dagger(x)|0\rangle_m = (-i)^2\left\{\frac{\delta\mathcal{Z}[J_m]}{\delta J_j^*(x)\delta J_i(y)}\right\}_{J=0}, \qquad i,j=1,2\,. \tag{32}$$

In this case, rewriting the partition function in terms of flavor currents, we get:

$$\mathcal{Z}[J_\varphi] \;=\; e^{-i\int d^4x \int d^4y\, J_\varphi^\dagger(x)U\Delta_F(x-y)U^T J_\varphi(y)}\,. \tag{33}$$

Now we can define the *flavor propagator* as:

$$\begin{aligned}\Delta_F^f(x-y) &= U\Delta_F(x-y)U^T \qquad\qquad (34)\\ &= \begin{bmatrix} \Delta_F^1(x-y)\cos^2\theta + \Delta_F^2(x-y)\sin^2\theta & (\Delta_F^2(x-y)-\Delta_F^1(x-y))\cos\theta\sin\theta \\ (\Delta_F^2(x-y)-\Delta_F^1(x-y))\cos\theta\sin\theta & \Delta_F^2(x-y)\cos^2\theta + \Delta_F^1(x-y)\sin^2\theta \end{bmatrix}.\end{aligned}$$

Therefore

$$\mathcal{Z}[J_\varphi] \;=\; e^{-i\int d^4x \int d^4y\, J_\varphi^\dagger(x)\,\Delta_F^f(x-y)\,J_\varphi(y)}\,. \tag{35}$$

This is indeed the generating functional for the flavor Green's functions on the mass vacuum:

$${}_m\langle 0|\varphi_\alpha(y)\varphi_\beta^\dagger(x)|0\rangle_m \;=\; (-i)^2\left\{\frac{\delta\mathcal{Z}[J_\varphi]}{\delta J_\beta^*(x)\delta J_\alpha(y)}\right\}_{J=0}, \qquad \alpha,\beta = A,B\,. \tag{36}$$

How can we find the Green's functions on the flavor vacuum? Here we follow the approach presented in Ref. [26] for the evaluation for the Green's functions on exotic vacua. Flavor fields correlation functions on the flavor vacuum will be given by:

$${}_f\langle 0(t_+)|\varphi_\alpha(y)\varphi_\beta^\dagger(x)|0(t_-)\rangle_f \;=\; (-i)^2\left\{\frac{\delta\mathcal{Z}_f[J_\varphi]}{\delta J_\beta^*(x)\delta J_\alpha(y)}\right\}_{J=0}, \qquad \alpha,\beta = A,B\,, \tag{37}$$

and mass fields correlations functions on the flavor vacua are:

$${}_f\langle 0(t_+)|\varphi_i(y)\varphi_j^\dagger(x)|0(t_-)\rangle_f \;=\; (-i)^2\left\{\frac{\delta\mathcal{Z}_f[J_m]}{\delta J_j^*(x)\delta J_i(y)}\right\}_{J=0}, \qquad i,j=1,2\,, \tag{38}$$

where, taking into account the form of mixing generator (6) and using the Gaussian decomposition (14):

$$\begin{aligned}\mathcal{Z}_f[J_\varphi] = \Bigg\{ & e^{-i\tan\theta S_+\left[\frac{\delta}{\delta J_\phi},\frac{\delta}{\delta J_\pi},\frac{\delta}{\delta J_\phi^\dagger},\frac{\delta}{\delta J_\pi^\dagger}\right](\tau_+)}\, e^{if(\theta)S_3\left[\frac{\delta}{\delta J_\phi},\frac{\delta}{\delta J_\pi},\frac{\delta}{\delta J_\phi^\dagger},\frac{\delta}{\delta J_\pi^\dagger}\right](\tau_+)} \\ \times\; & e^{i\tan\theta S_-\left[\frac{\delta}{\delta J_\phi},\frac{\delta}{\delta J_\pi},\frac{\delta}{\delta J_\phi^\dagger},\frac{\delta}{\delta J_\pi^\dagger}\right](\tau_+)}\, e^{i\tan\theta S_+\left[\frac{\delta}{\delta J_\phi},\frac{\delta}{\delta J_\pi},\frac{\delta}{\delta J_\phi^\dagger},\frac{\delta}{\delta J_\pi^\dagger}\right](\tau_-)}\, e^{-if(\theta)S_3\left[\frac{\delta}{\delta J_\phi},\frac{\delta}{\delta J_\pi},\frac{\delta}{\delta J_\phi^\dagger},\frac{\delta}{\delta J_\pi^\dagger}\right](\tau_-)} \\ \times\; & e^{-i\tan\theta S_-\left[\frac{\delta}{\delta J_\phi},\frac{\delta}{\delta J_\pi},\frac{\delta}{\delta J_\phi^\dagger},\frac{\delta}{\delta J_\pi^\dagger}\right](\tau_-)}\, \mathcal{Z}[J_\varphi, J_\pi]\,\}_{\mathbf{J}_\pi=0} \Bigg\}\,. \end{aligned} \tag{39}$$

[1] This fact can be proved (cf. Ref [26]) by noticing that in the standard derivation of the generating functional for the Green's functions (see e.g. [27], one makes use of the resolution of unity $\sum_n |n\rangle\langle n| = \mathbb{1}$ in terms of Hamiltonian eigenstates, which, in this case, can be constructed only on $|0\rangle_m$.)

IOP Conf. Series: Journal of Physics: Conf. Series **880** (2017) 012051 doi:10.1088/1742-6596/880/1/012051

Here $f(\theta)$ has to be determined by fixing a particular operator ordering and $\mathbf{J}_\pi = 0$ is a shorthand notation for $J_\pi = J_\pi^\dagger = 0$. This definition can be understood by means of the definition (11) and by regarding the generator $\hat{G}_\theta(t)$ as an extra interaction term. Mass and flavor field Green's functions evaluated on the flavor vacuum at time t are thus given, by

$$
{}_f\langle 0(t)|\varphi_\alpha(y)\varphi_\beta^\dagger(x)|0(t)\rangle_f \;=\; \lim_{\substack{t_+\to t\\ t_-\to t}} (-i)^2 \left\{\frac{\delta \mathcal{Z}_f[J_\varphi]}{\delta J_\beta^*(x)\delta J_\alpha(y)}\right\}_{J_\varphi=0}, \qquad \alpha,\beta = A,B\,, \tag{40}
$$

$$
{}_f\langle 0(t)|\varphi_i(y)\varphi_j^\dagger(x)|0(t)\rangle_f \;=\; \lim_{\substack{t_+\to t\\ t_-\to t}} (-i)^2 \left\{\frac{\delta \mathcal{Z}_f[J_\varphi]}{\delta J_j^*(x)\delta J_i(y)}\right\}_{J_m=0}, \qquad i,j = 1,2\,. \tag{41}
$$

The evaluation of Eq. (39) is long and cumbersome. However, it is possible to perform calculations relatively easily under the assumption that $\theta \ll 1$. With this we keep only the first order terms in θ. Since

$$
\log\cos\theta \;\approx\; -\theta^2/2\,, \tag{42}
$$

we can neglect the terms containing $S_3(t)$ in the leading order in θ. We thus have:

$$
\mathcal{Z}_f[J_\varphi] \;\approx\; \mathcal{Z}[J_\varphi] \;-\; i\theta\left[(S_+(\tau_+) \;-\; S_-(\tau_+) \;+\; S_-(\tau_-) \;-\; S_+(\tau_-))\,\mathcal{Z}[J_\varphi, J_\pi]\right]_{\mathbf{J}_\pi=0}\,. \tag{43}
$$

To explicitly evaluate Eq. (43) we consider each term in the sum on the RHS separately, namely

$$
\begin{aligned}
&S_+(\tau_+)\mathcal{Z}[J_\varphi, J_\pi] \\
&\quad = i\int \mathrm{d}^3\mathbf{x}\left[\frac{\delta}{\delta J_{\pi\,A}(\tau_+,\mathbf{x})}\frac{\delta}{\delta J^*_{\phi\,B}(\tau_+,\mathbf{x})} - \frac{\delta}{\delta J_{\phi\,A}(\tau_+,\mathbf{x})}\frac{\delta}{\delta J^*_{\pi\,B}(\tau_+,\mathbf{x})}\right]\mathcal{Z}[J_\varphi, J_\pi]\,,
\end{aligned} \tag{44}
$$

$$
\begin{aligned}
&S_+(\tau_-)\mathcal{Z}[J_\varphi, J_\pi] \\
&\quad = i\int \mathrm{d}^3\mathbf{x}\left[\frac{\delta}{\delta J_{\pi\,A}(\tau_-,\mathbf{x})}\frac{\delta}{\delta J^*_{\phi\,B}(\tau_-,\mathbf{x})} - \frac{\delta}{\delta J_{\phi\,A}(\tau_-,\mathbf{x})}\frac{\delta}{\delta J^*_{\pi\,B}(\tau_-,\mathbf{x})}\right]\mathcal{Z}[J_\varphi, J_\pi]\,,
\end{aligned} \tag{45}
$$

$$
\begin{aligned}
&S_-(\tau_+)\mathcal{Z}[J_\varphi, J_\pi] \\
&\quad = -i\int \mathrm{d}^3\mathbf{x}\left[\frac{\delta}{\delta J^*_{\pi\,A}(\tau_+,\mathbf{x})}\frac{\delta}{\delta J_{\phi\,B}(\tau_+,\mathbf{x})} - \frac{\delta}{\delta J^*_{\phi\,A}(\tau_+,\mathbf{x})}\frac{\delta}{\delta J_{\pi\,B}(\tau_+,\mathbf{x})}\right]\mathcal{Z}[J_\varphi, J_\pi]\,,
\end{aligned} \tag{46}
$$

$$
\begin{aligned}
&S_-(\tau_-)\mathcal{Z}[J_\varphi, J_\pi] \\
&\quad = -i\int \mathrm{d}^3\mathbf{x}\left[\frac{\delta}{\delta J^*_{\pi\,A}(\tau_-,\mathbf{x})}\frac{\delta}{\delta J_{\phi\,B}(\tau_-,\mathbf{x})} - \frac{\delta}{\delta J^*_{\phi\,A}(\tau_-,\mathbf{x})}\frac{\delta}{\delta J_{\pi\,B}(\tau_-,\mathbf{x})}\right]\mathcal{Z}[J_\varphi, J_\pi]\,.
\end{aligned} \tag{47}
$$

First we consider the expression (44). This explicitly reads as

$$
\begin{aligned}
\{S_+(\tau_+)\mathcal{Z}[J_\varphi, J_\pi]\}_{\mathbf{J}_\pi=0} = i\int \mathrm{d}^3\mathbf{x}_+\int \mathrm{d}^4x\int \mathrm{d}^4y\,[\,& J_\varphi^{*\,\alpha}(x)\partial_{\tau_+}\Delta^{\alpha A}(x-x_+)\Delta^{B\beta}(x_+-y)J_\varphi^\beta(y) \\
-\;& J_\varphi^{*\,\alpha}(x)\Delta^{\alpha A}(x-x_+)\partial_{\tau^+}\Delta^{B\beta}(x_+-y)J_\varphi^\beta(y)\,]\,,
\end{aligned} \tag{48}
$$

where

$$
\alpha,\beta = A,B\,, \qquad \tilde{x}_+ \;\equiv\; (\tau_+,\mathbf{x}_+)\,, \qquad \tilde{x}_- \;\equiv\; (\tau_-,\mathbf{x}_+)\,, \tag{49}
$$

IOP Conf. Series: Journal of Physics: Conf. Series **880** (2017) 012051 doi:10.1088/1742-6596/880/1/012051

where we explicitly wrote components of the flavor matrix (34), and a sum over repeated index is understood. In the same way we have

$$\{S_+(\tau_-)\mathcal{Z}[J_\varphi, J_\pi]\}_{\mathbf{J}_\pi=0} = i\int d^3\mathbf{x}_- \int d^4x \int d^4y\, [\, J^{*\,\alpha}_\varphi(x)\partial_{\tau_-}\Delta^{\alpha A}(x-x_-)\Delta^{B\beta}(x_- - y)J^\beta_\varphi(y)$$
$$- \; J^{*\,\alpha}_\varphi(x)\Delta^{\alpha A}(x-x_-)\partial_{\tau+}\Delta^{B\beta}(x_- - y)J^\beta_\varphi(y)\,]\,\mathcal{Z}[J]\,, \tag{50}$$

$$\{S_-(\tau_+)\mathcal{Z}[J_\varphi, J_\pi]\}_{\mathbf{J}_\pi=0} = -i\int d^3\mathbf{x}_+ \int d^4x \int d^4y\, [\, J^{*\,\beta}_\varphi(x)\Delta^{\beta B}(x-x_+)\partial_{\tau_+}\Delta^{\alpha A}(x_+ - y)J^\alpha_\varphi(y)$$
$$- \; J^{*\,\beta}_\varphi(x)\partial_{\tau_+}\Delta^{\beta B}(x-x_+)\Delta^{A\alpha}(x_+ - y)J^\alpha_\varphi(y)\,]\,\mathcal{Z}[J]\,, \tag{51}$$

$$\{S_-(\tau_-)\mathcal{Z}[J_\varphi, J_\pi]\}_{\mathbf{J}_\pi=0} = -i\int d^3\mathbf{x}_- \int d^4x \int d^4y\, [\, J^{*\,\beta}_\varphi(x)\Delta^{B\beta}(x-x_-)\partial_{\tau_-}\Delta^{\alpha A}(x_- - y)J^\alpha_\varphi(y)$$
$$- \; J^{*\,\beta}_\varphi(x)\partial_{\tau_+}\Delta^{\beta B}(x-x_-)\Delta^{A\alpha}(x_- - y)J^\alpha_\varphi(y)\,]\,\mathcal{Z}[J]\,. \tag{52}$$

The result can be written in a shorthand notation as

$$\mathcal{Z}_f[J_\varphi] \;\approx\; \mathcal{Z}[J_\varphi]\Big\{1-\theta\int d^3\mathbf{x}_\pm \int d^4x \int d^4y\, [\, -\, J^{*\,\alpha}_\varphi(x)\partial_{\tau_+}\Delta^{\alpha A}(x-x_+)\Delta^{B\beta}(x_+ - y)J^\beta_\varphi(y)$$
$$+\; J^{*\,\alpha}_\varphi(x)\Delta^{\alpha A}(x-x_+)\partial_{\tau+}\Delta^{B\beta}(x_+ - y)J^\beta_\varphi(y) + J^\alpha_\varphi(x)\partial_{\tau_-}\Delta^{\alpha A}(x-x_-)\Delta^{B\beta}(x_- - y)J^{*\,\beta}_\varphi(y)$$
$$-\; J^\alpha_\varphi(x)\Delta^{\alpha A}(x-x_-)\partial_{\tau^-}\Delta^{B\beta}(x_- - y)J^{*\,\beta}_\varphi(y) - J^\beta_\varphi(x)\Delta^{B\beta}(x-x_+)\partial_{\tau_+}\Delta^{\alpha A}(x_+ - y)J^{*\,\alpha}_\varphi(y)$$
$$+\; J^\beta_\varphi(x)\partial_{\tau_+}\Delta^{\beta B}(x-x_+)\Delta^{A\alpha}(x_+ - y)J^{*\,\alpha}_\varphi(y) + J^{*\,\beta}_\varphi(x)\Delta^{B\beta}(x-x_-)\partial_{\tau_-}\Delta^{\alpha A}(x_- - y)J^\alpha_\varphi(y)$$
$$-\; J^{*\,\beta}_\varphi(x)\partial_{\tau_+}\Delta^{\beta B}(x-x_-)\Delta^{A\alpha}(x_- - y)J^\alpha_\varphi(y)\,]\Big\}\,. \tag{53}$$

If $\theta = 0$ this reduces, correctly, to Eq. (33), i.e. the standard definition[2].

4. Conclusions

In this paper we have shown how the conventional definition of generating functional of Green's functions has to be modified in order to take properly into account the rich vacuum structure associated to mixing transformation in QFT. Although we performed our calculations in the approximation of a small mixing angle, the partition function obtained (53) is evidently structurally richer than its standard counterpart given by Eq. (35). This extra information is encoded in the presence of two-point mass-vacuum correlations and their temporal changes.

It should be stressed that we have not touched other subtle issues that are relevant in relativistic QFT. Clearly, we have not discussed a renormalization issues which would need to be included in more realistic models. Second, our Hamiltonian-based computations were not manifestly Lorentz invariant. This should be no problem, say in the condensed matter framework, but is a clearly disadvantageous for particle physics applications. So, in the context of mixing, a Lagrangian formulation till needs to be done. For some related discussion see, e.g. [28, 29, 30].

This work should be a starting point in an exact (non-perturbative) FI-based study of meson oscillations. Moreover, the extension to the fermionic case is required to the study of flavor oscillations and to a complete description of the dynamical generation of fermionic mixing. An exact calculation will reveal if the flavor Green's functions derived in this way are in agreement with those derived in Ref. [15].

[2] This choice affects the vacuum only. Here is understood that we choose currents independently.

Acknowledgments
P.J. was supported by the Czech Science Foundation Grant No. 17-33812L

References

[1] Cabibbo N 1963 *Phys. Rev. Lett.* **10** 531
[2] Glashow S L, Iliopoulos J and Maiani L 1970 *Phys. Rev.* D **2** 1285
[3] Kobayashi M and Maskawa T 1973 *Prog.Theor.Phys.* **49** 652
[4] Gell Mann M and Pais A 1955 *Phys. Rev.* **95** 1387
[5] Christenson J H, Cronin J W, Fitch W R and Turlay R 1964 *Phys. Rev. Lett.* **13** 138
[6] KTeV Collaboration 1999; *Phys. Rev. Lett.* **83** 22
[7] Pontecorvo B 1957 *Sov. Phys. JETP* **6** 429
[8] Pontecorvo B 1957 *Sov. Phys. JETP* **7** 172-02
[9] Bilenky S and Pontecorvo B 1957 *Phys. Rep.* **41** 225
[10] Davis R 1994 *Prog. Part. Nucl. Phys.* **32** 13
[11] Smy M B (Super-Kamiokande) 2003 *Nucl. Phys. Proc. Suppl.* **118** 25 (2003) Hallin A L et al. 2003 *Nucl. Phys. Proc. Suppl.* **118** 3; Eguchi K et al. (KamLAND) 2003 *Phys. Rev. Lett.* **90** 021802; Hampel W et al. (GALLEX) 1999 *Phys. Lett.* B **447** 127; Altmann M et al. (GNO) 2000 *Phys. Lett.* B **490** 16
[12] Giunti C 2004 *Preprint* arXiv:hep-ph/0401244
[13] Blasone M and Vitiello G 1995 *Ann. Phys.* **244** 283
[14] Blasone M, Jizba P and Vitiello G 2011 *Quantum Field Theory and its Macroscopic Manifestations* (London: World Scientific & ICP)
[15] Blasone M, Henning P A and Vitiello G 1999 *Phys. Lett.* B **451** 140
[16] Blasone M, Gargiulo M V and Vitiello G 2016 *Phys. Lett.* B **716** 104
[17] Blasone M, Lambiase G and Luciano G G 2015 *J. Phys. Conf. Ser.* **631** no.1 012053
[18] Blasone M, Jizba P, Lambiase G and Mavromatos N E 2014 *J. Phys.* **538** 012003
[19] Blasone M, Jizba P and Smaldone L 2015 *Il Nuovo Cimento* **38 C** 201
[20] Umezawa H, Takahashi Y and Kamefuchi S 1964 *Ann. Phys.* **26** 336
[21] Umezawa H, Matsumoto H and Tachiki M 1982 *Thermo Field Dynamics And Condensed States* (Amsterdam: North-Holland)
[22] Haag R 1996 *Local Quantum Physics: Fields, Particles, Algebras* (Berlin: Springer-Verlag)
[23] Blasone M, Capolupo A, Romei O and Vitiello G 2001 *Phys. Rev.* D **63** 125015
[24] Tung W K, *Group Theory in Physics* 1980 (Singapore: World Scientific)
[25] Fujii K, Habe C and Yabuki T 1999 *Phys. Rev.* D **59** 113003
[26] Blasone M, Jizba P and Smaldone L 2017 *in preparation*
[27] Greiner W and Reinhardt J 1996 *Field Quantization* (Berlin Heidelberg: Springer-Verlag)
[28] Giunti C 2004 *Am. J. Phys.* **72** 699
[29] Blasone M, Magueijo J and Pires Pacheco P 2005 *Braz. J. Phys.* **35** 447
[30] Blasone M, Di Mauro M and Lambiase G 2005 *Acta Phys. Polon.* B **36** 3255

Covariant electromagnetic field lines

Y Hadad[1], E Cohen[2], I Kaminer[3] and A C Elitzur[4]

[1] Department of Mathematics, University of Arizona, Tucson, AZ, 85721, USA
[2] H.H. Wills Physics Laboratory, University of Bristol, Tyndall Avenue, Bristol, BS8 1TL, U.K
[3] Department of Physics, Massachusetts Institute of Technology, 77 Massachusetts Avenue, Cambridge, Massachusetts 02139, USA
[4] Iyar, The Israeli Institute for Advanced Research, POB 651, Zichron Ya'akov 3095303, Israel

E-mail: yaron.hadad@gmail.com

Abstract. Faraday introduced electric field lines as a powerful tool for understanding the electric force, and these field lines are still used today in classrooms and textbooks teaching the basics of electromagnetism within the electrostatic limit. However, despite attempts at generalizing this concept beyond the electrostatic limit, such a fully relativistic field line theory still appears to be missing. In this work, we propose such a theory and define covariant electromagnetic field lines that naturally extend electric field lines to relativistic systems and general electromagnetic fields. We derive a closed-form formula for the field lines curvature in the vicinity of a charge, and show that it is related to the world line of the charge. This demonstrates how the kinematics of a charge can be derived from the geometry of the electromagnetic field lines. Such a theory may also provide new tools in modeling and analyzing electromagnetic phenomena, and may entail new insights regarding long-standing problems such as radiation-reaction and self-force. In particular, the electromagnetic field lines curvature has the attractive property of being non-singular everywhere, thus eliminating all self-field singularities without using renormalization techniques.

1. Introduction

Mathematical formalism is usually preferred over pictorial explanations in physics. It is the equations that enable us to predict the evolution of physical systems with nearly any desired precision. When we think of atoms as tiny solar systems, or draw spacetime as two-dimensional we are well aware that these pictorial methods are limited, leaving it to mathematics to give the "right" description. The compatibility between the two methods is indeed a source of wonder [1].

The visual aspect, however, goes far beyond being an intuitive aid to the formalism. Physical reality, after all, is often richer and more complex than what mathematics captures. Moreover, since mathematics is based on abstract definitions, in general it is compatible with many possible realities. This leaves open the question why this particular set of assumptions rather than another is compatible with our universe. Gödel's theorem [2], sometimes phrased as "truth transcends theoremhood" [3], stresses this incompleteness.

Indeed, geometry-based physical models not only preceded mathematical ones, but often kept guiding their development. The evolution of general relativity from the special one is a case in point. It was highly pictorial almost naive thought-experiments which enabled the two theories to generalize their insights from inertial to non-inertial motions.

This is also the case of electromagnetism, and the nature of the electromagnetic field in particular. It was Faraday's highly pictorial description of the field lines [4] as carriers of the electromagnetic field through their tension and pressure which later led to Maxwell's concise mathematical theory yielding the new predictions of this unique field representation [5]. Intriguingly, Maxwell himself kept heavily employing Faraday's "strings" and "rubber bands" alongside with his abstract equations, stressing time and again their mechanical properties.

In the current paper we resume this line of investigation by exploring the notion of fully relativistic electromagnetic field lines. We ascribe the field-lines mechanical properties and prove quantitatively that their curvature is not only the source of the charge's acceleration but also its origin.

In [6] the authors found a relationship between the curvature of electric field lines and the acceleration of a non-relativistic charge (Fig. 1). While external fields and acceleration of charges bend the electric field lines, non-relativistic electric charges always accelerate in the direction of straight electric field lines. Another advantage of the field lines curvature framework is its lack of singularities. It is well known that the total electric field is singular nearby point charges due to the notorious self field problem [7]. However, the field lines curvature is well-defined and everywhere non-singular, and provides a potential route to study electrodynamics without the use of renormalization. There are reasons to believe that similar results hold in a more general setting, and this paper is attempting to convey this.

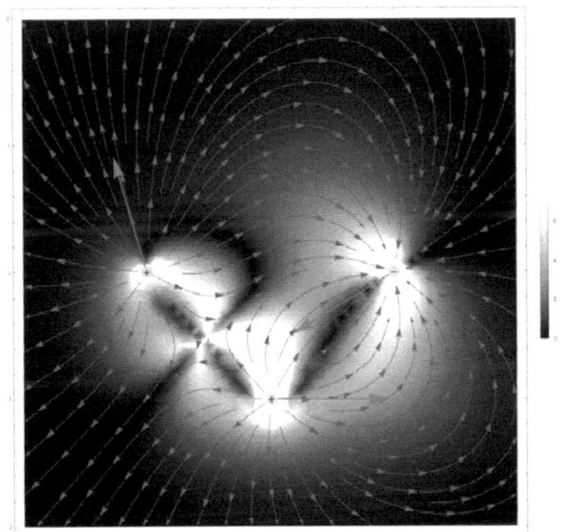

Figure 1. Electromagnetic field lines are represented by curves, whose curvature is computed explicitly below in Eq. (22). Bright colors in the figure represent high scalar curvature and black areas have zero electromagnetic field lines curvature. The red arrows are the acceleration of the charges, and the figure shows one of the main conclusions of this work - charges always accelerate in the direction of straight electromagnetic field lines. It is also proven that while the total electromagnetic field is singular due to the self fields of the charges, the electromagnetic field lines curvature is always non-singular (even for point charges).

Geometrically, the four-acceleration of a particle is the curvature vector of its world line [8]. This is an equivalence between the kinematics of a charge and the geometry of its world line, which links acceleration (of a particle) and curvature (of its world line). This paper defines a covariant notion of field lines for the electromagnetic tensor in Minkowski spacetime. We study the curvature of the electromagnetic field lines and show that it is closely related to the four-acceleration of the charge. This gives another equivalence, relating the kinematics of charges to the geometry of field lines. Therefore, at least in the framework of the classical theory of electrodynamics there is a sequence of three equivalences between four-acceleration, curvature of world lines and curvature of electromagnetic field lines.

Classical field theories suffer from the problem of self-field singularities when point-like particles are introduced. This is the notorious *self-force (or self-field) problem.* For example, Maxwell's equations predict the electromagnetic field to be singular at the location of point-like electrons [7]. One of the consequences of this work is that the curvature of the electromagnetic field lines is well-defined and non-singular, even in the vicinity of a point charge. This means that the framework studied in this work does not suffer from the problem of self-force. In other words, the electromagnetic field lines interpretation may shed light on how charges interact with their own electromagnetic field.

In this conference proceedings paper we summarize some of our recent results regarding the relativistic electromagnetic curvature. Many derivations and further examples and implications were omitted for brevity. A more detailed analysis can be found in [9].

The outline of the paper is as follows: in section 2 we motivate the equivalence between field lines curvature and acceleration. This is done by studying the special case of non-relativistic charges in an external electric field. In section 3 we provide a covariant definition of electromagnetic field lines. Section 4 studies the curvature of covariant field lines and its relationship to four-acceleration. Section 5 summarizes the results of this work.

2. Acceleration implies curvature

Consider a charge in empty (flat) space. Such a charge generates an electric field, which is commonly depicted by electric field lines. If the charge's velocity changes abruptly from zero to a constant velocity, it will be emitting an electromagnetic wave. For a sufficiently far observer the charge still appears to be at rest and its field-lines isotropic outside of the forward light-cone of the event of instantaneous acceleration. Inside the light-cone the charge is already moving, hence the field-lines appear denser in the direction perpendicular to the direction of motion. To ensure continuity of the field-lines, they appear to be partially "broken-and-redrawn" on the light-cone's surface, as shown in Fig. 2a.

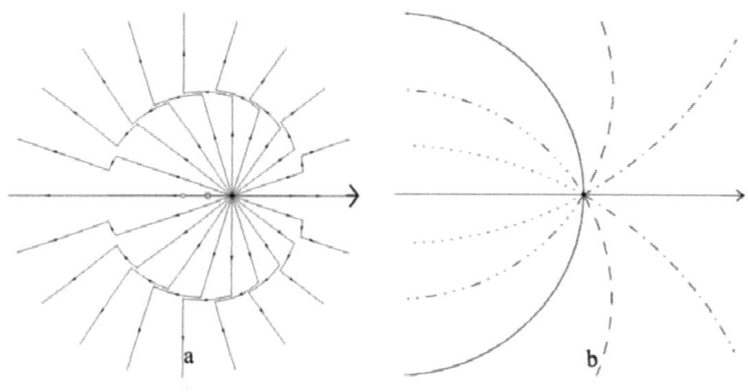

Figure 2. (a) The electric field lines generated by a charge after an instantaneous velocity change. The field lines appear to be partially "broken-and-redrawn" on the light-cone due to the instantaneous acceleration of the charge. (b) Electric field lines of a constantly accelerating charge bend. The charge always accelerates in the direction of straight field lines.

When the velocity of the charge changes at a constant rate (constant acceleration) the field lines appear to be "pealing off" the charge (see Fig. 2b). The only field-lines that do not curve are the two that are directed along the charge's acceleration vector. The lines remain straight because they are on the only axis on which symmetric constraints act from both sides.

This implies that as a charge accelerates, the electric field lines around the charge curve. At least for purely electric field lines we may say that charge acceleration implies that the field lines are (locally) curved, a fact that was proven in [6].

The goals of the rest of this paper are to (a) extend the notion of electric field lines to covariant electromagnetic field lines, (b) study the curvature of electromagnetic field lines, and (c) show that the electromagnetic field lines curvature is fundamentally related to the notion of charge acceleration.

3. Covariant field lines

Electric (or magnetic) field lines are defined to be curves in three-dimensional space which are everywhere tangent to the electric (magnetic) field, while their density is proportional to the magnitude of the field. Since the tangent is defined using the derivative (a first order operator), a point in space with a non-singular electric (magnetic) field has a unique field line that passes through it.

Studying the electromagnetic field lines in a fully relativistic manner requires a covariant definition of field lines that generalizes the notions of electric and magnetic field lines. It was previously argued [10] that field lines are intrinsically non-relativistic for two reasons. First, since electric and magnetic fields transform into each other under Lorentz transformations one cannot expect to find a covariant notion of a pure electric field line or a pure magnetic field line. Secondly, points on electric (or magnetic) field line all have the same time coordinate and are therefore inherently non-local.

In Minkowski spacetime the electromagnetic field is represented by the Faraday tensor $F^{\alpha\beta}$. The Faraday tensor is a covariant tensor that treats the electric and magnetic fields as a single physical object. One would expect that the first issue raised in [10] can be addressed by studying electromagnetic field lines using the Faraday electromagnetic tensor. Therefore, we are not expecting to split the electric field lines and magnetic field lines, but maintaining them as a single entity. The second issue can be overcome by refraining to use the notion of an absolute time in the definition of the electromagnetic field lines. Instead, electromagnetic field lines must be defined relatively to the observer who is measuring them.

The electromagnetic field $F^{\alpha\beta}$ as measured by an observer with four-velocity u^α can be reduced to a (four-)vector field. Consider an electromagnetic field line $e^\alpha(\lambda)$ with a parameterization variable λ. We require the electromagnetic field line to satisfy *the electromagnetic field line equation*

$$\frac{de^\alpha}{d\lambda} = E^\alpha, \tag{1}$$

where

$$E^\alpha(x^\beta) = F^{\alpha\gamma}(x^\beta)u_\gamma(\tau_{\text{ret}}), \tag{2}$$

and τ_{ret} is the retarded time from the observer to the spacetime event x^β, namely, the time it takes a ray of light emitted from the observer to reach x^β. Here and in the rest of this work we use a spacetime metric with signature $(-+++)$. We also employ the following convention: space-like four-vectors with no indices are to be interpreted as their Lorentz norm ($E = \sqrt{E^\beta E_\beta}$). Note that this is well-defined as space-like vectors have a positive Minkowski norm.

Let κ^α denote the curvature vector of the electromagnetic field line. In general, the curvature vector of a curve is defined to be the curve's acceleration with respect to unit-length parameterization. We may therefore compute the curvature of an electromagnetic field line as a curve in Minkowski spacetime by calculating

$$\kappa^\alpha = \frac{d^2e^\alpha}{ds^2}, \tag{3}$$

where s is a unit-length parameterization satisfying

$$\frac{de^\alpha}{ds} = \frac{E^\alpha}{E}. \tag{4}$$

By using Eq. (4) the curvature vector may be written explicitly as

$$\kappa^\alpha = \kappa_1^\alpha - \kappa_2^\alpha, \tag{5}$$

where we used the chain-rule $\frac{d}{ds} = \frac{dx^\gamma}{ds}\frac{\partial}{\partial x^\gamma}$, and introduce the notation

$$\kappa_1^\alpha = \frac{1}{E^2} E^\beta \frac{\partial E^\alpha}{\partial x^\beta}, \tag{6}$$

and

$$\kappa_2^\alpha = \frac{1}{E^4}(E_\gamma E^\beta \frac{\partial E^\gamma}{\partial x^\beta})E^\alpha \tag{7}$$

which will be used in the next section.

Many questions may rise regarding the relationship between the definition of covariant electromagnetic field lines (4) and field lines of electric or magnetic fields in the electrostatic limit. Unfortunately this is beyond the scope of this paper and such questions will be answered in details elsewhere [9]. The reader can easily verify that for a charge at rest, the spatial components of Eq. (4), the spatial components of the curvature four-vector (5), and the scalar curvature $\kappa = \sqrt{\kappa^\alpha \kappa_\alpha}$ restore their definition in this electrostatic limit[1]. Therefore, the form (4) of covariant electromagnetic field lines naturally extends its non-relativistic form used in [6].

Now, we have a covariant notion of electromagnetic field lines that passes the concerns raised in [10]. Having introduced this natural generalization, we proceed to studying its implications.

4. Covariant field lines curvature

Consider a charge q with mass m traveling along a world line $z^\alpha(\tau)$ with proper time τ. The charge generates an electromagnetic field on its forward light cone. A point x^α is on the light cone of the charge if

$$k^\alpha k_\alpha = 0 \tag{8}$$

where the null vector k^α is

$$k^\alpha = x^\alpha - z^\alpha(\tau_{\text{ret}}). \tag{9}$$

The (self) electromagnetic field produced by the charge is given by the retarded Liénard-Wiechert equation [11]

$$F_{\text{self}}^{\alpha\beta} = \frac{q}{R^2}\left(U^\alpha k^\beta - U^\beta k^\alpha\right), \tag{10}$$

where R is the retarded distance

$$R = -k^\alpha u_\alpha \tag{11}$$

and U^α is the Synge vector [12]

$$U^\alpha = Bu^\alpha + a^\alpha. \tag{12}$$

B is the so-called Plebański invariant [13]

$$B = \frac{1-W}{R} \tag{13}$$

[1] One may study another form of (dual) electromagnetic field lines, by considering the dual Faraday tensor. Here the spatial components of the dual electromagnetic field lines for a charge which is at rest give the magnetic field lines. The study of the dual electromagnetic field lines is straightforward and will not be discussed here.

and W is the contraction

$$W = -k^\alpha a_\alpha. \tag{14}$$

The four-velocity and four-acceleration used in Eqs. (11), (12) and (14) are the retarded four-velocity $u^\alpha = u^\alpha(\tau_{\text{ret}})$ and the retarded four-acceleration $a^\alpha = a^\alpha(\tau_{\text{ret}})$ respectively.

We shall consider first the problem of a free charge in empty space. Since q is a free charge, the only contribution to the field lines is the self field produced by the charge q itself. Substituting Eq. (10) into Eq. (2) gives

$$E^\alpha_{\text{self}} = -\frac{q}{R}U^\alpha + \frac{q}{R^3}(1-W)k^\alpha, \tag{15}$$

with squared Minkowski norm

$$(E_{\text{self}})^2 = \frac{q^2}{R^4}(1 - W^2 + a^2R^2). \tag{16}$$

We use Eqs. (15) and (16) in the curvature field vector Eq. (5)

$$\kappa^\alpha = \kappa_1^\alpha - \kappa_2^\alpha. \tag{17}$$

under the assumption that *the field lines are measured by the charge itself.* The first term in (17) defined in Eq. (6) gives

$$\kappa_1^\alpha = \frac{1}{\Delta}\left\{\left[\frac{-2(1-W)}{R^2} + \frac{V}{R} - a^2\right]P^\alpha + \frac{W(1-W)}{R^2}k^\alpha + (1-W)a^\alpha + R\dot{a}^\alpha\right\}, \tag{18}$$

and the second term which was defined in Eq. (7) gives

$$\kappa_2^\alpha = \frac{1}{\Delta^2}\left[(1-W)(-2+W^2) + WRV - R^2a^2 - R^3 a\cdot\dot{a}\right]\left(\frac{1-W}{R^2}P^\alpha - a^\alpha\right). \tag{19}$$

Here we defined the projection four-vector

$$P^\alpha = k^\alpha - Ru^\alpha, \tag{20}$$

the scalar Δ to be

$$\Delta = 1 - W^2 + a^2R^2 \tag{21}$$

and $V = -k^\alpha \dot{a}_\alpha$ to be the projection of the jerk $\dot{a}^\alpha = \frac{da^\alpha}{d\tau}$ on the null-vector k^α. The projection four-vector P^α defined in Eq. (20) has convenient properties that will be studied elsewhere [9].

Eq. (17) together with Eqs. (18) and (19) give a closed-form expression for the curvature of the covariant electromagnetic field lines produced by a free charge in Minkowski spacetime. The expression is ready to be evaluated at any event in spacetime explicitly. Studying Eq. (17) reveals a couple of interesting properties the electromagnetic field lines curvature possesses.

As is well-known due to the self-field problem, the electromagnetic field tensor (10) is singular at the position of the charge and behaves like $\frac{1}{R^2}$ in the limit $R \to 0$. Despite that fact, *the electromagnetic field lines curvature is non-singular in the vicinity of the charge.* In order to see this we recall from Eq. (10) that the electromagnetic (self) field at each event x^α is determined by the four-position of the charge at the retarded four-position $z^\alpha(\tau_{\text{ret}})$. This means that we need to study the limit $x^\alpha \to z^\alpha$ along the light-cone, and we do so by letting $k^\alpha = \varepsilon\hat{k}^\alpha$, where $\hat{k}^\alpha$ is of order unity and ε is a small parameter.

The leading order expansion of the curvature four-vector (17) in ε gives

$$\kappa^\alpha = -a^\alpha - \frac{\hat{W}}{\hat{R}^2}\hat{k}^\alpha + \frac{2\hat{W}}{\hat{R}}u^\alpha + O(\varepsilon), \tag{22}$$

where the hats denote normalized quantities of order one in ε, namely $k^\alpha = \varepsilon\hat{k}^\alpha$, $W = \varepsilon\hat{W}$ and $R = \varepsilon\hat{R}$. Eq. (22) shows that the field lines curvature does not contain any singular terms in ε.

It may appear that Eq. (22) is singular due to the denominators containing the normalized retarded position $\hat{R}$, but this is not the case. Let us explicitly write the normalized null-vector $\hat{k}^\alpha = (1, \hat{\vec{k}})$ and the four-velocity to be $u^\alpha = \gamma(1, \vec{v})$, where $\hat{\vec{k}}$ is a unit spatial vector, $\vec{v}$ is the spatial velocity of the charge in units where the speed of light is unity $c = 1$, and $\gamma = \frac{1}{\sqrt{1-\vec{v}^2}}$ is the relativistic Lorentz factor. We now see that the normalized retarded distance is always non-zero

$$\hat{R} = \gamma(1 - \hat{\vec{k}} \cdot \vec{v}) \neq 0 \tag{23}$$

as the charge is not massless and is moving slower than the speed of light in vacuum. Therefore we may conclude that the covariant electromagnetic field lines curvature (17) is everywhere non-singular even for point charges. This is a non-trivial property as quantities that are dependent on the self field of point charges are typically singular.

Contracting Eq. (22) with the normalized null-vector $\hat{k}_\alpha$ gives

$$\hat{k}_\alpha \kappa^\alpha = -\hat{W} \tag{24}$$

which vanishes when $\hat{W} = -\hat{k} \cdot a = 0$ and defines a two-dimensional surface in spacetime that intersects the four-position of the charge. On this surface Eq. (22) reveals that the four-curvature is just the negative four-acceleration

$$\kappa^\alpha = -a^\alpha. \tag{25}$$

Eq. (25) implies that the four-acceleration of the charge is precisely (minus) the curvature of the electromagnetic field lines that it is producing. By virtue of the perpendicularity condition (24), this relationship holds whenever the field line curvature itself is perpendicular to the light cone. Eq. (25) also implies that we may interchangeably translate a property related to the kinematics of the charge (acceleration) and the geometry of the field lines (curvature).

We can observe other properties of Eq. (17) by considering the motion of the charge in its instantaneous rest frame. The four-velocity now is $u^\alpha = (1, 0, 0, 0)$ and the four-acceleration is $a^\alpha = (0, \vec{a})$, where $\vec{a}$ is the spatial acceleration of the charge. The electromagnetic field lines four-curvature (25) is

$$\kappa^\alpha = |\vec{a}|(-\cos\theta, -\hat{a} + \hat{k}\cos\theta). \tag{26}$$

The scalar curvature is given by the squared norm of the curvature four-vector,

$$\kappa^2 = -\vec{a}^2 \cos(2\theta), \tag{27}$$

where θ is the angle between the spatial acceleration $\vec{a}$ and the retarded position of the particle $\vec{k} = \vec{x} - \vec{z}(\tau_{\text{ret}})$. If we separate the four-curvature κ^α to its temporal part (0-component) and spatial part (components $1, 2, 3$) by writing $\kappa^\alpha = (\kappa^0, \vec{\kappa})$, we see that the *scalar spatial curvature* is

$$|\vec{\kappa}| = |\vec{a}||\sin\theta|. \tag{28}$$

Therefore, locally the field lines of an accelerating charge ($\vec{a} \neq \vec{0}$) are spatially straight only in the direction of the charge acceleration $\theta = 0$. In this case the temporal component of the field lines curvature yields the magnitude of the acceleration

$$\kappa^0 = -|\vec{a}|. \tag{29}$$

In other words, nearby a freely accelerating charge all of its kinematical properties are determined by the field lines curvature. This generalizes the result derived in [6].

The next obvious step for this work is to study the curvature of arbitrary covariant electromagnetic field lines. In the electrostatic limits, this was done in [6], where the local behavior of the electric field lines curvature and the dynamics of the charge were shown to be related. A similar analysis can be provided in this fully relativistic case.

The total electromagnetic field can be written as a superposition

$$F^{\alpha\beta} = F^{\alpha\beta}_{\text{ext}} + F^{\alpha\beta}_{\text{self}} \tag{30}$$

of the external field $F^{\alpha\beta}_{\text{ext}}$ and the self field $F^{\alpha\beta}_{\text{self}}$ produced by the charge as defined in (10). Even when there is an external electromagnetic, in the leading order the covariant electromagnetic field lines curvature satisfies Eq. (22). This implies that also in the most general case, the electromagnetic field lines curvature is a non-singular physical object. The next order correction to (22) is directly related to the external electromagnetic field $F^{\alpha\beta}_{\text{ext}}$, and reveals great insights on the relationship between curvature of electromagnetic field lines and the full dynamics of accelerating electric charges. This result will be studied in details in [9].

5. Summary

In this paper we returned to the origins of the electromagnetic theory as outlined by Faraday and Maxwell. Following the founding fathers we brought the field lines back into the heart of discussion, and also elevated them to be fully covariant objects.

To summarize our main results: we defined a notion of covariant electromagnetic field lines (4) and calculated their four-curvature at an arbitrary event x^α for a free charge. We derived an explicit closed-form formula for the four-curvature of self fields (22). The field lines curvature was shown to be everywhere non-singular. Most importantly, the field lines curvature seems to completely determine the kinematics of the electric charges. Specifically and as depicted in Fig. 1, *charges always accelerate in the direction of straight electromagnetic field lines.* Further results are presented in [9].

Acknowledgments

E.C. was supported by ERC AdG NLST. I.K. was supported by a Marie Curie Grant No. 328853-MC-BSiCS.

References

[1] Wigner E P 1960 *Commun. Pur. Appl. Math.* **13** 1-14
[2] Gödel K 1931 *Monatsh. f. Mathematik und Physik* **38** 173-198
[3] Hofstadter D R 1980 *Gödel, Escher, Bach* (New York: Basic Books)
[4] Faraday M 1846 *Philos. Mag.* **28** 345-350
[5] Belcher J W and Olbert S 2003 *Am. J. Phys.* **71** 220-228
[6] Hadad Y, Cohen E, Kaminer I and Elitzur A C 2016 *J. Phys.: Conf. Ser.* **626** 012024
[7] Gralla S E, Harte A I and Wald R M 2009 *Phys. Rev. D* **80** 024031
[8] Pauli W 1981 *Theory of relativity* (New York: Dover Publications)
[9] Hadad Y, Cohen E, Kaminer I and Elitzur A C 2017 *On the relationship between electromagnetic field lines curvature and acceleration* (forthcoming)
[10] Van Enk S J 2013 *J. Phys. A: Math. Theor.* **46** 175204
[11] Rohrlich F 1965 *Classical charged particles* (Reading: Addison Wesley)
[12] Synge J L 1970 *Annali di Matematica* **84** 33-59.
[13] Plebański J 1972 *The structure of the field of a point charge* Mexico: Cinvestav

The Standard Model Algebra - a summary

Ovidiu Cristinel Stoica

Department of Theoretical Physics, National Institute of Physics and Nuclear Engineering – Horia Hulubei, Bucharest, Romania

E-mail: `cristi.stoica@theory.nipne.ro`

Abstract. A generation of leptons and quarks and the gauge symmetries of the Standard Model can be obtained from the Clifford algebra $\mathbb{C}\ell_6$. An instance of $\mathbb{C}\ell_6$ is implicitly generated by the Dirac algebra combined with the electroweak symmetry, while the color symmetry gives another instance of $\mathbb{C}\ell_6$ with a Witt decomposition. The minimal mathematical model proposed here results by identifying the two instances of $\mathbb{C}\ell_6$. The left ideal decomposition generated by the Witt decomposition represents the leptons and quarks, and their antiparticles. The $SU(3)_c$ and $U(1)_{em}$ symmetries of the SM are the symmetries of this ideal decomposition. The patterns of electric charges, colors, chirality, weak isospins, and hypercharges, follow from this, without predicting additional particles or forces, or proton decay. The electroweak symmetry is present in its broken form, due to the geometry. The predicted Weinberg angle is given by $\sin^2\theta_W = 0.25$. The model shares common features with previously known models, particularly with Chisholm and Farwell, 1996, Trayling and Baylis, 2004, and Furey, 2016.

1. Introduction

The approach [1] summarized here unifies leptons and quarks of a generation, and the gauge symmetries from the SM, in the Clifford algebra $\mathbb{C}\ell_6$, whose matrix form is (1).

		$\mathbf{1}_c$	$\mathbf{3}_c$			$\overline{\mathbf{1}}_c$	$\overline{\mathbf{3}}_c$		
Dirac, Lorentz	$\mathbf{1}_w$	ν_{R1}	u^r_{R1}	u^y_{R1}	u^b_{R1}	$\bar{e}_{L1}$	$\bar{d}^r_{L1}$	$\bar{d}^y_{L1}$	$\bar{d}^b_{L1}$
		ν_{R2}	u^r_{R2}	u^y_{R2}	u^b_{R2}	$\bar{e}_{L2}$	$\bar{d}^r_{L2}$	$\bar{d}^y_{L2}$	$\bar{d}^b_{L2}$
	$\mathbf{2}_w$	ν_{L1}	u^r_{L1}	u^y_{L1}	u^b_{L1}	$\bar{e}_{R1}$	$\bar{d}^r_{R1}$	$\bar{d}^y_{R1}$	$\bar{d}^b_{R1}$
		ν_{L2}	u^r_{L2}	u^y_{L2}	u^b_{L2}	$\bar{e}_{R2}$	$\bar{d}^r_{R2}$	$\bar{d}^y_{R2}$	$\bar{d}^b_{R2}$
		e_{L1}	d^r_{L1}	d^y_{L1}	d^b_{L1}	$\bar{\nu}_{R1}$	$\bar{u}^r_{R1}$	$\bar{u}^y_{R1}$	$\bar{u}^b_{R1}$
Dirac, Lorentz		e_{L2}	d^r_{L2}	d^y_{L2}	d^b_{L2}	$\bar{\nu}_{R2}$	$\bar{u}^r_{R2}$	$\bar{u}^y_{R2}$	$\bar{u}^b_{R2}$
	$\mathbf{1}_w$	e_{R1}	d^r_{R1}	d^y_{R1}	d^b_{R1}	$\bar{\nu}_{L1}$	$\bar{u}^r_{L1}$	$\bar{u}^y_{L1}$	$\bar{u}^b_{L1}$
		e_{R2}	d^r_{R2}	d^y_{R2}	d^b_{R2}	$\bar{\nu}_{L2}$	$\bar{u}^r_{L2}$	$\bar{u}^y_{L2}$	$\bar{u}^b_{L2}$

(1)

The group $SU(3)_c$ permutes the columns according to the representations $\mathbf{1}_c$, $\mathbf{3}_c$, $\overline{\mathbf{1}}_c$, and $\overline{\mathbf{3}}_c$. The rows are permuted by the $SU(2)_L$ group according to the representations $\mathbf{1}_w$ and $\mathbf{2}_w$, as well as by the Lorentz group and the Dirac algebra.

In the Standard Model Algebra (SMA), the Clifford algebra $\mathbb{C}\ell\left(\chi^\dagger \oplus \chi\right)$ is algebraically generated by a three-dimensional Hermitian space χ, and is naturally split into left ideals because of the decomposition $\chi^\dagger \oplus \chi$. In a basis adapted to the ideal decomposition, each column contains two 4-spinors associated to different flavors, as illustrated in (1). Each ideal is indexed with an electric charge which is multiple of $\frac{1}{3}$ partially representing the charge of the upper particle, and its color is determined by the ideal to which belongs, having thus associated the correct representation of $\mathrm{SU}(3)_\mathrm{c}$.

The earliest Grand Unified Theories (GUT) are based on the group SU(5) [2], and on SO(10) [3, 4]. Both these models require additional spontaneous symmetry breaking, and predict new interactions, hence new bosons, and proton decay. These predictions are not confirmed. More modern approaches either extend them with supersymmetry, or use different, larger groups. The SU(5) and SO(10) models predict a Weinberg angle θ_W satisfying $\sin^2\theta_W = \frac{3}{8} = 0.375$.

There are common ideas shared by the SMA model with previously known approaches. In previous models, particles of two distinct flavors were also combined into 8-spinor ideals: in a unified spin gauge theory of electroweak interactions and gravity based on $C\ell_{1,6} \cong C\ell_{1,3} \otimes C\ell_{0,3}$ [5], and in the model [6, 7, 8] based on $C\ell_7 \cong C\ell_3 \otimes C\ell_4$. This $C\ell_7$ model uses three space dimensions, time being a scalar, and four extra dimensions related to the Higgs boson. It predicts a Weinberg angle given by $\sin^2\theta_W = 0.375$. Remarkably, the condition to preserve the current and to leave right-handed neutrino sterile results in the full symmetries of the SM. Some of the differences are that the SMA model uses different structures, which lead to the algebra $\mathbb{C}\ell(\chi^\dagger \oplus \chi) \cong \mathbb{C}\ell_6$, it contains the Dirac algebra $C\ell_{1,3} \otimes \mathbb{C}$, and $\sin^2\theta_W = 0.25$. The ideals in the $C\ell_{1,6}$ and $C\ell_7$ models are obtained from primitive idempotents. The SMA is decomposed using a Witt decomposition $\chi^\dagger \oplus \chi$ and the exterior algebra $\bigwedge^k \chi$ contained within the minimal right ideal $\mathfrak{q}\mathfrak{q}^\dagger \mathbb{C}\ell_6 = \mathfrak{q}\mathfrak{q}^\dagger \bigwedge^k \chi$, into left ideals $\mathbb{C}\ell_6 \mathfrak{q}\mathfrak{q}^\dagger \mathfrak{q}_K$, where $\mathbb{K} \subseteq \{1, 2, 3\}$ (notations of section 2). This emphasizes the colors and charges, and allow the coupling of minimal left ideals of the same charge and different colors into a larger ideal.

Furey uses the Witt decomposition for $\mathbb{C}\ell_6$ in a model based on octonions [9, 10], to represent colors and charges of up- and down-type particles by $\mathfrak{q}^\dagger_K \mathfrak{q}\mathfrak{q}^\dagger$ and $\mathfrak{q}_K \mathfrak{q}^\dagger \mathfrak{q}$, on the minimal left ideals $\mathbb{C}\ell_6 \mathfrak{q}\mathfrak{q}^\dagger$ and $\mathbb{C}\ell_6 \mathfrak{q}^\dagger \mathfrak{q}$. They are united into a single irreducible representation of $\mathbb{C}\ell_6 \otimes_\mathbb{C} \mathbb{C}\ell_2$ obtained by using the octonion algebra. To represent the complete particles, with spin and chirality, Furey proposes including the quaternion algebra, resulting in a representation of leptons and quarks as spinors of an algebra isomorphic to $\mathbb{C}\ell_{12}$. On the other hand, in the SMA model, the complete particles are contained in the $\mathbb{C}\ell_6$ ideals classified by the elements $\mathfrak{q}\mathfrak{q}^\dagger \mathfrak{q}_K$. However, the $\mathrm{SU}(3)_\mathrm{c}$ and $\mathrm{U}(1)_\mathrm{em}$ symmetries in the SMA are identical to those obtained previously by Furey [10] as the unitary spin transformations preserving the Witt decomposition of $\mathbb{C}\ell_6$, improving by this previous results based on octonions and Clifford algebras [11, 12, 13], and to those from [8].

2. The Standard Model Algebra

The representation of the group $\mathrm{SU}(3)_\mathrm{c}$ corresponding to leptons is $\mathbf{1}_c$, the one corresponding to quarks is $\mathbf{3}_c$, and for their antiparticles, $\overline{\mathbf{1}}_c$ and respectively $\overline{\mathbf{3}}_c$ (1). The group $\mathrm{U}(1)_\mathrm{em}$ acts by multiplication with $e^{i\frac{1}{k}\varphi}$, where $\varphi \in \mathbb{R}$, and k is an integer satisfying $-3 \leq k \leq 3$. Thus, the representations of both symmetry groups can be seen naturally as representations on the exterior algebras $\bigwedge^\bullet \chi$ and $\bigwedge^\bullet \overline{\chi}$, where χ is a three-dimensional complex vector space endowed with a Hermitian inner product $\mathfrak{h}$, and $\overline{\chi}$ is its complex conjugate. Due to the presence of $\mathfrak{h}$, the following identifications hold: $\chi^\dagger \cong \overline{\chi}$, and also $\chi \cong \overline{\chi}^\dagger$.

The space $\chi^\dagger \oplus \chi$ is naturally endowed with the *inner product* (see *e.g.* [14])

$$\langle u_1^\dagger + u_2, u_3^\dagger + u_4 \rangle := \frac{1}{2}\left(u_1^\dagger(u_4) + u_3^\dagger(u_2)\right) \in \mathbb{C}, \tag{2}$$

where $u_2, u_4 \in \chi$ and $u_1^\dagger, u_3^\dagger \in \chi^\dagger$. The inner product $\langle u_1^\dagger + u_2, u_3^\dagger + u_4 \rangle$ vanishes on χ, so it is not the same as the Hermitian inner product $\mathfrak{h}$.

The *Standard Model Algebra* is the Clifford algebra defined by the inner product (2),

$$\mathcal{A}_{\mathrm{SM}} := \mathbb{C}\ell(\chi^\dagger \oplus \chi) \cong \mathbb{C}\ell_6 \cong \mathbf{M}_{\mathbb{C}}(8), \tag{3}$$

with the distinguished *Witt decomposition* $\chi^\dagger \oplus \chi$, and endowed with the Hermitian inner product $\mathfrak{h}$ on χ and $\chi^\dagger$. Let us fix an $\mathfrak{h}$-orthonormal basis $\mathfrak{q}_1, \mathfrak{q}_2, \mathfrak{q}_3$ of χ. Then, the following anticommutation relations hold

$$\begin{array}{l} \{\mathfrak{q}_j, \mathfrak{q}_k\} = 0, \\ \{\mathfrak{q}^\dagger_j, \mathfrak{q}^\dagger_k\} = 0, \\ \{\mathfrak{q}_j, \mathfrak{q}^\dagger_k\} = \delta_{jk} \end{array} \tag{4}$$

for $j, k \in \{1, 2, 3\}$ (see [15, 10]), which amounts to say that $(\mathfrak{q}^\dagger_1, \mathfrak{q}^\dagger_2, \mathfrak{q}^\dagger_3, \mathfrak{q}_1, \mathfrak{q}_2, \mathfrak{q}_3)$ is a *Witt basis* of $\chi^\dagger \oplus \chi$. From $(\mathfrak{q}^\dagger_1, \mathfrak{q}^\dagger_2, \mathfrak{q}^\dagger_3, \mathfrak{q}_1, \mathfrak{q}_2, \mathfrak{q}_3)$ one gets an orthonormal basis on $\chi^\dagger \oplus \chi$, by

$$\begin{cases} \mathfrak{e}_j = \mathfrak{q}_j + \mathfrak{q}^\dagger_j \\ \tilde{\mathfrak{e}}_j = i\left(\mathfrak{q}^\dagger_j - \mathfrak{q}_j\right), \end{cases} \tag{5}$$

where $j \in \{1, 2, 3\}$. The basis elements satisfy $\mathfrak{e}_j^2 = 1$, $\tilde{\mathfrak{e}}_j^2 = 1$, $\mathfrak{e}_j^\dagger = \mathfrak{e}_j$, and $\tilde{\mathfrak{e}}_j^\dagger = \tilde{\mathfrak{e}}_j$. One can go back to the Witt basis with

$$\begin{cases} \mathfrak{q}_j = \frac{1}{2}\left(\mathfrak{e}_j + i\tilde{\mathfrak{e}}_j\right) \\ \mathfrak{q}^\dagger_j = \frac{1}{2}\left(\mathfrak{e}_j - i\tilde{\mathfrak{e}}_j\right). \end{cases} \tag{6}$$

As a Clifford algebra, the algebra $\mathcal{A}_{\mathrm{SM}}$ is graded, $\mathcal{A}_{\mathrm{SM}} = \bigoplus_{k=0}^{6} \mathcal{A}^k_{\mathrm{SM}}$. Then,

$$\mathcal{A}^k_{\mathrm{SM}} = \mathbb{C}\ell^k(\chi^\dagger \oplus \chi) = \bigwedge{}^k \left(\chi^\dagger \oplus \chi\right). \tag{7}$$

We define

$$\begin{cases} \mathfrak{e} := & \mathfrak{e}_1\mathfrak{e}_2\mathfrak{e}_3, \\ \tilde{\mathfrak{e}} = & \tilde{\mathfrak{e}}_1\tilde{\mathfrak{e}}_2\tilde{\mathfrak{e}}_3. \end{cases} \tag{8}$$

It follows that $\mathfrak{e}^2 = -1$, $\tilde{\mathfrak{e}}^2 = -1$, $\mathfrak{e}\tilde{\mathfrak{e}} = -\tilde{\mathfrak{e}}\mathfrak{e}$, and $(\mathfrak{e}\tilde{\mathfrak{e}})^2 = -1$.

The elements

$$\begin{cases} \mathfrak{q} := & \mathfrak{q}_1\mathfrak{q}_2\mathfrak{q}_3, \\ \mathfrak{q}^\dagger = & \mathfrak{q}^\dagger_3\mathfrak{q}^\dagger_2\mathfrak{q}^\dagger_1, \end{cases} \tag{9}$$

are *nilpotent*, that is, $\mathfrak{q}^2 = 0$ and $\mathfrak{q}^{\dagger 2} = 0$. We denote $\mathfrak{p} := \mathfrak{q}\mathfrak{q}^\dagger$ and $\mathfrak{p}' := \mathfrak{q}^\dagger\mathfrak{q}$. Then, $\mathfrak{p}$ and $\mathfrak{p}'$ are *idempotent* elements, that is, $(\mathfrak{p})^2 = \mathfrak{p}$ and $(\mathfrak{p}')^2 = \mathfrak{p}'$.

Each of the nilpotent elements $\mathfrak{q}$ and $\mathfrak{q}^\dagger$, and each of the (primitive) idempotent elements $\mathfrak{p}$ and $\mathfrak{p}'$, define minimal left and right ideals of the algebra $\mathcal{A}_{\mathrm{SM}}$ [16, 17]. Via the representation of the Clifford algebra $\mathcal{A}_{\mathrm{SM}}$ as an endomorphism algebra $\mathrm{End}_{\mathbb{C}}(\mathbb{C}^8) \cong \mathbf{M}_{\mathbb{C}}(8)$, idempotent elements are represented as projectors.

The Clifford product is given by

$$(u^\dagger + v)\omega\mathfrak{p} = (u^\dagger \wedge \omega)\mathfrak{p} + (i_v\omega)\mathfrak{p} \in \bigwedge{}^\bullet \chi^\dagger \mathfrak{p}, \tag{10}$$

where $u^\dagger + v \in \chi^\dagger \oplus \chi$ and $\omega\mathfrak{p} \in \bigwedge^\bullet \chi^\dagger \mathfrak{p}$, and $i_v\omega$ is the *interior product*, defined for any $\omega \in \bigwedge^k \chi^\dagger$ by

$$(i_v\omega)(u_1, \ldots, u_k) = \begin{cases} \omega(v, u_1, \ldots, u_{k-1}), & \text{for } k \in \{1, 2, 3\}, \text{ and} \\ 0 & \text{for } k = 0. \end{cases} \tag{11}$$

The elements $\mathfrak{q}_j$ and $\mathfrak{q}^\dagger_j$ act like *annihilation operators*, and respectively *creation operators* on $\bigwedge^\bullet \chi^\dagger \mathfrak{p}$:

$$\begin{cases} \mathfrak{q}^\dagger_j(\omega\mathfrak{p}) = (\mathfrak{q}^\dagger_j \wedge \omega)\mathfrak{p}, \\ \mathfrak{q}_j(\omega\mathfrak{p}) = (i_{\mathfrak{q}_j}\omega)\mathfrak{p}. \end{cases} \tag{12}$$

We fix the following basis for the minimal left ideal $\bigwedge^\bullet \chi^\dagger \mathfrak{p}$:

$$(1\ \mathfrak{p}, \mathfrak{q}^\dagger_{23}\ \mathfrak{p}, \mathfrak{q}^\dagger_{31}\ \mathfrak{p}, \mathfrak{q}^\dagger_{12}\ \mathfrak{p}, \mathfrak{q}^\dagger_{321}\ \mathfrak{p}, \mathfrak{q}^\dagger_{1}\ \mathfrak{p}, \mathfrak{q}^\dagger_{2}\ \mathfrak{p}, \mathfrak{q}^\dagger_{3}\ \mathfrak{p}). \tag{13}$$

We now determine the matrix representation of $\mathfrak{q}_j$, $\mathfrak{q}^\dagger_j$, $\mathfrak{e}_j$, and $\tilde{\mathfrak{e}}_j$ in the basis (13).

In terms of the *Pauli matrices* σ_j, $j \in \{1,2,3\}$, and the matrices $\sigma_\pm = \frac{1}{2}(\sigma_1 \pm i\sigma_2)$, $\sigma_3^\pm = \frac{1}{2}(1 \pm \sigma_3)$, we obtain

$$\mathfrak{q}^\dagger_1 = \begin{pmatrix} 0 & 0 & 0 & 0 \\ 0 & 0 & 0 & -i\sigma_2 \\ -i\sigma_2 & 0 & 0 & 0 \\ 0 & 0 & 0 & 0 \end{pmatrix}, \mathfrak{q}^\dagger_2 = \begin{pmatrix} 0 & 0 & 0 & \sigma_3^- \\ 0 & 0 & -\sigma_3^- & 0 \\ 0 & -\sigma_3^+ & 0 & 0 \\ \sigma_3^+ & 0 & 0 & 0 \end{pmatrix}, \mathfrak{q}^\dagger_3 = \begin{pmatrix} 0 & 0 & 0 & -\sigma_- \\ 0 & 0 & \sigma_+ & 0 \\ 0 & -\sigma_+ & 0 & 0 \\ \sigma_- & 0 & 0 & 0 \end{pmatrix}. \tag{14}$$

$$\mathfrak{q}_1 = \begin{pmatrix} 0 & 0 & i\sigma_2 & 0 \\ 0 & 0 & 0 & 0 \\ 0 & 0 & 0 & 0 \\ 0 & i\sigma_2 & 0 & 0 \end{pmatrix}, \mathfrak{q}_2 = \begin{pmatrix} 0 & 0 & 0 & \sigma_3^+ \\ 0 & 0 & -\sigma_3^+ & 0 \\ 0 & -\sigma_3^- & 0 & 0 \\ \sigma_3^- & 0 & 0 & 0 \end{pmatrix}, \mathfrak{q}_3 = \begin{pmatrix} 0 & 0 & 0 & \sigma_+ \\ 0 & 0 & -\sigma_- & 0 \\ 0 & \sigma_- & 0 & 0 \\ -\sigma_+ & 0 & 0 & 0 \end{pmatrix}. \tag{15}$$

$$\mathfrak{q}^\dagger = \begin{pmatrix} 0 & 0 & 0 & 0 \\ 0 & 0 & 0 & 0 \\ \sigma_3^+ & 0 & 0 & 0 \\ 0 & 0 & 0 & 0 \end{pmatrix}, \mathfrak{q} = \begin{pmatrix} 0 & 0 & \sigma_3^+ & 0 \\ 0 & 0 & 0 & 0 \\ 0 & 0 & 0 & 0 \\ 0 & 0 & 0 & 0 \end{pmatrix}, \mathfrak{p} = \begin{pmatrix} \sigma_3^+ & 0 & 0 & 0 \\ 0 & 0 & 0 & 0 \\ 0 & 0 & 0 & 0 \\ 0 & 0 & 0 & 0 \end{pmatrix}, \mathfrak{p}' = \begin{pmatrix} 0 & 0 & 0 & 0 \\ 0 & 0 & 0 & 0 \\ 0 & 0 & \sigma_3^+ & 0 \\ 0 & 0 & 0 & 0 \end{pmatrix}. \tag{16}$$

By (5) one gets

$$\mathfrak{e}_1 = \begin{pmatrix} 0 & 0 & i\sigma_2 & 0 \\ 0 & 0 & 0 & -i\sigma_2 \\ -i\sigma_2 & 0 & 0 & 0 \\ 0 & i\sigma_2 & 0 & 0 \end{pmatrix}, \mathfrak{e}_2 = \begin{pmatrix} 0 & 0 & 0 & 1_2 \\ 0 & 0 & -1_2 & 0 \\ 0 & -1_2 & 0 & 0 \\ 1_2 & 0 & 0 & 0 \end{pmatrix}, \mathfrak{e}_3 = \begin{pmatrix} 0 & 0 & 0 & i\sigma_2 \\ 0 & 0 & i\sigma_2 & 0 \\ 0 & -i\sigma_2 & 0 & 0 \\ -i\sigma_2 & 0 & 0 & 0 \end{pmatrix}, \tag{17}$$

$$\tilde{\mathfrak{e}}_1 = \begin{pmatrix} 0 & 0 & \sigma_2 & 0 \\ 0 & 0 & 0 & \sigma_2 \\ \sigma_2 & 0 & 0 & 0 \\ 0 & \sigma_2 & 0 & 0 \end{pmatrix}, \tilde{\mathfrak{e}}_2 = \begin{pmatrix} 0 & 0 & 0 & -i\sigma_3 \\ 0 & 0 & i\sigma_3 & 0 \\ 0 & -i\sigma_3 & 0 & 0 \\ i\sigma_3 & 0 & 0 & 0 \end{pmatrix}, \tilde{\mathfrak{e}}_3 = \begin{pmatrix} 0 & 0 & 0 & -i\sigma_1 \\ 0 & 0 & i\sigma_1 & 0 \\ 0 & -i\sigma_1 & 0 & 0 \\ i\sigma_1 & 0 & 0 & 0 \end{pmatrix}, \tag{18}$$

$$\mathfrak{e} = \begin{pmatrix} 0_4 & 1_4 \\ -1_4 & 0_4 \end{pmatrix}, \tilde{\mathfrak{e}} = i\begin{pmatrix} 0_4 & 1_4 \\ 1_4 & 0_4 \end{pmatrix}, \mathfrak{e}\tilde{\mathfrak{e}} = i\begin{pmatrix} 1_4 & 0_4 \\ 0_4 & -1_4 \end{pmatrix}. \tag{19}$$

The Witt decomposition $\mathcal{A}^1_{\mathrm{SM}} = \chi^\dagger \oplus \chi$ induces the following decomposition of $\mathcal{A}_{\mathrm{SM}}$ as a direct sum of left ideals

$$\mathcal{A}_{\mathrm{SM}} = \bigoplus_{k=0}^{3} \left(\bigwedge{}^\bullet \chi^\dagger\right) \mathfrak{p} \bigwedge{}^k \chi. \tag{20}$$

The ideals correspond to internal degrees of freedom in $\bigwedge^k \chi$, similar to those of leptons and quarks.

3. The weak symmetry

Each minimal left ideal $\mathcal{A}_{\mathrm{SM}}\mathfrak{p}\mathfrak{q}_K$, $K \subseteq \{1,2,3\}$, has complex dimension eight. We cannot identify these ideals with four-spinor spaces. But we can identify them with pairs of particles whose left chiral components interact weakly. Hence, the action of the Dirac algebra on each minimal left ideal is reducible. The irreducible decomposition of each ideal seen as a representation of the Dirac algebra is given by the projectors $\frac{1}{2}(1 \pm i\mathfrak{e}\tilde{\mathfrak{e}})$.

The reducible representation of the Dirac algebra on a minimal left ideal will be taken to be the direct sum of two irreducible representations,

$$\Gamma^\mu = \begin{pmatrix} \tilde{\gamma}^\mu & 0 \\ 0 & \gamma^\mu \end{pmatrix}, \tag{21}$$

where

$$\gamma^0 = \begin{pmatrix} 0 & 1_2 \\ 1_2 & 0 \end{pmatrix}, \gamma^j = \begin{pmatrix} 0 & \sigma_j \\ -\sigma_j & 0 \end{pmatrix}, \tilde{\gamma}^0 = \begin{pmatrix} 0 & 1_2 \\ 1_2 & 0 \end{pmatrix}, \tilde{\gamma}^j = \begin{pmatrix} 0 & -\sigma_j \\ \sigma_j & 0 \end{pmatrix} \tag{22}$$

are the Weyl representation and a variation of it, chosen to match some notations that will appear later.

The chirality operator is

$$\Gamma^5 = -i\mathfrak{e}_1\tilde{\mathfrak{e}}_1 = \begin{pmatrix} \Gamma^5 & 0 \\ 0 & \gamma^5 \end{pmatrix} = \begin{pmatrix} 1_2 & 0 & 0 & 0 \\ 0 & -1_2 & 0 & 0 \\ 0 & 0 & -1_2 & 0 \\ 0 & 0 & 0 & 1_2 \end{pmatrix}. \tag{23}$$

We will work in the ideal $\mathcal{A}_{\mathrm{SM}}\mathfrak{p}$, the others being obtained from it by multiplying at right with $\mathfrak{q}_K$.

To find on the ideal $\mathcal{A}_{\mathrm{SM}}\mathfrak{p}$ a basis adapted to the weak symmetry, we will use the elements

$$\omega_{\mathrm{u}} = \begin{pmatrix} 0 & 1_2 & 0 & 0 \\ 0 & 0 & 0 & 0 \\ 0 & 0 & 0 & -1_2 \\ 0 & 0 & 0 & 0 \end{pmatrix}, \omega_{\mathrm{d}} = \begin{pmatrix} 0 & 0 & -1_2 & 0 \\ 0 & 0 & 0 & -1_2 \\ 0 & 0 & 0 & 0 \\ 0 & 0 & 0 & 0 \end{pmatrix}, \omega_{\circ} = \begin{pmatrix} \sigma_+ & 0 & 0 & 0 \\ 0 & -\sigma_+ & 0 & 0 \\ 0 & 0 & -\sigma_+ & 0 \\ 0 & 0 & 0 & \sigma_+ \end{pmatrix}. \tag{24}$$

Then,

$$(\omega_{\mathrm{u}}, \omega_{\mathrm{d}}, \omega_{\circ}, \omega^\dagger_{\mathrm{u}}, \omega^\dagger_{\mathrm{d}}, \omega^\dagger_{\circ}) \tag{25}$$

is a Witt basis of the vector space $\mathscr{N}^\dagger \oplus \mathscr{N}$, where

$$\begin{cases} \mathscr{N} := \mathrm{span}_{\mathbb{C}}(\omega_{\mathrm{u}}, \omega_{\mathrm{d}}, \omega_{\circ}), \\ \mathscr{N}^\dagger := \mathrm{span}_{\mathbb{C}}(\omega^\dagger_{\mathrm{u}}, \omega^\dagger_{\mathrm{d}}, \omega^\dagger_{\circ}). \end{cases} \tag{26}$$

We define

$$\begin{cases} \mathfrak{u}_j = \omega_j + \omega^\dagger_j \\ \mathfrak{u}'_j = i(\omega^\dagger_j - \omega_j), \end{cases} \tag{27}$$

where $j \in \{\mathrm{u}, \mathrm{d}, \circ\}$. They satisfy $\mathfrak{u}_j^2 = 1$, $\mathfrak{u}_j'^2 = 1$ and have the matrix form

$$\mathfrak{u}_{\mathrm{u}} = \begin{pmatrix} 0 & 1_2 & 0 & 0 \\ 1_2 & 0 & 0 & 0 \\ 0 & 0 & 0 & -1_2 \\ 0 & 0 & -1_2 & 0 \end{pmatrix}, \mathfrak{u}_{\mathrm{d}} = \begin{pmatrix} 0 & 0 & -1_2 & 0 \\ 0 & 0 & 0 & -1_2 \\ -1_2 & 0 & 0 & 0 \\ 0 & -1_2 & 0 & 0 \end{pmatrix}, \mathfrak{u}_{\circ} = \begin{pmatrix} \sigma_1 & 0 & 0 & 0 \\ 0 & -\sigma_1 & 0 & 0 \\ 0 & 0 & -\sigma_1 & 0 \\ 0 & 0 & 0 & \sigma_1 \end{pmatrix}, \tag{28}$$

$$\mathfrak{u}'_{\mathrm{u}} = \begin{pmatrix} 0 & -i1_2 & 0 & 0 \\ i1_2 & 0 & 0 & 0 \\ 0 & 0 & 0 & i1_2 \\ 0 & 0 & -i1_2 & 0 \end{pmatrix}, \mathfrak{u}'_{\mathrm{d}} = \begin{pmatrix} 0 & 0 & i1_2 & 0 \\ 0 & 0 & 0 & i1_2 \\ -i1_2 & 0 & 0 & 0 \\ 0 & -i1_2 & 0 & 0 \end{pmatrix}, \mathfrak{u}'_{\circ} = \begin{pmatrix} \sigma_2 & 0 & 0 & 0 \\ 0 & -\sigma_2 & 0 & 0 \\ 0 & 0 & -\sigma_2 & 0 \\ 0 & 0 & 0 & \sigma_2 \end{pmatrix}. \tag{29}$$

We define the nilpotent elements

$$\begin{cases} \omega := & \omega_{\mathrm{u}}\omega_{\mathrm{d}}\omega_{\circ}, \\ \omega^{\dagger} = & \omega^{\dagger}_{\circ}\omega^{\dagger}_{\mathrm{d}}\omega^{\dagger}_{\mathrm{u}}. \end{cases} \tag{30}$$

Then, the idempotent $\omega\omega^{\dagger}$ is equal to $\mathfrak{p}$, and

$$\mathcal{A}_{\mathrm{SM}}\mathfrak{p} = \bigwedge{}^{\bullet}\mathscr{N}^{\dagger}\mathfrak{p}. \tag{31}$$

Since (25) is a Witt basis, one can identify $\mathscr{N}^{\dagger}$ with the dual of $\mathscr{N}$, and obtain a Hermitian inner product $h_{\mathscr{N}}$ on $\mathscr{N}$ and $\mathscr{N}^{\dagger}$ similar to $\mathfrak{h}$.

Then, ω_j and $\omega^{\dagger}_j$ act like ladder operators on this ideal $\bigwedge{}^{\bullet}\mathscr{N}^{\dagger}\mathfrak{p}$, just like in (12), on the basis

$$\left(1\ \mathfrak{p}, \omega^{\dagger}_{\circ}\ \mathfrak{p}, \omega^{\dagger}_{\mathrm{u}}\ \mathfrak{p}, \omega^{\dagger}_{\mathrm{u}}\omega^{\dagger}_{\circ}\ \mathfrak{p}, \omega^{\dagger}_{\mathrm{d}}\ \mathfrak{p}, \omega^{\dagger}_{\mathrm{d}}\omega^{\dagger}_{\circ}\ \mathfrak{p}, \omega^{\dagger}_{\mathrm{d}}\omega^{\dagger}_{\mathrm{u}}\ \mathfrak{p}, \omega^{\dagger}_{\mathrm{d}}\omega^{\dagger}_{\mathrm{u}}\omega^{\dagger}_{\circ}\ \mathfrak{p}\right). \tag{32}$$

While the spaces $\bigwedge{}^{\bullet}\mathscr{N}^{\dagger}$ and $\bigwedge{}^{\bullet}\chi^{\dagger}$ are different, the ideals $\bigwedge{}^{\bullet}\mathscr{N}^{\dagger}\mathfrak{p}$ and $\bigwedge{}^{\bullet}\chi^{\dagger}\mathfrak{p}$ are equal.

Let us define the two-dimensional Hermitian vector space $\mathbb{W}_w$

$$\left(\mathbb{W}_w := \mathrm{span}_{\mathbb{C}}\left(\omega^{\dagger}_{\mathrm{u}}, \omega^{\dagger}_{\mathrm{d}}\right), h_{\mathrm{w}} := h_{\mathscr{N}}|_{\mathbb{W}_w}\right). \tag{33}$$

Then, the singlet representations of the action of $SU(2)_{\mathrm{L}}$ are $\bigwedge^0\mathbb{W}_w = \mathrm{span}_{\mathbb{C}}(1)$ and $\bigwedge^2\mathbb{W}_w = \mathrm{span}_{\mathbb{C}}(\omega^{\dagger}_{\mathrm{u}}\omega^{\dagger}_{\mathrm{d}})$, and the doublet representation is $\bigwedge^1\mathbb{W}_w = \mathrm{span}_{\mathbb{C}}(\omega^{\dagger}_{\mathrm{u}}, \omega^{\dagger}_{\mathrm{d}})$. The additional element $\omega^{\dagger}_{\circ}$ is required by the fact that the chiral spaces have two dimensions.

Let us define $\mathbb{W}_{0\mathrm{R}} := \mathrm{span}_{\mathbb{C}}\left(\mathfrak{p}, \omega^{\dagger}_{\circ}\mathfrak{p}\right)$. Then, we define the right-handed and left-handed up singlet spaces $\mathbb{W}_{0\mathrm{R}} := 1\,\mathrm{span}_{\mathbb{C}}\left(\mathfrak{p}, \omega^{\dagger}_{\circ}\mathfrak{p}\right)$ and $\mathbb{W}_{0\mathrm{L}} := \omega^{\dagger}_{\mathrm{u}}\mathbb{W}_{0\mathrm{R}}$, and the right-handed and left-handed down singlet spaces $\mathbb{W}_{1\mathrm{R}} := \omega^{\dagger}_{\mathrm{u}}\omega^{\dagger}_{\mathrm{d}}\mathbb{W}_{0\mathrm{R}}$ and $\mathbb{W}_{1\mathrm{L}} := \omega^{\dagger}_{\mathrm{d}}\mathbb{W}_{0\mathrm{R}}$. The spin representation of the weak group $SU(2)_{\mathrm{L}}$ is given by the generators

$$\begin{cases} \widetilde{T}_1 := \mathfrak{u}_{\mathrm{u}}\mathfrak{u}'_{\mathrm{d}} - \mathfrak{u}'_{\mathrm{u}}\mathfrak{u}_{\mathrm{d}} \\ \widetilde{T}_2 := \mathfrak{u}_{\mathrm{u}}\mathfrak{u}_{\mathrm{d}} + \mathfrak{u}'_{\mathrm{u}}\mathfrak{u}'_{\mathrm{d}} \\ \widetilde{T}_3 := \mathfrak{u}_{\mathrm{u}}\mathfrak{u}'_{\mathrm{u}} - \mathfrak{u}_{\mathrm{d}}\mathfrak{u}'_{\mathrm{d}}, \end{cases} \tag{34}$$

whose matrix form in the basis (32) is

$$\widetilde{T}_1 = 2i\begin{pmatrix} 0 & 0 & 0 & 0 \\ 0 & 0 & 1_2 & 0 \\ 0 & 1_2 & 0 & 0 \\ 0 & 0 & 0 & 0 \end{pmatrix}, \widetilde{T}_2 = 2\begin{pmatrix} 0 & 0 & 0 & 0 \\ 0 & 0 & -1_2 & 0 \\ 0 & 1_2 & 0 & 0 \\ 0 & 0 & 0 & 0 \end{pmatrix}, \widetilde{T}_3 = 2i\begin{pmatrix} 0 & 0 & 0 & 0 \\ 0 & -1_2 & 0 & 0 \\ 0 & 0 & 1_2 & 0 \\ 0 & 0 & 0 & 0 \end{pmatrix}. \tag{35}$$

The proof that their adjoint action gives exactly the desired singlet and doublet representations can be found in [1].

4. The color and electromagnetic symmetries

We will see now that the algebra $\mathcal{A}_{\mathrm{SM}}$, which has the symmetries of the gauge groups $\mathrm{SU}(3)_{\mathrm{c}}$ and $\mathrm{U}(1)_{\mathrm{em}}$, contains the generators of these groups. The groups $\mathrm{SU}(3)_{\mathrm{c}}$ and $\mathrm{U}(1)_{\mathrm{em}}$ are obtained as subgroups of $\mathrm{SO}(\chi^{\dagger} \oplus \chi)$, which is double-covered by the group $\mathrm{Spin}(\chi^{\dagger} \oplus \chi)$, The fact that finite dimensional Lie groups can be expressed as spin groups was explored in detail in [18].

The Lie algebra $\left(\mathcal{A}^2_{\mathrm{SM}}, [\cdot,\cdot]\right)$, where $[a,b] := ab - ba$ for any $a, b \in \mathcal{A}^2_{\mathrm{SM}}$, is just $\mathfrak{so}(\chi^{\dagger} \oplus \chi) \cong \mathfrak{so}(6,\mathbb{C}) \cong \mathfrak{spin}(\chi^{\dagger} \oplus \chi) \cong \mathfrak{spin}(6,\mathbb{C})$ corresponding to the inner product (2). The Lie algebra $\mathfrak{su}(\chi, \mathfrak{h})$ of the color group $\mathrm{SU}(3)_{\mathrm{c}}$ is a Lie subalgebra of $\left(\mathcal{A}^2_{\mathrm{SM}}, [\cdot,\cdot]\right)$, characterized by preserving the ideal decomposition (20). A set of generators for the group $\mathrm{SU}(3)_{\mathrm{c}}$ is

$$\begin{array}{lll} \widetilde{\lambda}_1 = \mathfrak{e}_1\tilde{\mathfrak{e}}_2 - \tilde{\mathfrak{e}}_1\mathfrak{e}_2, & \widetilde{\lambda}_2 = \mathfrak{e}_1\mathfrak{e}_2 + \tilde{\mathfrak{e}}_1\tilde{\mathfrak{e}}_2, & \widetilde{\lambda}_3 = \mathfrak{e}_1\tilde{\mathfrak{e}}_1 - \mathfrak{e}_2\tilde{\mathfrak{e}}_2, \\ \widetilde{\lambda}_4 = \mathfrak{e}_1\tilde{\mathfrak{e}}_3 - \tilde{\mathfrak{e}}_1\mathfrak{e}_3, & \widetilde{\lambda}_5 = \mathfrak{e}_1\mathfrak{e}_3 + \tilde{\mathfrak{e}}_1\tilde{\mathfrak{e}}_3, & \\ \widetilde{\lambda}_6 = \mathfrak{e}_2\tilde{\mathfrak{e}}_3 - \tilde{\mathfrak{e}}_2\mathfrak{e}_3, & \widetilde{\lambda}_7 = \mathfrak{e}_2\mathfrak{e}_3 + \tilde{\mathfrak{e}}_2\tilde{\mathfrak{e}}_3, & \widetilde{\lambda}_8 = \frac{1}{\sqrt{3}}(\mathfrak{e}_1\tilde{\mathfrak{e}}_1 + \mathfrak{e}_2\tilde{\mathfrak{e}}_2 - 2\mathfrak{e}_3\tilde{\mathfrak{e}}_3). \end{array} \tag{36}$$

The generators $\widetilde{\lambda}_j$ act on the ideals $\mathcal{A}_{\mathrm{SM}}\mathfrak{p}\bigwedge^k\chi$ with $k=0$ and $k=3$ like the representation $\mathbf{1}_c$ and $\overline{\mathbf{1}}_c$, and on those with $k=1$ or $k=2$ like $\mathbf{3}_c$ and $\overline{\mathbf{3}}_c$ [1], similar to the *Gell-Mann matrices*.

The generator of the $\mathrm{U}(1)_{\mathrm{em}}$ gauge transformations is

$$Q = \mathfrak{e}_1\tilde{\mathfrak{e}}_1 + \mathfrak{e}_2\tilde{\mathfrak{e}}_2 + \mathfrak{e}_3\tilde{\mathfrak{e}}_3. \tag{37}$$

Its adjoint action on $\mathfrak{p}\bigwedge^k\chi$ corresponds to the correct electric charges, given by $\frac{k}{3}e$, where e is the electron charge, and $k \in \{\pm 0, \pm 1, \pm 2, \pm 3\}$, with the identification $\bigwedge^{-k}\chi = \bigwedge^k\chi^{\dagger} = \bigwedge^k\overline{\chi}$.

From (36) and (37) we see that the color and electromagnetic forces are unified into an *electrocolor* symmetry $\mathrm{U}(3)_{ec}$.

In addition to acting on $\mathfrak{p}\bigwedge^k\chi$, the symmetry generated by (37) also transforms $\omega^{\dagger}_{\mathrm{d}}\mathfrak{p}$. Since

$$\omega^{\dagger}_{\mathrm{d}}\mathfrak{p} = \mathfrak{u}_{\mathrm{d}}\mathfrak{p} = i\tilde{\mathfrak{e}}\mathfrak{p} = \mathfrak{q}^{\dagger}\mathfrak{p}, \tag{38}$$

it follows that the charge of $\mathfrak{q}^{\dagger}\mathfrak{p}$ is equal to -1. This is essential to account for the fact that in each minimal left ideal actually live two particles, whose charge differs exactly by this value.

A proof that unitary spin transformations preserving a Witt decomposition in $\mathbb{C}\ell_6$ give these symmetries, and a set of generators constructed from the $\mathfrak{q}_j$ and $\mathfrak{q}^{\dagger}_j$ but equivalent to (36), was given by Furey [9]. Based on the algebra $C\ell_7$, [8] proposed generators of $\mathrm{SU}(3)_{\mathrm{c}}$ are equivalent to (36) via to the isomorphisms $C\ell_7 \cong \mathbf{M}_{\mathbb{C}}(8) \cong \mathbb{C}\ell_6$.

5. The electroweak symmetry

While in the standard electroweak theory the electromagnetic and weak interactions arise by spontaneous symmetry breaking of the electroweak symmetry, in the SMA they are already "broken" by the geometry, and appear more fundamental than the electroweak symmetry. The *hypercharge* Y is given by the *Gell-Mann–Nishijima formula*, in terms of the electric charge and the weak isospin

$$Y = 2(Q - T_3). \tag{39}$$

This reversed position about the electroweak symmetry breaking may be considered to be due to the construction of the SMA. However, since the split of $\mathbb{W}_w\mathfrak{p}$ is determined by $i\mathfrak{e}\tilde{\mathfrak{e}}$, it follows that it is an invariant property of $\mathcal{A}_{\mathrm{SM}}$. Hence, the spontaneous symmetry breaking mechanism seems not to be needed, but the Higgs boson is still required to give masses to particles. Moreover, the electromagnetic and weak interactions are unified in the SMA too, but they are distinguished by the geometry. In [1] it is shown that in the SMA the *Weinberg angle*, which characterizes the electroweak symmetry breaking, is $\theta_W = \frac{\pi}{6}$, being given by

$$\sin^2\theta_W = \frac{1}{4} = 0.25. \tag{40}$$

The experimental estimations range between ~ 0.223 and ~ 0.24, depending on the utilized scheme [19]. In particular, CODATA gives a value of 0.23129(5) [20]. It is hoped that by taking into account the higher order perturbative corrections, the value of 0.375 can be accommodated with the experimental value.

6. Leptons, quarks, and symmetries

The Witt decomposition $\mathcal{A}^1_{\mathrm{SM}} = \chi^\dagger \oplus \chi$, together with the operator Γ^5, induce a decomposition of the algebra $\mathcal{A}_{\mathrm{SM}}$ into chiral spaces,

$$\mathcal{A}_{\mathrm{SM}} = \bigoplus_{k=0}^{3} \left(\mathbb{W}_{0\mathrm{R}} \oplus \mathbb{W}_{0\mathrm{L}} \oplus \mathbb{W}_{1\mathrm{L}} \oplus \mathbb{W}_{1\mathrm{R}}\right) \bigwedge{}^k \chi. \tag{41}$$

In the basis (13), the chiral spaces are arranged in columns, as shown in (1).

Since

$$\left(\mathbb{W}_0 \oplus \mathbb{W}_1\right) \bigwedge{}^k \chi = \left(\overline{\mathbb{W}_1} \oplus \overline{\mathbb{W}_0}\right) \mathfrak{p}' \bigwedge{}^{3-k} \overline{\chi}, \tag{42}$$

the minimal left ideals contain both particles and their antiparticles, and the antiparticles have opposite chirality.

Any element of the $\mathcal{A}_{\mathrm{SM}}$ is a linear combination of elements of the form

$$\omega_\circ^{\dagger\,a}\, \omega_{\mathrm{u}}^{\dagger\,b}\, \mathfrak{q}^{\dagger c}\, \mathfrak{p}\, \mathfrak{q}_K, \tag{43}$$

where $a, b, c \in \{0, 1\}$, $\left(\omega^\dagger_{\mathrm{u}}\right)^0 = \left(\omega^\dagger_\circ\right)^0 = \left(\mathfrak{q}^\dagger\right)^0 = 1$ by convention, and $K \subset \{1, 2, 3\}$. The elements of the form (43) are cells in the matrix representation of the SMA. The factors present before $\mathfrak{p}$ identify the row, and those after it, the column.

The action of the group $\mathrm{SU}(3)_{\mathrm{c}}$ permutes the columns according to the representations $\mathbf{1}_c$, $\mathbf{3}_c$, $\overline{\mathbf{1}}_c$, and $\overline{\mathbf{3}}_c$. The color of each column is determined by the representation of $\mathrm{SU}(3)_{\mathrm{c}}$ to which it belongs. Each ideal has associated an electric charge which is multiple of $\frac{1}{3}$, but the lower half of each column has in addition to the upper half an extra charge equal to -1, due to the presence of the factor $\mathfrak{q}^\dagger$, since $\mathfrak{q}^\dagger \mathfrak{p} = \omega^\dagger_{\mathrm{d}} \mathfrak{p}$.

The Dirac algebra, the Lorentz group, and the weak group $\mathrm{SU}(2)_{\mathrm{L}}$ act by permuting the rows. The action of $\mathrm{SU}(2)_{\mathrm{L}}$ corresponds to $\mathbf{1}_w$ and $\mathbf{2}_w$, according to the chirality.

Therefore, the leptons, quarks, their antiparticles, and the gauge symmetries of the SM are reproduced by the Standard Model Algebra.

7. Future plans

The results presented here are concerned only the spaces in which the fields associated to particles are represented at each point. Future developments need to explore the consequences of these results to the field equations and the Lagrangian. It is at that level where masses become relevant, and we see if the SMA has implications on the Higgs field, the generations, the neutrino and quark mixing matrices, and the nature of the neutrino. In particular, is the neutrino a Majorana or a Dirac spinor? Hopefully, the geometric properties of the SMA may constrain other parameters of the SM, in addition to the Weinberg angle. Also, the geometric nature of this approach, and the fact that it is based on a higher-dimensional Clifford algebra, may have something to say about the geometry of spacetime.

References

[1] Stoica O C 2017 *Preprint arXiv:1702.04336*
[2] Georgi H and Glashow S 1974 *Phys. Rev. Lett.* **32** 438
[3] Georgi H 1975 *AIP (Am. Inst. Phys.) Conf. Proc., no. 23, pp. 575-582* (Harvard Univ., Cambridge, MA)
[4] Fritzsch H and Minkowski P 1975 *Ann. of Phys.* **93** 193–266

[5] Chisholm J and Farwell R 1996 *Clifford (Geometric) Algebras: With Applications to Physics, Mathematics, and Engineering* ed Baylis W E (Boston, MA: Birkhäuser Boston) pp 365–388 ISBN 978-1-4612-4104-1 URL `http://dx.doi.org/10.1007/978-1-4612-4104-1_27`

[6] Trayling G 1999 *Preprint arXiv:hep-th/9912231*

[7] Trayling G and Baylis W 2001 *J. Phys. A: Math. Theor.* **34** 3309

[8] Trayling G and Baylis W 2004 *Clifford Algebras: Applications to Mathematics, Physics, and Engineering* ed Abłamowicz R (Boston, MA: Birkhäuser Boston) pp 547–558 ISBN 978-1-4612-2044-2 URL `http://dx.doi.org/10.1007/978-1-4612-2044-2_34`

[9] Furey C 2015 *Phys. Lett B* **742** 195–199

[10] Furey C 2016 *Preprint arXiv:1611.09182*

[11] Günaydin M and Gürsey F 1974 *Phys. Rev. D* **9** 3387

[12] Barducci A, Buccella F, Casalbuoni R, Lusanna L and Sorace E 1977 *Phys. Lett. B* **67** 344–346

[13] Casalbuoni R and Gatto R 1979 *Phys. Lett. B* **88** 306–310

[14] Gualtieri M 2004 *Arxiv preprint math/0401221* arXiv:math/0401221 (*Preprint* `0401221`)

[15] Baez, J and Huerta, J 2010 *American Mathematical Society* **47** 483–552

[16] Chevalley C 1997 *The algebraic theory of spinors and Clifford algebras (Collected works)* vol 2 (Springer)

[17] Crumeyrolle A 1990 Orthogonal and symplectic Clifford algebras. Spinor structures

[18] Doran C, Hestenes D, Sommen F and Van Acker N 1993 *J. Math. Phys.* **34** 3642–3669

[19] Erler J and Freitas A 2015 *Particle Data Group* http://pdg.lbl.gov/2016/reviews/rpp2016-rev-standard-model.pdf

[20] Mohr P and Newe D 2016 *Particle Data Group* http://pdg.lbl.gov/2016/reviews/rpp2016-rev-phys-constants.pdf

New class of entropy-power-based uncertainty relations

P Jizba[1], **A Hayes**[2] **and JA Dunningham**[2]

[1] FNSPE, Czech Technical University in Prague, Břehová 7, 115 19 Praha 1, Czech Republic
[2] Department of Physics and Astronomy, University of Sussex, Brighton BN1 9QH, UK

E-mail: p.jizba@fjfi.cvut.cz; Anthony.Hayes@sussex.ac.uk; J.Dunningham@sussex.ac.uk

Abstract. We use the concept of entropy power to introduce a new one-parameter class of information-theoretic uncertainty relations. This class constitutes an infinite hierarchy of uncertainty relations, which allows to determine the shape of the underlying information-distribution function by measuring the relevant entropy powers. The efficiency of such uncertainty relations in quantum mechanics is illustrated with two examples: superpositions of two squeezed states and the Lévy-type heavy-tailed wave function. Improvement over both the variance-based and Shannon entropy based uncertainty relations is demonstrated in both these cases.

1. Introduction

In his 1948 seminal paper, Shannon laid down the foundations of modern communication theory currently known as information theory [1]. He was instrumental in pointing out that, in contrast to discrete signals or messages where information is quantified via (Shannon's) entropy, the case with continuous variables is less satisfactory. The continuous version of Shannon's entropy (SE) — the so-called differential entropy — may take negative values [1, 2] and so it does not enjoys the same status as its discrete-variable counterpart. To solve a range of communication theoretic problems related to continuous cases Shannon shifted the emphasis from the differential entropy to entropy power (EP). The EP represents the variance of a would-be Gaussian random variable with the same differential entropy as the random variable under investigation. In this connection we remind that the concept of EP was used by Shannon [1] in order to formulate information theory for continuous random variables. Since then, the EP has proved to be essential in number of information-theoretic applications ranging from interference channels to secrecy capacity [3, 4, 5, 6]. It has also lead to new advances in information parametric statistics [7] and network information theory. Apart from its rôle in information theory, the EP has found wide use in pure mathematics, namely in theory of inequalities [8] and mathematical statistics [9, 10, 11].

Information theory now extends far beyond the realm of communications and the same principles and concepts can be employed in applications that include statistical physics, biological science and quantum mechanics [12]. In this paper we focus on the application of the EP to quantum-mechanical uncertainty relations (URs). In essence, quantum-mechanical URs place fundamental limits on the accuracy with which one is able to know the values of different physical quantities. In the 1920s, Kennard and independently Robertson and Schrödinger reformulated original Heisenberg's UR in terms of variances of the observables [13, 14, 15]. In 1959, Stam [16] conjectured that the EP could be used to obtain Heisenberg's UR. This conjecture was bolstered in [17] by showing that the usual Schrödinger–Robertson variance-based URs (VURs) [14, 15] can be derived from entropic URs. VURs are useful and widely applied but have two major restrictions: Firstly, the product of the conjugate variances is a single number and so can only give partial information about the underlying states; secondly, variances are only useful concepts for well-behaved bell-like distributions. For heavy-tailed or multimodal distributions, the variances can be large or even infinite, making VURs ill-suited or even useless.

Here we show that Stam conjectured UR is just a member of a one-parameter class of EP-based inequalities, all of which stem from yet another important information measure, namely the differential Rényi entropy (RE) [18, 19, 20] and the ensuing Rényi entropy power (REP). We prove that this class constitutes an infinite hierarchy of higher-order cumulant URs, which allows one in principle to reconstruct the underlying information-distribution function in a process akin to quantum state tomography [21] using EPs in the place of the usual measurements.

For a better understanding of the upcoming results, we begin in Section 2 by introducing the concept of Rényi's EP. By virtue of the Beckner–Babenko theorem we derive in Section 3 the one-parameter hierarchy of entropy-power-based URs. After this we introduce, in Section 4 the concepts of information distribution and show how cumulants of the information distribution can be obtained from the knowledge of EPs. With cummulants at hand one is (in principle) able to reconstruct the underlying information distribution by means of the reconstruction theorem. Section 5 is devoted to the extension of REPURs to the quantum-mechanical setting. In Section 6 we illustrate the inner workings of the REPURs obtained on the two quantum-mechanical examples. Various remarks and generalizations are addressed in the concluding section.

2. Differential entropies and entropy powers

Let $\boldsymbol{X}$ be a random vector in $\mathbb{R}^d$ with the probability density function (PDF), $\wp$. The *differential entropy* $H(\boldsymbol{X})$ of $\boldsymbol{X}$ is defined as [1]

$$H(\boldsymbol{X}) \;=\; -\int_{\mathbb{R}^d} \wp(\boldsymbol{x}) \log_2 \wp(\boldsymbol{x})\, d\boldsymbol{x}\,. \tag{1}$$

The discrete version of (1) is nothing but the SE [1] and in such a case it represents an average number of binary questions that are needed to reveal the value of $\boldsymbol{X}$. Strictly speaking, $H(\boldsymbol{X})$ defined in (1) is not a proper entropy but rather an information gain [2, 18].

The entropy power $N(\boldsymbol{X})$ of $\boldsymbol{X}$ is the unique number such that [1, 22]

$$H(\boldsymbol{X}) \;=\; H(\boldsymbol{X}_G)\,, \tag{2}$$

where $\boldsymbol{X}_G$ is a Gaussian random vector with zero mean and variance equal to $N(\boldsymbol{X})$, i.e., $\boldsymbol{X}_G \sim \mathcal{N}(\boldsymbol{0}, N(\boldsymbol{X})\mathbf{1}_{d\times d})$. Eq.(2) can be equivalently rewritten in the form

$$H(\boldsymbol{X}) \;=\; H\Big(\sqrt{N(\boldsymbol{X})}\cdot \boldsymbol{G}\Big)\,, \tag{3}$$

with $\boldsymbol{G}$ representing a Gaussian random vector with zero mean and unit covariance matrix. The solution of (3) has the form [1]

$$N(\boldsymbol{X}) \;=\; \frac{2^{\frac{2}{d}H(\boldsymbol{X})}}{2\pi e}\,. \tag{4}$$

IOP Conf. Series: Journal of Physics: Conf. Series **880** (2017) 012054 doi:10.1088/1742-6596/880/1/012054

When the Shannon differential entropy is measured in *nats* then for the EP we have

$$N(\boldsymbol{X}) \;=\; \frac{1}{2\pi e}\exp\left(\frac{2}{d}\,H(\boldsymbol{X})\right). \tag{5}$$

Let us now assume that $\boldsymbol{X}_1$ and $\boldsymbol{X}_2$ are two independent random vectors of finite variance. The entropy power (4) [as well as (5)] satisfies the following important *entropy power inequality* [16, 20]

$$N(\boldsymbol{X}_1+\boldsymbol{X}_2) \;\geq\; N(\boldsymbol{X}_1) \;+\; N(\boldsymbol{X}_2)\,, \tag{6}$$

where the equality holds if and only if $\boldsymbol{X}_1$ and $\boldsymbol{X}_2$ are multivariate normal random variables with proportional covariance matrices [1]. In general, inequality (6) does not hold when $\boldsymbol{X}_1$ and $\boldsymbol{X}_2$ are discrete random variables and the differential entropy $H\left(\boldsymbol{X}\right)$ is replaced with the (discrete) Shannon entropy.

The differential RE $I_p(\boldsymbol{X})$ of $\boldsymbol{X}$ is defined as [2, 18, 19]

$$I_p(\boldsymbol{X}) \;=\; \frac{1}{(1-p)}\log_2\left(\int_M d\boldsymbol{x}\,\wp^p(\boldsymbol{x})\right), \tag{7}$$

where the index $p \in \mathbb{R}^+$. By virtue of L'Hospital's rule we obtain that $\lim_{p\to 1} I_p(\boldsymbol{X}) = H(\boldsymbol{X})$. As was the case with H, I_p is also additive for independent events [2]. The p-th Rényi entropy power $N_p(\boldsymbol{X})$ is defined as the solution of the equation

$$I_p\left(\boldsymbol{X}\right) \;=\; I_p\left(\sqrt{N_p(\boldsymbol{X})}\cdot \boldsymbol{G}\right), \tag{8}$$

where $\boldsymbol{G}$ represents a Gaussian random vector with zero mean and unit covariance matrix. This equation was studied in [23, 24] where it was shown that the only class of solution of (8) is

$$N_p(\boldsymbol{X}) \;=\; \frac{1}{2\pi}p^{-p'/p}\exp\left(\frac{2}{d}\,I_p(\boldsymbol{X})\right), \tag{9}$$

with $1/p + 1/p' = 1$ and $p \in \mathbb{R}^+$. So, p' and p are Hölder conjugates. In addition, one can check that $\lim_{p\to 1_+} N_p(\boldsymbol{X}) = N(\boldsymbol{X})$. For explicitness sake we consider here *nats* as units of information. We might also note in passing that from (9) follows that $N_p(\sigma\boldsymbol{G}) = \sigma^2$, i.e. for Gaussian processes the EP is simply the variance σ^2. When $\boldsymbol{X}_G$ represents a random Gaussian vector of zero mean and covariance matrix $\mathbb{K}_{ij}$, then $N_p(\boldsymbol{X}_G) = [\det(\mathbb{K}_{ij})]^{1/d} \equiv |\mathbb{K}|^{1/d}$. Importantly, since the REs are measurable qunatities [25, 26], the associated REPs are experimentally accessible. For some recent applications of the REs in quantum theory see, e.g., Refs. [27, 28, 29].

3. Entropy-power-based Uncertainty Relations

In the following we will need the so-called Beckner–Babenko theorem [30, 31].

Beckner–Babenko Theorem: Let

$$f^{(2)}(\boldsymbol{x}) = \int_{\mathbb{R}^d} e^{2\pi i \boldsymbol{x}.\boldsymbol{y}}\, f^{(1)}(\boldsymbol{y})\, d\boldsymbol{y}\,,$$

then for $p \in [1,2]$

$$|(p')^{d/2}|^{1/p'}\|f^{(2)}\|_{p'} \;\leq\; |p^{d/2}|^{1/p}\|f^{(1)}\|_p\,, \tag{10}$$

IOP Conf. Series: Journal of Physics: Conf. Series **880** (2017) 012054 doi:10.1088/1742-6596/880/1/012054

where p and p' are the Hölder conjugates and

$$\|g\|_p \equiv \left(\int_{\mathbb{R}^d} |g(\boldsymbol{y})|^p \, d\boldsymbol{y} \right)^{1/p} , \tag{11}$$

for any $g \in L^p(\mathbb{R}^d)$. Of course, the role of $f^{(1)}$ and $f^{(2)}$ may be interchanged in the inequality (10). An elementary proof can be found, e.g., in [23]. Inequality (10) is saturated only for Gaussian functions [31, 32].

Anticipating quantum-mechanical applications we define the square-root density likelihood as $\sqrt{\wp(\boldsymbol{y})} \equiv |f(\boldsymbol{y})|$. After simple algebra we recast (10) in the form [23]

$$\left(\int_{\mathbb{R}^d} [\wp^{(2)}(\boldsymbol{y})]^{(1+t)} \, d\boldsymbol{y} \right)^{1/t} \left(\int_{\mathbb{R}^d} [\wp^{(1)}(\boldsymbol{x})]^{(1+r)} \, d\boldsymbol{x} \right)^{1/r} \leq [2(1+t)]^d \, |t/r|^{d/2r} . \tag{12}$$

Here, $r = p/2-1$ and $t = p'/2-1$. Because $1/p+1/p' = 1$ we have the constraint $t = -r/(2r+1)$. Since $p \in [1,2]$ one has $r \in [-1/2, 0]$ and $t \in [0, \infty)$. Eq. (12) can be equivalently written as

$$I_{1+t}(\wp^{(2)}) \; + \; I_{1+r}(\wp^{(1)}) \; \geq \; \frac{1}{r} \log_2 [2(1+r)]^{d/2} \; + \; \frac{1}{t} \log_2 [2(1+t)]^{d/2} . \tag{13}$$

Note, in particular, that in the limit $t \to 0_+$ and $r \to 0_-$ this boils down to

$$H(\wp^{(2)}) \; + \; H(\wp^{(1)}) \; \geq \; \log_2 \left(\frac{e}{2} \right)^d , \tag{14}$$

which is just the classical Hirschman conjecture for Shannon's differential entropies [17, 33]. However, the semidefiniteness of I_p makes the URs (13) unsatisfactory. Fortunately, we can use the positiveness of REPs and rewrite (13) as

$$N_{1+t}(\wp^{(2)}) N_{1+r}(\wp^{(1)}) \; \equiv \; N_{p/2}(\boldsymbol{X}) N_{q/2}(\boldsymbol{Y}) \; \geq \; \frac{1}{16\pi^2} , \tag{15}$$

where $q \equiv p'$ and the REs involved are measured in bits. This is a one-parameter family of inequalities (p and q are the Hölder conjugates). In contrast to (13), the r.h.s. of (15) represents a *universal* lower bound independent of t and r. Note that when $\boldsymbol{X}$ is Gaussian, then $\boldsymbol{Y}$ is also Gaussian and (15) reduces to

$$|\mathbb{K}_{\boldsymbol{X}}|^{1/d} |\mathbb{K}_{\boldsymbol{Y}}|^{1/d} \; = \; \frac{1}{16\pi^2} . \tag{16}$$

Here, $|\mathbb{K}_{\boldsymbol{X}}|$ and $|\mathbb{K}_{\boldsymbol{Y}}|$ are determinants of the respective covariance matrices. The equality follows from the saturation of the inequality (10) by Gaussian functions.

By assuming that a PDF has a finite covariance matrix $(\mathbb{K}_{\boldsymbol{X}})_{ij}$ then

$$N(\boldsymbol{X}) \; \leq \; |\mathbb{K}_{\boldsymbol{X}}|^{1/d} \; \leq \; \sigma^2_{\boldsymbol{X}} , \tag{17}$$

with equality in the first inequality if and only if $\boldsymbol{X}$ is a Gaussian vector, and in the second if and only if $\boldsymbol{X}$ has covariance matrix that is proportional to the identity matrix. The proof of (17) is based on the non-negativity of the Kullback–Leibler divergence and can be found, e.g. in [23, 34, 35]. Inequality (17) immediately gives

$$\sigma^2_{\boldsymbol{X}} \sigma^2_{\boldsymbol{Y}} \; \geq \; |\mathbb{K}_{\boldsymbol{X}}|^{1/d} |\mathbb{K}_{\boldsymbol{Y}}|^{1/d} \geq \; N(\boldsymbol{X}) N(\boldsymbol{Y}) \; \geq \; \frac{1}{16\pi^2} , \tag{18}$$

which saturates only for Gaussian (respective white) random vectors $\boldsymbol{X}$ and $\boldsymbol{Y}$. Note, that when $(\mathbb{K}_{\boldsymbol{X}})_{ij}$ and $(\mathbb{K}_{\boldsymbol{Y}})_{ij}$ exist then (18) automatically implies the conventional Robertson–Schrödinger VUR. Since the VUR is implied by the Shannon EPUR alone, a question arises; in what sense is the general set of inequalities (15) more informative than the special case $r = t = 0$?

4. Reconstruction theorem

To guide our intuition and to show the conceptual underpinning for REPURs (12) we first note that the differential RE can be written as ($\mathbb{E}[\cdots]$ denotes the mean value)

$$I_p(\boldsymbol{X}) \;=\; \frac{1}{(1-p)}\log_2 \mathbb{E}\left[2^{(1-p)i_{\boldsymbol{X}}}\right]. \tag{19}$$

Here $i_{\boldsymbol{X}}(\boldsymbol{x}) \equiv -\log_2 \wp(\boldsymbol{x})$ is the information in $\boldsymbol{x}$ (with respect to the PDF $\wp(\boldsymbol{x})$). From (19), the differential RE can be viewed as a reparametrized version of the *cumulant generating function* of the information random variable $i_{\boldsymbol{X}}(\boldsymbol{X})$. The ensuing cumulant expansion is

$$pI_{1-p}(\boldsymbol{X}) \;=\; \log_2 e \sum_{n=1}^{\infty}\frac{\kappa_n(\boldsymbol{X})}{n!}\left(\frac{p}{\log_2 e}\right)^n \;=\; pH(\boldsymbol{X}) \;+\; \log_2 e\sum_{n=2}^{\infty}\frac{\kappa_n(\boldsymbol{X})}{n!}\left(\frac{p}{\log_2 e}\right)^n, \tag{20}$$

where $\kappa_n(\boldsymbol{X}) \equiv \kappa_n(i_{\boldsymbol{X}})$ denotes the n-th cumulant of $i_{\boldsymbol{X}}(\boldsymbol{X})$ (in units of *bits*^{n}). From (20) it follows that REPs can be written in terms of κ_n's. In fact, N_p's of order $p > 0$ can, in principle, uniquely determine the underlying information PDF (for the proof see Ref. [36]). So, the REPURs of different orders provide additional structural constraints between $\wp^{(1)}$ and $\wp^{(2)}$ which cannot be seen with the VUR or Shannon entropy UR alone. In this connection we list some further salient results [36]:

a) Only Gaussian PDFs saturate all REPURs. REPURs with $r = -1/2$ can be saturated with a wider class of PDFs. b) When $\wp(\boldsymbol{x})$ is close to (or equimeasurable with) a Gaussian PDF then only N_p's with p's in a neighborhood of 1 are needed. The closer the shape is to the Gaussian PDF, the smaller neighborhood of 1 needed. c) For heavy-tailed PDFs, we need to know all REPURs to estimate the shape of the underlying information PDF. d) The non-linear nature of the RE emphasizes the more probable parts of the PDF (typically the middle parts) for Rényi's index $p > 1$ while for $p < 1$ the less probable parts of the PDF (typically the tails) are accentuated. So, when the accentuated parts in $|\psi|^2$ and $|\hat{\psi}|^2$ are close to Gaussian PDF sectors, the associated REPUR will approach its lower bound. In the asymptotic regime when $r = -1/2$, the saturation of the REPUR means that the peak of $\wp^{(1)}$ and tails of $\wp^{(2)}$ are Gaussian, though both $\wp^{(1)}$ and $\wp^{(2)}$ might be non-Gaussian.

5. REPUR in Quantum Mechanics

Let us consider state vectors that are Fourier transform duals – the most prominent example being *phase quadratures* with the configuration and momentum space wave functions as special examples. In the latter case there is a reciprocal relation between $\psi(\mathbf{x})$ and $\hat{\psi}(\mathbf{p})$, namely

$$\psi(\boldsymbol{x}) \;=\; \int_{\mathbb{R}^d} e^{i\boldsymbol{p}\cdot\boldsymbol{x}/\hbar}\,\hat{\psi}(\boldsymbol{p})\,\frac{d\boldsymbol{p}}{(2\pi\hbar)^{d/2}}. \tag{21}$$

The Riesz–Fischer equality [37] guarantees mutual normalization $\|\psi\|_2 = \|\hat{\psi}\|_2 = 1$. Let us define

$$f^{(2)}(\boldsymbol{x}) \;=\; (2\pi\hbar)^{d/4}\psi(\sqrt{2\pi\hbar}\,\boldsymbol{x}), \qquad f^{(1)}(\boldsymbol{p}) \;=\; (2\pi\hbar)^{d/4}\hat{\psi}(\sqrt{2\pi\hbar}\,\boldsymbol{p}). \tag{22}$$

The factor $(2\pi\hbar)^{d/4}$ ensures that the new functions are normalized (in sense of $\|\ldots\|_2$) to unity. With these we have the same structure of the Fourier transform as in the Beckner–Babenko theorem. Consequently we can write the associated RE-based URs (13) in the form

$$I_{1+t}(|\psi|^2) \;+\; I_{1+r}(|\hat{\psi}|^2) \;\geq\; \frac{1}{r}\log_2\left(\frac{1+r}{\pi\hbar}\right)^{d/2} \;+\; \frac{1}{t}\log_2\left(\frac{1+t}{\pi\hbar}\right)^{d/2}, \tag{23}$$

where we have employed of the identity

$$I_p(|f^{(1)}|^2) \;=\; I_p(|\hat{\psi}|^2) - \frac{d}{2}\log_2(2\pi\hbar)\,, \tag{24}$$

(and similarly for $f^{(2)}$). Relation (24) just states that the scaled PDFs $|f^{(1)}|^2$ and $|f^{(2)}|^2$ obtained from (22) are less peaked (and hence less informative) than the original PDFs $|\hat{\psi}|^2$ and $|\psi|^2$, respectively. Consequently, we increase our ignorance when passing from $\hat{\psi}$ to $f^{(1)}$, and from ψ to $f^{(2)}$. In terms of the REP we can recast (23) into the form [cf. Eq. (15)]

$$N_{1+t}(|\psi|^2)N_{1+r}(|\hat{\psi}|^2) \;\geq\; \frac{\hbar^2}{4}\,. \tag{25}$$

This is akin to the Robertson–Schrödinger VUR, but is now a family of relations parametrised by t each having the same universal lower bound $\hbar^2/4$. It should be noted that the familiar VUR follows directly from Shannon's EPUR alone since

$$\sigma_x^2\sigma_p^2 \;\geq\; N_1(|\psi|^2)N_1(|\hat{\psi}|^2) \;\geq\; \frac{\hbar^2}{4}\,. \tag{26}$$

In the special case of Gaussian PDFs, the whole family reduces to the single relation

$$\sigma_x^2\sigma_p^2 \;=\; N_{1+t}(|\psi_G|^2)N_{1+r}(|\hat{\psi}_G|^2) \;=\; \frac{\hbar^2}{4}\,, \tag{27}$$

which is nothing but the familiar coherent-state VUR.

6. Applications in Quantum Mechanics

As a first example we consider an optical state that is pertinent to quantum metrology [38]. It consists of a superposition of a vacuum $|0\rangle$ and a squeezed vacuum $|z_\zeta\rangle$ which has the form $|\psi_\zeta\rangle = \mathcal{N}\left(|0\rangle + |z_\zeta\rangle\right)$, with the normalization factor $\mathcal{N} = 1/\sqrt{2+2(\cosh\zeta)^{-1/2}}$, and

$$|z_\zeta\rangle \;=\; \sum_{m=0}^{\infty}(-1)^m\frac{\sqrt{(2m)!}}{2^m m!}\left[\frac{(\tanh\zeta)^m}{\sqrt{\cosh\zeta}}\right]|2m\rangle\,, \tag{28}$$

where $|2m\rangle$ are even-number energy eigenstates and $\zeta \in \mathbb{R}$ is the squeezing parameter. If we rewrite $|\psi_\zeta\rangle$ in the basis of the eigenstates of the position and momentum *quadrature operators*

$$\hat{X} \;=\; \sqrt{\frac{\hbar}{2\omega}}(\hat{a} + \hat{a}^\dagger),\quad \hat{P} \;=\; -i\sqrt{\frac{\hbar\omega}{2}}(\hat{a} - \hat{a}^\dagger)\,, \tag{29}$$

(ω is the optical frequency and $\hat{a}$ and $\hat{a}^\dagger$ are the photon ladder operators), we get for the PDFs

$$|\psi_\zeta(x)|^2 \;=\; \sqrt{\frac{\omega}{\pi\hbar}}\left|\exp\left(-\frac{\omega x^2}{2\hbar}\right) + e^{\zeta/2}\exp\left(-\frac{\omega e^{2\zeta}x^2}{2\hbar}\right)\right|^2,$$

$$|\hat{\psi}_\zeta(p)|^2 \;=\; \frac{1}{\sqrt{\pi\hbar\omega}}\left|\exp\left(-\frac{p^2}{2\hbar\omega}\right) + e^{-\zeta/2}\exp\left(-\frac{e^{-2\zeta}p^2}{2\hbar\omega}\right)\right|^2. \tag{30}$$

Here the normalization factor $\mathcal{N}^2$ was dropped out. Relations (30) can be now employed to calculate the product $N_{1+t}(x)N_{1+r}(p)$ for different values of t. What we find is that the lower bound $\hbar^2/4$ is saturated for both $N_\infty(x)N_{1/2}(p)$ and $N_{1/2}(x)N_\infty(p)$ irrespective of the squeezing parameter ζ. From our foregoing analysis of REPURs this is easy to understand because the infinite and half indices of the EPs focus on the peak and tails of the PDF, respectively and from (30) we see that both the x and p PDFs are Gaussian in the tails as well as at the peaks (i.e., at $x = p = 0$). A REPUR is saturated only when the RE-accentuated sectors in both dual PDFs are Gaussian [36]. On the other hand, it is also clear that both PDFs (30) as a whole are highly non-Gaussian. We would therefore not expect REPURs with different indices to saturate the bound.

It is also interesting to compute the VUR for the state $|\psi_\zeta\rangle$. The associated variances read

$$\langle(\Delta X)^2\rangle_\zeta = \mathcal{N}^2\frac{\hbar}{\omega}\left[\frac{1}{2}(1+e^{-2\zeta}) + \sqrt{\text{sech}\zeta}\,(1-\tanh\zeta)\right],$$
$$\langle(\Delta P)^2\rangle_\zeta = \mathcal{N}^2\hbar\omega\left[\frac{1}{2}(1+e^{2\zeta}) + \sqrt{\text{sech}\zeta}\,(1+\tanh\zeta)\right]. \tag{31}$$

Note that for $\zeta = 0$, we have $\langle(\Delta X)^2\rangle_0\langle(\Delta P)^2\rangle_0 = \hbar^2/4$, i.e. the VUR is saturated. This is not surprising because the vacuum $|\psi_0\rangle = |0\rangle$ is the Glauber coherent state. However, as the squeezing parameter ζ is increased the product blows up rapidly, which makes the VUR uninformative. So the set of REPURs outperforms both the Shannon EPUR and the VUR by providing more information on the structural features of $|\psi_\zeta\rangle$ via the related PDFs (e.g., Gaussian peaks and tails in p-x quadratures).

We note that the conventional VUR does not pose any restriction on the variance of the observable whose conjugate observable has a PDF with infinite covariance matrix. So, such a state is maximally uncertain. In contrast to this, the set of related REPURs brings considerably more information about the structure of these states. To illuminate this point we discuss in our second example a power-law tail wave packet (PLTWP). PTLWPs are emblematic examples of quantum states with anomalous behavior during their temporal evolution [39]. For definiteness we will consider the PLTWP of the form

$$\psi(x) = \sqrt{\frac{\gamma}{\pi}}\sqrt{\frac{1}{\gamma^2+(x-m)^2}} \quad\Rightarrow\quad |\psi(x)|^2 = \frac{\gamma}{\pi}\frac{1}{\gamma^2+(x-m)^2}. \tag{32}$$

The latter is the Cauchy PDF with a scale parameter γ and median m. The Fourier transform gives

$$\hat{\psi}(p) = e^{-imp/\hbar}\sqrt{\frac{2\gamma}{\pi^2\hbar}}\,K_0(\gamma|p|/\hbar) \quad\Rightarrow\quad |\hat{\psi}(p)|^2 = \frac{2\gamma}{\pi^2\hbar}\,K_0^2(\gamma|p|/\hbar), \tag{33}$$

where K_0 is the modified Bessel function. With these we can directly write two emblematic REPURs

$$N_1(|\hat{\psi}|^2)N_1(|\psi|^2) = 0.0052\,\hbar^2\pi^4 > \hbar^2/4, \tag{34}$$
$$N_{1/2}(|\hat{\psi}|^2)N_\infty(|\psi|^2) = \frac{\hbar^2}{4}. \tag{35}$$

Note also that $\langle(\Delta p)^2\rangle_\psi = \hbar^2\pi/16c^2$ and $\langle(\Delta x)^2\rangle_\psi \to \infty$ (the latter behavior is typical for many PLTWPs), and so the Schrödinger–Robertson's VUR is completely uninformative. What can we conclude from (34)–(35)? First, the REPUR (35) is saturated. This implies that the peak part of $|\psi|^2$ and the tail part of $|\hat{\psi}|^2$ are Gaussian (as can be directly checked). Shannon's EPUR (34) implies: a) the involved PDFs are not Gaussian, b) in contrast to other REPURs it quantifies only shape structures of PDFs but is γ insensitive [36], c) from (14) [cf. also (24)] the lower bound of Hirschman's UR is $\log_2(\pi\hbar e)$ while (34) gives $\log_2(\pi\hbar e) + 0.5141$, so one could still gain 0.5141 bits of information should the system by prepared in a Gaussian state. Finally, we note that $N_\infty(|\hat{\psi}|^2) = 0$ and $N_{1/2}(|\psi|^2) \to \infty$, hence the related REPUR is indeterminate. This behavior is easy to understand. For a strongly leptokurtic PDF (such as $|\psi|^2$) $N_{1/2}$ emphasizes the very flat power-law tails of $|\psi|^2$, and hence $N_{1/2}$ represents the variance of a very flat Gaussian PDF. Similarly, N_∞ accentuates only the peak part of $|\hat{\psi}|^2$ that is sharply (almost δ-function) peaked, and so N_∞ represents the variance of the Gaussian PDF with zero spread. In principle it is possible to deduce from REPs also the scaling characteristics for PLTWPs. This fact was discussed in Ref. [36].

7. Conclusions and perspectives

Here we have introduced a new one-parameter class of REP-based URs for pairs of observables in an infinite-dimensional Hilbert space. On the conceptual side, the infinite tower of inequalities obtained possess a clear advantage over the single VUR by revealing the finer structure of the underlying PDFs further to their standard deviations. This was demonstrated on two relevant quantum mechanical examples.

On application side the proposed REPURs could be directly relevant, e.g., in

a) *quantum-state diagnostics, quantum metrology and measurement*:

- whenever Heisenberg (variance based)URs are inconvenient (uninformative), e.g., when state probability distributions (PDFs) are multimodal or do not have finite variances (heavy tailed PDFs).
- respective members in the tower of URs represent higher-order statistics (i.e., beyond simple variances) restrictions between conjugated-state PDFs. So by knowing one entropy power one can restrict the shape of the conjugated-state PDF via reconstruction theorem
- the more entropy powers are measured or controlled the more structural details about conjugated-state PDF can be obtained (number and height of peaks, tails asymptotic behavior, etc.). This is done via information scan and reconstruction theorem
- designing quantum states (or better their associated PDFs) via the reconstruction theorem. This may give a lot of insight into the structure of quantum states spectrum and so enable states to be tailored for specific metrology tasks.

b) *foundational and conceptual issues*:

- black hole mass-temperature (Hawking) relation can be obtained from the Heisenberg (variance-based) UR via Susskind–Adler mechanism. Mass-temperature relation could thus get corrections when the higher-order statistics restrictions are taken into account.
- origin and distribution of the structure in the universe is attributed to seeding quantum inhomogeneities formed during and after inflation. Usual picture is that these originate from density fluctuations controlled by Heisenberg URs. If the higher-order statistics URs are employed, the power spectrum of the density fluctuations will get corrections. This might better fit the experimentally observed spectrum (the Planck satellite and WMAP data).
- on more speculative vein, one can study the quantum vacuum structure (e.g., in QCD) when the higher-order correlations in energy fluctuations are taken into account.
- REPURs succinctly grasp the otherwise cumbersome higher-order cumulant URs but at the same time both the form and intuition behind the entropy-power based URs are very close to the usual Heisenberg UR.

Most of these issues are presently under active investigation.

Acknowledgments

It is pleasure to acknowledge helpful feedbacks from R Rivers and I Białynicki-Birula. P.J. was supported by the Czech Science Foundation Grant No. 17-33812L. J.D. acknowledges support from DSTL and the UK EPSRC through the NQIT Quantum Technology Hub (Grant No. EP/M013243/1).

References

[1] Shannon C E 1948 *A mathematical theory of communication Bell Syst. Tech. J.* **27** 379; 623
[2] Jizba P and Arimitsu T 2004 *Ann. Phys.* **312** 17
[3] Liu R, Liu T, Poor H V and Shamai S 2010 *IEEE Transactions on Information Theory* **56** 1865
[4] Costa M H M 1985 *IEEE Transactions on Information Theory* **31** 607
[5] Polyanskiy Y and Wu Y 2015 *Preprint* arXiv:1504.04419
[6] Bagherikaram G, Motahari A S and Khandani A K 2013 *IEEE Transactions on Information Theory* **59** 2673
[7] Frieden B R 2004 *Science from Fisher Information: A Unification* (Cambridge: Cambridge Univ. Press)
[8] Courtade T A 2016 *Preprint* arXiv:1602.03033
[9] Barron A 1986 *Ann. Probab.* **14** 336
[10] Artstein S, Ball K M, Barthe F and Naor A 2004 *Probabil. Theory Rel. Fields* **129** 381
[11] Johnson O and Barron A 2004 *Probabil. Theory Rel. Fields* **129** 391
[12] see, e.g., Roederer J 2005 *Information and Its Role in Nature* (Berlin: Springer)
[13] Kennard E H 1927 *Zeitschrift für Physik* **44** 326
[14] Schrödinger E 1930 *Sitzungsber. Preuss. Acad. Wiss.* **24** 296
[15] Robertson H P 1929 *Phys. Rev.* **34** 163
[16] Stam A 1959 *Inform. Contr.* **2** 101
[17] Białynicki-Birula I and Mycielski J 1975 *Commun. Math. Phys.* **44** 129
[18] Rényi A 1970 *Probability Theory* (Amsterdam: North-Holland)
[19] Rényi A 1976 *Selected Papers of Alfred Rényi* Vol.2 (Budapest: Akadémia Kiado)
[20] Cover T and Thomas J A 1991 *Elements of Information Theory* (London: Wiley)
[21] Lvovsky A I and Raymer M G 2009 *Rev. Mod. Phys.* **81** 299
[22] Costa M H 1985 *IEEE Trans. Inform. Theory* **31** 751
[23] Jizba P., Dunningham J A and Joo J 2015 *Annals of Physics* **355** 87
[24] Gardner R J 2002 *Bull. Amer. Math. Soc.* **39** 355
[25] Campbell L L 1965 *Inf. Control* **8** 423
[26] Jizba P and Arimitsu T 2004 *Phys. Rev.* **E69** 026128
[27] Bacco D et all 2013 *Nat. Commun.* **4** 2363
[28] Müller-Lennert M et all 2013 *J. Math. Phys.* **54** 122203
[29] Coles P J et all 2012 *Phys. Rev. Lett.* **108** 210405
[30] Beckner W 1975 *Ann. of Math.* **102** 159
[31] Babenko K I 1962 *Amer. Math. Soc. Transl.* **44** 115
[32] Lieb E H 1990 *Invent. Math.* **102** 179
[33] Hirschman I I 1957 *Am. J. Math.* **79** 152
[34] Dembo A and Cover T M 1991 *IEEE Trans. Inf. Theory* **37** 1501
[35] Rioul O 2011 *IEEE Trans. Inf. Theory* **57** 33
[36] Jizba P, Ma Y, Hayes A and Dunningham J A 2016 *Phys. Rev.* E **93** 060104
[37] Hardy G, Littlewood J E and Pólya G 1952 *Inequalities* (Cambridge: Cambridge Univ. Press)
[38] Knott P A, Cooling J P, Hayes A, Proctor T J and Dunningham J A 2016 *Phys. Rev.* A **93** 033859
[39] Lillo F and Mantegna R N 2000 *Phys. Rev. Lett.* **84** 1061

Operator algebra in the space of images

Enrico Celeghini

Dipartimento di Fisica, Università di Firenze, I50019 Sesto Fiorentino, Firenze, Italy
Depatamento de Fisica Téorica, Universidad de Valladolid, E-47005, Valladolid, Spain

E-mail: celeghini@fi.infn.it

Abstract.
A consistent description of images on the disk and of their transformations is given as elements of a vector space and of an operators algebra. The vector space of images on the disk $\mathbb{D}$ is the Hilbert space $L^2(\mathbb{D})$ that has as a basis the Zernike functions. To construct the operator algebra that transforms the images, $L^2(\mathbb{D})$ must be complemented and the full rigged Hilbert space $RHS(\mathbb{D})$ considered. Only this rigged Hilbert space allows indeed to write the operators of different cardinality we need to build the ladder operators on the Zernike functions that by inspection, belong to the representation $D^+_{1/2} \otimes D^+_{1/2}$ of the algebra $su(1,1) \oplus su(1,1)$. Consequently the transformations of images are operators contained inside the universal enveloping algebra $UEA[su(1,1) \oplus su(1,1)]$. Because of limited precision of experimental measures, physical states can be always described by vectors of the Schwartz space $\mathbb{S}(\mathbb{D})$, dense in the $L^2(\mathbb{D})$ space where the manipulation of images is performed.

1. Introduction

Images are essential in our life and in our research. The problem is that the information contained in an image is too large in order that the human mind can manage it and we have to isolate the limited relevant information. To our opinion, this work to clean the signal from spurious or forgettable elements and highlight the interesting information could be improved combining computer action and human ability of synthesis.

The only approach in this elaboration of images that has obtained up to now relevant results is the cleaning-up realized by adaptive optics where an auxiliary photoreceptor measures the wave front deformations introduced by the medium and acts on the instrument to induce the opposite effect. Adaptive optics removes indeed the spurious phases in the complex function $f(r,\theta)$ that represents the optical signal on the disk allowing to obtain the cleaned image $|f(r,\theta)|^2$.

We are proposing here that $|f(r,\theta)|^2$ can be consider not necessarily the final result of the process but possibly an intermediate step that can be further elaborated by means of an action we can call soft adaptive optics.

While hard adaptive optics acts only on the perturbations of the phase introduced by the medium, this soft elaboration of the numerical image $|f(r,\theta)|^2$ operates on a set of pixels to obtain another set of pixels, independently from the cause of the distortion (wind in atmosphere, diffraction in lents, defects of the apparatus ...) and the particular instrument of measure.

More, while hard adaptive optics can be applied only to the cleaning of images, the proposed soft approach converts images into images and can be employed everywhere a transformation of images can play a role, like -for instance- laser physics, microscopic images, radioastronomy or in general instrumental improvement.

The construction of a theory of operators that act in the space of images defined on the disk, transforming images into images, is the subject of this paper. The action on images/functions defined on a rectangle is similar and will be discussed elsewhere [1].

Characteristics of soft adaptive optics are:
(i) It can be applied both on and of-line, so it can be used later with particular accuracy when convenient.
(ii) Consuming only time machine, it has not mechanical moving parts and
(iii) it allows to consider together images from different sources like optical and radio images.

The relevant mathematical points are:
- Images on the disk belong to the Hilbert space $L^2(\mathbb{D})$ with basis the Zernike functions [2].
- To introduce together continuous and discrete operators we need to construct the ladder operators on the Zernike functions we have to complement $L^2(\mathbb{D})$ considering the rigged Hilbert space on the disk $(RHS(\mathbb{D}))$ i.e. $\{\mathbb{S}(\mathbb{D}) \subset L^2(\mathbb{D}) \subset \mathbb{S}(\mathbb{D})^x\}$.
- The Hilbert space $L^2(\mathbb{D})$ is isomorphic to the unitary irreducible representation $D^+_{1/2}\otimes D^+_{1/2}$ of the Lie group $SU(1,1)\otimes SU(1,1)$.
- $L^2(\mathbb{D})$ operators belong to the Universal Enveloping Algebra $UEA[su(1,1)\oplus su(1,1)]$ and
- can be computed algebraically or by means of first order differential operators.

The mathematical theory of rigged Hilbert spaces can be found in [3, 4, 5]. A physically relevant -basis dependent- approach is discussed in [6, 7, 8], centered on the two points:
(i) In physics every measured quantity has a limited precision so that it can be described by an element of the Schwartz space $\mathbb{S}(\mathbb{D})$ dense into the Hilbert space $L^2(\mathbb{D})$.
(ii) The inclusion map moves basis vectors from the dual of $\mathbb{S}(\mathbb{D})$ -$\mathbb{S}(\mathbb{D})^x$- into the dual of $L^2(\mathbb{D})$, $L^2(\mathbb{D})^x \equiv L^2(\mathbb{D})$.

In sect.2 we introduce the Zernike functions. In sect.3 we sketch the functional analysis of $RHS(\mathbb{D})$. In sect.4 we introduce the operators, their algebraic structure and their realization. In sect.5 we give a description of the practical modus operandi in images manipulation.

2. Zernike functions

Let us start from the radial Zernike polynomials on the circle, $R_n^m(r)$[2], where n is a natural number and m an integer, such that $0 \le m \le n$ with $n-m$ even. $\{R_n^m(r)\}$ are real polynomials defined for $0 \le r \le 1$ such that $R_n^m(1) = 1$ and are usually given in explicit form or in terms of hypergeometric functions. We prefer to define them by means of their second order differential equation, fundamental in our discussion:

$$\left[(1-r^2)\frac{d^2}{dr^2} - (3r - \frac{1}{r})\frac{d}{dr} + n(n+2) - \frac{m^2}{r^2}\right] R_n^m(r) = 0.$$

For m fix, $\{R_n^m(r)\}$ satisfy

$$\int_0^1 R_n^m(r)\, R_{n'}^m(r)\, r\, dr = \frac{\delta_{n\,n'}}{2(n+1)}\,, \qquad \sum_{n=m}^{\infty} (n+1)\, R_n^m(r)\, R_n^m(r') = \frac{\delta(r-r')}{2r}\,.$$

As the interest in optics is focused on real functions defined on the disk, starting from $R_n^m(r)$ it is usual to introduce the functions

$$\mathcal{Z}_n^{-m}(r,\theta) := R_n^m(r)\,\sin(m\theta) \qquad \mathcal{Z}_n^{m}(r,\theta) := R_n^m(r)\,\cos(m\theta)$$

and also to resume the two indices n and m in a unique sequential index [9].

However this paper is centered on symmetry and the objects "invariant in form" respect to rotations [10] are those defined in the complex space by Born and Wolf [2]:

$$Z_n^m(r,\theta) := R_n^{|m|}(r)\, e^{\mathbf{i}m\theta}$$

with n and m integer, $n-m$ even, but $-n \le m \le n$. The symmetry can be further improved writing n and m in function of two natural numbers k and l [11]

$$n = k + l \qquad m = k - l$$

and introducing a multiplicative factor. The final objects we consider here are thus

$$V_{k,l}(r,\theta) \;:=\; \sqrt{k+l+1}\;\, R_{k+l}^{|k-l|}(r)\, e^{\mathbf{i}(k-l)\theta} \qquad (k,l = 0,1,\ldots)$$

that have the symmetries

$$V_{l,k}(r,\theta) = V_{k,l}(r,\theta)^* = V_{k,l}(r,-\theta)$$

and satisfy

$$\frac{d^2 V_{k,l}(r,\theta)}{dr^2} = \frac{1}{1-r^2}\left[\left(3r - \frac{1}{r}\right)\frac{d}{dr} - (k+l)(k+l+2) + \frac{(k-l)^2}{r^2}\right] V_{k,l}(r,\theta)\,. \tag{1}$$

They are orthonormal on the unit disk $\mathbb{D}$:

$$\int_0^{2\pi} d\theta \int_0^1 dr\; r\; V_{k,l}(r,\theta)\; V_{k',l'}(r,\theta)^* \;=\; \delta_{k,k'}\,\delta_{l,l'} \tag{2}$$

so that they are a basis in the Hilbert space $L^2(\mathbb{D})$ of complex square integrable functions defined on the unit disk $\mathbb{D}$.

They also satisfy

$$\sum_{k,l=0}^{\infty} V_{k,l}(r,\theta)\; V_{k,l}(r',\theta')^* \;=\; \frac{1}{r}\,\delta(r-r')\,\delta(\theta-\theta') \tag{3}$$

exhibiting that the space of images is not a simple Hilbert space but of a more complex structure, a RHS. As explained in in sect.3 and, in more detail, in [8], the sum (3) is indeed outside $L^2(\mathbb{D})$: in fact it converges to a tempered distribution in the topology of $\mathbb{S}(\mathbb{D})^x$, a space contained in $RHS(\mathbb{D})$ but larger of $L^2(\mathbb{D})$. However this and the related convergence problems are irrelevant in applications where, because of experimental errors, only elements of $\mathbb{S}(\mathbb{D}) \subset L^2(\mathbb{D})$ are considered.

As the set $\{V_{k,l}(r,\theta)\}$ is a basis in $L^2(\mathbb{D})$, every function $f(r,\theta) \in L^2(\mathbb{D})$ can be written

$$f(r,\theta) = \sum_{k,l} f_{k,l}\; V_{k,l}(r,\theta)$$

where [4]

$$f_{k,l} := \frac{1}{\pi}\int_0^{2\pi} d\theta \int_0^1 dr\; r\;\, f(r,\theta)\, V_{k,l}(r,\theta)^* \qquad \text{with } \sum_{k,l} |f_{k,l}|^2 < \infty. \tag{4}$$

The completeness determines the inner product and the Parceval identity (5)

$$\frac{1}{\pi}\int_0^{2\pi} d\theta \int_0^1 dr\, r\; f(r,\theta)\; g(r,\theta)^* \;=\; \sum_{k,l} f_{k,l}\; {g_{k,l}}^*\,,$$

$$\frac{1}{\pi}\int_0^{2\pi} d\theta \int_0^1 dr\, r\; |f(r,\theta)|^2 \;=\; \sum_{k,l} |f_{k,l}|^{\,2}. \tag{5}$$

Now we can introduce in $\{V_{k,l}(r,\theta)\}$ the operators R, D_R, Θ, D_Θ, K and L such that:

$$R\; V_{k,l}(r,\theta) := r\, V_{k,l}(r,\theta)\;, \qquad D_R\; V_{k,l}(r,\theta) := \frac{dV_{k,l}(r,\theta)}{dr}\;,$$

$$\Theta\; V_{k,l}(r,\theta) := \theta\, V_{k,l}(r,\theta)\;, \qquad D_\Theta\; V_{k,l}(r,\theta) := \frac{dV_{k,l}(r,\theta)}{d\theta}\;,$$

$$K\; V_{k,l}(r,\theta) := k\, V_{k,l}(r,\theta)\;, \qquad L\; V_{k,l}(r,\theta) := l\; V_{k,l}(r,\theta)\;,$$

that allow to write eq.(1) as an operatorial identity on the whole space $\{V_{k,l}(r,\theta)\}$

$$D_R^2 = \frac{1}{1-R^2}\left[(3R-\frac{1}{R})D_R - (K+L)(K+L+2) + \frac{1}{R^2}(K-L)^2\right] \tag{6}$$

i.e. for each $f(r,\theta) \in \{V_{k,l}(r,\theta)\}$ the operator D_R^2 is not independent but related to a first order differential operator.

3. Functional analysis of functions on the disk

The general discussion of $RHS(\mathbb{D})$ will be presented in [8]. We limit ourselves here to sketch the fundamentals statements in the Zernike basis.

The Rigged Hilbert space on the disk, $RHS(\mathbb{D})$, is the Gel'fand triple [4]

$$\mathbb{S}(\mathbb{D}) \subset L^2(\mathbb{D}) \subset \mathbb{S}(\mathbb{D})^x\,.$$

$\mathbb{S}(\mathbb{D})$ is a Schwartz space, sub-space of $L^1(\mathbb{D})$, defined as

$$\mathbb{S}(\mathbb{D}) := \;\{\, f(r,\theta) \in C^\infty(\mathbb{D}) \mid \|f(r,\theta)\|_{\alpha,\beta,\gamma,\delta} < \infty \quad \forall\, \alpha,\, \beta,\, \gamma,\, \delta < \infty\}$$

$$\|f(r,\theta)\|_{\alpha,\beta,\gamma,\delta} \;:=\; sup_{\{0\le r<1,\, 0\le\theta<2\pi\}}\; |\, r^\alpha\, \theta^\beta\, {D_r}^\gamma\, {D_\theta}^\delta f(r,\theta)\,|\,.$$

$\mathbb{S}(\mathbb{D})^x$ (that includes the temperated distributions) is the dual of $\mathbb{S}(\mathbb{D})$, i.e. the space of the applications from $\mathbb{S}(\mathbb{D})$ into the field of complex numbers $\mathbf{C}$

$$\langle\, \mathbb{S}(\mathbb{D})^x \,|\, \mathbb{S}(\mathbb{D})\,\rangle \in \mathbf{C}.$$

$L^2(\mathbb{D})$ is a Hilbert space. It is the completion of $\mathbb{S}(\mathbb{D})$ i.e. it contains every Cauchy sequence of $\mathbb{S}(\mathbb{D})$ and $\mathbb{S}(\mathbb{D})$ is dense in $L^2(\mathbb{D})$.

The Riesz representation theorem states that $L^2(\mathbb{D})$ is isomorphic to its dual $L^2(\mathbb{D})^x$. $f(r,\theta) \in L^2(\mathbb{D})$ if eq.(4) is satisfied and $f(r,\theta) \in \mathbb{S}(\mathbb{D})$ if [4]

$$\sum_{k,l} (k+1)^2(l+1)^2\, |f_{k,l}|^2 < \infty \tag{7}$$

The inclusion map allows to move $\langle V_{k,l}(r,\theta)|$ from $\mathbb{S}(\mathbb{D})^x$ to $L^2(\mathbb{D})^x \equiv L^2(\mathbb{D})$. So, for all $V_{k,l}(r,\theta)$,

$$\langle V_{k,l}(r,\theta)|V_{k',l'}(r',\theta')\rangle_{\mathbb{S}(\mathbb{D})^x\otimes\mathbb{S}(\mathbb{D})} \;=\; \langle V_{k,l}(r,\theta)|V_{k',l'}(r',\theta')\rangle_{L^2(\mathbb{D})^x\otimes L^2(\mathbb{D})}$$

i.e., $RHS(\mathbb{D}) \equiv L^2(\mathbb{D})$ for $f(r,\theta) \in \mathbb{S}(\mathbb{D})$. This is exactly the situation we have in applications where, as every measure is subject to an error of measure, every experimental $f(r,\theta)$ can be considered to fulfill eq.(7) and thus such that $f(r,\theta) \in \mathbb{S}(\mathbb{D})$. So that for a physicist the only difference between a RHS and a Hilbert space is that the RHS allows to consider together operators of different cardinality (R, Θ, D_R and D_Θ from one side and K and L from the other) while Hilbert space not.

IOP Conf. Series: Journal of Physics: Conf. Series **880** (2017) 012055 doi:10.1088/1742-6596/880/1/012055

4. Operators and algebra

The fundamental benefit of $RHS(\mathbb{D})$ compared to $L^2(\mathbb{D})$ is indeed that in $RHS(\mathbb{D})$ we can give a rigorous description of all (discrete and continuous) operators we need to realize the algebra of rising and lowering operators.

Let us indeed introduce in $\{V_{k,l}(r,\theta)\}$ the recurrence operators:

$$\begin{aligned} A_+ \, V_{k,l}(r,\theta) &= (k+1) \; V_{k+1,l}(r,\theta), \\ A_- \, V_{k,l}(r,\theta) &= k \; V_{k-1,l}(r,\theta), \\ B_+ \, V_{k,l}(r,\theta) &= (l+1) \; V_{k,l+1}(r,\theta), \\ B_- \, V_{k,l}(r,\theta) &= l \; V_{k,l-1}(r,\theta) \, . \end{aligned} \tag{8}$$

They, by inspection, are functions of R, D_R, Θ, K and L :

$$\begin{aligned} A_+ &:= \frac{e^{+\mathrm{i}\Theta}}{2} \left[-(1-R^2)D_R + R(K+L+2) + \frac{1}{R}(K-L) \right] \sqrt{\frac{K+L+2}{K+L+1}}, \\ A_- &:= \frac{e^{-\mathrm{i}\Theta}}{2} \left[+(1-R^2)D_R + R(K+L) + \frac{1}{R}(K-L) \right] \sqrt{\frac{K+L}{K+L+1}}, \\ B_+ &:= \frac{e^{-\mathrm{i}\Theta}}{2} \left[-(1-R^2)D_R + R(K+L+2) - \frac{1}{R}(K-L) \right] \sqrt{\frac{K+L+2}{K+L+1}}, \\ B_- &:= \frac{e^{+\mathrm{i}\Theta}}{2} \left[+(1-R^2)D_R + R(K+L) - \frac{1}{R}(K-L) \right] \sqrt{\frac{K+L}{K+L+1}} \, . \end{aligned}$$

The discrete operators K and L are diagonal on $\{V_{k,l}(r,\theta)\}$ but, as the parameters k and l are modified by the action of $A_\pm$ and $B_\pm$, do not commute with $A_\pm$ and $B_\pm$:

$$[K, A_\pm] = \pm A_\pm \, , \qquad [L, B_\pm] = \pm B_\pm \, , \tag{9}$$

exhibiting that not only eq.(3) but also the operator structure needs the $RHS(\mathbb{D})$.

Now as $A_\pm$, $B_\pm$, K and L are operators, we can apply them iteratively in $\{V_{k,l}(r,\theta)\}$ and, in particular, we can calculate their commutators

$$[A_+, A_-] \, V_{k,l}(r,\theta) = -2(k+1/2) \, V_{k,l}(r,\theta) \, , \qquad [K, A_\pm] \, V_{k,l}(r,\theta) = \pm \, V_{k\pm1,l}(r,\theta) \, .$$

So, defining $A_3 := K + 1/2$, we see that $\{A_+, A_3, A_-\}$ are on $\{V_{k,l}(r,\theta)\}$ a differential realization of the Lie algebra $su(1,1)$:

$$[A_+, \, A_-] \; = \; -2A_3 \, , \qquad [A_3, \, A_\pm] \; = \; \pm A_\pm$$

and, analogously, as

$$[B_+, B_-] \, V_{k,l}(r,\theta) = -2(l+1/2) \, V_{k,l}(r,\theta) \, , \qquad [L, B_\pm] V_{k,l}(r,\theta) = \pm \, V_{k,l\pm1}(r,\theta) \, ,$$

$\{B_+, \; B_3 := L + 1/2 \, , \; B_-\}$ are on $\{V_{k,l}(r,\theta)\}$ one other algebra $su(1,1)$

$$[B_+, \, B_-] \; = \; -2B_3 \qquad [B_3, \, B_\pm] \; = \; \pm B_\pm \, .$$

Finally, as A_i and B_j commute in $\{V_{k,l}(r,\theta)\}$, we can complete the algebra with

$$[A_i, B_j] = 0,$$

defining thus a differential realization of the 6 dimensional Lie algebra $su(1,1) \oplus su(1,1)$.

The explicit calculation of the $su(1,1)$ Casimir invariants gives

$$C_A\, V_{k,l}(r,\theta) = \left[\frac{1}{2}\{A_+, A_-\} - A_3^2\right] V_{k,l}(r,\theta) = \frac{1}{4} V_{k,l}(r,\theta),$$

$$C_B\, V_{k,l}(r,\theta) = \left[\frac{1}{2}\{B_+, B_-\} - B_3^2\right] V_{k,l}(r,\theta) = \frac{1}{4} V_{k,l}(r,\theta).$$

As the Casimir of the discrete series D_j^+ of $su(1,1)$ is $j(1-j)$ with $j = 1/2, 1, ..$ [12], $\{V_{k,l}(r,\theta)\}$ is isomorphic to the representation $D_{1/2}^+ \otimes D_{1/2}^+$ of the group $SU(1,1) \otimes SU(1,1)$.

Now we can move from the algebra $su(1,1) \oplus su(1,1)$ to the associated universal enveloping algebra (UEA) i.e. the algebra UEA$[su(1,1) \oplus su(1,1)]$ that has, as a basis, the ordered monomials $A_+^{\alpha_1} A_3^{\alpha_2} A_-^{\alpha_3} B_+^{\beta_1} B_3^{\beta_2} B_-^{\beta_3}$ (where α_i and β_j are natural numbers) submitted to the $su(1,1) \oplus su(1,1)$ commutation relations. Every operator $\mathcal{O} \in$ UEA$[su(1,1) \oplus su(1,1)]$ can be written

$$\mathcal{O} = \sum_{\bar{\alpha},\bar{\beta}} \mathcal{O}_{\bar{\alpha},\bar{\beta}} = \sum_{\bar{\alpha},\bar{\beta}} c_{\bar{\alpha},\bar{\beta}}\, A_+^{\alpha_1} A_3^{\alpha_2} A_-^{\alpha_3} B_+^{\beta_1} B_3^{\beta_2} B_-^{\beta_3}, \tag{10}$$

where $c_{\bar{\alpha},\bar{\beta}}$ are constants depending from $\bar{\alpha}$ and $\bar{\beta}$.

Of course, as $\{V_{k,l}(r,\theta)\}$ is a differential representation of the algebra $su(1,1) \oplus su(1,1)$, it is also a differential representation of $UEA[su(1,1) \oplus su(1,1)]$.

Because the representation $D_{1/2}^+ \otimes D_{1/2}^+$ is unitary and irreducible the vector space of unitary operators acting on the space $L^2(\mathbb{D})$ is isomorphic to the set of operators acting on $D_{1/2}^+ \otimes D_{1/2}^+$ i.e. all invertible transformations of one image in another image can be written in the form (10), belong to $UEA[su(1,1) \oplus su(1,1)]$ and are differential operators in $L^2(\mathbb{D})$.

$\mathcal{O}$, in principle, can be a differential operator of higher order but, by means of iterated use of eqs.(6, 9), can be always reduced in the space $\{V_{k,l}(r,\theta)\}$ to an equivalent first order differential operators. Let us stress that in mathematics the UEA contains the closure of polynomials (i.e. the series) that in physical applications, because of the measure errors, can be excluded.

Thus, the operators that relate two arbitrary images are polynomials of the UEA and can be computed algebraically by means of iterated applications of eqs.(8) or analytically as first order differential operators.

5. Applications to images

Starting from $f(r,\theta) \in L^2(\mathbb{D})$, images on the disk $\mathbb{D}$ are described as $|f(r,\theta)|^2$ that, disregarding the phases, we relate to $|f(r,\theta)| \in L^2(\mathbb{D})$. By means of eq.(4) we write its components in terms of basic vectors:

$$f_{k,l} = \frac{1}{\pi} \int_0^{2\pi} d\theta \int_0^1 dr\; r\; |f(r,\theta)|\, V_{k,l}(r,\theta)^*$$

where, as $|f(r,\theta)|$ is real, $f_{l,k} = (f_{k,l})^*$.

In applications we are limited to values of k and l such that the Parceval identity, eq.(5), is satisfied in the approximation consistent with the experimental errors in $|f(r,\theta)|$, writing the finite sum:

$$|f(r,\theta)| \approx \sum_{k,l=0}^{k_M,l_M} f_{k,l}\, V_{k,l}(r,\theta) \tag{11}$$

where, as $R_{k+l}^{|k-l|}(r)$ are polynomials, k_M and l_M are quite smaller if $|f(r,\theta)|$ is a smooth function.

IOP Conf. Series: Journal of Physics: Conf. Series **880** (2017) 012055 doi:10.1088/1742-6596/880/1/012055

Now, as discussed before, any operator $\mathcal{O}$ that perform the transformation from the starting image $|f(r,\theta)|$ to a new arbitrary chosen image $|g(r,\theta)|$ can be written in the form eq.(10) with finite sums in $\bar{\alpha}$ and $\bar{\beta}$.

At this point we have two computational options. Following the algebraic approach, we can continue writing

$$\mathcal{O}_{\bar{\alpha},\bar{\beta}}\,|f(r,\theta)| \;=\; \sum_{k,l}^{k_M,l_M} f_{kl}\; c_{\bar{\alpha},\bar{\beta}}\;\; A_+^{\alpha_1} A_3^{\alpha_2} A_-^{\alpha_3} B_+^{\beta_1} B_3^{\beta_2} B_-^{\beta_3}\; V_{k,l}(r,\theta)$$

and, with iterated applications of eqs.(8), calculating

$$A_+^{\alpha_1} A_3^{\alpha_2} A_-^{\alpha_3} B_+^{\beta_1} B_3^{\beta_2} B_-^{\beta_3}\; V_{k,l}(r,\theta).$$

We obtain, in this way, the coefficients $g_{k,l}$ that satisfy

$$A_+^{\alpha_1} A_3^{\alpha_2} A_-^{\alpha_3} B_+^{\beta_1} B_3^{\beta_2} B_-^{\beta_3}\; V_{k,l}(r,\theta) \;=\; g_{k+\alpha_1-\alpha_3,l+\beta_1-\beta_3}\; V_{k+\alpha_1-\alpha_3,\,l+\beta_1-\beta_3}(r,\theta)$$

and we have

$$\mathcal{O}_{\bar{\alpha},\bar{\beta}}\,|f(r,\theta)| = \sum_{k,l} f_{kl}\; c_{\bar{\alpha},\bar{\beta}}\; g_{k+\alpha_1-\alpha_3,\,l+\beta_1-\beta_3}\; V_{k+\alpha_1-\alpha_3,\,l+\beta_1-\beta_3}(r,\theta)\;,$$

that allows to write our final result:

$$\mathcal{O}\;|f(r,\theta)| = |g(r,\theta)|\;.$$

In the alternative analytical approach, the operator $\mathcal{O}$ can be described in terms of it differential expression: combining in $\mathcal{O}$ iterated applications of eq.(6) and eq.(9) all dependence from the operators K and L can be moved to the right and all power of D_R higher of one reduces to one, obtaining

$$\mathcal{O} = \sum_i \;[f_i(R) D_R + h_i(R)]\; g_i(K,L)\;. \tag{12}$$

Applying this first order differential operator $\mathcal{O}$ -as written in eq.(12)- to $\sum f_{k,l}\; V_{k,l}(r,\theta)$ -as defined in eq.(11)- we have again:

$$|g(r,\theta)| = \sum_{k,l} f_{k,l}\;\mathcal{O}\,V_{k,l}(r,\theta).$$

References

[1] Celeghini E 2017 *to be published*
[2] Born M and Wolf E 2005 *Principles of Optics* (Cambridge, Un Press, 7th ed)
[3] Böhm A 1978 *The Rigged Hilbert Space and Quantum Mechanics* (Berlin, Springer-Verlag)
[4] Reed M and Simon B 1980 *Methods of Modern Mathematical Physics 1, 2* (London, Acad Press)
[5] Gadella M and Gómez F 2002 *Foundations of Physics* **32** 815
[6] Celeghini E 2015 *J of Phys: Conf Series* **626** 012047
[7] Celeghini E, Gadella M and del Olmo M A 2016 *J Math Phys* **57** 072105
[8] Celeghini E, Gadella M and del Olmo M A 2017 *to be published*
[9] Noll R J 1976 *J Opt Soc Am* **66** 207
[10] Bhatia A B and Wolf E 1954 *Proc Cambridge Soc* **50** 40
[11] Dunkl C F 2001 *Enciclopaedia of Mathematics* **Suppl 3** (Dordrecht, Kluwert) 454
[12] Bargmann V 1954 *Ann of Math* **59** 1

Schrödinger equation in (pre-Planckian?) space–time early universe, with minimal distance, and minimum effective graviton mass

Andrew W Beckwith

Physics Department, Chongqing University, College of Physics, Chongqing University Huxi Campus, No. 44 Daxuechen Nanlu, Shapinba District, Chongqing 401331, People's Republic of China

E-mail: Rwill9955b@gmail.com; abeckwith@uh.edu

Abstract. We review Vuille's generalized Schrödinger equation with its Ricci scalar inclusion, in curved space–time. This has a simplified version in the pre-Planckian regime, which leads to comparing a resultant admissible wave function with Bohmian reformulations of quantum physics, a radial distance given by a modified Poisson's equation and a minimal graviton mass. Finally, we look if Bohmian mechanics has a role in our formulation.

Keywords. Ricci tensor, Schrödinger equation, modified Poisson's equation, massive gravity, inflaton physics, Bohmian mechanics.

1. Introduction, organization of the paper

In Section 2, we develop an initial Schrödinger equation for curved space–time. In Eq. (5), there is a simple modified planar wave with a restricted curvature term in the phase. Later in Section 2, using a modified Poisson's equation, we present a scheme for a minimal radial component to the phase term in the modified Schrödinger equation. Section 3 introduces an early-universe modified version of the Heisenberg uncertainty principle (HUP), and Section 4 shows how this HUP affects the radial contribution in Eq. (5). Section 5 concludes with reflections on LIGO, and the Compton wavelength that need to be taken into consideration.

2. Vuille's treatment of Schrödinger equation for curved space–time

Here, we bring up [1] and a reset of the Schrödinger equation in early space–time, with curvature. From [1], we start with

$$ih\partial_t \Psi = -\frac{h^2}{2m}\nabla^2\Psi + mc^2\Psi \quad (1)$$

and apply several substitutions:

$$\nabla^2 \rightarrow \left(\frac{p^\alpha p^\beta}{m^2c^2} - g^{\alpha\beta}\right)\nabla^\alpha\nabla^\beta$$
$$p^\alpha = ih \cdot g^{\alpha\beta}\nabla_\alpha$$

IOP Conf. Series: Journal of Physics: Conf. Series **880** (2017) 012056 doi:10.1088/1742-6596/880/1/012056

$$p^{\beta} = ih \cdot \nabla^{\beta}$$
$$mc^2 = \frac{g^{\alpha\beta} p^{\alpha} p^{\beta}}{m}. \tag{2}$$

Then, after more derivation and by using the Ricci tensor, R_a^c, Vuille [1] obtained [2]

$$\nabla^a \nabla_a \left(\nabla^b \nabla_b + \frac{m^2 c^2}{h^2} \right) \Psi + \nabla^a R_a^c \nabla_c \Psi = 0. \tag{3}$$

3. Simplifying Eq. (3) in (pre-Planckian?) space–time

We will rewrite Eq. (3), with the result that, in pre-Planckian space–time,

$$\left(\partial_r^4 + \frac{m^2 c^2}{h^2} \partial_r^2 \right) \Psi + R_r^r \partial_r^2 \Psi = 0. \tag{4}$$

The Ricci tensor, R_a^c, in this setting becomes a constant, and is part of how the wave function evolves. Our candidate wave functional takes the form

$$\Psi \propto \Psi_{\text{initial}} e^{\aleph r} \approx \Psi_{\text{initial}} \exp \left(i \sqrt{\frac{m^2 c^2}{h^2} + R_r^r} \cdot r \right). \tag{5}$$

This wave function, as a spatial variable, r, which we will delineate by the treatment involving Eq. (6) below. To do this we use a modified Poisson's equation as given below.

4. What is important about the modified Poisson's equation [2]?

We first refer to two necessary and sufficient conditions for the onset of a massive graviton given in [2] and combine them with Padmanabhan's work [3]. We will redo the reference calculations in [2] with

$$\left\{ \nabla^2 + \left[\lambda^{-2} = \left(\frac{m_{\text{grav}} c}{h} \right)^2 \right] \right\} \left\{ U = \frac{Gm}{r} \cdot \exp \left[\left(-\frac{r}{\lambda} \right) = \left(\frac{r \cdot m_{\text{grav}} c}{h} \right) \right] \right\} = -4\pi G \rho. \tag{6}$$

Here, we will be using in the pre-Planckian potential as the inputs from the data usually associated with [3]:

$$a \approx a_{\min} t^{\gamma} \Leftrightarrow \phi \approx \sqrt{\frac{\gamma}{4\pi G}} \cdot \ln \left[\sqrt{\frac{8\pi G V_0}{\gamma \cdot (3\gamma - 1)}} \cdot t \right] \Leftrightarrow V \approx V_0 \exp \left\{ -\sqrt{\frac{16\pi G}{\gamma}} \cdot \phi(t) \right\}. \tag{7}$$

In other words, we will be using the inflation given by

$$\phi \approx \sqrt{\frac{\gamma}{4\pi G}} \cdot \ln \left[\sqrt{\frac{8\pi G V_0}{\gamma \cdot (3\gamma - 1)}} \cdot t \right]. \tag{8}$$

If this is accurate, then our approximation is to call the potential in Eq. (7) the same as U in Eq. (6). Then, with rearrangements, we arrive at

$$\left[\frac{d^2}{dr^2} + \left(\frac{m_{\text{grav}} c}{h} \right)^2 \right] \cdot \left[\frac{r^{-1} \alpha \cdot (3\alpha - 1)}{32\pi^2} \right] = G \cdot \rho. \tag{9}$$

Then, after algebra, the quadratic equation this engenders is

$$m_{\text{grav}}^2 \approx \left[\frac{32\pi^2 r \cdot h^2 \cdot G \cdot \rho}{c^2 \cdot \alpha \cdot (3\alpha - 1)} - \frac{16\pi^2 r^{-1} \cdot h^2 \cdot G}{c^2 \cdot \alpha \cdot (3\alpha - 1)} \right] \Rightarrow r^2 - \frac{r \cdot m_{\text{grav}}^2}{\frac{32\pi^2 h^2 \cdot G \cdot \rho}{c^2 \cdot \alpha \cdot (3\alpha - 1)}} - \frac{1}{2\rho} = 0. \tag{10}$$

A candidate for the density function comes next, and how we obtain a critical value for r.

IOP Conf. Series: Journal of Physics: Conf. Series **880** (2017) 012056 doi:10.1088/1742-6596/880/1/012056

5. Density function inserted into Eq. (10)

In [4] we assume

$$V_{\text{pre-Planckian}} \approx \left(\Delta E \approx \frac{h}{\delta t \cdot a_{\min}^2 \phi_{\inf}} \right). \tag{11}$$

As far as applications to [3],

$$r^2 - \frac{r^4 \cdot m_{\text{grav}}^2 a_{\min}^2 \cdot \delta t \cdot \sqrt{\frac{\gamma}{4\pi G}} \cdot \ln \left[\sqrt{\frac{8\pi G V_0}{\gamma \cdot (3\gamma - 1)}} \cdot t \right]}{\frac{32\pi^2 h^2 \cdot G}{c^2 \cdot \alpha \cdot (3\alpha - 1)}} - \frac{r^3 a_{\min}^2 \delta t \cdot \sqrt{\frac{\gamma}{4\pi G}} \cdot \ln \left[\sqrt{\frac{8\pi G V_0}{\gamma \cdot (3\gamma - 1)}} \cdot t \right]}{2} = 0. \tag{12}$$

Then, if we use Eq. (12) and look at extremely small times in the inflaton, the above becomes

$$r^2 + \frac{\frac{32\pi^2 \cdot h^2 \cdot G}{c^2 \cdot \alpha \cdot (3\alpha - 1)}}{m_{\text{grav}}^2} \cdot r + \frac{\frac{32\pi^2 h^2 \cdot G}{c^2 \cdot \alpha \cdot (3\alpha - 1)}}{m_{\text{grav}}^2 a_{\min}^2 \cdot \delta t \cdot |\phi|} = 0. \tag{13}$$

If r is very small, we can eliminate the graviton mass; if r is not so small, our answer will contain the graviton mass. We claim that, warts and all, this is a first-order approximation of the distance, from the moment of creation, for definitive acquisition for a 'mass' to a massive graviton. Also, it is a definitive restraint.

6. Conclusion. An initial radial condition depending upon $m_{\text{grav}} \neq 0$?

To incorporate some sort of measurement protocol, Eq. (13) would have to be the absolute magnitude. This has yet to be determined in our discussions with members of the Chongqing University HFGW, but what is noticeable is that the inflaton equation as given by Padmanabhan [3] hopefully will not be incommensurate with the physics of the Corda Criteria given in the gravity's breath document [5]. Also, [6] gives $N_{\text{e-foldings}} \geq 65$. Furthermore, we should keep in mind the physics incorporated in [7, 8]. That is, as to the work of LIGO. it is important to keep in mind that, in addition, [9] has confirmed that a subsequent analysis of the event GW150914 by the LSC constrained the graviton Compton wavelength of those alternative theories of gravity in which the graviton is massive and placed a 90% confidence on the upper bound of e^{-13} km for the Compton wavelength of the graviton. We assume that this may be changed considerably to a much smaller value. This will in its own way subsequently add additional rigor to the analysis of the graviton mass stated in Eq. (10) and is in accord with [10] as well as the Compton wavelength lower bound as given in Eq. (6) above. Finally, we leave open the possibility of refining Eq. (3) and Eq. (4) based on later review of [11]. The implications of having a radial value dependent upon graviton mass, if r not so small, are astounding and need clarification. That plus a review of [11] for possible refinement of Eq. (3) and Eq. (4) will be controversial. This needs much study.

Finally, we should consider, for future work, a Bohmian interpretation of Eq. (5) while comparing the results of a nonzero volume, $\Delta V^{(3)}$, in a (pre-Planckian) wave functional for Eq. (14) as given by Rubakov [12]. We then get

$$\frac{M}{2}\frac{\mathrm{d}q}{\mathrm{d}\tau} = \frac{Mc}{2} = V(q) \Rightarrow \Psi_{\text{initial}} \propto \Psi_0^+ \approx \exp\left(-\int_0^{q^+} M\sqrt{c}\,\mathrm{d}q \right) \xrightarrow[c=1]{} \Psi_0^+ \approx \mathrm{e}^{-Mq^+}. \tag{14}$$

Below, we outline the De Broglie–Bohmian path of a wave functional according to [13]. The second part of our conclusion regards examining if we can compare a (pre-Planckian?) wave function against the construction given in [13]. According to [13], we would have

$$\Psi(x,t) = \Psi(x_0,t_0) \exp\left\{ \left[\frac{i}{h} \int_{x_0,t_0}^{x,t} \left(\frac{\nabla S_{\text{action}}^2}{2m} - (Q+V) \right) \mathrm{d}t \right] - \int_{x_0,t_0}^{x,t} \frac{\nabla S_{\text{action}}^2}{2m} \mathrm{d}t \right\}. \tag{15}$$

IOP Conf. Series: Journal of Physics: Conf. Series **880** (2017) 012056 doi:10.1088/1742-6596/880/1/012056

Here, S_{action} would be the same as Eq. (16), given below, In order to do this, it may be useful to look at the classical degeneracy argument for forming Ψ_{initial}, and the reference by Rubakov [12] may be useful for this. That is, using a false vacuum analogy, we can write, if q is a generalized space–time unit of "length," and we examine a quartic potential

$$V(q) = -\frac{\mu^2 \cdot q^2}{2} + \frac{\widetilde{\lambda} q^4}{4} \ \&\ S_{\text{action}} = \int_{q_0^-}^{q_0^+} \sqrt{2MV(q)}\,\mathrm{d}q$$

$$\Rightarrow \Psi_{\text{initial}} \approx \Psi_0^+ \approx \exp\left(- \int_0^{q^+} \sqrt{2MV(q)}\,\mathrm{d}q \right), \quad (16)$$

Whereas we have V as given by Eq. (15), and then we have, by [13],

$$\begin{aligned} R &= \operatorname{Re} \Psi(x,t) \\ Q &= -\frac{h^2 \nabla^2 R}{2mR} \\ \nabla &\xrightarrow[\text{pre-Planckian}]{} \partial_r . \end{aligned} \quad (17)$$

Here, if we use $h = c = 1$ and have $\Psi(x_0, t_0)$ as given by Eq. (15), we find

$$Q = -\frac{h^2 \nabla^2 R}{2mR} \xrightarrow[\text{pre-Planckian}]{} \frac{1}{2M} \cdot (m_g^2 + R_r^r). \quad (18)$$

Here, M would be given by Eq. (19) as given below: If we use Ng infinite quantum statistics, [14], and a nonzero massive graviton mass (massive gravity) [10],

$$E = M = [S_{\text{entropy}} \approx n_{\text{grav}}] \cdot m_{\text{grav}}. \quad (19)$$

That is, if V were given by Eq. (18), we would find to a point that

$$\Psi(x,t) \approx \Psi(x_0,t_0) \exp\left\{ \frac{i}{h} \int_{x_0,t_0}^{x,t} \left[\frac{\nabla S_{\text{action}}^2}{2m} - (Q+V) \right] \mathrm{d}t \right\}. \quad (20)$$

That is, the term $\Psi(x_0, t_0)$ would likely be the same, but, interestingly enough, $\int_{x_0,t_0}^{x,t} \frac{\nabla S_{\text{action}}^2}{2m}\,\mathrm{d}t$ would likely be almost zero, not contributing at all. We will be examining if the following true. This has to be confirmed rigorously.

$$\int_{x_0,t_0}^{x,t} \frac{\nabla S_{\text{action}}^2}{2m}\,\mathrm{d}t \xrightarrow[pre-Planckian]{} \epsilon^+ \approx 0. \quad (21)$$

This value, and if it exists or is not viable, will be contrasted with a radial distance value of approximately

$$r = -\frac{1}{2} \cdot \frac{32\pi^2 h^2 \cdot G}{c^2 \cdot \alpha \cdot (3\alpha - 1)} \cdot \frac{1}{2 \cdot m_{\text{grav}}^2}$$
$$\cdot \left\{ 1 - \sqrt{1 + 16c \cdot (\Delta t) \cdot \left[\sqrt{2} V_0 \left(\frac{\gamma(3\gamma - 1)}{8\pi G V_0} \right)^{\frac{1}{\sqrt{2}}} \frac{(c \cdot m_{\text{grav}})^2 c^{1+\sqrt{2}}}{(\Delta t)^{1+\sqrt{2}}} \right]^2} \right\}. \quad (22)$$

See the appendix for why this was chosen.

This is the future work of our coming investigations: linking Bohmian mechanics with Eq. (5). If Eq. (21) holds, then Eq. (20) is very close to Eq. (5) and Bohmian mechanics, with some semiclassical interpretations is close to our results based on [1].

Acknowledgements
This work is supported in part by National Nature Science Foundation of China grant No. 11375279

References

[1] Vuille C 1999 Schrödinger's equation in general relativity. *General Relativity and Relativistic Astrophysics, Eighth Canadian Conference* ed C P Burgess and R C Myers (New York: Melville) p 243
[2] Poisson E and Will C 2014 *Gravity, Newtonian, Post Newtonian, Relativistic* (Cambridge, England: Cambridge University Press)
[3] Padmanabhan T 2005 Understanding our universe, current status and open issues. *100 Years of Relativity, Space–Time Structure: Einstein and Beyond* ed A Ashatekar (Singapore: World Scientific) p 175 arxiv:0503107 [gr-qc]
[4] Beckwith A 2016 Gedanken experiment for refining the Unruh metric tensor uncertainty principle via Schwarzschild geometry and Planckian space–time with initial nonzero entropy and applying the Riemannian–Penrose inequality and initial kinetic energy for a lower bound to graviton mass (massive gravity). *J High Energy Phys., Gravit. Cosmology* **2** 106 doi:10.4236/jhepgc.2016.21012
[5] Corda C 2012 Gravity's primordial breath. *Elect. J. Theo. Phys.* **9** 1 `http://www.ejtp.com/articles/ejtpv9i26p1.pdf`
[6] Freese K 1992 Natural inflation. *Particles, Strings, and Cosmology* ed P Nath and S Recucroft (Singapore: World Scientific) p 408
[7] Abbott B, the LIGO Scientific Collaboration and the Virgo Collaboration 2016 Observation of gravitational waves from a binary black hole merger *Phys. Rev. Lett.* **116** 061102
[8] Abbott B, the LIGO Scientific Collaboration and the Virgo Collaboration 2016 GW151226: Observation of gravitational waves from a 22-solar-mass binary black hole coalescence. *Phys. Rev. Lett.* **116** 241103
[9] . Abbott B et al. 2016 Tests of general relativity with GW150914 *Preprint* arxiv:1602.03841
[10] Goldhaber A and Nieto M 2010 Photon and graviton mass limits *Rev. Mod. Phys.* **82** 939 (*Preprint* arxiv:0809.1003)
[11] Carmelli M 1982 *Classical Fields, General Relativity and Gauge Theory* (New York: Wiley Interscience)
[12] Rubakov V 2002 *Classical Theory of Gauge Fields* (Princeton, NJ: Princeton University Press)
[13] Licata I and Fiscaletti D 2014 *Quantum Potential: Physics, Geometry, and Algebra* (New York: Springer-Verlag)
[14] Ng Y J 2008 Spacetime foam: From entropy and holography to infinite statistics and nonlocality. *Entropy* 10 441 doi:10.3390/e10040441

Appendix

Fixing a closed form of the radial distance, based upon [3] and [4]:

$$F_{\text{grav}} = -\nabla V_{\text{potential}} = \sqrt{2}V_0 \left[\frac{\gamma(3\gamma-1)}{8\pi G V_0}\right]^{\frac{1}{\sqrt{2}}} \frac{c^2}{r^{1+\sqrt{2}}}. \tag{A1}$$

Now, the time derivative of energy is [4]

$$\frac{\mathrm{d}E}{\mathrm{d}t} = v_{\text{velocity}} \cdot (m_{\text{mass}} \cdot a_{\text{accel}} - F_{\text{force}}). \tag{A2}$$

Then, if the derivative $\frac{\mathrm{d}E}{\mathrm{d}t}$ becomes instead incremental,

$$\Delta E \Delta t = c(\Delta t)^2 \cdot \left\{ m_{\text{mass}} \cdot a_{\text{accel}} - \sqrt{2}V_0 \left[\frac{\gamma(3\gamma-1)}{8\pi G V_0}\right]^{\frac{1}{\sqrt{2}}} \frac{c^2}{r^{1+\sqrt{2}}} \right\}. \tag{A3}$$

Then, using $\Delta E \approx \frac{h}{\delta t \cdot a_{\min}^2 \phi_{\inf}}$,

$$\frac{h}{a_{\min}^2 \sqrt{\frac{\gamma}{4\pi G}} \cdot \ln\left[\sqrt{\frac{8\pi G V_0}{\gamma\cdot(3\gamma-1)}} \cdot \delta t\right]} = c(\Delta t)^2 \left\{ m_{\text{mass}} \cdot a_{\text{accel}} - \sqrt{2}V_0 \left[\frac{\gamma(3\gamma-1)}{8\pi G V_0}\right]^{\frac{1}{\sqrt{2}}} \frac{c^2}{r^{1+\sqrt{2}}} \right\}. \tag{A4}$$

Now, set $\delta t = (\Delta t)$, and assume in the pre-Planckian regime that $m_{\text{mass}} \cdot a_{\text{accel}} \approx 0$ to get

$$\frac{h}{(\Delta t) \cdot a_{\min}^2 \sqrt{\frac{\gamma}{4\pi G}} \cdot \ln\left[\sqrt{\frac{8\pi G V_0}{\gamma \cdot (3\gamma-1)}} \cdot (\Delta t)\right]} = c(\Delta t)\left\{-\sqrt{2}V_0\left[\frac{\gamma(3\gamma-1)}{8\pi G V_0}\right]^{\frac{1}{\sqrt{2}}} \frac{c^2}{r^{1+\sqrt{2}}}\right\}. \tag{A5}$$

Hence, after algebra, we get

$$r = -\frac{1}{2} \cdot \frac{32\pi^2 h^2 \cdot G}{c^2 \cdot \alpha \cdot (3\alpha - 1)} \cdot \frac{1}{2 \cdot m_{\text{grav}}^2} \cdot \left\{1 - \sqrt{1 + 16c \cdot (\Delta t) \cdot \left[\sqrt{2}V_0 \left(\frac{\gamma(3\gamma-1)}{8\pi G V_0}\right)^{\frac{1}{\sqrt{2}}} \frac{(c \cdot m_{\text{grav}})^2}{r^{1+\sqrt{2}}}\right]^2}\right\}. \tag{A6}$$

which is further simplified by removing the spatial component on the right-hand side of Eq. (20), which is done via

$$r = -\frac{1}{2} \cdot \frac{32\pi^2 h^2 \cdot G}{c^2 \cdot \alpha \cdot (3\alpha - 1)} \cdot \frac{1}{2 \cdot m_{\text{grav}}^2} \cdot \left\{1 - \sqrt{1 + 16c \cdot (\Delta t) \cdot \left[\sqrt{2}V_0 \left(\frac{\gamma(3\gamma-1)}{8\pi G V_0}\right)^{\frac{1}{\sqrt{2}}} \frac{(c \cdot m_{\text{grav}})^2 c^{1+\sqrt{2}}}{(\Delta t)^{1+\sqrt{2}}}\right]^2}\right\}. \tag{A7}$$

Role of acceleration in the expansion of the universe and its influence on an early-universe modified version of the Heisenberg uncertainty principle

Andrew W Beckwith

Physics Department, Chongqing University, College of Physics, Chongqing University Huxi Campus, No. 44 Daxuechen Nanlu, Shapinba District, Chongqing 401331, People's Republic of China

E-mail: Rwill9955b@gmail.com; abeckwith@uh.edu

Abstract. From first principles, we examine what adding acceleration does, and does not do, as to an early-universe modified version of the Heisenberg uncertainty principle. In doing so, we examine a Friedmann equation for the evolution of the scale factor, using two cases explicitly—when the acceleration of the expansion of the scale factor is kept in and when it is out—and the intermediate case of when the acceleration and scale factors are important but not dominant. In doing so we tie this discussion into earlier work done on the HUP.

1. Introduction

We will be examining a Friedmann equation for the evolution of the scale factor, using two cases explicitly—when the acceleration of the expansion of the scale factor is kept in and when it is out—and the intermediate cases of when the acceleration and scale factors are important but not dominant. In doing so, we will be tying this discussion into earlier work done on the early-universe modified version of the Heisenberg uncertainty principle (HUP), but from the context of how the acceleration term will affect the HUP, and making sense of [1]

$$\left\langle (\delta g_{uv})^2 \left(\widehat{T}_{uv}\right)^2 \right\rangle \geq \frac{h^2}{V^2} \xrightarrow[uv \to tt]{} \left\langle (\delta g_{it})^2 \left(\widehat{T}_{it}\right)^2 \right\rangle \geq \frac{h^2}{V^2} \quad \& \quad \delta g_{rr} \approx \delta g_{\theta\theta} \approx \delta g_{\phi\phi} \approx 0^+. \tag{1}$$

Namely, we will be working with [1]

$$\delta t \Delta E = \frac{h}{\delta g_{tt}} = \frac{h}{a^2(t) \cdot \phi} \ll h$$
$$\Leftrightarrow S_{\text{initial}}(\text{with}[\delta g_{tt}]) = (\delta g_{tt})^{-3} S_{\text{initial}}(\text{without}[\delta g_{tt}]) \gg S_{\text{initial}}(\text{without}[\delta g_{tt}]). \tag{2}$$

That is, the fluctuation of $\delta g_{tt} \ll 1$ dramatically boosts initial entropy. It is not what it would be if $\delta g_{tt} \approx 1$. The next question to ask would be how could one actually have [1]

$$\delta g_{tt} \approx a^2(t) \cdot \phi \xrightarrow[\phi \text{ very large}]{} 1? \tag{3}$$

IOP Conf. Series: Journal of Physics: Conf. Series **880** (2017) 012057 doi:10.1088/1742-6596/880/1/012057

In short, we would require an enormous 'inflaton' style ϕ-valued scalar function and $a^2(t) \approx 10^{-110}$. How could ϕ be initially quite large? Within Planck time, the following holds for mass, as a lower bound [1].

$$m_{\text{graviton}} \geq \frac{2h^2}{(\delta g_{tt})^2 l_P^2} \cdot \frac{E-V}{\Delta T_{tt}^2} \tag{4}$$

Here, [1]

$$K.E. \approx (E-V) \approx \dot{\phi}^2 \propto a^{-6}. \tag{5}$$

Then [1],

$$\dot{\phi} \approx a^{-3} \Leftrightarrow \phi \approx t \cdot a^{-3} + H.O.T. \tag{6}$$

The question to ask, now, is about the acceleration of the scale factor, due to time. This will be the subject of our inquiry in the next section of this paper.

2. How could anyone get the acceleration of the universe factored into our scale factor?

Begin by looking at material from pages 483–485 of [2]:

$$\left(\frac{\dot{a}}{a}\right)^3 - \frac{3}{2}\cdot\left(\frac{\dot{a}}{a}\right)^2 - 2\cdot\left(\frac{\ddot{a}}{a}\right)\cdot\left(\frac{\dot{a}}{a}\right) + \left[\frac{-1}{64\pi G \cdot t^2} + \frac{\Lambda}{2}\right] = 0. \tag{7}$$

Then, consider two cases of what to do with the ratio $\frac{\dot{a}}{a}$ and solve the above as a cubic equation.

2.1. What if $\frac{\ddot{a}}{a}$ offers a vanishingly small contribution (low acceleration)?

$$\left(\frac{\dot{a}}{a}\right)^3 - \frac{3}{2}\cdot\left(\frac{\dot{a}}{a}\right)^2 + \left[\frac{-1}{64\pi G \cdot t^2} + \frac{\Lambda}{2}\right] \approx 0. \tag{8}$$

Then, using the idea of a *repressed cubic*, we will have the following solution for $\frac{\dot{a}}{a}$ [3].

$$\frac{\dot{a}}{a} = \text{Solution} = \xi \tag{9}$$

2.1.1. Solutions for Eq. (8), in reduced cubic form.

$$\xi = A + B, \frac{\sqrt{-3}}{2}\cdot(A-B) - \left(\frac{A+B}{2}\right), -\frac{\sqrt{-3}}{2}\cdot(A-B) - \left(\frac{A+B}{2}\right) \tag{10}$$

$$A = \left[\left(\frac{-1}{128\pi G \cdot t^2} + \frac{\Lambda}{4}\right) + \sqrt{\frac{1}{4}\cdot\left(\frac{-1}{64\pi G \cdot t^2} + \frac{\Lambda}{2}\right)^2 + \frac{1}{8}}\right]^{\frac{1}{3}}$$

$$B = -\left[\left(\frac{1}{128\pi G \cdot t^2} - \frac{\Lambda}{4}\right) + \sqrt{\frac{1}{4}\cdot\left(\frac{-1}{64\pi G \cdot t^2} + \frac{\Lambda}{2}\right)^2 + \frac{1}{8}}\right]^{\frac{1}{3}}. \tag{11}$$

Then

$$\Theta = \frac{1}{8}\cdot\left[2\cdot\left(\frac{-1}{64\pi G \cdot t^2} + \frac{\Lambda}{2}\right)^2 - 1\right]. \tag{12}$$

$$\begin{aligned} &\Theta > 0 \Rightarrow \xi \text{ has one real and two imaginary roots.}\\ &\Theta = 0 \Rightarrow \xi \text{ has three real roots, two of which are equal.}\\ &\Theta < 0 \Rightarrow \xi \text{ has three real roots, all of which are unique.} \end{aligned} \tag{13}$$

The situation to watch is when the time, t, is extremely small. Then, one must work with the situation where $\Theta > 0 \Rightarrow \xi$ has one real and two imaginary roots. That is, the situation is then dominated with one real root and two imaginary roots. We will comment on value of what happens to $\frac{\dot{a}}{a}$ = Solution = ξ if there is one real and two imaginary roots? What would be a possible constraint if we had, for undimensioned units,

$$\left[2 \cdot \left(\frac{-1}{64\pi G \cdot t^2} + \frac{\Lambda}{2}\right)^2 - 1\right] \approx 0 \Leftrightarrow \left(\frac{-1}{64\pi G \cdot t^2} + \frac{\Lambda}{2}\right) \approx \frac{1}{\sqrt{2}} \Leftrightarrow \Lambda \approx \sqrt{2} + \frac{1}{32\pi G \cdot t^2}? \tag{14}$$

That is, for the case that one uses undimensioned units, we would have, then,

$$\Theta \leq 0 \Leftrightarrow \Lambda \geq \sqrt{2} + \frac{1}{32\pi G \cdot t^2}. \tag{15}$$

How likely is this to happen in the pre-Planckian regime? Not likely. Secondly, we get

$$\Theta \leq 0 \Rightarrow \xi \text{ has three real roots.} \tag{16}$$

So if we neglect having the acceleration of the scale factor, by abandoning $\frac{\ddot{a}}{a}$ acceleration, we get a weird family of solutions for Eq. (8).

2.1.2. Solutions for Eq. (7), in cubic form, gained by not abandoning $\frac{\ddot{a}}{a}$. Following [2, 3], look first at

$$\begin{aligned} \widetilde{a}_1 &= -2 \cdot \left[\left(\frac{\ddot{a}}{a}\right) + \frac{3}{2}\right] \\ \widetilde{b}_1 &= -\frac{1}{4} \cdot \left[1 + 4 \cdot \left(\frac{\ddot{a}}{a}\right) + \frac{1}{256\pi G \cdot t^2} - \frac{\Lambda}{8}\right]. \end{aligned} \tag{17}$$

Our approximation is to set $\frac{\ddot{a}}{a}$ as a nonzero constant. In that case, set $\frac{\ddot{a}}{a}$ as a nondimensional but very large quantity. Then, a solution exists as given as for a reduced cubic version of Eq. (7), which can be given by

$$\xi = A_1 + B_1, \frac{\sqrt{-3}}{2} \cdot (A_1 - B_1) - \left(\frac{A_1 + B_1}{2}\right), -\frac{\sqrt{-3}}{2} \cdot (A_1 - B_1) - \left(\frac{A_1 + B_1}{2}\right) \tag{18}$$

and

$$\begin{aligned} A_1 &= \sqrt[3]{-\frac{\widetilde{b}_1}{2} + \sqrt{\left(\frac{\widetilde{b}_1}{2}\right)^2 + \frac{\widetilde{a}_1}{27}}} \\ B_1 &= -\sqrt[3]{\frac{\widetilde{b}_1}{2} + \sqrt{\left(\frac{\widetilde{b}_1}{2}\right)^2 + \frac{\widetilde{a}_1}{27}}} \end{aligned} \tag{19}$$

When $\frac{\ddot{a}}{a}$ is set as a nondimensional constant quantity and possibly quite large, then

$$\frac{\dot{a}}{a} = \text{Solution} = \xi_1 + \frac{1}{2}. \tag{20}$$

If so, then

$$\Theta = \frac{(\widetilde{b}_1)^2}{4} + \frac{(\widetilde{a}_1)^3}{27}, \tag{21}$$

If $\frac{\ddot{a}}{a}$ is constant and very large, the results of the sign of Eq. (21) are as follows.

$$\begin{aligned}&\Theta > 0 \Rightarrow \xi \text{ has one real and two imaginary roots.}\\&\Theta = 0 \Rightarrow \xi \text{ has three real roots, two of which are equal.}\\&\Theta < 0 \Rightarrow \xi \text{ has three real roots, all of which are unique.}\end{aligned} \tag{22}$$

Here, with very large constant initial $\frac{\ddot{a}}{a}$, we have that the third outcome is by far most likely to happen, in contrast to what would happen in the situation with $\frac{\ddot{a}}{a} = 0$ [2], and all of this will make sense of the transition to the "normal" HUP in which we will be considering [1]

$$\delta g_{tt} \approx a^2(t) \cdot \phi \xrightarrow[\phi \text{ very large}]{} 1. \tag{23}$$

In short, we would require an enormous inflaton style ϕ-valued scalar function, and $a^2(t) \approx 10^{-110}$. We will next discuss the implications of this point in the next section, of a nonzero smallest-scale factor. Secondly, the fact that we are working with a massive graviton will be given some credence when we obtain a lower bound, as will come up in our derivation of modification of the values [4]

$$\left\langle (\delta g_{uv})^2 \left(\widehat{T}_{uv}\right)^2 \right\rangle \geq \frac{h^2}{V^2} \xrightarrow[uv \to tt]{} \left\langle (\delta g_{it})^2 \left(\widehat{T}_{it}\right)^2 \right\rangle \geq \frac{h^2}{V^2} \quad \& \quad \delta g_{rr} \approx \delta g_{\theta\theta} \approx \delta g_{\phi\phi} \approx 0^+. \tag{24}$$

The reasons for saying this set of values for the variation of the non-g_{tt} metric will be presented in the third section: It is due to the smallness of the square of the scale factor in the vicinity of the Planck time interval.

3. Nonzero scale factor, initially, and what this tells us physically.

We start with a configuration from Unruh [5, 6].

$$\Delta l \cdot \Delta p \geq \frac{h}{2} \tag{25}$$

We will be using the approximation given by Unruh [5, 6], of a generalization we will write as

$$\begin{aligned}(\Delta l)_{ij} &= \frac{\delta g_{ij}}{g_{ig}} \cdot \frac{l}{2}\\(\Delta p)_{ij} &= \Delta T_{ij} \cdot \delta t \cdot \Delta A.\end{aligned} \tag{26}$$

We use the following from the Roberson–Walker metric [7].

$$\begin{aligned}g_{tt} &= 1\\g_{rr} &= -\frac{a^2(t)}{1 - k \cdot r^2}\\g_{\theta\theta} &= -a^2(t) \cdot r^2\\g_{\phi\phi} &= -a^2(t) \cdot \sin^2\theta \, d\phi^2\end{aligned} \tag{27}$$

Following Unruh [5, 6], write then, an uncertainty-of-metric tensor, with the following inputs:

$$a^2(t) \approx 10^{-110}, r \equiv l_p \approx 10^{-35}\,\text{m}. \tag{28}$$

Then, if $\Delta T_{tt} \approx \Delta\rho$,

$$V^{(4)} = \delta t \cdot \Delta A \cdot r$$
$$g_{rr} \cdot \Delta T_{tt} \cdot \delta t \cdot \Delta A \cdot \frac{r}{2} \geq \frac{h}{2}$$
$$\Leftrightarrow \delta g_{tt} \cdot \Delta T_{tt} \geq \frac{h}{V^{(4)}}. \tag{29}$$

Eq. (29) is such that we can extract, up to a point, the HUP in time and energy, with one very large caveat: we must use the fluid approximation of space–time [6] $T_{ii} = \text{diag}(\rho, -p, -p, -p)$.

$$\Delta T_{tt} \approx \Delta\rho \approx \frac{\Delta E}{V^{(3)}} \tag{30}$$

Then, Eq. (29) and Eq. (30) together yield

$$\delta t \Delta E \geq \frac{h}{\delta g_{tt}} \neq \frac{h}{2}, \text{ unless } \delta g_{tt} \approx \mathrm{O}(1). \tag{31}$$

How likely is $\delta g_{tt} \approx \mathrm{O}(1)$? Not going to happen. Why? The homogeneity of the early universe will keep

$$\delta g_{tt} \neq g_{tt} = 1. \tag{32}$$

In fact, we have that, from Giovannini [7], if ϕ is a scalar function and $a^2(t) \approx 10^{-110}$, then if

$$\delta g_{tt} \approx a^2(t) \cdot \phi \ll 1, \tag{33}$$

there is no way that Eq. (31) is going to come close to $\delta t \Delta E \geq \frac{h}{2}$.

4. How we can justify writing very small $\delta g_{rr} \approx \delta g_{\theta\theta} \approx \delta g_{\phi\phi} \approx 0^+$ values?

To begin this process, we break it down into the coordinates rr, $\theta\theta$, and $\phi\phi$. We will use the fluid approximation, $T_{ii} = \text{diag}(\rho, -p, -p, -p)$ [8], with [1]

$$\delta g_{rr} T_{rr} \geq -\left|\frac{h \cdot a^2(t) \cdot r^2}{V^{(4)}}\right| \xrightarrow[a \to 0]{} 0 \tag{34}$$
$$\delta g_{rr} T_{\theta\theta} \geq -\left|\frac{h \cdot a^2(t)}{V^{(4)}\left(1 - k \cdot r^2\right)}\right| \xrightarrow[a \to 0]{} 0 \tag{35}$$
$$\delta g_{rr} T_{\phi\phi} \geq -\left|\frac{h \cdot a^2(t) \cdot \sin^2\theta \cdot \mathrm{d}\phi^2}{V^{(4)}}\right| \xrightarrow[a \to 0]{} 0 \tag{36}$$

If, as an example, we have negative pressure, with T_{rr}, $T_{\theta\theta}$, and $T_{\phi\phi} < 0$, and $p = -\rho$, then the only choice we have is to set $\delta g_{rr} \approx \delta g_{\theta\theta} \approx \delta g_{\phi\phi} \approx 0^+$, since there is no way that $p = -\rho$ is zero valued. Having said this, the value of δg_{rr} being nonzero will be part of how we will be looking at a nonzero lower bound of the graviton mass.

5. Conclusion

To solidify the approach given here in terms of early universe GR theory, we refer to Einstein spaces, via [9], as well as make certain of the stress-energy tensor [10] as we can write it as a modified Einstein field equation. With, then, $\aleph$ as a constant,

$$R_{ij} = \aleph g_{ij}. \tag{37}$$

Here, the term in the left hand side of the metric tensor is a constant. So then, we can write, with R also a constant [10],

$$T_{ij} = -\frac{2}{\sqrt{-g}} \frac{\delta S}{\delta g_{ij}} = -\frac{1}{8\pi} \cdot [\aleph - R + \Lambda] \cdot g_{ij}. \tag{38}$$

The terms, if we use the fluid approximation, will then tend to a constant energy term on the RHS of Eq. (37) as well as restricting i, and j to t and t. If so, we have via an Einstein space, which may be more fully developed, in the pre-Planckian regime, a tight justification of Eq. (31) to Eq. (33). We also have the possibility of tying this in with Barbour [11]. This has a tie in with the lower bound to the graviton mass as brought up in Eq. (4), Eq. (5), and Eq. (6), which should be developed more fully later.

Acknowledgments
This work is supported in part by National Nature Science Foundation of China grant No. 11375279

References

[1] Beckwith A 2016 Gedanken experiment for refining the Unruh metric tensor uncertainty principle via Schwarzschild geometry and Planckian space–time with initial nonzero entropy and applying the Riemannian–Penrose inequality and initial kinetic energy for a lower bound to graviton mass (massive gravity). *J High Energy Phys., Gravit. Cosmology* **2** 106 doi:10.4236/jhepgc.2016.21012
[2] Rinstrom H 2003 *On the Topology and Future Stability of the Universe* (Oxford, England: Oxford University Press)
[3] Beyer W 1990 *CRC Standard Mathematical Tables,* 28th ed. (Boca Raton, FL: CRC Press)
[4] Downes T G, Milburn G J and Caves C M 2012 Optimal quantum estimation for gravitation *Preprint* arXiv:1108.5220 [gr-qc]
[5] Unruh W G 1986 Why study quantum theory? *Can. J. Phys.* **64** 128 doi:10.1139/p86-019
[6] Unruh W G 1986 Erratum: Why study quantum gravity? *Can. J. Phys.* **64** 128
[7] Giovannini M 2008 *A Primer on the Physics of the Cosmic Microwave Background* (Hackensack, NJ: World Press Scientific)
[8] Kolb E W and Turner M S *The Early Universe* (Redwood City, CA: Addison-Wesley)
[9] Petrov A Z 1969 *Einstein Spaces,* Oxford, England: Pergamon Press
[10] Gorbunov D and Rubakov V 2011 *Introduction to the Theory of the Early Universe, Cosmological Perturbations and Inflationary Theory* (Singapore: World Scientific)
[11] Barbour J 2009 The nature of time *Preprint* arXiv:0903.3489 [gr-qc]

Frequency-based redshift for cosmological observation and Hubble diagram from the 4-D spherical model in comparison with observed supernovae

Shigeto Nagao
Ishibashi, Ikeda, Osaka, 563-0032 Japan
E-mail: snagao@lilac.plala.or.jp

Abstract. According to the formerly reported 4-D spherical model of the universe, factors on Hubble diagrams are discussed. The observed redshift is not the prolongation of wavelength from that of the source at the emission but from the wavelength of spectrum of the present atom of the same element. It is equal to the redshift based on the shift of frequency from the time of emission. We demonstrate that the K-correction corresponds to conversion of the light propagated distance (luminosity distance) to the proper distance at present (present distance). Comparison of the graph of the present distance times $1 + z$ versus the frequency-based redshift with the reported Hubble diagrams from the Supernova Cosmology Project, which were time-dilated by $1 + z$ and K-corrected, showed an excellent fit for the Present Time (the radius of 4-D sphere) being c.a. 0.7 of its maximum.

1. Introduction

The author previously reported the light propagated distance and the redshift derived from the 4-D spherical model of the universe [1]. However, in actual cosmological observation, the redshift is based on the wavelength prolongation from the spectrum of the present atom but not from the spectrum at the time of emission at the light source. Under varying light speed along with the cosmological expansion, the redshift based on wavelength is different from that based on frequency. In this paper, we will discuss the two kinds of redshift, and provide the Hubble diagram expected from the 4-D spherical model. Subsequently, we will induce the proper distance at present (present distance) as a function of redshift or time of emission, and compare its Hubble diagram with those reported by the Supernova Cosmology Project (SCP).

According to the 4-D spherical model [1–4], the space energy spreads with expansion in a 3-D surface of a 4-D sphere as shown in Fig.1. Light and any other energy we detect in the 3-D space is a vibration of the intrinsic space energy in the 3-D space. The radius x of the 4-D sphere for the space energy distribution is our Observed Time (shown by Time or T), which we feel passing constantly and commonly for all of us [5]. Listed below definitions and abbreviations of key terms for the model.

x (Radius of universe): Radius of the 4D sphere, equal to the Observed Time (Time or T)
CU (Cosmic Unit): Unit of x and T, being one at its maximum.
(Time-related variables are expressed in the CU in this article.)
T_E (Time of Emission): Time when the light was emitted.
T_P (Present Time): Present Time of universe, when the light reaches us.
T_{ER} (Relative Time of Emission): Relative ratio of T_E to T_P. $T_{ER} \equiv T_E/T_P$
T_B (Back in Time): Back in Time from present when the light was emitted. $T_B \equiv T_P - T_E$
T_{BR} (Relative Back in Time): Relative ratio of T_B to T_P. $T_{BR} \equiv T_B/T_P$
T_C (Time Clear): Time when the space became transparent to light.
T_{CR} (Relative Time Clear): Relative ratio of T_C to T_P. $T_{CR} \equiv T_C/T_P$

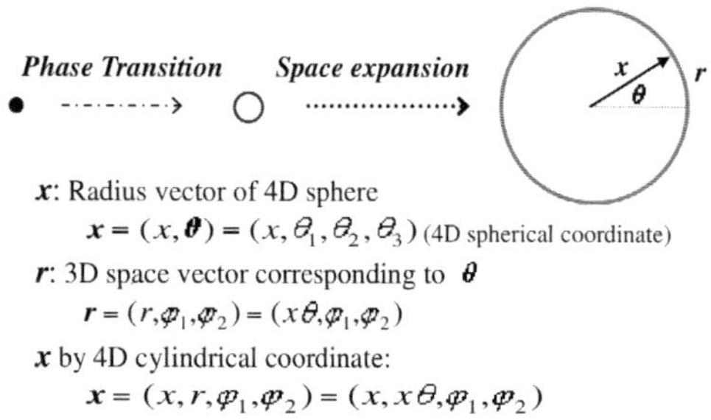

Figure 1. 4-D spherical model of the universe. The space energy spreads with expansion in a 3-D surface of a 4-D sphere.

2. Redshift for cosmological observation

Under the varying light speed $C(T)$, redshift based on wavelength is different from that based on frequency. Light of $\nu_0(T_E), \lambda_0(T_E)$ released at T_E reaches us now at T_P exhibiting $\nu(T_P), \lambda(T_P)$. $\nu_0(T_P), \lambda_0(T_P)$ is the spectrum of the present atom. $\nu_0(T_E)$ is equal to $\nu_0(T_P)$ because the electric orbital energy gap of the atom is same for both.

$$C(T_E) = \nu_0(T_E) * \lambda_0(T_E) \quad \rightarrow \quad C(T_P) = \nu(T_P) * \lambda(T_P) \tag{1}$$

$$Present\ atom\ spectrum: \quad C(T_P) = \nu_0(T_P) * \lambda_0(T_P) \tag{2}$$

The observed redshift is not the prolongation of wavelength from that of the source at the emission but from the wavelength of spectrum of the present atom of the same element. It is equal to the redshift based on the shift of frequency from the time of emission.

$$z + 1 = \frac{\lambda(T_P)}{\lambda_0(T_P)} = \frac{\nu_0(T_P)}{\nu(T_P)} = \frac{\nu_0(T_E)}{\nu(T_P)} = z_\nu + 1 \tag{3}$$

Wavelength-based redshift z_λ: Wavelength prolongation from T_E to T_P is the space expansion ratio n times the light speed ratio $C(T_P)/C(T_E)$ as formerly reported [1].

$$z_\lambda + 1 \equiv \frac{\lambda(T_P)}{\lambda_0(T_E)} = n \times \frac{C(T_P)}{C(T_E)} = \frac{T_P}{T_E} \times \frac{C(T_P)}{C(T_E)} = \frac{1}{T_{ER}} \times \frac{C(T_P)}{C(T_E)} \tag{4}$$

Frequency-based redshift z_ν (observed redshift):

$$z_\nu + 1 \equiv \frac{\nu_0(T_E)}{\nu(T_P)} = \frac{\lambda(T_P)}{\lambda_0(T_P)} = \frac{\lambda(T_P)\lambda_0(T_E)}{\lambda_0(T_E)\lambda_0(T_P)} = n \times \frac{C(T_P)}{C(T_E)} \times \frac{C(T_E)}{C(T_P)} = n = \frac{1}{T_{ER}} \tag{5}$$

Hereafter we use simply z for z_ν.

3. Factors affecting the brightness of a star

3.1. Factor by wavelength prolongation

Along with the space expansion, the wavelength is stretched by $n \equiv T_P/T_E = 1/T_{ER}$. While the energy of a single photon $h\nu$ decreases to $1/n$ times, the number of photons increases to n-fold to offset and preserve the total energy. As far as we use the luminosity as energy per unit time, we do not have to consider the factor by wavelength prolongation.

3.2. Factor by scattering

I previously proposed the following light speed equation, where x is the radius of 4-D sphere equal to our Observed Time that we feel passing constantly and commonly for all of us [1].

$$C(x) = K * f_D * f_{EM} = K * \frac{1}{x\sqrt{1-x}} * \left(1 - \frac{T_C^3}{x^3}\right) \tag{6}$$

The Electromagnetic interaction factor f_{EM} is effect of scattering. There are plural Time Clear T_C values. One is for the Cosmic Microwave Background (CMB) radiation [6] about 380 thousands years after the Big Bang when the universe cleared up after completion of atom formation from plasma. Later, hydrogen atoms started to form stars and radiate light, which re-ionized interstellar hydrogen atoms [7]. Then, the space became opaque again. Once the number of stars reached roughly plateau, the factor f_{EM} immediately increased and approached to one due to rapid decrease of the star density and the hydrogen atom density in the space. The third type of T_C-like is local one for light propagation in substances such as glass, water and air on the earth, and dust and gas in galaxies, whereas the optical path in the substance replaces the radius x in the f_{EM} formula. T_C of a substance and its density ρ give the absorbance. If the density is constant, the transmittance becomes as follows, where x is the optical depth.

$$I(x) = I_0\left(1 - \int (T_C\rho)^3\, dx\right) = I_0(1 - kx) \tag{7}$$

Stellar light emitted during the reionization period, said to be roughly from 150 million years to one billion years after the Big Bang [7], does not reach us or is darken by scattering. For stellar light emitted later, corresponding to $T_{ER} > 0.072$ or $z < 12.8$, the factor of scattering by reionization is negligible. Observed one is extinction due to absorption and scattering by dust and gas [8]. Atmospheric extinction on the earth is depending on location and altitude of the observing device. Galactic extinctions by our galaxy or host galaxy also dim the brightness a little, depending on directions of light lines. Light of shorter wavelength is dimmed more than light of longer wavelength, resulting in reddening of observed light. Corrections in brightness for extinctions are made from color excess functions of individual objects using respective standard extinction curves.

3.3. Factor by variation of the space energy density

Along with the space expansion, the density of the space energy (medium) decreases and causes light speed alteration. I formerly proposed the energy density factor of light speed f_D [1,3] as

$$f_D = \frac{1}{x\sqrt{1-x}}\ . \tag{8}$$

As an example of varying density of a wave medium, think of sound propagation. The sound propagates faster in a higher density medium. The intensity of volume is greater in a medium of higher density for same distance, that is, the sound propagates louder in the water than in the air. This phenomenon should be same for variation of the space energy density.

The brightness is usually expressed as energy per unit time, that is, luminosity as energy per unit time and flux as energy per unit time per unit area. The flux is shown as follows, where L is the luminosity and r is the distance from the light source.

$$F = \frac{L}{4\pi r^2} \tag{9}$$

Under constant light speed, the luminosity would be invariant by time. However, it varies depending on the light speed at the time of detection because it is a per unit-time value of energy, while the total energy from the source is invariant. Therefore, the luminosity is a function of the time of detection $T = x$.

$$F(x) = \frac{L(C(x))}{4\pi r^2} = \frac{L(x)}{4\pi r^2} \tag{10}$$

For our observation, the variable x is fixed to the Present Time T_P. $L = L(T_P)$ and $F = F(T_P)$. Therefore, we do not have to consider influence of light speed variation by space expansion for our observation of brightness whenever light was emitted.

4. Magnitude of flux and distance

Light of luminosity L emitted at T_E reaches us now at T_P. We observe now the flux $F(T_E)$ after correction of observational biases. $LD(T_E)$ is the light propagated distance equal to the luminosity distance. Define the magnitude of the flux as follows.

$$F(T_E) = \frac{L}{4\pi LD(T_E)^2} \tag{11}$$

$$m(T_E) \equiv -2.5\lg F(T_E) \tag{12}$$

Define the relative magnitude to the reference of the same luminosity exhibiting redshift 0.05 as the "magnitude of LD to $z = 0.05$, $DM_{0.05}$". It is a distance modulus. $z = 0.05$ corresponds to $T_{ER} = 1/(1+z) = 1/1.05$.

$$\begin{aligned} DM_{0.05} &\equiv m(T_{ER}) - m(1/1.05) \\ &= 5\lg LD(T_{ER}) - 5\lg LD(1/1.05) \end{aligned} \tag{13}$$

From eq.6 of light speed, the light propagated distance is given as follows [1], where f_{EM} is negligible if $T_{ER} > 0.072$ or $z < 12.8$.

$$\begin{aligned} LD(T_E) &= \int_{T_E}^{T_P} C(x)dx = \int_{T_E}^{T_P} \frac{K}{x\sqrt{1-x}}\left(1 - \frac{T_C^3}{x^3}\right)dx \approx \int_{T_E}^{T_P} \frac{K}{x\sqrt{1-x}}dx \\ &= K\left(\log\left(\frac{1-\sqrt{1-T_P}}{1+\sqrt{1-T_P}}\right) - \log\left(\frac{1-\sqrt{1-T_E}}{1+\sqrt{1-T_E}}\right)\right) \end{aligned} \tag{14}$$

$$LD(T_{ER}) = K\log\left(\frac{1-\sqrt{1-T_P}}{1+\sqrt{1-T_P}} \cdot \frac{1+\sqrt{1-T_P T_{ER}}}{1-\sqrt{1-T_P T_{ER}}}\right) \tag{15}$$

We get the following equation for the magnitude of LD to $z = 0.05$.

$$\begin{aligned} DM_{0.05}(T_{ER}) &= 5\lg\left(\log\left(\frac{1-\sqrt{1-T_P}}{1+\sqrt{1-T_P}} \cdot \frac{1+\sqrt{1-T_P T_{ER}}}{1-\sqrt{1-T_P T_{ER}}}\right)\right) \\ &\quad -5\lg\left(\log\left(\frac{1-\sqrt{1-T_P}}{1+\sqrt{1-T_P}} \cdot \frac{1+\sqrt{1-T_P/1.05}}{1-\sqrt{1-T_P/1.05}}\right)\right) \end{aligned} \tag{16}$$

Fig.2 shows graphs of $DM_{0.05}(T_{ER})$ with parameter T_{ER} for T_P values 0.6, 0.7 and 0.8 in dual logarithmic scale. (a) is versus the frequency-based redshift $z = 1/T_{ER} - 1$. (b) is versus the Relative Back in Time $T_{BR} = 1 - T_{ER}$, which times T_P signifies the light travel time since the emission. The dotted line in black is a reference for constant light speed based on the distance $LD_C(T_{ER}) = c(T_P - T_E) = c \cdot T_P(1 - T_{ER})$.

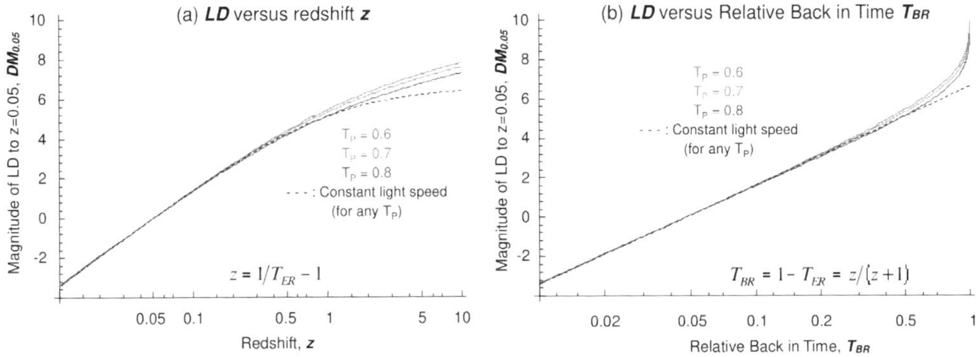

Figure 2. Magnitude of light propagated distance LD to $z = 0.05$ versus (a) redshift or (b) Relative Back in Time. T_P is Present Time. The dotted line in black is a reference for constant light speed.

5. Hubble diagram

According to the 4-D spherical model, the expansion speed of the 3-D space by our Observed Time is constant for a given angle θ as shown by

$$\frac{dr}{dT} = \frac{dr}{dx} = \frac{d(x\theta)}{dx} = \theta \ . \tag{17}$$

r is a distance between any two points in the 3-D space. θ is the angle between the two positions from the center of 4-D sphere. For variable θ at a given x, the recessional velocity is in proportion to θ and the distance $r = \theta \cdot x$. This is the Hubble's law. The said velocity is of the proper distance. Refer the proper distance at present to as the "present distance, PD".

5.1. LD-PD conversion

Take a case shown in Fig.3. Light emitted at A_E (3-D location A, time T_E) reaches us at B_P (B, T_P). The light propagated distance LD is the 3-D space component of the line $A_E B_P$, equal to the line CB_P. n is the space expansion ratio given by $n = A_P B_P / A_E B_E = T_P / T_E = 1/T_{ER}$. As we see in Fig.3, the ratio of CB_P to $A_E B_E$ is

$$\frac{CB_P}{A_E B_E} = 1 + \frac{1}{2}(n-1) = \frac{n+1}{2} \ . \tag{18}$$

The ratio of converting the light propagated distance LD to the present distance PD becomes as follows.

$$\frac{PD}{LD} = \frac{A_P B_P}{A_E B_E} \cdot \frac{A_E B_E}{CB_P} = n \cdot \frac{2}{n+1} = \frac{2(z+1)}{z+2} = \frac{2}{1+T_{ER}} \tag{19}$$

We get the following "magnitude of PD to $z = 0.05$, $DM_{0.05}^{PD}$".

$$\begin{aligned} DM_{0.05}^{PD}(T_{ER}) &= DM_{0.05}(T_{ER}) + 5\lg\left(2/\left(1+T_{ER}\right)\right) - 5\lg\left(2/\left(1+1/1.05\right)\right) \\ &= DM_{0.05}(T_{ER}) - 5\lg\left(1+T_{ER}\right) + 5\lg 2.05 - 5\lg 1.05 \end{aligned} \tag{20}$$

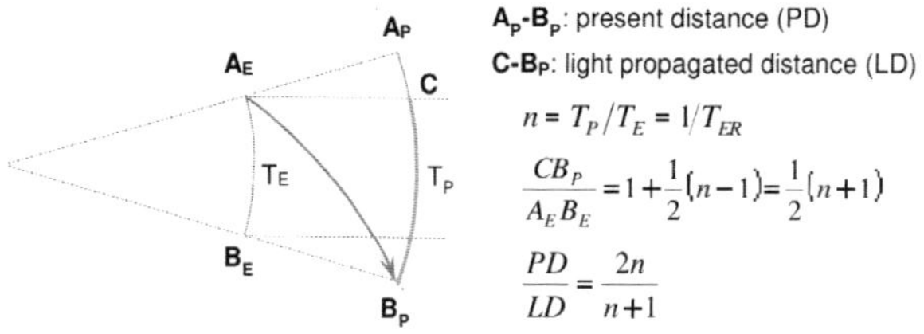

Figure 3. Light line in the 4-D coordinate. Light emitted at $A_E(A, T_E)$ reaches us at $B_P(B, T_P)$. The light propagated distance LD is the 3-D space component of the line $A_E B_P$ equal to the line CB_P.

5.2. Time dilation

Any model of universe including space expansion requires the so-called time dilation [9]. Take a time interval of two events of light emission. Light emits at location A, time T_E , and then reaches B at T_P. Another light dispatches A at $T_E + \Delta t$ and arrives at B at $T_E + \Delta t'$. Due to the space expansion by $1 + z$ from T_E to T_P, the time interval of the events increases by $1 + z$ for light detection at T_P.

$$\Delta t' = (1 + z)\Delta t \tag{21}$$

This time dilation was verified by observation on the light curve width of supernovae. Observed supernovae exhibited a uniform light curve width after divided by $1+z$ along with normalization of maximum luminosities [10, 11].

However, this is not a real time dilation but the distance $c\Delta t$ expands during the light travels from A to B. In Hubble diagrams including those by the SCP [12], the luminosity distance is multiplied by $1 + z$ as the time dilation. A local light-path near T_E expands by $1 + z$ at T_P. However, the expansion ratio of later path becomes smaller and approaches one for path around T_P. The expansion ratio of the whole light propagated distance to the proper distance at T_E is given by

$$\frac{CB_P}{A_E B_E} = \frac{n+1}{2} = 1 + \frac{z}{2} \tag{22}$$

, which is not $1 + z$. The dilation $1 + z$ is not the LD-PD conversion, either. The time duration $T_P - T_E$ is not dilated whereas is given as a function of the redshift. What does $(1 + z)LD$ signify? Provide tentatively that the light speed has been constant from the emission to the present. The light propagated distance LD_C from T_E to T_P is given as follows.

$$LD_C = c(T_P - T_E) = c \cdot T_P(1 - T_{ER}) = c \cdot T_P \cdot z/(1 + z) \tag{23}$$

Multiply the both sides by $1 + z$, we get the following equation.

$$(1 + z)LD_C = c \cdot T_P \cdot z \equiv k \cdot z \tag{24}$$

The time duration of light traveling is dilated by $1+z$, and becomes $T_P \cdot z$ equal to the duration when light travels from the present location A at T_P to our location B in future at $T_P(1 + z)$. An important feature of this modification is that the $(1 + z)$-dilated LD becomes proportional to the redshift z if light speed has been constant from the emission. This could be a reason why time-dilated luminosity distance versus redshift graphs are often used as a Hubble diagram, in addition to the time interval dilation seen on the light curve width of supernovae [10, 11].

5.3. K-correction

The K-correction converts a measurement of an object at a redshift z to an equivalent measurement in the rest frame of the object at $z = 0$ [13–16]. $f(\lambda)$ is the flux density by spectral energy distribution (SED) at wavelength λ of an object 10 parsecs away (energy per unit time per unit area per unit wavelength). $S_i(\lambda)$ is the effective transmission of band i. $Z_i(\lambda)$

is the flux density of a zero-magnitude standard star. Virtually consider a stationary universe without expansion. The apparent magnitude in band x becomes as follows, where DM^0 is the distance modulus defined by $DM^0 = 5\lg\left(D_L^0/10pc\right)$, where D_L^0 is the luminosity distance.

$$m_x^0 = M_x + DM^0 = -2.5\lg\left(\frac{\int f(\lambda)S_x(\lambda)d\lambda}{\int Z_x(\lambda)S_x(\lambda)d\lambda}\right) + DM^0 \tag{25}$$

In a space-expansion model of the universe, the SED is redshifted at the observer. If the SED is redshifted by z in the above case with the same flux density function and distance, the magnitude in band y becomes as follows because the wavelength is dilated by $1+z$.

$$m_y = -2.5\lg\left(\frac{\frac{1}{1+z}\int f\left(\frac{\lambda}{1+z}\right)S_y(\lambda)d\lambda}{\int Z_y(\lambda)S_y(\lambda)d\lambda}\right) + DM^0 \tag{26}$$

The flux in the numerator is divided by $1+z$ because of the dilation of wavelength width for the flux density. The difference of the two magnitudes is the K-correction given as follows.

$$\begin{aligned} K_{xy} &\equiv m_y - m_x^0 = -2.5\lg\left(\frac{\frac{1}{1+z}\int f\left(\frac{\lambda}{1+z}\right)S_y(\lambda)d\lambda}{\int Z_y(\lambda)S_y(\lambda)d\lambda}\right) + 2.5\lg\left(\frac{\int f(\lambda)S_x(\lambda)d\lambda}{\int Z_x(\lambda)S_x(\lambda)d\lambda}\right) \\ &= -2.5\lg\left(\frac{\int Z_x(\lambda)S_x(\lambda)d\lambda}{\int Z_y(\lambda)S_y(\lambda)d\lambda}\right) + 2.5\lg(1+z) + 2.5\lg\left(\frac{\int f(\lambda)S_x(\lambda)d\lambda}{\int f(\frac{\lambda}{1+z})S_y(\lambda)d\lambda}\right) \end{aligned} \tag{27}$$

This formula is exactly the energy-based K-correction, which the SCP applied [14]. We can rewrite the definition of K-correction as

$$m_y = M_x + DM^0 + K_{xy}\ . \tag{28}$$

If we actually observe m_y with redshift z, it gives the following formula with the absolute magnitude in band y and the luminosity distance in the observed frame.

$$m_y = M_y + DM(z)\ . \tag{29}$$

Accordingly, we get the following relation for the K-correction.

$$K_{xy} = M_y - M_x + DM(z) - DM^0 \tag{30}$$

As shown in eq.30, the K-correction amounts to the sum of difference in absolute magnitude and that in distance modulus between the redshifted and rest frames. The difference $M_y - M_x$ is the cross-filter adjustment on absolute magnitude between x and y bands, an observational bias. The rest part

$$K_{xy}^D \equiv DM(z) - DM^0 \tag{31}$$

is the theoretical part of K-correction in distance. The luminosity distance $D_L(z)$ in the observed frame is equal to the light propagated distance $LD(z)$. The luminosity distance D_L^0 in the rest frame corresponds to the present distance PD of the light source. We can conclude that the theoretical part of K-correction for frame conversion exactly expresses the LD-PD conversion.

IOP Conf. Series: Journal of Physics: Conf. Series **880** (2017) 012058 doi:10.1088/1742-6596/880/1/012058

6. Comparison with Hubble diagrams from the Supernova Cosmology Project

The reported graphs of Hubble diagram by the SCP applied the time dilation by $1+z$ and K-corrections [12, 17, 18]. For directly comparing the expected theoretical curves from the 4-D spherical model with the reported Hubble diagrams by the SCP, we multiply the PD by $1+z$ because the theoretical part of K-correction corresponds to the LD-PD conversion.

$$(1+z)\cdot PD = (1+z)\cdot \frac{2(z+1)}{z+2}\cdot LD \tag{32}$$

$$\frac{1}{T_{ER}}\cdot PD = \frac{1}{T_{ER}}\cdot \frac{2}{1+T_{ER}}\cdot LD \tag{33}$$

Refer the magnitude of this quantity to as the "adjusted magnitude of PD to $z = 0.05$, $DM^{adj}_{0.05}$".

$$\begin{aligned} DM^{adj}_{0.05}(T_{ER}) &= 5\,(\lg LD(T_{ER}) + \lg(1/T_{ER}) + \lg(2/(1+T_{ER})) - \lg LD(1/1.05) - \lg 1.05 - \lg(2*1.05/2.05)) \\ &= 5\,(\lg LD(T_{ER}) - \lg(T_{ER}) - \lg(1+T_{ER}) - \lg LD(1/1.05) - 2\lg 1.05 + \lg 2.05) \end{aligned} \tag{34}$$

It corresponds to the time-dilated by $1+z$ and K-corrected magnitude for a common luminosity like type Ia supernovae in real observation, while respective zero points are different. Substituting eq.15 for LD in eq.34, we get the following final equation for comparison with observed data.

$$\begin{aligned} DM^{adj}_{0.05}(T_{ER}) &= 5\lg\left(\log\left(\frac{1-\sqrt{1-T_P}}{1+\sqrt{1-T_P}}\cdot\frac{1+\sqrt{1-T_P T_{ER}}}{1-\sqrt{1-T_P T_{ER}}}\right)\right) - 5\lg(T_{ER}) - 5\lg(1+T_{ER}) \\ &\quad -5\lg\left(\log\left(\frac{1-\sqrt{1-T_P}}{1+\sqrt{1-T_P}}\cdot\frac{1+\sqrt{1-T_P/1.05}}{1-\sqrt{1-T_P/1.05}}\right)\right) - 10\lg 1.05 + 5\lg 2.05 \end{aligned} \tag{35}$$

As a reference, take $DM^{adj}_{0.05}$ based on LD_C presuming that the light speed had been constant since the emission to the present. From $LD_C = c\cdot T_P(1-T_{ER})$, it becomes as follows.

$$\begin{aligned} DM^{adj}_{0.05}(T_{ER}) &= 5\lg(1-T_{ER}) - 5\lg(1-1/1.05) - 5\lg(T_{ER}) - 5\lg(1+T_{ER}) - 10\lg 1.05 + 5\lg 2.05 \\ &= 5\lg(1-T_{ER}) + 5\lg 21 - 5\lg(T_{ER}) - 5\lg(1+T_{ER}) - 10\lg 1.05 + 5\lg 2.05 \end{aligned} \tag{36}$$

Fig.4 shows the graphs of the adjusted magnitude of PD to $z = 0.05$, $DM^{adj}_{0.05}$ versus the redshift $z = 1/T_{ER} - 1$, respectively in a logarithmic scale and a uniform scale for z. The dotted line in black is the reference $DM^{adj}_{0.05}$ for constant light speed.

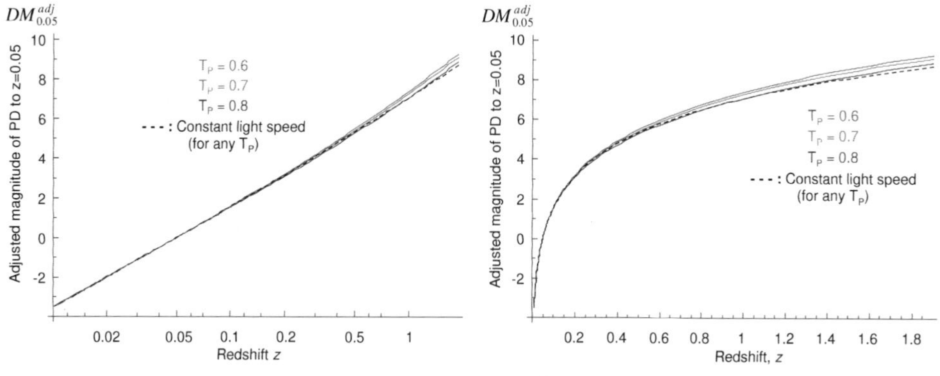

Figure 4. Adjusted magnitude of present distance PD to $z = 0.05$ versus redshift. z in logarithmic (left) and uniform (right) scales. The dotted line in black is a reference for constant light speed.

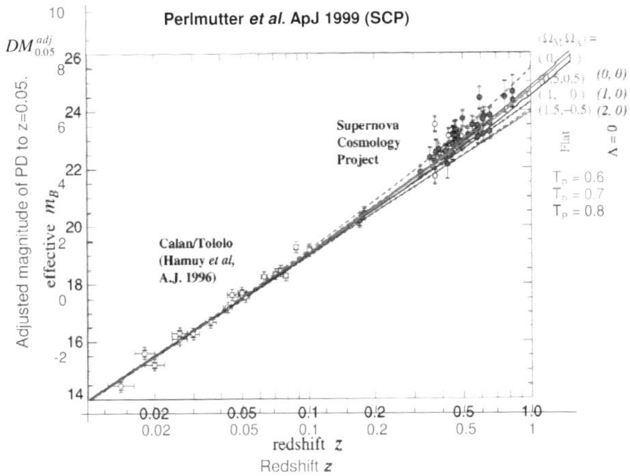

Figure 5. Superimposition of adjusted magnitude of present distance PD to $z = 0.05$ versus redshift graphs on the reported Hubble diagram from SCP (1999): z in a logarithmic scale.

Fig.5 is the superimposed results of Fig.4 on the Hubble diagram in a logarithmic scale for z, which Perlmutter et al reported in 1999 [12]. The vertical axis of their reported graph is the effective rest-frame B magnitude m_B^{eff}, which is defined as $m_B^{eff} = m_B + \alpha(s-1)$ and $m_B = m_R - K_{BR} - A_R$, with time dilation by $1+z$. The term $\alpha(s-1)$ is the correction for stretch factor s. The stretch factor s is a ratio of each light-curve width to the mean after divided by $1+z$ for each supernova. The average of s is one and the mean $\alpha(s-1)$ is zero. A_R is the galactic extinction in band R. K_{BR} is the K-correction from observed frame in band R to rest-frame in band B.

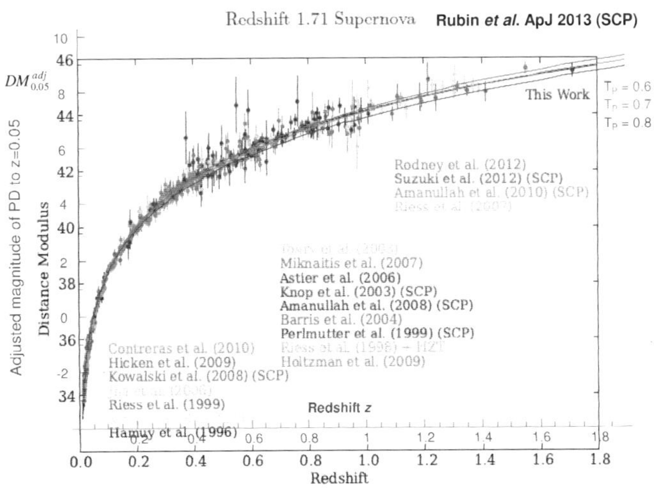

Figure 6. Superimposition of adjusted magnitude of present distance PD to $z = 0.05$ versus redshift graphs on the reported Hubble diagram from SCP (2013): z in a uniform scale.

Fig.6 is the superimposition of Fig.4 on the latest Hubble diagram from the SCP in a uniform scale for z, which Rubin et al reported in 2013 [18]. The vertical axis is the distance modulus, which is time-dilated and K-corrected m_B minus the absolute B-magnitude of supernovae Ia with other observational corrections for light curve width (stretch factor), color and host galaxy mass (extinctions). We find excellent fits to the observed data of supernovae for T_P around 0.7. In Fig.6, the graph for $T_P = 0.7$ very closely overlaps the line of flat $\Omega_m = 0.27$ ΛCDM universe that Rubin et al concluded as the best fit [18].

7. Conclusion

Before this article, I proposed the 4-D spherical model claiming a medium for light propagation called the space energy [1–4]. Light speed depends on the space energy density varying along with the space expansion. I demonstrated that Michelson-Morley type experiments [19–21] are incapable to detect a location shift of interference fringe while its brightness should vary [1]. Then, I proposed the acceleration factor of stationary waves for equation of motion instead of the Lorentz factor [2]. I also formerly reported the light speed equation, the light propagated distance and the redshift based on wavelength [1].

Under varying light speed, redshift based on wavelength is different from that based on frequency. Our observed redshift is the frequency-based redshift for comparison with initial value at the time of emission. In this article, we presented the Hubble diagram expected from the 4-D spherical model. The light propagated distance LD is the integration of the light speed by time from the emission to the present, and is equal to the luminosity distance. We showed the equation for converting the LD to the present distance PD (proper distance at present), which corresponds to the frame conversion by the K-correction. We compared the graph of magnitude of the PD times $1 + z$ versus the frequency-based redshift with the reported Hubble diagrams from the SCP, which were on the time-dilated by $1 + z$ and K-corrected luminosity distance versus the redshift. The superimposition of the graph from the model on the observed plots revealed an excellent fit for the Present Time being around 0.7 of its maximum.

References

[1] Nagao S 2015 Light propagated distance and redshift of a distant star *J. Phys.: Conf. Ser.* **626** 012068
[2] Nagao S 2013 Acceleration factor for propagation of a stationary wave in its wave medium: movement of energy in the 3-D space *J. Phys.: Conf. Ser.* **442** 012067
[3] Nagao S 2011 *J. Phys.: Conf. Ser.* **306** 012073
[4] Nagao S 2009 *J. Phys.: Conf. Ser.* **174** 012069
[5] Features of time *MiTiempo* (accessed 24 October 2016) http://www3.plala.or.jp/MiTiempo/Time-summary.pdf
[6] Cosmic microwave background *Wikipedia* (accessed 10 February 2016) https://en.wikipedia.org/wiki/Cosmic_microwave_background
[7] Reionization *Wikipedia* (accessed 10 February 2016) https://en.wikipedia.org/wiki/Reionization
[8] Extinction (astronomy) *Wikipedia* (accessed 10 February 2016) https://en.wikipedia.org/wiki/Extinction_(astronomy)
[9] Kirshner R P 2004 *PNAS* **101** 8
[10] Goldhaber G *et al* 2001 *Astrophys. J.* **558** 359
[11] Blondin S *et al* 2008 *Astrophys. J.* **682** 724
[12] Perlmutter S *et al* 1999 *Astrophys. J.* **517** 565
[13] Oke J B and Sandage A 1968 *Astrophys. J.* **154** 21
[14] Nugent P *et al* 2002 *PASP* **114** 803
[15] de Lapparent V *et al* 2003 *A&A* **404** 831
[16] Chilingarian I V *et al* 2010 *MNRAS* **405** 1409
[17] Suzuki N *et al* 2012 *Astrophys. J.* **746** 85
[18] Rubin D *et al* 2013 *Astrophys. J.* **763** 35
[19] Michelson A A 1881 *American Journal of Science* **22** 120
[20] Michelson A A and Morley E W 1887 *American Journal of Science* **34** 333
[21] Herrmann S *et al* 2009 *Physical Review D* **80** 105011 (arXiv:1002.1284)

Condensates as components of dark matter and dark energy

Antonio Capolupo

Dipartimento di Fisica E.R.Caianiello and INFN Gruppo Collegato di Salerno, Universitá di Salerno, Fisciano (SA) - 84084, Italy

E-mail: capolupo@sa.infn.it

Abstract. We report on recent results according to which the vacuum condensate characterizing many physical systems can give contributions to dark energy and dark matter of the universe. In particular, it is shown that thermal states of the intercluster medium, the vacuum energy of fields in curved space-time and of mixed neutrinos contribute to the dark matter. The vacuum condensates generated by the mixing of axions and photons and the one produced by superpartners of neutrinos may represent a component of the dark energy.

1. Introduction

According to many experimental data, the universe is made up approximately of 68% of dark energy [1]-[5] and about of 27% of dark matter. The rest is ordinary matter. Although different models have been proposed to explain the dark energy [6]- [15] and the dark matter problem [16]- [19], however, the solution to such a problem represents still a very big challenge.

Here it is reported the recent result [20] according to which the vacuum condensate energy characterizing many physical systems [21]-[40], can give contributions to the dark sector of the universe. It is shown that non trivial contributions to the dark matter can be originated by the thermal vacuum of the hot plasma present at the center of a galaxy cluster (intracluster medium), by the vacuum fluctuations of fields in curved space [41] and by the flavor neutrino vacuum [20]. Moreover, it is shown that the condensate of mixed boson [32, 33], such as axions and axion like particles (ALPs) in their interaction with photons, and the condensate of superpartners of mixed neutrinos, can give contributions to the dark energy with the state equation of the cosmological constant [20]. The mixing of unstable particles particles like kaons, B^0, D^0 mesons and $\eta - \eta'$ is expected do not contribute to the energy on large scale.

The origin of the non trivial vacuum energy contributions of the systems described above is due to the fact that the physical vacuum of these systems is a condensate of couples of particles and antiparticles which generate a positive value of the zero point energy. The formal analogy among the phenomena generating condensates could allow to simulate the systems here analyzed by means of phenomena reproducible in table top experiments such as the superconductivity, the Casimir effect and the Schwinger effect.

In Sec.II, the Bogoliubov transformations in QFT are introduced and the condensate structure of the transformed vacuum is presented. The general form of the energy momentum tensor of the vacuum condensates is presented in Sec.III. In Secs.IV, V, VI, the contributions given to the

IOP Conf. Series: Journal of Physics: Conf. Series **880** (2017) 012059 doi:10.1088/1742-6596/880/1/012059

energy of the universe by thermal states, by fields in curved space and by particle mixing are analyzed. Sec.VII is devoted to the conclusions.

2. Vacuum condensate induced by Bogoliubov transformations

Many systems such as the Hawking-Unruh [21, 22] and the Schwinger effects [23], the BCS theory of superconductivity [24], thermal fields in finite temperature QFT [26], the particle mixing [31]-[36], quantum fields in curved spacetimes [40] are described by the Bogoliubov transformations in the QFT context [27]. For bosons, general Bogliubov transformations are

$$\begin{aligned}\tilde{a}_{\mathbf{k}}(\xi,t) &= U^B_{\mathbf{k}}(\xi)\,a_{\mathbf{k}}(t) - V^B_{-\mathbf{k}}(\xi)\,a^\dagger_{-\mathbf{k}}(t)\,,\\ \tilde{a}^\dagger_{-\mathbf{k}}(\xi,t) &= U^{B*}_{-\mathbf{k}}(\xi)\,a^\dagger_{-\mathbf{k}}(t) - V^{B*}_{\mathbf{k}}(\xi)\,a_{\mathbf{k}}(t)\,,\end{aligned} \tag{1}$$

with $a_{\mathbf{k}}(t) = a_{\mathbf{k}}e^{-i\omega_k t}$, such that $a_{\mathbf{k}}|0\rangle_B = 0$ and $\omega_{\mathbf{k}} = \sqrt{k^2+m^2}$. The coefficients for bosons satisfy the conditions $U^B_{\mathbf{k}} = U^B_{-\mathbf{k}}$, $V^B_{\mathbf{k}} = V^B_{-\mathbf{k}}$, $|U^B_{\mathbf{k}}|^2 - |V^B_{\mathbf{k}}|^2 = 1$, and similar relations hold for fermions. The parameter ξ depends on the particular system one considers. For example, ξ is related to the acceleration of the observer in the Unruh effect case. The Bogoliubov transformations (1) can be written in terms of the generator $J(\xi,t)$ as: $a_{\mathbf{k}}(\xi,t) = J^{-1}(\xi,t)\,a_{\mathbf{k}}(t)J(\xi,t)$, with $J^{-1}(\xi) = J(-\xi)$, and similar transformations for fermions. The vacua $|0(\xi,t)\rangle$ annihilated by $a_{\mathbf{k}}(\xi,t)$ are related to the original ones $|0\rangle$ by $|0(\xi,t)\rangle = J^{-1}(\xi,t)|0\rangle$. This is a unitary operation at finite volume, but in QFT, in which $\mathbf{k}$ assumes a continuous infinity of values, is not a unitary transformation any more [28]. Then the vacuum $|0(\xi,t)\rangle$ cannot be expressed as a superposition of vectors in the Fock space built over $|0\rangle$ and the two Fock spaces are unitarily inequivalent [26]-[27]. This fact leads to the problem of the right choice of the Fock space and of the physical vacua associated with the particles which appear in observations. For systems characterized by Bogoliubov transformations, the physical vacua are the transformed ones $|0(\xi,t)\rangle$ [28]. Such vacua have a condensate structure, indeed one has $\langle 0(\xi,t)|a^\dagger_{\mathbf{k}}a_{\mathbf{k}}|0(\xi,t)\rangle = |V_{\mathbf{k}}|^2$, and similar for fermions. Such a property generate a non zero energy momentum tensor for $|0(\xi,t)\rangle$ which is responsible of the contributions to the dark sector of the universe which will be analyzed in the following and produces the spontaneous supersymmetry breaking in all the systems characterized by the presence of condensates [42]-[46].

3. Energy-momentum tensor of vacuum condensate

One computes the expectation value of the energy momentum tensor densities for scalar and for Majorana fields $T^\lambda_{\mu\nu}(x)$, ($\lambda = B, F$, with $B=$ bosons and $F=$ fermions) on the transformed vacuum $|0(\xi,t)\rangle_\lambda$,

$$\Xi^\lambda_{\mu\nu}(x) \equiv_\lambda \langle 0(\xi,t)| : T^\lambda_{\mu\nu}(x) : |0(\xi,t)\rangle_\lambda =_\lambda \langle 0(\xi,t)|T^\alpha_{\mu\nu}(x)|0(\xi,t)\rangle_\lambda -_\lambda \langle 0|T^\lambda_{\mu\nu}(x)|0\rangle_\lambda\,. \tag{2}$$

Here, : ... :, is the normal ordering with respect to the original vacuum $|0\rangle_\lambda$. Since $\langle 0(\xi,t)| : T^\lambda_{i,j}(x) : |0(\xi,t)\rangle = 0$, for $i \neq j$, the condensates induced by Bogoliubov transformations behave as a perfect fluid and the energy density and pressure of boson and fermion condensates are

$$\rho^\lambda = \langle 0(\xi,t)| : T^\lambda_{00}(x) : |0(\xi,t)\rangle\,, \tag{3}$$
$$p^\lambda = \langle 0(\xi,t)| : T^\lambda_{jj}(x) : |0(\xi,t)\rangle\,, \tag{4}$$

respectively.

For bosons, one has

$$\rho_B = \frac{1}{2}\langle 0(\xi,t)| : \Big[\pi^2(x) + \Big(\vec{\nabla}\phi(x)\Big)^2 + m^2\phi^2(x)\Big] : |0(\xi,t)\rangle\,; \tag{5}$$
$$p_B = \langle 0(\xi,t)| : \Big([\partial_j\phi(x)]^2 + \frac{1}{2}\Big[\pi^2(x) - \Big(\vec{\nabla}\phi(x)\Big)^2 - m^2\phi^2(x)\Big]\Big) : |0(\xi,t)\rangle\,. \tag{6}$$

In the particular case of the isotropy of the momenta, $k_1 = k_2 = k_3$, one has, $[\partial_j \phi(x)]^2 = \frac{1}{3}\left[\vec{\nabla}\phi(x)\right]^2$, then the pressure can be written as

$$p_B = \frac{1}{2}\langle 0(\xi,t)| : \left[\pi^2(x) - \frac{1}{3}\left(\vec{\nabla}\phi(x)\right)^2 - m^2\phi^2(x)\right] : |0(\xi,t)\rangle \,,$$

and the energy density and pressure can be expressed as [20]

$$\rho_B = \frac{1}{2\pi^2}\int_0^\infty dk k^2 \omega_k |V_k^B|^2 \,, \tag{7}$$

$$p_B = \frac{1}{6\pi^2}\int_0^\infty dk k^2 \left[\frac{k^2}{\omega_k}|V_k^B|^2 - \left(\frac{k^2}{\omega_k} + \frac{3m^2}{2\omega_k}\right)|U_k^B||V_k^B|\cos(\omega_k t)\right]. \tag{8}$$

The state equation is

$$w_B = \frac{1}{3}\frac{\int d^3\mathbf{k}\frac{k^2}{\omega_k}|V_k^B|^2}{\int d^3\mathbf{k}\,\omega_k|V_k^B|^2} - \frac{1}{3}\frac{\int d^3\mathbf{k}\left(\frac{k^2}{\omega_k} + \frac{3m^2}{2\omega_k}\right)U_k^B V_k^B\cos(\omega_k t)}{\int d^3\mathbf{k}\,\omega_k|V_k^B|^2}. \tag{9}$$

For fermions, the energy density and the pressure are

$$\rho_F = \frac{1}{2}\langle 0(\xi,t)| : \left[-i\bar{\psi}\gamma_j\partial^j\psi + m\bar{\psi}\psi\right] : |0(\xi,t)\rangle \,; \tag{10}$$

$$p_F = \langle 0(\xi,t)| : \left(\frac{i}{2}\bar{\psi}\gamma_j\overleftrightarrow{\partial_j}\psi\right) : |0(\xi,t)\rangle \,. \tag{11}$$

and explicitly one has [20]

$$\rho_F = \frac{1}{\pi^2}\int_0^\infty dk k^2 \omega_k |V_k^F|^2 \,, \tag{12}$$

$$p_F = \frac{1}{3\pi^2}\int_0^\infty dk \frac{k^4}{\omega_k}|V_k^F|^2 \,, \tag{13}$$

$$w_F = \frac{1}{3}\frac{\int d^3\mathbf{k}\frac{k^2}{\omega_k}|V_k^F|^2}{\int d^3\mathbf{k}\,\omega_k|V_k^F|^2}. \tag{14}$$

The non-zero values of ρ_λ and p_λ are due to the condensate structure of the physical vacuum. Notice also that, being $J^{-1}(\xi,t) = J^\dagger(\xi,t) = J(-\xi,t)$, one can write

$${}_\lambda\langle 0(\xi,t)| : T^\lambda_{\mu\nu}(x) : |0(\xi,t)\rangle_\lambda = {}_\lambda\langle 0|J_\lambda^{-1}(-\xi,t) : T^\lambda_{\mu\nu}(x) : J_\lambda(-\xi,t)|0\rangle_\lambda \,.$$

In the following we use the notation $\Theta(-\xi,x) = J^{-1}(-\xi,t)\Theta(x)J(-\xi,t)$ to denote the operators transformed by the generator $J(-\xi,t)$. Moreover, it will be used the condition, $\vec{\nabla}J(\xi,t) = 0$, satisfied by the generators $J's$ of the thermal states and of the particle mixing phenomenon.

Eqs.(7), (8) and (12), (13), hold for many phenomena. The particular system is specified by the explicit form of the Bogoliubov coefficients.

4. Contributions given by thermal states, Hawking and Unruh effects

In the framework of the Thermo Field Dynamics (TFD) [26]–[27], the physical vacuum of systems at non-zero temperature is the thermal vacuum state $|0(\xi(\beta))\rangle_\lambda$, with $\lambda = B, F$, $\beta \equiv 1/(k_B T)$ and k_B the Boltzmann constant. In such a formalism, the thermal statistical average is given by $\mathcal{N}_{\chi_{\mathbf{k}}}(\xi) =_\lambda \langle 0(\xi(\beta))|N_{\chi_{\mathbf{k}}}|0(\xi(\beta))\rangle_\lambda$, with $N_{\chi_{\mathbf{k}}} = \chi^\dagger_{\mathbf{k}}\chi_{\mathbf{k}}$, $(\chi = a, \alpha)$ the number operator [26]–[27].

The Bogoliubov coefficients are given by $U^T_{\mathbf{k}} = \sqrt{\frac{e^{\beta\omega_{\mathbf{k}}}}{e^{\beta\omega_{\mathbf{k}}}\pm 1}}$ and $V^T_{\mathbf{k}} = \sqrt{\frac{1}{e^{\beta\omega_{\mathbf{k}}}\pm 1}}$, with $-$ for bosons and $+$ for fermions, $\beta = 1/k_B T$ and $\omega_{\mathbf{k}} = \sqrt{k^2+m^2}$ [26]–[27]. Such coefficients, used in Eqs.(7), (8) and (12), (13) give the contributions of the thermal vacuum states to the energy and pressure. The results one finds are the following: for the cosmic microwave background temperature, $T = 2.72K$, only photons and particles with masses of order of $(10^{-3} - 10^{-4})eV$ contribute to the energy radiation with $\rho \simeq 10^{-51}GeV^4$ and state equations, $w = 1/3$ [47]. Negligible contributions to vacuum energy are given by the thermal states describing the Unruh and of the Hawking effects [47, 48]. On the other hand, the thermal vacuum of the hot plasma filling the center of galaxy clusters, which has temperatures of order of $(10 \div 100) \times 10^6 K$, has an energy of $(10^{-48} - 10^{-47})GeV^4$ and a state equation $w = 0.01$. Such values are in agreement with the ones on the dark matter [20].

5. Contribution of fields in curved background

A vacuum condensed structure characterizes also fields in curved spaces [40]. In these cases, the energy momentum tensor depends on the particular metric considered. Here one analyzes the spatially flat Friedmann Robertson-Walker metric, $ds^2 = dt^2 - a^2(t)d\mathbf{x}^2 = a^2(\eta)(d\eta^2 - d\mathbf{x}^2)$, where a is the scale factor, t is the comoving time, η is the conformal time, $\eta(t) = \int_{t_0}^t \frac{dt}{a(t)}$, with t_0 arbitrary constant and considers the boson case. Assuming a cutoff on the momenta much smaller than the comoving mass of the field, $K \ll ma$ and assuming that $m \gg H$, with H Hubble constant, the energy density and pressure of the vacuum condensate, become [41]

$$\begin{aligned} \rho_{curv} &= \frac{1}{8\pi^2}\int_0^K dk k^2 \left(\frac{2m}{a^3} + \frac{9H^2}{4ma^3} + \frac{k^2}{ma^5}\right), \\ p_{curv} &= \frac{1}{8\pi^2}\int_0^K dk k^2 \left(\frac{9H^2}{4ma^3} - \frac{k^2}{3ma^5}\right). \end{aligned} \tag{15}$$

The state equation, in the infrared regime, is then the one of the dark matter, $w_{curv} \simeq 0$. Numerical values of ρ_{curv}, compatible with the ones of dark matter can be found when, $\frac{mK^3}{a^3} \sim 10^{-45}GeV^4$.

6. Contributions given by particle mixing

The phenomenon of particle mixing is represented by the neutrino and quark mixing in fermion sector and by the axion-photon mixing and the mixing of kaons, B^0, D^0, and $\eta - \eta'$ systems, in boson sector. For two fields, it is expressed, both for fermions and bosons, as

$$\begin{aligned} \varphi_1(\theta, x) &= \varphi_1(x)\cos(\theta) + \varphi_2(x)\sin(\theta), \\ \varphi_2(\theta, x) &= -\varphi_1(x)\sin(\theta) + \varphi_2(x)\cos(\theta), \end{aligned} \tag{16}$$

where, θ is the mixing angle, $\varphi_i(\theta, x)$ are the mixed fields and $\varphi_i(x)$ are the free fields, with $i = 1, 2$.

The mixing transformations (16) can be written as $\varphi_i(\theta, x) \equiv J^{-1}(\theta, t)\varphi_i(x)J(\theta, t)$, where $i = 1, 2$, and $J(\theta, t)$ is the transformation generator [31, 32].

The physical vacuum where particle oscillations appears is $|0(\theta,t)\rangle \equiv J^{-1}(\theta,t)\,|0\rangle_{1,2}$. It has a condensate structure [31, 32]:

$$\langle 0(\theta,t)|\chi^{r\dagger}_{\mathbf{k},i}\chi^{r}_{\mathbf{k},i}|0(\theta,t)\rangle = \sin^2\theta\;|\Upsilon^{\lambda}_{\mathbf{k}}|^2, \tag{17}$$

where $\lambda = B, F$, $i = 1, 2$. The Bogoliubov coefficient $\Upsilon^{\lambda}_{\mathbf{k}}$ is for boson and fermion [31, 32]

$$|\Upsilon^{B}_{\mathbf{k}}| = \frac{1}{2}\left(\sqrt{\frac{\Omega_{k,1}}{\Omega_{k,2}}} - \sqrt{\frac{\Omega_{k,2}}{\Omega_{k,1}}}\right), \tag{18}$$

$$|\Upsilon^{F}_{\mathbf{k}}| = \frac{(\Omega_{k,1}+m_1)-(\Omega_{k,2}+m_2)}{2\sqrt{\Omega_{k,1}\Omega_{k,2}(\Omega_{k,1}+m_1)(\Omega_{k,2}+m_2)}}\,|\mathbf{k}|, \tag{19}$$

respectively, with $|\Sigma^{B}_{\mathbf{k}}|^2 - |\Upsilon^{B}_{\mathbf{k}}|^2 = 1$ and $|\Sigma^{F}_{\mathbf{k}}|^2 + |\Upsilon^{F}_{\mathbf{k}}|^2 = 1$, $\Omega_{k,i}$ energies of the free fields, $i = 1, 2$.

6.1. Boson mixing

For mixed bosons, the kinetic and gradient terms of mixed vacuum are equal to zero [20], then the energy density and pressure become

$$\rho^{B}_{mix} = \langle 0| : \sum_i m_i^2\phi_i^2(-\theta,x) : |0\rangle\,, \tag{20}$$

$$p^{B}_{mix} = -\langle 0| : \sum_i m_i^2\phi_i^2(-\theta,x) : |0\rangle\,, \tag{21}$$

and the state equation coincides with the one of the the cosmological constant, $w^{B}_{mix} = -1$, independently on the choice of the cut-off on the momenta. By setting $\Delta m^2 = |m_2^2 - m_1^2|$, the energy density of the boson mixed vacuum is

$$\rho^{B}_{mix} = \frac{\Delta m^2\sin^2\theta}{8\pi^2}\int_0^K dk k^2\left(\frac{1}{\omega_{k,1}} - \frac{1}{\omega_{k,2}}\right), \tag{22}$$

where K is the cut-off on the momenta.

- In the case of the axion-photon mixing, one has a value compatible with the estimated upper bound on the dark energy, $\rho^{axion}_{mix} = 2.3\times10^{-47}GeV^4$, for magnetic field strength $B \in [10^6 - 10^{17}]G$, axion mass $m_a \simeq 2\times10^{-2}eV$, $\sin^2_a\theta \sim 10^{-2}$ and a Planck scale cut-off, $K \sim 10^{19}GeV$.

- In the case superpartners of the neutrinos, considering masses $m_1 = 10^{-3}eV$ and $m_2 = 9\times10^{-3}eV$, such that $\Delta m^2 = 8\times10^{-5}eV^2$ and $\sin^2\theta = 0.3$, one obtains, $\rho^{B}_{mix} = 7\times10^{-47}GeV^4$ for a cut-off on the momenta $K = 10eV$, and $\rho^{B}_{mix} = 6.9\times10^{-46}GeV^4$ for a cut-off of order of the Planck scale, $10^{19}GeV$. Smaller values of the mixing angle lead to values which are compatible with the estimated value of the dark energy also in the case in which the cut-off is $K = 10^{19}GeV$, indeed ρ^{B}_{mix} depends linearly by $\sin^2\theta$ [20].

6.2. Fermion mixing

For fermion mixing, energy density and pressure become

$$\rho^{F}_{mix} = -\langle 0| : \sum_i\left[\psi_i^\dagger(-\theta,x)\gamma_0\gamma^j\partial_j\psi_i(-\theta,x) + m\psi_i^\dagger(-\theta,x)\gamma_0\psi_i(-\theta,x)\right] : |0\rangle\,; \tag{23}$$

$$p^{F}_{mix} = i\langle 0| : \sum_i\left[\psi_i^\dagger(-\theta,x)\gamma_0\gamma_j\partial_j\psi_i(-\theta,x)\right] : |0\rangle\,, \tag{24}$$

with $\psi_i(-\theta, x)$ flavor neutrino fields or the quark fields. Being

$$\langle 0| : \sum_i \bar{\psi}_i(-\theta,x)\gamma^j\partial_j\psi_i(-\theta,x) : |0\rangle \; = \langle 0| : \sum_i \left[\psi_i^\dagger(-\theta,x)\gamma_0\gamma_j\partial_j\psi_i(-\theta,x)\right] : |0\rangle \; = 0\,,$$

one has

$$\rho^F_{mix} \;=\; -\langle 0| : \sum_i \left[m_i\psi_i^\dagger(-\theta,x)\gamma_0\psi_i(-\theta,x)\right] : |0\rangle, \tag{25}$$

$$p^F_{mix} \;=\; 0\,. \tag{26}$$

The state equation is then $w^F_{mix} = 0$, which is the one of the dark matter. The energy density of the fermion mixed vacuum is

$$\rho^F_{mix} = \frac{\Delta m \sin^2\theta}{2\pi^2}\int_0^K dk k^2\left(\frac{m_2}{\omega_{k,2}} - \frac{m_1}{\omega_{k,1}}\right).$$

For masses of order of $10^{-3}eV$, such that $\Delta m^2 \simeq 8\times 10^{-5}eV^2$, one has $\rho^F_{mix} = 4\times 10^{-47}GeV^4$ for a cut-off on the momenta $K = m_1 + m_2$. Such a value is in agreement with the estimated upper bound of the dark matter. For K of order of the Plank scale one has $\rho^F_{mix} \sim \times 10^{-46}GeV^4$ [20]. Notice that for quarks, the confinement inside the hadrons inhibits the gravitational interaction of the quark vacuum condensate, thus we suppose that hadronic quark condensate should not contribute to the dark matter of the universe.

7. Conclusions

The vacuum condensates generated by many phenomena can give contributions to the dark energy and to the dark matter of the universe. In particular, the thermal states of the intercluster medium, the vacuum of fields in curved space-time and the flavor vacuum of neutrinos have a behavior similar to one of dark matter, while the axion-photon mixing reproduces the behavior and the estimated value of the dark energy.

Acknowledgments

Partial financial support from MIUR and INFN is acknowledged.

References

[1] de Bernardis P et al. 2000 *Nature* **404** 955
[2] Spergel D N et al. 2003 *ApJS* **148** 175
[3] Dodelson S et al. 2002 *ApJ* **572** 140
[4] Szalay A S et al. 2003 *ApJ* **591** 1
[5] Riess A G et al. 2004 *ApJ* **607** 665
[6] Amendola L and Tsujikawa S 2010 *Dark Energy: Theory and Observations* (Cambridge University Press)
[7] Nojiri S and Odintsov S D 2011 *Phys. Rep.* **505** 59
[8] Silvestri A and Trodden M 2009 *Rep. Prog. Phys.* **72** 096901
[9] Frieman J A, Turner M S and Huterer D 2008 *Annu. Rev. Astron. Astrophys.* **46** 385
[10] Durrer R and Maartens R 2008 *Gen. Relat. Grav.* **40** 301
[11] Sotiriou T P and Faraoni V 2010 *Rev. Mod. Phys.* **82** 451
[12] Schützhold R 2002 *Phys. Rev. Lett.* **89** 081302
[13] Thomas E C, Urban F R and Zhitnitsky A R 2009 *J. High Energy Phys.* **08** 043
[14] Klinkhamer F R and Volovik G E 2009 *Phys. Rev. D* **80** 083001
[15] Alexander S, Biswas T and Calcagni G 2010 *Phys. Rev. D* **81** 043511
[16] Bertone G, Hooper D and Silk J 2005 *Phys. Rept.* **405** 279
[17] Derraro S, Schmidt F and Hu W 2011 *Phys. Rev. D* **83** 063503
[18] Lombriser L et al. 2012 *Phys. Rev. D* **85** 102001

[19] Schmidt F, Vikhlinin A and Hu W 2009 *Phys. Rev. D* **80** 083505
[20] Capolupo A 2016 *Adv. High Energy Phys.* **2016** 8089142
[21] Hawking S W 1975 *Commun. Math. Phys.* **43** 199 [1976 Erratum-ibid. **46** 206]
[22] Unruh W G 1976 *Phys. Rev. D* **14** 870
[23] Schwinger J S 1951 *Phys. Rev.* **82** 664
[24] Bardeen J, Cooper L N and Schrieffer J R 1957 *Phys. Rev.* **108** 1175
[25] Iorio A 2011 *Annals Phys.* **326** 1334
[26] Takahasi Y and Umezawa H 1975 *Collect. Phenom.* **2** 55
[27] Umezawa H 1993 *Advanced field theory: Micro, macro, and thermal physics,* (New York, USA: AIP)
[28] Umezawa H, Matsumoto H and Tachiki M 1982 *Thermo field dynamics and condensed states* (North-Holland Publishing Company)
[29] Casimir H B G and Polder D 1948 *Phys. Rev.* **73** 360
[30] Casimir H B G 1948 *Proc. Kon. Nederland. Akad. Wetensch. B* **51** 793
[31] Blasone M, Capolupo A and Vitiello G 2002 *Phys. Rev. D* **66** 025033 and references therein
[32] Blasone M, Capolupo A, Romei O and Vitiello G 2001 *Phys. Rev. D* **63** 125015
[33] Capolupo A, Ji C-R, Mishchenko Y and Vitiello G 2004 *Phys. Lett. B* **594** 135-140
[34] Capolupo A, Capozziello S and Vitiello G 2007 *Phys. Lett. A* **363** 53
[35] Capolupo A, Capozziello S and Vitiello G 2009 *Phys. Lett. A* **373** 601
[36] Capolupo A,Capozziello S and Vitiello G 2008 *Int. J. Mod. Phys. A* **23** 4979
[37] Blasone M, Capolupo A, Capozziello S and Vitiello G 2008 *Nucl. Instrum. Meth. A* **588** 272
[38] Blasone M, Capolupo A and Vitiello G 2010 *Prog. Part. Nucl. Phys.* **64** 451
[39] Blasone M, Capolupo A, Capozziello S, Carloni S and Vitiello G 2004 *Phys. Lett. A* **323** 182
[40] Birrell N D and Davies P C W 1984 *Quantum Fields in Curved Space* (Cambridge, UK: Cambridge University Press)
[41] Albareti F D, Cembranos J A R and Maroto A L 2014 *Phys. Rev. D* **90** 123509
[42] Capolupo A and Di Mauro M and Iorio A 2011 *Phys. Lett. A* **375** 3415
[43] Capolupo A and Di Mauro M 2012 *Phys. Lett. A* **376** 2830; *do.* 2013 *Acta Phys. Polon. B* **44** 81
[44] Capolupo A and Vitiello G 2013 *Adv. High Energy Phys.* **2013** 850395
[45] Capolupo A and Di Mauro M 2015 *Adv. High Energy Phys.* **2015** 929362
[46] Mavromatos N E, Sarkar S and Tarantino W 2011 *Phys. Rev. D* **84** 044050
[47] Capolupo A, Lambiase G and Vitiello G 2016 *Adv. High Energy Phys.* **2016** 3127597
[48] Capolupo A, Lambiase G and Vitiello G 2017 Proceedings DICE2016 *J. Phys.: Conf. Series*, in press
[49] Parker L and Fulling S A 1974 *Phys. Rev. D* **9** 341
[50] Fulling S A and Parker L 1974 *Ann. Phys. (N.Y.) D* **87** 176

IOP Conf. Series: Journal of Physics: Conf. Series **880** (2017) 012060 doi:10.1088/1742-6596/880/1/012060

Thermal vacuum, cosmic microwave radiation, neutrino masses and fractal-like self-similar structure

Antonio Capolupo, Gaetano Lambiase and Giuseppe Vitiello

Dipartimento di Fisica E.R.Caianiello and INFN gruppo collegato di Salerno, Universitá di Salerno, Fisciano (SA) - 84084, Italy

E-mail: capolupo@sa.infn.it , lambiase@sa.infn.it , vitiello@sa.infn.it

Abstract. The behavior of thermal vacuum condensates of scalar and fermion fields is analyzed and it is shown that the condensate of Maxwell fields reproduces the characteristics of the cosmic microwave radiation. By studying fermion thermal states with the temperature of the cosmic neutrino background, we derive a value of the sum of the active neutrino masses which is compatible with its estimated lower bound. Moreover, we reveal the fractal self-similar structure of the thermal radiation and we relate it to the coherent structure of the thermal vacuum.

1. Introduction

In the context of the Thermo Field Dynamic [1, 2], we analyze the possible links existing among the thermal vacuum states, the cosmic microwave background (CMB) radiation and the neutrino masses [3].

The CMB radiation is a thermal radiation, left over from an early stage in the expansion of the universe [4], filling almost uniformly the observable universe. Such a radiation has a thermal black body spectrum corresponding to the temperature of $T_\gamma = 2.72548 \pm 0.00057$ K. The anisotropy is very small [5]-[7].

Together with the CMB, there is an indirect evidence of the universe's background particle radiation composed of neutrinos, called, cosmic neutrino background (CNB). The today estimated temperature for the CNB is roughly $1.95K$ [7]-[12].

Neutrinos play a significant role in the understanding of many phenomena of high energy physics and astrophysics [13]-[24]. The discovery of neutrino oscillations has shown that neutrinos have mass. Oscillations do not yield a value for the mass, but do set a lower limit of order of $0.06eV$ on the sum of the three neutrino masses $\sum m_\nu$ [25]. The upper limit of $\sum m_\nu$ is $2eV$.

In this paper we report the results of Ref.[3], in which we have shown that the thermal condensate derived in thermal field theories, behave as a perfect fluid, gives a contribution to the radiation of the universe and reproduces the behavior of CMB and CNB. Moreover, by exploiting the CMB and CNB energy contribution, we derive an upper bound on the absolute mass of the lightest neutrino. The absolute masses of the other neutrinos and the sum of the active neutrino masses are obtained by assuming the hierarchical neutrino model and using the experimental differences between the squared masses Δm^2_{12} and Δm^2_{23}. We find a value of $\sum m_\nu$ of the order of $0.06eV$, which is in agreement with the estimated lower bound on the neutrino masses. A result which does not depend on whether the neutrino is a Dirac or Majorana particle. Moreover, we study the fractal self-similar structure of the thermal states. It might be interesting to investigate about a possible relation between such fractal self-similarity properties and some aspects of holographic cosmology (see e.g.[26]).

The structure of the paper is the following, in Section II, we present the Thermo Field Dynamic formalism and derive the general expressions of energy density and pressure of the vacuum condensate. In Section III, we show the explicit results for Maxwell and scalar field. The results for neutrinos are reported in Section IV. The fractal structure of the thermal vacuum is studied in Section V and the conclusions are in Section VI.

2. Thermal vacuum condensate

We describe the observed black body spectrum of the CMB at $T = T_\gamma$, by using the Thermo Field Dynamics (TFD) formalism [1, 2] in quantum field theory (QFT). Thus we introduce the thermal vacuum state $|0(\theta)\rangle$, with $\theta = \theta(\beta)$, $\beta \equiv 1/(k_B T)$ and k_B the Boltzmann constat. The thermal vacuum $|0(\theta)\rangle$ is defined in such a way that the thermal statistical average $\mathcal{N}_{a_{\mathbf{k}}}(\theta)$ is given by the expectation value of the number operator $N_{a_{\mathbf{k}}} = a^\dagger_{\mathbf{k}} a_{\mathbf{k}}$, with $a_{\mathbf{k}}$ denoting the bosonic CMB modes, on $|0(\theta)\rangle$, i.e. $\mathcal{N}_{a_{\mathbf{k}}}(\theta) = \langle 0(\theta)|N_{a_{\mathbf{k}}}|0(\theta)\rangle$.

The boson modes $a_{\mathbf{k}}$ have usual canonical commutation relations (CCR) and $|0(\theta)\rangle$ is given by

$$|0(\theta)\rangle = \prod_{\mathbf{k}} \frac{1}{\cosh\theta_k} \exp\left(\tanh\theta_k \; a^\dagger_{\mathbf{k}} b^\dagger_{\mathbf{k}}\right)|0\rangle, \qquad \theta = \theta(\beta). \tag{1}$$

$|0(\theta)\rangle$ is a two-mode generalized $SU(1,1)$ squeezed coherent state [27, 28] at finite temperature, condensate of pairs of $a_{\mathbf{k}}$ and $b_{\mathbf{k}}$ quanta, with $b_{\mathbf{k}}$ auxiliary boson mode, commuting with $a_{\mathbf{k}}$. The quantum $b_{\mathbf{k}}$ is introduced in order to produce the trace operation in the computing thermal averages of the observable quantities. In Eq.(1), $|0\rangle$ is the vacuum for $a_{\mathbf{k}}$ and $b_{\mathbf{k}}$. One has that $\langle 0(\theta)|0(\theta)\rangle = 1, \ \forall\ \theta$, and in the infinite volume limit (for $\int d^3\kappa\ \theta_\kappa$ finite and positive) $\langle 0(\theta(\beta))|0\rangle \to 0$ as $V \to \infty, \quad \forall\ \beta$. Moreover, the thermal vacuum $|0(\theta)\rangle$ provides a representation of the CCR defined at each β and unitarily inequivalent to any other representation $\{|0(\beta')\rangle,\ \forall\ \beta' \neq \beta\}$ in the infinite volume limit. Indeed one has $\langle 0(\theta(\beta))|0(\theta(\beta'))\rangle \to 0$ as $V \to \infty$, $\forall\ \beta$ and $\beta', \beta' \neq \beta$. This means that in thermal equilibrium (β constant in time t) at a given temperature T, the system sits in the representation of the CCR corresponding to such a T. In thermal non-equilibrium conditions, with temperature changing in time, i.e. $\beta = \beta(t)$, the system evolves in time through unitarily inequivalent representations of the CCR and it is known that such a time evolution is controlled by the entropy operator [1, 2].

It is crucial to note that the annihilation operators for the state $|0(\theta)\rangle$, denoted with $A_{\mathbf{k}}(\theta_k)$ and $B_{\mathbf{k}}(\theta_k)$, are different from $a_{\mathbf{k}}$ and $b_{\mathbf{k}}$ which annihilate the state $|0\rangle$. The relation among the two sets of annihilators is provided by the Bogoliubov transformation

$$\begin{aligned} A_{\mathbf{k}}(\theta_k) &= a_{\mathbf{k}} \cosh\theta_k - b^\dagger_{\mathbf{k}} \sinh\theta_k, \\ B_{\mathbf{k}}(\theta_k) &= b_{\mathbf{k}} \cosh\theta_k - a^\dagger_{\mathbf{k}} \sinh\theta_k\,. \end{aligned}$$

The condensate density is given by the expectation value of the number operator $a^\dagger_{\mathbf{k}} a_{\mathbf{k}}$ on the thermal vacuum, $\mathcal{N}_{a_{\mathbf{k}}}(\theta) = \langle 0(\theta)|a^\dagger_{\mathbf{k}} a_{\mathbf{k}}|0(\theta)\rangle = \sinh^2\theta_k$. In the boson case, minimization of the free energy leads to the Bose-Einstein distribution function for $a_{\mathbf{k}}$

$$\mathcal{N}^B_{a_{\mathbf{k}}}(\theta) = \sinh^2\theta_k = \frac{1}{e^{\beta\Omega_k} - 1}\,. \tag{2}$$

Thus, we can reproduce the observed black body spectrum of the CMB by computing the expectation value $\mathcal{N}^B_{a_{\mathbf{k}}}(\theta)$, at $T = T_\gamma$.

For fermion fields we can obtain in a similar way the corresponding thermal $SU(2)$ vacuum and Bogoliubov transformations [1, 2]. We then obtain the Fermi-Dirac distribution function for fermions:

$$\mathcal{N}^F_{a_{\mathbf{k}}}(\theta) = \sin^2\theta_k = \frac{1}{e^{\beta\Omega_k}+1}\,. \tag{3}$$

In the following, we compute the energy momentum tensor $T^{\mu\nu}$ of the thermal vacuum for Maxwell, for massive boson and fermion fields and we show that particle with masses much higher than the CMB temperature, i.e. $m \gg 2\times 10^{-4}eV$, give negligible contribution to the energy of the universe. Next we derive the absolute mass of the neutrinos [3].

Notice that the vacuum condensate is homogenous and isotropic and behaves as a perfect fluid, indeed the off-diagonal terms of $T^{\mu\nu}$ on the vacuum state are zero for any field, i.e. $\langle 0(\theta)|T^{ij}(x)|0(\theta)\rangle = 0$, for $i \neq j$. Then, the energy density and pressure of the vacuum condensates (2), (3) at a given time (we consider the red shift z), are represented by the $(0,0)$ and (j,j) components of $T^{\mu\nu}$ of the vacuum $|0(\theta,z)\rangle$,

$$\rho(z) = g_{00}\langle 0(\theta,z)| : T^{00}(x) : |0(\theta,z)\rangle\,, \tag{4}$$

$$p(z) = g_{jj}\langle 0(\theta,z)| : T^{jj}(x) : |0(\theta,z)\rangle\,, \tag{5}$$

with : ... :, normal ordering with respect to $|0\rangle$.

3. Thermal vacuum, CMB radiation and boson fields

In the photon fields case, Eqs.(4) and (5) become

$$\rho_\gamma(z) = \frac{\pi^2\, k_B^4\,(1+z)^4 T_\gamma^4}{15\,\hbar^3\, c^3}\,, \tag{6}$$

$$p_\gamma(z) = \frac{\pi^2\, k_B^4\,(1+z)^4 T_\gamma^4}{45\,\hbar^3\, c^3}\,, \tag{7}$$

and the state equation is the one of the radiation $w_\gamma(z) = p_\gamma(z)/\rho_\gamma(z) = 1/3$. These results coincides with the ones obtained by solving the Boltzmann equation for the distribution function of photons in thermal equilibrium [12].

By considering the present CMB temperature and the present red shift, $z = 0$, one obtains a value of the thermal vacuum energy density $\rho_\gamma = 2\times 10^{-51} GeV^4$, coinciding with the energy density of the CMB [12].

Let us now study the energy momentum tensor of the thermal vacuum $|0(\theta,z)\rangle$ for massive boson and fermion fields. For bosons, at any epoch, one has [3]

$$\rho_B(z) = \frac{1}{2\pi^2}\int_0^\infty dk k^2 \frac{\Omega_k}{exp\left(\frac{\Omega_k}{k_B T_\gamma(1+z)}\right)-1}\,, \tag{8}$$

$$p_B(z) = \frac{1}{6\pi^2}\int_0^\infty dk k^2 \Bigg[\frac{k^2}{\Omega_k}\frac{1}{exp\left(\frac{\Omega_k}{k_B T_\gamma(1+z)}\right)-1} - \left(\frac{k^2}{\Omega_k}+\frac{3m^2}{2\Omega_k}\right)\frac{exp\left(\frac{\Omega_k}{2k_B T_\gamma(1+z)}\right)}{exp\left(\frac{\Omega_k}{k_B T_\gamma(1+z)}\right)-1}\cos(\Omega_k t)\Bigg]\,. \tag{9}$$

In particular, at the present epoch, $z = 0$ one obtains $\rho_B \simeq 9 \times 10^{-52} GeV^4$ for masses less or equal than the CMB temperature $m \leq T_\gamma$, i.e. $m \leq 2.3 \times 10^{-4} eV$. The maximum value of ρ_B is achieved for masses $m \ll 10^{-4} eV$. In this case, the energy density is $\rho_B \simeq 10^{-51} GeV^4$. Negligible values of ρ_B are obtained for boson masses $m \gg 10^{-3} eV$. Therefore, the thermal vacuum contribution of bosons to the energy of the universe is completely negligible with respect to ρ_γ and with respect to the critical density of the present universe $\rho_{cr} \simeq 4.5 \times 10^{-47} GeV^4$. Only hypothetical particles as axions with masses $m_a \in (10^{-3} - 10^{-6}) eV$, could generate non trivial contributions [3].

4. Thermal vacuum and neutrino masses

The thermal vacuum contribution of fermion fields to the vacuum energy density and to the pressure, are

$$\rho_F(z) = \frac{1}{\pi^2} \int_0^\infty dk k^2 \frac{\Omega_k}{exp\left(\frac{\Omega_k}{k_B T_\gamma (1+z)}\right) + 1}, \tag{10}$$

$$p_F(z) = \frac{1}{3\pi^2} \pi \int_0^\infty dk \frac{k^4}{\Omega_k} \frac{1}{exp\left(\frac{\Omega_k}{k_B T_\gamma (1+z)}\right) + 1}. \tag{11}$$

At the present epoch, $z = 0$, and for masses $m \leq T_\gamma$, one has the maximum value of ρ_F at $T = T_\gamma$, which is $\rho_F \sim 1.6 \times 10^{-51} GeV^4$. In this case, the state equation is $w_F \sim 1/3$. We note that only the thermal vacuum condensate of sub-eV massive particles, like the neutrinos, is relevant, since the condensate of more heavy fermions give negligible contributions to the energy of universe.

We consider now the relic neutrino temperature $T_\nu = 1.95K$ and we study the energy density of the thermal vacuum for cosmic neutrino background (CNB). We use massive Dirac neutrinos. Let $|0(\theta, z)\rangle$ denote the vacuum condensate of the neutrino modes with temperature T_ν. We derive the absolute neutrino masses from the plausible value of the CNB energy.

In Eqs.(10) and (11), if one considers $T = T_\nu$ and the sum on the three neutrino fields with masses m_i, one has only small variations from the above results. Indeed, for $z = 0$ and $m_i \sim 10^{-4} eV$, we find the maximum value of the energy density $\rho_\nu \sim 10^{-51} GeV^4$, and the state equation is $w_\nu \sim 1/3$. We assume $\rho_\nu \leq \rho_\gamma$, and the mass of the lighter neutrino $m_{\nu,1} \sim 10^{-3} eV$ [3]. Considering the hierarchical neutrino model, according to which $\Delta m_{12}^2 = 8 \times 10^{-5} eV^2$ and $\Delta m_{23}^2 = 2.7 \times 10^{-3} eV^2$, one can derive the following values of the other neutrino masses: $m_{\nu,2} = 9 \times 10^{-3} eV$ and $m_{\nu,3} = 5.3 \times 10^{-2} eV$. Then, the sum of the three neutrino masses is $\sum m_\nu = 6 \times 10^{-2} eV$, which is in agreement with its estimated lower bound. Notice that the contributions to the energy density ρ_ν given by ν_2 and ν_3 are negligible.

5. Fractal structure of the thermal vacuum

In this Section we show the fractal self-similar structure of the thermal vacuum $|0(\theta)\rangle$ considered above for the CMB and CNB [3]. Consider the boson case. In full generality, we consider time-dependent θ: $\theta = \theta(t)$. We use the notation $|0(t)\rangle \equiv |0(\theta(t))\rangle$. As shown in [29], $|0(t)\rangle$ provides the quantum representation of the system of couples of damped/amplified oscillators

$$m\ddot{x} + \gamma\dot{x} + kx = 0, \tag{12}$$

$$m\ddot{y} - \gamma\dot{y} + ky = 0, \tag{13}$$

$$L = m\dot{x}\dot{y} + \frac{\gamma}{2}(x\dot{y} - \dot{x}y) - kx\,y, \tag{14}$$

where "dot" denotes time derivative, m, γ and κ are positive real constants and L is the Lagrangian from which Eqs. (12) and (13) are derived. The state $|0(t)\rangle$ is indeed obtained by quantizing the system represented by Eqs. (12) - (14). Therefore, we take into account the canonical commutation relations $[\,x, p_x\,] = i\,\hbar = [\,y, p_y\,]$, $[\,x, y\,] = 0 = [\,p_x, p_y\,]$ and the corresponding annihilators

$$\alpha \;\equiv\; \left(\frac{1}{2\hbar\Omega}\right)^{\frac{1}{2}} \left(\frac{p_x}{\sqrt{m}} - i\sqrt{m}\Omega x\right); \tag{15}$$

$$\widetilde{\alpha} \;\equiv\; \left(\frac{1}{2\hbar\Omega}\right)^{\frac{1}{2}} \left(\frac{p_y}{\sqrt{m}} - i\sqrt{m}\Omega y\right); \tag{16}$$

satisfying the relations $[\,\alpha, \alpha^\dagger\,] = 1 = [\,\widetilde{\alpha}, \widetilde{\alpha}^\dagger\,]$, $\quad [\,\alpha, \widetilde{\alpha}\,] = 0 = [\,\alpha, \widetilde{\alpha}^\dagger\,]$. Moreover, we consider the transformations $a \equiv (1/\sqrt{2})(\alpha + \widetilde{\alpha})$, $b \equiv (1/\sqrt{2})(\alpha - \widetilde{\alpha})$. As shown in [29], the proper quantization setting is the one of the quantum field theory (QFT) and the Hamiltonian H of the system is $H = H_0 + H_I$, with $H_0 = \sum_{\mathbf{k}} \hbar\Omega_k(a_{\mathbf{k}}^\dagger a_{\mathbf{k}} - b_{\mathbf{k}}^\dagger b_{\mathbf{k}})$, and $H_I = i\sum_{\mathbf{k}} \hbar\Gamma_k(a_{\mathbf{k}}^\dagger b_{\mathbf{k}}^\dagger - a_{\mathbf{k}} b_{\mathbf{k}})$. In this case, the parameter $\theta_\kappa(t)$ of the thermal vacuum is $\theta_\kappa(t) = \Gamma_k\, t \equiv (\gamma_k/2m)\, t$ for each κ-mode [29]. The time evolution of the vacuum $|0\rangle$ for $a_{\mathbf{k}}$ and $b_{\mathbf{k}}$ is driven by H_I and given by $|0(\theta(t))\rangle = e^{-it\frac{H}{\hbar}}|0\rangle = e^{-it\frac{H_I}{\hbar}}|0\rangle$, which leads to the explicit expression for $|0(\theta(t))\rangle$ similar to the one in Eq. (1) for time dependent β. The free energy for the $a-$modes is defined by

$$\mathcal{F}_a \equiv \langle 0(t)| \left(H_a - \frac{1}{\beta} S_a\right) |0(t)\rangle\,, \tag{17}$$

where H_a is the free Hamiltonian for the $a-$modes, $H_a = \sum_{\mathbf{k}} \hbar\Omega_k\, a_{\mathbf{k}}^\dagger a_{\mathbf{k}}$, and S_a is the entropy:

$$S_a \equiv -\sum_{\mathbf{k}} \left\{ a_{\mathbf{k}}^\dagger a_{\mathbf{k}} \ln \sinh^2(\theta) - a_{\mathbf{k}} a_{\mathbf{k}}^\dagger \ln \cosh^2(\theta) \right\}\,. \tag{18}$$

For $\beta(t)$ slowly varying in time, the Bose-Einstein distribution function Eq. (2) is derived by minimizing $\mathcal{F}_a$. Therefore, $|0(t)\rangle$ provides a finite temperature representation of the CCR which is equivalent to the Thermo Field Dynamics representation $\{|0(\beta)\rangle\}$, with time dependent β.

We now remark that Eqs. (12) and (13) describe a system which has self-similarity properties. We indeed note that solutions of Eqs. (12) and (13) are the parametric time evolution of clockwise and anti-clockwise logarithmic spirals. Indeed, by defining $x(t) \equiv [z_1(t) + z_2^*(-t)]/2$ and $y(t) \equiv [z_1^*(-t) + z_2(t)]/2$, Eqs. (12) and (13) can be rewritten as [30]

$$m\,\ddot{z}_1 \;+\; \gamma\,\dot{z}_1 \;+\; \kappa\,z_1 \;=\; 0, \tag{19}$$

$$m\,\ddot{z}_2 \;-\; \gamma\,\dot{z}_2 \;+\; \kappa\,z_2 \;=\; 0, \tag{20}$$

whose solutions are the clockwise and anti-clockwise logarithmic spirals $z_1(t) = r_0\, e^{-i\Omega t}\, e^{-\Gamma t}$ and $z_2(t) = r_0\, e^{+i\Omega t}\, e^{+\Gamma t}$, respectively, with $\Gamma \equiv \gamma/2m$ and $\Omega^2 = (1/m)(\kappa - \gamma^2/4m)$, $\kappa > \gamma^2/4m$ [30, 31]. Therefore, Eqs. (12) and (13), whose quantum representation is given by $|0(t)\rangle$, describe the self-similar fractal structure of their logarithmic spiral solutions [32, 33]. The link between $SU(1,1)$ coherent states, fractal-like self-similarity and CMB is thus established [3]. A similar result may be obtained in the fermion case for CNB by considering that the $su(2)$ algebra contracts in the infinite volume limit to the $e(2)$ algebra (as it can be seen by considering the Holstein-Primakoff non-linear boson realization [34, 35, 36]), which is isomorph to the Weyl-Heisenberg algebra.

6. Conclusion

We have shown that the structure of CMB is characterized by the arrow of time and by coherent, thermal and fractal-like self-similar properties. The observed black body spectrum of CMB is confirmed by the results here presented. Moreover, by analyzing the thermal vacuum energy for neutrinos and considering the neutrino hierarchical model, we have derived the value of the neutrino masses which are in agreement with estimated lower bounds. An interesting question to ask is on a possible relation between the fractal self-similarity structures considered in this paper and the recently reported observations on the holographic structure of the Universe [26].

Acknowledgments
Partial financial support from MIUR and INFN is acknowledged.

References
[1] Takahashi Y, Umezawa H 1975 *Collect. Phenom.* **2** 55; reprinted 1996 in *Int. J. Mod. Phys.* B **10** 1755
[2] Umezawa H 1993 *Advanced field theory: Micro, macro, and thermal physics* (New York: AIP)
[3] Capolupo A, Lambiase G and Vitiello G 2016 *Adv. High Energy Phys.* **2016** 3127597
[4] Gawiser E and Silk J 2000 *Phys. Rep.* **333** 245
[5] Fixsen D J 2009 *APJ* **707** 916
[6] Bennett C L et al. 1996 *APJL* **464** L1 L4
[7] de Bernardis P et al. 2000 *Nature* **404** 955
[8] Weinberg S 2008 *Cosmology* (Oxford: OUP)
[9] Bashinsky S and Seljak U 2004 *Phys. Rev. D* **69** 083002
[10] Mangano G et al. 2005 *Nucl. Phys. B* **729** 221
[11] Komatsu E 2011 *APJ, Supp. Series* **192** 18
[12] Kolb E and Turner M 1994 *The Early Universe* (Frontiers in Physics)
[13] Capolupo A, Capozziello S and Vitiello G 2007 *Phys. Lett. A* **363** 53
[14] Capolupo A, Capozziello S and Vitiello G 2009 *Phys. Lett. A* **373** 601
[15] Capolupo A, Capozziello S and Vitiello G 2008 *Int. J. Mod. Phys. A* **23** 4979
[16] Blasone M, Capolupo A, Capozziello S, Carloni S and Vitiello G 2004 *Phys. Lett. A* **323** 182
[17] Blasone M, Capolupo A, Capozziello S and Vitiello G 2008 *Nucl. Instrum. Meth. A* **588** 272
[18] Blasone M, Capolupo A and Vitiello G 2010 *Prog. Part. Nucl. Phys.* **64** 451
[19] Capolupo A, Di Mauro M and Iorio A 2011 *Phys. Lett. A* **375** 3415
[20] Capolupo A and Di Mauro M 2015 *Adv. in High Energy Phys.*, **2015** 929362
[21] Capolupo A and Vitiello G 2013 *Adv. in High Energy Phys.*, **2013** 850395
[22] Capolupo A and Di Mauro M 2012 *Phys. Lett. A* **376** 2830
[23] Capolupo A and Di Mauro M 2013 *Acta Phys. Polonica B* **44** 81
[24] Capolupo A 2016 *Adv. High Energy Phys.* **2016** 8089142
[25] Gonzalez-Garcia M C, Maltoni M, Salvado J and Schwetz T 2012 *J. High Energy Phys.* **12** 123
[26] Afshordi N et al. 2017 *Phys. Rev. Lett.* **118** 041301
[27] Perelomov A 1986 *Generalized Coherent States and Their Applications* (Berlin: Springer-Verlag)
[28] Klauder J R and Skagerstam B 1985 *Coherent States* (Singapore: World Scientific)
[29] Celeghini E, Rasetti M and Vitiello G 1992 *Ann. Phys.* **215** 156
[30] Vitiello G 2012 *Phys. Lett. A* **376** 2527
[31] Vitiello G 2014 *Systems* **2** 203
[32] Peitgen H O, Jürgens H and Saupe D 1986 *Chaos and Fractals: New Frontiers of Science* (Berlin: Springer-Verlag)
[33] Andronov A A, Vitt A A and Khaikin S E 1966 *Theory of Oscillators* (Mineola, NY, USA: Dover Publications, Inc.)
[34] Holstein T and Primakoff H 1940 *Phys. Rev.* **58** 1098
[35] Shah M N, Umezawa H and Vitiello G 1974 *Phys. Rev. B* **10** 4724
[36] De Concini C and Vitiello G 1976 *Nuc. Phys. B* **116** 141

IOP Conf. Series: Journal of Physics: Conf. Series **880** (2017) 012061 doi:10.1088/1742-6596/880/1/012061

Phase space picture of neutrino mixing and oscillations

Massimo Blasone[1,2], **Maria Vittoria Gargiulo**[1,2], **and Giuseppe Vitiello**[1,2]

[1] Dipartimento di Fisica "E.R.Caianiello", Università degli Studi di Salerno, Via Giovanni Paolo II, 132, I-84084 Fisciano (SA), Italy

[2] INFN Sezione di Napoli, Gruppo collegato di Salerno, Italy

E-mail: blasone@sa.infn.it, mgargiulo@unisa.it, vitiello@sa.infn.it

Abstract. We consider a simple classical model in phase-space resembling the quantum one used for the description of neutrino oscillations and investigate the possibility of defining an analogue to the mixing transformation and to the oscillation formula in terms of generalized coordinates.

1. Introduction

The phenomenon of neutrino oscillations [1] is one of the most intriguing and fascinating ones in the context of particle physics. It is indeed a beautiful manifestation of quantum mechanics at work on macroscopic scales and it is a continuous source for novel physical insights and deeper analysis. In fact, in recent years many interesting features emerged in the study of neutrino mixing and oscillations in the context of Quantum Field Theory [2]: it was found indeed that the mixing transformations at level of fields contain Bogoliubov transformations [3, 4], which produce dramatic changes in the structure of the vacuum inducing a condensate of particle-antiparticle pairs, reflecting into additional terms in the oscillation formulas [5, 6]. Another interesting feature is represented by the geometric phase(s) associated to neutrino oscillations, which are present already at QM level and could provide important information if detectable [7]. Finally, it has been shown that the phenomenon of neutrino mixing and oscillations can be equivalently understood and described in quantum information language: time dependent (single–particle) entanglement among neutrino flavor states is generated by the time evolution of the system and could be in principle used for quantum information tasks [8, 9].

In view of these developments, it is definitely interesting any attempt to reformulate such a phenomenon in a new language and from alternative points of view. In this report, we consider the correspondence of neutrino mixing transformations with transformations in (classical) phase space. This can be interesting since it can help to better understand a phenomenon which has a specific quantum nature by means of some classical system analogue for it. Also, the (classical) phase space picture could be at some point implemented with non-commutativity in order to fully reproduce the quantum mechanical framework. Here we make a first step in this direction, defining a consistent classical phase space picture of neutrino mixing and oscillations leading to the same oscillation formulas obtained in the quantum formulation.

Our analysis is limited to the case of oscillations between two flavors only. This report is structured as follows. In Section 2 we summarize the essential features of neutrino mixing and oscillations which then we rephrase for a system of two coupled oscillators. In Section 3 and 4 we present the mixing transformation and the oscillation formula in terms of generalized coordinates. Finally, in Section 5 we draw our conclusions.

2. Flavor mixing and oscillations

In this section we briefly review the simplest description for neutrino oscillations and we rephrase this in terms of a system of coupled oscillators.

2.1. Neutrino oscillations

Pontecorvo mixing transformations [1] are written as a rotation of the states with definite masses m_1, m_2, $|\nu_1\rangle$, $|\nu_2\rangle$, into those with definite flavors $|\nu_e\rangle$ and $|\nu_\mu\rangle$ as:

$$|\nu_e\rangle = \cos\theta\,|\nu_1\rangle + \sin\theta\,|\nu_2\rangle, \tag{1}$$

$$|\nu_\mu\rangle = \cos\theta\,|\nu_2\rangle - \sin\theta\,|\nu_1\rangle. \tag{2}$$

Neutrino oscillations arise from time evolution of the state $|\nu_e\rangle$ in Eq.(1) which gives:

$$|\nu_e(t)\rangle = \cos\theta\, e^{-i\omega_1 t}\,|\nu_1\rangle \,+\, \sin\theta\, e^{-i\omega_2 t}\,|\nu_2\rangle. \tag{3}$$

When computing the probability amplitude of finding a given neutrino, *e.g.* ν_e, at a time t from its generation, one obtains flavor oscillations:

$$P_{\nu_e\to\nu_e}(t) = |\langle\nu_e|\nu_e(t)\rangle|^2 = 1 - \sin^2 2\theta\,\sin^2\left(\frac{\Delta\omega}{2}t\right) = 1 - P_{\nu_e\to\nu_\mu}(t). \tag{4}$$

One can also verify that flavor conservation holds:

$$|\langle\nu_e|\nu_e(t)\rangle|^2 \,+\, |\langle\nu_\mu|\nu_e(t)\rangle|^2 = 1. \tag{5}$$

2.2. Coupled oscillators

Since we are interested in the mixing transformation properties and in the above treatment no reference appears to the fermionic or bosonic nature of the particles, we treat such particles as bosons and consider the simplest case of two harmonic oscillators with different frequencies. The (normal ordered) Hamiltonian for such system is

$$:H(\hat{a}_{1,2};\hat{a}^\dagger_{1,2}) := \omega_1\,\hat{a}_1^\dagger\hat{a}_1 + \omega_2\,\hat{a}_2^\dagger\hat{a}_2\,, \tag{6}$$

with $|a_1\rangle = \hat{a}_1^\dagger|0\rangle_1$, $|a_2\rangle = \hat{a}_2^\dagger|0\rangle_2$, $[\hat{a}_i, \hat{a}_j^\dagger] = \delta_{ij}$. We use $\hbar = 1$. Eqs.(1),(2) can be seen as arising by the application to the vacuum state $|0\rangle_{1,2} = |0\rangle_1 \otimes |0\rangle_2$ of the following flavor operators (adopting the conventional neutrino terminology):

$$\hat{a}_e^\dagger = \cos\theta\,\hat{a}_1^\dagger + \sin\theta\,\hat{a}_2^\dagger, \qquad \hat{a}_\mu^\dagger = \cos\theta\,\hat{a}_2^\dagger - \sin\theta\,\hat{a}_1^\dagger. \tag{7}$$

The transformation in Eq.(7) is a canonical one. The Hamiltonian in Eq.(6) is written in terms of the flavor ladder operators as:

$$:H(\hat{a}_{e,\mu};\hat{a}^\dagger_{e,\mu}) := \omega_e\,\hat{a}_e^\dagger\hat{a}_e + \omega_\mu\,\hat{a}_\mu^\dagger\hat{a}_\mu + \omega_{e\mu}\,(\hat{a}_e^\dagger\hat{a}_\mu + \hat{a}_\mu^\dagger\hat{a}_e), \tag{8}$$

with

$$\begin{aligned} \omega_e &= \omega_1\cos^2\theta + \omega_2\sin^2\theta, \\ \omega_\mu &= \omega_2\cos^2\theta + \omega_1\sin^2\theta, \\ \omega_{e\mu} &= (\omega_2 - \omega_1)\sin\theta\cos\theta. \end{aligned} \tag{9}$$

Thus we also have $\omega_1+\omega_2 = \omega_e+\omega_\mu$. We remark that the hermiticity of $H(\hat{a}_{e,\mu};\hat{a}^\dagger_{e,\mu})$ is preserved, it is not spoiled by the non–diagonal mixing terms.

In order to discuss flavor oscillations within the classical phase–space picture of Section 3, we consider here the Heisenberg picture. Thus we introduce the flavor number operator $\hat{N}_e(t) = \hat{a}^\dagger_e(t)\hat{a}_e(t)$ on the one–particle state $|a_e\rangle$ using - cf. Eq.(7):

$$\hat{a}_e(t) = \cos\theta\,\hat{a}_1\,e^{-i\omega_1 t} + \sin\theta\,\hat{a}_2\,e^{-i\omega_2 t}, \qquad \hat{a}_\mu(t) = \cos\theta\,\hat{a}_2\,e^{-i\omega_2 t} - \sin\theta\,\hat{a}_1\,e^{-i\omega_1 t}. \tag{10}$$

and h.c. We obtain the expectation value:

$$\begin{aligned}\langle a_e|\hat{N}_e(t)|a_e\rangle &= \langle a_e|\,(\cos\theta\,\hat{a}^\dagger_1\,e^{i\omega_1 t} + \sin\theta\,\hat{a}^\dagger_2\,e^{i\omega_2 t})(\cos\theta\,\hat{a}_1\,e^{-i\omega_1 t} + \sin\theta\,\hat{a}_2\,e^{-i\omega_2 t})\,|a_e\rangle \\ &= 1 - 2\sin^2 2\theta\,\sin^2\Big(\frac{\omega_2-\omega_1}{2}t\Big). \end{aligned}\tag{11}$$

In the same way we obtain

$$\langle a_e|\hat{N}_\mu(t)|a_e\rangle = 2\sin^2 2\theta\,\sin^2\Big(\frac{\omega_2-\omega_1}{2}t\Big) \tag{12}$$

where $\hat{N}_\mu(t) = \hat{a}^\dagger_\mu(t)\hat{a}_\mu(t)$.

3. Classical phase–space picture

We now consider a classical analogue of the above system, *i.e.* two harmonic oscillators, with the same mass m but different frequencies $\omega_1 \neq \omega_2$. The total Hamiltonian is:

$$H(Q_{1,2};P_{1,2}) = \frac{P_1^2}{2m} + \frac{\omega_1^2 m}{2}Q_1^2 + \frac{P_2^2}{2m} + \frac{\omega_2^2 m}{2}Q_2^2. \tag{13}$$

Defining

$$\alpha_i = \sqrt{\frac{m\omega_i}{2}}\left(Q_i + i\frac{P_i}{m\omega_i}\right), \qquad \alpha_i^* = \sqrt{\frac{m\omega_i}{2}}\left(Q_i - i\frac{P_i}{m\omega_i}\right), \tag{14}$$

one can rewrite the Hamiltonian in Eq.(13) as

$$H(\alpha_{1,2};\alpha^*_{1,2}) = \omega_1\,|\alpha_1|^2 + \omega_2\,|\alpha_2|^2. \tag{15}$$

where α_i are c–number counterparts of the above ladder operators and may be thought as eigenvalues of a coherent state corrisponding to the annihilation operator $\hat{a}_i$. In analogy with Eq.(7) we write the following transformations:

$$\alpha_e = \cos\theta\,\alpha_1 + \sin\theta\,\alpha_2, \qquad \alpha_\mu = \cos\theta\,\alpha_2 - \sin\theta\,\alpha_1. \tag{16}$$

and c.c., and substitute them in Eq.(15), obtaining the Hamiltonian in the "flavor" variables:

$$H(\alpha_{e,\mu};\alpha^*_{e,\mu}) = \omega_e|\alpha_e|^2 + \omega_\mu|\alpha_\mu|^2 + \omega_{e\mu}(\alpha_e^*\alpha_\mu + \alpha_\mu^*\alpha_e). \tag{17}$$

where ω_e, ω_μ and $\omega_{e\mu}$ are given in Eq.(8). Eq.(15) and Eq.(17) represent alternative forms of the Hamiltonian Eq.(13) in terms of $(\alpha_1,\alpha_2,\alpha_1^*,\alpha_2^*)$, $(\alpha_e,\alpha_\mu,\alpha_e^*,\alpha_\mu^*)$, respectively. It is now interesting to ask whether the same Hamiltonian can be written in terms of other canonical variables (Q_e,Q_μ,P_e,P_μ) in such a way to close the chain depicted in the following graph:

$$\begin{bmatrix} H(Q_{1,2};P_{1,2}) & \overset{4}{\longleftrightarrow} & H(\alpha_{1,2};\alpha^*_{1,2}) \\ \updownarrow 1 & & \updownarrow 3 \\ H(Q_{e,\mu};P_{e,\mu}) & \overset{2}{\longleftrightarrow} & H(\alpha_{e,\mu};\alpha^*_{e,\mu}) \end{bmatrix} \tag{18}$$

In analogy with Eq.(14) we define ($\sigma = e, \mu$):

$$\alpha_\sigma = \sqrt{\frac{m\omega_\sigma}{2}}\left(Q_\sigma + i\frac{P_\sigma}{m\omega_\sigma}\right), \quad \alpha^*_\sigma = \sqrt{\frac{m\omega_\sigma}{2}}\left(Q_\sigma - i\frac{P_\sigma}{m\omega_\sigma}\right). \tag{19}$$

Inspired by Eq.(7) and Eq.(16), we introduce the following canonical transformations:

$$Q_e = \cos\theta\sqrt{\frac{\omega_1}{\omega_e}}Q_1 + \sin\theta\sqrt{\frac{\omega_2}{\omega_e}}Q_2, \qquad P_e = \cos\theta\sqrt{\frac{\omega_e}{\omega_1}}P_1 + \sin\theta\sqrt{\frac{\omega_e}{\omega_2}}P_2, \tag{20}$$

$$Q_\mu = \cos\theta\sqrt{\frac{\omega_2}{\omega_\mu}}Q_2 - \sin\theta\sqrt{\frac{\omega_1}{\omega_\mu}}Q_1, \qquad P_\mu = \cos\theta\sqrt{\frac{\omega_\mu}{\omega_2}}P_2 - \sin\theta\sqrt{\frac{\omega_\mu}{\omega_1}}P_1. \tag{21}$$

They indeed guarantee the conservation of the Poisson brackets. For example:

$$\{Q_e, P_e\} = \sum_{i=1}^{2}\left(\frac{\partial Q_e}{\partial Q_i}\frac{\partial P_e}{\partial P_i} - \frac{\partial Q_e}{\partial P_i}\frac{\partial P_e}{\partial Q_i}\right) = \cos^2\theta\sqrt{\frac{\omega_1}{\omega_e}}\sqrt{\frac{\omega_e}{\omega_1}} + \sin^2\theta\sqrt{\frac{\omega_2}{\omega_e}}\sqrt{\frac{\omega_e}{\omega_2}} = 1. \tag{22}$$

The generator of the canonical (point) transformation in Eqs.(20),(21) is of the F_2–type [10]:

$$F_2(Q_{1,2}, P_{e,\mu}) = \left(\cos\theta\sqrt{\frac{\omega_1}{\omega_e}}Q_1 + \sin\theta\sqrt{\frac{\omega_2}{\omega_e}}Q_2\right)P_e + \left(\cos\theta\sqrt{\frac{\omega_2}{\omega_\mu}}Q_2 - \sin\theta\sqrt{\frac{\omega_1}{\omega_\mu}}Q_1\right)P_\mu;$$

$$P_{1,2} = \frac{\partial F_2}{\partial Q_{1,2}}, \qquad Q_{e,\mu} = \frac{\partial F_2}{\partial P_{e,\mu}}.$$

In order for the chain in Eq.(18) to close, we invert the transformations in Eqs.(20),(21) and substitute them in Eq.(13). We obtain

$$H(Q_{e,\mu};P_{e,\mu}) = \frac{P_e^2}{2m} + \frac{P_\mu^2}{2m} + \frac{m\omega_e^2}{2}Q_e^2 + \frac{m\omega_\mu^2}{2}Q_\mu^2 + \omega_{e\mu}\Big(\frac{P_e}{\sqrt{m\omega_e}}\frac{P_\mu}{\sqrt{m\omega_\mu}} + (\sqrt{m\omega_e}Q_e)(\sqrt{m\omega_\mu}Q_\mu)\Big). \tag{23}$$

We also verify that, in turn, substituting Eq.(19) in Eq.(17), we obtain again Eq.(23). Consistently with the canonicity of the used transformations, $H(Q_{e,\mu};P_{e,\mu})$ turns out to be a real quantity; the imaginary terms proportional to $Q_e P_\mu$ and $Q_\mu P_e$ cancel. From Eq.(23) one can derive the equations of motion for these variables:

$$\dot{Q}_e = \frac{P_e}{m} + \omega_{e\mu}\frac{P_\mu}{m\sqrt{\omega_e\omega_\mu}}, \quad \dot{P}_e = -m\omega_e^2 - \omega_{e\mu}m\sqrt{\omega_e\omega_\mu}Q_\mu, \tag{24}$$

$$\dot{Q}_\mu = \frac{P_\mu}{m} + \omega_{e\mu}\frac{P_e}{m\sqrt{\omega_e\omega_\mu}}, \quad \dot{P}_\mu = -m\omega_\mu^2 - \omega_{e\mu}m\sqrt{\omega_e\omega_\mu}Q_e, \tag{25}$$

which can be rewritten in the following form:

$$\begin{cases} \ddot{Q}_e = -(\omega_e^2 + \omega_{e\mu}^2)Q_e - \omega_{e\mu}Q_\mu\left(\frac{\omega_e\omega_\mu + \omega_\mu^2}{\sqrt{\omega_e\omega_\mu}}\right) \\ \ddot{Q}_\mu = -(\omega_\mu^2 + \omega_{e\mu}^2)Q_\mu - \omega_{e\mu}Q_e\left(\frac{\omega_e\omega_\mu + \omega_e^2}{\sqrt{\omega_e\omega_\mu}}\right) \end{cases} \tag{26}$$

Summarizing, we have found the analogue, at classical level, of Pontecorvo mixing transformations. These are given by the the canonical (classical) transformations Eqs.(20),(21).

IOP Conf. Series: Journal of Physics: Conf. Series **880** (2017) 012061 doi:10.1088/1742-6596/880/1/012061

4. Oscillations in phase space

We wonder whether it is possible to obtain the oscillation formulas Eqs.(11),(12) within the classical model in terms of the Q_σ, P_σ variables. We know that the (real) solutions of Hamilton's equations for the Hamiltonian Eq.(13) are:

$$Q_1(t) = A_1\, e^{i\omega_1 t} + A_1^*\, e^{-i\omega_1 t}, \qquad Q_2(t) = A_2\, e^{i\omega_2 t} + A_2^*\, e^{-i\omega_2 t}. \tag{27}$$

and

$$P_1(t) = i\omega_1 A_1 e^{i\omega_1 t} - i\omega_1 A_1^* e^{-i\omega_1 t}, \qquad P_2(t) = i\omega_2 A_2 e^{i\omega_2 t} - i\omega_2 A_2^* e^{-i\omega_2 t}. \tag{28}$$

We now set $m = 1$ for simplicity and consider the following quantities:

$$\alpha_1(t) = \frac{1}{2}\left(\sqrt{\omega_1}Q_1(t) + i\frac{1}{\sqrt{\omega_1}}P_1(t)\right) \;=\; A_1^*\sqrt{\omega_1}e^{-i\omega_1 t} \tag{29}$$

$$\alpha_2(t) = \frac{1}{2}\left(\sqrt{\omega_2}Q_2(t) + i\frac{1}{\sqrt{\omega_2}}P_2(t)\right) \;=\; A_2^*\sqrt{\omega_2}e^{-i\omega_2 t} \tag{30}$$

and c.c.. Using Eqs.(20),(21) the quantity $\alpha_e(t)$ and $\alpha_\mu(t)$ take the following form:

$$\alpha_e(t) = \frac{1}{2}\left(\sqrt{\omega_e}Q_e(t) + i\frac{1}{\sqrt{\omega_e}}P_e(t)\right) \;=\; \cos\theta A_1^*\sqrt{\omega_1}e^{-i\omega_1 t} + \sin\theta A_2^*\sqrt{\omega_2}e^{-i\omega_2 t}, \tag{31}$$

$$\alpha_\mu(t) = \frac{1}{2}\left(\sqrt{\omega_\mu}Q_\mu(t) + i\frac{1}{\sqrt{\omega_\mu}}P_\mu(t)\right) \;=\; \cos\theta A_2^*\sqrt{\omega_2}e^{-i\omega_2 t} - \sin\theta A_1^*\sqrt{\omega_1}e^{-i\omega_1 t}. \tag{32}$$

Setting as initial conditions

$$e: \qquad A_1 = e^{i\gamma}\frac{\cos\theta}{\sqrt{\omega_1}}, \quad A_2 = e^{i\gamma}\frac{\sin\theta}{\sqrt{\omega_2}} \tag{33}$$

with γ arbitrary real, we obtain

$$|\alpha_e(t)|^2\Big|_e = 1 - 2\sin^2 2\theta\, \sin^2\left(\frac{\omega_2 - \omega_1}{2}t\right), \tag{34}$$

which shows that already at a classical level we can reproduce flavor oscillation formulas. By the same reasoning, we have:

$$|\alpha_\mu(t)|^2\Big|_e = 2\sin^2 2\theta\, \sin^2\left(\frac{\omega_2 - \omega_1}{2}t\right). \tag{35}$$

Choosing as initial conditions:

$$\mu: \qquad A_1 = e^{i\gamma}\frac{-\sin\theta}{\sqrt{\omega_1}}, \quad A_2 = e^{i\gamma}\frac{\cos\theta}{\sqrt{\omega_2}}, \tag{36}$$

we obtain similar results as above, cf. Eqs.(34),(35), by exchanging $e \leftrightarrow \mu$.

$$|\alpha_\mu(t)|^2\Big|_\mu = 1 - 2\sin^2 2\theta\, \sin^2\left(\frac{\omega_2 - \omega_1}{2}t\right), \tag{37}$$

and

$$|\alpha_e(t)|^2\Big|_\mu = 2\sin^2 2\theta\, \sin^2\left(\frac{\omega_2 - \omega_1}{2}t\right). \tag{38}$$

We note that, in the above derivation, the quantities $|\alpha_\sigma(t)|^2\Big|_\rho$ behave exactly as the expectation values of the operators $\hat{N}_\sigma(t)$ on the state $|a_\rho\rangle$.

On the same line of reasoning, we can consider the "expectation value" of the Hamiltonian Eq.(17) on the "flavor state", corresponding to the initial conditions in Eq.(33):

$$H(\alpha_{e,\mu};\alpha^*_{e,\mu})\Big|_e = \omega_e + (\omega_\mu - \omega_e)|\alpha_\mu(t)|^2\Big|_e + \omega_{e\mu}\Big(\alpha^*_e(t)\alpha_\mu(t) + \alpha^*_\mu(t)\alpha_e(t)\Big)\Big|_e, \tag{39}$$

where we used $|\alpha_e(t)|^2\Big|_e = 1 - |\alpha_\mu(t)|^2\Big|_e$.

Following Ref.[11], we can regard the "free" Hamiltonians $\tilde{H}_e = \omega_e\,|\alpha_e(t)|^2$ and $\tilde{H}_\mu = \omega_\mu\,|\alpha_\mu(t)|^2$ as free energies F_σ, and write:

$$H = \sum_{\sigma=e,\mu}\Big(F_\sigma(t) + TS_\sigma(t)\Big), \tag{40}$$

where we make the identification $T = \tan 2\theta$ and

$$S_\sigma = \frac{1}{4}\delta\omega\Big(\alpha^*_e(t)\alpha_\mu(t) + \alpha^*_\mu(t)\alpha_e(t)\Big), \tag{41}$$

with $\delta\omega = \frac{2}{\tan 2\theta}\,\omega_{e\mu}$. We have:

$$S_e(t)\Big|_e = -\frac{1}{4}\delta\omega \sin 4\theta \sin^2\Big(\frac{\omega_2 - \omega_1}{2}t\Big). \tag{42}$$

The other results are summarized in *Table 1*, from which we see how the energetic balance is recovered.

Table 1. Energetic balance.

init.cond.	H	F_e	F_μ	$TS_e = TS_\mu$
e	ω_e	$\omega_e\,|\alpha_e(t)|^2\Big|_e$	$\omega_\mu\,|\alpha_\mu(t)|^2\Big|_e$	$\frac{1}{2}\delta\omega|\alpha_\mu(t)|^2\Big|_e$
μ	ω_μ	$\omega_e\,|\alpha_e(t)|^2\Big|_\mu$	$\omega_\mu\,|\alpha_\mu(t)|^2\Big|_\mu$	$-\frac{1}{2}\delta\omega|\alpha_e(t)|^2\Big|_\mu$

Note finally, that the integral of the entropy expectation value over an oscillation cycle, is only dependent on the mixing angle:

$$\int_0^\tau S_e(t)\Big|_e dt = \pi\cos^2 2\theta \sin 2\theta, \tag{43}$$

where the period is $\tau = \frac{2\pi}{\omega_2-\omega_1}$. Such a quantity, being independent of dynamical parameters, can be related to other geometric invariants as geometric phase for oscillating neutrinos [7].

5. Conclusions

In this paper we have considered a simple classical model in phase-space resembling the quantum one used for the description of neutrino oscillations. We have found that it is indeed possible to obtain flavor oscillation formulas by considering classical analogues of the (time–dependent) flavor number operators and (initial) neutrino states. By resorting to previous results [11], we have given a thermodynamical interpretation of the phenomenon of flavor oscillations in terms of energy fluxes between two open subsystems.

One interesting output of the present work is a form for the classical hamiltonian describing flavor oscillations. This can be useful for identifying classical (hamiltonian) systems analogue for this phenomenon. Another possible direction of investigation which is suggested by the above hamiltonian form Eq.(23) is the mapping of such system into an equivalent non–commutative one. On such aspects, work is in progress.

References

[1] Bilenky S M and Pontecorvo B 1978 *Phys. Rept.* **41** 225
[2] Blasone M and Vitiello G 1995 *Ann. Phys.* **244** 283
[3] Blasone M, Gargiulo M V and Vitiello G 2016 *Phys. Lett.* B **761** 104
[4] Blasone M, Gargiulo M V and Vitiello G 2015 *J. Phys.: Conf. Series* **626** 012026
[5] Blasone M, Henning P A and Vitiello G 1999 *Phys. Lett.* B **451** 140
[6] Blasone M, Jizba P and Vitiello G 2001 *Phys. Lett.* B **517** 471
[7] Blasone M, Henning P A and Vitiello G 1999 *Phys. Lett.* B **466** 262
[8] Blasone M, Dell'Anno F, De Siena S and Illuminati F 2009 *EPL* **85** 50002
[9] Blasone M, Dell'Anno F, De Siena S and Illuminati F 2010 *J. Phys.: Conf. Series* **237** 012007
[10] Goldstein H, Poole C and Safko J 2000 *Classical Mechanics* (Boston: Addison Wesley)
[11] Blasone M, Di Mauro M and Vitiello G 2011 *Phys. Lett.* B **697** 238
[12] Blasone M, Jizba P and Vitiello G 2011 *Quantum Field Theory and its Macroscopic Manifestations* (London: World Scientific & ICP)
[13] Perelomov A 1986 *Generalized Coherent States and Their Applications* (Berlin: Springer–Verlag)
[14] Umezawa H 1993 *Advanced Field Theory: Micro, Macro and Thermal Physics* (New York: AIP)
[15] Blasone M, Jizba P, Lambiase G and Mavromatos N E 2014 *J.Phys.: Conf. Series* **538** 012003

IOP Conf. Series: Journal of Physics: Conf. Series **880** (2017) 012062 doi:10.1088/1742-6596/880/1/012062

Thermodynamic instabilities and statistical effects in protoneutron stars

G Gervino[1,3], A Lavagno[2,3], D Pigato[2]
[1]Dipartimento di Fisica, Università di Torino, I-10126 Torino, Italy
[2]Department of Applied Science and Technology, Politecnico di Torino, I-10129 Torino, Italy
[3]INFN, Sezione di Torino, I-10125 Torino, Italy

Abstract.
We investigate the physical properties of the protoneutron stars in the framework of a relativistic mean-field model and we study the finite-temperature equation of state in β-stable matter at fixed entropy per baryon, in the presence of hyperons, Δ-isobar resonances and trapped neutrinos. In this context, we study the possible presence of thermodynamic instabilities and a phase transition from nucleonic matter to resonance-dominated Δ-matter can take place. Such a phase transition is characterized by both mechanical instability (fluctuations on the baryon density) and by chemical-diffusive instability (fluctuations on the isospin concentration) in asymmetric nuclear matter. We show that such statistical effects could play a crucial role in the structure and in the evolution of the protoneutron stars.

1. Introduction

A protoneutron star (PNS) is formed in a stellar remnant after a successful core-collapse supernova explosion of a star with a mass smaller than about 20 solar masses and in the first seconds of its evolution it is a very hot (temperature of up to 50 MeV), lepton rich and β-stable object and a lepton concentration typical of the pre-supernova matter [1].

The knowledge of the nuclear EOS of dense matter at finite temperature plays a crucial role in the determination of the structure and in the macrophysical evolution of the PNS [2]. The processes related to strong interaction should in principle be described by quantum chromodynamics. However, in the energy density range reached in the compact stars, strongly non-perturbative effects in the complex theory of QCD are not negligible [3]. In the absence of a converging method to approach QCD at finite density one often turns to effective and phenomenological model investigations.

In this article, we study a hadronic equation of state (EOS) at finite temperature and density by means of a relativistic mean-field model with the inclusion $\Delta(1232)$-isobars [4, 5, 6] and by requiring the Gibbs conditions on the global conservation of baryon number and net electric charge. Transport model calculations and experimental results indicate that an excited state of baryonic matter is dominated by the Δ resonance at the energies from the BNL Alternating Gradien Synchrotron (AGS) to RHIC [7]. Moreover, in the framework of the nonlinear Walecka model, it has been predicted that a phase transition from nucleonic matter to Δ-excited nuclear matter can take place and the occurrence of this transition sensibly depends on the Δ-meson coupling constants [8, 9, 10, 11, 12].

2. Hadronic equation of state and stability conditions

In this section, we start by introducing the hadronic equation of state (EOS) in the framework of a relativistic mean-field theory. In this investigation we include all the baryon octet in order to reproduce the chemical composition of the PNS at high baryon chemical potential. We also take into account of leptons particle by fixing the lepton fraction $Y_L = Y_e + Y_{\nu_e} = (\rho_e + \rho_{\nu_e})/\rho_B$, where ρ_e, ρ_{ν_e} and ρ_B are the electron, neutrino and baryon number densities, respectively. This is because, in the first stage of PNS evolution, electrons and neutrinos are trapped inside the stellar matter and, therefore, the lepton number must be conserved until neutrinos escape out of the PNS [13].

The Lagrangian density can be written in term of the hadronic [4, 14] plus leptonic component, as follow:

$$\begin{aligned}
\mathcal{L}_{tot} = \mathcal{L}_H + \mathcal{L}_l = \sum_B \bar{\psi}_B[i\gamma_\mu \partial^\mu - (M_B - g_{\sigma B}\sigma) \\
-g_{\omega B}\gamma_\mu \omega^\mu - g_{\rho B}\gamma^\mu \vec{\tau} \cdot \vec{\rho}_\mu]\psi_B + \frac{1}{2}(\partial_\mu \sigma \partial^\mu \sigma - m_\sigma^2 \sigma^2) \\
-U(\sigma) + \frac{1}{2}m_\omega^2 \omega_\mu \omega^\mu + \frac{1}{2}m_\rho^2 \vec{\rho}_\mu \cdot \vec{\rho}^{\,\mu} - \frac{1}{4}F_{\mu\nu}F^{\mu\nu} \\
-\frac{1}{4}\vec{G}_{\mu\nu}\vec{G}^{\mu\nu} + \sum_l \bar{\psi}_l[i\gamma_\mu \partial^\mu - m_l]\psi_l \,, \qquad (1)
\end{aligned}$$

where the sums over B and l are over the baryon octet and lepton particles, respectively. The field strength tensors for the vector mesons are given by the usual expressions $F_{\mu\nu} \equiv \partial_\mu \omega_\nu - \partial_\nu \omega_\mu$, $\vec{G}_{\mu\nu} \equiv \partial_\mu \vec{\rho}_\nu - \partial_\nu \vec{\rho}_\mu$, and $U(\sigma)$ is a nonlinear potential of σ meson

$$U(\sigma) = \frac{1}{3}a\sigma^3 + \frac{1}{4}b\sigma^4 \,, \qquad (2)$$

usually introduced to achieve a reasonable compression modulus for equilibrium nuclear matter.

The field equations in a mean field approximation are

$$\begin{aligned}
(i\gamma_\mu \partial^\mu - (M - g_{\sigma B}\sigma) - g_{\omega B}\gamma^0 \omega - g_{\rho B}\gamma^0 \tau_3 \rho)\psi = 0 \,, \qquad (3) \\
m_\sigma^2 \sigma + a\sigma^2 + b\sigma^3 = g_{\sigma B} < \bar{\psi}\psi >= g_{\sigma B}\rho_S \,, \qquad (4) \\
m_\omega^2 \omega = g_{\omega B} < \bar{\psi}\gamma^0 \psi >= g_{\omega B}\rho_B \,, \qquad (5) \\
m_\rho^2 \rho = g_{\rho B} < \bar{\psi}\gamma^0 \tau_3 \psi >= g_{\rho B}\rho_I \,, \qquad (6)
\end{aligned}$$

where $\sigma = \langle\sigma\rangle$, $\omega = \langle\omega^0\rangle$ and $\rho = \langle\rho_3^0\rangle$ are the nonvanishing expectation values of mesons fields, ρ_I is the total isospin density, ρ_B and ρ_S are the baryon density and the baryon scalar density, respectively. They are given by

$$\rho_B = 2\sum_{i=B} \int \frac{\mathrm{d}^3k}{(2\pi)^3}[n_i(k) - \overline{n}_i(k)] \,, \qquad (7)$$

$$\rho_S = 2\sum_{i=B} \int \frac{\mathrm{d}^3k}{(2\pi)^3} \frac{M_i^*}{E_i^*}[n_i(k) + \overline{n}_i(k)] \,, \qquad (8)$$

where $n_i(k)$ and $\overline{n}_i(k)$ are the fermion particle and antiparticle distributions.

The nucleon effective energy is defined as $E_i{}^*(k) = \sqrt{k^2 + M_i{}^{*2}}$, where $M_i{}^* = M_i - g_{\sigma B}\sigma$. The effective chemical potentials μ_i^* are given in terms of the meson fields as follows

$$\mu_i^* = \mu_i - g_{\omega B}\omega - \tau_{3iB}g_{\rho B}\rho \,, \qquad (9)$$

IOP Conf. Series: Journal of Physics: Conf. Series **880** (2017) 012062 doi:10.1088/1742-6596/880/1/012062

where μ_i are the thermodynamical chemical potentials ($\mu_i = \partial\epsilon/\partial\rho_i$).

The further conditions, required for the β-stable chemical equilibrium and charge neutrality, can be written as

$$\mu_\Lambda = \mu_{\Sigma^0} = \mu_{\Xi^0} = \mu_n \,, \tag{10}$$
$$\mu_{\Sigma^-} = \mu_{\Xi^-} = \mu_n + \mu_e \,, \tag{11}$$
$$\mu_p = \mu_{\Sigma^+} = \mu_n - \mu_e \,; \tag{12}$$
$$\rho_p + \rho_{\Sigma^+} - \rho_{\Sigma^-} - \rho_{\Xi^-} - \rho_e = 0 \,. \tag{13}$$

In the case of trapped neutrinos, the new equalities are obtained by the replacement of $\mu_e \to \mu_e - \mu_{\nu_e}$. The total entropy per baryon is calculated using $s = (S_B + S_l)/(T\rho_B)$, where $S_B = P_B + \epsilon_B - \sum_{i=B} \mu_i\rho_i$ and $S_l = P_l + \epsilon_l - \sum_{i=l} \mu_i\rho_i$, and the sums are extended over all the baryons and leptons species.

The thermodynamical quantities can be obtained from the thermodynamic potential in the standard way. More explicitly, the baryon pressure P_B and the energy density ϵ_B can be written as

$$\begin{aligned} P_B = \; & \frac{2}{3}\sum_i \int \frac{\mathrm{d}^3k}{(2\pi)^3}\frac{k^2}{E_i^*(k)}[n_i(k) + \overline{n}_i(k)] - \frac{1}{2}m_\sigma^2\sigma^2 \\ & -U(\sigma) + \frac{1}{2}m_\omega^2\omega^2 + \frac{1}{2}m_\rho^2\rho^2 \,, \end{aligned} \tag{14}$$
$$\begin{aligned} \epsilon_B = \; & 2\sum_i \int \frac{\mathrm{d}^3k}{(2\pi)^3} E_i^*(k)[n_i(k) + \overline{n}_i(k)] + \frac{1}{2}m_\sigma^2\sigma^2 \\ & +U(\sigma) + \frac{1}{2}m_\omega^2\omega^2 + \frac{1}{2}m_\rho^2\rho^2 \,. \end{aligned} \tag{15}$$

Here and in the following, we focus our investigation by considering the so-called GM3 [4] and the SFHo parameter sets [15]. The implementation of hyperon degrees of freedom comes from determination of the corresponding meson-hyperon coupling constants that have been fitted to hypernuclear properties.

Concerning the stability conditions, we are dealing with the study of a multi-component system at finite temperature and density with two conserved charges: baryon number and electric charge. For such a system, the Helmholtz free energy density F can be written as [16]

$$F(T, \rho_B, \rho_C) = -P(T, \mu_B, \mu_C) + \mu_B\rho_B + \mu_C\rho_C \,, \tag{16}$$

with

$$\mu_B = \left(\frac{\partial F}{\partial \rho_B}\right)_{T,\rho_C} , \quad \mu_C = \left(\frac{\partial F}{\partial \rho_C}\right)_{T,\rho_B} . \tag{17}$$

In a system with N different particles, the particle chemical potentials are expressed as the linear combination of the two independent chemical potentials μ_B and μ_C and, as a consequence, $\sum_{i=1}^{N} \mu_i\rho_i = \mu_B\rho_B + \mu_C\rho_C$.

Assuming the presence of two phases (denoted as I and II, respectively), the system is stable against the separation in two phases if the free energy of a single phase is lower than the free energy in all two phases configuration. The phase coexistence is given by the Gibbs conditions

$$\mu_B^I = \mu_B^{II} \,, \qquad \mu_C^I = \mu_C^{II} \,, \tag{18}$$
$$P^I(T, \mu_B, \mu_C) = P^{II}(T, \mu_B, \mu_C) \,. \tag{19}$$

Therefore, at a given baryon density ρ_B and at a given net electric charge density $\rho_C = y\,\rho_B$ (with $y = Z/A$), the chemical potentials μ_B are μ_C are univocally determined. An important feature of this conditions is that, unlike the case of a single conserved charge, the pressure in the mixed phase is not constant and, although the total ρ_B and ρ_C are fixed, baryon and charge densities can be different in the two phases. For such a system in thermal equilibrium, the possible phase transition can be characterized by mechanical (fluctuations in the baryon density) and chemical instabilities (fluctuations in the electric charge density) [9, 16]. As usual the condition of the mechanical stability implies

$$\rho_B \left(\frac{\partial P}{\partial \rho_B}\right)_{T,\rho_C} > 0\,. \tag{20}$$

By introducing the notation $\mu_{i,j} = (\partial\mu_i/\partial\rho_j)_{T,P}$ (with $i,j = B,C$), the chemical stability for a process at constant P and T can be expressed with the following conditions [9]

$$\rho_B\,\mu_{B,B} + \rho_C\,\mu_{C,B} = 0\,, \tag{21}$$

$$\rho_B\,\mu_{B,C} + \rho_C\,\mu_{C,C} = 0\,. \tag{22}$$

Whenever the above stability conditions are not respected, the system becomes unstable and the phase transition takes place. The coexistence line of a system with one conserved charge becomes in this case a two dimensional surface in (T,P,y) space, enclosing the region where mechanical and diffusive instabilities occur.

By increasing the temperature and the baryon density during the high energy heavy ion collisions ($T \approx 50$ MeV and $\rho_B \geq \rho_0$), a multi-particle system with Δ-isobar and pion degrees of freedom may take place.

In analogy with the liquid-gas case, we are going to investigate the existence of a possible phase transition in the nuclear medium by studying the presence of instabilities (mechanical and/or chemical) in the system. The chemical stability condition is satisfied if [9]

$$\left(\frac{\partial\mu_C}{\partial y}\right)_{T,P} > 0 \;\; \text{or} \;\; \begin{cases} \left(\frac{\partial\mu_B}{\partial y}\right)_{T,P} < 0\,, & \text{if } y > 0\,, \\ \left(\frac{\partial\mu_B}{\partial y}\right)_{T,P} > 0\,, & \text{if } y < 0\,. \end{cases} \tag{23}$$

In the Fig. 1, we report the baryon and electric charge chemical potential isobars as a function of y, at fixed temperature $T = 50$ MeV and $x_{\sigma\Delta} \equiv g_{\sigma\Delta}/g_{\sigma N} = 1.3$ (the ratio related to the scalar σ meson-Δ coupling constants) in the GM3 parameters set [4].

From the analysis of the above chemical potential isobars, we are able to construct the binodal surface relative to the nucleon-Δ matter phase transition. In Fig. 2, we show the binodal section at $T = 50$ MeV and $x_{\sigma\Delta} = 1.3$.

The right branch (at lower density) corresponds to the initial phase (I), where the dominant component of the system is given by nucleons. The left branch (II) is related to the final phase at higher densities, where the system is composed primarily by Δ-isobar degrees of freedom (Δ-dominant phase). In presence of Δ-isobars the phase coexistence region results very different from what obtained in the liquid-gas case, in particular it extends up to regions of negative electric charge fraction and the mixed phase region ends in a point of maximum asymmetry with $y = -1$ (corresponding to a system with almost all Δ^--particles, being antiparticles and pions contribution almost negligible in this regime). We analyze the phase evolution of the system during the isothermal compression from an arbitrary initial point A, indicated in Fig. 2. In this point the system becomes unstable and starts to be energetically favorable the separation into two phases, therefore an infinitesimal Δ-dominant phase appears in B, at the same temperature and pressure. Let us observe that, although in B the electric charge fraction is substantially

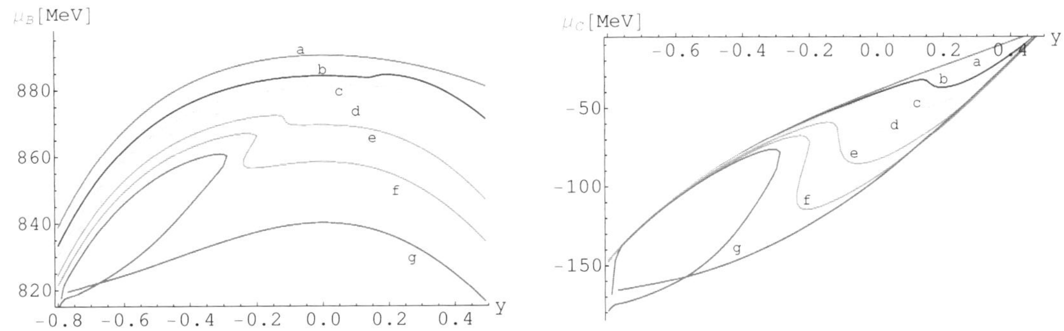

Figure 1. Baryon (right panel) and electric charge (left panel) chemical potential isobars as a function of y at $T = 50$ MeV and $x_{\sigma\Delta} = 1.3$. The curves labeled a through g have pressure P=9,7,6,5,4,3,2 MeV/fm^3, respectively.

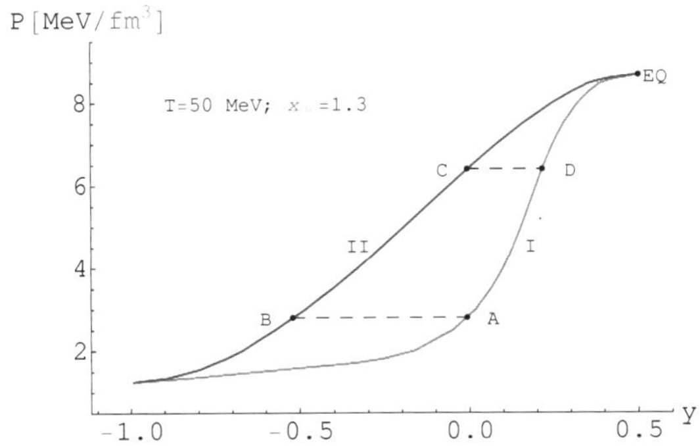

Figure 2. Binodal section at $T = 50$ MeV and $x_{\sigma\Delta} = 1.3$.

negative, the relative Δ^- abundance must be weighed on the low volume fraction occupied by the phase II near the point B. During the phase transition, each phase evolves towards a configuration with increasing y, in contrast to the liquid-gas case, where each phase evolves through a configuration with a decreasing value of y (with the exception of the gas phase after the maximum asymmetry point). We will see in the next section how the presence of such features are relevant in the structure and in the evolution of the PNS.

3. Results and discussion

We investigate the relevance Δ-isobar degrees of freedom and the stability conditions in the bulk properties of compact star and PNS.

Let us start by considering β-stable and electric-charge neutral nuclear matter at $T = 0$. In Fig. 3, we report the mass-radius relations in absence (no Δ) and in presence of Δ-isobars with different scalar coupling ratios ($x_{\sigma\Delta} = 1.0$ and $x_{\sigma\Delta} = 1.2$) in the GM3 model [4]. The presence of Δ-isobar degrees of freedom smooths the equation of state enlarging the effect of thermodynamic instabilities and reduces the maximum gravitational mass. On the other hand very compact object with smaller radii can be formed.

In Fig. 4, we show the temperature as a function of the baryon density (in units of the saturation nuclear density), in absence (np) and in presence (npH) of hyperons in the SFHo model [15, 17]. We limit our analysis in the first two phases: in the left panel, the first leptonic rich state ($s = 1, Y_L = 0.4$) and, in the right panel, the maximum heating phase ($s = 2, Y_{\nu_e} = 0$). Indeed in the cold-catalyzed phase ($s = 0, Y_{\nu_e} = 0$), the temperature is very low (fews MeV), and the above statistical effects due to thermodynamical instabilities

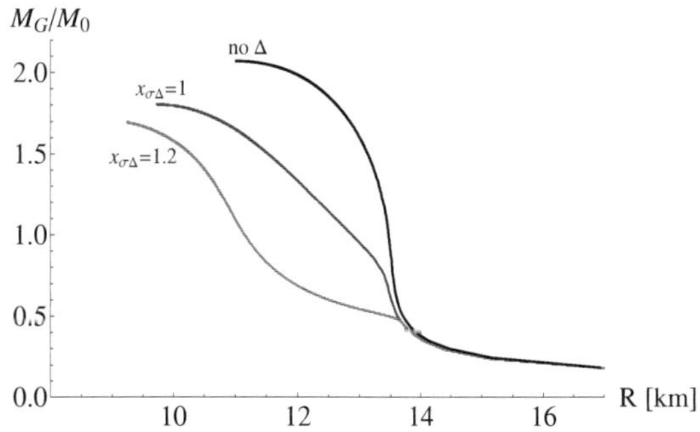

Figure 3. Mass-radius relations in absence (no Δ) and in presence of Δ-isobars with different scalar coupling ratios ($x_{\sigma\Delta} = 1.0$ and $x_{\sigma\Delta} = 1.2$).

and Delta-isobar formation may be neglected. In both previous cases, we observe a reduction in temperature in presence of hyperons and Delta-isobar degrees of freedom. Note also that, when hyperons are present, for $s = 1$ and $Y_L = 0.4$, the system evolves in a quasi isothermal configuration above $\rho_B = (2.5 \div 6)\,\rho_0$. The different behavior in the stellar temperature have important consequences in the PNS evolution and in its particles concentration. Finite temperature properties of matter at high density influence the diffusion of neutrinos, being the neutrino mean free paths strongly temperature dependent [13]. In particular, neutrino opacity is very sensitive to the inner temperature (in general proportional to T^2) and, therefore, this would affect sensibly the cooling of the PNS.

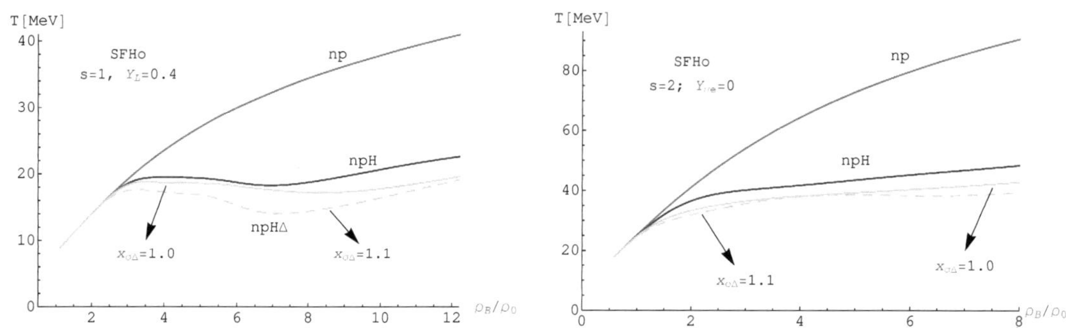

Figure 4. Temperature as a function of the baryon density (in units of the saturation nuclear density ρ_0) in the SFHo model and for different value of the Delta coupling constant $x_{\sigma\Delta}$. Left panel: entropy per baryon and neutrino fraction $s = 1, Y_L = 0.4$, right panel: $s = 2, Y_{\nu_e} = 0$. The labels np and npH stand for nucleons and nucleons plus hyperons.

In Fig. 5, we show the variation of the maximum baryonic mass in units of solar mass $M_\odot$ as a function of the central baryon density ρ_c, for pure nucleonic (np) and hyperonic plus Delta-isobars (npH) stars in the first leptonic rich state (left panel, $s = 1, Y_L = 0.4$) and in the maximum heating phase (right panel, $s = 2, Y_{\nu_e} = 0$). For a comparison, in the figure we have considered the two models GM3 and SFHo model in presence of Δ-isobars with different scalar coupling ratios ($x_{\sigma\Delta} = 1.0$ and $x_{\sigma\Delta} = 1.1$).

Let us note the strong reduction of the maximum baryonic mass with the introduction of hyperons and Delta-isobar degrees of freedom. This effect is remarkable stronger in the maximum heating phase ($s = 2, Y_{\nu_e} = 0$) and for a greater value of the $x_{\sigma\Delta}$ coupling due also to the presence of thermodynamical instabilities conditions.

In the presence of hyperons, when the stellar core contains non-leptonic negative charges,

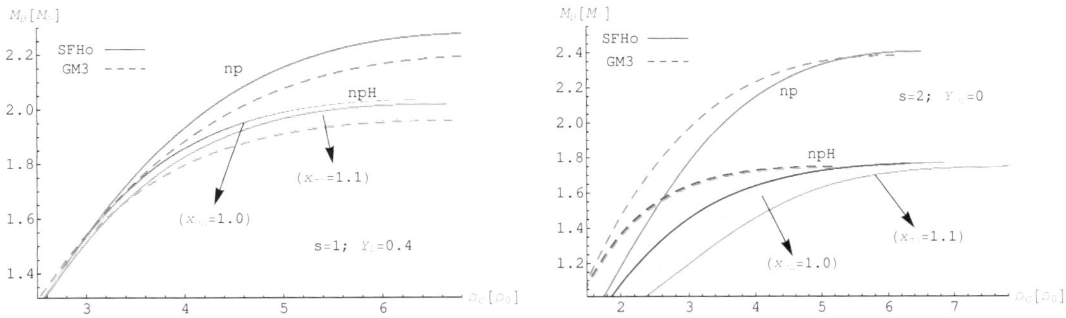

Figure 5. Maximum baryonic mass M_B in units of solar mass $M_\odot$ as a function of the central baryon density ρ_c (in units of the nuclear saturation density ρ_0) for nucleons (np) and hyperons stars (npH) stars in the case $s = 1$ and $Y_L = 0.4$ (left panel) and during the maximum heating phase $s = 2$ and $Y_\nu = 0$ (right panel).

the maximum masses of neutrino-trapped stars result to be significantly larger than for low temperatures and for lepton poor matter. Hence, there exists a window of initial masses for which the star becomes unstable to gravitational collapse during deleptonization and a black hole can take place [13]. We can see that the formation of such a metastable phase strongly depends on the presence of statistical effects and the window of metastability grows with the value $x_{\sigma\Delta}$.

References

[1] Prakash M et al. 2001 *Lect. Notes Phys.* **578** 364
[2] Lavagno A and Pigato D 2011 *Eur. Phys. J* A **47** 52
[3] Bonanno L, Drago A and Lavagno A 2007 *Phys. Rev. Lett.* **99** 242301
[4] Glendenning N K 1992 *Phys. Rev.* D **46** 1274
[5] Lavagno A 2010 *Phys. Rev.* C **81** 044909
[6] Lavagno A and Pigato D 2012 *J. Phys. G: Nucl. Part. Phys.* **39** 125106
[7] Hofmann M, Mattiello R, Sorge H, Stocker H and Greiner W 1995 *Phys. Rev.* C **51** 2095
[8] Li Z, Mao G, Zhuo Y and Greiner W 1997 *Phys. Rev.* C **56** 1570
[9] Lavagno A and Pigato D 2012 *Phys. Rev.* C **86** 024917
[10] Lavagno A 2013 *Eur. Phys. J.* A **49** 102
[11] Drago A, Lavagno A, Pagliara G and Pigato D 2014 *Phys. Rev.* C **90** 065809
[12] Drago A, Lavagno A and Pagliara G 2014 *Phys. Rev.* D **89** 043014
[13] Prakash M *et al.* 1997 *Phys. Rep.* **280** 1
[14] Di Toro M *et al.* 2006 *Nucl. Phys.* A **775** 102
[15] Steiner A W, Hempel M and Fischer T 2013 *Astrophys. J.* **774** 17
[16] Müller H and Serot B D 1995 *Phys. Rev.* C **52** 2072
[17] Drago A *et al.* 2016 *Phys. Rev.* D **93** 103001

Dirac bi-spinor entanglement under local noise and its simulation by Jaynes-Cummings interactions

Victor A. S. V. Bittencourt and Alex E. Bernardini

Departamento de Física, Universidade Federal de São Carlos, PO Box 676, 13565-905, São Carlos, SP, Brasil.

E-mail: vbittencourt@df.ufscar.br

Abstract. A description of the effects of the local noise on the quantum entanglement constraining the internal degrees of freedom of Dirac bi-spinor structures driven by arbitrary Poincaré invariant potentials is proposed. Given that the Dirac equation dynamics including external potentials can be simulated by a suitable four level trapped ion setup, quantum entanglement of two-qubit ionic states with quantum numbers related to the total angular momentum and to its projection onto the direction of the external magnetic field (used for lift the ions degeneracy), are recovered by means of a suitable ansatz. This formalism allows the inclusion of noise effects, which leads to disentanglement in the four level trapped ion quantum system. Our results indicate the role of interactions in bi-spinor entanglement, as well as the description of disentanglement in ionic states under local noises. For a state prepared initially in one of the ionic levels, local noise induces entanglement sudden death followed by sudden revivals driven by the noiseless dynamics of the state. Residual quantum correlations are observed in the intervals where such state is separable. Schrödinger cat and Werner states partially loose their initial entanglement content due to the interaction with the noisy environment but presenting entanglement oscillations without sudden death. Because Dirac equation describes low energy excitations of mono layer and bi-layer graphene, the formalism can also be applied to compute, for instance, electron-hole or electron/electron entanglement in various circumstances.

1. Introduction

Quantum correlations has been in the core of recent developments exploring the interface between quantum and classical physics [1, 2, 3, 4]. Quantum entanglement and other quantum correlations are widely considered for the engineering of quantum information protocols, in particular for quantum cryptography [5, 6] and quantum computing processes [7]. The non-local coherence generated by entanglement is essential to various applications of such quantum information/computing tasks in physical systems [7].

One platform suitable for implementing and characterizing quantum correlations is the ion-trap technology [8, 9], which has provided a phenomenological access to generate and manipulate quantum correlational properties of the trapped ions [10, 11, 12, 13]. Moreover, the trapped ion setup has also been engendered for simulation protocols, such as for simulating open quantum systems and quantum phase transitions [14, 15, 16]. Simulating Dirac equation structures is another example of a quantum system engendered by trapped ion dynamics. Under suitable conditions, the ionic (anti)Jaynes-Cummings ((A)JC) Hamiltonian dynamics can be mapped onto the structure of the Dirac equation, reproducing a series of relativistic-like quantum effects [17, 18, 19]. On the other hand, the two-*qubit* intrinsic entanglement present on the

$SU(2) \otimes SU(2)$ bi-spinor structure of Dirac equation solutions is also encoded in the ionic system simulating the Dirac dynamics by quantum numbers related to the total angular momentum and to its projection onto the direction of the external magnetic field (which is applied for lift the degeneracy of the ions internal levels) [13, 20, 21].

A quantum system not isolated has its correlational properties affected by the coupling with its environment. Fluctuations of the external magnetic field acts as an environment, causing entanglement degradation and a consequent decoherence [22, 23, 24]. To include such environment effects on the trapped ion two-*qubit* structure considered in [13], as well as to elucidate how the local decoherece is related to local and non-local disentanglement, a noise model for the environment can be adopted to describe more realistic setups [25].

This letter is concerned with the effects of a local noise model, where each subsystem is individually affected by the environment, in the informational content of internal states of a four level trapped ion. The local noise is included via Kraus operators and the dynamics of an arbitrary ionic state is recovered through the connection between the (A)JC dynamics and the Dirac equation structure. Once the Dirac-like spinor density matrix is constructed, the quantum entanglement can be quantified through negativity [26, 27]. Due to the coupling with the environment, an initial pure state evolves into a separable mixed state, which can display quantum correlations other than entanglement. In particular, the geometric discord [28] shall be adopted as a measure of such quantum correlations. For quantum states constrained to a specific ionic level, sudden deaths and quantum revivals can be quantified by the intrinsic entanglement. Moreover, those states exhibit residual nonclassical correlations for intervals were the state is separable. On the other hand, when the maximally entangled (Werner and Schrödinger cat) quantum states – prepared as superposition of such internal levels – are considered, coupling to environment causes a partial loss of their initially entangled properties, but no entanglement sudden death is present. In the three states considered, entanglement oscillations, caused by their noiseless dynamics, are observed.

2. Trapped ion framework

The trapped ion framework can be formulated by a Hamiltonian $\hat{H}_{RJC}$ which, in the rotating wave approximation [29], includes three interactions, the (anti) Jaynes-Cummings and the carrier interactions. The (A)JC interactions are given by the Hamiltonian

$$\hat{H}_j^{(A)JC} = \hbar\, \eta_j \tilde{\Omega}_j \left(\hat{\sigma}^{+(-)} a_j e^{+(-)i\phi_{r(b)}} + \hat{\sigma}^{-(+)} a_j^{\dagger} e^{-(+)i\phi_{r(b)}} \right) + \hbar\, \delta_j \hat{\sigma}_z, \tag{1}$$

with $j = x, y, z$, where the phases $\phi_{r\,(b)}$ describe red(blue)-sideband excitations, $\tilde{\Omega}_j$ are the Rabi frequencies, $\hat{\sigma}^{+\,(-)}$ are the internal level raising(lowering) operators, δ_j is an emergent detuning frequency between the external field and the two-level system, and $\eta_j = k\sqrt{\hbar\,/2\tilde{m}\nu_j}$ is the Lamb-Dicke parameter (where $\tilde{m}$ is the ion mass and k is the wave number of the external field). The JC interaction corresponds to a de-excitation of the internal level accompanied by an excitation of the vibrational degree of freedom, while the AJC interaction accomplished an excitation of the internal level with an excitation of the vibrational degree of freedom. The carrier interaction is given by

$$\hat{H}_j^C = \hbar\, \Omega_j (\hat{\sigma}^+ e^{i\phi} + \hat{\sigma}^- e^{-i\phi}), \tag{2}$$

which excites the internal levels without changing the vibrational state of the ion. From the experimental perspective, the above described dynamics is observed, for example, in the hyperfine levels $(2s^2\, S_{1/2})$ of alkali ions.

From now on, we consider a trapped ion with four internal levels labeled $\{|a\rangle, |b\rangle, |c\rangle, |d\rangle\}$. The combination of JC, AJC and carrier interactions acting on such system can be mapped into a Dirac Hamiltonian including tensor and pseudotensor external fields [17, 18, 30, 10, 11],

$$\hat{H}_D = \hat{\beta}\, mc^2 + c\,\hat{\boldsymbol{\alpha}}\cdot\boldsymbol{p} + \hat{\beta}\hat{\boldsymbol{\Sigma}}\cdot(\kappa\,\boldsymbol{\mathcal{E}}) + i\hat{\beta}\hat{\boldsymbol{\alpha}}\cdot\left(\mu\,\frac{\boldsymbol{\mathcal{E}}}{c}\right), \tag{3}$$

from which one firstly notices that the Dirac mass and kinetic term reads

$$\hat{\beta}mc^2 \rightarrow 2\hbar\,\delta(\hat{\sigma}_z^{ad} + \hat{\sigma}_z^{bc}), \tag{4}$$

$$c\,\hat{\boldsymbol{\alpha}}\cdot\boldsymbol{p} \rightarrow 2\eta\Delta_x\tilde{\Omega}(\hat{\sigma}_x^{ad} + \hat{\sigma}_x^{bc})p_x + 2\eta\Delta_y\tilde{\Omega}(\hat{\sigma}_y^{ad} - \hat{\sigma}_y^{bc})p_y + 2\eta\Delta_z\tilde{\Omega}(\hat{\sigma}_x^{ac} - \hat{\sigma}_x^{bd})p_z, \tag{5}$$

such that, for example, $\hat{\sigma}_z^{ad} + \hat{\sigma}_z^{bc} \equiv \hat{\beta}$, where the upper indices denote the involved internal levels [30, 10, 11, 13]. The momenta p_j are given by $p_j \rightarrow \frac{i\hbar}{2\Delta_j}\,(a_j^\dagger - a_j\,)$, where $\Delta_j = \sqrt{\hbar/2\tilde{m}\nu_j}$ is the delocalization width of the ground state wave function.

The tensor and pseudotensor external potentials are mapped in terms of the carrier interactions (2) with frequencies $\Omega_j^{(1)}$ and $\Omega_j^{(2)}$:

$$\hat{\beta}\hat{\boldsymbol{\Sigma}}\cdot(\kappa\,\boldsymbol{\mathcal{E}}) \rightarrow 2\hbar\,\Omega_x^{(1)}\,(\,\hat{\sigma}_x^{ab} - \hat{\sigma}_x^{cd}\,) + 2\hbar\,\Omega_y^{(1)}\,(\,\hat{\sigma}_y^{ab} - \hat{\sigma}_y^{cd}\,) + 2\hbar\,\Omega_z^{(1)}\,(\,\hat{\sigma}_z^{ab} - \hat{\sigma}_z^{cd}\,), \tag{6a}$$

$$i\hat{\beta}\hat{\boldsymbol{\alpha}}\cdot\left(\mu\,\frac{\boldsymbol{\mathcal{E}}}{c}\right) \rightarrow 2\hbar\,\Omega_x^{(2)}\,(\,-\hat{\sigma}_y^{ad} - \hat{\sigma}_y^{bc}\,) + 2\hbar\,\Omega_y^{(2)}\,(\,\hat{\sigma}_x^{bc} - \hat{\sigma}_x^{ad}\,) + 2\hbar\,\Omega_z^{(2)}\,(\,\hat{\sigma}_y^{bd} - \hat{\sigma}_y^{ac}\,). \tag{6b}$$

The Dirac dynamics driven by (3) is reproduced in the trapped ion setup through the maps (4), (5) and (6) and the relations between Dirac and ionic parameters are identified as

$$\frac{\mu\,\mathcal{E}_j}{c} = 2\hbar\,\Omega_j^{(2)}, \qquad \kappa\,\mathcal{E}_j = 2\hbar\,\Omega_j^{(1)}, \qquad c = 2\eta\Delta\tilde{\Omega}, \qquad mc^2 = 2\hbar\,\delta, \tag{7}$$

which set a one-to-one correspondence between (3) and the sum of the interactions (4), (5) and (6). Accordingly, the bi-spinor eigenstates of (3) $|\psi_{n,s}\rangle$ (with $n,\ s = 0,\ 1$), are described as linear combinations of the internal ionic levels,

$$|\psi_{n,s}\rangle \rightarrow M_{n,s}^a|a\rangle + M_{n,s}^b|b\rangle + M_{n,s}^c|c\rangle + M_{n,s}^d|d\rangle. \tag{8}$$

Each of the ionic state $|i\rangle$ is characterized by two quantum number: the total angular momentum F, and the projection M of the angular momentum onto an external magnetic field used for lift the degeneracy of the internal ionic levels. Such structure allows the assignment of a two-*qubit* correspondence given by

$$|a\rangle \equiv |0\,0\rangle, \qquad |b\rangle \equiv |0\,1\rangle, \qquad |c\rangle \equiv |1\,0\rangle, \qquad |d\rangle \equiv |1\,1\rangle. \tag{9}$$

According to the analysis of the classes of Poincaré invariant Dirac-like interactions [20, 13] the Hamiltonian eigenstates $|\psi_{n,s}\rangle$ indeed exhibit a naturally *spin-parity* entangled structure [31] which can be straightforwardly computed from stationary pure states, $\varrho_{n,s} = |\psi_{n,s}\rangle\langle\psi_{n,s}|$, given by [20]

$$\begin{aligned}\varrho_{n,s} &= \frac{1}{4}\left[I + \frac{(-1)^n}{|\lambda_{n,s}|}\hat{H}_D\right]\left[I + \frac{(-1)^s}{\sqrt{g_2}}\hat{\mathcal{O}}\right],\\ \hat{\mathcal{O}} &= \frac{1}{2}\left(\hat{H}_D^2 - \hat{I}\right) = m\,\kappa\,\hat{\boldsymbol{\Sigma}}\cdot\boldsymbol{\mathcal{E}} + \mu\,\hat{\beta}\,\hat{\boldsymbol{\Sigma}}\cdot\,(\,\boldsymbol{p}\times\boldsymbol{\mathcal{E}}\,) - i\kappa\,\hat{\beta}\,\hat{\boldsymbol{\alpha}}\cdot(\,\boldsymbol{p}\times\boldsymbol{\mathcal{E}}\,),\end{aligned} \tag{10}$$

where the parameter g_2 is evaluated as $g_2 = \frac{1}{16}\,\mathrm{Tr}\left[\left(\hat{H}_D^2 - \frac{1}{4}\,\mathrm{Tr}[\hat{H}_D^2]\right)^2\right]$ and $\lambda_{n,s}$ is identified as the mean energy of $\varrho_{n,s}$, $\lambda_{n,s} = \mathrm{Tr}[\,\hat{H}_D\,\varrho_{n,s}\,]$. Considering a Dirac-like one-dimensional propagation along the x axis, with the electric field lying in the xy-plane, such that $\boldsymbol{\mathcal{E}} = \mathcal{E}(\cos\theta\,\boldsymbol{i} + \sin\theta\,\boldsymbol{j}\,)$, with $\boldsymbol{p} = p\,\boldsymbol{i}$, where $\boldsymbol{i}, \boldsymbol{j}, \boldsymbol{k}$ define an orthonormal basis, one has $\boldsymbol{p}\times\boldsymbol{\mathcal{E}} = p\,\mathcal{E}\sin\theta\,\boldsymbol{k}$. The arbitrary choice of θ does not qualitatively affect the results, thus, one shall adopt $\theta = \pi/4$, such that the corresponding expressions for g_2 and $\lambda_{n,s}$ read

$$g_2 = \mathcal{E}^2\left[m^2\kappa^2 + \frac{1}{2}(\mu^2+\kappa^2)p^2\right], \tag{11a}$$

$$\lambda_{n,s} = (-1)^n\left[p^2 + m^2 + (\kappa^2+\mu^2)\mathcal{E}^2 + 2(-1)^s\mathcal{E}\sqrt{m^2\kappa^2 + \frac{1}{2}(\mu^2+\kappa^2)\,p^2}\,\right]^{1/2}. \tag{11b}$$

To recover the dynamics of an internal ionic state, the Dirac bi-spinor basis, $\{|\,\psi_{n,s}\,\rangle\}$, shall be related to the ionic state basis, $\{|\,i\,\rangle\}$. The temporal evolution of a single internal level is thus reconstructed by using the completeness relation $\sum_{n,s=0}^{1}\varrho_{n,s} = \hat{I}$ as to have (for $|j(t=0)\rangle \equiv |j\rangle$)

$$\rho_j(t) = |\,j(t)\,\rangle\langle\,j(t)\,| = e^{-i\hat{H}_D t}|j\rangle\langle j|e^{i\hat{H}_D t} = \sum_{n,s=0}^{1}\sum_{m,l=0}^{1} e^{-i(\lambda_{n,s}-\lambda_{m,l})t}\,\varrho_{n,s}\,\rho_j(0)\,\varrho_{m,l}. \tag{12}$$

A typical four level quantum oscillation pattern is present in this time evolution. For instance, a state initially in $|j\rangle$ evolves into a generic superposition, with conversion probability to a particular ionic level $|k\rangle \neq |j\rangle$ obtained through (12) [13].

The assignment (9) suggests the identification of two subsystems - $\mathcal{S}_F$, related to the total angular momentum quantum number F, and $\mathcal{S}_M$ associated to the projection of the angular momentum onto the direction of the projection M of the external magnetic field. In this framework, a state initially prepared in the internal state $|j\rangle$ will evolve to a superposition between the four internal states and shall exhibit quantum entanglement (and possibly other quantum correlations in the case of mixed states) between the two subsystems $\mathcal{S}_F$ and $\mathcal{S}_M$.

To describe more realistic setups, one is inclined to consider environment effects on the trapped ion dynamics. The inclusion of such open systems effects can be accomplished by an effective Hamiltonian $\hat{H}_{Env}$ describing environment coupling as to have a total Hamiltonian given by

$$\hat{H} = \hat{H}_j^{(A)JC} + \hat{H}_{Env}. \tag{13}$$

In particular, the influence of an environment is included through the influence of two *local* noises, generated by random fluctuations acting separately on each qubit which is given by the Hamiltonian [25]:

$$\hat{H}_{\mathrm{Env}}^{\mathrm{Loc}} = -\frac{1}{2}\mu\left(b_1(t)\,(\hat{\sigma}_z^{(1)}\otimes\hat{I}^{(2)}) + b_2(t)\,(\hat{I}^{(1)}\otimes\hat{\sigma}_z^{(2)})\right). \tag{14}$$

In this model, separate fluctuations act on the corresponding subsystem and the fields $b_i(t)$ are characterized by Markovian conditions

$$\langle b_i(t)\rangle = 0, \qquad \langle b_i(t)b_i(t')\rangle = \frac{\Gamma_i}{\mu^2}\delta(t-t'), \tag{15}$$

were $\langle\dots\rangle$ represent the ensemble average and Γ_i are the phase relaxations due to the local interactions. This setup was previously considered for study phonon decoherence [32], for protocols to measure magnetic field gradient [33] in the trapped ion framework, and for investigating the influence of noise on transport properties in chains of trapped ions [34]. Here it is supposed that the fluctuations on the degeneracy lifting magnetic field affect each degree of freedom individually. Other often considered noise model is the *global* noise [25] in which the random fluctuations act equally on both qubits. The influence of such environment will be investigated in a future issue.

3. Dynamics and Results

The temporal evolution driven by the total Hamiltonian $\hat{H}$ (13), can be recovered by decoupling it into two pieces through the interaction picture:

$$\tilde{\rho}(t) = e^{i\int_0^t \hat{H}_{\mathrm{Env}}(s)ds}\rho(0)e^{-i\int_0^t \hat{H}_{\mathrm{Env}}ds}, \tag{16}$$

with $\rho(0)$ being the initial state and $\tilde{\rho}(t)$ describing the time-evolved density matrix in the interaction picture, which is a solution of the corresponding master equation given in terms of the Kraus representation [7, 35, 36]. The inclusion of noise effects through Kraus representation is thus implemented by taking the statistical mean of Eq. (16) [25]. The behavior of $\tilde{\rho}(t)$ under the influence of the local noise described by the Hamiltonian (14) is then expressed in a compact way as

$$\tilde{\rho}(t) = \mathcal{E}_D(\rho(0)) = \sum_{\mu,\nu=1}^{2} E_\mu^\dagger(t)F_\nu^\dagger(t)\rho(0)F_\nu(t)E_\mu(t), \tag{17}$$

where the Kraus operators describing the local interaction (14) are given by:

$$\begin{aligned} E_1 &= \mathrm{diag}\{1,\gamma_1\}\otimes\hat{I}^{(2)},\ E_2 = \mathrm{diag}\{0,\omega_1\}\otimes\hat{I}^{(2)}, \\ F_1 &= \hat{I}^{(1)}\otimes\mathrm{diag}\{1,\gamma_2\},\ F_2 = \hat{I}^{(1)}\otimes\mathrm{diag}\{0,\omega_2\}, \end{aligned} \tag{18}$$

with $\gamma_i = e^{-\Gamma_i t/2}$ and $\omega_i = \sqrt{1-e^{-\Gamma_i t}}$. For simplicity we adopted equal phase relaxations on both subsystems, i.e. $\Gamma_1 = \Gamma_2 = \Gamma$.

After the above described procedure, to recover Schrödinger picture density matrix one uses the completeness relation for the eigenstates $\{\varrho_{n,s}\}$ as to have

$$\rho(t) = e^{i\hat{H}_D t}\tilde{\rho}(t)e^{-i\hat{H}_D t} = \sum_{n,s=0}^{1}\sum_{m,l=0}^{1}\sum_{\mu=1}^{3} e^{-i(\lambda_{n,s}-\lambda_{m,l})t}\,\varrho_{n,s}\,D_\mu^\dagger(t)\,\rho(0)\,D_\mu(t)\,\varrho_{m,l}, \tag{19}$$

from which any observable can be evaluated for any given initial state $\rho(0)$.

From now on, the disentangling properties of ionic states under local noise effects can be investigated. In particular we shall consider the evolution of the internal ionic state $|a(0)\rangle = |a\rangle$, and the evolution of the Schrödinger cat state $\rho_C(0) = |\psi_C\rangle\langle\psi_C|$ and of the Werner state $\rho_W(0) = |\psi_W\rangle\langle\psi_W|$, which are written in terms of internal ionic states as $|\psi_C\rangle = \frac{|a\rangle+|d\rangle}{\sqrt{2}}$ and $|\psi_W\rangle = \frac{|b\rangle+|c\rangle}{\sqrt{2}}$. The time evolutions of these states are recovered by Eq. (19) from which all correlational properties can be computed. Accordingly to the Peres criterion, any separable state must have all eigenvalues of its partial transpose density matrix positive [26]. The negativity of a state ϱ is thus defined by [27]

$$\mathcal{N}[\varrho] = ||\,\varrho_A^T\,|| - 1 = \sum_i |\mu_i| - 1, \tag{20}$$

where $||\,\varrho_1^T\,|| = \sum_i |\mu_i|$ stands for the trace norm of the partial transposed matrix ϱ_1^T, whose eigenvalues are $\{\mu_i\}$. Roughly speaking, negativity measures the extent to which the partial transpose fails to be positive.

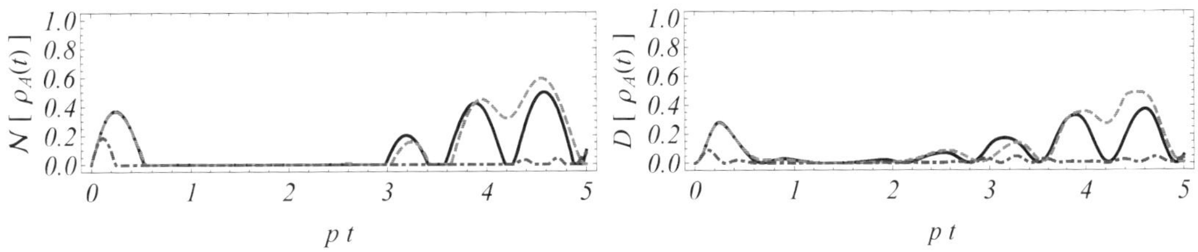

Figure 1. Entanglement (left plot) and quantum correlations (right plot) of the state $|a(t)\rangle$ as function of pt for $\kappa = \mu = 1$, $\mathcal{E}/p = 1$, $\Gamma/p = 1/2$ and for $m/p = 0$ (solid curves), 1 (dashed curves), 10 (dot-dashed curves). The influence of the local noise causes entanglement sudden death, followed by sudden revivals and quantum oscillations generated by the noiseless evolution of the state. Moreover, in the periods where the mixed state is separable, only residual quantum correlations are present, as depicted in the right plot.

Separable mixed states can exhibit quantum correlations other than entanglement [37], and the characterization of such quantum correlations is still an open problem. The quantum discord, defined as the difference between total correlations among two subsystems before and after the action of perfect local measurements in one of them, is a quantifier for nonclassical correlations [38]. The evaluation of quantum discord requires an extremization procedure over all possible sets of local measurements in one of the subsystems, being a NP-hard problem with no analytical solution or even efficient algorithm (unless $P = NP$) [39]. To avoid such computational issue, we shall adopt the geometric discord $D[\varrho]$, as a measure for quantum correlations [40]. Such quantity is defined as the minimum distance between the state ρ and the set of states with zero quantum discord. Given the Fano decomposition of a state ϱ

$$\varrho = \frac{1}{4}\left[\hat{I} + (\hat{\boldsymbol{\sigma}}^{(1)} \otimes \hat{I}^{(2)}) \cdot \boldsymbol{a}^{(1)} + (\hat{I}^{(1)} \otimes \hat{\boldsymbol{\sigma}}^{(2)}) \cdot \boldsymbol{a}^{(2)} + \sum_{i,j=1}^{3} t_{ij}\hat{\sigma}_i^{(1)} \otimes \hat{\sigma}_j^{(2)}\right],$$

geometric discord associated with the i subsystem of the state is analytically computed as [28]

$$D[\varrho]_{1(2)} = \frac{1}{4}\left(a_{1(2)}^2 + \parallel T \parallel^2 - k_{max}\right), \tag{21}$$

where $\parallel T \parallel^2 = Tr[TT^T]$, T being the matrix composed by the t_{ij} coefficients, and k_{max} is the largest eigenvalue of $\boldsymbol{a}_{1(2)}\boldsymbol{a}_{1(2)}^T + TT^T$. It was proved that $D[\varrho] = 0$ is a necessary and sufficient condition for vanishing quantum discord [28], and that the geometric discord and negativity satisfy the inequality $(\mathcal{N}[\varrho])^2 \leq D[\varrho]$ [41].

The correlational properties of the state $\rho_A(t) = |a(t)\rangle\langle a(t)|$ are depicted in Fig. 1 which shows the negativity (left plot) and the geometric discord (right plot) as function of pt for $\kappa = \mu = 1$, $\mathcal{E}/p = 1$, $\Gamma/p = 1/2$ and for $m/p = 0$ (solid curves), 1 (dashed curves), 10 (dot-dashed curves). The noiseless evolution drives the state to a superposition among all the internal states for $t > 0$, generating intrinsic entanglement. Nevertheless the influence of the local noise suddenly makes the entanglement vanish in an effect known as entanglement sudden death [42], which is followed by sudden revivals in a oscillation pattern without definite frequency. In the intervals where the state is separable, only residual quantum correlations persists, as depicted in the left plot of Fig. 1 and, similar to the negativity, geometric discord oscillates without a definite frequency. Moreover, large values of m/p corresponds to a stronger influence of the environment on the correlations.

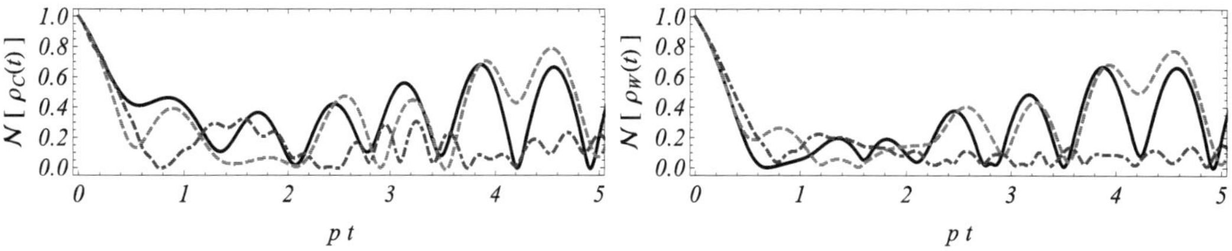

Figure 2. Werner (right plot) and cat (left plot) state entanglement under local noise as function of pt. The parameters and plot styles are in correspondence with those of Fig. 1. The local noise dissipates the entanglement in both states, but for low values of m/p entanglement oscillations are still present. For $t \gg 1/\Gamma$, both Werner and cat states are pure, being mixed only for intermediate values of pt.

Under local noise, both Werner and Schrödinger cat states suffer decoherence, as depicted in Fig. 2. Entanglement is partially dissipated, nevertheless, different from the previous case, no entanglement sudden death is observed and negativity oscillates without a definite periodicity. Even for $t \gg 1/\Gamma$, the states exhibit entanglement. Intermediate values of m/p characterize a stronger influence of the local noise on the entangling properties of such states, as depicted in the blue curves of Fig. 2, which correspond to $m/p = 10$. The Werner and Schrödinger cat states are, qualitatively, equally affected by the local noise, and both are pure states for $t \gg 1/\Gamma$.

4. Conclusions

The issue investigated throughout this letter is concerned with the inclusion of local noise effects, described by stochastic fields acting on four ionic levels, in the context of a Dirac-like dynamics simulated by the trapped ion systems. The entanglement properties were quantified in terms of negativity and the quantum correlations were quantified by means of the geometric discord, where the noiseless dynamics of ionic states was recovered through the correspondence between JC, AJC and carrier interactions through the inclusion of tensor and pseudotensor Dirac structures into the simulated Hamiltonian. The noise was introduced by means of the Kraus representation formalism, such that quantum correlations of states prepared as internal levels of the trapped ion were then investigated. The quantum correlational properties exhibited by a state initially on a single level, and by Werner and Schrödinger cat states formed by superpositions between two internal ionic levels were finally obtained.

Under local noise, for Werner and Schrödinger cat states, the entanglement is partially lost, but entanglement oscillations are still present due to the pureness of both states for $t \gg 1/\Gamma$. The mass parameter on the Dirac Hamiltonian actively suppresses the entanglement content of the states. For a state initially in the internal level a, entanglement shows sudden deaths and sudden revivals, which are driven by the coupled dynamics with the environment. For such state, the noise also suppress quantum correlations other than entanglement, and in the intervals where the state is separable it exhibits residual quantum correlations.

One concludes by noticing that such trapped ion platforms can work to implement more complex quantum simulations [43] where, from the beginning, the only observable that can be measured (through fluorescence techniques) still is $\hat{\sigma}_z$, related to issues that deserve some further investigations.

Acknowledgments

Acknowledgments - The work of AEB was supported by the Brazilian Agency FAPESP (grant 15/05903-4). The work of VASVB was supported by the Brazilian Agency CNPq (grant 140900/2014-4).

IOP Conf. Series: Journal of Physics: Conf. Series **880** (2017) 012063 doi:10.1088/1742-6596/880/1/012063

References

[1] Zurek W H 1991 *Phys. Today* **44** (10) 36; 1981 *Phys. Rev.* **D 24** 1516; 1982 *Phys. Rev.* **D 26** 1862
[2] Joos E and Zeh H D 1985 *Z. Phys.* **B 59** 223
[3] Braun D, Haake F and Strunz W T 2001 *Phys. Rev. Lett.* **86** 2913
[4] Anastopoulos C and Hu B L 2000 *Phys. Rev.* **A 62** 033821
[5] Shore B W 1990 *The Theory of Coherent Atomic Excitation*, vols. 1 and 2 (New York: Wiley), Chap. 23
[6] Gisin N, Ribordy G, Tittel W and Zbinden W 2002 *Rev. Mod. Phys.* **74** 145
[7] Nielsen M A and Chuang I L 2000 *Quantum Computation and Quantum Information* (England: Cambridge University Press)
[8] Gessner M, Ramm M, Pruttivarasin T, Buchleitner A, Breuer H P and Haeffner H 2014 *Nature Physics* **10** 105
[9] Ospelkaus C, Warring U, Colombe Y, Brown K R, Amini J M, Leibfried D and Wineland D J 2011 *Nature* **476** 181
[10] Bermudez A, Martin-Delgado M A and Solano E 2007 *Phys. Rev.* **A 76**, 041801(R)
[11] Tenev T G, Ivanov P A and Vitanov N V 2013 *Phys. Rev.* **A 87** 022103
[12] Lamata L, Casanova J, Gerritsma R, Roos C F, García-Ripoll J J and Solano E 2011 *New Journal of Physics* **13** 095003
[13] Bittencourt V A S V, Bernardini A E, Blasone M 2016 *Phys. Rev.* **A 93** 053823
[14] Schindler P, Müller M, Nigg D, Barreiro J T, Martinez E A, Hennrich M, Monz T, Diehl S, Zoller P and Blatt R 2013 *Nature Physics* **9** 361
[15] Barreiro J T, Müller M, Schindler P, Nigg D, Monz T, Chwalla M, Hennrich M, Roos C F, Zoller P and Blatt R 2011 *Nature* **470** 486
[16] Islam R *et al* 2011 *Nature Commun.* **2** 377
[17] Lamata L, León J, Schätz T and Solano E 2007 *Phys. Rev. Lett.* **98** 253005
[18] Casanova J, García-Ripoll J J, Gerritsma R, Roos C F and Solano E 2010 *Phys. Rev.* **A 82** 020101(R)
[19] Lee T E, Alvarez-Rodriguez U, Cheng X H, Lamata L and Solano E 2015 *Phys. Rev.* **A 92** 032129
[20] Bittencourt V A S V and Bernardini A E 2016 *Annals of Physics* **364** 182
[21] Bittencourt V A S V, Mizrahi S S and Bernardini A E 2015 *Annals of Physics* **355** 35
[22] Leibfried D, Blatt R, Monroe C and Wineland D 2003 *Rev. Mod. Phys.* **75** 281
[23] Kielpinski D, Monroe C and Wineland D J 2002 *Nature* **417** 709
[24] Wineland D J, Monroe C, Itano W M, Leibfried D, King B E and Meekhof D M 1998 *J. Res. Natl. Inst. Stand. Technol.* **103** 259
[25] Yu T and Eberly J H 2006 *Opt. Commun.* **264** 393; 2004 *Phys. Rev. Lett.* **93** 140404; 2003 *Phys. Rev.* **B 68** 165322
[26] Peres A 1995 *Phys. Rev. Lett.* **77** 8
[27] Vidal G and Werner R F 2002 *Phys. Rev.* **A 65** 032314
[28] Dakić B, Vedral V and Brukner C 2010 *Phys. Rev. Lett.* **105** 190502
[29] Liebfried D, Blatt R, Monroe C and Wineland D 2003 *Rev. Mod. Phys.* **75** 281
[30] Gerritsma R, Lanyon B P, Kirchmair G, Zähringer F, Hempel C, Casanova J, García-Ripoll J J, Solano E, Blatt R and Roos C F 2011 *Phys. Rev. Lett.* **106** 060503
[31] Bernardini A E and Mizrahi S S 2014 *Physica Scripta* **89** 075105
[32] Yu T and Eberly J H 2002 *Phys. Rev.* **B 66** 193306
[33] Ng H T and Kim K 2014 *Optics Communication* **331** 353-358
[34] Cormick C and Schmiegelow C T 2016 *Phys. Rev.* **A 94** 053406
[35] Kraus K 1983 *States, Effects, and Operations: Fundamental Notions in Quantum Theory* (Berlin: Springer-Verlag)
[36] Wódkiewicz K 2001 *Opt. Express* **8** 145
[37] Henderson L and Vedral V 2001 *J. Phys. A: Math. Gen.* **34** 6899-6905
[38] Ollivier H and Zurek W H 2001 *Phys. Rev. Lett.* **88** 017901
[39] Huang Y 2014 *New Journal of Physics* **16** 033027
[40] Dakić B, Vedral V and Brukner C 2010 *Phys. Rev. Lett.* **105** 190502
[41] Girolami D and Adesso G 2011 *Phys. Rev.* **A 84** 052110
[42] Diósi L 2003 *Lect. Notes Phys.* **622** 157-163
[43] Gerritsma R, Kirchmais G, Zahringer F, Solando E, Blatt R and Ross C F 2010 *Nature* **463** 68

IOP Conf. Series: Journal of Physics: Conf. Series **880** (2017) 012064 doi:10.1088/1742-6596/880/1/012064

Effect of boundaries on vacuum field fluctuations and radiation-mediated interactions between atoms

F Armata[1], S Butera[2], G Fiscelli[3], R Incardone[4,5], V Notararigo[6], R Palacino[3], R Passante[3,7], L Rizzuto[3,7], S Spagnolo[3]

[1] QOLS, Blackett Laboratory, Imperial College London, London SW7 2AZ, UK
[2] SUPA, Institute of Photonics and Quantum Sciences, Heriot-Watt University, Edinburgh, UK
[3] Dipartimento di Fisica e Chimica, Università degli Studi di Palermo, Via Archirafi 36, I-90123 Palermo, Italy
[4] 4th Institute for Theoretical Physics, Universität Stuttgart, 70569 Stuttgart, Germany
[5] Max Planck Institute for Intelligent Systems, 70569 Stuttgart, Germany
[6] Department of Physics and Astronomy, University College London, Gower Street, London WC1E 6BT, UK
[7] INFN, Laboratori Nazionali del Sud, I-95123 Catania, Italy

E-mail: lucia.rizzuto@unipa.it

Abstract. In this paper we discuss and review several aspects of the effect of boundary conditions and structured environments on dispersion and resonance interactions involving atoms or molecules, as well as on vacuum field fluctuations. We first consider the case of a perfect mirror, which is free to move around an equilibrium position and whose mechanical degrees of freedom are treated quantum mechanically. We investigate how the quantum fluctuations of the mirror's position affect vacuum field fluctuations for both a one-dimensional scalar and electromagnetic field, showing that the effect is particularly significant in the proximity of the moving mirror. This result can be also relevant for possible gravitational effects, since the field energy density couples to gravity. We stress that this interaction-induced modification of the vacuum field fluctuations can be probed through the Casimir-Polder interaction with a polarizable body, thus allowing to detect the effect of the mirror's quantum position fluctuations. We then consider the effect of an environment such as an isotropic photonic crystal or a metallic waveguide, on the resonance interaction between two entangled identical atoms, one excited and the other in the ground state. We discuss the strong dependence of the resonance interaction with the relative position of the atomic transition frequency with the gap of the photonic crystal in the former case, and with the cut-off frequency of waveguide in the latter.

1. Introduction

A peculiar aspect of the quantum theory is the presence of unavoidable quantum fluctuations of the fields, even in the ground state of the system [1]. Field fluctuations have striking consequences in quantum electrodynamics; for example: the Lamb shift, the anomalous magnetic moment of the electron, Casimir and Casimir-Polder forces, vacuum polarization [2, 3]. In quantum field theory, the vacuum state is not *void*, but contains virtual particles or quanta which give a sort of structure to the empty space [4] and have observable effects on matter. On the other hand, the presence of matter, i.e. microscopic objects such as atoms or molecules as well as macroscopic objects, can change the physical features of vacuum fluctuations and ultimately of the vacuum state. Thus, macroscopic objects as fixed or moving metallic or dielectric plates or structured materials such as photonic crystals or photonic crystals waveguides can modify vacuum field fluctuations and the properties of the vacuum [5, 6], yielding striking observable effects such as the Casimir effect, that is a force between two neutral conducting plates in the vacuum, and Casimir-Polder forces [7, 8, 9]. Moreover, this allows us to modify and tailor radiative processes that are related to vacuum fluctuations, for example spontaneous emission [10, 11, 12], radiative corrections [13] and radiation-mediated interaction between atoms and molecules (for example, van der Waals/Casimir-Polder [14, 15, 16, 17] and resonance interactions [18, 19]).

In this paper we shall discuss and review some aspects related to the role of fixed or moving boundaries on vacuum field energy densities and Casimir interactions, as well as the influence of a photonic bandgap environment or a metallic waveguide on the resonance (radiation-mediated) interaction between two entangled identical atoms, one excited and the other in the ground state. Specifically, we will first consider vacuum field energy densities near a boundary that is free to move and whose mechanical degrees of freedom are treated quantum mechanically, thus subjected to quantum fluctuations of its position. Compared to the case of a fixed configuration, the effect of the boundary's position fluctuations is mainly concentrated in the very proximity of the boundary itself [20, 21] and yields a smearing out of the otherwise divergent field energy density at the position of the boundary [22, 23]. This fact not only affects the Casimir-Polder force on a polarizable body placed near the wall, but, in principle, can be relevant from a gravitational point of view, because the energy density couples to the gravitational field. We will then consider a photonic bandgap environment, in particular a photonic crystal, and discuss how it can alter the resonance interaction between two atoms embedded in the crystal [19, 24]. Finally, we briefly mention the case of the resonance interaction between two atoms placed inside a metallic cylindrical waveguide.

This paper is organized as follows. In section 2 we consider a one-dimensional cavity with a mobile wall described quantum mechanically and discuss how quantum fluctuations of its position affect the field energy density inside the cavity and the Casimir-Polder interaction energy with a polarizable body placed in the proximity of the fluctuating wall, as well as the observability of this effect. In section 3 we discuss the effect of an external environment such as a photonic crystal or a metallic cylindric waveguide on the resonance interaction between two entangled atoms, mediated by the radiation field. Last Section is devoted to our conclusive remarks.

2. Vacuum fluctuations near a fluctuating wall and Casimir-Polder interactions

We first consider a one-dimensional massless scalar field confined between two perfectly reflecting mirrors placed at $x = 0$ and $x = L(t)$. The mirror at $x = 0$ has a fixed position, while we allow the other mirror of mass M to move in space, assuming it bounded by a harmonic potential of frequency ω_{osc} around an equilibrium position L_0. We assume Dirichlet boundary conditions for the scalar field operator at the boundaries, i.e. $\phi(0,t) = \phi(L(t),t) = 0$. The mechanical degrees of freedom of the moving mirror are treated quantum mechanically; this permits us to include quantum position fluctuations of the mirror. We find that their effect is particularly relevant for small mirror's masses, such as those attainable in modern optomechanics experiments [25]. Since the mobile mirror can oscillate, an interaction between the cavity field modes and the mirror's mechanical degrees of freedom (phonons) is generated, as well as an effective interaction between field modes, mediated by the moving mirror. Our system is described by the following effective Hamiltonian, valid for a small displacement of the moving mirror from its equilibrium position [26, 27]

$$H = \hbar\omega_{osc}b^\dagger b + \hbar\sum_k \omega_k a_k^\dagger a_k - \sum_{kj} C_{kj}\Big\{\left(b + b^\dagger\right)\mathcal{N}\Big[\left(a_k + a_k^\dagger\right)\left(a_j + a_j^\dagger\right)\Big]\Big\}, \qquad (1)$$

where a_k, $a_k^\dagger$ are annihilation and creation operators of the cavity field modes relative to the equilibrium position L_0 with wavenumber k, b, $b^\dagger$ are annihilation and creation operators relative to the mobile wall (mechanical excitations), $\mathcal{N}$ is the normal ordering operator, and

$$C_{kj} = (-1)^{k+j} \left(\frac{\hbar}{2}\right)^{3/2} \frac{1}{L_0\sqrt{M}} \sqrt{\frac{\omega_k \omega_j}{\omega_{osc}}} \tag{2}$$

is the effective coupling constant between the moving mirror and the field. In our 1D model, the field modes are equally spaced in frequency and the allowed frequencies are $\omega_j = ck_j$, where $k_j = j\pi/L_0$, with j an integer number.

We now consider the field energy density inside the cavity. The corresponding operator is

$$\mathcal{H}(x) = \frac{1}{2}\left[\frac{1}{c^2}\dot{\phi}^2(x) + \left(\frac{d\phi(x)}{dx}\right)^2\right]. \tag{3}$$

After subtraction of the (divergent) energy density that would be present even in absence of the wall, in the case of fixed walls its average value on the ground state $|\{0_k\}\rangle$ of the field, is $\langle\{0_k\}|\mathcal{H}|\{0_k\}\rangle = -\pi c\hbar/(24L_0^2)$, yielding a constant value inside the cavity. At the position of the two fixed walls, extra divergent terms are present, as in the case of the vacuum electromagnetic field fluctuations [22, 28, 29]. In the case of fluctuating boundaries, such divergences can be smeared out by an average on the probability distribution function of the mirror's position [22, 23]. We now consider how the field energy density changes when we allow one of the mirrors to move around its equilibrium position, using the Hamiltonian (1). Due to the interaction term in (1), the bare ground state $|\{0_k\}, 0\rangle$ of the field-mobile wall system is not an eigenstate of the total Hamiltonian. The true (dressed) ground state can be obtained by first-order perturbation theory, and it contains an admixture of states with a mechanical excitation of the mirror and pairs of virtual excitations of the cavity field. At the lowest order in the mirror-field interaction, the dressed ground state is

$$|g\rangle = |\{0_k\}, 0\rangle + \sum_{kj} D_{kj} |\{1_k, 1_j\}, 1\rangle, \tag{4}$$

where

$$D_{kj} = (-1)^{k+j} \frac{1}{L_0} \sqrt{\frac{\hbar\omega_k\omega_j}{8M\omega_{osc}}} \left(\frac{1}{\omega_{osc} + \omega_k + \omega_j}\right). \tag{5}$$

In Eq. (4), the elements inside the curly bracket in the state vectors refer to field excitations, while the other element refers to phonons, that is excitations of the wall's mechanical degrees of freedom.

We can now evaluate the average value of the *renormalized* energy density on the dressed state (4), namely after subtraction of its space-independent value in the absence of the walls given above, obtaining [20]

$$\langle g|\mathcal{H}(x)|g\rangle_R = \frac{\hbar}{2L_0^3 M\omega_{osc}} \sum_{j\ell r} (-1)^{\ell+r} \cos[(q_\ell - q_r)x] \frac{\omega_\ell \omega_j \omega_r}{(\omega_{osc} + \omega_\ell + \omega_j)(\omega_{osc} + \omega_r + \omega_j)}, \tag{6}$$

where $q_\ell = \ell\pi/L_0$, with ℓ an integer number. To evaluate (6), it is necessary to introduce an upper cutoff frequency ω_{cut}, related to a sort of plasma frequency of the cavity walls. To simplify the numerical calculation, we use a sharp cutoff at the frequency ω_{cut}, thus considering a finite number of field modes in the cavity.

Using the dressed ground state (4) we can also evaluate the expectation value of the renormalized field correlation function for the scalar field, that in a successive part of this Section

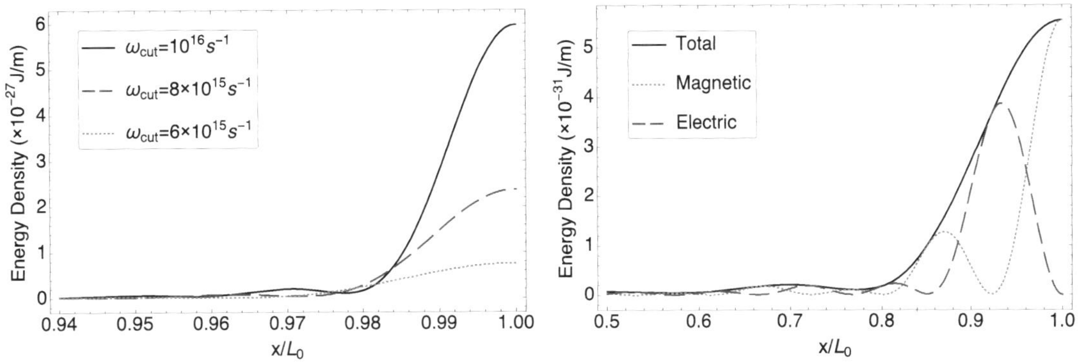

(a) (b)

Figure 1. Change of the renormalized field energy density, compared to the static walls case, in the proximity of the moving mirror. In a) the three curves differ for the value of the cutoff frequency: $\omega_{cut} = 6 \cdot 10^{15}\,\mathrm{s}^{-1}$ (red short-dashed line) , $\omega_{cut} = 8 \cdot 10^{15}\,\mathrm{s}^{-1}$ (blue long-dashed line) and $\omega_{cut} = 10^{16}\,\mathrm{s}^{-1}$ (black continuous line). The field energy density change is concentrated near the mobile wall and grows when the cut-off frequency is increased. b) shows the change of the field energy density (black continuous line), and its electric (blue dashed line) and magnetic (red dotted line) components, due to the motion of the wall. The cutoff frequency has been set to $\omega_{cut} = 10^{15}\mathrm{s}^{-1}$. In all plots we have used: $\omega_{osc} = 10^5\,\mathrm{s}^{-1}$, $L_0 = 10\,\mu\mathrm{m}$, $M = 10^{-11}\,\mathrm{kg}$.

we will use to extend our results for the scalar energy density to the case of the one-dimensional electromagnetic field. Writing explicitly space and time components, the renormalized scalar field propagator on the dressed vacuum state (4) is given by [21]

$$\begin{aligned} G_R(x,t;x',t') &= \langle g|\phi(x,t)\phi(x',t')|g\rangle_{bc} - \langle\{0_r\}|\phi(x,t)\phi(x',t')|\{0_r\}\rangle_{un} \\ &= \left(\sum_p \frac{\hbar c^2}{L_0\omega_p} e^{-i\omega_p(t-t')}\sin(k_p x)\sin(k_p x') - \int \frac{dp}{2\pi}\frac{\hbar c^2}{2\omega_p} e^{-i\omega_p(t-t')} e^{ik_p(x-x')}\right) \\ &+ 8\sum_{j\ell r} \frac{\hbar c^2}{L_0(\omega_j\omega_r)^{1/2}} D_{j\ell}D_{\ell r}\left[\cos(\omega_j t - \omega_r t')\right]\sin(k_j x)\sin(k_r x'), \end{aligned} \quad (7)$$

where subscripts bc and un refer respectively to the bounded (Dirichlet) and unbounded case. The quantity in the second line of (7) coincides with the renormalized propagator of a fixed-wall cavity, while the term in the third line is the change consequent to quantum motion of the mobile wall around its equilibrium position.

Figure 1(a) shows the change of the renormalized energy density of the field (6) in the proximity of the mobile wall, for three different cut-off frequencies, specifically $\omega_{cut} = 10^{16}\,\mathrm{s}^{-1}$, $\omega_{\mathrm{cut}} = 8 \cdot 10^{15}\,\mathrm{s}^{-1}$, $\omega_{\mathrm{cut}} = 6 \cdot 10^{15}\,\mathrm{s}^{-1}$. This figure clearly shows that the motion of the mobile wall mainly affects the field energy density in its very proximity and is more relevant increasing the cut-off frequency. The values for the physical parameters used are $\omega_{osc} = 10^5\,\mathrm{s}^{-1}$, $L_0 = 10\,\mu\mathrm{m}$ and $M = 10^{-11}\,\mathrm{kg}$, which are the typical values of commercial microelectromechanical systems (MEMS). It should be noted that the effect scales as M^{-1}. Although in the limit of infinite upper cut-off frequency the energy density diverges at the boundaries [28], it is possible to shows that averaging over the probability distribution of the position of the mobile wall around its equilibrium position at L_0, smears out the divergence [22, 23].

Using the known relation between the electromagnetic vacuum stress tensor and the renormalized scalar propagator [30], and the possibility to express the electromagnetic field propagator in terms of that for the scalar field [31], our results can be generalized to the case of the one-dimensional electromagnetic field. By assuming a one-dimensional model with the cavity length along the x axis, magnetic and electric fields respectively along y and z, the renormalized electric and magnetic energy densities result,

$$\langle E_z^2(x)\rangle = \frac{1}{c^2}\lim_{(x',t')\to(x,t)} \partial_t\partial_{t'}\langle G_R(x,t;x',t')\rangle = \langle E_z^2(x)\rangle_0 + \langle E_z^2(x)\rangle_1, \tag{8}$$

$$\langle B_y^2(x)\rangle = \lim_{(x',t')\to(x,t)} \partial_x\partial_{x'}\langle G_R(x,t;x',t')\rangle = \langle B_y^2(x)\rangle_0 + \langle B_y^2(x)\rangle_1, \tag{9}$$

where the subscript 0 indicates the fixed-wall contribution, while 1 the modification induced by the quantum fluctuactions of the boundary.

Using (7), (8) and (9), we obtain an analytical expression of the average value of the contribution of the wall's motion to the field energy density in the cavity, $[\langle E_z^2(x)\rangle_0+\langle E_z^2(x)\rangle_1+\langle B_y^2(x)\rangle_0+\langle B_y^2(x)\rangle_1]/2$, after introducing an upper cut-off frequency. Figure 1(b) shows the energy density as a function of the position in the proximity of the mobile wall, as well as its electric and magnetic components, using $\omega_{osc} = 10^5\,\mathrm{s}^{-1}$, $L_0 = 10\,\mu\mathrm{m}$, $M = 10^{-11}\,\mathrm{kg}$ and $\omega_{cut} = 8\cdot 10^{15}\,\mathrm{s}^{-1}$. As it can be inferred from the figure, also in the electromagnetic case, the changes are significant mainly close to the mobile wall. This is indeed expected because the pairs of virtual photons emitted by the mobile cavity wall remain confined close to the wall.

The electric and magnetic energy densities can be probed by using a non-dispersive electrically and magnetically polarizable body placed in the cavity at position x_0, whose interaction energy with the field fluctuations is given by [32]

$$\Delta E_{int} = -\frac{1}{2}\alpha_E\langle E^2(x_0)\rangle - \frac{1}{2}\alpha_M\langle B^2(x_0)\rangle, \tag{10}$$

where α_E and α_M are respectively the static electric and magnetic polarizability of the probe. Therefore the changes of the electric and magnetic energy densities induced by the wall motion can be measured through an appropriate polarizable body, allowing to detect the effect we have obtained. These results can be also generalized to a massless scalar field in a three-dimensional cavity with a mobile wall [21].

3. Resonance interaction between two entangled atoms in an external environment

We now briefly review some recent results we have obtained on the resonance interaction (mediated by the electromagnetic radiation field) between two identical atoms in the presence of an external environment, specifically a photonic crystal. In the multipolar coupling scheme and in the dipole approximation, our system of two identical atoms A and B interacting with the quantum electromagnetic field is described by the following Hamiltonian [2]

$$H = H_A + H_B + \sum_{\mathbf{k}j} \hbar\omega_k a^\dagger_{\mathbf{k}j}a_{\mathbf{k}j} - \boldsymbol{\mu}_A\cdot\mathbf{E}(\mathbf{r}_A) - \boldsymbol{\mu}_B\cdot\mathbf{E}(\mathbf{r}_B), \tag{11}$$

where H_A and H_B are the Hamiltonian of atoms A and B with dipole moment operators $\boldsymbol{\mu}_A$ and $\boldsymbol{\mu}_B$ and position $\mathbf{r}_A$ and $\mathbf{r}_B$, respectively. We use the two-level approximation with one atom in the excited state $|e\rangle$ and the other in the ground $|g\rangle$, whose energy difference is $\hbar\omega_a$: we consider the atoms in their symmetric or antisymmetric state and the field in the vacuum

$$|\psi\rangle_\pm = \frac{1}{\sqrt{2}}\left(|g_A, e_B, \{0_{\mathbf{k}j}\}\rangle \pm |e_A, g_B, \{0_{\mathbf{k}j}\rangle\}\right), \tag{12}$$

The excitation is thus shared between the two atoms. These are also called superradiant (symmetric) and subradiant (antisymmetric) states, and the effect of a perfectly conducting plate on their spontaneous decay rate has been recently considered as a function of the distance of the two atoms from the plate [33]. Due to the interaction term in (11), the state (12) is not an eigenstate of the Hamiltonian and we can evaluate the resulting energy shift by second-order perturbation theory, assuming a weak coupling regime. In general the expression of the second-order energy shift depends on the distance between the atoms and gives rise to the so-called resonance interaction energy, thus to a force between the atoms. In the vacuum space, the resonance interaction, due to the exchange of one real or virtual photon between the atoms, is given by [19, 34]

$$\Delta E_{\pm}^{vac}(r) = \mp(\mu_A^{ge})_i(\mu_B^{eg})_j \left(-\nabla^2\delta_{ij} + \nabla_i\nabla_j\right) \frac{\cos(k_a r)}{r}$$
$$= \pm\frac{(\mu_A^{ge})_i(\mu_B^{eg})_j}{r^3} \left[(\delta_{ij} - 3\hat{\mathbf{r}}_i\hat{\mathbf{r}}_j)(\cos(k_a r) + k_a r\sin(k_a r)) - (\delta_{ij} - \hat{\mathbf{r}}_i\hat{\mathbf{r}}_j)\, k_a^2 r^2 \cos(k_a r)\right], \quad (13)$$

where $k_a = \omega_a/c$ and $\mathbf{r} = \mathbf{r}_B - \mathbf{r}_A$ is the distance between the two atoms, and the upper/lower sign refers to the symmetric/antisymmetric state, respectively. In vacuum space, the resonance interaction exhibits a behaviour as r^{-3} in the *near* zone $r \ll k_a^{-1}$ and as r^{-1} in the *far* zone $r \gg k_a^{-1}$. It is thus a very long range interaction which can be relatively intense, as it is a second-order effect, contrarily to the fourth-order van der Waals and Casimir-Polder dispersion interactions between two atoms. However, in order to observe this interaction, it is necessary to preserve the fragile entangled state (12) for a sufficiently long time, and this can be a formidable task from the experimental point of view.

The resonance interaction energy, similarly to other radiative processes, can be changed and tailored through external boundaries or a structured environment. An important example of a structured environment is a photonic crystal, that is a periodic array of dielectric slabs with different refractive index yielding a photonic bandgap, where the photonic density of states vanishes [12, 18, 35, 36]. Photonic crystals are a powerful tool to manipulate and tailor many radiative processes, even in the strong-coupling regime [37]. In our analysis we have considered both the one-dimensional case and the isotropic three-dimensional case, where, in the latter, the one-dimensional dispersion relation is assumed to be valid, regardless of the photon propagation direction. In both cases the photonic density of states outside the forbidden gap, and in the proximity of its edges, can be approximated by a quadratic dispersion relation (effective mass approximation), $\omega_k = \omega_g \mp A(k - k_0)^2$, where ω_g can be the lower gap frequency ω_l or the upper one ω_u, occurring at the gap wavenumber k_0, A is a positive constant and the minus or plus sign applies respectively below the lower edge or above the upper edge of the gap. The quantities $\omega_g = \omega_l\,(\omega_u), k_0, A$ depend on the physical properties of the crystal, for example periodicity, dielectric constants, etc. This dispersion relation yields a change of the photon density of states and thus a possible modification of any radiation-mediated physical process: in the one-dimensional and isotropic three-dimensional cases, the density of states vanishes inside the bandgap and has a strong peak in the regions external to the gap and in the proximity of its edges (see [38], for example).

We now report some of our results on the resonance interaction between two two-level quantum emitters (atoms or quantum dots) embedded in an isotropic three-dimensional photonic crystal, in two specific cases: (i) when their transition frequency is outside the gap and not far from its upper edge [19]; (ii) when it is inside the forbidden gap and in the proximity of its lower edge [24]. We use in both cases the effective mass approximation with the appropriate interpolation $\omega_l = Ak_0^2$, which is necessary to obtain the correct linear dispersion relation for very low-frequency photons, i.e. having wavelength larger than the periodicity of the crystal: in this case, indeed, the photon propagation is essentially the same as in free space. Both results have been obtained by second-order perturbation theory.

In the case (i), we obtain

$$\Delta E_{\pm} = \pm \frac{\omega_u}{2\sqrt{A(\omega_a - \omega_u)}} (\mu_A^{ge})_i (\mu_B^{eg})_j \left(-\nabla^2 \delta_{ij} + \nabla_i \nabla_j \right) \frac{\cos(k_0 r)}{k_0 r}. \quad (14)$$

Comparison of (14) with (13) shows that the interaction energy scales with the distance as r^{-1} asymptotically ($r \gg k_a^{-1} \sim k_0^{-1}$), as in the vacuum space; however in the photonic crystal case there is the extra numerical factor $\omega_u/(2\sqrt{A(\omega_a - \omega_u)})$. When the transition frequency is close to the edge of the gap, but not so close to require a nonperturbative calculation, this factor can provide a significant increase of the resonance interactions energy and force, up to a factor 10^3 for realistic values of the parameters involved [19]. It must be said, however, that in such situations also the spontaneous emission rate is increased, making the correlated state more fragile and difficult to preserve in time. Similar results are obtained when ω_a is below the lower edge ω_l of the gap.

On the other hand, in the case (ii), that is when the atomic transition frequency ω_a is inside the gap and in the proximity of its lower edge ω_l, an explicit evaluation of the resonance interaction energy yields a scaling with the distance as r^{-2} in the far zone, rather than the r^{-1} behaviour obtained in the vacuum space. Using typical values for the parameters of a photonic crystal and an optical transition frequency of the two identical atoms, we find that, although the interaction energy can be reduced asymptotically with respect to the free-space case, it is still significant and much stronger than dispersion interactions between atoms [24]. However, contrarily to the case (i), spontaneous emission is strongly suppressed by the presence of the photonic crystal, making more stable the entangled state (12), thus improving the possibility to detect the elusive resonance interatomic interaction.

Another structured environment we have recently investigated to control the resonance interatomic interaction is a cylindric metallic waveguide, when the two entangled atoms are placed on the axis of the guide. In this case the main difference is the presence of a lower cut-off frequency of the guide, directly related to its diameter. This yields an asymptotical exponential behavior of the interaction with the distance when the atomic transition frequency is below the waveguide's cut-off frequency [39].

All the results reported in this section prove that a structured environment, for example a photonic crystal or a waveguide, allows to control and tailor the resonance interaction between two entangled atoms, permitting to increase or reduce the strength of the interaction, as well as controlling the collective spontaneous emission rate.

Conclusions

In this paper we have reviewed some recent results related to the effect of boundaries on several radiative processes, in particular a fluctuating reflecting plate or a photonic crystal. We have first considered a one-dimensional model of a massless scalar field in a cavity with a mobile wall, whose degrees of freedom have been treated quantum mechanically, thus allowing the possibility of quantum fluctuations of its position. We have then extended this model to the case of the one-dimensional electromagnetic field. We have discussed how the position fluctuations of the wall affect the field energy density in the cavity and the related Casimir interaction energies with a polarizable body. We have found significant changes in the very proximity of the mobile wall, which are particularly relevant for masses of the wall as low as those currently realizable in state-of-the-art optomechanics experiments. Successively, we have considered the resonance interaction energy between two atoms or quantum dots embedded in a photonic bandgap material such as a photonic crystal, showing how the presence of the crystal significantly changes the features of the interaction, depending on the position of the atomic transition frequency with respect to the edges of the bandgap. We have also discussed how this setup is promising in the perspective of the direct observation of the elusive quantum resonance interaction between atoms. Finally, we have briefly mentioned preliminary results obtained for the resonance interaction when the two quantum emitters are placed within a cylindrical metallic waveguide.

Acknowledgments
F.A. acknowledges the Marie Curie Actions of the EUs 7th Framework Program under REA [grant number 317232] for their financial support. S.B. acknowledges support from the EPSRC CM-CDT Grant No. EP/403673X/1.

References
[1] Milonni P W 1994 *The Quantum Vacuum: An Introduction to Quantum Electrodynamics* (San Diego, CA: Academic Press)
[2] Compagno G, Passante R and Persico F 1995 *Atom-Field Interactions and Dressed Atoms* (Cambridge: Cambridge University Press)
[3] Buhmann S Y 2012 *Dispersion Forces I and II* (Heidelberg: Springer)
[4] Aitchison I J R 1985 *Contemp. Phys.* **26** 333
[5] Simpson W M R and Leonhardt U 2015 *Forces of the Quantum Vacuum. An Introduction to Casimir Physics* (Singapore: World Scientific Publishing Co.)
[6] Lamoreaux S K 2005 *Rep. Prog. Phys.* **68** 201
[7] Casimir H B G 1948 *Proc. K. Ned. Akad. Wet.*, Proc. K. Ned. Akad. Wet. **51** 793
[8] Casimir H B G and Polder D 1948 *Phys. Rev.* **73** 360
[9] Milonni P W 2007 *Phys. Scr.* **76** C167
[10] Purcell E M 1946 *Phys. Rev.* **69** 674
[11] Kleppner D 1981 *Phys. Rev. Lett.* **47** 233
[12] Yablonovitch E 1987 *Phys. Rev. Lett.* **58** 2059
[13] Wang X H, Kivshar Y S and Gu B Y 2004 *Phys. Rev. Lett.* **93** 073901
[14] Haakh H R and Scheel S 2015 *Phys. Rev. A* **91** 052707
[15] Hood J D, Goban A, Asenjo-Garcia A, Lu M, Yu S P, Chang D E and Kimble H J 2016 *Proc. Natl. Acad. Sci.* **113** 10507
[16] Messina R, Passante R, Rizzuto L, Spagnolo S and Vasile R 2008 *J. Phys. A: Math. Theor.* **41** 164031
[17] Shahmoon E, Mazets I and Kurizki G 2014 *Proc. Natl. Acad. Sci.* **111** 10485
[18] El-Ganainy R and John S 2013 *New J. Phys.* **15** 083033
[19] Incardone R, Fukuta T, Tanaka S, Petrosky T, Rizzuto L and Passante R 2014 *Phys. Rev. A* **89** 062117
[20] Butera S and Passante R 2013 *Phys. Rev. Lett.* **111** 060403
[21] Armata F and Passante R 2015 *Phys. Rev. D* **91** 025012
[22] Ford L H and Svaiter N F 1998 *Phys. Rev. D* **58** 065007
[23] Bartolo N, Butera S, Lattuca M, Passante R, Rizzuto L and Spagnolo S 2015 *J. Phys. Condens. Matter* **27** 214015
[24] Notararigo V, Passante R and Rizzuto L 2016 *Preprint* quant-ph/1512.05253
[25] Aspelmeyer M, Kippenberg T J and Marquardt F 2014 *Rev. Mod. Phys.* **86** 1391
[26] Law C K 1994 *Phys. Rev. A* **49** 433
[27] Law C K 1995 *Phys. Rev. A* **51** 2537
[28] Sopova V and Ford L H 2002 *Phys. Rev. D* **66** 045026
[29] Bartolo N and Passante R 2012 *Phys. Rev. A* **86** 012122
[30] Deutsch D and Candelas P 1979 *Phys. Rev. D* **20** 3063
[31] Brown L S and Maclay G J 1969 *Phys. Rev.* **184** 1272
[32] Passante R, Power E A and Thirunamachandran T 1998 *Phys. Lett. A* **249** 77
[33] Palacino R, Passante R, Rizzuto L, Barcellona P and Buhmann S Y 2017 *preprint submitted to J. Phys. B*
[34] Salam A 2010 *Molecular Quantum Electrodynamics* (Hoboken, NJ: John Wiley & Sons)
[35] John S 1987 *Phys. Rev. Lett.* **58** 2486
[36] Angelakis D G, Knight P L and Paspalakis E 2004 *Contemp. Phys.* **45** 303
[37] Liu Y and Houck A A 2017 *Nature Phys.* **13** 48
[38] Lambropoulos P, Nikolopoulos G M, Nielsen T R and Bay S 2000 *Rep. Prog. Phys.* **63** 455
[39] Fiscelli G, Rizzuto L and Passante R 2017 *Preprint in preparation*

Violation of Bell's inequality: Must the Einstein locality really be abandoned?

Kurt Jung
Fachbereich Physik, Technische Universität Kaiserslautern,
Erwin-Schrödinger-Str. 56, D-67663 Kaiserslautern, Germany

E-mail: jung@physik.uni-kl.de

Abstract.
Since John Bell has established his famous inequality and several independent experiments have confirmed the distinct polarization correlation of entangled photons predicted by quantum mechanics it is evident that quantum mechanics cannot be explained by local realistic theories. Actually, the observed polarization correlation can be deduced from wave optical considerations. The correlation has its origin in the phase coupling of the two circularly polarized wave packets leaving the photon source simultaneously. The experimental results violate Bell's inequality although no non-local interactions have to be assumed. In consequence the principle of locality remains valid in the scope of quantum mechanics. However, the principle of realism has to be replaced by the less stringent principle of contextuality.

1. Introduction

In 1964 John Bell [1] presented an inequality allowing physicists to discriminate between quantum mechanics and local realistic theories. Bell showed that spin correlations of entangled particles derived from local realistic theories must be less pronounced than predicted by quantum mechanics. When experiments especially those with entangled photons confirmed the quantum mechanical results [2, 3, 4] local realistic theories were ruled out.

When A. Khrennikov [5] discusses the dilemma whether locality or objectivity should be abandoned he comes to the conclusion that quantum mechanics is nonobjective in the sense that the values of physical observables cannot be assigned to a quantum system before the measurement process has taken place. In Khrennikov's opinion objectivity has to be replaced by contextuality.

B.-G. Englert concluded in his article on quantum theory [6] from considerations on statistical operators that quantum mechanics does not require any non-local action in order to reproduce the polarization correlations of entangled photons. Therefore it might be wise to query the concept of realism or the closely related concept of objectivity. Neither Khrennikov nor Englert propose a mechanism leading to the observed polarization correlations.

The best evidence for the violation of Bell's inequality has been achieved by the measurements of Aspect *et al* [2, 3] and Weihs *et al* [4]. In the experiments of Aspect and coworkers the linear polarizations of the entangled photons are correlated. In contrast, in the experiment of Weihs *et al* the linear polarizations are anti-correlated. In this article mainly the experiments of Weihs *et al* will be discussed. The experiments of Aspect *et al* [2, 3] will be only shortly touched upon at the end.

With respect to the experiment of Weihs *et al* not only the experimentally approved polarization correlations on singlet states will be deduced from wave optical considerations but also polarization correlations of entangled photon pairs in triplet configuration. When future experiments with down conversion sources will confirm the predicted results for triplet states it will be a strong argument in favour of the wave optical considerations presented in this article.

2. Symmetries of photon source and detector systems

By means of an optically nonlinear BBO crystal Weihs *et al* [4] convert ultraviolet photons into entangled pairs of green photons. The two electromagnetic wave packets associated with the photons are simultaneously emitted in (nearly) the same direction. The senses of rotation of the photons leaving such a parametric down conversion source are statistically distributed. Photon pairs in triplet configuration are always rotating in the same sense and photon pairs in singlet configuration are always rotating in opposite sense.

After the photons have passed a half wave plate in order to exchange the vertically polarized ordinary and the horizontally polarized extra-ordinary beam the two wave packets traverse separate compensating BBO plates which can be tilted with respect to a horizontal axis perpendicular to the propagation direction of the photons. The compensating plates serve three purposes. They cancel to a great extend the undesired phase shift between ordinary and extra-ordinary beam summed up in the primary birefringent BBO plate.

By tilting the compensating plates the phase of ordinary and extra-ordinary beam can be shifted with respect to each other. The relative phase of the two linearly polarized waves has to be fine-tuned to be $\pm\,\pi/2$ thus forming circularly polarized waves. If the probability of finding single photons with a given linear polarization does no longer depend on the polarization direction the pulses are really circularly polarized. This fine-tuning has to be done in both photon channels.

By a larger change of the tilting angle the sense of rotation of the photons passing through the compensating plate can be reverted. If the sense of rotation is reverted in one of the two photon channels a triplet configuration is converted into a singlet configuration or vice versa.

When the coordinate systems of the two observers are suitably chosen the polarization correlation of entangled photons exclusively depends on $\alpha - \beta$ where α and β are the polarization angles looked for by Alice and Bob. If the circular polarizations of the two photons are opposite to each other (singlet state) the coordinate systems of the two observers must have opposite handedness. If the circular polarizations of the two photons are equal (triplet state) the coordinate systems of the observers must have equal handedness. That means the symmetries of source and detector systems have to match. In figures 1 and 2 appropriate coordinate systems are shown for studying singlet and triplet states.

In the figures the observer Bob, the corresponding $\boldsymbol{z}$-axis and the angle β are distinguished by a superscript indicating that the coordinate systems of Bob are different when the photon pairs are in singlet (S) or triplet (T) configuration. One has to bear in mind that the angles β^S and β^T are opposite to each other, i.e. $\beta^S = -\beta^T$. In the past mainly entangled photon pairs in singlet configuration have been studied with the coordinate system depicted in figure 1. Therefore the angle β traditionally used in the literature is identical with β^S.

The singlet case has been exhaustively investigated by Weihs *et al* [4] using the above described down conversion source. Although the triplet configuration has only fragmentarily studied by Gregor Weihs [7] with the same down conversion source it will be discussed at first because the argumentation is simpler than in the singlet case which will be treated afterwards.

3. Polarization correlation of photon pairs in triplet configuration

Circularly polarized waves can be understood as the superposition of two equally large linearly polarized waves. The phase shift of these two linearly polarized waves with orthogonal

IOP Conf. Series: Journal of Physics: Conf. Series **880** (2017) 012065 doi:10.1088/1742-6596/880/1/012065

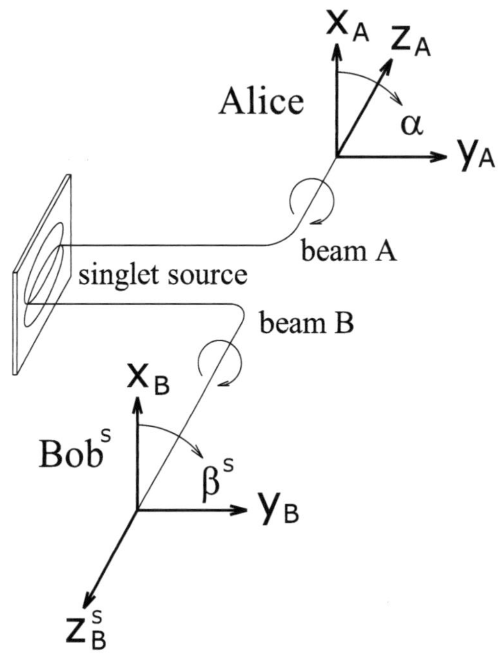

Figure 1. Coordinate systems of two observers with a down conversion source emitting photon pairs in singlet configuration. The coordinate systems of Alice and Bob are right-handed and left-handed, respectively. The angles α and β^S indicate the orientation of the linear polarization looked for by Alice and Bob^S. The thin lines between source and observers indicate optical fibres.

polarization directions must be $\pi/2$. The absolute orientation of the linearly polarized waves is not relevant as long as they remain orthogonal to each other.

The fragmentation of the circularly polarized wave into two equally sized linearly polarized waves is a convenient method to separate photons into two commensurate disjunct groups. Photons initially embedded in one of the two linearly polarized wave components will always stay in this wave component. The two groups of photons do not intermix. This conservation law even holds when the polarization character of the waves is temporarily changed by the use of birefringent plates or electro-optical modulators. Thus photons detected by Alice or Bob behind polarizing filters were already contained in the corresponding linearly polarized wave components just at the source.

In the following it is assumed that the Wollaston prisma of Alice is orientated in direction α. Thus one of Alice's detectors looks for photons polarized in the directions α and $\alpha + 180°$. The other one looks for photons polarized in the directions $\alpha - 90°$ and $\alpha + 90°$.

One can assume that the light source contains two sets of linear oscillators. The oscillators A_α and $A_{\alpha+90°}$ send circularly polarized light towards Alice. The oscillators B^T_α and $B^T_{\alpha+90°}$ send circularly polarized light to Bob^T. The subscripts indicate the orientation of the linear oscillators.

In the experiments with the BBO down conversion source the linear polarizations are anti-correlated. That means at the source the electric fields of the two clockwisely or anti-clockwisely rotating wave packets are always orthogonal to each other or in other words the circularly polarized wave packets are phase shifted by $\Delta\varphi = +\pi/2$ or $-\pi/2$.

Four configurations of the simultaneously emitted wave packets have to be distinguished.

IOP Conf. Series: Journal of Physics: Conf. Series **880** (2017) 012065 doi:10.1088/1742-6596/880/1/012065

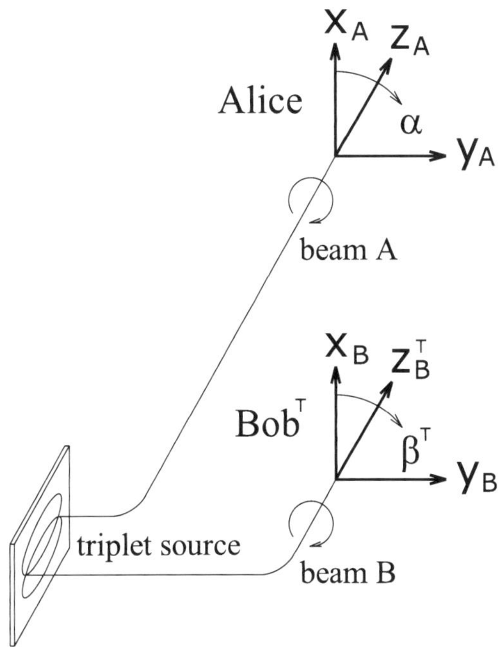

Figure 2. Coordinate systems of two observers with a down conversion source emitting photon pairs in triplet configuration. The coordinate systems of Alice and Bob are both right-handed. The angles α and β^T indicate the orientation of the linear polarization looked for by Alice and Bob^T. The thin lines between source and observers indicate optical fibres.

The time dependences of the four linear oscillators are given in table 1 on the premise that the time dependence of oscillator A_α is arbitrarily chosen to be $+\cos\omega t$. All four cases lead to the same polarization correlation because the transmission probability of the analyzers is periodic with a period of 180°. Thus the signs of the time dependences are not relevant.

Linearly polarized wave packets simultaneously emitted from the oscillators A_α and $B^T_{\alpha+90°}$ are equal or opposite in phase and thus induce a maximum coincidence rate. The same is true for the wave packets simultaneously emitted by the oscillators $A_{\alpha+90°}$ and B^T_α. The coincidence rates in the two channels with matching (or opposite) phases must be equal because of symmetry reasons. Each channel provides half of the total coincidence rate I_0 obtained when both polarizing filters are removed. The channel A_α and B^T_α and the channel $A_{\alpha+90°}$ and $B^T_{\alpha+90°}$ do not provide coincidences because the two simultaneously emitted wave packets have a phase shift of $\pi/2$ at the source and thus are orthogonal to each other.

As has been explained above the orientation of the linearly polarized waves can be arbitrarily chosen. Therefore the coincidence rates found in the four possible coincidence channels are for each angle α

$$\begin{aligned} &I(\alpha, \beta^T\!= \alpha) = 0 \\ &I(\alpha, \beta^T\!= \alpha + 90°) = I_0/2 \\ &I(\alpha + 90°, \beta^T\!= \alpha + 90°) = 0 \\ &I(\alpha + 90°, \beta^T\!= \alpha) = I_0/2. \end{aligned} \tag{1}$$

For $\beta^T = \alpha \pm 90°$ the photon pairs are fully correlated. Assuming both detectors have perfect efficiency the correlation holds for each individual pair of entangled photons. If one of the photons is detected at Alice by the detector looking for photons polarized in the direction α the entangled photon will be detected at Bob^T by the detector looking for the polarization direction parallel to $\beta^T = \alpha + 90°$.

The analyzer on Alice's side oriented along α lets pass 100% of the linearly polarized wave emitted by oscillator A_α. If the analyzer at Bob^T is oriented along $\beta^T \neq \alpha \pm 90°$ it lets pass only a fraction

$$\cos^2(\alpha - \beta^T \pm 90°) = \sin^2(\alpha - \beta^T) \tag{2}$$

of the wave packets send out by the linear oscillator $B^T_{\alpha+90°}$. Bob's detector also lets pass linearly polarized light emitted by oscillator B_α. But this light does not provide additonal coincidences because the oscillations of A_α and B_α are orthogonal to each other. Therefore the coincidence rate is

$$I(\alpha, \beta^T) = I_0 \sin^2(\alpha - \beta^T)/2\,. \tag{3}$$

The complementary coincidence rate is consequently

$$I(\alpha, \beta^T \pm 90°) = I_0 \sin^2(\alpha - \beta^T \mp 90°)/2 = I_0 \cos^2(\alpha - \beta^T)/2. \tag{4}$$

The two coincidence rates $I(\alpha, \beta^T)$ and $I(\alpha, \beta^T + 90°)$ add up to $I_0/2$.

The considerations are symmetric with respect to the role of Alice and Bob^T. If the two sets of oscillators are oriented along β^T and $\beta^T + 90°$ the same polarization correlation is obtained.

The polarization correlation can be derived more formally by determining the overlap integral of the two linearly polarized waves looked for by Alice and Bob^T just at the source.

The time dependences of the linear oscillators A_α, $A_{\alpha+90°}$ and B^T_β are given in table 2 again on the premise that the time dependence of oscillator A_α is chosen to be $+\cos\omega t$. The time dependences of B^T_β in the four rows can be summarized in the form

$$\pm \sin(\omega t \pm (\alpha - \beta^T)). \tag{5}$$

The overlap integral of the two normalized functions $\sqrt{2}\cos\omega t$ and $\pm\sqrt{2}\sin(\omega t \pm (\alpha - \beta^T))$ is

$$\begin{aligned}
&\pm\ \frac{\omega}{2\pi}\int_0^{2\pi/\omega} 2\cos\omega t \sin\left(\omega t \pm\ (\alpha - \beta^T)\right) dt = \\
&\pm\,(\frac{\omega}{2\pi}\int_0^{2\pi/\omega} 2\cos\omega t \sin\omega t\, dt\ \ \cos\left(\alpha - \beta^T\right) \\
&\pm\ \frac{\omega}{2\pi}\int_0^{2\pi/\omega} 2\cos\omega t \cos\omega t\, dt\ \ \sin\left(\alpha - \beta^T\right)) = \pm\ \sin\left(\alpha - \beta^T\right).
\end{aligned} \tag{6}$$

Table 1. Time dependences of the linear oscillators A_α, $A_{\alpha+90°}$, B^T_α, and $B^T_{\alpha+90°}$ simultaneously emitting two circularly polarized waves in (nearly) the same direction towards Alice and Bob^T. The two circularly polarized photons form a triplet pair. At the source the polarization directions of the two waves are always perpendicular to each other. The two circularly polarized wave packets are twisted by $\Delta\varphi = +90°$ or $-90°$ with respect to each other.

	Rotation	$\Delta\varphi$	A_α	$A_{\alpha+90°}$	B^T_α	$B^T_{\alpha+90°}$
A	clockwise	$+90°$	$+\cos\omega t$	$+\sin\omega t$	$+\sin\omega t$	$-\cos\omega t$
B	countercl.	$+90°$	$+\cos\omega t$	$-\sin\omega t$	$-\sin\omega t$	$-\cos\omega t$
C	clockwise	$-90°$	$+\cos\omega t$	$+\sin\omega t$	$-\sin\omega t$	$+cos\omega t$
D	countercl.	$-90°$	$+\cos\omega t$	$-\sin\omega t$	$+\sin\omega t$	$+\cos\omega t$

The coincidence rate is proportional to the square of the overlap amplitude of the two wave packets looked for by Alice and Bob^T just at the source. Thus the coincidence rate is

$$I(\alpha, \beta^T) = I_0 \sin^2(\alpha - \beta^T)/2. \tag{7}$$

in agreement with equation (3).

The predicted results should be carefully tested in future experiments. This means among others that quarter wave plates in front and at the rear of the electro-optical modulators should be introduced. By means of a Poincaré sphere Gregor Weihs describes at large in his doctor thesis [7] that an electro-optical modulator only twists a linear polarization into another linear polarization when the modulator is sandwiched between two quarter wave plates with appropriate orientations of the optical axes. In the experiment of Weihs *et al* [4] these quarter wave plates have been omitted. Only for the angles $0°, 90°, 180°$ and $270°$ Alice and Bob really look for linearly polarized light. For the pretended angles $45°$, $135°$, $225°$ and $315°$ the observers look for circularly polarized waves. For all other angles Alice and Bob look for elliptically polarized light.

In the triplet configuration the entangled photons are anti-correlated with respect to the linear polarization and correlated with respect to the circular polarization. Therefore the omission of the quarter wave plates would lead to fallacious angular dependences of the coincidence rate.

4. Polarization correlation of photon pairs in singlet configuration

The argumentation to derive the polarization correlation in the singlet case is less obvious than in the triplet case because the electric fields of the associated wave packets rotate in opposite directions. In fact, it is not so difficult to define a constant phase as could have been assumed at first sight. In the experiment of Weihs *et al* [4] this counter-rotation has been achieved by switching the tilting angle of the compensating plate on Bob's side and thus reverting the sense of rotation of the photons propagating towards Bob. A constant phase difference of the two counter-rotating waves can be defined when the sense of rotation of the photons on Bob's side is once more reverted by going over from the right-handed coordinate system of Bob^T (Fig. 1) to the left-handed coordinate system of Bob^S (Fig. 2).

The two reversions cancel each other when the associated mirror planes coincide. In the experiment of Weihs *et al* [4] this condition is fulfilled because both mirror planes are oriented along $\beta = 0°$.

Consequently the argumentation used in the previous section can be applied analogously. Thus the coincidence rate is

$$I(\alpha, \beta^S) = I_0 \sin^2(\alpha - \beta^S)/2 \tag{8}$$

Table 2. Time dependences of the linear oscillators A_α, $A_{\alpha+90°}$ emitting circularly polarized waves towards Alice. The oscillators B^T_β and $B^T_{\beta+90°}$ send circularly polarized waves towards Bob. The time dependence of the oscillator B^T_β is given in the last column. The four rows for clockwise and anticlockwise rotation and the phase shift $\Delta\varphi = +90°$ or $-90°$ are listed in the same order as in table I

	Rotation	$\Delta\varphi$	A_α	$A_{\alpha+90°}$	B^T_β
A	clockwise	$+90°$	$+\cos\omega t$	$+\sin\omega t$	$+\sin(\omega t + (\alpha - \beta^T))$
B	countercl.	$+90°$	$+\cos\omega t$	$-\sin\omega t$	$-\sin(\omega t - (\alpha - \beta^T))$
C	clockwise	$-90°$	$+\cos\omega t$	$+\sin\omega t$	$-\sin(\omega t + (\alpha - \beta^T))$
D	countercl.	$-90°$	$+\cos\omega t$	$-\sin\omega t$	$+\sin(\omega t - (\alpha - \beta^T))$

as has been confirmed by Weihs *et al* [4].

As has already been mentioned in the previous section, the quarter wave plates before and behind the electro-optical modulators have been omitted in the experiment of Weihs *et al* [4]. This experimental laxness has no consequences for the coincidence rates because in the singlet configuration the entangled photons are not only anti-correlated with respect to the linear polarization but also with respect to the circular polarization. In fact, for the sake of clarity the missing quarter wave plates should be introduced in future experiments in order to really prove the intended polarization correlations.

5. Polarization correlation of entangled photons emitted by doubly excited atoms

Aspect, Grangier and Roger could show in 1981 that polarization correlations of entangled photons significantly violate Bell's inequality. They studied circularly polarized photon pairs emitted by doubly excited calcium-40 atoms in a $(4p^2\ {}^1S_0) \rightarrow (4s\,4p\ {}^1P_1) \rightarrow (4s^2\ {}^1S_0)$ decay cascade. The short living intermediate ${}^1P_1$ state has a life time of $5\,ns$. Only entangled photon pairs leaving the atom in opposite directions have been detected by the two observers. The two photons form a singlet state because the two wave packets are rotating in opposite directions with respect to their propagation directions.

A constant phase shift can be defined when the rotational frequencies of the two wave packets are equal. The two photons ($\lambda_1 = 551.3\ nm, \lambda_2 = 422.7\ nm$) emitted in a decay cascade of calcium-40 have different energies. That means carrier and rotational frequencies cannot be equal for both photons. This particular feature should be systematically tested in future experiments.

The wave packets emitted by doubly excited atoms are in phase ($\Delta\varphi = 0°$) or opposite in phase ($\Delta\varphi = 180°$). That means the electric fields of the two wave packets are always parallel or antiparallel to each other at the source. In the wave optical model presented above the oscillators A_α and B_α are in phase. The same is true for the oscillators $A_{\alpha+90°}$ and $B_{\alpha+90°}$. In consequence the coincidence rate is maximal when the two detectors are parallel and zero if the two detectors are orthogonal to each other. Thus the polarization correlation is given by

$$I(\alpha, \beta^S) = I_0 \cos^2(\alpha - \beta^S)/2 \tag{9}$$

The photons are correlated with respect to their linear polarization, but anti-correlated with respect to their circular polarization.

6. Concluding remarks

The derivation of the polarization correlation is based on the fact that circularly polarized wave packets associated with entangled photon pairs are phase shifted at the source. For such waves the information about the phase relation of the two associated wave packets at the source can be deduced from the difference angle of the polarization directions looked for by the two observers.

Elementary processes providing circularly polarized entangled photon pairs are generally characterized by a typical phase difference of the two simultaneously emitted wave packets. If the phase difference at the source is 0° (or 180°) the linear polarizations are correlated [2, 3]. If the phase difference is 90° the linear polarizations are anti-correlated [4, 7].

When the polarization character of the wave packets is not changed the distances between source and observers have no influence on the experimentally found polarization correlations. Usually in optical fibers the polarization state is changed by spurious birefringent effects. Therefore Weihs *et al* [4, 7] have to recover the original polarization state by manually compensating the undesired phase shifts in both photon channels.

Initiated by the violation of Bell's inequality physicists investigating polarization correlations of entangled photons mostly assume that their must be a superluminal information transfer between the two observers. Salart *et al* [8] try to find a lower limit for the speed of spooky action at a distance. Shalm *et al* [9, 10] even claim to have proven that spooky action at a distance is really real.

Following the wave optical considerations presented above the polarization correlations can be explained without assuming non-local interactions. Therefore it is no longer necessary to perform further experiments with enlarged distances [8] or with ultra-high efficiency detectors [9, 10] in order to close possible loopholes.

The contradictory statements on the occurence of non-local interactions rest on different assumptions. Bell's inequality is derived from a purely particle based consideration. On the simplest guess it is purely accidental at which output of a Wollaston prism a circularly polarized photon is detected. Thus the coincidence rate is $I_0/4$ for all angles α and β.

Particle based considerations generally lead to wrong conclusions when quantum mechanical results depend on phase differences. For example the intensity ratio of the photons detected at the two outputs of a Mach-Zehnder interferometer can only be derived from the relative phase of the two partial waves reaching the second semi-permeable mirror. In the case of interferometers obviously only wave based considerations lead to proper results.

The wave optical explanation given above is purely wave based. The information contained in phase differences of distinct wave components is maintained. The polarization correlation determined by Alice and Bob is equal to the polarization correlation just at the source because the two photon groups contained in the two linearly polarized wave components do not intermix. Thus the principle of locality remains valid.

It is fully accidental through which output of a Wollaston prism the first photon leaves. For the second photon, however, the probability is well defined. There is no further evidence for any accidental behaviour. For $\alpha = \beta$ the strict correlation even holds for each pair of entangled photons. Thereby it is not relevant which of the two photons is detected first.

On the other hand the principle of realism has to be reconsidered. The linear polarization component of a circularly polarized photon is not defined before the measurement has taken place. Nonetheless the polarization correlation of entangled photons measured at distant locations is well defined and agrees with the predictions of quantum mechanics. Obviously the principle of realism has to be replaced by the principle of contextuality. That means the polarization correlation is only well defined when the polarization directions looked for by the two observers are both fixed. But this assertion is trivial. If coincidence rates do not depend on α as well as on β it would make no sense to measure coincidences at all.

References

[1] Bell J S 1964 *Physics* **1** 195
[2] Aspect A, Grangier P and Roger G 1981 *Phys. Rev. Lett.* **47** 460
[3] Aspect A, Grangier P and Roger G 1982 *Phys. Rev. Lett.* **49** 91
[4] Weihs G, Jennewein J, Simon C, Weinfurter H and Zeilinger A 1998 *Phys. Rev. Lett.* **81** 5039
[5] Khrennikov A 2012 *Int. J. Theor. Physics* **51** 2488
[6] Englert B-G 2013 *Eur. Phys. J.* D **67** 238
[7] Weihs G 1999 *Ein Experiment zum Test der Bellschen Ungleichung unter Einsteinscher Lokalität* Innsbruck (The doctor thesis is electronically available at the webpage of the Universität Innsbruck www.uibk.ac.at/exphys/photonik/people/gwdiss.pdf)
[8] Salart D, Baas A, Branciard C, Gisin N, Zbinden H 2008 *Nature* **454** 861
[9] Giustina M, Shalm L K, Zeilinger A *et al* 2015 *Phys. Rev. Lett.* **115** 250401
[10] Shalm L K, Jennewein T, Kwiat P G *et al* 2015 *Phys. Rev. Lett.* **115** 250402

Is direct measurement of time possible?

Thomas Reynolds

Department of Computer Science, University at Albany, 1400 Washington Ave., Albany, NY 12222 USA

E-mail: tracm2@acm.org

Abstract. Is direct measurement of time possible? The answer to this question may depend upon how one understands time. Is time an essential constituent of physical reality? Or is what scientists are talking about when they use the symbol 't' or the word 'time' an human cultural construct, as the Chief of the USA NIST Divisions of Time and Frequency and of Quantum Physics has suggested. Few aspects of physics do not reference activity to time, but many discussions within either view of time seem to use one same, largely traditional, language of time. Briefly considering the question of measurement, including from a formal measure-theoretic point of view, clarifies the situation.

1. Introduction

There is a thread in physics which raises questions about whether time as traditionally understood actually is or is not an essential aspect of physical reality, a fundamental constituent of the universe. There are a variety of interpretations of how the universe may be "timeless", but discussions of time metrology are not a prominent feature of them.[1]

Modern standardized time metrology involves exquisitely complex physical apparatus.[2] But one might think that an aspect of the world considered so fundamental as time would be accessible more directly than that. In the historical development of the concept and use of time, simple direct observation of readily apparent features of the world were considered sufficient to tell time.[3] Those features even now perhaps have not been entirely superseded—the present standard definition of the "second" is the current iteration of a sequence of refinements tracing back to pre-atomic science and folk culture.[4]

With the rise of science and technology since the European Renaissance and especially Galileo, the uses of time changed along with the increasingly fine scale at which time metrology operates.[5] As a comparison, matter, too, is ordinarily understood to be a fundamental aspect of the world albeit of a different kind, and the atomic theory of matter has existed since at least the Ancient Greeks.[6] But conceptions of the essential nature of matter were profoundly changed when increasingly refined experiments accessed increasingly smaller physical scales. The understanding of atoms came to include component entities which at first were not directly observable. Milliken's 1909-11 oil drop experiments produced an early direct or nearly so measurement of electrons, and his 1916 paper on the photoelectric effect continued with direct or nearly so measurements relating the atomic electron and the light photon. Now, based on complexly mediated experimental techniques and interpretations, the Standard Model presents a systematization of additional sub-component entities, and even the dynamics of atomic components and processes are becoming observable.[7] However, modern time metrology, despite its subtlety and fineness, has not found any substructure of time.

So if time metrology, as sophisticated in complexity and in fineness of scale as it now is, has not produced a clear answer to the question of whether nature is timeless or not, perhaps going back to basics will help. Consider the question: "Is Direct Measurement of Time Possible?".

2. "Direct"

By "direct measurement" is meant measurement made with systems the components and values of which do not involve chains of subcomponents and of reasoning and interpretation to provide results. A canonical simple example for measuring space is a measuring tape. A child can assess its surroundings and an adult can build a house using only a measuring tape to identify spatial extents. It is a spatial object which one can directly handle and move at will, and can reuse in different spatial locations and situations without alterations. Other objects are directly compared to it. One merely extends the tape and holds it directly beside some other spatial object to tell directly the length of the other object. No theoretical interpretation is necessary, nothing intervenes between measuring tool and measured object. One holds the tool and sees the value, whether measuring an object or an empty spatial extent.

To make the contrast explicit, a toy example of non-direct would be measuring a spatial object (a) visible in a mirror, by walking around the room until (b) some other object—a chair, say—exactly obscures the mirror image, then (c) measure the chair, and finally (d) use trigonometric relationships to calculate the original object's length from the chair's size and from relative positions of things.

If a time dimension exists, it is not like a spatial dimension. No formally organized direct physical manipulations, the usual concomitants of measurement, are available. One cannot stand beside a temporal object and see all of it, nor cut it in half, nor separately manipulate halves, nor move pieces around, nor assemble a temporal object from various temporal pieces. There are no Cuisenaire® rods showing a temporal parsing which matches a number system, like those spatial objects for children do.[8] Though numbers are used as time values, the only inherent physical operation is increase; and there is no control over that.

Further, in the case of time actually there are no recognizably temporal objects of any kind to which to have a direct relation. Neither is there anything like an empty extent of time. Although various spatial objects have been used during the historical development of the time concept as markers within various conceptions of time, they are merely spatial, even if motion is involved.[9]

As a source of indirect but pertinent evidence about the issue, consider humans, the dominant species on Earth. Humans evolved able to live in climates from hot tropics to icy Arctic. They evolved able to kill mastodons to eat, to kill lions to avoid being eaten, to kill three-dimensionally spatially motile birds in flight, and to kill each other, with only wood and stone tools. That human species as well as myriad "lesser" creatures have evolved binocular parallax vision and skin sensory receptors, and are able directly to sense spatial distances and lengths.[10] In the battling competitions of "nature red in tooth and claw", one would expect that over many millions of years some organ would have evolved to sense time directly, and would be known as such. But there seems to be no such evolved feature. This is a modern statement of Poincaré's point: "*We have not a direct intuition of the equality of two intervals of time.* The persons who believe they possess this intuition are dupes of an illusion."[11]

3. Formal Measurement Theory

The actual doing of measurement can be very complex. But in measure-theoretic terms, the formal essence of measurement is matching members of one set, considered to be the measured, to members of another set, considered to be the measure.[12]

More thoroughly stated, consider a set A of empirical phenomena distinguishable from all other phenomena, and a set of relations R over the elements of A. Together (A,R) form a relational system of which A is the domain. (A,R) is taken as data, found rather than constructed in the same way that a collection of people and their relative physical heights is considered to be found data for the purpose of measuring them. There is another relational system (B,S), with domain B and relations S, which is a cultural construct. That is, the choosing and pairing of B and S to be (B,S) and the use of (B,S) for measurement is an agreed arrangement specified—a shared choice made—by people and as such is an human cultural construct, a tool to be used for a purpose, in this case for measurement. A system of measurement is a mapping from (A,R) to (B,S), and also is a cultural construct.

To be considered science, and certainly to be a physical science, (B,S) is generally required to be a mathematical relational system, with B numbers and S some n-ary mathematical relationships. In terms of these basics, much of the activity of mathematical physics consists of identifying appropriate number systems to be candidate Bs and suitable mathematical relationships S_i among the members of each B_i. And experimental science's focus is to find whether an A of interest is partitionable according to some R in a way mappable to some (B,S). Note the distinction that whatever may be the raw uninterpreted result of an experimental run, or whatever mathematics may be involved in arriving at final relations among numbers, only the final output number for a phenomenon, comparable to other final output numbers, is the measure or more properly the "measured value".

A measuring tape gives the $(A,R) \longmapsto (B,S)$ mapping directly, including its visible final value output. In quantum physics for example the situation is less clear. Wigner's classic 1963 paper "The Problem of Measurement" is concerned with how theory is involved in the data collection and interpretation complexities of experiment, and how that relates to measured results. Formal measurement theory as discussed here was first proposed in 1957, and Wigner's article shows no obvious awareness of it. So it is an interesting exercise, which will not be done here, to go through Wigner's article and identify what relates to (A,R), what to (B,S), and what to mapping.[13]

When experiment gives satisfactory results, A and and its characterization in terms of R are talked about as though they are reality. By the scientific method, measurability justifies vocabulary. "Heavenly spheres "used to be considered real, now they are poetry and orbits instead are reality.

4. Time

> "For that is what time is: a number of change in respect of the before and after. So time is not change but the way in which change has a number." Time "measures change". (Aristotle, **Physics**)[14]
>
> "[T]ime is an abstraction, at which we arrive by means of the changes of things. ... [Absolute time] is an idle metaphysical conception." (Mach, **The Science of Mechanics**)[15]
>
> "My own personal opinion is that time is a human construct." "[B]ut what time really is, is a question that I can't answer for you." (Tom O'Brian, Chief of United States NIST Quantum Physics Division and of Time and Frequency Division)[16]

In dim prehistory, awareness that things change and some changes are vital and should be kept track of—eg when planting season begins—was a shared experience, and that brought vocabulary. The primitive origins of modern timekeeping, recognizing divisions within natural cycles and counting repetitions of natural events, required more vocabulary.[17] By now, time vocabulary is traditional and well-established, including its implicit reification of time as a thing real in the same sense the rest of the physical world is real. But traditional vocabulary can mislead. Textbook and even theoretical discussions still use time vocabulary, including "spacetime", in ways reifying an understanding that time is self-substantial and has various characteristics. Yet experimental practice in science uses time values simply as a more or less arbitrary uniform comparator in describing dynamic processes.

Historically, replicable scientific timekeeping became possible with the Galilean pendulum, and the low friction pivot isolating its periodicity from local disruptive influences. In the attoworld of modern science, atomic clocks provide highly granular timekeeping standards, with continual improvements based on increasingly sophisticated understanding of light phenomena.[18] These modern types are founded on light frequencies, which are believed to be constant when isolated in the same way that—and because—light speed is constant. In the USA the NIST Division responsible for this is named "Time and Frequency", but frequency is more basic. The system depends technologically on accurately comparing and counting atomic transition cycles and light frequency cycles.[19]

But time is not "measured". Rather some atomic transition cycle is chosen as a canonical unit of activity—informally, as a "tick"—; some number of sequential ticks is observed in coordination with some number of cycles of light; and that sequential existence is said to mark out what is to be considered one second of time. That is, seconds of "time" are generated, and are asserted as a reference to which to compare activity of other things. This "time" is merely a simple counting number, a culturally agreed upon reference counter usable anywhere, with zero being arbitrarily assignable at any point, exhibiting a mathematical elegance or pure numericality suiting it to be the ubiquitous 'dt' of 'dx/dt'.

5. Conclusion

The answer to the title question seems to be that the question itself is wrong. "Time" is not an entity to which the operations of measurement, direct or indirect, apply. Instead it is a widely shared cultural construct, based in social activity in which the system of standard time values provides an universal, neutral, non-perturbing reference parameter for use measuring dynamics of other things. The effect is not unlike Newton's absolute time but without reifying it as "absolute" or as "true", keeping only the "mathematical" and "abstract" and using technology and social organization to refine "vulgar measures" into the functional analogue of an absolute.[20] The idea of time can be reduced to pure counting and as such is an epitome of abstraction, limpid in comparisons, available and effective as parameterization.

Formally, very often the measurement task in physical science is to find a mapping, not (A,R) $\longmapsto$ (B,S) but ((A,R),(C,T)) $\longmapsto$ (B,S), where C is a set of clock timestamps and T is relations among them. Deciding how (A,R) and (C,T) are to be put together and compared is an aspect of experimental design. But (C,T) is supposed to be uniform in all experiments or at least relatable from experiment to experiment. The reason why the title's question is wrong is that ((C,T),(C,T)) $\longmapsto$ (B,S) is undefined—the measuring tape cannot be combined with nor compared to itself to measure its accuracy.

The traditional time vocabulary is well-established because in many social contexts "now", "then", "until", "before", "after", etc are all useful, efficient references to complex configurations of spatial objects. However, in science it would be desirable to remove the realness or reification presumption while keeping the aptness for parameterization, leaving all reification to apply only to physical clocks. For example it could be possible to clarify the scientific time vocabulary by making small adjustments. One low-effort start might be to substitute the phrase 'time value' many places where 'time' ordinarily would be used; use 'timekeeping' rather than 'time' in many discussions in Special Relativity; and in almost all contexts, to stop referring to "measuring time", perhaps using the not unusual phrase 'counting time'. Change is possible—in scientific publication equal use of 'she' and 'he' for the neutral personal reference has become common.

References

[1] Rovelli C 1991 Time in quantum gravity: an hypothesis *Phys. Rev. D* **42** n 2 pp 442-456
Barbour J 1999 *The End of Time* (Oxford: Oxford UP)
Prati E 2011 Generalized clocks in timeless canonical formalism *J. Phys.: Conf. Series* **306** 012013
Anderson E 2012 The problem of time in quantum gravity *Preprint* gr-qc/arXiv:1009.2157v3
Reynolds T 2015 Spacetime representation of electromagnetic radiation in a (2+1)-dimensional universe *J. Phys.: Conf. Series* **626** 012072

[2] Diddams S A, Bergquist J C, Jefferts S R and Oates C W Standards of time and frequency at the outset of the 21st Century *Science* **306** 19.Nov.2004 pp 1318-1324

[3] Nilsson M P 1920 *Primitive Time Reckoning* (Lund, Sweden: C.W.K. Gleerup)

[4] Lombardi M A 2012 The evolution of time measurement, part 4: the atomic second *IEEE Instrumentation & Measurement Mag.* **15** n 1 Feb 2012 p 47-51

[5] Drake S 1975 The role of music in Galileo's experiments *Sci Am* **232** June 1975 98-104
Dohrn-vanRossum G 1996 *History of the Hour: Clocks and Modern Temporal Orders* tr T Dunlap (Chicago: U Chicago P)
Sullivan D B 2001 Time and frequency measurement at NIST: the first 100 years *Proc. 2001 IEEE Frequency Control Symp.* pp 4-17

[6] Berryman S 2016 Ancient atomism *Stanford Encyclopedia of Philosophy* accessed 15.Feb.2017 at plato.stanford.edu/entries/atomism-ancient/

[7] Millikan R A 1917 *The Electron* (Chicago: U Chicago P)
Millikan R A 1916 A direct photoelectric determination of Planck's 'h' *Phys. Rev.* **7** n 3 pp 355-390
Anonymous 2017 Standard model *Wikipedia* accessed 15.Feb.2017 at en.wikipedia.org/wiki/Standard_Model
de Oteyza D G et al 2013 Direct imaging of covalent bond structure in single-molecule chemical reactions *Science* **340** n 6139 21.June.2013 pp 1434-1437
Pazourek R, Nagele S and Burgdörfer J 2015 Attosecond chronoscopy of photoemission *Rev. Mod. Phys.* **87** July-Sept 2015 pp 765-801

[8] Anonymous 2017 Cuisenaire rods *Wikipedia* accessed 15.Feb.2017 at en.wikipedia.org/wiki/Cuisenaire_rods

[9] Nilsson M P 1920 *Primitive Time Reckoning* (Lund, Sweden: C.W.K. Gleerup)

[10] Anonymous 2017 Stereopsis *Wikipedia* accessed 14.Feb.17 at en.wikipedia.org/wiki/Stereopsis
Anonymous 2017 Somatosensory system *Wikipedia* accessed 14.Feb.17 at en.wikipedia.org/wiki/Somatosensory_system

[11] Poincaré H 1898 The measure of time *The Foundations of Science* tr G B Halsted (New York: Science Press, 1913) pp 222-234 emphasis in original

[12] Scott D and Suppes P 1958 Aspects of theories of measurement *J Symbolic Logic* **23** n 2 pp 113-128

[13] Wigner E P 1963 The problem of measurement *Am J Phys* **31** n 6 pp 6-15

[14] Aristotle *Physics* BkIV Ch11 ln219b1, BkIV Ch12 ln221a1

[15] Mach E 1960 *The Science of Mechanics* tr T J McCormack (La Salle, IL: Open Court) 6th ed with revisions through 9th German ed (1st German ed publ. 1883) p 273

[16] O'Brian T 2014 New clock may end time as we know it *National Public Radio interview* accessed 15.Feb.2017 at www.npr.org/2014/11/03/361069820/new-clock-may-end-time-as-we-know-it

[17] Nilsson M P 1920 *Primitive Time Reckoning* (Lund, Sweden: C.W.K. Gleerup)

[18] Poli N, Oates C W, Gill P and Tino G M 2013 Optical atomic clocks *Rivista Nuovo Cimento* **36** n 12 p 555-624
Ludlow A D and Ye J 2015 Progress on the optical lattice clock *Comptes Rendues Phys.* **16** p 499-505

[19] Lombardi M A, Heavner T P and Jefferts S R 2007 NIST Primary Frequency Standards and the Realization of the SI Second *Measure* **2** n 4 Dec 2007 pp 74-89

[20] Newton I 1687 Scholium *Philosophiae Naturalis Principia Mathematica* tr Motte A (New York: Daniel Adee, 1st American edition 1846) pp 77-82

Does universality of free-fall apply to the motion of quantum probes?

Luigi Seveso[1], Valerio Peri[1,2], Matteo G. A. Paris[1,3]
[1]Quantum Technology Lab, Dipartimento di Fisica, Università degli Studi di Milano, I-20133 Milano, Italy
[2]Institute for Theoretical Physics, ETH Zurich, 8093 Zürich, Switzerland
[3]INFN, Sezione di Milano, I-20133 Milano, Italy

E-mail: `luigi.seveso@unimi.it`, `periv@student.ethz.ch`, `matteo.paris@fisica.unimi.it`

Abstract. Can quantum-mechanical particles propagating on a fixed spacetime background be approximated as test bodies satisfying the weak equivalence principle? We ultimately answer the question in the negative but find that, when universality of free-fall is assessed locally, a nontrivial agreement between quantum mechanics and the weak equivalence principle exists. Implications for mass sensing by quantum probes are discussed in some details.

1. Introduction

The weak equivalence principle (WEP) is arguably the most basic property of the classical gravitational field. It states that the trajectories of a particle subject only to gravitational forces are completely determined by the local spacetime geometry [1, 2]. Strictly speaking, the WEP applies only to test-bodies, i.e. particles which are sufficiently light so that their back-action on the external gravitational field is negligible, and sufficiently small so that the effects of tidal forces on them are inconsequential. For such probes, the WEP states that their trajectories under gravity are independent of the internal properties, most remarkably their masses.

In general relativity, the WEP is directly encoded into the specific form taken by the action functional for a free point-like material particle. The action $S(a, b)$ for the propagation of a particle of mass m between the two events a and b is given by

$$S(a, b) = -mc \int_a^b ds \ , \tag{1}$$

where $ds = \sqrt{g_{\mu\nu} dx^\mu dx^\nu}$ is the metric element. Since the mass appears as a multiplicative factor, it does not enter the equations of motions. As a consequence, particles of different masses follow the same trajectories given the same initial conditions. This is the content of the WEP, as iconically represented by the Leaning Tower of Pisa experiment, supposedly performed by Galileo Galilei in 1589 [3, 4], as well as its repetition on the moon by astronaut David Scott during the Apollo 15 mission [5].

The classical gravitational theory thus undisputably predicts that test-bodies move universally when in free-fall. However, it is a logically different question whether the behavior of real physical particles may, in fact, be approximated with that of test-bodies. The question

IOP Conf. Series: Journal of Physics: Conf. Series **880** (2017) 012067 doi:10.1088/1742-6596/880/1/012067

is especially relevant when the particle's propagation exhibits quantum features [6, 7, 8, 9, 10], i.e. if quantumness of the particle probing an external gravitational field is taken into account. Is the WEP satisfied in the quantum regime? In this paper, we address the conceptual aspects of the problem and, in particular, we devote our analysis to

1. **discussion of the status of the WEP within quantum mechanics:** in Section 2, we review some standard objections against the possibility of extending the WEP of classical physics to quantum particles;
2. **formulation of a WEP for quantum probes:** while with classical particles one can sensibly talk about trajectories, quantum particles are described by wavefunctions. Therefore, a suitable reformulation of the WEP that applies also to quantum probes is in order. Our proposal, based on quantifying the information extractable from position measurements, is described in details in Section 3;
3. **results for uniform and non-uniform gravitational fields:** in Section 4, we discuss to what extent quantum probes obey the WEP according to our information-theoretic formulation, pointing out in particular the differences between uniform and non-uniform gravitational fields.

Besides the foundational aspects, our analysis may be of interest to address experiments involving gravity and particles with an inherently quantum-mechanical behaviour, e.g. Bose-Einstein condensates and neutron interferometry.

2. Quantum mechanics and the WEP

Let us here review the status of the WEP in classical and quantum physics, with a focus on their non-relativistic limits. The Newtonian theory of gravity implies that a test-body of mass m in an external gravitational potential $\varphi(\mathbf{x})$ obeys the equation of motion

$$m\,\ddot{\mathbf{x}} = -m\,\nabla\varphi(\mathbf{x})\,. \tag{2}$$

The mass simplifies and, as a result, the solutions of (2) do not depend on m.

Moving to the quantum case, one may incorporate gravity at the semiclassical level by starting from the relativistic Klein-Gordon equation for a massive spinless boson in a fixed metric background [11]. Then, upon taking the non-relativistic limit, one recovers a Schrödinger-like equation with the gravitational potential energy $m\,\varphi(\mathbf{x})$ of the probe contributing an extra term to the Hamiltonian,

$$i\hbar\,\partial_t\psi(\mathbf{x},t) = -\frac{\hbar^2}{2m}\,\Delta\psi(\mathbf{x},t) + m\,\varphi(\mathbf{x})\,\psi(\mathbf{x},t)\,. \tag{3}$$

The particle's mass appears explicitly through the ratio $\hbar/m$. The solutions of (3) thus depend parametrically on the same ratio. It follows that quantumness of the state describing the probing particle is directly linked to the appearance of mass-dependent effects in its propagation. In other words, the dynamical evolution of the wavefunction for a particle falling in a gravitational field contains non-trivial information about its mass. This is because we may take repeated measurements on identically prepared probes and, since the quantum state of the system is mass-dependent, the statistics of the measurements would also be, in general, mass-dependent. At a first sight, such a conclusion appears to be at odd with the essence of the WEP, which dictates that no information about the mass of a freely-falling particle is available to an observer monitoring its trajectories. As we will see, the WEP is indeed clashing with quantum mechanics in the general case of a nonuniform field, whereas a nontrivial agreement between quantum mechanics and the WEP exists for uniform fields, i.e. locally.

2.1. Quantifying information on a parameter: quantum parameter estimation theory

Information about the value of a certain parameter may be extracted from the system under investigation by performing measurements. This intuitive concept may be made more precise in the framework of quantum parameter estimation theory. In order for this contribution to be self-contained, we present a brief overview of its fundamentals. A more extensive review can be found in [12], together with the more in-depth references [13, 14, 15].

Quantum parameter estimation theory provides the tools to quantify the precision achievable in any estimation procedure aimed at inferring the value of a certain physical parameter λ, taking into account the limitations imposed by quantum mechanics. As a matter of fact, there are many quantities in quantum systems which do not formally correspond to proper physical observables (e.g. coupling constants). As such, they are not directly measurable, even in principle. In these cases, one has to resort to an indirect estimation strategy, i.e. one measures some observable of the system, which is experimentally accessible, and whose outcomes' statistics depends on the parameter, and then extracts an estimate through a suitable processing of the data. Therefore, there are two distinct steps involved in any estimation task: 1. the choice of the measurement to be performed and 2. the choice of data processing, i.e. the choice of how to elaborate statistically the experimental data to extract an estimate of the parameter.

The last step is usually fulfilled by a locally unbiased estimator $\hat{\lambda}$, which is a function of the measurement outcomes such that its expectation value equals the true value of the parameter. That is, $\mathbb{E}_\lambda(\hat{\lambda}) = \lambda$, where $\mathbb{E}_\lambda$ denotes the expectation value taken with respect to the probability distribution $p_\lambda(x)$ of the experimental data $x = \{x_1, x_2, \dots, x_N\}$.

Typically, one further requires the estimator to minimize the expectation value of a suitable loss function. One of the most natural choices is to minimize the expectation value of the square of the deviations of the estimator with respect to the true value of the parameter, $\mathbb{E}_\lambda[(\hat{\lambda} - \lambda)^2]$, which for locally unbiased estimator coincides with the variance $\mathrm{Var}(\hat{\lambda})$ of $\hat{\lambda}$. It is then a well-known result of classical statistics (that goes under the name of Crámer-Rao theorem [16]) that the variance of any locally unbiased estimator after N repeated measurements is bounded from below by the inverse of the Fisher information $F_X(\lambda)$ (multiplied by N),

$$\mathrm{Var}(\hat{\lambda}) \geq \frac{1}{N\, F_X(\lambda)} \,. \tag{4}$$

The bound is saturable, at least in the asymptotic limit $N \to \infty$ of a large number of measurements, and the corresponding optimal estimator is said to be efficient. The Fisher information thus quantifies, for a fixed choice of measurement scheme, the maximum amount of information which can be extracted on some unknown parameter. Explicitly, it is defined as the expectation value of the logarithmic derivative squared of the probability distribution of the data,

$$F_X(\lambda) = \mathbb{E}[(\partial_\lambda \ln p_\lambda)^2] \,. \tag{5}$$

Regarding the first optimization step, i.e. the choice of the measurement, one is interested in achieving the best possible estimation precision, which corresponds to the maximum possible Fisher information. The Fisher information can in fact be maximized over the choice of the measurement scheme, which yields as an upper-bound the quantum Fisher information $J(\lambda)$,

$$J(\lambda) = \max_{\{X\}} F_X(\lambda) \,, \tag{6}$$

where the maximum is taken with respect to measurements of all possible observables X [17]. Closed-form expressions for $J(\lambda)$ are available. The quantum Fisher information encodes the ultimate bound to the achievable precision for any quantum-limited measurement.

2.2. A quantum Galilean experiment

Returning to the case of a freely-falling quantum probe under gravity, the reasoning below (3) suggests that the Fisher information will in general be non-zero when measurements of position are employed to estimate its mass. The same could also be true for other kinds of measurements, e.g. a measurement of momentum. However, since the WEP concerns the trajectories of particles, it is the Fisher information for position measurements to be relevant. Thus, from now on the parameter to be estimated is the mass of a probe in a gravitational field, and the probability $p_m(\mathbf{x})$ of measuring the particle at position $\mathbf{x}$ is the modulus square of the wavefunction $\psi(\mathbf{x}, t)$ which is a solution of equation (3) for appropriate initial conditions.

In order to quantitatively illustrate some of the observations made above, let us work out a simple example. More general examples of quantum measurement strategies aimed at mass estimation under gravity, and their corresponding sensitivities, can be found in [18]. Let us consider a quantum free-fall experiment á-la Galileo with a quantum probe in a uniform gravitational field $\mathbf{g} = -g\,\hat{\mathbf{i}}$ ($\hat{\mathbf{i}}$ is the unit vector along the x axis). The initial preparation of the probe is described by the Gaussian wavefunction

$$\psi(x,\,0) = \frac{1}{(\pi\,\Delta x^2)^{1/4}}\,e^{-\frac{1}{2}\left(\frac{x-h}{\Delta x}\right)^2}\,, \tag{7}$$

which in the classical limit describes a particle localized over the scale Δx around $x = h$. Under time evolution (setting $\hbar = 1$), we have

$$\psi(x,\,t) = \frac{1}{(\pi\,\Delta x^2)^{1/4}}\,e^{-img^2t^3/6}\,e^{-imgxt}\,\frac{1}{\sqrt{1+it/m\,\Delta x^2}}\,e^{-\frac{1}{2(1+it/m\,\Delta x^2)}\left(\frac{x-h}{\Delta x}\right)^2}\,. \tag{8}$$

One can then compute [18] the Fisher information on m which can be extracted from the system through measurements of position, and compare it with the ultimate quantum limit set by the quantum Fisher information,

$$F_X(m) = \frac{2}{m^2}\left[1+\left(\frac{m\,\Delta x^2}{t}\right)^2\right]^{-2}\,, \qquad J(m) = \frac{t^2}{2m^4\Delta x^4} + 2g^2t^2\Delta x^2 + \frac{2g^2t^4}{m^2\,\Delta x^2}\,. \tag{9}$$

The quantum Fisher information grows like t^4 for large interrogation times. In contrast, the Fisher information saturates to a constant value, which means that by monitoring the trajectory of a freely-falling quantum probe one cannot arbitrarily improve the precision.

This last remark is valid when the probe is subject only to gravitational forces. If this is not the case, a measurement of position can perform significantly better. Let us consider the following setup, which reproduces the basic physics of neutron interferometry experiments [19]. A beam of particles with de Broglie wavelength $\lambda\!\!\!^{-}$ is split in two and propagates along the two arms of an interferometer, with one arm being located higher by an amount h in the Earth's gravitational field. The beam propagating in the upper arm acquires an extra phase due to gravity, $\varphi_g = -m^2ghL\lambda\!\!\!^{-}/\hbar^2$, where L is the length of the horizontal arm of the interferometer. Then the two beams are recombined on a beamsplitter and a coarse-grained position measurement is made, i.e. it is measured whether the particle leaves the interferometer through one port or the other. The Fisher information on the particle's mass that can be extracted in this way is equal to $4\varphi_g^2/m^2$. It scales quadratically with L and therefore likewise with the interrogation time of the experiment. Thus, if the particle probing the field is not in pure free-fall, monitoring of its trajectory allows to estimate its mass to any desired level of accuracy by increasing the interrogation time.[1]

[1] However, a single measurement would not suffice. It is still necessary that the Crámer-Rao inequality be saturated, which in general requires a number of repetitions sufficiently large to be close to the asymptotic regime.

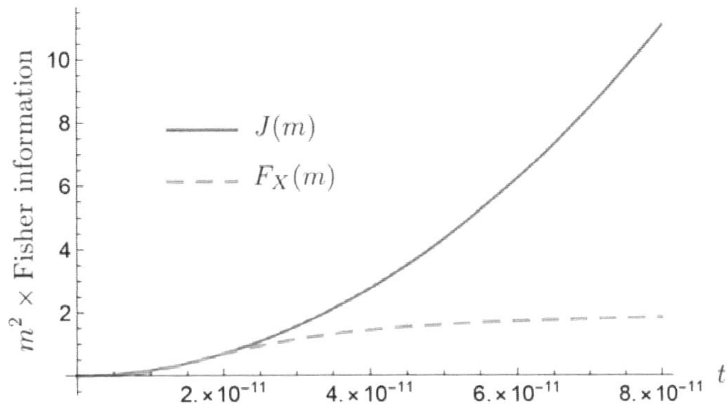

Figure 1. Comparison between the quantum Fisher information $J(m)$ (*solid*) and the position Fisher information $F_X(m)$ (*dashed*) for the mass of a freely-falling Gaussian probe in a uniform field. Both quantities are multiplied by m^2 to make them adimensional. For the purposes of this figure, the mass m is taken to be the neutron's mass $m \simeq 1.7 \times 10^{-27}$ kg and the time t is reported in seconds.

3. Information theory-inspired WEP for quantum probes?

The trajectories of classical test-bodies in free-fall carry no information about their masses. Vice versa, measurements of position on quantum probes allow to extract nonvanishing information. In particular, even in the point-like limit when $\Delta x \to 0$, $F_X(m)$ does not go to zero. The fact that the Fisher information does not vanish may be seen as evidence that a quantum probe cannot be approximated by a test-body.

However, we are going to argue that such a conclusion is premature. In fact, the solutions of the Schrödinger equation depend on the ratio $\hbar/m$ also in the free case. Therefore, the propagation of a wavepacket is subject to mass-dependent dispersive effects even in the absence of any gravitational field. As a consequence, one would generically obtain a nonzero Fisher information by monitoring the probe's trajectory.

In order to assess the validity of the test-body approximation for quantum probes, one should focus only on the information which is explicitly due to the gravitational coupling. In other words, one may split the Fisher information as

$$F_X(m) = F_X(m)|_{free} + F_X(m)|_{grav} , \tag{10}$$

where $F_X(m)|_{free}$ is the Fisher information one would have if no gravitational field were present (for the same initial preparation of the probe), and $F_X(m)|_{grav}$ is the Fisher information which is introduced by the external field. We will therefore say that quantum probes satisfy the WEP if the introduction of a gravitational field does not allow to extract any information $F_X(m)|_{grav}$, in addition to the amount $F_X(m)|_{free}$ which would already be present with no gravity. Vice versa, if it turns out that $F_X(m)|_{grav}$ is in general nonzero, then we will conclude that quantum probes cannot behave even in principle as test-particles.

Such formulation of the WEP is sufficiently general to apply to properly quantum particles. It has an operative flavour (as it is based on the results of physical measurements), and is inspired by information theory concepts. In the next section, we discuss to what degrees it is actually satisfied by quantum mechanics.

4. Results for uniform and non-uniform gravitational fields

We tackle first the case of a uniform gravitational field. Let $\psi(\mathbf{x}, 0)$ denote the initial preparation of the probe at time $t = 0$ and let $\mathbf{g}$ be a uniform gravitational field corresponding to a potential

$\varphi(\mathbf{x})$. The state of the probe is then evolved according to the Hamiltonian $H = \mathbf{p}^2/2m + m\varphi(\mathbf{x})$. The resulting state at general time t is

$$\psi(\mathbf{x}, 0) = e^{-img^2t^3/6}\, e^{-imt\varphi(\mathbf{x})}\, \psi_{free}(\mathbf{x} - \mathbf{g}t^2/2,\, t)\ , \tag{11}$$

where $\psi_{free}(\mathbf{x},\, t)$ is the time-evolved wavefunction in the absence of any gravitational field and given the same initial condition. It is then a simple matter [20] to check that the Fisher information for the state (11) satisfies $F_X(m) = F_X(m)|_{free}$, i.e. $F_X(m)|_{grav} = 0$. Thus, at least locally, no clash between quantum mechanics and the WEP is present.

More interesting is the case of a non-uniform gravitational field. In this case it is not possible in general to obtain exact solutions of the Schrödinger equation (3); however one can obtain approximate solutions by solving (3) through a BCH expansion [21] of the time-evolution operator, neglecting terms which are second order in the gravity gradient $\nabla\mathbf{g}$ and its derivatives [20]. Then, one finds that the probability distribution of position measurements is modified in a mass-dependent way by the introduction of an external gravitational field. More precisely,

$$|\psi(\mathbf{x},\, t)|^2 \sim |\psi_{free}(\mathbf{x} - \mathbf{g}t^2/2 + \mathbf{d}_m,\, t)|^2\ , \tag{12}$$

where $\mathbf{d}_m$ is a displacement which violates universality, i.e. it depends explicitly on the mass m of the probing particle,[2]

$$\mathbf{d}_m = \frac{t^2}{2}[\mathbf{x}\cdot\nabla\mathbf{g}(\mathbf{x}) - \mathbf{g}(\mathbf{x})] + \frac{t^3}{3m}\mathbf{p}\cdot\nabla\mathbf{g}(\mathbf{x}) + \frac{5t^4}{48}\nabla\mathbf{g}^2(\mathbf{x})\ . \tag{13}$$

Notice that the mass-dependent terms vanishes identically for a uniform field. More generally, we expect our formulation of the WEP to be satisfied when such a term can be neglected compared to the term $-\mathbf{g}t^2/2$, which would imply

$$\left|t^3\,\mathbf{p}\cdot\nabla\mathbf{g}/m\right| \ll |\mathbf{g}t^2|\ . \tag{14}$$

By substituting $\mathbf{p}$ with its classical expression $m\mathbf{g}t$ and approximating the gravity gradient as $\sim \mathbf{g}/\ell$, where ℓ is the characteristic scale of variation of the gravitational field, one finds that (14) is equivalent to the condition

$$|\mathbf{g}t^2| \ll \ell\ , \tag{15}$$

i.e. as long as the probe does not have the time to explore regions of the order of the curvature scale the WEP appproximately holds. In general, however, gravity gradients can encode non-trivial information about the mass of a particle into its wavefunction, so that $F_X(m)|_{grav} \neq 0$.

5. Conclusions

In conclusion, we have discussed the use of quantum parameter estimation theory to analyze the interplay between quantum mechanics and the WEP. We have shown that, besides being a fundamental tool for the design of sensing experiments, the Fisher information may also provide a way to discuss whether quantum probes can be approximated as test-bodies. We found that, in general, this is not the case: the evolution of the wavefunction of a quantum probe is sensitive in a mass-dependent way to gravity gradients. This, in turn, means that the introduction of a gravitational field may improve precision in estimating the mass of a freely-falling particle by monitoring its trajectory. Nonetheless, locally, i.e. in the limit of a uniform field, our formulation of the WEP does hold.

[2] A few words on notation. For example, the writing $\mathbf{x}\cdot\nabla\mathbf{g}(\mathbf{x})$ stands for the three-dimensional vector whose j^{th} component is $x_i\partial_i g_j$, with repeated indices summed over. There is no ambiguity over contractions of indices since $x_i\partial_i g_j = x_i\partial_j g_i$. The same considerations apply for other analogous expressions in the text.

IOP Conf. Series: Journal of Physics: Conf. Series **880** (2017) 012067 doi:10.1088/1742-6596/880/1/012067

Let us emphasize that for classical probes, universality of free-fall holds in a stronger sense: a probe, devoid of any internal structure as considered here, but subject only to classical physics, would satisfy our formulation of the WEP irrespective of whether the external field is uniform or not. On the other hand, if the classical probe has an internal structure, e.g. it is spatially extended, then it is known [22] that it does not in general move geodesically. Our results show that a quantum probe, even if in principle devoid of any internal structure, behaves instead in analogy with a non-pointlike classical probe, because of the extended character of its wavefunction.

Acknowledgments
This work has been supported by EU through the Collaborative Project QuProCS (Grant Agreement 641277) and by UniMI through the H2020 Transition Grant 15-6-3008000-625.

References
[1] Will C M 1993 *Theory and experiment in gravitational physics* (Cambridge: Cambridge University Press)
[2] Di Casola E, Liberati S and Sonego S 2015 *Am. J. Phys.* **83** 39–46
[3] Adler C G and Coulter B L 1978 *Am. J. Phys.* **46** 199–201
[4] Segre M 1989 *Stud. Hist. Phil. Sci. A* **20** 435 – 451
[5] Allen J 1972 Summary of scientific results *Apollo 15 Preliminary Science Report* (NASA SP-289) pp 2–11
[6] Greenberger D 1968 *Ann. Phys.* **47** 116–126
[7] Anandan J 1980 *Found. Phys.* **10** 601–629
[8] Sonego S 1995 *Phys. Lett. A* **208** 1–7
[9] Chryssomalakos C and Sudarsky D 2003 *Gen. Rel. Grav.* **35** 605–617
[10] Okon E and Callender C 2011 *Eur. J. Philos. Sci.* **1** 133–145
[11] Kiefer C and Singh T P 1991 *Phys. Rev. D* **44** 1067
[12] Paris M G A 2009 *Int. J. Quantum Inf.* **7** 125–137
[13] Helstrom C W 1976 *Quantum detection and estimation theory* (Cambridge: Academic Press)
[14] Holevo A S 2003 *Statistical structure of quantum theory* vol 67 (Berlin: Springer Science & Business Media)
[15] Hayashi M 2006 *Quantum information* (Berlin: Springer)
[16] Cramér H 2016 *Mathematical Methods of Statistics (PMS-9)* vol 9 (Princeton: Princeton University Press)
[17] Braunstein S L and Caves C M 1994 *Phys. Rev. Lett.* **72** 3439
[18] Seveso L, Peri V and Paris M G A 2017 *J. Phys. A: Math. Theor*
[19] Colella R, Overhauser A W and Werner S A 1975 *Phys. Rev. Lett.* **34** 1472
[20] Seveso L and Paris M G 2017 *Ann. Phys.* **380** 213–223
[21] Serre J P 2009 *Lie algebras and Lie groups: 1964 lectures given at Harvard University* (Berlin: Springer)
[22] Geroch R and Jang P S 1975 *J. Math. Phys.* **16** 65–67